Table of Atomic Masses*

Element	Symbol	Atomic Number	Atomic Mass	Element	Symbol	Atomic Number	Atomic Mass	Element	Symbol	Atomic Number	Atomic Mass
Actinium	Ac	89	[227]§	Hafnium	Hf	72	178.5	Promethium	Pm	61	[145]
Aluminum	Al	13	26.98	Hassium	Hs	108	[265]	Protactinium	Pa	91	[231]
Americium	Am	95	[243]	Helium	He	2	4.003	Radium	Ra	88	226
Antimony	Sb	51	121.8	Holmium	Ho	67	164.9	Radon	Rn	86	[222]
Argon	Ar	18	39.95	Hydrogen	H	1	1.008	Rhenium	Re	75	186.2
Arsenic	As	33	74.92	Indium	In	49	114.8	Rhodium	Rh	45	102.9
Astatine	At	85	[210]	Iodine	I	53	126.9	Rubidium	Rb	37	85.47
Barium	Ba	56	137.3	Iridium	Ir	77	192.2	Ruthenium	Ru	44	101.1
Berkelium	Bk	97	[247]	Iron	Fe	26	55.85	Rutherfordium	Rf	104	[261]
Beryllium	Be	4	9.012	Krypton	Kr	36	83.80	Samarium	Sm	62	150.4
Bismuth	Bi	83	209.0	Lanthanum	La	57	138.9	Scandium	Sc	21	44.96
Bohrium	Bh	107	[264]	Lawrencium	Lr	103	[260]	Seaborgium	Sg	106	[263]
Boron	B	5	10.81	Lead	Pb	82	207.2	Selenium	Se	34	78.96
Bromine	Br	35	79.90	Lithium	Li	3	6.941	Silicon	Si	14	28.09
Cadmium	Cd	48	112.4	Lutetium	Lu	71	175.0	Silver	Ag	47	107.9
Calcium	Ca	20	40.08	Magnesium	Mg	12	24.31	Sodium	Na	11	22.99
Californium	Cf	98	[251]	Manganese	Mn	25	54.94	Strontium	Sr	38	87.62
Carbon	C	6	12.01	Meitnerium	Mt	109	[268]	Sulfur	S	16	32.07
Cerium	Ce	58	140.1	Mendelevium	Md	101	[258]	Tantalum	Ta	73	180.9
Cesium	Cs	55	132.90	Mercury	Hg	80	200.6	Technetium	Tc	43	[98]
Chlorine	Cl	17	35.45	Molybdenum	Mo	42	95.94	Tellurium	Te	52	127.6
Chromium	Cr	24	52.00	Neodymium	Nd	60	144.2	Terbium	Tb	65	158.9
Cobalt	Co	27	58.93	Neon	Ne	10	20.18	Thallium	Tl	81	204.4
Copper	Cu	29	63.55	Neptunium	Np	93	[237]	Thorium	Th	90	232.0
Curium	Cm	96	[247]	Nickel	Ni	28	58.69	Thulium	Tm	69	168.9
Dubnium	Db	105	[262]	Niobium	Nb	41	92.91	Tin	Sn	50	118.7
Dysprosium	Dy	66	162.5	Nitrogen	N	7	14.01	Titanium	Ti	22	47.88
Einsteinium	Es	99	[252]	Nobelium	No	102	[259]	Tungsten	W	74	183.9
Erbium	Er	68	167.3	Osmium	Os	76	190.2	Uranium	U	92	238.0
Europium	Eu	63	152.0	Oxygen	O	8	16.00	Vanadium	V	23	50.94
Fermium	Fm	100	[257]	Palladium	Pd	46	106.4	Xenon	Xe	54	131.3
Fluorine	F	9	19.00	Phosphorus	P	15	30.97	Ytterbium	Yb	70	173.0
Francium	Fr	87	[223]	Platinum	Pt	78	195.1	Yttrium	Y	39	88.91
Gadolinium	Gd	64	157.3	Plutonium	Pu	94	[244]	Zinc	Zn	30	65.38
Gallium	Ga	31	69.72	Polonium	Po	84	[209]	Zirconium	Zr	40	91.22
Germanium	Ge	32	72.59	Potassium	K	19	39.10				
Gold	Au	79	197.0	Praseodymium	Pr	59	140.9				

*The values given here are to four significant figures.

§A value given in brackets denotes the mass of the longest-lived isotope.

Table of Atomic Masses*

Element	Symbol	Atomic Number	Atomic Mass	Element	Symbol	Atomic Number	Atomic Mass	Element	Symbol	Atomic Number	Atomic Mass
Actinium	Ac	89	[227][§]	Hafnium	Hf	72	178.5	Promethium	Pm	61	[145]
Aluminum	Al	13	26.98	Hassium	Hs	108	[265]	Protactinium	Pa	91	[231]
Americium	Am	95	[243]	Helium	He	2	4.003	Radium	Ra	88	226
Antimony	Sb	51	121.8	Holmium	Ho	67	164.9	Radon	Rn	86	[222]
Argon	Ar	18	39.95	Hydrogen	H	1	1.008	Rhenium	Re	75	186.2
Arsenic	As	33	74.92	Indium	In	49	114.8	Rhodium	Rh	45	102.9
Astatine	At	85	[210]	Iodine	I	53	126.9	Rubidium	Rb	37	85.47
Barium	Ba	56	137.3	Iridium	Ir	77	192.2	Ruthenium	Ru	44	101.1
Berkelium	Bk	97	[247]	Iron	Fe	26	55.85	Rutherfordium	Rf	104	[261]
Beryllium	Be	4	9.012	Krypton	Kr	36	83.80	Samarium	Sm	62	150.4
Bismuth	Bi	83	209.0	Lanthanum	La	57	138.9	Scandium	Sc	21	44.96
Bohrium	Bh	107	[264]	Lawrencium	Lr	103	[260]	Seaborgium	Sg	106	[263]
Boron	B	5	10.81	Lead	Pb	82	207.2	Selenium	Se	34	78.96
Bromine	Br	35	79.90	Lithium	Li	3	6.941	Silicon	Si	14	28.09
Cadmium	Cd	48	112.4	Lutetium	Lu	71	175.0	Silver	Ag	47	107.9
Calcium	Ca	20	40.08	Magnesium	Mg	12	24.31	Sodium	Na	11	22.99
Californium	Cf	98	[251]	Manganese	Mn	25	54.94	Strontium	Sr	38	87.62
Carbon	C	6	12.01	Meitnerium	Mt	109	[268]	Sulfur	S	16	32.07
Cerium	Ce	58	140.1	Mendelevium	Md	101	[258]	Tantalum	Ta	73	180.9
Cesium	Cs	55	132.90	Mercury	Hg	80	200.6	Technetium	Tc	43	[98]
Chlorine	Cl	17	35.45	Molybdenum	Mo	42	95.94	Tellurium	Te	52	127.6
Chromium	Cr	24	52.00	Neodymium	Nd	60	144.2	Terbium	Tb	65	158.9
Cobalt	Co	27	58.93	Neon	Ne	10	20.18	Thallium	Tl	81	204.4
Copper	Cu	29	63.55	Neptunium	Np	93	[237]	Thorium	Th	90	232.0
Curium	Cm	96	[247]	Nickel	Ni	28	58.69	Thulium	Tm	69	168.9
Dubnium	Db	105	[262]	Niobium	Nb	41	92.91	Tin	Sn	50	118.7
Dysprosium	Dy	66	162.5	Nitrogen	N	7	14.01	Titanium	Ti	22	47.88
Einsteinium	Es	99	[252]	Nobelium	No	102	[259]	Tungsten	W	74	183.9
Erbium	Er	68	167.3	Osmium	Os	76	190.2	Uranium	U	92	238.0
Europium	Eu	63	152.0	Oxygen	O	8	16.00	Vanadium	V	23	50.94
Fermium	Fm	100	[257]	Palladium	Pd	46	106.4	Xenon	Xe	54	131.3
Fluorine	F	9	19.00	Phosphorus	P	15	30.97	Ytterbium	Yb	70	173.0
Francium	Fr	87	[223]	Platinum	Pt	78	195.1	Yttrium	Y	39	88.91
Gadolinium	Gd	64	157.3	Plutonium	Pu	94	[244]	Zinc	Zn	30	65.38
Gallium	Ga	31	69.72	Polonium	Po	84	[209]	Zirconium	Zr	40	91.22
Germanium	Ge	32	72.59	Potassium	K	19	39.10				
Gold	Au	79	197.0	Praseodymium	Pr	59	140.9				

*The values given here are to four significant figures.

[§]A value given in brackets denotes the mass of the longest-lived isotope.

Introductory Chemistry

Steven S. Zumdahl
University of Illinois

Fifth Edition

Houghton Mifflin Company
Boston New York

To David Scott Hendrie

Vice President and Publisher: Charles Hartford
Executive Editor: Richard Stratton
Development Editors: Sara Wise, Bess Deck
Editorial Associate: Rosemary Mack
Senior Project Editor: Cathy Labresh Brooks
Senior Production/Design Coordinator: Jill Haber
Senior Manufacturing Coordinator: Marie Barnes
Senior Marketing Manager: Katherine Greig
Marketing Associate: Alexandra Shaw

Cover photograph © 2002 Charles Krebs/Getty Images

Photo Credits appear on the page following the Index/Glossary.

Printed in the U.S.A.

Library of Congress Catalog Card Number: 2003101239

ISBNs:

Case: 0-618-30501-7
Paper: 0-618-30503-3

23456789-DOW-07 06 05 04 03

Brief Contents

Contents

7 Reactions in Aqueous Solutions 172

8 Chemical Composition 210

9 Chemical Quantities 246

18 Radioactivity and Nuclear Energy 550

APPENDIX A1

Preface

In revising *Introductory Chemistry* we have rededicated ourselves to the goals we pursued in the first four editions: to make chemistry interesting, accessible, and understandable to the beginning student. For this edition, we have added unprecedented instructor support and additional student study support to further these goals in today's chemistry classroom.

Learning chemistry can be very rewarding. And even the novice, we believe, can relate the macroscopic world of chemistry—the observation of color changes and precipitate formation—to the microscopic world of ions and molecules. To achieve that goal, instructors are making a sincere attempt to provide more interesting and more effective ways to learn chemistry, and we hope that *Introductory Chemistry* will be perceived as a part of that effort. In this text we have presented concepts in a clear and sensible manner using language and analogies that students can relate to. We have also written the book in a way that supports active learning. In particular, the In-Class Discussion Questions, found at the end of each chapter, provide excellent material for collaborative work by students. In addition, we have connected chemistry to real-life experience at every opportunity, from chapter opening discussions of chemical applications to "Chemistry in Focus" features throughout the book. We are convinced that this approach will foster enthusiasm and real understanding as the student uses this text.

New to this Edition

Building on the success of previous editions of *Introductory Chemistry,* the following changes have been made to further enhance the text:

- **The Instructor's Toolkit** To assist instructors, coordinators, and the many adjunct professors teaching introductory chemistry we have created a new, comprehensive and integrated instructor support package. We have consolidated print and electronic resources to help instructors save time and to prepare effectively for class. (For more details of each resource see Supplements to the Text). The toolkit contains:

  A new *Instructor's Annotated Edition* with a wrap-around margin that appears as a "frame" around the student text page and consolidates much of the support previously supplied in separate manuals

  Complete Solutions Guide with detailed solutions

  HM ClassPrep, HM Testing, and *Chemistry Animations and Videos CD-ROMs* all conveniently located in one package

  Enhanced *Instructor web site*

  Updated printed *Test Bank*

- **Enhanced student support package** We have developed a range of student materials to address the needs of the increasing number of students who are less well prepared for this course. These resources have been designed to help students improve their basic math skills and to support good study habits. (For more details of each resource see Supplements to the Text). The following resources are automatically packaged with the text:

  Student web site including quizzes, interactive molecules, flashcards, problems linked to the text, and "Chemistry in Focus" boxes

SMARTHINKING™, live online tutoring by qualified chemistry professors

New *Math Review CD-ROM* (plus math tips in the text itself)

Student Study Card with formulas and study reminders

- A **new art program** enhances the visual impact of the text while clarifying important concepts. Illustrations have been updated to make it easier for students to connect molecular-level activity to macroscopic phenomena. Students can more readily see the connection between abstract chemical concepts and real-life situations, motivating them to learn the material.
- **Chapter 3 (Matter and Energy)** We have expanded our coverage of energy by adding a new section (3.6) that explains the concept of energy transfer as heat. In addition, we have explained exothermic and endothermic processes. Also, in this section, we give a full explanation of the concept of temperature as it relates to the microscopic motions of the particles in a system.
- **Chapter 10 (Modern Atomic Theory)** In Chapter 10 we have expanded our treatment of the atom by adding a section on the historical development of the structure of the atom. This material on Rutherford's work will give students a better perspective on how the current model of the atom was conceived. We have also greatly expanded our treatment of light with new emphasis on the properties of waves and the dual nature of light.
- We have replaced approximately one-third of the "Chemistry in Focus" boxes with new up-to-date topics.
- We have replaced many end-of-chapter questions and problems. For the instructor, answers to all the Self-Check and end-of-chapter exercises now appear in the wrap-around margins of the Instructor's Annotated Edition. In the student text, answers to Self-Checks and to even-numbered exercises are still provided at the back of the book.
- Content is now provided for WebCT and Blackboard systems. All content from the student web site, along with HMTesting content and assets from the instructor web site, is now available for professors using either of these popular course management systems.

Emphasis on Reaction Chemistry

We continue to emphasize chemical reactions early in the book, leaving the more abstract material on orbitals for later chapters. In a course in which many students encounter chemistry for the first time, it seems especially important that we present the chemical nature of matter before we discuss the theoretical intricacies of atoms and orbitals. Reactions are inherently interesting to students and can help us draw them to chemistry. In particular, reactions can form the basis for fascinating classroom demonstrations and laboratory experiments.

We have therefore chosen to emphasize reactions before going on to the details of atomic structure. Relying only on very simple ideas about the atom, Chapters 6 and 7 represent a thorough treatment of chemical reactions, including how to recognize a chemical change and what a chemical equation means. The properties of aqueous solutions are discussed in detail, and careful attention is given to precipitation and acid–base reactions. In addition, a simple treatment of oxidation–reduction reactions is given. These chapters should provide a solid foundation, relatively early in the course, for reaction-based laboratory experiments.

For instructors who feel that it is desirable to introduce orbitals early in the course, prior to chemical reactions, the chapters on atomic theory and bonding (Chapters 10 and 11) can be covered directly after Chapter 4. Chapter 5 deals solely with nomenclature and can be used wherever it is needed in a particular course.

Development of Problem-Solving Skills

Problem-solving is a high priority in chemical education. We all want our students to acquire problem-solving skills. Fostering the development of such skills has been a central focus of the earlier editions of this text and we have maintained this approach in this edition.

In the first chapters we spend considerable time guiding students to an understanding of the importance of learning chemistry. At the same time, we explain that the complexities that can make chemistry frustrating at times can also provide the opportunity to develop the problem-solving skills that are beneficial in any profession. Learning to think like a chemist is useful to everyone. To emphasize this idea, we apply scientific thinking to some real-life problems in Chapter 1.

One reason chemistry can be challenging for beginning students is that they often do not possess the required mathematical skills. Thus we have paid careful attention to such fundamental mathematical skills as using scientific notation, rounding off to the correct number of significant figures, and rearranging equations to solve for a particular quantity. And we have meticulously followed the rules we have set down, so as not to confuse students.

Attitude plays a crucial role in achieving success in problem solving. Students must learn that a systematic, thoughtful approach to problems is better than brute force memorization. We foster this attitude early in the book, using temperature conversions as a vehicle in Chapter 2. Throughout the book we encourage an approach that starts with trying to represent the essence of the problem using symbols and/or diagrams, and ends with thinking about whether the answer makes sense. We approach new concepts by carefully working through the material before we give mathematical formulas or overall strategies. We encourage a thoughtful step-by-step approach rather than the premature use of algorithms. Once we have provided the necessary foundation, we highlight important rules and processes in skill development boxes so that students can locate them easily.

Many of the worked examples are followed by Self-Check exercises, which provide additional practice. The Self-Check exercises are keyed to end-of-chapter exercises to offer another opportunity for students to practice a particular problem-solving skill or understand a particular concept.

We have expanded the number of end-of-chapter exercises. As in the first four editions, the end-of-chapter exercises are arranged in "matched pairs," meaning that both problems in the pair explore similar topics. An Additional Problems section includes further practice in chapter concepts as well as more challenging problems. Cumulative reviews, which appear after every few chapters, test concepts from the preceding chapter block. Answers for all even-numbered exercises appear in a special section at the end of the student edition.

Handling the Language of Chemistry and Applications

We have gone to great lengths to make this book "student friendly" and have received enthusiastic feedback from students who have used it.

As in the earlier editions, we present a systematic and thorough treatment of chemical nomenclature. Once this framework is established, students can progress through the book comfortably.

Along with chemical reactions, applications form an important part of descriptive chemistry. Because students are interested in chemistry's impact on their lives, we have included many new "Chemistry in Focus" boxes, which describe current applications of chemistry. These special interest boxes cover such topics as using light as a sex attractant, the effects of asteroid impacts on the earth, plants that help control arsenic pollution, and the science behind common household materials such as broccoli, stainless steel, and carbonated beverages.

Visual Impact of Chemistry

Responding to instructors' requests to include graphic illustrations of chemical reactions, phenomena, and processes, our full-color design enables color to be used functionally, thoughtfully, and consistently to help students understand chemistry and to make the subject more inviting to them. We have included only those photos that illustrate a chemical reaction or phenomenon or that make a connection between chemistry and the real world. Many new photos enhance the fifth edition.

Choices of Coverage

For the convenience of instructors, four versions of the fifth edition are available: two paperback versions and two hardbound versions. *Basic Chemistry,* Fifth Edition, a paperback text, provides basic coverage of chemical concepts and applications through acid–base chemistry and has fifteen chapters. *Introductory Chemistry,* Fifth Edition, available in hardcover and paperback, expands the coverage to eighteen chapters with the addition of equilibrium, oxidation–reduction reactions and electrochemistry, radioactivity, and nuclear energy. Finally, *Introductory Chemistry: A Foundation,* Fifth Edition, a hardbound text, has twenty chapters, with the final two chapters providing a brief introduction to organic and biological chemistry.

Supplements for the Text

A main focus of this revision is to provide instructors and students with an unparalleled level of support. We considered all aspects of the ancillary package and how we could make them better suited to student and instructor needs.

For the Student

Student web site (chemistry.college.hmco.com/students)

Generic user name and password packaged automatically with all new texts. Includes Houghton Mifflin's ACE self-quizzing, interactive molecules, flashcards of key terms and concepts, and "Chemistry in Focus" boxes.

SMARTHINKING™ is our free, **live online tutoring service.** A passkey that is valid for twelve months from initial sign-on is packaged automatically with all new texts. It allows access to support from a real professor or study resources from wherever students are, whenever they need help. With SMARTHINKING students can:

- Connect immediately to live help Sunday-Thursday, 2 P.M.–5 P.M. and 9 P.M.–1 A.M. EST
- Submit a question to get a response from an e-structor, usually within 24 hours
- Use the whiteboard with full scientific notation and graphics
- Preschedule time with an e-structor
- View past online sessions, questions, or essays in an archive on their personal academic homepage
- View their tutoring schedule
- Work on other projects while waiting for help

Math Review CD-ROM Packaged automatically with all new texts, this CD includes review and practice material for key math skills that students need in order to succeed in the course.

Student Study Card A laminated, two-sided study card with chemical formulas/reminders is included automatically with all new texts for students to use as a quick study aid.

***Introductory Chemistry Interactive* CD-ROM with Printed Media Activities package** This package is available for bundling with copies of Zumdahl, *Introductory Chemistry,* Fifh Edition, and is also available for sale separately. The handy guide provides information and access to technology resources available with the fifth edition. Package includes:

> ***Introductory Chemistry Interactive* Student CD-ROM** Highly interactive, offering topic summaries, conceptual enhancement, dynamic visualization, and more. Cross-platform (Macintosh/Windows) CD.
>
> **Printed Media Activities booklet** These activities direct students to the technology resources provided to explore topics, learn concepts, and solve problems.

Study Guide by Donald DeCoste of the University of Illinois contains Chapter Discussions and Learning Review (practice chapter tests).

Solutions Guide by James F. Hall, University of Massachusetts—Lowell contains detailed solutions for the even-numbered end-of-chapter questions and exercises and Cumulative Review exercises.

Introductory Chemistry in the Laboratory by James F. Hall, University of Massachusetts—Lowell, contains experiments organized according to the topical presentation in the text. Annotations in the Instructors Annotated Edition indicate where the experiments from this manual are relevant to chapter content. The lab manual has been updated and revised for this edition.

For the Instructor

Instructors Annotated Edition The new Instructors Annotated Edition gathers a wealth of teaching support in one convenient package.

The IAE contains all 20 chapters (the full contents *of Introductory Chemistry: A Foundation,* Fifth Edition). Annotations in the new wrap-around margins of the IAE include:

- **Answers to Self-Check exercises,** at point-of-use.
- **Answers to all end-of-chapter questions and exercises,** at point-of-use
- **Additional Examples** with answers to supplement worked-out examples in the text
- **Using Technology** Information about incorporating animations and video clips from the electronic support materials in lecture
- **Teaching Support** Suggestions for specific lecture/instruction methods, activities, and in-class demonstrations to help convey concepts
- **Overview** An overview of the chapter's learning objectives
- **Teaching Tips** Guidelines for highlighting critical information in the chapter
- **Misconceptions** Tips on where students may have trouble or confusion with a topic
- **Demonstrations** Detailed instructions for in-class demonstrations and activities. (These are similar to material in Teaching Support, and may be referenced in Teaching Support annotations.)
- **Laboratory Experiments** Information on which labs in the *Laboratory Manual* are relevant to chapter content
- **Background Information** Explanations of conventions used in the text
- **Icons** mark material correlations between the main text and the electronic support materials, the *Test Bank,* the *Laboratory Manual,* and the figures available in the overhead transparency package.
- **Historical Notes** Biographical or other historical information about science and scientists

Complete Solutions Guide by James F. Hall, University of Massachusetts—Lowell, contains detailed solutions for all end-of-chapter questions and exercises and Cumulative Review exercises.

HMClass Prep with *HMTesting* Version 6.0 and *Chemistry Animations and Videos* CD-ROM package

This package is a 2-CD set containing the electronic lecture support materials for instructors. The set includes both the *HMClass Prep* with *HMTesting* CD-ROM and the *Chemistry Animations and Videos* CD listed below. It allows instructors to access both lecture aids and testing software in one place. The two CDs cannot be ordered separately from each other.

***HMClass Prep* with *HMTesting* Version V.6.0 and CD-ROM (Cross-platform—Macintosh/Windows)** *HMClass Prep* includes everything an instructor will need to develop lectures: Teaching note annotations from the IAE, PowerPoint slides with all text figures, tables, and many photos, as well as the *Test Bank* (MS Word files of the printed *Test Bank*).

HM Testing Version 6.0 combines a flexible test-editing program with a comprehensive gradebook function for easy administration and tracking. *HM Testing* enables instructors to administer tests via network server or the Web. The *HM Testing* database can produce multiple-choice, true/false, fill-in-the-blank, and essay tests. Questions can be customized based on the chapter being covered, the question format, level of difficulty, and specific topics. *HM Testing* provides for the utmost security in accessing both test questions and grades.

Chemistry Animations and Videos V.2.0 CD-ROM (Cross-platform—Macintosh/Windows) Includes animations and video files that may be integrated into PowerPoint. Video files are selected from Video Series C. References to materials on this CD appear under the Using Technology annotations in the IAE and are marked with the CD icon.

Test Bank by Steven S. Zumdahl and Donald DeCoste, provides over 1600 multiple-choice, true-false, short-answer, matching, and completion questions. Approximately 300 questions from the previous edition have been replaced.

Instructor web site (chemistry.college.hmco.com/instructors) includes PowerPoint slides with all text figures, tables, and many photos; a list of lecture demos; discussion questions; and teaching notes from the textbook margins. User name and password required.

WebCT ePack and Blackboard Course Cartridge All content from the student web site, along with *HMTesting* content, is available for professors using either of these popular course management systems.

Transparencies package includes nearly 100 full-color acetates of selected illustrations from the text. Icons in the IAE margins indicate the figures that are available as acetates; all images are also available PowerPoint-ready on the *HMClassPrep* CD.

DVD contains all the video clips of lecture demonstrations from Video Series C, plus the animations of important chemical processes and concepts that are also available on the *Chemistry Animations and Videos* V.2.0 CD-ROM. The DVD is designed to provide another presentation option for instructors with DVD presentation technology.

Instructor's Guide for Introductory Chemistry in the Laboratory by James F. Hall includes general notes about each experiment, estimated completion time, materials required, and answers to both pre- and post-laboratory questions. Annotations in the IAE indicate where the experiments from this manual are relevant to chapter content. The lab manual has been updated and revised for this edition.

Houghton Mifflin Videotape Series A, B, C and D provide over 100 lecture demonstrations performed by John Luoma, Cleveland State University; John J. Fortman and Rubin Battino, Wright State University; Patricia L. Samuel; and Paul Kelter, University of North Carolina—Greensboro. Series C demonstrations appear on the *Chemistry Animations and Videos* V.2.0 CD-ROM and DVD as well.

We have worked hard to make this book and its supplements clear, interesting, and accurate. We would appreciate any comments that would make the book more useful to students and instructors.

Acknowledgments

A book such as this one depends on the expertise and dedication of many talented people. Richard Stratton, Executive Editor, is the ideal editor. He is knowledgeable, organized, considerate, and always supportive. Not only is he a great pleasure to work with, but he has outstanding taste in restaurants. Sara Wise and Bess Deck, development editors, have done an excellent job of helping to plan this revision and organizing its execution. They have been a tremendous help. Cathy Brooks, Senior Project Editor, is a truly outstanding project editor. It is a very secure feeling to know that she will always get it right. Sharon Donahue, Photo Researcher, has a real flair for choosing the right photos and I very much appreciate her efforts. Many thanks also to Jill Haber, Senior Production/Design Coordinator, who managed the design and typesetting of the book, and to Henry Rachlin, Designer, for a beautiful interior design.

I especially appreciate the efforts of Jim Hall of the University of Lowell who has been of tremendous help with the end-of-chapter questions and exercises and the Cumulative Review exercises. I also very much appreciate the help of my colleague Don DeCoste from the University of Illinois. Don designed the In-Class Discussion Questions and has had a big impact on my own teaching philosophy. As always I am especially grateful for the love and support of my wife Susan who truly understands the stresses and rewards of being an author. Finally I send my love and appreciation to my wonderful family: Mom and Dad, Scott, Whitney, Jessica, Joshua, Moriah and David, and Leslie and Steve and Tyler and Sunshine.

My sincerest appreciation goes to the following reviewers who examined the fourth edition in preparation for the revision:

Judith Chamberlain
Inver Hills Community College

Dorothy N. Eseonu
Virginia Union University, Richmond

Terry Gleason
Saddleback College

William H. Hersh
Queens College

Beth A. Landis
University of the Incarnate Word

Estelle Lebeau
Central Michigan University

Robert E. Loffredo
Truett-McConnell College

Dr. Scott Luaders
Quincy University

Kathryn A. Lysko, Ph.D.
Immaculata College

C. Michael McCallum
University of the Pacific

William A. Meena
Rock Valley College

Kathy Mitchell
St. Petersburg College

Alice J. Monroe
St. Petersburg College

Martha K. Newchurch
Nicholls State University

David A. Nyquist, Ph.D.
University of North Florida

James A. Petrich
San Antonio College

Dr. Mary Sohn
Florida Tech.

In addition, I want to thank the accuracy reviewers of the text and the supplements: David W. Shinn, Ph.D.; Ghassan M. Saed, Ph.D.; Kathy Mitchell; William A. Meena, Rock Valley College; and C. Michael McCallum, University of the Pacific.

Features of *Introductory Chemistry*, Fifth Edition

The fifth edition of *Introductory Chemistry* retains all of the qualities that have made it a trusted and authoritative first-year text. Its hallmark abilities to make chemistry interesting, accessible, and understandable to the beginning student are enhanced through superior teaching and learning support.

Unparalleled Learning Support

The Student Support Package comes automatically with all new texts. It includes the following resources to help students develop math and study skills.

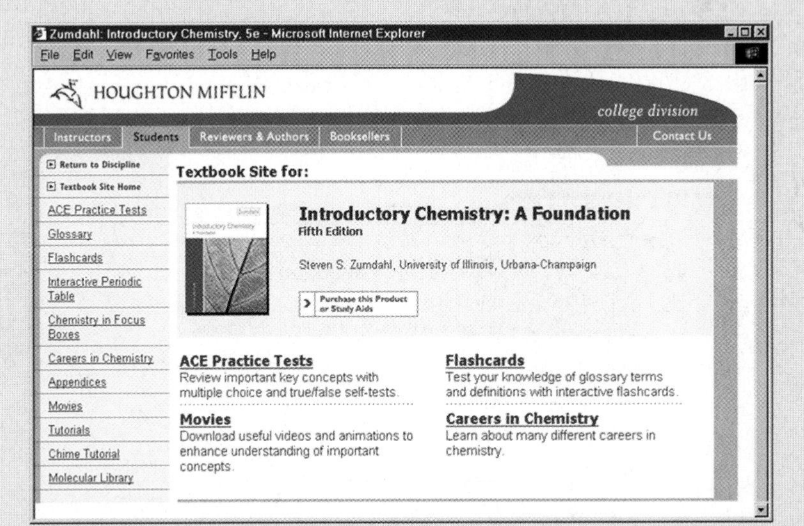

- The **Student web site** provides a full suite of interactive learning tools such as **ACE self-quizzing exercises** and **flashcards** to help students learn important concepts and terms. Generic user name and password packaged automatically with all new texts.

- **SMARTHINKING™** is **live, online tutoring.** A passkey offers twelve months of online, text-specific tutoring when students need it. Students can work one-on-one with an online tutor using a state-of-the-art whiteboard; submit a question anytime and receive a response, usually within 24 hours; and access additional study resources at any time.

- The **Student Study Card** is a two-sided quick study aid with chemical formulas and reminders.

Study Card to Accompany Zumdahl's *Introductory Chemistry* Series

Measurements and Calculations

Table 2.2 The Commonly Used Prefixes in the Metric System

Prefix	Symbol	Meaning	Power of 10 for Scientific Notation
mega	M	1,000,000	10^6
kilo	k	1000	10^3
deci	d	0.1	10^{-1}
centi	c	0.01	10^{-2}
milli	m	0.001	10^{-3}
micro	μ	0.000001	10^{-6}
nano	n	0.000000001	10^{-9}

Table 2.6 Some Examples of Commonly Used Units

length	A dime is 1 mm thick. A quarter is 2.5 cm in diameter. The average height of an adult man is 1.8 m.
mass	A nickel has a mass of about 5 g. A 120-lb woman has a mass of about 55 kg.
volume	A 12-oz can of soda has a volume of about 360 mL. A half gallon of milk is equal to about 2 L of milk.

Gases

STP: 0°C, 1 atm
Volume of 1 mole of ideal gas at STP = 22.4 L
$PV = nRT$ (Ideal Gas Law)
$R = 0.08206$ L atm/K mol
Process at constant n and T: $P_1V_1 = P_2V_2$ (Boyle's Law)
Process at constant n and P: $V_1/T_1 = V_2/T_2$ (Charles's Law)
Process at constant T and P: $V_1/n_1 = V_2/n_2$ (Avogadro's Law)

Types of Crystalline Solids

Crystalline solids: Ionic solids (Components are ions.), Molecular solids (Components are molecules.), Atomic solids (Components are atoms.)

Figure 13.13 The classes of crystalline solids.

The **Math Review CD-ROM** includes review and practice material for key math skills that students need in order to succeed in the course.

Unparalleled Teaching Support

We provide a comprehensive instructor support package to help save time in preparing for class.

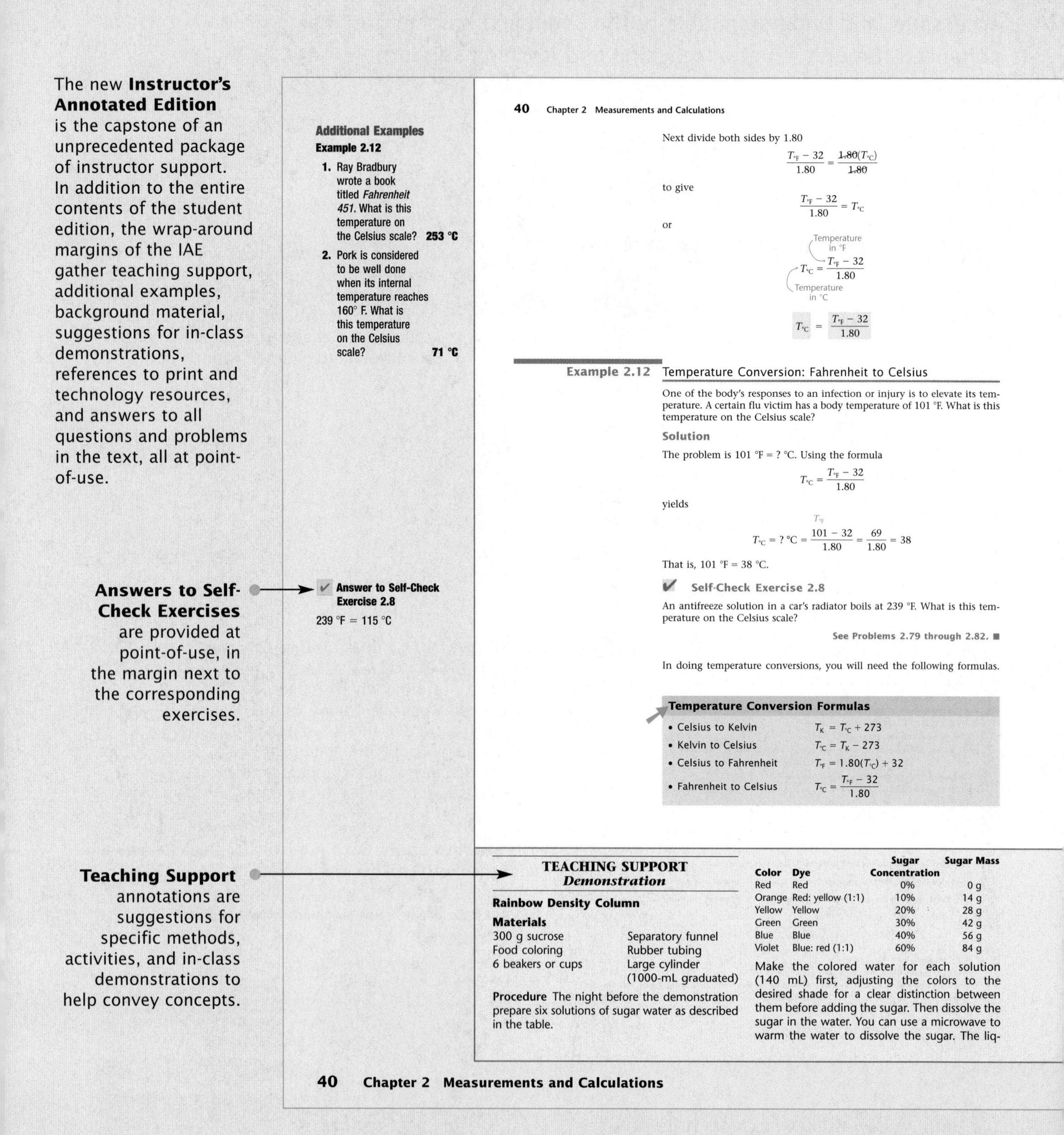

The new **Instructor's Annotated Edition** is the capstone of an unprecedented package of instructor support. In addition to the entire contents of the student edition, the wrap-around margins of the IAE gather teaching support, additional examples, background material, suggestions for in-class demonstrations, references to print and technology resources, and answers to all questions and problems in the text, all at point-of-use.

Answers to Self-Check Exercises are provided at point-of-use, in the margin next to the corresponding exercises.

Teaching Support annotations are suggestions for specific methods, activities, and in-class demonstrations to help convey concepts.

Additional Examples

Example 2.12

1. Ray Bradbury wrote a book titled *Fahrenheit 451*. What is this temperature on the Celsius scale? **253 °C**
2. Pork is considered to be well done when its internal temperature reaches 160° F. What is this temperature on the Celsius scale? **71 °C**

Next divide both sides by 1.80

$$\frac{T_{°F} - 32}{1.80} = \frac{1.80(T_{°C})}{1.80}$$

to give

$$\frac{T_{°F} - 32}{1.80} = T_{°C}$$

or

$$T_{°C} = \frac{T_{°F} - 32}{1.80}$$

Temperature in °F

Temperature in °C

$$T_{°C} = \frac{T_{°F} - 32}{1.80}$$

Example 2.12 Temperature Conversion: Fahrenheit to Celsius

One of the body's responses to an infection or injury is to elevate its temperature. A certain flu victim has a body temperature of 101 °F. What is this temperature on the Celsius scale?

Solution

The problem is 101 °F = ? °C. Using the formula

$$T_{°C} = \frac{T_{°F} - 32}{1.80}$$

yields

$$T_{°C} = ?\ °C = \frac{101 - 32}{1.80} = \frac{69}{1.80} = 38$$

That is, 101 °F = 38 °C.

Self-Check Exercise 2.8

An antifreeze solution in a car's radiator boils at 239 °F. What is this temperature on the Celsius scale?

See Problems 2.79 through 2.82. ■

Answer to Self-Check Exercise 2.8

239 °F = 115 °C

In doing temperature conversions, you will need the following formulas.

Temperature Conversion Formulas

- Celsius to Kelvin $T_K = T_{°C} + 273$
- Kelvin to Celsius $T_{°C} = T_K - 273$
- Celsius to Fahrenheit $T_{°F} = 1.80(T_{°C}) + 32$
- Fahrenheit to Celsius $T_{°C} = \frac{T_{°F} - 32}{1.80}$

TEACHING SUPPORT
Demonstration

Rainbow Density Column

Materials

300 g sucrose	Separatory funnel
Food coloring	Rubber tubing
6 beakers or cups	Large cylinder (1000-mL graduated)

Procedure The night before the demonstration prepare six solutions of sugar water as described in the table.

Color	Dye	Sugar Concentration	Sugar Mass
Red	Red	0%	0 g
Orange	Red: yellow (1:1)	10%	14 g
Yellow	Yellow	20%	28 g
Green	Green	30%	42 g
Blue	Blue	40%	56 g
Violet	Blue: red (1:1)	60%	84 g

Make the colored water for each solution (140 mL) first, adjusting the colors to the desired shade for a clear distinction between them before adding the sugar. Then dissolve the sugar in the water. You can use a microwave to warm the water to dissolve the sugar. The liq-

Other components of the instructor support package include the **Complete Solutions Guide, chemistry animations and video clips, Instructor web site,** and complete print and electronic testing support. Content for **WebCT** and **Blackboard** users, **transparencies, lab materials,** and more are also available. (For more detail on each of these materials, see the supplements section in the Preface.)

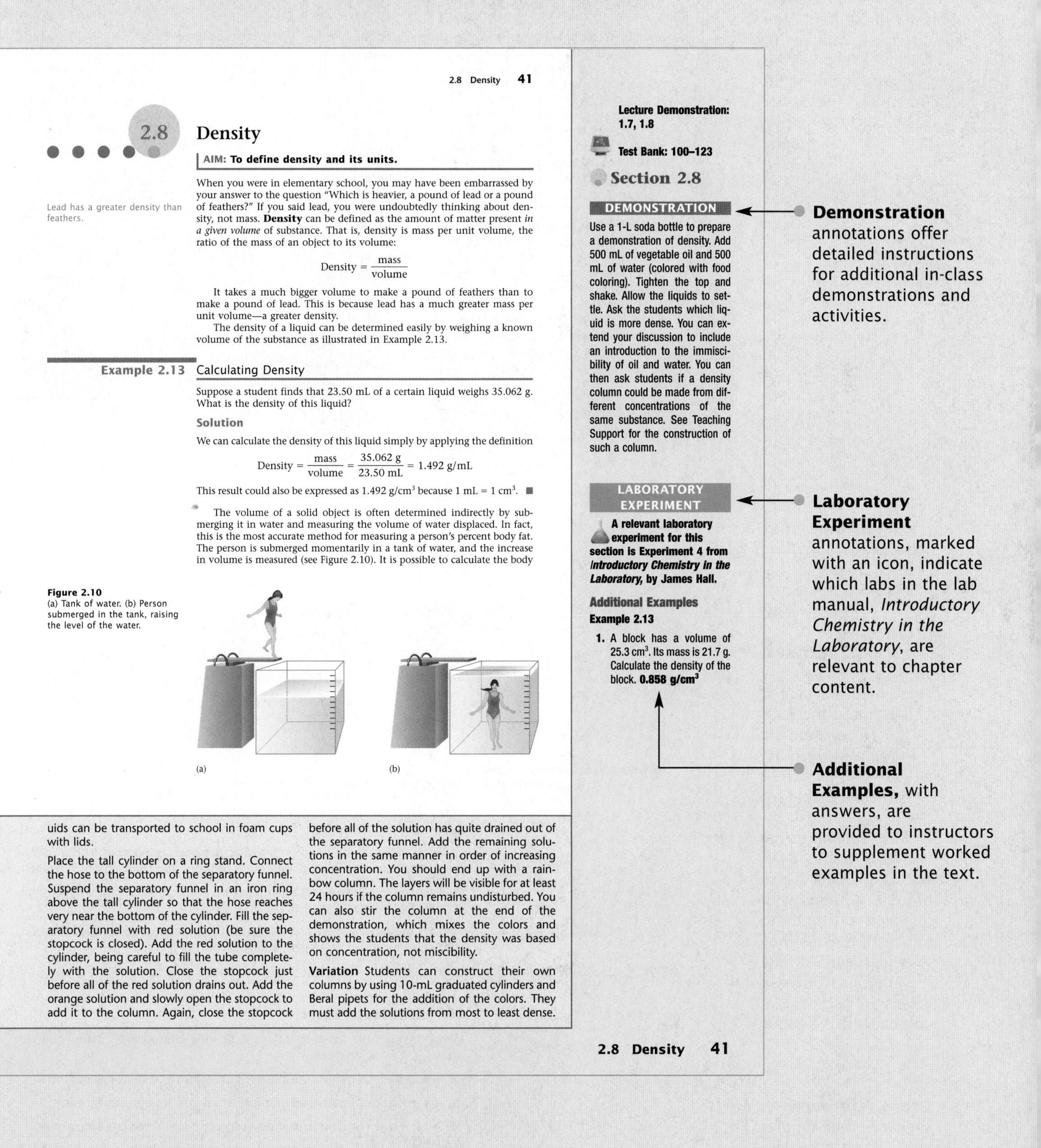

2.8 Density **41**

2.8 Density

AIM: To define density and its units.

Lead has a greater density than feathers.

When you were in elementary school, you may have been embarrassed by your answer to the question "Which is heavier, a pound of lead or a pound of feathers?" If you said lead, you were undoubtedly thinking about density, not mass. **Density** can be defined as the amount of matter present *in a given volume* of substance. That is, density is mass per unit volume, the ratio of the mass of an object to its volume:

$$\text{Density} = \frac{\text{mass}}{\text{volume}}$$

It takes a much bigger volume to make a pound of feathers than to make a pound of lead. This is because lead has a much greater mass per unit volume—a greater density.

The density of a liquid can be determined easily by weighing a known volume of the substance as illustrated in Example 2.13.

Example 2.13 Calculating Density

Suppose a student finds that 23.50 mL of a certain liquid weighs 35.062 g. What is the density of this liquid?

Solution

We can calculate the density of this liquid simply by applying the definition

$$\text{Density} = \frac{\text{mass}}{\text{volume}} = \frac{35.062\text{ g}}{23.50\text{ mL}} = 1.492\text{ g/mL}$$

This result could also be expressed as 1.492 g/cm^3 because 1 mL = 1 cm^3. ■

The volume of a solid object is often determined indirectly by submerging it in water and measuring the volume of water displaced. In fact, this is the most accurate method for measuring a person's percent body fat. The person is submerged momentarily in a tank of water, and the increase in volume is measured (see Figure 2.10). It is possible to calculate the body

Figure 2.10
(a) Tank of water. (b) Person submerged in the tank, raising the level of the water.

(a) (b)

Lecture Demonstration: 1.7, 1.8

Test Bank: 100–123

Section 2.8

DEMONSTRATION

Use a 1-L soda bottle to prepare a demonstration of density. Add 500 mL of vegetable oil and 500 mL of water (colored with food coloring). Tighten the top and shake. Allow the liquids to settle. Ask the students which liquid is more dense. You can extend your discussion to include an introduction to the immiscibility of oil and water. You can then ask students if a density column could be made from different concentrations of the same substance. See Teaching Support for the construction of such a column.

LABORATORY EXPERIMENT

A relevant laboratory experiment for this section is Experiment 4 from *Introductory Chemistry in the Laboratory*, by James Hall.

Additional Examples

Example 2.13

1. A block has a volume of 25.3 cm^3. Its mass is 21.7 g. Calculate the density of the block. **0.858 g/cm^3**

uids can be transported to school in foam cups with lids.

Place the tall cylinder on a ring stand. Connect the hose to the bottom of the separatory funnel. Suspend the separatory funnel in an iron ring above the tall cylinder so that the hose reaches very near the bottom of the cylinder. Fill the separatory funnel with red solution (be sure the stopcock is closed). Add the red solution to the cylinder, being careful to fill the tube completely with the solution. Close the stopcock just before all of the red solution drains out. Add the orange solution and slowly open the stopcock to add it to the column. Again, close the stopcock before all of the solution has quite drained out of the separatory funnel. Add the remaining solutions in the same manner in order of increasing concentration. You should end up with a rainbow column. The layers will be visible for at least 24 hours if the column remains undisturbed. You can also stir the column at the end of the demonstration, which mixes the colors and shows the students that the density was based on concentration, not miscibility.

Variation Students can construct their own columns by using 10-mL graduated cylinders and Beral pipets for the addition of the colors. They must add the solutions from most to least dense.

2.8 Density **41**

Demonstration annotations offer detailed instructions for additional in-class demonstrations and activities.

Laboratory Experiment annotations, marked with an icon, indicate which labs in the lab manual, *Introductory Chemistry in the Laboratory*, are relevant to chapter content.

Additional Examples, with answers, are provided to instructors to supplement worked examples in the text.

Motivation and Problem Solving

The **Chemistry in Focus** boxes describe current applications of chemistry. These special interest boxes help students see how chemistry applies to their lives by covering such topics as fat substitutes, the ozone layer, and alternative fuels.

CHEMISTRY IN FOCUS

Putting the Brakes on Arsenic

The toxicity of arsenic is well known. Indeed, arsenic has often been the poison of choice in classic plays and films—rent *Arsenic and Old Lace* sometime. Contrary to its treatment in the aforementioned movie, arsenic poisoning is a serious, contemporary problem. For example, the World Health Organization estimates that 77 million people in Bangladesh are at risk from drinking water that contains large amounts of naturally occurring arsenic. Recently, the Environmental Protection Agency announced more stringent standards for arsenic in U.S. public drinking water supplies. Studies show that prolonged exposure to arsenic can lead to a higher risk of bladder, lung, and skin cancers as well as other ailments, although the levels of arsenic that induce these symptoms remain in dispute in the scientific community.

Cleaning up arsenic-contaminated soil and water poses a significant problem. One approach is to find plants that will leach arsenic from the soil. Such a plant, the brake fern, recently has been shown to have a voracious appetite for arsenic. Research led by Lenna Ma, a chemist at the University of Florida in Gainesville, has shown that the brake fern accumulates arsenic at a rate 200 times that of the average plant. The arsenic, which becomes concentrated in fronds that grow up to 5 feet long, can be easily harvested and hauled away. Researchers are now investigating the best way to dispose of the plants so the arsenic can be isolated. The fern (*Pteris vittata*) looks promising for putting the brakes on arsenic pollution.

Lenna Ma and *Pteris vittata*—called the brake fern.

Important rules and steps appear in colored boxes so that students can locate them easily.

Many **Examples,** titled for easy reference, are located throughout the text. They model a thoughtful, step-by-step approach to solving problems.

Most examples are followed by **Self-Check Exercises,** which provide students with the opportunity to practice the skills they have just learned. The solutions to these exercises are provided at the back of the book.

Cross-references to similar end-of-chapter exercises are provided.

subscript 1 is always understood and not written). Following are some general rules for writing formulas:

Rules for Writing Formulas

1. Each atom present is represented by its element symbol.
2. The number of each type of atom is indicated by a subscript written to the right of the element symbol.
3. When only one atom of a given type is present, the subscript 1 is not written.

Example 4.1 Writing Formulas of Compounds

Write the formula for each of the following compounds, listing the elements in the order given.

a. Each molecule of a compound that has been implicated in the formation of acid rain contains one atom of sulfur and three atoms of oxygen.
b. Each molecule of a certain compound contains two atoms of nitrogen and five atoms of oxygen.
c. Each molecule of glucose, a type of sugar, contains six atoms of carbon, twelve atoms of hydrogen, and six atoms of oxygen.

Self-Check Exercise 4.1

Write the formula for each of the following compounds, listing the elements in the order given.

a. A molecule contains four phosphorus atoms and ten oxygen atoms.
b. A molecule contains one uranium atom and six fluorine atoms.
c. A molecule contains one aluminum atom and three chlorine atoms.

See Problems 4.19 and 4.20. ■

Improved Visualization

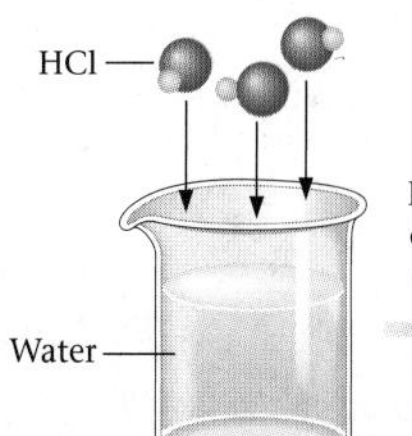

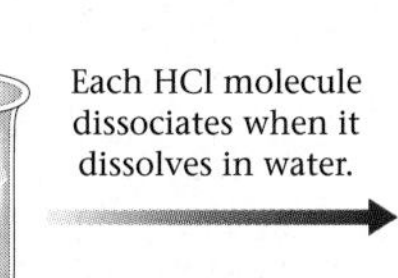

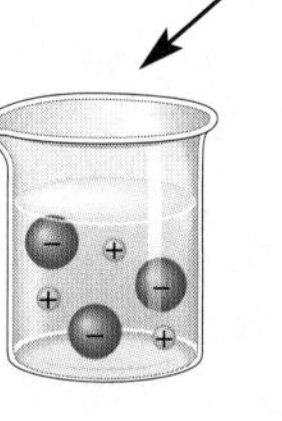

Figure 7.5
When gaseous HCl is dissolved in water, each molecule dissociates to produce H^+ and Cl^- ions. That is, HCl behaves as a strong electrolyte.

The **illustrations and photos** provide visual representation of chemical reactions, phenomena, and processes to help students understand chemistry and to make it more inviting to them.

Some of the more common elements are highlighted in **periodic table icons** to remind students about the position of selected elements and to help them become more familiar with the periodic table.

H Group 1   Cl Group 7

The Arrhenius definition of an acid: a substance that produces H^+ ions in aqueous solution.

$$HCl \xrightarrow{H_2O} H^+(aq) + Cl^-(aq)$$
$$HNO_3 \xrightarrow{H_2O} H^+(aq) + NO_3^-(aq)$$
$$H_2SO_4 \xrightarrow{H_2O} H^+(aq) + HSO_4^-(aq)$$

Arrhenius proposed that an **acid** *is a substance that produces H^+ ions (protons) when it is dissolved in water.*

Studies show that when HCl, HNO_3, and H_2SO_4 are placed in water, *virtually every molecule* dissociates to give ions. This means that when 100 molecules of HCl are dissolved in water, 100 H^+ ions and 100 Cl^- ions are produced. Virtually no HCl molecules exist in aqueous solution (see Figure 7.5). Because these substances are strong electrolytes that produce H^+ ions, they are called **strong acids.**

Arrhenius also found that *aqueous solutions that exhibit basic behavior* always contain hydroxide ions. He defined a **base** as a *substance that produces hydroxide ions (OH^-) in water.* The base most commonly used in the chemical laboratory is sodium hydroxide, NaOH, which contains Na^+ and OH^- ions and is very soluble in water. Sodium hydroxide, like all ionic substances, produces separated cations and anions when it is dissolved in water.

$$NaOH(s) \xrightarrow{H_2O} Na^+(aq) + OH^-(aq)$$

Although dissolved sodium hydroxide is usually represented as NaOH(*aq*), you should remember that the solution really contains separated Na^+ and OH^- ions. In fact, for every 100 units of NaOH dissolved in water, 100 Na^+ and 100 OH^- ions are produced.

Potassium hydroxide (KOH) has properties markedly similar to those of sodium hydroxide. It is very soluble in water and produces separated ions.

$$KOH(s) \xrightarrow{H_2O} K^+(aq) + OH^-(aq)$$

The marsh marigold is a beautiful but poisonous plant. Its toxicity results partly from the presence of erucic acid.

Revised drawings make artwork more realistic and easier to read. The art program helps students understand the molecular basis for chemical phenomena.

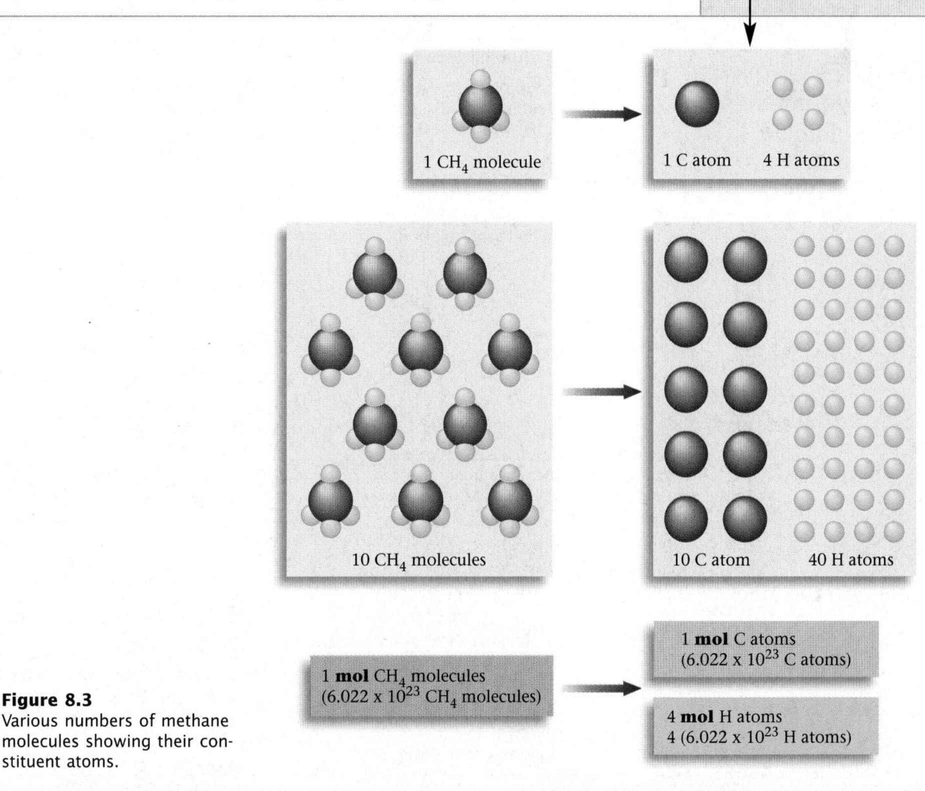

Figure 8.3
Various numbers of methane molecules showing their constituent atoms.

End-of-Chapter Material

Key Terms are printed in bold type and are defined within the chapter where they first appear. They are also grouped at the end of the chapter and in the Index/Glossary at the back of the text.

Each chapter has a **Summary** section to reinforce key concepts in the chapter.

In-Class Discussion Questions promote collaborative learning in class.

Questions and Problems are keyed to chapter sections. They are arranged in matched pairs and the even-numbered exercise is answered in the back of the text. **Additional Problems** are not keyed to any section or topic; they incorporate material from more than one section to provide an additional level of challenge for students.

Cumulative Reviews follow every two to three chapters, combining questions and problems from the chapters covered.

Chapter 5 Review

Key Terms

binary compound (5.1)
binary ionic compound (5.2)
polyatomic ion (5.5)
oxyanion (5.5)
acid (5.6)

Summary

1. Binary compounds can be named systematically by following a set of relatively simple rules. For compounds containing both a metal and a nonmetal, the metal is always named first, followed by a name derived from the root name for the nonmetal. For compounds containing a metal that can form more than one cation (Type II), we use a Roman numeral to specify the cation's charge. In binary compounds containing only nonmetals (Type III), prefixes are used to specify the numbers of atoms.
2. Polyatomic ions are charged entities composed of several atoms bound together. These have special names that must be memorized. Naming ionic compounds that contain polyatomic ions is very similar to naming binary ionic compounds.
3. The names of acids (molecules with one or more H^+ ions attached to an anion) depend on whether the anion contains oxygen.

In-Class Discussion Questions

These questions are designed to be considered by groups of students in class. Often these questions work well for introducing a particular topic in class.

1. Evaluate each of the following as an acceptable systematic name for water.
   a. dihydrogen oxide
   b. hydroxide hydride
   c. hydrogen hydroxide
   d. oxygen dihydride
2. Why do we call $Ba(NO_3)_2$ barium nitrate but call $Fe(NO_3)_2$ iron(II) nitrate?
3. Why is calcium dichloride not an acceptable name for $CaCl_2$?
4. What is the difference between sulfuric acid and hydrosulfuric acid?
5. Although we never use the systematic name for ammonia, NH_3, what do you think this name would be? Support your answer.

Questions and Problems

All even-numbered exercises have answers in the back of this book and solutions in the *Solutions Guide*.

5.1 Naming Compounds

QUESTIONS

1. Why is it necessary to have a *system* for the naming of chemical compounds?
2. What is a *binary* chemical compound? What are the two major *types* of binary chemical compounds? Give three examples of each type of binary compound.

5.2 Naming Compounds That Contain a Metal and a Nonmetal (Types I and II)

QUESTIONS

3. In ionic compounds, what general names are used to describe the *positive* ions and the *negative* ions?
4. In naming ionic compounds, we always name the ________ first.
5. A simple ________ ion has the same name as its parent element.
6. When we write the formula for an ionic compound, we are merely indicating the relative numbers of each type of ion in the compound, *not* the presence of "molecules" in the compound with that formula. Explain.
7. For a metallic element that forms two stable cations, the ending ________ is sometimes used for the cation with the lower charge.
8. We indicate the charge of a metallic element that forms more than one cation by adding a ________ after the name of the cation.
9. Give the name of each of the following simple binary ionic compounds.
   a. NaI
   b. CaF_2
   c. Al_2S_3
   d. $CaBr_2$
   e. SrO
   f. AgCl
   g. CsI
   h. Li_2O
10. Give the name of each of the following simple binary ionic compounds.
   a. NaBr
   b. CaS
   c. AlI_3
   d. Cs_2O
   e. MgF_2
   f. $SrCl_2$
   g. Li_2O
   h. BaI_2

Cumulative Review for Chapters 4–5

QUESTIONS

1. What is an element? How many elements are presently known? How many of these occur naturally and how many are man-made? Which elements are most abundant on the earth?
2. Without consulting any reference, write the name and symbol for as many elements as you can. How many could you name? How many symbols did you write correctly?
3. Why do the symbols for some elements seem to bear no relationship to the name for the element? Give several examples and explain.
4. Without consulting your textbook or notes, state as many points as you can of Dalton's atomic theory. Explain in your own words each point of the theory.
5. What is a compound? What is meant by the *law of* ... odic table be used to predict what ion an element's atoms will form?
12. What are some general physical properties of ionic compounds such as sodium chloride? How do we know that substances such as sodium chloride consist of positively and negatively charged particles? Since ionic compounds are made up of electrically charged particles, why doesn't such a compound have an overall electric charge? Can an ionic compound consist only of cations or anions (but not both)? Why not?
13. What principle do we use in writing the formula of an ionic compound such as NaCl or MgI_2? How do we know that *two* iodide ions are needed for each magnesium ion, whereas only one chloride ion is needed per sodium ion?
14. When writing the name of an ionic compound, which is named first, the anion or the cation? Give an ex-

Introductory Chemistry

1 Chemistry: An Introduction

Chemistry deals with the natural world.
▼

Did you ever see a fireworks display on July Fourth and wonder how it's possible to produce those beautiful, intricate designs in the air? Have you read about dinosaurs—how they ruled the earth for millions of years and then suddenly disappeared? Although the extinction happened 65 million years ago and may seem unimportant, could the same thing happen to us? Have you ever wondered why an ice cube (pure water) floats in a glass of water (also pure water)? Did you know that the "lead" in your pencil is made of the same substance (carbon) as the diamond in an engagement ring? Did you ever wonder how a corn plant or a palm tree grows seemingly by magic, or why leaves turn beautiful colors in autumn? Do you know how the battery works to start your car or run your calculator? Surely some of these things and many others in the world around you have intrigued you. The fact is that we can explain all of these things in convincing ways using the models of chemistry and the related physical and life sciences.

Fireworks are a beautiful illustration of chemistry in action.

1.1 Chemistry: An Introduction

AIM: To understand the importance of learning chemistry.

Although chemistry might seem to have little to do with dinosaurs, knowledge of chemistry was the tool that enabled paleontologist Luis W. Alvarez and his coworkers from the University of California at Berkeley to "crack the case" of the disappearing dinosaurs. The key was the relatively high level of iridium found in the sediment that represents the boundary between the earth's Cretaceous (K) and Tertiary (T) periods—the time when the dinosaurs disappeared virtually overnight (on the geological scale). The Berkeley researchers knew that meteorites also have unusually high iridium content (relative to the earth's composition), which led them to suggest that a large meteorite impacted the earth 65 million years ago, causing the climatic changes that wiped out the dinosaurs.

A knowledge of chemistry is useful to almost everyone—chemistry occurs all around us all of the time, and an understanding of chemistry is useful to doctors, lawyers, mechanics, business people, firefighters, and poets among others. Chemistry is important—there is no doubt about that. It lies at the heart of our efforts to produce new materials that make our lives safer and easier, to produce new sources of energy that are abundant and nonpolluting, and to understand and control the many diseases that threaten us and our food supplies. Even if your future career does not require the daily use of chemical principles, your life will be greatly influenced by chemistry.

Although a strong case can be made that the use of chemistry has greatly enriched all of our lives, there is also a dark side to the story. Our society has used its knowledge of chemistry to kill and destroy. It is important to understand that the principles of chemistry are inherently neither good nor bad—it's what we do with this knowledge that really matters. Although humans are clever, resourceful, and concerned about others, they also can be greedy, selfish, and ignorant. In addition, we tend to be shortsighted; we concentrate too much on the present and do not think enough about the long-range implications of our actions. This type of thinking has already caused us a great deal of trouble—severe environmental damage has occurred on many fronts. We cannot place all the responsibility on the chemical companies, because everyone has contributed to these problems. However, it is less important to lay blame than to figure out how to solve these problems. An important part of the answer must rely on chemistry.

One of the "hottest" fields in the chemical sciences is environmental chemistry—an area that involves studying our environmental ills and finding creative ways to address them. For example, meet Bart Eklund, who works in the atmospheric chemistry field for Radian Corporation in Austin, Texas. Bart's interest in a career in environmental science was fostered by two environmental chemistry courses and two ecology courses he took as an undergraduate. His original plan to gain several years of industrial experience and then to return to school for a graduate degree changed when he discovered that professional advancement with a B.S. degree was possible in the environmental research field. The multidisciplinary nature of environmental problems has allowed Bart to pursue his interest in several fields at the same time. You might say that he specializes in being a generalist.

Bart Eklund checking air quality at a hazardous waste site.

The environmental consulting field appeals to Bart for a number of reasons: the chance to define and solve a number of research problems; the simultaneous work on a number of diverse projects; the mix of desk, field, and laboratory work; the travel; and the opportunity to perform rewarding work that has a positive effect on people's lives.

Among his career highlights are the following:

- Spending a winter month doing air sampling in the Grand Tetons, where he also met his wife and learned to ski;
- Driving sampling pipes by hand into the rocky ground of Death Valley Monument in California;
- Working regularly with experts in their fields and with people who enjoy what they do;
- Doing vigorous work in 100 °F weather while wearing a rubberized suit, double gloves, and a respirator; and
- Getting to work in and see Alaska, Yosemite Park, Niagara Falls, Hong Kong, the People's Republic of China, Mesa Verde, New York City, and dozens of other interesting places.

Bart Eklund's career demonstrates how chemists are helping to solve our environmental problems. It is how we use our chemical knowledge that makes all the difference.

An example that shows how technical knowledge can be a "double-edged sword" is the case of chlorofluorocarbons (CFCs). When the compound CCl_2F_2 (originally called Freon-12) was first synthesized, it was hailed as a near-miracle substance. Because of its noncorrosive nature and its unusual ability to resist decomposition, Freon-12 was rapidly applied in refrigeration and air-conditioning systems, cleaning applications, the blowing of foams used for insulation and packing materials, and many other ways. For years everything seemed fine—the CFCs actually replaced more dangerous materials, such as the ammonia formerly used in refrigeration systems. The CFCs were definitely viewed as "good guys." But then a problem was discovered—the ozone in the upper atmosphere that protects us from the high-energy radiation of the sun began to decline. What was happening to cause the destruction of the vital ozone?

Much to everyone's amazement, the culprits turned out to be the seemingly beneficial CFCs. Inevitably, large quantities of CFCs had leaked into the atmosphere but nobody was very worried about this development because these compounds seemed totally benign. In fact, the great stability of the CFCs (a tremendous advantage for their various applications) was in the end a great disadvantage when they were released into the environment. Professor F. S. Rowland and his colleagues at the University of California at Irvine demonstrated that the CFCs eventually drifted to high altitudes in the atmosphere, where the energy of the sun stripped off chlorine atoms. These chlorine atoms in turn promoted the decomposition of the ozone in the upper atmosphere. (We will discuss this in more detail in Chapter 12.) Thus a substance that possessed many advantages in earth-bound applications turned against us in the atmosphere. Who could have guessed it would turn out this way?

The good news is that the U.S. chemical industry is leading the way to find environmentally safe alternatives to CFCs, and the levels of CFCs in the atmosphere are already dropping.

The saga of the CFCs demonstrates that we can respond relatively quickly to a serious environmental problem if we decide to do so. Also, it is important to understand that chemical manufacturers have a new attitude about the environment—they are now among the leaders in finding ways to address our environmental ills. The industries that apply the chemical sciences are now determined to be part of the solution rather than part of the problem.

A chemist in the laboratory.

As you can see, learning chemistry is both interesting and important. A chemistry course can do more than simply help you learn the principles of chemistry, however. A major by-product of your study of chemistry is that you will become a better problem solver. One reason chemistry has the reputation of being "tough" is that it often deals with rather complicated systems that require some effort to figure out. Although this might at first seem like a disadvantage, you can turn it to your advantage if you have the right attitude. Recruiters for companies of all types maintain that one of the first things they look for in a prospective employee is the ability to solve problems. We will spend a good deal of time solving various types of problems in this book by using a systematic, logical approach that will serve you well in solving any kind of problem in any field. Keep this broader goal in mind as you learn to solve the specific problems connected with chemistry.

Although learning chemistry is often not easy, it's never impossible. In fact, anyone who is interested, patient, and willing to work can learn the fundamentals of chemistry. In this book we will try very hard to help you

CHEMISTRY IN FOCUS

Dr. Ruth—Cotton Hero

Dr. Ruth Rogan Benerito may have saved the cotton industry in the United States. In the 1960s, synthetic fibers posed a serious competitive threat to cotton, primarily because of wrinkling. Synthetic fibers such as polyester can be formulated to be highly resistant to wrinkles both in the laundering process and in wearing. On the other hand, 1960s' cotton fabrics wrinkled easily—white cotton shirts had to be ironed to look good. This requirement put cotton at a serious disadvantage and endangered an industry very important to the economic health of the South.

During the 1960s Ruth Benerito worked as a scientist for the Department of Agriculture, where she was instrumental in developing the chemical treatment of cotton to make it wrinkle resistant. In so doing she enabled cotton to remain a preeminent fiber in the market—a place it continues to hold today.

Recently Dr. Benerito, who is now 86 years old and long retired, was honored with the Lemelson–MIT Lifetime Achievement Award for Inventions. Dr. Benerito, who holds 55 patents, including the one for wrinkle-free cotton awarded in 1969, began her career when women were not expected to enter scientific fields. However, her mother, who was an artist, adamantly encouraged her to be anything she wanted to be.

Dr. Benerito graduated from high school at 14 and attended Newcomb College, the women's college associated with Tulane University. She majored in chemistry with minors in physics and math. At that time she was one of only two women allowed to take the physical chemistry course at Tulane. She earned her B.S. degree in 1935 at age 19 and subsequently earned a master's degree at Tulane and a Ph.D. at the University of Chicago.

In 1953 Dr. Benerito began working in the Agriculture Department's Southern Regional Research Center in New Orleans, where she mainly worked on cotton and cotton-related products. She also invented a special method for intravenous feeding in long-term medical patients.

Since her retirement in 1986, she has continued to tutor science students to keep busy. Everyone who knows Dr. Benerito describes her as a class act.

Ruth Benerito, the inventor of easy-care cotton.

understand what chemistry is and how it works and to point out how chemistry applies to the things going on in your life.

Our sincere hope is that this text will motivate you to learn chemistry, make its concepts understandable to you, and demonstrate how interesting and vital the study of chemistry is.

What Is Chemistry?

AIM: To define chemistry.

Chemical and physical changes will be discussed in Chapter 3.

Chemistry can be defined as *the science that deals with the materials of the universe and the changes that these materials undergo.* Chemists are involved in activities as diverse as examining the fundamental particles of matter, looking for molecules in space, synthesizing and formulating new materials of all types, using bacteria to produce such chemicals as insulin, and inventing new diagnostic methods for early detection of disease.

Chemistry is often called the central science—and with good reason. Most of the phenomena that occur in the world around us involve chemical changes, changes where one or more substances become different substances. Here are some examples of chemical changes:

Wood burns in air, forming water, carbon dioxide, and other substances.

A plant grows by assembling simple substances into more complex substances.

The steel in a car rusts.

Eggs, flour, sugar, and baking powder are mixed and baked to yield a cake.

The definition of the term *chemistry* is learned and stored in the brain.

Grape juice ferments to form wine.

Emissions from a power plant lead to the formation of acid rain.

As we proceed, you will see how the concepts of chemistry allow us to understand the nature of these and other changes and thus help us manipulate natural materials to our benefit.

The launch of the space shuttle gives clear indications that chemical reactions are occurring.

Solving Problems Using a Scientific Approach

AIM: To understand scientific thinking.

One of the most important things we do in everyday life is solve problems. In fact, most of the decisions you make each day can be described as solving problems.

It's 8:30 A.M. on Friday. Which is the best way to drive to school to avoid traffic congestion?

You have two tests on Monday. Should you divide your study time equally or allot more time to one than to the other?

Your car stalls at a busy intersection and your little brother is with you. What should you do next?

These are everyday problems of the type we all face. What process do we use to solve them? You may not have thought about it before, but there are several steps that almost everyone uses to solve problems:

1. Recognize the problem and state it clearly. Some information becomes known, or something happens that requires action. In science we call this step *making an observation.*

CHEMISTRY IN FOCUS

A Mystifying Problem

To illustrate how science helps us solve problems, consider a true story about two people, David and Susan (not their real names). Several years ago David and Susan were healthy 40-year-olds living in California, where David was serving in the Air Force. Gradually Susan became quite ill, showing flu-like symptoms including nausea and severe muscle pains. Even her personality changed: she became uncharacteristically grumpy. She seemed like a totally different person from the healthy, happy woman of a few months earlier. Following her doctor's orders, she rested and drank a lot of fluids, including large quantities of coffee and orange juice from her favorite mug, part of a 200-piece set of pottery dishes recently purchased in Italy. However, she just got sicker, developing extreme abdominal cramps and severe anemia.

During this time David also became ill and exhibited symptoms much like Susan's: weight loss, excruciating pain in his back and arms, and uncharacteristic fits of temper. The disease became so debilitating that he retired early from the Air Force and the couple moved to Seattle. For a short time their health improved, but after they unpacked all their belongings (including those pottery dishes), their health began to deteriorate again. Susan's body became so sensitive that she could not tolerate the weight of a blanket. She was near death. What was wrong? The doctors didn't know, but one suggested she might have porphyria, a rare blood disease.

Desperate, David began to search the medical literature himself. One day while he was reading about porphyria, a phrase jumped off the page: "Lead poisoning can sometimes be confused with porphyria." Could the problem be lead poisoning?

We have described a very serious problem with life-or-death implications. What should David do next? Overlooking for a moment the obvious response of calling the couple's doctor immediately to discuss the possibility of lead poisoning, could David solve the problem via scientific thinking? Let's use the three steps described in Section 1.3 to attack the problem one part at a time. This is important: usually we solve complex problems by breaking them down into manageable parts. We can then assemble the solution to the overall problem from the answers we have found "piecemeal."

In this case there are many parts to the overall problem:

What is the disease?

Where is it coming from?

Can it be cured?

Let's attack "What is the disease?" first.

Observation: David and Susan are ill with the symptoms described. Is the disease lead poisoning?

Hypothesis: The disease is lead poisoning.

Experiment: If the disease is lead poisoning, the symptoms must match those known to characterize lead poisoning. Look up the symptoms of lead poisoning. David did this and found that they matched the couple's symptoms almost exactly.

This discovery points to lead poisoning as the source of their problem, but David needed more evidence.

Observation: Lead poisoning results from high levels of lead in the bloodstream.

Hypothesis: The couple have high levels of lead in their blood.

Experiment: Perform a blood analysis. Susan arranged for such an analysis, and the results

2. Propose *possible* solutions to the problem or *possible* explanations for the observation. In scientific language, suggesting such a possibility is called *formulating a hypothesis.*

3. Decide which of the solutions is the best or decide whether the explanation proposed is reasonable. To do this we search our memory for any pertinent information or we seek new information. In science we call searching for new information *performing an experiment.*

Italian pottery.

showed high lead levels for both David and Susan.

This confirms that lead poisoning is probably the cause of the trouble, but the overall problem is still not solved. David and Susan are likely to die unless they find out where the lead is coming from.

Observation: There is lead in the couple's blood.

Hypothesis: The lead is in their food or drink when they buy it.

Experiment: Find out whether anyone else who shopped at the same store was getting sick (no one was). Also note that moving to a new area did not solve the problem.

Observation: The food they buy is free of lead.

Hypothesis: The dishes they use are the source of the lead poisoning.

Experiment: Find out whether their dishes contain lead. David and Susan learned that lead compounds are often used to put a shiny finish on pottery objects. And laboratory analysis of their Italian pottery dishes showed that lead was present in the glaze.

Observation: Lead is present in their dishes, so the dishes are a possible source of their lead poisoning.

Hypothesis: The lead is leaching into their food.

Experiment: Place a beverage, such as orange juice, in one of the cups and then analyze the beverage for lead. The results showed high levels of lead in drinks that had had contact with the pottery cups.

After many applications of the scientific method, the problem is solved. We can summarize the answer to the problem (David and Susan's illness) as follows: the Italian pottery they used for everyday dishes contained a lead glaze that contaminated their food and drink with lead. This lead accumulated in their bodies to the point where it interfered seriously with normal functions and produced severe symptoms. This overall explanation, which summarizes the hypotheses that agree with the experimental results, is called a *theory* in science. This explanation accounts for the results of all the experiments performed.*

We could continue to use the scientific method to study other aspects of this problem, such as

What types of food or drink leach the most lead from the dishes?

Do all pottery dishes with lead glazes produce lead poisoning?

As we answer questions using the scientific method, other questions naturally arise. By repeating the three steps over and over, we can come to understand a given phenomenon thoroughly.

*"David" and "Susan" recovered from their lead poisoning and are now publicizing the dangers of using lead-glazed pottery. This happy outcome is the answer to the third part of their overall problem, "Can the disease be cured?" They simply stopped eating from that pottery!

As we will discover in the next section, scientists use these same procedures to study what happens in the world around us. The important point here is that scientific thinking can help you in all parts of your life. It's worthwhile to learn how to think scientifically—whether you want to be a scientist, an auto mechanic, a doctor, a politician, or a poet!

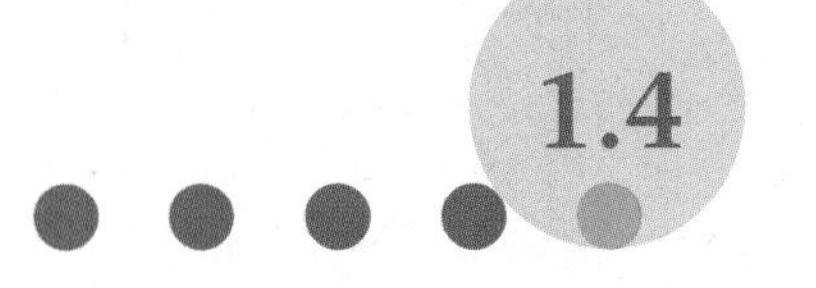

The Scientific Method

AIM: To describe the method scientists use to study nature.

In the last section we began to see how the methods of science are used to solve problems. In this section we will further examine this approach.

Science is a framework for gaining and organizing knowledge. Science is not simply a set of facts but also a plan of action—a *procedure* for processing and understanding certain types of information. Although scientific thinking is useful in all aspects of life, in this text we will use it to understand how the natural world operates. The process that lies at the center of scientific inquiry is called the **scientific method.** As we saw in the previous section, it consists of the following steps:

Quantitative observations involve a number. Qualitative ones do not.

Steps in the Scientific Method

1. *State the problem and collect data (make observations).* Observations may be *qualitative* (the sky is blue; water is a liquid) or *quantitative* (water boils at 100 °C; a certain chemistry book weighs 4.5 pounds). A qualitative observation does not involve a number. A quantitative observation is called a **measurement** and does involve a number (and a unit, such as pounds or inches). We will discuss measurements in detail in Chapter 2.
2. *Formulate hypotheses.* A hypothesis is a *possible* explanation for the observation.
3. *Perform experiments.* An experiment is something we do to test the hypothesis. We gather new information that allows us to decide whether the hypothesis is supported by the new information we have learned from the experiment. Experiments always produce new observations, and this brings us back to the beginning of the process again.

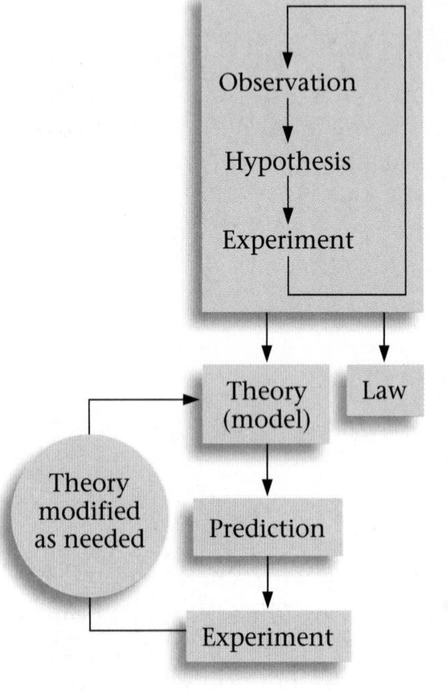

Figure 1.1
The various parts of the scientific method.

To explain the behavior of a given part of nature, we repeat these steps many times. Gradually we accumulate the knowledge necessary to understand what is going on.

Once we have a set of hypotheses that agrees with our various observations, we assemble them into a theory that is often called a *model.* A **theory** (model) is a set of tested hypotheses that gives an overall explanation of some part of nature (see Figure 1.1).

It is important to distinguish between observations and theories. An observation is something that is witnessed and can be recorded. A theory is an *interpretation*—a possible explanation of *why* nature behaves in a particular way. Theories inevitably change as more information becomes available. For example, the motions of the sun and stars have remained virtually the same over the thousands of years during which humans have been observing them, but our explanations—our theories—have changed greatly since ancient times.

The point is that we don't stop asking questions just because we have devised a theory that seems to account satisfactorily for some aspect of natural behavior. We continue doing experiments to refine our theories. We generally do this by using the theory to make a prediction and then doing an experiment (making a new observation) to see whether the results bear out this prediction.

Law: A summary of observed behavior.
Theory: An explanation of behavior.

Always remember that theories (models) are human inventions. They represent our attempts to explain observed natural behavior in terms of our human experiences. We must continue to do experiments and refine our theories to be consistent with new knowledge if we hope to approach a more nearly complete understanding of nature.

As we observe nature, we often see that the same observation applies to many different systems. For example, studies of innumerable chemical changes have shown that the total mass of the materials involved is the same before and after the change. We often formulate such generally observed behavior into a statement called a **natural law.** The observation that the total mass of materials is not affected by a chemical change in those materials is called the law of conservation of mass.

You must recognize the difference between a law and a theory. A law is a summary of observed (measurable) behavior, whereas a theory is an explanation of behavior. *A law tells what happens; a theory (model) is our attempt to explain why it happens.*

In this section, we have described the scientific method (which is summarized in Figure 1.1) as it might ideally be applied. However, it is important to remember that science does not always progress smoothly and efficiently. Scientists are human. They have prejudices; they misinterpret data; they can become emotionally attached to their theories and thus lose objectivity; and they play politics. Science is affected by profit motives, budgets, fads, wars, and religious beliefs. Galileo, for example, was forced to recant his astronomical observations in the face of strong religious resistance. Lavoisier, the father of modern chemistry, was beheaded because of his political affiliations. And great progress in the chemistry of nitrogen fertilizers resulted from the desire to produce explosives to fight wars. The progress of science is often slowed more by the frailties of humans and their institutions than by the limitations of scientific measuring devices. The scientific method is only as effective as the humans using it. It does not automatically lead to progress.

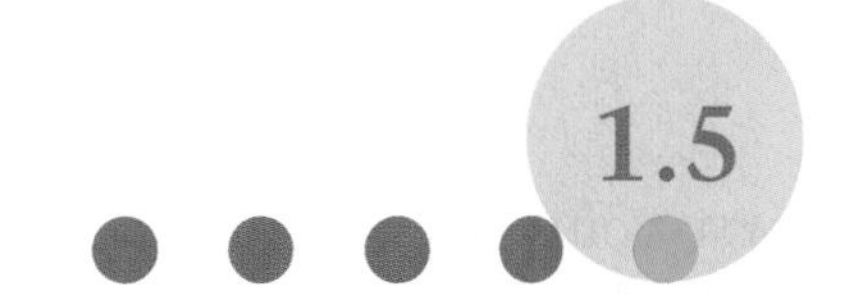

Learning Chemistry

AIM: To develop successful strategies for learning chemistry.

Chemistry courses have a universal reputation for being difficult. There are some good reasons for this. For one thing, the language of chemistry is unfamiliar in the beginning; many terms and definitions need to be memorized. As with any language, *you must know the vocabulary* before you can communicate effectively. We will try to help you by pointing out those things that need to be memorized.

But memorization is only the beginning. Don't stop there or your experience with chemistry will be frustrating. Be willing to do some thinking, and learn to trust yourself to figure things out. To solve a typical chemistry

CHEMISTRY IN FOCUS

Chemistry: An Important Component of Your Education

What is the purpose of education? Because you are spending considerable time, energy, and money to pursue an education, this is an important question.

Some people seem to equate education with the storage of facts in the brain. These people apparently believe that education simply means memorizing the answers to all of life's present and future problems. Although this is clearly unreasonable, many students seem to behave as though this were their guiding principle. These students want to memorize lists of facts and to reproduce them on tests. They regard as unfair any exam questions that require some original thought or some processing of information. Indeed, it might be tempting to reduce education to a simple filling up with facts, because that approach can produce short-term satisfaction for both student and teacher. And of course, storing facts in the brain *is* important. You cannot function without knowing that red means stop, electricity is hazardous, ice is slippery, and so on.

However, mere recall of abstract information, without the ability to process it, makes you little better than a talking encyclopedia. Former students always seem to bring the same message when they return to campus. The characteristics that are most important to their success are a knowledge of the fundamentals of their fields, the ability to recognize and solve problems, and the ability to communicate effectively. They also emphasize the importance of a high level of motivation.

How does studying chemistry help you achieve these characteristics? The fact that chemical systems are complicated is really a blessing, though one that is well disguised. Studying chemistry will not by itself make you a good problem solver, but it can help you develop a positive, aggressive attitude toward problem solving and can help boost your confidence. Learning to "think like a chemist" can be valuable to anyone in any field. In fact, the chemical industry is heavily populated at all levels and in all areas by chemists and chemical engineers. People who were trained as chemical professionals often excel not only in chemical research and production but also in the areas of personnel, marketing, sales, development, finance, and management. The point is that much of what you learn in this course can be applied to any field of endeavor. So be careful not to take too narrow a view of this course. Try to look beyond short-term frustration to long-term benefits. It may not be easy to learn to be a good problem solver, but it's well worth the effort.

Students pondering the structure of a molecule.

problem, you must sort through the given information and decide what is really crucial.

It is important to realize that chemical systems tend to be complicated—there are typically many components—and we must make approximations in describing them. Therefore, trial and error play a major role in solving chemical problems. In tackling a complicated system, a practicing chemist really does not expect to be right the first time he or she analyzes the problem. The usual practice is to make several simplifying assumptions and then give it a try. If the answer obtained doesn't make sense, the chemist adjusts the assumptions, using feedback from the first attempt, and tries again. The point is this: in dealing with chemical systems, do not expect to understand immediately everything that is going on. In fact, it is typical (even for an experienced chemist) *not* to understand at first. Make an attempt to solve the problem and then analyze the feedback. *It is no disaster to make a mistake as long as you learn from it.*

The only way to develop your confidence as a problem solver is to practice solving problems. To help you, this book contains examples worked out in detail. Follow these through carefully, making sure you understand each step. These examples are usually followed by a similar exercise (called a self-check exercise) that you should try on your own (detailed solutions of the self-check exercises are given at the end of each chapter). Use the self-check exercises to test whether you are understanding the material as you go along.

There are questions and problems at the end of each chapter. The questions review the basic concepts of the chapter and give you an opportunity to check whether you properly understand the vocabulary introduced. Some of the problems are really just exercises that are very similar to examples done in the chapter. If you understand the material in the chapter, you should be able to do these exercises in a straightforward way. Other problems require more creativity. These contain a knowledge gap—some unfamiliar territory that you must cross—and call for thought and patience on your part. For this course to be really useful to you, it is important to go beyond the questions and exercises. Life offers us many exercises, routine events that we deal with rather automatically, but the real challenges in life are true problems. This course can help you become a more creative problem solver.

As you do homework, be sure to use the problems correctly. If you cannot do a particular problem, do not immediately look at the solution. Review the relevant material in the text and then try the problem again. Don't be afraid to struggle with a problem. Looking at the solution as soon as you get stuck short-circuits the learning process.

Learning chemistry takes time. Use all the resources available to you and study on a regular basis. Don't expect too much of yourself too soon. You may not understand everything at first, and you may not be able to do many of the problems the first time you try them. This is normal. It doesn't mean you can't learn chemistry. Just remember to keep working and to keep learning from your mistakes, and you will make steady progress.

CHAPTER 1 REVIEW

IN-CLASS DISCUSSION QUESTIONS

These questions are designed to be considered by groups of students in class. Often these questions work well for introducing a particular topic in class.

1. Discuss how a hypothesis can become a theory. Can a theory become a law? Explain.
2. Make five qualitative and five quantitative observations about the room in which you now sit.
3. List as many chemical reactions you can think of that are part of your everyday life. Explain.
4. Differentiate between a "theory" and a "scientific theory."
5. Describe three situations when you used the scientific method (outside of school) in the past month.
6. Scientific models do not describe reality. They are simplifications and therefore incorrect at some level. So why are models useful?

7. Theories should inspire questions. Discuss a scientific theory you know and the questions it brings up.
8. Describe how you would set up an experiment to test the relationship between completion of assigned homework and the final grade you receive in the course.
9. If all scientists use the scientific method to try to arrive at a better understanding of the world, why do so many debates arise among scientists?

Questions and Problems

All even-numbered exercises have answers in the back of this book and solutions in the *Solutions Guide.*

1.1 Chemistry: An Introduction

QUESTIONS

1. Chemistry is an intimidating academic subject for many students. You are not alone if you are afraid of not doing well in this course! Why do you suppose the study of chemistry is so intimidating for many students? What about having to take a chemistry course bothers you? Make a list of your concerns and bring them to class for discussion with your fellow students and your instructor.
2. The first paragraphs in this chapter ask you if you have ever wondered how and why various things in our everyday lives happen the way they do. For your next class meeting, make a list of five similar chemistry-related things for discussion with your instructor and the other students in your class.
3. This section presents several ways our day-to-day lives have been enriched by chemistry. List three materials or processes involving chemistry that you feel have contributed to such an enrichment, and explain your choices.
4. The text admits that there has also been a "dark side" to our use of chemicals and chemical processes, and uses the example of chlorofluorocarbons (CFCs) to explain this. List three additional improper or unfortunate uses of chemicals or chemical processes, and explain your reasoning.

1.2 What Is Chemistry?

QUESTIONS

5. This textbook provides a specific definition of chemistry: the study of the materials of which the universe is made and the transformations that these materials undergo. Obviously, such a general definition has to be very broad and nonspecific. From your point of view at this time, how would *you* define chemistry? In your mind, what are "chemicals"? What do "chemists" do?
6. We use chemical reactions in our everyday lives, too, not just in the science laboratory. Give at least five examples of chemical transformations that you use in your daily activities. Indicate what the "chemical" is in each of your examples and how you recognize that a chemical change has taken place.

1.3 Solving Problems Using a Scientific Approach

QUESTIONS

7. For the "Chemistry in Focus" discussion of lead poisoning given in this section, discuss how David and Susan analyzed the situation, arriving at the theory that the lead glaze on the pottery was responsible for their symptoms.
8. Being a scientist is very much like being a detective. Detectives such as Sherlock Holmes or Miss Marple perform a very systematic analysis of a crime to solve it, much like a scientist does when addressing a scientific investigation. What are the steps that scientists (or detectives) use to solve problems?

1.4 The Scientific Method

QUESTIONS

9. What are the three operations involved in applying the scientific method? How does the scientific method help us to understand our observations of nature?
10. Which of the following observations are *qualitative* and which are *quantitative?*
    a. I wear a size 7¾ hat.
    b. My favorite shirt is blue.
    c. Robins like to eat worms after a rainstorm.
    d. The weather bureau reported that hailstones measuring half an inch in diameter fell during last night's storm.
    e. Washington and Baltimore are two very interesting cities to visit.
    f. Washington and Baltimore are only 40 miles apart.
11. Several words are used in this section which students sometimes may find hard to distinguish. Write your *own* definitions of the following terms, and bring them to class for discussion with your instructor and fellow students: *theory, experiment, natural law, hypothesis.*
12. What is a *natural law?* Give examples of such laws. How does a law differ from a theory?
13. Although science *should* lead to solutions to problems that are completely independent of outside forces, very often in history scientific investigations have been influenced by prejudice, profit motives, fads, wars, religious beliefs, and other forces. Your textbook mentions the case of Galileo having to change his theories about astronomy based on intervention by religious authorities. Can you give three additional examples of how scientific investigations have been similarly influenced by nonscientific forces?

1.5 Learning Chemistry

QUESTIONS

14. In some academic subjects, it may be possible to receive a good grade primarily by memorizing facts. Why is chemistry not one of these subjects?

15. Why is the ability to solve problems important in the study of chemistry? Why is it that the *method* used to attack a problem is as important as the answer to the problem itself?

16. Students approaching the study of chemistry must learn certain basic facts (such as the names and symbols of the most common elements), but it is much more important that they learn to think critically and to go beyond the specific examples discussed in class or in the textbook. Explain how learning to do this might be helpful in any career, even one far removed from chemistry.

2 Measurements and Calculations

A variety of chemical glassware.

As we pointed out in Chapter 1, making observations is a key part of the scientific process. Sometimes observations are *qualitative* ("the substance is a yellow solid") and sometimes they are *quantitative* ("the substance weighs 4.3 grams"). A quantitative observation is called a **measurement.** Measurements are very important in our daily lives. For example, we pay for gasoline by the gallon so the gas pump must accurately measure the gas delivered to our fuel tank. The efficiency of the modern automobile engine depends on various measurements, including the amount of oxygen in the exhaust gases, the temperature of the coolant, and the pressure of the lubricating oil. In addition, cars with traction control systems have devices to measure and compare the rates of rotation of all four wheels. As we will see in the "Chemistry in Focus" discussion in this chapter, measuring devices have become very sophisticated in dealing with our fast-moving and complicated society.

As we will discuss in this chapter, a measurement always consists of two parts: a number and a unit. Both parts are necessary to make the measurement meaningful. For example, suppose a friend tells you that she saw a bug 5 long. This statement is meaningless as it stands. Five what? If it's 5 millimeters, the bug is quite small. If it's 5 centimeters, the bug is quite large. If it's 5 meters, run for cover!

The point is that for a measurement to be meaningful, it must consist of both a number and a unit that tells us the scale being used.

In this chapter we will consider the characteristics of measurements and the calculations that involve measurements.

A gas pump measures the amount of gasoline delivered.

Scientific Notation

AIM: To show how very large or very small numbers can be expressed as the product of a number between 1 and 10 and a power of 10.

A measurement must always consist of a number *and* a unit.

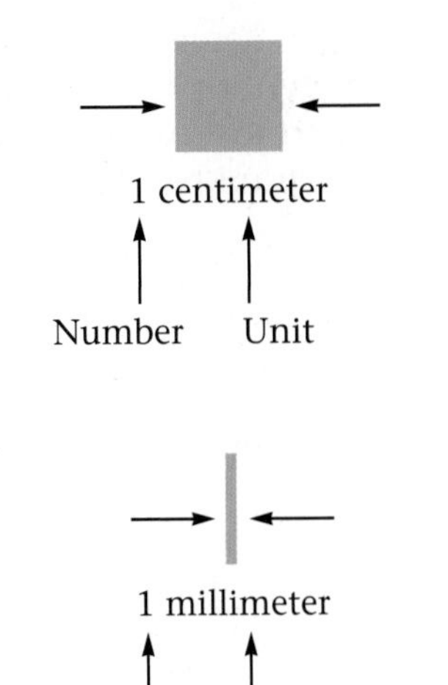

The numbers associated with scientific measurements are often very large or very small. For example, the distance from the earth to the sun is approximately 93,000,000 (93 million) miles. Written out, this number is rather bulky. Scientific notation is a method for making very large or very small numbers more compact and easier to write.

To see how this is done, consider the number 125, which can be written as the product

$$125 = 1.25 \times 100$$

Because $100 = 10 \times 10 = 10^2$, we can write

$$125 = 1.25 \times 100 = 1.25 \times 10^2$$

Similarly, the number 1700 can be written

$$1700 = 1.7 \times 1000$$

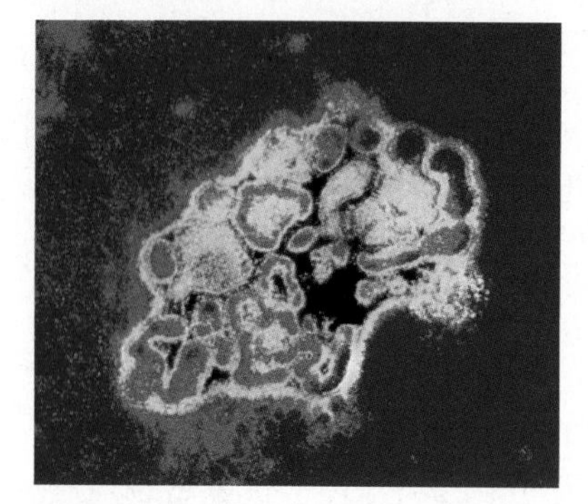

When describing very small distances, such as the diameter of a swine flu virus (shown here magnified 16,537 times), it is convenient to use scientific notation.

and because $1000 = 10 \times 10 \times 10 = 10^3$, we can write

$$1700 = 1.7 \times 1000 = 1.7 \times 10^3$$

Scientific notation simply expresses a number as *a product of a number between 1 and 10 and the appropriate power of 10.* For example, the number 93,000,000 can be expressed as

$$93{,}000{,}000 = 9.3 \times 10{,}000{,}000 = \underbrace{9.3}_{\text{Number between 1 and 10}} \times \underbrace{10^7}_{\text{Appropriate power of 10 } (10{,}000{,}000 = 10^7)}$$

The easiest way to determine the appropriate power of 10 for scientific notation is to start with the number being represented and count the number of places the decimal point must be moved to obtain a number between 1 and 10. For example, for the number

$$9\ 3\ 0\ 0\ 0\ 0\ 0\ 0$$
7 6 5 4 3 2 1

Keep one digit to the left of the decimal point.

we must move the decimal point seven places to the left to get 9.3 (a number between 1 and 10). To compensate for every move of the decimal point to the left, we must multiply by 10. That is, each time we move the decimal point to the left, we make the number smaller by one power of 10. So for each move of the decimal point to the left, we must multiply by 10 to restore the number to its original magnitude. Thus moving the decimal point seven places to the left means we must multiply 9.3 by 10 seven times, which equals 10^7:

$$93{,}000{,}000 = 9.3 \times 10^7$$

We moved the decimal point seven places to the left, so we need 10^7 to compensate.

Moving the decimal point to the left requires a positive exponent.

Remember: whenever the decimal point is moved to the *left,* the exponent of 10 is *positive.*

We can represent numbers smaller than 1 by using the same convention, but in this case the power of 10 is negative. For example, for the number 0.010 we must move the decimal point two places to the right to obtain a number between 1 and 10:

$$0\,.\,0\ 1\ 0$$
1 2

Moving the decimal point to the right requires a negative exponent.

This requires an exponent of −2, so $0.010 = 1.0 \times 10^{-2}$. Remember: whenever the decimal point is moved to the *right,* the exponent of 10 is *negative.*

Next consider the number 0.000167. In this case we must move the decimal point four places to the right to obtain 1.67 (a number between 1 and 10):

$$0\,.\,0\ 0\ 0\ 1\ 6\ 7$$
1 2 3 4

Read the Appendix if you need a further discussion of exponents and scientific notation.

Moving the decimal point four places to the right requires an exponent of −4. Therefore,

$$0.000167 = 1.67 \times 10^{-4}$$

We moved the decimal point four places to the right.

We summarize these procedures below.

$100 = 1.0 \times 10^2$
$0.010 = 1.0 \times 10^{-2}$

Left Is Positive; remember LIP.

Using Scientific Notation

- Any number can be represented as the product of a number between 1 and 10 and a power of 10 (either positive or negative).
- The power of 10 depends on the number of places the decimal point is moved and in which direction. The *number of places* the decimal point is moved determines the *power of 10.* The *direction* of the move determines whether the power of 10 is *positive* or *negative.* If the decimal point is moved to the left, the power of 10 is positive; if the decimal point is moved to the right, the power of 10 is negative.

Example 2.1 Scientific Notation: Powers of 10 (Positive)

Represent the following numbers in scientific notation.

a. 238,000

b. 1,500,000

Solution

a. First we move the decimal point until we have a number between 1 and 10, in this case 2.38.

2 3 8 0 0 0
5 4 3 2 1 The decimal point was moved five places to the left.

Because we moved the decimal point five places to the left, the power of 10 is positive 5. Thus $238{,}000 = 2.38 \times 10^5$.

b. 1 5 0 0 0 0 0
6 5 4 3 2 1 The decimal point was moved six places to the left, so the power of 10 is 6.

Thus $1{,}500{,}000 = 1.5 \times 10^6$. ■

Example 2.2 Scientific Notation: Powers of 10 (Negative)

Represent the following numbers in scientific notation.

a. 0.00043

b. 0.089

Solution

a. First we move the decimal point until we have a number between 1 and 10, in this case 4.3.

0 . 0 0 0 4 3
1 2 3 4 The decimal point was moved four places to the right.

Because we moved the decimal point four places to the right, the power of 10 is negative 4. Thus $0.00043 = 4.3 \times 10^{-4}$.

b. 0.089

1 2 The power of 10 is negative 2 because the decimal point was moved two places to the right.

Thus $0.089 = 8.9 \times 10^{-2}$.

✔ Self-Check Exercise 2.1

Write the numbers 357 and 0.0055 in scientific notation. If you are having difficulty with scientific notation at this point, reread the Appendix.

See Problems 2.7 through 2.12. ■

2.2 Units

AIM: To learn the English, metric, and SI systems of measurement.

The **units** part of a measurement tells us what *scale* or *standard* is being used to represent the results of the measurement. From the earliest days of civilization, trade has required common units. For example, if a farmer from one region wanted to trade some of his grain for the gold of a miner who lived in another region, the two people had to have common standards (units) for measuring the amount of the grain and the weight of the gold.

The need for common units also applies to scientists, who measure quantities such as mass, length, time, and temperature. If every scientist had her or his own personal set of units, complete chaos would result. Unfortunately, although standard systems of units did arise, different systems were adopted in different parts of the world. The two most widely used systems are the **English system** used in the United States and the **metric system** used in most of the rest of the industrialized world.

The metric system has long been preferred for most scientific work. In 1960 an international agreement set up a comprehensive system of units called the **International System** (*le Système Internationale* in French), or **SI.** The SI units are based on the metric system and units derived from the metric system. The most important fundamental SI units are listed in Table 2.1. Later in this chapter we will discuss how to manipulate some of these units.

Because the fundamental units are not always a convenient size, the SI system uses prefixes to change the size of the unit. The most commonly used prefixes are listed in Table 2.2. Although the fundamental unit for length is the meter (m), we can also use the decimeter (dm), which represents one-tenth (0.1) of a meter; the centimeter (cm), which represents one one-hundredth (0.01) of a meter; the millimeter (mm), which represents one one-thousandth (0.001) of a meter; and so on. For example, it's much more convenient to specify the diameter of a certain contact lens as 1.0 cm than as 1.0×10^{-2} m.

Table 2.1 Some Fundamental SI Units

Physical Quantity	Name of Unit	Abbreviation
mass	kilogram	kg
length	meter	m
time	second	s
temperature	kelvin	K

CHEMISTRY IN FOCUS

Critical Units!

How important are conversions from one unit to another? If you ask the National Aeronautic and Space Administration (NASA), very important! In 1999 NASA lost a $125 million Mars Climate Orbiter because of a failure to convert from English to metric units.

The problem arose because two teams working on the Mars mission were using different sets of units. NASA's scientists at the Jet Propulsion Laboratory in Pasadena, California, assumed that the thrust data for the rockets on the Orbiter they received from Lockheed Martin Astronautics in Denver, which built the spacecraft, were in metric units. In reality, the units were English. As a result the Orbiter dipped 100 kilometers lower into the Mars atmosphere than planned and the friction from the atmosphere caused the craft to burn up.

NASA's mistake refueled the controversy over whether Congress should require the United States to switch to the metric system. About 95% of the world now uses the metric system, and the United States is slowly switching from English to metric. For example, the automobile industry has adopted metric fasteners and we buy our soda in two-liter bottles.

Units can be very important. In fact, they can mean the difference between life and death on some occasions. In 1983, for example, a Canadian jetliner almost ran out of fuel when someone pumped 22,300 pounds of fuel into the aircraft instead of 22,300 kilograms. Remember to watch your units!

Artist's conception of the lost Mars Climate Orbiter.

Table 2.2 The Commonly Used Prefixes in the Metric System

Prefix	Symbol	Meaning	Power of 10 for Scientific Notation
mega	M	1,000,000	10^6
kilo	k	1000	10^3
deci	d	0.1	10^{-1}
centi	c	0.01	10^{-2}
milli	m	0.001	10^{-3}
micro	μ	0.000001	10^{-6}
nano	n	0.000000001	10^{-9}

2.3 Measurements of Length, Volume, and Mass

AIM: To understand the metric system for measuring length, volume, and mass.

The fundamental SI unit of length is the **meter,** which is a little longer than a yard (1 meter = 39.37 inches). In the metric system fractions of a

Table 2.3 The Metric System for Measuring Length

Unit	Symbol	Meter Equivalent
kilometer	km	1000 m or 10^3 m
meter	m	1 m or 1 m
decimeter	dm	0.1 m or 10^{-1} m
centimeter	cm	0.01 m or 10^{-2} m
millimeter	mm	0.001 m or 10^{-3} m
micrometer	μm	0.000001 m or 10^{-6} m
nanometer	nm	0.000000001 m or 10^{-9} m

The meter was originally defined, in the eighteenth century, as one ten-millionth of the distance from the equator to the North Pole and then, in the late nineteenth century, as the distance between two parallel marks on a special metal bar stored in a vault in Paris. More recently, for accuracy and convenience, a definition expressed in terms of light waves has been adopted.

meter or multiples of a meter can be expressed by powers of 10, as summarized in Table 2.3.

The English and metric systems are compared on the ruler shown in Figure 2.1. Note that

$$1 \text{ inch} = 2.54 \text{ centimeters}$$

Other English–metric equivalences are given in Section 2.6.

Volume is the amount of three-dimensional space occupied by a substance. The fundamental unit of volume in the SI system is based on the volume of a cube that measures 1 meter in each of the three directions. That is, each edge of the cube is 1 meter in length. The volume of this cube is

$$1 \text{ m} \times 1 \text{ m} \times 1 \text{ m} = (1 \text{ m})^3 = 1 \text{ m}^3$$

or, in words, one cubic meter.

In Figure 2.2 this cube is divided into 1000 smaller cubes. Each of these small cubes represents a volume of 1 dm^3, which is commonly called the **liter** (rhymes with meter and is slightly larger than a quart) and abbreviated L.

The cube with a volume of 1 dm^3 (1 liter) can in turn be broken into 1000 smaller cubes each representing a volume of 1 cm^3. This means that each liter contains 1000 cm^3. One cubic centimeter is called a **milliliter** (abbreviated mL), a unit of volume used very commonly in chemistry. This relationship is summarized in Table 2.4.

The *graduated cylinder* (see Figure 2.3), commonly used in chemical laboratories for measuring the volumes of liquids, is marked off in convenient units of volume (usually milliliters). The graduated cylinder is filled to the desired volume with the liquid, which then can be poured out.

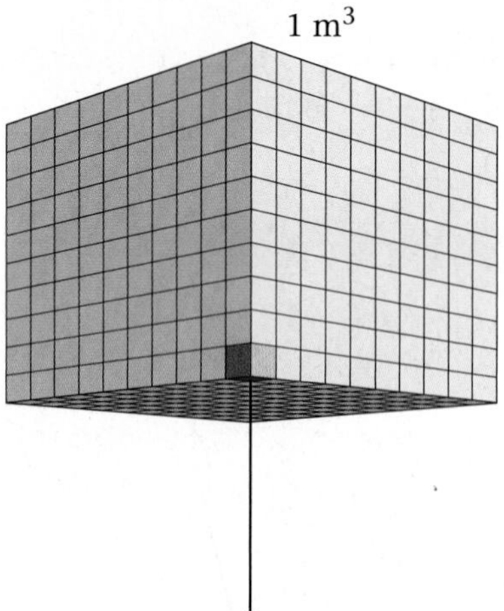

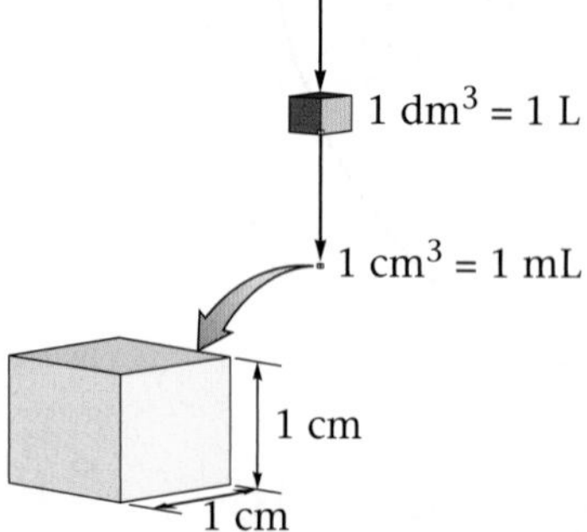

Figure 2.2
The largest drawing represents a cube that has sides 1 m in length and a volume of 1 m^3. The middle-size cube has sides 1 dm in length and a volume of 1 dm^3, or 1 L. The smallest cube has sides 1 cm in length and a volume of 1 cm^3, or 1 mL.

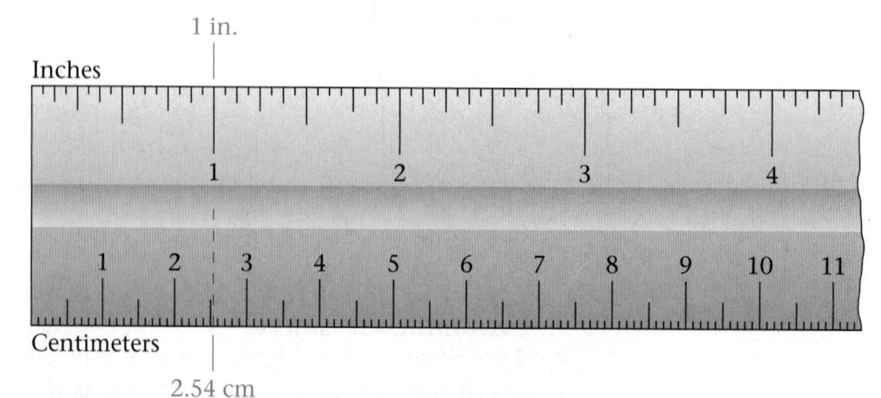

Figure 2.1
Comparison of English and metric units for length on a ruler.

CHEMISTRY IN FOCUS

Measurement: Past, Present, and Future

Measurement lies at the heart of doing science. We obtain the data for formulating laws and testing theories by doing measurements. Measurements also have very practical importance; they tell us if our drinking water is safe, whether we are anemic, and the exact amount of gasoline we put in our cars at the filling station.

Although the fundamental measuring devices we consider in this chapter are still widely used, new measuring techniques are being developed every day to meet the challenges of our increasingly sophisticated world. For example, engines in modern automobiles have oxygen sensors that analyze the oxygen content in the exhaust gases. This information is sent to the computer that controls the engine functions so that instantaneous adjustments can be made in spark timing and air–fuel mixtures to provide efficient power with minimum air pollution.

As another example, consider airline safety: How do we rapidly, conveniently, and accurately determine whether a given piece of baggage contains an explosive device? A thorough hand-search of each piece of luggage is out of the question. Scientists are now developing a screening procedure that bombards the luggage with high-energy particles that cause any substance present to emit radiation characteristic of that substance. This radiation is monitored to identify luggage with unusually large quantities of nitrogen, because most chemical explosives are based on compounds containing nitrogen.

A pollution control officer measuring the oxygen content of river water.

Scientists are also examining the natural world to find supersensitive detectors because many organisms are sensitive to tiny amounts of chemicals in their environments—recall, for example, the sensitive noses of bloodhounds. One of these natural measuring devices uses the sensory hairs from Hawaiian red swimming crabs, which are connected to electrical analyzers and used to detect hormones down to levels of 10^{-8} g/L. Likewise, tissues from pineapple cores can be used to detect tiny amounts of hydrogen peroxide.

These types of advances in measuring devices have led to an unexpected problem: detecting all kinds of substances in our food and drinking water scares us. Although these substances were always there, we didn't worry so much when we couldn't detect them. Now that we know they are present what should we do about them? How can we assess whether these trace substances are harmful or benign? Risk assessment has become much more complicated as our sophistication in taking measurements has increased.

mL 100 90 80 70 60 50 40 30 20 10

Figure 2.3
A 100-mL graduated cylinder.

Table 2.4 The Relationship of the Liter and Milliliter

Unit	Symbol	Equivalence
liter	L	1 L = 1000 mL
milliliter	mL	$\frac{1}{1000}$ L = 10^{-3} L = 1 mL

Another important measurable quantity is **mass,** which can be defined as the quantity of matter present in an object. The fundamental SI unit of mass is the **kilogram.** Because the metric system, which existed before the SI system, used the gram as the fundamental unit, the prefixes for the various mass units are based on the **gram,** as shown in Table 2.5.

In the laboratory we determine the mass of an object by using a balance. A balance compares the mass of the object to a set of standard masses

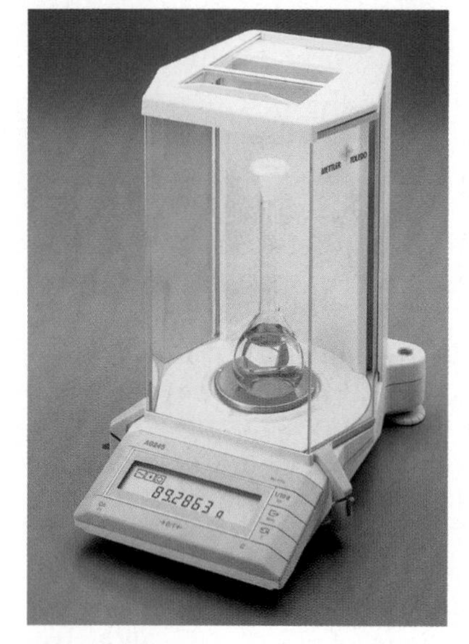

Figure 2.4
An electronic analytical balance used in chemistry labs.

Table 2.5 The Most Commonly Used Metric Units for Mass

Unit	Symbol	Gram Equivalent
kilogram	kg	1000 g $= 10^3$ g = 1 kg
gram	g	1 g
milligram	mg	0.001 g $= 10^{-3}$ g = 1 mg

Table 2.6 Some Examples of Commonly Used Units

length	A dime is 1 mm thick. A quarter is 2.5 cm in diameter. The average height of an adult man is 1.8 m.
mass	A nickel has a mass of about 5 g. A 120-lb woman has a mass of about 55 kg.
volume	A 12-oz can of soda has a volume of about 360 mL. A half gallon of milk is equal to about 2 L of milk.

("weights"). For example, the mass of an object can be determined by using a single-pan balance (Figure 2.4).

To help you get a feeling for the common units of length, volume, and mass, some familiar objects are described in Table 2.6.

Uncertainty in Measurement

AIMS: To understand how uncertainty in a measurement arises. To learn to indicate a measurement's uncertainty by using significant figures.

A student performing a titration in the laboratory.

Whenever a measurement is made with a device such as a ruler or a graduated cylinder, an estimate is required. We can illustrate this by measuring the pin shown in Figure 2.5a. We can see from the ruler that the pin is a little longer than 2.8 cm and a little shorter than 2.9 cm. Because there are no graduations on the ruler between 2.8 and 2.9, we must estimate the pin's length between 2.8 and 2.9 cm. We do this by *imagining* that the distance between 2.8 and 2.9 is broken into 10 equal divisions (Figure 2.5b) and estimating to which division the end of the pin reaches. The end of the pin appears to come about halfway between 2.8 and 2.9, which corresponds to 5 of our 10 imaginary divisions. So we estimate the pin's length as 2.85 cm. The result of our measurement is that the pin is approximately 2.85 cm in length, but we had to rely on a visual estimate, so it might actually be 2.84 or 2.86 cm.

Because the last number is based on a visual estimate, it may be different when another person makes the same measurement. For example, if five different people measured the pin, the results might be

Person	Result of Measurement
1	2.85 cm
2	2.84 cm
3	2.86 cm
4	2.85 cm
5	2.86 cm

Figure 2.5
Measuring a pin. (a) The length is between 2.8 cm and 2.9 cm. (b) Imagine that the distance between 2.8 and 2.9 is divided into 10 equal parts. The end of the pin occurs after about 5 of these divisions.

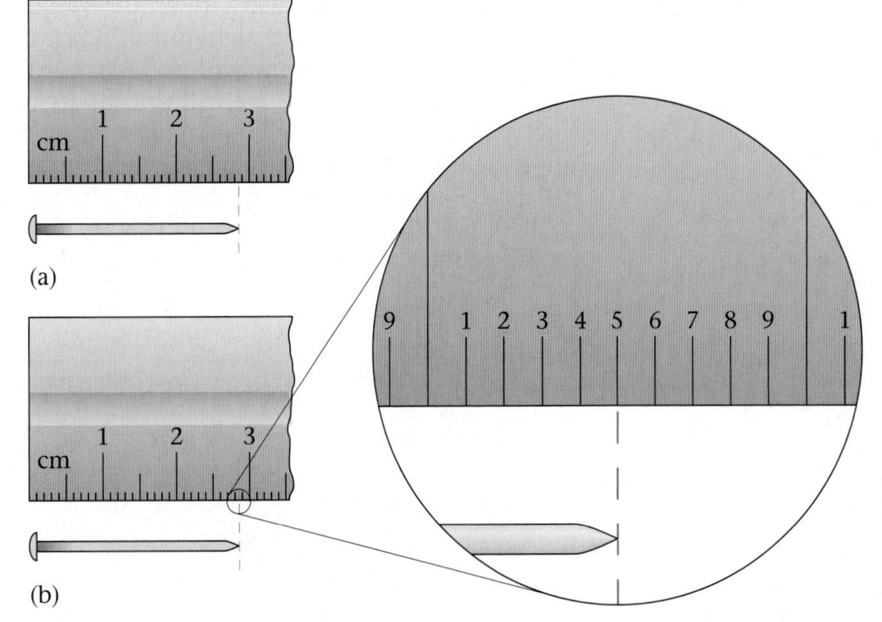

Note that the first two digits in each measurement are the same regardless of who made the measurement; these are called the *certain* numbers of the measurement. However, the third digit is estimated and can vary; it is called an *uncertain* number. When one is making a measurement, the custom is to record all of the certain numbers plus the *first* uncertain number. It would not make any sense to try to measure the pin to the third decimal place (thousandths of a centimeter), because this ruler requires an estimate of even the second decimal place (hundredths of a centimeter).

Every measurement has some degree of uncertainty.

It is very important to realize that *a measurement always has some degree of uncertainty.* The uncertainty of a measurement depends on the measuring device. For example, if the ruler in Figure 2.5 had marks indicating hundredths of a centimeter, the uncertainty in the measurement of the pin would occur in the thousandths place rather than the hundredths place, but some uncertainty would still exist.

The numbers recorded in a measurement (all the certain numbers plus the first uncertain number) are called **significant figures.** The number of significant figures for a given measurement is determined by the inherent uncertainty of the measuring device. For example, the ruler used to measure the pin can give results only to hundredths of a centimeter. Thus, when we record the significant figures for a measurement, we automatically give information about the uncertainty in a measurement. The uncertainty in the last number (the estimated number) is usually assumed to be ± 1 unless otherwise indicated. For example, the measurement 1.86 kilograms can be interpreted as 1.86 ± 0.01 kilograms, where the symbol $\pm$ means plus or minus. That is, it could be 1.86 kg − 0.01 kg = 1.85 kg or 1.86 kg + 0.01 kg = 1.87 kg.

2.5 Significant Figures

AIM: To learn to determine the number of significant figures in a calculated result.

We have seen that any measurement involves an estimate and thus is uncertain to some extent. We signify the degree of certainty for a particular measurement by the number of significant figures we record.

Because doing chemistry requires many types of calculations, we must consider what happens when we do arithmetic with numbers that contain uncertainties. It is important that we know the degree of uncertainty in the final result. Although we will not discuss the process here, mathematicians have studied how uncertainty accumulates and have designed a set of rules to determine how many significant figures the result of a calculation should have. You should follow these rules whenever you carry out a calculation. The first thing we need to do is learn how to count the significant figures in a given number. To do this we use the following rules:

Rules for Counting Significant Figures

1. *Nonzero integers.* Nonzero integers *always* count as significant figures. For example, the number 1457 has four nonzero integers, all of which count as significant figures.

2. *Zeros.* There are three classes of zeros:
   a. *Leading zeros* are zeros that *precede* all of the nonzero digits. They *never* count as significant figures. For example, in the number 0.0025, the three zeros simply indicate the position of the decimal point. The number has only two significant figures, the 2 and the 5.
   b. *Captive zeros* are zeros that fall *between* nonzero digits. They *always* count as significant figures. For example, the number 1.008 has four significant figures.
   c. *Trailing zeros* are zeros at the *right end* of the number. They are significant only if the number is written with a decimal point. The number one hundred written as 100 has only one significant figure, but written as 100., it has three significant figures.

3. *Exact numbers.* Often calculations involve numbers that were not obtained using measuring devices but were determined by counting: 10 experiments, 3 apples, 8 molecules. Such numbers are called *exact numbers.* They can be assumed to have an unlimited number of significant figures. Exact numbers can also arise from definitions. For example, 1 inch is defined as *exactly* 2.54 centimeters. Thus in the statement 1 in. = 2.54 cm, neither 2.54 nor 1 limits the number of significant figures when it is used in a calculation.

Leading zeros are never significant figures.

Captive zeros are always significant figures.

Trailing zeros are sometimes significant figures.

Exact numbers never limit the number of significant figures in a calculation.

Rules for counting significant figures also apply to numbers written in scientific notation. For example, the number 100. can also be written as 1.00×10^2, and both versions have three significant figures. Scientific notation offers two major advantages: the number of significant figures can be indicated easily, and fewer zeros are needed to write a very large or a very small number. For example, the number 0.000060 is much more conveniently represented as 6.0×10^{-5}, and the number has two significant figures, written in either form.

Significant figures are easily indicated by scientific notation.

Example 2.3 Counting Significant Figures

Give the number of significant figures for each of the following measurements.

a. A sample of orange juice contains 0.0108 g of vitamin C.

b. A forensic chemist in a crime lab weighs a single hair and records its mass as 0.0050060 g.

c. The distance between two points was found to be 5.030×10^3 ft.

d. In yesterday's bicycle race, 110 riders started but only 60 finished.

Solution

a. The number contains three significant figures. The zeros to the left of the 1 are leading zeros and are not significant, but the remaining zero (a captive zero) is significant.

b. The number contains five significant figures. The leading zeros (to the left of the 5) are not significant. The captive zeros between the 5 and the 6 are significant, and the trailing zero to the right of the 6 is significant because the number contains a decimal point.

c. This number has four significant figures. Both zeros in 5.030 are significant.

d. Both numbers are exact (they were obtained by counting the riders). Thus these numbers have an unlimited number of significant figures.

✔ Self-Check Exercise 2.2

Give the number of significant figures for each of the following measurements.

a. 0.00100 m

b. 2.0800×10^2 L

c. 480 Corvettes

See Problems 2.37 and 2.38. ■

Rounding Off Numbers

When you perform a calculation on your calculator, the number of digits displayed is usually greater than the number of significant figures that the result should possess. So you must "round off" the number (reduce it to fewer digits). The rules for **rounding off** follow.

These rules reflect the way calculators round off.

Rules for Rounding Off

1. If the digit to be removed
   a. is less than 5, the preceding digit stays the same. For example, 1.33 rounds to 1.3.
   b. is equal to or greater than 5, the preceding digit is increased by 1. For example, 1.36 rounds to 1.4, and 3.15 rounds to 3.2.
2. In a series of calculations, carry the extra digits through to the final result and *then* round off.* This means that you should carry all of the digits that show on your calculator until you arrive at the final number (the answer) and then round off, using the procedures in rule 1.

* This practice will not be followed in the worked-out examples in this text, because we want to show the correct number of significant figures in each step of the example.

We need to make one more point about rounding off to the correct number of significant figures. Suppose the number 4.348 needs to be rounded to two significant figures. In doing this, we look *only* at the *first number* to the right of the 3:

4.348
↑
Look at this number to round off to two significant figures.

The number is rounded to 4.3 because 4 is less than 5. It is incorrect to round sequentially. For example, do *not* round the 4 to 5 to give 4.35 and then round the 3 to 4 to give 4.4.

When rounding off, *use only the first number to the right of the last significant figure.*

Do not round off sequentially. The number 6.8347 rounded to three significant figures is 6.83, not 6.84.

Determining Significant Figures in Calculations

Next we will learn how to determine the correct number of significant figures in the result of a calculation. To do this we will use the following rules.

Rules for Using Significant Figures in Calculations

1. For *multiplication* or *division*, the number of significant figures in the result is the same as that in the measurement with the *smallest number* of significant figures. We say this measurement is *limiting*, because it limits the number of significant figures in the result. For example, consider this calculation:

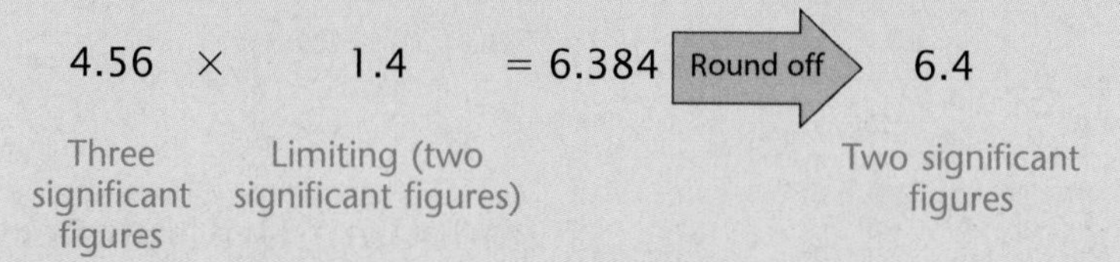

Because 1.4 has only two significant figures, it limits the result to two significant figures. Thus the product is correctly written as 6.4, which has two significant figures. Consider another example. In the division $\frac{8.315}{298}$, how many significant figures should appear in the answer? Because 8.315 has four significant figures, the number 298 (with three significant figures) limits the result. The calculation is correctly represented as

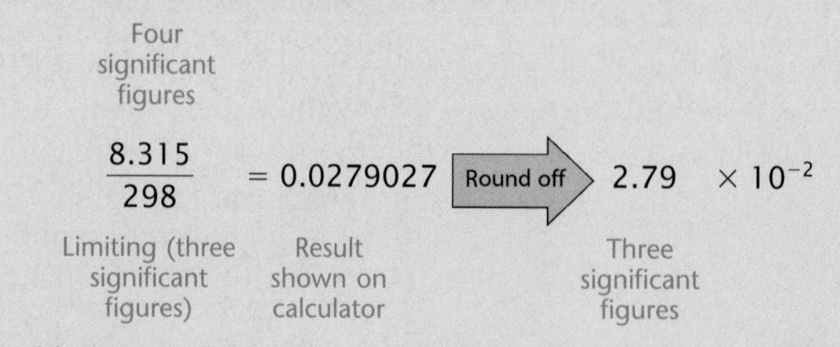

2. For *addition* or *subtraction*, the limiting term is the one with the smallest number of decimal places. For example, consider the following sum:

If you need help in using your calculator, see the Appendix.

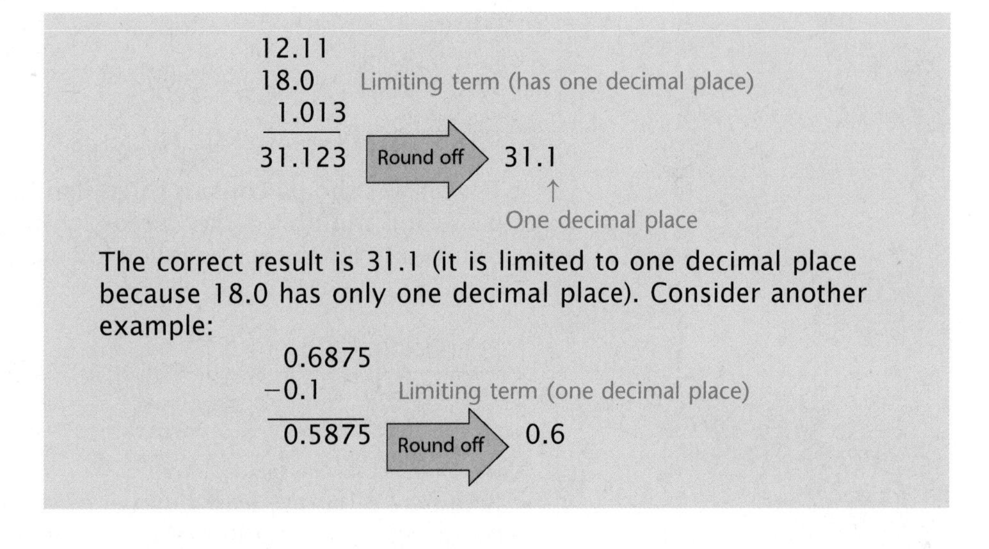

Note that *for multiplication and division, significant figures are counted. For addition and subtraction, the decimal places are counted.*

Now we will put together the things you have learned about significant figures by considering some mathematical operations in the following examples.

Example 2.4 Counting Significant Figures in Calculations

Without performing the calculations, tell how many significant figures each answer should contain.

a. 5.19
   1.9
   0.842

b. $1081 - 7.25$

c. 2.3×3.14

d. the total cost of 3 boxes of candy at $2.50 a box

Solution

a. The answer will have one digit after the decimal place. The limiting number is 1.9, which has one decimal place, so the answer has two significant figures.

b. The answer will have no digits after the decimal point. The number 1081 has no digits to the right of the decimal point and limits the result, so the answer has four significant figures.

c. The answer will have two significant figures because the number 2.3 has only two significant figures (3.14 has three).

d. The answer will have three significant figures. The limiting factor is 2.50 because 3 (boxes of candy) is an exact number. ■

Example 2.5 Calculations Using Significant Figures

Carry out the following mathematical operations and give each result to the correct number of significant figures.

a. 5.18×0.0208

b. $(3.60 \times 10^{-3}) \times (8.123) \div 4.3$

c. $21 + 13.8 + 130.36$

d. $116.8 - 0.33$

e. $(1.33 \times 2.8) + 8.41$

Solution

Limiting terms — Round to this digit.

a. $5.18 \times 0.0208 = 0.107744 \Rightarrow 0.108$

The answer should contain three significant figures because each number being multiplied has three significant figures (Rule 1). The 7 is rounded to 8 because the following digit is greater than 5.

Round to this digit.

b. $\dfrac{(3.60 \times 10^{-3})(8.123)}{4.3} = 6.8006 \times 10^{-3} \Rightarrow 6.8 \times 10^{-3}$

Limiting term (4.3)

Because 4.3 has the least number of significant figures (two), the result should have two significant figures (Rule 1).

c.
$$\begin{array}{r} 21 \\ 13.8 \\ 130.36 \\ \hline 165.16 \end{array} \Rightarrow 165$$

In this case 21 is limiting (there are no digits after the decimal point). Thus the answer must have no digits after the decimal point, in accordance with the rule for addition (Rule 2).

d.
$$\begin{array}{r} 116.8 \\ -\ 0.33 \\ \hline 116.47 \end{array} \Rightarrow 116.5$$

Because 116.8 has only one decimal place, the answer must have only one decimal place (Rule 2). The 4 is rounded up to 5 because the digit to the right (7) is greater than 5.

e. $1.33 \times 2.8 = 3.724 \Rightarrow 3.7$

$$\begin{array}{r} 3.7 \\ +\ 8.41 \\ \hline 12.11 \end{array} \Rightarrow 12.1$$

3.7 ← Limiting term

Note that in this case we multiplied and then rounded the result to the correct number of significant figures before we performed the addition so that we would know the correct number of decimal places.

✔ Self-Check Exercise 2.3

Give the answer for each calculation to the correct number of significant figures.

a. 12.6×0.53

b. $(12.6 \times 0.53) - 4.59$

c. $(25.36 - 4.15) \div 2.317$

See Problems 2.51 through 2.56. ■

Problem Solving and Dimensional Analysis

AIM: To learn how dimensional analysis can be used to solve various types of problems.

Suppose that the boss at the store where you work on weekends asks you to pick up 2 dozen doughnuts on the way to work. However, you find that the doughnut shop sells by the doughnut. How many doughnuts do you need?

This "problem" is an example of something you encounter all the time: converting from one unit of measurement to another. Examples of this occur in cooking (The recipe calls for 3 cups of cream, which is sold in pints. How many pints do I buy?); traveling (The purse costs 250 pesos. How much is that in dollars?); sports (A recent Tour de France bicycle race was 3215 kilometers long. How many miles is that?); and many other areas.

How do we convert from one unit of measurement to another? Let's explore this process by using the doughnut problem.

2 dozen doughnuts = ? individual doughnuts

where ? represents a number you don't know yet. The essential information you must have is the definition of a dozen:

1 dozen = 12

You can use this information to make the needed conversion as follows:

$$2 \text{ dozen doughnuts} \times \frac{12}{1 \text{ dozen}} = 24 \text{ doughnuts}$$

You need to buy 24 doughnuts.

Note two important things about this process.

1. The factor $\frac{12}{1 \text{ dozen}}$ is a conversion factor based on the definition of the term *dozen.* This conversion factor is a ratio of the two parts of the definition of a dozen given above.
2. The unit dozen itself cancels.

Now let's generalize a bit. To change from one unit to another we will use a conversion factor.

$$\text{Unit}_1 \times \text{conversion factor} = \text{Unit}_2$$

The **conversion factor** is a ratio of the two parts of the statement that relates the two units. We will see this in more detail on the following pages.

Earlier in this chapter we considered a pin that measured 2.85 cm in length. What is the length of the pin in inches? We can represent this problem as

$$2.85 \text{ cm} \rightarrow ? \text{ in.}$$

The question mark stands for the number we want to find. To solve this problem, we must know the relationship between inches and centimeters. In Table 2.7, which gives several equivalents between the English and metric systems, we find the relationship

2.54 cm = 1 in.

This is called an **equivalence statement.** In other words, 2.54 cm and 1 in. stand for *exactly the same distance.* (See Figure 2.1.) The respective numbers are different because they refer to different *scales* (*units*) of distance.

The equivalence statement 2.54 cm = 1 in. can lead to either of two conversion factors:

$$\frac{2.54 \text{ cm}}{1 \text{ in.}} \quad \text{or} \quad \frac{1 \text{ in.}}{2.54 \text{ cm}}$$

Note that these *conversion factors* are *ratios of the two parts of the equivalence statement* that relates the two units. Which of the two possible conversion factors do we need? Recall our problem:

2.85 cm = ? in.

Table 2.7 English–Metric and English–English Equivalents

Length	1 m = 1.094 yd
	2.54 cm = 1 in.
	1 mi = 5280. ft
	1 mi = 1760. yd
Mass	1 kg = 2.205 lb
	453.6 g = 1 lb
Volume	1 L = 1.06 qt
	1 ft^3 = 28.32 L

That is, we want to convert from units of centimeters to inches:

$$2.85 \text{ cm} \times \text{conversion factor} = ? \text{ in.}$$

We choose a conversion factor that cancels the units we want to discard and leaves the units we want in the result. Thus we do the conversion as follows:

Units cancel just as numbers do.

$$2.85 \text{ cm} \times \frac{1 \text{ in.}}{2.54 \text{ cm}} = \frac{2.85 \text{ in.}}{2.54} = 1.12 \text{ in.}$$

Note two important facts about this conversion:

1. The centimeter units cancel to give inches for the result. This is exactly what we had wanted to accomplish. Using the other conversion factor $\left(2.85 \text{ cm} \times \frac{2.54 \text{ cm}}{1 \text{ in.}}\right)$ would not work because the units would not cancel to give inches in the result.

When you finish a calculation, always check to make sure that the answer makes sense.

2. As the units changed from centimeters to inches, the number changed from 2.85 to 1.12. Thus 2.85 cm has exactly the same value (is the same length) as 1.12 in. Notice that in this conversion, the number decreased from 2.85 to 1.12. This makes sense because the inch is a larger unit of length than the centimeter is. That is, it takes fewer inches to make the same length in centimeters.

The result in the foregoing conversion has three significant figures as required. Caution: Noting that the term 1 appears in the conversion, you might think that because this number appears to have only one significant figure, the result should have only one significant figure. That is, the answer should be given as 1 in. rather than 1.12 in. However, in the equivalence statement 1 in. = 2.54 cm, the 1 is an exact number (by definition). In other words, exactly 1 in. equals 2.54 cm. Therefore, the 1 does not limit the number of significant digits in the result.

When exact numbers are used in a calculation, they never limit the number of significant digits.

We have seen how to convert from centimeters to inches. What about the reverse conversion? For example, if a pencil is 7.00 in. long, what is its length in centimeters? In this case, the conversion we want to make is

$$7.00 \text{ in.} \rightarrow ? \text{ cm}$$

What conversion factor do we need to make this conversion?

Remember that two conversion factors can be derived from each equivalence statement. In this case, the equivalence statement 2.54 cm = 1 in. gives

$$\frac{2.54 \text{ cm}}{1 \text{ in.}} \quad \text{or} \quad \frac{1 \text{ in.}}{2.54 \text{ cm}}$$

Again, we choose which to use by looking at the *direction* of the required change. For us to change from inches to centimeters, the inches must cancel. Thus the factor

$$\frac{2.54 \text{ cm}}{1 \text{ in.}}$$

is used, and the conversion is done as follows:

$$7.00 \text{ in.} \times \frac{2.54 \text{ cm}}{1 \text{ in.}} = (7.00)(2.54) \text{ cm} = 17.8 \text{ cm}$$

Consider the direction of the required change in order to select the correct conversion factor.

Here the inch units cancel, leaving centimeters as required.

Note that in this conversion, the number increased (from 7.00 to 17.8). This makes sense because the centimeter is a smaller unit of length than the inch. That is, it takes more centimeters to make the same length in inches. *Always take a moment to think about whether your answer makes sense.* This will help you avoid errors.

Changing from one unit to another via conversion factors (based on the equivalence statements between the units) is often called **dimensional analysis.** We will use this method throughout our study of chemistry.

We can now state some general steps for doing conversions by dimensional analysis.

Converting from One Unit to Another

STEP 1 To convert from one unit to another, use the equivalence statement that relates the two units. The conversion factor needed is a ratio of the two parts of the equivalence statement.

STEP 2 Choose the appropriate conversion factor by looking at the direction of the required change (make sure the unwanted units cancel).

STEP 3 Multiply the quantity to be converted by the conversion factor to give the quantity with the desired units.

STEP 4 Check that you have the correct number of significant figures.

STEP 5 Ask whether your answer makes sense.

We will now illustrate this procedure in Example 2.6.

Example 2.6 Conversion Factors: One-Step Problems

An Italian bicycle has its frame size given as 62 cm. What is the frame size in inches?

Solution

We can represent the problem as

$$62 \text{ cm} = ? \text{ in.}$$

In this problem we want to convert from centimeters to inches.

$$62 \text{ cm} \times \text{conversion factor} = ? \text{ in.}$$

STEP 1 To convert from centimeters to inches, we need the equivalence statement 1 in. = 2.54 cm. This leads to two conversion factors:

$$\frac{1 \text{ in.}}{2.54 \text{ cm}} \quad \text{and} \quad \frac{2.54 \text{ cm}}{1 \text{ in.}}$$

STEP 2 In this case, the direction we want is

$$\text{Centimeters} \rightarrow \text{inches}$$

so we need the conversion factor $\frac{1 \text{ in.}}{2.54 \text{ cm}}$. We know this is the one we want because using it will make the units of centimeters cancel, leaving units of inches.

STEP 3 The conversion is carried out as follows:

$$62 \text{ cm} \times \frac{1 \text{ in.}}{2.54 \text{ cm}} = 24 \text{ in.}$$

STEP 4 The result is limited to two significant figures by the number 62. The centimeters cancel, leaving inches as required.

STEP 5 Note that the number decreased in this conversion. This makes sense; the inch is a larger unit of length than the centimeter.

✔ Self-Check Exercise 2.4

Wine is often bottled in 0.750-L containers. Using the appropriate equivalence statement from Table 2.7, calculate the volume of such a wine bottle in quarts.

See Problems 2.63 and 2.64. ■

Next we will consider a conversion that requires several steps.

Example 2.7 Conversion Factors: Multiple-Step Problems

The length of the marathon race is approximately 26.2 mi. What is this distance in kilometers?

Solution

The problem before us can be represented as follows:

$$26.2 \text{ mi} = ? \text{ km}$$

We could accomplish this conversion in several different ways, but because Table 2.7 gives the equivalence statements 1 mi = 1760 yd and 1 m = 1.094 yd, we will proceed as follows:

$$\text{Miles} \rightarrow \text{yards} \rightarrow \text{meters} \rightarrow \text{kilometers}$$

This process will be carried out one conversion at a time to make sure everything is clear.

MILES → YARDS: We convert from miles to yards using the conversion factor $\frac{1760 \text{ yd}}{1 \text{ mi}}$.

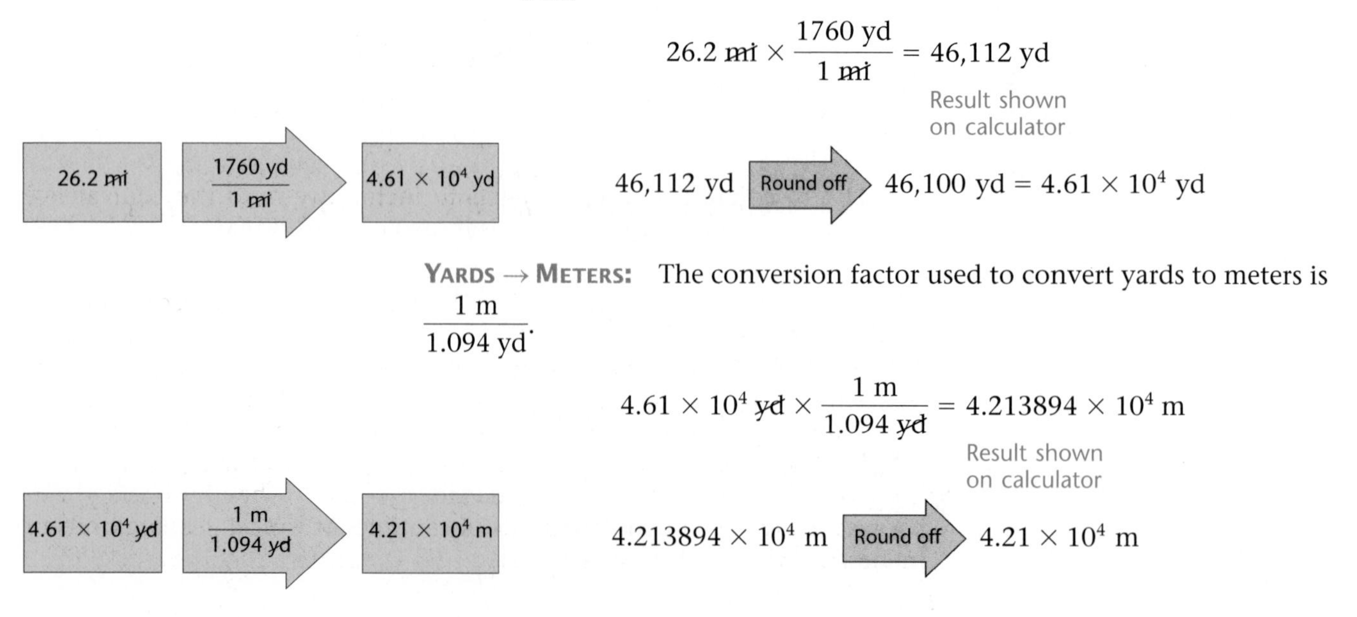

$$26.2 \text{ mi} \times \frac{1760 \text{ yd}}{1 \text{ mi}} = 46{,}112 \text{ yd}$$

$$46{,}112 \text{ yd} \xrightarrow{\text{Round off}} 46{,}100 \text{ yd} = 4.61 \times 10^4 \text{ yd}$$

YARDS → METERS: The conversion factor used to convert yards to meters is $\frac{1 \text{ m}}{1.094 \text{ yd}}$.

$$4.61 \times 10^4 \text{ yd} \times \frac{1 \text{ m}}{1.094 \text{ yd}} = 4.213894 \times 10^4 \text{ m}$$

$$4.213894 \times 10^4 \text{ m} \xrightarrow{\text{Round off}} 4.21 \times 10^4 \text{ m}$$

METERS → KILOMETERS: Because 1000 m = 1 km, or 10^3 m = 1 km, we convert from meters to kilometers as follows:

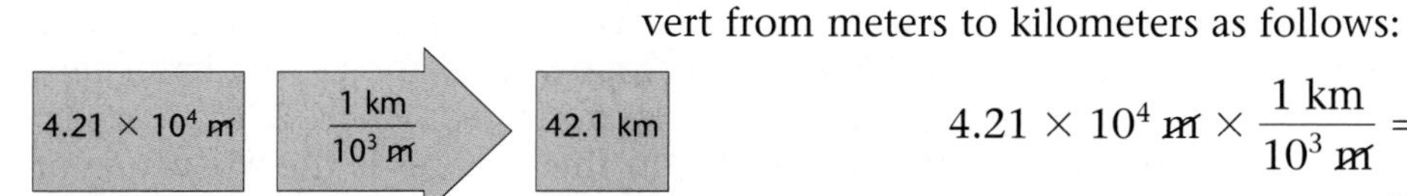

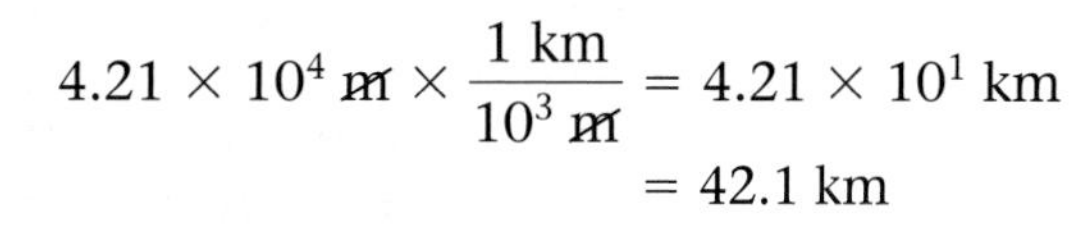

$$4.21 \times 10^4 \text{ m} \times \frac{1 \text{ km}}{10^3 \text{ m}} = 4.21 \times 10^1 \text{ km}$$
$$= 42.1 \text{ km}$$

Thus the marathon (26.2 mi) is 42.1 km.

Once you feel comfortable with the conversion process, you can combine the steps. For the above conversion, the combined expression is

$$\text{miles} \rightarrow \text{yards} \rightarrow \text{meters} \rightarrow \text{kilometers}$$
$$26.2 \text{ mi} \times \frac{1760 \text{ yd}}{1 \text{ mi}} \times \frac{1 \text{ m}}{1.094 \text{ yd}} \times \frac{1 \text{ km}}{10^3 \text{ m}} = 42.1 \text{ km}$$

Note that the units cancel to give the required kilometers and that the result has three significant figures.

Remember that we are rounding off at the end of each step to show the correct number of significant figures. However, in doing a multistep calculation, *you* should retain the extra numbers that show on your calculator and round off only at the end of the calculation.

✔ Self-Check Exercise 2.5

Racing cars at the Indianapolis Motor Speedway now routinely travel around the track at an average speed of 225 mi/h. What is this speed in kilometers per hour?

See Problems 2.69 and 2.70. ■

RECAP: Whenever you work problems, remember the following points:

1. Always include the units (a measurement always has two parts: a number *and* a unit).
2. Cancel units as you carry out the calculations.
3. Check that your final answer has the correct units. If it doesn't, you have done something wrong.
4. Check that your final answer has the correct number of significant figures.
5. Think about whether your answer makes sense.

Units provide a very valuable check on the validity of your solution. Always use them.

2.7 Temperature Conversions: An Approach to Problem Solving

AIMS: To learn the three temperature scales. To learn to convert from one scale to another. To continue to develop problem-solving skills.

When the doctor tells you your temperature is 102 degrees and the weatherperson on TV says it will be 75 degrees tomorrow, they are using the **Fahrenheit scale.** Water boils at 212 °F and freezes at 32 °F, and normal body temperature is 98.6 °F (where °F signifies "Fahrenheit degrees"). This temperature scale is widely used in the United States and Great Britain, and it is the scale employed in most of the engineering sciences. Another temperature scale, used in Canada and Europe and in the physical and life sciences in most countries, is the **Celsius scale.** In keeping with the metric system, which is based on powers of 10, the freezing and boiling points of water on the Celsius scale are assigned as 0 °C and 100 °C, respectively. On both the Fahrenheit and the Celsius scales, the unit of temperature is called a degree, and the symbol for it is followed by the capital letter representing the scale on which the units are measured: °C or °F.

Although 373 K is often stated as 373 degrees Kelvin, it is more correct to say 373 kelvins.

Still another temperature scale used in the sciences is the **absolute** or **Kelvin scale.** On this scale water freezes at 273 K and boils at 373 K. On the Kelvin scale, the unit of temperature is called a kelvin and is symbolized by K. Thus, on the three scales, the boiling point of water is stated as 212 Fahrenheit degrees (212 °F), 100 Celsius degrees (100 °C), and 373 kelvins (373 K).

The three temperature scales are compared in Figures 2.6 and 2.7. There are several important facts you should note.

1. The size of each temperature unit (each degree) is the same for the Celsius and Kelvin scales. This follows from the fact that the *difference* between the boiling and freezing points of water is 100 units on both of these scales.
2. The Fahrenheit degree is smaller than the Celsius and Kelvin unit. Note that on the Fahrenheit scale there are 180 Fahrenheit degrees between the boiling and freezing points of water, as compared with 100 units on the other two scales.
3. The zero points are different on all three scales.

In your study of chemistry, you will sometimes need to convert from one temperature scale to another. We will consider in some detail how this is done. In addition to learning how to change temperature scales, you should also use this section as an opportunity to further develop your skills in problem solving.

Converting Between the Kelvin and Celsius Scales

It is relatively simple to convert between the Celsius and Kelvin scales because the temperature unit is the same size; only the zero points are different. Because 0 °C corresponds to 273 K, converting from Celsius to Kelvin

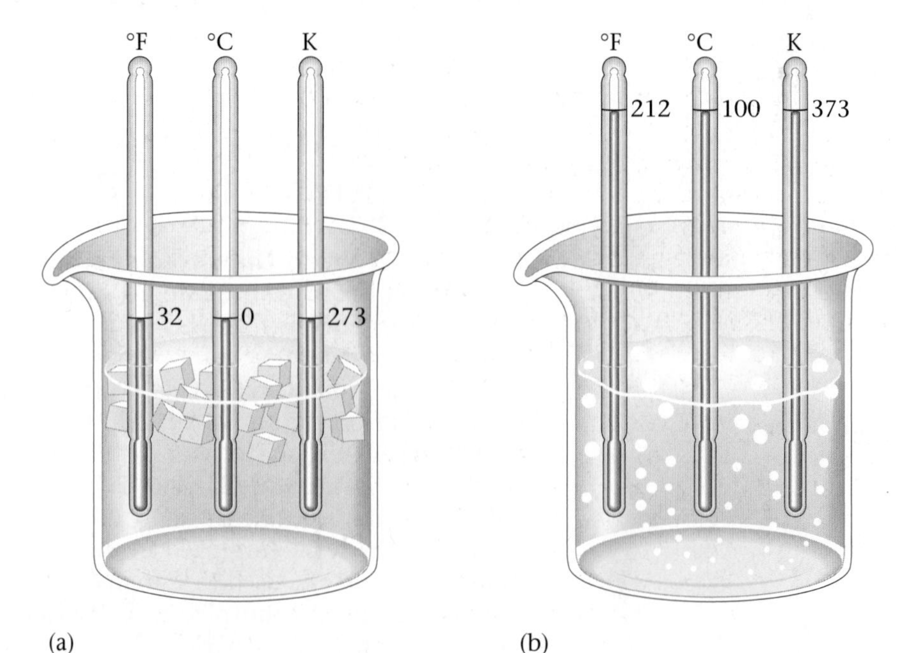

Figure 2.6
Thermometers based on the three temperature scales in (a) ice water and (b) boiling water.

Figure 2.7
The three major temperature scales.

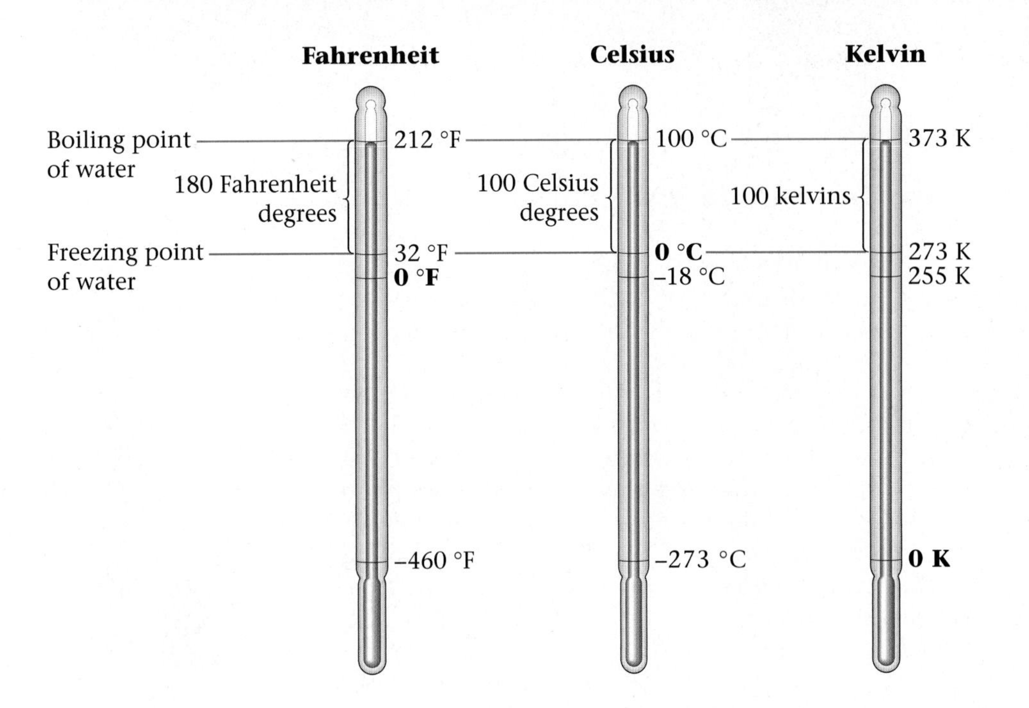

requires that we add 273 to the Celsius temperature. We will illustrate this procedure in Example 2.8.

Example 2.8 Temperature Conversion: Celsius to Kelvin

Boiling points will be discussed further in Chapter 13.

The boiling point of water at the top of Mt. Everest is 70. °C. Convert this temperature to the Kelvin scale. (The decimal point after the temperature reading indicates that the trailing zero is significant.)

Solution

This problem asks us to find 70. °C in units of kelvins. We can represent this problem simply as

$$70.\ °C = ?\ K$$

In solving problems, it is often helpful to draw a diagram that depicts what the words are telling you.

In doing problems, it is often helpful to draw a diagram in which we try to represent the words in the problem with a picture. This problem can be diagramed as shown in Figure 2.8a.

Figure 2.8
Converting 70. °C to units measured on the Kelvin scale.
(a) We know 0 °C = 273 K. We want to know 70. °C = ? K.
(b) There are 70 degrees on the Celsius scale between 0 °C and 70. °C. Because units on these two scales are the same size, there are also 70 kelvins in this same distance on the Kelvin scale.

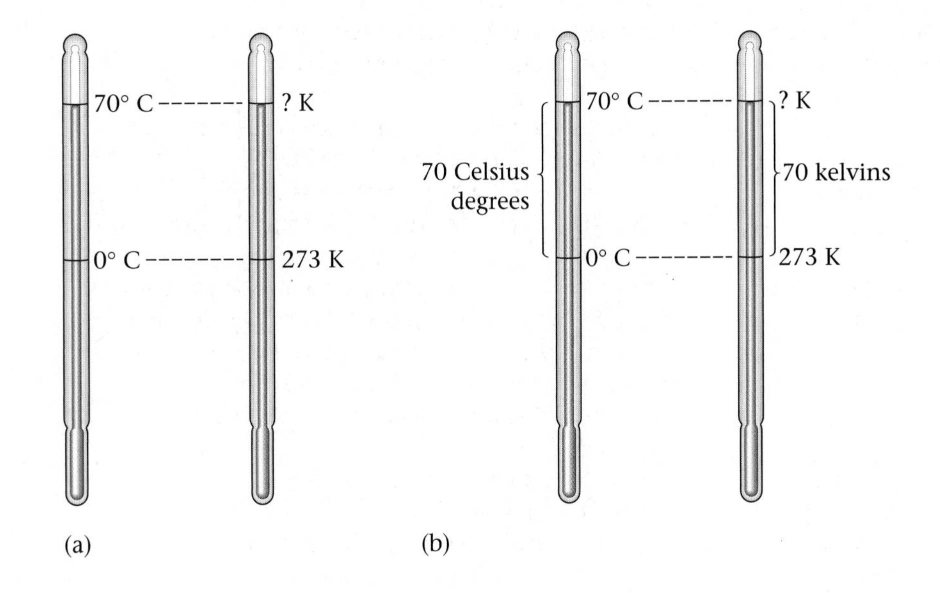

CHEMISTRY IN FOCUS

Tiny Thermometers

Can you imagine a thermometer that has a diameter equal to one one-hundredth of a human hair? Such a device has actually been produced by scientists Yihica Gao and Yoshio Bando of the National Institute for Materials Science in Tsukuba, Japan. The thermometer they constructed is so tiny that it must be read using a powerful electron microscope.

It turns out that the tiny thermometers were produced by accident. The Japanese scientists were actually trying to make tiny (nanoscale) gallium nitride wires. However, when they examined the results of their experiment, they discovered tiny tubes of carbon atoms that were filled with elemental gallium. Because gallium is a liquid over an unusually large temperature range, it makes a perfect working liquid for a thermometer. Just as in mercury thermometers, which have mostly been phased out because of the toxicity of mercury, the tiny gallium expands as the temperature increases. The gallium moves up the tube as the temperature increases.

These miniscule thermometers are not useful in the normal macroscopic world—they can't even be seen with the naked eye. However, they should be valuable for monitoring temperatures from 50 °C to 500 °C in materials in the nanoscale world.

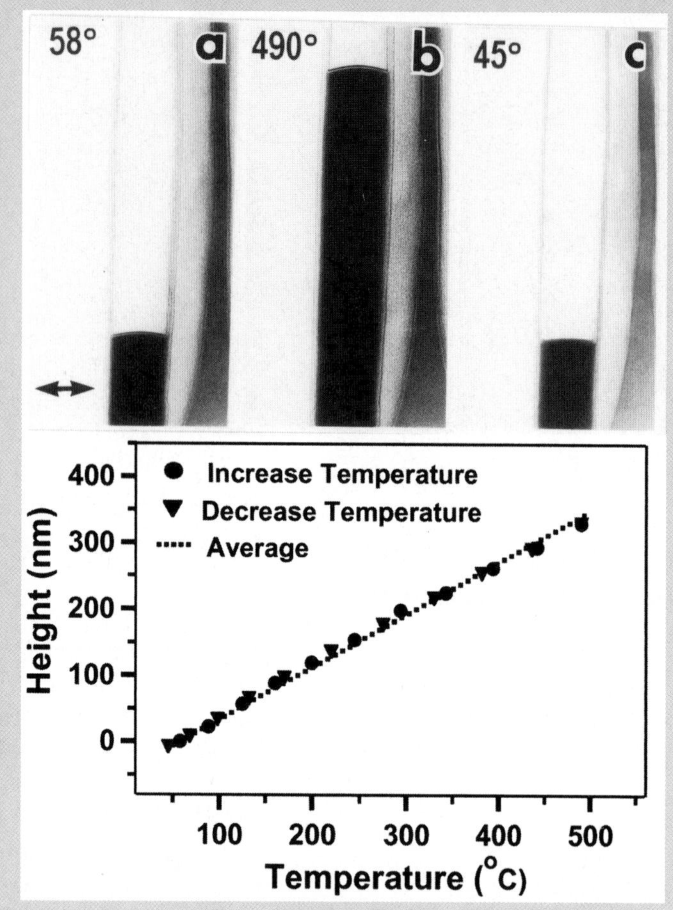

Liquid gallium expands within a carbon nanotube as the temperature increases (left to right).

In this picture we have shown what we want to find: "What temperature (in kelvins) is the same as 70. °C?" We also know from Figure 2.7 that 0 °C represents the same temperature as 273 K. How many degrees above 0 °C is 70. °C? The answer, of course, is 70. Thus we must add 70. to 0 °C to reach 70. °C. Because degrees are the *same size* on both the Celsius scale and the Kelvin scale (see Figure 2.8b), we must also add 70. to 273 K (same temperature as 0 °C) to reach ? K. That is,

$$? \text{ K} = 273 + 70. = 343 \text{ K}$$

Thus 70. °C corresponds to 343 K.

Note that to convert from the Celsius to the Kelvin scale, we simply add the temperature in °C to 273. That is,

$$T_{°C} + 273 = T_K$$

Temperature in Celsius degrees — Temperature in kelvins

Using this formula to solve the present problem gives

$$70. + 273 = 343$$

(with units of kelvins, K), which is the correct answer. ■

We can summarize what we learned in Example 2.8 as follows: to convert from the Celsius to the Kelvin scale, we can use the formula

$$T_{°C} + 273 = T_K$$

Temperature in Celsius degrees — Temperature in kelvins

Example 2.9 Temperature Conversion: Kelvin to Celsius

Liquid nitrogen boils at 77 K. What is the boiling point of nitrogen on the Celsius scale?

Solution

0 °C --------- 273 K

? °C --------- 77 K

The problem to be solved here is 77 K = ? °C. Let's explore this question by examining the picture to the left representing the two temperature scales. One key point is to recognize that 0 °C = 273 K. Also note that the difference between 273 K and 77 K is 196 kelvins (273 − 77 = 196). That is, 77 K is 196 kelvins below 273 K. The degree size is the same on these two temperature scales, so 77 K must correspond to 196 Celsius degrees below zero or −196 °C. Thus 77 K = ? °C = −196 °C.

We can also solve this problem by using the formula

$$T_{°C} + 273 = T_K$$

However, in this case we want to solve for the Celsius temperature, $T_{°C}$. That is, we want to isolate $T_{°C}$ on one side of the equals sign. To do this we use an important general principle: doing *the same thing on both sides of the equals sign* preserves the equality. In other words, it's always okay to perform the same operation on both sides of the equals sign.

To isolate $T_{°C}$ we need to subtract 273 from both sides:

$$T_{°C} + 273 - 273 = T_K - 273$$

↑ ↑ Sum is zero

to give

$$T_{°C} = T_K - 273$$

Using this equation to solve the problem, we have

$$T_{°C} = T_K - 273 = 77 - 273 = -196$$

So, as before, we have shown that

$$77 \text{ K} = -196 \text{ °C}$$

Self-Check Exercise 2.6

Which temperature is colder, 172 K or −75 °C?

See Problems 2.77 and 2.78. ■

In summary, because the Kelvin and Celsius scales have the same size unit, to switch from one scale to the other we must simply account for the different zero points. We must add 273 to the Celsius temperature to obtain the temperature on the Kelvin scale:

$$T_K = T_{°C} + 273$$

To convert from the Kelvin scale to the Celsius scale, we must subtract 273 from the Kelvin temperature:

$$T_{°C} = T_K - 273$$

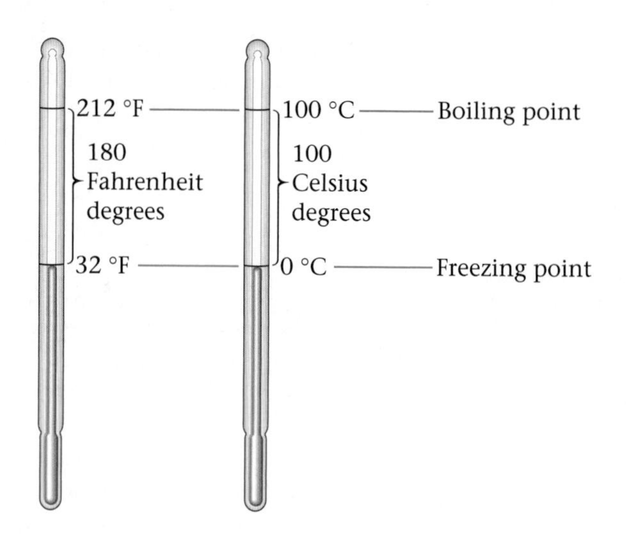

Figure 2.9
Comparison of the Celsius and Fahrenheit scales.

Converting Between the Fahrenheit and Celsius Scales

The conversion between the Fahrenheit and Celsius temperature scales requires two adjustments:

1. For the different size units
2. For the different zero points

To see how to adjust for the different unit sizes, consider the diagram in Figure 2.9. Note that because 212 °F = 100 °C and 32 °F = 0 °C,

$$212 - 32 = 180 \text{ Fahrenheit degrees} = 100 - 0 = 100 \text{ Celsius degrees}$$

Thus

$$180. \text{ Fahrenheit degrees} = 100. \text{ Celsius degrees}$$

Dividing both sides of this equation by 100. gives

Remember, it's okay to do the same thing to both sides of the equation.

$$\frac{180.}{100.} \text{ Fahrenheit degrees} = \frac{\cancel{100.}}{\cancel{100.}} \text{ Celsius degrees}$$

or

$$1.80 \text{ Fahrenheit degrees} = 1.00 \text{ Celsius degree}$$

The factor 1.80 is used to convert from one degree size to the other.

Next we have to account for the fact that 0 °C is *not* the same as 0 °F. In fact, 32 °F = 0 °C. Although we will not show how to derive it, the equation to convert a temperature in Celsius degrees to the Fahrenheit scale is

$$T_{°F} = 1.80(T_{°C}) + 32$$

Temperature in °F   Temperature in °C

In this equation the term $1.80(T_{°C})$ adjusts for the difference in degree size between the two scales. The 32 in the equation accounts for the different zero points. We will now show how to use this equation.

Example 2.10 Temperature Conversion: Celsius to Fahrenheit

On a summer day the temperature in the laboratory, as measured on a lab thermometer, is 28 °C. Express this temperature on the Fahrenheit scale.

Solution

This problem can be represented as 28 °C = ? °F. We will solve it using the formula

$$T_{°F} = 1.80(T_{°C}) + 32$$

In this case,

$$T_{°F} = ?\ °F = 1.80(\underset{T_{°C}}{28}) + 32 = \underset{\text{Rounds off to 50}}{50.4} + 32$$

$$= 50. + 32 = 82$$

Thus 28 °C = 82 °F. ■

Note that 28 °C is approximately equal to 82 °F. Because the numbers are just reversed, this is an easy reference point to remember for the two scales.

Example 2.11 Temperature Conversion: Celsius to Fahrenheit

Express the temperature −40. °C on the Fahrenheit scale.

Solution

We can express this problem as −40. °C = ? °F. To solve it we will use the formula

$$T_{°F} = 1.80(T_{°C}) + 32$$

In this case,

$$T_{°F} = ?\ °F = 1.80(\underset{T_{°C}}{-40.}) + 32$$
$$= -72 + 32 = -40$$

So −40 °C = −40 °F. This is a very interesting result and is another useful reference point.

✔ Self-Check Exercise 2.7

Hot tubs are often maintained at 41 °C. What is this temperature in Fahrenheit degrees?

See Problems 2.79 through 2.82. ■

To convert from Celsius to Fahrenheit, we have used the equation

$$T_{°F} = 1.80(T_{°C}) + 32$$

To convert a Fahrenheit temperature to Celsius, we need to rearrange this equation to isolate Celsius degrees ($T_{°C}$). Remember we can always do the same operation to both sides of the equation. First subtract 32 from each side:

$$T_{°F} - 32 = 1.80(T_{°C}) + \underbrace{32 - 32}_{\text{Sum is zero}}$$

to give

$$T_{°F} - 32 = 1.80(T_{°C})$$

Next divide both sides by 1.80

$$\frac{T_{°F} - 32}{1.80} = \frac{\cancel{1.80}(T_{°C})}{\cancel{1.80}}$$

to give

$$\frac{T_{°F} - 32}{1.80} = T_{°C}$$

or

Temperature in °F

$$T_{°C} = \frac{T_{°F} - 32}{1.80}$$

Temperature in °C

$$T_{°C} = \frac{T_{°F} - 32}{1.80}$$

Example 2.12 Temperature Conversion: Fahrenheit to Celsius

One of the body's responses to an infection or injury is to elevate its temperature. A certain flu victim has a body temperature of 101 °F. What is this temperature on the Celsius scale?

Solution

The problem is 101 °F = ? °C. Using the formula

$$T_{°C} = \frac{T_{°F} - 32}{1.80}$$

yields

$$T_{°C} = ?\ °C = \frac{\overset{T_{°F}}{101} - 32}{1.80} = \frac{69}{1.80} = 38$$

That is, 101 °F = 38 °C.

✔ Self-Check Exercise 2.8

An antifreeze solution in a car's radiator boils at 239 °F. What is this temperature on the Celsius scale?

See Problems 2.79 through 2.82. ■

In doing temperature conversions, you will need the following formulas.

Temperature Conversion Formulas

- Celsius to Kelvin $T_K = T_{°C} + 273$
- Kelvin to Celsius $T_{°C} = T_K - 273$
- Celsius to Fahrenheit $T_{°F} = 1.80(T_{°C}) + 32$
- Fahrenheit to Celsius $T_{°C} = \dfrac{T_{°F} - 32}{1.80}$

Density

AIM: To define density and its units.

Lead has a greater density than feathers.

When you were in elementary school, you may have been embarrassed by your answer to the question "Which is heavier, a pound of lead or a pound of feathers?" If you said lead, you were undoubtedly thinking about density, not mass. **Density** can be defined as the amount of matter present *in a given volume* of substance. That is, density is mass per unit volume, the ratio of the mass of an object to its volume:

$$\text{Density} = \frac{\text{mass}}{\text{volume}}$$

It takes a much bigger volume to make a pound of feathers than to make a pound of lead. This is because lead has a much greater mass per unit volume—a greater density.

The density of a liquid can be determined easily by weighing a known volume of the substance as illustrated in Example 2.13.

Example 2.13 Calculating Density

Suppose a student finds that 23.50 mL of a certain liquid weighs 35.062 g. What is the density of this liquid?

Solution

We can calculate the density of this liquid simply by applying the definition

$$\text{Density} = \frac{\text{mass}}{\text{volume}} = \frac{35.062 \text{ g}}{23.50 \text{ mL}} = 1.492 \text{ g/mL}$$

This result could also be expressed as 1.492 g/cm^3 because 1 mL = 1 cm^3. ■

The volume of a solid object is often determined indirectly by submerging it in water and measuring the volume of water displaced. In fact, this is the most accurate method for measuring a person's percent body fat. The person is submerged momentarily in a tank of water, and the increase in volume is measured (see Figure 2.10). It is possible to calculate the body

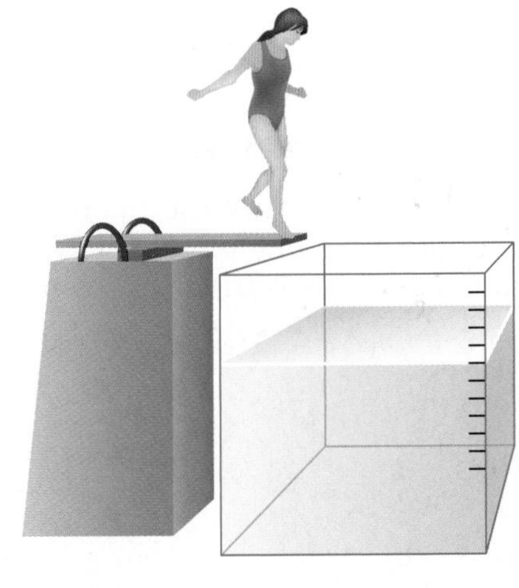

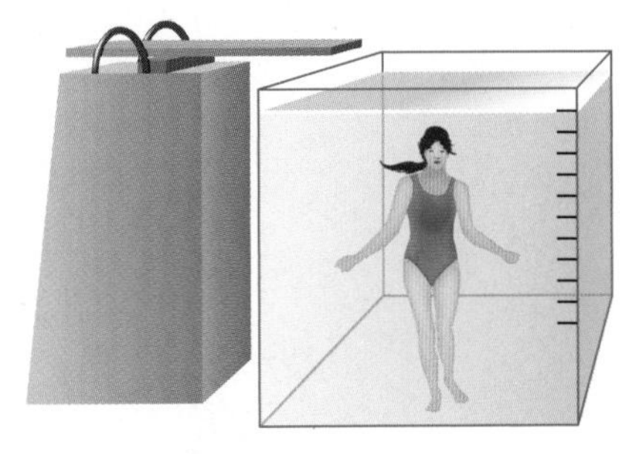

Figure 2.10
(a) Tank of water. (b) Person submerged in the tank, raising the level of the water.

density by using the person's weight (mass) and the volume of the person's body determined by submersion. Fat, muscle, and bone have different densities (fat is less dense than muscle tissue, for example), so the fraction of the person's body that is fat can be calculated. The more muscle and the less fat a person has, the higher his or her body density. For example, a muscular person weighing 150 lb has a smaller body volume (and thus a higher density) than a fat person weighing 150 lb.

Example 2.14 Determining Density

The most common units for density are g/mL = g/cm^3.

At a local pawn shop a student finds a medallion that the shop owner insists is pure platinum. However, the student suspects that the medallion may actually be silver and thus much less valuable. The student buys the medallion only after the shop owner agrees to refund the price if the medallion is returned within two days. The student, a chemistry major, then takes the medallion to her lab and measures its density as follows. She first weighs the medallion and finds its mass to be 55.64 g. She then places some water in a graduated cylinder and reads the volume as 75.2 mL. Next she drops the medallion into the cylinder and reads the new volume as 77.8 mL. Is the medallion platinum (density = 21.4 g/cm^3) or silver (density = 10.5 g/cm^3)?

Solution

The densities of platinum and silver differ so much that the measured density of the medallion will show which metal is present. Because by definition

$$\text{Density} = \frac{\text{mass}}{\text{volume}}$$

to calculate the density of the medallion, we need its mass and its volume. The mass of the medallion is 55.64 g. The volume of the medallion can be obtained by taking the difference between the volume readings of the water in the graduated cylinder before and after the medallion was added.

$$\text{Volume of medallion} = 77.8 \text{ mL} - 75.2 \text{ mL} = 2.6 \text{ mL}$$

The volume appeared to increase by 2.6 mL when the medallion was added, so 2.6 mL represents the volume of the medallion. Now we can use the measured mass and volume of the medallion to determine its density:

$$\text{Density of medallion} = \frac{\text{mass}}{\text{volume}} = \frac{55.64 \text{ g}}{2.6 \text{ mL}} = 21 \text{ g/mL}$$

or

$$= 21 \text{ g/cm}^3$$

The medallion is really platinum.

✔ Self-Check Exercise 2.9

A student wants to identify the main component in a commercial liquid cleaner. He finds that 35.8 mL of the cleaner weighs 28.1 g. Of the following possibilities, which is the main component of the cleaner?

Substance	Density, g/cm^3
chloroform	1.483
diethyl ether	0.714
isopropyl alcohol	0.785
toluene	0.867

See Problems 2.93 and 2.94. ■

Example 2.15 Using Density in Calculations

Mercury has a density of 13.6 g/mL. What volume of mercury must be taken to obtain 225 g of the metal?

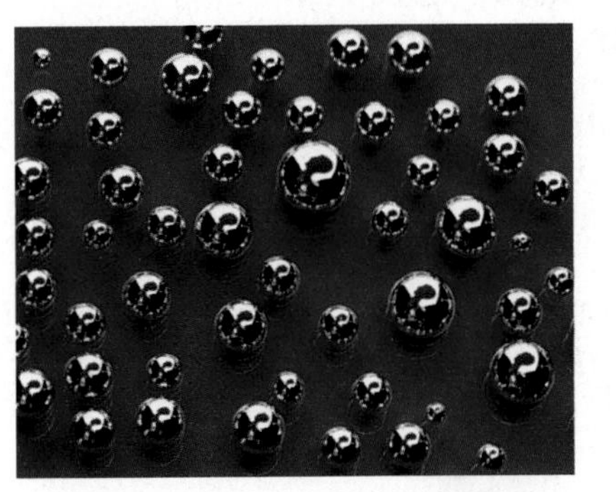

Spherical droplets of mercury, a very dense liquid.

Solution

To solve this problem, start with the definition of density,

$$\text{Density} = \frac{\text{mass}}{\text{volume}}$$

and then rearrange this equation to isolate the required quantity. In this case we want to find the volume. Remember that we maintain an equality when we do the same thing to both sides. For example, if we multiply *both sides* of the density definition by volume,

$$\text{Volume} \times \text{density} = \frac{\text{mass}}{\cancel{\text{volume}}} \times \cancel{\text{volume}}$$

volume cancels on the right, leaving

$$\text{Volume} \times \text{density} = \text{mass}$$

We want the volume, so we now divide both sides by density,

$$\frac{\text{Volume} \times \cancel{\text{density}}}{\cancel{\text{density}}} = \frac{\text{mass}}{\text{density}}$$

to give

$$\text{Volume} = \frac{\text{mass}}{\text{density}}$$

Now we can solve the problem by substituting the given numbers:

$$\text{Volume} = \frac{225\ \text{g}}{13.6\ \text{g/mL}} = 16.5\ \text{mL}$$

We must take 16.5 mL of mercury to obtain an amount that has a mass of 225 g. ■

The densities of various common substances are given in Table 2.8.

Table 2.8 Densities of Various Common Substances at 20 °C

Substance	Physical State	Density (g/cm^3)
oxygen	gas	0.00133*
hydrogen	gas	0.000084*
ethanol	liquid	0.785
benzene	liquid	0.880
water	liquid	1.000
magnesium	solid	1.74
salt (sodium chloride)	solid	2.16
aluminum	solid	2.70
iron	solid	7.87
copper	solid	8.96
silver	solid	10.5
lead	solid	11.34
mercury	liquid	13.6
gold	solid	19.32

*At 1 atmosphere pressure

Besides being a tool for the identification of substances, density has many other uses. For example, the liquid in your car's lead storage battery (a solution of sulfuric acid) changes density because the sulfuric acid is consumed as the battery discharges. In a fully charged battery, the density of the solution is about 1.30 g/cm^3. When the density falls below 1.20 g/cm^3, the battery has to be recharged. Density measurement is also used to determine the amount of antifreeze, and thus the level of protection against freezing, in the cooling system of a car. Water and antifreeze have different densities, so the measured density of the mixture tells us how much of each is present. The device used to test the density of the solution—a hydrometer—is shown in Figure 2.11.

In certain situations, the term *specific gravity* is used to describe the density of a liquid. **Specific gravity** is defined as the ratio of the density of a given liquid to the density of water at 4 °C. Because it is a ratio of densities, specific gravity has no units.

Figure 2.11
A hydrometer being used to determine the density of the antifreeze solution in a car's radiator.

Chapter 2 Review

Key Terms

measurement (p. 15)
scientific notation (2.1)
units (2.2)
English system (2.2)
metric system (2.2)
SI units (2.2)
volume (2.3)
mass (2.3)
significant figures (2.4)
rounding off (2.5)
conversion factor (2.6)
equivalence statement (2.6)
dimensional analysis (2.6)
Fahrenheit scale (2.7)
Celsius scale (2.7)
Kelvin (absolute) scale (2.7)
density (2.8)
specific gravity (2.8)

Summary

1. A quantitative observation is called a measurement and always consists of a number and a unit.
2. We can conveniently express very large or very small numbers using scientific notation, which represents the number as a number between 1 and 10 multiplied by 10 raised to a power.
3. Units give a scale on which to represent the results of a measurement. The three systems discussed are the English, metric, and SI systems. The metric and SI systems use prefixes (Table 2.2) to change the size of the units.
4. The mass of an object represents the quantity of matter in that object.
5. All measurements have a degree of uncertainty, which is reflected in the number of significant figures used to express them. Various rules are used to round off to the correct number of significant figures in a calculated result.
6. We can convert from one system of units to another by a method called dimensional analysis, in which conversion factors are used.

7. Temperature can be measured on three different scales: Fahrenheit, Celsius, and Kelvin. We can readily convert among these scales.

8. Density is the amount of matter present in a given volume (mass per unit volume). That is,

$$\text{Density} = \frac{\text{mass}}{\text{volume}}$$

In-Class Discussion Questions

These questions are designed to be considered by groups of students in class. Often these questions work well for introducing a particular topic in class.

1. a. There are 365 days/year, 24 hours/day, 12 months/year, and 60 minutes/hour. How many minutes are there in one month?
   b. There are 24 hours/day, 60 minutes/hour, 7 days/week, and 4 weeks/month. How many minutes are there in one month?
   c. Why are these answers different? Which (if either) is more correct and why?

2. You go to a convenience store to buy candy and find the owner to be rather odd. He allows you to buy pieces only in multiples of four, and to buy four, you need \$0.23. He allows you only to use 3 pennies and 2 dimes. You have a bunch of pennies and dimes, and instead of counting them, you decide to weigh them. You have 636.3 g of pennies, and each penny weighs an average of 3.03 g. Each dime weighs an average of 2.29 g. Each piece of candy weighs an average of 10.23 g.
   a. How many pennies do you have?
   b. How many dimes do you need to buy as much candy as possible?
   c. How much would all of your dimes weigh?
   d. How many pieces of candy could you buy (based on the number of dimes from part b)?
   e. How much would this candy weigh?
   f. How many pieces of candy could you buy with twice as many dimes?

3. When a marble is dropped into a beaker of water, it sinks to the bottom. Which of the following is the best explanation?
   a. The surface area of the marble is not large enough to be held up by the surface tension of the water.
   b. The mass of the marble is greater than that of the water.
   c. The marble weighs more than an equivalent volume of the water.
   d. The force from dropping the marble breaks the surface tension of the water.
   e. The marble has greater mass and volume than the water.

   Explain each choice. That is, for choices you did not pick, explain why you feel they are wrong, and justify the choice you did pick.

4. Consider water in each graduated cylinder as shown:

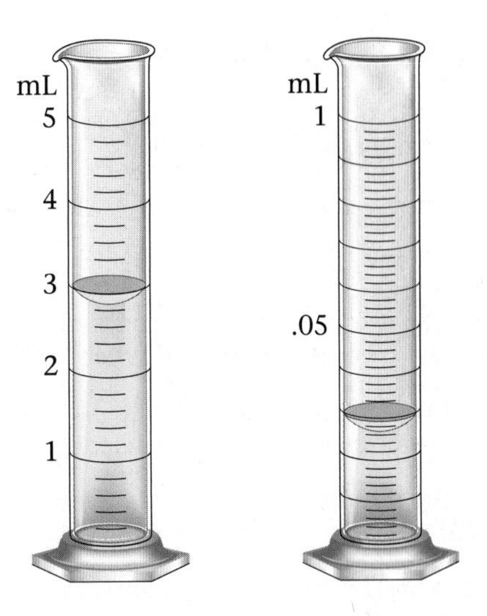

You add both samples of water to a beaker. How would you write the number describing the total volume? What limits the precision of this number?

5. What data would you need to estimate the money you would spend on gasoline to drive your car from New York to Chicago? Provide estimates of values and a sample calculation.

6. For each of the following numbers, indicate which zeros are significant and explain. Do not merely cite the rule that applies, but explain the rule.
   a. 10.020 b. 0.002050 c. 190 d. 270

7. Consider the addition of "15.4" to "28." What would a mathematician say the answer is? What would a scientist say? Justify the scientist's answer, not merely citing the rule, but explaining it.

8. Consider multiplying "26.2" by "16.43." What would a mathematician say the answer is? What would a scientist say? Justify the scientist's answer, not merely citing the rule, but explaining it.

9. In lab you report a measured volume of 128.7 mL of water. Using significant figures as a measure of the error, what range of answers does your reported volume imply? Explain.

10. Sketch two pieces of glassware: one that can measure volume to the thousandths place, and one that can measure volume only to the ones place.

11. Oil floats on water but is "thicker" than water. Why do you think this fact is true?

12. Show how converting numbers to scientific notation can help you decide which digits are significant.

13. You are driving 65 mph and take your eyes off the road "just for a second." How many feet do you travel in this time?

14. You have a 1.0 cm^3 sample of lead and a 1.0 cm^3 sample of glass. You drop each in a separate beaker of water. How do the volumes of water that are displaced by the samples compare? Explain.

Questions and Problems

All even-numbered exercises have answers in the back of this book and solutions in the *Solutions Guide.*

2.1 Scientific Notation

QUESTIONS

1. What is a *measurement?* Why does a measurement always consist of two parts, and what are those parts?
2. Write, in your own words, the steps involved in expressing a number in standard scientific notation.
3. When the number 1.521×10^3 is written in ordinary decimal notation, it is expressed as ______.
4. When expressed in standard scientific notation, numbers greater than 1 will have (positive/negative) exponents, whereas numbers less than 1 will have (positive/negative) exponents.

PROBLEMS

5. Write each of the following as an "ordinary" decimal number.
   a. 6.235×10^{-2}
   b. 7.229×10^3
   c. 5.001×10^{-6}
   d. 8.621×10^4
6. For each of the following numbers, if the number is rewritten in standard scientific notation, will the exponent of the power of 10 be positive or negative?
   a. 1,942,200
   b. 15
   c. 0.151
   d. 0.0000000721
7. Express each of the following numbers in *standard* scientific notation.
   a. 0.005219
   b. 5219
   c. 6,199,291
   d. 0.1973
   e. 93,000,000
   f. 72.41×10^{-2}
   g. 0.007241×10^{-5}
   h. 1.00
8. Express each of the following numbers in *standard* scientific notation.
   a. 9,367,421
   b. 7241
   c. 0.0005519
   d. 5.408
   e. 6.24×10^2
   f. 6319×10^{-2}
   g. 0.000000007215
   h. 0.721
9. Express each of the following as an "ordinary" decimal number.
   a. 6.442×10^3
   b. 5.991×10^{-5}
   c. 2.001×10^4
   d. 1.997×10^{-3}
   e. 7.871×10^{-1}
   f. 1.001×10^1
   g. 9.721×10^{-4}
   h. 2.015×10^6
   i. 5.583×10^{-2}
   j. 4.227×10^{-6}
   k. 9.734×10^3
   l. 1.000×10^1
10. Express each of the following as an "ordinary" decimal number.
    a. 7.327×10^{-4}
    b. 1.51×10^2
    c. 1×10^0
    d. 5.399×10^{-4}
    e. 0.221×10^3
    f. 7.83×10^{-2}
    g. 1218×10^{-4}
    h. 2.918×10^{-4}
    i. 7.251×10^3
    j. 1.911×10^{-9}
    k. 9.951×10^2
    l. 9.951×10^{-2}
11. Write each of the following numbers in *standard* scientific notation.
    a. 4381×10^{-4}
    b. $98{,}784 \times 10^4$
    c. 78.21×10^2
    d. 9.871×10^{-4}
    e. 0.009871×10^7
    f. $42{,}221 \times 10^4$
    g. 0.00008951×10^6
    h. $0.00008951 \times 10^{-6}$
12. Write each of the following numbers in *standard* scientific notation.
    a. 8714×10^2
    b. 0.0006591×10^8
    c. 0.0231×10^4
    d. $0.00000001519 \times 10^{-6}$
    e. 12.94×10^{-3}
    f. 1.921×10^{-4}
    g. 1
    h. $7{,}354{,}211 \times 10^8$
13. Write each of the following numbers in *standard* scientific notation.
    a. 1/1033
    b. $1/10^5$
    c. $1/10^{-7}$
    d. 1/0.0002
    e. 1/3,093,000
    f. $1/10^{-4}$
    g. $1/10^9$
    h. 1/0.000015
14. Write each of the following numbers in *standard* scientific notation.
    a. 1/0.00032
    b. $10^3/10^{-3}$
    c. $10^3/10^3$
    d. 1/55,000
    e. $(10^5)(10^4)(10^{-4})/(10^{-2})$
    f. $43.2/(4.32 \times 10^{-5})$
    g. $(4.32 \times 10^{-5})/432$
    h. $1/(10^5)(10^{-6})$

2.2 Units

QUESTIONS

15. What are the names for the fundamental units of mass, distance, and time in the metric system (SI)?
16. Although the standard of mass is, strictly speaking, the kilogram, in chemistry we more commonly make measurements of mass in terms of how many ______ a sample weighs.
17. For each of the following descriptions, identify the power of 10 being indicated by a *prefix* in the measurement.
    a. For my lunch, I had a cheeseburger that contained 1/8 of a *kilo*gram of beef.
    b. The sign on the highway I just passed said "Toronto 50 *kilo*meters."
    c. The hard drive on my new computer has a storage capacity of 40 *giga*bytes.
    d. My favorite radio station broadcasts at 93.7 *mega*hertz on the FM dial.
    e. The liquid medication I have to give my dog says it contains 2 *milli*grams of active ingredient per cubic *centi*meter.
    f. I just won the lottery for "*mega*bucks"!
18. Give the metric prefix that corresponds to each of the following:
    a. 1,000,000
    b. 10^{-3}
    c. 10^{-9}
    d. 10^6
    e. 10^{-2}
    f. 0.000001

2.3 Measurements of Length, Volume, and Mass

QUESTIONS

Students often have trouble relating measurements in the metric system to the English system they have grown up with. Give the approximate English system equivalents for each of the following metric system descriptions in exercises 19–22.

19. I just purchased *12 meters* of fabric at the sewing store for making new curtains.

20. My wife asked me to pick up *half a kilogram* of hamburger for dinner on the way home.

21. Many people purchase their soda in *two-liter* containers to save money.

22. My French cookbook says I need *250 milliliters* of wine to make this sauce.

23. Which is the greater distance, 50 miles or 100 kilometers?

24. Which weighs more, a pound of hamburger or a kilogram of hamburger?

25. The length 52.2 mm can also be expressed as ________ cm.

26. Who is taller, a man who is 1.62 m tall or a woman who is 5 ft 6 in. tall?

27. The fundamental SI unit of length is the meter. However, we often deal with larger or smaller lengths or distances for which multiples or fractions of the fundamental unit are more useful. For each of the following situations, suggest what fraction or multiple of the meter might be the most appropriate measurement.
    a. the distance between Chicago and Saint Louis
    b. the size of your bedroom
    c. the dimensions of this textbook
    d. the thickness of a hair

28. Which metric unit of length or distance is most comparable in scale to each of the following English system units for making measurements?
    a. an inch
    b. a yard
    c. a mile

29. The unit of volume in the metric system is the liter, which consists of 1000 milliliters. How many liters or milliliters is each of the following common English system measurements approximately equivalent to?
    a. a gallon of gasoline
    b. a pint of milk
    c. a cup of water

30. Which metric system unit is most appropriate for measuring the distance between two cities?
    a. meters
    b. millimeters
    c. centimeters
    d. kilometers

For exercises 31 and 32 some examples of simple approximate metric–English equivalents are given in Table 2.6.

31. What is the value in dollars of a stack of dimes that is 10 cm high?

32. How many quarters would have to be lined up in a row to reach a length of 1 meter?

2.4 Uncertainty in Measurement

QUESTIONS

33. If you were to measure the width of this page using a ruler, and you used the ruler to the limits of precision permitted by the scale on the ruler, the last digit you would write down for the measurement would be *uncertain* no matter how careful you were. Explain.

34. No matter how careful an experimenter may be, a measurement always has some degree of ________.

35. For the pin shown in Figure 2.5, why is the third figure determined for the length of the pin uncertain? Considering that the third figure is uncertain, explain why the length of the pin is indicated as 2.85 cm rather than, for example, 2.83 or 2.87 cm.

36. Why can the length of the pin shown in Figure 2.5 not be recorded as 2.850 cm?

2.5 Significant Figures

QUESTIONS

37. Indicate the number of significant figures in each of the following:
    a. 1000
    b. 1000.
    c. 100.01
    d. 1×10^2
    e. 0.0000000001

38. Indicate the number of significant figures implied in each of the following statements:
    a. The population of the United States is 250 million.
    b. One hour is equivalent to 60 minutes.
    c. There are 5280 feet in 1 mile.
    d. Jet airliners fly at 500 mi/h.
    e. The "Daytona 500" is a 500-mile race.

Rounding Off Numbers

QUESTIONS

39. When we are rounding off a number, if the number to the right of the digit to be rounded is less than 5, the digit should ________.

40. When performing a chain of several calculations, we round off the (final/intermediate) answer(s).

41. Round off each of the following numbers to three significant digits and express the result in standard scientific notation.
    a. 93,101,000
    b. 2.9881×10^{-6}
    c. 0.000048814
    d. 7896×10^{6}
    e. 0.004921×10^{-4}

42. Express each of the following numbers in standard scientific notation, rounding off each to three significant digits.

a. 422.65×10^{-3}
b. 71.246×10^{5}
c. 0.00044515
d. 22.9987×10^{-5}
e. 9.7222×10^{5}

43. Round off each of the following numbers to the indicated number of significant digits.
a. 102.4005 to five digits
b. 15.9995 to three digits
c. 1.6385 to four digits
d. 7.355 to three digits

44. Round off each of the following numbers to the indicated number of significant digits and write the answer in standard scientific notation.
a. 0.00034159 to three digits
b. 103.351×10^{2} to four digits
c. 17.9915 to five digits
d. 3.365×10^{5} to three digits

Determining Significant Figures in Calculations

QUESTIONS

45. When numbers are multiplied, the answer should contain the same number of significant figures as the multiplier with the ______ number of significant figures.

46. When numbers are added or subtracted, the limiting term is the one with the smallest number of ______ places.

47. When the calculation (0.005215)(0.08212)(273.2)/(4.1) is performed, the answer should be reported to ______ significant digits.

48. Try this with your calculator: Enter 2 ÷ 3 and press the = sign. What does your calculator say is the answer? What would be wrong with that answer if the 2 and 3 were experimentally determined numbers?

49. How many digits after the decimal point should be reported when the calculation (199.0354 + 43.09 + 121.2) is performed?

50. How many digits after the decimal point should be reported when the calculation (10,434 − 9.3344) is performed?

PROBLEMS

Note: See the Appendix for help in doing mathematical operations with numbers that contain exponents.

51. Evaluate each of the following, and write the answer to the correct number of significant figures.
a. 97.381 + 4.2502 + 0.99195
b. 171.5 + 72.915 − 8.23
c. 1.00914 + 0.87104 + 1.2012
d. 21.901 − 13.21 − 4.0215

52. Evaluate each of the following and write the answer to the appropriate number of significant figures.
a. 212.2 + 26.7 + 402.09
b. 1.0028 + 0.221 + 0.10337
c. 52.331 + 26.01 − 0.9981
d. $2.01 \times 10^{2} + 3.014 \times 10^{3}$

53. *Without actually performing the calculations indicated,* tell to how many significant digits the answer to the calculation should be expressed.
a. (10.12 + 17.381 + 18.2)/(1.41)
b. $(0.173)(6.022 \times 10^{23})$
c. (2.51)(0.08202)(298)/(765.2)
d. $\dfrac{(9.77732)(47.112)}{(273.24)(0.2)}$

54. *Without actually performing the calculations indicated,* tell to how many significant digits the answer to the calculation should be expressed.
a. $\dfrac{(9.7871)(2)}{(0.00182)(43.21)}$
b. (67.41 + 0.32 + 1.98)/(18.225)
c. $(2.001 \times 10^{-3})(4.7 \times 10^{-6})(68.224 \times 10^{-2})$
d. (72.15)(63.9)[1.98 + 4.8981]

55. Evaluate each of the following, and write the answer to the appropriate number of significant figures.
a. $(2.3232 + 0.2034 - 0.16) \times (4.0 \times 10^{3})$
b. $(1.34 \times 10^{2} + 3.2 \times 10^{1})/(3.32 \times 10^{-6})$
c. $(4.3 \times 10^{6})/(4.334 + 44.0002 - 0.9820)$
d. $(2.043 \times 10^{-2})^{3}$

56. Evaluate each of the following and write the answer to the appropriate number of significant figures.
a. (2.0944 + 0.0003233 + 12.22)/(7.001)
b. $(1.42 \times 10^{2} + 1.021 \times 10^{3})/(3.1 \times 10^{-1})$
c. $(9.762 \times 10^{-3})/(1.43 \times 10^{2} + 4.51 \times 10^{1})$
d. $(6.1982 \times 10^{-4})^{2}$

2.6 Problem Solving and Dimensional Analysis

QUESTIONS

57. A ______ represents a ratio based on an equivalence statement between two measurements.

58. How many significant figures are understood for the numbers in the following definition: 1 mi = 5280 ft?

59. Given that 1 mi = 1760 yd, determine what conversion factor is appropriate to convert 1849 yd to miles; to convert 2.781 mi to yards.

60. Given that 1 L = 1000 mL, determine what conversion factor is appropriate to convert 2.75 L to milliliters; to convert 255 mL to liters.

For exercises 61 and 62, apples cost $0.79 per pound.

61. What conversion factor is appropriate to express the cost of 5.3 lb of apples?

62. What conversion factor could be used to determine how many pounds of apples could be bought for $2.00?

PROBLEMS

Note: Appropriate equivalence statements for various units are found inside the back cover of this book.

63. Perform each of the following conversions, being sure to set up clearly the appropriate conversion factor in each case.
a. 32 seconds to minutes
b. 2.4 lb to kilograms
c. 2.4 lb to grams
d. 3150 ft to miles
e. 14.2 in. to feet
f. 22.4 g to kilograms
g. 9.72 mg to grams
h. 2.91 m to yards

64. Perform each of the following conversions, being sure to set up clearly the appropriate conversion factor in each case.
a. 2.23 m to yards
b. 46.2 yd to meters
c. 292 cm to inches
d. 881.2 in. to centimeters
e. 1043 km to miles
f. 445.5 mi to kilometers
g. 36.2 m to kilometers
h. 0.501 km to centimeters

65. Perform each of the following conversions, being sure to set up clearly the appropriate conversion factor in each case. The inside cover of this book provides equivalence statements in addition to those contained in this chapter.
a. 17.3 L to cubic feet
b. 17.3 L to milliliters
c. 8.75 L to gallons
d. 762 g to ounces
e. 1.00 g to atomic mass units
f. 1.00 L to pints
g. 64.5 g to kilograms
h. 72.1 mL to liters

66. Perform each of the following conversions, being sure to set up clearly the appropriate conversion factor in each case.
a. 254.3 g to kilograms
b. 2.75 kg to grams
c. 2.75 kg to pounds
d. 2.75 kg to ounces
e. 534.1 g to pounds
f. 1.75 lb to grams
g. 8.7 oz to grams
h. 45.9 g to ounces

67. If $1.00 is equivalent to 1.74 German marks, what is $20.00 worth in marks? What is the value in dollars of a 100-mark bill?

68. Boston and New York City are 190 miles apart. What is this distance in kilometers? in meters? in feet?

69. The United States has high-speed trains running between Boston and New York capable of speeds up to 160 mi/h. Are these trains faster or slower than the fastest trains in the United Kingdom, which reach speeds of 225 km/h?

70. The radius of an atom is on the order of 10^{-10} m. What is this radius in centimeters? in inches? in nanometers?

2.7 Temperature Conversions

QUESTIONS

71. The temperature scale used in everyday life in most of the world except the United States is the ______ scale.

72. The ______ point of water is at 32° on the Fahrenheit temperature scale.

73. The normal boiling point of water is ______ °F, or ______ °C.

74. The freezing point of water is ______ K.

75. On both the Celsius and Kelvin temperature scales, there are ______ degrees between the normal freezing and boiling points of water.

76. On which temperature scale (°F, °C, or K) does 1 degree represent the smallest change in temperature?

PROBLEMS

77. Make the following temperature conversions:
a. 22.5 °C to kelvins
b. 444.9 K to °C
c. 0 °C to kelvins
d. 298.1 K to °C

78. Make the following temperature conversions:
a. −210 °C to kelvins
b. 275 K to °C
c. 778 K to °C
d. 778 °C to kelvins

79. Convert the following Fahrenheit temperatures to Celsius degrees.
a. a chilly morning in early autumn, 45 °F
b. a hot, dry day in the Arizona desert, 115 °F
c. the temperature in winter when my car won't start, −10 °F
d. the surface of a star, 10,000 °F

80. Convert the following Celsius temperatures to Fahrenheit degrees.
a. the boiling temperature of ethyl alcohol, 78.1 °C
b. a hot day at the beach on a Greek isle, 40. °C
c. the lowest possible temperature, −273 °C
d. the body temperature of a person with hypothermia, 32 °C

81. Perform the indicated temperature conversions.
a. 25.1 °F to °C
b. 25.1 °C to °F
c. 25.1 K to °C
d. 25.1 K to °F

82. Perform the indicated temperature conversions.
a. 275 K to °C
b. 82 °F to °C
c. −21 °C to °F
d. −40 °F to °C (Notice anything unusual about your answer?)

2.8 Density

QUESTIONS

83. What does the *density* of a substance represent?

84. The most common units for density are ______.

85. A kilogram of lead occupies a much smaller volume than a kilogram of water, because ______ has a much higher density.

86. If a solid block of glass, with a volume of exactly 100 $in.^3$, is placed in a basin of water that is full to the brim, then ______ of water will overflow from the basin.

87. Typically, gases have very (high/low) densities compared to those of solids and liquids (see Table 2.8).

88. What property of density makes it useful as an aid in identifying substances?

89. Referring to Table 2.8, which substance listed is most dense? Which substance is least dense? For the two substances you have identified, for which one would a 1.00-g sample occupy the larger volume?

90. Referring to Table 2.8, determine whether copper, silver, lead, or mercury is the least dense.

PROBLEMS

91. For the masses and volumes indicated, calculate the density in grams per cubic centimeter.
   a. mass = 452.1 g; volume = 292 cm^3
   b. mass = 0.14 lb; volume = 125 mL
   c. mass = 1.01 kg; volume = 1000 cm^3
   d. mass = 225 mg; volume = 2.51 mL

92. For the masses and volumes indicated, calculate the density in grams per cubic centimeter.
   a. mass = 122.4 g; volume = 5.5 cm^3
   b. mass = 19,302 g; volume = 0.57 m^3
   c. mass = 0.0175 kg; volume = 18.2 mL
   d. mass = 2.49 g; volume = 0.12 m^3

93. If ethanol (grain alcohol) has a density of 0.785 g/mL, calculate the volume of 82.5 g of ethanol.

94. Mercury, the liquid metal used in thermometers, is very dense at 13.6 g/cm^3. What would be the mass of 125 mL of mercury?

95. A cube of metal weighs 1.45 kg and displaces 542 mL of water when immersed. Calculate the density of the metal.

96. A material will float on the surface of a liquid if the material has a density less than that of the liquid. Given that the density of water is approximately 1.0 g/mL under many conditions, will a block of material having a volume of 1.2×10^4 $in.^3$ and weighing 3.5 lb float or sink when placed in a reservoir of water?

97. Iron has density 7.87 g/cm^3. If 52.4 g of iron is added to 75.0 mL of water in a graduated cylinder, to what volume reading will the water level in the cylinder rise?

98. The density of pure silver is 10.5 g/cm^3 at 20 °C. If 5.25 g of pure silver pellets is added to a graduated cylinder containing 11.2 mL of water, to what volume level will the water in the cylinder rise?

99. Use the information in Table 2.8 to calculate the volume of 50.0 g of each of the following substances.
   a. sodium chloride
   b. mercury
   c. benzene
   d. silver

100. Use the information in Table 2.8 to calculate the mass of 50.0 cm^3 of each of the following substances.
   a. gold
   b. iron
   c. lead
   d. aluminum

ADDITIONAL PROBLEMS

101. Indicate the number of significant digits in the answer when each of the following expressions is evaluated (you do *not* have to evaluate the expression).
   a. (6.25)/(74.1143)
   b. $(1.45)(0.08431)(6.022 \times 10^{23})$
   c. (4.75512)(9.74441)/(3.14)

102. Express each of the following as an "ordinary" decimal number.
   a. 3.011×10^{23}
   b. 5.091×10^9
   c. 7.2×10^2
   d. 1.234×10^5
   e. 4.32002×10^{-4}
   f. 3.001×10^{-2}
   g. 2.9901×10^{-7}
   h. 4.2×10^{-1}

103. Write each of the following numbers in standard scientific notation, rounding off the numbers to three significant digits.
   a. 424.6174
   b. 0.00078145
   c. 26,755
   d. 0.0006535
   e. 72.5654

104. Which unit of length in the metric system would be most appropriate in size for measuring each of the following items?
   a. the dimensions of this page
   b. the size of the room in which you are sitting
   c. the distance from New York to London
   d. the diameter of a baseball
   e. the diameter of a common pin

105. Make the following conversions.
   a. 1.25 in. to feet and to centimeters
   b. 2.12 qt to gallons and to liters
   c. 2640 ft to miles and to kilometers
   d. 1.254 kg lead to its volume in cubic centimeters
   e. 250. mL ethanol to its mass in grams
   f. 3.5 $in.^3$ of mercury to its volume in milliliters and its mass in kilograms

106. On the planet Xgnu, the most common units of length are the blim (for long distances) and the kryll (for shorter distances). Because the Xgnuese have 14 fingers, it is not perhaps surprising that 1400 kryll = 1 blim.
   a. Two cities on Xgnu are 36.2 blim apart. What is this distance in kryll?

b. The average Xgnuese is 170 kryll tall. What is this height in blims?
c. This book is presently being used at Xgnu University. The area of the cover of this book is 72.5 square krylls. What is its area in square blims?

107. You pass a road sign saying "New York 110 km." If you drive at a constant speed of 100. km/h, how long should it take you to reach New York?

108. At the mall, you decide to try on a pair of French jeans. Naturally, the waist size of the jeans is given in centimeters. What does a waist measurement of 52 cm correspond to in inches?

109. Suppose your car is rated at 45 mi/gal for highway use and 38 mi/gal for city driving. If you wanted to write your friend in Spain about your car's mileage, what ratings in kilometers per liter would you report?

110. You are in Paris, and you want to buy some peaches for lunch. The sign in the fruit stand indicates that peaches are 11.5 francs per kilogram. Given that there are approximately 5 francs to the dollar, calculate what a pound of peaches will cost in dollars.

111. For a pharmacist dispensing pills or capsules, it is often easier to weigh the medication to be dispensed rather than to count the individual pills. If a single antibiotic capsule weighs 0.65 g, and a pharmacist weighs out 15.6 g of capsules, how many capsules have been dispensed?

112. On the planet Xgnu, the natives have 14 fingers. On the official Xgnuese temperature scale (°X), the boiling point of water (under an atmospheric pressure similar to Earth's) is 140 °X, whereas it freezes at 14 °X. Derive the relationship between °X and °C.

113. For a material to float on the surface of water, the material must have a density less than that of water (1.0 g/mL) and must not react with the water or dissolve in it. A spherical ball has a radius of 0.50 cm and weighs 2.0 g. Will this ball float or sink when placed in water? (*Note:* Volume of a sphere = $\frac{4}{3}\pi r^3$.)

114. A gas cylinder having a volume of 10.5 L contains 36.8 g of gas. What is the density of the gas?

115. Using Table 2.8, calculate the volume of 25.0 g of each of the following:
a. hydrogen gas (at 1 atmosphere pressure)
b. mercury
c. lead
d. water

116. Ethanol and benzene dissolve in each other. When 100. mL of ethanol is dissolved in 1.00 L of benzene, what is the mass of the mixture? (See Table 2.8.)

117. When 2891 is written in scientific notation, the exponent indicating the power of 10 is ______.

118. For each of the following numbers, if the number is rewritten in scientific notation, will the exponent of the power of 10 be positive, negative, or zero?
a. $1/10^3$
b. 0.00045
c. 52,550
d. 7.21
e. 1/3

119. For each of the following numbers, if the number is rewritten in scientific notation, will the exponent be positive, negative, or zero?
a. 4,915,442
b. 1/1000
c. 0.001
d. 3.75

120. For each of the following numbers, by how many places does the decimal point have to be moved to express the number in standard scientific notation? In each case, is the exponent positive or negative?
a. 102
b. 0.00000000003489
c. 2500
d. 0.00003489
e. 398,000
f. 1
g. 0.3489
h. 0.0000003489

121. For each of the following numbers, by how many places must the decimal point be moved to express the number in standard scientific notation? In each case, will the exponent be positive, negative, or zero?
a. 55,651
b. 0.000008991
c. 2.04
d. 883,541
e. 0.09814

122. For each of the following numbers, by how many places must the decimal point be moved to express the number in standard scientific notation? In each case, will the exponent be positive, negative, or zero?
a. 72.471
b. 0.008941
c. 9.9914
d. 6519
e. 0.000000008715

123. Express each of the following numbers in scientific (exponential) notation.
a. 529
b. 240,000,000
c. 301,000,000,000,000,000
d. 78,444
e. 0.0003442
f. 0.000000000902
g. 0.043
h. 0.0821

124. Express each of the following as an "ordinary" decimal number.
a. 2.98×10^{-5}
b. 4.358×10^{9}
c. 1.9928×10^{-6}
d. 6.02×10^{23}
e. 1.01×10^{-1}
f. 7.87×10^{-3}
g. 9.87×10^{7}
h. 3.7899×10^{2}
i. 1.093×10^{-1}
j. 2.9004×10^{0}
k. 3.9×10^{-4}
l. 1.904×10^{-8}

125. Write each of the following numbers in *standard* scientific notation.
a. 102.3×10^{-5}
b. 32.03×10^{-3}
c. 59933×10^{2}
d. 599.33×10^{4}
e. 5993.3×10^{3}
f. 2054×10^{-1}
g. $32,000,000 \times 10^{-6}$
h. 59.933×10^{5}

126. Write each of the following numbers in *standard* scientific notation. See the Appendix if you need help multiplying or dividing numbers with exponents.

a. $1/10^2$
b. $1/10^{-2}$
c. $55/10^3$
d. $(3.1 \times 10^6)/10^{-3}$
e. $(10^6)^{1/2}$
f. $(10^6)(10^4)/(10^2)$
g. $1/0.0034$
h. $3.453/10^{-4}$

127. The fundamental unit of length or distance in the metric system is the ______.

128. The SI unit of temperature is the ______.

129. Which distance is farther, 100 km or 50 mi?

130. The unit of volume corresponding to 1/1000 of a liter is referred to as 1 milliliter, or 1 cubic ______.

131. The volume 0.250 L could also be expressed as ______ mL.

132. The distance 10.5 cm could also be expressed as ______ m.

133. Would an automobile moving at a constant speed of 100 km/h violate a 65-mph speed limit?

134. Which weighs more, 100 g of water or 1 kg of water?

135. Which weighs more, 4.25 grams of gold or 425 milligrams of gold?

136. The length 100 mm can also be expressed as ______ cm.

137. When a measurement is made, the certain numbers plus the first uncertain number are called the ______ of the measurement.

138. In the measurement of the length of the pin indicated in Figure 2.5, what are the *certain* numbers in the measurement shown?

139. Indicate the number of significant figures in each of the following:

a. This book contains over 500 pages.
b. A mile is just over 5000 ft.
c. A liter is equivalent to 1.059 qt.
d. The population of the United States is approaching 250 million.
e. A kilogram is 1000 g.
f. The Boeing 747 cruises at around 600 mi/h.

140. Round off each of the following numbers to three significant digits.

a. 0.00042557
b. 4.0235×10^{-5}
c. 5,991,556
d. 399.85
e. 0.0059998

141. Round off each of the following numbers to the indicated number of significant digits.

a. 0.75555 to four digits
b. 292.5 to three digits
c. 17.005 to four digits
d. 432.965 to five digits

142. Evaluate each of the following, and write the answer to the appropriate number of significant figures.

a. $149.2 + 0.034 + 2000.34$
b. $1.0322 \times 10^3 + 4.34 \times 10^3$
c. $4.03 \times 10^{-2} - 2.044 \times 10^{-3}$
d. $2.094 \times 10^5 - 1.073 \times 10^6$

143. Evaluate each of the following, and write the answer to the appropriate number of significant figures.

a. $(0.0432)(2.909)(4.43 \times 10^8)$
b. $(0.8922)/[(0.00932)(4.03 \times 10^2)]$
c. $(3.923 \times 10^2)(2.94)(4.093 \times 10^{-3})$
d. $(4.9211)(0.04434)/[(0.000934)(2.892 \times 10^{-7})]$

144. Evaluate each of the following, and write the answer to the appropriate number of significant figures.

a. $(2.9932 \times 10^4)[2.4443 \times 10^2 + 1.0032 \times 10^1]$
b. $[2.34 \times 10^2 + 2.443 \times 10^{-1}]/(0.0323)$
c. $(4.38 \times 10^{-3})^2$
d. $(5.9938 \times 10^{-6})^{1/2}$

145. Given that 1 L = 1000 cm^3, determine what conversion factor is appropriate to convert 350 cm^3 to liters; to convert 0.200 L to cubic centimeters.

146. Given that 12 months = 1 year, determine what conversion factor is appropriate to convert 72 months to years; to convert 3.5 years to months.

147. Perform each of the following conversions, being sure to set up clearly the appropriate conversion factor in each case.

a. 8.43 cm to millimeters
b. 2.41×10^2 cm to meters
c. 294.5 nm to centimeters
d. 404.5 m to kilometers
e. 1.445×10^4 m to kilometers
f. 42.2 mm to centimeters
g. 235.3 m to millimeters
h. 903.3 nm to micrometers

148. Perform each of the following conversions, being sure to set up clearly the appropriate conversion factor(s) in each case.

a. 908 oz to kilograms
b. 12.8 L to gallons
c. 125 mL to quarts
d. 2.89 gal to milliliters
e. 4.48 lb to grams
f. 550 mL to quarts

149. The mean distance from the earth to the sun is 9.3×10^7 mi. What is this distance in kilometers? in centimeters?

150. Given that one gross = 144 items, how many pencils are contained in 6 gross?

151. Convert the following temperatures to kelvins.

a. 0 °C
b. 25 °C
c. 37 °C
d. 100 °C
e. −175 °C
f. 212 °C

152. Carry out the indicated temperature conversions.

a. 175 °F to kelvins
b. 255 K to Celsius degrees
c. −45 °F to Celsius degrees
d. 125 °C to Fahrenheit degrees

153. For the masses and volumes indicated, calculate the density in grams per cubic centimeter.

a. mass = 234 g; volume = 2.2 cm^3
b. mass = 2.34 kg; volume = 2.2 m^3
c. mass = 1.2 lb; volume = 2.1 ft^3
d. mass = 4.3 ton; volume = 54.2 yd^3

154. A sample of a liquid solvent has density 0.915 g/mL. What is the mass of 85.5 mL of the liquid?

155. An organic solvent has density 1.31 g/mL. What volume is occupied by 50.0 g of the liquid?

156. A solid metal sphere has a volume of 4.2 ft^3. The mass of the sphere is 155 lb. Find the density of the metal sphere in grams per cubic centimeter.

157. A sample containing 33.42 g of metal pellets is poured into a graduated cylinder initially containing 12.7 mL of water, causing the water level in the cylinder to rise to 21.6 mL. Calculate the density of the metal.

158. Convert the following temperatures to Fahrenheit degrees.

a. −5 °C
b. 273 K
c. −196 °C
d. 0 K
e. 86 °C
f. −273 °C

Matter and Energy

Grand Prismatic Spring in Yellowstone National Park.

As you look around you, you must wonder about the properties of matter. How do plants grow and why are they green? Why is the sun hot? Why does a hot dog get hot in a microwave oven? Why does wood burn whereas rocks do not? What is a flame? How does soap work? Why does soda fizz when you open the bottle? When iron rusts, what's happening? And why doesn't aluminum rust? How does a cold pack for an athletic injury, which is stored for weeks or months at room temperature, suddenly get cold when you need it? How does a hair permanent work?

Why does soda fizz when you open the bottle?

The answers to these and endless other questions lie in the domain of chemistry. In this chapter we begin to explore the nature of matter: how it is organized and how and why it changes. We will also consider the energy that accompanies these changes.

Matter

AIM: To learn about matter and its three states.

Matter, the "stuff" of which the universe is composed, has two characteristics: it has mass and it occupies space. Matter comes in a great variety of forms: the stars, the air that you are breathing, the gasoline that you put in your car, the chair on which you are sitting, the turkey in the sandwich you may have had for lunch, the tissues in your brain that enable you to read and comprehend this sentence, and so on.

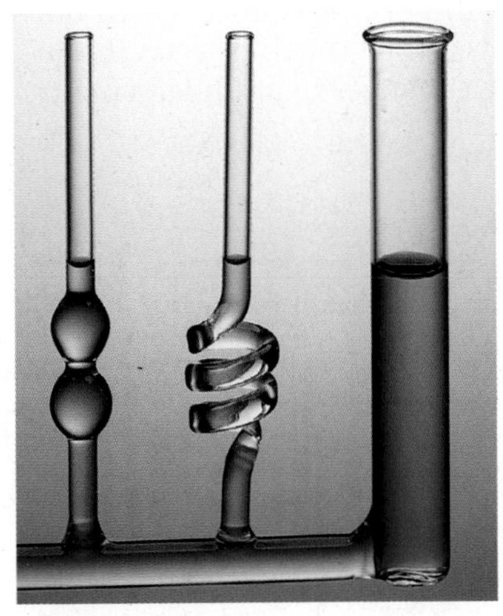

Figure 3.1
Liquid water takes the shape of its container.

To try to understand the nature of matter, we classify it in various ways. For example, wood, bone, and steel share certain characteristics. These things are all rigid; they have definite shapes that are difficult to change. On the other hand, water and gasoline, for example, take the shape of any container into which they are poured (see Figure 3.1). Even so, 1 L of water has a volume of 1 L whether it is in a pail or a beaker. In contrast, air takes the shape of its container and fills any container uniformly.

The substances we have just described illustrate the three **states of matter: solid, liquid,** and **gas.** These are defined and illustrated in Table 3.1. The state of a given sample of matter depends on the strength of the forces among the particles contained in the matter; the stronger these forces, the more rigid the matter. We will discuss this in more detail in the next section.

Table 3.1 The Three States of Matter

State	Definition	Examples
solid	rigid; has a fixed shape and volume	ice cube, diamond, iron bar
liquid	has a definite volume but takes the shape of its container	gasoline, water, alcohol, blood
gas	has no fixed volume or shape; takes the shape and volume of its container	air, helium, oxygen

How does this lush vegetation grow in a tropical rain forest, and why is it green?

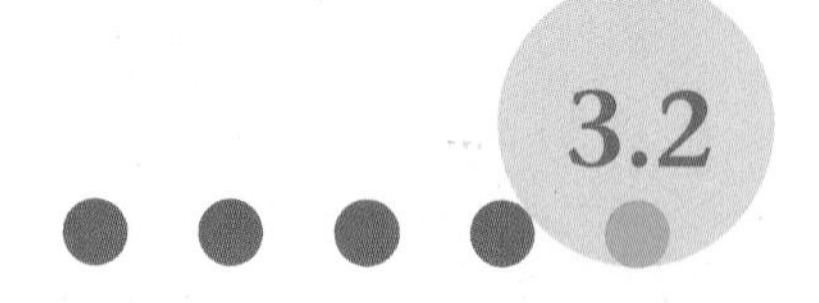

Physical and Chemical Properties and Changes

AIMS: To learn to distinguish between physical and chemical properties. To learn to distinguish between physical and chemical changes.

When you see a friend, you immediately respond and call him or her by name. We can recognize a friend because each person has unique characteristics or properties. The person may be thin and tall, may have blonde hair and blue eyes, and so on. The characteristics just mentioned are examples of **physical properties.** Substances also have physical properties. Typical physical properties of a substance include odor, color, volume, state (gas, liquid, or solid), density, melting point, and boiling point. We can also describe a pure substance in terms of its **chemical properties,** which refer to its ability to form new substances. An example of a chemical change is wood burning in a fireplace, giving off heat and gases and leaving a residue of ashes. In this process, the wood is changed to several new substances. Other examples of chemical changes include the rusting of the steel in our cars, the digestion of food in our stomachs, and the growth of grass in our yards. In a chemical change a given substance changes to a fundamentally different substance or substances.

Example 3.1 Identifying Physical and Chemical Properties

Classify each of the following as a physical or a chemical property.

a. The boiling point of a certain alcohol is 78 °C.

b. Diamond is very hard.

c. Sugar ferments to form alcohol.

d. A metal wire conducts an electric current.

Solution

Items (a), (b), and (d) are physical properties; they describe inherent characteristics of each substance, and no change in composition occurs. A metal wire has the same composition before and after an electric current has passed through it. Item (c) is a chemical property of sugar. Fermentation of sugars involves the formation of a new substance (alcohol).

Self-Check Exercise 3.1

Gallium metal has such a low melting point (30 °C) that it melts from the heat of a hand.

Which of the following are physical properties and which are chemical properties?

a. Gallium metal melts in your hand.

b. Platinum does not react with oxygen at room temperature.

c. This page is white.

d. The copper sheets that form the "skin" of the Statue of Liberty have acquired a greenish coating over the years.

See Problems 3.11 through 3.14. ■

The letters indicate atoms and the lines indicate attachments (bonds) between atoms.

The purpose here is to give an overview. Don't worry about the precise definitions of *atom* and *molecule* now. We will explore these concepts more fully in Chapter 4.

Matter can undergo changes in both its physical and its chemical properties. To illustrate the fundamental differences between physical and chemical changes, we will consider water. As we will see in much more detail in later chapters, a sample of water contains a very large number of individual units (called molecules), each made up of two atoms of hydrogen and one atom of oxygen—the familiar H_2O. This molecule can be represented as

where the letters stand for atoms and the lines show attachments (called bonds) between atoms, and the molecular model (on the right) represents water in a more three-dimensional fashion.

What is really occurring when water undergoes the following changes?

We will describe these changes of state precisely in Chapter 13, but you already know something about these processes because you have observed them many times.

When ice melts, the rigid solid becomes a mobile liquid that takes the shape of its container. Continued heating brings the liquid to a boil, and the water becomes a gas or vapor that seems to disappear into "thin air." The changes that occur as the substance goes from solid to liquid to gas are represented in Figure 3.2. In ice the water molecules are locked into fixed positions. In the liquid the molecules are still very close together, but some motion is occurring; the positions of the molecules are no longer fixed as they are in ice. In the gaseous state the molecules are much farther apart and move randomly, hitting each other and the walls of the container.

The most important thing about all these changes is that the water molecules are still intact. The motions of individual molecules and the distances between them change, but *H_2O molecules are still present.* These changes of state are **physical changes** because they do not affect the composition of the substance. In each state we still have water (H_2O), not some other substance.

An iron pyrite crystal (gold color) on a white quartz crystal.

Now suppose we run an electric current through water as illustrated in Figure 3.3. Something very different happens. The water disappears and is

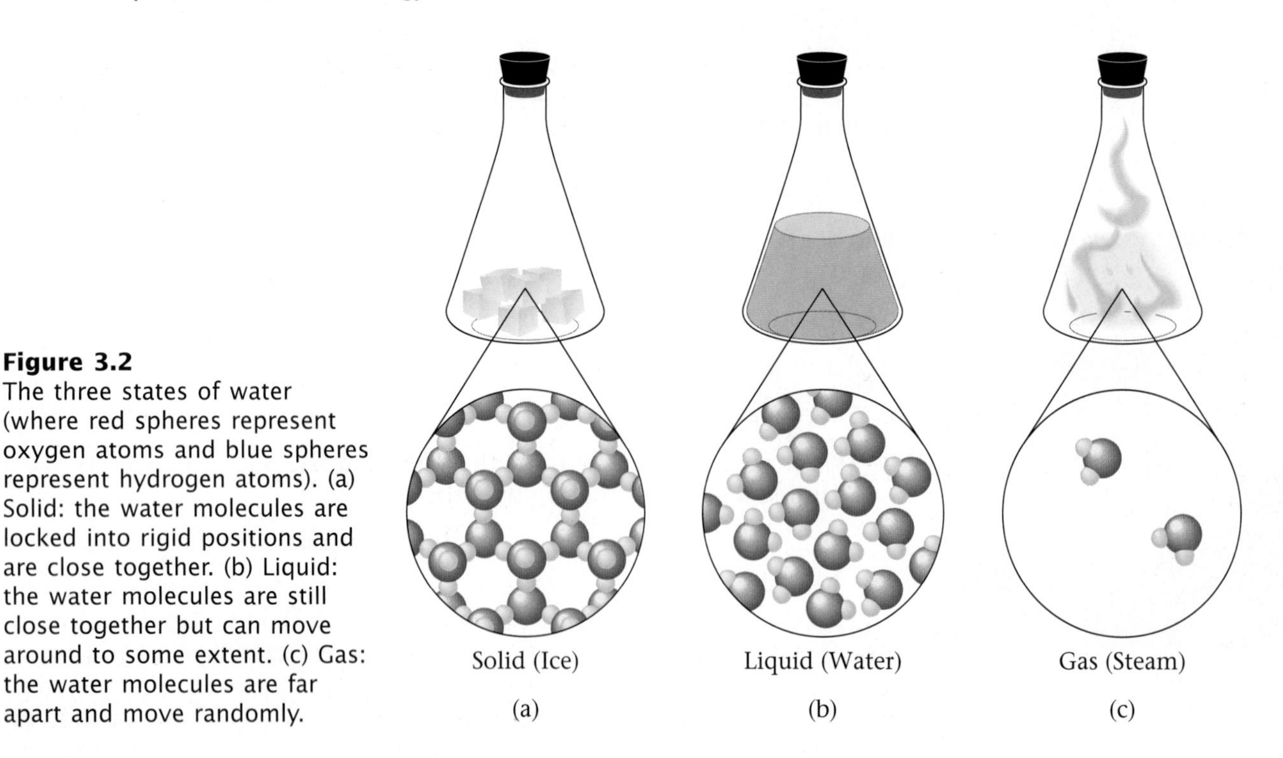

Figure 3.2
The three states of water (where red spheres represent oxygen atoms and blue spheres represent hydrogen atoms). (a) Solid: the water molecules are locked into rigid positions and are close together. (b) Liquid: the water molecules are still close together but can move around to some extent. (c) Gas: the water molecules are far apart and move randomly.

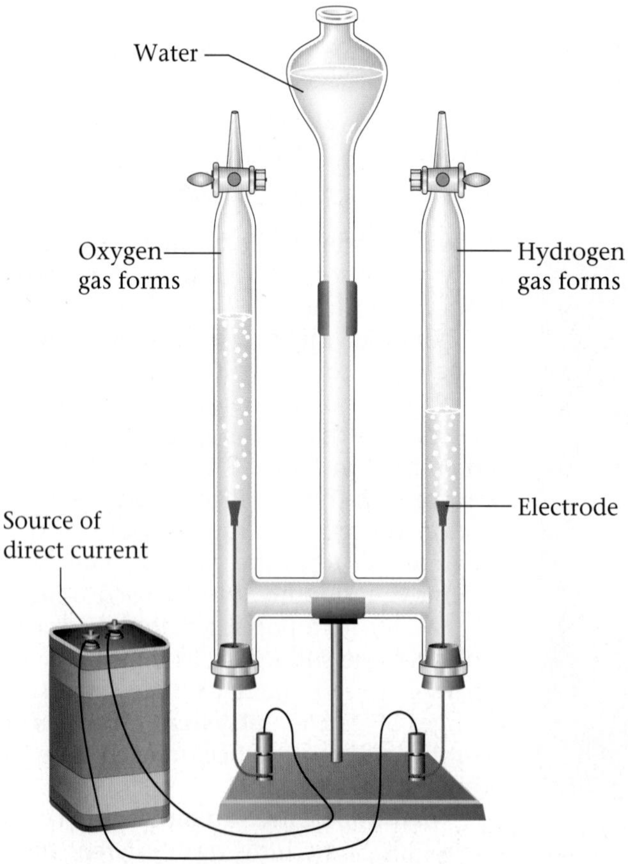

Water
Oxygen gas forms
Hydrogen gas forms
Electrode
Source of direct current

Figure 3.3
Electrolysis, the decomposition of water by an electric current, is a chemical process.

replaced by two new gaseous substances, hydrogen and oxygen. An electric current actually causes the water molecules to come apart—the water *decomposes* to hydrogen and oxygen. We can represent this process as follows:

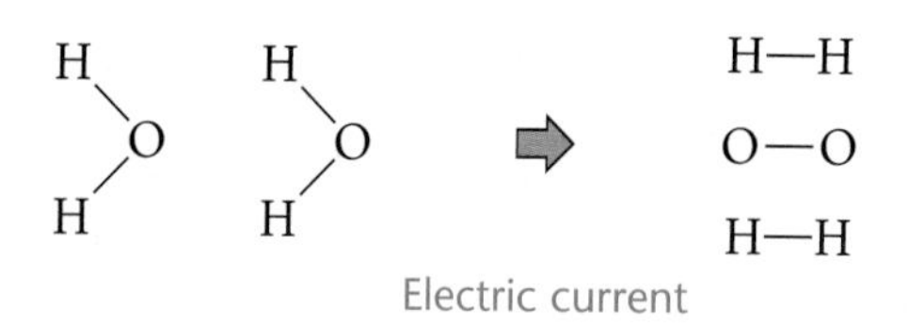

This is a **chemical change** because water (consisting of H_2O molecules) has changed into different substances: hydrogen (containing H_2 molecules) and oxygen (containing O_2 molecules). Thus in this process, the H_2O molecules have been replaced by O_2 and H_2 molecules. Let us summarize:

Physical and Chemical Changes

1. A *physical change* involves a change in one or more physical properties, but no change in the fundamental components that make up the substance. The most common physical changes are changes of state: solid ⇔ liquid ⇔ gas.
2. A *chemical change* involves a change in the fundamental components of the substance; a given substance changes into a different substance or substances. Chemical changes are called **reactions:** silver tarnishes by reacting with substances in the air; a plant forms a leaf by combining various substances from the air and soil; and so on.

Example 3.2 Identifying Physical and Chemical Changes

Oxygen combines with the chemicals in wood to produce flames. Is a physical or chemical change taking place?

Classify each of the following as a physical or a chemical change.

a. Iron metal is melted.

b. Iron combines with oxygen to form rust.

c. Wood burns in air.

d. A rock is broken into small pieces.

Solution

a. Melted iron is just liquid iron and could cool again to the solid state. This is a physical change.

b. When iron combines with oxygen, it forms a different substance (rust) that contains iron and oxygen. This is a chemical change because a different substance forms.

c. Wood burns to form different substances (as we will see later, they include carbon dioxide and water). After the fire, the wood is no longer in its original form. This is a chemical change.

d. When the rock is broken up, all the smaller pieces have the same composition as the whole rock. Each new piece differs from the original only in size and shape. This is a physical change.

Self-Check Exercise 3.2

Classify each of the following as a chemical change, a physical change, or a combination of the two.

a. Milk turns sour.

b. Wax is melted over a flame and then catches fire and burns.

See Problems 3.17 and 3.18. ■

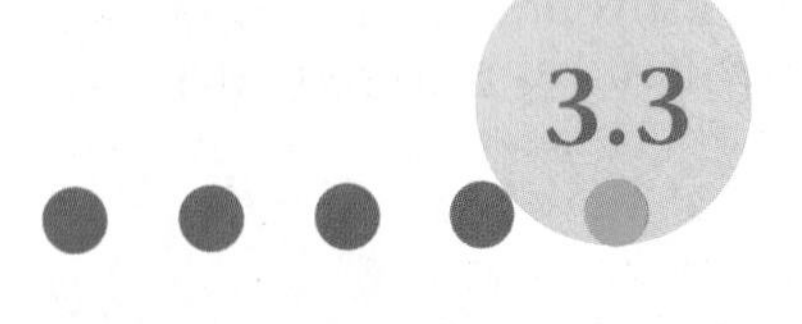

Elements and Compounds

AIM: To understand the definitions of elements and compounds.

Element: a substance that cannot be broken down into other substances by chemical methods.

As we examine the chemical changes of matter, we encounter a series of fundamental substances called **elements.** Elements cannot be broken down into other substances by chemical means. Examples of elements are iron, aluminum, oxygen, and hydrogen. All of the matter in the world around us contains elements. The elements sometimes are found in an isolated state, but more often they are combined with other elements. Most substances contain several elements combined together.

Compound: a substance composed of a given combination of elements that can be broken down into those elements by chemical methods.

The atoms of certain elements have special affinities for each other. They bind together in special ways to form **compounds,** substances that have the same composition no matter where we find them. Because compounds are made of elements, they can be broken down into elements through chemical changes:

Compounds ➡ Elements
Chemical changes

Water is an example of a compound. Pure water always has the same composition (the same relative amounts of hydrogen and oxygen) because it consists of H_2O molecules. Water can be broken down into the elements hydrogen and oxygen by chemical means, such as by the use of an electric current (see Figure 3.3).

As we will discuss in more detail in Chapter 4, each element is made up of a particular kind of atom: a pure sample of the element aluminum contains only aluminum atoms, elemental copper contains only copper atoms, and so on. Thus an element contains only one kind of atom; a sample of iron contains many atoms, but they are all iron atoms. Samples of certain pure elements do contain molecules; for example, hydrogen gas contains H—H (usually written H_2) molecules, and oxygen gas contains O—O (O_2) molecules. However, any pure sample of an element contains only atoms of that element, *never* any atoms of any other element.

A compound *always* contains atoms of *different* elements. For example, water contains hydrogen atoms and oxygen atoms, and there are always exactly twice as many hydrogen atoms as oxygen atoms because water consists of H—O—H molecules. A different compound, carbon dioxide, consists of CO_2 molecules and so contains carbon atoms and oxygen atoms (always in the ratio 1:2).

A compound, although it contains more than one type of atom, *always has the same composition*—that is, the same combination of atoms. The properties of a compound are typically very different from those of the elements it contains. For example, the properties of water are quite different from the properties of pure hydrogen and pure oxygen.

3.4 Mixtures and Pure Substances

AIM: To learn to distinguish between mixtures and pure substances.

Virtually all of the matter around us consists of mixtures of substances. For example, if you closely observe a sample of soil, you will see that it has many types of components, including tiny grains of sand and remnants of plants. The air we breathe is a complex mixture of such gases as oxygen, nitrogen, carbon dioxide, and water vapor. Even the sparkling water from a drinking fountain contains many substances besides water.

A **mixture** can be defined as something that has variable composition. For example, wood is a mixture (its composition varies greatly depending on the tree from which it originates); wine is a mixture (it can be red or pale yellow, sweet or dry); coffee is a mixture (it can be strong, weak, or bitter); and, although it looks very pure, water pumped from deep in the earth is a mixture (it contains dissolved minerals and gases).

A **pure substance,** on the other hand, will always have the same composition. Pure substances are either elements or compounds. For example, pure water is a compound containing individual H_2O molecules. However, as we find it in nature, liquid water always contains other substances in addition to pure water—it is a mixture. This is obvious from the different tastes, smells, and colors of water samples obtained from various locations. However, if we take great pains to purify samples of water from various sources (such as oceans, lakes, rivers, and the earth's interior), we always end up with the same pure substance—water, which is made up only of H_2O molecules. Pure water always has the same physical and chemical properties and is always made of molecules containing hydrogen and oxygen in exactly the same proportions, regardless of the original source of the water. The properties of a pure substance make it possible to identify that substance conclusively.

Although we say we can separate mixtures into pure substances, it is virtually impossible to separate mixtures into totally pure substances. No matter how hard we try, some impurities (components of the original mixture) remain in each of the "pure substances."

Mixtures can be separated into pure substances: elements and/or compounds.

For example, the mixture known as air can be separated into oxygen (element), nitrogen (element), water (compound), carbon dioxide (compound), argon (element), and other pure substances.

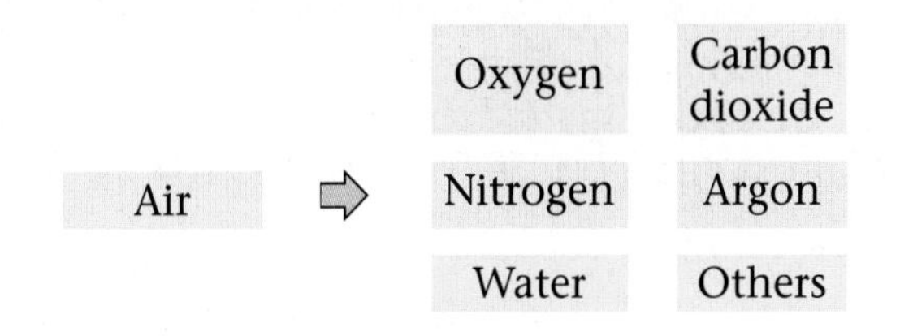

A solution is a homogeneous mixture.

Mixtures can be classified as either homogeneous or heterogeneous. A **homogeneous mixture** is *the same throughout.* For example, when we dissolve some salt in water and stir well, all regions of the resulting mixture

CHEMISTRY IN FOCUS

Instant Geology

Geologic changes on the earth are usually incredibly slow, taking place over millions of years. Tectonic plates grind past each other at rates of centimeters per year, slowly raising mountains; those mountains are then slowly ground away by wind and water. However, we do observe some exceptions to this slow-paced geology. The most dramatic geologic changes have occurred in milliseconds, when large asteroids and comets crashed into the earth at speeds of thousands of kilometers per hour. Impacts of this type literally vaporize rocks and the superheated gas that is created blasts pulverized matter into the atmosphere.

One such collision apparently wiped out the dinosaurs 65 million years ago. At that time a large asteroid hit the earth at a point that is now located along the coast of the Yucatan peninsula, creating toxic gases and throwing millions of tons of pulverized earth into the atmosphere. These events ultimately led to the demise of the dinosaurs and many other creatures. The crater of this impact is deeply buried and was identified only recently.

One impact crater that remains intact is Arizona's Meteor Crater, found 20 km west of Winslow, Arizona. This crater, measuring 200 m deep and 1.2 km in diameter, was formed about 50,000 years ago by a meteorite 45 m in diameter. The impact of this relatively small object had an energy roughly equivalent to 20 millions tons of TNT—about the same as a hydrogen bomb. However, most impact craters are not so obvious.

How do scientists find the ancient impact craters that are mostly disguised by normal geologic changes? It turns out that extraterrestrial impacts leave behind all kinds of clues. The huge seismic vibrations that accompany the impact of a large object create temperatures high enough to cause changes in minerals like no other geologic process. For example, when rock is heated to a high temperature, the magnetic fields in the minerals become disrupted and realign to match the rocks' current environment. Also, impacts can cause gravitational anomalies due to the lower density of the pulverized rock in the crater.

Thick sheets of rocks melted by the impact line the floors of most large craters. Sometimes the craters can yield mineral riches. For example, the 200-meter-thick impact melt in the ancient crater surrounding Sudbury, Ontario, has yielded more than $1 billion per year in ores containing nickel, platinum, and copper. Analysis shows that these minerals—which are from Earth, not the extraterrestrial object—separated from the surrounding material when the rocks were melted by the impact. Radioactive dating indicates that the Sudbury impact occurred nearly 2 billion years ago and produced a crater about 300 km in diameter. However, the crater has been greatly changed by subsequent geologic changes, including ice-age glaciers that have scraped away 4 km of the earth's surface at Sudbury.

So far geologists have identified about 200 impact craters on the earth. Many more remain to be discovered. This is one field where geology and chemistry come together to yield new information about our planet's history.

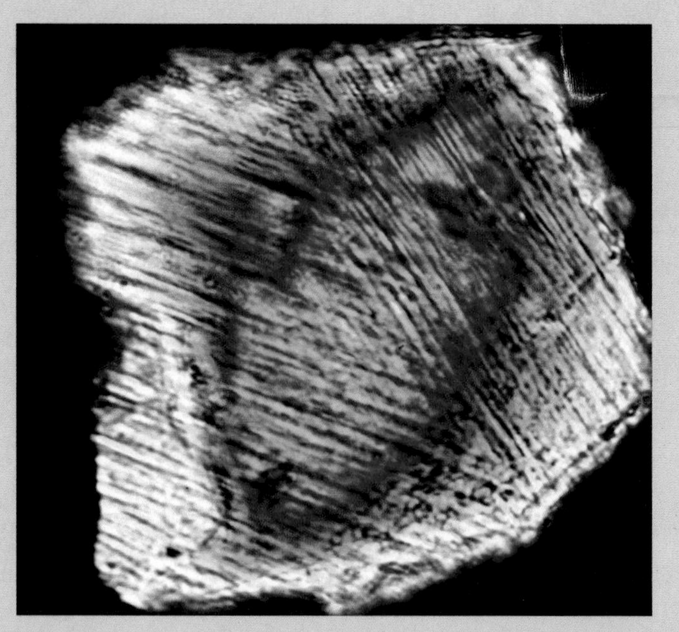

Note the fractures along multiple directions in this grain of quartz, which was shocked when an extraterrestrial impact produced intense seismic vibrations.

Figure 3.4
When table salt is stirred into water (left), a homogeneous mixture called a solution forms (right).

Figure 3.5
Sand and water do not mix to form a uniform mixture. After the mixture is stirred, the sand settles back to the bottom.

have the same properties. A homogeneous mixture is also called a **solution.** Of course, different amounts of salt and water can be mixed to form various solutions, but a homogeneous mixture (a solution) does not vary in composition from one region to another (see Figure 3.4).

Coffee is a solution that has variable composition. It can be strong or weak.

The air around you is a solution—it is a homogeneous mixture of gases. Solid solutions also exist. Brass is a homogeneous mixture of the metals copper and zinc.

A **heterogeneous mixture** contains regions that have different properties from those of other regions. For example, when we pour sand into water, the resulting mixture has one region containing water and another, very different region containing mostly sand (see Figure 3.5).

Example 3.3 Distinguishing Between Mixtures and Pure Substances

Identify each of the following as a pure substance, a homogeneous mixture, or a heterogeneous mixture.

a. gasoline
b. a stream with gravel at the bottom
c. air
d. brass
e. copper metal

Solution

a. Gasoline is a homogeneous mixture containing many compounds.

b. A stream with gravel on the bottom is a heterogeneous mixture.

c. Air is a homogeneous mixture of elements and compounds.

d. Brass is a homogeneous mixture containing the elements copper and zinc. Brass is not a pure substance because the relative amounts of copper and zinc are different in different brass samples.

e. Copper metal is a pure substance (an element).

✔ Self-Check Exercise 3.3

Classify each of the following as a pure substance, a homogeneous mixture, or a heterogeneous mixture.

a. wine
b. the oxygen and helium in a scuba tank
c. oil and vinegar salad dressing
d. common salt (sodium chloride)

See Problems 3.29 through 3.32. ■

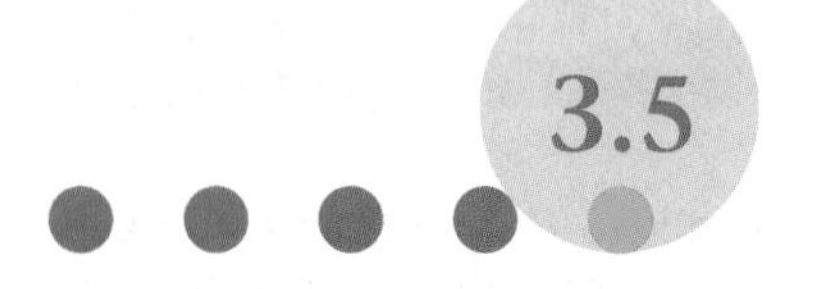

Separation of Mixtures

AIM: To learn two methods of separating mixtures.

We have seen that the matter found in nature is typically a mixture of pure substances. For example, seawater is water containing dissolved minerals. We can separate the water from the minerals by boiling, which changes the water to steam (gaseous water) and leaves the minerals behind as solids. If we collect and cool the steam, it condenses to pure water. This separation process, called **distillation**, is shown in Figure 3.6.

The separation of a mixture sometimes occurs in the natural environment and can be to our benefit (see photo on p. 65).

When we carry out the distillation of salt water, water is changed from the liquid state to the gaseous state and then back to the liquid state. These

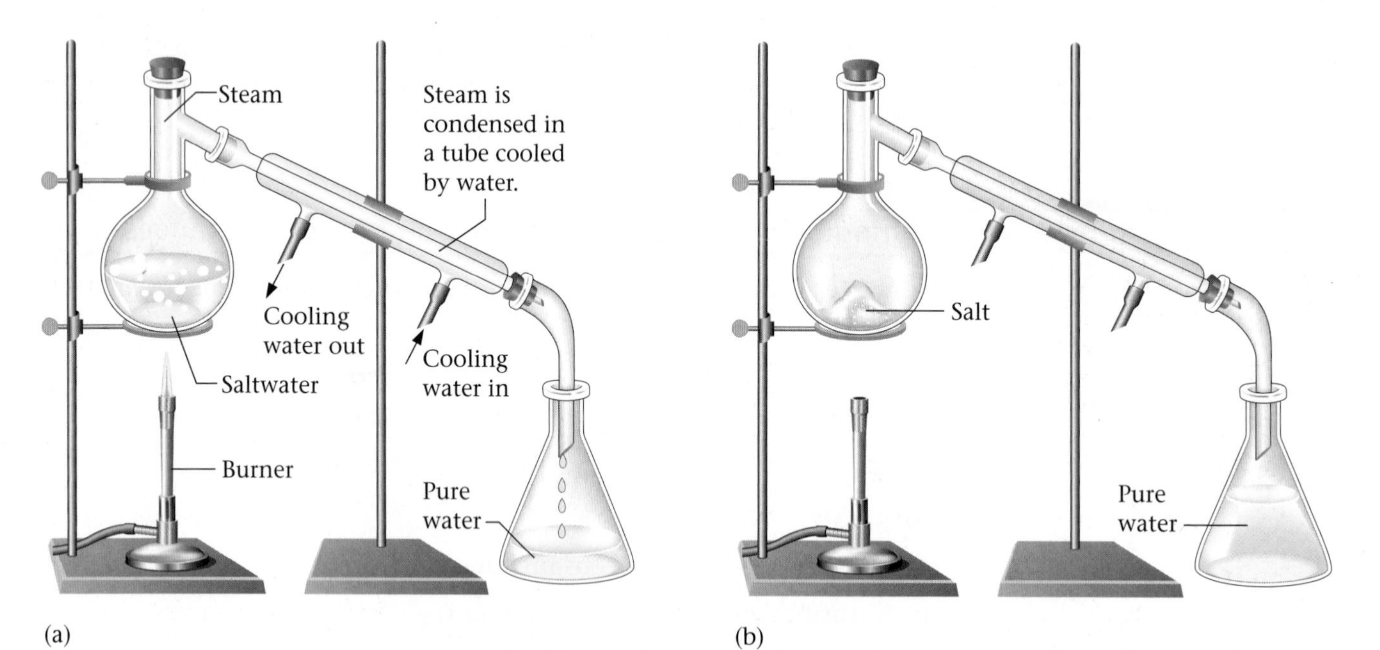

Figure 3.6
Distillation of a solution consisting of salt dissolved in water. (a) When the solution is boiled, steam (gaseous water) is driven off. If this steam is collected and cooled, it condenses to form pure water, which drips into the collection flask as shown. (b) After all of the water has been boiled off, the salt remains in the original flask and the water is in the collection flask.

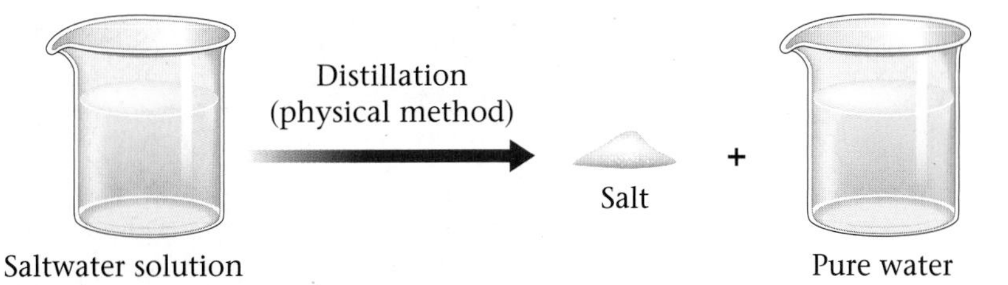

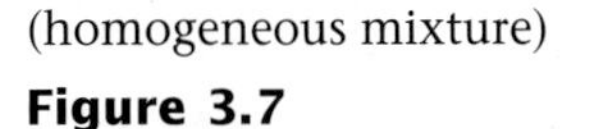
Saltwater solution
(homogeneous mixture)

Figure 3.7

No chemical change occurs when salt water is distilled.

Mixture of solid and liquid

Stirring rod

Funnel

Filter paper traps solid

Filtrate (liquid component of the mixture)

Figure 3.8
Filtration separates a liquid from a solid. The liquid passes through the filter paper, but the solid particles are trapped.

changes of state are examples of physical changes. We are separating a mixture of substances, but we are not changing the composition of the individual substances. We can represent this as shown in Figure 3.7.

Suppose we scooped up some sand with our sample of seawater. This sample is a heterogeneous mixture, because it contains an undissolved solid as well as the saltwater solution. We can separate out the sand by simple **filtration.** We pour the mixture onto a mesh, such as a filter paper, which allows the liquid to pass through and leaves the solid behind (see Figure 3.8). The salt can then be separated from the water by distillation. The total separation process is represented in Figure 3.9. All the changes involved are physical changes.

We can summarize the description of matter given in this chapter with the diagram shown in Figure 3.10. Note that a given sample of matter can be a pure substance (either an element or a compound) or, more commonly, a mixture (homogeneous or heterogeneous). We have seen that all matter exists as elements or can be broken down into elements, the most fundamental substances we have encountered up to this point. We will have more to say about the nature of elements in the next chapter.

When water from the Great Salt Lake evaporates (changes to a gas and escapes), the salt is left behind. This is one commercial source of salt.

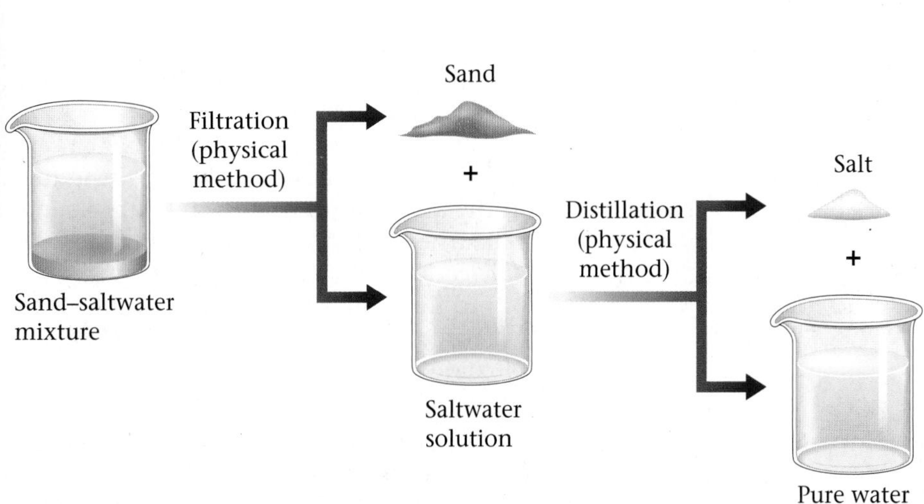

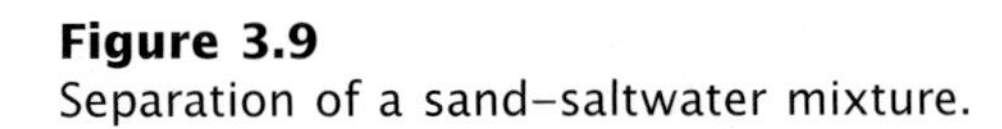
Figure 3.9
Separation of a sand–saltwater mixture.

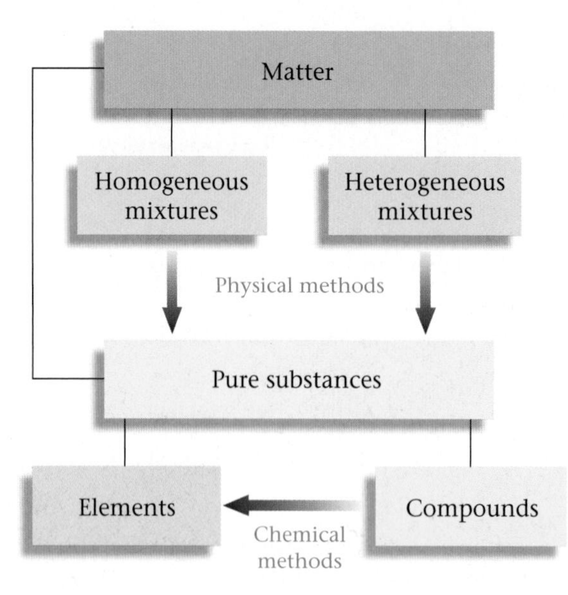

Figure 3.10
The organization of matter.

Energy, Temperature, and Heat

AIMS: To understand energy and its effect on matter. To understand the meanings of temperature and heat.

Energy is a familiar term. We speak of solar energy, nuclear energy, and energy from coal and gasoline, and when we're tired, we say that we have run out of energy. Energy allows us to "do things"—to work, to drive to school, and to cook eggs. A common definition of **energy** is the capacity to do work.

One way we use energy is to change the temperature of a substance. For example, we often heat water using the energy provided by a stove or Bunsen burner. As we will see in more detail later, the temperature of a substance reflects the random motions of the components of that substance. For example, in ice the components are water molecules that vibrate randomly about their fixed positions in the solid, as represented in Figure 3.11. When the solid is heated to higher temperatures, the random vibrations become more energetic. Finally, at the melting point of ice, the molecules are vibrating energetically enough to break loose from their positions, and rigid ice changes to liquid water.

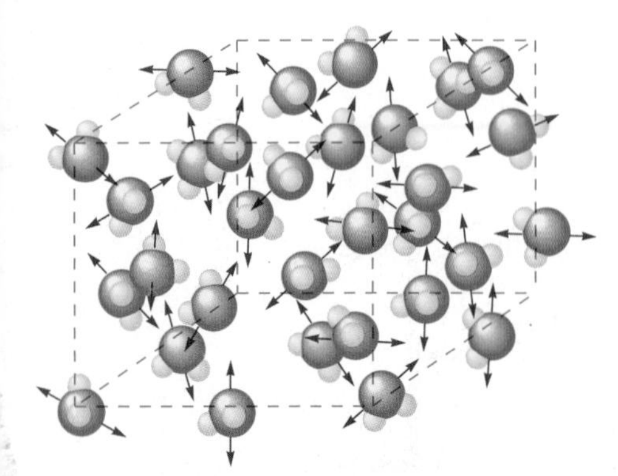

Figure 3.11
In ice, the water molecules vibrate randomly about their positions in the solid. Their motions are represented by arrows.

When we use a Bunsen burner on an ice cube, some of the energy of the reaction between natural gas (the fuel for the burner) and the oxygen in the air is transferred to the ice, causing it to melt. This flow of energy is called heat. **Heat** can be defined as *a flow of energy due to a temperature difference.*

To illustrate further the concepts of heat and temperature, consider an experiment in which we place 1.00 kg of hot water (90. °C) next to 1.00 kg of cold water (10. °C) in an insulated box. The water samples are separated from each other by a thin metal plate (see Figure 3.12). You know what will happen: the hot water will cool down and the cold water will warm up. But what, exactly, is happening?

Because the H_2O molecules in the hot water are moving faster than those in the cold water (see Figure 3.13), energy will be transferred through the metal wall from the hot water to the cold water. This energy transfer

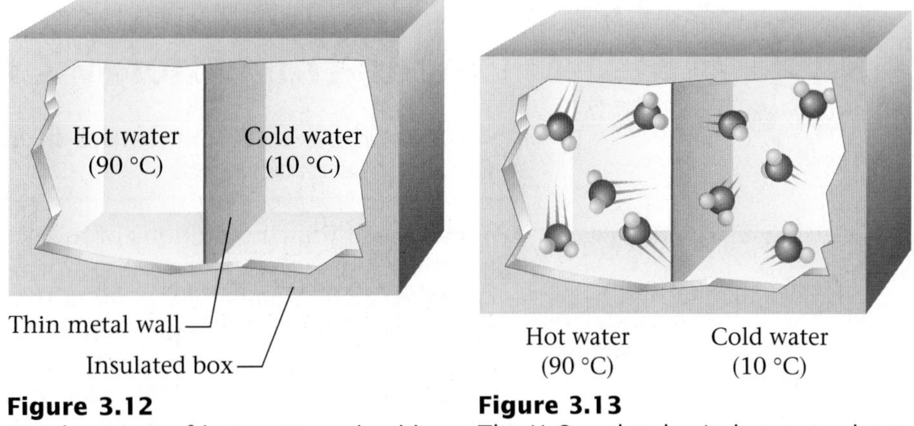

Figure 3.12
Equal masses of hot water and cold water separated by a thin metal wall in an insulated box.

Figure 3.13
The H_2O molecules in hot water have much greater random motions than the H_2O molecules in cold water.

A burning match releases energy.

will cause the H_2O molecules in the hot water to slow down and the H_2O molecules in the cold water to speed up. This transfer of energy from the hot water to the cold water is called heat. What will eventually happen? The two water samples will reach the same temperature (see Figure 3.14), which is the average of the original temperatures:

$$T_{\text{final}} = \frac{T_{\substack{\text{hot}\\\text{initial}}} + T_{\substack{\text{cold}\\\text{initial}}}}{2} = \frac{90\ ^\circ\text{C} + 10\ ^\circ\text{C}}{2} = 50\ ^\circ\text{C}$$

When a process results in the evolution of heat, it is said to be **exothermic** (*exo-* is a prefix meaning "out of"). For example, when a match is struck, energy is produced as heat; it is an exothermic process. Processes that absorb energy are said to be **endothermic.** Melting ice to form liquid water is a common endothermic process. Energy from a source such as a Bunsen burner is required to melt the ice. Energy is put into the ice to melt it.

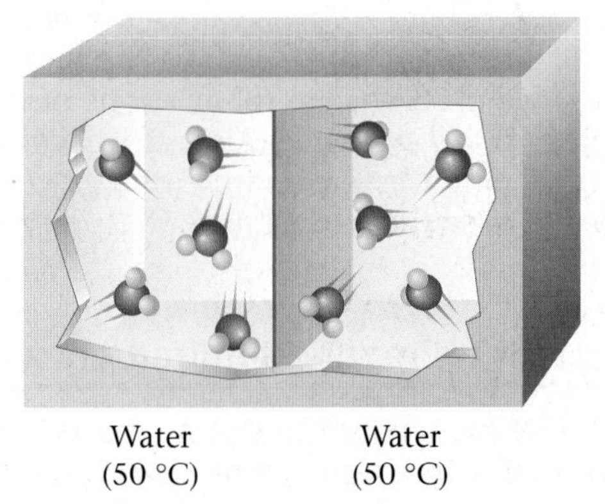

Figure 3.14
The water samples now have the same temperature (50 °C) and have the same random motions.

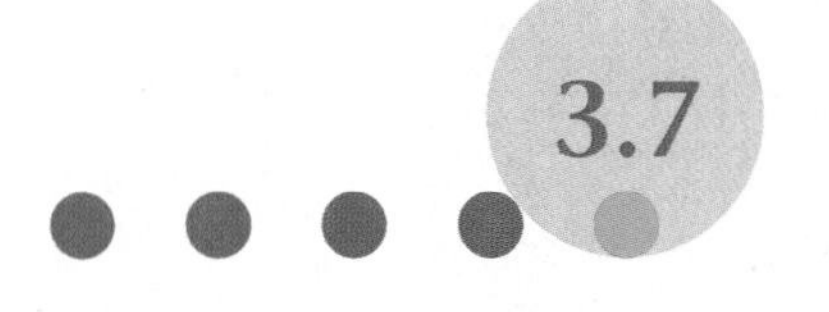

Calculating Energy Changes

AIM: To learn to calculate the quantities of energy required for various processes.

Gases will be discussed fully in Chapter 12.

Calorie: the energy (heat) required to raise the temperature of 1 g of water by 1 °C.

The abbreviations for calorie and joule are cal and J.

We have seen that the motions of molecules in a substance increase as we raise the temperature of that substance. *The amount of energy (heat) required to raise the temperature of one gram of water by one Celsius degree* is called a **calorie** in the metric system of units. The unit of energy in the SI system is called the **joule** (pronounced "jewel"). We can convert between joules and calories by using the definition that exactly 1 calorie = 4.184 joules, which leads to the equivalence statement

$$1 \text{ cal} = 4.184 \text{ J}$$

Example 3.4 Converting Calories to Joules

Express 60.1 cal of energy in units of joules.

Solution

By definition 1 cal = 4.184 J, so the conversion factor needed is $\frac{4.184 \text{ J}}{1 \text{ cal}}$, and the result is

$$60.1 \cancel{\text{cal}} \times \frac{4.184 \text{ J}}{1 \cancel{\text{cal}}} = 251 \text{ J}$$

Note that the 1 in the denominator is an exact number by definition and so does not limit the number of significant figures.

✔ Self-Check Exercise 3.4

How many calories of energy correspond to 28.4 J?

See Problems 3.45 through 3.51. ■

Now think about heating a substance from one temperature to another. How does the amount of substance heated affect the energy required? In 2 g of water there are twice as many molecules as in 1 g of water. It takes twice as much energy to change the temperature of 2 g of water by 1 °C, because we must change the motions of twice as many molecules in a 2-g sample as in a 1-g sample. Also, as we would expect, it takes twice as much energy to raise the temperature of a given sample of water by 2 degrees as it does to raise the temperature by 1 degree.

Example 3.5 Calculating Energy Requirements

Determine the amount of energy (heat) in joules required to raise the temperature of 7.40 g water from 29.0 °C to 46.0 °C.

Solution

In solving any kind of problem, it is often useful to draw a diagram that represents the situation. In this case, we have 7.40 g of water that is to be heated from 29.0 °C to 46.0 °C.

Nature Has Hot Plants

The exotic-looking voodoo lily is a beautiful and seductive plant. The voodoo lily features an elaborate reproductive mechanism—a purple spike that can reach nearly 3 feet in length and is cloaked by a hoodlike leaf. But approach to the plant reveals bad news—it smells terrible!

Despite its antisocial odor, this putrid plant has fascinated biologists for many years because of its ability to generate heat. At the peak of its metabolic activity, the plant's blossom can be as much as 15 °C above its surrounding temperature. To generate this much heat, the metabolic rate of the plant must be close to that of a flying hummingbird!

What's the purpose of this intense heat production? For a plant faced with limited food supplies in the very competitive tropical climate where it grows, heat production seems like a great waste of energy. The answer to this mystery is that the voodoo lily is pollinated mainly by carrion-loving insects. Thus the lily prepares a malodorous mixture of chemicals characteristic of rotting meat, which it then "cooks" off into the surrounding air to attract flesh-feeding beetles and flies. Then, once the insects enter the pollination chamber, the high temperatures there (as high as 110 °F) cause the insects to remain very active to better carry out their pollination duties.

The voodoo lily is only one of many thermogenic (heat-producing) plants. These plants are of special interest to biologists because they provide opportunities to study metabolic reactions that are quite subtle in "normal" plants.

A voodoo lily.

7.40 g water, T = 29.0 °C ⇨ 7.40 g water, T = 46.0 °C

? energy

Our task is to determine how much energy is required to accomplish this.

From the discussion in the text, we know that 4.184 J of energy is required to raise the temperature of *one* gram of water by *one* Celsius degree.

1.00 g water, T = 29.0 °C ⇨ 1.00 g water, T = 30.0 °C

4.184 J

Because in our case we have 7.40 g of water instead of 1.00 g, it will take 7.40×4.184 J to raise the temperature by one degree.

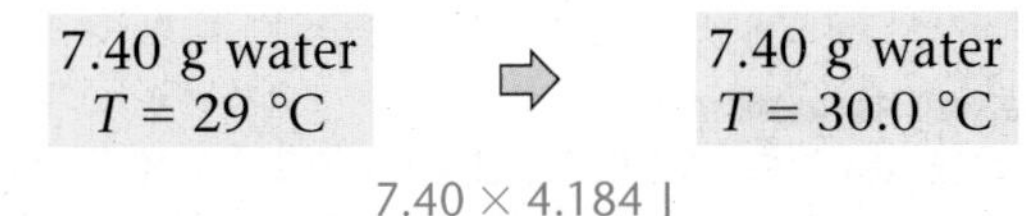

7.40 × 4.184 J

However, we want to raise the temperature of our sample of water by more than 1 °C. In fact, the temperature change required is from 29.0 °C to 46.0 °C. This is a change of 17.0 °C (46.0 °C − 29.0 °C = 17.0 °C). Thus we will have to supply 17.0 times the energy necessary to raise the temperature of 7.40 g of water by 1 °C.

7.40 g water $T = 29.0$ °C	⇨	7.40 g water $T = 46.0$ °C

$17.0 \times 7.40 \times 4.184$ J

This calculation is summarized as follows:

$$4.184 \frac{\text{J}}{\text{g °C}} \times 7.40 \text{ g} \times 17.0 \text{ °C} = 526 \text{ J}$$

Energy per gram of water per degree of temperature × Actual grams of water × Actual temperature change = Energy required

The result you will get on your calculator is $4.184 \times 7.40 \times 17.0 = 526.3472$, which rounds off to 526.

We have shown that 526 J of energy (as heat) is required to raise the temperature of 7.40 g of water from 29.0 °C to 46.0 °C. Note that because 4.184 J of energy is required to heat 1 g of water by 1 °C, the units are J/g °C (joules per gram per Celsius degree).

✔ Self-Check Exercise 3.5

Calculate the joules of energy required to heat 454 g of water from 5.4 °C to 98.6 °C. ■

So far we have seen that the energy (heat) required to change the temperature of a substance depends on

1. The amount of substance being heated (number of grams)
2. The temperature change (number of degrees)

There is another important factor: the identity of the substance.

Different substances respond differently to being heated. We have seen that 4.184 J of energy raises the temperature of 1 g of water 1 °C. In contrast, this same amount of energy applied to 1 g of gold raises its temperature by approximately 32 °C! The point is that some substances require relatively large amounts of energy to change their temperatures, whereas others require relatively little. Chemists describe this difference by saying that substances have different heat capacities. *The amount of energy required to change the temperature of one gram of a substance by one Celsius degree* is called its **specific heat capacity** or, more commonly, its **specific heat.** The specific heat capacities for several substances are listed in Table 3.2. You can see from the table that the specific heat capacity for water is very high compared to those of the other substances listed. This is why lakes and seas are much slower to respond to cooling or heating than are the surrounding land masses.

Table 3.2 The Specific Heat Capacities of Some Common Substances

Substance	Specific Heat Capacity (J/g °C)
water (*l*)* (liquid)	4.184
water (*s*) (ice)	2.03
water (*g*) (steam)	2.0
aluminum (*s*)	0.89
iron (*s*)	0.45
mercury (*l*)	0.14
carbon (*s*)	0.71
silver (*s*)	0.24
gold (*s*)	0.13

*The symbols (*s*), (*l*), and (*g*) indicate the solid, liquid, and gaseous states, respectively.

Example 3.6 Calculations Involving Specific Heat Capacity

a. What quantity of energy (in joules) is required to heat a piece of iron weighing 1.3 g from 25 °C to 46 °C?

CHEMISTRY IN FOCUS

Firewalking: Magic or Science?

For millennia people have been amazed at the ability of Eastern mystics to walk across beds of glowing coals without any apparent discomfort. Even in the United States, thousands of people have performed feats of firewalking as part of motivational seminars. How can this be possible? Do firewalkers have supernatural powers?

Actually, there are good scientific explanations of why firewalking is possible. First, human tissue is mainly composed of water, which has a relatively large specific heat capacity. This means that a large amount of energy must be transferred from the coals to change significantly the temperature of the feet. During the brief contact between feet and coals involved in firewalking, there is relatively little time for energy flow, so the feet do not reach a high enough temperature to cause damage.

Also, although the surface of the coals has a very high temperature, the red-hot layer is very thin. Therefore, the quantity of energy available to heat the feet is smaller than might be expected.

A Hindu firewalking ceremony in the Fiji Islands.

Thus, although firewalking is impressive, there are several scientific reasons why anyone with the proper training should be able to do it on a properly prepared bed of coals. (Don't try this on your own!)

b. What is the answer in calories?

Solution

a. It is helpful to draw the following diagram to represent the problem.

1.3 g iron $T = 25$ °C	⇨	1.3 g iron $T = 46$ °C
	? joules	

From Table 3.2 we see that the specific heat capacity of iron is 0.45 J/g °C. That is, it takes 0.45 J to raise the temperature of a 1-g piece of iron by 1 °C.

1.0 g iron $T = 25$ °C	⇨	1.0 g iron $T = 26$ °C
	0.45 J	

In this case our sample is 1.3 g, so 1.3×0.45 J is required for *each* degree of temperature increase.

Because the temperature increase is 21 °C (46 °C − 25 °C = 21 °C), the total amount of energy required is

The result you will get on your calculator is 0.45 × 1.3 × 21 = 12.285, which rounds off to 12.

$$0.45 \frac{J}{g\ ^\circ C} \times 1.3\ g \times 21\ ^\circ C = 12\ J$$

1.3 g iron $T = 25\ ^\circ C$	⇨	1.3 g iron $T = 46\ ^\circ C$

21 × 1.3 × 0.45 J

Note that the final units are joules, as they should be.

b. To calculate this energy in calories, we can use the definition 1 cal = 4.184 J to construct the appropriate conversion factor. We want to change from joules to calories, so cal must be in the numerator and J in the denominator where it cancels:

$$12\ \cancel{J} \times \frac{1 cal}{4.184\ \cancel{J}} = 2.9\ cal$$

Remember that 1 in this case is an exact number by definition and therefore does not limit the number of significant figures (the number 12 is limiting here).

✔ Self-Check Exercise 3.6

A 5.63-g sample of solid gold is heated from 21 °C to 32 °C. How much energy (in joules and calories) is required?

See Problems 3.53 through 3.60. ■

Note that in Example 3.6, to calculate the energy (heat) required, we took the product of the specific heat capacity, the sample size in grams, and the change in temperature in Celsius degrees.

Energy (heat) required (Q)	=	Specific heat capacity (*s*)	×	Mass (*m*) in grams of sample	×	Change in temperature (ΔT) in °C

The symbol Δ (the Greek letter delta) is shorthand for "change in."

We can represent this by the following equation:

$$Q = s \times m \times \Delta T$$

where

Q = energy (heat) required
s = specific heat capacity
m = mass of the sample in grams
ΔT = change in temperature in Celsius degrees

This equation always applies when a substance is being heated (or cooled) and no change of state occurs. Before you begin to use this equation, however, make sure you understand where it comes from.

Example 3.7 Specific Heat Capacity Calculations: Using the Equation

A 1.6-g sample of a metal that has the appearance of gold requires 5.8 J of energy to change its temperature from 23 °C to 41 °C. Is the metal pure gold?

Solution

We can represent the data given in this problem by the following diagram:

1.6 g metal $T = 23\ °C$	⇨	1.6 g metal $T = 41\ °C$
	5.8 J	

$$\Delta T = 41\ °C - 23\ °C = 18\ °C$$

Using the data given, we can calculate the value of the specific heat capacity for the metal and compare this value to the one for gold given in Table 3.2. We know that

$$Q = s \times m \times \Delta T$$

or, pictorially,

1.6 g metal $T = 23\ °C$	⇨	1.6 g metal $T = 41\ °C$
	5.8 J × 1.6 × 18	

When we divide both sides of the equation

$$Q = s \times m \times \Delta T$$

by $m \times \Delta T$, we get

$$\frac{Q}{m \times \Delta T} = s$$

Thus using the data given, we can calculate the value of s. In this case,

$$Q = \text{energy (heat) required} = 5.8\ J$$
$$m = \text{mass of the sample} = 1.6\ g$$
$$\Delta T = \text{change in temperature} = 18\ °C\ (41\ °C - 23\ °C = 18\ °C)$$

Thus

$$s = \frac{Q}{m \times \Delta T} = \frac{5.8\ J}{(1.6\ g)(18\ °C)} = 0.20\ J/g\ °C$$

The result you will get on your calculator is 5.8/(1.6)(18) = 0.2013889, which rounds off to 0.20.

From Table 3.2, the specific heat capacity for gold is 0.13 J/g °C. Thus the metal must not be pure gold.

✔ Self-Check Exercise 3.7

A 2.8-g sample of pure metal requires 10.1 J of energy to change its temperature from 21 °C to 36 °C. What is this metal? (Use Table 3.2.)

See Problems 3.61 and 3.62. ■

Chapter 3 Review

Key Terms

matter (3.1)
states of matter (3.1)
solid (3.1)
liquid (3.1)
gas (3.1)
physical properties (3.2)
chemical properties (3.2)
physical change (3.2)
chemical change (3.2)
reaction (3.2)
element (3.3)
compound (3.3)
mixture (3.4)
pure substance (3.4)
homogeneous mixture (3.4)
solution (3.4)
heterogeneous mixture (3.4)
distillation (3.5)
filtration (3.5)
energy (3.6)
heat (3.6)
exothermic (3.6)
endothermic (3.6)
calorie (3.7)
joule (3.7)
specific heat capacity (3.7)
specific heat (3.7)

SUMMARY

1. Matter can exist in three states—solid, liquid, and gas—and can be described in terms of its physical and chemical properties. Chemical properties describe a substance's ability to undergo a change to a different substance. Physical properties are the characteristics a substance exhibits as long as no chemical change occurs.

2. A physical change involves a change in one or more physical properties, but no change in composition. A chemical change transforms a substance into a new substance or substances.

3. A mixture has variable composition. A homogeneous mixture has the same properties throughout; a heterogeneous mixture does not. A pure substance always has the same composition. We can physically separate mixtures of pure substances by distillation and filtration.

4. Pure substances are of two types: elements, which cannot be broken down chemically into simpler substances, and compounds, which can be broken down chemically into elements.

5. Energy can be defined as the capacity to do work or generate heat. An exothermic process releases energy as heat. An endothermic process absorbs energy as heat.

6. Specific heat capacity is the amount of energy required to change the temperature of one gram of a substance by one Celsius degree. Each substance has a characteristic specific heat capacity. The energy (heat) required to change the temperature of a substance depends on three factors: the amount of substance, the temperature change, and the specific heat capacity of the substance.

IN-CLASS DISCUSSION QUESTIONS

These questions are designed to be considered by groups of students in class. Often these questions work well for introducing a particular topic in class.

1. Objects in the same container eventually reach the same temperature. When you go into a room and touch a piece of metal in that room, it feels colder than a piece of plastic. Explain.

2. When water boils, you can see bubbles rising to the surface of the water. Of what are these bubbles made?
   a. air
   b. hydrogen and oxygen gas
   c. oxygen gas
   d. water vapor
   e. carbon dioxide gas

3. If you place a glass rod over a burning candle, the glass turns black. What is happening to each of the following (physical change, chemical change, both, or neither) as the candle burns? Explain.
   a. the wax
   b. the wick
   c. the glass rod

4. Which characteristics of a solid, a liquid, and a gas do each of the following have, and in which category would you classify each? Explain.
   a. a bowl of pudding
   b. a bucketful of sand

5. The boiling of water is a
   a. physical change because the water disappears.
   b. physical change because the gaseous water is chemically the same as the liquid.
   c. chemical change because heat is needed for the process to occur.
   d. chemical change because hydrogen and oxygen gases are formed from water.
   e. chemical and physical change.

   Explain your answer.

6. Is there a difference between a homogeneous mixture of hydrogen and oxygen in a 2:1 ratio and a sample of water vapor? Explain.

7. Sketch a magnified view (showing atoms and/or molecules) of each of the following and explain why is the specified type of mixture:
   a. a heterogeneous mixture of two different compounds
   b. a homogeneous mixture of an element and a compound

8. Are all physical changes accompanied by chemical changes? Are all chemical changes accompanied by physical changes? Explain.

9. Why would a chemist find fault with the phrase "pure orange juice"?

10. Are separations of mixtures physical or chemical changes? Explain.

QUESTIONS AND PROBLEMS

All even-numbered exercises have answers in the back of this book and solutions in the *Solutions Guide*.

3.1 Matter

QUESTIONS

1. How, in general, do we define *matter?*

2. What are the three *physical states* in which matter may exist?

3. Solids and liquids are virtually incompressible, whereas ________ are very compressible.

4. _______ have definite volumes, but are able to take on the shape of their containers.

5. Compare and contrast the ease at which molecules are able to move relative to each other in the three states of matter.

6. Matter in the _______ state has no shape and fills completely whatever container holds it.

7. What similarities are there between the liquid and gaseous states of matter? What differences are there between these two states?

8. How is the rigidity of a sample of matter affected by the strength of the forces among the particles in the sample?

9. Consider three 10-g samples of water: one as ice, one as a liquid, and one as vapor. How do the volumes of these three samples compare with one another? How is this difference in volume related to the physical state involved?

10. In a sample of a gaseous substance, more than 99% of the overall volume of the sample is empty space. How is this fact reflected in the properties of a gaseous substance, compared with the properties of a liquid or solid substance?

3.2 Physical and Chemical Properties and Changes

QUESTIONS

11. Aluminum is a silver-colored metal, which can easily be rolled or hammered into a thin foil. Are these characteristics of aluminum physical or chemical properties?

12. When copper metal is heated in nitric acid, the copper dissolves to form a deep blue solution, and a brown gas is evolved from the acid. These characteristics are examples of (physical/chemical) changes.

(For exercises 13–14). Aqueous solutions of the substance nickel(II) sulfate are bright green in color. If an aqueous solution of barium chloride is added to an aqueous solution of nickel(II) sulfate, a white precipitate of barium sulfate forms.

13. From the information above, indicate one *physical* property of nickel(II) sulfate in solution.

14. From the information above, indicate one *chemical* property of nickel(II) sulfate in solution.

15. Choose a chemical substance with which you are familiar, and give an example of a *chemical change* that might take place to the substance.

16. What are the most common *physical changes* possible for a sample of matter?

17. Classify the following as *physical* or *chemical* changes.
   a. Mothballs gradually vaporize in a closet.
   b. A French chef making a sauce with brandy is able to burn off the alcohol from the brandy, leaving just the brandy flavoring.
   c. Hydrofluoric acid attacks glass, and is used to etch calibration marks on glass laboratory utensils.
   d. Calcium chloride lowers the temperature at which water freezes, and can be used to melt ice on city sidewalks and roadways.
   e. An antacid tablet fizzes and releases carbon dioxide gas when it comes in contact with hydrochloric acid in the stomach.
   f. Baking soda fizzes if mixed with vinegar.
   g. Chemistry majors usually get holes in the cotton jeans they wear to lab because of the acids used in many experiments.
   h. Whole milk curdles if vinegar is added to it.
   i. A piece of rubber stretches when you pull on it.
   j. Rubbing alcohol evaporates quickly from the skin.
   k. Acetone is used to dissolve and remove nail polish.

18. Classify the following as *physical* or *chemical* changes/properties.
   a. A shirt scorches when you leave the iron on one spot too long.
   b. The tires on your car seem to be getting flat in very cold weather.
   c. Your grandmother's silver tea set gets black with tarnish over time.
   d. A bottle of wine left open turns to vinegar.
   e. Spray-on oven cleaner converts grease in the oven into a soapy material.
   f. An ordinary flashlight battery begins to leak with age and can't be recharged.
   g. Acids produced by bacteria in plaque cause teeth to decay.
   h. Sugar will char if overheated while making homemade candy.
   i. Hydrogen peroxide fizzes when applied to a wound.
   j. Dry ice "evaporates" without melting as time passes.
   k. Chlorine laundry bleaches will sometimes change the color of brightly colored clothing.

3.3 Elements and Compounds

QUESTIONS

19. Give definitions for the terms *element* and *compound*. Give five examples of each type of substance.

20. A pure sample of a(n) _______ contains only one kind of atom.

21. Certain elements have special affinities for other elements. This causes them to bind together in special ways to form _______.

22. _______ can be broken down into the component elements by chemical changes.

23. The composition of a given pure compound is always ______ no matter what the source of the compound.

24. How do the properties of a compound, in general, compare to the properties of the elements that constitute it? Give an example of a common compound and the elements of which it is composed to illustrate your answer.

3.4 Mixtures and Pure Substances

QUESTIONS

25. Give definitions for the terms *pure substance* and *mixture.* Give five examples of each type of material.

26. How does the composition of a mixture compare to the composition of a pure substance? Which material has a fixed composition and which has a variable composition?

27. What does it mean to say that a solution is a *homogeneous mixture?*

28. Give three examples of heterogeneous *mixtures* and three examples of *solutions* that you might use in everyday life.

29. Classify the following as *mixtures* or as *pure substances.*
   a. the air you are breathing
   b. the soda you are drinking while reading this book
   c. the water with which you just watered your lawn
   d. the diamond in the ring that your fiancé just presented to you

30. Classify the following as *mixtures* or as *pure substances.*
   a. a multivitamin tablet
   b. the blue liquid in your car's windshield washer
   c. a Spanish omelet
   d. distilled water

31. Classify the following mixtures as *heterogeneous* or *homogeneous.*
   a. soil
   b. mayonnaise
   c. Italian salad dressing
   d. the wood from which the desk you are studying on is made
   e. sand at the beach

32. Classify the following mixtures as *homogeneous* or *heterogeneous.*
   a. gasoline
   b. a jar of jelly beans
   c. chunky peanut butter
   d. margarine
   e. the paper this question is printed on

3.5 Separation of Mixtures

QUESTIONS

33. Describe how the process of *distillation* could be used to separate a solution into its component substances. Give an example.

34. Describe how the process of *filtration* could be used to separate a mixture into its components. Give an example.

35. In a common laboratory experiment in general chemistry, students are asked to determine the relative amounts of benzoic acid and charcoal in a solid mixture. Benzoic acid is relatively soluble in hot water, but charcoal is not. Devise a method for separating the two components of this mixture.

36. Describe the process of distillation depicted in Figure 3.6. Does the separation of the components of a mixture by distillation represent a chemical or a physical change?

3.6 Energy, Temperature, and Heat

QUESTIONS

37. How do we define *energy?*

38. Two units of energy that are used very commonly are the *calorie* and the *joule.* Which unit represents the larger amount of energy?

39. Describe what happens to the molecules in a sample of ice as the sample is slowly heated until it liquefies and then vaporizes.

40. Consider a sample of *steam* (water in the gaseous state) at 150 °C. Describe what happens to the molecules in the sample as the sample is slowly cooled until it liquefies and then solidifies.

3.7 Calculating Energy Changes

QUESTIONS

41. Metallic substances tend to have (higher/lower) specific heat capacities than nonmetallic substances.

42. The quantity of energy required to change the temperature of a sample is calculated by taking the product of the mass of the sample, the specific heat capacity of the sample, and the ______ change undergone by the sample.

43. If it takes 654 J of energy to warm a 5.51-g sample of water, how much energy would be required to warm 55.1 g of water by the same amount?

44. If it takes 526 J of energy to warm 7.40 g of water by 17 °C, how much heat would be needed to warm 7.40 g of water by 55 °C?

45. Convert the following numbers of calories or kilocalories into joules and kilojoules. (Remember: kilo means 1000.)
 a. 75.2 kcal c. 1.41×10^3 cal
 b. 75.2 cal d. 1.41 kcal

46. Convert the following numbers of kilojoules into kilocalories. (Remember: kilo means 1000.)
 a. 462.4 kJ c. 1.014 kJ
 b. 18.28 kJ d. 190.5 kJ

47. Convert the following numbers of calories into kilocalories.
 a. 7518 cal c. 1 cal
 b. 7.518×10^3 cal d. 655,200 cal

48. Convert the following numbers of kilocalories into calories.
 a. 12.30 kcal c. 940,000 kcal
 b. 290.4 kcal d. 4201 kcal

49. Convert the following numbers of joules (J) into kilojoules (kJ). (Remember: kilo means 1000.)
 a. 243,000 J c. 0.251 J
 b. 4.184 J d. 450.3 J

50. Perform the indicated conversions.
 a. 76.52 cal into kilojoules
 b. 7.824 kJ into kilocalories
 c. 489.4 J into calories
 d. 1.598×10^4 J into kilocalories

51. Perform the indicated conversions.
 a. 89.74 kJ into kilocalories
 b. 1.756×10^4 J into kilojoules
 c. 1.756×10^4 J into kilocalories
 d. 1.00 kJ into calories

52. If 72.4 kJ of heat is applied to a 952-g block of metal, the temperature of the metal increases by 10.7 °C. Calculate the specific heat capacity of the metal in J/g °C.

53. Calculate the energy required in joules and calories to heat 29.2 g of aluminum from 27.2 °C to 41.5 °C. See Table 3.2.

54. A particular sample of iron requires 562 J to raise its temperature from 25.0 °C to 50.0 °C. What must be the mass of the sample of iron? See Table 3.2.

55. If 100. J of heat energy is applied to a 25-g sample of mercury, by how many degrees will the temperature of the mercury sample increase? See Table 3.2.

56. Calculate the quantity of heat required to raise the temperature of a 852.5-g sample of iron from 40.1 °C to 75.5 °C.

57. The specific heat capacity of silver is 0.24 J/g °C. Express this in terms of calories per gram per Celsius degree.

58. The specific heat capacity of gold is 0.13 J/g °C. Calculate the specific heat capacity of gold in cal/g °C.

59. Suppose you have samples of gold, iron, and aluminum, all of the same mass. If the same quantity of heat energy is applied in turn to each of the three samples, which sample of metal will end up at the highest temperature? Which will end up at the lowest temperature?

60. If the temperatures of separate 25.0-g samples of gold, mercury, and carbon are to be raised by 20. °C, how much heat (in joules) must be applied to each substance?

61. A 5.00-g sample of one of the substances listed in Table 3.2 was heated from 25.2 °C to 55.1 °C, requiring 133 J to do so. What substance was it?

62. A 35.2 g sample of metal Z requires 1251 J of energy to heat the sample by 25.0 °C. Calculate the specific heat capacity of metal Z.

Additional Problems

63. If solid iron pellets and sulfur powder are poured into a container at room temperature, a simple ________ has been made. If the iron and sulfur are heated until a chemical reaction takes place between them, a(n) ________ will form.

64. Pure substance X is melted, and the liquid is placed in an electrolysis apparatus such as that shown in Figure 3.3. When an electric current is passed through the liquid, a brown solid forms in one chamber and a white solid forms in the other chamber. Is substance X a compound or an element?

65. If a piece of hard white blackboard chalk is heated strongly in a flame, the mass of the piece of chalk will decrease, and eventually the chalk will crumble into a fine white dust. Does this change suggest that the chalk is composed of an element or a compound?

66. During a very cold winter, the temperature may remain below freezing for extended periods. However, fallen snow can still disappear, even though it cannot melt. This is possible because a solid can vaporize directly, without passing through the liquid state. Is this process (sublimation) a physical or a chemical change?

67. Perform the indicated conversions.
 a. 4.52 cal to kilocalories
 b. 5.27 kcal to joules
 c. 852,000 cal to kilojoules
 d. 352.4 kcal to kilojoules
 e. 5.72 kJ to calories
 f. 4.52×10^3 J to kilojoules

68. Calculate the amount of energy required (in joules) to heat 2.5 kg of water from 18.5 °C to 55.0 °C.

69. If 10. J of heat is applied to 5.0-g samples of each of the substances listed in Table 3.2, which substance's temperature will increase the most? Which substance's temperature will increase the least?

70. A 5-g sample of aluminum and a 5-g sample of iron are heated in a boiling water bath in separate test tubes. The test tubes are then placed together into a beaker containing ice. Which metal will lose the most heat in cooling down?

71. Hydrogen gives off 120. J/g of energy when burned in oxygen, and methane gives off 50. J/g under the same circumstances. If a mixture of 5.0 g of hydrogen and 10. g of methane is burned, and the heat released is transferred to 500. g of water at 25 °C, what final temperature will be reached by the water?

72. A 5.00-g sample of aluminum pellets and a 10.00-g sample of iron pellets are placed together in a dry test tube, and the test tube is heated in a boiling water bath to 100. °C. The mixture of hot iron and aluminum is then poured into 97.3 g of water at 22.5 °C. To what final temperature is the water heated by the metals?

73. A 50.0-g sample of water at 100. °C is poured into a 50.0-g sample of water at 25 °C. What will be the final temperature of the water?

74. A 25.0-g sample of pure iron at 85 °C is dropped into 75 g of water at 20. °C. What is the final temperature of the water–iron mixture?

75. Discuss the similarities and differences between a liquid and a gas.

76. In gaseous substances, the individual molecules are relatively (close/far apart) and are moving freely, rapidly, and randomly.

77. The fact that solutions of potassium chromate are bright yellow is an example of a _______ property.

78. The fact that the substance copper(II) sulfate pentahydrate combines with ammonia in solution to form a new compound is an example of a _______ property.

(For exercises 79–80) Solutions containing copper(II) ions are bright blue in color. When sodium hydroxide is added to such a solution, a solid material forms that is colored a much paler shade of blue than the original solution of copper(II) ions.

79. The fact that a solution containing copper(II) ions is bright blue is a _______ property.

80. The fact that a reaction takes place when sodium hydroxide is added to a solution of copper(II) ions is a _______ property.

81. The processes of melting and evaporation involve changes in the _______ of a substance.

82. _______ is the process of making a chemical reaction take place by passage of an electric current through a substance or solution.

83. Classify the following as *physical* or *chemical* properties/changes.
   a. Milk curdles if a few drops of lemon juice are added.
   b. Butter turns rancid if left exposed at room temperature.
   c. Salad dressing separates into layers after standing.
   d. Milk of magnesia neutralizes stomach acid.
   e. The steel in a car has rust spots.
   f. A person is asphyxiated by breathing carbon monoxide.
   g. Sulfuric acid spilled on a laboratory notebook page causes the paper to char and disintegrate.
   h. Sweat cools the body as it evaporates from the skin.
   i. Aspirin reduces fever.
   j. Oil feels slippery.
   k. Alcohol burns, forming carbon dioxide and water.

84. Classify the following mixtures as *homogeneous* or as *heterogeneous*.
   a. the freshman class at your school
   b. salsa
   c. mashed potatoes
   d. cream of tomato soup
   e. cream of mushroom soup

85. Classify the following mixtures as *homogeneous* or as *heterogeneous*.
   a. potting soil
   b. white wine
   c. your sock drawer
   d. window glass
   e. granite

86. If it takes 4.5 J of energy to warm 5.0 g of aluminum from 25 °C to a certain higher temperature, then it will take _______ J to warm 10. g of aluminum over the same temperature interval.

87. If it takes 103 J of energy to warm a certain mass of iron from 25 °C to 50. °C, then it will take _______ J to warm the same mass of iron from 25 °C to 75 °C.

88. Convert the following numbers of calories/kilocalories into joules (J).
   a. 44.21 cal
   b. 162.4 cal
   c. 3.721×10^3 cal
   d. 146.2 kcal

89. Convert the following numbers of joules/kilojoules into kilocalories.
   a. 52.18 kJ
   b. 4.298 J
   c. 5.433×10^3 J
   d. 455.9 kJ

90. Perform the following conversions.
   a. 5.442×10^4 J to kilojoules
   b. 5.442×10^4 J to calories
   c. 352.6 kcal to kilojoules
   d. 17.24 kJ to kilocalories

91. Calculate the energy required to heat 25.0 g of gold from 20.0 °C to 75.0 °C. Express your answer in joules, kilojoules, calories, and kilocalories.

92. Calculate the energy required (in joules) to heat 75 g of water from 25 °C to 39 °C.

93. If a 37.5-lb sample of aluminum is warmed from 22.1 °F to 85.2 °F, how much energy does the aluminum absorb?

94. For each of the substances listed in Table 3.2, calculate the quantity of heat required to heat 150. g of the substance by 11.2 °C.

95. Suppose you had 10.0-g samples of each of the substances listed in Table 3.2 and that 1.00 kJ of heat is applied to each of these samples. By what amount would the temperature of each sample be raised?

96. A 125-g sample of an unknown metal requires 1.351 kJ of energy to heat it from 25.0 °C to 112.1 °C. Calculate the specific heat capacity of the unknown metal.

CUMULATIVE REVIEW FOR CHAPTERS 1–3

QUESTIONS

1. In the exercises for Chapter 1 of this text, you were asked to give your *own* definition of what chemistry represents. After having completed a few more chapters in this book, has your definition changed? Do you have a better appreciation for what chemists do? Explain.

2. Early on in this text, some aspects of the best way to go about learning chemistry were presented. In *beginning* your study of chemistry, you may initially have approached studying chemistry as you would any of your other academic subjects (taking notes in class, reading the text, memorizing facts, and so on). Discuss why the ability to sort through and analyze facts and the ability to propose and solve problems are so much more important in learning chemistry.

3. You have learned the basic way in which scientists analyze problems, propose models to explain the systems under consideration, and then experiment to test their models. Suppose you have a sample of a liquid material. You are not sure whether the liquid is a pure *compound* (for example, water or alcohol) or a *solution*. How could you apply the scientific method to study the liquid and to determine which type of material the liquid is?

4. Many college students would not choose to take a chemistry course if it were not required for their major. Do you have a better appreciation of *why* chemistry is a required course for your own particular major or career choice? Discuss.

5. In Chapter 2 of this text, you were introduced to the International System *(SI)* of measurements. What are the basic units of this system for mass, distance, time, and temperature? What are some of the common multiples and subdivisions of these basic units? Why do you suppose the metric system is used practically everywhere in the world except the United States? Why do you suppose the United States is reluctant to adopt this system? Do you think the United States *should* adopt this system? Why or why not?

6. Most people think of science as being a specific, exact discipline, with a "correct" answer for every problem. Yet you were introduced to the concept of *uncertainty* in scientific measurements. What is meant by "uncertainty"? How does uncertainty creep into measurements? How is uncertainty *indicated* in scientific measurements? Can uncertainty ever be completely eliminated in experiments? Explain.

7. After studying a few chapters of this text, and perhaps having done a few lab experiments and taken a few quizzes in chemistry, you are probably sick of hearing the term *significant figures*. Most chemistry teachers make a big deal about significant figures. Why is reporting the correct number of significant figures so important in science? Summarize the rules for deciding whether a figure in a calculation is "significant." Summarize the rules for rounding off numbers. Summarize the rules for doing arithmetic with the correct number of significant figures.

8. This chemistry course may have been the first time you have encountered the method of *dimensional analysis* in problem solving. Explain what are meant by a *conversion factor* and an *equivalence statement*. Give an everyday example of how you might use dimensional analysis to solve a simple problem.

9. You have learned about several temperature scales so far in this text. Describe the Fahrenheit, Celsius, and Kelvin temperature scales. How are these scales defined? Why were they defined this way? Which of these temperature scales is the most fundamental? Why?

10. What is *matter?* What is matter composed of? What are some of the different types of matter? How do these types of matter differ and how are they the same?

11. What is the difference between a chemical property and a physical property? Give examples of each. What is the difference between a chemical change and a physical change? Give examples of each.

12. What is an *element* and what is a *compound?* Give examples of each. What does it mean to say that a compound has a *constant composition?* Would samples of a particular compound here and in another part of the world have the same composition and properties?

13. What is a mixture? What is a solution? How do mixtures differ from pure substances? What are some of the techniques by which mixtures can be resolved into their components?

14. How is the concept of *energy* defined? What are some common *units* of energy, and how are these units defined? What is meant by the *specific heat capacity* of a material? How is specific heat capacity used to measure energy changes in processes?

PROBLEMS

15. For each of the following, make the indicated conversion.
    a. 144,400 to standard scientific notation
    b. 3.489×10^{-3} to ordinary decimal notation
    c. 125.2×10^{3} to standard scientific notation
    d. 5.22×10^{2} to ordinary decimal notation
    e. 0.001781×10^{-3} to standard scientific notation
    f. 25.1×10^{0} to ordinary decimal notation

16. For each of the following, make the indicated conversion.
   a. 2.4 L to quarts
   b. 785.2 g to pounds
   c. 785.2 g to kilograms
   d. 7521 m to miles
   e. 25.2 cm to meters
   f. 451 mL to liters
   g. 6.45 m to feet
   h. 125.5 cm to inches
   i. 6.29 inches to feet
   j. 4.62 ounces to pounds

17. Evaluate each of the following mathematical expressions, being sure to express the answer to the correct number of significant figures.
   a. $10.20 + 4.1 + 26.001 + 2.4$
   b. $(1.091 - 0.991) + 1.2$
   c. $(4.06 + 5.1)(2.032 - 1.02)$
   d. $(67.21)(1.003)(2.4)$
   e. $[(7.815 + 2.01)(4.5)]/(1.9001)$
   f. $(1.67 \times 10^{-9})(1.1 \times 10^{-4})$
   g. $(4.02 \times 10^{-4})(2.91 \times 10^{3})/(9.102 \times 10^{-1})$
   h. $(1.04 \times 10^{2} + 2.1 \times 10^{1})/(4.51 \times 10^{3})$
   i. $(1.51 \times 10^{-3})^{2}/(1.074 \times 10^{-7})$
   j. $(1.89 \times 10^{2})/[(7.01 \times 10^{-3})(4.1433 \times 10^{4})]$

18. Make the indicated temperature changes.
   a. −40 °C to Fahrenheit degrees
   b. 577 K to Celsius degrees
   c. 225 °F to Celsius degrees
   d. −158 °C to kelvins
   e. 25.9 °F to Celsius degrees
   f. 25.9 °C to Fahrenheit degrees

19. Given the following mass, volume, and density information, calculate the missing quantity.
   a. mass = 121.4 g; volume = 42.4 cm^3; density = ? g/cm^3
   b. mass = 0.721 lb; volume = 241 cm^3; density = ? g/cm^3
   c. mass = ? g; volume = 124.1 mL; density = 0.821 g/mL
   d. mass = ? g; volume = 4.51 L; density = 1.15 g/cm^3
   e. mass = 142.4 g; volume = ? mL; density = 0.915 g/mL
   f. mass = 4.2 lb; volume = ? cm^3; density = 3.75 g/cm^3

20. For each of the following amounts of energy, perform the indicated conversion.
   a. 425 J to kilojoules
   b. 425 J to calories
   c. 425 J to kilocalories
   d. 78.5 kcal to joules
   e. 78.5 kcal to kilojoules
   f. 78.5 kcal to calories

21. Calculate the *mass* (in grams) of each of the following substances that could be warmed over the indicated temperature range by application of exactly 1.0 kJ of energy.
   a. water, from 15 °C to 42 °C
   b. iron, from 25 °C to 125 °C
   c. carbon, from −10 °C to 47 °C
   d. gold, from 56 °F to 75 °F
   e. silver, from 289 K to 385 K
   f. aluminum, from −10 °C to 85 °F

4 Chemical Foundations: Elements, Atoms, and Ions

The new Guggenheim Museum in Bilbao, Spain. The museum's signature feature is a roof clad in titanium that forms a "metallic flower."

The chemical elements are very important to each of us in our daily lives. Although certain elements are present in our bodies in tiny amounts, they can have a profound impact on our health and behavior. As we will see in this chapter, lithium can be a miracle treatment for someone with manic-depressive disease and our cobalt levels can have a remarkable impact on whether we behave violently.

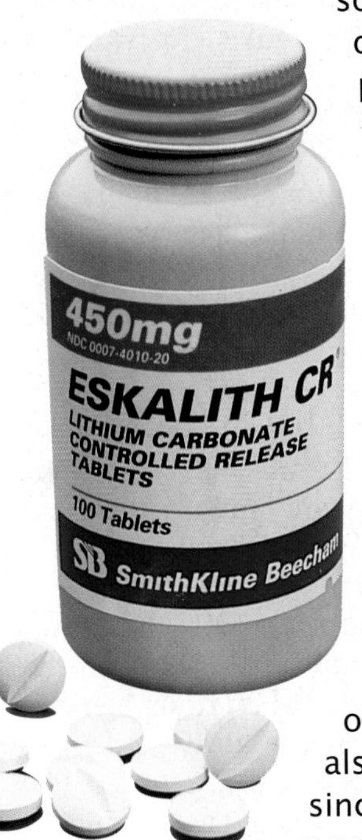

Lithium is administered in the form of lithium carbonate pills.

Since ancient times, humans have used chemical changes to their advantage. The processing of ores to produce metals for ornaments and tools and the use of embalming fluids are two applications of chemistry that were used before 1000 B.C.

The Greeks were the first to try to explain why chemical changes occur. By about 400 B.C. they had proposed that all matter was composed of four fundamental substances: fire, earth, water, and air.

The next 2000 years of chemical history were dominated by alchemy. Some alchemists were mystics and fakes who were obsessed with the idea of turning cheap metals into gold. However, many alchemists were sincere scientists and this period saw important events: the elements mercury, sulfur, and antimony were discovered, and alchemists learned how to prepare acids.

The first scientist to recognize the importance of careful measurements was the Irishman Robert Boyle (1627–1691). Boyle is best known for his pioneering work on the properties of gases, but his most important contribution to science was probably his insistence that science should be firmly grounded in experiments. For example, Boyle held no preconceived notions about how many elements there might be. His definition of the term *element* was based on experiments: a substance was an element unless it could be broken down into two or more simpler substances. For example, air could not be an element as the Greeks believed, because it could be broken down into many pure substances.

As Boyle's experimental definition of an element became generally accepted, the list of known elements grew, and the Greek system of four elements died. But although Boyle was an excellent scientist, he was not always right. For some reason he ignored his own definition of an element and clung to the alchemists' views that metals were not true elements and that a way would be found eventually to change one metal into another.

Robert Boyle at 62 years of age.

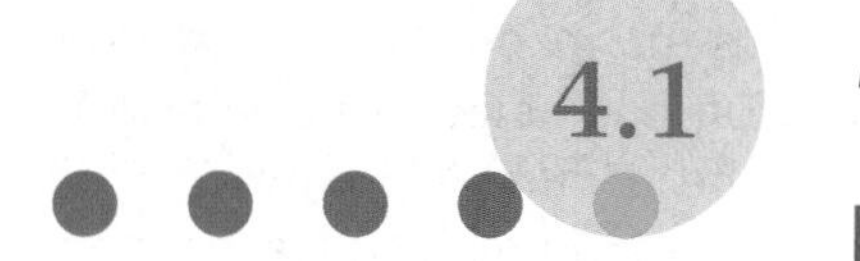

The Elements

AIMS: To learn about the relative abundances of the elements. To learn the names of some elements.

In studying the materials of the earth (and other parts of the universe), scientists have found that all matter can be broken down chemically into about one hundred different elements. At first it might seem amazing that the millions of known substances are composed of so few fundamental elements. Fortunately for those trying to understand and systematize it, nature often uses a relatively small number of fundamental units to assemble even extremely complex materials. For example, proteins, a group of substances that serve the human body in almost uncountable ways, are all made by linking together a few fundamental units to form huge molecules. A nonchemical example is the English language, where hundreds of thousands of words are constructed from only 26 letters. If you take apart the thousands of words in an English dictionary, you will find only these 26 fundamental components. In much the same way, when we take apart all of the substances in the world around us, we find only about one hundred fundamental building blocks—the elements. Compounds are made by combining atoms of the various elements, just as words are constructed from the 26 letters of the alphabet. And just as you had to learn the letters of the alphabet before you learned to read and write, you need to learn the names and symbols of the chemical elements before you can read and write chemistry.

Presently 114 different elements are known,* 88 of which occur naturally. (The rest have been made in laboratories.) The elements vary tremendously in abundance. In fact, only 9 elements account for most of the compounds found in the earth's crust. In Table 4.1, the elements are listed in order of their abundance (mass percent) in the earth's crust, oceans, and atmosphere. Note that nearly half of the mass is accounted for by oxygen alone. Also note that the 9 most abundant elements account for over 98% of the total mass.

Table 4.1 Distribution (Mass Percent) of the 18 Most Abundant Elements in the Earth's Crust, Oceans, and Atmosphere

Element	Mass Percent	Element	Mass Percent
oxygen	49.2	titanium	0.58
silicon	25.7	chlorine	0.19
aluminum	7.50	phosphorus	0.11
iron	4.71	manganese	0.09
calcium	3.39	carbon	0.08
sodium	2.63	sulfur	0.06
potassium	2.40	barium	0.04
magnesium	1.93	nitrogen	0.03
hydrogen	0.87	fluorine	0.03
		all others	0.49

* This number changes as new elements are made in particle accelerators.

Footprints in the sand of the Namib Desert in Namibia.

Oxygen, in addition to accounting for about 20% of the earth's atmosphere (where it occurs as O_2 molecules), is also found in virtually all the rocks, sand, and soil on the earth's crust. In these latter materials, oxygen is not present as O_2 molecules but exists in compounds that usually contain silicon and aluminum atoms. The familiar substances of the geological world, such as rocks and sand, contain large groups of silicon and oxygen atoms bound together to form huge clusters.

The list of elements found in living matter is very different from that for the earth's crust. Table 4.2 shows the distribution of elements in the human body. Oxygen, carbon, hydrogen, and nitrogen form the basis for all biologically important molecules. Some elements found in the body (called trace elements) are crucial for life, even though they are present in relatively small amounts. For example, chromium helps the body use sugars to provide energy.

One more general comment is important at this point. As we have seen, elements are fundamental to understanding chemistry. However, students are often confused by the many different ways that chemists use the term *element.* Sometimes when we say *element,* we mean a single atom of that element. We might call this the microscopic form of an element. Other times when we use the term *element,* we mean a sample of the element large enough to weigh on a balance. Such a sample contains many, many atoms of the element, and we might call this the macroscopic form of the element. There is yet a further complication. As we will see in more detail in Section 4.9 the macroscopic forms of several elements contain molecules rather than individual atoms as the fundamental components. For example, chemists know that oxygen gas consists of molecules with two oxygen atoms connected together (represented as O—O or more commonly as O_2). Thus when we refer to the element oxygen we might mean a single atom of oxygen, a single O_2 molecule, or a macroscopic sample containing many O_2 molecules. Finally, we often use the term *element* in a generic fashion. When we say the human body contains the element sodium or lithium, we do not mean that free elemental sodium or lithium is present. Rather, we mean that atoms of these elements are present in some form. In this text we will try to make clear what we mean when we use the term *element* in a particular case.

Table 4.2 Abundance of Elements in the Human Body

Major Elements	Mass Percent	Trace Elements (in alphabetical order)
oxygen	65.0	arsenic
carbon	18.0	chromium
hydrogen	10.0	cobalt
nitrogen	3.0	copper
calcium	1.4	fluorine
phosphorus	1.0	iodine
magnesium	0.50	manganese
potassium	0.34	molybdenum
sulfur	0.26	nickel
sodium	0.14	selenium
chlorine	0.14	silicon
iron	0.004	vanadium
zinc	0.003	

CHEMISTRY IN FOCUS

Trace Elements: Small but Crucial

We all know that certain chemical elements, such as calcium, carbon, nitrogen, phosphorus, and iron, are essential for humans to live. However, many other elements that are present in tiny amounts in the human body are also essential to life. Examples are chromium, cobalt, iodine, manganese, and copper. Chromium assists in the metabolism of sugars, cobalt is present in vitamin B_{12}, iodine is necessary for the proper functioning of the thyroid gland, manganese appears to play a role in maintaining the proper calcium levels in bones, and copper is involved in the production of red blood cells.

It is becoming clear that certain of the trace elements are very important in determining human behavior. For example, lithium (administered as lithium carbonate) has been a miracle drug for some people afflicted with manic-depressive syndrome, a disease that produces oscillatory behavior between inappropriate "highs" and the blackest of depressions. Although its exact function remains unknown, lithium seems to moderate the levels of neurotransmitters (compounds that are essential to nerve function), thus relieving some of the extreme emotions in sufferers of manic-depressive disease.

In addition, a chemist named William Walsh has done some very interesting studies on the inmates of Stateville Prison in Illinois. By analyzing the trace elements in the hair of prisoners, he has found intriguing relationships between the behavior of the inmates and their trace element profile. For example, Walsh found an inverse relationship between the level of cobalt in the prisoner's body and the degree of violence in his behavior.

Besides the levels of trace elements in our bodies, our exposure to various substances in our water, our food, and the air we breathe also has great importance for our health. For example, many scientists are concerned about our exposure to aluminum, through aluminum compounds used in water purification, baked goods (sodium aluminum phosphate is a common leavening agent), and cheese (so that it melts easily when cooked), and the aluminum that dissolves from our cookware and utensils. The effects of exposure to low levels of aluminum on humans are not presently clear, but there are some indications that we should limit our intake of this element.

Another example of low-level exposure to an element is the fluoride placed in many water supplies and toothpastes to control tooth decay by making tooth enamel more resistant to dissolving. However, the exposure of large numbers of people to fluoride is quite controversial—many people think it is harmful.

The chemistry of trace elements is fascinating and important. Keep your eye on the news for further developments.

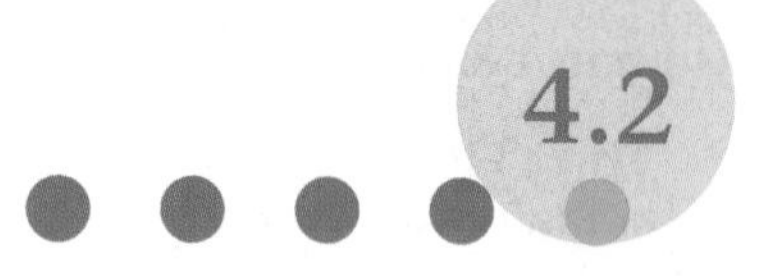

4.2 Symbols for the Elements

AIM: To learn the symbols of some elements.

The names of the chemical elements have come from many sources. Often an element's name is derived from a Greek, Latin, or German word that describes some property of the element. For example, gold was originally called *aurum,* a Latin word meaning "shining dawn," and lead was known as *plumbum,* which means "heavy." The names for chlorine and iodine come from Greek words describing their colors, and the name for bromine comes from a Greek word meaning "stench." In addition, it is very common for an element to be named for the place where it was discovered. You can guess where the elements francium, germanium, californium,* and americium* were first found. Some of the heaviest elements are named after famous scientists—for example, einsteinium* and nobelium.*

We often use abbreviations to simplify the written word. For example, it is much easier to put MA on an envelope than to write out Massachu-

* These elements are made artificially. They do not occur naturally.

setts, and we often write USA instead of United States of America. Likewise, chemists have invented a set of abbreviations or **element symbols** for the chemical elements. These symbols usually consist of the first letter or the first two letters of the element names. The first letter is always capitalized, and the second is not. Examples include:

fluorine	F	neon	Ne
oxygen	O	silicon	Si
carbon	C		

In the symbol for an element, only the first letter is capitalized.

Sometimes, however, the two letters used are not the first two letters in the name. For example,

zinc	Zn	cadmium	Cd
chlorine	Cl	platinum	Pt

Various forms of the element gold.

The symbols for some other elements are based on the original Latin or Greek name.

Current Name	Original Name	Symbol
gold	aurum	Au
lead	plumbum	Pb
sodium	natrium	Na
iron	ferrum	Fe

A list of the most common elements and their symbols is given in Table 4.3. You can also see the elements represented on a table in the inside front cover of this text. We will explain the form of this table (which is called the periodic table) in later chapters.

Table 4.3 The Names and Symbols of the Most Common Elements

Element	Symbol	Element	Symbol
aluminum	Al	lithium	Li
antimony (stibium)*	Sb	magnesium	Mg
argon	Ar	manganese	Mn
arsenic	As	mercury (hydrargyrum)	Hg
barium	Ba	neon	Ne
bismuth	Bi	nickel	Ni
boron	B	nitrogen	N
bromine	Br	oxygen	O
cadmium	Cd	phosphorus	P
calcium	Ca	platinum	Pt
carbon	C	potassium (kalium)	K
chlorine	Cl	radium	Ra
chromium	Cr	silicon	Si
cobalt	Co	silver (argentium)	Ag
copper (cuprum)	Cu	sodium (natrium)	Na
fluorine	F	strontium	Sr
gold (aurum)	Au	sulfur	S
helium	He	tin (stannum)	Sn
hydrogen	H	titanium	Ti
iodine	I	tungsten (wolfram)	W
iron (ferrum)	Fe	uranium	U
lead (plumbum)	Pb	zinc	Zn

*Where appropriate, the original name is shown in parentheses so that you can see where some of the symbols came from.

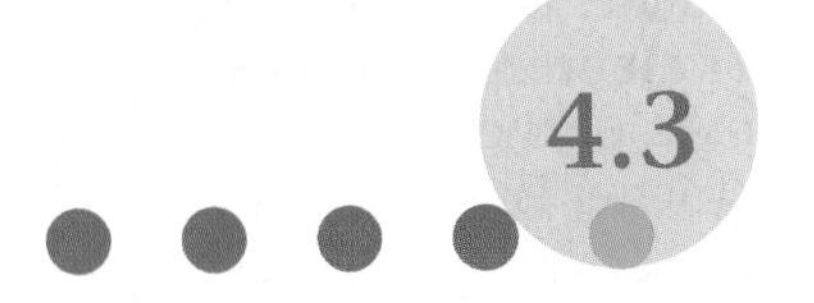

Dalton's Atomic Theory

AIMS: To learn about Dalton's theory of atoms. To understand and illustrate the law of constant composition.

Figure 4.1
John Dalton (1766–1844) was an English scientist who made his living as a teacher in Manchester. Although Dalton is best known for his atomic theory, he made contributions in many other areas, including meteorology (he recorded daily weather conditions for 46 years, producing a total of 200,000 data entries). A rather shy man, Dalton was colorblind to red (a special handicap for a chemist) and suffered from lead poisoning contracted from drinking stout (strong beer or ale) that had been drawn through lead pipes.

As scientists of the eighteenth century studied the nature of materials, several things became clear:

1. Most natural materials are mixtures of pure substances.
2. Pure substances are either elements or combinations of elements called compounds.
3. A given compound always contains the same proportions (by mass) of the elements. For example, water *always* contains 8 g of oxygen for every 1 g of hydrogen, and carbon dioxide *always* contains 2.7 g of oxygen for every 1 g of carbon. This principle became known as the **law of constant composition.** It means that a given compound always has the same composition, regardless of where it comes from.

John Dalton (Figure 4.1), an English scientist and teacher, was aware of these observations, and in about 1808 he offered an explanation for them that became known as **Dalton's atomic theory.** The main ideas of this theory (model) can be stated as follows:

Dalton's Atomic Theory

1. Elements are made of tiny particles called **atoms.**
2. All atoms of a given element are identical.
3. The atoms of a given element are different from those of any other element.
4. Atoms of one element can combine with atoms of other elements to form compounds. A given compound always has the same relative numbers and types of atoms.
5. Atoms are indivisible in chemical processes. That is, atoms are not created or destroyed in chemical reactions. A chemical reaction simply changes the way the atoms are grouped together.

Dalton's model successfully explained important observations such as the law of constant composition. This law makes sense because if a compound always contains the same relative numbers of atoms, it will always contain the same proportions by mass of the various elements.

Like most new ideas, Dalton's model was not accepted immediately. However, Dalton was convinced he was right and *used his model to predict* how a given pair of elements might combine to form more than one com-

N O
NO

O N O
NO_2

N N O
N_2O

Figure 4.2
Dalton pictured compounds as collections of atoms. Here NO, NO_2, and N_2O are represented. Note that the number of atoms of each type in a molecule is given by a subscript, except that the number 1 is always assumed and never written.

CHEMISTRY IN FOCUS

No Laughing Matter

Sometimes solving one problem leads to another. One such example involves the catalytic converters now required on all automobiles sold in much of the world. The purpose of these converters is to remove harmful pollutants such as CO and NO_2 from automobile exhausts. The good news is that these devices are quite effective and have led to much cleaner air in congested areas. The bad news is that these devices produce significant amounts of nitrous oxide, N_2O, commonly known as laughing gas because when inhaled it produces relaxation and mild inebriation. It was long used by dentists to make their patients more tolerant of some painful dental procedures.

The problem with N_2O is not that it is an air pollutant but that it is a "greenhouse gas." Certain molecules, such as CO_2, CH_4, N_2O, and others, strongly absorb infrared light ("heat radiation"), which causes the earth's atmosphere to retain more of its heat energy. Human activities have significantly increased the concentrations of these gases in the atmosphere. Mounting evidence suggests that the earth is warming as a result, leading to possible dramatic climatic changes.

A recent study by the Environmental Protection Agency (EPA) indicates that N_2O now accounts for over 7% of the greenhouse gases in the atmosphere and that automobiles equipped with catalytic converters produce nearly half of this N_2O. Ironically, N_2O is not regulated, because the Clean Air Act of 1970 was written to control smog—not greenhouse gases. The United States and other industrialized nations are now negotiating to find ways to control global warming but no agreement is now in place.

The N_2O situation illustrates just how complex environmental issues are. Clean may not necessarily be "green."

pound. For example, nitrogen and oxygen might form a compound containing one atom of nitrogen and one atom of oxygen (written NO), a compound containing two atoms of nitrogen and one atom of oxygen (written N_2O), a compound containing one atom of nitrogen and two atoms of oxygen (written NO_2), and so on (Figure 4.2). When the existence of these substances was verified, it was a triumph for Dalton's model. Because Dalton was able to predict correctly the formation of multiple compounds between two elements, his atomic theory became widely accepted.

Formulas of Compounds

AIM: To learn how a formula describes a compound's composition.

Here, *relative* refers to ratios.

A **compound** is a distinct substance that is composed of the atoms of two or more elements and always contains exactly the same relative masses of those elements. In light of Dalton's atomic theory, this simply means that a compound always contains the same relative *numbers* of atoms of each element. For example, water always contains two hydrogen atoms for each oxygen atom.

The types of atoms and the number of each type in each unit (molecule) of a given compound are conveniently expressed by a **chemical formula.** In a chemical formula the atoms are indicated by the element symbols, and the number of each type of atom is indicated by a subscript, a number that appears to the right of and below the symbol for the element. The formula for water is written H_2O, indicating that each molecule of water contains two atoms of hydrogen and one atom of oxygen (the subscript 1 is always understood and not written). Following are some general rules for writing formulas:

Rules for Writing Formulas

1. Each atom present is represented by its element symbol.
2. The number of each type of atom is indicated by a subscript written to the right of the element symbol.
3. When only one atom of a given type is present, the subscript 1 is not written.

Example 4.1 Writing Formulas of Compounds

Write the formula for each of the following compounds, listing the elements in the order given.

a. Each molecule of a compound that has been implicated in the formation of acid rain contains one atom of sulfur and three atoms of oxygen.

b. Each molecule of a certain compound contains two atoms of nitrogen and five atoms of oxygen.

c. Each molecule of glucose, a type of sugar, contains six atoms of carbon, twelve atoms of hydrogen, and six atoms of oxygen.

Solution

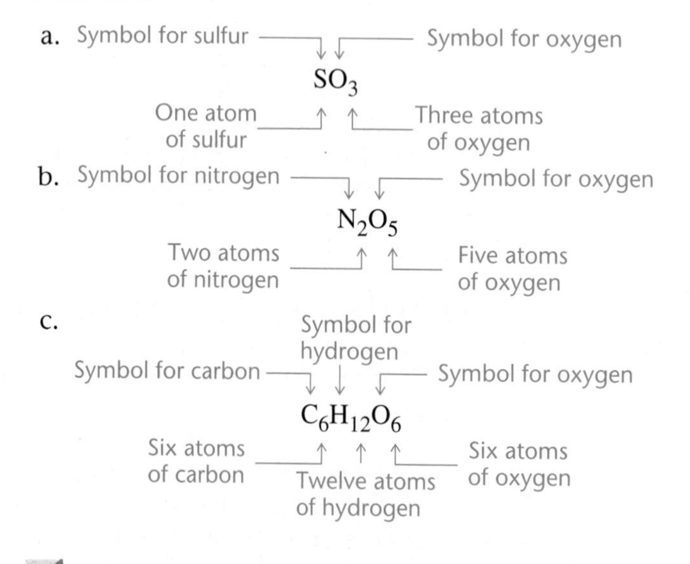

✔ Self-Check Exercise 4.1

Write the formula for each of the following compounds, listing the elements in the order given.

a. A molecule contains four phosphorus atoms and ten oxygen atoms.

b. A molecule contains one uranium atom and six fluorine atoms.

c. A molecule contains one aluminum atom and three chlorine atoms.

See Problems 4.19 and 4.20. ■

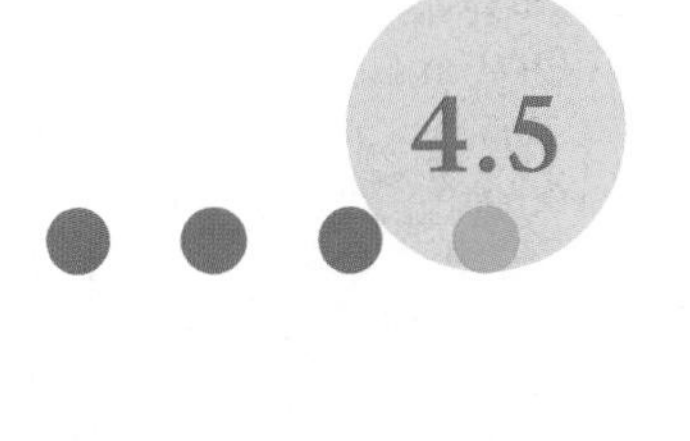

The Structure of the Atom

AIMS: To learn about the internal parts of an atom. To understand Rutherford's experiment to characterize the atom's structure.

Spherical cloud of positive charge

Electrons

Figure 4.3
One of the early models of the atom was the plum pudding model, in which the electrons were pictured as embedded in a positively charged spherical cloud, much as raisins are distributed in an old-fashioned plum pudding.

Dalton's atomic theory, proposed in about 1808, provided such a convincing explanation for the composition of compounds that it became generally accepted. Scientists came to believe that *elements consist of atoms* and that *compounds are a specific collection of atoms* bound together in some way. But what is an atom like? It might be a tiny ball of matter that is the same throughout with no internal structure—like a ball bearing. Or the atom might be composed of parts—it might be made up of a number of subatomic particles. But if the atom contains parts, there should be some way to break up the atom into its components.

Many scientists pondered the nature of the atom during the 1800s, but it was not until almost 1900 that convincing evidence became available that the atom has a number of different parts.

A physicist in England named J. J. Thomson showed in the late 1890s that the atoms of any element can be made to emit tiny negative particles. (He knew they had a negative charge because he could show that they were repelled by the negative part of an electric field.) Thus he concluded that all types of atoms must contain these negative particles, which are now called **electrons.**

On the basis of his results, Thomson wondered what an atom must be like. Although atoms contain these tiny negative particles, he also knew that whole atoms are not negatively *or* positively charged. Thus he concluded that the atom must also contain positive particles that balance exactly the negative charge carried by the electrons, giving the atom a zero overall charge.

Another scientist pondering the structure of the atom was William Thomson (better known as Lord Kelvin and no relation to J. J. Thomson). Lord Kelvin got the idea (which might have occurred to him during dinner) that the atom might be something like plum pudding (a pudding with raisins randomly distributed throughout). Kelvin reasoned that the atom might be thought of as a uniform "pudding" of positive charge with enough negative electrons scattered within to counterbalance that positive charge (see Figure 4.3). Thus the plum pudding model of the atom came into being.

Some historians credit J. J. Thomson for the plum pudding model.

If you had taken this course in 1910, the plum pudding model would have been the only picture of the atom described. However, our ideas about the atom were changed dramatically in 1911 by a physicist named Ernest Rutherford (Figure 4.4), who learned physics in J. J. Thomson's laboratory in the late 1890s. By 1911 Rutherford had become a distinguished scientist with many important discoveries to his credit. One of his main areas of interest involved alpha particles (α particles), positively charged particles with a mass approximately 7500 times that of an electron. In studying the flight of these particles through air, Rutherford found that some of the α particles were deflected by something in the air. Puzzled by this, he designed an experiment that involved directing α particles toward a thin metal foil. Surrounding the foil was a detector coated with a substance

Figure 4.4
Ernest Rutherford (1871–1937) was born on a farm in New Zealand. In 1895 he placed second in a scholarship competition to attend Cambridge University but was awarded the scholarship when the winner decided to stay home and get married. Rutherford was an intense, hard-driving person who became a master at designing just the right experiment to test a given idea. He was awarded the Nobel Prize in chemistry in 1908.

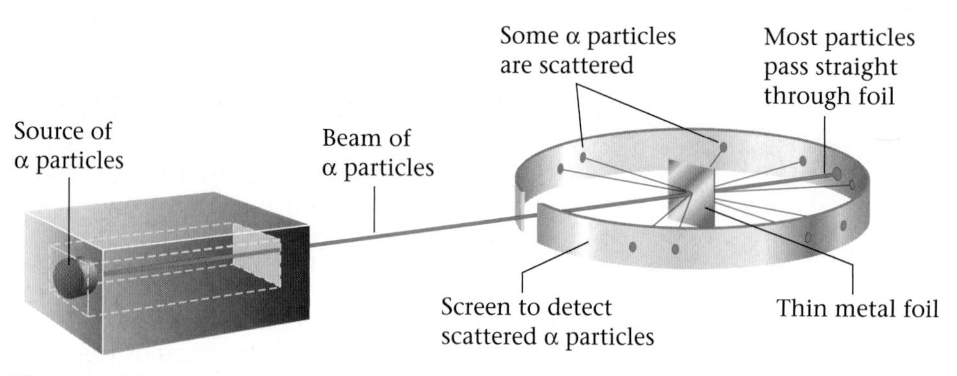

Figure 4.5
Rutherford's experiment on α-particle bombardment of metal foil.

that produced tiny flashes wherever it was hit by an α particle (Figure 4.5). The results of the experiment were very different from those Rutherford anticipated. Although most of the α particles passed straight through the foil, some of the particles were deflected at large angles, as shown in Figure 4.5, and some were reflected backward.

This outcome was a great surprise to Rutherford. (He described this result as comparable to shooting a gun at a piece of paper and having the bullet bounce back.) Rutherford knew that if the plum pudding model of the atom was correct, the massive α particles would crash through the thin foil like cannonballs through paper (as shown in Figure 4.6a). So he expected the α particles to travel through the foil experiencing, at most, very minor deflections of their paths.

Rutherford concluded from these results that the plum pudding model for the atom could not be correct. The large deflections of the α particles could be caused only by a center of concentrated positive charge that would repel the positively charged α particles, as illustrated in Figure 4.6b. Most of the α particles passed directly through the foil because the atom is mostly open space. The deflected α particles were those that had a "close encounter" with the positive center of the atom, and the few reflected α particles were those that scored a "direct hit" on the positive center. In Rutherford's mind

One of Rutherford's coworkers in this experiment was an undergraduate named Ernest Marsden who, like Rutherford, was from New Zealand.

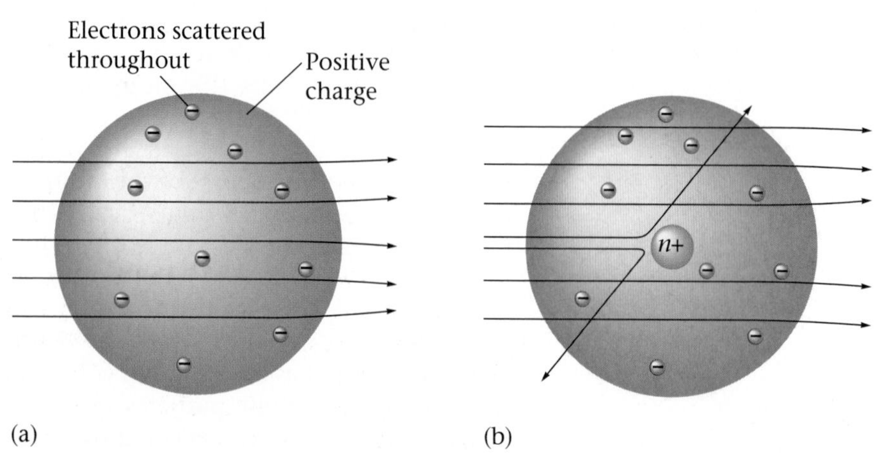

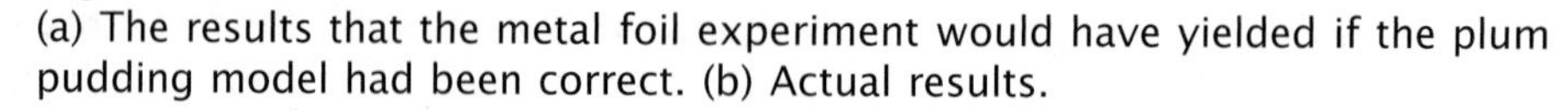

Figure 4.6
(a) The results that the metal foil experiment would have yielded if the plum pudding model had been correct. (b) Actual results.

CHEMISTRY IN FOCUS

Glowing Tubes for Signs, Television Sets, and Computers

J. J. Thomson discovered that atoms contain electrons by using a device called a cathode ray tube (often abbreviated CRT today). When he did these experiments, he could not have imagined that he was making television sets and computer monitors possible. A cathode ray tube is a sealed glass tube that contains a gas and has separated metal plates connected to external wires (Figure 4.7). When a source of electrical energy is applied to the metal plates, a glowing beam is produced (Figure 4.8). Thomson became convinced that the glowing gas was caused by a stream of negatively charged particles coming from the metal plate. In addition, because Thomson always got the same kind of negative particles no matter what metal he used, he concluded that all types of atoms must contain these same negative particles (we now call them electrons).

Figure 4.8
A CRT being used to display computer graphics.

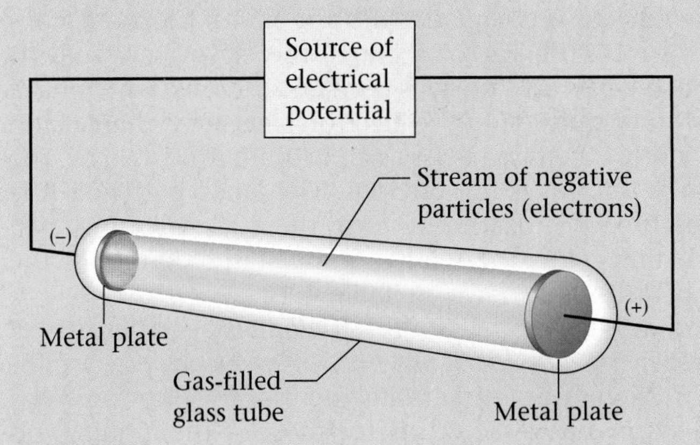

Figure 4.7
Schematic of a cathode ray tube. A stream of electrons passes between the electrodes. The fast-moving particles excite the gas in the tube, causing a glow between the plates.

Thomson's cathode ray tube has many modern applications. For example, "neon" signs consist of small-diameter cathode ray tubes containing different kinds of gases to produce various colors. For example, if the gas in the tube is neon, the tube glows with a red–orange color; if argon is present, a blue glow appears. The presence of krypton gives an intense white light.

A television picture tube or computer monitor is also fundamentally a cathode ray tube. In this case the electrons are directed onto a screen containing chemical compounds that glow when struck by fast-moving electrons. The use of various compounds that emit different colors when they are struck by the electrons makes color pictures possible on the screens of these CRTs.

these results could be explained only in terms of a **nuclear atom**—an atom with a dense center of positive charge (the **nucleus**) around which tiny electrons moved in a space that was otherwise empty.

He concluded that the nucleus must have a positive charge to balance the negative charge of the electrons and that it must be small and dense. What was it made of? By 1919 Rutherford concluded that the nucleus of an atom contained what he called protons. A **proton** has the same magnitude (size) of charge as the electron, but its charge is *positive*. We say that the proton has a charge of 1+ and the electron a charge of 1−.

If the atom were expanded to the size of a huge stadium, the nucleus would be only about as big as a fly at the center.

Rutherford reasoned that the hydrogen atom has a single proton at its center and one electron moving through space at a relatively large distance from the proton (the hydrogen nucleus). He also reasoned that other atoms must have nuclei (the plural of *nucleus*) composed of many protons bound together somehow. In addition, Rutherford and a coworker, James Chadwick, were able to show in 1932 that most nuclei also contain a neutral particle that they named the **neutron.** A neutron is slightly more massive than a proton but has no charge.

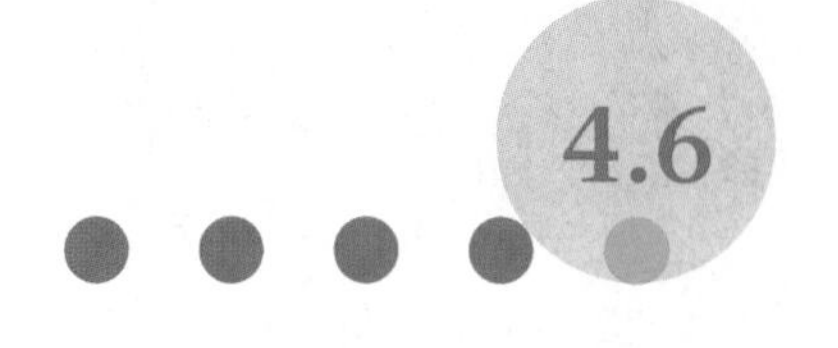

Introduction to the Modern Concept of Atomic Structure

AIM: To understand some important features of subatomic particles.

In the years since Thomson and Rutherford, a great deal has been learned about atomic structure. The simplest view of the atom is that it consists of a tiny nucleus (about 10^{-13} cm in diameter) and electrons that move about the nucleus at an average distance of about 10^{-8} cm from it (Figure 4.9). To visualize how small the nucleus is compared with the size of the atom, consider that if the nucleus were the size of a grape, the electrons would be about one *mile* away on average. The nucleus contains protons, which have a positive charge equal in magnitude to the electron's negative charge, and neutrons, which have almost the same mass as a proton but no charge. The neutrons' function in the nucleus is not obvious. They may help hold the protons (which repel each other) together to form the nucleus, but we will not be concerned with that here. The relative masses and charges of the electron, proton, and neutron are shown in Table 4.4.

In this model the atom is called a nuclear atom because the positive charge is localized in a small, compact structure (the nucleus) and not spread out uniformly, as in the plum pudding view.

An important question arises at this point: *"If all atoms are composed of these same components, why do different atoms have different chemical properties?"* The answer lies in the number and arrangement of the electrons. The space in which the electrons move accounts for most of the atomic volume. The electrons are the parts of atoms that "intermingle" when atoms combine to form molecules. Therefore, the number of electrons a given atom possesses greatly affects the way it can interact with other atoms. As a result, atoms of different elements, which have different numbers of electrons, show different chemical behavior. Although the atoms of different elements also differ in their numbers of protons, it is the number of electrons that really determines chemical behavior. We will discuss how this happens in later chapters.

The *chemistry* of an atom arises from its electrons.

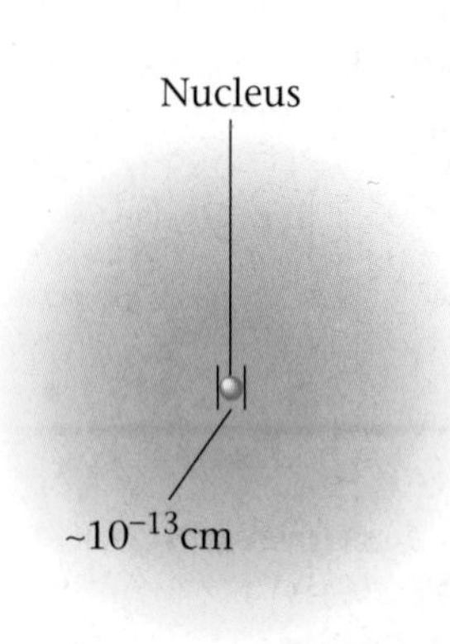

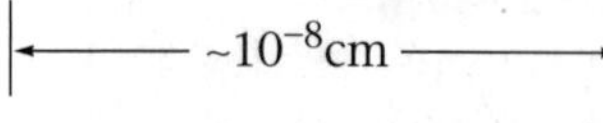

Figure 4.9
A nuclear atom viewed in cross section. (The symbol ~ means approximately.) This drawing does not show the actual scale. The nucleus is actually *much* smaller compared with the size of an atom.

Table 4.4 The Mass and Charge of the Electron, Proton, and Neutron

Particle	Relative Mass*	Relative Charge
electron	1	1−
proton	1836	1+
neutron	1839	none

*The electron is arbitrarily assigned a mass of 1 for comparison.

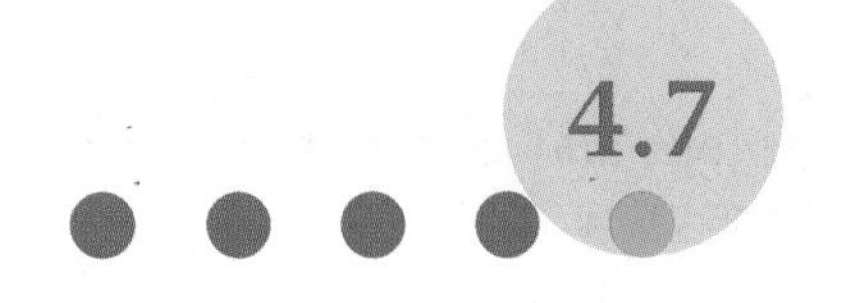

4.7 Isotopes

AIMS: To learn about the terms isotope, atomic number, and mass number. To understand the use of the symbol $^{A}_{Z}X$ to describe a given atom.

We have seen that an atom has a nucleus with a positive charge due to its protons and has electrons in the space surrounding the nucleus at relatively large distances from it.

All atoms of the same element have the same number of protons (the element's atomic number) and the same number of electrons.

In a free atom, the positive and negative charges always balance to yield a net zero charge.

As an example, consider a sodium atom, which has 11 protons in its nucleus. Because an atom has no overall charge, the number of electrons must equal the number of protons. Therefore, a sodium atom has 11 electrons in the space around its nucleus. It is *always* true that a sodium atom has 11 protons and 11 electrons. However, each sodium atom also has neutrons in its nucleus, and different types of sodium atoms exist that have different numbers of neutrons.

When Dalton stated his atomic theory in the early 1800s, he assumed all of the atoms of a given element were identical. This idea persisted for over a hundred years, until James Chadwick discovered that the nuclei of most atoms contain neutrons as well as protons. (This is a good example of how a theory changes as new observations are made.) After the discovery of the neutron, Dalton's statement that all atoms of a given element are identical had to be changed to "All atoms of the same element contain the same number of protons and electrons, but atoms of a given element may have different numbers of neutrons."

Atomic number: the number of protons. Mass number: the sum of protons and neutrons.

To illustrate this idea, consider the sodium atoms represented in Figure 4.10. These atoms are **isotopes,** or *atoms with the same number of protons but different numbers of neutrons.* The number of protons in a nucleus is called the atom's **atomic number.** The *sum* of the number of neutrons and the number of protons in a given nucleus is called the atom's **mass number.** To specify which of the isotopes of an element we are talking about, we use the symbol

$$^{A}_{Z}X$$

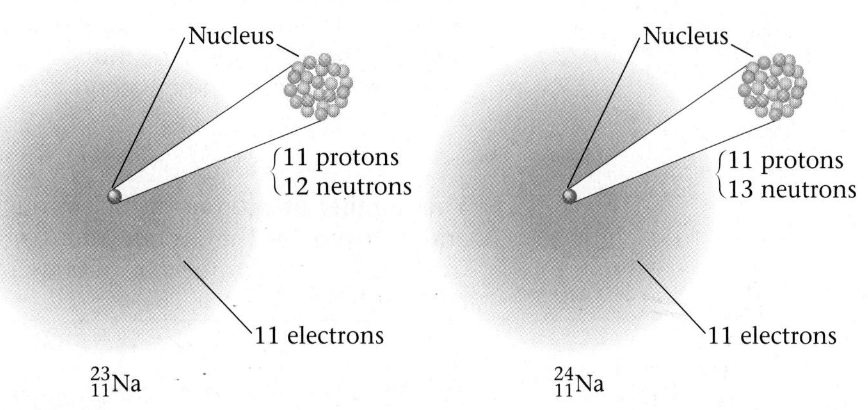

Figure 4.10
Two isotopes of sodium. Both have 11 protons and 11 electrons, but they differ in the number of neutrons in their nuclei.

where

X = the symbol of the element
A = the mass number (number of protons and neutrons)
Z = the atomic number (number of protons)

For example, the symbol for one particular type of sodium atom is written

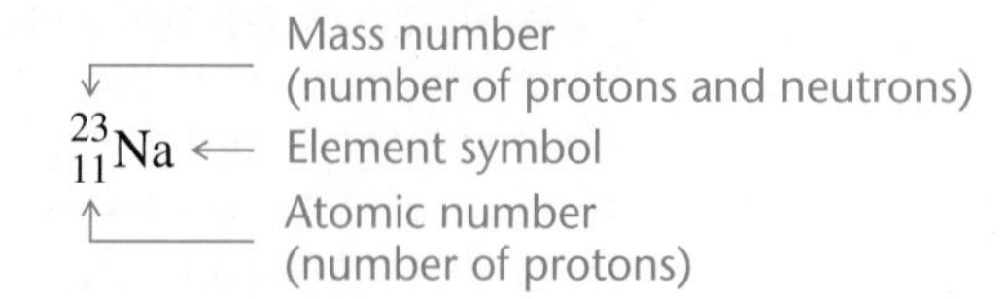

The particular atom represented here is called sodium-23, because it has a mass number of 23. Let's specify the number of each type of subatomic particle. From the atomic number 11 we know that the nucleus contains 11 protons. And because the number of electrons is equal to the number of protons, we know that this atom contains 11 electrons. How many neutrons are present? We can calculate the number of neutrons from the definition of the mass number

$$\text{Mass number} = \text{number of protons} + \text{number of neutrons}$$

or, in symbols,

$$A = Z + \text{number of neutrons}$$

We can isolate (solve for) the number of neutrons by subtracting Z from both sides of the equation

$$A - Z = Z - Z + \text{number of neutrons}$$
$$A - Z = \text{number of neutrons}$$

This is a general result. You can always determine the number of neutrons present in a given atom by subtracting the atomic number from the mass number. In this case ($^{23}_{11}Na$), we know that $A = 23$ and $Z = 11$. Thus

$$A - Z = 23 - 11 = 12 = \text{number of neutrons}$$

In summary, sodium-23 has 11 electrons, 11 protons, and 12 neutrons.

Example 4.2 Interpreting Symbols for Isotopes

In nature, elements are usually found as a mixture of isotopes. Three isotopes of elemental carbon are $^{12}_{6}C$ (carbon-12), $^{13}_{6}C$ (carbon-13), and $^{14}_{6}C$ (carbon-14). Determine the number of each of the three types of subatomic particles in each of these carbon atoms.

Solution

The number of protons and electrons is the same in each of the isotopes and is given by the atomic number of carbon, 6. The number of neutrons can be determined by subtracting the atomic number (Z) from the mass number:

$$A - Z = \text{number of neutrons}$$

The numbers of neutrons in the three isotopes of carbon are

$$^{12}_{6}C\text{: number of neutrons} = A - Z = 12 - 6 = 6$$
$$^{13}_{6}C\text{: number of neutrons} = 13 - 6 = 7$$
$$^{14}_{6}C\text{: number of neutrons} = 14 - 6 = 8$$

CHEMISTRY IN FOCUS

Isotope Tales

The atoms of a given element typically consist of several isotopes—atoms with the same number of protons but different numbers of neutrons. It turns out that the ratio of isotopes found in nature can be very useful in natural detective work. One reason is that the ratio of isotopes of elements found in living animals and humans reflect their diets. For example, African elephants that feed on grasses have a different $^{13}C/^{12}C$ ratio in their tissues than elephants that primarily eat tree leaves. This difference arises because grasses have a different growth pattern than leaves do, resulting in different amounts of ^{13}C and ^{12}C being incorporated from the CO_2 in the air. Because leaf-eating and grass-eating elephants live in different areas of Africa, the observed differences in the $^{13}C/^{12}C$ isotope ratios in elephant ivory samples have enabled authorities to identify the sources of illegal samples of ivory.

Another case of isotope detective work involves the tomb of King Midas, who ruled the kingdom Phyrgia in the eighth century B.C. Analysis of nitrogen isotopes in the king's decayed casket has revealed details about the king's diet. Scientists have learned that the $^{15}N/^{14}N$ ratios of carnivores are higher than those of herbivores, which in turn are higher than those of plants. It turns out that the organism responsible for decay of the king's wooden casket has an unusually large requirement for nitrogen. The source of this nitrogen was the body of the dead king. Because the decayed wood under his now-decomposed body showed a high $^{15}N/^{14}N$ ratio, researchers feel sure that the king's diet was rich in meat.

Ancient Anasazi Indian cliff dwellings.

A third case of historical isotope detective work concerns the Pueblo ancestor people (commonly called the Anasazi), who lived in what is now northwestern New Mexico between A.D. 900 and 1150. The center of their civilization, Chaco Canyon, was a thriving cultural center boasting dwellings made of hand-hewn sandstone and more than 200,000 logs. The sources of the logs have always been controversial. Many theories have been advanced concerning the distances over which the logs were hauled. Recent research by Nathan B. English, a geochemist at the University of Arizona in Tucson, has used the distribution of strontium isotopes in the wood to identify the probable sources of the logs. This effort has enabled scientists to understand more clearly the Anasazi building practices.

These stories illustrate how isotopes can serve as valuable sources of biologic and historic information.

In summary,

Symbol	Number of Protons	Number of Electrons	Number of Neutrons
$^{12}_{6}C$	6	6	6
$^{13}_{6}C$	6	6	7
$^{14}_{6}C$	6	6	8

✔ Self-Check Exercise 4.2

Give the number of protons, neutrons, and electrons in the atom symbolized by $^{90}_{38}Sr$. Strontium-90 occurs in fallout from nuclear testing. It can accumulate in bone marrow and may cause leukemia and bone cancer.

See Problems 4.39 and 4.40. ■

✔ Self-Check Exercise 4.3

Give the number of protons, neutrons, and electrons in the atom symbolized by $^{201}_{80}Hg$.

See Problems 4.39 and 4.40. ■

Example 4.3 Writing Symbols for Isotopes

Magnesium burns in air to give a bright white flame.

Write the symbol for the magnesium atom (atomic number 12) with a mass number of 24. How many electrons and how many neutrons does this atom have?

Solution

The atomic number 12 means the atom has 12 protons. The element magnesium is symbolized by Mg. The atom is represented as

$$^{24}_{12}Mg$$

and is called magnesium-24. Because the atom has 12 protons, it must also have 12 electrons. The mass number gives the total number of protons and neutrons, which means that this atom has 12 neutrons ($24 - 12 = 12$). ■

Example 4.4 Calculating Mass Number

Write the symbol for the silver atom ($Z = 47$) that has 61 neutrons.

Solution

The element symbol is $^{A}_{Z}Ag$, where we know that $Z = 47$. We can find A from its definition, $A = Z +$ number of neutrons. In this case,

$$A = 47 + 61 = 108$$

The complete symbol for this atom is $^{108}_{47}Ag$.

✔ Self-Check Exercise 4.4

Give the symbol for the phosphorus atom ($Z = 15$) that contains 17 neutrons.

See Problems 4.41 and 4.42. ■

4.8 Introduction to the Periodic Table

AIMS: To learn about various features of the periodic table. To learn some of the properties of metals, nonmetals, and metalloids.

In any room where chemistry is taught or practiced, you are almost certain to find a chart called the **periodic table** hanging on the wall. This chart shows all of the known elements and gives a good deal of information about each. As our study of chemistry progresses, the usefulness of the periodic table will become more obvious. This section will simply introduce it.

Alkali metals | Alkaline earth metals | Transition metals | Halogens | Noble gases

1 1A	2 2A	3	4	5	6	7	8	9	10	11	12	13 3A	14 4A	15 5A	16 6A	17 7A	18 8A
1 H																	2 He
3 Li	4 Be											5 B	6 C	7 N	8 O	9 F	10 Ne
11 Na	12 Mg											13 Al	14 Si	15 P	16 S	17 Cl	18 Ar
19 K	20 Ca	21 Sc	22 Ti	23 V	24 Cr	25 Mn	26 Fe	27 Co	28 Ni	29 Cu	30 Zn	31 Ga	32 Ge	33 As	34 Se	35 Br	36 Kr
37 Rb	38 Sr	39 Y	40 Zr	41 Nb	42 Mo	43 Tc	44 Ru	45 Rh	46 Pd	47 Ag	48 Cd	49 In	50 Sn	51 Sb	52 Te	53 I	54 Xe
55 Cs	56 Ba	57 La*	72 Hf	73 Ta	74 W	75 Re	76 Os	77 Ir	78 Pt	79 Au	80 Hg	81 Tl	82 Pb	83 Bi	84 Po	85 At	86 Rn
87 Fr	88 Ra	89 Ac†	104 Rf	105 Db	106 Sg	107 Bh	108 Hs	109 Mt	110 Uun	111 Uuu	112 Uub		114 Uuq				

*Lanthanides	58 Ce	59 Pr	60 Nd	61 Pm	62 Sm	63 Eu	64 Gd	65 Tb	66 Dy	67 Ho	68 Er	69 Tm	70 Yb	71 Lu
†Actinides	90 Th	91 Pa	92 U	93 Np	94 Pu	95 Am	96 Cm	97 Bk	98 Cf	99 Es	100 Fm	101 Md	102 No	103 Lr

Figure 4.11
The periodic table.

A simple version of the periodic table is shown in Figure 4.11. Note that each box of this table contains a number written over one, two, or three letters. The letters are the symbols for the elements. The number shown above each symbol is the atomic number (the number of protons and also the number of electrons) for that element. For example, carbon (C) has atomic number 6:

6 C

Lead (Pb) has atomic number 82:

82 Pb

Notice that elements 110 through 112, and element 114, have unusual three-letter designations beginning with U. These are abbreviations for the systematic names of the atomic numbers of these elements. "Regular" names for these elements will be chosen eventually by the scientific community.

Note that the elements are listed on the periodic table in order of increasing atomic number. They are also arranged in specific horizontal rows and vertical columns. The elements were first arranged in this way in 1869 by Dmitri Mendeleev, a Russian scientist. Mendeleev arranged the elements in this way because of similarities in the chemical properties of various "families" of elements. For example, fluorine and chlorine are reactive gases that form similar compounds. It was also known that sodium and potassium behave very similarly. Thus the name *periodic table* refers to the fact that as we increase the atomic numbers, every so often an element occurs with properties similar to those of an earlier (lower-atomic-number) element. For example, the elements

Mendeleev actually arranged the elements in order of increasing atomic mass rather than atomic number.

9 F
17 Cl
35 Br
53 I
85 At

Throughout the text, we will highlight the location of various elements by presenting a small version of the periodic table.

all show similar chemical behavior and so are listed vertically, as a "family" of elements.

These families of elements with similar chemical properties that lie in the same vertical column on the periodic table are called **groups.** Groups are often referred to by the number over the column (see Figure 4.11). Note that the group numbers are accompanied by the letter A on the periodic table in Figure 4.11 and the one inside the front cover of the text. For simplicity we will delete the A's when we refer to groups in the text. Many of the groups have special names. For example, the first column of elements (Group 1) has the name **alkali metals.** The Group 2 elements are called the **alkaline earth metals,** the Group 7 elements are the **halogens,** and the elements in Group 8 are called the **noble gases.** A large collection of elements that spans many vertical columns consists of the **transition metals.**

There's another convention recommended by the International Union of Pure and Applied Chemistry for group designations that uses numbers 1 through 18 and includes the transition metals (see Fig. 4.11). Do not confuse that system with the one used in this text, where only the representative elements have group numbers (1 through 8).

Most of the elements are **metals.** Metals have the following characteristic physical properties:

Physical Properties of Metals

1. Efficient conduction of heat and electricity
2. Malleability (they can be hammered into thin sheets)
3. Ductility (they can be pulled into wires)
4. A lustrous (shiny) appearance

Nonmetals sometimes have one or more metallic properties. For example, solid iodine is lustrous, and graphite (a form of pure carbon) conducts electricity.

For example, copper is a typical metal. It is lustrous (although it tarnishes readily); it is an excellent conductor of electricity (it is widely used in electrical wires); and it is readily formed into various shapes, such as pipes for water systems. Copper is one of the transition metals—the metals shown in the center of the periodic table. Iron, aluminum, and gold are other familiar elements that have metallic properties. All of the elements shown to the left of and below the heavy "stair-step" black line in Figure 4.11 are classified as metals, except for hydrogen (Figure 4.12).

The relatively small number of elements that appear in the upper-right corner of the periodic table (to the right of the heavy line in Figures 4.11 and 4.12) are called **nonmetals.** Nonmetals generally lack those properties that characterize metals and show much more variation in their properties than metals do. Whereas almost all metals are solids at normal temperatures, many nonmetals (such as nitrogen, oxygen, chlorine, and neon) are gaseous and one (bromine) is a liquid. Several nonmetals (such as carbon, phosphorus, and sulfur) are also solids.

The elements that lie close to the "stair-step" line as shown in blue in Figure 4.12 often show a mixture of metallic and nonmetallic properties. These elements, which are called **metalloids** or **semimetals,** include silicon, germanium, arsenic, antimony, and tellurium.

As we continue our study of chemistry, we will see that the periodic table is a valuable tool for organizing accumulated knowledge and that it helps us predict the properties we expect a given element to exhibit. We will also develop a model for atomic structure that will explain why there are groups of elements with similar chemical properties.

Example 4.5 Interpreting the Periodic Table

For each of the following elements, use the periodic table in the front of the book to give the symbol and atomic number and to specify whether the element is a metal or a nonmetal. Also give the named family to which the element belongs (if any).

a. iodine
c. gold
b. magnesium
d. lithium

Indonesian men carrying chunks of elemental sulfur in baskets.

Solution

a. Iodine (symbol I) is element 53 (its atomic number is 53). Iodine lies to the right of the stair-step line in Figure 4.12 and thus is a nonmetal. Iodine is a member of Group 7, the family of halogens.

b. Magnesium (symbol Mg) is element 12 (atomic number 12). Magnesium is a metal and is a member of the alkaline earth metal family (Group 2).

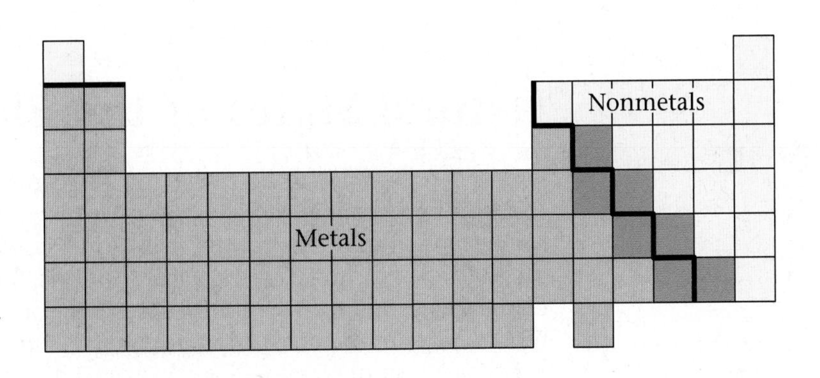

Figure 4.12
The elements classified as metals and as nonmetals.

CHEMISTRY IN FOCUS

Putting the Brakes on Arsenic

The toxicity of arsenic is well known. Indeed, arsenic has often been the poison of choice in classic plays and films—rent *Arsenic and Old Lace* sometime. Contrary to its treatment in the aforementioned movie, arsenic poisoning is a serious, contemporary problem. For example, the World Health Organization estimates that 77 million people in Bangladesh are at risk from drinking water that contains large amounts of naturally occurring arsenic. Recently, the Environmental Protection Agency announced more stringent standards for arsenic in U.S. public drinking water supplies. Studies show that prolonged exposure to arsenic can lead to a higher risk of bladder, lung, and skin cancers as well as other ailments, although the levels of arsenic that induce these symptoms remain in dispute in the scientific community.

Cleaning up arsenic-contaminated soil and water poses a significant problem. One approach is to find plants that will leach arsenic from the soil. Such a plant, the brake fern, recently has been shown to have a voracious appetite for arsenic. Research led by Lenna Ma, a chemist at the University of Florida in Gainesville, has shown that the brake fern accumulates arsenic at a rate 200 times that of the average plant. The arsenic, which becomes concentrated in fronds that grow up to 5 feet long, can be easily harvested and hauled away. Researchers are now investigating the best way to dispose of the plants so the arsenic can be isolated. The fern (*Pteris vittata*) looks promising for putting the brakes on arsenic pollution.

Lenna Ma and *Pteris vittata*—called the brake fern.

c. Gold (symbol Au) is element 79 (atomic number 79). Gold is a metal and is not a member of a named vertical family. It is classed as a transition metal.

d. Lithium (symbol Li) is element 3 (atomic number 3). Lithium is a metal in the alkali metal family (Group 1).

✔ Self-Check Exercise 4.5

Give the symbol and atomic number for each of the following elements. Also indicate whether each element is a metal or a nonmetal and whether it is a member of a named family.

a. argon
b. chlorine
c. barium
d. cesium

See Problems 4.53 and 4.54. ■

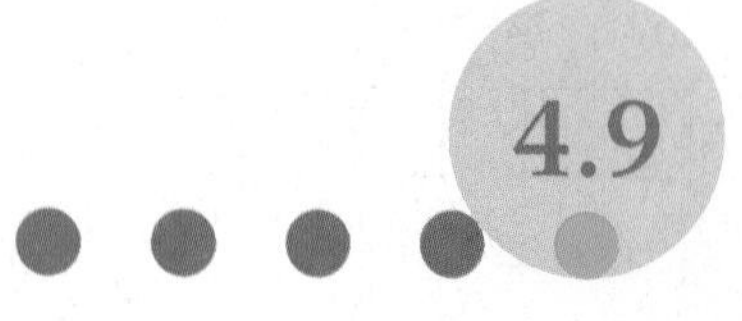

4.9 Natural States of the Elements

AIM: To learn the natures of the common elements.

As we have noted, the matter around us consists mainly of mixtures. Most often these mixtures contain compounds, in which atoms from different elements are bound together. Most elements are quite reactive: their atoms tend to combine with those of other elements to form compounds. Thus

A gold nugget weighing 13 lb, 7 oz, which came to be called Tom's Baby, was found by Tom Grove near Breckenridge, Colorado, on July 23, 1887.

we do not often find elements in nature in pure form—uncombined with other elements. However, there are notable exceptions. The gold nuggets found at Sutter's Mill in California that launched the Gold Rush in 1849 are virtually pure elemental gold. And platinum and silver are often found in nearly pure form.

Gold, silver, and platinum are members of a class of metals called *noble metals* because they are relatively unreactive. (The term *noble* implies a class set apart.)

Other elements that appear in nature in the uncombined state are the elements in Group 8: helium, neon, argon, krypton, xenon, and radon. Because the atoms of these elements do not combine readily with those of other elements, we call them the *noble gases*. For example, helium gas is found in uncombined form in underground deposits with natural gas.

When we take a sample of air (the mixture of gases that constitute the earth's atmosphere) and separate it into its components, we find several pure elements present. One of these is argon. Argon gas consists of a collection of separate argon atoms, as shown in Figure 4.13.

Recall that a molecule is a collection of atoms that behaves as a unit. Molecules are always electrically neutral (zero charge).

Air also contains nitrogen gas and oxygen gas. When we examine these two gases, however, we find that they do not contain single atoms, as argon does, but instead contain **diatomic molecules:** molecules made up of *two atoms,* as represented in Figure 4.14. In fact, any sample of elemental oxygen gas at normal temperatures contains O_2 molecules. Likewise, nitrogen gas contains N_2 molecules.

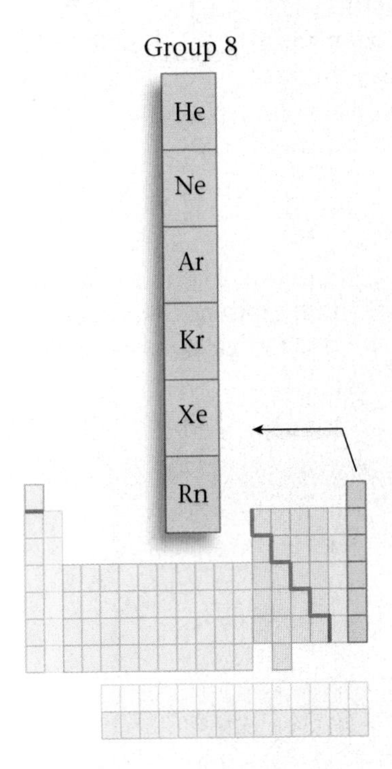

Hydrogen is another element that forms diatomic molecules. Though virtually all of the hydrogen found on earth is present in compounds with other elements (such as with oxygen in water), when hydrogen is prepared as a free element it contains diatomic H_2 molecules. For example, an electric current can be used to decompose water (see Figure 4.15 on p. 104 and Figure 3.3 on p. 58) into elemental hydrogen and oxygen containing H_2 and O_2 molecules, respectively.

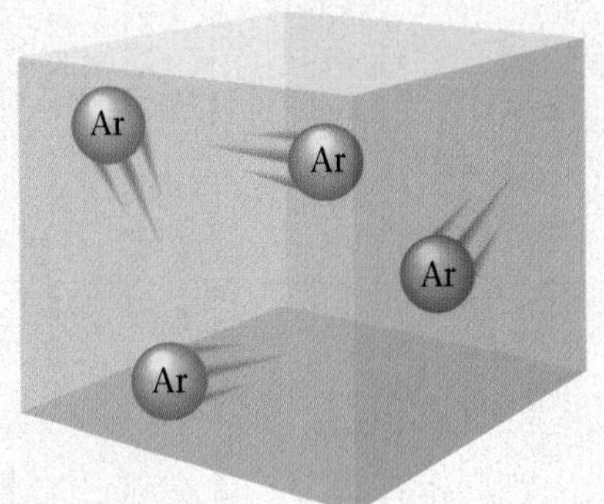

Figure 4.13
Argon gas consists of a collection of separate argon atoms.

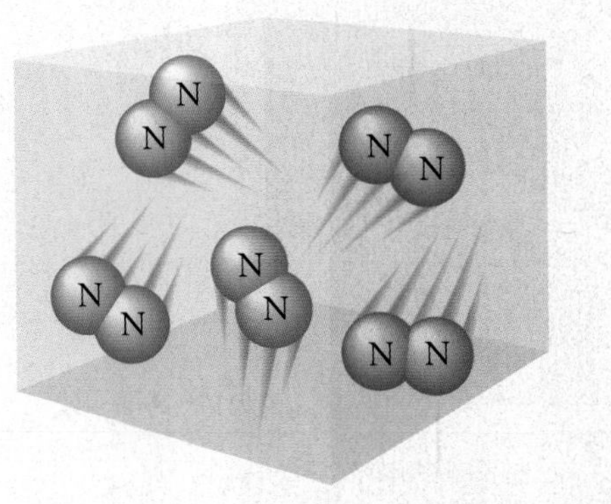

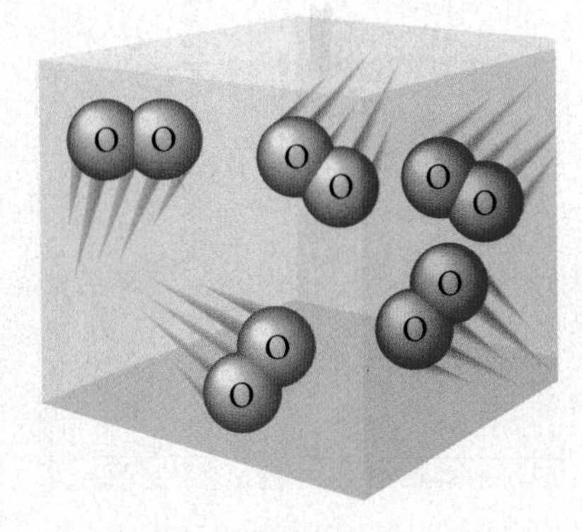

(a) (b)

Figure 4.14
Gaseous nitrogen and oxygen contain diatomic (two-atom) molecules. (a) Nitrogen gas contains N—N (N_2) molecules. (b) Oxygen gas contains O—O (O_2) molecules.

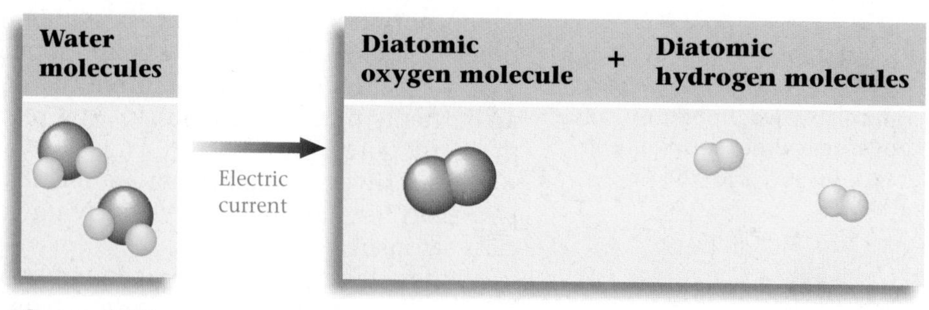

Figure 4.15
The decomposition of two water molecules (H_2O) to form two hydrogen molecules (H_2) and an oxygen molecule (O_2). Note that only the grouping of the atoms changes in this process; no atoms are created or destroyed. There must be the same number of H atoms and O atoms before and after the process. Thus the decomposition of two H_2O molecules (containing four H atoms and two O atoms) yields one O_2 molecule (containing two O atoms) and two H_2 molecules (containing a total of four H atoms).

The only elemental hydrogen found naturally on earth occurs in the exhaust gases of volcanoes.

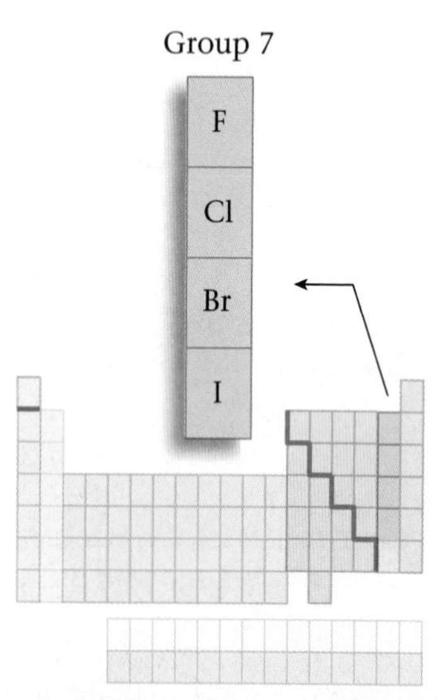

Several other elements, in addition to hydrogen, nitrogen, and oxygen, exist as diatomic molecules. For example, when sodium chloride is melted and subjected to an electric current, chlorine gas is produced (along with sodium metal). This chemical change is represented in Figure 4.16. Chlorine gas is a pale green gas that contains Cl_2 molecules.

Chlorine is a member of Group 7, the halogen family. All the elemental forms of the Group 7 elements contain diatomic molecules. Fluorine is a pale yellow gas containing F_2 molecules. Bromine is a brown liquid made up of Br_2 molecules. Iodine is a lustrous, purple solid that contains I_2 molecules.

Table 4.5 lists the elements that contain diatomic molecules in their pure, elemental forms.

So far we have seen that several elements are gaseous in their elemental forms at normal temperatures (~25 °C). The noble gases (the Group 8 elements) contain individual atoms, whereas several other gaseous elements contain diatomic molecules (H_2, N_2, O_2, F_2, and Cl_2).

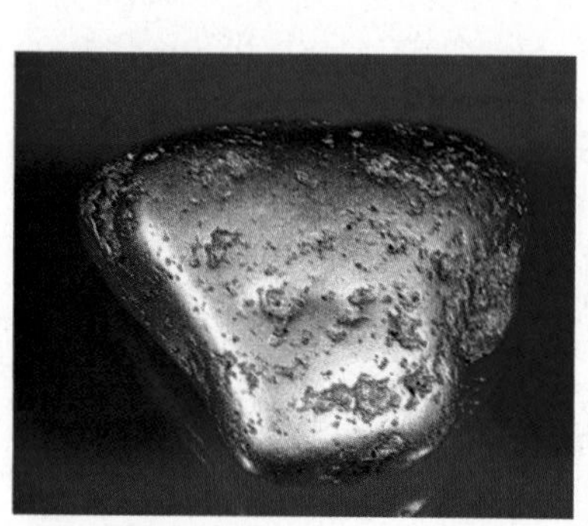

Platinum is a noble metal used in jewelry and in many industrial processes.

(a)

(b)

Figure 4.16
(a) Sodium chloride (common table salt) can be decomposed to the elements (b) sodium metal (on the left) and chlorine gas.

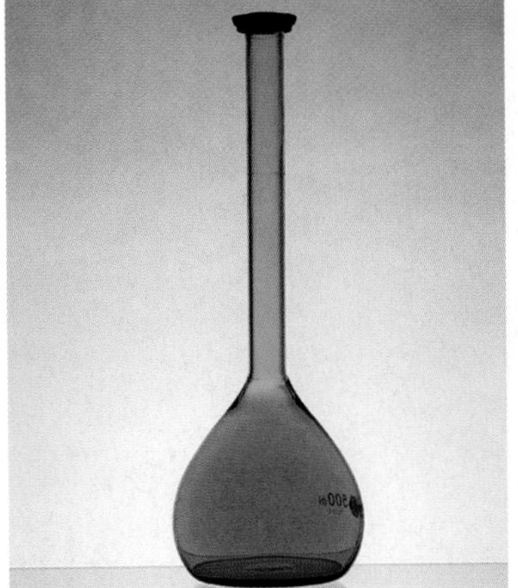

Liquid bromine in a flask with bromine vapor.

Table 4.5 Elements That Exist as Diatomic Molecules in Their Elemental Forms

Element Present	Elemental State at 25 °C	Molecule
hydrogen	colorless gas	H_2
nitrogen	colorless gas	N_2
oxygen	pale blue gas	O_2
fluorine	pale yellow gas	F_2
chlorine	pale green gas	Cl_2
bromine	reddish brown liquid	Br_2
iodine	lustrous, dark purple solid	I_2

~ means "approximately."

Graphite and diamond, two forms of carbon.

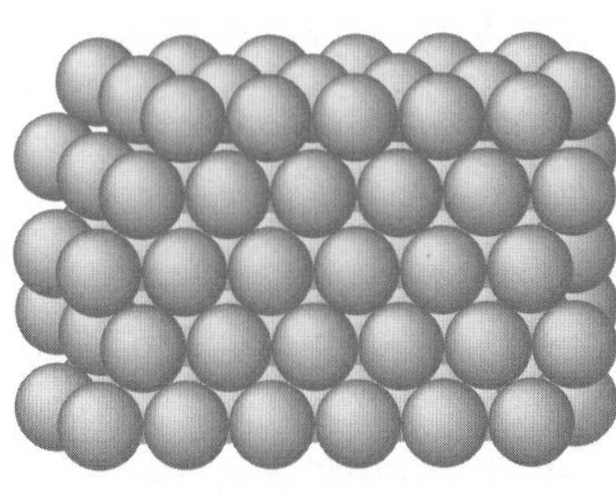

Figure 4.17
In solid metals, the spherical atoms are packed closely together.

Only two elements are liquids in their elemental forms at 25 °C: the nonmetal bromine (containing Br_2 molecules) and the metal mercury. The metals gallium and cesium almost qualify in this category; they are solids at 25 °C, but both melt at ~30 °C.

The other elements are solids in their elemental forms at 25 °C. For metals these solids contain large numbers of atoms packed together much like marbles in a jar (see Figure 4.17).

The structures of solid nonmetallic elements are more varied than those of metals. In fact, different forms of the same element often occur. For example, solid carbon occurs in three forms. Different forms of a given element are called *allotropes*. The three allotropes of carbon are the familiar diamond and graphite forms plus a form that has only recently been discovered called *buckminsterfullerene*. These elemental forms have very different properties because of their different structures (see Figure 4.18). Diamond is the hardest natural substance known and is often used for industrial cutting tools. Diamonds are also valued as gemstones. Graphite, on the other hand, is a rather soft material useful for writing (pencil "lead" is really graphite) and (in the form of a powder) for lubricating locks. The rather odd name given to buckminsterfullerene comes from the structure of the C_{60} molecules that comprise it. The soccer-ball-like structure contains five- and six-member rings reminiscent of the structure of geodesic domes suggested by the late industrial designer Buckminster Fuller. Other "fullerenes" containing molecules with more than 60 carbon atoms have also recently been discovered, leading to a new area of chemistry.

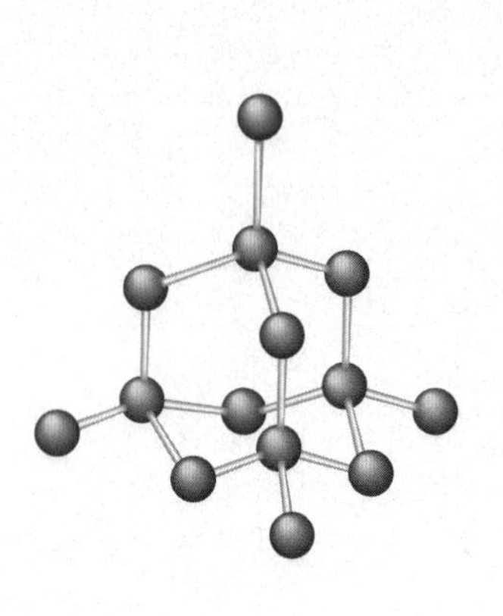

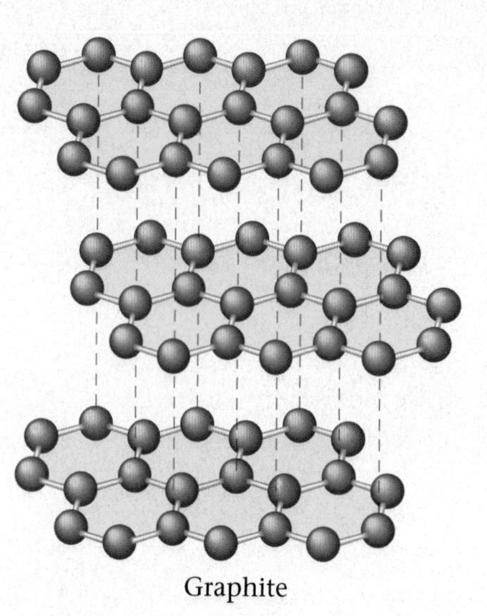

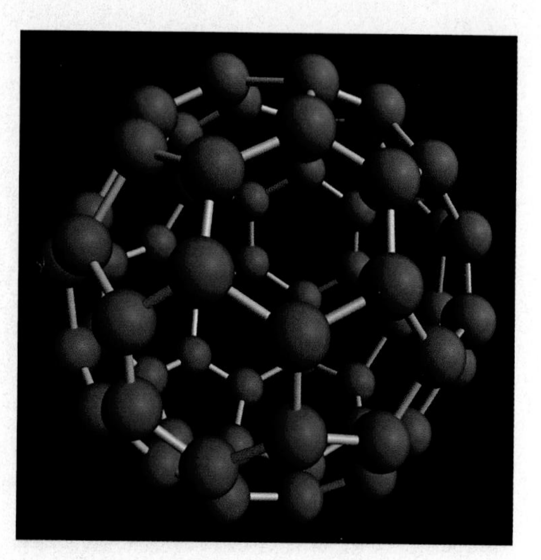

Figure 4.18
The three solid elemental forms of carbon (allotropes): (a) diamond, (b) graphite, and (c) buckminsterfullerene. The representations of diamond and graphite are fragments of much larger structures that extend in all directions from the parts shown here. Buckminsterfullerene contains C_{60} molecules, one of which is shown.

4.10 Ions

AIMS: To understand the formation of ions from their parent atoms, and learn to name them. To learn how the periodic table can help predict which ion a given element forms.

We have seen that an atom has a certain number of protons in its nucleus and an equal number of electrons in the space around the nucleus. This results in an exact balance of positive and negative charges. We say that an atom is a neutral entity—it has *zero net charge.*

We can produce a charged entity, called an **ion,** by taking a neutral atom and adding or removing one or more electrons. For example, a sodium atom ($Z = 11$) has eleven protons in its nucleus and eleven electrons outside its nucleus.

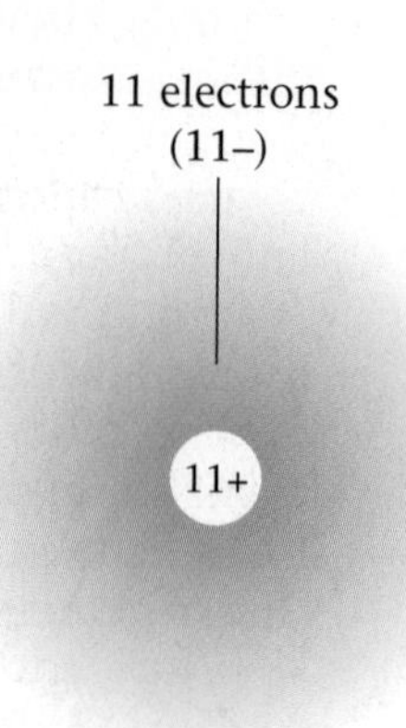

Neutral sodium atom (Na)

An ion has a net positive or negative charge.

If one of the electrons is lost, there will be eleven positive charges but only ten negative charges. This gives an ion with a net positive one (1+) charge: (11+) + (10−) = 1+. We can represent this process as follows:

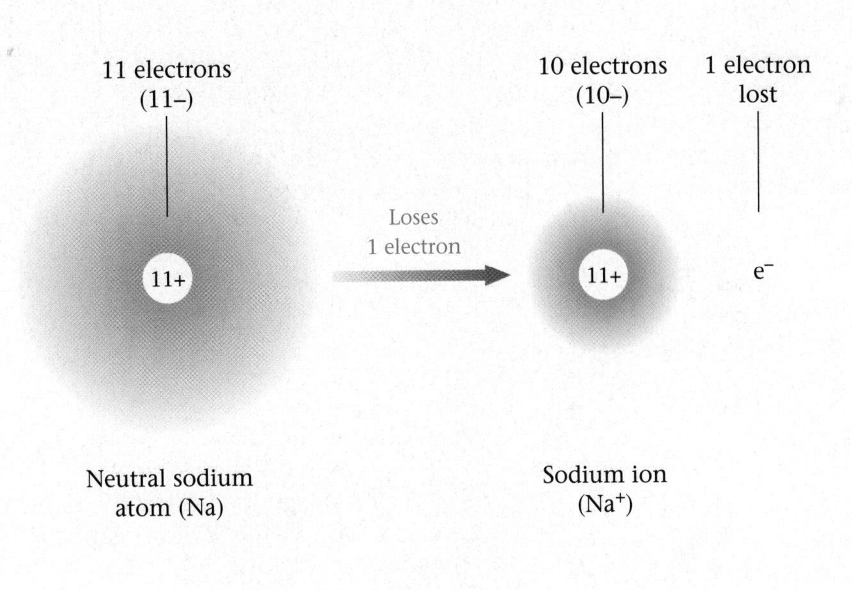

or, in shorthand form, as

$$Na \rightarrow Na^+ + e^-$$

where Na represents the neutral sodium atom, Na^+ represents the 1+ ion formed, and e^- represents an electron.

A positive ion, called a **cation** (pronounced *cat' eye on*), is produced when one or more electrons are *lost* from a neutral atom. We have seen that sodium loses one electron to become a 1+ cation. Some atoms lose more than one electron. For example, a magnesium atom typically loses two electrons to form a 2+ cation:

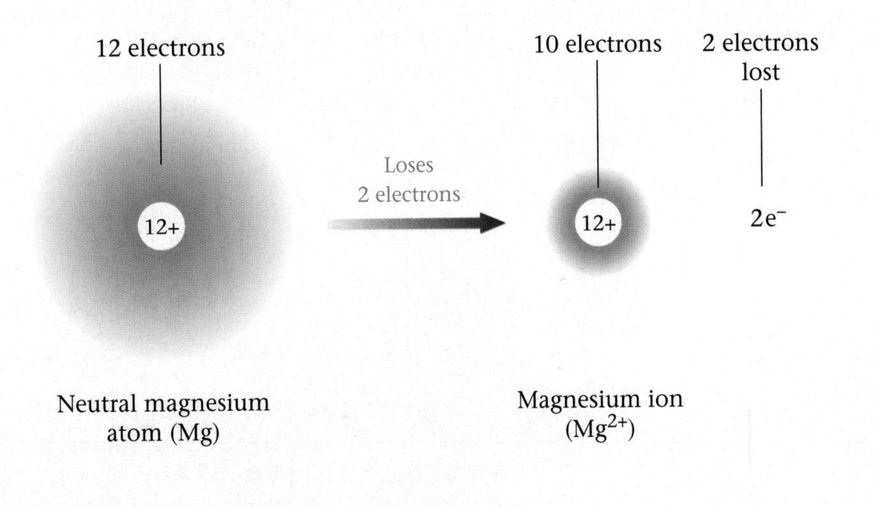

We usually represent this process as follows:

$$Mg \rightarrow Mg^{2+} + 2e^-$$

Aluminum forms a 3+ cation by losing three electrons:

Note the size decreases dramatically when an atom loses one or more electrons to form a positive ion.

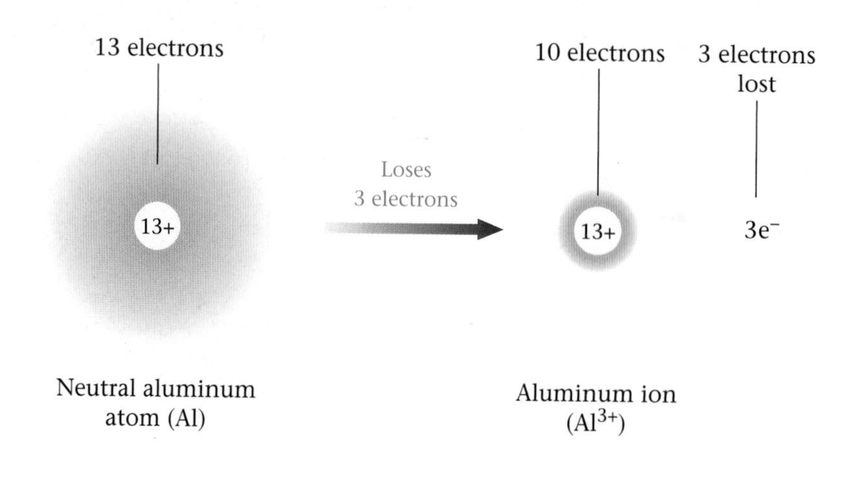

or

$$Al \rightarrow Al^{3+} + 3e^-$$

A cation is named using the name of the parent atom. Thus Na^+ is called the sodium ion (or sodium cation), Mg^{2+} is called the magnesium ion (or magnesium cation), and Al^{3+} is called the aluminum ion (or aluminum cation).

When electrons are *gained* by a neutral atom, an ion with a negative charge is formed. A negatively charged ion is called an **anion** (pronounced *an' ion*). An atom that gains one extra electron forms an anion with a 1− charge. An example of an atom that forms a 1− anion is the chlorine atom, which has seventeen protons and seventeen electrons:

Note the size increases dramatically when an atom gains one or more electrons to form a negative ion.

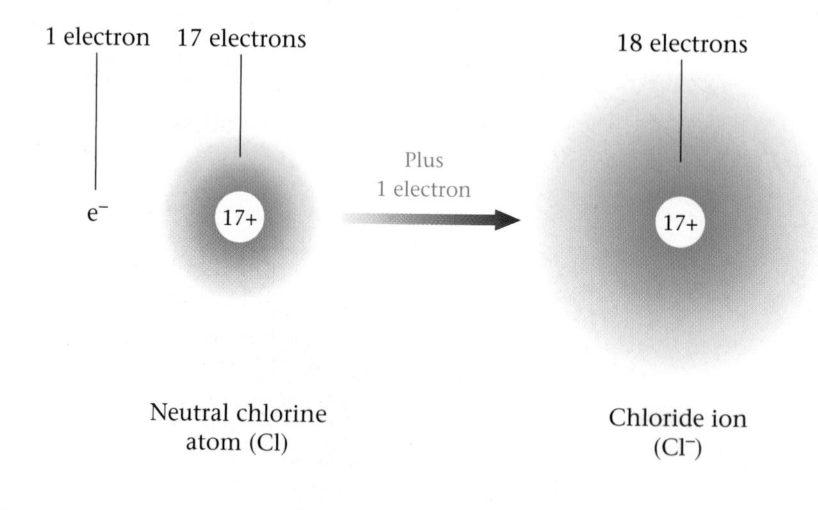

or

$$Cl + e^- \rightarrow Cl^-$$

Note that the anion formed by chlorine has eighteen electrons but only seventeen protons, so the net charge is (18−) + (17+) = 1−. Unlike a cation, which is named for the parent atom, an anion is named by taking the root name of the atom and changing the ending. For example, the Cl^- anion produced from the Cl (chlorine) atom is called the *chloride* ion (or chloride anion). Notice that the word *chloride* is obtained from the root of the atom name (*chlor-*) plus the suffix *-ide*. Other atoms that add one electron to form 1− ions include

The name of an anion is obtained by adding *-ide* to the root of the atom name.

fluorine	$F + e^- \rightarrow F^-$	(*fluor*ide ion)
bromine	$Br + e^- \rightarrow Br^-$	(*brom*ide ion)
iodine	$I + e^- \rightarrow I^-$	(*iod*ide ion)

Note that the name of each of these anions is obtained by adding *-ide* to the root of the atom name.

Some atoms can add two electrons to form 2− anions. Examples include

oxygen	$O + 2e^- \rightarrow O^{2-}$	(*ox*ide ion)
sulfur	$S + 2e^- \rightarrow S^{2-}$	(*sulf*ide ion)

Note that the names for these anions are derived in the same way as those for the 1− anions.

It is important to recognize that ions are always formed by removing electrons from an atom (to form cations) or adding electrons to an atom (to form anions). *Ions are never formed by changing the number of protons* in an atom's nucleus.

It is essential to understand that isolated atoms do not form ions on their own. Most commonly, ions are formed when metallic elements combine with nonmetallic elements. As we will discuss in detail in Chapter 7, when metals and nonmetals react, the metal atoms tend to lose one or more electrons, which are in turn gained by the atoms of the nonmetal. Thus reactions between metals and nonmetals tend to form compounds that contain metal cations and nonmetal anions. We will have more to say about these compounds in Section 4.11.

Ion Charges and the Periodic Table

For Groups 1, 2, and 3, the charges of the cations equal the group numbers.

We find the periodic table very useful when we want to know what type of ion is formed by a given atom. Figure 4.19 shows the types of ions formed by atoms in several of the groups on the periodic table. Note that the Group 1 metals all form 1+ ions (M^+), the Group 2 metals all form 2^+ ions (M^{2+}), and the Group 3 metals form 3^+ ions (M^{3+}). Thus for Groups 1 through 3 the charges of the cations formed are identical to the group numbers.

In contrast to the Group 1, 2, and 3 metals, most of the many *transition metals* form cations with various positive charges. For these elements there is no easy way to predict the charge of the cation that will be formed.

Note that metals always form positive ions. This tendency to lose electrons is a fundamental characteristic of metals. Nonmetals, on the other hand, form negative ions by gaining electrons. Note that the Group 7 atoms all gain one electron to form 1− ions and that all the nonmetals in Group 6 gain two electrons to form 2− ions.

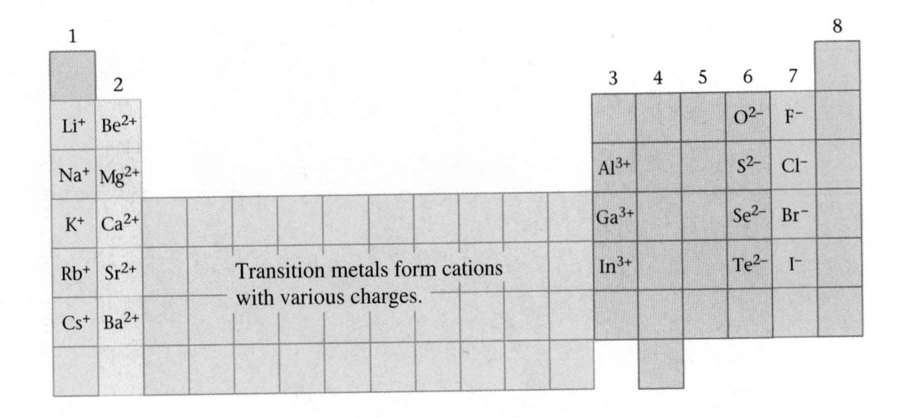

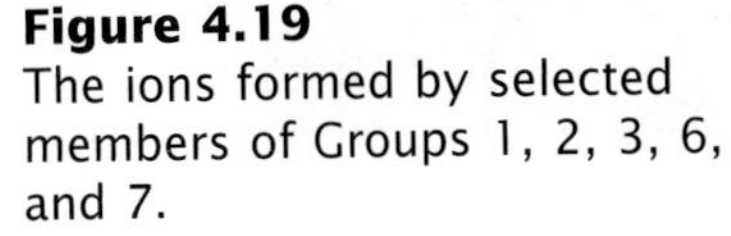
Figure 4.19
The ions formed by selected members of Groups 1, 2, 3, 6, and 7.

At this point you should memorize the relationships between the group number and the type of ion formed, as shown in Figure 4.19. You will understand why these relationships exist after we further discuss the theory of the atom in Chapter 10.

4.11 Compounds That Contain Ions

AIM: To learn how ions combine to form neutral compounds.

Melting means that the solid, where the ions are locked into place, is changed to a liquid, where the ions can move.

Chemists have good reasons to believe that many chemical compounds contain ions. For instance, consider some of the properties of common table salt, sodium chloride (NaCl). It must be heated to about 800 °C to melt and to almost 1500 °C to boil (compare to water, which boils at 100 °C). As a solid, salt will not conduct an electric current, but when melted it is a very good conductor. Pure water does not conduct electricity (does not allow an electric current to flow), but when salt is dissolved in water, the resulting solution readily conducts electricity (see Figure 4.20).

Chemists have come to realize that we can best explain these properties of sodium chloride (NaCl) by picturing it as containing Na^+ ions and Cl^- ions packed together as shown in Figure 4.21. Because the positive and negative charges attract each other very strongly, it must be heated to a very high temperature (800 °C) before it melts.

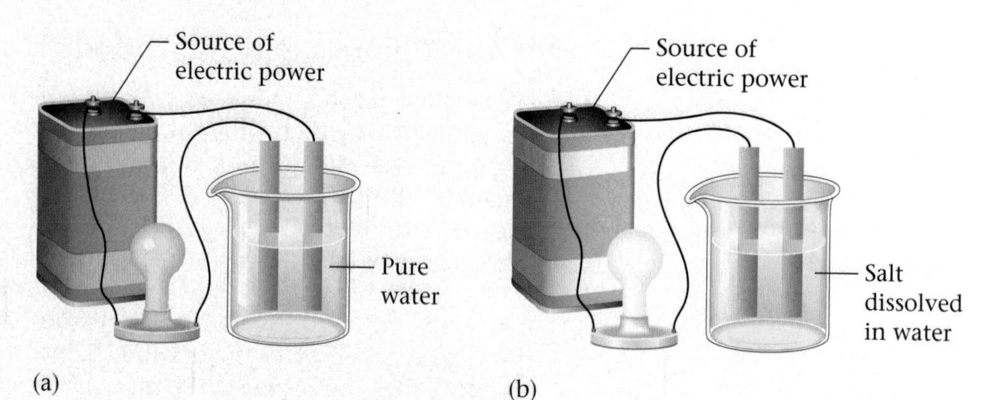

Figure 4.20
(a) Pure water does not conduct a current, so the circuit is not complete and the bulb does not light. (b) Water containing dissolved salt conducts electricity and the bulb lights.

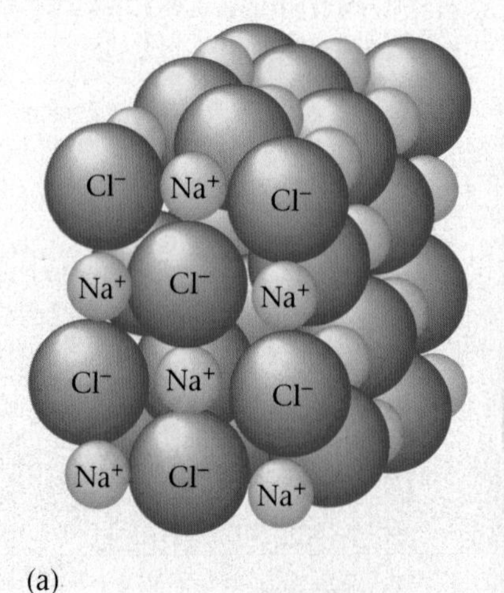

Figure 4.21
(a) The arrangement of sodium ions (Na^+) and chloride ions (Cl^-) in the ionic compound sodium chloride. (b) Solid sodium chloride highly magnified.

To explore further the significance of the electrical conductivity results, we need to discuss briefly the nature of electric currents. An electric current can travel along a metal wire because *electrons are free to move* through the wire; the moving electrons carry the current. In ionic substances the ions carry the current. Thus substances that contain ions can conduct an electric current *only if the ions can move*—the current travels by the movement of the charged ions. In solid NaCl the ions are tightly held and cannot move, but when the solid is melted and changed to a liquid, the structure is disrupted and the ions can move. As a result, an electric current can travel through the melted salt.

A substance containing ions that can move can conduct an electric current.

The same reasoning applies to NaCl dissolved in water. When the solid dissolves, the ions are dispersed throughout the water and can move around in the water, allowing it to conduct a current.

Dissolving NaCl causes the ions to be randomly dispersed in the water, allowing them to move freely. Dissolving is not the same as melting, but both processes free the ions to move.

Thus, we recognize substances that contain ions by their characteristic properties. They often have very high melting points, and they conduct an electric current when melted or when dissolved in water.

Many substances contain ions. In fact, whenever a compound forms between a metal and a nonmetal, it can be expected to contain ions. We call these substances **ionic compounds.**

One fact very important to remember is that *a chemical compound must have a net charge of zero.* This means that if a compound contains ions, then

An ionic compound cannot contain only anions or only cations, because the net charge of a compound must be zero.

The net charge of a compound (zero) is the sum of the positive and negative charges.

1. There must be both positive ions (cations) and negative ions (anions) present.
2. The numbers of cations and anions must be such that the net charge is zero.

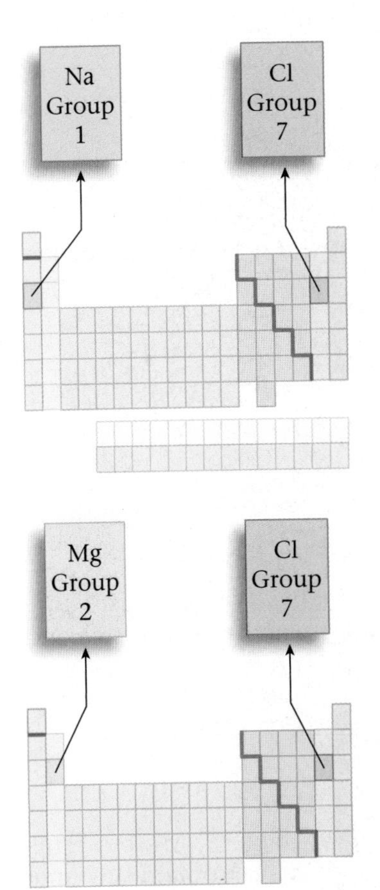

For example, note that the formula for sodium chloride is written NaCl, indicating one of each type of these elements. This makes sense because sodium chloride contains Na^+ ions and Cl^- ions. Each sodium ion has a 1+ charge and each chloride ion has a 1− charge, so they must occur in equal numbers to give a net charge of zero.

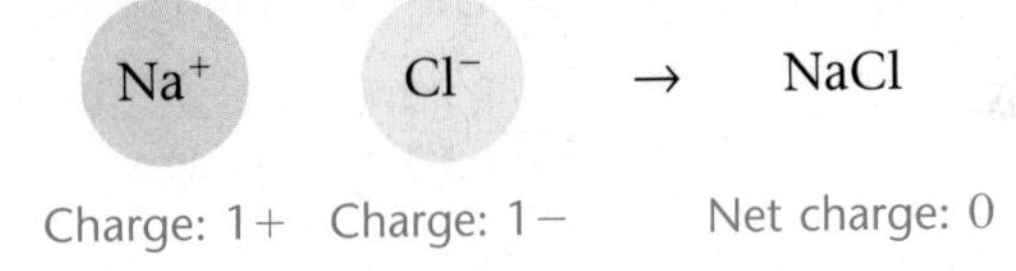

And for *any* ionic compound,

$$\begin{matrix}\text{Total charge} \\ \text{of cations}\end{matrix} + \begin{matrix}\text{Total charge} \\ \text{of anions}\end{matrix} = \begin{matrix}\text{Zero} \\ \text{net charge}\end{matrix}$$

Consider an ionic compound that contains the ions Mg^{2+} and Cl^-. What combination of these ions will give a net charge of zero? To balance the 2+ charge on Mg^{2+}, we will need two Cl^- ions to give a net charge of zero.

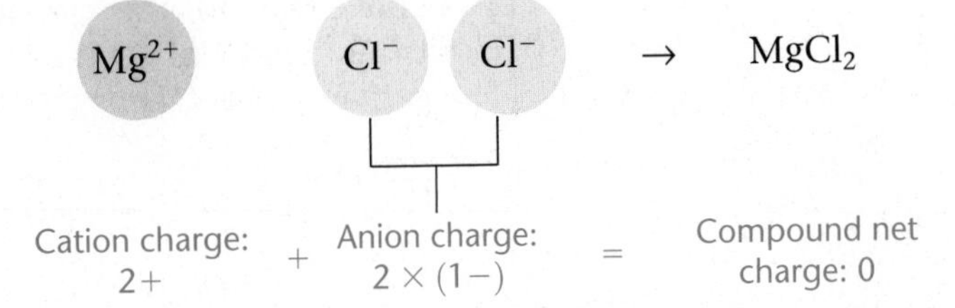

This means that the formula of the compound must be $MgCl_2$. Remember that subscripts are used to give the relative numbers of atoms (or ions).

Now consider an ionic compound that contains the ions Ba^{2+} and O^{2-}. What is the correct formula? These ions have charges of the same size (but

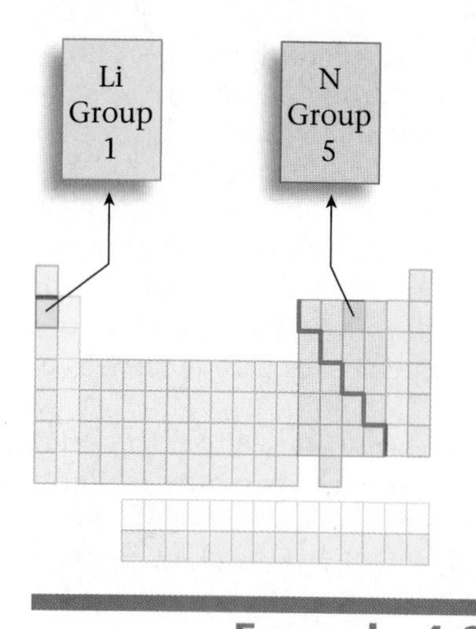

opposite sign), so they must occur in equal numbers to give a net charge of zero. The formula of the compound is BaO, because (2+) + (2−) = 0.

Similarly, the formula of a compound that contains the ions Li^+ and N^{3-} is Li_3N, because three Li^+ cations are needed to balance the charge of the N^{3-} anion.

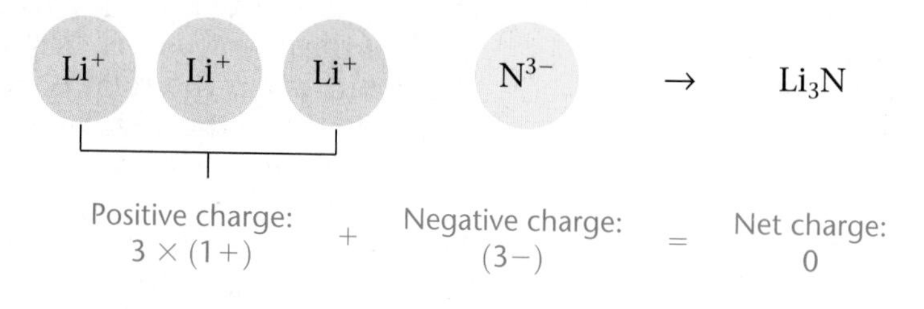

Example 4.6 Writing Formulas for Ionic Compounds

The pairs of ions contained in several ionic compounds are listed below. Give the formula for each compound.

a. Ca^{2+} and Cl^- b. Na^+ and S^{2-} c. Ca^{2+} and P^{3-}

The subscript 1 in a formula is not written.

Solution

a. Ca^{2+} has a 2+ charge, so two Cl^- ions (each with the charge 1−) will be needed.

The formula is $CaCl_2$.

b. In this case S^{2-}, with its 2− charge, requires two Na^+ ions to produce a zero net charge.

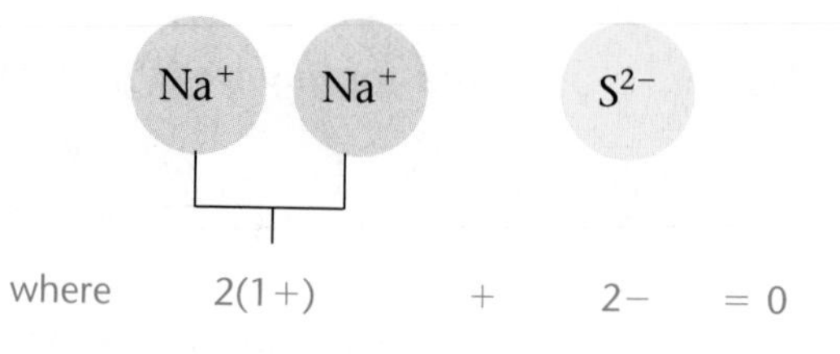

The formula is Na_2S.

c. We have the ions Ca^{2+} (charge 2+) and P^{3-} (charge 3−). We must figure out how many of each are needed to balance exactly the positive and negative charges. Let's try two Ca^{2+} and one P^{3-}.

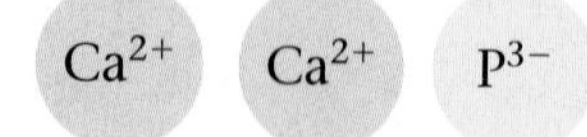

The resulting net charge is 2(2+) + (3−) = (4+) + (3−) = 1−. This doesn't work because the net charge is not zero. We can obtain the same total positive and total negative charges by having three Ca^{2+} ions and two P^{3-} ions.

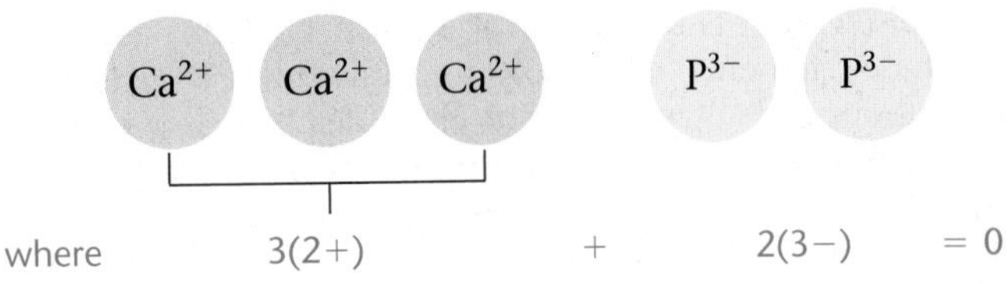

Thus the formula must be Ca_3P_2.

Self-Check Exercise 4.6

Give the formulas for the compounds that contain the following pairs of ions.

a. K^+ and I^- b. Mg^{2+} and N^{3-} c. Al^{3+} and O^{2-}

See Problems 4.83 and 4.84. ■

Chapter 4 Review

Key Terms

element symbols (4.2)
law of constant composition (4.3)
Dalton's atomic theory (4.3)
atom (4.3)
compound (4.4)
chemical formula (4.4)
electron (4.5)
nuclear atom (4.5)
nucleus (4.5)
proton (4.5)
neutron (4.5)
isotopes (4.7)
atomic number, Z (4.7)
mass number, A (4.7)
periodic table (4.8)
groups (4.8)
alkali metals (4.8)
alkaline earth metals (4.8)
halogens (4.8)
noble gases (4.8)
transition metals (4.8)
metals (4.8)
nonmetals (4.8)
metalloids (semimetals) (4.8)
diatomic molecule (4.9)
ion (4.10)
cation (4.10)
anion (4.10)
ionic compound (4.11)

Summary

1. Of the more than 100 different elements now known, only 9 account for about 98% of the total mass of the earth's crust, oceans, and atmosphere. In the human body, oxygen, carbon, hydrogen, and nitrogen are the most abundant elements.

2. Elements are represented by symbols that usually consist of the first one or two letters of the element's name. Sometimes, however, the symbol is taken from the element's original Latin or Greek name.

3. The law of constant composition states that a given compound always contains the same proportions by mass of the elements of which it is composed.

4. Dalton accounted for this law with his atomic theory. He postulated that all elements are composed of atoms; that all atoms of a given element are identical, but that atoms of different elements are different; that chemical compounds are formed when atoms combine; and that atoms are not created or destroyed in chemical reactions.

5. A compound can be represented by a chemical formula that uses the symbol for each type of atom and gives the number of each type of atom that appears in a molecule of the compound.

6. Atoms consist of a nucleus containing protons and neutrons, surrounded by electrons that occupy a large volume relative to the size of the nucleus. Electrons have a relatively small mass (1/1836 of the proton mass) and a negative charge. Protons have a positive charge equal in magnitude (but opposite in sign) to that of the electron. A neutron has a slightly greater mass than the proton but no charge.

7. Isotopes are atoms with the same number of protons but different numbers of neutrons.

8. The periodic table displays the elements in rows and columns in order of increasing atomic number. Elements that have similar chemical properties fall into vertical columns called groups. Most of the elements are metals. These occur on the left-hand side of the periodic table; the nonmetals appear on the right-hand side.

9. Each chemical element is composed of a given type of atom. These elements may exist as individual atoms or as groups of like atoms. For example, the noble gases contain single, separated atoms. However, elements such as oxygen, nitrogen, and chlorine exist as diatomic (two-atom) molecules.

10. When an atom loses one or more electrons, it forms a positive ion called a cation. This behavior is characteristic of metals. When an atom gains one or more electrons, it becomes a negatively charged ion called an anion. This behavior is characteristic of nonmetals. Oppositely charged ions form ionic compounds. A compound is always neutral overall—it has zero net charge.

11. The elements in Groups 1 and 2 on the periodic table form 1+ and 2+ cations, respectively. Group 7 atoms can gain one electron to form 1− ions. Group 6 atoms form 2− ions.

In-Class Discussion Questions

These questions are designed to be considered by groups of students in class. Often these questions work well for introducing a particular topic in class.

1. Knowing the number of protons in the atom of a neutral element enables you to determine which of the following?
   a. the number of neutrons in the atom of the neutral element
   b. the number of electrons in the atom of the neutral element
   c. the name of the element
   d. two of the above
   e. none of the above

   Explain.

2. The average mass of a carbon atom is 12.011. Assuming you could pick up one carbon atom, what is the chance that you would randomly get one with a mass of 12.011?
   a. 0%
   b. 0.011%
   c. about 12%
   d. 12.011%
   e. greater than 50%
   f. none of the above

   Explain.

3. How is an ion formed?
   a. by either adding or subtracting protons from the atom
   b. by either adding or subtracting neutrons from the atom
   c. by either adding or subtracting electrons from the atom
   d. all of the above
   e. two of the above

   Explain.

4. The formula of water, H_2O, suggests which of the following?
   a. There is twice as much mass of hydrogen as oxygen in each molecule.
   b. There are two hydrogen atoms and one oxygen atom per water molecule.
   c. There is twice as much mass of oxygen as hydrogen in each molecule.
   d. There are two oxygen atoms and one hydrogen atom per water molecule.
   e. Two of the above.

   Explain.

5. The vitamin niacin (nicotinic acid, $C_6H_5NO_2$) can be isolated from a variety of natural sources, such as liver, yeast, milk, and whole grain. It also can be synthesized from commercially available materials. Which source of nicotinic acid, from a nutritional view, is best for use in a multivitamin tablet? Why?

6. One of the best indications of a useful theory is that it raises more questions for further experimentation than it originally answered. Does this apply to Dalton's atomic theory? Give examples.

7. Dalton assumed that all atoms of the same element are identical in all their properties. Explain why this assumption is not valid.

8. How does Dalton's atomic theory account for the law of constant composition?

9. Which of the following is true about the state of an individual atom?
   a. An individual atom should be considered to be a solid.
   b. An individual atom should be considered to be a liquid.
   c. An individual atom should be considered to be a gas.
   d. The state of the atom depends on which element it is.
   e. An individual atom cannot be considered to be a solid, liquid, or gas.

   For choices you did not pick, explain what you feel is wrong with them, and justify the choice you did pick.

10. These questions concern the work of J. J. Thomson:
    a. From Thomson's work, which particles do you think he would feel are most important in the formation of compounds (chemical changes) and why?
    b. Of the remaining two subatomic particles, which do you place second in importance for forming compounds and why?
    c. Come up with three models that explain Thomson's findings and evaluate them. To be complete you should include Thomson's findings.

11. Heat is applied to an ice cube until only steam is present. Draw a sketch of this process, assuming you can see it at an extremely high level of magnification.

What happens to the size of the molecules? What happens to the total mass of the sample?

12. What makes a carbon atom different from a nitrogen atom? How are they alike?

13. Hundreds of years ago, alchemists tried to turn lead into gold. Is this possible? If not, why not? If yes, how would you do it?

14. Compare Dalton's atom with Thomson's atom and Rutherford's atom.

15. Identify the following:
    a. the heaviest noble gas
    b. the transition metal that has 25 electrons as a 2+ ion
    c. the halogen in the fourth period

16. Models are always simplifications. List at least two observations that Dalton's model does not explain.

17. Chlorine has two prominent isotopes, ^{37}Cl and ^{35}Cl. Which is more abundant? How do you know?

Questions and Problems

All even-numbered exercises have answers in the back of this book and solutions in the *Solutions Guide*.

4.1 The Elements

QUESTIONS

1. The ancient Greeks believed that all matter was composed of four fundamental substances: earth, air, fire, and water. How does this early conception of matter compare with our modern theories about matter?

2. Who was the first scientist generally accredited with putting the study of chemistry on a firm experimental basis?

3. In addition to his important work on the properties of gases, what other valuable contributions did Robert Boyle make to the development of the study of chemistry?

4. How many elements are presently known? How many of these elements occur naturally, and how many are synthesized artificially? What are the most common elements present on the earth?

5. What is the most abundant element on the earth in terms of mass? Where are several places where this element is commonly found?

6. What are the most abundant elements found in living creatures? Are these elements also the most abundant elements found in the nonliving world?

4.2 Symbols for the Elements

Note: Refer to the tables on the inside front cover when appropriate.

QUESTIONS

7. The letters *C*, *S*, and *T* have been very popular when naming the elements, and there are ten or more elements whose names begin with each of these letters. Without looking in your textbook, see if you can list the symbol and name of five elements for each letter.

8. In some cases, the symbol of an element does not seem to bear any relationship to the name we use for the element. Generally, the symbol for such an element is based on its name in another language. Give the symbols and names for five examples of such elements.

9. Match the name in column 1 with the chemical symbol in column 2.

Column 1	Column 2
a. hydrogen	1. He
b. cobalt	2. H
c. potassium	3. Na
d. bromine	4. So
e. barium	5. Ag
f. sulfur	6. S
g. silver	7. B
h. sodium	8. Ba
i. helium	9. Br
j. carbon	10. Co
	11. C
	12. K
	13. Po
	14. Ne

10. Match the chemical symbol in column 1 with the name in column 2.

Column 1	Column 2
a. Al	1. silicon
b. Ne	2. gold
c. Ni	3. sulfur
d. Si	4. aluminum
e. Cu	5. copper
f. Fe	6. neon
g. F	7. iron
h. Au	8. silver
i. Zn	9. nickel
j. U	10. magnesium
	11. barium
	12. fluorine
	13. uranium
	14. zinc

11. Use the periodic table inside the front cover of this book to look up the symbol or name for each of the following elements.

Symbol	Name
_______	tungsten
_______	germanium
Pd	_______
_______	platinum
_______	zirconium
Ir	_______

12. Use the periodic table found inside the front cover of this book to look up the symbol or name for each of the following less common elements.

Symbol	Name
Ru	_______
_______	thallium
No	_______
_______	thorium
_______	plutonium
Os	_______

13. For each of the following chemical symbols, give the name of the corresponding element.
    a. K c. P e. N g. Ne
    b. Ge d. C f. Na h. I

14. Several elements have chemical symbols beginning with the letter *B*. For each of the following chemical symbols, give the name of the corresponding element.
    a. Br c. Be e. Ba
    b. B d. Bi

4.3 Dalton's Atomic Theory

QUESTIONS

15. Indicate whether each of the following statements is true or false. If a statement is false, correct the statement so that it becomes true.
    a. Most materials occur in nature as pure substances.
    b. A given compound usually contains the same relative number of atoms of its various elements.
    c. Atoms are made up of tiny particles called molecules.

16. State the main idea of the *law of constant composition* in your own words, using an example to illustrate your understanding of this important idea.

4.4 Formulas of Compounds

QUESTIONS

17. What is a compound?

18. A given compound always contains the same relative masses of its constituent elements. How is this related to the relative numbers of each kind of atom present?

19. Write the formula for each of the following substances, listing the elements in the order given.
    a. a molecule containing three carbon atoms and eight hydrogen atoms
    b. a compound containing two nitrogen atoms for every oxygen atom
    c. a compound containing half as many barium atoms as iodine atoms
    d. a compound containing aluminum atoms and also three times as many chlorine atoms as there are aluminum atoms
    e. a sugar whose molecules contain 12 carbon atoms, 22 hydrogen atoms, and 11 oxygen atoms
    f. a compound that contains twice as many potassium atoms as carbon atoms, and three times as many oxygen atoms as carbon atoms

20. Write the formula for each of the following substances, listing the elements in the order given.
    a. a molecule containing six carbon atoms and six hydrogen atoms
    b. a molecule containing two nitrogen atoms and four oxygen atoms
    c. a compound containing half as many calcium atoms as chlorine atoms
    d. a compound containing one iron atom for every three bromine atoms
    e. a compound containing equal numbers of sodium and nitrogen atoms, but three times as many oxygen atoms as there are sodium atoms
    f. a compound containing three calcium atoms for every two nitrogen atoms

4.5 The Structure of the Atom

QUESTIONS

21. Scientists J. J. Thomson and William Thomson (Lord Kelvin) made numerous contributions to our understanding of the atom's structure.
    a. What subatomic particle did J. J. Thomson discover, and what did this lead him to postulate about the nature of the atom?
    b. William Thomson postulated what became known as the "plum pudding" model of the atom's structure. What did this model suggest?

22. Indicate whether each of the following statements is true or false. If false, correct the statement so that it becomes true.
    a. Rutherford's bombardment experiments with metal foil suggested that the alpha particles were being deflected by coming near a large, negatively charged atomic nucleus.
    b. The proton and the electron have similar masses but opposite electrical charges.
    c. Most atoms also contain neutrons, which are slightly heavier than protons but carry no charge.

4.6 Introduction to the Modern Concept of Atomic Structure

QUESTIONS

23. Where are neutrons found in an atom? Are neutrons positively charged, negatively charged, or electrically uncharged?

24. What two common types of particles are found in the nucleus of the atom? What are the relative charges of these particles? What are the relative masses of these particles?

25. Do the proton and the neutron have exactly the same mass? How do the masses of the proton and the neutron compare to the mass of the electron? Which particles make the greatest contribution to the mass of an atom? Which particles make the greatest contribution to the chemical properties of an atom?

26. The proton and the (electron/neutron) have almost equal masses. The proton and the (electron/neutron) have charges that are equal in magnitude but opposite in nature.

27. An average atomic nucleus has a diameter of about ________ m.

28. Which particles in an atom are most responsible for the chemical properties of the atom? Where are these particles located in the atom?

4.7 Isotopes

QUESTIONS

29. Explain what do we mean when we say that a particular element consists of several *isotopes?*

30. True or false? The mass number of a nucleus represents the number of protons in the nucleus.

31. For an atom, the number of protons and electrons is (different/the same).

32. The ________ number represents the sum of the number of protons and neutrons in a nucleus.

33. Was Dalton's original atomic theory consistent with the discovery that many elements consist of several isotopes? Why or why not?

34. Are all atoms of the same element identical? If not, how can they differ?

35. For each of the following elements, use the periodic table on the inside cover of this book to write the element's atomic number.
    a. Ni
    b. copper
    c. Se
    d. cadmium
    e. S
    f. silicon
    g. V
    h. xenon

36. For each of the following elements, use the periodic table on the inside cover of this book to write the element's atomic number.
    a. Ge
    b. zinc
    c. Cr
    d. tungsten
    e. Sr
    f. cobalt
    g. Be
    h. lithium

37. Write the atomic symbol ($^{A}_{Z}X$) for each of the isotopes described below.
    a. $Z = 8$, number of neutrons = 9
    b. the isotope of chlorine in which $A = 37$
    c. $Z = 27$, $A = 60$
    d. number of protons = 26, number of neutrons = 31
    e. the isotope of I with a mass number of 131
    f. $Z = 3$, number of neutrons = 4

38. Write the atomic symbol $^{A}_{Z}X$ for each of the isotopes described below.
    a. number of protons = 27, number of neutrons = 31
    b. the isotope of boron with mass number 10
    c. $Z = 12$, $A = 23$
    d. atomic number 53, number of neutrons = 79
    e. $Z = 9$, number of neutrons = 10
    f. number of protons = 29, mass number 65

39. How many protons and neutrons are contained in the nucleus of each of the following atoms? Assuming each atom is uncharged, how many electrons are present?
    a. $^{244}_{94}Pu$
    b. $^{241}_{95}Am$
    c. $^{227}_{89}Ac$
    d. $^{133}_{55}Cs$
    e. $^{193}_{77}Ir$
    f. $^{56}_{25}Mn$

40. How many protons and neutrons are contained in the nucleus of each of the following atoms? Assuming each atom is uncharged, how many electrons are present?
    a. $^{235}_{92}U$
    b. $^{13}_{6}C$
    c. $^{57}_{26}Fe$
    d. $^{208}_{82}Pb$
    e. $^{86}_{37}Rb$
    f. $^{41}_{20}Ca$

41. Complete the following table.

Name	Symbol	Atomic Number	Mass Number	Neutrons
sodium	________	11	23	________
nitrogen	$^{15}_{7}N$	________	________	________
________	$^{136}_{56}Ba$	________	________	________
lithium	________	________	________	6
boron	________	5	11	________

42. Complete the following table.

Name	Neutrons	Atomic Number	Mass Number	Symbol
nitrogen	6	______	______	______
______	______	7	14	______
lead	______	______	206	______
______	31	26	______	______
______	______	______	______	$^{84}_{36}Kr$

4.8 Introduction to the Periodic Table

QUESTIONS

43. On the basis of what property are the elements arranged in order on the periodic table? Why?

44. In which direction on the periodic table, horizontal or vertical, are elements with similar chemical properties aligned? What are families of elements with similar chemical properties called?

45. List the characteristic physical properties that distinguish the metallic elements from the nonmetallic elements.

46. Where are the metallic elements found on the periodic table? Are there more metallic elements or nonmetallic elements?

47. Most, but not all, metallic elements are solids under ordinary laboratory conditions. Which metallic elements are *not* solids?

48. List five nonmetallic elements that exist as gaseous substances under ordinary conditions. Do any metallic elements ordinarily occur as gases?

49. Under ordinary conditions, only a few pure elements occur as liquids. Give an example of a metallic and a nonmetallic element that ordinarily occur as liquids.

50. What is a *metalloid?* Where are the metalloids found on the periodic table?

51. Write the number and name (if any) of the group (family) to which each of the following elements belongs.
    a. cesium
    b. Ra
    c. Rn
    d. chlorine
    e. strontium
    f. Xe
    g. Rb

52. Write the number and name (if any) of the group (family) to which each of the following elements belongs.
    a. K
    b. aluminum
    c. He
    d. silicon
    e. sulfur
    f. Kr
    g. selenium

53. For each of the following elements, use the tables on the inside cover of this book to give the chemical symbol, atomic number, and group number of each element, and to specify whether each element is a metal, nonmetal, or metalloid.
    a. strontium
    b. iodine
    c. silicon
    d. cesium
    e. sulfur

54. For each of the following elements, use the tables on the inside cover of this book to give the chemical symbol, atomic number, and group number of each element, and to specify whether the element is a metal, nonmetal, or metalloid.
    a. lithium
    b. arsenic
    c. radon
    d. radium
    e. germanium

4.9 Natural States of the Elements

QUESTIONS

55. Most substances are composed of ______ rather than elemental substances.

56. Are most of the chemical elements found in nature in the elemental form or combined in compounds? Why?

57. The noble gas present in relatively large concentrations in the atmosphere is ______.

58. Why are the elements of Group 8 referred to as the noble or inert gas elements?

59. Molecules of nitrogen gas and oxygen gas are said to be ______, which means they consist of pairs of atoms.

60. Give three examples of gaseous elements that exist as diatomic molecules. Give three examples of gaseous elements that exist as monatomic species.

61. A simple way to generate elemental hydrogen gas is to pass ______ through water.

62. If sodium chloride (table salt) is melted and then subjected to an electric current, elemental ______ gas is produced, along with sodium metal.

63. Most of the elements are solids at room temperature. Give three examples of elements that are *liquids* at room temperature, and three examples of elements that are *gases* at room temperature.

64. The two most common elemental forms of carbon are graphite and ______.

4.10 Ions

QUESTIONS

65. An isolated atom has a net charge of ______.

66. How are ions produced from an atom? Does the formation of a simple ion involve changes to the atom's nucleus?

67. A simple ion with a 3+ charge (for example, Al^{3+}) results when an atom (gains/loses) ________ electrons.

68. An ion that contains more protons than electrons will be (positively/negatively) charged.

69. Positive ions are called ________, whereas negative ions are called ________.

70. Simple negative ions formed from single atoms are given names that end in ________.

71. Based on their location in the periodic table, give the symbols for three elements that would be expected to form positive ions in their reactions.

72. The tendency to *gain* electrons is a fundamental property of the ________ elements.

73. How many electrons are contained in each of the following ions?
a. Fe^{2+} d. Co^{3+} g. Cr^{3+}
b. Ca^{2+} e. S^{2-} h. K^{+}
c. Co^{2+} f. Cl^{-}

74. Consider the following ions. For the positive ions listed, predict the formula of the simplest compound that would be expected to be formed between the given ion and the sulfide ion, S^{2-}. For the negative ions listed, predict the formula of the simplest compound that would be expected to be formed between the given ion and the aluminum ion, Al^{3+}.
a. Fe^{3+} d. I^{-} g. Mn^{2+}
b. Fe^{2+} e. Rb^{+} h. Sn^{4+}
c. O^{2-} f. P^{3-}

75. For the following processes that show the formation of ions, use the periodic table to indicate the number of electrons and protons present in both the *ion* and the *neutral atom* from which the ion is made.
a. $Ca \rightarrow Ca^{2+} + 2e^{-}$ d. $Fe \rightarrow Fe^{3+} + 3e^{-}$
b. $P + 3e^{-} \rightarrow P^{3-}$ e. $Al \rightarrow Al^{3+} + 3e^{-}$
c. $Br + e^{-} \rightarrow Br^{-}$ f. $N + 3e^{-} \rightarrow N^{3-}$

76. For the following processes that show the formation of ions, fill in the number of electrons that must be lost or gained to complete the process.
a. $Co \rightarrow Co^{2+} +$ ________ e^{-}
b. $N +$ ________ $e^{-} \rightarrow N^{3-}$
c. $Sn \rightarrow Sn^{2+} +$ ________ e^{-}
d. $Sn \rightarrow Sn^{4+} +$ ________ e^{-}
e. $Rb \rightarrow Rb^{+} +$ ________ e^{-}
f. $S +$ ________ $e^{-} \rightarrow S^{2-}$

77. For each of the following atomic numbers, use the periodic table to write the formula (including the charge) for the simple *ion* that the element is most likely to form.
a. 53 c. 55 e. 9
b. 38 d. 88 f. 13

78. On the basis of its location in the periodic table, indicate what simple ion each of the following elements is most likely to form.
a. Tl ($Z = 81$) d. As ($Z = 33$)
b. Se ($Z = 34$) e. Fr ($Z = 87$)
c. Ba ($Z = 56$) f. Cs ($Z = 55$)

4.11 Compounds That Contain Ions

QUESTIONS

79. List some properties of a substance that would lead you to believe it consists of ions. How do these properties differ from those of nonionic compounds?

80. Why does a solution of sodium chloride in water conduct an electric current, whereas a solution of sugar in water does not?

81. Why does an ionic compound conduct an electric current when the compound is melted but not when it is in the solid state?

82. Why must the total number of positive charges in an ionic compound equal the total number of negative charges?

83. For the following pairs of ions, use the concept that a chemical compound must have a net charge of zero to predict the formula of the simplest compound that the ions are most likely to form.
a. Fe^{3+} and P^{3-} e. Mg^{2+} and O^{2-}
b. Fe^{3+} and S^{2-} f. Mg^{2+} and N^{3-}
c. Fe^{3+} and Cl^{-} g. Na^{+} and P^{3-}
d. Mg^{2+} and Cl^{-} h. Na^{+} and S^{2-}

84. For the following pairs of ions, use the concept that a chemical compound must have a net charge of zero to predict the simplest compound that the ions are most likely to form.
a. Co^{2+} and O^{2-} d. Ba^{2+} and N^{3-}
b. Co^{3+} and S^{2-} e. Ca^{2+} and C^{4-}
c. Al^{3+} and Cl^{-} f. K^{+} and N^{3-}

ADDITIONAL PROBLEMS

85. For each of the following elements, give the chemical symbol and atomic number.
a. astatine d. strontium g. argon
b. xenon e. lead h. cesium
c. radium f. selenium

86. Give the group number (if any) in the periodic table for the elements listed in problem 85. If the group has a family name, give that name.

87. List the names, symbols, and atomic numbers of the top four elements in Groups 1, 2, 6, and 7.

88. List the names, symbols, and atomic numbers of the top four elements in Groups 3, 5, and 8.

89. What is the difference between the atomic number and the mass number of an element? Can atoms of

two different elements have the same atomic number? Could they have the same mass number? Why or why not?

90. Which subatomic particles contribute most to the atom's mass? Which subatomic particles determine the atom's chemical properties?

91. Is it possible for the same two elements to form more than one compound? Is this consistent with Dalton's atomic theory? Give an example.

92. Carbohydrates, a class of compounds containing the elements carbon, hydrogen, and oxygen, were originally thought to contain one water molecule (H_2O) for each carbon atom present. The carbohydrate glucose contains six carbon atoms. Write a general formula showing the relative numbers of each type of atom present in glucose.

93. When iron rusts in moist air, the product is typically a mixture of two iron–oxygen compounds. In one compound, there is an equal number of iron and oxygen atoms. In the other compound, there are three oxygen atoms for every two iron atoms. Write the formulas for the two iron oxides.

94. How many protons and neutrons are contained in the nucleus of each of the following atoms? For an atom of the element, how many electrons are present?

a. $^{63}_{29}Cu$ b. $^{80}_{35}Br$ c. $^{24}_{12}Mg$

95. Though the common isotope of aluminum has a mass number of 27, isotopes of aluminum have been isolated (or prepared in nuclear reactors) with mass numbers of 24, 25, 26, 28, 29, and 30. How many neutrons are present in each of these isotopes? Why are they all considered aluminum atoms, even though they differ greatly in mass? Write the atomic symbol for each isotope.

96. The principal goal of alchemists was to convert cheaper, more common metals into gold. Considering that gold had no particular practical uses (for example, it was too soft to be used for weapons), why do you think early civilizations placed such emphasis on the value of gold?

97. How did Robert Boyle define an element?

98. Give the chemical symbol for each of the following elements.

a. iodine d. iron
b. silicon e. copper
c. tungsten f. cobalt

99. Give the chemical symbol for each of the following elements.

a. calcium d. lead
b. potassium e. platinum
c. cesium f. gold

100. Give the chemical symbol for each of the following elements.

a. bromine d. vanadium
b. bismuth e. fluorine
c. mercury f. calcium

101. Give the chemical symbol for each of the following elements.

a. silver d. antimony
b. aluminum e. tin
c. cadmium f. arsenic

102. For each of the following chemical symbols, give the name of the corresponding element.

a. Os e. U
b. Zr f. Mn
c. Rb g. Ni
d. Rn h. Br

103. For each of the following chemical symbols, give the name of the corresponding element.

a. Te e. Cs
b. Pd f. Bi
c. Zn g. F
d. Si h. Ti

104. Write the simplest formula for each of the following substances, listing the elements in the order given.

a. a molecule containing one carbon atom and two oxygen atoms
b. a compound containing one aluminum atom for every three chlorine atoms
c. perchloric acid, which contains one hydrogen atom, one chlorine atom, and four oxygen atoms
d. a molecule containing one sulfur atom and six chlorine atoms

105. For each of the following atomic numbers, write the name and chemical symbol of the corresponding element. (Refer to Figure 4.11.)

a. 7 e. 22
b. 10 f. 18
c. 11 g. 36
d. 28 h. 54

106. Write the atomic symbol (^A_ZX) for each of the isotopes described below.

a. $Z = 6$, number of neutrons $= 7$
b. the isotope of carbon with a mass number of 13
c. $Z = 6$, $A = 13$
d. $Z = 19$, $A = 44$
e. the isotope of calcium with a mass number of 41
f. the isotope with 19 protons and 16 neutrons

107. How many protons and neutrons are contained in the nucleus of each of the following atoms? In an atom of each element, how many electrons are present?

a. $^{41}_{22}Ti$ c. $^{76}_{32}Ge$ e. $^{75}_{33}As$

b. $^{64}_{30}Zn$ d. $^{86}_{36}Kr$ f. $^{41}_{19}K$

108. Complete the following table.

Symbol	Protons	Neutrons	Mass Number
$^{41}_{20}Ca$	______	______	______
______	25	30	______
______	47	______	109
$^{45}_{21}Sc$	______	______	______

109. For each of the following elements, use the table on the inside front cover of the book to give the chemical symbol and atomic number and to specify whether the element is a metal or a nonmetal. Also give the named family to which the element belongs (if any).

a. carbon c. radon

b. selenium d. beryllium

5 Nomenclature

Mineral towers in Mono Lake, California.

When chemistry was an infant science, there was no system for naming compounds. Names such as sugar of lead, blue vitriol, quicklime, Epsom salts, milk of magnesia, gypsum, and laughing gas were coined by early chemists. Such names are called *common names.* As our knowledge of chemistry grew, it became clear that using common names for compounds was not practical. More than four million chemical compounds are currently known. Memorizing common names for all these compounds would be impossible.

The solution, of course, is a *system* for naming compounds in which the name tells something about the composition of the compound. After learning the system, you should be able to name a compound when you are given its formula. And, conversely, you should be able to construct a compound's formula, given its name. In the next few sections we will specify the most important rules for naming compounds other than organic compounds (those based on chains of carbon atoms).

An artist using plaster of Paris, a gypsum plaster.

5.1 Naming Compounds

AIM: To understand why it is necessary to have a system for naming compounds.

We will begin by discussing the system for naming **binary compounds**—compounds composed of two elements. We can divide binary compounds into two broad classes:

1. Compounds that contain a metal and a nonmetal
2. Compounds that contain two nonmetals

We will describe how to name compounds in each of these classes in the next several sections. Then, in succeeding sections, we will describe the systems used for naming more complex compounds.

CHEMISTRY IN FOCUS

Sugar of Lead

In ancient Roman society it was common to boil wine in a lead-lined vessel, driving off much of the water to produce a very sweet, viscous syrup called *sapa.* This syrup was commonly used as a sweetener for many types of food and drink.

We now realize that a major component of this syrup was lead acetate, $Pb(C_2H_3O_2)_2$. This compound has a very sweet taste—hence its original name, sugar of lead.

Many historians believe that the fall of the Roman Empire was due at least in part to lead poisoning, which causes lethargy and mental malfunctions. One major source of this lead was the sapa syrup. In addition, the Romans' highly advanced plumbing system employed lead water pipes, which allowed lead to be leached into their drinking water.

Sadly, this story is more relevant to today's society than you might think. Lead-based solder was widely used for many years to connect the copper pipes in water systems in homes and commercial buildings. There is evidence that dangerous amounts of lead can be leached from these soldered joints into drinking water. In fact, large quantities of lead have been found in the water that some drinking fountains and water coolers dispense. In response to these problems, the U.S. Congress has passed a law banning lead from the solder used in plumbing systems for drinking water.

An ancient painting showing Romans drinking wine.

5.2 Naming Compounds That Contain a Metal and a Nonmetal (Types I and II)

AIM: To learn to name binary compounds of a metal and a nonmetal.

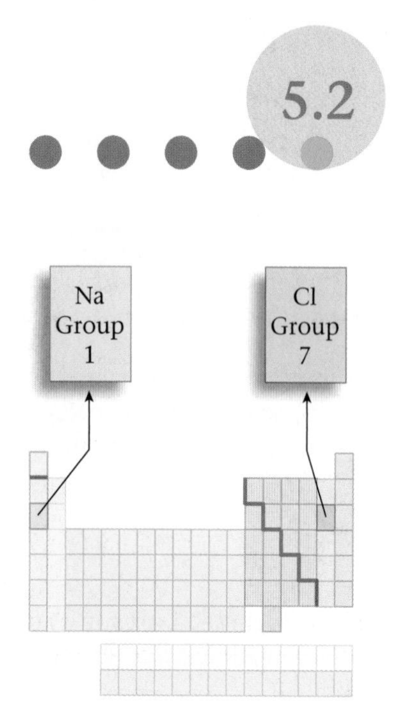

As we saw in Section 4.11, when a metal such as sodium combines with a nonmetal such as chlorine, the resulting compound contains ions. The metal loses one or more electrons to become a cation, and the nonmetal gains one or more electrons to form an anion. The resulting substance is called a **binary ionic compound.** Binary ionic compounds contain a positive ion (cation), which is always written first in the formula, and a negative ion (anion). *To name these compounds we simply name the ions.*

In this section we will consider binary ionic compounds of two types based on the cations they contain. Certain metal atoms form only one cation. For example, the Na atom always forms Na^+, *never* Na^{2+} or Na^{3+}. Likewise, Cs always forms Cs^+, Ca always forms Ca^{2+}, and Al always forms Al^{3+}. We will call compounds that contain this type of metal atom Type I binary compounds and the cations they contain Type I cations. Examples of Type I cations are Na^+, Ca^{2+}, Cs^+, and Al^{3+}.

Table 5.1 Common Simple Cations and Anions

Cation	Name	Anion	Name*
H^+	hydrogen	H^-	hydride
Li^+	lithium	F^-	fluoride
Na^+	sodium	Cl^-	chloride
K^+	potassium	Br^-	bromide
Cs^+	cesium	I^-	iodide
Be^{2+}	beryllium	O^{2-}	oxide
Mg^{2+}	magnesium	S^{2-}	sulfide
Ca^{2+}	calcium		
Ba^{2+}	barium		
Al^{3+}	aluminum		
Ag^+	silver		
Zn^{2+}	zinc		

*The root is given in color.

Other metal atoms can form two or more cations. For example, Cr can form Cr^{2+} and Cr^{3+} and Cu can form Cu^+ and Cu^{2+}. We will call such ions Type II cations and their compounds Type II binary compounds.

In summary:

Type I compounds: The metal present forms only one type of cation.

Type II compounds: The metal present can form two (or more) cations that have different charges.

Some common cations and anions and their names are listed in Table 5.1. You should memorize these. They are an essential part of your chemical vocabulary.

Type I Binary Ionic Compounds

The following rules apply for Type I ionic compounds:

Rules for Naming Type I Ionic Compounds

1. The cation is always named first and the anion second.
2. A simple cation (obtained from a single atom) takes its name from the name of the element. For example, Na^+ is called sodium in the names of compounds containing this ion.
3. A simple anion (obtained from a single atom) is named by taking the first part of the element name (the root) and adding *-ide.* Thus the Cl^- ion is called chloride.

A simple cation has the same name as its parent element.

We will illustrate these rules by naming a few compounds. For example, the compound NaI is called sodium iodide. It contains Na^+ (the sodium cation, named for the parent metal) and I^- (iodide: the root of iodine plus *-ide*). Similarly, the compound CaO is called calcium oxide because it contains Ca^{2+} (the calcium cation) and O^{2-} (the oxide anion).

The rules for naming binary compounds are also illustrated by the following examples:

Compound	Ions Present	Name
NaCl	Na^+, Cl^-	sodium chloride
KI	K^+, I^-	potassium iodide
CaS	Ca^{2+}, S^{2-}	calcium sulfide
CsBr	Cs^+, Br^-	cesium bromide
MgO	Mg^{2+}, O^{2-}	magnesium oxide

It is important to note that in the *formulas* of ionic compounds, simple ions are represented by the element symbol: Cl means Cl^-, Na means Na^+, and so on. However, when *individual ions* are shown, the charge is always included. Thus the formula of potassium bromide is written KBr, but when the potassium and bromide ions are shown individually, they are written K^+ and Br^-.

Example 5.1 Naming Type I Binary Compounds

Name each binary compound.

a. CsF b. $AlCl_3$ c. MgI_2

Solution

We will name these compounds by systematically following the rules given above.

a. CsF

STEP 1 Identify the cation and anion. Cs is in Group 1, so we know it will form the 1+ ion Cs^+. Because F is in Group 7, it forms the 1− ion F^-.

STEP 2 Name the cation. Cs^+ is simply called cesium, the same as the element name.

STEP 3 Name the anion. F^- is called fluoride: we use the root name of the element plus *-ide*.

STEP 4 Name the compound by combining the names of the individual ions. The name for CsF is cesium fluoride. (Remember that the name of the cation is always given first.)

b.

Compound	Ions Present	Ion Names	Comments
$AlCl_3$ (Cation)	Al^{3+}	aluminum	Al (Group 3) always forms Al^{3+}.
$AlCl_3$ (Anion)	Cl^-	chloride	Cl (Group 7) always forms Cl^-.

The name of $AlCl_3$ is aluminum chloride.

c.

Compound	Ions Present	Ion Names	Comments
MgI_2 (Cation)	Mg^{2+}	magnesium	Mg (Group 2) always forms Mg^{2+}.
MgI_2 (Anion)	I^-	iodide	I (Group 7) gains one electron to form I^-.

The name of MgI_2 is magnesium iodide.

✔ Self-Check Exercise 5.1

Name the following compounds.

a. Rb_2O b. SrI_2 c. K_2S

See Problems 5.9 and 5.10. ■

Example 5.1 reminds us of three things:

1. Compounds formed from metals and nonmetals are ionic.
2. In an ionic compound the cation is always named first.
3. The *net* charge on an ionic compound is always zero. Thus, in CsF, one of each type of ion (Cs^+ and F^-) is required: (1+) + (1−) = 0 charge. In $AlCl_3$, however, three Cl^- ions are needed to balance the charge of Al^{3+}: (3+) + 3 (1−) = 0 charge. In MgI_2, two I^- ions are needed for each Mg^{2+} ion: (2+) + 2(1−) = 0 charge.

Type II Binary Ionic Compounds

So far we have considered binary ionic compounds (Type I) containing metals that always give the same cation. For example, sodium always forms the Na^+ ion, calcium always forms the Ca^{2+} ion, and aluminum always forms the Al^{3+} ion. As we said in the previous section, we can predict with certainty that each Group 1 metal will give a 1+ cation and each Group 2 metal will give a 2+ cation. Aluminum always forms Al^{3+}.

Type II binary ionic compounds contain a metal that can form more than one type of cation.

However, there are many metals that can form more than one type of cation. For example, lead (Pb) can form Pb^{2+} or Pb^{4+} in ionic compounds. Also, iron (Fe) can produce Fe^{2+} or Fe^{3+}, chromium (Cr) can produce Cr^{2+} or Cr^{3+}, gold (Au) can produce Au^+ or Au^{3+}, and so on. This means that if we saw the name gold chloride, we wouldn't know whether it referred to the compound AuCl (containing Au^+ and Cl^-) or the compound $AuCl_3$ (containing Au^{3+} and three Cl^- ions). Therefore, we need a way of specifying which cation is present in compounds containing metals that can form more than one type of cation.

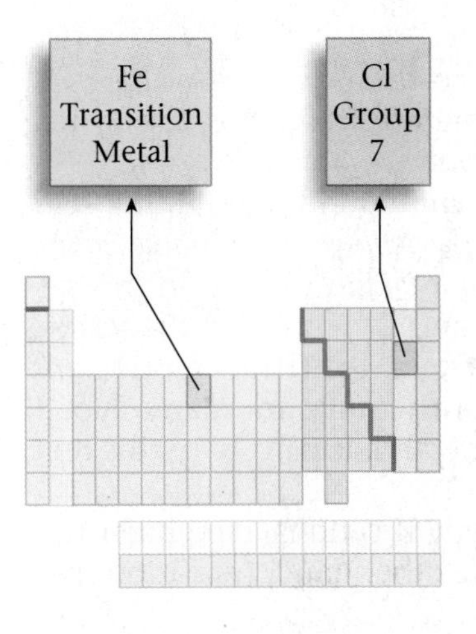

Chemists have decided to deal with this situation by using a Roman numeral to specify the charge on the cation. To see how this works, consider the compound $FeCl_2$. Iron can form Fe^{2+} or Fe^{3+}, so we must first decide which of these cations is present. We can determine the charge on the iron cation, because we know it must just balance the charge on the two 1− anions (the chloride ions). Thus if we represent the charges as

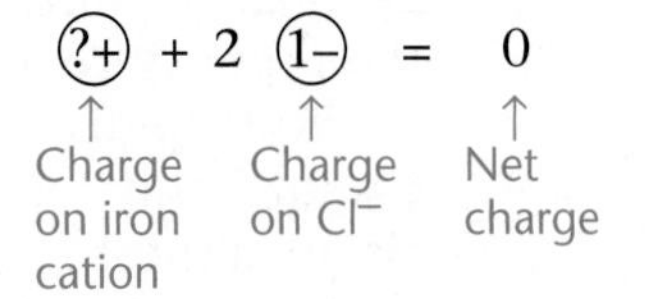

we know that ? must represent 2 because

$$(2+) + 2(1-) = 0$$

$FeCl_3$ must contain Fe^{3+} to balance the charge of three Cl^- ions.

The compound $FeCl_2$, then, contains one Fe^{2+} ion and two Cl^- ions. We call this compound iron(II) chloride, where the II tells the charge of the iron cation. That is, Fe^{2+} is called iron(II). Likewise, Fe^{3+} is called iron(III). And $FeCl_3$, which contains one Fe^{3+} ion and three Cl^- ions, is called iron(III) chloride. Remember that the Roman numeral tells the *charge* on the ion, not the number of ions present in the compound.

Copper(II) sulfate crystals.

Table 5.2 Common Type II Cations

Ion	Systematic Name	Older Name
Fe^{3+}	iron(III)	ferric
Fe^{2+}	iron(II)	ferrous
Cu^{2+}	copper(II)	cupric
Cu^{+}	copper(I)	cuprous
Co^{3+}	cobalt(III)	cobaltic
Co^{2+}	cobalt(II)	cobaltous
Sn^{4+}	tin(IV)	stannic
Sn^{2+}	tin(II)	stannous
Pb^{4+}	lead(IV)	plumbic
Pb^{2+}	lead(II)	plumbous
Hg^{2+}	mercury(II)	mercuric
Hg_2^{2+}*	mercury(I)	mercurous

*Mercury(I) ions always occur bound together in pairs to form Hg_2^{2+}.

Note that in the above examples the Roman numeral for the cation turned out to be the same as the subscript needed for the anion (to balance the charge). This is often not the case. For example, consider the compound PbO_2. Since the oxide ion is O^{2-}, for PbO_2 we have:

$$(?+) \quad + \quad 2\,(2-) \quad = \quad 0$$

Charge on lead ion; (4–) Charge on two O^{2-} ions; Net charge

Thus the charge on the lead ion must be 4+ to balance the 4− charge of the two oxide ions. The name of PbO_2 is therefore lead(IV) oxide, where the IV indicates the presence of the Pb^{4+} cation.

There is another system for naming ionic compounds containing metals that form two cations. *The ion with the higher charge has a name ending in* -ic, *and the one with the lower charge has a name ending in* -ous. In this system, for example, Fe^{3+} is called the ferric ion, and Fe^{2+} is called the ferrous ion. The names for $FeCl_3$ and $FeCl_2$, in this system, are ferric chloride and ferrous chloride, respectively. Table 5.2 gives both names for many Type II cations. We will use the system of Roman numerals exclusively in this text; the other system is falling into disuse.

To help distinguish between Type I and Type II cations, remember that Group 1 and 2 metals are always Type I. On the other hand, transition metals are almost always Type II.

Rules for Naming Type II Ionic Compounds

1. The cation is always named first and the anion second.
2. Because the cation can assume more than one charge, the charge is specified by a Roman numeral in parentheses.

Example 5.2 Naming Type II Binary Compounds

Give the systematic name of each of the following compounds.

a. $CuCl$ b. HgO c. Fe_2O_3 d. MnO_2 e. $PbCl_4$

Solution

All these compounds include a metal that can form more than one type of cation; thus we must first determine the charge on each cation. We do this by recognizing that a compound must be electrically neutral; that is, the positive and negative charges must balance exactly. We will use the known charge on the anion to determine the charge of the cation.

a. In $CuCl$ we recognize the anion as Cl^-. To determine the charge on the copper cation, we invoke the principle of charge balance.

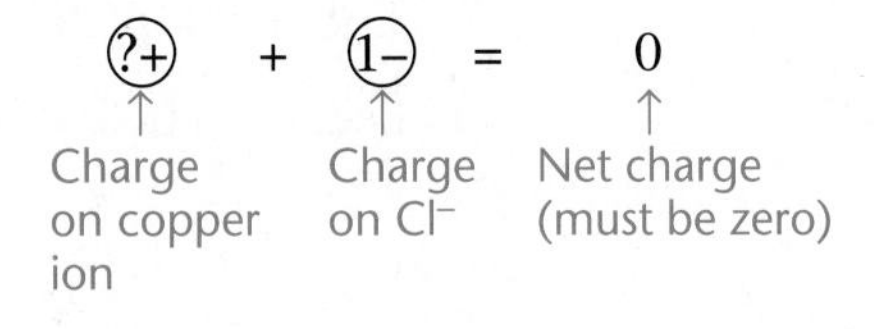

In this case, ?+ must be 1+ because (1+) + (1−) = 0. Thus the copper cation must be Cu^+. Now we can name the compound by using the regular steps.

Compound	Ions Present	Ion Names	Comments
CuCl — Cation	Cu^+	copper(I)	Copper forms other cations (it is a transition metal), so we must include the I to specify its charge.
CuCl — Anion	Cl^-	chloride	

The name of $CuCl$ is copper(I) chloride.

b. In HgO we recognize the O^{2-} anion. To yield zero net charge, the cation must be Hg^{2+}.

Compound	Ions Present	Ion Names	Comments
HgO — Cation	Hg^{2+}	mercury(II)	The II is necessary to specify the charge.
HgO — Anion	O^{2-}	oxide	

The name of HgO is mercury(II) oxide.

c. Because Fe_2O_3 contains three O^{2-} anions, the charge on the iron cation must be 3+.

$$2(3+) + 3(2-) = 0$$

↑ Fe^{3+} ↑ O^{2-} ↑ Net charge

Compound	Ions Present	Ion Names	Comments
Fe_2O_3 — Cation	Fe^{3+}	iron(III)	Iron is a transition metal and requires a III to specify the charge on the cation.
Fe_2O_3 — Anion	O^{2-}	oxide	

The name of Fe_2O_3 is iron(III) oxide.

d. MnO_2 contains two O^{2-} anions, so the charge on the manganese cation is 4+.

$$\underset{Mn^{4+}}{(4+)} + \underset{O^{2-}}{2(2-)} = \underset{\text{Net charge}}{0}$$

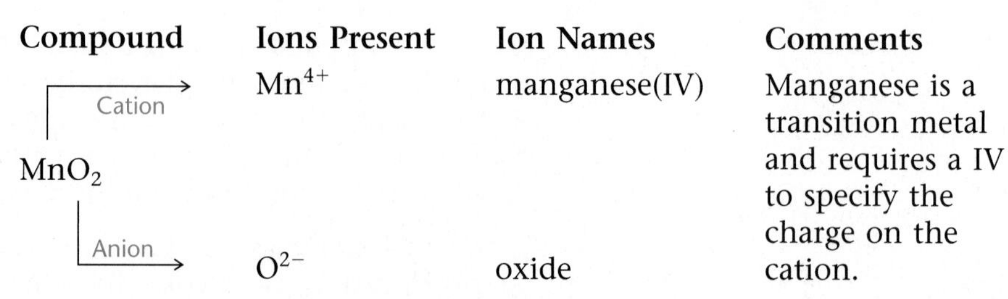

Compound	Ions Present	Ion Names	Comments
MnO_2 (Cation)	Mn^{4+}	manganese(IV)	Manganese is a transition metal and requires a IV to specify the charge on the cation.
MnO_2 (Anion)	O^{2-}	oxide	

The name of MnO_2 is manganese(IV) oxide.

e. Because $PbCl_4$ contains four Cl^- anions, the charge on the lead cation is 4+.

$$\underset{Pb^{4+}}{(4+)} + \underset{Cl^{-}}{4(1-)} = \underset{\text{Net charge}}{0}$$

Compound	Ions Present	Ion Names	Comments
$PbCl_4$ (Cation)	Pb^{4+}	lead(IV)	Lead forms both Pb^{2+} and Pb^{4+}, so a Roman numeral is required.
$PbCl_4$ (Anion)	Cl^-	chloride	

The name for $PbCl_4$ is lead(IV) chloride. ■

Sometimes transition metals form only one ion, such as silver, which forms Ag^+; zinc, which forms Zn^{2+}; and cadmium, which forms Cd^{2+}. In these cases, chemists do not use a Roman numeral, although it is not "wrong" to do so.

The use of a Roman numeral in a systematic name for a compound is required only in cases where more than one ionic compound forms between a given pair of elements. This occurs most often for compounds that contain transition metals, which frequently form more than one cation. *Metals that form only one cation do not need to be identified by a Roman numeral.* Common metals that do not require Roman numerals are the Group 1 elements, which form only 1+ ions; the Group 2 elements, which form only 2+ ions; and such Group 3 metals as aluminum and gallium, which form only 3+ ions.

As shown in Example 5.2, when a metal ion that forms more than one type of cation is present, the charge on the metal ion must be determined by balancing the positive and negative charges of the compound. To do this, you must be able to recognize the common anions and you must know their charges (see Table 5.1).

Example 5.3 Naming Binary Ionic Compounds: Summary

Give the systematic name of each of the following compounds.

a. $CoBr_2$ c. Al_2O_3

b. $CaCl_2$ d. $CrCl_3$

Solution

Compound	Ions and Names	Compound Name	Comments
a. $CoBr_2$	Co^{2+} cobalt(II); Br^- bromide	cobalt(II) bromide	Cobalt is a transition metal; the name of the compound must have a Roman numeral. The two Br^- ions must be balanced by a Co^{2+} cation.
b. $CaCl_2$	Ca^{2+} calcium; Cl^- chloride	calcium chloride	Calcium, a Group 2 metal, forms only the Ca^{2+} ion. A Roman numeral is not necessary.
c. Al_2O_3	Al^{3+} aluminum; O^{2-} oxide	aluminum oxide	Aluminum forms only Al^{3+}. A Roman numeral is not necessary.
d. $CrCl_3$	Cr^{3+} chromium(III); Cl^- chloride	chromium(III) chloride	Chromium is a transition metal. The name of the compound must have a Roman numeral. $CrCl_3$ contains Cr^{3+}.

Self-Check Exercise 5.2

Give the names of the following compounds.

a. $PbBr_2$ and $PbBr_4$ b. FeS and Fe_2S_3 c. $AlBr_3$ d. Na_2S e. $CoCl_3$

See Problems 5.9, 5.10, and 5.13 through 5.16. ■

The following flow chart is useful when you are naming binary ionic compounds:

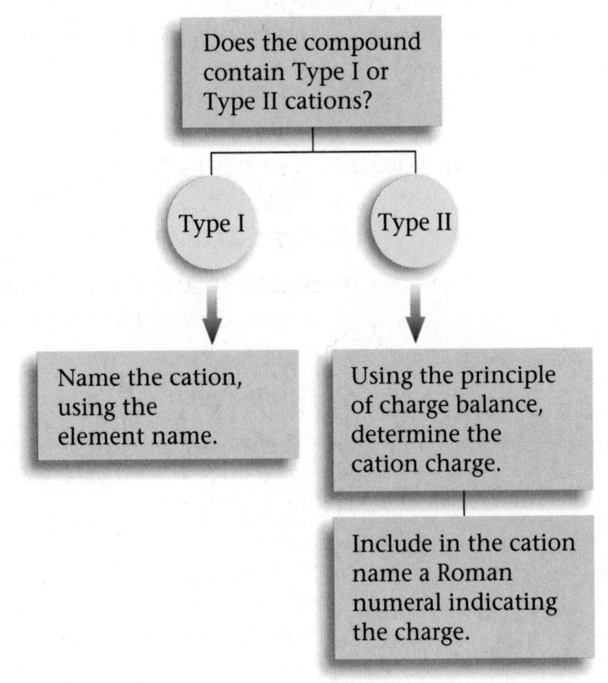

Naming Binary Compounds That Contain Only Nonmetals (Type III)

AIM: To learn how to name binary compounds containing only nonmetals.

Binary compounds that contain only nonmetals are named in accordance with a system similar in some ways to the rules for naming binary ionic compounds, but there are important differences. *Type III binary compounds contain only nonmetals.* The following rules cover the naming of these compounds.

Table 5.3 Prefixes Used to Indicate Numbers in Chemical Names

Prefix	Number Indicated
mono-	1
di-	2
tri-	3
tetra-	4
penta-	5
hexa-	6
hepta-	7
octa-	8

Rules for Naming Type III Binary Compounds

1. The first element in the formula is named first, and the full element name is used.
2. The second element is named as though it were an anion.
3. Prefixes are used to denote the numbers of atoms present. These prefixes are given in Table 5.3.
4. The prefix *mono-* is never used for naming the first element. For example, CO is called carbon monoxide, *not* monocarbon monoxide.

We will illustrate the application of these rules in Example 5.4.

Example 5.4 Naming Type III Binary Compounds

Name the following binary compounds, which contain two nonmetals (Type III).

a. BF_3 b. NO c. N_2O_5

Solution

a. BF_3

RULE 1 Name the first element, using the full element name: boron.

RULE 2 Name the second element as though it were an anion: fluoride.

RULES 3 AND 4 Use prefixes to denote numbers of atoms. One boron atom: do not use *mono-* in first position. Three fluorine atoms: use the prefix *tri-*.

The name of BF_3 is boron trifluoride.

b.

Compound	Individual Names	Prefixes	Comments
NO	nitrogen	none	*Mono-* is not used
	oxide	*mono-*	for the first element.

The name for NO is nitrogen monoxide. Note that the second *o* in *mono-* has been dropped for easier pronunciation. The *common* name for NO, which is often used by chemists, is nitric oxide.

A piece of copper metal about to be placed in nitric acid (left). Copper reacts with nitric acid to produce colorless NO, which immediately reacts with the oxygen in the air to form reddish-brown NO_2 gas (right).

c. **Compound**	**Individual Names**	**Prefixes**	**Comments**
N_2O_5	nitrogen	*di-*	two N atoms
	oxide	*penta-*	five O atoms

The name for N_2O_5 is dinitrogen pentoxide. The *a* in *penta-* has been dropped for easier pronunciation.

Self-Check Exercise 5.3

Name the following compounds.

a. CCl_4 b. NO_2 c. IF_5

See Problems 5.17 and 5.18. ■

The previous examples illustrate that, to avoid awkward pronunciation, we often drop the final *o* or *a* of the prefix when the second element is oxygen. For example, N_2O_4 is called dinitrogen tetroxide, *not* dinitrogen tetr*a*oxide, and CO is called carbon monoxide, *not* carbon mon*o*oxide.

Water and ammonia are always referred to by their common names.

Some compounds are always referred to by their common names. The two best examples are water and ammonia. The systematic names for H_2O and NH_3 are never used.

To make sure you understand the procedures for naming binary nonmetallic compounds (Type III), study Example 5.5 and then do Self-Check Exercise 5.4.

Example 5.5 Naming Type III Binary Compounds: Summary

Name each of the following compounds.

a. PCl_5 c. SF_6 e. SO_2

b. P_4O_6 d. SO_3 f. N_2O_3

Solution

	Compound	Name
a.	PCl_5	phosphorus pentachloride
b.	P_4O_6	tetraphosphorus hexoxide
c.	SF_6	sulfur hexafluoride
d.	SO_3	sulfur trioxide
e.	SO_2	sulfur dioxide
f.	N_2O_3	dinitrogen trioxide

✔ Self-Check Exercise 5.4

Name the following compounds.

a. SiO_2

b. O_2F_2

c. XeF_6

See Problems 5.17 and 5.18. ■

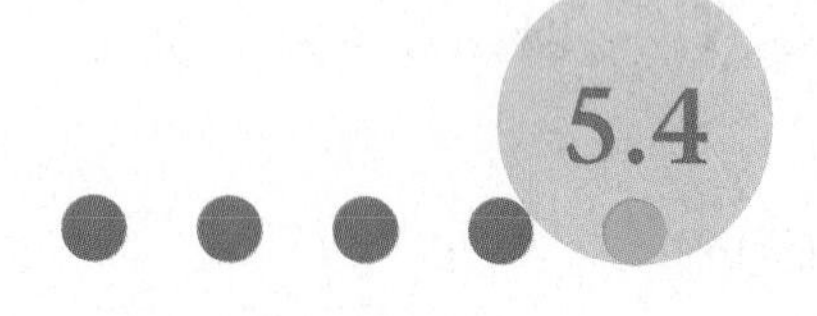

Naming Binary Compounds: A Review

AIM: To review the naming of Type I, Type II, and Type III binary compounds.

Because different rules apply for naming various types of binary compounds, we will now consider an overall strategy to use for these compounds. We have considered three types of binary compounds, and naming each of them requires different procedures.

Type I: Ionic compounds with metals that always form a cation with the same charge

Type II: Ionic compounds with metals (usually transition metals) that form cations with various charges

Type III: Compounds that contain only nonmetals

In trying to determine which type of compound you are naming, use the periodic table to help identify metals and nonmetals and to determine which elements are transition metals.

The flow chart given in Figure 5.1 should help you as you name binary compounds of the various types.

Example 5.6 Naming Binary Compounds: Summary

Name the following binary compounds.

a. CuO e. K_2S

b. SrO f. OF_2

c. B_2O_3 g. NH_3

d. $TiCl_4$

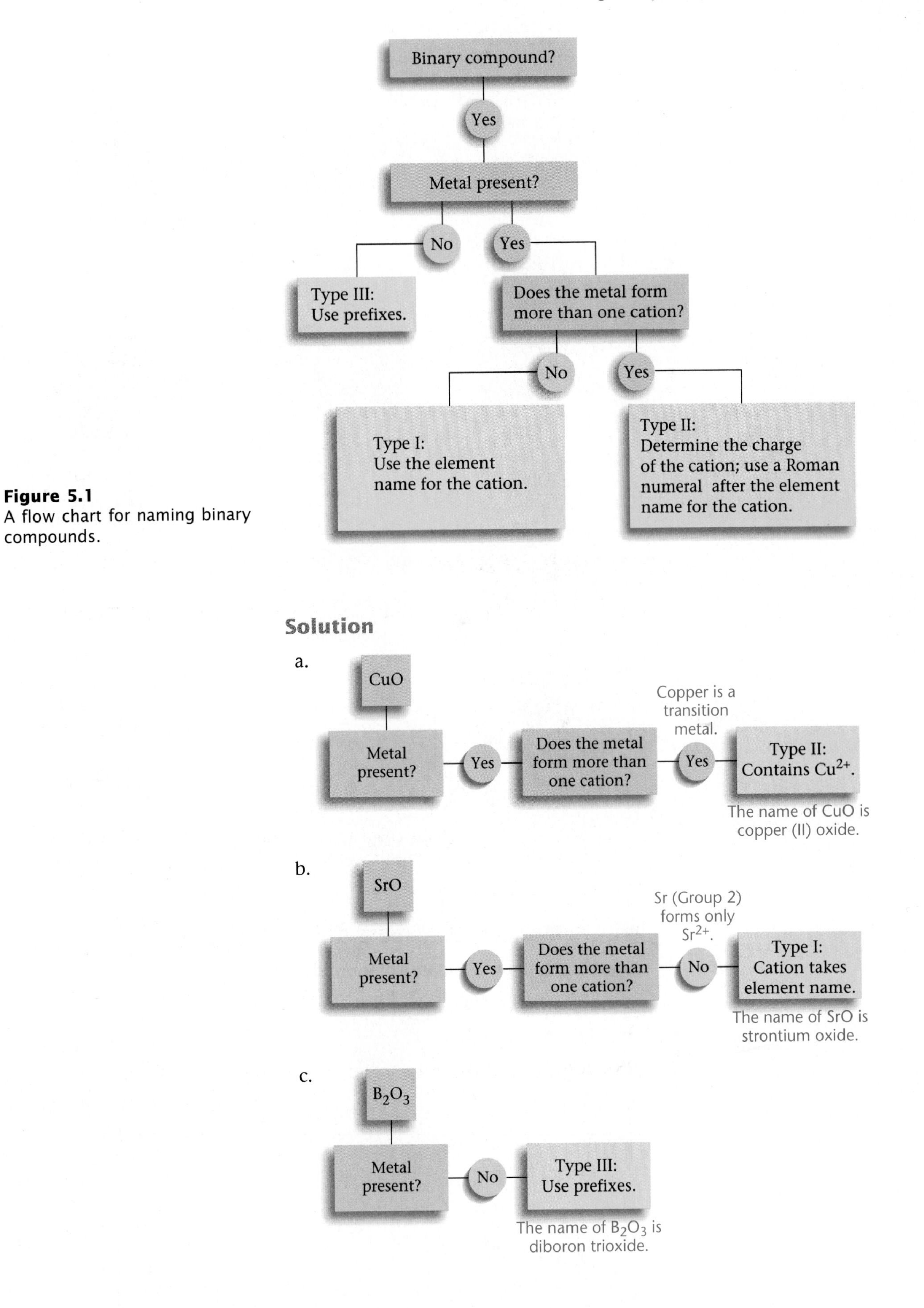

Figure 5.1
A flow chart for naming binary compounds.

Solution

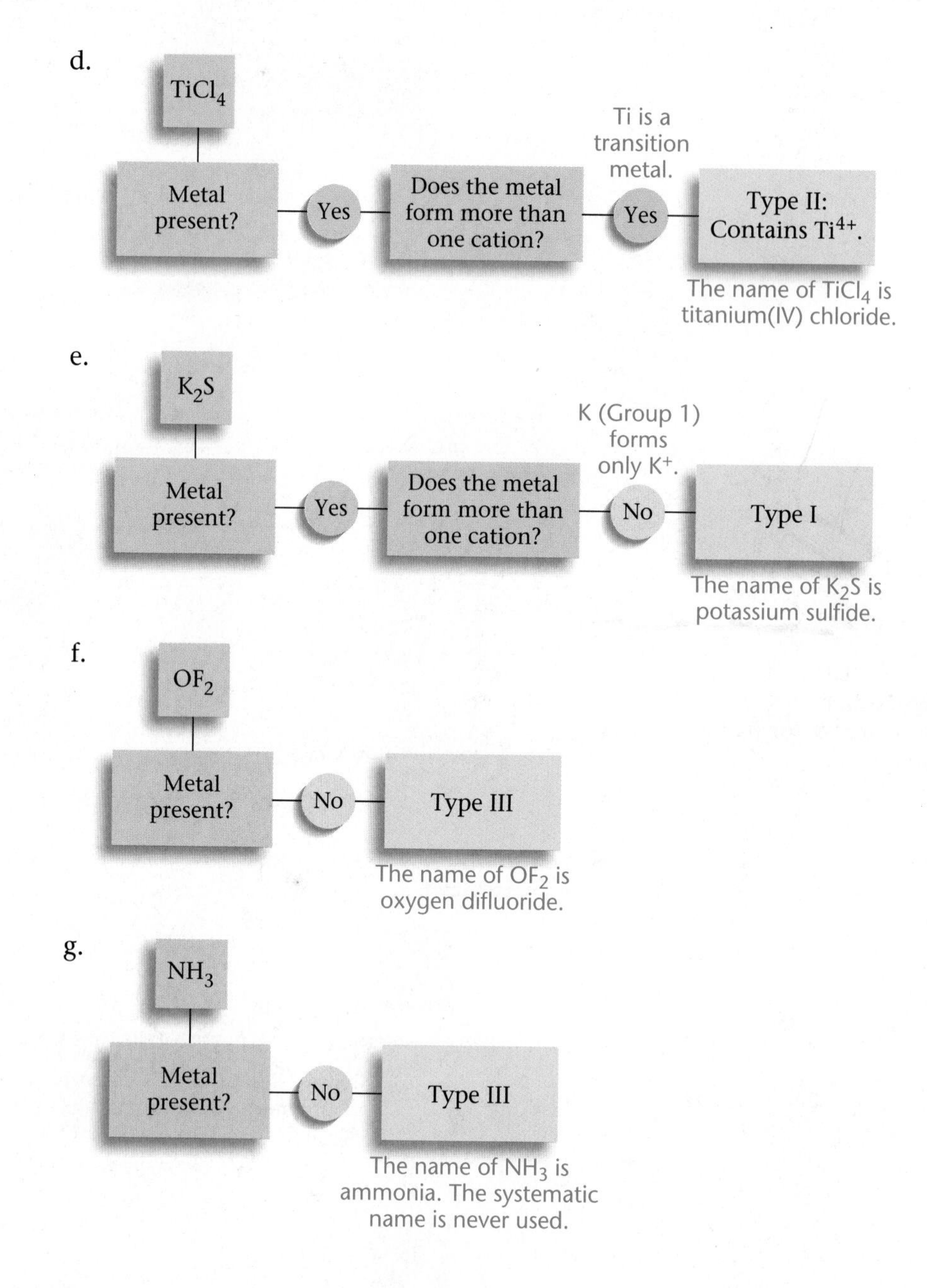

✔ Self-Check Exercise 5.5

Name the following binary compounds.

a. ClF_3 d. MnO_2

b. VF_5 e. MgO

c. CuCl f. H_2O

See Problems 5.19 and 5.20. ■

Naming Compounds That Contain Polyatomic Ions

AIM: To learn the names of common polyatomic ions and how to use them in naming compounds.

Ionic compounds containing polyatomic ions are not binary compounds, because they contain more than two elements.

The names and charges of polyatomic ions must be memorized. They are an important part of the vocabulary of chemistry.

A type of ionic compound that we have not yet considered is exemplified by ammonium nitrate, NH_4NO_3, which contains the **polyatomic ions** NH_4^+ and NO_3^-. As their name suggests, polyatomic ions are charged entities composed of several atoms bound together. Polyatomic ions are assigned special names that you *must memorize* in order to name the compounds containing them. The most important polyatomic ions and their names are listed in Table 5.4.

Note in Table 5.4 that several series of polyatomic anions exist that contain an atom of a given element and different numbers of oxygen atoms. These anions are called **oxyanions.** When there are two members in such a series, the name of the one with the smaller number of oxygen atoms ends in *-ite,* and the name of the one with the larger number ends in *-ate.* For example, SO_3^{2-} is sulfite and SO_4^{2-} is sulfate. When more than two oxyanions make up a series, *hypo-* (less than) and *per-* (more than) are used as prefixes to name the members of the series with the fewest and the most oxygen atoms, respectively. The best example involves the oxyanions containing chlorine:

Note that the SO_3^{2-} anion has very different properties from SO_3 (sulfur trioxide), a pungent, toxic gas.

ClO^-	*hypo*chlor*ite*
ClO_2^-	chlor*ite*
ClO_3^-	chlor*ate*
ClO_4^-	*per*chlor*ate*

Except for hydroxide and cyanide, the names of polyatomic ions do not have an *-ide* ending.

Naming ionic compounds that contain polyatomic ions is very similar to naming binary ionic compounds. For example, the compound NaOH is called sodium hydroxide, because it contains the Na^+ (sodium) cation and the OH^- (hydroxide) anion. To name these compounds, *you must learn to recognize the common polyatomic ions.* That is, you must learn the *composition* and *charge* of each of the ions in Table 5.4. Then when you see

Table 5.4 Names of Common Polyatomic Ions

Ion	Name	Ion	Name
NH_4^+	ammonium	CO_3^{2-}	carbonate
NO_2^-	nitrite	HCO_3^-	hydrogen carbonate (bicarbonate is a widely used common name)
NO_3^-	nitrate		
SO_3^{2-}	sulfite		
SO_4^{2-}	sulfate	ClO^-	hypochlorite
HSO_4^-	hydrogen sulfate (bisulfate is a widely used common name)	ClO_2^-	chlorite
		ClO_3^-	chlorate
		ClO_4^-	perchlorate
OH^-	hydroxide	$C_2H_3O_2^-$	acetate
CN^-	cyanide	MnO_4^-	permanganate
PO_4^{3-}	phosphate	$Cr_2O_7^{2-}$	dichromate
HPO_4^{2-}	hydrogen phosphate	CrO_4^{2-}	chromate
$H_2PO_4^-$	dihydrogen phosphate	O_2^{2-}	peroxide

the formula $NH_4C_2H_3O_2$, you should immediately recognize its two "parts":

$$\underbrace{NH_4}_{NH_4^+}\ \underbrace{C_2H_3O_2}_{C_2H_3O_2^-}$$

The correct name is ammonium acetate.

Remember that when a metal is present that forms more than one cation, a Roman numeral is required to specify the cation charge, just as in naming Type II binary ionic compounds. For example, the compound $FeSO_4$ is called iron(II) sulfate, because it contains Fe^{2+} (to balance the 2^- charge on SO_4^{2-}). Note that to determine the charge on the iron cation, you must know that sulfate has a 2^- charge.

Example 5.7 Naming Compounds That Contain Polyatomic Ions

Give the systematic name of each of the following compounds.

a. Na_2SO_4 c. $Fe(NO_3)_3$ e. Na_2SO_3

b. KH_2PO_4 d. $Mn(OH)_2$ f. NH_4ClO_3

Solution

Compound	Ions Present	Ion Names	Compound Name
a. Na_2SO_4	two Na^+ SO_4^{2-}	sodium sulfate	sodium sulfate
b. KH_2PO_4	K^+ $H_2PO_4^-$	potassium dihydrogen phosphate	potassium dihydrogen phosphate
c. $Fe(NO_3)_3$	Fe^{3+} three NO_3^-	iron(III) nitrate	iron(III) nitrate
d. $Mn(OH)_2$	Mn^{2+} two OH^-	manganese(II) hydroxide	manganese(II) hydroxide
e. Na_2SO_3	two Na^+ SO_3^{2-}	sodium sulfite	sodium sulfite
f. NH_4ClO_3	NH_4^+ ClO_3^-	ammonium chlorate	ammonium chlorate

✔ Self-Check Exercise 5.6

Name each of the following compounds.

a. $Ca(OH)_2$

b. Na_3PO_4

c. $KMnO_4$

d. $(NH_4)_2Cr_2O_7$

e. $Co(ClO_4)_2$

f. $KClO_3$

g. $Cu(NO_2)_2$

See Problems 5.35 and 5.36. ■

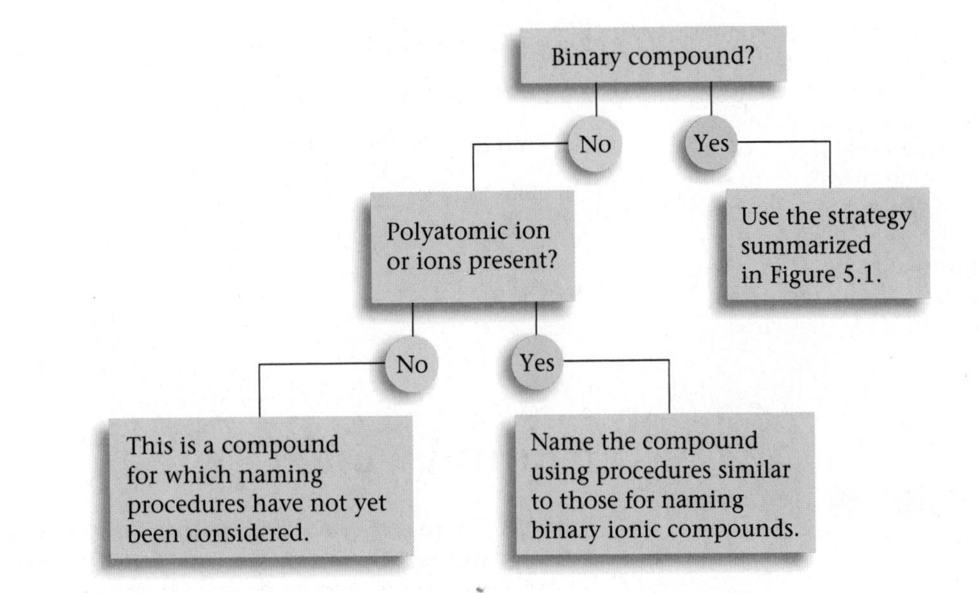

Figure 5.2
Overall strategy for naming chemical compounds.

Example 5.7 illustrates that when more than one polyatomic ion appears in a chemical formula, parentheses are used to enclose the ion and a subscript is written after the closing parenthesis. Other examples are $(NH_4)_2SO_4$ and $Fe_3(PO_4)_2$.

In naming chemical compounds, use the strategy summarized in Figure 5.2. If the compound being considered is binary, use the procedure summarized in Figure 5.1. If the compound has more than two elements, ask yourself whether it has any polyatomic ions. Use Table 5.4 to help you recognize these ions until you have committed them to memory. If a polyatomic ion is present, name the compound using procedures very similar to those for naming binary ionic compounds.

Example 5.8 Summary of Naming Binary Compounds and Compounds That Contain Polyatomic Ions

Name the following compounds.

a. Na_2CO_3

b. $FeBr_3$

c. $CsClO_4$

d. PCl_3

e. $CuSO_4$

Solution

Compound	Name	Comments
a. Na_2CO_3	sodium carbonate	Contains $2Na^+$ and CO_3^{2-}.
b. $FeBr_3$	iron(III) bromide	Contains Fe^{3+} and $3Br^-$.
c. $CsClO_4$	cesium perchlorate	Contains Cs^+ and ClO_4^-.
d. PCl_3	phosphorus trichloride	Type III binary compound (both P and Cl are nonmetals).
e. $CuSO_4$	copper(II) sulfate	Contains Cu^{2+} and SO_4^{2-}.

Although we have emphasized that a Roman numeral is required in the name of a compound that contains a transition metal ion, certain transition metals form only one ion. Common examples are zinc (forms only Zn^{2+}) and silver (forms only Ag^{+}). For these cases the Roman numeral is omitted from the name.

✔ Self-Check Exercise 5.7

Name the following compounds.

a. $NaHCO_3$

b. $BaSO_4$

c. $CsClO_4$

d. BrF_5

e. NaBr

f. KOCl

g. $Zn_3(PO_4)_2$

See Problems 5.29 through 5.36. ■

5.6 Naming Acids

AIMS: To learn how the anion composition determines the acid's name. To learn names for common acids.

When dissolved in water, certain molecules produce H^+ ions (protons). These substances, which are called **acids,** were first recognized by the sour taste of their solutions. For example, citric acid is responsible for the tartness of lemons and limes. Acids will be discussed in detail later. Here we simply present the rules for naming acids.

An acid can be viewed as a molecule with one or more H^+ ions attached to an anion. The rules for naming acids depend on whether the anion contains oxygen.

Rules for Naming Acids

1. If the *anion does not contain oxygen*, the acid is named with the prefix *hydro-* and the suffix *-ic* attached to the root name for the element. For example, when gaseous HCl (hydrogen chloride) is dissolved in water, it forms hydrochloric acid. Similarly, hydrogen cyanide (HCN) and dihydrogen sulfide (H_2S) dissolved in water are called hydrocyanic acid and hydrosulfuric acid, respectively.

2. When the *anion contains oxygen*, the acid name is formed from the root name of the central element of the anion or the anion name, with a suffix of *-ic* or *-ous*. When the anion name ends in *-ate*, the suffix *-ic* is used.

   For example,

Acid	Anion	Name
H_2SO_4	SO_4^{2-} (sulfate)	sulfuric acid
H_3PO_4	PO_4^{3-} (phosphate)	phosphoric acid
$HC_2H_3O_2$	$C_2H_3O_2^-$ (acetate)	acetic acid

   When the anion name ends in *-ite*, the suffix *-ous* is used in the acid name.

   For example,

Acid	Anion	Name
H_2SO_3	SO_3^{2-} (sulfite)	sulfurous acid
HNO_2	NO_2^- (nitrite)	nitrous acid

Table 5.5 Names of Acids That Do Not Contain Oxygen

Acid	Name
HF	hydrofluoric acid
HCl	hydrochloric acid
HBr	hydrobromic acid
HI	hydroiodic acid
HCN	hydrocyanic acid
H_2S	hydrosulfuric acid

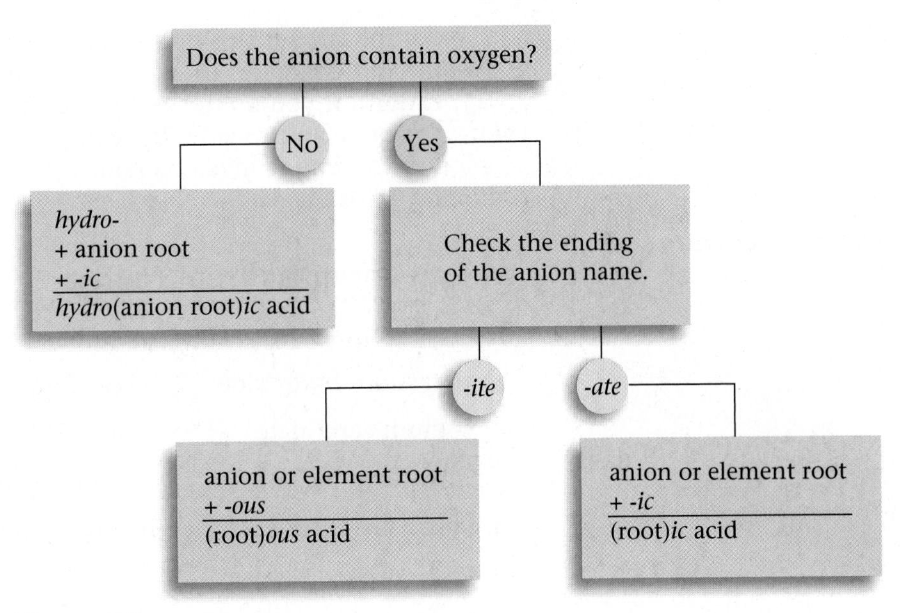

Figure 5.3
A flow chart for naming acids. The acid is considered as one or more H^+ ions attached to an anion.

Table 5.6 Names of Some Oxygen-Containing Acids

Acid	Name
HNO_3	nitric acid
HNO_2	nitrous acid
H_2SO_4	sulfuric acid
H_2SO_3	sulfurous acid
H_3PO_4	phosphoric acid
$HC_2H_3O_2$	acetic acid

The application of Rule 2 can be seen in the names of the acids of the oxyanions of chlorine below. The rules for naming acids are given in schematic form in Figure 5.3. The names of the most important acids are given in Table 5.5 and Table 5.6. These should be memorized.

Acid	Anion	Name
$HClO_4$	perchlor*ate*	perchlor*ic* acid
$HClO_3$	chlor*ate*	chlor*ic* acid
$HClO_2$	chlor*ite*	chlor*ous* acid
HClO	hypochlor*ite*	hypochlor*ous* acid

5.7 Writing Formulas from Names

AIM: To learn to write the formula of a compound, given its name.

So far we have started with the chemical formula of a compound and decided on its systematic name. Being able to reverse the process is also important. Often a laboratory procedure describes a compound by name, but the label on the bottle in the lab shows only the formula of the chemical it contains. It is essential that you are able to get the formula of a compound from its name. In fact, you already know enough about compounds to do this. For example, given the name calcium hydroxide, you can write the formula as $Ca(OH)_2$ because you know that calcium forms only Ca^{2+} ions and that, since hydroxide is OH^-, two of these anions are required to give a neutral compound. Similarly, the name iron(II) oxide implies the formula FeO, because the Roman numeral II indicates the presence of the cation Fe^{2+} and the oxide ion is O^{2-}.

We emphasize at this point that it is essential to learn the name, composition, and charge of each of the common polyatomic anions (and the NH_4^+ cation). If you do not recognize these ions by formula and by name, you will not be able to write the compound's name given its formula or the compound's formula given its name. You must also learn the names of the common acids.

Example 5.9 Writing Formulas from Names

Give the formula for each of the following compounds.

a. potassium hydroxide
b. sodium carbonate
c. nitric acid
d. cobalt(III) nitrate
e. calcium chloride
f. lead(IV) oxide
g. dinitrogen pentoxide
h. ammonium perchlorate

Solution

Name	Formula	Comments
a. potassium hydroxide	KOH	Contains K^+ and OH^-.
b. sodium carbonate	Na_2CO_3	We need two Na^+ to balance CO_3^{2-}.
c. nitric acid	HNO_3	Common strong acid; memorize.
d. cobalt(III) nitrate	$Co(NO_3)_3$	Cobalt(III) means Co^{3+}; we need three NO_3^- to balance Co^{3+}.
e. calcium chloride	$CaCl_2$	We need two Cl^- to balance Ca^{2+}; Ca (Group 2) always forms Ca^{2+}.
f. lead(IV) oxide	PbO_2	Lead(IV) means Pb^{4+}; we need two O^{2-} to balance Pb^{4+}.
g. dinitrogen pentoxide	N_2O_5	*di-* means two; *pent(a)-* means five.
h. ammonium perchlorate	NH_4ClO_4	Contains NH_4^+ and ClO_4^-.

✔ Self-Check Exercise 5.8

Write the formula for each of the following compounds.

a. ammonium sulfate
b. vanadium(V) fluoride
c. disulfur dichloride
d. rubidium peroxide
e. aluminum oxide

See Problems 5.41 through 5.46. ■

CHAPTER 5 REVIEW

KEY TERMS

binary compound (5.1)
binary ionic compound (5.2)
polyatomic ion (5.5)
oxyanion (5.5)
acid (5.6)

SUMMARY

1. Binary compounds can be named systematically by following a set of relatively simple rules. For compounds containing both a metal and a nonmetal, the metal is always named first, followed by a name derived from the root name for the nonmetal. For compounds containing a metal that can form more than one cation (Type II), we use a Roman numeral to specify the cation's charge. In binary compounds containing only nonmetals (Type III), prefixes are used to specify the numbers of atoms.

2. Polyatomic ions are charged entities composed of several atoms bound together. These have special names that must be memorized. Naming ionic compounds that contain polyatomic ions is very similar to naming binary ionic compounds.

3. The names of acids (molecules with one or more H^+ ions attached to an anion) depend on whether the anion contains oxygen.

IN-CLASS DISCUSSION QUESTIONS

These questions are designed to be considered by groups of students in class. Often these questions work well for introducing a particular topic in class.

1. Evaluate each of the following as an acceptable systematic name for water.
   a. dihydrogen oxide
   b. hydroxide hydride
   c. hydrogen hydroxide
   d. oxygen dihydride

2. Why do we call $Ba(NO_3)_2$ barium nitrate but call $Fe(NO_3)_2$ iron(II) nitrate?

3. Why is calcium dichloride not an acceptable name for $CaCl_2$?

4. What is the difference between sulfuric acid and hydrosulfuric acid?

5. Although we never use the systematic name for ammonia, NH_3, what do you think this name would be? Support your answer.

QUESTIONS AND PROBLEMS

All even-numbered exercises have answers in the back of this book and solutions in the *Solutions Guide*.

5.1 Naming Compounds

QUESTIONS

1. Why is it necessary to have a *system* for the naming of chemical compounds?

2. What is a *binary* chemical compound? What are the two major *types* of binary chemical compounds? Give three examples of each type of binary compound.

5.2 Naming Compounds That Contain a Metal and a Nonmetal (Types I and II)

QUESTIONS

3. In ionic compounds, what general names are used to describe the *positive* ions and the *negative* ions?

4. In naming ionic compounds, we always name the ________ first.

5. A simple ________ ion has the same name as its parent element.

6. When we write the formula for an ionic compound, we are merely indicating the relative numbers of each type of ion in the compound, *not* the presence of "molecules" in the compound with that formula. Explain.

7. For a metallic element that forms two stable cations, the ending ________ is sometimes used for the cation with the lower charge.

8. We indicate the charge of a metallic element that forms more than one cation by adding a ________ after the name of the cation.

9. Give the name of each of the following simple binary ionic compounds.
   a. NaI
   b. CaF_2
   c. Al_2S_3
   d. $CaBr_2$
   e. SrO
   f. AgCl
   g. CsI
   h. Li_2O

10. Give the name of each of the following simple binary ionic compounds.
    a. NaBr
    b. CaS
    c. AlI_3
    d. Cs_2O
    e. MgF_2
    f. $SrCl_2$
    g. Li_2O
    h. BaI_2

11. Identify each case in which the name is incorrect. Give the correct name.
    a. BaH_2, barium hydroxide
    b. Na_2O, disodium oxide
    c. $SnCl_4$, tin(IV) chloride
    d. SiO_2, silver dioxide
    e. $FeBr_3$, iron(III) bromide

12. Identify each case in which the formula is incorrect. Give the correct formula for the indicated name.
    a. silver oxide, SiO_2
    b. sodium sulfide, Na_2S
    c. barium oxide, B_2O_3
    d. calcium nitride, Ca_3N_2
    e. lithium bromide, Li_3B

13. Write the name of each of the following ionic substances, using the system that includes a Roman numeral to specify the charge of the cation.
    a. $SnBr_2$ d. Cr_2O_3
    b. SnI_4 e. Hg_2I_2
    c. CrO f. HgI_2

14. Write the name of each of the following ionic substances, using the system that includes a Roman numeral to specify the charge of the cation.
    a. SnO d. Cu_2S
    b. SnO_2 e. $FeCl_3$
    c. CuS f. FeI_2

15. Write the name of each of the following ionic substances, using *-ous* or *-ic* endings to indicate the charge of the cation.
    a. $CoCl_2$ d. SnO_2
    b. $CrBr_3$ e. Fe_2O_3
    c. PbO f. $FeCl_3$

16. Write the name for each of the following substances, using the *-ous*/*-ic* notation.
    a. SnS d. CoI_2
    b. SnS_2 e. $HgBr_2$
    c. $CoCl_3$ f. Hg_2Br_2

5.3 Naming Binary Compounds That Contain Only Nonmetals (Type III)

QUESTIONS

17. Name each of the following binary compounds of nonmetallic elements.
    a. IF_5 c. SeO e. NI_3
    b. $AsCl_3$ d. XeF_4 f. B_2O_3

18. Write the name for each of the following binary compounds of nonmetallic elements.
    a. XeF_6 c. P_2O_5 e. $AsBr_3$
    b. $SiCl_4$ d. H_2O f. Cl_4

5.4 Naming Binary Compounds: A Review

QUESTIONS

19. Name each of the following binary compounds, using the periodic table to determine whether the compound is likely to be ionic (containing a metal and a nonmetal) or nonionic (containing only nonmetals).
    a. SnO_2 d. Fe_2S_3
    b. CaH_2 e. OCl_2
    c. $SiBr_4$ f. XeF_4

20. Name each of the following binary compounds, using the periodic table to determine whether the compound is likely to be ionic (containing a metal and a nonmetal) or nonionic (containing only nonmetals).
    a. B_2H_6 d. Ag_2S
    b. Ca_3N_2 e. $CuCl_2$
    c. CBr_4 f. ClF

21. Name each of the following binary compounds, using the periodic table to determine whether the compound is likely to be ionic (containing a metal and a nonmetal) or nonionic (containing only nonmetals).
    a. MgS d. $ClBr$
    b. $AlCl_3$ e. Li_2O
    c. PH_3 f. P_4O_{10}

22. Name each of the following binary compounds, using the periodic table to determine whether the compound is likely to be ionic (containing a metal and a nonmetal) or nonionic (containing only nonmetals).
    a. BaF_2 d. Rb_2O
    b. RaO e. As_2O_5
    c. N_2O f. Ca_3N_2

5.5 Naming Compounds That Contain Polyatomic Ions

QUESTIONS

23. What is a *polyatomic* ion? Give examples of five common polyatomic ions.

24. What is an *oxyanion?* How is a series of oxyanions named? (Give an example.)

25. For the oxyanions of sulfur, the ending *-ite* is used for SO_3^{2-} to indicate that it contains ______ than does SO_4^{2-}.

26. In naming oxyanions, when there are more than two members in the series for a given element, what prefixes are used to indicate the oxyanions in the series with the *fewest* and the *most* oxygen atoms?

27. Complete the following list by filling in the missing names or formulas of the oxyanions of chlorine.

ClO_4^-	______
______	hypochlorite
ClO_3^-	______
______	chlorite

28. A series of four oxyanions of bromine, comparable to the series for chlorine discussed in the text, also exists. Write the formulas and names for the four oxyanions of bromine.

29. Write the formula for each of the following phosphorus-containing ions, including the overall charge of the ion.
    a. phosphide
    b. phosphate
    c. phosphite
    d. hydrogen phosphate

30. Write the formula for each of the following nitrogen-containing polyatomic ions, including the overall charge of the ion.
    a. nitrate
    b. nitrite
    c. ammonium
    d. cyanide

31. Write the formula for each of the following chlorine-containing ions, including the overall charge of the ion.
    a. chloride
    b. hypochlorite
    c. chlorate
    d. perchlorate

32. Write the formula for each of the following carbon-containing ions, including the overall charge of the ions.
    a. carbonate
    b. hydrogen carbonate
    c. acetate
    d. cyanide

33. Give the name of each of the following polyatomic anions.
    a. MnO_4^-
    b. O_2^{2-}
    c. CrO_4^{2-}
    d. $Cr_2O_7^{2-}$
    e. NO_3^-
    f. SO_3^{2-}

34. Give the name of each of the following polyatomic ions.
    a. NH_4^+
    b. $H_2PO_4^-$
    c. SO_4^{2-}
    d. HSO_3^-
    e. ClO_4^-
    f. IO_3^-

35. Name each of the following compounds, which contain polyatomic ions.
    a. NH_4NO_2
    b. $Ba(OH)_2$
    c. K_2O_2
    d. $Al(HSO_4)_3$
    e. AgCN
    f. $CaHPO_4$

36. Name each of the following compounds, which contain polyatomic ions.
    a. $(NH_4)_2SO_4$
    b. $KClO_4$
    c. $Fe_2(SO_4)_3$
    d. $Ca_3(PO_4)_2$
    e. $Ca(OH)_2$
    f. K_2CO_3

5.6 Naming Acids

QUESTIONS

37. Give a simple definition of an *acid*.

38. Many acids contain the element ________ in addition to hydrogen.

39. Name each of the following acids.
    a. HCl
    b. H_2SO_4
    c. HNO_3
    d. HI
    e. HNO_2
    f. $HClO_3$
    g. HBr
    h. HF
    i. $HC_2H_3O_2$

40. Name each of the following acids.
    a. HOCl
    b. H_2SO_3
    c. $HBrO_3$
    d. HOI
    e. $HBrO_4$
    f. H_2S
    g. H_2Se
    h. H_3PO_3

5.7 Writing Formulas from Names

PROBLEMS

41. Write the formula for each of the following simple binary ionic compounds.
    a. radium oxide
    b. silver sulfide
    c. rubidium iodide
    d. silver iodide
    e. calcium hydride
    f. magnesium phosphide
    g. cesium bromide
    h. barium nitride

42. Write the formula for each of these ionic substances.
    a. plumbic oxide
    b. stannous bromide
    c. cupric sulfide
    d. cuprous iodide
    e. mercurous chloride
    f. chromic fluoride

43. Write the formula for each of the following binary compounds of nonmetallic elements.
    a. phosphorus triiodide
    b. silicon tetrachloride
    c. dinitrogen pentoxide
    d. iodine monobromide
    e. diboron trioxide
    f. nitrogen trichloride
    g. carbon monoxide

44. Write the formula for each of the following binary compounds of nonmetallic elements.
    a. dinitrogen oxide
    b. nitrogen dioxide
    c. dinitrogen tetraoxide (tetroxide)
    d. sulfur hexafluoride
    e. phosphorus tribromide
    f. carbon tetraiodide
    g. oxygen dichloride

45. Write the formula for each of the following compounds that contain polyatomic ions. Be sure to enclose the polyatomic ion in parentheses if more than one such ion is needed to balance the oppositely charged ion(s).
    a. ammonium nitrate
    b. magnesium acetate
    c. calcium peroxide
    d. potassium hydrogen sulfate
    e. iron(II) sulfate
    f. potassium hydrogen carbonate
    g. cobalt(II) sulfate
    h. lithium perchlorate

46. Write the formula for each of the following compounds that contain polyatomic ions. Be sure to enclose the polyatomic ion in parentheses if more than one such ion is needed.
    a. barium sulfite
    b. calcium dihydrogen phosphate
    c. ammonium perchlorate
    d. sodium permanganate
    e. iron(III) sulfate
    f. cobalt(II) carbonate
    g. nickelous hydroxide
    h. zinc chromate

47. Write the formula for each of the following acids.
    a. hydrosulfuric acid
    b. perbromic acid
    c. acetic acid
    d. hydrobromic acid
    e. chlorous acid
    f. hydroselenic acid
    g. sulfurous acid
    h. perchloric acid

48. Write the formula for each of the following acids.
    a. hydrocyanic acid
    b. nitric acid
    c. sulfuric acid
    d. phosphoric acid
    e. hypochlorous acid
    f. hydrobromic acid
    g. bromous acid
    h. hydrofluoric acid

49. Write the formula for each of the following substances.
    a. sodium peroxide
    b. calcium chlorate
    c. rubidium hydroxide
    d. zinc nitrate
    e. ammonium dichromate
    f. hydrosulfuric acid
    g. calcium bromide
    h. hypochlorous acid
    i. potassium sulfate
    j. nitric acid
    k. barium acetate
    l. lithium sulfite

50. Write the formula for each of the following substances.
    a. magnesium hydrogen sulfate
    b. cesium perchlorate
    c. iron(II) oxide
    d. hydrotelluric acid
    e. strontium nitrate
    f. tin(IV) acetate
    g. manganese(II) sulfate
    h. dinitrogen tetroxide
    i. sodium hydrogen phosphate
    j. lithium peroxide
    k. nitrous acid
    l. cobalt(III) nitrate

ADDITIONAL PROBLEMS

51. Although sucrose (table sugar) and sodium chloride (table salt) are very similar in appearance, sodium chloride melts at over 800 °C, whereas table sugar melts at around 185 °C. What kinds of forces give an ionic compound such a high resistance to melting?

52. Before an electrocardiogram (ECG) is recorded for a cardiac patient, the ECG leads are usually coated with a moist paste containing sodium chloride. What property of an ionic substance such as NaCl is being made use of here?

53. What is a *binary* compound? What is a *polyatomic* anion? What is an *oxyanion?*

54. On some periodic tables, hydrogen is listed both as a member of Group 1 and as a member of Group 7. Write an equation showing the formation of H^+ ion and an equation showing the formation of H^- ion.

55. In general, oxyacids contain a nonmetal, oxygen, and what other element?

56. Complete the following list by filling in the missing oxyanion or oxyacid for each pair.

ClO_4^-	______
______	HIO_3
ClO^-	______
BrO_2^-	______
______	$HClO_2$

57. Name the following compounds.
    a. $Ca(C_2H_3O_2)_2$
    b. PCl_3
    c. $Cu(MnO_4)_2$
    d. $Fe_2(CO_3)_3$
    e. $LiHCO_3$
    f. Cr_2S_3
    g. $Ca(CN)_2$

58. Name the following compounds.
    a. $AuBr_3$
    b. $Co(CN)_3$
    c. $MgHPO_4$
    d. B_2H_6
    e. NH_3
    f. Ag_2SO_4
    g. $Be(OH)_2$

59. Name the following compounds.
    a. $HClO_3$
    b. $CoCl_3$
    c. B_2O_3
    d. H_2O
    e. $HC_2H_3O_2$
    f. $Fe(NO_3)_3$
    g. $CuSO_4$

60. Name the following compounds.
    a. $(NH_4)_2CO_3$
    b. NH_4HCO_3
    c. $Ca_3(PO_4)_2$
    d. H_2SO_3
    e. MnO_2
    f. HIO_3
    g. KH

61. Most metallic elements form *oxides*, and often the oxide is the most common compound of the element that is found in the earth's crust. Write the formulas for the oxides of the following metallic elements.
    a. potassium
    b. magnesium
    c. iron(II)
    d. iron(III)
    e. zinc(II)
    f. lead(II)
    g. aluminum

62. Consider a hypothetical simple ion M^{4+}. Determine the formula of the compound this ion would form with each of the following anions.
a. acetate
b. permanganate
c. oxide
d. hydrogen phosphate
e. hydroxide
f. nitrite

63. Consider a hypothetical element M, which is capable of forming stable simple cations that have charges of 1+, 2+, and 3+, respectively. Write the formulas of the compounds formed by the various M cations with each of the following anions.
a. chromate
b. dichromate
c. sulfide
d. bromide
e. bicarbonate
f. hydrogen phosphate

64. Consider the hypothetical metallic element M, which is capable of forming stable simple cations that have charges of 1+, 2+, and 3+, respectively. Consider also the nonmetallic elements D, E, and F, which form anions that have charges of 1−, 2−, and 3−, respectively. Write the formulas of all possible compounds between metal M and nonmetals D, E, and F.

65. Complete Table 5.A (on page 149) by writing the names and formulas for the ionic compounds formed when the cations listed across the top combine with the anions shown in the left-hand column.

66. Complete Table 5.B (on page 149) by writing the formulas for the ionic compounds formed when the anions listed across the top combine with the cations shown in the left-hand column.

67. The noble metals gold, silver, and platinum are often used in fashioning jewelry because they are relatively ________.

68. The noble gas ________ is frequently found in underground deposits of natural gas.

69. The elements of Group 7 (fluorine, chlorine, bromine, and iodine) consist of molecules containing ________ atom(s).

70. Under what physical state at room temperature do each of the halogen elements exist?

71. When an atom gains two electrons, the ion formed has a charge of ________.

72. An ion with one more electron than it has protons has a ________ charge.

73. An atom that has lost three electrons will have a charge of ________.

74. An atom that has gained one electron has a charge of ________.

75. For each of the negative ions listed in column 1, use the periodic table to find in column 2 the total number of electrons the ion contains. A given answer may be used more than once.

Column 1	Column 2
[1] Se^{2-}	[a] 18
[2] S^{2-}	[b] 35
[3] P^{3-}	[c] 52
[4] O^{2-}	[d] 34
[5] N^{3-}	[e] 36
[6] I^-	[f] 54
[7] F^-	[g] 10
[8] Cl^-	[h] 9
[9] Br^-	[i] 53
[10] At^-	[j] 86

76. For each of the following processes that show the formation of ions, complete the process by indicating the number of electrons that must be gained or lost to form the ion. Indicate the total number of electrons in the ion, and in the atom from which it was made.
a. $Al \rightarrow Al^{3+}$
b. $S \rightarrow S^{2-}$
c. $Cu \rightarrow Cu^+$
d. $F \rightarrow F^-$
e. $Zn \rightarrow Zn^{2+}$
f. $P \rightarrow P^{3-}$

77. For each of the following atomic numbers, use the periodic table to write the formula (including the charge) for the simple *ion* that the element is most likely to form.
a. 36
b. 31
c. 52
d. 81
e. 35
f. 87

78. For the following pairs of ions, use the principle of electrical neutrality to predict the formula of the binary compound that the ions are most likely to form.
a. Na^+ and S^{2-}
b. K^+ and Cl^-
c. Ba^{2+} and O^{2-}
d. Mg^{2+} and Se^{2-}
e. Cu^{2+} and Br^-
f. Al^{3+} and I^-
g. Al^{3+} and O^{2-}
h. Ca^{2+} and N^{3-}

79. Give the name of each of the following simple binary ionic compounds.
a. BeO
b. MgI_2
c. Na_2S
d. Al_2O_3
e. HCl
f. LiF
g. Ag_2S
h. CaH_2

80. In which of the following pairs is the name incorrect?
a. SiI_4, silver iodide
b. $CoCl_2$, copper(II) chloride
c. CaH_2, hydrocalcinic acid
d. $Zn(C_2H_3O_2)_2$, zinc acetate
e. PH_3, phosphoric trihydride

81. Write the name of each of the following ionic substances, using the system that includes a Roman numeral to specify the charge of the cation.
a. $FeBr_2$
b. CoS
c. Co_2S_3
d. SnO_2
e. Hg_2Cl_2
f. $HgCl_2$

Table 5.A

Ions	Fe^{2+}	Al^{3+}	Na^+	Ca^{2+}	NH_4^+	Fe^{3+}	Ni^{2+}	Hg_2^{2+}	Hg^{2+}
CO_3^{2-}	___	___	___	___	___	___	___	___	___
BrO_3^-	___	___	___	___	___	___	___	___	___
$C_2H_3O_2^-$	___	___	___	___	___	___	___	___	___
OH^-	___	___	___	___	___	___	___	___	___
HCO_3^-	___	___	___	___	___	___	___	___	___
PO_4^{3-}	___	___	___	___	___	___	___	___	___
SO_3^{2-}	___	___	___	___	___	___	___	___	___
ClO_4^-	___	___	___	___	___	___	___	___	___
SO_4^{2-}	___	___	___	___	___	___	___	___	___
O^{2-}	___	___	___	___	___	___	___	___	___
Cl^-	___	___	___	___	___	___	___	___	___

Table 5.B

Ions	nitrate	sulfate	hydrogen sulfate	dihydrogen phosphate	oxide	chloride
calcium	___	___	___	___	___	___
strontium	___	___	___	___	___	___
ammonium	___	___	___	___	___	___
aluminum	___	___	___	___	___	___
iron(III)	___	___	___	___	___	___
nickel(II)	___	___	___	___	___	___
silver(I)	___	___	___	___	___	___
gold(III)	___	___	___	___	___	___
potassium	___	___	___	___	___	___
mercury(II)	___	___	___	___	___	___
barium	___	___	___	___	___	___

82. Write the name of each of the following ionic substances, using *-ous* or *-ic* to indicate the charge of the cation.
a. $CoBr_3$
b. PbI_4
c. Fe_2O_3
d. FeS
e. $SnCl_4$
f. SnO

83. Name each of the following binary compounds.
a. XeF_6
b. OF_2
c. AsI_3
d. N_2O_4
e. Cl_2O
f. SF_6

84. Name each of the following compounds.
a. $Fe(C_2H_3O_2)_3$
b. BrF
c. K_2O_2
d. $SiBr_4$
e. $Cu(MnO_4)_2$
f. $CaCrO_4$

85. Which oxyanion of nitrogen contains a larger number of oxygen atoms, the nitr*ate* ion or the nitr*ite* ion?

86. Write the formula for each of the following carbon-containing polyatomic ions, including the overall charge of the ion.
a. carbonate
b. hydrogen carbonate
c. acetate
d. cyanide

87. Write the formula for each of the following chromium-containing ions, including the overall charge of the ion.
a. chromous
b. chromate
c. chromic
d. dichromate

88. Give the name of each of the following polyatomic anions.
a. CO_3^{2-}
b. ClO_3^-
c. SO_4^{2-}
d. PO_4^{3-}
e. ClO_4^-
f. MnO_4^-

89. Name each of the following compounds, which contain polyatomic ions.
a. LiH_2PO_4
b. $Cu(CN)_2$
c. $Pb(NO_3)_2$
d. Na_2HPO_4
e. $NaClO_2$
f. $Co_2(SO_4)_3$

90. Write the formula for each of the following simple binary ionic compounds.

a. calcium chloride
b. silver(I) oxide (usually called silver oxide)
c. aluminum sulfide
d. beryllium bromide
e. hydrosulfuric acid
f. potassium hydride
g. magnesium iodide
h. cesium fluoride

91. Write the formula for each of the following binary compounds of nonmetallic elements.
a. sulfur dioxide
b. dinitrogen monoxide
c. xenon tetrafluoride
d. tetraphosphorus decoxide
e. phosphorus pentachloride
f. sulfur hexafluoride
g. nitrogen dioxide

92. Write the formula of each of the following ionic substances.
a. sodium dihydrogen phosphate
b. lithium perchlorate
c. copper(II) hydrogen carbonate
d. potassium acetate
e. barium peroxide
f. cesium sulfite

93. Write the formula for each of the following compounds, which contain polyatomic ions. Be sure to enclose the polyatomic ion in parentheses if more than one such ion is needed to balance the oppositely charged ion(s).
a. silver(I) perchlorate (usually called silver perchlorate)
b. cobalt(III) hydroxide
c. sodium hypochlorite
d. potassium dichromate
e. ammonium nitrite
f. ferric hydroxide
g. ammonium hydrogen carbonate
h. potassium perbromate

Cumulative Review for Chapters 4–5

QUESTIONS

1. What is an element? How many elements are presently known? How many of these occur naturally and how many are man-made? Which elements are most abundant on the earth?

2. Without consulting any reference, write the name and symbol for as many elements as you can. How many could you name? How many symbols did you write correctly?

3. Why do the symbols for some elements seem to bear no relationship to the name for the element? Give several examples and explain.

4. Without consulting your textbook or notes, state as many points as you can of Dalton's atomic theory. Explain in your own words each point of the theory.

5. What is a compound? What is meant by the *law of constant composition* for compounds and why is this law so important to our study of chemistry?

6. What is meant by a *nuclear atom?* Describe the points of Rutherford's model for the nuclear atom and how he tested this model. Based on his experiments, how did Rutherford envision the structure of the atom? How did Rutherford's model of the atom's structure differ from Kelvin's "plum pudding" model?

7. What are the three fundamental particles that compose all atoms? Indicate the electrical charge and relative mass of each of these particles. Where is each type of particle found in the atom? Which of these particles is responsible for the chemical behavior of a given type of atom? Why?

8. What are *isotopes?* To what do the *atomic number* and the *mass number* of an isotope refer? How are specific isotopes indicated symbolically (give an example and explain)? Do the isotopes of a given element have the same chemical and physical properties? Explain.

9. Describe the periodic table of the elements. How are the elements arranged in the table? What significance is there in the way the elements are arranged into vertical groups? Which general area of the periodic table contains the metallic elements? Which general area contains the nonmetallic elements? Give the names of some of the families of elements in the periodic table.

10. Are most elements found in nature in the elemental or the combined form? Why? Name several elements that are usually found in the elemental form.

11. What are *ions?* How are ions formed from atoms? Do isolated atoms form ions spontaneously? To what do the terms *cation* and *anion* refer? In terms of subatomic particles, how is an ion related to the atom from which it is formed? Does the nucleus of an atom change when the atom is converted into an ion? How can the periodic table be used to predict what ion an element's atoms will form?

12. What are some general physical properties of ionic compounds such as sodium chloride? How do we know that substances such as sodium chloride consist of positively and negatively charged particles? Since ionic compounds are made up of electrically charged particles, why doesn't such a compound have an overall electric charge? Can an ionic compound consist only of cations or anions (but not both)? Why not?

13. What principle do we use in writing the formula of an ionic compound such as NaCl or MgI_2? How do we know that *two* iodide ions are needed for each magnesium ion, whereas only one chloride ion is needed per sodium ion?

14. When writing the name of an ionic compound, which is named first, the anion or the cation? Give an example. What ending is added to the root name of an element to show that it is a simple anion in a Type I ionic compound? Give an example. What *two* systems are used to show the charge of the cation in a Type II ionic compound? Give examples of each system for the same compound. What general type of element is involved in Type II compounds?

15. Describe the system used to name Type III binary compounds (compounds of nonmetallic elements). Give several examples illustrating the method. How does this system differ from that used for ionic compounds? How is the system for Type III compounds similar to those for ionic compounds?

16. What is a *polyatomic* ion? Without consulting a reference, list the formulas and names of at least ten polyatomic ions. When writing the overall formula of an ionic compound involving polyatomic ions, why are parentheses used around the formula of a polyatomic ion when more than one such ion is present? Give an example.

17. What is an *oxyanion?* What special system is used in a series of related oxyanions that indicates the relative number of oxygen atoms in each ion? Give examples.

18. What is an *acid?* How are acids that do *not* contain oxygen named? Give several examples. Describe the naming system for the oxyacids. Give examples of a series of oxyacids illustrating this system.

PROBLEMS

19. Write the *symbol* and *atomic number* for each of the following elements: magnesium, tin, lead, sodium, hydrogen, chlorine, silver, potassium, calcium, bromine, neon, aluminum, gold, mercury, and iodine.

20. Write the name, atomic number, and group number for each of the following elements.

a. Br e. B i. Mg
b. Li f. Cl j. Al
c. Sr g. Xe
d. K h. C

21. Write the *name* and *chemical symbol* corresponding to each of the following atomic numbers:
a. 19 f. 6 k. 29
b. 12 g. 15 l. 35
c. 36 h. 20 m. 2
d. 92 i. 79 n. 8
e. 1 j. 82

22. Indicate the number of protons, neutrons, and electrons in isolated atoms having the following nuclear symbols:
a. ${}^{4}_{2}He$ d. ${}^{41}_{20}Ca$ g. ${}^{235}_{92}U$
b. ${}^{37}_{17}Cl$ e. ${}^{40}_{20}Ca$ h. ${}^{1}_{1}H$
c. ${}^{79}_{35}Br$ f. ${}^{238}_{92}U$

23. What simple ion does each of the following elements most commonly form?
a. Mg f. Ba j. Ca
b. F g. Na k. S
c. Ag h. Br l. Li
d. Al i. K m. Cl
e. O

24. For each of the following simple ions, indicate the number of protons and electrons the ion contains.
a. Mg^{2+} e. Ni^{2+} i. S^{2-}
b. Fe^{2+} f. Zn^{2+} j. Rb^{+}
c. Fe^{3+} g. Co^{3+} k. Se^{2-}
d. F^{-} h. N^{3-} l. K^{+}

25. Using the ions indicated in Problem 24, write the formulas and give the names for all possible simple ionic compounds involving these ions.

26. Give the name of each of the following binary ionic compounds.
a. $FeCl_3$ d. Fe_2O_3 g. MnO_2
b. Cu_2S e. AuI_3 h. CuO
c. $CoBr_2$ f. Cr_2S_3 i. NiS

27. Which of the following formulas are incorrect? Why?
a. $SiNa_4$ d. AgOH g. $Fe(OH)_4$
b. $CaBr_3$ e. NH_4CO_3 h. $CaCl_2$
c. $NaCl_4$ f. Al_3S_2

28. Give the name of each of the following polyatomic ions.
a. NH_4^{+} e. NO_2^{-} h. ClO_4^{-}
b. SO_3^{2-} f. CN^{-} i. ClO^{-}
c. NO_3^{-} g. OH^{-} j. PO_4^{3-}
d. SO_4^{2-}

29. Using the negative polyatomic ions listed in Table 5.4, write formulas for each of their sodium and calcium compounds.

30. Give the name of each of the following compounds.
a. B_2O_3 d. N_2O_4 g. SF_6
b. NO_2 e. P_2O_5 h. N_2O_3
c. PCl_5 f. ICl

31. Write formulas for each of the following compounds.
a. potassium sulfide i. magnesium perchlorate
b. sodium hydride j. diphosphorus pentoxide
c. hydrochloric acid k. nitric acid
d. dinitrogen tetroxide l. silver sulfite
e. aluminum nitrate m. copper(II) bromide
f. calcium sulfate n. barium phosphate
g. hydrosulfuric acid o. gold(III) chloride
h. ammonium acetate p. manganous chloride

6 Chemical Reactions: An Introduction

Lightning over the town of Tamworth in New South Wales, Australia.

Chemistry is about change. Grass grows. Steel rusts. Hair is bleached, dyed, "permed," or straightened. Natural gas burns to heat houses. Nylon is produced for jackets, swimsuits, and pantyhose. Water is decomposed to hydrogen and oxygen gas by an electric current. Grape juice ferments in the production of wine. The bombardier beetle concocts a toxic spray to shoot at its enemies (see "Chemistry in Focus," p. 161).

These are just a few examples of chemical changes that affect each of us. Chemical reactions are the heart and soul of chemistry, and in this chapter we will discuss the fundamental ideas about chemical reactions.

Nylon jackets are sturdy and dry quickly. These characteristics make them ideal for athletic wear.

Production of plastic film for use in containers such as soft drink bottles *(left)*. Nylon being drawn from the boundary between two solutions containing different reactants *(right)*.

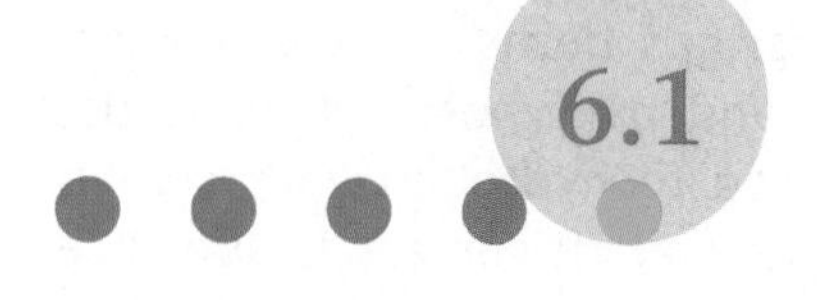

6.1 Evidence for a Chemical Reaction

AIM: To learn the signals that show a chemical reaction has occurred.

Energy and chemical reactions will be discussed in more detail in Chapter 7.

How do we know when a chemical reaction has occurred? That is, what are the clues that a chemical change has taken place? A glance back at the processes in the introduction suggests that *chemical reactions often give a visual signal.* Steel changes from a smooth, shiny material to a reddish-brown, flaky substance when it rusts. Hair changes color when it is bleached. Solid nylon is formed when two particular liquid solutions are brought into contact. A blue flame appears when natural gas reacts with oxygen. Chemical reactions, then, often give *visual* clues: a color changes, a solid forms, bubbles are produced (see Figure 6.1), a flame occurs, and so on. However, reactions are not always visible. Sometimes the only signal that a reaction is occurring is a change in temperature as heat is produced or absorbed (see Figure 6.2).

Table 6.1 summarizes common clues to the occurrence of a chemical reaction, and Figure 6.3 gives some examples of reactions that show these clues.

Figure 6.1
Bubbles of hydrogen and oxygen gas form when an electric current is used to decompose water.

Table 6.1 Some Clues That a Chemical Reaction Has Occurred

1. The color changes.
2. A solid forms.
3. Bubbles form.
4. Heat and/or a flame is produced, or heat is absorbed.

(a)

(b)

Figure 6.2
(a) An injured girl wearing a cold pack to help prevent swelling. The pack is activated by breaking an ampule; this initiates a chemical reaction that absorbs heat rapidly, lowering the temperature of the area to which the pack is applied. (b) A hot pack used to warm hands and feet in winter. When the package is opened, oxygen from the air penetrates a bag containing solid chemicals. The resulting reaction produces heat for several hours.

(a)

(b) (c) (d)

Figure 6.3
(a) When colorless hydrochloric acid is added to a red solution of cobalt(II) nitrate, the solution turns blue, a sign that a chemical reaction has taken place. (b) A solid forms when a solution of sodium dichromate is added to a solution of lead nitrate. (c) Bubbles of hydrogen gas form when calcium metal reacts with water. (d) Methane gas reacts with oxygen to produce a flame in a bunsen burner.

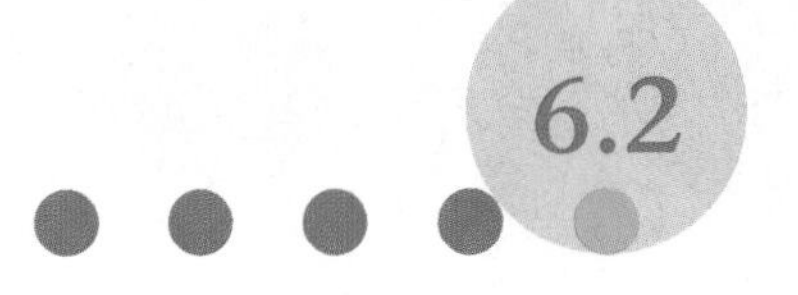

6.2 Chemical Equations

AIM: To learn to identify the characteristics of a chemical reaction and the information given by a chemical equation.

Chemists have learned that a chemical change always involves a rearrangement of the ways in which the atoms are grouped. For example, when the methane, CH_4, in natural gas combines with oxygen, O_2, in the air and burns, carbon dioxide, CO_2, and water, H_2O, are formed. A chemical change such as this is called a **chemical reaction.** We represent a chemical reaction by writing a **chemical equation** in which the chemicals present before the reaction (the **reactants**) are shown to the left of an arrow and the chemicals formed by the reaction (the **products**) are shown to the right of an arrow. The arrow indicates the direction of the change and is read as "yields" or "produces":

$$\text{Reactants} \rightarrow \text{Products}$$

For the reaction of methane with oxygen, we have

$$\underbrace{\overset{\text{Methane}}{CH_4} + \overset{\text{Oxygen}}{O_2}}_{\text{Reactants}} \rightarrow \underbrace{\overset{\text{Carbon dioxide}}{CO_2} + \overset{\text{Water}}{H_2O}}_{\text{Products}}$$

Note from this equation that the products contain the same atoms as the reactants but that the atoms are associated in different ways. That is, a *chemical reaction involves changing the ways the atoms are grouped.*

It is important to recognize that **in a chemical reaction, atoms are neither created nor destroyed.** *All atoms present in the reactants must be accounted for among the products.* In other words, there must be the same number of each type of atom on the product side as on the reactant side of the arrow. Making sure that the equation for a reaction obeys this rule is called **balancing the chemical equation** for a reaction.

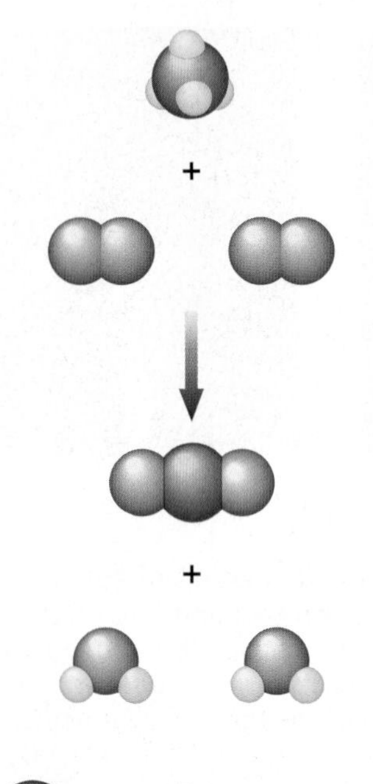

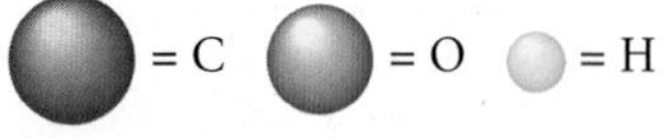

Figure 6.4
The reaction between methane and oxygen to give water and carbon dioxide. Note that there are four oxygen atoms in the products *and* in the reactants; none has been gained or lost in the reaction. Similarly, there are four hydrogen atoms and one carbon atom in the reactants *and* in the products. The reaction simply changes the way the atoms are grouped.

The equation that we have shown for the reaction between CH_4 and O_2 is not balanced. We can see that it is not balanced by taking the reactants and products apart.

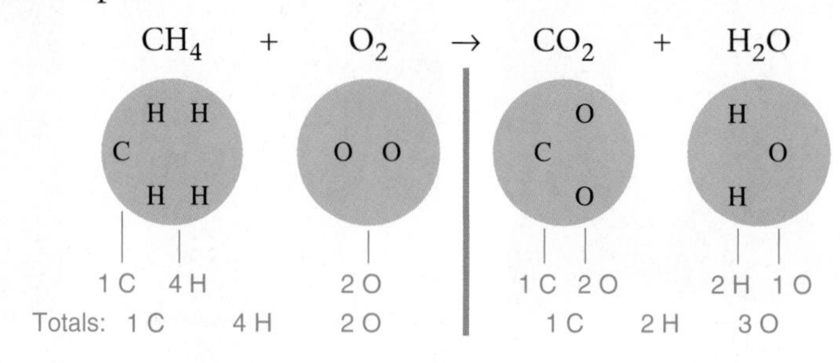

The reaction cannot happen this way because, as it stands, this equation states that one oxygen atom is created and two hydrogen atoms are destroyed. A reaction is only rearrangement of the way the atoms are grouped; atoms are not created or destroyed. The total number of each type of atom must be the same on both sides of the arrow. We can fix the imbalance in this equation by involving one more O_2 molecules on the left and by showing the production of one more H_2O molecules on the right.

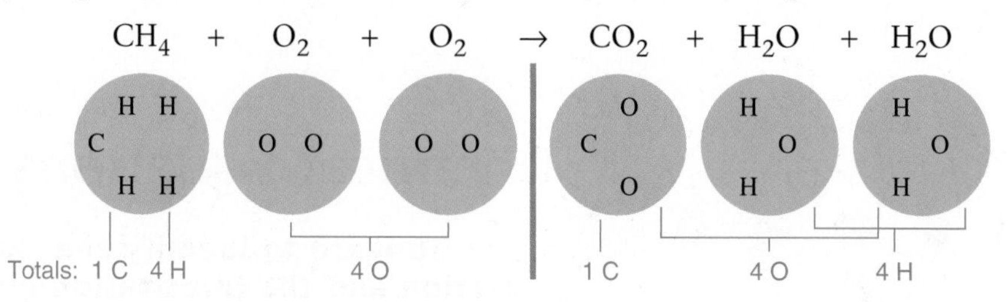

This *balanced chemical equation* shows the actual numbers of molecules involved in this reaction (see Figure 6.4).

When we write the balanced equation for a reaction, we group like molecules together. Thus

$$CH_2 + \boxed{O_2 + O_2} \rightarrow CO_2 + \boxed{H_2O + H_2O}$$

is written

$$CH_4 + \boxed{2O_2} \rightarrow CO_2 + \boxed{2H_2O}$$

The chemical equation for a reaction provides us with two important types of information:

1. The identities of the reactants and products
2. The relative numbers of each

Physical States

Besides specifying the compounds involved in the reaction, we often indicate in the equation the *physical states* of the reactants and products by using the following symbols:

Symbol	State
(*s*)	solid
(*l*)	liquid
(*g*)	gas
(*aq*)	dissolved in water (in aqueous solution)

(a)

(b) (c)

Figure 6.5
The reactants (a) potassium metal (stored in mineral oil to prevent oxidation) and (b) water. (c) The reaction of potassium with water. The flame occurs because the hydrogen gas, $H_2(g)$, produced by the reaction burns in air (reacts with $O_2(g)$) at the high temperatures caused by the reaction.

For example, when solid potassium reacts with liquid water, the products are hydrogen gas and potassium hydroxide; the latter remains dissolved in the water. From this information about the reactants and products, we can write the equation for the reaction. Solid potassium is represented by $K(s)$; liquid water is written as $H_2O(l)$; hydrogen gas contains diatomic molecules and is represented as $H_2(g)$; potassium hydroxide dissolved in water is written as $KOH(aq)$. So the *unbalanced* equation for the reaction is

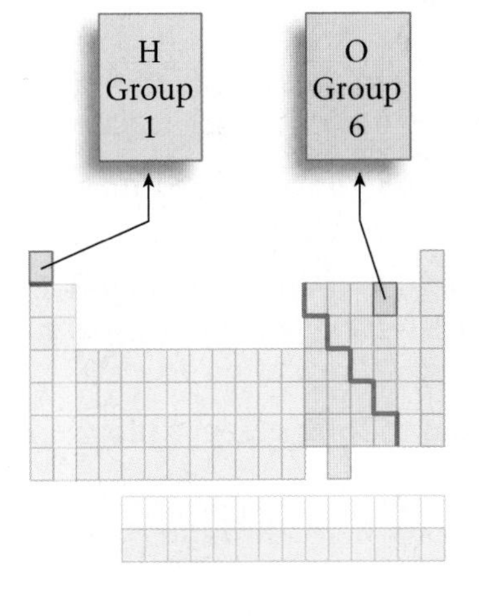

Solid potassium		Water		Hydrogen gas		Potassium hydroxide dissolved in water
$K(s)$	$+$	$H_2O(l)$	$\rightarrow$	$H_2(g)$	$+$	$KOH(aq)$

This reaction is shown in Figure 6.5.

The hydrogen gas produced in this reaction then reacts with the oxygen gas in the air, producing gaseous water and a flame. The *unbalanced* equation for this second reaction is

$$H_2(g) + O_2(g) \rightarrow H_2O(g)$$

Both of these reactions produce a great deal of heat. In Example 6.1 we will practice writing the unbalanced equations for reactions. Then, in the next section, we will discuss systematic procedures for balancing equations.

Example 6.1 Chemical Equations: Recognizing Reactants and Products

Write the *unbalanced* chemical equation for each of the following reactions.

a. Solid mercury(II) oxide decomposes to produce liquid mercury metal and gaseous oxygen.

b. Solid carbon reacts with gaseous oxygen to form gaseous carbon dioxide.

c. Solid zinc is added to an aqueous solution containing dissolved hydrogen chloride to produce gaseous hydrogen that bubbles out of the solution and zinc chloride that remains dissolved in the water.

Zinc metal reacts with hydrochloric acid to produce bubbles of hydrogen gas.

Solution

a. In this case we have only one reactant, mercury(II) oxide. The name mercury(II) oxide means that the Hg^{2+} cation is present, so one O^{2-} ion is required for a zero net charge. Thus the formula is HgO, which is written HgO(*s*) in this case because it is given as a solid. The products are liquid mercury, written Hg(*l*), and gaseous oxygen, written $O_2(g)$. (Remember that oxygen exists as a diatomic molecule under normal conditions.) The unbalanced equation is

$$\underbrace{HgO(s)}_{\text{Reactant}} \rightarrow \underbrace{Hg(l) + O_2(g)}_{\text{Products}}$$

b. In this case, solid carbon, written C(*s*), reacts with oxygen gas, $O_2(g)$, to form gaseous carbon dioxide, which is written $CO_2(g)$. The equation (which happens to be balanced) is

$$\underbrace{C(s) + O_2(g)}_{\text{Reactants}} \rightarrow \underbrace{CO_2(g)}_{\text{Product}}$$

c. In this reaction solid zinc, Zn(*s*), is added to an aqueous solution of hydrogen chloride, which is written HCl(*aq*) and called hydrochloric acid. These are the reactants. The products of the reaction are gaseous hydrogen, $H_2(g)$, and aqueous zinc chloride. The name zinc chloride means that the Zn^{2+} ion is present, so two Cl^- ions are needed to achieve a zero net charge. Thus zinc chloride dissolved in water is written $ZnCl_2(aq)$. The unbalanced equation for the reaction is

$$\underbrace{Zn(s) + HCl(aq)}_{\text{Reactants}} \rightarrow \underbrace{H_2(g) + ZnCl_2(aq)}_{\text{Products}}$$

Because Zn forms only the Zn^{2+} ion, a Roman numeral is usually not used. Thus $ZnCl_2$ is commonly called zinc chloride.

✔ Self-Check Exercise 6.1

Identify the reactants and products and write the *unbalanced* equation (including symbols for states) for each of the following chemical reactions.

a. Solid magnesium metal reacts with liquid water to form solid magnesium hydroxide and hydrogen gas.

b. Solid ammonium dichromate (review Table 5.4 if this compound is unfamiliar) decomposes to solid chromium(III) oxide, gaseous nitrogen, and gaseous water.

c. Gaseous ammonia reacts with gaseous oxygen to form gaseous nitrogen monoxide and gaseous water.

See Problems 6.13 through 6.34. ■

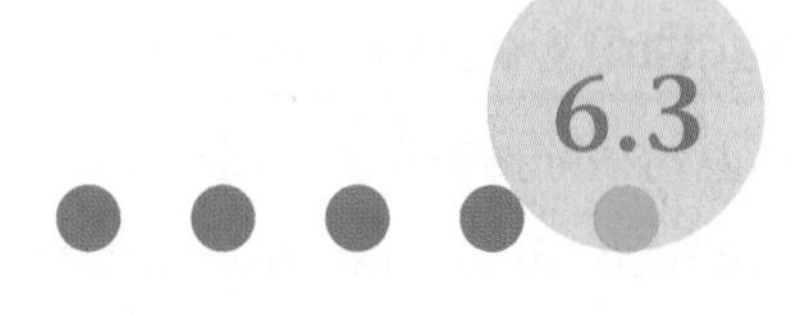

Balancing Chemical Equations

AIM: To learn how to write a balanced equation for a chemical reaction.

As we saw in the previous section, an unbalanced chemical equation is not an accurate representation of the reaction that occurs. Whenever you see an equation for a reaction, you should ask yourself whether it is balanced. The principle that lies at the heart of the balancing process is that **atoms are conserved in a chemical reaction.** That is, atoms are neither created nor destroyed. They are just grouped differently. The same number of each type of atom is found among the reactants and among the products.

Chemists determine the identity of the reactants and products of a reaction by experimental observation. For example, when methane (natural gas) is burned in the presence of sufficient oxygen gas, the products are always carbon dioxide and water. **The identities (formulas) of the compounds must never be changed in balancing a chemical equation.** In other words, the subscripts in a formula cannot be changed, nor can atoms be added to or subtracted from a formula.

Trial and error is often useful for solving problems. It's okay to make a few wrong turns before you get to the right answer.

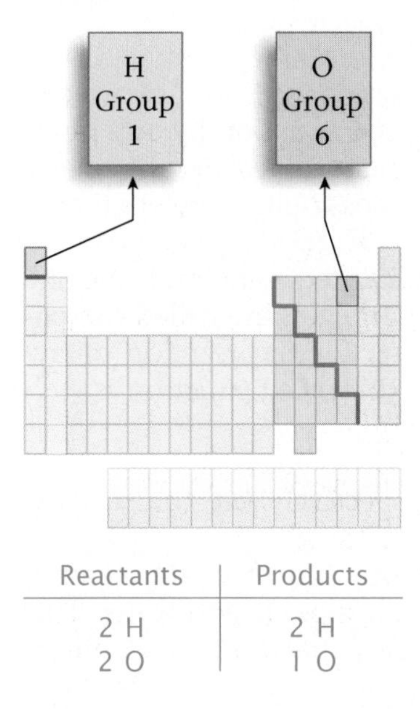

Most chemical equations can be balanced by trial and error—that is, by inspection. Keep trying until you find the numbers of reactants and products that give the same number of each type of atom on both sides of the arrow. For example, consider the reaction of hydrogen gas and oxygen gas to form liquid water. First, we write the unbalanced equation from the description of the reaction.

$$H_2(g) + O_2(g) \rightarrow H_2O(l)$$

We can see that this equation is unbalanced by counting the atoms on both sides of the arrow.

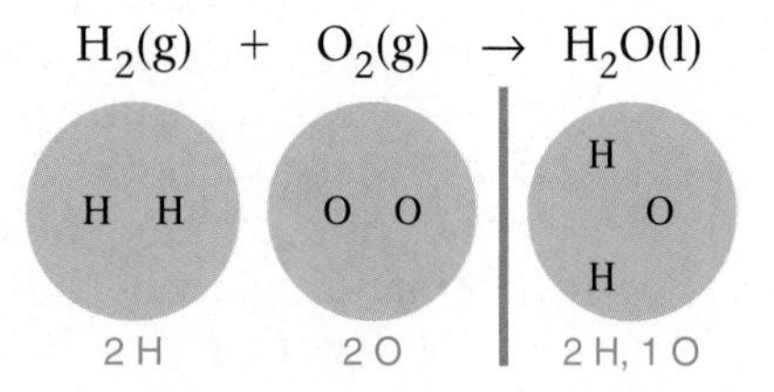

Reactants	Products
2 H	2 H
2 O	1 O

We have one more oxygen atom in the reactants than in the products. Because we cannot create or destroy atoms and because we *cannot change the formulas* of the reactants or products, we must balance the equation by adding more molecules of reactants and/or products. In this case we need one more oxygen atom on the right, so we add another water molecule (which contains one O atom). Then we count all of the atoms again.

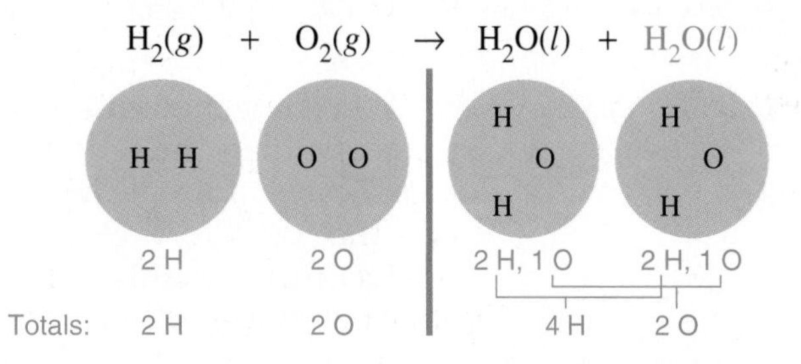

Reactants	Products
2 H	4 H
2 O	2 O

We have balanced the oxygen atoms, but now the hydrogen atoms have become unbalanced. There are more hydrogen atoms on the right than on the left. We can solve this problem by adding another hydrogen molecule (H_2) to the reactant side.

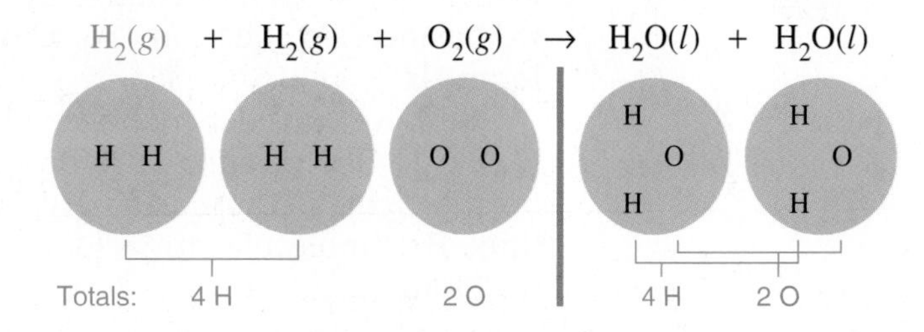

Reactants	Products
4 H	4 H
2 O	2 O

The equation is now balanced. We have the same numbers of hydrogen and oxygen atoms represented on both sides of the arrow. Collecting like molecules, we write the balanced equation as

$$2H_2(g) + O_2(g) \rightarrow 2H_2O(l)$$

Consider next what happens if we multiply every part of this balanced equation by 2:

$$2 \times [2H_2(g) + O_2(g) \rightarrow 2H_2O(l)]$$

to give

$$4H_2(g) + 2O_2(g) \rightarrow 4H_2O(l)$$

This equation is balanced (count the atoms to verify this). In fact, we can multiply or divide *all parts* of the original balanced equation by any number to give a new balanced equation. Thus each chemical reaction has many possible balanced equations. Is one of the many possibilities preferred over the others? Yes.

The accepted convention is that the "best" balanced equation is the one with the *smallest integers (whole numbers)*. These integers are called the **coefficients** for the balanced equation. Therefore, for the reaction of hydrogen and oxygen to form water, the "correct" balanced equation is

$$2H_2(g) + O_2(g) \rightarrow 2H_2O(l)$$

The coefficients 2, 1 (never written), and 2, respectively, are the smallest *integers* that give a balanced equation for this reaction.

Next we will balance the equation for the reaction of liquid ethanol, C_2H_5OH, with oxygen gas to form gaseous carbon dioxide and water. This reaction, among many others, occurs in engines that burn a gasoline–ethanol mixture called gasohol.

The first step in obtaining the balanced equation for a reaction is always to identify the reactants and products from the description given for the reaction. In this case we are told that liquid ethanol, $C_2H_5OH(l)$, reacts with gaseous oxygen, $O_2(g)$, to produce gaseous carbon dioxide, $CO_2(g)$, and gaseous water, $H_2O(g)$. Therefore, the unbalanced equation is

$C_2H_5OH(l)$	+	$O_2(g)$	$\rightarrow$	$CO_2(g)$	+	$H_2O(g)$
Liquid ethanol		Gaseous oxygen		Gaseous carbon dioxide		Gaseous water

In balancing equations, start by looking at the most complicated molecule.

When one molecule in an equation is more complicated (contains more elements) than the others, it is best to start with that molecule. The most complicated molecule here is C_2H_5OH, so we begin by considering the products that contain the atoms in C_2H_5OH. We start with carbon. The only product that contains carbon is CO_2. Because C_2H_5OH contains two carbon atoms, we place a 2 before the CO_2 to balance the carbon atoms.

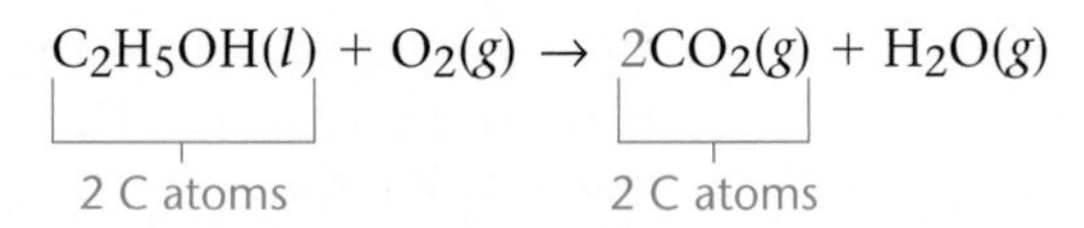

C_2H_5OH

2 C, 6H, 1 O

Remember, we cannot change the formula of any reactant or product when we balance an equation. We can only place coefficients in front of the formulas.

Next we consider hydrogen. The only product containing hydrogen is H_2O. C_2H_5OH contains six hydrogen atoms, so we need six hydrogen atoms on the right. Because each H_2O contains two hydrogen atoms, we need three H_2O molecules to yield six hydrogen atoms. So we place a 3 before the H_2O.

$$C_2H_5OH(l) + O_2(g) \rightarrow 2CO_2(g) + 3H_2O(g)$$

(5 + 1) H → 6 H (reactant side); (3 × 2) H → 6 H (product side)

CHEMISTRY IN FOCUS

The Beetle That Shoots Straight

If someone said to you, "Name something that protects itself by spraying its enemies," your answer would almost certainly be "a skunk." Of course, you would be correct, but there is another correct answer—the bombardier beetle. When threatened, this beetle shoots a boiling stream of toxic chemicals at its enemy. How does this clever beetle accomplish this? Obviously, the boiling mixture cannot be stored inside the beetle's body all the time. Instead, when endangered, the beetle mixes chemicals that produce the hot spray. The chemicals involved are stored in two compartments. One compartment contains the chemicals hydrogen peroxide (H_2O_2) and methylhydroquinone ($C_7H_8O_2$). The key reaction is the decomposition of hydrogen peroxide to form oxygen gas and water:

$$2H_2O_2(aq) \rightarrow 2H_2O(l) + O_2(g)$$

Hydrogen peroxide also reacts with the hydroquinones to produce other compounds that become part of the toxic spray.

However, none of these reactions occurs very fast unless certain enzymes are present. (Enzymes are natural substances that speed up biological reactions by means we will not discuss here.) When the beetle mixes the hydrogen peroxide and hydroquinones with the enzyme, the decomposition of H_2O_2 occurs rapidly, producing a hot mixture pressurized by the formation of oxygen gas. When the gas pressure becomes high enough, the hot spray is ejected in one long stream or in short bursts. The beetle has a highly accurate aim and can shoot several attackers with one batch of spray.

A bombardier beetle defending itself.

O—C—O	H—O—H
O—C—O	H—O—H
	H—O—H
4 O atoms	3 O atoms

Finally, we count the oxygen atoms. On the left we have three oxygen atoms (one in C_2H_5OH and two in O_2), and on the right we have seven oxygen atoms (four in $2CO_2$ and three in $3H_2O$). We can correct this imbalance if we have three O_2 molecules on the left. That is, we place a coefficient of 3 before the O_2 to produce the balanced equation.

$$C_2H_5OH(l) + 3O_2(g) \rightarrow 2CO_2(g) + 3H_2O(g)$$

1 O, (3 × 2) O → 7 O; (2 × 2) O, 3 O → 7 O

At this point you may have a question: why did we choose O_2 on the left when we balanced the oxygen atoms? Why not use C_2H_5OH, which has an oxygen atom? The answer is that if we had changed the coefficient in front of C_2H_5OH, we would have unbalanced the hydrogen and carbon atoms. Now we count all of the atoms as a check to make sure the equation is balanced.

$$C_2H_5OH(l) + 3O_2(g) \rightarrow 2CO_2(g) + 3H_2O(g)$$

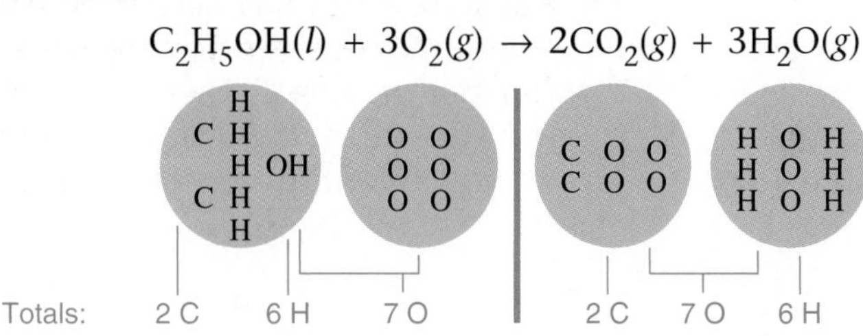

Reactants	Products
2 C	2 C
6 H	6 H
7 O	7 O

The equation is now balanced. We have the same numbers of all types of atoms on both sides of the arrow. Notice that these coefficients are the smallest integers that give a balanced equation.

The process of writing and balancing the equation for a chemical reaction consists of several steps:

How to Write and Balance Equations

STEP 1 Read the description of the chemical reaction. What are the reactants, the products, and their states? Write the appropriate formulas.

STEP 2 Write the *unbalanced* equation that summarizes the information from step 1.

STEP 3 Balance the equation by inspection, starting with the most complicated molecule. Proceed element by element to determine what coefficients are necessary so that the same number of each type of atom appears on both the reactant side and the product side. Do not change the identities (formulas) of any of the reactants or products.

STEP 4 Check to see that the coefficients used give the same number of each type of atom on both sides of the arrow. (Note that an "atom" may be present in an element, a compound, or an ion.) Also check to see that the coefficients used are the smallest integers that give the balanced equation. This can be done by determining whether all coefficients can be divided by the same integer to give a set of smaller *integer* coefficients.

Example 6.2 Balancing Chemical Equations I

For the following reaction, write the unbalanced equation and then balance the equation: solid potassium reacts with liquid water to form gaseous hydrogen and potassium hydroxide that dissolves in the water.

Solution

STEP 1 From the description given for the reaction, we know that the reactants are solid potassium, $K(s)$, and liquid water, $H_2O(l)$. The products are gaseous hydrogen, $H_2(g)$, and dissolved potassium hydroxide, $KOH(aq)$.

STEP 2 The unbalanced equation for the reaction is

$$K(s) + H_2O(l) \rightarrow H_2(g) + KOH(aq)$$

STEP 3 Although none of the reactants or products is very complicated, we will start with KOH because it contains the most elements (three). We will arbitrarily consider hydrogen first. Note that on the reactant side of the equation in step 2, there are two hydrogen atoms but on the product side there are three. If we place a coefficient of 2 in front of both H_2O and KOH, we now have four H atoms on each side.

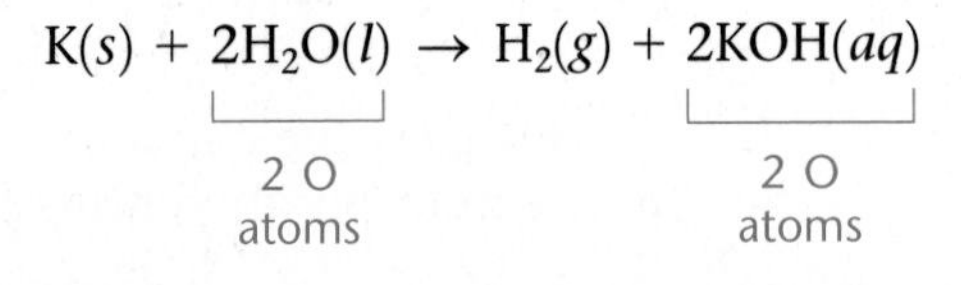

Also note that the oxygen atoms balance.

$$K(s) + 2H_2O(l) \rightarrow H_2(g) + 2KOH(aq)$$

2 O atoms (reactants) — 2 O atoms (products)

However, the K atoms do not balance; we have one on the left and two on the right. We can fix this easily by placing a coefficient of 2 in front of K(s) to give the balanced equation:

$$2K(s) + 2H_2O(l) \rightarrow H_2(g) + 2KOH(aq)$$

STEP 4

CHECK: There are 2 K, 4 H, and 2 O on both sides of the arrow, and the coefficients are the smallest integers that give a balanced equation. We know this because we cannot divide through by a given integer to give a set of smaller *integer* (whole-number) coefficients. For example, if we divide all of the coefficients by 2, we get

Reactants	Products
2 K	2 K
4 H	4 H
2 O	2 O

$$K(s) + H_2O(l) \rightarrow \tfrac{1}{2}H_2(g) + KOH(aq)$$

This is not acceptable because the coefficient for H_2 is not an integer. ■

Example 6.3 Balancing Chemical Equations II

Under appropriate conditions at 1000 °C, ammonia gas reacts with oxygen gas to produce gaseous nitrogen monoxide (common name, nitric oxide) and gaseous water. Write the unbalanced and balanced equations for this reaction.

Solution

STEP 1 The reactants are gaseous ammonia, $NH_3(g)$, and gaseous oxygen, $O_2(g)$. The products are gaseous nitrogen monoxide, $NO(g)$, and gaseous water, $H_2O(g)$.

STEP 2 The unbalanced equation for the reaction is

$$NH_3(g) + O_2(g) \rightarrow NO(g) + H_2O(g)$$

Reactants	Products
1 N	1 N
3 H	2 H
2 O	2 O

STEP 3 In this equation there is no molecule that is obviously the most complicated. Three molecules contain two elements, so we arbitrarily start with NH_3. We arbitrarily begin by looking at hydrogen. A coefficient of 2 for NH_3 and a coefficient of 3 for H_2O give six atoms of hydrogen on both sides.

$$2NH_3(g) + O_2(g) \rightarrow NO(g) + 3H_2O(g)$$

6 H (reactants) — 6 H (products)

We can balance the nitrogen by giving NO a coefficient of 2.

$$2NH_3(g) + O_2(g) \rightarrow 2NO(g) + 3H_2O(g)$$

2 N (reactants) — 2 N (products)

$\frac{5}{2} = 2\frac{1}{2}$

O—O
O—O
O+O
$2\frac{1}{2}\ O_2$ contains 5 O atoms

Finally, we note that there are two atoms of oxygen on the left and five on the right. The oxygen can be balanced with a coefficient of $\frac{5}{2}$ for O_2, because $\frac{5}{2} \times O_2$ gives five oxygen atoms.

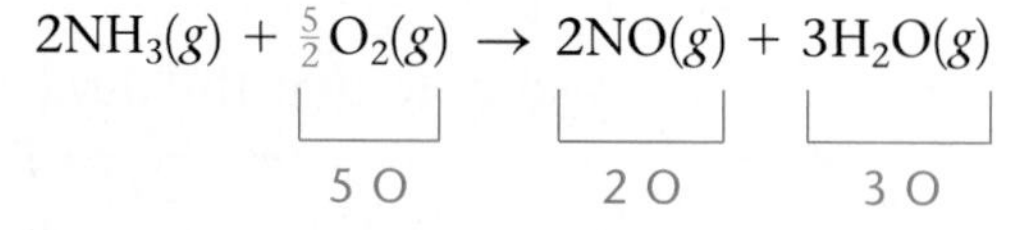

$$2NH_3(g) + \tfrac{5}{2}O_2(g) \rightarrow 2NO(g) + 3H_2O(g)$$

However, the convention is to have integer (whole-number) coefficients, so we multiply the entire equation by 2.

$$2 \times [2NH_3(g) + \tfrac{5}{2}O_2(g) \rightarrow 2NO(g) + 3H_2O(g)]$$

or

$$2 \times 2NH_3(g) + 2 \times \tfrac{5}{2}O_2(g) \rightarrow 2 \times 2NO(g) + 2 \times 3H_2O(g)$$
$$4NH_3(g) + 5O_2(g) \rightarrow 4NO(g) + 6H_2O(g)$$

Step 4

Check: There are 4 N, 12 H, and 10 O atoms on both sides, so the equation is balanced. These coefficients are the smallest integers that give a balanced equation. That is, we cannot divide all coefficients by the same integer and obtain a smaller set of *integers*.

Reactants	Products
4 N	4 N
12 H	12 H
10 O	10 O

✔ Self-Check Exercise 6.2

Propane, C_3H_8, a liquid at 25 °C under high pressure, is often used for gas grills and as a fuel in rural areas where there is no natural gas pipeline. When liquid propane is released from its storage tank, it changes to propane gas that reacts with oxygen gas (it "burns") to give gaseous carbon dioxide and gaseous water. Write and balance the equation for this reaction.

Hint: This description of a chemical process contains many words, some of which are crucial to solving the problem and some of which are not. First sort out the important information and use symbols to represent it.

See Problems 6.37 through 6.44. ■

Example 6.4 Balancing Chemical Equations III

Decorations on glass are produced by etching with hydrofluoric acid.

Glass is sometimes decorated by etching patterns on its surface. Etching occurs when hydrofluoric acid (an aqueous solution of HF) reacts with the silicon dioxide in the glass to form gaseous silicon tetrafluoride and liquid water. Write and balance the equation for this reaction.

Solution

Step 1 From the description of the reaction we can identify the reactants:

hydrofluoric acid	$HF(aq)$
solid silicon dioxide	$SiO_2(s)$

and the products:

gaseous silicon tetrafluoride	$SiF_4(g)$
liquid water	$H_2O(l)$

Step 2 The unbalanced equation is

$$SiO_2(s) + HF(aq) \rightarrow SiF_4(g) + H_2O(l)$$

Reactants	Products
1 Si	1 Si
1 H	2 H
1 F	4 F
2 O	1 O

STEP 3 There is no clear choice here for the most complicated molecule. We arbitrarily start with the elements in SiF_4. The silicon is balanced (one atom on each side), but the fluorine is not. To balance the fluorine, we need a coefficient of 4 before the HF.

$$SiO_2(s) + 4HF(aq) \rightarrow SiF_4(g) + H_2O(l)$$

Reactants	Products
1 Si	1 Si
4 H	2 H
4 F	4 F
2 O	1 O

Hydrogen and oxygen are not balanced. Because we have four hydrogen atoms on the left and two on the right, we place a 2 before the H_2O:

$$SiO_2(s) + 4HF(aq) \rightarrow SiF_4(g) + 2H_2O(l)$$

This balances the hydrogen *and* the oxygen (two atoms on each side).

Reactants	Products
1 Si	1 Si
4 H	4 H
4 F	4 F
2 O	2 O

STEP 4

CHECK: $SiO_2(s) + 4HF(aq) \rightarrow SiF_4(g) + 2H_2O(l)$

Totals: 1 Si, 2 O, 4 H, 4 F → 1 Si, 4 F, 4 H, 2 O

All atoms check, so the equation is balanced.

✔ Self-Check Exercise 6.3

If you are having trouble writing formulas from names, review the appropriate sections of Chapter 5. It is very important that you are able to do this.

Give the balanced equation for each of the following reactions.

a. When solid ammonium nitrite is heated, it produces nitrogen gas and water vapor.

b. Gaseous nitrogen monoxide (common name, nitric oxide) decomposes to produce dinitrogen monoxide gas (common name, nitrous oxide) and nitrogen dioxide gas.

c. Liquid nitric acid decomposes to reddish-brown nitrogen dioxide gas, liquid water, and oxygen gas. (This is why bottles of nitric acid become yellow upon standing.)

See Problems 6.37 through 6.44. ■

CHAPTER 6 REVIEW

KEY TERMS

chemical reaction (6.2)
chemical equation (6.2)
reactant (6.2)
product (6.2)
balancing a chemical equation (6.2)
coefficient (6.3)

SUMMARY

1. Chemical reactions usually give some kind of visual signal—a color changes, a solid forms, bubbles form, heat and/or flame is produced.

2. A chemical equation represents a chemical reaction. Reactants are shown on the left side of an arrow and products on the right. In a chemical reaction, atoms are neither created nor destroyed; they are merely rearranged. A balanced chemical equation gives the relative numbers of reactant and product molecules.

3. A chemical equation for a reaction can be balanced by using a systematic approach. First identify the reactants and products and write the formulas. Next write the unbalanced equation. Then balance by trial and error, starting with the most complicated molecule(s). Finally, check to be sure the equation is balanced.

In-Class Discussion Questions

These questions are designed to be considered by groups of students in class. Often these questions work well for introducing a particular topic in class.

1. The following are actual student responses to the question: Why is it necessary to balance chemical equations?
   a. The chemicals will not react until you have added the correct ratios.
   b. The correct products will not form unless the right amounts of reactants have been added.
   c. A certain number of products cannot form without a certain number of reactants.
   d. The balanced equation tells you how much reactant you need, and allows you to predict how much product you will make.
   e. A ratio must be established for the reaction to occur as written.

   Justify the best choice, and for choices you did not pick, explain what is wrong with them.
2. What information do we get from a formula? From an equation?
3. Given the equation for the reaction: $N_2 + H_2 \rightarrow NH_3$, draw a molecular diagram that represents the reaction (make sure it is balanced).
4. What do the subscripts in a chemical formula represent? What do the coefficients in a balanced chemical equation represent?
5. Can the subscripts in a chemical formula be fractions? Explain.
6. Can the coefficients in a balanced chemical equation be fractions? Explain.
7. Changing the subscripts of chemicals can mathematically balance the equations. Why is this unacceptable?
8. Table 6.1 lists some clues that a chemical reaction has occurred. However, these events do not necessarily prove the existence of a chemical change. Give an example for each of the clues that is not a chemical reaction but a physical change.
9. Use molecular-level drawings to show the difference between physical and chemical changes.

Questions and Problems

All even-numbered exercises have answers in the back of this book and solutions in the *Solutions Guide.*

6.1 Evidence for a Chemical Reaction

QUESTIONS

1. How do we *know* when a chemical reaction is taking place? Can you think of an example of how each of the five senses (sight, hearing, taste, touch, smell) might be used in detecting when a chemical reaction has taken place?
2. When home-made biscuits are baked, baking powder is added to make the biscuits "rise." What evidence suggests that baking powder works through means of a chemical reaction?
3. If you have had a clogged sink drain at your home, you have undoubtedly tried using a commercial drain cleaner to dissolve the clog. What evidence is there that such drain cleaners work by chemical reaction?
4. Small cuts and abrasions on the skin are frequently cleaned using hydrogen peroxide solution. What evidence is there that treating a wound with hydrogen peroxide causes a chemical reaction to take place?
5. You have probably had the unpleasant experience of discovering that a flashlight battery has gotten old and begun to leak. Is there evidence that this change is due to a chemical reaction?
6. If a bottle of wine is left open to the air, eventually it turns into vinegar. Is there evidence that this change represents a chemical reaction?

6.2 Chemical Equations

QUESTIONS

7. We write chemical equations to describe a reaction in "shorthand." Where are the *reactants* found in an equation? (left/right) Where are the products? (left/right) What does the arrow (→) signify?
8. In an ordinary chemical reaction, ________ are neither created nor destroyed.
9. In a chemical reaction, the total number of atoms present after the reaction is complete is (larger than/smaller than/the same as) the total number of atoms present before the reaction began.
10. Balancing an equation for a reaction ensures that the number of each type of atom is ________ on both sides of the equation.
11. How do we indicate the physical state of a substance when we write a chemical equation for a reaction?
12. In a chemical equation for a reaction, the notation "(*aq*)" after a substance's formula means that the substance is dissolved in ________.

PROBLEMS

Note: In some of the following problems you will need to write a chemical formula from the name of the compound. Review Chapter 5 if you are having trouble.

13. Pennies in the United States now consist of a zinc disk that is coated with a thin layer of copper. If a penny is scratched and then soaked in hydrochloric acid, it is possible to dissolve the zinc disk, leaving only a thin, hollow shell of copper. Write an unbalanced chemical equation to illustrate the reaction of zinc

metal with hydrochloric acid, which produces dissolved zinc chloride and evolves gaseous hydrogen.

14. Liquefied propane gas is often used for cooking in suburban areas away from natural gas lines. Propane (C_3H_8) burns in oxygen gas, producing carbon dioxide gas, water vapor, and heat. Write the unbalanced chemical equation for this process.

15. If a sample of pure hydrogen gas is ignited very carefully, the hydrogen burns gently, combining with the oxygen gas of the air to form water vapor. Write the unbalanced chemical equation for this reaction.

16. If a patient experiencing digestive problems needs a series of gastrointestinal X rays taken, they frequently must drink what is called a "barium cocktail." It consists of a thick suspension of barium sulfate in water, which is used because barium sulfate is opaque to X rays, and the thick suspension coats the walls of the digestive tract and makes them more visible on the X rays. Barium sulfate may be prepared in the laboratory by mixing a solution of barium chloride with a solution of sodium sulfate, which produces solid barium sulfate suspended in a solution of sodium chloride. Write the unbalanced chemical equation for this process.

17. Although silver is not easily attacked by acids (and for that reason it has been used to make jewelry and household items), it will dissolve in concentrated nitric acid, producing brown NO gas and leaving a solution of silver nitrate in water. Write the unbalanced chemical equation for this process.

18. Your family may have a "gas grill" for outdoor cooking. Gas grills typically use bottled propane gas (C_3H_8), which burns in air (oxygen) to produce carbon dioxide gas and water vapor. Write the unbalanced chemical equation for this process. Gas grills should never be used indoors, however, because if the supply of oxygen is restricted, the products of the reaction tend to be water vapor and toxic carbon monoxide, instead of nontoxic carbon dioxide. Write the unbalanced chemical equation for this process.

19. Elemental boron is produced in one industrial process by heating diboron trioxide with magnesium metal, also producing magnesium oxide as a by-product. Write the unbalanced chemical equation for this process.

20. Many over-the-counter antacid tablets are now formulated using calcium carbonate as the active ingredient, which enables such tablets to also be used as dietary calcium supplements. As an antacid for gastric hyperacidity, calcium carbonate reacts by combining with hydrochloric acid found in the stomach, producing a solution of calcium chloride, converting the stomach acid to water, and releasing carbon dioxide gas (which the person suffering from stomach problems may feel as a "burp"). Write the unbalanced chemical equation for this process.

21. Phosphorus trichloride is used in the manufacture of certain pesticides, and may be synthesized by direct combination of its constituent elements. Write the unbalanced chemical equation for this process.

22. Pure silicon, which is needed in the manufacturing of electronic components, may be prepared by heating silicon dioxide (sand) with carbon at high temperatures, releasing carbon monoxide gas. Write the unbalanced chemical equation for this process.

23. Nitrous oxide gas (systematic name: dinitrogen monoxide) is used by some dental practitioners as an anesthetic. Nitrous oxide (and water vapor as by-product) can be produced in small quantities in the laboratory by careful heating of ammonium nitrate. Write the unbalanced chemical equation for this reaction.

24. Hydrogen sulfide gas is responsible for the odor of rotten eggs. Hydrogen sulfide burns in air, producing sulfur dioxide gas and water vapor. Write the unbalanced chemical equation for this process.

25. Acetylene gas (C_2H_2) is often used by plumbers, welders, and glass blowers because it burns in oxygen with an intensely hot flame. The products of the combustion of acetylene are carbon dioxide and water vapor. Write the unbalanced chemical equation for this process.

26. If ferric oxide is heated strongly in a stream of carbon monoxide gas, it produces molten elemental iron and carbon dioxide gas. Write the unbalanced chemical equation for this process.

27. The Group 2 metals (Ba, Ca, Sr) can be produced in the elemental state by the reaction of their oxides with aluminum metal at high temperatures, also producing solid aluminum oxide as a by-product. Write the unbalanced chemical equations for the reactions of barium oxide, calcium oxide, and strontium oxide with aluminum.

28. Ozone gas is a form of elemental oxygen containing molecules with *three* oxygen atoms, O_3. Ozone is produced from atmospheric oxygen gas, O_2, by the high-energy outbursts found in lightning storms. Write the unbalanced equation for the formation of ozone gas from oxygen gas.

29. Carbon tetrachloride was widely used for many years as a solvent until its harmful properties became well established. Carbon tetrachloride may be prepared by the reaction of natural gas (methane, CH_4) and elemental chlorine gas in the presence of ultraviolet light. Write the unbalanced chemical equation for this process.

30. Ammonium nitrate is used as a "high-nitrogen" fertilizer, despite the fact that it is quite explosive if not handled carefully. Ammonium nitrate can be synthesized by the reaction of ammonia gas and

nitric acid. Write the unbalanced chemical equation for this process.

31. Calcium oxide is sometimes very challenging to store in the chemistry laboratory. This compound reacts with moisture in the air and is converted to calcium hydroxide. If a bottle of calcium oxide is left on the shelf too long, it gradually absorbs moisture from the humidity in the laboratory. Eventually the bottle cracks and spills the calcium hydroxide that has been produced. Write the unbalanced chemical equation for this process.

32. Although they were formerly called the inert gases, the heavier elements of Group 8 do form relatively stable compounds. For example, at high temperatures in the presence of an appropriate catalyst, xenon gas will combine directly with fluorine gas to produce solid xenon tetrafluoride. Write the unbalanced chemical equation for this process.

33. Ammonium nitrate is a high explosive if not handled carefully, breaking down into nitrogen gas, oxygen gas, and water vapor. The expansion of the three gases produced yields the explosive force in this case. Write the unbalanced chemical equation for this process.

34. Hydrogen peroxide, H_2O_2, has many applications. For example, it is used as an antiseptic and also as a bleach during hair-coloring. Hydrogen peroxide may be prepared by the reaction of barium peroxide with sulfuric acid, which produces a precipitate of barium sulfate (which may be easily filtered) and a concentrated solution of hydrogen peroxide. Write the unbalanced chemical equation for this process.

6.3 Balancing Chemical Equations

QUESTIONS

35. When balancing chemical equations, why is it permitted to change the numbers *in front of* a chemical formula (coefficients) but not a number *within* a chemical formula (subscripts)?

36. After balancing a chemical equation, we ordinarily make sure that the coefficients are the smallest ________ possible.

PROBLEMS

37. Balance each of the following chemical equations.
   a. $FeCl_3(aq) + KOH(aq) \rightarrow Fe(OH)_3(s) + KCl(aq)$
   b. $Pb(C_2H_3O_2)_2(aq) + KI(aq) \rightarrow PbI_2(s) + KC_2H_3O_2(aq)$
   c. $P_4O_{10}(s) + H_2O(l) \rightarrow H_3PO_4(aq)$
   d. $Li_2O(s) + H_2O(l) \rightarrow LiOH(aq)$
   e. $MnO_2(s) + C(s) \rightarrow Mn(s) + CO_2(g)$
   f. $Sb(s) + Cl_2(g) \rightarrow SbCl_3(s)$
   g. $CH_4(g) + H_2O(g) \rightarrow CO(g) + H_2(g)$
   h. $FeS(s) + HCl(aq) \rightarrow FeCl_2(aq) + H_2S(g)$

38. Balance each of the following chemical equations.
   a. $H_2O_2(aq) \rightarrow H_2O(l) + O_2(g)$
   b. $Ag(s) + H_2S(g) \rightarrow Ag_2S(s) + H_2(g)$
   c. $FeO(s) + C(s) \rightarrow Fe(l) + CO_2(g)$
   d. $Cl_2(g) + KI(aq) \rightarrow KCl(aq) + I_2(s)$
   e. $Na_2B_4O_7(s) + H_2SO_4(aq) + H_2O(l) \rightarrow H_3BO_3(s) + Na_2SO_4(aq)$
   f. $CaC_2(s) + H_2O(l) \rightarrow Ca(OH)_2(s) + C_2H_2(g)$
   g. $NaCl(s) + H_2SO_4(l) \rightarrow HCl(g) + Na_2SO_4(s)$
   h. $SiO_2(s) + C(s) \rightarrow Si(l) + CO(g)$

39. Balance each of the following chemical equations.
   a. $Br_2(l) + KI(aq) \rightarrow KBr(aq) + I_2(s)$
   b. $K_2O_2(s) + H_2O(l) \rightarrow KOH(aq) + O_2(g)$
   c. $LiOH(s) + CO_2(g) \rightarrow Li_2CO_3(s) + H_2O(g)$
   d. $K_2CO_3(s) + HNO_3(aq) \rightarrow KNO_3(aq) + H_2O(l) + CO_2(g)$
   e. $LiAlH_4(s) + AlCl_3(s) \rightarrow AlH_3(s) + LiCl(s)$
   f. $Mg(s) + H_2S(g) \rightarrow MgS(s) + H_2(g)$
   g. $Na_2SO_4(s) + C(s) \rightarrow Na_2S(s) + CO_2(g)$
   h. $NaCl(s) + H_2SO_4(l) \rightarrow Na_2SO_4(s) + HCl(g)$

40. Balance each of the following chemical equations.
   a. $CaF_2(s) + H_2SO_4(l) \rightarrow CaSO_4(s) + HF(g)$
   b. $KBr(s) + H_3PO_4(aq) \rightarrow K_3PO_4(aq) + HBr(g)$
   c. $TiCl_4(l) + Na(s) \rightarrow NaCl(s) + Ti(s)$
   d. $K_2CO_3(s) \rightarrow K_2O(s) + CO_2(g)$
   e. $KO_2(s) + H_2O(l) \rightarrow KOH(aq) + O_2(g)$
   f. $Na_2O_2(s) + H_2O(g) + CO_2(g) \rightarrow NaHCO_3(s) + O_2(g)$
   g. $KNO_2(s) + C(s) \rightarrow K_2CO_3(s) + CO(g) + N_2(g)$
   h. $BaO(s) + Al(s) \rightarrow Al_2O_3(s) + Ba(s)$

41. Balance each of the following chemical equations.
   a. $Li(s) + Cl_2(g) \rightarrow LiCl(s)$
   b. $Ba(s) + N_2(g) \rightarrow Ba_3N_2(s)$
   c. $NaHCO_3(s) \rightarrow Na_2CO_3(s) + CO_2(g) + H_2O(g)$
   d. $Al(s) + HCl(aq) \rightarrow AlCl_3(aq) + H_2(g)$
   e. $NiS(s) + O_2(g) \rightarrow NiO(s) + SO_2(g)$
   f. $CaH_2(s) + H_2O(l) \rightarrow Ca(OH)_2(s) + H_2(g)$
   g. $H_2(g) + CO(g) \rightarrow CH_3OH(l)$
   h. $B_2O_3(s) + C(s) \rightarrow B_4C_3(s) + CO_2(g)$

42. Balance each of the following chemical equations.
   a. $Ag_2S(s) + H_2O(l) \rightarrow Ag(s) + H_2S(g) + O_2(g)$
   b. $CaO(s) + SO_3(g) \rightarrow CaSO_4(s)$
   c. $CS_2(g) + Cl_2(g) \rightarrow CCl_4(l) + S_2Cl_2(g)$
   d. $Al(s) + F_2(g) \rightarrow AlF_3(s)$
   e. $NH_3(g) + O_2(g) \rightarrow NO(g) + H_2O(l)$
   f. $PI_3(s) + H_2O(l) \rightarrow H_3PO_3(aq) + HI(g)$
   g. $C(s) + S_8(s) \rightarrow CS_2(l)$
   h. $CaH_2(s) + H_2O(l) \rightarrow Ca(OH)_2(aq) + H_2(g)$

43. Balance each of the following chemical equations.
   a. $KO_2(s) + H_2O(l) \rightarrow KOH(aq) + O_2(g) + H_2O_2(aq)$
   b. $Fe_2O_3(s) + HNO_3(aq) \rightarrow Fe(NO_3)_3(aq) + H_2O(l)$
   c. $NH_3(g) + O_2(g) \rightarrow NO(g) + H_2O(g)$
   d. $PCl_5(l) + H_2O(l) \rightarrow H_3PO_4(aq) + HCl(g)$
   e. $C_2H_5OH(l) + O_2(g) \rightarrow CO_2(g) + H_2O(l)$
   f. $CaO(s) + C(s) \rightarrow CaC_2(s) + CO_2(g)$
   g. $MoS_2(s) + O_2(g) \rightarrow MoO_3(s) + SO_2(g)$
   h. $FeCO_3(s) + H_2CO_3(aq) \rightarrow Fe(HCO_3)_2(aq)$

44. Balance each of the following chemical equations.
a. $Ba(NO_3)_2(aq) + Na_2CrO_4(aq) \rightarrow BaCrO_4(s) + NaNO_3(aq)$
b. $PbCl_2(aq) + K_2SO_4(aq) \rightarrow PbSO_4(s) + KCl(aq)$
c. $C_2H_5OH(l) + O_2(g) \rightarrow CO_2(g) + H_2O(l)$
d. $CaC_2(s) + H_2O(l) \rightarrow Ca(OH)_2(s) + C_2H_2(g)$
e. $Sr(s) + HNO_3(aq) \rightarrow Sr(NO_3)_2(aq) + H_2(g)$
f. $BaO_2(s) + H_2SO_4(aq) \rightarrow BaSO_4(s) + H_2O_2(aq)$
g. $AsI_3(s) \rightarrow As(s) + I_2(s)$
h. $CuSO_4(aq) + KI(s) \rightarrow CuI(s) + I_2(s) + K_2SO_4(aq)$

Additional Problems

45. Sodium hydrogen carbonate, more commonly known as baking soda, can be produced by bubbling carbon dioxide gas through an ice-cold aqueous solution containing large amounts of sodium chloride and ammonia. Because sodium hydrogen carbonate is not very soluble in cold water, it precipitates, whereas the other product of the reaction, ammonium chloride, remains dissolved. Write the unbalanced chemical equation for the process. (*Hint:* Water is a reactant.)

46. Many ships are built with aluminum superstructures to save weight. Aluminum, however, burns in oxygen if there is a sufficiently hot ignition source, which has led to several tragedies at sea. Write the unbalanced chemical equation for the reaction of aluminum with oxygen, producing aluminum oxide as product.

47. Crude gunpowders often contain a mixture of potassium nitrate and charcoal (carbon). When such a mixture is heated until reaction occurs, a solid residue of potassium carbonate is produced. The explosive force of the gunpowder comes from the fact that two gases are also produced (carbon monoxide and nitrogen), which increase in volume with great force and speed. Write the unbalanced chemical equation for the process.

48. The sugar sucrose, which is present in many fruits and vegetables, reacts in the presence of certain yeast enzymes to produce ethyl alcohol (ethanol) and carbon dioxide gas. Balance the following equation for this reaction of sucrose.

$$C_{12}H_{22}O_{11}(aq) + H_2O(l) \rightarrow C_2H_5OH(aq) + CO_2(g)$$

49. Methanol (methyl alcohol), CH_3OH, is a very important industrial chemical. Formerly, methanol was prepared by heating wood to high temperatures in the absence of air. The complex compounds present in wood are degraded by this process into a charcoal residue and a volatile portion that is rich in methanol. Today, methanol is instead synthesized from carbon monoxide and elemental hydrogen. Write the balanced chemical equation for this latter process.

50. The Hall process is an important method by which pure aluminum is prepared from its oxide (alumina, Al_2O_3) by indirect reaction with graphite (carbon). Balance the following equation, which is a simplified representation of this process.

$$Al_2O_3(s) + C(s) \rightarrow Al(s) + CO_2(g)$$

51. Iron oxide ores, commonly a mixture of FeO and Fe_2O_3, are given the general formula Fe_3O_4. They yield elemental iron when heated to a very high temperature with either carbon monoxide or elemental hydrogen. Balance the following equations for these processes.

$$Fe_3O_4(s) + H_2(g) \rightarrow Fe(s) + H_2O(g)$$
$$Fe_3O_4(s) + CO(g) \rightarrow Fe(s) + CO_2(g)$$

52. The elements of Group 1 all react with sulfur to form the metal sulfides. Write balanced chemical equations for the reactions of the Group 1 elements with sulfur.

53. When steel wool (iron) is heated in pure oxygen gas, the steel wool bursts into flame and a fine powder consisting of a mixture of iron oxides (FeO and Fe_2O_3) forms. Write *separate* unbalanced equations for the reaction of iron with oxygen to give each of these products.

54. One method of producing hydrogen peroxide is to add barium peroxide to water. A precipitate of barium oxide forms, which may then be filtered off to leave a solution of hydrogen peroxide. Write the balanced chemical equation for this process.

55. When elemental boron, B, is burned in oxygen gas, the product is diboron trioxide. If the diboron trioxide is then reacted with a measured quantity of water, it reacts with the water to form what is commonly known as boric acid, $B(OH)_3$. Write a balanced chemical equation for each of these processes.

56. A common experiment in introductory chemistry courses involves heating a weighed mixture of potassium chlorate, $KClO_3$, and potassium chloride. Potassium chlorate decomposes when heated, producing potassium chloride and evolving oxygen gas. By measuring the volume of oxygen gas produced in this experiment, students can calculate the relative percentage of $KClO_3$ and KCl in the original mixture. Write the balanced chemical equation for this process.

57. A common demonstration in chemistry courses involves adding a tiny speck of manganese(IV) oxide to a concentrated hydrogen peroxide, H_2O_2, solution. Hydrogen peroxide is unstable, and it decomposes quite spectacularly under these conditions to produce oxygen gas and steam (water vapor). Manganese(IV) oxide is a catalyst for the decomposition of hydrogen peroxide and is not consumed in the reaction. Write the balanced equation for the decomposition reaction of hydrogen peroxide.

58. The benches in many undergraduate chemistry laboratories are often covered by a film of white dust. This may be due to poor housekeeping, but the dust is usually ammonium chloride, produced by the gaseous reaction in the laboratory of hydrogen chloride and ammonia; most labs have aqueous solutions of these common reagents. Write the balanced chemical equation for the reaction of gaseous ammonia and hydrogen chloride to form solid ammonium chloride.

59. Glass is a mixture of several compounds, but a major constituent of most glass is calcium silicate, $CaSiO_3$. Glass can be etched by treatment with hydrogen fluoride: HF attacks the calcium silicate of the glass, producing gaseous and water-soluble products (which can be removed by washing the glass). For example, the volumetric glassware in chemistry laboratories is often graduated by using this process. Balance the following equation for the reaction of hydrogen fluoride with calcium silicate.

$$CaSiO_3(s) + HF(g) \rightarrow CaF_2(aq) + SiF_4(g) + H_2O(l)$$

60. You decided to toast some English muffins, but left the muffins in the toaster too long so that they began to char. Is there evidence that a chemical reaction has taken place?

61. If you had a "sour stomach," you might try an over-the-counter antacid tablet to relieve the problem. Can you think of evidence that the action of such an antacid is a chemical reaction?

62. When iron wire is heated in the presence of sulfur, the iron soon begins to glow, and a chunky, blue-black mass of iron(II) sulfide is formed. Write the unbalanced chemical equation for this reaction.

63. When finely divided solid sodium is dropped into a flask containing chlorine gas, an explosion occurs and a fine powder of sodium chloride is deposited on the walls of the flask. Write the unbalanced chemical equation for this process.

64. If aqueous solutions of potassium chromate and barium chloride are mixed, a bright yellow solid (barium chromate) forms and settles out of the mixture, leaving potassium chloride in solution. Write a balanced chemical equation for this process.

65. When hydrogen sulfide, H_2S, gas is bubbled through a solution of lead(II) nitrate, $Pb(NO_3)_2$, a black precipitate of lead(II) sulfide, PbS, forms, and nitric acid, HNO_3, is produced. Write the unbalanced chemical equation for this reaction.

66. If an electric current is passed through aqueous solutions of sodium chloride, sodium bromide, and sodium iodide, the elemental halogens are produced at one electrode in each case, with hydrogen gas being evolved at the other electrode. If the liquid is then evaporated from the mixture, a residue of sodium hydroxide remains. Write balanced chemical equations for these electrolysis reactions.

67. When a strip of magnesium metal is heated in oxygen, it bursts into an intensely white flame and produces a finely powdered dust of magnesium oxide. Write the unbalanced chemical equation for this process.

68. When small amounts of acetylene gas are needed, a common process is to react calcium carbide with water. Acetylene gas is evolved rapidly from this combination even at room temperature, leaving a residue of calcium hydroxide. Write the balanced chemical equation for this process.

69. When solid red phosphorus, P_4, is burned in air, the phosphorus combines with oxygen, producing a choking cloud of tetraphosphorus decoxide. Write the unbalanced chemical equation for this reaction.

70. When copper(II) oxide is boiled in an aqueous solution of sulfuric acid, a strikingly blue solution of copper(II) sulfate forms along with additional water. Write the unbalanced chemical equation for this reaction.

71. When lead(II) sulfide is heated to high temperatures in a stream of pure oxygen gas, solid lead(II) oxide forms with the release of gaseous sulfur dioxide. Write the unbalanced chemical equation for this reaction.

72. When sodium sulfite is boiled with sulfur, the sulfite ions, SO_3^{2-}, are converted to thiosulfate ions, $S_2O_3^{2-}$, resulting in a solution of sodium thiosulfate, $Na_2S_2O_3$. Write the unbalanced chemical equation for this reaction.

73. Balance each of the following chemical equations.
   a. $Cl_2(g) + KBr(aq) \rightarrow Br_2(l) + KCl(aq)$
   b. $Cr(s) + O_2(g) \rightarrow Cr_2O_3(s)$
   c. $P_4(s) + H_2(g) \rightarrow PH_3(g)$
   d. $Al(s) + H_2SO_4(aq) \rightarrow Al_2(SO_4)_3(aq) + H_2(g)$
   e. $PCl_3(l) + H_2O(l) \rightarrow H_3PO_3(aq) + HCl(aq)$
   f. $SO_2(g) + O_2(g) \rightarrow SO_3(g)$
   g. $C_7H_{16}(l) + O_2(g) \rightarrow CO_2(g) + H_2O(g)$
   h. $C_2H_6(g) + O_2(g) \rightarrow CO_2(g) + H_2O(g)$

74. Balance each of the following chemical equations.
   a. $ZnCl_2(aq) + Na_2CO_3(aq) \rightarrow ZnCO_3(s) + NaCl(aq)$
   b. $Al(s) + H_2SO_4(aq) \rightarrow Al_2(SO_4)_3(aq) + H_2(g)$
   c. $Mn(s) + S(s) \rightarrow MnS_2(s)$
   d. $C_5H_{12}(l) + O_2(g) \rightarrow CO_2(g) + H_2O(g)$
   e. $H_2O(l) + Br_2(l) \rightarrow HBr(aq) + HOBr(aq)$
   f. $MnS_2(s) + O_2(g) \rightarrow MnO_2(s) + SO_2(g)$
   g. $PbCl_2(aq) + K_2CrO_4(aq) \rightarrow PbCrO_4(s) + KCl(aq)$
   h. $AgNO_3(aq) + H_2SO_4(aq) \rightarrow Ag_2SO_4(s) + HNO_3(aq)$

75. Balance each of the following chemical equations.
   a. $SiCl_4(l) + Mg(s) \rightarrow Si(s) + MgCl_2(s)$
   b. $NO(g) + Cl_2(g) \rightarrow NOCl(g)$
   c. $MnO_2(s) + Al(s) \rightarrow Mn(s) + Al_2O_3(s)$
   d. $Cr(s) + S_8(s) \rightarrow Cr_2S_3(s)$
   e. $NH_3(g) + F_2(g) \rightarrow NH_4F(s) + NF_3(g)$
   f. $Ag_2S(s) + H_2(g) \rightarrow Ag(s) + H_2S(g)$
   g. $O_2(g) \rightarrow O_3(g)$
   h. $Na_2SO_3(aq) + S_8(s) \rightarrow Na_2S_2O_3(aq)$

76. Balance each of the following chemical equations.

a. $Pb(NO_3)_2(aq) + K_2CrO_4(aq) \rightarrow PbCrO_4(s) + KNO_3(aq)$

b. $BaCl_2(aq) + Na_2SO_4(aq) \rightarrow BaSO_4(s) + NaCl(aq)$

c. $CH_3OH(l) + O_2(g) \rightarrow CO_2(g) + H_2O(g)$

d. $Na_2CO_3(aq) + S(s) + SO_2(g) \rightarrow CO_2(g) + Na_2S_2O_3(aq)$

e. $Cu(s) + H_2SO_4(aq) \rightarrow CuSO_4(aq) + SO_2(g) + H_2O(l)$

f. $MnO_2(s) + HCl(aq) \rightarrow MnCl_2(aq) + Cl_2(g) + H_2O(l)$

g. $As_2O_3(s) + KI(aq) + HCl(aq) \rightarrow AsI_3(s) + KCl(aq) + H_2O(l)$

h. $Na_2S_2O_3(aq) + I_2(aq) \rightarrow Na_2S_4O_6(aq) + NaI(aq)$

7 Reactions in Aqueous Solutions

Sodium reacting with water.

The chemical reactions that are most important to us occur in water—in aqueous solutions. Virtually all of the chemical reactions that keep each of us alive and well take place in the aqueous medium present in our bodies. For example, the oxygen you breathe dissolves in your blood, where it associates with the hemoglobin in the red blood cells. While attached to the hemoglobin it is transported to your cells, where it reacts with fuel (from the food you eat) to provide energy for living. However, the reaction between oxygen and fuel is not direct—the cells are not tiny furnaces. Instead, electrons are transferred from the fuel to a series of molecules that pass them along (this is called the respiratory chain) until they eventually reach oxygen. Many other reactions are also crucial to our health and well-being. You will see numerous examples of these as you continue your study of chemistry.

In this chapter we will study some common types of reactions that take place in water, and we will become familiar with some of the driving forces that make these reactions occur. We will also learn how to predict the products for these reactions and how to write various equations to describe them.

Developing a photo involves several aqueous chemical reactions.

7.1 Predicting Whether a Reaction Will Occur

AIM: To learn about some of the factors that cause reactions to occur.

In this text we have already seen many chemical reactions. Now let's consider an important question: Why does a chemical reaction occur? What causes reactants to "want" to form products? As chemists have studied reactions, they have recognized several "tendencies" in reactants that drive them to form products. That is, there are several "driving forces" that pull reactants toward products—changes that tend to make reactions go in the direction of the arrow. The most common of these driving forces are

1. Formation of a solid
2. Formation of water
3. Transfer of electrons
4. Formation of a gas

When two or more chemicals are brought together, if any of these things can occur, a chemical change (a reaction) is likely to take place. Accordingly, when we are confronted with a set of reactants and want to predict whether a reaction will occur and what products might form, we will consider these driving forces. They will help us organize our thoughts as we encounter new reactions.

7.2

Reactions in Which a Solid Forms

AIM: To learn to identify the solid that forms in a precipitation reaction.

One driving force for a chemical reaction is the formation of a solid, a process called **precipitation.** The solid that forms is called a **precipitate,** and the reaction is known as a **precipitation reaction.** For example, when an aqueous (water) solution of potassium chromate, $K_2CrO_4(aq)$, which is yellow, is added to a colorless aqueous solution containing barium nitrate, $Ba(NO_3)_2(aq)$, a yellow solid forms (see Figure 7.1). The fact that a solid forms tells us that a reaction—a chemical change—has occurred. That is, we have a situation where

$$\text{Reactants} \rightarrow \text{Products}$$

Figure 7.1
The precipitation reaction that occurs when yellow potassium chromate, $K_2CrO_4(aq)$, is mixed with a colorless barium nitrate solution, $Ba(NO_3)_2(aq)$.

What is the equation that describes this chemical change? To write the equation, we must decipher the identities of the reactants and products. The reactants have already been described: $K_2CrO_4(aq)$ and $Ba(NO_3)_2(aq)$. Is there some way in which we can predict the identities of the products? What is the yellow solid? The best way to predict the identity of this solid is to first *consider what products are possible.* To do this we need to know what chemical species are present in the solution that results when the reactant solutions are mixed. First, let's think about the nature of each reactant in an aqueous solution.

What Happens When an Ionic Compound Dissolves in Water?

The designation $Ba(NO_3)_2(aq)$ means that barium nitrate (a white solid) has been dissolved in water. Note from its formula that barium nitrate contains the Ba^{2+} and NO_3^- ions. *In virtually every case when a solid containing ions dissolves in water, the ions separate* and move around independently. That is, $Ba(NO_3)_2(aq)$ does not contain $Ba(NO_3)_2$ units. Rather, it contains separated Ba^{2+} and NO_3^- ions. In the solution there are two NO_3^- ions for every Ba^{2+} ion. Chemists know that separated ions are present in this solution because it is an excellent conductor of electricity (see Figure 7.2). Pure water does not conduct an electric current. Ions must be present in water for a current to flow.

When each unit of a substance that dissolves in water produces separated ions, the substance is called a **strong electrolyte.** Barium nitrate is a strong electrolyte in water, because each $Ba(NO_3)_2$ unit produces the separated ions (Ba^{2+}, NO_3^-, NO_3^-).

Similarly, aqueous K_2CrO_4 also behaves as a strong electrolyte. Potassium chromate contains the K^+ and CrO_4^{2-} ions, so an aqueous solution of potassium chromate (which is prepared by dissolving solid K_2CrO_4 in water) contains these separated ions. That is, $K_2CrO_4(aq)$ does not contain K_2CrO_4 units but instead contains K^+ cations and CrO_4^{2-} anions, which move around independently. (There are two K^+ ions for each CrO_4^{2-} ion.)

The idea introduced here is very important: when ionic compounds dissolve, the *resulting solution contains the separated ions.* Therefore, we can

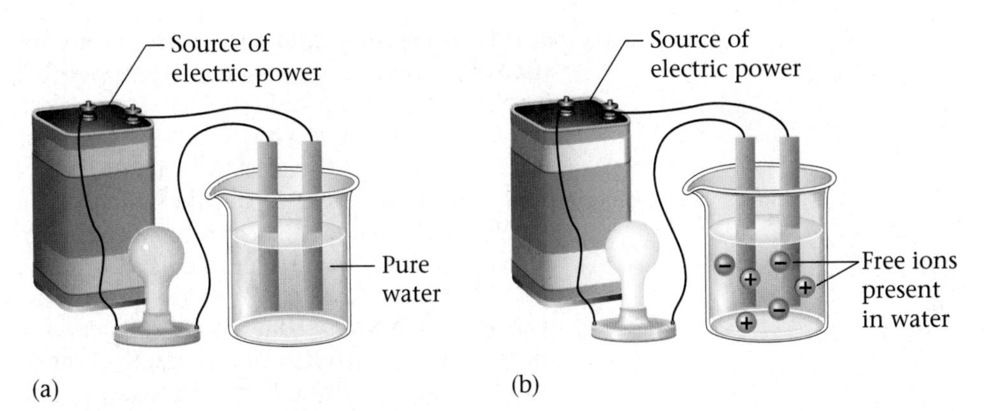

Figure 7.2
Electrical conductivity of aqueous solutions. (a) Pure water does not conduct an electric current. The lamp does not light. (b) When an ionic compound is dissolved in water, current flows and the lamp lights. The result of this experiment is strong evidence that ionic compounds dissolved in water exist in the form of separated ions.

represent the mixing of $K_2CrO_4(aq)$ and $Ba(NO_3)_2(aq)$ in two ways. We usually write these reactants as

$$K_2CrO_4(aq) + Ba(NO_3)_2(aq) \rightarrow \text{Products}$$

However, a more accurate representation of the situation is:

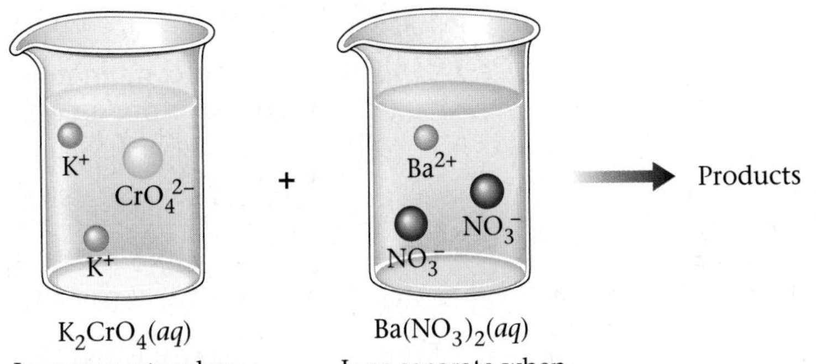

We can express this information in equation form as follows:

$$\underbrace{2K^+(aq) + CrO_4^{2-}(aq)}_{\text{The ions in } K_2CrO_4(aq)} + \underbrace{Ba^{2+}(aq) + 2NO_3^-(aq)}_{\text{The ions in } Ba(NO_3)_2(aq)} \rightarrow \text{Products}$$

Thus the *mixed solution* contains four types of ions: K^+, CrO_4^{2-}, Ba^{2+}, and NO_3^-. Now that we know what the reactants are, we can make some educated guesses about the possible products.

How to Decide What Products Form

Which of these ions combine to form the yellow solid observed when the original solutions are mixed? This is not an easy question to answer. Even an experienced chemist is not sure what will happen in a new reaction. The chemist tries to think of the various possibilities, considers the likelihood of each possibility, and then makes a prediction (an educated guess). Only after identifying each product experimentally can the chemist be sure what reaction actually has taken place. However, an educated guess is very useful because it indicates what kinds of products are most likely. It gives us a place to start. So the best way to proceed is first to think of the various possibilities and then to decide which of them is most likely.

What are the possible products of the reaction between $K_2CrO_4(aq)$ and $Ba(NO_3)_2(aq)$ or, more accurately, what reaction can occur among the ions K^+, CrO_4^{2-}, Ba^{2+}, and NO_3^-? We already know some things that will help us decide. We know that a *solid compound must have a zero net charge.* This means that the product of our reaction must contain *both anions and cations* (negative and positive ions). For example, K^+ and Ba^{2+} could not combine to form the solid because such a solid would have a positive charge. Similarly, CrO_4^{2-} and NO_3^- could not combine to form a solid because that solid would have a negative charge.

Something else that will help us is an observation that chemists have made by examining many compounds: *most ionic materials contain only two types of ions*—one type of cation and one type of anion. This idea is illustrated by the following compounds (among many others):

Compound	Cation	Anion
$NaCl$	Na^+	Cl^-
KOH	K^+	OH^-
Na_2SO_4	Na^+	SO_4^{2-}
NH_4Cl	NH_4^+	Cl^-
Na_2CO_3	Na^+	CO_3^{2-}

All the possible combinations of a cation and an anion to form uncharged compounds from among the ions K^+, CrO_4^{2-}, Ba^{2+}, and NO_3^- are shown below:

	NO_3^-	CrO_4^{2-}
K^+	KNO_3	K_2CrO_4
Ba^{2+}	$Ba(NO_3)_2$	$BaCrO_4$

So the list of compounds that *might* make up the solid are

K_2CrO_4	$BaCrO_4$
KNO_3	$Ba(NO_3)_2$

Which of these possibilities is most likely to represent the yellow solid? We know it's not K_2CrO_4 or $Ba(NO_3)_2$; these are the reactants. They were present (dissolved) in the separate solutions that were mixed initially. The only real possibilities are KNO_3 and $BaCrO_4$. To decide which of these is more likely to represent the yellow solid, we need more facts. An experienced chemist, for example, knows that KNO_3 is a white solid. On the other hand, the CrO_4^{2-} ion is yellow. Therefore, the yellow solid most likely is $BaCrO_4$.

We have determined that one product of the reaction between $K_2CrO_4(aq)$ and $Ba(NO_3)_2(aq)$ is $BaCrO_4(s)$, but what happened to the K^+ and NO_3^- ions? The answer is that these ions are left dissolved in the solution. That is, KNO_3 does not form a solid when the K^+ and NO_3^- ions are present in water. In other words, if we took the white solid $KNO_3(s)$ and put it in water, it would totally dissolve (the white solid would "disappear," yielding a colorless solution). So when we mix $K_2CrO_4(aq)$ and $Ba(NO_3)_2(aq)$, $BaCrO_4(s)$ forms but KNO_3 is left behind in solution [we write it as $KNO_3(aq)$]. (If we poured the mixture through a filter to remove the solid

$BaCrO_4$ and then evaporated all of the water, we would obtain the white solid KNO_3.)

After all this thinking, we can finally write the unbalanced equation for the precipitation reaction.

$$K_2CrO_4(aq) + Ba(NO_3)_2(aq) \rightarrow BaCrO_4(s) + KNO_3(aq)$$

We can represent this reaction in pictures as follows:

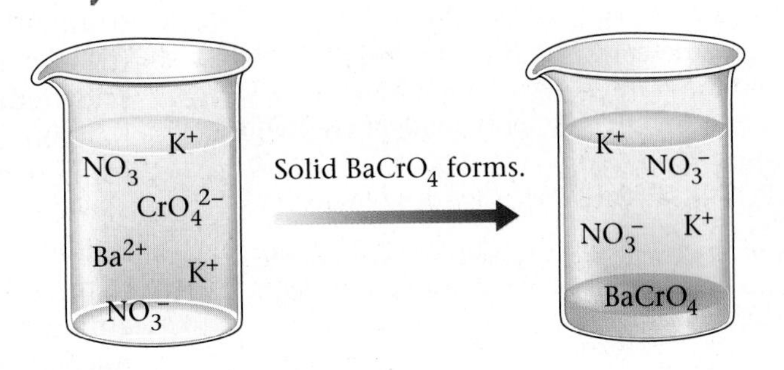

Note that the K^+ and NO_3^- ions are not involved in the chemical change. They remain dispersed in the water before and after the reaction.

Using Solubility Rules

In the example considered above we were finally able to identify the products of the reaction by using two types of chemical knowledge:

1. Knowledge of facts
2. Knowledge of concepts

For example, knowing the colors of the various compounds proved very helpful. This represents factual knowledge. Awareness of the concept that solids always have a net charge of zero was also essential. These two kinds of knowledge allowed us to make a good guess about the identity of the solid that formed. As you continue to study chemistry, you will see that a balance of factual and conceptual knowledge is always required. You must both *memorize* important facts and *understand* crucial concepts in order to succeed.

Solids must contain both anions and cations in the relative numbers necessary to produce zero net charge.

In the present case we are dealing with a reaction in which an ionic solid forms—that is, a process in which ions that are dissolved in water combine to give a solid. We know that for a solid to form, both positive and negative ions must be present in relative numbers that give zero net charge. However, oppositely charged ions in water do not always react to form a solid, as we have seen for K^+ and NO_3^-. In addition, Na^+ and Cl^- can coexist in water in very large numbers with no formation of solid NaCl. In other words, when solid NaCl (common salt) is placed in water, it dissolves—the white solid "disappears" as the Na^+ and Cl^- ions are dispersed throughout the water. (You probably have observed this phenomenon in preparing salt water to cook food.) The following two statements, then, are really saying the same thing.

1. Solid NaCl is very soluble in water.
2. Solid NaCl does not form when one solution containing Na^+ is mixed with another solution containing Cl^-.

To predict whether a given pair of dissolved ions will form a solid when mixed, we must know some facts about the solubilities of various types of

Table 7.1 General Rules for Solubility of Ionic Compounds (Salts) in Water at 25 °C

1. Most nitrate (NO_3^-) salts are soluble.
2. Most salts of Na^+, K^+, and NH_4^+ are soluble.
3. Most chloride salts are soluble. Notable exceptions are $AgCl$, $PbCl_2$, and Hg_2Cl_2.
4. Most sulfate salts are soluble. Notable exceptions are $BaSO_4$, $PbSO_4$, and $CaSO_4$.
5. Most hydroxide compounds are only slightly soluble.* The important exceptions are $NaOH$ and KOH. $Ba(OH)_2$ and $Ca(OH)_2$ are only moderately soluble.
6. Most sulfide (S^{2-}), carbonate (CO_3^{2-}), and phosphate (PO_4^{3-}) salts are only slightly soluble.*

*The terms *insoluble* and *slightly soluble* really mean the same thing: such a tiny amount dissolves that it is not possible to detect it with the naked eye.

ionic compounds. In this text we will use the term **soluble solid** to mean a solid that readily dissolves in water; the solid "disappears" as the ions are dispersed in the water. The terms **insoluble solid** and **slightly soluble solid** are taken to mean the same thing: a solid where such a tiny amount dissolves in water that it is undetectable with the naked eye. The solubility information about common solids that is summarized in Table 7.1 is based on observations of the behavior of many compounds. This is factual knowledge that you will need to predict what will happen in chemical reactions where a solid might form. This information is summarized in Figure 7.3.

Notice that in Table 7.1 and Figure 7.3 the term *salt* is used to mean *ionic compound*. Many chemists use the terms *salt* and *ionic compound* interchangeably. In Example 7.1, we will illustrate how to use the solubility rules to predict the products of reactions among ions.

Example 7.1 Identifying Precipitates in Reactions Where a Solid Forms

$AgNO_3$ is usually called silver nitrate rather than silver(I) nitrate because silver forms only Ag^+.

When an aqueous solution of silver nitrate is added to an aqueous solution of potassium chloride, a white solid forms. Identify the white solid and write the balanced equation for the reaction that occurs.

Solution

First let's use the description of the reaction to represent what we know:

$$AgNO_3(aq) + KCl(aq) \rightarrow \text{White solid}$$

Remember, try to determine the essential facts from the words and represent these facts by symbols or diagrams. To answer the main question (What

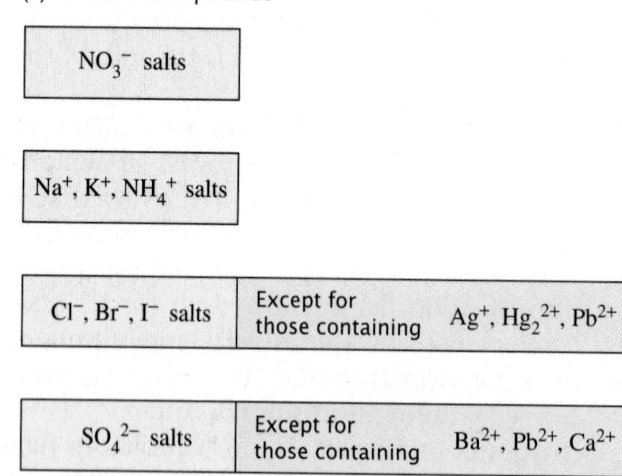

Figure 7.3
Solubilities of common compounds.

is the white solid?) we must establish what ions are present in the mixed solution. That is, we must know what the reactants are really like. Remember that *when ionic substances dissolve in water, the ions separate.* So we can write the equation

$$Ag^+(aq) + NO_3^-(aq) + K^+(aq) + Cl^-(aq) \rightarrow \text{Products}$$

Ions in $AgNO_3(aq)$ | Ions in $KCl(aq)$

or using pictures

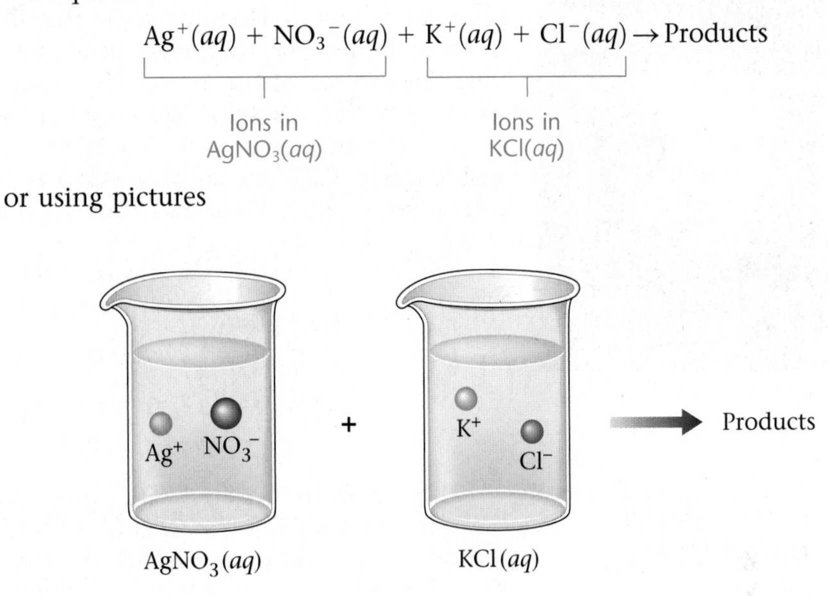

to represent the ions present in the mixed solution before any reaction occurs. In summary:

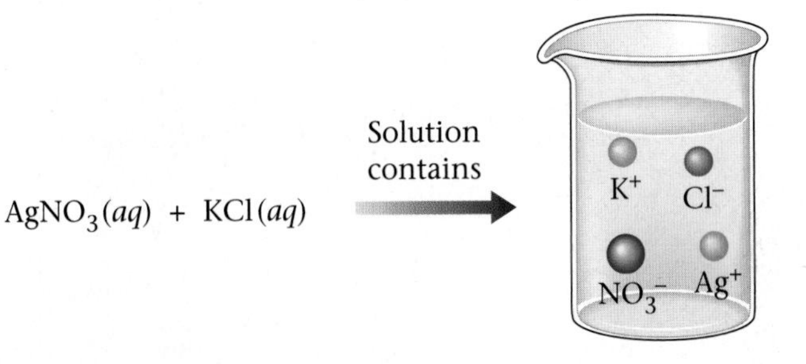

	NO_3^-	Cl^-
Ag+	$AgNO_3$	AgCl
K+	KNO_3	KCl

Now we will consider what solid *might* form from this collection of ions. Because the solid must contain both positive and negative ions, the possible compounds that can be assembled from this collection of ions are

$AgNO_3$ AgCl
KNO_3 KCl

$AgNO_3$ and KCl are the substances already dissolved in the reactant solutions, so we know that they do not represent the white solid product. We are left with two possibilities:

AgCl

KNO_3

Another way to obtain these two possibilities is by *ion interchange.* This means that in the reaction of $AgNO_3(aq)$ and $KCl(aq)$, we take the cation from one reactant and combine it with the anion of the other reactant.

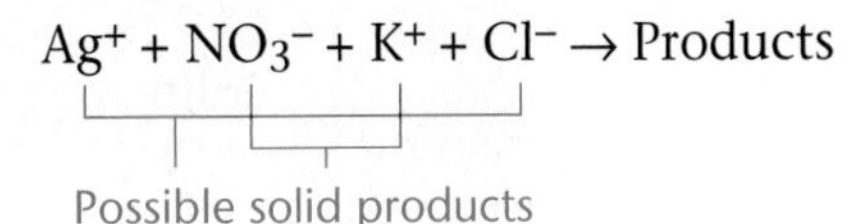

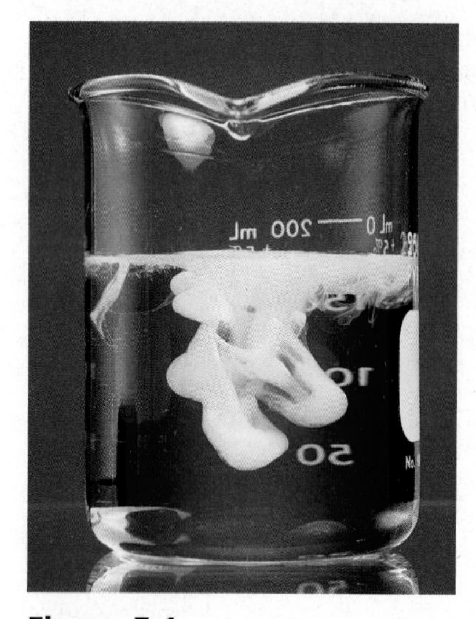

Figure 7.4
Precipitation of silver chloride occurs when solutions of silver nitrate and potassium chloride are mixed. The K^+ and NO_3^- ions remain in solution.

Ion interchange also leads to the following possible solids:

$AgCl$ or KNO_3

To decide whether $AgCl$ or KNO_3 is the white solid, we need the solubility rules (Table 7.1). Rule 2 states that most salts containing K^+ are soluble in water. Rule 1 says that most nitrate salts (those containing NO_3^-) are soluble. So the salt KNO_3 is water-soluble. That is, when K^+ and NO_3^- are mixed in water, a solid (KNO_3) does *not* form.

On the other hand, Rule 3 states that although most chloride salts (salts that contain Cl^-) are soluble, AgCl is an exception. That is, $AgCl(s)$ is insoluble in water. Thus the white solid must be AgCl. Now we can write

$$AgNO_3(aq) + KCl(aq) \rightarrow AgCl(s) + ?$$

What is the other product?

To form $AgCl(s)$, we have used the Ag^+ and Cl^- ions:

$$Ag^+(aq) + NO_3^-(aq) + K^+(aq) + Cl^-(aq) \rightarrow AgCl(s)$$

This leaves the K^+ and NO_3^- ions. What do they do? Nothing. Because KNO_3 is very soluble in water (Rules 1 and 2), the K^+ and NO_3^- ions remain separate in the water; the KNO_3 remains dissolved and we represent it as $KNO_3(aq)$. We can now write the full equation:

$$AgNO_3(aq) + KCl(aq) \rightarrow AgCl(s) + KNO_3(aq)$$

Figure 7.4 shows the precipitation of $AgCl(s)$ that occurs when this reaction takes place. In graphic form, the reaction is

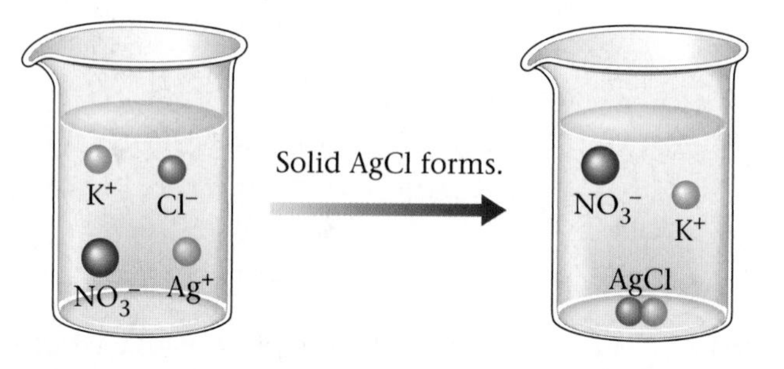

■

The following strategy is useful for predicting what will occur when two solutions containing dissolved salts are mixed.

How to Predict Precipitates When Solutions of Two Ionic Compounds Are Mixed

STEP 1 Write the reactants as they actually exist before any reaction occurs. Remember that when a salt dissolves, its ions separate.

STEP 2 Consider the various solids that could form. To do this, simply *exchange the anions* of the added salts.

STEP 3 Use the solubility rules (Table 7.1) to decide whether a solid forms and, if so, to predict the identity of the solid.

Example 7.2 Using Solubility Rules to Predict the Products of Reactions

Using the solubility rules in Table 7.1, predict what will happen when the following solutions are mixed. Write the balanced equation for any reaction that occurs.

a. $KNO_3(aq)$ and $BaCl_2(aq)$

b. $Na_2SO_4(aq)$ and $Pb(NO_3)_2(aq)$

c. $KOH(aq)$ and $Fe(NO_3)_3(aq)$

Solution (a)

STEP 1 $KNO_3(aq)$ represents an aqueous solution obtained by dissolving solid KNO_3 in water to give the ions $K^+(aq)$ and $NO_3^-(aq)$. Likewise, $BaCl_2(aq)$ is a solution formed by dissolving solid $BaCl_2$ in water to produce $Ba^{2+}(aq)$ and $Cl^-(aq)$. When these two solutions are mixed, the following ions will be present:

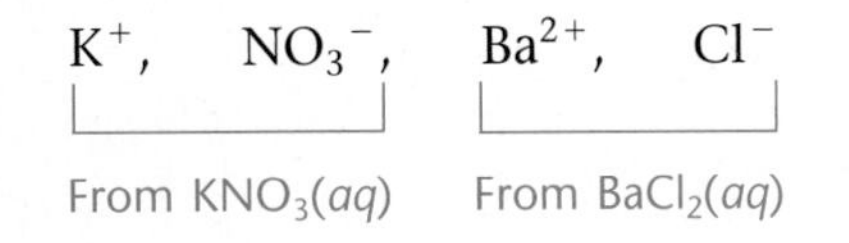

STEP 2 To get the possible products, we exchange the anions.

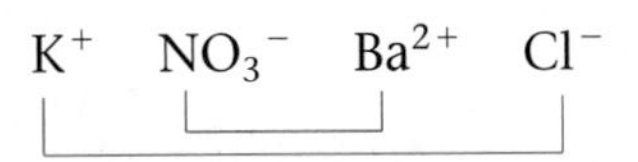

This yields the possibilities KCl and $Ba(NO_3)_2$. These are the solids that *might* form. Notice that two NO_3^- ions are needed to balance the 2+ charge on Ba^{2+}.

STEP 3 The rules listed in Table 7.1 indicate that both KCl and $Ba(NO_3)_2$ are soluble in water. So no precipitate forms when $KNO_3(aq)$ and $BaCl_2(aq)$ are mixed. All of the ions remain dissolved in the solution. This means that no reaction takes place. That is, no chemical change occurs.

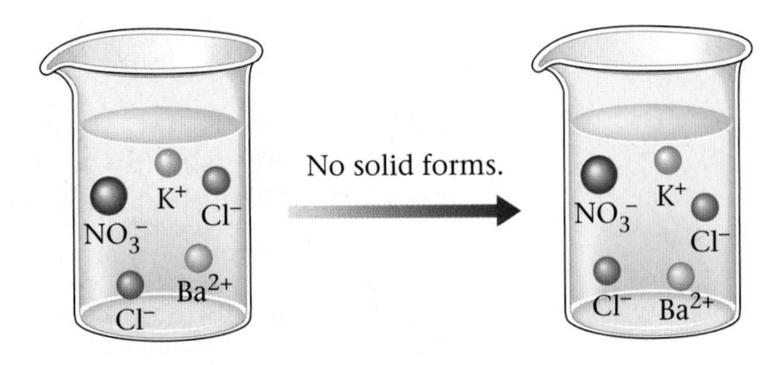

Solution (b)

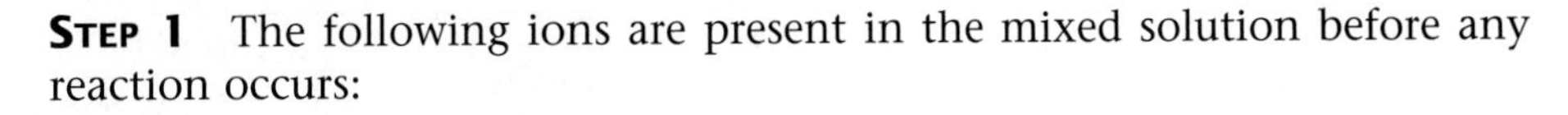

STEP 1 The following ions are present in the mixed solution before any reaction occurs:

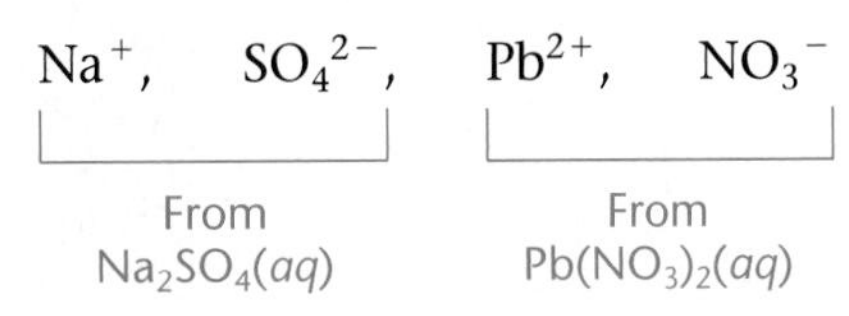

Step 2 Exchanging anions as follows:

$$Na^+ \quad SO_4^{2-} \quad Pb^{2+} \quad NO_3^-$$

yields the *possible* solid products $PbSO_4$ and $NaNO_3$.

Step 3 Using Table 7.1, we see that $NaNO_3$ is soluble in water (Rules 1 and 2) but that $PbSO_4$ is only slightly soluble (Rule 4). Thus, when these solutions are mixed, solid $PbSO_4$ forms. The balanced reaction is

$$Na_2SO_4(aq) + Pb(NO_3)_2(aq) \rightarrow PbSO_4(s) + 2NaNO_3(aq)$$

Remains dissolved

which can be represented as

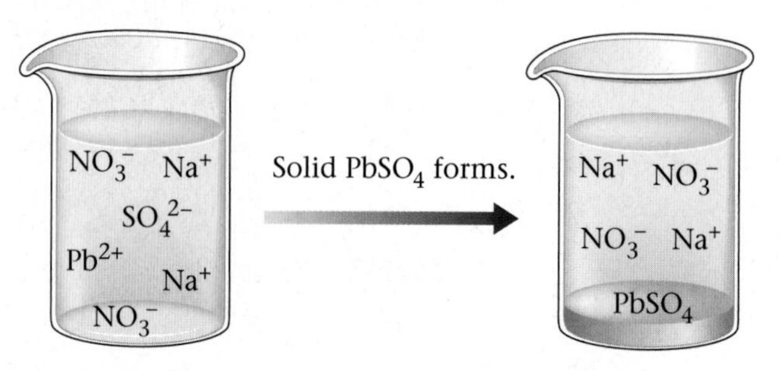

Solution (c)

Step 1 The ions present in the mixed solution before any reaction occurs are

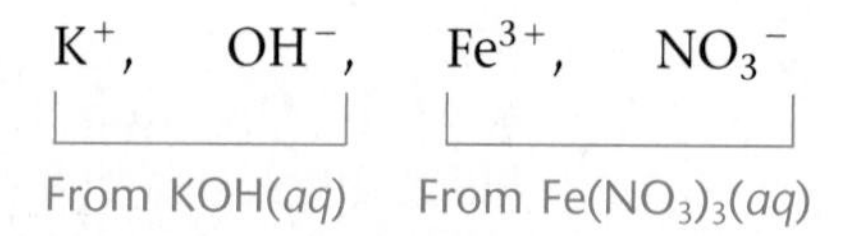

Step 2 Exchanging anions

$$K^+ \quad OH^- \quad Fe^{3+} \quad NO_3^-$$

yields the possible solid products KNO_3 and $Fe(OH)_3$.

Step 3 Rules 1 and 2 (Table 7.1) state that KNO_3 is soluble, whereas $Fe(OH)_3$ is only slightly soluble (Rule 5). Thus, when these solutions are mixed, solid $Fe(OH)_3$ forms. The balanced equation for the reaction is

$$3KOH(aq) + Fe(NO_3)_3(aq) \rightarrow Fe(OH)_3(s) + 3KNO_3(aq)$$

which can be represented as

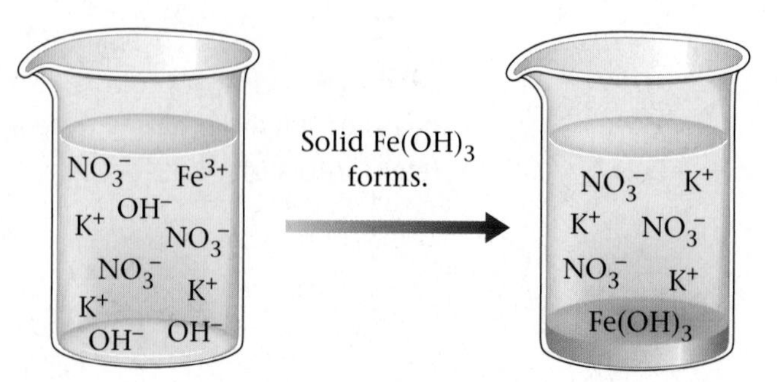

Self-Check Exercise 7.1

Predict whether a solid will form when the following pairs of solutions are mixed. If so, identify the solid and write the balanced equation for the reaction.

a. $Ba(NO_3)_2(aq)$ and $NaCl(aq)$

b. $Na_2S(aq)$ and $Cu(NO_3)_2(aq)$

c. $NH_4Cl(aq)$ and $Pb(NO_3)_2(aq)$

See Problems 7.17 and 7.18. ■

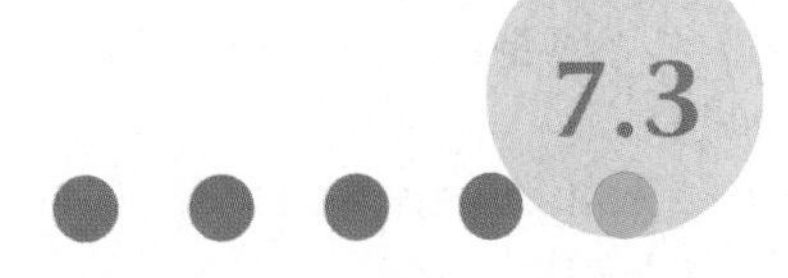

Describing Reactions in Aqueous Solutions

AIM: To learn to describe reactions in solutions by writing molecular, complete ionic, and net ionic equations.

Much important chemistry, including virtually all of the reactions that make life possible, occurs in aqueous solutions. We will now consider the types of equations used to represent reactions that occur in water. For example, as we saw earlier, when we mix aqueous potassium chromate with aqueous barium nitrate, a reaction occurs to form solid barium chromate and dissolved potassium nitrate. One way to represent this reaction is by the equation

$$K_2CrO_4(aq) + Ba(NO_3)_2(aq) \rightarrow BaCrO_4(s) + 2KNO_3(aq)$$

This is called the **molecular equation** for the reaction; it shows the complete formulas of all reactants and products. However, although this equation shows the reactants and products of the reaction, it does not give a very clear picture of what actually occurs in solution. As we have seen, aqueous solutions of potassium chromate, barium nitrate, and potassium nitrate contain the individual ions, not molecules as is implied by the molecular equation. Thus the **complete ionic equation,**

$$\underbrace{2K^+(aq) + CrO_4^{2-}(aq)}_{\text{Ions from } K_2CrO_4} + \underbrace{Ba^{2+}(aq) + 2NO_3^-(aq)}_{\text{Ions from } Ba(NO_3)_2} \rightarrow BaCrO_4(s) + 2K^+(aq) + 2NO_3^-(aq)$$

better represents the actual forms of the reactants and products in solution. *In a complete ionic equation, all substances that are strong electrolytes are represented as ions.* Notice that $BaCrO_4$ is not written as the separate ions, because it is present as a solid; it is not dissolved.

A strong electrolyte is a substance that completely breaks apart into ions when dissolved in water. The resulting solution readily conducts an electric current.

The complete ionic equation reveals that only some of the ions participate in the reaction. Notice that the K^+ and NO_3^- ions are present in solution both before and after the reaction. Ions such as these, which do not participate directly in a reaction in solution, are called **spectator ions.** The ions that participate in this reaction are the Ba^{2+} and CrO_4^{2-} ions, which combine to form solid $BaCrO_4$:

$$Ba^{2+}(aq) + CrO_4^{2-}(aq) \rightarrow BaCrO_4(s)$$

The net ionic equation includes only those components that undergo a change in the reaction.

This equation, called the **net ionic equation,** includes only those components that are directly involved in the reaction. Chemists usually write the net ionic equation for a reaction in solution, because it gives the actual forms of the reactants and products and includes only the species that undergo a change.

Types of Equations for Reactions in Aqueous Solutions

Three types of equations are used to describe reactions in solutions.

1. The *molecular equation* shows the overall reaction but not necessarily the actual forms of the reactants and products in solution.
2. The *complete ionic equation* represents all reactants and products that are strong electrolytes as ions. All reactants and products are included.
3. The *net ionic equation* includes only those components that undergo a change. Spectator ions are not included.

To make sure these ideas are clear, we will do another example. In Example 7.2 we considered the reaction between aqueous solutions of lead nitrate and sodium sulfate. The molecular equation for this reaction is

$$Pb(NO_3)_2(aq) + Na_2SO_4(aq) \rightarrow PbSO_4(s) + 2NaNO_3(aq)$$

Because any ionic compound that is dissolved in water is present as the separated ions, we can write the complete ionic equation as follows:

$$Pb^{2+}(aq) + 2NO_3^-(aq) + 2Na^+(aq) + SO_4^{2-}(aq) \rightarrow PbSO_4(s) + 2Na^+(aq) + 2NO_3^-(aq)$$

The $PbSO_4$ is not written as separate ions because it is present as a solid. The ions that take part in the chemical change are the Pb^{2+} and the SO_4^{2-} ions, which combine to form solid $PbSO_4$. Thus the net ionic equation is

$$Pb^{2+}(aq) + SO_4^{2-}(aq) \rightarrow PbSO_4(s)$$

The Na^+ and NO_3^- ions do not undergo any chemical change; they are spectator ions.

Example 7.3 Writing Equations for Reactions

For each of the following reactions, write the molecular equation, the complete ionic equation, and the net ionic equation.

Because silver is present as Ag^+ in all of its common ionic compounds, we usually delete the (I) when naming silver compounds.

a. Aqueous sodium chloride is added to aqueous silver nitrate to form solid silver chloride plus aqueous sodium nitrate.

b. Aqueous potassium hydroxide is mixed with aqueous iron(III) nitrate to form solid iron(III) hydroxide and aqueous potassium nitrate.

Solution

a. *Molecular equation:*

$$NaCl(aq) + AgNO_3(aq) \rightarrow AgCl(s) + NaNO_3(aq)$$

Complete ionic equation:

$$Na^+(aq) + Cl^-(aq) + Ag^+(aq) + NO_3^-(aq) \rightarrow AgCl(s) + Na^+(aq) + NO_3^-(aq)$$

Net ionic equation:

$$Cl^-(aq) + Ag^+(aq) \rightarrow AgCl(s)$$

b. *Molecular equation:*

$$3KOH(aq) + Fe(NO_3)_3(aq) \rightarrow Fe(OH)_3(s) + 3KNO_3(aq)$$

Complete ionic equation:

$$3K^+(aq) + 3OH^-(aq) + Fe^{3+}(aq) + 3NO_3^-(aq) \rightarrow Fe(OH)_3(s) + 3K^+(aq) + 3NO_3^-(aq)$$

Net ionic equation:

$$3OH^-(aq) + Fe^{3+}(aq) \rightarrow Fe(OH)_3(s)$$

✔ Self-Check Exercise 7.2

For each of the following reactions, write the molecular equation, the complete ionic equation, and the net ionic equation.

a. Aqueous sodium sulfide is mixed with aqueous copper(II) nitrate to produce solid copper(II) sulfide and aqueous sodium nitrate.

b. Aqueous ammonium chloride and aqueous lead(II) nitrate react to form solid lead(II) chloride and aqueous ammonium nitrate.

See Problems 7.25 through 7.30. ■

7.4 Reactions That Form Water: Acids and Bases

AIM: To learn the key characteristics of the reactions between strong acids and strong bases.

In this section we encounter two very important classes of compounds: acids and bases. Acids were first associated with the sour taste of citrus fruits. In fact, the word *acid* comes from the Latin word *acidus,* which means "sour." Vinegar tastes sour because it is a dilute solution of acetic acid; citric acid is responsible for the sour taste of a lemon. Bases, sometimes called *alkalis,* are characterized by their bitter taste and slippery feel, like wet soap. Most commercial preparations for unclogging drains are highly basic.

Don't taste chemicals!

Acids have been known for hundreds of years. For example, the *mineral acids* sulfuric acid, H_2SO_4, and nitric acid, HNO_3, so named because they were originally obtained by the treatment of minerals, were discovered around 1300. However, it was not until the late 1800s that the essential nature of acids was discovered by Svante Arrhenius, then a Swedish graduate student in physics.

Arrhenius, who was trying to discover why only certain solutions could conduct an electric current, found that conductivity arose from the presence of ions. In his studies of solutions, Arrhenius observed that when the substances HCl, HNO_3, and H_2SO_4 were dissolved in water, they behaved as strong electrolytes. He suggested that this was the result of ionization reactions in water.

The Nobel Prize in chemistry was awarded to Arrhenius in 1903 for his studies of solution conductivity.

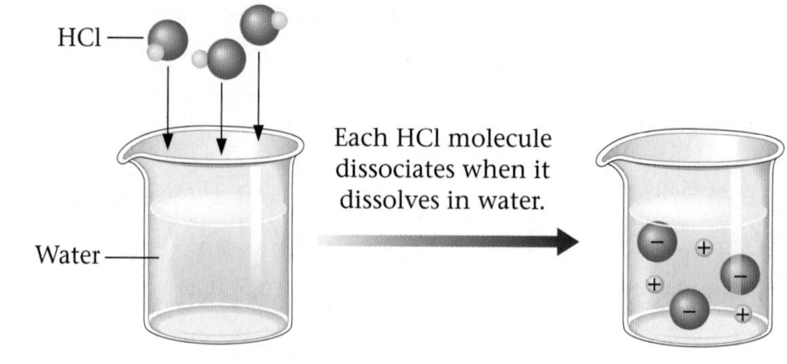

Figure 7.5
When gaseous HCl is dissolved in water, each molecule dissociates to produce H^+ and Cl^- ions. That is, HCl behaves as a strong electrolyte.

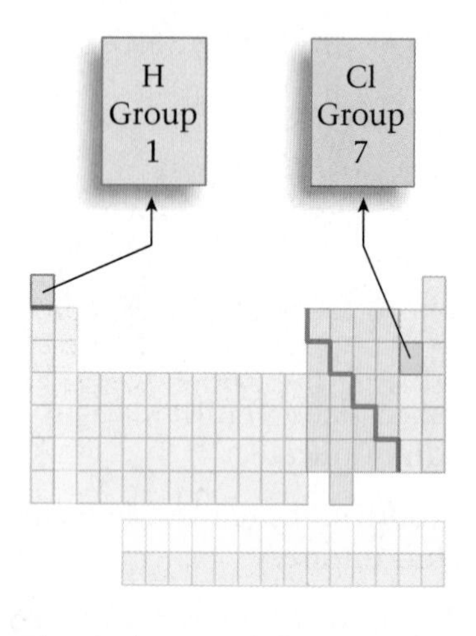

The Arrhenius definition of an acid: a substance that produces H^+ ions in aqueous solution.

$$HCl \xrightarrow{H_2O} H^+(aq) + Cl^-(aq)$$
$$HNO_3 \xrightarrow{H_2O} H^+(aq) + NO_3^-(aq)$$
$$H_2SO_4 \xrightarrow{H_2O} H^+(aq) + HSO_4^-(aq)$$

Arrhenius proposed that an **acid** *is a substance that produces H^+ ions (protons) when it is dissolved in water.*

Studies show that when HCl, HNO_3, and H_2SO_4 are placed in water, *virtually every molecule* dissociates to give ions. This means that when 100 molecules of HCl are dissolved in water, 100 H^+ ions and 100 Cl^- ions are produced. Virtually no HCl molecules exist in aqueous solution (see Figure 7.5). Because these substances are strong electrolytes that produce H^+ ions, they are called **strong acids.**

Arrhenius also found that *aqueous solutions that exhibit basic behavior* always contain hydroxide ions. He defined a **base** as a *substance that produces hydroxide ions (OH^-) in water.* The base most commonly used in the chemical laboratory is sodium hydroxide, NaOH, which contains Na^+ and OH^- ions and is very soluble in water. Sodium hydroxide, like all ionic substances, produces separated cations and anions when it is dissolved in water.

$$NaOH(s) \xrightarrow{H_2O} Na^+(aq) + OH^-(aq)$$

Although dissolved sodium hydroxide is usually represented as NaOH(*aq*), you should remember that the solution really contains separated Na^+ and OH^- ions. In fact, for every 100 units of NaOH dissolved in water, 100 Na^+ and 100 OH^- ions are produced.

Potassium hydroxide (KOH) has properties markedly similar to those of sodium hydroxide. It is very soluble in water and produces separated ions.

$$KOH(s) \xrightarrow{H_2O} K^+(aq) + OH^-(aq)$$

Because these hydroxide compounds are strong electrolytes that contain OH^- ions, they are called **strong bases.**

When strong acids and strong bases (hydroxides) are mixed, the fundamental chemical change that always occurs is that *H^+ ions react with OH^- ions to form water.*

$$H^+(aq) + OH^-(aq) \rightarrow H_2O(l)$$

The marsh marigold is a beautiful but poisonous plant. Its toxicity results partly from the presence of erucic acid.

Water is a very stable compound, as evidenced by the abundance of it on the earth's surface. Therefore, when substances that can form water are mixed, there is a strong tendency for the reaction to occur. In particular, the hydroxide ion OH^- has a high affinity for H^+ ions, because water is produced in the reaction between these ions.

The tendency to form water is the second of the driving forces for reactions that we mentioned in Section 7.1. Any compound that produces OH^- ions in water reacts vigorously to form H_2O with any compound that

Hydrochloric acid is an aqueous solution that contains dissolved hydrogen chloride. It is a strong electrolyte.

can furnish H^+ ions. For example, the reaction between hydrochloric acid and aqueous sodium hydroxide is represented by the following molecular equation:

$$HCl(aq) + NaOH(aq) \rightarrow H_2O(l) + NaCl(aq)$$

Because HCl, NaOH, and NaCl exist as completely separated ions in water, the complete ionic equation for this reaction is

$$H^+(aq) + Cl^-(aq) + Na^+(aq) + OH^-(aq) \rightarrow H_2O(l) + Na^+(aq) + Cl^-(aq)$$

Notice that the Cl^- and Na^+ are spectator ions (they undergo no changes), so the net ionic equation is

$$H^+(aq) + OH^-(aq) \rightarrow H_2O(l)$$

Thus the only chemical change that occurs when these solutions are mixed is that water is formed from H^+ and OH^- ions.

Example 7.4 Writing Equations for Acid–Base Reactions

Nitric acid is a strong acid. Write the molecular, complete ionic, and net ionic equations for the reaction of aqueous nitric acid and aqueous potassium hydroxide.

Solution

Molecular equation:

$$HNO_3(aq) + KOH(aq) \rightarrow H_2O(l) + KNO_3(aq)$$

Complete ionic equation:

$$H^+(aq) + NO_3^-(aq) + K^+(aq) + OH^-(aq) \rightarrow H_2O(l) + K^+(aq) + NO_3^-(aq)$$

Net ionic equation:

$$H^+(aq) + OH^-(aq) \rightarrow H_2O(l)$$

Note that K^+ and NO_3^- are spectator ions and that the formation of water is the driving force for this reaction. ■

Hydrochloric acid is an aqueous solution of HCl.

There are two important things to note as we examine the reaction of hydrochloric acid with aqueous sodium hydroxide and the reaction of nitric acid with aqueous potassium hydroxide.

1. The net ionic equation is the same in both cases; water is formed.

$$H^+(aq) + OH^-(aq) \rightarrow H_2O(l)$$

2. Besides water, which is *always a product* of the reaction of an acid with OH^-, the second product is an ionic compound, which might precipitate or remain dissolved, depending on its solubility.

$$HCl(aq) + NaOH(aq) \rightarrow H_2O(l) + NaCl(aq)$$

$$HNO_3(aq) + KOH(aq) \rightarrow H_2O(l) + KNO_3(aq)$$

Dissolved ionic compounds

This ionic compound is called a **salt.** In the first case the salt is sodium chloride, and in the second case the salt is potassium nitrate. We can obtain these soluble salts in solid form (both are white solids) by evaporating the water.

Summary of Strong Acids and Strong Bases

The following points about strong acids and strong bases are particularly important.

1. The common strong acids are aqueous solutions of HCl, HNO_3, and H_2SO_4.
2. A strong acid is a substance that completely dissociates (ionizes) in water. (Each molecule breaks up into an H^+ ion plus an anion.)
3. A strong base is a metal hydroxide compound that is very soluble in water. The most common strong bases are NaOH and KOH, which completely break up into separated ions (Na^+ and OH^- or K^+ and OH^-) when they are dissolved in water.
4. The net ionic equation for the reaction of a strong acid and a strong base (contains OH^-) is always the same: it shows the production of water.

$$H^+(aq) + OH^-(aq) \rightarrow H_2O(l)$$

5. In the reaction of a strong acid and a strong base, one product is always water and the other is always an ionic compound called a salt, which remains dissolved in the water. This salt can be obtained as a solid by evaporating the water.
6. The reaction of H^+ and OH^- is often called an acid–base reaction, where H^+ is the acidic ion and OH^- is the basic ion.

Both strong acids and strong bases are strong electrolytes.

Drano contains a strong base.

7.5 Reactions of Metals with Nonmetals (Oxidation–Reduction)

AIMS: To learn the general characteristics of a reaction between a metal and a nonmetal. To understand electron transfer as a driving force for a chemical reaction.

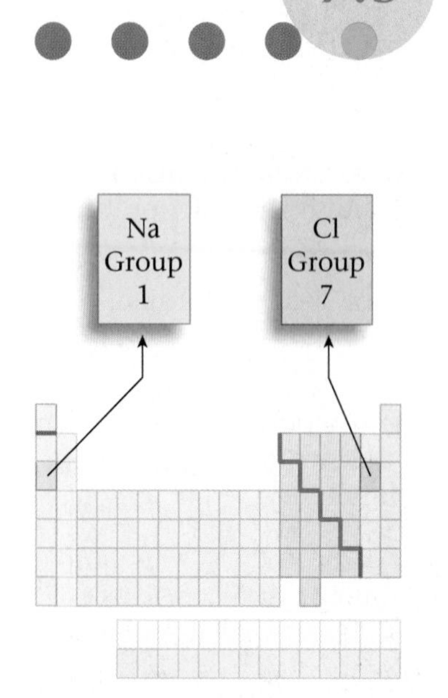

In Chapter 4 we spent considerable time discussing ionic compounds—compounds formed in the reaction of a metal and a nonmetal. A typical example is sodium chloride, formed by the reaction of sodium metal and chlorine gas:

$$2Na(s) + Cl_2(g) \rightarrow 2NaCl(s)$$

Let's examine what happens in this reaction. Sodium metal is composed of sodium atoms, each of which has a net charge of zero. (The positive charges of the eleven protons in its nucleus are exactly balanced by the negative charges on the eleven electrons.) Similarly, the chlorine molecule consists of two uncharged chlorine atoms (each has seventeen protons and seventeen electrons). However, in the product (sodium chloride), the sodium is present as Na^+ and the chlorine as Cl^-. By what process do the neutral atoms become ions? The answer is that one electron is transferred from each sodium atom to each chlorine atom.

$$Na + Cl \xrightarrow{e^-} Na^+ + Cl$$

After the electron transfer, each sodium has ten electrons and eleven protons (a net charge of 1+), and each chlorine has eighteen electrons and seventeen protons (a net charge of 1−).

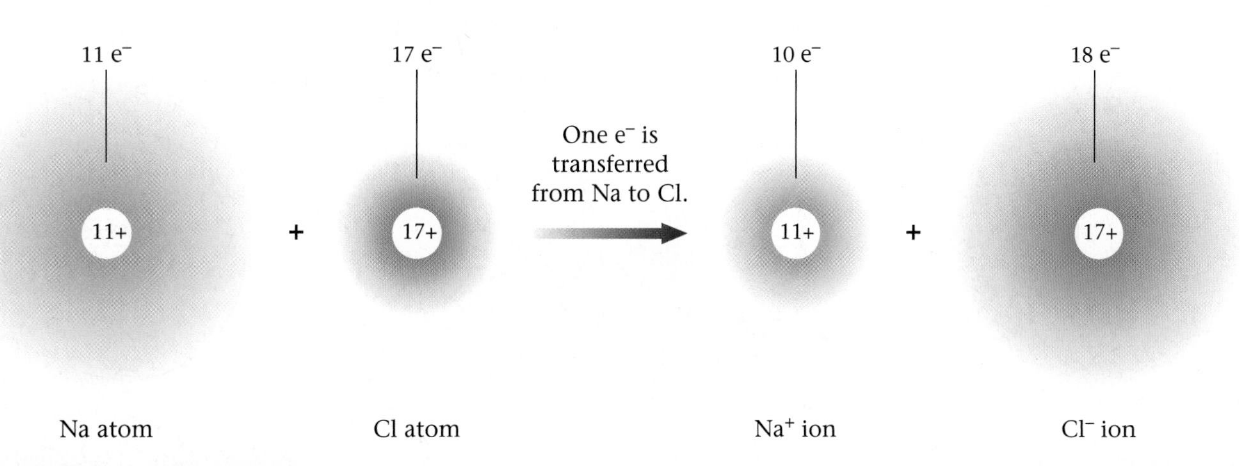

Figure 7.6
The thermite reaction gives off so much heat that the iron formed is molten.

Thus the reaction of a metal with a nonmetal to form an ionic compound involves the transfer of one or more electrons from the metal (which forms a cation) to the nonmetal (which forms an anion). This tendency to transfer electrons from metals to nonmetals is the third driving force for reactions that we listed in Section 7.1. A reaction that *involves a transfer of electrons* is called an **oxidation–reduction reaction.**

There are many examples of oxidation–reduction reactions in which a metal reacts with a nonmetal to form an ionic compound. Consider the reaction of magnesium metal with oxygen,

$$2Mg(s) + O_2(g) \rightarrow 2MgO(s)$$

which produces a bright, white light useful in camera flash units. Note that the reactants contain uncharged atoms, but the product contains ions:

$$MgO$$
Contains Mg^{2+}, O^{2-}

Therefore, in this reaction, each magnesium atom loses two electrons ($Mg \rightarrow Mg^{2+} + 2e^-$) and each oxygen atom gains two electrons ($O + 2e^- \rightarrow O^{2-}$). We might represent this reaction as follows:

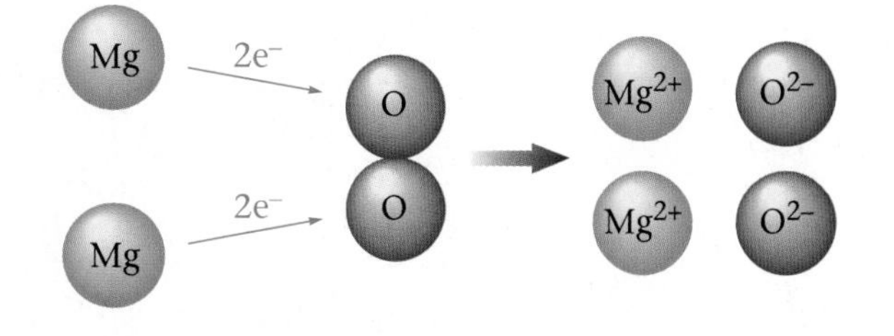

Another example is

$$2Al(s) + Fe_2O_3(s) \rightarrow 2Fe(s) + Al_2O_3(s)$$

which is a reaction (called the thermite reaction) that produces so much energy (heat) that the iron is initially formed as a liquid (see Figure 7.6). In this case the aluminum is originally present as the elemental metal (which contains uncharged Al atoms) and ends up in Al_2O_3, where it is present as Al^{3+} cations (the $2Al^{3+}$ ions just balance the charge of the $3O^{2-}$ ions). Therefore, in the reaction each aluminum atom loses three electrons.

This equation is read, "An aluminum atom yields an aluminum ion with a 3+ charge and three electrons."

$$Al \rightarrow Al^{3+} + 3e^-$$

The opposite process occurs with the iron, which is initially present as Fe^{3+} ions in Fe_2O_3 and ends up as uncharged atoms in the elemental iron. Thus each iron cation gains three electrons to form an uncharged atom:

$$Fe^{3+} + 3e^- \rightarrow Fe$$

We can represent this reaction in schematic form as follows:

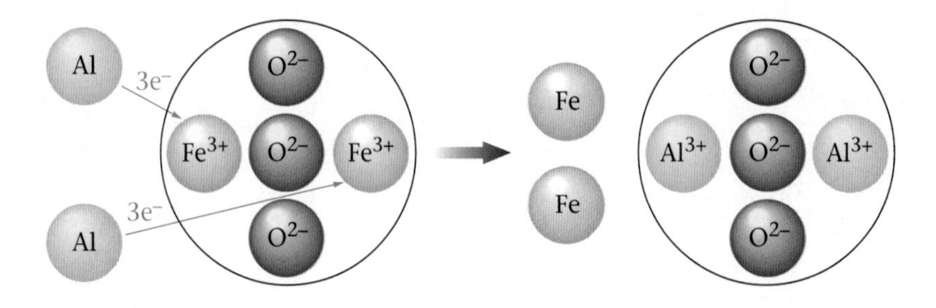

Example 7.5 Identifying Electron Transfer in Oxidation–Reduction Reactions

For each of the following reactions, show how electrons are gained and lost.

a. $2Al(s) + 3I_2(s) \rightarrow 2AlI_3(s)$ (This reaction is shown in Figure 7.7. Note the purple "smoke," which is excess I_2 being driven off by the heat.)

b. $2Cs(s) + F_2(g) \rightarrow 2CsF(s)$

Figure 7.7
When powdered aluminum and iodine (shown in the foreground) are mixed (and a little water added), they react vigorously.

Solution

a. In AlI_3 the ions are Al^{3+} and I^- (aluminum always forms Al^{3+}, and iodine always forms I^-). In Al(s) the aluminum is present as uncharged atoms. Thus aluminum goes from Al to Al^{3+} by losing three electrons ($Al \rightarrow Al^{3+} + 3e^-$). In I_2 each iodine atom is uncharged. Thus each iodine atom goes from I to I^- by gaining one electron ($I + e^- \rightarrow I^-$). A schematic for this reaction is

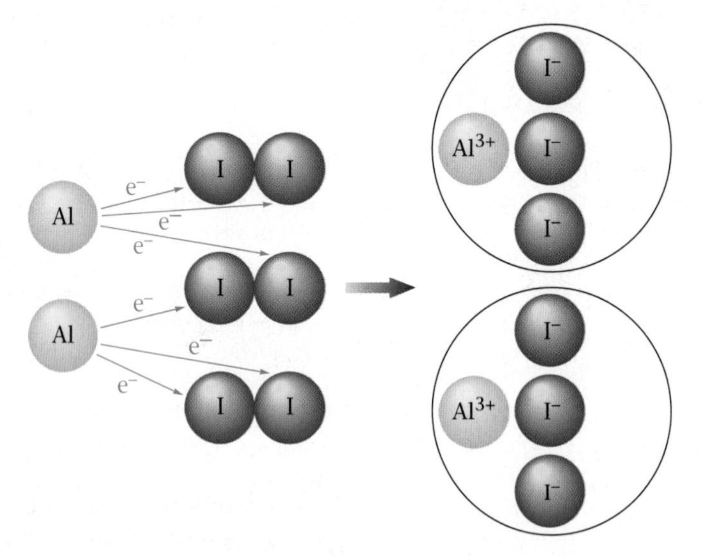

b. In CsF the ions present are Cs^+ and F^-. Cesium metal, Cs(s), contains uncharged cesium atoms, and fluorine gas, $F_2(g)$, contains uncharged fluorine atoms. Thus in the reaction each cesium atom loses one electron ($Cs \rightarrow Cs^+ + e^-$) and each fluorine atom gains one electron ($F + e^- \rightarrow F^-$). The schematic for this reaction is

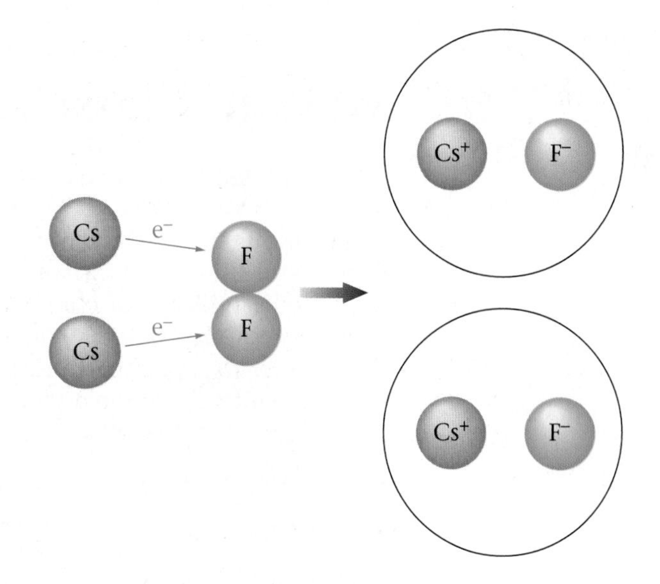

Self-Check Exercise 7.3

For each reaction, show how electrons are gained and lost.

a. $2Na(s) + Br_2(l) \rightarrow 2NaBr(s)$

b. $2Ca(s) + O_2(g) \rightarrow 2CaO(s)$

See Problems 7.47 and 7.48. ■

So far we have emphasized electron transfer (oxidation–reduction) reactions that involve a metal and a nonmetal. Electron transfer reactions can also take place between two nonmetals. We will not discuss these reactions in detail here. All we will say at this point is that one sure sign of an oxidation–reduction reaction between nonmetals is the presence of oxygen, $O_2(g)$, as a reactant or product. In fact, oxidation got its name from oxygen. Thus the reactions

$$CH_4(g) + 2O_2(g) \rightarrow CO_2(g) + 2H_2O(g)$$

and

$$2SO_2(g) + O_2(g) \rightarrow 2SO_3(g)$$

are electron transfer reactions, even though it is not obvious at this point.

We can summarize what we have learned about oxidation–reduction reactions as follows:

Characteristics of Oxidation–Reduction Reactions

1. When a metal reacts with a nonmetal, an ionic compound is formed. The ions are formed when the metal transfers one or more electrons to the nonmetal, the metal atom becoming a cation and the nonmetal atom becoming an anion. *Therefore, a metal–nonmetal reaction can always be assumed to be an oxidation–reduction reaction, which involves electron transfer.*
2. Two nonmetals can also undergo an oxidation–reduction reaction. At this point we can recognize these cases only by looking for O_2 as a reactant or product. When two nonmetals react, the compound formed is not ionic.

Ways to Classify Reactions

AIM: To learn various classification schemes for reactions.

So far in our study of chemistry we have seen many, many chemical reactions—and this is just Chapter 7. In the world around us and in our bodies, literally millions of chemical reactions are taking place. Obviously, we need a system for putting reactions into meaningful classes that will make them easier to remember and easier to understand.

In Chapter 7 we have so far considered the following "driving forces" for chemical reactions:

- Formation of a solid
- Formation of water
- Transfer of electrons

We will now discuss how to classify reactions involving these processes. For example, in the reaction

$$\underset{\text{Solution}}{K_2CrO_4(aq)} + \underset{\text{Solution}}{Ba(NO_3)_2(aq)} \rightarrow \underset{\text{Solid formed}}{BaCrO_4(s)} + \underset{\text{Solution}}{2KNO_3(aq)}$$

solid $BaCrO_4$ (a precipitate) is formed. Because the *formation of a solid when two solutions are mixed* is called *precipitation,* we call this a **precipitation reaction.**

Notice in this reaction that two anions (NO_3^- and CrO_4^{2-}) are simply exchanged. Note that CrO_4^{2-} was originally associated with K^+ in K_2CrO_4 and that NO_3^- was associated with Ba^{2+} in $Ba(NO_3)_2$. In the products these associations are reversed. Because of this double exchange, we sometimes call this reaction a double-exchange reaction or **double-displacement reaction.** We might represent such a reaction as

$$AB + CD \rightarrow AD + CB$$

So we can classify a reaction such as this one as a precipitation reaction or as a double-displacement reaction. Either name is correct, but the former is more commonly used by chemists.

In this chapter we have also considered reactions in which water is formed when a strong acid is mixed with a strong base. All of these reactions had the same net ionic equation:

$$H^+(aq) + OH^-(aq) \rightarrow H_2O(l)$$

The H^+ ion comes from a strong acid, such as $HCl(aq)$ or $HNO_3(aq)$, and the origin of the OH^- ion is a strong base, such as $NaOH(aq)$ or $KOH(aq)$. An example is

$$HCl(aq) + KOH(aq) \rightarrow H_2O(l) + KCl(aq)$$

We classify these reactions as **acid–base reactions.** You can recognize this as an acid–base reaction because it *involves an H^+ ion that ends up in the product water.*

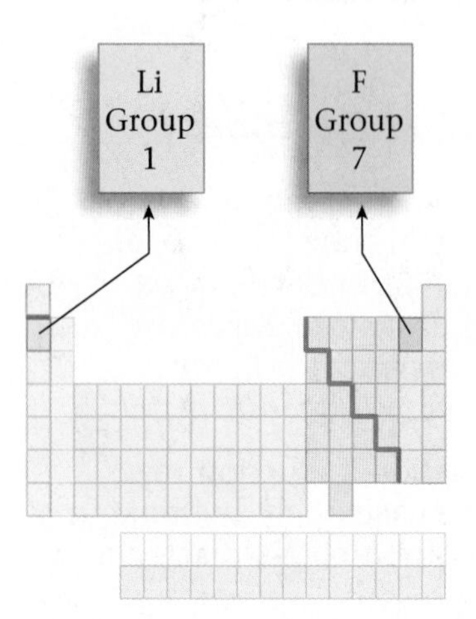

The third driving force is electron transfer. We see evidence of this driving force particularly in the "desire" of a metal to donate electrons to nonmetals. An example is

$$2Li(s) + F_2(g) \rightarrow 2LiF(s)$$

CHEMISTRY IN FOCUS

Do We Age by Oxidation?

People (especially those over age 30) seem obsessed about staying young, but the fountain of youth sought since the days of Ponce de Leon has proved elusive. The body inevitably seems to wear out after 70 or 80 years. Is this our destiny or can we find ways to combat aging?

Why do we age? No one knows for certain, but many scientists think that oxidation plays a major role. Although oxygen is essential for life, it can also have a detrimental effect. The oxygen molecule and other oxidizing substances in the body can extract single electrons from the large molecules that make up cell membranes (walls), thus causing them to become very reactive. In fact, these activated molecules can react with each other to change the properties of the cell membranes. If enough of these changes accumulate, the body's immune system comes to view the changed cell as "foreign" and destroys it. This action is particularly harmful to the organism if the cells involved are irreplaceable, such as nerve cells.

Because the human body is so complex, it is very difficult to pinpoint the cause or causes of aging. Scientists are therefore studying simpler life forms. For example, Rajundar Sohal and his coworkers at Southern Methodist University in Dallas are examining aging in common houseflies. Their work indicates that the accumulated damage from oxidation is linked to both the fly's vitality and its life expectancy. One study showed that flies that were forced to be sedentary (couldn't fly around) showed much less damage from oxidation (because of their lower oxygen consumption) and lived twice as long as flies that had normal activities.

Accumulated knowledge from various studies indicates that oxidation is probably a major cause of aging. If this is true, how can we protect ourselves? The best way to approach the answer to this question is to study the body's natural defenses against oxidation. A recent study by Russel J. Reiter of the Texas Health Science Center at San Antonio has shown that melatonin—a chemical secreted by the pineal gland in the brain (but only at night)—protects against oxidation. In addition, it has long been known that vitamin E is an antioxidant. Studies have shown that red blood cells deficient in vitamin E age much faster than cells with normal vitamin E levels. On the basis of this type of evidence many people take daily doses of vitamin E to ward off the effects of aging.

Oxidation is only one possible cause of aging. Research continues on many fronts to find out why we get "older" as time passes.

Foods that contain natural antioxidants.

where each lithium atom loses one electron to form Li^+, and each fluorine atom gains one electron to form the F^- ion. The process of electron transfer is also called oxidation–reduction. Thus we classify the above reaction as an **oxidation–reduction reaction.**

An additional driving force for chemical reactions that we have not yet discussed is *formation of a gas*. A reaction in aqueous solution that forms a gas (which escapes as bubbles) is pulled toward the products by this event. An example is the reaction

$$2HCl(aq) + Na_2CO_3(aq) \rightarrow CO_2(g) + H_2O(l) + NaCl(aq)$$

for which the net ionic equation is

$$2H^+(aq) + CO_3^{2-}(aq) \rightarrow CO_2(g) + H_2O(l)$$

Note that this reaction forms carbon dioxide gas as well as water, so it illustrates two of the driving forces that we have considered. Because this reaction involves H^+ that ends up in the product water, we classify it as an acid–base reaction.

Consider another reaction that forms a gas:

$$Zn(s) + 2HCl(aq) \rightarrow H_2(g) + ZnCl_2(aq)$$

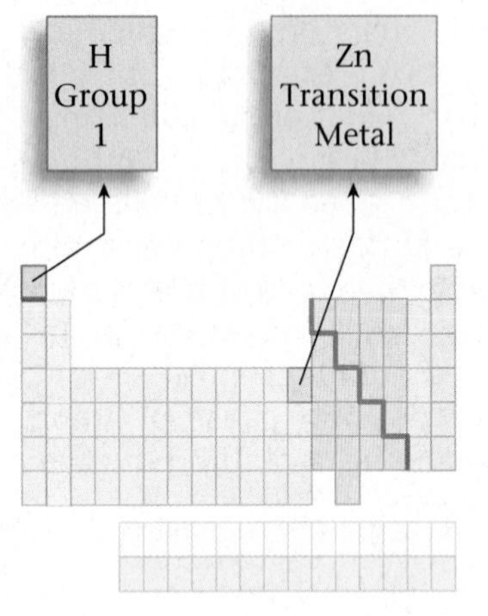

How might we classify this reaction? A careful look at the reactants and products shows the following:

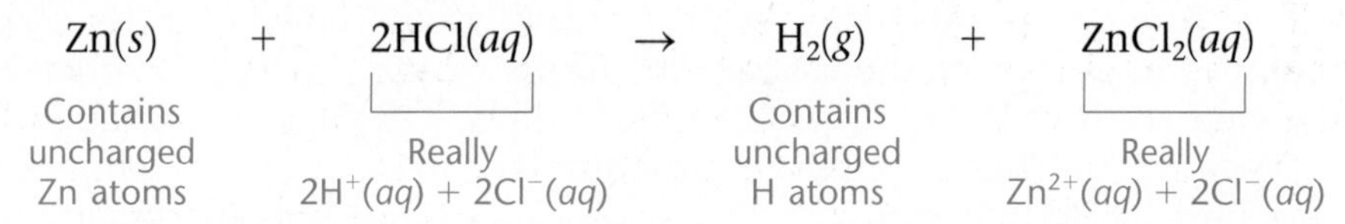

Note that in the reactant zinc metal, Zn exists as uncharged atoms, whereas in the product it exists as Zn^{2+}. Thus each Zn atom loses two electrons. Where have these electrons gone? They have been transferred to two H^+ ions to form H_2. The schematic for this reaction is

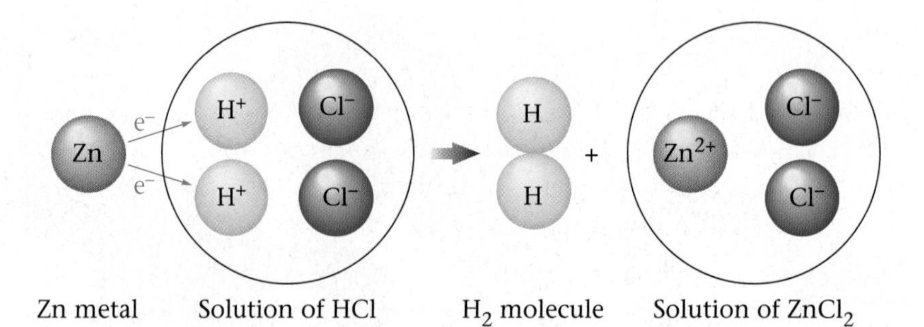

This is an electron transfer process, so the reaction can be classified as an oxidation–reduction reaction.

Another way this reaction is sometimes classified is based on the fact that a *single* type of anion (Cl^-) has been exchanged between H^+ and Zn^{2+}. That is, Cl^- is originally associated with H^+ in HCl and ends up associated with Zn^{2+} in the product $ZnCl_2$. We can call this a *single-replacement reaction* in contrast to double-displacement reactions, in which two types of anions are exchanged. We can represent a single replacement as

$$A + BC \rightarrow B + AC$$

CHEMISTRY IN FOCUS

Oxidation–Reduction Reactions Launch the Space Shuttle

Launching into space a vehicle that weighs millions of pounds requires unimaginable quantities of energy—all furnished by oxidation–reduction reactions.

Notice from Figure 7.8 that three cylindrical objects are attached to the shuttle orbiter. In the center is a tank about 28 feet in diameter and 154 feet long that contains liquid oxygen and liquid hydrogen (in separate compartments). These fuels are fed to the orbiter's rocket engines, where they react to form water and release a huge quantity of energy.

$$2H_2 + O_2 \rightarrow 2H_2O + \text{energy}$$

Note that we can recognize this reaction as an oxidation–reduction reaction because O_2 is a reactant.

Two solid-fuel rockets 12 feet in diameter and 150 feet long are also attached to the orbiter. Each rocket contains 1.1 million pounds of fuel: ammonium perchlorate (NH_4ClO_4) and powdered aluminum mixed with a binder ("glue"). Because the rockets are so large, they are built in segments and assembled at the launch site as shown in Figure 7.9. Each segment is filled with the syrupy propellant (Figure 7.10), which then solidifies to a consistency much like that of a hard rubber eraser.

The oxidation–reduction reaction between the ammonium perchlorate and the aluminum is represented as follows:

$$3NH_4ClO_4(s) + 3Al(s) \rightarrow Al_2O_3(s) + AlCl_3(s) + 3NO(g) + 6H_2O(g) + \text{energy}$$

It produces temperatures of about 5700 °F and 3.3 million pounds of thrust in each rocket.

Thus we can see that oxidation–reduction reactions furnish the energy to launch the space shuttle.

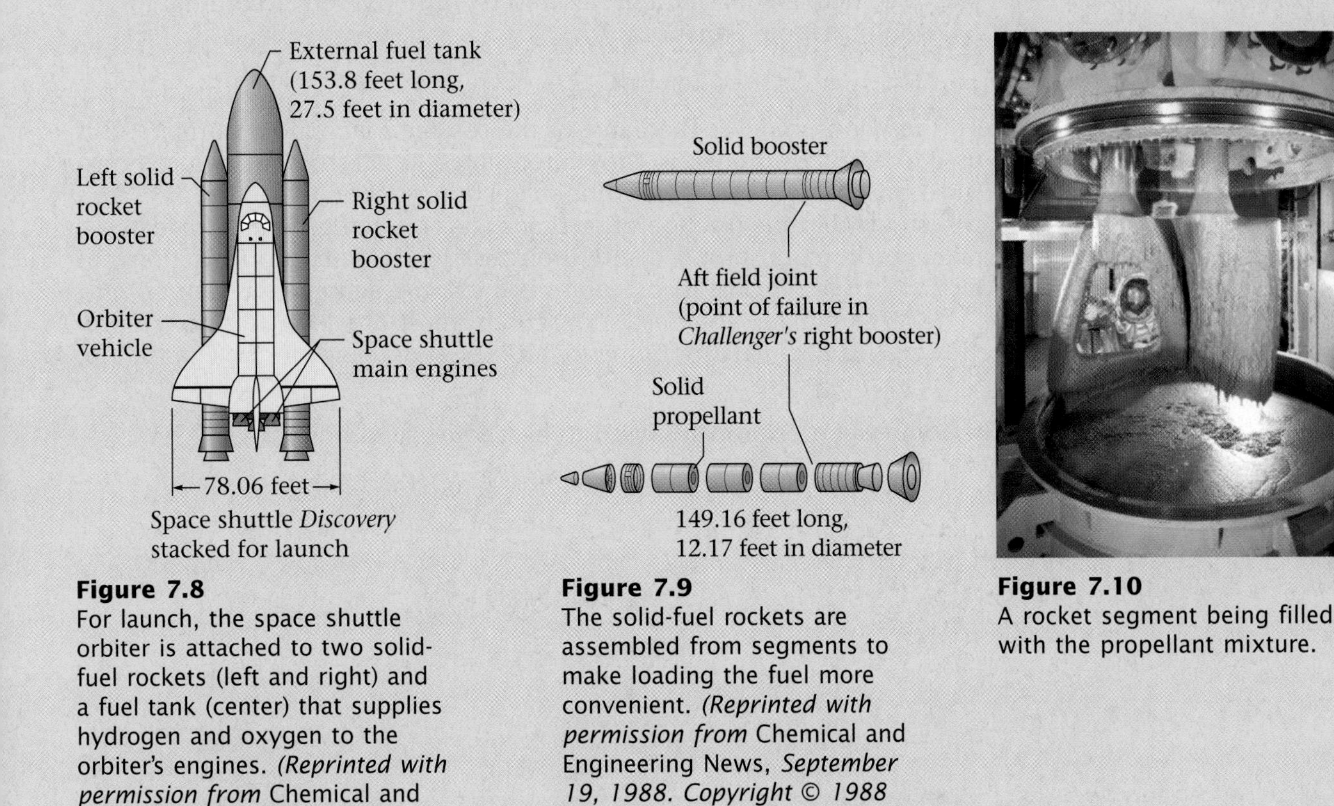

Figure 7.8
For launch, the space shuttle orbiter is attached to two solid-fuel rockets (left and right) and a fuel tank (center) that supplies hydrogen and oxygen to the orbiter's engines. *(Reprinted with permission from* Chemical and Engineering News, *September 19, 1988. Copyright © 1988 American Chemical Society.)*

Figure 7.9
The solid-fuel rockets are assembled from segments to make loading the fuel more convenient. *(Reprinted with permission from* Chemical and Engineering News, *September 19, 1988. Copyright © 1988 American Chemical Society.)*

Figure 7.10
A rocket segment being filled with the propellant mixture.

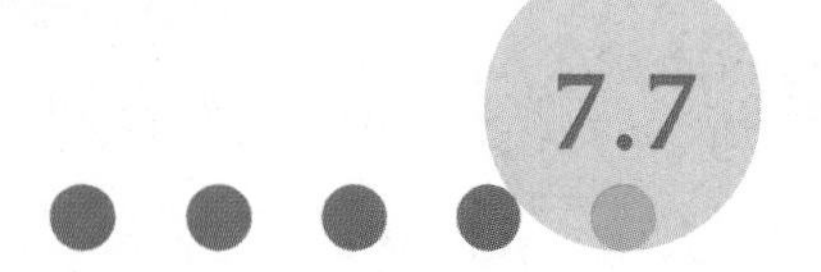

Other Ways to Classify Reactions

AIM: To consider additional classes of chemical reactions.

So far in this chapter we have classified chemical reactions in several ways. The most commonly used of these classifications are

- Precipitation reactions
- Acid–base reactions
- Oxidation–reduction reactions

However, there are still other ways to classify reactions that you may encounter in your future studies of chemistry. We will consider several of these in this section.

Combustion Reactions

Many chemical reactions that involve oxygen produce energy (heat) so rapidly that a flame results. Such reactions are called **combustion reactions.** We have considered some of these reactions previously. For example, the methane in natural gas reacts with oxygen according to the following balanced equation:

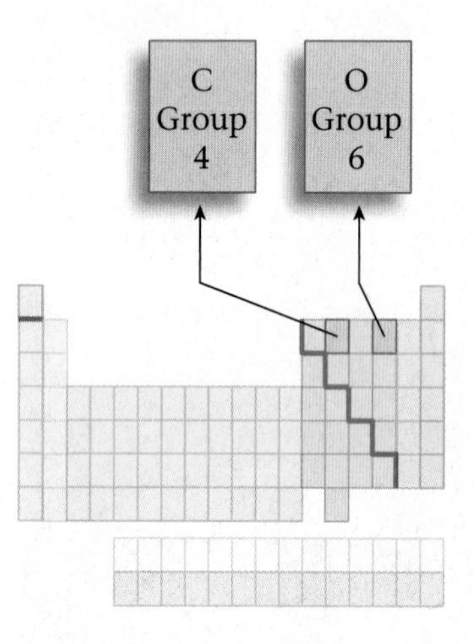

$$CH_4(g) + 2O_2(g) \rightarrow CO_2(g) + 2H_2O(g)$$

This reaction produces the flame of the common laboratory burner and is used to heat most homes in the United States. Recall that we originally classified this reaction as an oxidation–reduction reaction in Section 7.5. Thus we can say that the reaction of methane with oxygen is both an oxidation–reduction reaction and a combustion reaction. Combustion reactions, in fact, are a special class of oxidation–reduction reactions (see Figure 7.11).

There are many combustion reactions, most of which are used to provide heat or electricity for homes or businesses or energy for transportation. Some examples are:

- Combustion of propane (used to heat some rural homes)

$$C_3H_8(g) + 5O_2(g) \rightarrow 3CO_2(g) + 4H_2O(g)$$

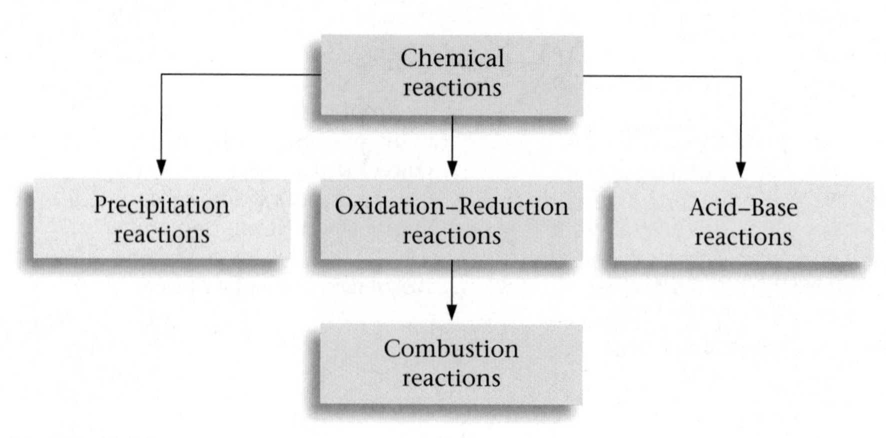

Figure 7.11
Classes of reactions. Combustion reactions are a special type of oxidation–reduction reaction.

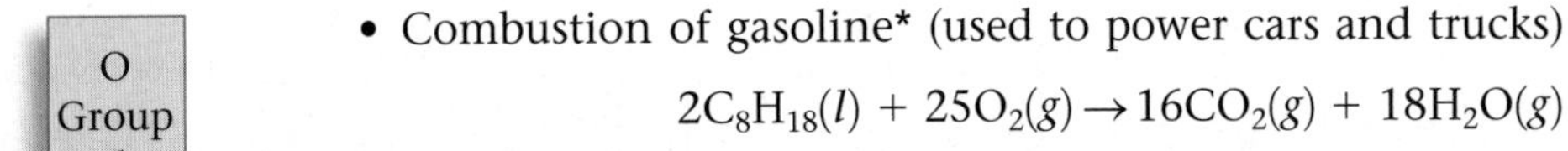

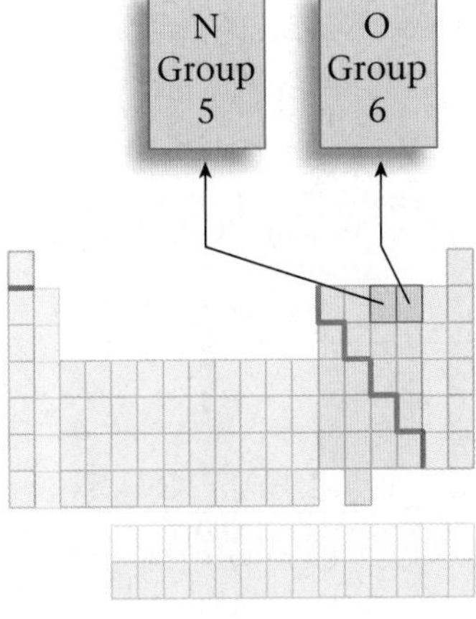

- Combustion of gasoline* (used to power cars and trucks)

$$2C_8H_{18}(l) + 25O_2(g) \rightarrow 16CO_2(g) + 18H_2O(g)$$

- Combustion of coal* (used to generate electricity)

$$C(s) + O_2(g) \rightarrow CO_2(g)$$

Synthesis (Combination) Reactions

One of the most important activities in chemistry is the synthesis of new compounds. Each of our lives has been greatly affected by synthetic compounds such as plastic, polyester, and aspirin. When a given compound is formed from simpler materials, we call this a **synthesis** (or **combination**) **reaction.**

In many cases synthesis reactions start with elements, as shown by the following examples:

- Synthesis of water $2H_2(g) + O_2(g) \rightarrow 2H_2O(l)$
- Synthesis of carbon dioxide $C(s) + O_2(g) \rightarrow CO_2(g)$
- Synthesis of nitrogen monoxide $N_2(g) + O_2(g) \rightarrow 2NO(g)$

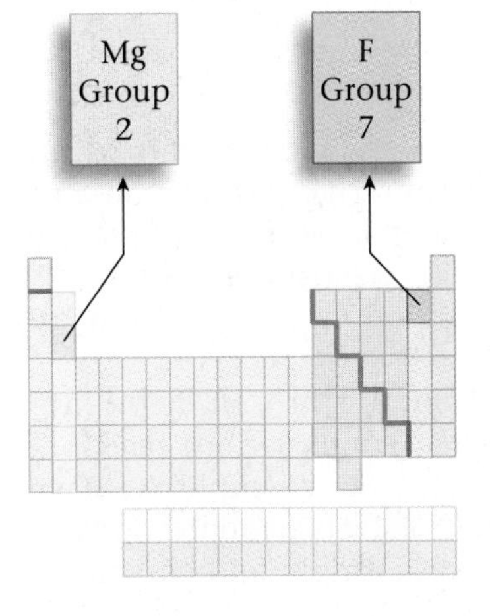

Notice that each of these reactions involves oxygen, so each can be classified as an oxidation–reduction reaction. The first two reactions are also commonly called combustion reactions because they produce flames. The reaction of hydrogen with oxygen to produce water, then, can be classified three ways: as an oxidation–reduction reaction, as a combustion reaction, and as a synthesis reaction.

There are also many synthesis reactions that do not involve oxygen:

- Synthesis of sodium chloride $2Na(s) + Cl_2(g) \rightarrow 2NaCl(s)$
- Synthesis of magnesium fluoride $Mg(s) + F_2(g) \rightarrow MgF_2(s)$

We have discussed the formation of sodium chloride before and have noted that it is an oxidation–reduction reaction; uncharged sodium atoms lose electrons to form Na^+ ions, and uncharged chlorine atoms gain electrons to form Cl^- ions. The synthesis of magnesium fluoride is also an oxidation–reduction reaction because Mg^{2+} and F^- ions are produced from the uncharged atoms.

We have seen that synthesis reactions in which the reactants are elements are oxidation–reduction reactions as well. In fact, we can think of these synthesis reactions as another subclass of the oxidation–reduction class of reactions.

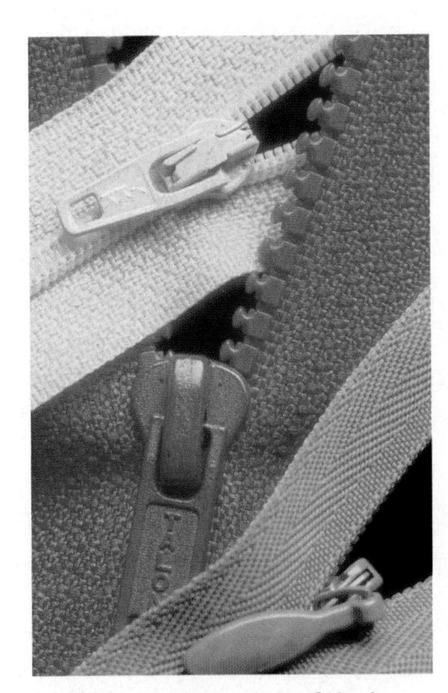

Formation of the colorful plastics used in these zippers is an example of a synthetic reaction.

Decomposition Reactions

In many cases a compound can be broken down into simpler compounds or all the way to the component elements. This is usually accomplished by heating or by the application of an electric current. Such reactions are called **decomposition reactions.** We have discussed decomposition reactions before, including

- Decomposition of water

$$2H_2O(l) \xrightarrow[\text{Electric current}]{} 2H_2(g) + O_2(g)$$

* This substance is really a complex mixture of compounds, but the reaction shown is representative of what takes place.

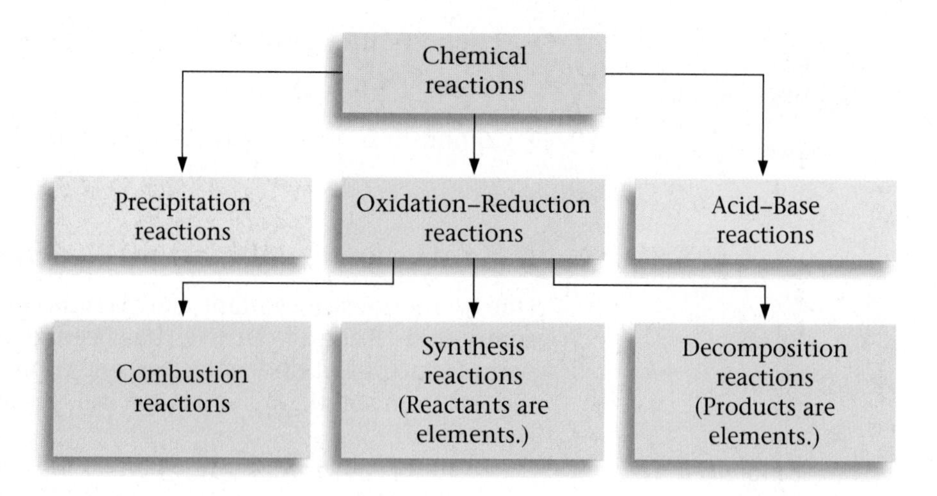

Figure 7.12
Summary of classes of reactions.

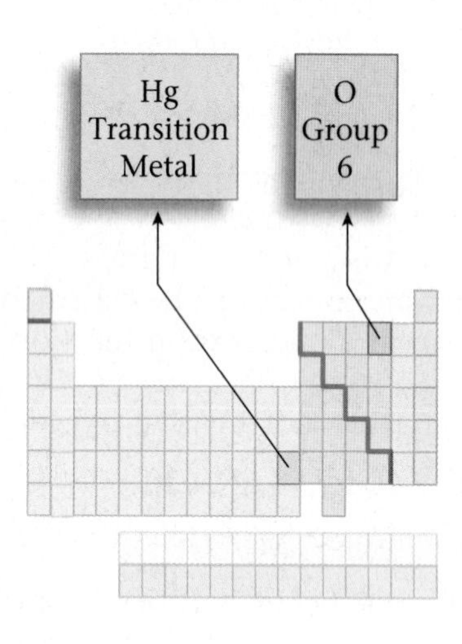

• Decomposition of mercury(II) oxide

$$2HgO(s) \xrightarrow[\text{Heat}]{} 2Hg(l) + O_2(g)$$

Because O_2 is involved in the first reaction, we recognize it as an oxidation–reduction reaction. In the second reaction, HgO, which contains Hg^{2+} and O^{2-} ions, is decomposed to the elements, which contain uncharged atoms. In this process each Hg^{2+} gains two electrons and each O^{2-} loses two electrons, so this is both a decomposition reaction and an oxidation–reduction reaction.

A decomposition reaction, in which a compound is broken down into its elements, is just the opposite of the synthesis (combination) reaction, in which elements combine to form the compound. For example, we have just discussed the synthesis of sodium chloride from its elements. Sodium chloride can be decomposed into its elements by melting it and passing an electric current through it:

$$2NaCl(l) \xrightarrow[\text{Electric current}]{} 2Na(l) + Cl_2(g)$$

There are other schemes for classifying reactions that we have not considered. However, we have covered many of the classifications that are commonly used by chemists as they pursue their science in laboratories and industrial plants.

It should be apparent that many important reactions can be classified as oxidation–reduction reactions. As shown in Figure 7.12, various types of reactions can be viewed as subclasses of the overall oxidation–reduction category.

Example 7.6 Classifying Reactions

Classify each of the following reactions in as many ways as possible.

a. $2K(s) + Cl_2(g) \rightarrow 2KCl(s)$

b. $Fe_2O_3(s) + 2Al(s) \rightarrow Al_2O_3(s) + 2Fe(s)$

c. $2Mg(s) + O_2(g) \rightarrow 2MgO(s)$

d. $HNO_3(aq) + NaOH(aq) \rightarrow H_2O(l) + NaNO_3(aq)$

e. $KBr(aq) + AgNO_3(aq) \rightarrow AgBr(s) + KNO_3(aq)$

f. $PbO_2(s) \rightarrow Pb(s) + O_2(g)$

Solution

a. This is both a synthesis reaction (elements combine to form a compound) and an oxidation–reduction reaction (uncharged potassium and chlorine atoms are changed to K^+ and Cl^- ions in KCl).

b. This is an oxidation–reduction reaction. Iron is present in $Fe_2O_3(s)$ as Fe^{3+} ions and in elemental iron, Fe(*s*), as uncharged atoms. So each Fe^{3+} must gain three electrons to form Fe. The reverse happens to aluminum, which is present initially as uncharged aluminum atoms, each of which loses three electrons to give Al^{3+} ions in Al_2O_3. Note that this reaction might also be called a single-replacement reaction because O is switched from Fe to Al.

c. This is both a synthesis reaction (elements combine to form a compound) and an oxidation–reduction reaction (each magnesium atom loses two electrons to give Mg^{2+} ions in MgO, and each oxygen atom gains two electrons to give O^{2-} in MgO).

d. This is an acid–base reaction. It might also be called a double-displacement reaction because NO_3^- and OH^- "switch partners."

e. This is a precipitation reaction that might also be called a double-displacement reaction in which the anions Br^- and NO_3^- are exchanged.

f. This is a decomposition reaction (a compound breaks down into elements). It also is an oxidation–reduction reaction, because the ions in PbO_2 (Pb^{4+} and O^{2-}) are changed to uncharged atoms in the elements Pb(*s*) and $O_2(g)$. That is, electrons are transferred from O^{2-} to Pb^{4+} in the reaction.

✔ Self-Check Exercise 7.4

Classify each of the following reactions in as many ways as possible.

a. $4NH_3(g) + 5O_2(g) \rightarrow 4NO(g) + 6H_2O(g)$

b. $S_8(s) + 8O_2(g) \rightarrow 8SO_2(g)$

c. $2Al(s) + 3Cl_2(g) \rightarrow 2AlCl_3(s)$

d. $2AlN(s) \rightarrow 2Al(s) + N_2(g)$

e. $BaCl_2(aq) + Na_2SO_4(aq) \rightarrow BaSO_4(s) + 2NaCl(aq)$

f. $2Cs(s) + Br_2(l) \rightarrow 2CsBr(s)$

g. $KOH(aq) + HCl(aq) \rightarrow H_2O(l) + KCl(aq)$

h. $2C_2H_2(g) + 5O_2(g) \rightarrow 4CO_2(g) + 2H_2O(l)$

See Problems 7.53 and 7.54. ■

Chapter 7 Review

Key Terms

precipitation (7.2)
precipitate (7.2)
precipitation reaction (7.2, 7.6)
strong electrolyte (7.2)
soluble solid (7.2)
insoluble (slightly soluble) solid (7.2)
molecular equation (7.3)
complete ionic equation (7.3)
spectator ions (7.3)
net ionic equation (7.3)
acid (7.4)
strong acid (7.4)
base (7.4)
strong base (7.4)
salt (7.4)
oxidation–reduction reaction (7.5, 7.6)
double-displacement reaction (7.6)
acid–base reaction (7.6)
combustion reaction (7.7)
synthesis (combination) reaction (7.7)
decomposition reaction (7.7)

Summary

1. Four driving forces that favor chemical change (chemical reaction) are formation of a solid, formation of water, transfer of electrons, and formation of a gas.
2. A reaction where a solid forms is called a precipitation reaction. General rules on solubility help predict whether a solid—and what solid—will form when two solutions are mixed.
3. Three types of equations are used to describe reactions in solution: (1) the molecular equation, which shows the complete formulas of all reactants and products; (2) the complete ionic equation, in which all reactants and products that are strong electrolytes are shown as ions; and (3) the net ionic equation, which includes only those components of the solution that undergo a change. Spectator ions (those ions that remain unchanged in a reaction) are not included in a net ionic equation.
4. A strong acid is a compound in which virtually every molecule dissociates in water to give an H^+ ion and an anion. Similarly, a strong base is a metal hydroxide compound that is soluble in water, giving OH^- ions and cations. The products of the reaction of a strong acid and a strong base are water and a salt.
5. Reactions of metals and nonmetals involve a transfer of electrons and are called oxidation–reduction reactions. A reaction between a nonmetal and oxygen is also an oxidation–reduction reaction. Combustion reactions involve oxygen and are a subgroup of oxidation–reduction reactions.
6. When a given compound is formed from simpler materials, such as elements, the reaction is called a synthesis or combination reaction. The reverse process, which occurs when a compound is broken down into its component elements, is called a decomposition reaction. These reactions are also subgroups of oxidation–reduction reactions.

In-Class Discussion Questions

These questions are designed to be considered by groups of students in class. Often these questions work well for introducing a particular topic in class.

1. You add an aqueous solution of lead nitrate to an aqueous solution of potassium iodide. Draw highly magnified views of each individual solution and the mixed solution, including any product that forms. Write the balanced equation for the reaction.
2. Assume a highly magnified view of a solution of HCl that allows you to "see" the HCl. Draw this magnified view. If you dropped in a piece of magnesium, the magnesium would disappear and hydrogen gas would be released. Represent this change using symbols for the elements, and write the balanced equation.
3. Consider exposed electrodes of a lightbulb in a solution of H_2SO_4 such that the lightbulb is on. You add a dilute solution and the bulb grows dim. Which of the following could be in the added solution?
   a. $Ba(OH)_2$ c. K_2SO_4
   b. $NaNO_3$ d. $Ca(NO_3)_2$
   For choices you did not pick, explain what you feel is wrong with them, and justify the choice you did pick.
4. Why is the formation of a solid evidence of a chemical reaction? Use a molecular-level drawing in your explanation.
5. Sketch molecular-level drawings to differentiate between two soluble compounds: one that is a strong electrolyte, and one that is not an electrolyte.
6. Mixing an aqueous solution of potassium nitrate with an aqueous solution of sodium chloride does not result in a chemical reaction. Why?
7. Why is the formation of water evidence of a chemical reaction? Use a molecular-level drawing in your explanation.

8. Use the Arrhenius definition of acids and bases to write the net ionic equation for the reaction of an acid with a base.

9. Why is the transfer of electrons evidence of a chemical reaction? Use a molecular-level drawing in your explanation.

10. Why is the formation of a gas evidence of a chemical reaction? Use a molecular-level drawing in your explanation.

Questions and Problems

All even-numbered exercises have answers in the back of this book and solutions in the *Solutions Guide.*

7.1 Predicting Whether a Reaction Will Occur

QUESTIONS

1. Why is water an important solvent? Although you have not yet studied water in detail, can you think of some properties of water that make it so important?

2. What is a "driving force"? What are some of the driving forces discussed in this section that tend to make reactions likely to occur? Can you think of any other possible driving forces?

7.2 Reactions in Which a Solid Forms

QUESTIONS

3. What do we mean by a *precipitation reaction?*

4. When two solutions of ionic substances are mixed and a precipitate forms, what is the net charge of the precipitate? Why?

5. Describe briefly what happens when an ionic substance is dissolved in water.

6. When an ionic substance dissolves, the resulting solution contains the separated ________.

7. What is meant by a *strong electrolyte?* Give two examples of substances that behave in solution as strong electrolytes.

8. How do chemists know that the ions behave independently of one another when an ionic solid is dissolved in water?

9. When aqueous solutions of sodium chloride, NaCl, and silver nitrate, $AgNO_3$, are mixed, a precipitate forms, but this precipitate is *not* sodium nitrate. What does this reaction tell you about the solubility of $NaNO_3$ in water?

10. What do we mean when we say that a solid is "slightly" soluble in water? Is there a practical difference between *slightly* soluble and *in*soluble?

11. On the basis of the general solubility rules given in Table 7.1, predict which of the following substances are likely to be soluble in water.
    a. aluminum nitrate
    b. magnesium chloride
    c. rubidium sulfate
    d. nickel(II) hydroxide
    e. lead(II) sulfide
    f. magnesium hydroxide
    g. iron(III) phosphate

12. On the basis of the general solubility rules given in Table 7.1, predict which of the following substances are likely to be soluble in water.
    a. zinc chloride
    b. lead(II) nitrate
    c. lead(II) sulfate
    d. sodium iodide
    e. cobalt(III) sulfide
    f. chromium(III) hydroxide
    g. magnesium carbonate
    h. ammonium carbonate

13. On the basis of the general solubility rules given in Table 7.1, for each of the following compounds, indicate why the compound is *not* likely to be soluble in water. Indicate which of the solubility rules covers each substance's particular situation.
    a. chromium(III) sulfide
    b. cobalt(II) hydroxide
    c. zinc phosphate
    d. mercurous chloride

14. On the basis of the general solubility rules given in Table 7.1, for each of the following compounds, indicate why the compound is *not* likely to be soluble in water. Indicate which of the solubility rules covers each substance's particular situation.
    a. calcium sulfate
    b. manganese(II) sulfide
    c. copper(II) hydroxide
    d. zinc carbonate

15. On the basis of the general solubility rules given in Table 7.1, predict the identity of the precipitate that forms when aqueous solutions of the following substances are mixed. If no precipitate is likely, indicate which rules apply.
    a. iron(III) chloride, $FeCl_3$, and phosphoric acid, H_3PO_4
    b. barium nitrate, $Ba(NO_3)_2$, and sodium sulfate, Na_2SO_4
    c. potassium chloride, KCl, and iron(II) sulfate, $FeSO_4$
    d. lead(II) nitrate, $Pb(NO_3)_2$, and hydrochloric acid, HCl
    e. calcium nitrate, $Ca(NO_3)_2$, and sodium chloride, NaCl
    f. ammonium sulfide, $(NH_4)_2S$, and copper(II) chloride, $CuCl_2$

16. On the basis of the general solubility rules given in Table 7.1, predict the identity of the precipitate that forms when aqueous solutions of the following substances are mixed. If no precipitate is likely, indicate which rules apply.
 a. sodium sulfate, Na_2SO_4, and calcium chloride, $CaCl_2$
 b. ammonium iodide, NH_4I, and silver nitrate, $AgNO_3$
 c. potassium phosphate, K_3PO_4, and lead(II) nitrate, $Pb(NO_3)_2$
 d. sodium hydroxide, NaOH, and iron(III) chloride, $FeCl_3$
 e. potassium sulfate, K_2SO_4, and sodium nitrate, $NaNO_3$
 f. sodium carbonate, Na_2CO_3, and barium nitrate, $Ba(NO_3)_2$

PROBLEMS

17. On the basis of the general solubility rules given in Table 7.1, write a balanced molecular equation for the precipitation reactions that take place when the following aqueous solutions are mixed. Underline the formula of the precipitate (solid) that forms. If no precipitation reaction is likely for the reactants given, explain why.
 a. ammonium chloride, NH_4Cl, and sulfuric acid, H_2SO_4
 b. potassium carbonate, K_2CO_3, and tin(IV) chloride, $SnCl_4$
 c. ammonium chloride, NH_4Cl, and lead(II) nitrate, $Pb(NO_3)_2$
 d. copper(II) sulfate, $CuSO_4$, and potassium hydroxide, KOH
 e. sodium phosphate, Na_3PO_4, and chromium(III) chloride, $CrCl_3$
 f. ammonium sulfide, $(NH_4)_2S$, and iron(III) chloride, $FeCl_3$

18. On the basis of the general solubility rules given in Table 7.1, write a balanced molecular equation for the precipitation reactions that take place when the following aqueous solutions are mixed. Underline the precipitate (solid) that forms. If no precipitation reaction is likely for the solutes given, so indicate.
 a. sodium sulfide, Na_2S, and copper(II) chloride, $CuCl_2$
 b. potassium phosphate, K_3PO_4, and aluminum chloride, $AlCl_3$
 c. sulfuric acid, H_2SO_4, and barium chloride, $BaCl_2$
 d. sodium hydroxide, NaOH, and iron(III) chloride, $FeCl_3$
 e. sodium chloride, NaCl, and mercurous nitrate, $Hg_2(NO_3)_2$
 f. potassium carbonate, K_2CO_3, and chromium(III) acetate, $Cr(C_2H_3O_2)_3$

19. Balance each of the following equations that describe precipitation reactions.
 a. $Na_2SO_4(aq) + CaCl_2(aq) \rightarrow CaSO_4(s) + NaCl(aq)$
 b. $Co(C_2H_3O_2)_2(aq) + Na_2S(aq) \rightarrow CoS(s) + NaC_2H_3O_2(aq)$
 c. $KOH(aq) + NiCl_2(aq) \rightarrow Ni(OH)_2(s) + KCl(aq)$

20. Balance each of the following equations that describe precipitation reactions.
 a. $AgNO_3(aq) + H_2SO_4(aq) \rightarrow Ag_2SO_4(s) + HNO_3(aq)$
 b. $Ca(NO_3)_2(aq) + H_2SO_4(aq) \rightarrow CaSO_4(s) + HNO_3(aq)$
 c. $Pb(NO_3)_2(aq) + H_2SO_4(aq) \rightarrow PbSO_4(s) + HNO_3(aq)$

21. For each of the following precipitation reactions, complete and balance the equation, indicating clearly which product is the precipitate.
 a. $Ba(NO_3)_2(aq) + (NH_4)_2SO_4(aq) \rightarrow$
 b. $CoCl_3(aq) + NaOH(aq) \rightarrow$
 c. $FeCl_3(aq) + (NH_4)_2S(aq) \rightarrow$

22. For each of the following precipitation reactions, complete and balance the equation, indicating clearly which product is the precipitate.
 a. $(NH_4)_2S(aq) + CoCl_2(aq) \rightarrow$
 b. $FeCl_3(aq) + NaOH(aq) \rightarrow$
 c. $CuSO_4(aq) + Na_2CO_3(aq) \rightarrow$

7.3 Describing Reactions in Aqueous Solutions

QUESTIONS

23. What is a net ionic equation? What species are shown in such an equation, and which species are not shown?

24. Ions that do not directly participate in a reaction in solution are called ______ ions.

PROBLEMS

25. Write balanced net ionic equations for the reactions that occur when the following aqueous solutions are mixed. If no reaction is likely to occur, so indicate.
 a. silver nitrate, $AgNO_3$, and potassium chloride, KCl
 b. nickel(II) sulfate, $NiSO_4$, and barium chloride, $BaCl_2$
 c. ammonium phosphate, $(NH_4)_3PO_4$, and calcium chloride, $CaCl_2$
 d. hydrofluoric acid, HF, and potassium sulfate, K_2SO_4
 e. calcium chloride, $CaCl_2$, and ammonium sulfate, $(NH_4)_2SO_4$
 f. lead(II) nitrate, $Pb(NO_3)_2$, and barium chloride, $BaCl_2$

26. Write balanced net ionic equations for the reactions that occur when the following aqueous solutions are mixed. If no reaction is likely to occur, so indicate.
 a. sodium sulfate and potassium chloride
 b. potassium sulfide and calcium nitrate
 c. sodium hydroxide and silver nitrate

d. sodium carbonate and iron(III) chloride
e. ammonium phosphate and aluminum chloride
f. barium nitrate and potassium chloride

27. A common analysis for the quantity of halide ions (Cl^-, Br^-, and I^-) in a sample is to precipitate and weigh the halide ions as their silver salts. For example, a given sample of seawater can be treated with dilute silver nitrate, $AgNO_3$, solution to precipitate the halides. The mixture of precipitated silver halides can then be filtered from the solution, dried, and weighed as an indication of the halide content of the original sample. Write the net ionic equations showing the precipitation of halide ions from seawater with silver nitrate.

28. The procedures and principles of qualitative analysis are covered in many introductory chemistry laboratory courses. In qualitative analysis, students learn to analyze mixtures of the common positive and negative ions, separating and confirming the presence of the particular ions in the mixture. One of the first steps in such an analysis is to treat the mixture with hydrochloric acid, which precipitates and removes silver ion, lead(II) ion, and mercury(I) ion from the aqueous mixture as the insoluble chloride salts. Write balanced net ionic equations for the precipitation reactions of these three cations with chloride ion.

29. Many plants are poisonous because their stems and leaves contain oxalic acid, $H_2C_2O_4$, or sodium oxalate, $Na_2C_2O_4$; when ingested, these substances cause swelling of the respiratory tract and suffocation. A standard analysis for determining the amount of oxalate ion, $C_2O_4^{2-}$, in a sample is to precipitate this species as calcium oxalate, which is insoluble in water. Write the net ionic equation for the reaction between sodium oxalate and calcium chloride, $CaCl_2$, in aqueous solution.

30. Another step in the qualitative analysis of cations (see problem 28) involves precipitating some of the metal ions as the insoluble sulfides (followed by subsequent treatment of the mixed sulfide precipitate to separate the individual ions). Write balanced net ionic equations for the reactions of Co(II), Co(III), Fe(II), and Fe(III) ions with sulfide ion, S^{2-}.

7.4 Reactions That Form Water: Acids and Bases

QUESTIONS

31. What is meant by a *strong acid?* Are the strong acids also strong *electrolytes?* Explain.

32. What is meant by a *strong base?* Are the strong bases also strong *electrolytes?* Explain.

33. The same net ionic process takes place when any strong acid reacts with any strong base. Write the equation for that process.

34. Write the formulas and names of three common strong acids and strong bases.

35. If 1000 NaOH units were dissolved in a sample of water, the NaOH would produce ______ Na^+ ions and ______ OH^- ions.

36. The ionic compound produced when a strong acid and a strong base react is called a(n) ______.

PROBLEMS

37. In addition to the three strong acids emphasized in the chapter (HCl, HNO_3, and H_2SO_4), hydrobromic acid, HBr, and perchloric acid, $HClO_4$, are strong acids. Write equations for the dissociation of each of these additional strong acids in water.

38. In addition to the strong bases NaOH and KOH discussed in this chapter, the hydroxide compounds of other Group 1 elements behave as strong bases when dissolved in water. Write equations for RbOH and CsOH that show which ions form when they dissolve in water.

39. What salt would form when each of the following strong acid/strong base reactions takes place?
a. $HCl(aq) + KOH(aq) \rightarrow$
b. $RbOH(aq) + HNO_3(aq) \rightarrow$
c. $HClO_4(aq) + NaOH(aq) \rightarrow$
d. $HBr(aq) + CaOH(aq) \rightarrow$

40. Below are the formulas of some salts. Such salts could form by the reaction of the appropriate strong acid with the appropriate strong base (with the other product of the reaction being, of course, water). For each salt, write an equation showing the formation for the salt from the reaction of the appropriate strong acid and strong base.
a. KCl
b. $NaClO_4$
c. $CsNO_3$
d. K_2SO_4

7.5 Reactions of Metals with Nonmetals (Oxidation–Reduction)

QUESTIONS

41. What is an oxidation–reduction reaction? What is transferred during such a reaction?

42. The reaction $2Na + Cl_2 \rightarrow 2NaCl$, like any reaction between a metal and a nonmetal, involves the ______ of electrons.

43. What do we mean when we say that the transfer of electrons can be the "driving force" for a reaction? Give an example of a reaction where this happens.

44. If atoms of a metallic element (such as sodium) react with atoms of a nonmetallic element (such as sulfur), which element loses electrons and which element gains them?

45. If potassium atoms were to react with atoms of the nonmetal sulfur, how many electrons would each

potassium atom lose? How many electrons would each sulfur atom gain? How many potassium atoms would have to react to provide enough electrons for one sulfur atom? What charges would the resulting potassium and sulfur ions have?

46. If a nitrogen molecule, N_2, were to react with a reactive metal such as potassium, what charge would the resulting nitride ions have? How many electrons would be gained by each nitrogen atom? How many electrons would be gained by each N_2 molecule?

PROBLEMS

47. For the reaction $Mg(s) + Cl_2(g) \rightarrow MgCl_2(s)$, illustrate how electrons are gained and lost during the reaction.

48. For the reaction $2Al(s) + 3Br_2(l) \rightarrow 2AlBr_3(s)$, show how electrons are gained and lost by the atoms.

49. Balance each of the following oxidation–reduction reactions. In each, indicate which substance is being oxidized and which is being reduced.
   a. $Na(s) + S(s) \rightarrow Na_2S(s)$
   b. $Mg(s) + O_2(g) \rightarrow MgO(s)$
   c. $Ca(s) + F_2(g) \rightarrow CaF_2(s)$
   d. $Fe(s) + Cl_2(g) \rightarrow FeCl_3(s)$

50. Balance each of the following oxidation–reduction chemical reactions.
   a. $Fe(s) + S(s) \rightarrow Fe_2S_3(s)$
   b. $Zn(s) + HNO_3(aq) \rightarrow Zn(NO_3)_2(aq) + H_2(g)$
   c. $Sn(s) + O_2(g) \rightarrow SnO(s)$
   d. $K(s) + H_2(g) \rightarrow KH(s)$
   e. $Cs(s) + H_2O(l) \rightarrow CsOH(s) + H_2(g)$

7.6 Ways to Classify Reactions

QUESTIONS

51. Distinguish between what we mean by a *single*-displacement reaction and a *double*-displacement reaction by giving two examples of each type.

52. Two "driving forces" for reactions discussed in this section are the formation of water in an acid–base reaction and the formation of a gaseous product. Write balanced chemical equations showing two examples of each type.

53. Identify each of the following unbalanced reaction equations as belonging to one or more of the following categories: precipitation, acid–base, or oxidation–reduction.
   a. $K_2SO_4(aq) + Ba(NO_3)_2(aq) \rightarrow BaSO_4(s) + KNO_3(aq)$
   b. $HCl(aq) + Zn(s) \rightarrow H_2(g) + ZnCl_2(aq)$
   c. $HCl(aq) + AgNO_3(aq) \rightarrow HNO_3(aq) + AgCl(s)$
   d. $HCl(aq) + KOH(aq) \rightarrow H_2O(l) + KCl(aq)$
   e. $Zn(s) + CuSO_4(aq) \rightarrow ZnSO_4(aq) + Cu(s)$
   f. $NaH_2PO_4(aq) + NaOH(aq) \rightarrow Na_3PO_4(aq) + H_2O(l)$
   g. $Ca(OH)_2(aq) + H_2SO_4(aq) \rightarrow CaSO_4(s) + H_2O(l)$
   h. $ZnCl_2(aq) + Mg(s) \rightarrow Zn(s) + MgCl_2(aq)$
   i. $BaCl_2(aq) + H_2SO_4(aq) \rightarrow BaSO_4(s) + HCl(aq)$

54. Identify each of the following unbalanced reaction equations as belonging to one or more of the following categories: precipitation, acid–base, or oxidation–reduction.
   a. $H_2O_2(aq) \rightarrow H_2O(l) + O_2(g)$
   b. $H_2SO_4(aq) + Cu(s) \rightarrow CuSO_4(aq) + H_2(g)$
   c. $H_2SO_4(aq) + NaOH(aq) \rightarrow Na_2SO_4(aq) + H_2O(l)$
   d. $H_2SO_4(aq) + Ba(OH)_2(aq) \rightarrow BaSO_4(s) + H_2O(l)$
   e. $AgNO_3(aq) + CuCl_2(aq) \rightarrow Cu(NO_3)_2(aq) + AgCl(s)$
   f. $KOH(aq) + CuSO_4(aq) \rightarrow Cu(OH)_2(s) + K_2SO_4(aq)$
   g. $Cl_2(g) + F_2(g) \rightarrow ClF(g)$
   h. $NO(g) + O_2(g) \rightarrow NO_2(g)$
   i. $Ca(OH)_2(s) + HNO_3(aq) \rightarrow Ca(NO_3)_2(aq) + H_2O(l)$

7.7 Other Ways to Classify Reactions

QUESTIONS

55. A reaction in which a compound reacts rapidly with elemental oxygen, usually with the release of heat or light, is referred to as a ________ reaction.

56. Reactions involving the combustion of fuel substances make up a subclass of ________ reactions.

57. What is a *synthesis* or *combination* reaction? Give an example. Can such reactions also be classified in other ways? Give an example of a synthesis reaction that is also a *combustion* reaction. Give an example of a synthesis reaction that is also an *oxidation–reduction* reaction, but which does not involve combustion.

58. What is a *decomposition* reaction? Give an example. Can such reactions also be classified in other ways?

PROBLEMS

59. Balance each of the following equations that describe combustion reactions.
   a. $C_2H_6(g) + O_2(g) \rightarrow CO_2(g) + H_2O(g)$
   b. $C_4H_{10}(g) + O_2(g) \rightarrow CO_2(g) + H_2O(g)$
   c. $C_6H_{14}(l) + O_2(g) \rightarrow CO_2(g) + H_2O(g)$

60. Balance each of the following equations that describe combustion reactions.
   a. $C_2H_5OH(l) + O_2(g) \rightarrow CO_2(g) + H_2O(g)$
   b. $C_6H_{14}(l) + O_2(g) \rightarrow CO_2(g) + H_2O(g)$
   c. $C_6H_{12}(l) + O_2(g) \rightarrow CO_2(g) + H_2O(g)$

61. By now, you are familiar with enough chemical compounds to begin to write your own chemical reaction equations. Write two examples of what we mean by a *combustion* reaction.

62. By now, you are familiar with enough chemical compounds to begin to write your own chemical reaction equations. Write two examples each of what we mean by a *synthesis* reaction and by a *decomposition* reaction.

63. Balance each of the following equations that describe synthesis reactions.
   a. $Ni(s) + CO(g) \rightarrow Ni(CO)_4(g)$

b. $Al(s) + S(s) \rightarrow Al_2S_3(s)$
c. $Na_2SO_3(aq) + S(s) \rightarrow Na_2S_2O_3(aq)$
d. $Fe(s) + Br_2(l) \rightarrow FeBr_3(s)$
e. $Na(s) + O_2(g) \rightarrow Na_2O_2(s)$

64. Balance each of the following equations that describe synthesis reactions.
a. $Co(s) + S(s) \rightarrow Co_2S_3(s)$
b. $NO(g) + O_2(g) \rightarrow NO_2(g)$
c. $FeO(s) + CO_2(g) \rightarrow FeCO_3(s)$
d. $Al(s) + F_2(g) \rightarrow AlF_3(s)$
e. $NH_3(g) + H_2CO_3(aq) \rightarrow (NH_4)_2CO_3(s)$

65. Balance each of the following equations that describe decomposition reactions.
a. $CaSO_4(s) \rightarrow CaO(s) + SO_3(g)$
b. $Li_2CO_3(s) \rightarrow Li_2O(s) + CO_2(g)$
c. $LiHCO_3(s) \rightarrow Li_2CO_3(s) + H_2O(g) + CO_2(g)$
d. $C_6H_6(l) \rightarrow C(s) + H_2(g)$
e. $PBr_3(l) \rightarrow P_4(s) + Br_2(l)$

66. Balance each of the following equations that describe decomposition reactions.
a. $NI_3(s) \rightarrow N_2(g) + I_2(s)$
b. $BaCO_3(s) \rightarrow BaO(s) + CO_2(g)$
c. $C_6H_{12}O_6(s) \rightarrow C(s) + H_2O(g)$
d. $Cu(NH_3)_4SO_4(s) \rightarrow CuSO_4(s) + NH_3(g)$
e. $NaN_3(s) \rightarrow Na_3N(s) + N_2(g)$

Additional Problems

67. Distinguish between the *molecular* equation, the *complete ionic* equation, and the *net ionic* equation for a reaction in solution. Which type of equation most clearly shows the species that actually react with one another?

68. Using the general solubility rules given in Table 7.1, name three reactants that would form precipitates with each of the following ions in aqueous solution. Write the net ionic equation for each of your suggestions.
a. chloride ion
b. calcium ion
c. iron(III) ion
d. sulfate ion
e. mercury(I) ion, Hg_2^{2+}
f. silver ion

69. Without first writing a full molecular or ionic equation, write the net ionic equations for any precipitation reactions that occur when aqueous solutions of the following compounds are mixed. If no reaction occurs, so indicate.
a. iron(III) nitrate and sodium carbonate
b. mercurous nitrate and sodium chloride
c. sodium nitrate and ruthenium nitrate
d. copper(II) sulfate and sodium sulfide
e. lithium chloride and lead(II) nitrate
f. calcium nitrate and lithium carbonate
g. gold(III) chloride and sodium hydroxide

70. Complete and balance each of the following molecular equations for strong acid/strong base reactions. Underline the formula of the *salt* produced in each reaction.
a. $HNO_3(aq) + KOH(aq) \rightarrow$
b. $H_2SO_4(aq) + Ba(OH)_2(aq) \rightarrow$
c. $HClO_4(aq) + NaOH(aq) \rightarrow$
d. $HCl(aq) + Ca(OH)_2(aq) \rightarrow$

71. For the cations listed in the left-hand column, give the formulas of the precipitates that would form with each of the anions in the right-hand column. If no precipitate is expected for a particular combination, so indicate.

Cations	Anions
Ag^+	$C_2H_3O_2^-$
Ba^{2+}	Cl^-
Ca^{2+}	CO_3^{2-}
Fe^{3+}	NO_3^-
Hg_2^{2+}	OH^-
Na^+	PO_4^{3-}
Ni^{2+}	S^{2-}
Pb^{2+}	SO_4^{2-}

72. On the basis of the general solubility rules given in Table 7.1, predict which of the following substances are likely to be soluble in water.
a. potassium hexacyanoferrate(III), $K_3Fe(CN)_6$
b. ammonium molybdate, $(NH_4)_2MoO_4$
c. osmium(II) carbonate, $OsCO_3$
d. gold(III) phosphate, $AuPO_4$
e. sodium hexanitrocobaltate(III), $Na_3Co(NO_2)_6$
f. barium carbonate, $BaCO_3$
g. iron(III) chloride, $FeCl_3$

73. On the basis of the general solubility rules given in Table 7.1, predict the identity of the precipitate that forms when aqueous solutions of the following substances are mixed. If no precipitate is likely, indicate why (which rules apply).
a. iron(III) chloride and sodium hydroxide
b. nickel(II) nitrate and ammonium sulfide
c. silver nitrate and potassium chloride
d. sodium carbonate and barium nitrate
e. potassium chloride and mercury(I) nitrate
f. barium nitrate and sulfuric acid

74. On the basis of the general solubility rules given in Table 7.1, write a balanced molecular equation for the precipitation reactions that take place when the following aqueous solutions are mixed. Underline the formula of the precipitate (solid) that forms. If no precipitation reaction is likely for the reactants given, so indicate.
a. silver nitrate and hydrochloric acid
b. copper(II) sulfate and ammonium carbonate
c. iron(II) sulfate and potassium carbonate
d. silver nitrate and potassium nitrate
e. lead(II) nitrate and lithium carbonate
f. tin(IV) chloride and sodium hydroxide

75. For each of the following *un*balanced molecular equations, write the corresponding *balanced net ionic equation* for the reaction.

a. $HCl(aq) + AgNO_3(aq) \rightarrow AgCl(s) + HNO_3(aq)$
b. $CaCl_2(aq) + Na_3PO_4(aq) \rightarrow Ca_3(PO_4)_2(s) + NaCl(aq)$
c. $Pb(NO_3)_2(aq) + BaCl_2(aq) \rightarrow PbCl_2(s) + Ba(NO_3)_2(aq)$
d. $FeCl_3(aq) + NaOH(aq) \rightarrow Fe(OH)_3(s) + NaCl(aq)$

76. Most sulfide compounds of the transition metals are insoluble in water. Many of these metal sulfides have striking and characteristic colors by which we can identify them. Therefore, in the analysis of mixtures of metal ions, it is very common to precipitate the metal ions by using dihydrogen sulfide (commonly called hydrogen sulfide), H_2S. Suppose you had a mixture of Fe^{2+}, Cr^{3+}, and Ni^{2+}. Write net ionic equations for the precipitation of these metal ions by the use of H_2S.

77. What strong acid and what strong base would react in aqueous solution to produce the following salts?
a. potassium perchlorate, $KClO_4$
b. cesium nitrate, $CsNO_3$
c. potassium chloride, KCl
d. sodium sulfate, Na_2SO_4

78. Using the general solubility rules given in Table 7.1, name three reactants that would form precipitates with each of the following ions in aqueous solutions. Write the balanced molecular equation for each of your suggested reactants.
a. sulfide ion
b. carbonate ion
c. hydroxide ion
d. phosphate ion

79. For the reaction $16Fe(s) + 3S_8(s) \rightarrow 8Fe_2S_3(s)$, show how electrons are gained and lost by the atoms.

80. Balance the equation for each of the following oxidation–reduction chemical reactions.
a. $Na(s) + O_2(g) \rightarrow Na_2O_2(s)$
b. $Fe(s) + H_2SO_4(aq) \rightarrow FeSO_4(aq) + H_2(g)$
c. $Al_2O_3(s) \rightarrow Al(s) + O_2(g)$
d. $Fe(s) + Br_2(l) \rightarrow FeBr_3(s)$
e. $Zn(s) + HNO_3(aq) \rightarrow Zn(NO_3)_2(aq) + H_2(g)$

81. Identify each of the following unbalanced reaction equations as belonging to one or more of the following categories: precipitation, acid–base, or oxidation–reduction.
a. $Fe(s) + H_2SO_4(aq) \rightarrow Fe_2(SO_4)_3(aq) + H_2(g)$
b. $HClO_4(aq) + RbOH(aq) \rightarrow RbClO_4(aq) + H_2O(l)$
c. $Ca(s) + O_2(g) \rightarrow CaO(s)$
d. $H_2SO_4(aq) + NaOH(aq) \rightarrow Na_2SO_4(aq) + H_2O(l)$
e. $Pb(NO_3)_2(aq) + Na_2CO_3(aq) \rightarrow PbCO_3(s) + NaNO_3(aq)$
f. $K_2SO_4(aq) + CaCl_2(aq) \rightarrow KCl(aq) + CaSO_4(s)$
g. $HNO_3(aq) + KOH(aq) \rightarrow KNO_3(aq) + H_2O(l)$
h. $Ni(C_2H_3O_2)_2(aq) + Na_2S(aq) \rightarrow NiS(s) + NaC_2H_3O_2(aq)$
i. $Ni(s) + Cl_2(g) \rightarrow NiCl_2(s)$

82. Complete and balance each of the following equations that describe combustion reactions.
a. $C_4H_{10}(l) + O_2(g) \rightarrow$
b. $C_4H_{10}O(l) + O_2(g) \rightarrow$
c. $C_4H_{10}O_2(l) + O_2(g) \rightarrow$

83. Balance each of the following equations that describe synthesis reactions.
a. $FeO(s) + O_2(g) \rightarrow Fe_2O_3(s)$
b. $CO(g) + O_2(g) \rightarrow CO_2(g)$
c. $H_2(g) + Cl_2(g) \rightarrow HCl(g)$
d. $K(s) + S_8(s) \rightarrow K_2S(s)$
e. $Na(s) + N_2(g) \rightarrow Na_3N(s)$

84. Balance each of the following equations that describe decomposition reactions.
a. $NaHCO_3(s) \rightarrow Na_2CO_3(s) + H_2O(g) + CO_2(g)$
b. $NaClO_3(s) \rightarrow NaCl(s) + O_2(g)$
c. $HgO(s) \rightarrow Hg(l) + O_2(g)$
d. $C_{12}H_{22}O_{11}(s) \rightarrow C(s) + H_2O(g)$
e. $H_2O_2(l) \rightarrow H_2O(l) + O_2(g)$

85. Write a balanced oxidation–reduction equation for the reaction of each of the metals in the left-hand column with each of the nonmetals in the right-hand column.

Ba	O_2
K	S
Mg	Cl_2
Rb	N_2
Ca	Br_2
Li	

86. Sulfuric acid, H_2SO_4, oxidizes many metallic elements. One of the effects of acid rain is that it produces sulfuric acid in the atmosphere, which then reacts with metals used in construction. Write balanced oxidation–reduction equations for the reaction of sulfuric acid with Fe, Zn, Mg, Co, and Ni.

87. Although the metals of Group 2 of the periodic table are not nearly as reactive as those of Group 1, many of the Group 2 metals will combine with common nonmetals, especially at elevated temperatures. Write balanced chemical equations for the reactions of Mg, Ca, Sr, and Ba with Cl_2, Br_2, and O_2.

88. For each of the following metals, how many electrons will the metal atoms lose when the metal reacts with a nonmetal?
a. sodium
b. potassium
c. magnesium
d. barium
e. aluminum

89. For each of the following nonmetals, how many electrons will each atom of the nonmetal gain in reacting with a metal?
a. oxygen
b. fluorine
c. nitrogen
d. chlorine
e. sulfur

90. There is much overlapping of the classification schemes for reactions discussed in this chapter. Give an example of a reaction that is, at the same time,

an oxidation–reduction reaction, a combustion reaction, and a synthesis reaction.

91. Classify the reactions represented by the following unbalanced equations by as many methods as possible. Balance the equations.
   a. $I_4O_9(s) \rightarrow I_2O_6(s) + I_2(s) + O_2(g)$
   b. $Mg(s) + AgNO_3(aq) \rightarrow Mg(NO_3)_2(aq) + Ag(s)$
   c. $SiCl_4(l) + Mg(s) \rightarrow MgCl_2(s) + Si(s)$
   d. $CuCl_2(aq) + AgNO_3(aq) \rightarrow Cu(NO_3)_2(aq) + AgCl(s)$
   e. $Al(s) + Br_2(l) \rightarrow AlBr_3(s)$

92. Classify the reactions represented by the following unbalanced equations by as many methods as possible. Balance the equations.
   a. $C_3H_8O(l) + O_2(g) \rightarrow CO_2(g) + H_2O(g)$
   b. $HCl(aq) + AgC_2H_3O_2(aq) \rightarrow AgCl(s) + HC_2H_3O_2(aq)$
   c. $HCl(aq) + Al(OH)_3(s) \rightarrow AlCl_3(aq) + H_2O(l)$
   d. $H_2O_2(aq) \rightarrow H_2O(l) + O_2(g)$
   e. $N_2H_4(l) + O_2(g) \rightarrow N_2(g) + H_2O(g)$

93. Corrosion of metals costs us billions of dollars annually, slowly destroying cars, bridges, and buildings. Corrosion of a metal involves the oxidation of the metal by the oxygen in the air, typically in the presence of moisture. Write a balanced equation for the reaction of each of the following metals with O_2: Zn, Al, Fe, Cr, and Ni.

94. Elemental chlorine, Cl_2, is very reactive, combining with most metallic substances. Write a balanced equation for the reaction of each of the following metals with Cl_2: Na, Al, Zn, Ca, and Fe.

CUMULATIVE REVIEW FOR CHAPTERS 6–7

QUESTIONS

1. What kind of *visual* evidence indicates that a chemical reaction has occurred? Give an example of each type of evidence you have mentioned. Do *all* reactions produce visual evidence that they have taken place?

2. What, in general terms, does a chemical equation indicate? What are the substances indicated to the left of the arrow called in a chemical equation? To the right of the arrow?

3. What does it mean to "balance" an equation? Why is it so important that equations be balanced? What does it mean to say that atoms must be *conserved* in a balanced chemical equation? How are the physical states of reactants and products indicated when writing chemical equations?

4. When balancing a chemical equation, why is it *not* permissible to adjust the subscripts in the formulas of the reactants and products? What would changing the subscripts within a formula do? What do the *coefficients* in a balanced chemical equation represent? Why is it acceptable to adjust a substance's coefficient but not permissible to adjust the subscripts within the substance's formula?

5. What is meant by the *driving force* for a reaction? Give some examples of driving forces that make reactants tend to form products. Write a balanced chemical equation illustrating each type of driving force you have named.

6. What is a *precipitation reaction?* What would you see if a precipitation reaction were to take place in a beaker? Write a balanced chemical equation illustrating a precipitation reaction.

7. Define the term *strong electrolyte.* What types of substances tend to be strong electrolytes? What does a solution of a strong electrolyte contain? Give a way to determine if a substance is a strong electrolyte.

8. Summarize the simple solubility rules for ionic compounds. How do we use these rules in determining the identity of the solid formed in a precipitation reaction? Give examples including balanced complete and net ionic equations.

9. In general terms, what are the *spectator ions* in a precipitation reaction? Why are the spectator ions not included in writing the net ionic equation for a precipitation reaction? Does this mean that the spectator ions do not have to be present in the solution?

10. Describe some physical and chemical properties of *acids* and *bases*. What is meant by a *strong* acid or base? Are strong acids and bases also strong electrolytes? Give several examples of strong acids and strong bases.

11. What is a *salt?* How are salts formed by acid–base reactions? Write chemical equations showing the formation of three different salts. What other product is formed when an aqueous acid reacts with an aqueous base? Write the net ionic equation for the formation of this substance.

12. What is essential in an oxidation–reduction reaction? What is *oxidation?* What is *reduction?* Can an oxidation reaction take place without a reduction reaction also taking place? Why? Write a balanced chemical equation illustrating an oxidation–reduction reaction between a metal and a nonmetal. Indicate which species is oxidized and which is reduced.

13. What is a *combustion* reaction? Are combustion reactions a unique type of reaction, or are they a special case of a more general type of reaction? Write an equation that illustrates a combustion reaction.

14. Give an example of a *synthesis* reaction and of a *decomposition* reaction. Are synthesis and decomposition reactions always also oxidation–reduction reactions? Explain.

15. List and define all the ways of classifying chemical reactions that have been discussed in the text. Give a balanced chemical equation as an example of each type of reaction, and show clearly how your example fits the definition you have given.

PROBLEMS

16. Potassium and sodium are highly reactive metals. Write balanced chemical equations for the reaction of each of these metals with the following substances, with the indicated products being produced.
    a. with water, producing aqueous sodium hydroxide or potassium hydroxide, and hydrogen gas
    b. with chlorine gas, producing sodium chloride or potassium chloride
    c. with phosphorus, producing sodium phosphide or potassium phosphide
    d. with nitrogen gas, producing sodium nitride or potassium nitride
    e. with hydrogen gas, producing sodium hydride or potassium hydride

17. Balance the following chemical equations.
    a. $FeCl_3(aq) + KOH(aq) \rightarrow Fe(OH)_3(s) + KCl(aq)$
    b. $AgC_2H_3O_2(aq) + HCl(aq) \rightarrow AgCl(s) + HC_2H_3O_2(aq)$
    c. $Na_2O_2(s) + H_2O(l) \rightarrow NaOH(aq) + O_2(g)$
    d. $SnO(s) + C(s) \rightarrow Sn(s) + CO_2(g)$
    e. $Fe(s) + Br_2(l) \rightarrow FeBr_3(s)$
    f. $Na_2S(s) + HCl(aq) \rightarrow NaCl(aq) + H_2S(g)$
    g. $K_2O(s) + H_2O(l) \rightarrow KOH(aq)$
    h. $H_2SO_4(l) + NaCl(s) \rightarrow Na_2SO_4(s) + HCl(g)$
    i. $N_2(g) + I_2(s) \rightarrow NI_3(s)$

18. The reagent shelf in a general chemistry lab contains aqueous solutions of the following substances: silver nitrate, sodium chloride, acetic acid, nitric acid, sulfuric acid, potassium chromate, barium nitrate, phosphoric acid, hydrochloric acid, lead nitrate, sodium hydroxide, and sodium carbonate. Suggest how you might prepare the following pure substances using these reagents and any normal laboratory equipment. If it is *not* possible to prepare a substance using these reagents, indicate why.
   a. $BaCrO_4(s)$
   b. $NaC_2H_3O_2(s)$
   c. $AgCl(s)$
   d. $PbSO_4(s)$
   e. $Na_2SO_4(s)$
   f. $BaCO_3(s)$

19. The common strong acids are HCl, HNO_3, and H_2SO_4, whereas NaOH and KOH are the common strong bases. Write the neutralization reaction equations for each of these strong acids with each of these strong bases in aqueous solution.

20. Classify each of the following chemical equations in as *many* ways as possible based on what you have learned. Balance each equation.
   a. $FeO(s) + HNO_3(aq) \rightarrow Fe(NO_3)_2(aq) + H_2O(l)$
   b. $Mg(s) + CO_2(g) + O_2(g) \rightarrow MgCO_3(s)$
   c. $NaOH(s) + CuSO_4(aq) \rightarrow Cu(OH)_2(s) + Na_2SO_4(aq)$
   d. $HI(aq) + KOH(aq) \rightarrow KI(aq) + H_2O(l)$
   e. $C_3H_8(g) + O_2(g) \rightarrow CO_2(g) + H_2O(g)$
   f. $Co(NH_3)_6Cl_2(s) \rightarrow CoCl_2(s) + NH_3(g)$
   g. $HCl(aq) + Pb(C_2H_3O_2)_2(aq) \rightarrow HC_2H_3O_2(aq) + PbCl_2(s)$
   h. $C_{12}H_{22}O_{11}(s) \rightarrow C(s) + H_2O(g)$
   i. $Al(s) + HNO_3(aq) \rightarrow Al(NO_3)_3(aq) + H_2(g)$
   j. $B(s) + O_2(g) \rightarrow B_2O_3(s)$

21. In Column 1 are listed some reactive metals; in Column 2 are listed some nonmetals. Write a balanced chemical equation for the combination/synthesis reaction of each element in Column 1 with each element in Column 2.

Column 1	**Column 2**
sodium, Na	fluorine gas, F_2
calcium, Ca	oxygen gas, O_2
aluminum, Al	sulfur, S
magnesium, Mg	chlorine gas, Cl_2

22. Give balanced equations for two examples of each of the following types of reactions.
   a. precipitation
   b. single-displacement
   c. combustion
   d. synthesis
   e. oxidation–reduction
   f. decomposition
   g. acid–base neutralization

8 Chemical Composition

These seashells from Sanibel Island, Florida, contain the mineral calcium carbonate.
▼

One very important chemical activity is the synthesis of new substances. Nylon, the artificial sweetener aspartame (Nutra-Sweet®), Kevlar used in bulletproof vests and the body parts of exotic cars, polyvinyl chloride (PVC) for plastic water pipes, Teflon, Nitinol (the alloy that remembers its shape even after being severely distorted), and so many other materials that make our lives easier—all originated in some chemist's laboratory. Some of the new materials have truly amazing properties such as the plastic that listens and talks, described in the "Chemistry in Focus" on page 213. When a chemist makes a new substance, the first order of business is to identify it. What is its composition? What is its chemical formula?

In this chapter we will learn to determine a compound's formula. Before we can do that, however, we need to think about counting atoms. How do we determine the number of each type of atom in a substance so that we can write its formula? Of course, atoms are too small to count individually. As we will see in this chapter, we typically count atoms by weighing them. So let us first consider the general principle of counting by weighing.

The new Enzo Ferrari has a body made of carbon fiber composite materials.

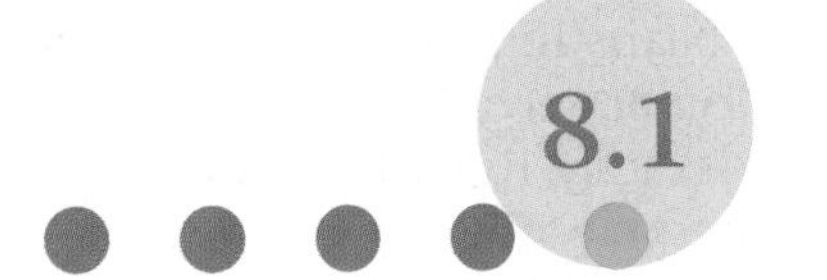

8.1 Counting by Weighing

AIM: To understand the concept of average mass and explore how counting can be done by weighing.

Suppose you work in a candy store that sells gourmet jelly beans by the bean. People come in and ask for 50 beans, 100 beans, 1000 beans, and so on, and you have to count them out—a tedious process at best. As a good problem solver, you try to come up with a better system. It occurs to you that it might be far more efficient to buy a scale and count the jelly beans by weighing them. How can you count jelly beans by weighing them? What information about the individual beans do you need to know?

Assume that all of the jelly beans are identical and that each has a mass of 5 g. If a customer asks for 1000 jelly beans, what mass of jelly beans would be required? Each bean has a mass of 5 g, so you would need 1000 beans $\times$ 5 g/bean, or 5000 g (5 kg). It takes just a few seconds to weigh out 5 kg of jelly beans. It would take much longer to count out 1000 of them.

In reality, jelly beans are not identical. For example, let's assume that you weigh 10 beans individually and get the following results:

Bean	Mass
1	5.1 g
2	5.2 g
3	5.0 g
4	4.8 g
5	4.9 g
6	5.0 g
7	5.0 g
8	5.1 g
9	4.9 g
10	5.0 g

Can we count these nonidentical beans by weighing? Yes. The key piece of information we need is the *average mass* of the jelly beans. Let's compute the average mass for our 10-bean sample.

$$\text{Average mass} = \frac{\text{total mass of beans}}{\text{number of beans}}$$

$$= \frac{5.1\text{ g} + 5.2\text{ g} + 5.0\text{ g} + 4.8\text{ g} + 4.9\text{ g} + 5.0\text{ g} + 5.0\text{ g} + 5.1\text{ g} + 4.9\text{ g} + 5.0\text{ g}}{10}$$

$$= \frac{50.0}{10} = 5.0\text{ g}$$

The average mass of a jelly bean is 5.0 g. Thus, to count out 1000 beans, we need to weigh out 5000 g of beans. This sample of beans, in which the beans have an average mass of 5.0 g, can be treated exactly like a sample where all of the beans are identical. Objects do not need to have identical masses in order to be counted by weighing. We simply need to know the average mass of the objects. For purposes of counting, the objects *behave as though they were all identical,* as though they each actually had the average mass.

Suppose a customer comes into the store and says, "I want to buy a bag of candy for each of my kids. One of them likes jelly beans and the other one likes mints. Please put a scoopful of jelly beans in a bag and a scoopful of mints in another bag." Then the customer recognizes a problem. "Wait! My kids will fight unless I bring home exactly the same number of candies for each one. Both bags must have the same number of pieces because they'll definitely count them and compare. But I'm really in a hurry, so we don't have time to count them here. Is there a simple way you can be sure the bags will contain the same number of candies?"

You need to solve this problem quickly. Suppose you know the average masses of the two kinds of candy:

Jelly beans: average mass = 5 g
Mints: average mass = 15 g

You fill the scoop with jelly beans and dump them onto the scale, which reads 500 g. Now the key question: What mass of mints do you need to give the same number of mints as there are jelly beans in 500 g of jelly beans? Comparing the average masses of the jelly beans (5 g) and mints (15 g), you realize that each mint has three times the mass of each jelly bean:

$$\frac{15\text{ g}}{5\text{ g}} = 3$$

CHEMISTRY IN FOCUS

Plastic That Talks and Listens!

Imagine a plastic so "smart" that it can be used to sense a baby's breath, measure the force of a karate punch, sense the presence of a person 100 ft away, or make a balloon that sings. There is a plastic film capable of doing all these things. It's called **polyvinylidene difluoride (PVDF),** which has the structure

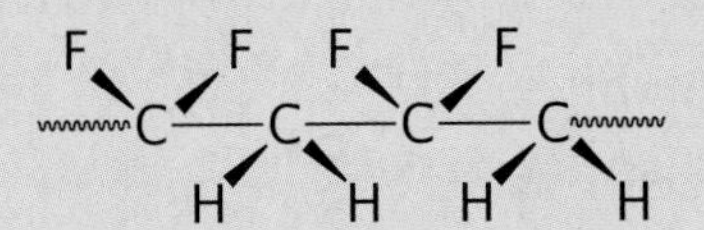

When this polymer is processed in a particular way, it becomes piezoelectric and pyroelectric. A *piezoelectric* substance produces an electric current when it is physically deformed or, alternatively, undergoes a deformation when a current is applied. A *pyroelectric* material is one that develops an electrical potential in response to a change in its temperature.

Because PVDF is piezoelectric, it can be used to construct a paper-thin microphone; it responds to sound by producing a current proportional to the deformation caused by the sound waves. A ribbon of PVDF plastic one-quarter of an inch wide could be strung along a hallway and used to listen to all the conversations going on as people walk through. On the other hand, electric pulses can be applied to the PVDF film to produce a speaker. A strip of PVDF film glued to the inside of a balloon can play any song stored on a microchip attached to the film—hence a balloon that can sing happy birthday at a party. The PVDF film also can be used to construct a sleep apnea monitor, which, when placed beside the mouth of a sleeping infant, will set off an alarm if the breathing stops, thus helping to prevent sudden infant death syndrome (SIDS). The same type of film is used by the U.S. Olympic karate team to measure the force of kicks and punches as the team trains. Also, gluing two strips of film together gives a material that curls in response to a current, creating an artificial muscle. In addition, because the PVDF film is pyroelectric, it responds to the infrared (heat) radiation emitted by a human as far away as 100 ft, making it useful for burglar alarm systems. Making the PVDF polymer piezoelectric and pyroelectric requires some very special processing, which makes it costly ($10 per square foot), but this seems a small price to pay for its near-magical properties.

This means that you must weigh out an amount of mints that is three times the mass of the jelly beans:

$$3 \times 500 \text{ g} = 1500 \text{ g}$$

You weigh out 1500 g of mints and put them in a bag. The customer leaves with your assurance that both the bag containing 500 g of jelly beans and the bag containing 1500 g of mints contain the same number of candies.

In solving this problem, you have discovered a principle that is very important in chemistry: two samples containing different types of components, A and B, both *contain the same number of components if the ratio of the sample masses is the same as the ratio of the masses of the individual components* of A and B.

Let's illustrate this rather intimidating statement by using the example we just discussed. The individual components have the masses 5 g (jelly beans) and 15 g (mints). Consider several cases.

- Each sample contains 1 component:

$$\text{Mass of mint} = 15 \text{ g}$$
$$\text{Mass of jelly bean} = 5 \text{ g}$$

- Each sample contains 10 components:

$$10 \cancel{\text{mints}} \times \frac{15 \text{ g}}{\cancel{\text{mint}}} = 150 \text{ g of mints}$$

$$10 \cancel{\text{jelly beans}} \times \frac{5 \text{ g}}{\cancel{\text{jelly bean}}} = 50 \text{ g of jelly beans}$$

- Each sample contains 100 components:

$$100 \cancel{\text{mints}} \times \frac{15 \text{ g}}{\cancel{\text{mint}}} = 1500 \text{ g of mints}$$

$$100 \cancel{\text{jelly beans}} \times \frac{5 \text{ g}}{\cancel{\text{jelly bean}}} = 500 \text{ g of jelly beans}$$

Note in each case that the ratio of the masses is always 3 to 1:

$$\frac{1500}{500} = \frac{150}{50} = \frac{15}{5} = \frac{3}{1}$$

which is the ratio of the masses of the individual components:

$$\frac{\text{Mass of mint}}{\text{Mass of jelly bean}} = \frac{15}{5} = \frac{3}{1}$$

Any two samples, one of mints and one of jelly beans, that have a *mass ratio* of 15/5 = 3/1 will contain the same number of components. And these same ideas apply also to atoms, as we will see in the next section.

8.2

Atomic Masses: Counting Atoms by Weighing

AIM: To understand atomic mass and its experimental determination.

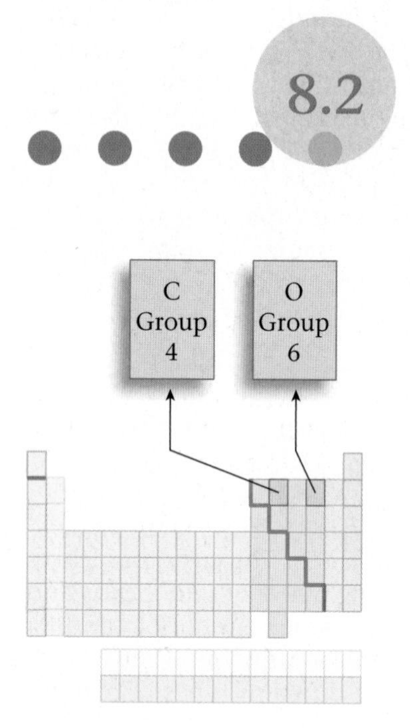

In Chapter 6 we considered the balanced equation for the reaction of solid carbon and gaseous oxygen to form gaseous carbon dioxide:

$$C(s) + O_2(g) \rightarrow CO_2(g)$$

Now suppose you have a small pile of solid carbon and want to know how many oxygen molecules are required to convert all of this carbon into carbon dioxide. The balanced equation tells us that one oxygen molecule is required for each carbon atom.

$$C(s) \quad + \quad O_2(g) \quad \rightarrow \quad CO_2(g)$$

1 atom reacts with 1 molecule to yield 1 molecule

To determine the number of oxygen molecules required, we must know how many carbon atoms are present in the pile of carbon. But individual atoms are far too small to see. We must learn to count atoms by weighing samples containing large numbers of them.

In the last section we saw that we can easily count things like jelly beans and mints by weighing. Exactly the same principles can be applied to counting atoms.

Because atoms are so tiny, the normal units of mass—the gram and the kilogram—are much too large to be convenient. For example, the mass of a single carbon atom is 1.99×10^{-23} g. To avoid using terms like 10^{-23} when describing the mass of an atom, scientists have defined a much smaller unit of mass called the **atomic mass unit,** which is abbreviated **amu.** In terms of grams,

$$1 \text{ amu} = 1.66 \times 10^{-24} \text{ g}$$

Now let's return to our problem of counting carbon atoms. To count carbon atoms by weighing, we need to know the mass of individual atoms, just as we needed to know the mass of the individual jelly beans. Recall from

Chapter 4 that the atoms of a given element exist as isotopes. The isotopes of carbon are $^{12}_{6}C$, $^{13}_{6}C$, and $^{14}_{6}C$. Any sample of carbon contains a mixture of these isotopes, always in the same proportions. Each of these isotopes has a slightly different mass. Therefore, just as with the nonidentical jelly beans, we need to use an average mass for the carbon atoms. The **average atomic mass** for carbon atoms is 12.01 amu. This means that any sample of carbon from nature *can be treated as though it were composed of identical carbon atoms,* each with a mass of 12.01 amu. Now that we know the average mass of the carbon atom, we can count carbon atoms by weighing samples of natural carbon. For example, what mass of natural carbon must we take to have 1000 carbon atoms present? Because 12.01 amu is the average mass,

Remember that 1000 is an exact number here.

$$\text{Mass of 1000 natural carbon atoms} = (1000\ \cancel{\text{atoms}})\left(12.01\ \frac{\text{amu}}{\cancel{\text{atom}}}\right)$$
$$= 12{,}010\ \text{amu} = 12.01 \times 10^3\ \text{amu}$$

Now let's assume that when we weigh the pile of natural carbon mentioned earlier, the result is 3.00×10^{20} amu. How many carbon atoms are present in this sample? We know that an average carbon atom has the mass 12.01 amu, so we can compute the number of carbon atoms by using the equivalence statement

$$1\ \text{carbon atom} = 12.01\ \text{amu}$$

to construct the appropriate conversion factor,

$$\frac{1\ \text{carbon atom}}{12.01\ \text{amu}}$$

The calculation is carried out as follows:

$$3.00 \times 10^{20}\ \cancel{\text{amu}} \times \frac{1\ \text{carbon atom}}{12.01\ \cancel{\text{amu}}} = 2.50 \times 10^{19}\ \text{carbon atoms}$$

The principles we have just discussed for carbon apply to all the other elements as well. All the elements as found in nature typically consist of a mixture of various isotopes. So to count the atoms in a sample of a given element by weighing, we must know the mass of the sample and the average mass for that element. Some average masses for common elements are listed in Table 8.1.

Table 8.1 Average Atomic Mass Values for Some Common Elements

Element	Average Atomic Mass (amu)
Hydrogen	1.008
Carbon	12.01
Nitrogen	14.01
Oxygen	16.00
Sodium	22.99
Aluminum	26.98

Example 8.1 Calculating Mass Using Atomic Mass Units (amu)

Calculate the mass, in amu, of a sample of aluminum that contains 75 atoms.

Solution

To solve this problem we use the average mass for an aluminum atom: 26.98 amu. We set up the equivalence statement:

$$1\ \text{Al atom} = 26.98\ \text{amu}$$

which gives the conversion factor we need:

$$75\ \cancel{\text{Al atoms}} \times \frac{26.98\ \text{amu}}{\cancel{\text{Al atom}}} = 2024\ \text{amu}$$

The 75 in this problem is an exact number—the number of atoms.

✔ Self-Check Exercise 8.1

Calculate the mass of a sample that contains 23 nitrogen atoms.

See Problems 8.5 and 8.8. ■

The opposite calculation can also be carried out. That is, if we know the mass of a sample, we can determine the number of atoms present. This procedure is illustrated in Example 8.2.

Example 8.2 Calculating the Number of Atoms from the Mass

Calculate the number of sodium atoms present in a sample that has a mass of 1172.49 amu.

Solution

We can solve this problem by using the average atomic mass for sodium (see Table 8.1) of 22.99 amu. The appropriate equivalence statement is

$$1 \text{ Na atom} = 22.99 \text{ amu}$$

which gives the conversion factor we need:

$$1172.49 \, \cancel{\text{amu}} \times \frac{1 \text{ Na atom}}{22.99 \, \cancel{\text{amu}}} = 51.00 \text{ Na atoms}$$

✔ Self-Check Exercise 8.2

Calculate the number of oxygen atoms in a sample that has a mass of 288 amu.

See Problems 8.6 and 8.7. ■

To summarize, we have seen that we can count atoms by weighing if we know the average atomic mass for that type of atom. This is one of the fundamental operations in chemistry, as we will see in the next section.

The average atomic mass for each element is listed in tables found inside the front cover of this book. Chemists often call these values the *atomic weights* for the elements, although this terminology is passing out of use.

8.3 The Mole

AIMS: To understand the mole concept and Avogadro's number. To learn to convert among moles, mass, and number of atoms in a given sample.

In the previous section we used atomic mass units for mass, but these are extremely small units. In the laboratory a much larger unit, the gram, is the convenient unit for mass. In this section we will learn to count atoms in samples with masses given in grams.

Let's assume we have a sample of aluminum that has a mass of 26.98 g. What mass of copper contains exactly the same number of atoms as this sample of aluminum?

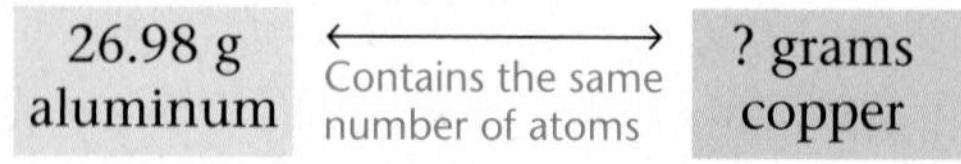

To answer this question, we need to know the average atomic masses for aluminum (26.98 amu) and copper (63.55 amu). Which atom has the greater atomic mass, aluminum or copper? The answer is copper. If we have 26.98 g of aluminum, do we need more or less than 26.98 g of copper to have the same number of copper atoms as aluminum atoms? We need more

Figure 8.1
All these samples of pure elements contain the *same number* (a mole) of atoms: 6.022×10^{23} atoms.

than 26.98 g of copper because each copper atom has a greater mass than each aluminum atom. Therefore, a given number of copper atoms will weigh more than an equal number of aluminum atoms. How much copper do we need? Because the average masses of aluminum and copper atoms are 26.98 amu and 63.55 amu, respectively, 26.98 g of aluminum and 63.55 g of copper contain exactly the same number of atoms. So we need 63.55 g of copper. As we saw in the first section when we were discussing candy, *samples in which the ratio of the masses is the same as the ratio of the masses of the individual atoms always contain the same number of atoms.* In the case just considered, the ratios are

$$\underbrace{\frac{26.98\text{ g}}{63.55\text{ g}}}_{\text{Ratio of sample masses}} = \underbrace{\frac{26.98\text{ amu}}{63.55\text{ amu}}}_{\text{Ratio of atomic masses}}$$

Therefore, 26.98 g of aluminum contains the same number of aluminum atoms as 63.55 g of copper contains copper atoms.

Now compare carbon (average atomic mass, 12.01 amu) and helium (average atomic mass, 4.003 amu). A sample of 12.01 g of carbon contains the same number of atoms as 4.003 g of helium. In fact, if we weigh out samples of all the elements such that each sample has a mass equal to that element's average atomic mass in grams, these samples all contain the same number of atoms (Figure 8.1). This number (the number of atoms present in all of these samples) assumes special importance in chemistry. It is called the mole, the unit all chemists use in describing numbers of atoms. The **mole** (abbreviated mol) can be defined as *the number equal to the number of carbon atoms in 12.01 grams of carbon.* Techniques for counting atoms very precisely have been used to determine this number to be 6.022×10^{23}. This number is called **Avogadro's number.** *One mole of something consists of 6.022×10^{23} units of that substance.* Just as a dozen eggs is 12 eggs, a mole of eggs is 6.022×10^{23} eggs. And a mole of water contains 6.022×10^{23} H_2O molecules.

Figure 8.2
One-mole samples of iron (nails), iodine crystals, liquid mercury, and powdered sulfur.

This definition of the mole is slightly different from the SI definition but is used because it is easier to understand at this point.

Avogadro's number (to four significant figures) is 6.022×10^{23}. One mole of *anything* is 6.022×10^{23} units of that substance.

The magnitude of the number 6.022×10^{23} is very difficult to imagine. To give you some idea, 1 mol of seconds represents a span of time 4 million times as long as the earth has already existed! One mole of marbles is enough to cover the entire earth to a depth of 50 miles! However, because atoms are so tiny, a mole of atoms or molecules is a perfectly manageable quantity to use in a reaction (Figure 8.2).

How do we use the mole in chemical calculations? Recall that Avogadro's number is defined such that a 12.01-g sample of carbon contains 6.022×10^{23} atoms. By the same token, because the average atomic mass of hydrogen is 1.008 amu (Table 8.1), 1.008 g of hydrogen contains 6.022×10^{23} hydrogen

Table 8.2 Comparison of 1-Mol Samples of Various Elements

Element	Number of Atoms Present	Mass of Sample (g)
Aluminum	6.022×10^{23}	26.98
Gold	6.022×10^{23}	196.97
Iron	6.022×10^{23}	55.85
Sulfur	6.022×10^{23}	32.07
Boron	6.022×10^{23}	10.81
Xenon	6.022×10^{23}	131.3

The mass of 1 mol of an element is equal to its average atomic mass in grams.

atoms. Similarly, 26.98 g of aluminum contains 6.022×10^{23} aluminum atoms. The point is that a sample of *any* element that weighs a number of grams equal to the average atomic mass of that element contains 6.022×10^{23} atoms (1 mol) of that element.

Table 8.2 shows the masses of several elements that contain 1 mol of atoms.

In summary, *a sample of an element with a mass equal to that element's average atomic mass expressed in grams contains 1 mol of atoms.*

To do chemical calculations, you *must* understand what the mole means and how to determine the number of moles in a given mass of a substance. However, before we do any calculations, let's be sure that the process of counting by weighing is clear. Consider the following "bag" of H atoms (symbolized by dots), which contains 1 mol (6.022×10^{23}) of H atoms and has a mass of 1.008 g. Assume the bag itself has no mass.

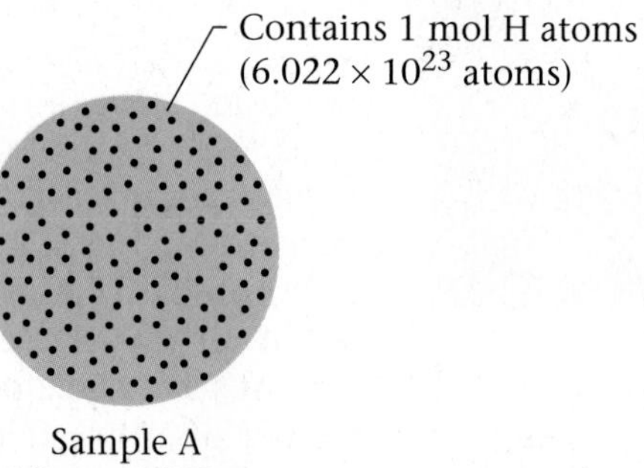

A 1-mol sample of graphite (a form of carbon) weighs 12.01 g.

Now consider another "bag" of hydrogen atoms in which the number of hydrogen atoms is unknown.

Contains an unknown number of H atoms

Sample B

We want to find out how many H atoms are present in sample ("bag") B. How can we do that? We can do it by weighing the sample. We find the mass of sample B to be 0.500 g.

How does this measured mass help us determine the number of atoms in sample B? We know that 1 mol of H atoms has a mass of 1.008 g. Sample B has a mass of 0.500 g, which is approximately half the mass of a mole of H atoms.

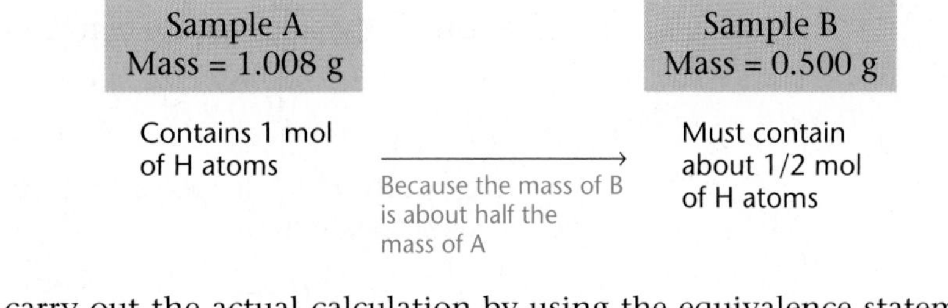

We carry out the actual calculation by using the equivalence statement

$$1 \text{ mol H atoms} = 1.008 \text{ g H}$$

to construct the conversion factor we need:

$$0.500\ \cancel{\text{g H}} \times \frac{1 \text{ mol H}}{1.008\ \cancel{\text{g H}}} = 0.496 \text{ mol H in sample B}$$

Let's summarize. We know the mass of 1 mol of H atoms, so we can determine the number of moles of H atoms in any other sample of pure hydrogen by weighing the sample and *comparing* its mass to 1.008 g (the mass of 1 mol of H atoms). We can follow this same process for any element, because we know the mass of 1 mol for each of the elements.

Also, because we know that 1 mol is 6.022×10^{23} units, once we know the *moles* of atoms present, we can easily determine the *number* of atoms present. In the case considered above, we have approximately 0.5 mol of H atoms in sample B. This means that about 1/2 of 6×10^{23}, or 3×10^{23}, H atoms are present. We carry out the actual calculation by using the equivalence statement

$$1 \text{ mol} = 6.022 \times 10^{23}$$

to determine the conversion factor we need:

$$0.496\ \cancel{\text{mol H atoms}} \times \frac{6.022 \times 10^{23} \text{ H atoms}}{1\ \cancel{\text{mol H atoms}}} = 2.99 \times 10^{23} \text{ H atoms in sample B}$$

These procedures are illustrated in Example 8.3.

Example 8.3 Calculating Moles and Number of Atoms

A bicycle with an aluminum frame.

Aluminum (Al), a metal with a high strength-to-weight ratio and a high resistance to corrosion, is often used for structures such as high-quality bicycle frames. Compute both the number of moles of atoms and the number of atoms in a 10.0-g sample of aluminum.

Solution

In this case we want to change from mass to moles of atoms:

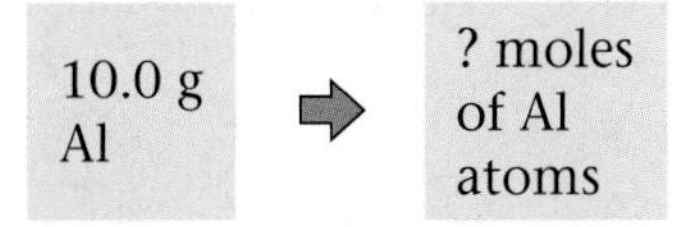

The mass of 1 mol (6.022×10^{23} atoms) of aluminum is 26.98 g. The sample we are considering has a mass of 10.0 g. Its mass is less than 26.98 g, so this sample contains less than 1 mol of aluminum atoms. We calculate the number of moles of aluminum atoms in 10.0 g by using the equivalence statement

$$1 \text{ mol Al} = 26.98 \text{ g Al}$$

to construct the appropriate conversion factor:

$$10.0 \text{ g Al} \times \frac{1 \text{ mol Al}}{26.98 \text{ g Al}} = 0.371 \text{ mol Al}$$

Next we convert from moles of atoms to the number of atoms, using the equivalence statement

$$6.022 \times 10^{23} \text{ Al atoms} = 1 \text{ mol Al atoms}$$

We have

$$0.371 \text{ mol Al} \times \frac{6.022 \times 10^{23} \text{ Al atoms}}{1 \text{ mol Al}} = 2.23 \times 10^{23} \text{ Al atoms}$$

We can summarize this calculation as follows:

$$10.0 \text{ g Al} \times \frac{1 \text{ mol}}{26.98 \text{ g}} \Rightarrow 0.371 \text{ mol Al}$$

$$0.371 \text{ mol Al atoms} \times \frac{6.022 \times 10^{23} \text{ Al atoms}}{\text{mol}} \Rightarrow 2.23 \times 10^{23} \text{ Al atoms}$$

■

Example 8.4 Calculating the Number of Atoms

A silicon chip used in an integrated circuit of a microcomputer has a mass of 5.68 mg. How many silicon (Si) atoms are present in this chip? The average atomic mass for silicon is 28.09 amu.

Solution

A silicon chip of the type used in electronic equipment.

Our strategy for doing this problem is to convert from milligrams of silicon to grams of silicon, then to moles of silicon, and finally to atoms of silicon:

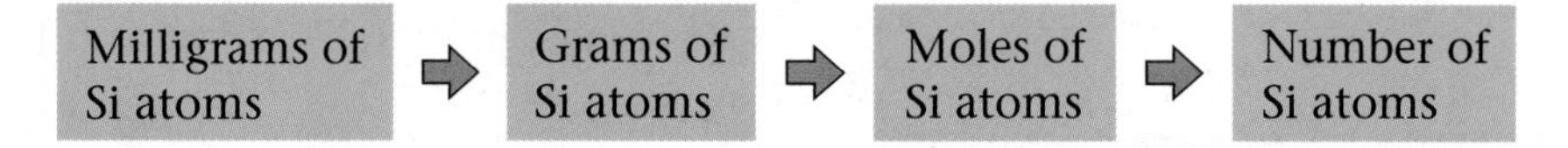

where each arrow in the schematic represents a conversion factor. Because 1 g = 1000 mg, we have

$$5.68 \text{ mg Si} \times \frac{1 \text{ g Si}}{1000 \text{ mg Si}} = 5.68 \times 10^{-3} \text{ g Si}$$

Next, because the average mass of silicon is 28.09 amu, we know that 1 mol of Si atoms weighs 28.09 g. This leads to the equivalence statement

$$1 \text{ mol Si atoms} = 28.09 \text{ g Si}$$

Thus,

$$5.68 \times 10^{-3} \text{ g Si} \times \frac{1 \text{ mol Si}}{28.09 \text{ g Si}} = 2.02 \times 10^{-4} \text{ mol Si}$$

Using the definition of a mole (1 mol = 6.022×10^{23}), we have

$$2.02 \times 10^{-4} \text{ mol Si} \times \frac{6.022 \times 10^{23} \text{ atoms}}{1 \text{ mol Si}} = 1.22 \times 10^{20} \text{ Si atoms}$$

We can summarize this calculation as follows:

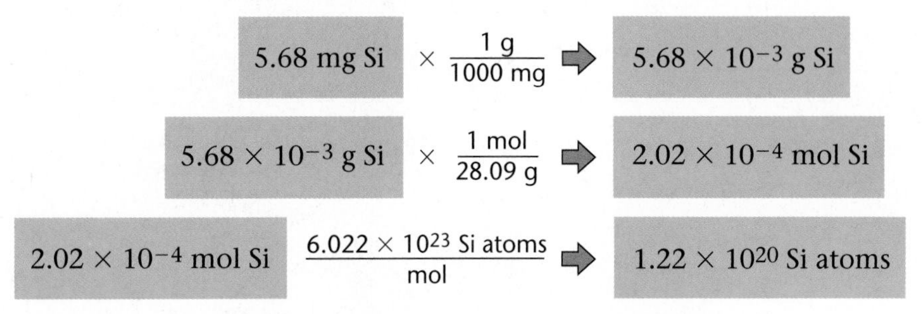

Problem Solving: Does the Answer Make Sense?

When you finish a problem, always think about the "reasonableness" of your answers. In Example 8.4, 5.68 mg of silicon is clearly much less than 1 mol of silicon (which has a mass of 28.09 g), so the final answer of 1.22×10^{20} atoms (compared to 6.022×10^{23} atoms in a mole) at least lies in the right direction. That is, 1.22×10^{20} atoms is a smaller number than 6.022×10^{23}. Also, always include the units as you perform calculations and make sure the correct units are obtained at the end. Paying careful attention to units and making this type of general check can help you detect errors such as an inverted conversion factor or a number that was incorrectly entered into your calculator.

✔ Self-Check Exercise 8.3

The values for the average masses of the atoms of the elements are listed inside the front cover of this book.

Chromium (Cr) is a metal that is added to steel to improve its resistance to corrosion (for example, to make stainless steel). Calculate both the number of moles in a sample of chromium containing 5.00×10^{20} atoms and the mass of the sample.

See Problems 8.19 through 8.24. ■

Molar Mass

AIMS: To understand the definition of molar mass. To learn to convert between moles and mass of a given sample of a chemical compound.

A chemical compound is, fundamentally, a collection of atoms. For example, methane (the major component of natural gas) consists of molecules each containing one carbon atom and four hydrogen atoms (CH_4). How can we calculate the mass of 1 mol of methane? That is, what is the mass of 6.022×10^{23} CH_4 molecules? Because each CH_4 molecule contains one carbon atom and four hydrogen atoms, 1 mol of CH_4 molecules consists of 1 mol of carbon atoms and 4 mol of hydrogen atoms (Figure 8.3). The mass of 1 mol of methane can be found by summing the masses of carbon and hydrogen present:

Note that when we say 1 mol of methane, we mean 1 mol of methane *molecules*.

Remember that the least number of decimal places limits the number of significant figures in addition.

$$\begin{aligned}
\text{Mass of 1 mol of C} &= 1 \times 12.01\ \text{g} = 12.01\ \ \text{g} \\
\text{Mass of 4 mol of H} &= 4 \times 1.008\ \text{g} = \underline{\ \ 4.032\ \text{g}} \\
\text{Mass of 1 mol of } CH_4 &= 16.04\ \ \text{g}
\end{aligned}$$

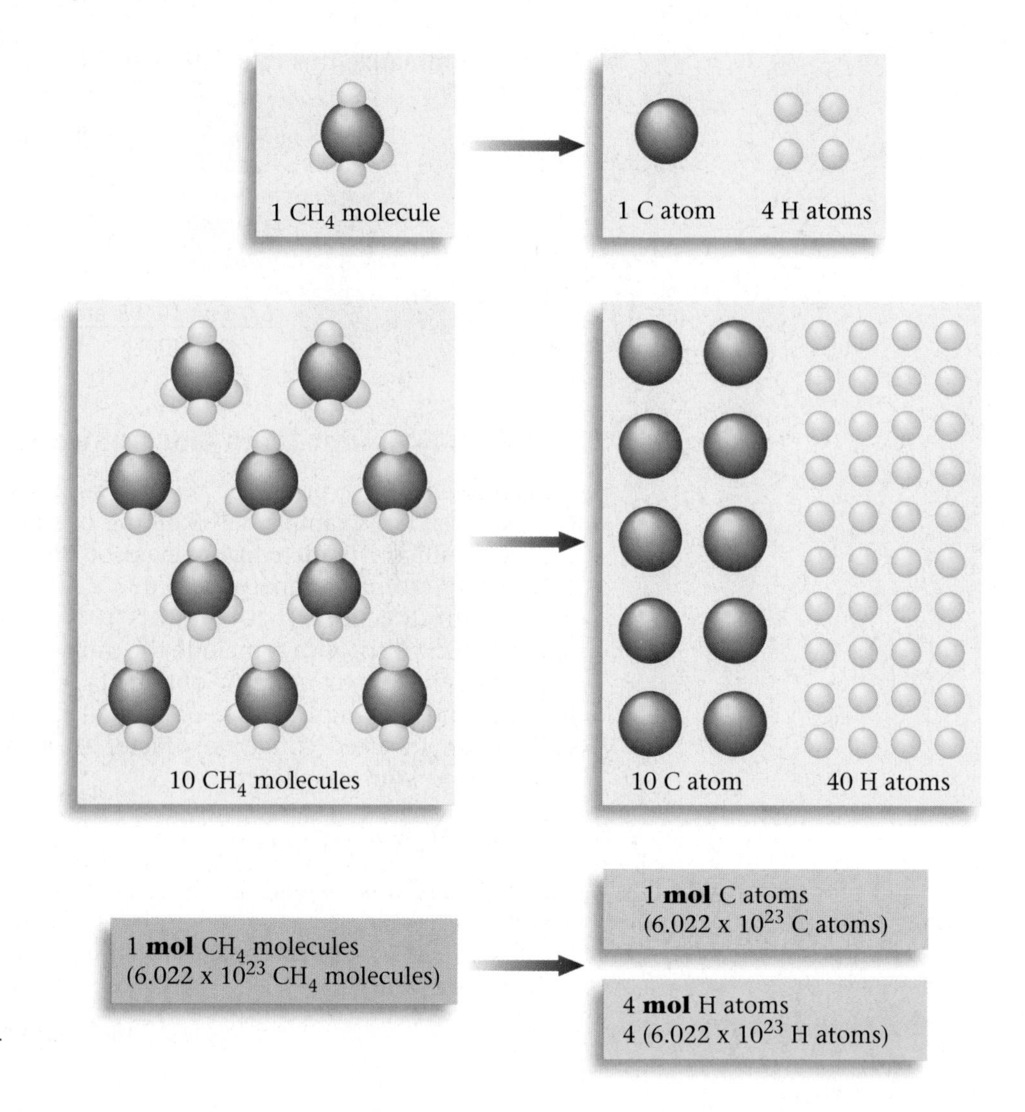

Figure 8.3
Various numbers of methane molecules showing their constituent atoms.

A substance's molar mass (in grams) is the mass of 1 mol of that substance.

The quantity 16.04 g is called the molar mass for methane: the mass of 1 mol of CH_4 molecules. The **molar mass*** of any substance is the *mass (in grams) of 1 mol of the substance.* The molar mass is obtained by summing the masses of the component atoms.

Example 8.5 Calculating Molar Mass

Calculate the molar mass of sulfur dioxide, a gas produced when sulfur-containing fuels are burned. Unless "scrubbed" from the exhaust, sulfur dioxide can react with moisture in the atmosphere to produce acid rain.

Solution

The formula for sulfur dioxide is SO_2. We need to compute the mass of 1 mol of SO_2 molecules—the molar mass for sulfur dioxide. We know that 1 mol of SO_2 molecules contains 1 mol of sulfur atoms and 2 mol of oxygen atoms:

* The term *molecular weight* was traditionally used instead of *molar mass*. The terms *molecular weight* and *molar mass* mean exactly the same thing. Because the term *molar mass* more accurately describes the concept, it will be used in this text.

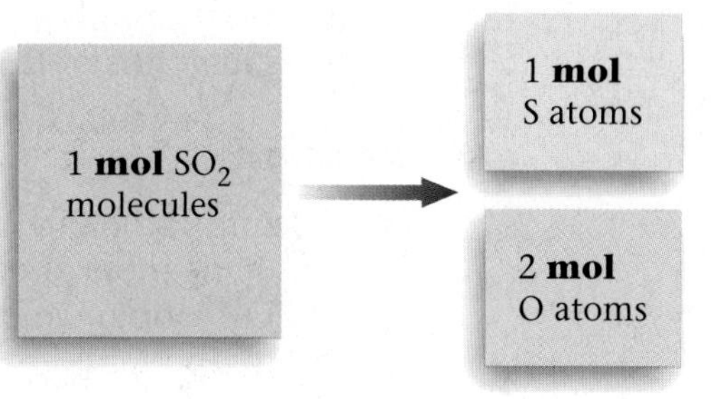

$$\begin{aligned}\text{Mass of 1 mol of S} &= 1 \times 32.07 &= 32.07\text{ g}\\ \text{Mass of 2 mol of O} &= 2 \times 16.00 &= \underline{32.00\text{ g}}\\ \text{Mass of 1 mol of } SO_2 & &= 64.07\text{ g} = \text{molar mass}\end{aligned}$$

The molar mass of SO_2 is 64.07 g. It represents the mass of 1 mol of SO_2 molecules.

Self-Check Exercise 8.4

Polyvinyl chloride (called PVC), which is widely used for floor coverings ("vinyl") and for plastic pipes in plumbing systems, is made from a molecule with the formula C_2H_3Cl. Calculate the molar mass of this substance.

See Problems 8.27 through 8.30. ■

Some substances exist as a collection of ions rather than as separate molecules. For example, ordinary table salt, sodium chloride (NaCl), is composed of an array of Na^+ and Cl^- ions. There are no NaCl molecules present. In some books the term **formula weight** is used instead of molar mass for ionic compounds. However, in this book we will apply the term *molar mass* to both ionic and molecular substances.

To calculate the molar mass for sodium chloride, we must realize that 1 mol of NaCl contains 1 mol of Na^+ ions and 1 mol of Cl^- ions.

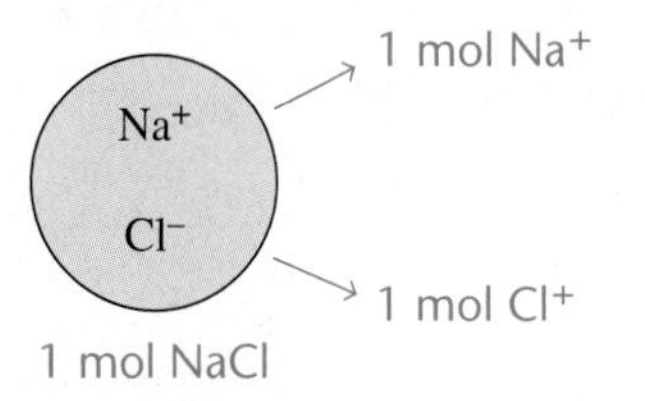

The mass of the electron is so small that Na^+ and Na have the same mass for our purposes, even though Na^+ has one electron less than Na. Also the mass of Cl virtually equals the mass of Cl^- even though it has one more electron than Cl.

Therefore, the molar mass (in grams) for sodium chloride represents the sum of the mass of 1 mol of sodium ions and the mass of 1 mol of chloride ions.

$$\begin{aligned}\text{Mass of 1 mol of } Na^+ &= 22.99\text{ g}\\ \text{Mass of 1 mol of } Cl^- &= \underline{35.45\text{ g}}\\ \text{Mass of 1 mol of NaCl} &= 58.44\text{ g} = \text{molar mass}\end{aligned}$$

The molar mass of NaCl is 58.44 g. It represents the mass of 1 mol of sodium chloride.

Example 8.6 Calculating Mass from Moles

Calcium carbonate, $CaCO_3$ (also called calcite), is the principal mineral found in limestone, marble, chalk, pearls, and the shells of marine animals such as clams.

a. Calculate the molar mass of calcium carbonate.

b. A certain sample of calcium carbonate contains 4.86 mol. What is the mass in grams of this sample?

Solution

a. Calcium carbonate is an ionic compound composed of Ca^{2+} and CO_3^{2-} ions. One mole of calcium carbonate contains 1 mol of Ca^{2+} and 1 mol of CO_3^{2-} ions. We calculate the molar mass by summing the masses of the components.

Mass of 1 mol of Ca^{2+} = 1 × 40.08 g = 40.08 g
Mass of 1 mol of CO_3^{2-} (contains 1 mol of C and 3 mol of O):
1 mol of C = 1 × 12.01 g = 12.01 g
3 mol of O = 3 × 16.00 g = 48.00 g
Mass of 1 mol of $CaCO_3$ = 100.09 g = molar mass

b. We determine the mass of 4.86 mol of $CaCO_3$ by using the molar mass.

$$4.86 \text{ mol } CaCO_3 \times \frac{100.09 \text{ g } CaCO_3}{1 \text{ mol } CaCO_3} = 486 \text{ g } CaCO_3$$

which can be diagrammed as follows:

$$4.86 \text{ mol } CaCO_3 \times \frac{100.09 \text{ g}}{\text{mol}} \Rightarrow 486 \text{ g } CaCO_3$$

Note that the sample under consideration contains nearly 5 mol and thus should have a mass of nearly 500 g, so our answer makes sense.

✔ Self-Check Exercise 8.5

For average atomic masses, look inside the front cover of this book.

Calculate the molar mass for sodium sulfate, Na_2SO_4. A sample of sodium sulfate with a mass of 300.0 g represents what number of moles of sodium sulfate?

See Problems 8.35 through 8.38. ■

In summary, the molar mass of a substance can be obtained by summing the masses of the component atoms. The molar mass (in grams) represents the mass of 1 mol of the substance. Once we know the molar mass of a compound, we can compute the number of moles present in a sample of known mass. The reverse, of course, is also true as illustrated in Example 8.7.

Example 8.7 Calculating Moles from Mass

Juglone, a dye known for centuries, is produced from the husks of black walnuts. It is also a natural herbicide (weed killer) that kills off competitive plants around the black walnut tree but does not affect grass and other noncompetitive plants. The formula for juglone is $C_{10}H_6O_3$.

a. Calculate the molar mass of juglone.

b. A sample of 1.56 g of pure juglone was extracted from black walnut husks. How many moles of juglone does this sample represent?

Black walnuts with and without their green hulls.

Solution

a. The molar mass is obtained by summing the masses of the component atoms. In 1 mol of juglone there are 10 mol of carbon atoms, 6 mol of hydrogen atoms, and 3 mol of oxygen atoms.

$$\begin{aligned}
\text{Mass of 10 mol of C} &= 10 \times 12.01\text{ g} = 120.1\text{ g}\\
\text{Mass of 6 mol of H} &= 6 \times 1.008\text{ g} = 6.048\text{ g}\\
\text{Mass of 3 mol of O} &= 3 \times 16.00\text{ g} = \underline{48.00\text{ g}}\\
\text{Mass of 1 mol of } C_{10}H_6O_3 &= 174.1\text{ g} = \text{molar mass}
\end{aligned}$$

b. The mass of 1 mol of this compound is 174.1 g, so 1.56 g is much less than a mole. We can determine the exact fraction of a mole by using the equivalence statement

$$1\text{ mol} = 174.1\text{ g juglone}$$

to derive the appropriate conversion factor:

$$1.56\ \cancel{\text{g juglone}} \times \frac{1\text{ mol juglone}}{174.1\ \cancel{\text{g juglone}}} = 0.00896\text{ mol juglone} = 8.96 \times 10^{-3}\text{ mol juglone}$$

$$\boxed{1.56\text{ g juglone}} \times \frac{1\text{ mol}}{174.1\text{ g}} \Rightarrow \boxed{8.96 \times 10^{-3}\text{ mol juglone}}$$

■

Example 8.8 Calculating Number of Molecules

Isopentyl acetate, $C_7H_{14}O_2$, the compound responsible for the scent of bananas, can be produced commercially. Interestingly, bees release about 1 μg (1×10^{-6} g) of this compound when they sting. This attracts other bees, which then join the attack. How many moles and how many molecules of isopentyl acetate are released in a typical bee sting?

Solution

We are given a mass of isopentyl acetate and want the number of molecules, so we must first compute the molar mass.

$$\begin{aligned}
7\ \cancel{\text{mol C}} \times 12.01\frac{\text{g}}{\cancel{\text{mol}}} &= 84.07\text{ g C}\\
14\ \cancel{\text{mol H}} \times 1.008\frac{\text{g}}{\cancel{\text{mol}}} &= 14.11\text{ g H}\\
2\ \cancel{\text{mol O}} \times 16.00\frac{\text{g}}{\cancel{\text{mol}}} &= \underline{32.00\text{ g O}}\\
\text{molar mass is } &130.18\text{ g}
\end{aligned}$$

This means that 1 mol of isopentyl acetate (6.022×10^{23} molecules) has a mass of 130.18 g.

Next we determine the number of moles of isopentyl acetate in 1 μg, which is 1×10^{-6} g. To do this, we use the equivalence statement

$$1\text{ mol isopentyl acetate} = 130.18\text{ g isopentyl acetate}$$

which yields the conversion factor we need:

$$1 \times 10^{-6}\ \cancel{\text{g } C_7H_{14}O_2} \times \frac{1\text{ mol } C_7H_{14}O_2}{130.18\ \cancel{\text{g } C_7H_{14}O_2}} = 8 \times 10^{-9}\text{ mol } C_7H_{14}O_2$$

Using the equivalence statement 1 mol = 6.022×10^{23} units, we can determine the number of molecules:

$$8 \times 10^{-9}\ \cancel{\text{mol } C_7H_{14}O_2} \times \frac{6.022 \times 10^{23}\text{ molecules}}{1\ \cancel{\text{mol } C_7H_{14}O_2}} = 5 \times 10^{15}\text{ molecules}$$

This very large number of molecules is released in each bee sting.

Self-Check Exercise 8.6

The substance Teflon, the slippery coating on many frying pans, is made from the C_2F_4 molecule. Calculate the number of C_2F_4 units present in 135 g of Teflon.

See Problems 8.39 and 8.40. ■

8.5 Percent Composition of Compounds

AIM: To learn to find the mass percent of an element in a given compound.

So far we have discussed the composition of compounds in terms of the numbers of constituent atoms. It is often useful to know a compound's composition in terms of the *masses* of its elements. We can obtain this information from the formula of the compound by comparing the mass of each element present in 1 mol of the compound to the total mass of 1 mol of the compound. The mass fraction for each element is calculated as follows:

$$\text{Mass fraction for a given element} = \frac{\text{mass of the element present in 1 mol of compound}}{\text{mass of 1 mol of compound}}$$

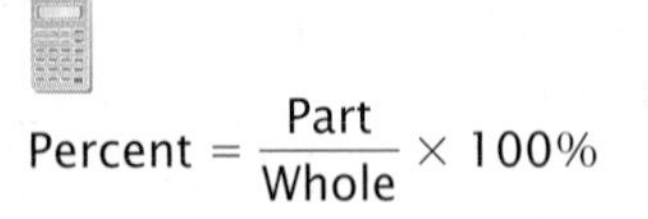

$$\text{Percent} = \frac{\text{Part}}{\text{Whole}} \times 100\%$$

The mass fraction is converted to *mass percent* by multiplying by 100%.

We will illustrate this concept using the compound ethanol, an alcohol obtained by fermenting the sugar in grapes, corn, and other fruits and grains. Ethanol is often added to gasoline as an octane enhancer to form a fuel called gasohol. The added ethanol has the effect of increasing the octane of the gasoline and also lowering the carbon monoxide in automobile exhaust.

The formula for ethanol is written C_2H_5OH, although you might expect it to be written simply as C_2H_6O.

Note from its formula that each molecule of ethanol contains two carbon atoms, six hydrogen atoms, and one oxygen atom. This means that each mole of ethanol contains 2 mol of carbon atoms, 6 mol of hydrogen atoms, and 1 mol of oxygen atoms. We calculate the mass of each element present and the molar mass for ethanol as follows:

$$\text{Mass of C} = 2\ \cancel{\text{mol}} \times 12.01\frac{\text{g}}{\cancel{\text{mol}}} = 24.02\text{ g}$$
$$\text{Mass of H} = 6\ \cancel{\text{mol}} \times 1.008\frac{\text{g}}{\cancel{\text{mol}}} = 6.048\text{ g}$$
$$\text{Mass of O} = 1\ \cancel{\text{mol}} \times 16.00\frac{\text{g}}{\cancel{\text{mol}}} = 16.00\text{ g}$$
$$\text{Mass of 1 mol of } C_2H_5OH = 46.07\text{ g} = \text{molar mass}$$

The **mass percent** (sometimes called the weight percent) of carbon in ethanol can be computed by comparing the mass of carbon in 1 mol of ethanol with the total mass of 1 mol of ethanol and multiplying the result by 100%.

$$\text{Mass percent of C} = \frac{\text{mass of C in 1 mol } C_2H_5OH}{\text{mass of 1 mol } C_2H_5OH} \times 100\%$$
$$= \frac{24.02 \text{ g}}{46.07 \text{ g}} \times 100\% = 52.14\%$$

That is, ethanol contains 52.14% by mass of carbon. The mass percents of hydrogen and oxygen in ethanol are obtained in a similar manner.

$$\text{Mass percent of H} = \frac{\text{mass of H in 1 mol } C_2H_5OH}{\text{mass of 1 mol } C_2H_5OH} \times 100\%$$
$$= \frac{6.048 \text{ g}}{46.07 \text{ g}} \times 100\% = 13.13\%$$
$$\text{Mass percent of O} = \frac{\text{mass of O in 1 mol } C_2H_5OH}{\text{mass of 1 mol } C_2H_5OH} \times 100\%$$
$$= \frac{16.00 \text{ g}}{46.07 \text{ g}} \times 100\% = 34.73\%$$

Sometimes, because of rounding-off effects, the sum of the mass percents in a compound is not exactly 100%.

The mass percentages of all the elements in a compound add up to 100%, although rounding-off effects may produce a small deviation. Adding up the percentages is a good way to check the calculations. In this case, the sum of the mass percents is 52.14% + 13.13% + 34.73% = 100.00%.

Example 8.9 Calculating Mass Percent

Carvone is a substance that occurs in two forms, both of which have the same molecular formula ($C_{10}H_{14}O$) and molar mass. One type of carvone gives caraway seeds their characteristic smell; the other is responsible for the smell of spearmint oil. Compute the mass percent of each element in carvone.

Solution

Because the formula for carvone is $C_{10}H_{14}O$, the masses of the various elements in 1 mol of carvone are

$$\text{Mass of C in 1 mol} = 10 \cancel{\text{mol}} \times 12.01 \frac{\text{g}}{\cancel{\text{mol}}} = 120.1 \text{ g}$$
$$\text{Mass of H in 1 mol} = 14 \cancel{\text{mol}} \times 1.008 \frac{\text{g}}{\cancel{\text{mol}}} = 14.11 \text{ g}$$
$$\text{Mass of O in 1 mol} = 1 \cancel{\text{mol}} \times 16.00 \frac{\text{g}}{\cancel{\text{mol}}} = \underline{16.00 \text{ g}}$$
$$\text{Mass of 1 mol of } C_{10}H_{14}O = 150.21 \text{ g}$$
$$\text{Molar mass} = 150.2 \text{ g}$$

(rounding to the correct number of significant figures)

The 120.1 limits the sum to one decimal place.

Next we find the fraction of the total mass contributed by each element and convert it to a percentage.

$$\text{Mass percent of C} = \frac{120.1 \text{ g C}}{150.2 \text{ g } C_{10}H_{14}O} \times 100\% = 79.96\%$$
$$\text{Mass percent of H} = \frac{14.11 \text{ g H}}{150.2 \text{ g } C_{10}H_{14}O} \times 100\% = 9.394\%$$
$$\text{Mass percent of O} = \frac{16.00 \text{ g O}}{150.2 \text{ g } C_{10}H_{14}O} \times 100\% = 10.65\%$$

CHECK: Add the individual mass percent values—they should total 100% within a small range due to rounding off. In this case, the percentages add up to 100.00%.

✔ Self-Check Exercise 8.7

Penicillin, an important antibiotic (antibacterial agent), was discovered accidentally by the Scottish bacteriologist Alexander Fleming in 1928, although he was never able to isolate it as a pure compound. This and similar antibiotics have saved millions of lives that would otherwise have been lost to infections. Penicillin, like many of the molecules produced by living systems, is a large molecule containing many atoms. One type of penicillin, penicillin F, has the formula $C_{14}H_{20}N_2SO_4$. Compute the mass percent of each element in this compound.

See Problems 8.45 through 8.50. ■

Formulas of Compounds

AIM: To understand the meaning of empirical formulas of compounds.

Assume that you have mixed two solutions, and a solid product (a precipitate) forms. How can you find out what the solid is? What is its formula? There are several possible approaches you can take to answering these questions. For example, we saw in Chapter 7 that we can usually predict the identity of a precipitate formed when two solutions are mixed in a reaction of this type if we know some facts about the solubilities of ionic compounds.

However, although an experienced chemist can often predict the product expected in a chemical reaction, the only sure way to identify the product is to perform experiments. Usually we compare the physical properties of the product to the properties of known compounds.

Sometimes a chemical reaction gives a product that has never been obtained before. In such a case, a chemist determines what compound has been formed by determining which elements are present and how much of each. These data can be used to obtain the formula of the compound. In Section 8.5 we used the formula of the compound to determine the mass of each element present in a mole of the compound. To obtain the formula of an unknown compound, we do the opposite. That is, we use the measured masses of the elements present to determine the formula.

Recall that the formula of a compound represents the relative numbers of the various types of atoms present. For example, the molecular formula CO_2 tells us that for each carbon atom there are two oxygen atoms in each molecule of carbon dioxide. So to determine the formula of a substance we need to count the atoms. As we have seen in this chapter, we can do this by weighing. Suppose we know that a compound contains only the elements carbon, hydrogen, and oxygen, and we weigh out a 0.2015-g sample for analysis. Using methods we will not discuss here, we find that this 0.2015-g sample of compound contains 0.0806 g of carbon, 0.01353 g of hydrogen, and 0.1074 g of oxygen. We have just learned how to convert these masses to numbers of atoms by using the atomic mass of each element. We begin by converting to moles.

Carbon

$$(0.0806 \cancel{\text{g C}}) \times \frac{1 \text{ mol C atoms}}{12.01 \cancel{\text{g C}}} = 0.00671 \text{ mol C atoms}$$

Hydrogen

$$(0.01353 \text{ \sout{g H}}) \times \frac{1 \text{ mol H atoms}}{1.008 \text{ \sout{g H}}} = 0.01342 \text{ mol H atoms}$$

Oxygen

$$(0.1074 \text{ \sout{g O}}) \times \frac{1 \text{ mol O atoms}}{16.00 \text{ \sout{g O}}} = 0.006713 \text{ mol O atoms}$$

Let's review what we have established. We now know that 0.2015 g of the compound contains 0.00671 mol of C atoms, 0.01342 mol of H atoms, and 0.006713 mol of O atoms. Because 1 mol is 6.022×10^{23}, these quantities can be converted to actual numbers of atoms.

Carbon

$$(0.00671 \text{ \sout{mol C atoms}}) \frac{(6.022 \times 10^{23} \text{ C atoms})}{1 \text{ \sout{mol C atoms}}} = 4.04 \times 10^{21} \text{ C atoms}$$

Hydrogen

$$(0.01342 \text{ \sout{mol H atoms}}) \frac{(6.022 \times 10^{23} \text{ H atoms})}{1 \text{ \sout{mol H atoms}}} = 8.08 \times 10^{21} \text{ H atoms}$$

Oxygen

$$(0.006713 \text{ \sout{mol O atoms}}) \frac{(6.022 \times 10^{23} \text{ O atoms})}{1 \text{ \sout{mol O atoms}}} = 4.043 \times 10^{21} \text{ O atoms}$$

These are the numbers of the various types of atoms *in 0.2015 g of compound.* What do these numbers tell us about the formula of the compound? Note the following:

1. The compound contains the same number of C and O atoms.

2. There are twice as many H atoms as C atoms or O atoms.

We can represent this information by the formula CH_2O, which expresses the *relative* numbers of C, H, and O atoms present. Is this the true formula for the compound? In other words, is the compound made up of CH_2O molecules? It may be. However, it might also be made up of $C_2H_4O_2$ molecules, $C_3H_6O_3$ molecules, $C_4H_8O_4$ molecules, $C_5H_{10}O_5$ molecules, $C_6H_{12}O_6$ molecules, and so on. Note that each of these molecules has the required 1:2:1 ratio of carbon to hydrogen to oxygen atoms (the ratio shown by experiment to be present in the compound).

When we break a compound down into its separate elements and "count" the atoms present, we learn only the ratio of atoms—we get only the *relative* numbers of atoms. The formula of a compound that expresses the smallest whole-number ratio of the atoms present is called the **empirical formula** or *simplest formula.* A compound that contains the molecules $C_4H_8O_4$ has the same empirical formula as a compound that contains $C_6H_{12}O_6$ molecules. The empirical formula for both is CH_2O. The actual formula of a compound—the one that gives the composition of the molecules that are present—is called the **molecular formula.** The sugar called glucose is made of molecules with the molecular formula $C_6H_{12}O_6$ (Figure 8.4). Note from the molecular formula for glucose that the empirical formula is CH_2O. We can represent the molecular formula as a multiple (by 6) of the empirical formula:

$$C_6H_{12}O_6 = (CH_2O)_6$$

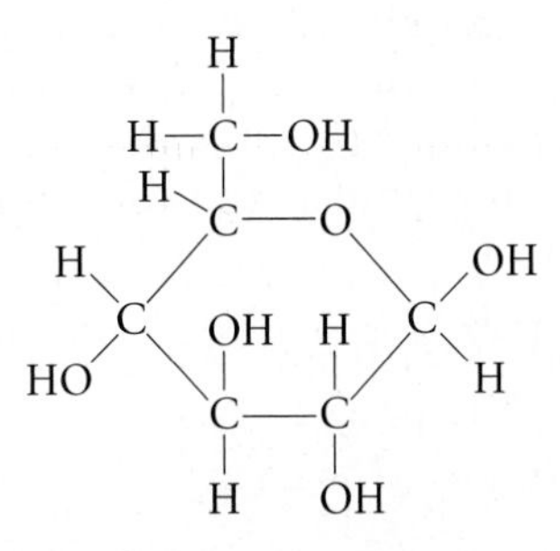

Figure 8.4
The glucose molecule. The molecular formula is $C_6H_{12}O_6$, as can be verified by counting the atoms. The empirical formula for glucose is CH_2O.

In the next section, we will explore in more detail how to calculate the empirical formula for a compound from the relative masses of the elements present. As we will see in Sections 8.7 and 8.8, we must know the molar mass of a compound to determine its molecular formula.

Example 8.10 Determining Empirical Formulas

In each case below, the molecular formula for a compound is given. Determine the empirical formula for each compound.

a. C_6H_6. This is the molecular formula for benzene, a liquid commonly used in industry as a starting material for many important products.

b. $C_{12}H_4Cl_4O_2$. This is the molecular formula for a substance commonly called dioxin, a powerful poison that sometimes occurs as a by-product in the production of other chemicals.

c. $C_6H_{16}N_2$. This is the molecular formula for one of the reactants used to produce nylon.

Solution

a. $C_6H_6 = (CH)_6$; CH is the empirical formula. Each subscript in the empirical formula is multiplied by 6 to obtain the molecular formula.

b. $C_{12}H_4Cl_4O_2$; $C_{12}H_4Cl_4O_2 = (C_6H_2Cl_2O)_2$; $C_6H_2Cl_2O$ is the empirical formula. Each subscript in the empirical formula is multiplied by 2 to obtain the molecular formula.

c. $C_6H_{16}N_2 = (C_3H_8N)_2$; C_3H_8N is the empirical formula. Each subscript in the empirical formula is multiplied by 2 to obtain the molecular formula. ■

8.7 Calculation of Empirical Formulas

AIM: To learn to calculate empirical formulas.

As we said in the previous section, one of the most important things we can learn about a new compound is its chemical formula. To calculate the empirical formula of a compound, we first determine the relative masses of the various elements that are present.

One way to do this is to measure the masses of elements that react to form the compound. For example, suppose we weigh out 0.2636 g of pure nickel metal into a crucible and heat this metal in the air so that the nickel can react with oxygen to form a nickel oxide compound. After the sample has cooled, we weigh it again and find its mass to be 0.3354 g. The gain in mass is due to the oxygen that reacts with the nickel to form the oxide. Therefore, the mass of oxygen present in the compound is the total mass of the product minus the mass of the nickel:

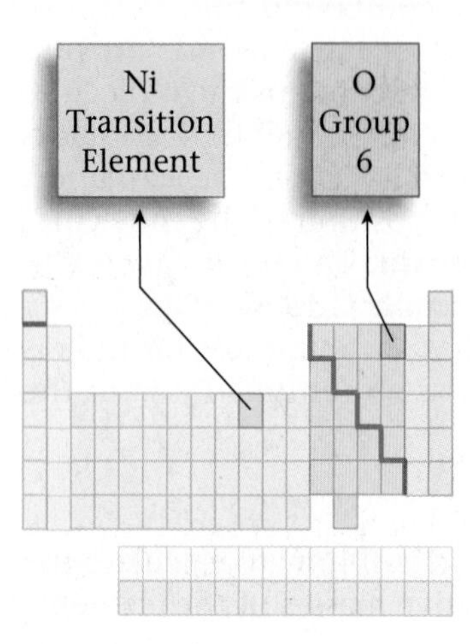

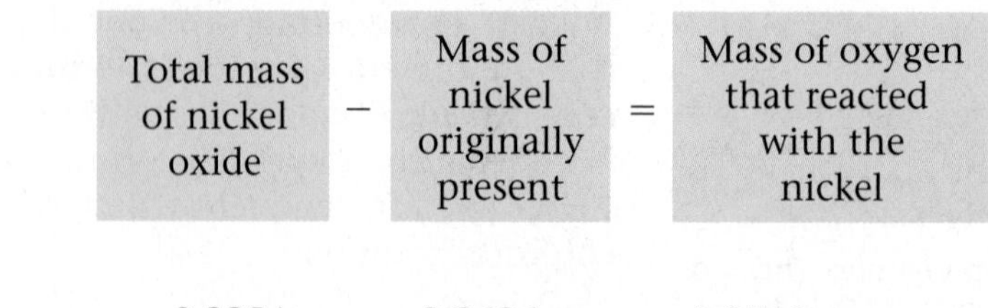

or

$$0.3354 \text{ g} - 0.2636 \text{ g} = 0.0718 \text{ g}$$

Note that the mass of nickel present in the compound is the nickel metal originally weighed out. So we know that the nickel oxide contains 0.2636 g of nickel and 0.0718 g of oxygen. What is the empirical formula of this compound?

To answer this question we must convert the masses to numbers of atoms, using atomic masses:

Four significant figures allowed.

$$0.2636 \text{ g Ni} \times \frac{1 \text{ mol Ni atoms}}{58.69 \text{ g Ni}} = 0.004491 \text{ mol Ni atoms}$$

Three significant figures allowed.

$$0.0718 \text{ g O} \times \frac{1 \text{ mol O atoms}}{16.00 \text{ g O}} = 0.00449 \text{ mol O atoms}$$

These mole quantities represent numbers of atoms (remember that a mole of atoms is 6.022×10^{23} atoms). It is clear from the moles of atoms that the compound contains an equal number of Ni and O atoms, so the formula is NiO. This is the *empirical formula;* it expresses the smallest whole-number (integer) ratio of atoms:

$$\frac{0.004491 \text{ mol Ni atoms}}{0.00449 \text{ mol O atoms}} = \frac{1 \text{ Ni}}{1 \text{ O}}$$

That is, this compound contains equal numbers of nickel atoms and oxygen atoms. We say the ratio of nickel atoms to oxygen atoms is 1:1 (1 to 1).

Example 8.11 Calculating Empirical Formulas

An oxide of aluminum is formed by the reaction of 4.151 g of aluminum with 3.692 g of oxygen. Calculate the empirical formula for this compound.

Solution

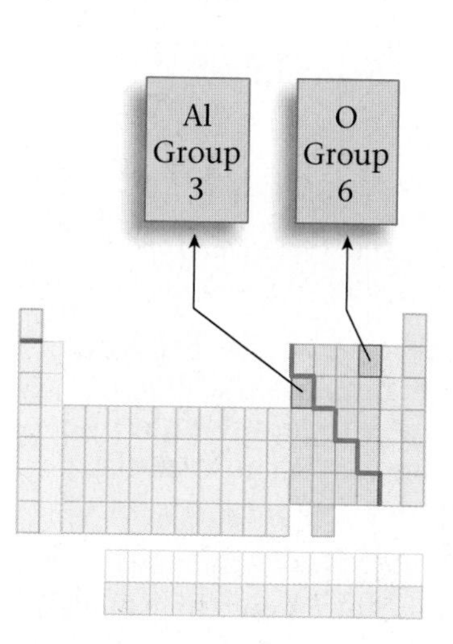

We know that the compound contains 4.151 g of aluminum and 3.692 g of oxygen. But we need to know the relative numbers of each type of atom to write the formula, so we must convert these masses to moles of atoms to get the empirical formula. We carry out the conversion by using the atomic masses of the elements.

$$4.151 \text{ g Al} \times \frac{1 \text{ mol Al}}{26.98 \text{ g Al}} = 0.1539 \text{ mol Al atoms}$$

$$3.692 \text{ g O} \times \frac{1 \text{ mol O}}{16.00 \text{ g O}} = 0.2308 \text{ mol O atoms}$$

Because chemical formulas use only whole numbers, we next find the integer (whole-number) ratio of the atoms. To do this we start by dividing both numbers by the smallest of the two. This converts the smallest number to 1.

$$\frac{0.1539 \text{ mol Al}}{0.1539} = 1.000 \text{ mol Al atoms}$$

$$\frac{0.2308 \text{ mol O}}{0.1539} = 1.500 \text{ mol O atoms}$$

Note that dividing both numbers of moles of atoms by the *same* number does not change the *relative* numbers of oxygen and aluminum atoms. That is,

$$\frac{0.2308 \text{ mol O}}{0.1539 \text{ mol Al}} = \frac{1.500 \text{ mol O}}{1.000 \text{ mol Al}}$$

Thus we know that the compound contains 1.500 mol of O atoms for every 1.000 mol of Al atoms, or, in terms of individual atoms, we could say that the compound contains 1.500 O atoms for every 1.000 Al atom. However,

We might express these data as:

$Al_{1.000\ mol}O_{1.500\ mol}$

or

$Al_{2.000\ mol}O_{3.000\ mol}$

or

Al_2O_3

because only *whole* atoms combine to form compounds, we must find a set of *whole numbers* to express the empirical formula. When we multiply both 1.000 and 1.500 by 2, we get the integers we need.

$$1.500 \text{ O} \times 2 = 3.000 = 3 \text{ O atoms}$$
$$1.000 \text{ Al} \times 2 = 2.000 = 2 \text{ Al atoms}$$

Therefore, this compound contains two Al atoms for every three O atoms, and the empirical formula is Al_2O_3. Note that the *ratio* of atoms in this compound is given by each of the following fractions:

$$\frac{0.2308 \text{ O}}{0.1539 \text{ Al}} = \frac{1.500 \text{ O}}{1.000 \text{ Al}} = \frac{\frac{3}{2}\text{O}}{1 \text{ Al}} = \frac{3 \text{ O}}{2 \text{ Al}}$$

The smallest whole-number ratio corresponds to the subscripts of the empirical formula, Al_2O_3. ■

Sometimes the relative numbers of moles you get when you calculate an empirical formula will turn out to be nonintegers, as was the case in Example 8.11. When this happens, you must convert to the appropriate whole numbers. This is done by multiplying all the numbers by the same small integer, which can be found by trial and error. The multiplier needed is almost always between 1 and 6. We will now summarize what we have learned about calculating empirical formulas.

Steps for Determining the Empirical Formula of a Compound

STEP 1 Obtain the mass of each element present (in grams).

STEP 2 Determine the number of moles of each type of atom present.

STEP 3 Divide the number of moles of each element by the smallest number of moles to convert the smallest number to 1. If all of the numbers so obtained are integers (whole numbers), these are the subscripts in the empirical formula. If one or more of these numbers are not integers, go on to step 4.

STEP 4 Multiply the numbers you derived in step 3 by the smallest integer that will convert all of them to whole numbers. This set of whole numbers represents the subscripts in the empirical formula.

Example 8.12 Calculating Empirical Formulas for Binary Compounds

When a 0.3546-g sample of vanadium metal is heated in air, it reacts with oxygen to achieve a final mass of 0.6330 g. Calculate the empirical formula of this vanadium oxide.

Solution

STEP 1 All the vanadium that was originally present will be found in the final compound, so we can calculate the mass of oxygen that reacted by taking the following difference:

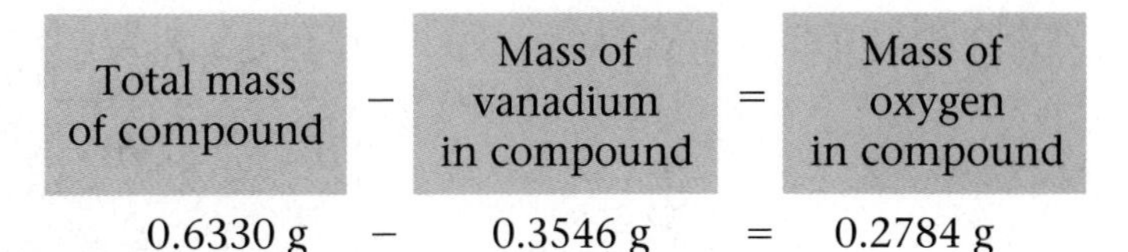

STEP 2 Using the atomic masses (50.94 for V and 16.00 for O), we obtain

$$0.3546\ \cancel{\text{g V}} \times \frac{1 \text{ mol V atoms}}{50.94\ \cancel{\text{g V}}} = 0.006961 \text{ mol V atoms}$$

$$0.2784\ \cancel{\text{g O}} \times \frac{1 \text{ mol O atoms}}{16.00\ \cancel{\text{g O}}} = 0.01740 \text{ mol O atoms}$$

STEP 3 Then we divide both numbers of moles by the smaller, 0.006961.

$$\frac{0.006961 \text{ mol V atoms}}{0.006961} = 1.000 \text{ mol V atoms}$$

$$\frac{0.01740 \text{ mol O atoms}}{0.006961} = 2.500 \text{ mol O atoms}$$

Because one of these numbers (2.500) is not an integer, we go on to step 4.

$V_{1.000}O_{2.500}$ becomes V_2O_5.

STEP 4 We note that $2 \times 2.500 = 5.000$ and $2 \times 1.000 = 2.000$, so we multiply both numbers by 2 to get integers.

$$2 \times 1.000 \text{ V} = 2.000 \text{ V} = 2 \text{ V}$$
$$2 \times 2.500 \text{ O} = 5.000 \text{ O} = 5 \text{ O}$$

This compound contains 2 V atoms for every 5 O atoms, and the empirical formula is V_2O_5.

✔ Self-Check Exercise 8.8

In a lab experiment it was observed that 0.6884 g of lead combines with 0.2356 g of chlorine to form a binary compound. Calculate the empirical formula of this compound.

See Problems 8.58, 8.61, 8.63, 8.65, and 8.66. ■

The same procedures we have used for binary compounds also apply to compounds containing three or more elements, as Example 8.13 illustrates.

Example 8.13 Calculating Empirical Formulas for Compounds Containing Three or More Elements

A sample of lead arsenate, an insecticide used against the potato beetle, contains 1.3813 g of lead, 0.00672 g of hydrogen, 0.4995 g of arsenic, and 0.4267 g of oxygen. Calculate the empirical formula for lead arsenate.

Solution

STEP 1 The compound contains 1.3813 g Pb, 0.00672 g H, 0.4995 g As, and 0.4267 g O.

STEP 2 We use the atomic masses of the elements present to calculate the moles of each.

$$1.3813\ \cancel{\text{g Pb}} \times \frac{1 \text{ mol Pb}}{207.2\ \cancel{\text{g Pb}}} = 0.006667 \text{ mol Pb}$$

Only three significant figures allowed.

$$0.00672\ \cancel{\text{g H}} \times \frac{1 \text{ mol H}}{1.008\ \cancel{\text{g H}}} = 0.00667 \text{ mol H}$$

$$0.4995 \text{ g As} \times \frac{1 \text{ mol As}}{74.92 \text{ g As}} = 0.006667 \text{ mol As}$$

$$0.4267 \text{ g O} \times \frac{1 \text{ mol O}}{16.00 \text{ g O}} = 0.02667 \text{ mol O}$$

STEP 3 Now we divide by the smallest number of moles.

$$\frac{0.006667 \text{ mol Pb}}{0.006667} = 1.000 \text{ mol Pb}$$

$$\frac{0.00667 \text{ mol H}}{0.006667} = 1.00 \text{ mol H}$$

$$\frac{0.006667 \text{ mol As}}{0.006667} = 1.000 \text{ mol As}$$

$$\frac{0.02667 \text{ mol O}}{0.006667} = 4.000 \text{ mol O}$$

The numbers of moles are all whole numbers, so the empirical formula is $PbHAsO_4$.

Self-Check Exercise 8.9

Sevin, the commercial name for an insecticide used to protect crops such as cotton, vegetables, and fruit, is made from carbamic acid. A chemist analyzing a sample of carbamic acid finds 0.8007 g of carbon, 0.9333 g of nitrogen, 0.2016 g of hydrogen, and 2.133 g of oxygen. Determine the empirical formula for carbamic acid.

See Problems 8.57 and 8.59. ■

When a compound is analyzed to determine the relative amounts of the elements present, the results are usually given in terms of percentages by masses of the various elements. In Section 8.5 we learned to calculate the percent composition of a compound from its formula. Now we will do the opposite. Given the percent composition, we will calculate the empirical formula.

Percent by mass for a given element means the grams of that element in 100 g of the compound.

To understand this procedure, you must understand the meaning of *percent.* Remember that percent means parts of a given component per 100 parts of the total mixture. For example, if a given compound is 15% carbon (by mass), the compound contains 15 g of carbon per 100 g of compound.

Calculation of the empirical formula of a compound when one is given its percent composition is illustrated in Example 8.14.

Example 8.14 Calculating Empirical Formulas from Percent Composition

Cisplatin, the common name for a platinum compound that is used to treat cancerous tumors, has the composition (mass percent) 65.02% platinum, 9.34% nitrogen, 2.02% hydrogen, and 23.63% chlorine. Calculate the empirical formula for cisplatin.

Solution

STEP 1 Determine how many grams of each element are present in 100 g of compound. Cisplatin is 65.02% platinum (by mass), which means there is 65.02 g of platinum (Pt) per 100.00 g of compound. Similarly, a 100.00-g sample of cisplatin contains 9.34 g of nitrogen (N), 2.02 g of hydrogen (H), and 26.63 g of chlorine (Cl).

If we have a 100.00-g sample of cisplatin, we have 65.02 g Pt, 9.34 g N, 2.02 g H, and 23.63 g Cl.

STEP 2 Determine the number of moles of each type of atom. We use the atomic masses to calculate moles.

$$65.02 \text{ g Pt} \times \frac{1 \text{ mol Pt}}{195.1 \text{ g Pt}} = 0.3333 \text{ mol Pt}$$
$$9.34 \text{ g N} \times \frac{1 \text{ mol N}}{14.01 \text{ g N}} = 0.667 \text{ mol N}$$
$$2.02 \text{ g H} \times \frac{1 \text{ mol H}}{1.008 \text{ g H}} = 2.00 \text{ mol H}$$
$$23.63 \text{ g Cl} \times \frac{1 \text{ mol Cl}}{35.45 \text{ g Cl}} = 0.6666 \text{ mol Cl}$$

STEP 3 Divide through by the smallest number of moles.

$$\frac{0.3333 \text{ mol Pt}}{0.3333} = 1.000 \text{ mol Pt}$$
$$\frac{0.667 \text{ mol N}}{0.3333} = 2.00 \text{ mol N}$$
$$\frac{2.00 \text{ mol H}}{0.3333} = 6.01 \text{ mol H}$$
$$\frac{0.6666 \text{ mol Cl}}{0.3333} = 2.000 \text{ mol Cl}$$

The empirical formula for cisplatin is $PtN_2H_6Cl_2$. Note that the number for hydrogen is slightly greater than 6 because of rounding-off effects.

✔ Self-Check Exercise 8.10

The most common form of nylon (Nylon-6) is 63.68% carbon, 12.38% nitrogen, 9.80% hydrogen, and 14.14% oxygen. Calculate the empirical formula for Nylon-6.

See Problems 8.67 through 8.74. ■

Note from Example 8.14 that once the percentages are converted to masses, this example is the same as earlier examples in which the masses were given directly.

Calculation of Molecular Formulas

AIM: To learn to calculate the molecular formula of a compound, given its empirical formula and molar mass.

If we know the composition of a compound in terms of the masses (or mass percentages) of the elements present, we can calculate the empirical formula but not the molecular formula. For reasons that will become clear as we consider Example 8.15, to obtain the molecular formula we must know the molar mass. In this section we will consider compounds where both the percent composition and the molar mass are known.

Example 8.15 Calculating Molecular Formulas

A white powder is analyzed and found to have an empirical formula of P_2O_5. The compound has a molar mass of 283.88 g. What is the compound's molecular formula?

Solution

To obtain the molecular formula, we must compare the empirical formula mass to the molar mass. The empirical formula mass for P_2O_5 is the mass of 1 mol of P_2O_5 units.

P Group 5 | O Group 6

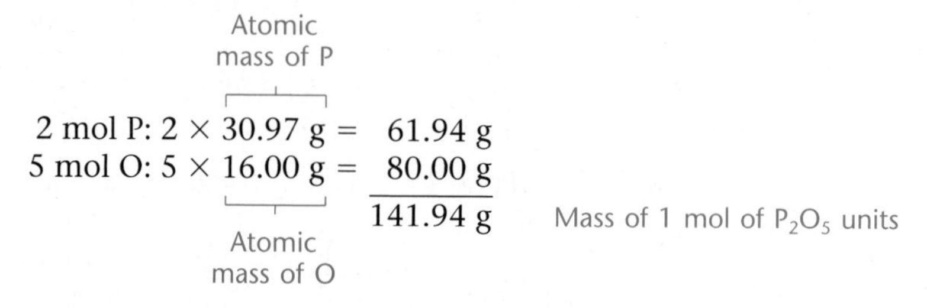

Recall that the molecular formula contains a whole number of empirical formula units. That is,

$$\text{Molecular formula} = (\text{empirical formula})_n$$

where n is a small whole number. Now, because

$$\text{Molecular formula} = n \times \text{empirical formula}$$

then

$$\text{Molar mass} = n \times \text{empirical formula mass}$$

Solving for n gives

$$n = \frac{\text{molar mass}}{\text{empirical formula mass}}$$

Thus, to determine the molecular formula, we first divide the molar mass by the empirical formula mass. This tells us how many empirical formula masses there are in one molar mass.

$$\frac{\text{Molar mass}}{\text{Empirical formula mass}} = \frac{283.88\text{ g}}{141.94\text{ g}} = 2$$

This result means that $n = 2$ for this compound, so the molecular formula consists of two empirical formula units, and the molecular formula is $(P_2O_5)_2$, or P_4O_{10}. The structure of this interesting compound is shown in Figure 8.5.

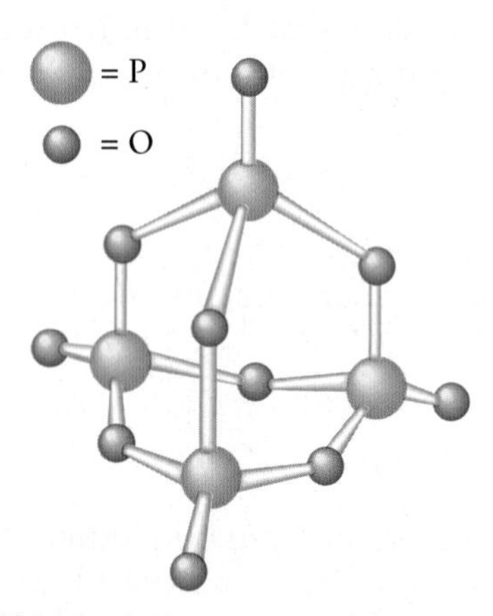

Figure 8.5
The structure of P_4O_{10} as a "ball-and-stick" model. This compound has a great affinity for water and is often used as a desiccant, or drying agent.

✔ Self-Check Exercise 8.11

A compound used as an additive for gasoline to help prevent engine knock shows the following percentage composition:

71.65% Cl   24.27% C   4.07% H

The molar mass is known to be 98.96 g. Determine the empirical formula and the molecular formula for this compound.

See Problems 8.81 and 8.82. ■

It is important to realize that the molecular formula is always an integer multiple of the empirical formula. For example, the sugar glucose (see

Figure 8.4) has the empirical formula CH_2O and the molecular formula $C_6H_{12}O_6$. In this case there are six empirical formula units in each glucose molecule:

$$(CH_2O)_6 = C_6H_{12}O_6$$

In general, we can represent the molecular formula in terms of the empirical formula as follows:

Molecular formula = (empirical formula)$_n$, where n is an integer.

$$(\text{Empirical formula})_n = \text{molecular formula}$$

where n is an integer. If $n = 1$, the molecular formula is the same as the empirical formula. For example, for carbon dioxide the empirical formula (CO_2) and the molecular formula (CO_2) are the same, so $n = 1$. On the other hand, for tetraphosphorus decoxide the empirical formula is P_2O_5 and the molecular formula is $P_4O_{10} = (P_2O_5)_2$. In this case $n = 2$.

CHAPTER 8 REVIEW

KEY TERMS

atomic mass unit (8.2)
average atomic mass (8.2)
mole (8.3)
Avogadro's number (8.3)
molar mass (8.4)
mass percent (8.5)
empirical formula (8.6)
molecular formula (8.6)

SUMMARY

1. We can count individual units by weighing if we know the average mass of the units. Thus, when we know the average mass of the atoms of an element as that element occurs in nature, we can calculate the number of atoms in any given sample of that element by weighing the sample.
2. A mole is a unit of measure equal to 6.022×10^{23}, which is called Avogadro's number. One mole of any substance contains 6.022×10^{23} units.
3. One mole of an element has a mass equal to the element's atomic mass expressed in grams. The molar mass of any compound is the mass (in grams) of 1 mol of the compound and is the sum of the masses of the component atoms.
4. Percent composition consists of the mass percent of each element in a compound:

$$\text{Mass percent} = \frac{\text{mass of a given element in 1 mol of compound}}{\text{mass of 1 mol of compound}} \times 100\%$$

5. The empirical formula of a compound is the simplest whole-number ratio of the atoms present in the compound; it can be derived from the percent composition of the compound. The molecular formula is the exact formula of the molecules present; it is always an integer multiple of the empirical formula. The following diagram summarizes these different ways of expressing the same information.

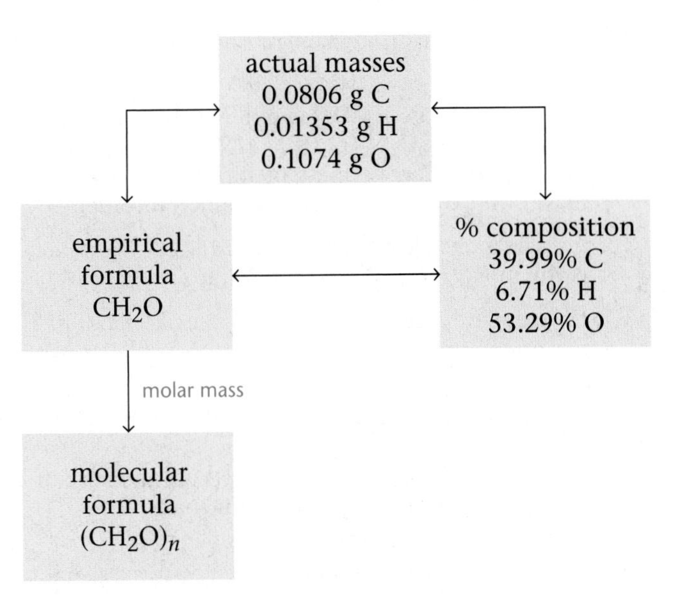

IN-CLASS DISCUSSION QUESTIONS

These questions are designed to be considered by groups of students in class. Often these questions work well for introducing a particular topic in class.

1. In chemistry, what is meant by the term *mole?* What is the importance of the mole concept?
2. What is the difference between the empirical and molecular formulas of a compound? Can they ever be the same? Explain.

3. You find a compound composed only of element X and hydrogen, and know that it is 91.33% element X by mass. Each molecule has 2.67 times as many H atoms as X atoms. What is element X?

4. What is the empirical formula of an oxide of nitrogen that contains 36.8% nitrogen?

5. A substance A_2B is 60% A by mass. Calculate the percent B (by mass) for AB_2.

6. Give the formula for calcium phosphate and then answer the following questions:
   a. Calculate the percent composition of each of the elements in this compound.
   b. If you knew that there was 50.0 g of phosphorus in your sample, how many grams of calcium phosphate would you have? How many moles of calcium phosphate would this be? How many formula units of calcium phosphate?

7. How would you find the number of "chalk molecules" it takes to write your name on the board? Explain what you would need to do, and provide a sample calculation.

8. A 0.821-mol sample of a substance composed of diatomic molecules has a mass of 131.3 g. Identify this molecule.

9. How many molecules of water are there in a 10.0-g sample of water? How many hydrogen atoms are there in this sample?

10. What is the mass (in grams) of one molecule of ammonia?

11. Consider separate 100.0-g samples of each of the following: NH_3, N_2O, N_2H_4, HCN, HNO_3. Arrange these samples from largest mass of nitrogen to smallest mass of nitrogen and prove/explain your order.

12. A single molecule has a mass of 8.25×10^{-23} g. What is the molar mass of this compound?

13. Differentiate between the terms *atomic mass* and *molar mass.*

Questions and Problems*

All even-numbered exercises have answers in the back of this book and solutions in the *Solutions Guide.*

8.1 Counting by Weighing

PROBLEMS

1. Merchants usually sell small nuts, washers, and bolts by weight (like jelly beans!) rather than by individually counting the items. Suppose a particular type of washer weighs 0.110 g on the average. What would 100 such washers weigh? How many washers would there be in 100. g of washers?

* The element symbols and formulas are given in some problems but not in others to help you learn this necessary "vocabulary."

2. A particular small laboratory cork weighs 1.63 g, whereas a rubber lab stopper of the same size weighs 4.31 g. How many corks would there be in 500. g of such corks? How many rubber stoppers would there be in 500. g of similar stoppers? How many grams of rubber stoppers would be needed to contain the same number of stoppers as there are corks in 1.00 kg of corks?

8.2 Atomic Masses: Counting Atoms by Weighing

QUESTIONS

3. Define the *amu.* What is one amu equivalent to in grams?

4. Why do we use the *average* atomic mass of the elements when performing calculations?

PROBLEMS

5. Using the average atomic masses for each of the following elements (see the table inside the front cover of this book), calculate the mass, in amu, of each of the following samples.
   a. 278 atoms of Li
   b. 1 million C atoms
   c. 5×10^{25} sodium atoms
   d. 1 atom of cadmium
   e. 6.022×10^{23} atoms of mercury

6. Using the average atomic masses for each of the following elements (see the table inside the front cover of this book), calculate the number of atoms present in each of the following samples.
   a. 52.00 amu of chromium
   b. 749.2 amu of arsenic
   c. 4274 amu of rubidium
   d. 2698 amu of aluminum
   e. 1900. amu of fluorine

7. What does an average magnesium atom weigh (in amu)? What would 345 magnesium atoms weigh? How many magnesium atoms are contained in a sample of magnesium that has a mass of 2.071×10^4 amu?

8. What does an average iodine atom weigh (in amu)? How many atoms of iodine are contained in a sample of iodine that has a mass of 7.043×10^4 amu? What would 451 iodine atoms weigh?

8.3 The Mole

QUESTIONS

9. In 26.98 g of aluminum, there are ________ aluminum atoms present.

10. In 78.20 g of potassium, there are ________ potassium atoms present.

PROBLEMS

11. Suppose you have a sample of sodium weighing 11.50 g. How many atoms of sodium are present in the sample? What mass of potassium would you need to have the same number of potassium atoms as there are sodium atoms in the sample of sodium?

12. What mass of iron contains the same number of atoms as 14.01 g of nitrogen?

13. What mass of hydrogen contains the same number of atoms as 7.00 g of nitrogen?

14. What mass of cobalt contains the same number of atoms as 57.0 g of fluorine?

15. If an average aluminum atom has mass 4.480×10^{-23} g, what is the average mass of a boron atom in grams?

16. Calculate the average mass in grams of 1 atom of oxygen.

17. Which has the smaller mass, 1 mol of He atoms or 4 mol of H atoms?

18. Which weighs more, 0.50 mol of oxygen atoms or 4 mol of hydrogen atoms?

19. Use the average atomic masses given inside the front cover of this book to calculate the number of *moles* of each element present in each of the following samples.
    a. 21.50 g of arsenic
    b. 9.105 g of phosphorus
    c. 0.05152 g of barium
    d. 43.15 g of carbon
    e. 26.02 g of chromium
    f. 1.951 g of platinum

20. Use the average atomic masses given inside the front cover of this book to calculate the number of *moles* of each element present in each of the following samples.
    a. 1.71×10^{-3} g of silver
    b. 280.9 mg of silicon
    c. 63.45 g of zinc
    d. 5.869 g of nickel
    e. 128.6 g of uranium
    f. 3.251 kg of lithium

21. Use the average atomic masses given inside the front cover of this book to calculate the mass in grams of each of the following samples.
    a. 0.251 mol of sodium
    b. 2.25 mol of helium
    c. 4.27×10^5 mol of iron
    d. 5.58 mol of copper
    e. 1.45×10^{-6} mol of lithium
    f. 6.25 mol of aluminum

22. Use the average atomic masses given inside the front cover of this book to calculate the mass in grams of each of the following samples.
    a. 1.76×10^{-3} mol of cesium
    b. 0.0125 mol of neon
    c. 5.29×10^3 mol of lead
    d. 0.00000122 mol of sodium
    e. 5.51 millimol of arsenic (1 millimol = 1/1000 mol)
    f. 8.72 mol of carbon

23. Using the average atomic masses given inside the front cover of the text, calculate the number of *atoms* present in each of the following samples.
    a. 1.50 g of silver, Ag
    b. 0.0015 mol of copper, Cu
    c. 0.0015 g of copper, Cu
    d. 2.00 kg of magnesium, Mg
    e. 2.34 oz of calcium, Ca
    f. 2.34 g of calcium, Ca
    g. 2.34 mol of calcium, Ca

24. Using the average atomic masses given inside the front cover of the text, calculate the indicated quantities.
    a. the number of cobalt atoms in 0.00103 g of cobalt
    b. the number of cobalt atoms in 0.00103 mol of cobalt
    c. the number of mol of cobalt in 2.75 g of cobalt
    d. the number of mol of cobalt represented by 5.99×10^{21} cobalt atoms
    e. the mass of 4.23 mol of cobalt
    f. the number of cobalt atoms in 4.23 mol of cobalt
    g. the number of cobalt atoms in 4.23 g of cobalt

8.4 Molar Mass

QUESTIONS

25. The ________ of a substance is the mass (in grams) of 1 mol of the substance.

26. The molar mass of a substance can be obtained by ________ the atomic weights of the component atoms.

PROBLEMS

27. Give the name and calculate the molar mass for each of the following substances.
    a. Cr_2O_3
    b. $Cu(NO_3)_2$
    c. P_4O_6
    d. Bi_2O_3
    e. CS_2
    f. H_2SO_3

28. Give the name and calculate the molar mass for each of the following substances.
    a. NO_2
    b. N_2O
    c. XeF_4
    d. NaOCl
    e. HNO_3
    f. $NaC_2H_3O_2$

29. Calculate the molar mass for each of the following substances.
 a. barium perchlorate
 b. magnesium sulfate
 c. lead(II) chloride
 d. copper(II) nitrate
 e. tin(IV) chloride
 f. phenol, C_6H_6O

30. Calculate the molar mass for each of the following substances.
 a. ammonium sulfide, $(NH_4)_2S$
 b. dichlorophenol, $C_6H_4OCl_2$
 c. barium hydride, BaH_2
 d. potassium dihydrogen phosphate, KH_2PO_4
 e. potassium hydrogen phosphate, K_2HPO_4
 f. potassium phosphate, K_3PO_4

31. Calculate the number of *moles* of the indicated substance present in each of the following samples.
 a. 21.4 mg of nitrogen dioxide
 b. 1.56 g of copper(II) nitrate
 c. 2.47 g of carbon disulfide
 d. 5.04 g of aluminum sulfate
 e. 2.99 g of lead(II) chloride
 f. 62.4 g of calcium carbonate

32. Calculate the number of *moles* of the indicated substance present in each of the following samples.
 a. 92.4 g of hydrogen bromide
 b. 5.34 mg of ferric chloride
 c. 2.21 kg of sulfuric acid
 d. 3.44 g of barium carbonate
 e. 2.89 g of aluminum chloride
 f. 7.21 g of lithium carbonate

33. Calculate the number of *moles* of the indicated substance in each of the following samples.
 a. 18.0 g of dextrose, $C_6H_{12}O_6$
 b. 21.94 g of nitrous oxide, N_2O
 c. 21.94 g of nitric oxide, NO
 d. 1.24 oz of gold(III) acetate, $Au(C_2H_3O_2)_3$
 e. 44.2 g of ammonium dichromate, $(NH_4)_2Cr_2O_7$

34. Calculate the number of *moles* of the indicated substance in each of the following samples.
 a. 4.26×10^{-3} g of sodium dihydrogen phosphate
 b. 521 g of copper(I) chloride
 c. 151 kg of iron
 d. 8.76 g of strontium fluoride
 e. 1.26×10^4 g of aluminum

35. Calculate the mass in grams of each of the following samples.
 a. 2.41 millimol of potassium nitrate (1 millimol = 1/1000 mol)
 b. 8.91 mol of ethanol, C_2H_5OH
 c. 0.0141 mol of calcium oxide
 d. 1.91 mol of gold(III) bromide
 e. 0.0000117 mol of water
 f. 2.68 mol of silver nitrate

36. Calculate the mass in grams of each of the following samples.
 a. 0.000471 mol of carbon monoxide
 b. 1.75×10^{-6} mol of gold(III) chloride
 c. 228 mol of iron(III) chloride
 d. 2.98 millimol of potassium phosphate (1 millimol = 1/1000 mol)
 e. 2.71×10^{-3} mol of lithium chloride
 f. 6.55 mol of ammonia

37. Calculate the mass in grams of each of the following samples.
 a. 0.251 mol of ethyl alcohol, C_2H_6O
 b. 1.26 mol of carbon dioxide
 c. 9.31×10^{-4} mol of gold(III) chloride
 d. 7.74 mol of sodium nitrate
 e. 0.000357 mol of iron

38. Calculate the mass in grams of each of the following samples.
 a. 1.27 mmol of carbon dioxide
 b. 4.12×10^3 mol of nitrogen trichloride
 c. 0.00451 mol of ammonium nitrate
 d. 18.0 mol of water
 e. 62.7 mol of copper(II) sulfate

39. Calculate the number of *molecules* present in each of the following samples.
 a. 4.75 mmol of phosphine, PH_3
 b. 4.75 g of phosphine, PH_3
 c. 1.25×10^{-2} g of lead(II) acetate, $Pb(CH_3CO_2)_2$
 d. 1.25×10^{-2} mol of lead(II) acetate, $Pb(CH_3CO_2)_2$
 e. a sample of benzene, C_6H_6, which contains a total of 5.40 mol of carbon

40. Calculate the number of *molecules* present in each of the following samples.
 a. 6.37 mol of carbon monoxide
 b. 6.37 g of carbon monoxide
 c. 2.62×10^{-6} g of water
 d. 2.62×10^{-6} mol of water
 e. 5.23 g of benzene, C_6H_6

41. Calculate the number of *moles* of carbon atoms present in each of the following samples.
 a. 1.271 g of ethanol, C_2H_5OH
 b. 3.982 g of 1,4-dichlorobenzene, $C_6H_4Cl_2$
 c. 0.4438 g of carbon suboxide, C_3O_2
 d. 2.910 g of methylene chloride, CH_2Cl_2

42. Calculate the number of *moles* of sulfur atoms present in each of the following samples.
 a. 2.01 g of sodium sulfate
 b. 2.01 g of sodium sulfite
 c. 2.01 g of sodium sulfide
 d. 2.01 g of sodium thiosulfate, $Na_2S_2O_3$

8.5 Percent Composition of Compounds

QUESTIONS

43. The mass fraction of an element present in a compound can be obtained by comparing the mass of the particular element present in 1 mol of the compound to the ________ mass of the compound.

44. The mass percentage of a given element in a compound must always be (greater/less) than 100%.

PROBLEMS

45. Calculate the percent by mass of each element in the following compounds.
    a. $HClO_3$
    b. UF_4
    c. CaH_2
    d. Ag_2S
    e. $NaHSO_3$
    f. MnO_2

46. Calculate the percent by mass of each element in the following compounds.
    a. Na_2S
    b. NH_4NO_2
    c. NH_4NO_3
    d. NH_2Cl
    e. PH_3
    f. H_3PO_3

47. Calculate the percent by mass of the element listed *first* in the formulas for each of the following compounds.
    a. methane, CH_4
    b. sodium nitrate, $NaNO_3$
    c. carbon monoxide, CO
    d. nitrogen dioxide, NO_2
    e. 1-octanol, $C_8H_{18}O$
    f. calcium phosphate, $Ca_3(PO_4)_2$
    g. 3-phenylphenol, $C_{12}H_{10}O$
    h. aluminum acetate, $Al(C_2H_3O_2)_3$

48. Calculate the percent by mass of the element listed *first* in the formulas for each of the following compounds.
    a. copper(II) bromide, $CuBr_2$
    b. copper(I) bromide, CuBr
    c. iron(II) chloride, $FeCl_2$
    d. iron(III) chloride, $FeCl_3$
    e. cobalt(II) iodide, CoI_2
    f. cobalt(III) iodide, CoI_3
    g. tin(II) oxide, SnO
    h. tin(IV) oxide, SnO_2

49. Calculate the percent by mass of the element listed *first* in the formulas for each of the following compounds.
    a. adipic acid, $C_6H_{10}O_4$
    b. ammonium nitrate, NH_4NO_3
    c. caffeine, $C_8H_{10}N_4O_2$
    d. chlorine dioxide, ClO_2
    e. cyclohexanol, $C_6H_{11}OH$
    f. dextrose, $C_6H_{12}O_6$
    g. eicosane, $C_{20}H_{42}$
    h. ethanol, C_2H_5OH

50. Calculate the percent by mass of the element listed *first* in the formulas for each of the following compounds.
    a. iron(III) chloride
    b. oxygen difluoride, OF_2
    c. benzene, C_6H_6
    d. ammonium perchlorate, NH_4ClO_4
    e. silver oxide
    f. cobalt(II) chloride
    g. dinitrogen tetroxide
    h. manganese(II) chloride

51. For each of the following samples of ionic substances, calculate the number of moles and mass of the positive ions present in each sample.
    a. 4.25 g of ammonium iodide, NH_4I
    b. 6.31 mol of ammonium sulfide, $(NH_4)_2S$
    c. 9.71 g of barium phosphide, Ba_3P_2
    d. 7.63 mol of calcium phosphate, $Ca_3(PO_4)_2$

52. For each of the following ionic substances, calculate the *percentage* of the overall molar mass of the compound that is represented by the *positive ions* the compound contains.
    a. ammonium chloride
    b. copper(II) sulfate
    c. gold(III) chloride
    d. silver nitrate

8.6 Formulas of Compounds

QUESTIONS

53. What experimental evidence about a new compound must be known before its formula can be determined?

54. What does the *empirical* formula of a compound represent? How does the *molecular* formula differ from the empirical formula?

55. Give the empirical formula that corresponds to each of the following molecular formulas.
    a. sodium peroxide, Na_2O_2
    b. terephthalic acid, $C_8H_6O_4$
    c. phenobarbital, $C_{12}H_{12}N_2O_3$
    d. 1,4-dichloro-2-butene, $C_4H_6Cl_2$

56. Which of the following pairs of compounds have the same *empirical* formula?
    a. acetylene, C_2H_2, and benzene, C_6H_6
    b. ethane, C_2H_6, and butane, C_4H_{10}
    c. nitrogen dioxide, NO_2, and dinitrogen tetroxide, N_2O_4
    d. diphenyl ether, $C_{12}H_{10}O$, and phenol, C_6H_5OH

8.7 Calculation of Empirical Formulas

PROBLEMS

57. A compound was analyzed and was found to contain the following percentages by mass: phosphorus, 90.10%; hydrogen 8.90%. Determine the empirical formula of the compound.

58. A compound was analyzed and was found to contain the following percentages by mass: hydrogen, 3.09%; phosphorus, 31.60%; oxygen, 65.31%. Determine the empirical formula of the compound.

59. A 0.5998-g sample of a new compound has been analyzed and found to contain the following masses of elements: carbon, 0.2322 g; hydrogen, 0.05848 g;

oxygen, 0.3091 g. Calculate the empirical formula of the compound.

60. A compound has the following percentages by mass: barium, 58.84%; sulfur, 13.74%; oxygen, 27.43%. Determine the empirical formula of the compound.

61. If a 1.271-g sample of aluminum metal is heated in a chlorine gas atmosphere, the mass of aluminum chloride produced is 6.280 g. Calculate the empirical formula of aluminum chloride.

62. Analysis of a certain compound yielded the following percentages of the elements by mass: nitrogen, 29.16%; hydrogen, 8.392%; carbon, 12.50%; oxygen, 49.95%. Determine the empirical formula of the compound.

63. When 3.269 g of zinc is heated in pure oxygen, the sample gains 0.800 g of oxygen in forming the oxide. Calculate the empirical formula of zinc oxide.

64. If cobalt metal is mixed with excess sulfur and heated strongly, a sulfide is produced that contains 55.06% cobalt by mass. Calculate the empirical formula of the sulfide.

65. If 2.461 g of metallic calcium is heated in a stream of chlorine gas, 4.353 g of Cl_2 is absorbed in forming the metal chloride. Calculate the empirical formula of calcium chloride.

66. If 10.00 g of copper metal is heated strongly in the air, the sample gains 2.52 g of oxygen in forming an oxide. Determine the empirical formula of this oxide.

67. A compound used in the nuclear industry has the following composition: uranium, 67.61%; fluorine, 32.39%. Determine the empirical formula of the compound.

68. A compound has the following percentages by mass: aluminum, 32.13%; fluorine, 67.87%. Calculate the empirical formula of the compound.

69. A compound has the following percentage composition by mass: copper, 33.88%; nitrogen, 14.94%; oxygen, 51.18%. Determine the empirical formula of the compound.

70. When lithium metal is heated strongly in an atmosphere of pure nitrogen, the product contains 59.78% Li and 40.22% N on a mass basis. Determine the empirical formula of the compound.

71. A compound has been analyzed and has been found to have the following composition: copper, 66.75%; phosphorus, 10.84%; oxygen, 22.41%. Determine the empirical formula of the compound.

72. A compound was analyzed and found to have the following percentage composition: aluminum, 15.77%; sulfur, 28.11%; oxygen, 56.12%. Calculate the empirical formula of the compound.

73. When 1.00 mg of lithium metal is reacted with fluorine gas (F_2), the resulting fluoride salt has a mass of 3.73 mg. Calculate the empirical formula of lithium fluoride.

74. Phosphorus and chlorine form two binary compounds, in which the percentages of phosphorus are 22.55% and 14.87%, respectively. Calculate the empirical formulas of the two binary phosphorus–chlorine compounds.

8.8 Calculation of Molecular Formulas

QUESTIONS

75. How does the *molecular* formula of a compound differ from the empirical formula? Can a compound's empirical and molecular formulas be the same? Explain.

76. What information do we need to determine the molecular formula of a compound if we know only the empirical formula?

PROBLEMS

77. A binary compound of boron and hydrogen has the following percentage composition: 78.14% boron, 21.86% hydrogen. If the molar mass of the compound is determined by experiment to be between 27 and 28 g, what are the empirical and molecular formulas of the compound?

78. A compound with empirical formula CH was found by experiment to have a molar mass of approximately 78 g. What is the molecular formula of the compound?

79. A compound with the empirical formula CH_2 was found to have a molar mass of approximately 84 g. What is the molecular formula of the compound?

80. A compound with the empirical formula CH_4O was found in a subsequent experiment to have a molar mass of approximately 192 g. What is the molecular formula of the compound?

81. A compound having an approximate molar mass of 165–170 g has the following percentage composition by mass: carbon, 42.87%; hydrogen, 3.598%; oxygen, 28.55%; nitrogen, 25.00%. Determine the empirical and molecular formulas of the compound.

82. A very toxic sodium compound consists of 46.91% Na, 24.51% C, and 28.59% N on a mass basis and has a molar mass of approximately 50 g/mol. Determine the empirical and molecular formulas of the compound.

Additional Problems

83. Use the periodic table inside the front cover of this text to determine the atomic mass (per mole) or molar mass of each of the substances in column 1, and find that mass in column 2.

Column 1	Column 2
(1) molybdenum	(a) 33.99 g
(2) lanthanum	(b) 79.9 g
(3) carbon tetrabromide	(c) 95.94 g
(4) mercury(II) oxide	(d) 125.84 g
(5) titanium(IV) oxide	(e) 138.9 g
(6) manganese(II) chloride	(f) 143.1 g
(7) phosphine, PH_3	(g) 156.7 g
(8) tin(II) fluoride	(h) 216.6 g
(9) lead(II) sulfide	(i) 239.3 g
(10) copper(I) oxide	(j) 331.6 g

84. Complete the following table.

Mass of Sample	Moles of Sample	Atoms in Sample
5.00 g Al	_______	_______
_______	0.00250 mol Fe	_______
_______	_______	2.6×10^{24} atoms Cu
0.00250 g Mg	_______	_______
_______	2.7×10^{-3} mol Na	_______
_______	_______	1.00×10^{4} atoms U

85. Complete the following table.

Mass of Sample	Moles of Sample	Molecules in Sample	Atoms in Sample
4.24 g C_6H_6	_______	_______	_______
_______	0.224 mol H_2O	_______	_______
_______	_______	2.71×10^{22} molecules CO_2	_______
_______	1.26 mol HCl	_______	_______
_______	_______	4.21×10^{24} molecules H_2O	_______
0.297 g CH_3OH	_______	_______	_______

86. Consider a hypothetical compound composed of elements X, Y, and Z with the empirical formula X_2YZ_3. Given that the atomic masses of X, Y, and Z are 41.2, 57.7, and 63.9, respectively, calculate the percentage composition by mass of the compound. If the molecular formula of the compound is found by molar mass determination to be actually $X_4Y_2Z_6$, what is the percentage of each element present? Explain your results.

87. A binary compound of magnesium and nitrogen is analyzed, and 1.2791 g of the compound is found to contain 0.9240 g of magnesium. When a second sample of this compound is treated with water and heated, the nitrogen is driven off as ammonia, leaving a compound that contains 60.31% magnesium and 39.69% oxygen by mass. Calculate the empirical formulas of the two magnesium compounds.

88. When a 2.118-g sample of copper is heated in an atmosphere in which the amount of oxygen present is restricted, the sample gains 0.2666 g of oxygen in forming a reddish-brown oxide. However, when 2.118 g of copper is heated in a stream of pure oxygen, the sample gains 0.5332 g of oxygen. Calculate the empirical formulas of the two oxides of copper.

89. Hydrogen gas reacts with each of the halogen elements to form the hydrogen halides (HF, HCl, HBr, HI). Calculate the percent by mass of hydrogen in each of these compounds.

90. Calculate the number of atoms of each element present in each of the following samples.
   a. 4.21 g of water
   b. 6.81 g of carbon dioxide
   c. 0.000221 g of benzene, C_6H_6
   d. 2.26 mol of $C_{12}H_{22}O_{11}$

91. Calculate the mass in grams of each of the following samples.
   a. 10,000,000,000 nitrogen molecules
   b. 2.49×10^{20} carbon dioxide molecules
   c. 7.0983 mol of sodium chloride
   d. 9.012×10^{-6} mol of 1,2-dichloroethane, $C_2H_4Cl_2$

92. Calculate the mass of carbon in grams, the percent carbon by mass, and the number of individual carbon atoms present in each of the following samples.
   a. 7.819 g of carbon suboxide, C_3O_2
   b. 1.53×10^{21} molecules of carbon monoxide
   c. 0.200 mol of phenol, C_6H_6O

93. Find the item in column 2 that best explains or completes the statement or question in column 1.

Column 1

(1) 1 amu
(2) 1008 amu
(3) mass of the "average" atom of an element
(4) number of carbon atoms in 12.01 g of carbon
(5) 6.022×10^{23} molecules
(6) total mass of all atoms in 1 mol of a compound
(7) smallest whole-number ratio of atoms present in a molecule
(8) formula showing actual number of atoms present in a molecule
(9) product formed when any carbon-containing compound is burned in O_2
(10) have the same empirical formulas, but different molecular formulas

Column 2

(a) 6.022×10^{23}
(b) atomic mass
(c) mass of 1000 hydrogen atoms
(d) benzene, C_6H_6, and acetylene, C_2H_2
(e) carbon dioxide
(f) empirical formula
(g) 1.66×10^{-24} g
(h) molecular formula
(i) molar mass
(j) 1 mol

94. Calculate the number of grams of iron that contain the same number of atoms as 2.24 g of cobalt.

95. Calculate the number of grams of cobalt that contain the same number of atoms as 2.24 g of iron.

96. Calculate the number of grams of mercury that contain the same number of atoms as 5.00 g of tellurium.

97. Calculate the number of grams of lithium that contain the same number of atoms as 1.00 kg of zirconium.

98. Given that the molar mass of carbon tetrachloride, CCl_4, is 153.8 g, calculate the mass in grams of 1 molecule of CCl_4.

99. Calculate the mass in grams of hydrogen present in 2.500 g of each of the following compounds.
a. benzene, C_6H_6
b. calcium hydride, CaH_2
c. ethyl alcohol, C_2H_5OH
d. serine, $C_3H_7O_3N$

100. Calculate the mass in grams of nitrogen present in 5.000 g of each of the following compounds.
a. glycine, $C_2H_5O_2N$
b. magnesium nitride, Mg_3N_2
c. calcium nitrate
d. dinitrogen tetroxide

101. A strikingly beautiful copper compound with the common name "blue vitriol" has the following elemental composition: 25.45% Cu, 12.84% S, 4.036% H, 57.67% O. Determine the empirical formula of the compound.

102. A magnesium salt has the following elemental composition: 16.39% Mg, 18.89% N, 64.72% O. Determine the empirical formula of the salt.

103. The mass 1.66×10^{-24} g is equivalent to 1 ________.

104. Although exact isotopic masses are known with great precision for most elements, we use the *average* mass of an element's atoms in most chemical calculations. Explain.

105. Using the average atomic masses given in Table 8.1, calculate the number of atoms present in each of the following samples.
a. 160,000 amu of oxygen
b. 8139.81 amu of nitrogen
c. 13,490 amu of aluminum
d. 5040 amu of hydrogen
e. 367,495.15 amu of sodium

106. If an average sodium atom weighs 22.99 amu, how many sodium atoms are contained in 1.98×10^{13} amu of sodium? What will 3.01×10^{23} sodium atoms weigh?

107. Using the average atomic masses given inside the front cover of this text, calculate how many *moles* of each element the following *masses* represent.
a. 1.5 mg of chromium
b. 2.0×10^{-3} g of strontium
c. 4.84×10^{4} g of boron
d. 3.6×10^{-6} μg of californium
e. 1.0 ton (2000 lb) of iron
f. 20.4 g of barium
g. 62.8 g of cobalt

108. Using the average atomic masses given inside the front cover of this text, calculate the *mass in grams* of each of the following samples.
a. 5.0 mol of potassium
b. 0.000305 mol of mercury
c. 2.31×10^{-5} mol of manganese
d. 10.5 mol of phosphorus
e. 4.9×10^{4} mol of iron
f. 125 mol of lithium
g. 0.01205 mol of fluorine

109. Using the average atomic masses given inside the front cover of this text, calculate the number of *atoms* present in each of the following samples.
a. 2.89 g of gold
b. 0.000259 mol of platinum
c. 0.000259 g of platinum
d. 2.0 lb of magnesium
e. 1.90 mL of liquid mercury (density = 13.6 g/mL)
f. 4.30 mol of tungsten
g. 4.30 g of tungsten

110. Calculate the molar mass for each of the following substances.
a. ferrous sulfate
b. mercuric iodide
c. stannic oxide
d. cobaltous chloride
e. cupric nitrate

111. Calculate the molar mass for each of the following substances.
a. adipic acid, $C_6H_{10}O_4$
b. caffeine, $C_8H_{10}N_4O_2$
c. eicosane, $C_{20}H_{42}$
d. cyclohexanol, $C_6H_{11}OH$
e. vinyl acetate, $C_4H_6O_2$
f. dextrose, $C_6H_{12}O_6$

112. Calculate the number of *moles* of the indicated substance present in each of the following samples.
a. 21.2 g of ammonium sulfide
b. 44.3 g of calcium nitrate
c. 4.35 g of dichlorine monoxide
d. 1.0 lb of ferric chloride
e. 1.0 kg of ferric chloride

113. Calculate the number of *moles* of the indicated substance present in each of the following samples.
a. 1.28 g of iron(II) sulfate
b. 5.14 mg of mercury(II) iodide
c. 9.21 μg of tin(IV) oxide
d. 1.26 lb of cobalt(II) chloride
e. 4.25 g of copper(II) nitrate

114. Calculate the mass in grams of each of the following samples.
a. 2.6×10^{-2} mol of copper(II) sulfate, $CuSO_4$
b. 3.05×10^3 mol of tetrafluoroethylene, C_2F_4
c. 7.83 mmol (1 mmol = 0.001 mol) of 1,4-pentadiene, C_5H_8
d. 6.30 mol of bismuth trichloride, $BiCl_3$
e. 12.2 mol of sucrose, $C_{12}H_{22}O_{11}$

115. Calculate the mass in grams of each of the following samples.
a. 3.09 mol of ammonium carbonate
b. 4.01×10^{-6} mol of sodium hydrogen carbonate
c. 88.02 mol of carbon dioxide
d. 1.29 mmol of silver nitrate
e. 0.0024 mol of chromium(III) chloride

116. Calculate the number of *molecules* present in each of the following samples.
a. 3.45 g of $C_6H_{12}O_6$
b. 3.45 mol of $C_6H_{12}O_6$
c. 25.0 g of ICl_5
d. 1.00 g of B_2H_6
e. 1.05 mmol of $Al(NO_3)_3$

117. Calculate the number of moles of hydrogen atoms present in each of the following samples.
a. 2.71 g of ammonia
b. 0.824 mol of water
c. 6.25 mg of sulfuric acid
d. 451 g of ammonium carbonate

118. Calculate the percent by mass of each element in the following compounds.
a. calcium phosphate
b. cadmium sulfate
c. iron(III) sulfate
d. manganese(II) chloride
e. ammonium carbonate
f. sodium hydrogen carbonate
g. carbon dioxide
h. silver(I) nitrate

119. Calculate the percent by mass of the element mentioned *first* in the formulas for each of the following compounds.
a. sodium azide, NaN_3
b. copper(II) sulfate, $CuSO_4$
c. gold(III) chloride, $AuCl_3$
d. silver nitrate, $AgNO_3$
e. rubidium sulfate, Rb_2SO_4
f. sodium chlorate, $NaClO_3$
g. nitrogen triiodide, NI_3
h. cesium bromide, CsBr

120. Calculate the percent by mass of the element mentioned *first* in the formulas for each of the following compounds.
a. iron(II) sulfate
b. silver(I) oxide
c. strontium chloride
d. vinyl acetate, $C_4H_6O_2$
e. methanol, CH_3OH
f. aluminum oxide
g. potassium chlorite
h. potassium chloride

121. A 1.2569-g sample of a new compound has been analyzed and found to contain the following masses of elements: carbon, 0.7238 g; hydrogen, 0.07088 g; nitrogen, 0.1407 g; oxygen, 0.3214 g. Calculate the empirical formula of the compound.

122. A 0.7221-g sample of a new compound has been analyzed and found to contain the following masses of elements: carbon, 0.2990 g; hydrogen, 0.05849 g; nitrogen, 0.2318 g; oxygen, 0.1328 g. Calculate the empirical formula of the compound.

123. When 2.004 g of calcium is heated in pure nitrogen gas, the sample gains 0.4670 g of nitrogen. Calculate the empirical formula of the calcium nitride formed.

124. When 4.01 g of mercury is strongly heated in air, the resulting oxide weighs 4.33 g. Calculate the empirical formula of the oxide.

125. When 1.00 g of metallic chromium is heated with elemental chlorine gas, 3.045 g of a chromium chloride salt results. Calculate the empirical formula of the compound.

126. When barium metal is heated in chlorine gas, a binary compound forms that consists of 65.95% Ba and 34.05% Cl by mass. Calculate the empirical formula of the compound.

9 Chemical Quantities

Environmental scientists testing pond water for industrial pollutants.

Suppose you work for a consumer advocate organization and you want to test a company's advertising claims about the effectiveness of its antacid. The company claims that its product neutralizes 10 times as much stomach acid per tablet as its nearest competitor. How would you test the validity of this claim?

Or suppose that after graduation you go to work for a chemical company that makes methanol (methyl alcohol), a substance used as a starting material for the manufacture of products such as antifreeze and aviation fuels and as a fuel in the cars that race in the Indianapolis 500 (see "Chemistry in Focus" on p. 257. You are working with an experienced chemist who is trying to improve the company's process for making methanol from the reaction of gaseous hydrogen with carbon monoxide gas. The first day on the job, you are instructed to order enough hydrogen and carbon monoxide to produce 6.0 kg of methanol in a test run. How would you determine how much carbon monoxide and hydrogen you should order?

Methanol is a starting material for some jet fuels.

More than 10 *billion* pounds of methanol are produced annually.

After you study this chapter, you will be able to answer these questions.

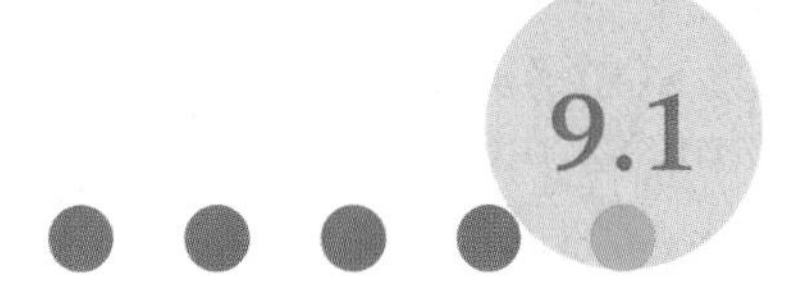

Information Given by Chemical Equations

AIM: To understand the molecular and mass information given in a balanced equation.

Reactions are what chemistry is really all about. Recall that chemical changes are really rearrangements of atom groupings that can be described by chemical equations. In this section we will review the meaning and usefulness of chemical equations by considering one of the processes mentioned in the introduction: the reaction between gaseous carbon monoxide and hydrogen to produce liquid methanol, $CH_3OH(l)$. The reactants and products are

$$\text{Unbalanced: } \underbrace{CO(g) + H_2(g)}_{\text{Reactants}} \rightarrow \underbrace{CH_3OH(l)}_{\text{Product}}$$

Because atoms are just rearranged (not created or destroyed) in a chemical reaction, we must always balance a chemical equation. That is, we must choose coefficients that give the same number of each type of atom on both

Table 9.1 Information Conveyed by the Balanced Equation for the Production of Methanol

$CO(g)$	+	$2H_2(g)$	→	$CH_3OH(l)$
1 molecule CO	+	2 molecules H_2	→	1 molecule CH_3OH
1 dozen CO molecules	+	2 dozen H_2 molecules	→	1 dozen CH_3OH molecules
6.022×10^{23} CO molecules	+	$2(6.022 \times 10^{23})$ H_2 molecules	→	6.022×10^{23} CH_3OH molecules
1 mol CO molecules	+	2 mol H_2 molecules	→	1 mol CH_3OH molecules

sides. Using the smallest set of integers that satisfies this condition gives the balanced equation

$$\text{Balanced: } CO(g) + 2H_2(g) \rightarrow CH_3OH(l)$$

Check: Reactants: 1 C, 1 O, 4 H; Products: 1 C, 1 O, 4 H
It is important to recognize that the coefficients in a balanced equation give the *relative* numbers of molecules. That is, we could multiply this balanced equation by any number and still have a balanced equation. For example, we could multiply by 12:

$$12[CO(g) + 2H_2(g) \rightarrow CH_3OH(l)]$$

to obtain

$$12CO(g) + 24H_2(g) \rightarrow 12CH_3OH(l)$$

This is still a balanced equation (check to be sure). Because 12 represents a dozen, we could even describe the reaction in terms of dozens:

$$1 \text{ dozen } CO(g) + 2 \text{ dozen } H_2(g) \rightarrow 1 \text{ dozen } CH_3OH(l)$$

We could also multiply the original equation by a very large number, such as 6.022×10^{23}:

$$6.022 \times 10^{23}[CO(g) + 2H_2(g) \rightarrow CH_3OH(l)]$$

which leads to the equation

$$6.022 \times 10^{23}\ CO(g) + 2(6.022 \times 10^{23})\ H_2(g) \rightarrow 6.022 \times 10^{23}\ CH_3OH(l)$$

One mole is 6.022×10^{23} units.

Just as 12 is called a dozen, chemists call 6.022×10^{23} a *mole* (abbreviated mol). Our equation, then, can be written in terms of moles:

$$1 \text{ mol } CO(g) + 2 \text{ mol } H_2(g) \rightarrow 1 \text{ mol } CH_3OH(l)$$

Various ways of interpreting this balanced chemical equation are given in Table 9.1.

Example 9.1 Relating Moles to Molecules in Chemical Equations

Propane, C_3H_8, is a fuel commonly used for cooking on gas grills and for heating in rural areas where natural gas is unavailable. Propane reacts with oxygen gas to produce heat and the products carbon dioxide and water. This combustion reaction is represented by the unbalanced equation

$$C_3H_8(g) + O_2(g) \rightarrow CO_2(g) + H_2O(g)$$

Give the balanced equation for this reaction, and state the meaning of the equation in terms of numbers of molecules and moles of molecules.

Propane is often used as a fuel for outdoor grills.

Solution

Using the techniques explained in Chapter 6, we can balance the equation.

$$C_3H_8(g) + 5O_2(g) \rightarrow 3CO_2(g) + 4H_2O(g)$$

CHECK: 3 C, 8 H, 10 O → 3 C, 8 H, 10 O

This equation can be interpreted in terms of molecules as follows:

1 molecule of C_3H_8 reacts with 5 molecules of O_2 to give

3 molecules of CO_2 plus 4 molecules of H_2O

or as follows in terms of moles (of molecules):

1 mol C_3H_8 reacts with 5 mol O_2 to give 3 mol

of CO_2 plus 4 mol H_2O ■

9.2 Mole–Mole Relationships

AIM: To learn to use a balanced equation to determine relationships between moles of reactants and moles of products.

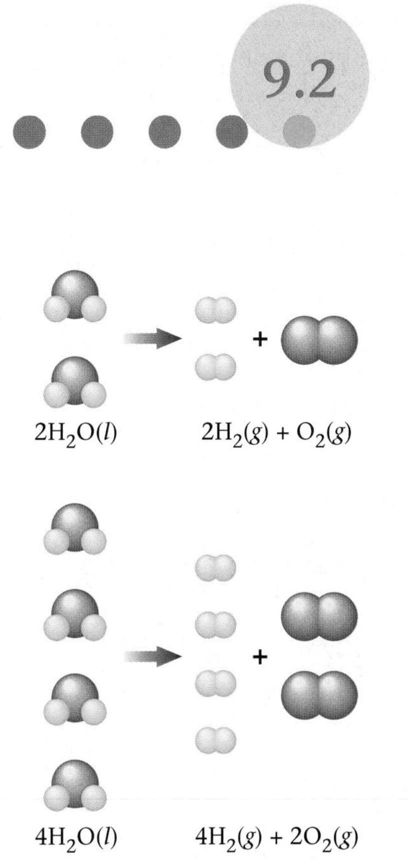

Now that we have discussed the meaning of a balanced chemical equation in terms of moles of reactants and products, we can use an equation to predict the moles of products that a given number of moles of reactants will yield. For example, consider the decomposition of water to give hydrogen and oxygen, which is represented by the following balanced equation:

$$2H_2O(l) \rightarrow 2H_2(g) + O_2(g)$$

This equation tells us that 2 mol of H_2O yields 2 mol of H_2 and 1 mol of O_2.

Now suppose that we have 4 mol of water. If we decompose 4 mol of water, how many moles of products do we get?

One way to answer this question is to multiply the entire equation by 2 (that will give us 4 mol of H_2O).

$$2[2H_2O(l) \rightarrow 2H_2(g) + O_2(g)]$$
$$4H_2O(l) \rightarrow 4H_2(g) + 2O_2(g)$$

Now we can state that

4 mol of H_2O yields 4 mol of H_2 plus 2 mol of O_2

which answers the question of how many moles of products we get with 4 mol of H_2O.

Next, suppose we decompose 5.8 mol of water. What numbers of moles of products are formed in this process? We could answer this question by rebalancing the chemical equation as follows: First, we divide *all coefficients* of the balanced equation

$$2H_2O(l) \rightarrow 2H_2(g) + O_2(g)$$

by 2, to give

$$H_2O(l) \rightarrow H_2(g) + \tfrac{1}{2}O_2(g)$$

Now, because we have 5.8 mol of H_2O, we multiply this equation by 5.8.

$$5.8[H_2O(l) \rightarrow H_2(g) + \tfrac{1}{2}O_2(g)]$$

This gives

This equation with noninteger coefficients makes sense only if the equation means moles (of molecules) of the various reactants and products.

$$5.8H_2O(l) \rightarrow 5.8H_2(g) + 5.8(\tfrac{1}{2})O_2(g)$$
$$5.8H_2O(l) \rightarrow 5.8H_2(g) + 2.9O_2(g)$$

(Verify that this is a balanced equation.) Now we can state that

5.8 mol of H_2O yields 5.8 mol of H_2 plus 2.9 mol of O_2

This procedure of rebalancing the equation to obtain the number of moles involved in a particular situation always works, but it can be cumbersome. In Example 9.2 we will develop a more convenient procedure, which uses conversion factors, or **mole ratios,** based on the balanced chemical equation.

Example 9.2 Determining Mole Ratios

What number of moles of O_2 will be produced by the decomposition of 5.8 mol of water?

Solution

Our problem can be diagrammed as follows:

5.8 mol H_2O → yields → ? mol O_2

To answer this question, we need to know the relationship between moles of H_2O and moles of O_2 in the balanced equation (conventional form):

$$2H_2O(l) \rightarrow 2H_2(g) + O_2(g)$$

From this equation we can state that

2 mol H_2O → yields → 1 mol O_2

The statement 2 mol H_2O = 1 mol O_2 is obviously not true in a literal sense, but it correctly expresses the chemical equivalence between H_2O and O_2.

which can be represented by the following equivalence statement:

$$2 \text{ mol } H_2O = 1 \text{ mol } O_2$$

We now want to use this equivalence statement to obtain the conversion factor (mole ratio) that we need. Because we want to go from moles of H_2O to moles of O_2, we need the mole ratio

$$\frac{1 \text{ mol } O_2}{2 \text{ mol } H_2O}$$

so that mol H_2O will cancel in the conversion from moles of H_2O to moles of O_2.

$$5.8 \cancel{\text{mol } H_2O} \times \frac{1 \text{ mol } O_2}{2 \cancel{\text{mol } H_2O}} = 2.9 \text{ mol } O_2$$

So if we decompose 5.8 mol of H_2O, we will get 2.9 mol of O_2. Note that this is the same answer we obtained earlier when we rebalanced the equation to give

$$5.8H_2O(l) \rightarrow 5.8H_2(g) + 2.9O_2(g)$$ ■

We saw in Example 9.2 that to determine the moles of a product that can be formed from a specified number of moles of a reactant, we can use the balanced equation to obtain the appropriate mole ratio. We will now extend these ideas in Example 9.3.

Example 9.3 Using Mole Ratios in Calculations

Calculate the number of moles of oxygen required to react exactly with 4.30 mol of propane, C_3H_8, in the reaction described by the following balanced equation:

$$C_3H_8(g) + 5O_2(g) \rightarrow 3CO_2(g) + 4H_2O(g)$$

Solution

In this case the problem can be stated as follows:

4.30 mol C_3H_8 requires ? mol O_2

To solve this problem, we need to consider the relationship between the reactants C_3H_8 and O_2. Using the balanced equation, we find that

1 mol of C_3H_8 requires 5 mol of O_2

which can be represented by the equivalence statement

$$1 \text{ mol } C_3H_8 = 5 \text{ mol } O_2$$

This leads to the required mole ratio

$$\frac{5 \text{ mol } O_2}{1 \text{ mol } C_3H_8}$$

for converting from moles of C_3H_8 to moles of O_2. We construct the conversion ratio this way so that mol C_3H_8 cancels:

$$4.30 \cancel{\text{mol } C_3H_8} \times \frac{5 \text{ mol } O_2}{1 \cancel{\text{mol } C_3H_8}} = 21.5 \text{ mol } O_2$$

We can now answer the original question:

4.30 mol of C_3H_8 requires 21.5 mol of O_2

✔ Self-Check Exercise 9.1

Calculate the moles of CO_2 formed when 4.30 mol of C_3H_8 reacts with the required 21.5 mol of O_2.

HINT: Use the moles of C_3H_8, and obtain the mole ratio between C_3H_8 and CO_2 from the balanced equation.

See Problems 9.15 and 9.16. ■

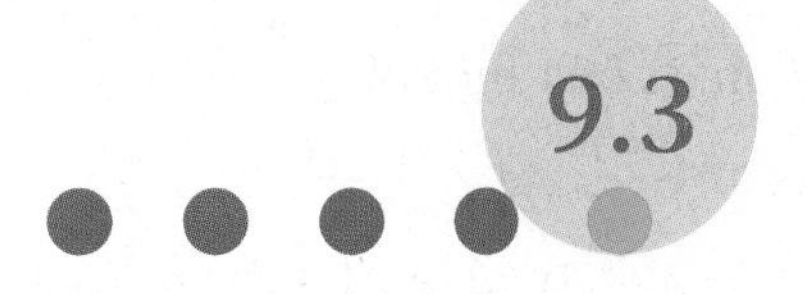

Mass Calculations

AIM: To learn to relate masses of reactants and products in a chemical reaction.

In the last section we saw how to use the balanced equation for a reaction to calculate the numbers of moles of reactants and products for a particular case. However, moles represent numbers of molecules, and we cannot count molecules directly. In chemistry we count by weighing. Therefore,

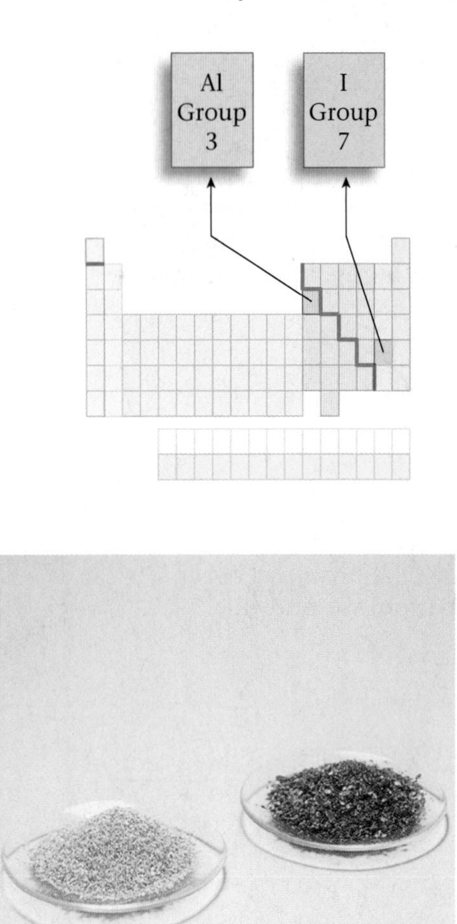

in this section we will review the procedures for converting between moles and masses and will see how these procedures are applied to chemical calculations.

To develop these procedures we will consider the reaction between powdered aluminum metal and finely ground iodine to produce aluminum iodide. The balanced equation for this vigorous chemical reaction is

$$2\text{Al}(s) + 3\text{I}_2(s) \rightarrow 2\text{AlI}_3(s)$$

Suppose we have 35.0 g of aluminum. What mass of I_2 should we weigh out to react exactly with this amount of aluminum?

To answer this question we need to think about what the balanced equation tells us. The equation states that

$$2 \text{ mol of Al requires } 3 \text{ mol of I}_2$$

which leads to the mole ratio

$$\frac{3 \text{ mol I}_2}{2 \text{ mol Al}}$$

We can use this ratio to calculate the moles of I_2 needed:

$$\text{Moles of Al present} \times \frac{3 \text{ mol I}_2}{2 \text{ mol Al}} = \text{moles of I}_2 \text{ required}$$

This leads us to the question: How many moles of Al are present? The problem states that we have 35.0 g of aluminum, so we must convert from grams to moles of aluminum. This is something we already know how to do. Using the table of average atomic masses inside the front cover of this book, we find the atomic mass of aluminum to be 26.98. This means that 1 mol of aluminum has a mass of 26.98 g. We can use the equivalence statement

$$1 \text{ mol Al} = 26.98 \text{ g}$$

to find the moles of Al in 35.0 g.

$$35.0 \cancel{\text{g Al}} \times \frac{1 \text{ mol Al}}{26.98 \cancel{\text{g Al}}} = 1.30 \text{ mol Al}$$

Now that we have moles of Al, we can find the moles of I_2 required.

$$1.30 \cancel{\text{mol Al}} \times \frac{3 \text{ mol I}_2}{2 \cancel{\text{mol Al}}} = 1.95 \text{ mol I}_2$$

We now know the *moles* of I_2 required to react with the 1.30 mol of Al (35.0 g). The next step is to convert 1.95 mol of I_2 to grams so we will know how much to weigh out. We do this by using the molar mass of I_2. The atomic mass of iodine is 126.9 g (for 1 mol of I atoms), so the molar mass of I_2 is

$$2 \times 126.9 \text{ g/mol} = 253.8 \text{ g/mol} = \text{mass of 1 mol of I}_2$$

Now we convert the 1.95 mol of I_2 to grams of I_2.

$$1.95 \cancel{\text{mol I}_2} \times \frac{253.8 \text{ g I}_2}{\cancel{\text{mol I}_2}} = 495 \text{ g I}_2$$

We have solved the problem. We need to weigh out 495 g of iodine (contains I_2 molecules) to react exactly with the 35.0 g of aluminum. We will further develop procedures for dealing with masses of reactants and products in Example 9.4.

Aluminum (*left*) and iodine (*right*) shown at the top, react vigorously to form aluminum iodide. The purple cloud results from excess iodine vaporized by the heat of the reaction.

Example 9.4 Using Mass–Mole Conversions with Mole Ratios

Propane, C_3H_8, when used as a fuel, reacts with oxygen to produce carbon dioxide and water according to the following unbalanced equation:

$$C_3H_8(g) + O_2(g) \rightarrow CO_2(g) + H_2O(g)$$

What mass of oxygen will be required to react exactly with 96.1 g of propane?

Solution

Always balance the equation for the reaction first.

To deal with the amounts of reactants and products, we first need the balanced equation for this reaction:

$$C_3H_8(g) + 5O_2(g) \rightarrow 3CO_2(g) + 4H_2O(g)$$

Remember that to show the correct significant figures in each step, we are rounding off after each calculation. In doing problems, you should carry extra numbers, rounding off only at the end.

Next, let's summarize what we know and what we want to find.

What we know:

- The balanced equation for the reaction
- The mass of propane available (96.1 g)

What we want to calculate:

- The mass of oxygen (O_2) required to react exactly with all the propane

Our problem, in schematic form, is

Using the ideas we developed when we discussed the aluminum–iodine reaction, we will proceed as follows:

1. We are given the number of grams of propane, so we must convert to moles of propane (C_3H_8).
2. Then we can use the coefficients in the balanced equation to determine the moles of oxygen (O_2) required.
3. Finally, we will use the molar mass of O_2 to calculate grams of oxygen.

We can sketch this strategy as follows:

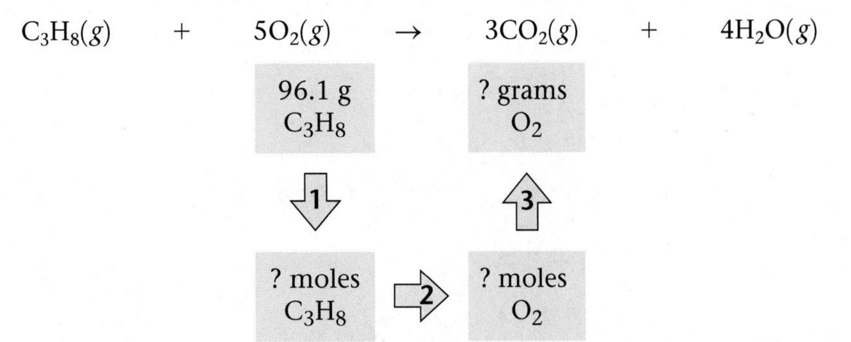

Thus the first question we must answer is, *How many moles of propane are present in 96.1 g of propane?* The molar mass of propane is 44.09 g ($3 \times 12.01 + 8 \times 1.008$). The moles of propane present can be calculated as follows:

$$96.1 \text{ g } \cancel{C_3H_8} \times \frac{1 \text{ mol } C_3H_8}{44.09 \text{ g } \cancel{C_3H_8}} = 2.18 \text{ mol } C_3H_8$$

Next we recognize that each mole of propane reacts with 5 mol of oxygen. This gives us the equivalence statement

$$1 \text{ mol } C_3H_8 = 5 \text{ mol } O_2$$

from which we construct the mole ratio

$$\frac{5 \text{ mol } O_2}{1 \text{ mol } C_3H_8}$$

that we need to convert from moles of propane molecules to moles of oxygen molecules.

$$2.18 \cancel{\text{mol } C_3H_8} \times \frac{5 \text{ mol } O_2}{1 \cancel{\text{mol } C_3H_8}} = 10.9 \text{ mol } O_2$$

Notice that the mole ratio is set up so that the moles of C_3H_8 cancel and the resulting units are moles of O_2.

Because the original question asked for the *mass* of oxygen needed to react with 96.1 g of propane, we must convert the 10.9 mol of O_2 to grams, using the molar mass of O_2 (32.00 = 2 × 16.00).

$$10.9 \cancel{\text{mol } O_2} \times \frac{32.0 \text{ g } O_2}{1 \cancel{\text{mol } O_2}} = 349 \text{ g } O_2$$

Therefore, 349 g of oxygen is required to burn 96.1 g of propane. We can summarize this problem by writing out a "conversion string" that shows how the problem was done.

$$\overset{1}{\downarrow} \qquad \overset{2}{\downarrow} \qquad \overset{3}{\downarrow}$$

$$96.1 \cancel{\text{g } C_3H_8} \times \frac{1 \cancel{\text{mol } C_3H_8}}{44.09 \cancel{\text{g } C_3H_8}} \times \frac{5 \cancel{\text{mol } O_2}}{1 \cancel{\text{mol } C_3H_8}} \times \frac{32.0 \text{ g } O_2}{1 \cancel{\text{mol } O_2}} \; 5 \; 349 \text{ g } O_2$$

Use units as a check to see that you have used the correct conversion factors (mole ratios).

This is a convenient way to make sure the final units are correct. The procedure we have followed is summarized below.

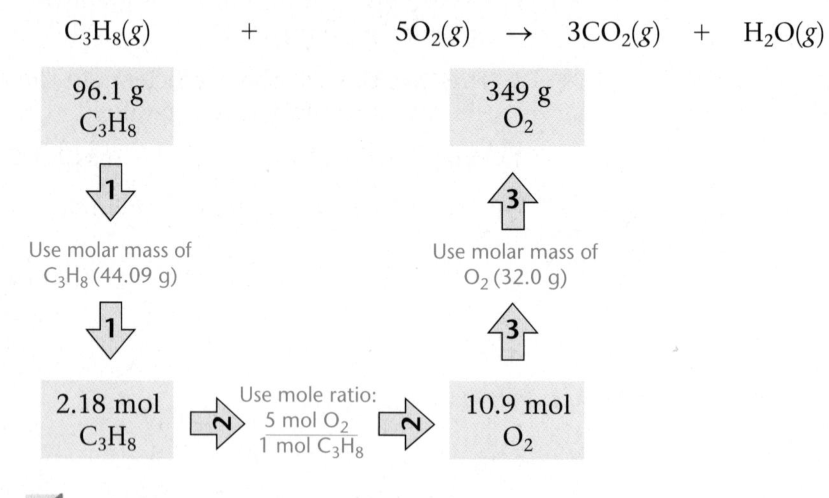

Self-Check Exercise 9.2

What mass of carbon dioxide is produced when 96.1 g of propane reacts with sufficient oxygen?

See Problems 9.23 through 9.26.

Self-Check Exercise 9.3

Calculate the mass of water formed by the complete reaction of 96.1 g of propane with oxygen.

See Problems 9.23 through 9.26. ■

So far in this chapter, we have spent considerable time "thinking through" the procedures for calculating the masses of reactants and products in chemical reactions. We can summarize these procedures in the following steps:

Steps for Calculating the Masses of Reactants and Products in Chemical Reactions

STEP 1 Balance the equation for the reaction.

STEP 2 Convert the masses of reactants or products to moles.

STEP 3 Use the balanced equation to set up the appropriate mole ratio(s).

STEP 4 Use the mole ratio(s) to calculate the number of moles of the desired reactant or product.

STEP 5 Convert from moles back to masses.

The process of using a chemical equation to calculate the relative masses of reactants and products involved in a reaction is called **stoichiometry** (pronounced stoý·kē·óm·ĕtry). Chemists say that the balanced equation for a chemical reaction describes the stoichiometry of the reaction.

We will now consider a few more examples that involve chemical stoichiometry. Because real-world examples often involve very large or very small masses of chemicals that are most conveniently expressed by using scientific notation, we will deal with such a case in Example 9.5.

Example 9.5 Stoichiometric Calculations: Using Scientific Notation

For a review of writing formulas of ionic compounds, see Chapter 5.

Solid lithium hydroxide is used in space vehicles to remove exhaled carbon dioxide from the living environment. The products are solid lithium carbonate and liquid water. What mass of gaseous carbon dioxide can 1.00×10^3 g of lithium hydroxide absorb?

Solution

STEP 1 Using the description of the reaction, we can write the unbalanced equation

$$LiOH(s) + CO_2(g) \rightarrow Li_2CO_3(s) + H_2O(l)$$

The balanced equation is

$$2LiOH(s) + CO_2(g) \rightarrow Li_2CO_3(s) + H_2O(l)$$

Check this for yourself.

STEP 2 We convert the given mass of LiOH to moles, using the molar mass of LiOH, which is 6.941 g + 16.00 g + 1.008 g = 23.95 g.

$$1.00 \times 10^3 \text{ } \cancel{\text{g LiOH}} \times \frac{1 \text{ mol LiOH}}{23.95 \text{ } \cancel{\text{g LiOH}}} = 41.8 \text{ mol LiOH}$$

STEP 3 The appropriate mole ratio is

$$\frac{1 \text{ mol } CO_2}{2 \text{ mol LiOH}}$$

Carrying extra significant figures and rounding off only at the end gives an answer of 919 g CO_2.

STEP 4 Using this mole ratio, we calculate the moles of CO_2 needed to react with the given mass of LiOH.

$$41.8\ \cancel{\text{mol LiOH}} \times \frac{1\ \text{mol CO}_2}{2\ \cancel{\text{mol LiOH}}} = 20.9\ \text{mol CO}_2$$

Astronaut Sidney M. Gutierrez changes the lithium hydroxide canisters on Space Shuttle *Columbia*. The lithium hydroxide is used to purge carbon dioxide from the air in the shuttle's cabin.

STEP 5 We calculate the mass of CO_2 by using its molar mass (44.01 g).

$$20.9\ \cancel{\text{mol CO}_2} \times \frac{44.01\ \text{g CO}_2}{1\ \cancel{\text{mol CO}_2}} = 920.\ \text{g CO}_2 = 9.20 \times 10^2\ \text{g CO}_2$$

Thus 1.00×10^3 g of LiOH(*s*) can absorb 920. g of $CO_2(g)$.

We can summarize this problem as follows:

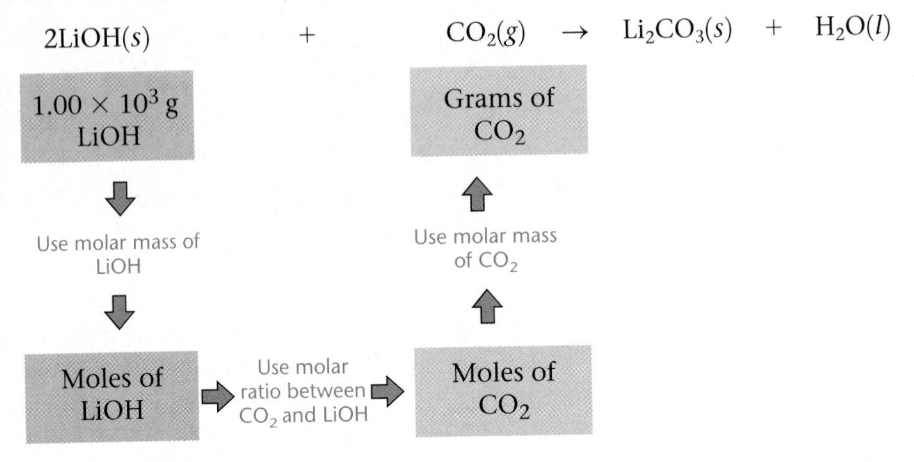

The conversion string is

$$1.00 \times 10^3\ \cancel{\text{g LiOH}} \times \frac{1\ \cancel{\text{mol LiOH}}}{23.95\ \cancel{\text{g LiOH}}} \times \frac{1\ \cancel{\text{mol CO}_2}}{2\ \cancel{\text{mol LiOH}}} \times \frac{44.01\ \text{g CO}_2}{1\ \cancel{\text{mol CO}_2}} = 9.20 \times 10^2\ \text{g CO}_2$$

✔ Self-Check Exercise 9.4

Hydrofluoric acid, an aqueous solution containing dissolved hydrogen fluoride, is used to etch glass by reacting with the silica, SiO_2, in the glass to produce gaseous silicon tetrafluoride and liquid water. The unbalanced equation is

$$HF(aq) + SiO_2(s) \rightarrow SiF_4(g) + H_2O(l)$$

a. Calculate the mass of hydrogen fluoride needed to react with 5.68 g of silica. *Hint:* Think carefully about this problem. What is the balanced equation for the reaction? What is given? What do you need to calculate? Sketch a map of the problem before you do the calculations.

b. Calculate the mass of water produced in the reaction described in part a.

See Problems 9.23 through 9.26. ■

Example 9.6 Stoichiometric Calculations: Comparing Two Reactions

Baking soda, $NaHCO_3$, is often used as an antacid. It neutralizes excess hydrochloric acid secreted by the stomach. The balanced equation for the reaction is

$$NaHCO_3(s) + HCl(aq) \rightarrow NaCl(aq) + H_2O(l) + CO_2(g)$$

CHEMISTRY IN FOCUS

Methyl Alcohol: Fuel with a Future?

Southern California is famous for many things, and among them, unfortunately, is smog. Smog is produced when pollutants in the air are trapped near the ground and are caused to react by sunlight. One step being considered by the state of California to help solve the smog problem is to replace gasoline with methyl alcohol (usually called methanol). One advantage of methanol is that it reacts more nearly completely than gasoline with oxygen in a car's engine, thus releasing lower amounts of unburned fuel into the atmosphere. Methanol also produces less carbon monoxide (CO) in the exhaust than does gasoline. Carbon monoxide not only is toxic itself but also encourages the formation of nitrogen dioxide by the reaction

$$CO(g) + O_2(g) + NO(g) \rightarrow CO_2(g) + NO_2(g)$$

Nitrogen dioxide is a reddish-brown gas that leads to ozone formation and acid rain.

Using methanol as a fuel is not a new idea. For example, it is the only fuel allowed in the open-wheeled race cars used in the Indianapolis 500 and in similar races. Methanol works very well in racing engines because it has outstanding antiknock characteristics, even at the tremendous speeds at which these engines operate.

The news about methanol is not all good, however. One problem is lower fuel mileage. Because it takes about twice as many gallons of methanol as gasoline to travel a given distance, a methanol-powered car's fuel tank must be twice the usual size. However, although costs vary greatly depending on market conditions, the cost of methanol averages about half that of gasoline, so the net cost is about the same for both fuels.

A second disadvantage of methanol is that its high affinity for water causes condensation from the air, which leads to increased corrosion of the fuel tank and fuel lines. This problem can be solved by using more expensive stainless steel for these parts.

Crew members change tires and add methanol fuel to a race car in the Indianapolis 500 during a pit stop.

The most serious problem with methanol may be its tendency to form formaldehyde, HCHO, when it is combusted. Formaldehyde has been implicated as a carcinogen (a substance that causes cancer). Formaldehyde can also lead to ozone formation in the air, which causes even more severe smog. Researchers are now working on catalytic converters for exhaust systems to help decompose the formaldehyde.

To test the feasibility of methanol as a motor fuel, California has operated several hundred vehicles on methanol since 1980. Because accessibility to methanol is limited, cars are now being prepared that can run on methanol or gasoline. These vehicles are being tested on a large scale in California. So if you live in southern California, in a few years your neighborhood "gas station" may actually be pumping methanol.

Milk of magnesia, which is an aqueous suspension of magnesium hydroxide, $Mg(OH)_2$, is also used as an antacid. The balanced equation for the reaction is

$$Mg(OH)_2(s) + 2HCl(aq) \rightarrow 2H_2O(l) + MgCl_2(aq)$$

Which antacid can consume the most stomach acid, 1.00 g of $NaHCO_3$ or 1.00 g of $Mg(OH)_2$?

Solution

Before we begin, let's think about the problem to be solved. The question we must ask for each antacid is, *How many moles of HCl will react with 1.00 g of each antacid?* The antacid that reacts with the larger number of moles of HCl is more effective because it will neutralize more moles of acid. A schematic for this procedure is

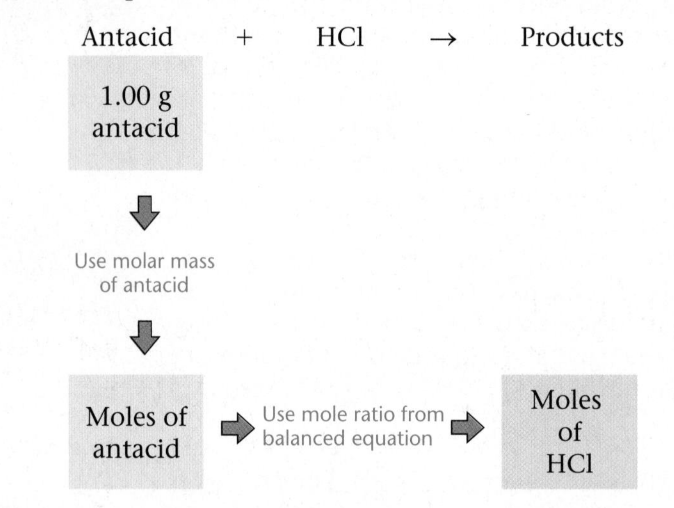

Notice that in this case we do not need to calculate how many grams of HCl react; we can answer the question with moles of HCl. We will now solve this problem for each antacid. Both of the equations are balanced, so we can proceed with the calculations.

Using the molar mass of $NaHCO_3$, which is 22.99 g + 1.008 g + 12.01 g + 3(16.00 g) = 84.01 g, we determine the moles of $NaHCO_3$ in 1.00 g of $NaHCO_3$.

$$1.00\ \cancel{\text{g NaHCO}_3} \times \frac{1\ \text{mol NaHCO}_3}{84.01\ \cancel{\text{g NaHCO}_3}} = 0.0119\ \text{mol NaHCO}_3$$
$$= 1.19 \times 10^{-2}\ \text{mol NaHCO}_3$$

Next we determine the moles of HCl, using the mole ratio $\frac{1\ \text{mol HCl}}{1\ \text{mol NaHCO}_3}$.

$$1.19 \times 10^{-2}\ \cancel{\text{mol NaHCO}_3} \times \frac{1\ \text{mol HCl}}{1\ \cancel{\text{mol NaHCO}_3}} = 1.19 \times 10^{-2}\ \text{mol HCl}$$

Thus 1.00 g of $NaHCO_3$ neutralizes 1.19×10^{-2} mol of HCl. We need to compare this to the number of moles of HCl that 1.00 g of $Mg(OH)_2$ neutralizes.

Using the molar mass of $Mg(OH)_2$, which is 24.31 g + 2(16.00 g) + 2(1.008 g) = 58.33 g, we determine the moles of $Mg(OH)_2$ in 1.00 g of $Mg(OH)_2$.

$$1.00\ \cancel{\text{g Mg(OH)}_2} \times \frac{1\ \text{mol Mg(OH)}_2}{58.33\ \cancel{\text{g Mg(OH)}_2}} = 0.0171\ \text{mol Mg(OH)}_2$$
$$= 1.71 \times 10^{-2}\ \text{mol Mg(OH)}_2$$

To determine the moles of HCl that react with this amount of $Mg(OH)_2$, we use the mole ratio $\frac{2\ \text{mol HCl}}{1\ \text{mol Mg(OH)}_2}$.

$$1.71 \times 10^{-2}\ \cancel{\text{mol Mg(OH)}_2} \times \frac{2\ \text{mol HCl}}{1\ \cancel{\text{mol Mg(OH)}_2}} = 3.42 \times 10^{-2}\ \text{mol HCl}$$

Therefore, 1.00 g of $Mg(OH)_2$ neutralizes 3.42×10^{-2} mol of HCl. We have already calculated that 1.00 g of $NaHCO_3$ neutralizes only 1.19×10^{-2} mol of HCl. Therefore, $Mg(OH)_2$ is a more effective antacid than $NaHCO_3$ on a mass basis.

✔ Self-Check Exercise 9.5

In Example 9.6 we answered one of the questions we posed in the introduction to this chapter. Now let's see if you can answer the other question posed there. Determine what mass of carbon monoxide and what mass of hydrogen are required to form 6.0 kg of methanol by the reaction

$$CO(g) + 2H_2(g) \rightarrow CH_3OH(l)$$

See Problem 9.39. ■

9.4 Calculations Involving a Limiting Reactant

AIMS: To learn to recognize the limiting reactant in a reaction. To learn to use the limiting reactant to do stoichiometric calculations.

Farmer Rodney Donala looks out over his corn fields in front of his 30,000-gallon tank (at right) of anhydrous ammonia, a liquid fertilizer.

Manufacturers of cars, bicycles, and appliances order parts in the same proportion as they are used in their products. For example, auto manufacturers order four times as many wheels as engines and bicycle manufacturers order twice as many pedals as seats. Likewise, when chemicals are mixed together so that they can undergo a reaction, they are often mixed in stoichiometric quantities—that is, in exactly the correct amounts so that all reactants "run out" (are used up) at the same time. To clarify this concept, we will consider the production of hydrogen for use in the manufacture of ammonia. Ammonia, a very important fertilizer itself and a starting material for other fertilizers, is made by combining nitrogen from the air with hydrogen. The hydrogen for this process is produced by the reaction of methane with water according to the balanced equation

$$CH_4(g) + H_2O(g) \rightarrow 3H_2(g) + CO(g)$$

Let's consider the question, *What mass of water is required to react exactly with 249 g of methane?* That is, how much water will just use up all of the 249 g of methane, leaving no methane or water remaining?

This problem requires the same strategies we developed in the previous section. Again, drawing a map of the problem is helpful.

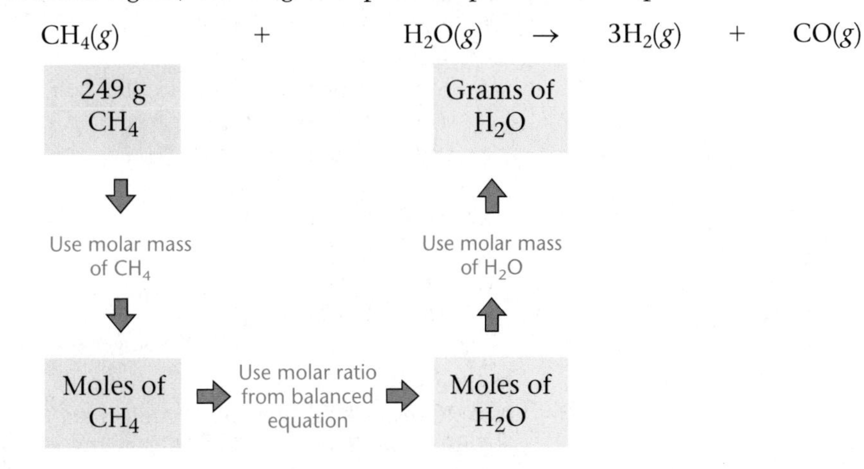

We first convert the mass of CH_4 to moles, using the molar mass of CH_4 (16.04 g/mol).

$$249\ \cancel{\text{g CH}_4} \times \frac{1\ \text{mol CH}_4}{16.04\ \cancel{\text{g CH}_4}} = 15.5\ \text{mol CH}_4$$

Because in the balanced equation 1 mol of CH_4 reacts with 1 mol of H_2O, we have

$$15.5\ \cancel{\text{mol CH}_4} \times \frac{1\ \text{mol H}_2\text{O}}{1\ \cancel{\text{mol CH}_4}} = 15.5\ \text{mol H}_2\text{O}$$

Therefore, 15.5 mol of H_2O will react exactly with the given mass of CH_4. Converting 15.5 mol of H_2O to grams of H_2O (molar mass = 18.02 g/mol) gives

$$15.5\ \cancel{\text{mol H}_2\text{O}} \times \frac{18.02\ \text{g H}_2\text{O}}{1\ \cancel{\text{mol H}_2\text{O}}} = 279\ \text{g H}_2\text{O}$$

This result means that if 249 g of methane is mixed with 279 g of water, both reactants will "run out" at the same time. The reactants have been mixed in stoichiometric quantities.

If, on the other hand, 249 g of methane is mixed with 300 g of water, the methane will be consumed before the water runs out. The water will be in *excess*. In this case, the quantity of products formed will be determined by the quantity of methane present. Once the methane is consumed, no more products can be formed, even though some water still remains. In this situation, because the amount of methane *limits* the amount of products that can be formed, it is called the **limiting reactant,** or **limiting reagent.** In any stoichiometry problem, where reactants are not mixed in stoichiometric quantities, it is essential to determine which reactant is limiting in order to calculate correctly the amounts of products that will be formed. This concept is illustrated in Figure 9.1. Note from this figure that because there are fewer water molecules than CH_4 molecules, the water is consumed first. After the water molecules are

The reactant that is consumed first limits the amounts of products that can form.

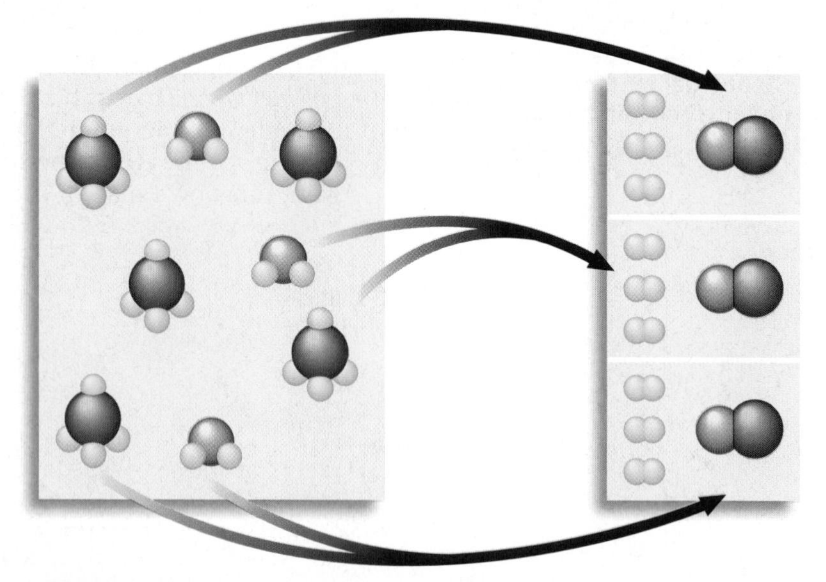

Figure 9.1
A mixture of $5CH_4$ and $3H_2O$ molecules undergoes the reaction $CH_4(g) + H_2O(g) \rightarrow 3H_2(g) + CO(g)$. Note that the H_2O molecules are used up first, leaving two CH_4 molecules unreacted.

gone, no more products can form. So in this case water is the limiting reactant.

You probably have been dealing with limiting-reactant problems for most of your life. For example, suppose a lemonade recipe calls for 1 cup of sugar for every 6 lemons. You have 12 lemons and 3 cups of sugar. Which ingredient is limiting, the lemons or the sugar?*

Example 9.7 Stoichiometric Calculations: Identifying the Limiting Reactant

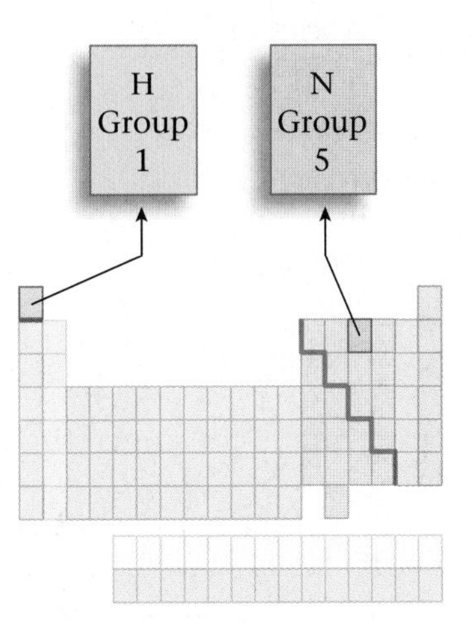

Suppose 25.0 kg (2.50×10^4 g) of nitrogen gas and 5.00 kg (5.00×10^3 g) of hydrogen gas are mixed and reacted to form ammonia. Calculate the mass of ammonia produced when this reaction is run to completion.

Solution

The unbalanced equation for this reaction is

$$N_2(g) + H_2(g) \rightarrow NH_3(g)$$

which leads to the balanced equation

$$N_2(g) + 3H_2(g) \rightarrow 2NH_3(g)$$

This problem is different from the others we have done so far in that we are mixing *specified amounts of two reactants* together. To know how much product forms, we must determine which reactant is consumed first. That is, we must determine which is the limiting reactant in this experiment. To do so we must add a step to our normal procedure. We can map this process as follows:

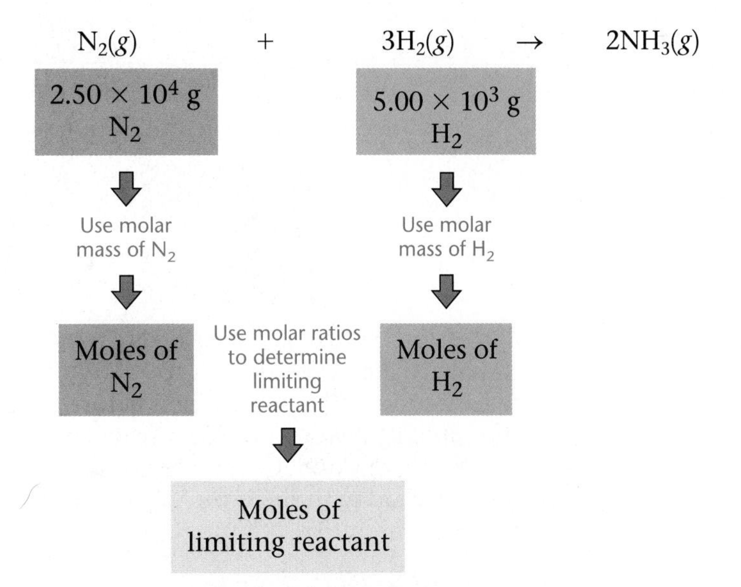

We will use the moles of the limiting reactant to calculate the moles and then the grams of the product.

* The ratio of lemons to sugar that the recipe calls for is 6 lemons to 1 cup of sugar. We can calculate the number of lemons required to "react with" the 3 cups of sugar as follows:

$$3 \text{ cups sugar} \times \frac{6 \text{ lemons}}{1 \text{ cup sugar}} = 18 \text{ lemons}$$

Thus 18 lemons would be required to use up 3 cups of sugar. However, we have only 12 lemons, so the lemons are limiting.

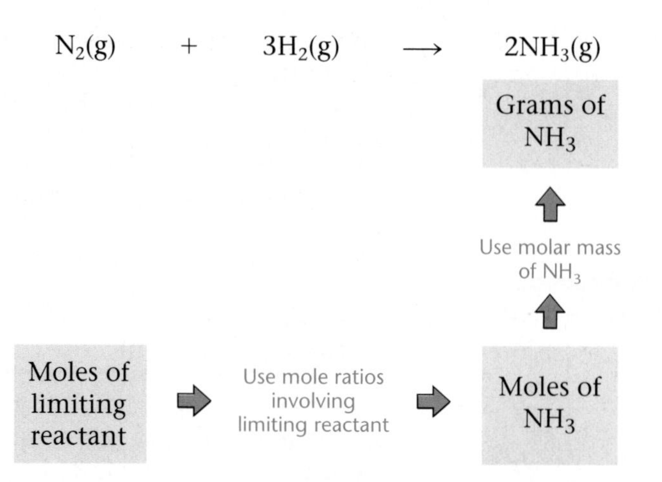

We first calculate the moles of the two reactants present:

$$2.50 \times 10^4 \text{ g N}_2 \times \frac{1 \text{ mol N}_2}{28.02 \text{ g N}_2} = 8.92 \times 10^2 \text{ mol N}_2$$

$$5.00 \times 10^3 \text{ g H}_2 \times \frac{1 \text{ mol H}_2}{2.016 \text{ g H}_2} = 2.48 \times 10^3 \text{ mol H}_2$$

Now we must determine which reactant is limiting (will be consumed first). We have 8.92×10^2 mol of N_2. Let's determine *how many moles of H_2 are required to react with this much N_2*. Because 1 mol of N_2 reacts with 3 mol of H_2, the number of moles of H_2 we need to react completely with 8.92×10^2 mol of N_2 is determined as follows:

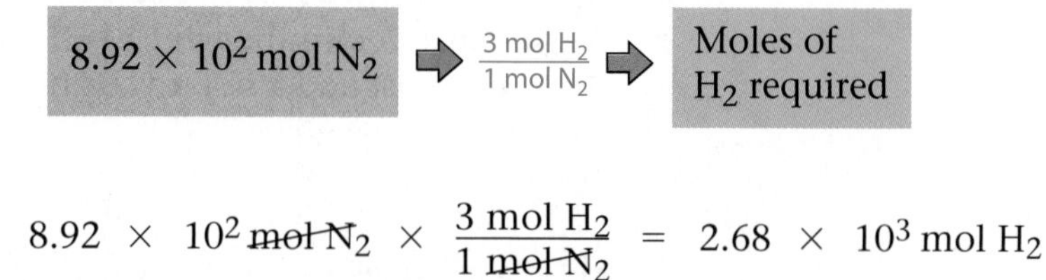

$$8.92 \times 10^2 \text{ mol N}_2 \times \frac{3 \text{ mol H}_2}{1 \text{ mol N}_2} = 2.68 \times 10^3 \text{ mol H}_2$$

Is N_2 or H_2 the limiting reactant? The answer comes from the comparison

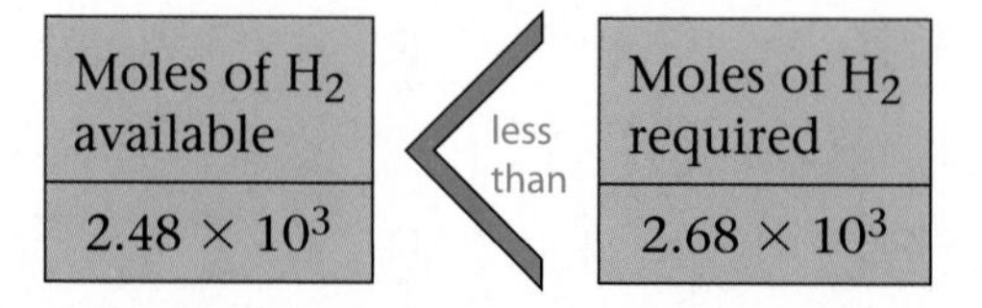

We see that 8.92×10^2 mol of N_2 requires 2.68×10^3 mol of H_2 to react completely. However, only 2.48×10^3 mol of H_2 is present. This means that the hydrogen will be consumed before the nitrogen runs out, so hydrogen is the *limiting reactant* in this particular situation.

Note that in our effort to determine the limiting reactant, we could have started instead with the given amount of hydrogen and calculated the moles of nitrogen required.

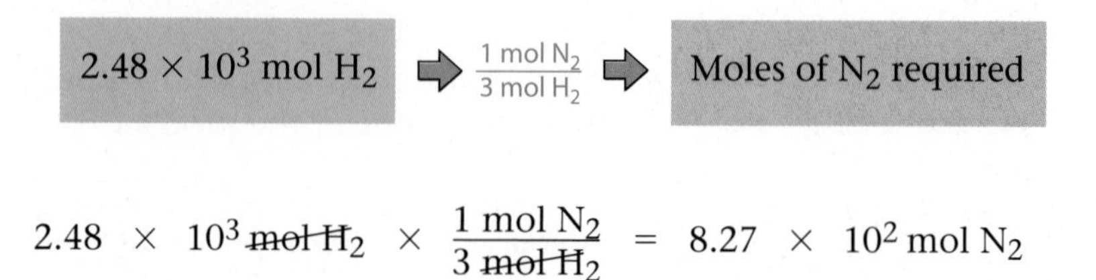

$$2.48 \times 10^3 \text{ mol H}_2 \times \frac{1 \text{ mol N}_2}{3 \text{ mol H}_2} = 8.27 \times 10^2 \text{ mol N}_2$$

Thus 2.48×10^3 mol of H_2 requires 8.27×10^2 mol of N_2. Because 8.92×10^2 mol of N_2 is actually present, the nitrogen is in excess.

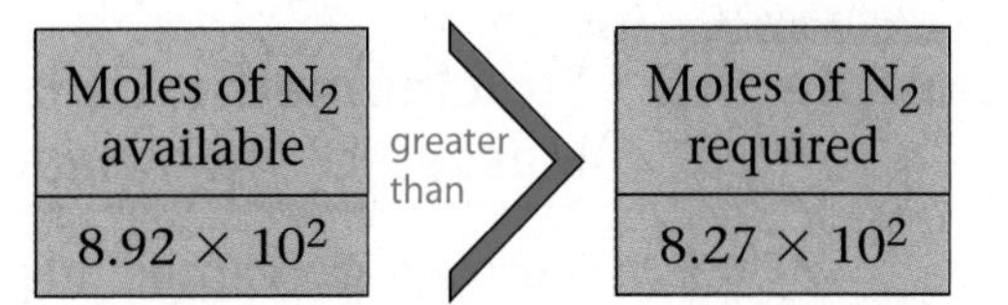

Always check to see which, if any, reactant is limiting when you are given the amounts of two or more reactants.

If nitrogen is in excess, hydrogen will "run out" first; again we find that hydrogen limits the amount of ammonia formed.

Because the moles of H_2 present are limiting, we must use this quantity to determine the moles of NH_3 that can form.

$$2.48 \times 10^3 \cancel{\text{mol H}_2} \times \frac{2 \text{ mol NH}_3}{3 \cancel{\text{mol H}_2}} = 1.65 \times 10^3 \text{ mol NH}_3$$

Next we convert moles of NH_3 to mass of NH_3.

$$1.65 \times 10^3 \cancel{\text{mol NH}_3} \times \frac{17.03 \text{ g NH}_3}{1 \cancel{\text{mol NH}_3}} = 2.81 \times 10^4 \text{ g NH}_3 = 28.1 \text{ kg NH}_3$$

Therefore, 25.0 kg of N_2 and 5.00 kg of H_2 can form 28.1 kg of NH_3. ■

The strategy used in Example 9.7 is summarized in Figure 9.2.

The following list summarizes the steps to take in solving stoichiometry problems in which the amounts of two (or more) reactants are given.

Steps for Solving Stoichiometry Problems Involving Limiting Reactants

STEP 1 Write and balance the equation for the reaction.

STEP 2 Convert known masses of reactants to moles.

STEP 3 Using the numbers of moles of reactants and the appropriate mole ratios, determine which reactant is limiting.

STEP 4 Using the amount of the limiting reactant and the appropriate mole ratios, compute the number of moles of the desired product.

STEP 5 Convert from moles of product to grams of product, using the molar mass (if this is required by the problem).

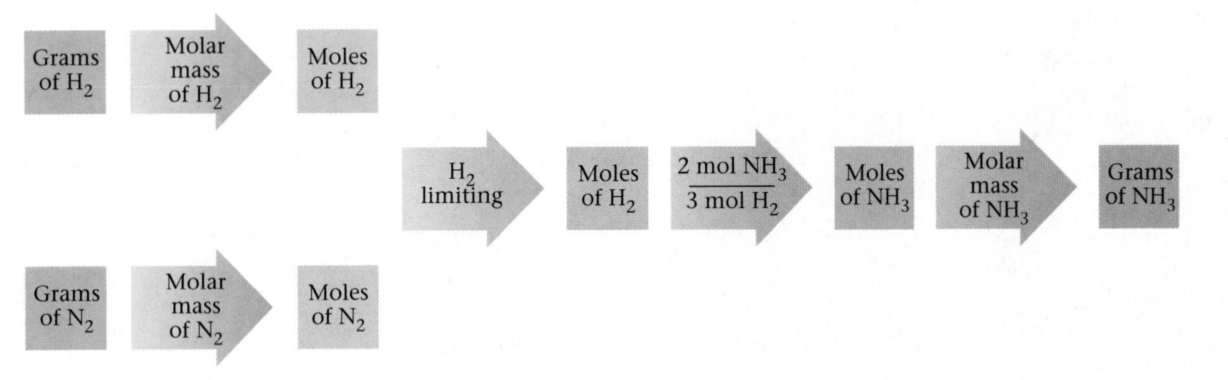

Figure 9.2
A map of the procedure used in Example 9.7.

Example 9.8 Stoichiometric Calculations: Reactions Involving the Masses of Two Reactants

Nitrogen gas can be prepared by passing gaseous ammonia over solid copper(II) oxide at high temperatures. The other products of the reaction are solid copper and water vapor. How many grams of N_2 are formed when 18.1 g of NH_3 are reacted with 90.4 g of CuO?

Copper(II) oxide reacting with ammonia in a heated tube.

Solution

STEP 1 From the description of the problem, we obtain the following balanced equation:

$$2NH_3(g) + 3CuO(s) \rightarrow N_2(g) + 3Cu(s) + 3H_2O(g)$$

STEP 2 Next, from the masses of reactants available we must compute the moles of NH_3 (molar mass = 17.03 g) and of CuO (molar mass = 79.55 g).

$$18.1 \cancel{\text{g NH}_3} \times \frac{1 \text{ mol NH}_3}{17.03 \cancel{\text{g NH}_3}} = 1.06 \text{ mol NH}_3$$

$$90.4 \cancel{\text{g CuO}} \times \frac{1 \text{ mol CuO}}{79.55 \cancel{\text{g CuO}}} = 1.14 \text{ mol CuO}$$

STEP 3 To determine which reactant is limiting, we use the mole ratio between CuO and NH_3.

$$1.06 \cancel{\text{mol NH}_3} \times \frac{3 \text{ mol CuO}}{2 \cancel{\text{mol NH}_3}} = 1.59 \text{ mol CuO}$$

Then we compare how much CuO we have with how much of it we need.

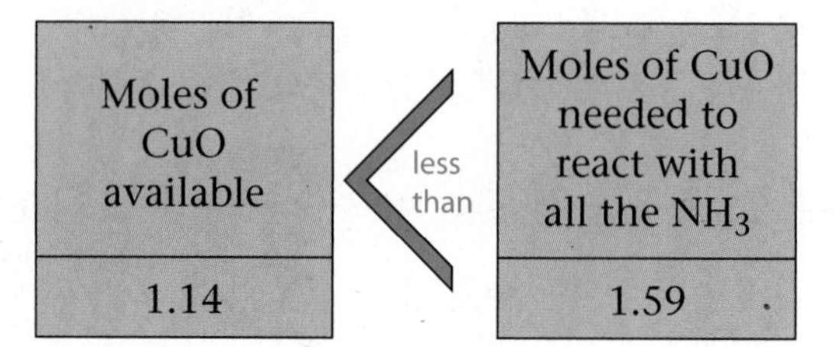

Therefore, 1.59 mol of CuO is required to react with 1.06 mol of NH_3, but only 1.14 mol of CuO is actually present. So the amount of CuO is limiting; CuO will run out before NH_3 does.

STEP 4 CuO is the limiting reactant, so we must use the amount of CuO in calculating the amount of N_2 formed. Using the mole ratio between CuO and N_2 from the balanced equation, we have

$$1.14 \cancel{\text{mol CuO}} \times \frac{1 \text{ mol N}_2}{3 \cancel{\text{mol CuO}}} = 0.380 \text{ mol N}_2$$

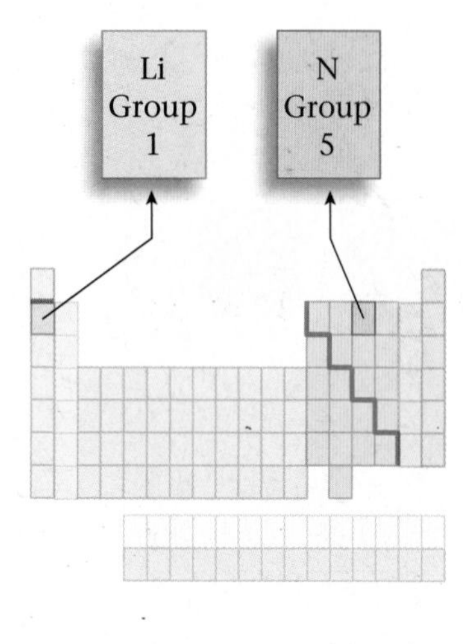

STEP 5 Using the molar mass of N_2 (28.02), we can now calculate the mass of N_2 produced.

$$0.380 \cancel{\text{mol N}_2} \times \frac{28.02 \text{ g N}_2}{1 \cancel{\text{mol N}_2}} = 10.6 \text{ g N}_2$$

✔ Self-Check Exercise 9.6

Lithium nitride, an ionic compound containing the Li^+ and N^{3-} ions, is prepared by the reaction of lithium metal and nitrogen gas. Calculate the mass of lithium nitride formed from 56.0 g of nitrogen gas and 56.0 g of lithium in the unbalanced reaction

$$Li(s) + N_2(g) \rightarrow Li_3N(s)$$

See Problems 9.51 through 9.54. ■

9.5 Percent Yield

AIM: To learn to calculate actual yield as a percentage of theoretical yield.

In the previous section we learned how to calculate the amount of products formed when specified amounts of reactants are mixed together. In doing these calculations, we used the fact that the amount of product is controlled by the limiting reactant. Products stop forming when one reactant runs out.

The amount of product calculated in this way is called the **theoretical yield** of that product. It is the amount of product predicted from the amounts of reactants used. For instance, in Example 9.8, 10.6 g of nitrogen represents the theoretical yield. This is the *maximum amount* of nitrogen that can be produced from the quantities of reactants used. Actually, however, the amount of product predicted (the theoretical yield) is seldom obtained. One reason for this is the presence of side reactions (other reactions that consume one or more of the reactants or products).

The *actual yield* of product, which is the amount of product *actually obtained,* is often compared to the theoretical yield. This comparison, usually expressed as a percent, is called the **percent yield.**

Percent yield is important as an indicator of the efficiency of a particular reaction.

$$\frac{\text{Actual yield}}{\text{Theoretical yield}} \times 100\% = \text{percent yield}$$

For example, *if* the reaction considered in Example 9.8 *actually* gave 6.63 g of nitrogen instead of the *predicted* 10.6 g, the percent yield of nitrogen would be

$$\frac{6.63\ \cancel{\text{g } N_2}}{10.6\ \cancel{\text{g } N_2}} \times 100\% = 62.5\%$$

Example 9.9 Stoichiometric Calculations: Determining Percent Yield

In Section 9.1, we saw that methanol can be produced by the reaction between carbon monoxide and hydrogen. Let's consider this process again. Suppose 68.5 kg (6.85×10^4 g) of $CO(g)$ is reacted with 8.60 kg (8.60×10^3 g) of $H_2(g)$.

a. Calculate the theoretical yield of methanol.

b. If 3.57×10^4 g of CH_3OH is actually produced, what is the percent yield of methanol?

Solution (a)

STEP 1 The balanced equation is

$$2H_2(g) + CO(g) \rightarrow CH_3OH(l)$$

STEP 2 Next we calculate the moles of reactants.

$$6.85 \times 10^4 \text{ g CO} \times \frac{1 \text{ mol CO}}{28.01 \text{ g CO}} = 2.45 \times 10^3 \text{ mol CO}$$

$$8.60 \times 10^3 \text{ g } H_2 \times \frac{1 \text{ mol } H_2}{2.016 \text{ g } H_2} = 4.27 \times 10^3 \text{ mol } H_2$$

STEP 3 Now we determine which reactant is limiting. Using the mole ratio between CO and H_2 from the balanced equation, we have

$$2.45 \times 10^3 \text{ mol CO} \times \frac{2 \text{ mol } H_2}{1 \text{ mol CO}} = 4.90 \times 10^3 \text{ mol } H_2$$

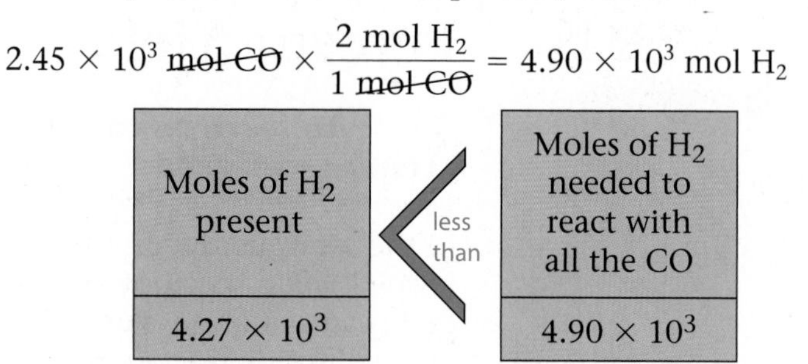

We see that 2.45×10^3 mol of CO requires 4.90×10^3 mol of H_2. Because only 4.27×10^3 mol of H_2 is actually present, *H_2 is limiting.*

STEP 4 We must therefore use the amount of H_2 and the mole ratio between H_2 and CH_3OH to determine the maximum amount of methanol that can be produced in the reaction.

$$4.27 \times 10^3 \text{ mol } H_2 \times \frac{1 \text{ mol } CH_3OH}{2 \text{ mol } H_2} = 2.14 \times 10^3 \text{ mol } CH_3OH$$

This represents the theoretical yield in moles.

STEP 5 Using the molar mass of CH_3OH (32.04 g), we can calculate the theoretical yield in grams.

$$2.14 \times 10^3 \text{ mol } CH_3OH \times \frac{32.04 \text{ g } CH_3OH}{1 \text{ mol } CH_3OH} = 6.86 \times 10^4 \text{ g } CH_3OH$$

So, from the amounts of reactants given, the maximum amount of CH_3OH that can be formed is 6.85×10^4 g. This is the *theoretical yield.*

Solution (b)

The percent yield is

$$\frac{\text{Actual yield (grams)}}{\text{Theoretical yield (grams)}} \times 100\% = \frac{3.57 \times 10^4 \text{ g } CH_3OH}{6.86 \times 10^4 \text{ g } CH_3OH} \times 100\%$$
$$= 52.0\%$$

✔ Self-Check Exercise 9.7

Titanium(IV) oxide is a white compound used as a coloring pigment. In fact, the page you are now reading is white because of the presence of this compound in the paper. Solid titanium(IV) oxide can be prepared by reacting gaseous titanium(IV) chloride with oxygen gas. A second product of this reaction is chlorine gas.

$$TiCl_4(g) + O_2(g) \rightarrow TiO_2(s) + Cl_2(g)$$

a. Suppose 6.71×10^3 g of titanium(IV) chloride is reacted with 2.45×10^3 g of oxygen. Calculate the maximum mass of titanium(IV) oxide that can form.

b. If the percent yield of TiO_2 is 75%, what mass is actually formed?

See Problems 9.63 and 9.64. ■

CHAPTER 9 REVIEW

KEY TERMS

mole ratio (9.2)
stoichiometry (9.3)
limiting reactant (limiting reagent) (9.4)
theoretical yield (9.5)
percent yield (9.5)

SUMMARY

1. A balanced equation relates the numbers of molecules of reactants and products. It can also be expressed in terms of the numbers of moles of reactants and products.
2. The process of using a chemical equation to calculate the relative amounts of reactants and products involved in the reaction is called doing stoichiometric calculations. To convert between moles of reactants and moles of products, we use mole ratios derived from the balanced equation.
3. Often reactants are not mixed in stoichiometric quantities (they do not "run out" at the same time). In that case, we must use the limiting reactant to calculate the amounts of products formed.
4. The actual yield of a reaction is usually less than its theoretical yield. The actual yield is often expressed as a percentage of the theoretical yield, which is called the percent yield.

IN-CLASS DISCUSSION QUESTIONS

These questions are designed to be considered by groups of students in class. Often these questions work well for introducing a particular topic in class.

1. Relate In-Class Discussion Question 2 from Chapter 2 to the concepts of chemical stoichiometry.
2. You are making cookies and are missing a key ingredient—eggs. You have plenty of the other ingredients, except that you have only 1.33 cups of butter and no eggs. You note that the recipe calls for 2 cups of butter and 3 eggs (plus the other ingredients) to make 6 dozen cookies. You telephone a friend and have him bring you some eggs.
   a. How many eggs do you need?
   b. If you use all the butter (and get enough eggs), how many cookies can you make?
   Unfortunately, your friend hangs up before you tell him how many eggs you need. When he arrives, he has a surprise for you—to save time he has broken the eggs in a bowl for you. You ask him how many he brought, and he replies, "All of them, but I spilled some on the way over." You weigh the eggs and find that they weigh 62.1 g. Assuming that an average egg weighs 34.21 g:
   c. How much butter is needed to react with all the eggs?
   d. How many cookies can you make?
   e. Which will you have left over, eggs or butter?
   f. How much is left over?
   g. Relate this question to the concepts of chemical stoichiometry.
3. Nitrogen (N_2) and hydrogen (H_2) react to form ammonia (NH_3). Consider the mixture of N_2 ( ) and H_2( ) in a closed container as illustrated below:

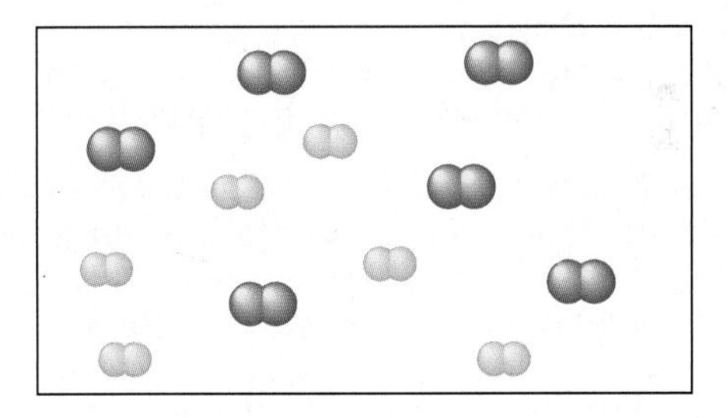

   Assuming the reaction goes to completion, draw a representation of the product mixture. Explain how you arrived at this representation.
4. Which of the following equations best represents the reaction for Question 3?
   a. $6N_2 + 6H_2 \rightarrow 4NH_3 + 4N_2$
   b. $N_2 + H_2 \rightarrow NH_3$
   c. $N + 3H \rightarrow NH_3$
   d. $N_2 + 3H_2 \rightarrow 2NH_3$
   e. $2N_2 + 6H_2 \rightarrow 4NH_3$
   For choices you did not pick, explain what you feel is wrong with them, and justify the choice you did pick.
5. You know that chemical A reacts with chemical B. You react 10.0 g A with 10.0 g B. What information do you need to know to determine the amount of product that will be produced? Explain.
6. If 10.0 g of hydrogen gas is reacted with 10.0 g of oxygen gas, what mass of water can be produced?

Questions 7 and 8 deal with the following situation: You react chemical A with chemical B to make one product. It takes 100 g A to react completely with 20 g B.

7. What is the mass of the product?
   a. Less than 20 g
   b. Between 20 g and 100 g
   c. Between 100 g and 120 g
   d. Exactly 120 g
   e. More than 120 g

8. What is true about the chemical properties of the product?
   a. The properties are more like those of chemical A.
   b. The properties are more like those of chemical B.
   c. The properties are equally like those of chemical A and chemical B.
   d. The properties are not necessarily like either those of A or B.
   e. The properties are more like those of A or more like those of B, but more information is needed.
   For choices you did not pick, explain what you feel is wrong with them, and justify the choice you did pick.

9. The limiting reactant in a reaction:
   a. has the lowest coefficient in a balanced equation.
   b. is the reactant for which you have the fewest number of moles.
   c. has the lowest ratio: moles available/coefficient in the balanced equation.
   d. has the lowest ratio: coefficient in the balanced equation/moles available.
   e. None of the above.
   For choices you did not pick, explain what you feel is wrong with them, and justify the choice you did pick.

10. Given the equation $3A + B \rightarrow C + D$, if 4 moles of A is reacted with 2 moles of B, which of the following is true?
    a. The limiting reactant is the one with the higher molar mass.
    b. A is the limiting reactant because you need 6 moles of A and have 4 moles.
    c. B is the limiting reactant because you have fewer moles of B than moles of A.
    d. B is the limiting reactant because three A molecules react with every one B molecule.
    e. Neither reactant is limiting.
    For choices you did not pick, explain what you feel is wrong with them, and justify the choice you did pick.

11. A kerosene lamp has a mass of 1.5 kg. You put 0.5 kg of kerosene in the lamp. You burn all of the kerosene until the lamp has a mass of 1.5 kg. What is the mass of the gases given off? Explain.

12. What happens to the weight of an iron bar when it rusts?
    a. There is no change because mass is always conserved.
    b. The weight increases.
    c. The weight increases, but if the rust is scraped off, the bar has the original weight.
    d. The weight decreases.
    Justify your choice and, for choices you did not pick, explain what is wrong with them. Explain what it means for something to rust.

13. You may have noticed that water sometimes drips from the exhaust pipe of a car as it is running. Is this evidence that at least a small amount of water is originally present in the gasoline? Explain.

14. You have a chemical in a sealed glass container filled with air. The system has a mass of 250.0 g. The chemical is ignited by means of a magnifying glass focusing sunlight on the reactant. After the chemical is completely burned, what is the mass of the system? Explain your answer.

15. Consider the equation $2A + B \rightarrow A_2B$. If you mix 1.0 mol of A and 1.0 mol of B, how many moles of A_2B can be produced?

16. Can the percent yield of a reaction ever be greater than 100%? Explain.

17. According to the law of conservation of mass, mass cannot be gained or destroyed in a chemical reaction. So why can't you simply add the masses of two reactants to determine the total mass of product produced?

Questions and Problems

All even-numbered exercises have answers in the back of this book and solutions in the *Solutions Guide.*

9.1 Information Given by Chemical Equations

QUESTIONS

1. What do the coefficients of a balanced chemical equation tell us about the proportions in which atoms and molecules react on an individual (microscopic) basis?

2. What do the coefficients of a balanced chemical equation tell us about the proportions in which substances react on a macroscopic (mole) basis?

3. Although *mass* is a property of matter we can conveniently measure in the laboratory, the coefficients of a balanced chemical equation are *not* directly interpreted on the basis of mass. Explain why.

4. Explain why, in the balanced chemical equation $C + O_2 \rightarrow CO_2$, we know that 1 g of C will *not* react exactly with 1 g of O_2.

PROBLEMS

5. For each of the following reactions, give the balanced equation for the reaction and state the meaning of the equation in terms of the numbers of *individual molecules* and in terms of *moles of molecules*.
   a. $PCl_3(l) + H_2O(l) \rightarrow H_3PO_3(aq) + HCl(g)$
   b. $XeF_2(g) + H_2O(l) \rightarrow Xe(g) + HF(g) + O_2(g)$
   c. $S(s) + HNO_3(aq) \rightarrow H_2SO_4(aq) + H_2O(l) + NO_2(g)$
   d. $NaHSO_3(s) \rightarrow Na_2SO_3(s) + SO_2(g) + H_2O(l)$

6. For each of the following reactions, give the balanced chemical equation for the reaction and state the meaning of the equation in terms of *individual molecules* and in terms of *moles* of molecules.
   a. $MnO_2(s) + Al(s) \rightarrow Mn(s) + Al_2O_3(s)$
   b. $B_2O_3(s) + CaF_2(s) \rightarrow BF_3(g) + CaO(s)$
   c. $NO_2(g) + H_2O(l) \rightarrow HNO_3(aq) + NO(g)$
   d. $C_6H_6(g) + H_2(g) \rightarrow C_6H_{12}(g)$

9.2 Mole–Mole Relationships

QUESTIONS

7. Consider the reaction represented by the chemical equation

$$KOH(s) + SO_2(g) \rightarrow KHSO_3(s)$$

Since the coefficients of the balanced chemical equation are all equal to 1, we know that exactly 1 g of KOH will react with exactly 1 g of SO_2. True or false? Explain.

8. True or false? For the reaction represented by the balanced chemical equation

$$N_2(g) + 3I_2(s) \rightarrow 2NI_3(s)$$

if you want to prepare 2 mol of $NI_3(s)$, then 1 g of $N_2(g)$ and 3 g of $I_2(s)$ will be needed.

9. Consider the balanced equation

$$CH_4(g) + 2O_2(g) \rightarrow CO_2(g) + 2H_2O(g)$$

What is the mole ratio that would enable you to calculate the number of moles of oxygen needed to react exactly with a given number of moles of $CH_4(g)$? What mole ratios would you use to calculate how many moles of each product form from a given number of moles of CH_4?

10. Consider the unbalanced chemical equation

$$CaH_2(s) + H_2O(l) \rightarrow Ca(OH)_2(aq) + H_2(g)$$

Balance the equation and then write the mole ratios that would allow you to calculate the number of moles of each product that would form for a given number of moles of water reacting.

PROBLEMS

11. For each of the following balanced reactions, calculate how many *moles of product* would be produced by complete conversion of 0.15 mol of the reactant indicated in boldface. State clearly the mole ratio used for the conversion.
    a. $\mathbf{2Mg}(s) + O_2(g) \rightarrow 2MgO(s)$
    b. $2Mg(s) + \mathbf{O_2}(g) \rightarrow 2MgO(s)$
    c. $\mathbf{4Fe}(s) + 3O_2(g) \rightarrow 2Fe_2O_3(s)$
    d. $4Fe(s) + \mathbf{3O_2}(g) \rightarrow 2Fe_2O_3(s)$

12. For each of the following unbalanced equations, calculate how many *moles of each product* would form if 0.250 mol of the reactant in boldface were to react completely.
    a. $\mathbf{C_{12}H_{22}O_{11}}(s) \rightarrow C(s) + H_2O(g)$
    b. $\mathbf{S}(s) + H_2SO_4(aq) \rightarrow SO_2(g) + H_2O(l)$
    c. $\mathbf{Bi_2O_3}(s) \rightarrow Bi(s) + O_2(g)$
    d. $\mathbf{H_2S}(g) + O_2(g) \rightarrow SO_2(g) + H_2O(l)$

13. For each of the following balanced reactions, calculate how many *moles of each product* would be produced by complete conversion of 1.25 mol of the reactant indicated in boldface. State clearly the mole ratio used for the conversion.
    a. $\mathbf{C_2H_5OH}(l) + 3O_2(g) \rightarrow 2CO_2(g) + 3H_2O(g)$
    b. $\mathbf{N_2}(g) + O_2(g) \rightarrow 2NO(g)$
    c. $\mathbf{2NaClO_2}(s) + Cl_2(g) \rightarrow 2ClO_2(g) + 2NaCl(s)$
    d. $\mathbf{3H_2}(g) + N_2(g) \rightarrow 2NH_3(g)$

14. For each of the following balanced chemical equations, calculate how many *moles* and how many *grams* of each product would be produced by the complete conversion of 0.50 mol of the reactant indicated in boldface. State clearly the mole ratio used for each conversion.
    a. $\mathbf{NH_3}(g) + HCl(g) \rightarrow NH_4Cl(s)$
    b. $CH_4(g) + \mathbf{4S}(s) \rightarrow CS_2(l) + 2H_2S(g)$
    c. $\mathbf{PCl_3} + 3H_2O(l) \rightarrow H_3PO_3(aq) + 3HCl(aq)$
    d. $\mathbf{NaOH}(s) + CO_2(g) \rightarrow NaHCO_3(s)$

15. For each of the following *unbalanced* equations, indicate how many *moles* of the *second reactant* would be required to react exactly with *0.275 mol* of the *first reactant.* State clearly the mole ratio used for the conversion.
    a. $Cl_2(g) + KI(aq) \rightarrow I_2(s) + KCl(aq)$
    b. $Co(s) + P_4(s) \rightarrow Co_3P_2(s)$
    c. $Zn(s) + HNO_3(aq) \rightarrow ZnNO_3(aq) + H_2(g)$
    d. $C_5H_{12}(l) + O_2(g) \rightarrow CO_2(g) + H_2O(g)$

16. For each of the following *unbalanced* equations, indicate how many *moles* of the *first product* are produced if *0.625 mol* of the *second product* forms. State clearly the mole ratio used for each conversion.
    a. $KO_2(s) + H_2O(l) \rightarrow O_2(g) + KOH(s)$
    b. $SeO_2(g) + H_2Se(g) \rightarrow Se(s) + H_2O(g)$
    c. $CH_3CH_2OH(l) + O_2(g) \rightarrow CH_3CHO(aq) + H_2O(l)$
    d. $Fe_2O_3(s) + Al(s) \rightarrow Fe(l) + Al_2O_3(s)$

9.3 Mass Calculations

QUESTIONS

17. What quantity serves as the conversion factor between the mass of a sample and how many moles the sample contains?

18. What does it mean to say that the balanced chemical equation for a reaction describes the *stoichiometry* of the reaction?

PROBLEMS

19. Using the average atomic masses given inside the front cover of this book, calculate how many *moles* of each substance the following masses represent.
    a. 12.7 g of hydrogen gas, H_2
    b. 5.2 g of calcium hydride, CaH_2
    c. 41.6 mg of potassium hydroxide, KOH
    d. 6.93 g of hydrogen sulfide, H_2S
    e. 94.7 g of water, H_2O
    f. 321 mg of lead
    g. 8.79 g of silver nitrate, $AgNO_3$

20. Using the average atomic masses given inside the front cover of this book, calculate how many *moles* of each substance the following masses represent.
    a. 1.47×10^{-3} g of iridium
    b. 8.95 g of lead(II) sulfide, PbS

c. 293 mg of copper(II) oxide, CuO
d. 91.4 g of iron(III) chloride, $FeCl_3$
e. 2.67 g of nitrogen gas, N_2
f. 89.2 g of carbon disulfide, CS_2
g. 1.43 kg of iron

21. Using the average atomic masses given inside the front cover of this book, calculate the *mass in grams* of each of the following samples.
a. 4.25 mol of oxygen gas, O_2
b. 1.27 millimol of platinum (1 millimol = 1/1000 mol)
c. 0.00101 mol of iron(II) sulfate, $FeSO_4$
d. 75.1 mol of calcium carbonate, $CaCO_3$
e. 1.35×10^{-4} mol of gold
f. 1.29 mol of hydrogen peroxide, H_2O_2
g. 6.14 mol of copper(II) sulfide, CuS

22. Using the average atomic masses given inside the front cover of this book, calculate the *mass in grams* of each of the following samples.
a. 0.624 mol of copper(I) iodide, CuI
b. 4.24 mol of bromine, Br_2
c. 0.000211 mol of xenon tetrafluoride, XeF_4
d. 9.11 mol of ethylene, C_2H_4
e. 1.21 millimol of ammonia, NH_3 (1 millimol = 1/1000 mol)
f. 4.25 mol of sodium hydroxide, $NaOH$
g. 1.27×10^{-6} mol of potassium iodide, KI

23. For each of the following *unbalanced* equations, calculate the *mass of each product* that could be produced by complete reaction of *1.55 g* of the reactant indicated in boldface.
a. $\mathbf{CS_2}(l) + O_2(g) \rightarrow CO_2(g) + SO_2(g)$
b. $\mathbf{NaNO_3}(s) \rightarrow NaNO_2(s) + O_2(g)$
c. $H_2(g) + \mathbf{MnO_2}(s) \rightarrow MnO(s) + H_2O(g)$
d. $\mathbf{Br_2}(l) + Cl_2(g) \rightarrow BrCl(g)$

24. For each of the following *unbalanced* equations, calculate how many *moles* of the *second* reactant would be required to react completely with exactly *25.0 g* of the *first* reactant. Indicate clearly the mole ratio used for each conversion.
a. $Mg(s) + CuCl_2(aq) \rightarrow MgCl_2(aq) + Cu(s)$
b. $AgNO_3(aq) + NiCl_2(aq) \rightarrow AgCl(s) + Ni(NO_3)_2(aq)$
c. $NaHSO_3(aq) + NaOH(aq) \rightarrow Na_2SO_3(aq) + H_2O(l)$
d. $KHCO_3(aq) + HCl(aq) \rightarrow KCl(aq) + H_2O(l) + CO_2(g)$

25. For each of the following *unbalanced* equations, calculate how many *grams of each product* would be produced by complete reaction of *12.5 g* of the reactant indicated in boldface. Indicate clearly the mole ratio used for the conversion.
a. $TiBr_4(g) + \mathbf{H_2}(g) \rightarrow Ti(s) + HBr(g)$
b. $\mathbf{SiH_4}(g) + NH_3(g) \rightarrow Si_3N_4(s) + H_2(g)$
c. $NO(g) + \mathbf{H_2}(g) \rightarrow N_2(g) + 2H_2O(l)$
d. $\mathbf{Cu_2S}(s) \rightarrow Cu(s) + S(g)$

26. For each of the following *balanced* equations, calculate how many *grams of each product* would be produced by complete reaction of *15.0 g* of the reactant indicated in boldface.
a. $\mathbf{2BCl_3}(s) + 3H_2(g) \rightarrow 2B(s) + 6HCl(g)$
b. $\mathbf{2Cu_2S}(s) + 3O_2(g) \rightarrow 2Cu_2O(s) + 2SO_2(g)$
c. $2Cu_2O(s) + \mathbf{Cu_2S}(s) \rightarrow 6Cu(s) + SO_2(g)$
d. $CaCO_3(s) + \mathbf{SiO_2}(s) \rightarrow CaSiO_3(s) + CO_2(g)$

27. Bottled propane is used in areas away from natural gas pipelines for cooking and heating, and is also the source of heat in most gas barbecue grills. Propane burns in oxygen according to the following balanced chemical equation:

$$C_3H_8(g) + 5O_2(g) \rightarrow 3CO_2(g) + 4H_2O(g)$$

Calculate the mass in grams of water vapor produced if 3.11 mol of propane is burned.

28. Given the information in Problem 27, calculate the mass of oxygen gas consumed in the burning of 25.0 lb of propane.

29. When elemental carbon is burned in the open atmosphere, with plenty of oxygen gas present, the product is carbon dioxide.

$$C(s) + O_2(g) \rightarrow CO_2(g)$$

However, when the amount of oxygen present during the burning of the carbon is restricted, carbon monoxide is more likely to result.

$$2C(s) + O_2(g) \rightarrow 2CO(g)$$

What mass of each product is expected when a 5.00-g sample of pure carbon is burned under each of these conditions?

30. The more reactive halogen elements are able to replace the less reactive halogens from their compounds. For example, if chlorine gas is bubbled through a potassium iodide solution, elemental iodine is produced.

$$Cl_2(g) + KI(aq) \rightarrow I_2(s) + KCl(aq)$$

Calculate the mass of iodine produced when 2.55 g of chlorine gas is bubbled through an excess amount of potassium iodide solution.

31. Although we usually think of substances as "burning" only in oxygen gas, the process of rapid oxidation to produce a flame may also take place in other strongly oxidizing gases. For example, when iron is heated and placed in pure chlorine gas, the iron "burns" according to the following (unbalanced) reaction:

$$Fe(s) + Cl_2(g) \rightarrow FeCl_3(s)$$

How many milligrams of iron(III) chloride result when 15.5 mg of iron is reacted with an excess of chlorine gas?

32. You have probably seen images of a chef preparing a "flaming" dessert or entrée. The flame is usually the result of the combustion of ethyl alcohol, C_2H_5OH, that has been added to the food (perhaps as cognac or rum).

$$C_2H_5OH(l) + O_2(g) \rightarrow CO_2(g) + H_2O(l)$$

If 25.0 g of ethyl alcohol is burned in air (excess oxygen), calculate the mass of carbon dioxide produced.

33. The halogen elements are so reactive that the halides of many metals can be prepared by the direct combination of the elements. For example, iron(III) chloride can be prepared by the following reaction:

$$2Fe(s) + 3Cl_2(g) \rightarrow 2FeCl_3(s)$$

Calculate the mass of $FeCl_3$ that is formed if 12.4 g of iron reacts completely.

34. When the hydroxide compound of many metals is heated, water is driven off and the oxide of the metal remains. For example, if cobalt(II) hydroxide is heated, cobalt(II) oxide is produced.

$$Co(OH)_2(s) \rightarrow CoO(s) + H_2O(g)$$

What mass of cobalt(II) oxide would remain if 5.75 g of cobalt(II) hydroxide were heated strongly?

35. With the news that calcium deficiencies in many women's diets may contribute to the development of osteoporosis (bone weakening), the use of dietary calcium supplements has become increasingly common. Many of these calcium supplements consist of nothing more than calcium carbonate, $CaCO_3$. When a calcium carbonate tablet is ingested, it dissolves by reaction with stomach acid, which contains hydrochloric acid, HCl. The unbalanced equation is

$$CaCO_3(s) + HCl(aq) \rightarrow CaCl_2(aq) + H_2O(l) + CO_2(g)$$

What mass of HCl is required to react with a tablet containing 500. mg of calcium carbonate?

36. A tried-and-true introductory chemistry experiment involves heating finely divided copper metal with sulfur to determine the proportions in which the elements react to form copper(II) sulfide. The experiment works well, because any excess sulfur beyond that required to react with the copper may be simply boiled away from the reaction container.

$$Cu(s) + S(s) \rightarrow CuS(s)$$

If 1.25 g of copper is heated with an excess of sulfur, how many grams of sulfur will react?

37. Ammonium nitrate has been used as a high explosive because it is unstable and decomposes into several gaseous substances. The rapid expansion of the gaseous substances produces the explosive force.

$$NH_4NO_3(s) \rightarrow N_2(g) + O_2(g) + H_2O(g)$$

Calculate the mass of each product gas if 1.25 g of ammonium nitrate reacts.

38. Solutions of sodium hydroxide cannot be kept for very long because they absorb carbon dioxide from the air, forming sodium carbonate. The unbalanced equation is

$$NaOH(aq) + CO_2(g) \rightarrow Na_2CO_3(aq) + H_2O(l)$$

Calculate the number of grams of carbon dioxide that can be absorbed by complete reaction with a solution that contains 5.00 g of sodium hydroxide.

39. Thionyl chloride, $SOCl_2$, is used as a very powerful drying agent in many synthetic chemistry experiments in which the presence of even small amounts of water would be detrimental. The unbalanced chemical equation is

$$SOCl_2(l) + H_2O(l) \rightarrow SO_2(g) + HCl(g)$$

Calculate the mass of water consumed by complete reaction of 35.0 g of $SOCl_2$.

40. Magnesium metal, which burns in oxygen with an intensely bright white flame, has been used in photographic flash units. The balanced equation for this reaction is

$$2Mg(s) + O_2(g) \rightarrow 2MgO(s)$$

How many grams of MgO(s) are produced by complete reaction of 1.25 g of magnesium metal?

9.4 Calculations Involving a Limiting Reactant

QUESTIONS

41. What is the *limiting reactant* for a process? Why does a reaction stop when the limiting reactant is consumed, even though there may be plenty of the other reactants present?

42. Explain how one determines which reactant in a process is the limiting reactant. Does this depend only on the masses of the reactant present? Is the mole ratio in which the reactants combine involved?

43. What is the *theoretical yield* for a reaction, and how does this quantity depend on the limiting reactant?

44. What does it mean to say a reactant is present "in excess" in a process? Can the *limiting reactant* be present in excess? Does the presence of an excess of a reactant affect the mass of products expected for a reaction?

PROBLEMS

45. For each of the following *unbalanced* reactions, suppose exactly 5.00 g of *each reactant* is taken. Determine which reactant is limiting, and also determine what mass of the excess reagent will remain after the limiting reactant is consumed.
   a. $Na_2B_4O_7(s) + H_2SO_4(aq) + H_2O(l) \rightarrow H_3BO_3(s) + Na_2SO_4(aq)$
   b. $CaC_2(s) + H_2O(l) \rightarrow Ca(OH)_2(s) + C_2H_2(g)$
   c. $NaCl(s) + H_2SO_4(l) \rightarrow HCl(g) + Na_2SO_4(s)$
   d. $SiO_2(s) + C(s) \rightarrow Si(l) + CO(g)$

46. For each of the following *unbalanced* chemical equations, suppose that exactly 5.00 g of *each* reactant is taken. Determine which reactant is limiting, and calculate what mass of each product is expected (assuming that the limiting reactant is completely consumed).
   a. $S(s) + H_2SO_4(aq) \rightarrow SO_2(g) + H_2O(l)$
   b. $MnO_2(s) + H_2SO_4(l) \rightarrow Mn(SO_4)_2(s) + H_2O(l)$
   c. $H_2S(g) + O_2(g) \rightarrow SO_2(g) + H_2O(l)$
   d. $AgNO_3(aq) + Al(s) \rightarrow Ag(s) + Al(NO_3)_3(aq)$

47. For each of the following *unbalanced* chemical equations, suppose 10.0 g of *each* reactant is taken. Show by calculation which reactant is the limiting reagent. Calculate the mass of each product that is expected.
    a. $C_3H_8(g) + O_2(g) \rightarrow CO_2(g) + H_2O(g)$
    b. $Al(s) + Cl_2(g) \rightarrow AlCl_3(s)$
    c. $NaOH(s) + CO_2(g) \rightarrow Na_2CO_3(s) + H_2O(l)$
    d. $NaHCO_3(s) + HCl(aq) \rightarrow NaCl(aq) + H_2O(l) + CO_2(g)$

48. For each of the following *unbalanced* equations, suppose that exactly 1.00 g of *each* reactant is taken. Determine which reactant is limiting, and calculate what mass of the product in boldface is expected (assuming that the limiting reactant is completely consumed).
    a. $CS_2(l) + O_2(g) \rightarrow \mathbf{CO_2}(g) + SO_2(g)$
    b. $NH_3(g) + CO_2(g) \rightarrow CN_2H_4O(s) + \mathbf{H_2O}(g)$
    c. $H_2(g) + MnO_2(s) \rightarrow MnO(s) + \mathbf{H_2O}(g)$
    d. $I_2(l) + Cl_2(g) \rightarrow \mathbf{ICl}(g)$

49. For each of the following *unbalanced* chemical equations, suppose 1.00 g of *each* reactant is taken. Show by calculation which reactant is limiting. Calculate the mass of each product that is expected.
    a. $UO_2(s) + HF(aq) \rightarrow UF_4(aq) + H_2O(l)$
    b. $NaNO_3(aq) + H_2SO_4(aq) \rightarrow Na_2SO_4(aq) + HNO_3(aq)$
    c. $Zn(s) + HCl(aq) \rightarrow ZnCl_2(aq) + H_2(g)$
    d. $B(OH)_3(s) + CH_3OH(l) \rightarrow B(OCH_3)_3(s) + H_2O(l)$

50. For each of the following *unbalanced* chemical equations, suppose 10.0 mg of *each* reactant is taken. Show by calculation which reactant is limiting. Calculate the mass of each product that is expected.
    a. $CO(g) + H_2(g) \rightarrow CH_3OH(l)$
    b. $Al(s) + I_2(s) \rightarrow AlI_3(s)$
    c. $Ca(OH)_2(aq) + HBr(aq) \rightarrow CaBr_2(aq) + H_2O(l)$
    d. $Cr(s) + H_3PO_4(aq) \rightarrow CrPO_4(s) + H_2(g)$

51. The more reactive halogen elements are able to replace the less reactive halogens from their compounds.

$$Cl_2(g) + NaI(aq) \rightarrow NaCl(aq) + I_2(s)$$
$$Br_2(l) + NaI(aq) \rightarrow NaBr(aq) + I_2(s)$$

Suppose separate solutions each containing 25.0 g of NaI are available. If 5.00 g of Cl_2 gas is bubbled into one NaI solution, and 5.00 g of liquid bromine is added to the other, calculate the number of grams of elemental iodine produced in each case.

52. An experiment that led to the formation of the new field of organic chemistry involved the synthesis of urea, CN_2H_4O, by the controlled reaction of ammonia and carbon dioxide:

$$2NH_3(g) + CO_2(g) \rightarrow CN_2H_4O(s) + H_2O(l)$$

What is the theoretical yield of urea when 100. g of ammonia is reacted with 100. g of carbon dioxide?

53. Lead(II) oxide from an ore can be reduced to elemental lead by heating in a furnace with carbon.

$$PbO(s) + C(s) \rightarrow Pb(l) + CO(g)$$

Calculate the expected yield of lead if 50.0 kg of lead oxide is heated with 50.0 kg of carbon.

54. If steel wool (iron) is heated until it glows and is placed in a bottle containing pure oxygen, the iron reacts spectacularly to produce iron(III) oxide.

$$Fe(s) + O_2(g) \rightarrow Fe_2O_3(s)$$

If 1.25 g of iron is heated and placed in a bottle containing 0.0204 mol of oxygen gas, what mass of iron(III) oxide is produced?

55. One method for chemical analysis involves finding some reagent that will precipitate the species of interest. The mass of the precipitate is then used to determine what mass of the species of interest was present in the original sample. For example, calcium ion can be precipitated from solution by addition of sodium oxalate. The balanced equation is

$$Ca^{2+}(aq) + Na_2C_2O_4(aq) \rightarrow CaC_2O_4(s) + 2Na^+(aq)$$

Suppose a solution is known to contain approximately 15 g of calcium ion. Show by calculation whether the addition of a solution containing 15 g of sodium oxalate will precipitate all of the calcium from the sample.

56. The copper(II) ion in a copper(II) sulfate solution reacts with potassium iodide to produce the triiodide ion, I_3^-. This reaction is commonly used to determine how much copper is present in a given sample.

$$CuSO_4(aq) + KI(aq) \rightarrow CuI(s) + KI_3(aq) + K_2SO_4(aq)$$

If 2.00 g of KI is added to a solution containing 0.525 g of $CuSO_4$, calculate the mass of each product produced.

57. Hydrogen peroxide is used as a cleaning agent in the treatment of cuts and abrasions for several reasons. It is an oxidizing agent that can directly kill many microorganisms; it decomposes upon contact with blood, releasing elemental oxygen gas (which inhibits the growth of anaerobic microorganisms); and it foams upon contact with blood, which provides a cleansing action. In the laboratory, small quantities of hydrogen peroxide can be prepared by the action of an acid on an alkaline earth metal peroxide, such as barium peroxide.

$$BaO_2(s) + 2HCl(aq) \rightarrow H_2O_2(aq) + BaCl_2(aq)$$

What amount of hydrogen peroxide should result when 1.50 g of barium peroxide is treated with 25.0 mL of hydrochloric acid solution containing 0.0272 g of HCl per mL?

58. Silicon carbide, SiC, is one of the hardest materials known. Surpassed in hardness only by diamond, it is sometimes known commercially as carborundum. Silicon carbide is used primarily as an abrasive for sandpaper and is manufactured by heating common sand (silicon dioxide, SiO_2) with carbon in a furnace.

$$SiO_2(s) + C(s) \rightarrow CO(g) + SiC(s)$$

What mass of silicon carbide should result when 1.0 kg of pure sand is heated with an excess of carbon?

9.5 Percent Yield

QUESTIONS

59. What is the *actual yield* of a reaction? What is the *percent yield* of a reaction? How do the actual yield and the percent yield differ from the theoretical yield?

60. The text explains that one reason why the actual yield for a reaction may be less than the theoretical yield is side reactions. Suggest some other reasons why the percent yield for a reaction might not be 100%.

61. According to his prelaboratory theoretical yield calculations, a student's experiment should have produced 1.44 g of magnesium oxide. When he weighed his product after reaction, only 1.23 g of magnesium oxide was present. What is the student's percent yield?

62. A student calculated the theoretical yield of barium sulfate in a precipitation experiment to be 1.352 g. When she filtered, dried, and weighed her precipitate, however, her yield was only 1.279 g. Calculate the student's percent yield.

PROBLEMS

63. The compound sodium thiosulfate pentahydrate, $Na_2S_2O_3 \cdot 5H_2O$, is important commercially to the photography business as "hypo," because it has the ability to dissolve unreacted silver salts from photographic film during development. Sodium thiosulfate pentahydrate can be produced by boiling elemental sulfur in an aqueous solution of sodium sulfite.

$$S_8(s) + Na_2SO_3(aq) + H_2O(l) \rightarrow Na_2S_2O_3 \cdot 5H_2O(s) \quad \text{(unbalanced)}$$

What is the theoretical yield of sodium thiosulfate pentahydrate when 3.25 g of sulfur is boiled with 13.1 g of sodium sulfite? Sodium thiosulfate pentahydrate is very soluble in water. What is the percent yield of the synthesis if a student doing this experiment is able to isolate (collect) only 5.26 g of the product?

64. Alkali metal hydroxides are sometimes used to "scrub" excess carbon dioxide from the air in closed spaces (such as submarines and spacecraft). For example, lithium hydroxide reacts with carbon dioxide according to the unbalanced chemical equation

$$LiOH(s) + CO_2(g) \rightarrow Li_2CO_3(s) + H_2O(g)$$

Suppose a lithium hydroxide canister contains 155 g of $LiOH(s)$. What mass of $CO_2(g)$ will the canister be able to absorb? If it is found that after 24 hours of use the canister has absorbed 102 g of carbon dioxide, what percentage of its capacity has been reached?

65. Although they were formerly called the inert gases, at least the heavier elements of Group 8 do form relatively stable compounds. For example, xenon combines directly with elemental fluorine at elevated temperatures in the presence of a nickel catalyst.

$$Xe(g) + 2F_2(g) \rightarrow XeF_4(s)$$

What is the theoretical mass of xenon tetrafluoride that should form when 130. g of xenon is reacted with 100. g of F_2? What is the percent yield if only 145 g of XeF_4 is actually isolated?

66. A common undergraduate laboratory analysis for the amount of sulfate ion in an unknown sample is to precipitate and weigh the sulfate ion as barium sulfate.

$$Ba^{2+}(aq) + SO_4^{2-}(aq) \rightarrow BaSO_4(s)$$

The precipitate produced, however, is very finely divided, and frequently some is lost during filtration before weighing. If a sample containing 1.12 g of sulfate ion is treated with 5.02 g of barium chloride, what is the theoretical yield of barium sulfate to be expected? If only 2.02 g of barium sulfate is actually collected, what is the percent yield?

ADDITIONAL PROBLEMS

67. Natural waters often contain relatively high levels of calcium ion, Ca^{2+}, and hydrogen carbonate ion (bicarbonate), HCO_3^-, from the leaching of minerals into the water. When such water is used commercially or in the home, heating of the water leads to the formation of solid calcium carbonate, $CaCO_3$, which forms a deposit ("scale") on the interior of boilers, pipes, and other plumbing fixtures.

$$Ca(HCO_3)_2(aq) \rightarrow CaCO_3(s) + CO_2(g) + H_2O(l)$$

If a sample of well water contains 2.0×10^{-3} mg of $Ca(HCO_3)_2$ per milliliter, what mass of $CaCO_3$ scale would 1.0 mL of this water be capable of depositing?

68. One process for the commercial production of baking soda (sodium hydrogen carbonate) involves the following reaction, in which the carbon dioxide is used in its solid form ("dry ice") both to serve as a source of reactant and to cool the reaction system to a temperature low enough for the sodium hydrogen carbonate to precipitate:

$$NaCl(aq) + NH_3(aq) + H_2O(l) + CO_2(s) \rightarrow NH_4Cl(aq) + NaHCO_3(s)$$

Because they are relatively cheap, sodium chloride and water are typically present in excess. What is the expected yield of $NaHCO_3$ when one performs such a synthesis using 10.0 g of ammonia and 15.0 g of dry ice, with an excess of NaCl and water?

69. A favorite demonstration among chemistry instructors, to show that the properties of a compound

differ from those of its constituent elements, involves iron filings and powdered sulfur. If the instructor takes samples of iron and sulfur and just mixes them together, the two elements can be separated from one another with a magnet (iron is attracted to a magnet, sulfur is not). If the instructor then combines and *heats* the mixture of iron and sulfur, a reaction takes place and the elements combine to form iron(II) sulfide (which is not attracted by a magnet).

$$Fe(s) + S(s) \rightarrow FeS(s)$$

Suppose 5.25 g of iron filings is combined with 12.7 g of sulfur. What is the theoretical yield of iron(II) sulfide?

70. When the sugar glucose, $C_6H_{12}O_6$, is burned in air, carbon dioxide and water vapor are produced. Write the balanced chemical equation for this process, and calculate the theoretical yield of carbon dioxide when 1.00 g of glucose is burned completely.

71. When elemental copper is strongly heated with sulfur, a mixture of CuS and Cu_2S is produced, with CuS predominating.

$$Cu(s) + S(s) \rightarrow CuS(s)$$
$$2Cu(s) + S(s) \rightarrow Cu_2S(s)$$

What is the theoretical yield of CuS when 31.8 g of Cu(*s*) is heated with 50.0 g of S? (Assume only CuS is produced in the reaction.) What is the percent yield of CuS if only 40.0 g of CuS can be isolated from the mixture?

72. Barium chloride solutions are used in chemical analysis for the quantitative precipitation of sulfate ion from solution.

$$Ba^{2+}(aq) + SO_4^{2-}(aq) \rightarrow BaSO_4(s)$$

Suppose a solution is known to contain on the order of 150 mg of sulfate ion. What mass of barium chloride should be added to guarantee precipitation of all the sulfate ion?

73. The traditional method of analysis for the amount of chloride ion present in a sample was to dissolve the sample in water and then slowly to add a solution of silver nitrate. Silver chloride is very insoluble in water, and by adding a slight excess of silver nitrate, it is possible effectively to remove all chloride ion from the sample.

$$Ag^+(aq) + Cl^-(aq) \rightarrow AgCl(s)$$

Suppose a 1.054-g sample is known to contain 10.3% chloride ion by mass. What mass of silver nitrate must be used to completely precipitate the chloride ion from the sample? What mass of silver chloride will be obtained?

74. For each of the following reactions, give the balanced equation for the reaction and state the meaning of the equation in terms of numbers of *individual molecules* and in terms of *moles* of molecules.
   a. $UO_2(s) + HF(aq) \rightarrow UF_4(aq) + H_2O(l)$
   b. $NaC_2H_3O_2(aq) + H_2SO_4(aq) \rightarrow Na_2SO_4(aq) + HC_2H_3O_2(aq)$
   c. $Mg(s) + HCl(aq) \rightarrow MgCl_2(aq) + H_2(g)$
   d. $B_2O_3(s) + H_2O(l) \rightarrow B(OH)_3(aq)$

75. True or false? For the reaction represented by the balanced chemical equation

$$Mg(OH)_2(aq) + 2HCl(aq) \rightarrow 2H_2O(l) + MgCl_2(aq)$$

for 0.40 mol of $Mg(OH)_2$, 0.20 mol of HCl will be needed.

76. Consider the balanced equation

$$C_3H_8(g) + 5O_2(g) \rightarrow 3CO_2(g) + 4H_2O(g)$$

What mole ratio enables you to calculate the number of moles of oxygen needed to react exactly with a given number of moles of $C_3H_8(g)$? What mole ratios enable you to calculate how many moles of each product form from a given number of moles of C_3H_8?

77. For each of the following balanced reactions, calculate how many *moles of each product* would be produced by complete conversion of *0.50 mol* of the reactant indicated in boldface. Indicate clearly the mole ratio used for the conversion.
   a. $\mathbf{2H_2O_2}(l) \rightarrow 2H_2O(l) + O_2(g)$
   b. $\mathbf{2KClO_3}(s) \rightarrow 2KCl(s) + 3O_2(g)$
   c. $\mathbf{2Al}(s) + 6HCl(aq) \rightarrow 2AlCl_3(aq) + 3H_2(g)$
   d. $\mathbf{C_3H_8}(g) + 5O_2(g) \rightarrow 3CO_2(g) + 4H_2O(g)$

78. For each of the following balanced equations, indicate how many *moles of the product* could be produced by complete reaction of *1.00 g* of the reactant indicated in boldface. Indicate clearly the mole ratio used for the conversion.
   a. $\mathbf{NH_3}(g) + HCl(g) \rightarrow NH_4Cl(s)$
   b. $\mathbf{CaO}(s) + CO_2(g) \rightarrow CaCO_3(s)$
   c. $\mathbf{4Na}(s) + O_2(g) \rightarrow 2Na_2O(s)$
   d. $\mathbf{2P}(s) + 3Cl_2(g) \rightarrow 2PCl_3(l)$

79. Using the average atomic masses given inside the front cover of the text, calculate how many *moles* of each substance the following masses represent.
   a. 4.21 g of copper(II) sulfate
   b. 7.94 g of barium nitrate
   c. 1.24 mg of water
   d. 9.79 g of tungsten
   e. 1.45 lb of sulfur
   f. 4.65 g of ethyl alcohol, C_2H_5OH
   g. 12.01 g of carbon

80. Using the average atomic masses given inside the front cover of the text, calculate the *mass in grams* of each of the following samples.
   a. 5.0 mol of nitric acid
   b. 0.000305 mol of mercury
   c. 2.31×10^{-5} mol of potassium chromate
   d. 10.5 mol of aluminum chloride
   e. 4.9×10^4 mol of sulfur hexafluoride
   f. 125 mol of ammonia
   g. 0.01205 mol of sodium peroxide

81. For each of the following *incomplete* and *unbalanced* equations, indicate how many *moles* of the *second reactant* would be required to react completely with *0.145 mol* of the *first reactant.*
   a. $BaCl_2(aq) + H_2SO_4 \rightarrow$
   b. $AgNO_3(aq) + NaCl(aq) \rightarrow$
   c. $Pb(NO_3)_2(aq) + Na_2CO_3(aq) \rightarrow$
   d. $C_3H_8(g) + O_2(g) \rightarrow$

82. One step in the commercial production of sulfuric acid, H_2SO_4, involves the conversion of sulfur dioxide, SO_2, into sulfur trioxide, SO_3.

$$2SO_2(g) + O_2(g) \rightarrow 2SO_3(g)$$

If 150 kg of SO_2 reacts completely, what mass of SO_3 should result?

83. Many metals occur naturally as sulfide compounds; examples include ZnS and CoS. Air pollution often accompanies the processing of these ores, because toxic sulfur dioxide is released as the ore is converted from the sulfide to the oxide by roasting (smelting). For example, consider the unbalanced equation for the roasting reaction for zinc:

$$ZnS(s) + O_2(g) \rightarrow ZnO(s) + SO_2(g)$$

How many kilograms of sulfur dioxide are produced when 1.0×10^2 kg of ZnS is roasted in excess oxygen by this process?

84. If sodium peroxide is added to water, elemental oxygen gas is generated:

$$Na_2O_2(s) + H_2O(l) \rightarrow NaOH(aq) + O_2(g)$$

Suppose 3.25 g of sodium peroxide is added to a large excess of water. What mass of oxygen gas will be produced?

85. When elemental copper is placed in a solution of silver nitrate, the following oxidation–reduction reaction takes place, forming elemental silver:

$$Cu(s) + 2AgNO_3(aq) \rightarrow Cu(NO_3)_2(aq) + 2Ag(s)$$

What mass of copper is required to remove all the silver from a silver nitrate solution containing 1.95 mg of silver nitrate?

86. When small quantities of elemental hydrogen gas are needed for laboratory work, the hydrogen is often generated by chemical reaction of a metal with acid. For example, zinc reacts with hydrochloric acid, releasing gaseous elemental hydrogen:

$$Zn(s) + 2HCl(aq) \rightarrow ZnCl_2(aq) + H_2(g)$$

What mass of hydrogen gas is produced when 2.50 g of zinc is reacted with excess aqueous hydrochloric acid?

87. The gaseous hydrocarbon acetylene, C_2H_2, is used in welders' torches because of the large amount of heat released when acetylene burns with oxygen.

$$2C_2H_2(g) + 5O_2(g) \rightarrow 4CO_2(g) + 2H_2O(g)$$

How many grams of oxygen gas are needed for the complete combustion of 150 g of acetylene?

88. For each of the following *unbalanced* chemical equations, suppose exactly 5.0 g of each reactant is taken. Determine which reactant is limiting, and calculate what mass of each product is expected, assuming that the limiting reactant is completely consumed.
   a. $Na(s) + Br_2(l) \rightarrow NaBr(s)$
   b. $Zn(s) + CuSO_4(aq) \rightarrow ZnSO_4(aq) + Cu(s)$
   c. $NH_4Cl(aq) + NaOH(aq) \rightarrow NH_3(g) + H_2O(l) + NaCl(aq)$
   d. $Fe_2O_3(s) + CO(g) \rightarrow Fe(s) + CO_2(g)$

89. For each of the following *unbalanced* chemical equations, suppose 25.0 g of each reactant is taken. Show by calculation which reactant is limiting. Calculate the theoretical yield in grams of the product in boldface.
   a. $C_2H_5OH(l) + O_2(g) \rightarrow \mathbf{CO_2}(g) + H_2O(l)$
   b. $N_2(g) + O_2(g) \rightarrow \mathbf{NO}(g)$
   c. $NaClO_2(aq) + Cl_2(g) \rightarrow ClO_2(g) + \mathbf{NaCl}(aq)$
   d. $H_2(g) + N_2(g) \rightarrow \mathbf{NH_3}(g)$

90. Hydrazine, N_2H_4, emits a large quantity of energy when it reacts with oxygen, which has led to hydrazine's use as a fuel for rockets:

$$N_2H_4(l) + O_2(g) \rightarrow N_2(g) + 2H_2O(g)$$

How many moles of each of the gaseous products are produced when 20.0 g of pure hydrazine is ignited in the presence of 20.0 g of pure oxygen? How many grams of each product are produced?

91. Although elemental chlorine, Cl_2, is added to drinking water supplies primarily to kill microorganisms, another beneficial reaction that also takes place removes sulfides (which would impart unpleasant odors or tastes to the water). For example, the noxious-smelling gas hydrogen sulfide (its odor resembles that of rotten eggs) is removed from water by chlorine by the following reaction:

$$H_2S(aq) + Cl_2(aq) \rightarrow HCl(aq) + S_8(s) \quad \text{(unbalanced)}$$

What mass of sulfur is removed from the water when 50. L of water containing 1.5×10^{-5} g of H_2S per liter is treated with 1.0 g of $Cl_2(g)$?

92. Before going to lab, a student read in his lab manual that the percent yield for a difficult reaction to be studied was likely to be only 40.% of the theoretical yield. The student's prelab stoichiometric calculations predict that the theoretical yield should be 12.5 g. What is the student's actual yield likely to be?

CUMULATIVE REVIEW FOR CHAPTERS 8–9

QUESTIONS

1. What does the average *atomic mass* of an element represent? What unit is used for average atomic mass? Express the atomic mass unit in grams. Why is the average atomic mass for an element typically *not* a whole number?

2. Perhaps the most important concept in introductory chemistry concerns what a *mole* of a substance represents. The mole concept will come up again and again in later chapters in this book. What does one mole of a substance represent on a microscopic, atomic basis? What does one mole of a substance represent on a macroscopic, mass basis? Why have chemists defined the mole in this manner?

3. How do we know that 16.00 g of oxygen contains the same number of atoms as does 12.01 g of carbon, and that 22.99 g of sodium contains the same number of atoms as each of these? How do we know that 106.0 g of Na_2CO_3 contains the same number of carbon atoms as does 12.01 g of carbon, but three times as many oxygen atoms as in 16.00 g of oxygen, and twice as many sodium atoms as in 22.99 g of sodium?

4. Define *molar mass*. Using H_3PO_4 as an example, calculate the molar mass from the atomic masses of the elements.

5. What is meant by the *percentage composition* by mass for a compound? Describe in general terms how this information is obtained by experiment for new compounds. How can this information be calculated for known compounds?

6. Define, compare, and contrast what are meant by the *empirical* and *molecular* formulas for a substance. What does each of these formulas tell us about a compound? What information must be known for a compound before the molecular formula can be determined? Why is the molecular formula an *integer multiple* of the empirical formula?

7. When chemistry teachers prepare an exam question on determining the empirical formula of a compound, they usually take a known compound and calculate the percentage composition of the compound from the formula. They then give students this percentage composition data and have the students calculate the original formula. Using a compound of *your* choice, first use the molecular formula of the compound to calculate the percentage composition of the compound. Then use this percentage composition data to calculate the empirical formula of the compound.

8. Rather than giving students straight percentage composition data for determining the empirical formula of a compound (see Question 7), sometimes chemistry teachers will try to emphasize the experimental nature of formula determination by converting the percentage composition data into actual experimental masses. For example, the compound CH_4 contains 74.87% carbon by mass. Rather than giving students the data in this form, a teacher might instead say, "When 1.000 g of a compound was analyzed, it was found to contain 0.7487 g of carbon, with the remainder consisting of hydrogen." Using the compound you chose for Question 7, and the percentage composition data you calculated, reword your data as suggested in this problem in terms of actual "experimental" masses. Then from these masses, calculate the empirical formula of your compound.

9. Balanced chemical equations give us information in terms of individual molecules reacting in the proportions indicated by the coefficients, and also in terms of macroscopic amounts (that is, moles). Write a balanced chemical equation of your choice, and interpret in words the meaning of the equation on the molecular and macroscopic levels.

10. Consider the *unbalanced* equation for the combustion of ethyl alcohol, C_2H_5OH:

$$C_2H_5OH(l) + O_2(g) \rightarrow CO_2(g) + H_2O(g)$$

For a given amount of ethyl alcohol, write the mole ratios that would enable you to calculate the number of moles of each product, as well as the number of moles of O_2 that would be required. Show how these mole ratios would be applied if 0.65 mol of ethyl alcohol is combusted.

11. In the practice of chemistry one of the most important calculations concerns the masses of products expected when particular masses of reactants are used in an experiment. For example, chemists judge the practicality and efficiency of a reaction by seeing how close the amount of product actually obtained is to the expected amount. Using a balanced chemical equation and an amount of starting material of your choice, summarize and illustrate the various steps needed in such a calculation for the expected amount of product.

12. What is meant by a *limiting reactant* in a particular reaction? In what way is the reaction "limited"? What does it mean to say that one or more of the reactants are present *in excess?* What happens to a reaction when the limiting reactant is used up?

13. For a balanced chemical equation of your choice, and using 25.0 g of each of the reactants in your equation, illustrate and explain how you would determine which reactant is the limiting reactant. Indicate *clearly* in your discussion how the choice of limiting reactant follows from your calculations.

14. What do we mean by the *theoretical yield* for a reaction? What is meant by the *actual yield?* Why might the actual yield for an experiment be *less* than the theoretical yield? Can the actual yield be *more* than the theoretical yield?

PROBLEMS

15. Consider 2.50-g samples of each of the following substances. Calculate the number of moles of each substance present in the sample.
    a. $Hg(l)$
    b. $CS_2(l)$
    c. $Al(NO_3)_3(s)$
    d. $C_3H_8(g)$
    e. $F_2(g)$
    f. $CaCO_3(s)$
    g. $CO(g)$
    h. $CO_2(g)$
    i. $CaCl_2(s)$
    j. $S_8(s)$

16. For the compounds in Question 15, calculate the percentage by mass of the element whose symbol occurs first in the compound's formula.

17. A compound was analyzed and was found to have the following percentage composition by mass: magnesium, 28.83%; carbon, 14.24%; oxygen, 56.93%. Calculate the empirical formula of the compound.

18. For each of the following balanced equations, calculate how many grams of each product would be formed if 25.0 g of the reactant listed first reacts completely with the second.
    a. $2AgNO_3(aq) + CaSO_4(aq) \rightarrow Ag_2SO_4(s) + Ca(NO_3)_2(aq)$
    b. $2Al(s) + 6HNO_3(aq) \rightarrow 2Al(NO_3)_3(aq) + 3H_2(g)$
    c. $H_3PO_4(aq) + 3NaOH(aq) \rightarrow Na_3PO_4(aq) + 3H_2O(l)$
    d. $CaO(s) + 2HCl(aq) \rightarrow CaCl_2(aq) + H_2O(l)$

19. For the reactions in Question 18, calculate the mass of each product formed if 12.5 g of the first reactant is combined with 10.0 g of the second reactant. Indicate which substance is the limiting reactant for each case.

20. Depending on the concentration of oxygen gas present, when carbon is burned, either of two oxides may result.

    $$2C(s) + O_2(g) \rightarrow 2CO(g)$$
    (restricted amount of oxygen)
    $$C(s) + O_2(g) \rightarrow CO_2(g)$$
    (unrestricted amount of oxygen)

    Suppose that experiments are performed in which duplicate 5.00-g samples of carbon are burned under both conditions. Calculate the theoretical yield of product for each experiment.

10 Modern Atomic Theory

The flames of metal salts are often brightly colored.

The concept of atoms is a very useful one. It explains many important observations, such as why compounds always have the same composition (a specific compound always contains the same types and numbers of atoms) and how chemical reactions occur (they involve a rearrangement of atoms).

Once chemists came to "believe" in atoms, a logical question followed: What are atoms like? What is the structure of an atom? In Chapter 4 we learned to picture the atom with a positively charged nucleus composed of protons and neutrons at its center and electrons moving around the nucleus in a space very large compared to the size of the nucleus.

In this chapter we will look at atomic structure in more detail. In particular, we will develop a picture of the electron arrangements in atoms—a picture that allows us to account for the chemistry of the various elements. Recall from our discussion of the periodic table in Chapter 4 that, although atoms exhibit a great variety of characteristics, certain elements can be grouped together because they behave similarly. For example, fluorine, chlorine, bromine, and iodine (the halogens) show great chemical similarities. Likewise, lithium, sodium, potassium, rubidium, and cesium (the alkali metals) exhibit many similar properties, and the noble gases (helium, neon, argon, krypton, xenon, and radon) are all very nonreactive. Although the members of each of these groups of elements show great similarity *within* the group, the differences in behavior *between* groups are striking. In this chapter we will see that it is the way the electrons are arranged in various atoms that accounts for these facts.

A neon sign celebrating Route 66.

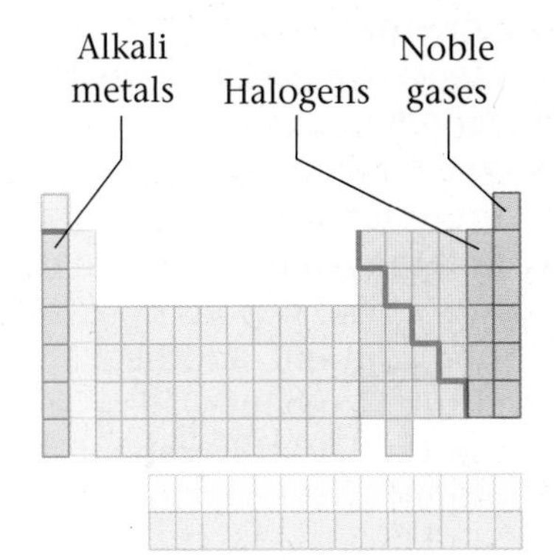

10.1 Rutherford's Atom

AIM: To describe Rutherford's model of the atom.

Remember that in Chapter 4 we discussed the idea that an atom has a small positive core (called the nucleus) with negatively charged electrons moving around the nucleus in some way (Figure 10.1). This concept of a *nuclear*

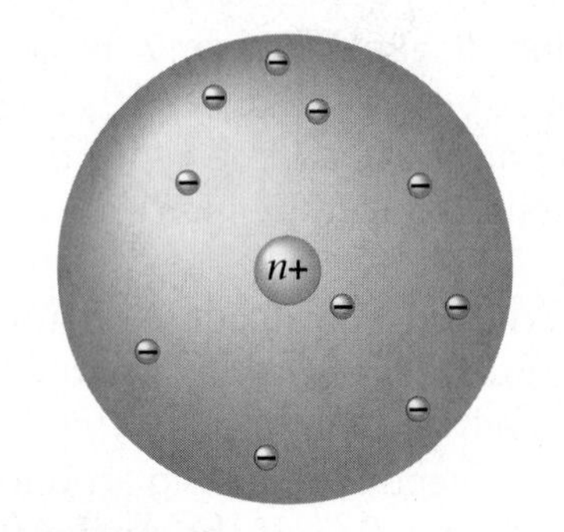

Figure 10.1
The Rutherford atom. The nuclear charge ($n+$) is balanced by the presence of n electrons moving in some way around the nucleus.

atom resulted from Ernest Rutherford's experiments in which he bombarded metal foil with α particles (see Section 4.5). Rutherford and his coworkers were able to show that the nucleus of the atom is composed of positively charged particles called *protons* and neutral particles called *neutrons*. Rutherford also found that the nucleus is apparently very small compared to the size of the entire atom. The electrons account for the rest of the atom.

A major question left unanswered by Rutherford's work was, What are the electrons doing? That is, how are the electrons arranged and how do they move? Rutherford suggested that electrons might revolve around the nucleus like the planets revolve around the sun in our solar system. He couldn't explain, however, why the negative electrons aren't attracted into the positive nucleus, causing the atom to collapse.

At this point it became clear that more observations of the properties of atoms were needed to understand the structure of the atom more fully. To help us understand these observations, we need to discuss the nature of light and how it transmits energy.

10.2 Electromagnetic Radiation

AIM: To explore the nature of electromagnetic radiation.

If you hold your hand a few inches from a brightly glowing light bulb, what do you feel? Your hand gets warm. The "light" from the bulb somehow transmits energy to your hand. The same thing happens if you move close to the glowing embers of wood in a fireplace—you receive energy that makes you feel warm. The energy you feel from the sun is a similar example.

Figure 10.2
A seagull floating on the ocean moves up and down as waves pass.

In all three of these instances, energy is being transmitted from one place to another by light—more properly called **electromagnetic radiation.** Many kinds of electromagnetic radiation exist, including the X rays used to make images of bones, the "white" light from a light bulb, the microwaves used to cook hot dogs and other food, and the radio waves that transmit voices and music. How do these various types of electromagnetic radiation differ from one another? To answer this question we need to talk about waves. To explore the characteristics of waves, let's think about ocean waves. In Figure 10.2 a seagull is shown floating on the ocean and being raised and lowered by the motion of the water surface as waves pass by. Notice that the gull just moves up and down as the waves pass—it is not moved forward. A particular wave is characterized by three properties: *wavelength, frequency,* and *speed.*

The **wavelength** (symbolized by the Greek letter lambda, λ) is the distance between two consecutive wave peaks (see Figure 10.3). The **frequency** of the wave (symbolized by the Greek letter nu, ν) indicates how many wave peaks pass a certain point per given time period. This idea can best be understood by thinking about how many times the seagull in Figure 10.2 goes up and down per minute. The *speed* of a wave indicates how fast a given peak travels through the water.

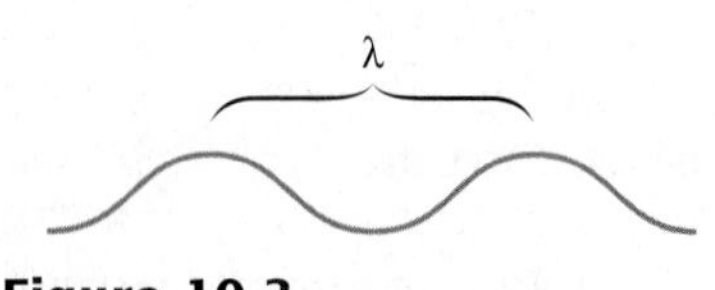

Figure 10.3
The wavelength of a wave is the distance between peaks.

Although it is more difficult to picture than water waves, light (electromagnetic radiation) also travels as waves. The various types of electromagnetic radiation (X rays, microwaves, and so on) differ in their wavelengths.

CHEMISTRY IN FOCUS

Light as a Sex Attractant

Parrots, which are renowned for their vibrant colors, apparently have a secret weapon that enhances their colorful appearance—a phenomenon called fluorescence. Fluorescence occurs when a substance absorbs ultraviolet (UV) light, which is invisible to the human eye, and converts it to visible light. This phenomenon is widely used in interior lighting in which long tubes are coated with a fluorescent substance. The fluorescent coating absorbs UV light (produced in the interior of the tube) and emits intense white light, which consists of all wavelengths of visible light.

Interestingly, scientists have recently shown that parrots have fluorescent feathers that are used to attract the opposite sex. Note in the accompanying photos that a bridgerigar parrot has certain feathers that produce fluorescence. Kathryn E. Arnold of the University of Glasgow in Scotland examined the skins of 700 Australian parrots from museum collections and found that the feathers that showed fluorescence were always display feathers—ones that were fluffed or waggled during courtship. To test her theory that fluorescence is a significant aspect of parrot romance, Arnold studied the behavior of a parrot toward birds of the opposite sex. In some cases, the potential mate had a UV-blocking substance applied to its feathers, blocking its fluorescence. Arnold's study revealed that parrots always preferred partners that showed fluorescence over those in which the fluorescence was blocked. Perhaps on your next date you might consider wearing a shirt with some fluorescent decoration!

The back and front of a bridgerigar parrot. In the photo at the right, the same parrot is seen under ultraviolet light.

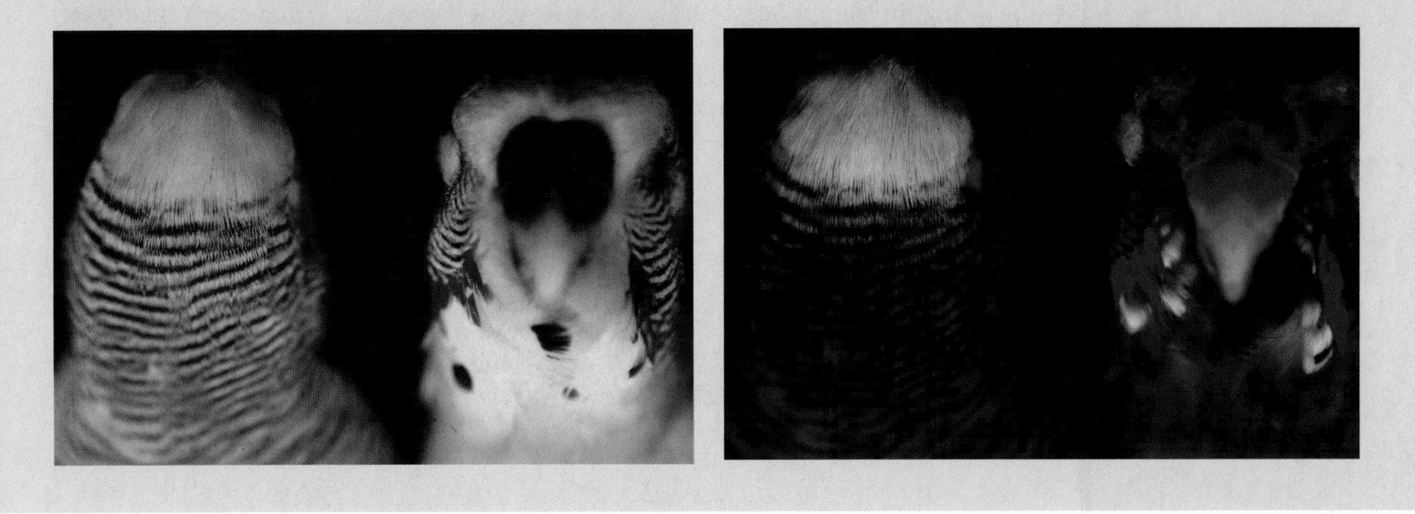

The classes of electromagnetic radiation are shown in Figure 10.4. Notice that X rays have very short wavelengths, whereas radiowaves have very long wavelengths.

Radiation provides an important means of energy transfer. For example, the energy from the sun reaches the earth mainly in the forms of visible and ultraviolet radiation. The glowing coals of a fireplace transmit heat energy by infrared radiation. In a microwave oven, the water molecules in food absorb microwave radiation, which increases their motions; this energy is then transferred to other types of molecules by collisions, increasing the food's temperature.

Thus we visualize electromagnetic radiation ("light") as a wave that carries energy through space. Sometimes, however, light doesn't behave as

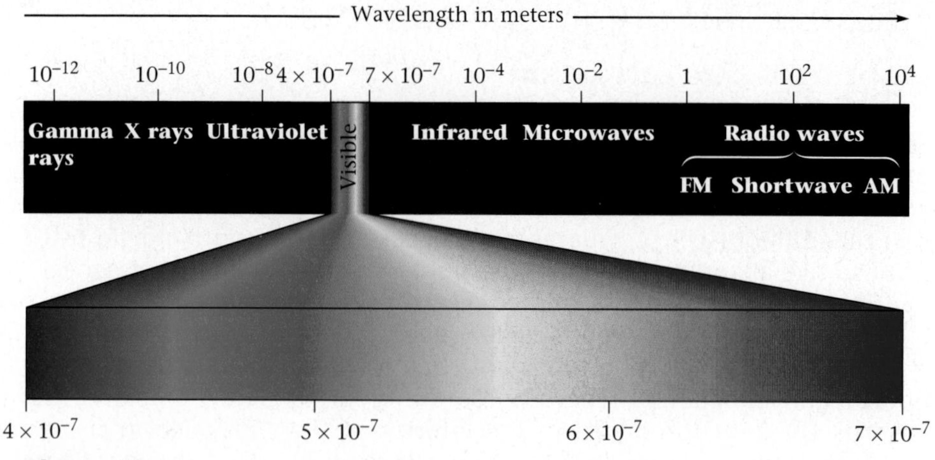

Figure 10.4
The different wavelengths of electromagnetic radiation.

though it were a wave. That is, electromagnetic radiation can sometimes have properties that are characteristic of particles. (You will learn more about this idea in later courses.) Another way to think of a beam of light traveling through space, then, is as a stream of tiny packets of energy called **photons.**

What is the exact nature of light? Does it consist of waves or is it a stream of particles of energy? It seems to be both (see Figure 10.5). This situation is often referred to as the wave–particle nature of light.

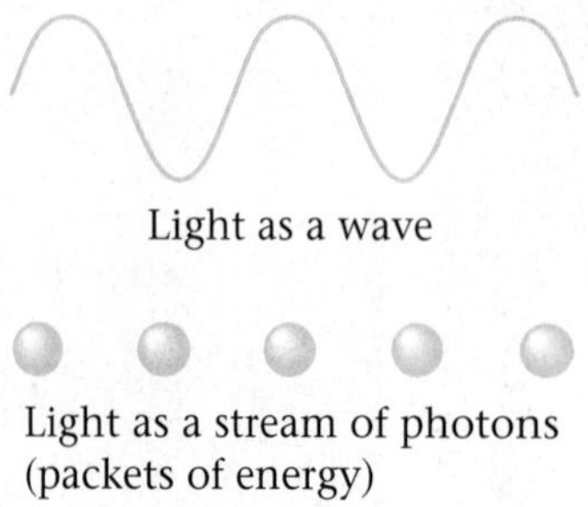

Different wavelengths of electromagnetic radiation carry different amounts of energy. For example, the photons that correspond to red light carry less energy than the photons that correspond to blue light. In general, the longer the wavelength of light, the lower the energy of its photons (see Figure 10.6).

Light as a wave

Light as a stream of photons (packets of energy)

Figure 10.5
Electromagnetic radiation (a beam of light) can be pictured in two ways: as a wave and as a stream of individual packets of energy called photons.

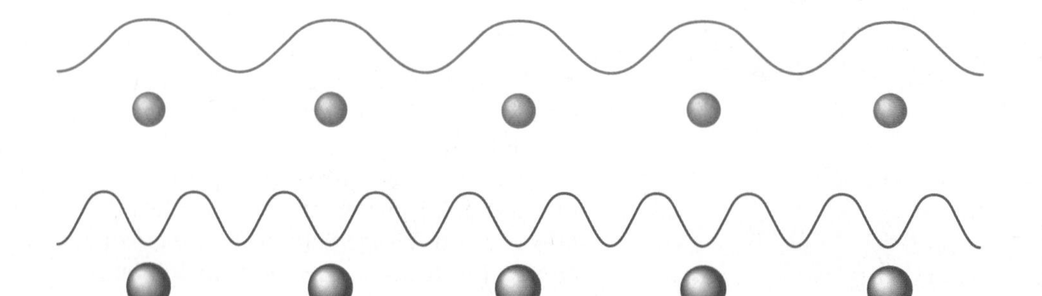

Figure 10.6
A photon of red light (relatively long wavelength) carries less energy than a photon of blue light (relatively short wavelength) does.

CHEMISTRY IN FOCUS

Atmospheric Effects

The gaseous atmosphere of the earth is crucial to life in many different ways. One of the most important characteristics of the atmosphere is the way its molecules absorb radiation from the sun.

If it weren't for the protective nature of the atmosphere, the sun would "fry" us with its high-energy radiation. We are protected by the atmospheric ozone, a form of oxygen consisting of O_3 molecules, which absorbs high-energy radiation and thus prevents it from reaching the earth. This explains why we are so concerned that chemicals released into the atmosphere are destroying this high-altitude ozone.

The atmosphere also plays a central role in controlling the earth's temperature, a phenomenon called the *greenhouse effect.* The atmospheric gases CO_2, H_2O, CH_4, N_2O, and others do not absorb light in the visible region. Therefore, the visible light from the sun passes through the atmosphere to warm the earth. In turn, the earth radiates this energy back toward space as infrared radiation. (For example, think of the heat radiated from black asphalt on a hot summer day.) But the gases listed earlier are strong *absorbers* of *infrared* waves, and they reradiate some of this energy back toward the earth as shown in Figure 10.7. Thus these gases act as an insulating blanket keeping the earth much warmer than it would be without them. (If these gases were not present, all of the heat the earth radiates would be lost into space.)

However, there is a problem. When we burn fossil fuels (coal, petroleum, and natural gas), one of the products is CO_2. Because we use such huge quantities of fossil fuels, the CO_2 content in the atmosphere is increasing gradually but significantly. This should cause the earth to get warmer, eventually changing the weather patterns on the earth's surface and melting the polar ice caps, which would flood many low-lying areas.

Because the natural forces that control the earth's temperature are not very well understood at this point, it is difficult to decide whether the greenhouse warming has already started. But many scientists think it has. For example, the 1980s and 1990s were among the warmest years the earth has experienced since people started keeping records.

The greenhouse effect is something we must watch closely. Controlling it may mean lowering our dependence on fossil fuels and increasing our reliance on nuclear, solar, or other power sources. In recent years, the trend has been in the opposite direction.

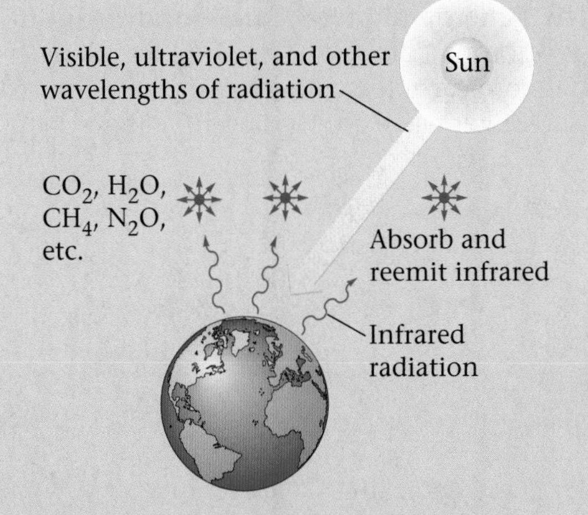

Figure 10.7
Certain of the gases in the earth's atmosphere reflect back some of the infrared (heat) radiation produced by the earth. This keeps the earth warmer than it would be otherwise.

A composite satellite image of the earth's biomass constructed from the radiation given off by living matter over a multiyear period.

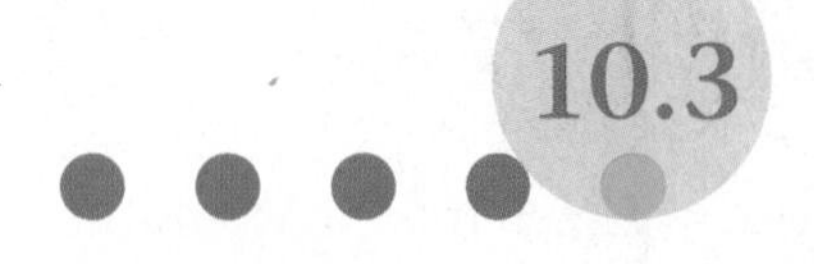

Emission of Energy by Atoms

AIM: To see how atoms emit light.

Consider the results of the experiment shown on the left. This experiment is run by dissolving compounds containing the Li^+ ion, the Cu^{2+} ion, and the Na^+ ion in separate dishes containing methyl alcohol (with a little water added to help dissolve the compounds). The solutions are then set on fire. Notice the brilliant colors that result. The solution containing Li^+ gives a beautiful, deep-red color, while the Cu^{2+} solution burns green. Notice that the Na^+ solution burns with a yellow–orange color, a color that should look familiar to you from the lights used in many parking lots. The color of these "sodium vapor lights" arises from the same source (the sodium atom) as the color of the burning solution containing Na^+ ions.

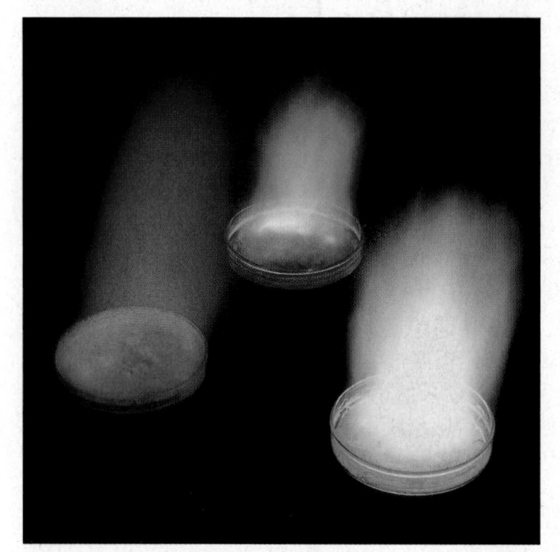

When salts containing Li^+, Cu^{2+}, and Na^+ dissolved in methyl alcohol are set on fire, brilliant colors result: Li^+, red; Cu^{2+}, green; and Na^+, yellow.

As we will see in more detail in the next section, the colors of these flames result from atoms in these solutions releasing energy by emitting visible light of specific wavelengths (that is, specific colors). The heat from the flame causes the atoms to absorb energy—we say that the atoms become *excited*. Some of this excess energy is then released in the form of light. The atom moves to a lower energy state as it emits a photon of light.

Lithium emits red light because its energy change corresponds to photons of red light (see Figure 10.8). Copper emits green light because it undergoes a different energy change than lithium; the energy change for copper corresponds to the energy of a photon of green light. Likewise, the energy change for sodium corresponds to a photon with a yellow–orange color.

To summarize, we have the following situation. When atoms receive energy from some source—they become excited—they can release this energy by emitting light. The emitted energy is carried away by a photon. Thus the energy of the photon corresponds exactly to the energy change experienced by the emitting atom. High-energy photons correspond to short-wavelength light and low-energy photons correspond to long-wavelength light. The photons of red light therefore carry less energy than the photons of blue light because red light has a longer wavelength than blue light does.

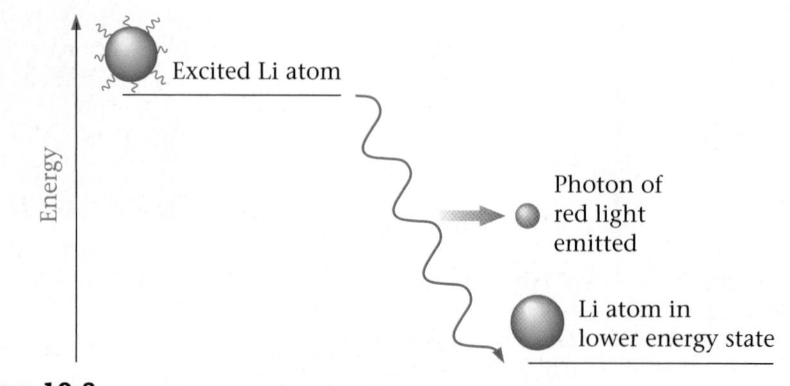

Figure 10.8
An excited lithium atom emitting a photon of red light to drop to a lower energy state.

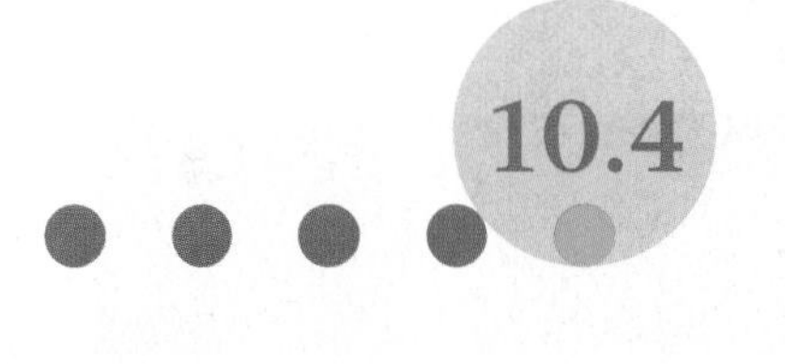

The Energy Levels of Hydrogen

AIM: To understand how the emission spectrum of hydrogen demonstrates the quantized nature of energy.

An atom can lose energy by *emitting* a photon.

As we learned in the last section, an atom with excess energy is said to be in an *excited state*. An excited atom can release some or all of its excess energy by emitting a photon (a "particle" of electromagnetic radiation) and thus move to a lower energy state. The lowest possible energy state of an atom is called its *ground state*.

We can learn a great deal about the energy states of hydrogen atoms by observing the photons they emit. To understand the significance of this, you need to remember that the *different wavelengths of light carry different amounts of energy per photon*. Recall that a beam of red light has lower-energy photons than a beam of blue light.

Each photon of blue light carries a larger quantity of energy than a photon of red light.

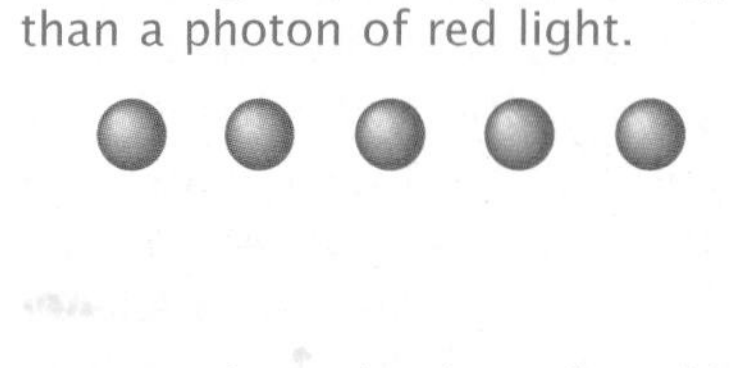

When a hydrogen atom absorbs energy from some outside source, it uses this energy to enter an excited state. It can release this excess energy (go back to a lower state) by emitting a photon of light (Figure 10.9). We can picture this process in terms of the energy-level diagram shown in Figure 10.10. The important point here is that *the energy contained in the photon corresponds to the change in energy that the atom experiences* in going from the excited state to the lower state.

A particular color (wavelength) of light carries a particular amount of energy per photon.

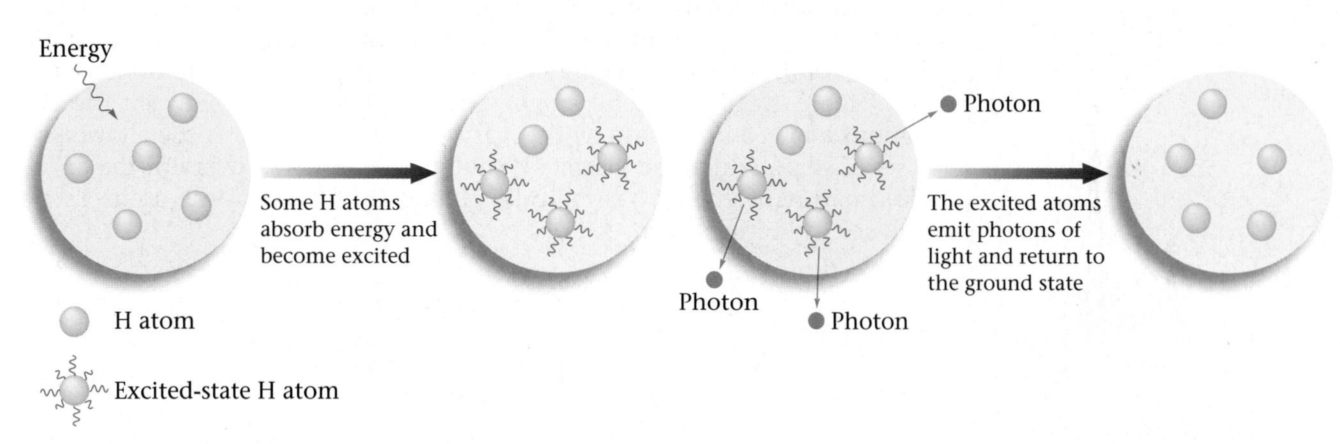

Figure 10.9
(a) A sample of H atoms receives energy from an external source, which causes some of the atoms to become excited (to possess excess energy).
(b) The excited atoms (H) can release the excess energy by emitting photons. The energy of each emitted photon corresponds exactly to the energy lost by each excited atom.

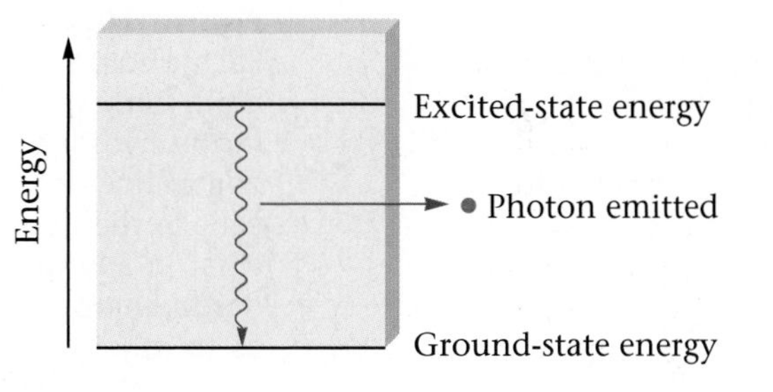

Figure 10.10
When an excited H atom returns to a lower energy level, it emits a photon that contains the energy released by the atom. Thus the energy of the photon corresponds to the difference in energy between the two states.

Figure 10.11
When excited hydrogen atoms return to lower energy states, they emit photons of certain energies, and thus certain colors. Shown here are the colors and wavelengths (in nanometers) of the photons in the visible region that are emitted by excited hydrogen atoms.

Consider the following experiment. Suppose we take a sample of H atoms and put a lot of energy into the system (as represented in Figure 10.9). When we study the photons of visible light emitted, we see only certain colors (Figure 10.11). That is, *only certain types of photons* are produced. We don't see all colors, which would add up to give "white light"; we see only selected colors. This is a very significant result. Let's discuss carefully what it means.

Because only certain photons are emitted, we know that only certain energy changes are occurring (Figure 10.12). This means that the hydrogen atom must have *certain discrete energy levels* (Figure 10.13). Excited hydrogen atoms *always* emit photons with the same discrete colors (wavelengths)—those shown in Figure 10.11. They *never* emit photons with energies (colors) in between those shown. So we can conclude that all hydrogen atoms have the same set of discrete energy levels. We say the energy levels of hydrogen are **quantized.** That is, only *certain values are allowed.* Scientists have found that the energy levels of *all* atoms are quantized.

The quantized nature of the energy levels in atoms was a surprise when scientists discovered it. It had been assumed previously that an atom could exist at any energy level. That is, everyone had assumed that atoms could have a continuous set of energy levels rather than only certain discrete values (Figure 10.14). A useful analogy here is the contrast between the elevations allowed by a ramp, which vary continuously, and those allowed by a set of steps, which are discrete (Figure 10.15). The discovery of the quantized nature of energy has radically changed our view of the atom, as we will see in the next few sections.

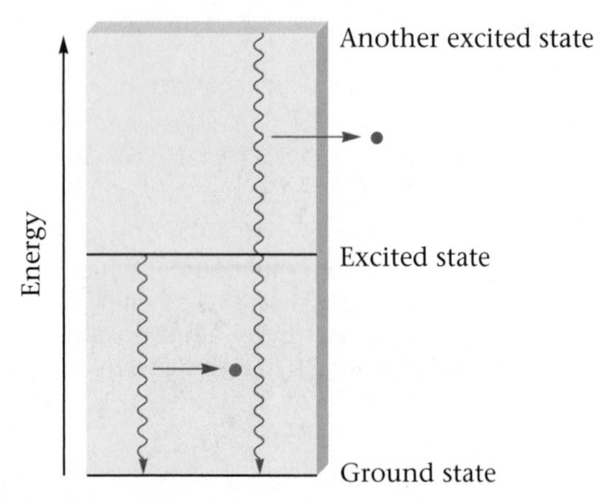

Figure 10.12
Hydrogen atoms have several excited-state energy levels. The color of the photon emitted depends on the energy change that produces it. A larger energy change may correspond to a blue photon, whereas a smaller change may produce a red photon.

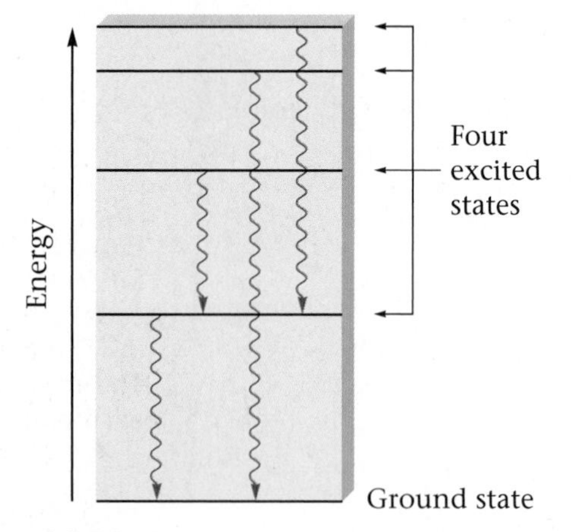

Figure 10.13
Each photon emitted by an excited hydrogen atom corresponds to a particular energy change in the hydrogen atom. In this diagram the horizontal lines represent discrete energy levels present in the hydrogen atom. A given H atom can exist in any of these energy states and can undergo energy changes to the ground state as well as to other excited states.

Figure 10.14
(a) Continuous energy levels. Any energy value is allowed. (b) Discrete (quantized) energy levels. Only certain energy states are allowed.

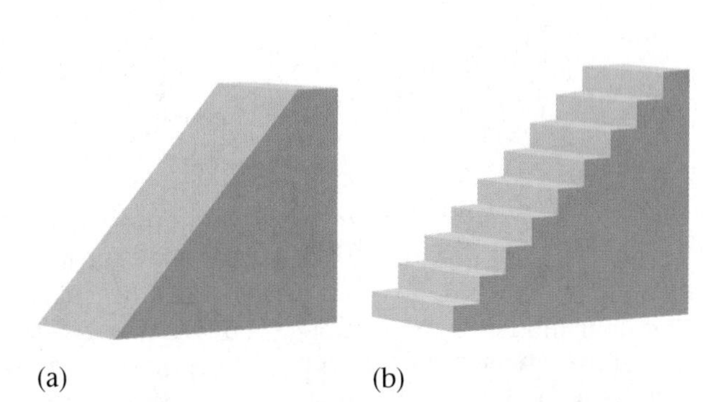

Figure 10.15
The difference between continuous and quantized energy levels can be illustrated by comparing a flight of stairs with a ramp. (a) A ramp varies continuously in elevation. (b) A flight of stairs allows only certain elevations; the elevations are quantized.

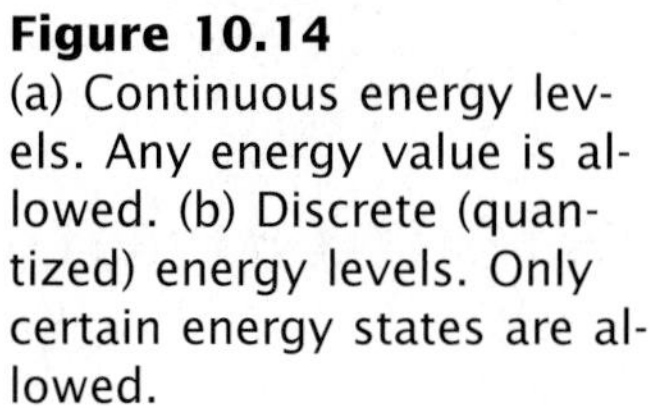

10.5 The Bohr Model of the Atom

AIM: To learn about Bohr's model of the hydrogen atom.

In 1911 at the age of twenty-five, Niels Bohr (Figure 10.16) received his Ph.D. in physics. He was convinced that the atom could be pictured as a small positive nucleus with electrons orbiting around it.

Over the next two years, Bohr constructed a model of the hydrogen atom with quantized energy levels that agreed with the hydrogen emission results we have just discussed. Bohr pictured the electron moving in circu-

Figure 10.16
Niels Hendrik David Bohr (1885–1962) as a boy lived in the shadow of his younger brother Harald, who played on the 1908 Danish Olympic Soccer Team and later became a distinguished mathematician. In school, Bohr received his poorest marks in composition and struggled with writing during his entire life. In fact, he wrote so poorly that he was forced to dictate his Ph.D. thesis to his mother. He is one of the very few people who felt the need to write rough drafts of postcards. Nevertheless, Bohr was a brilliant physicist. After receiving his Ph.D. in Denmark, he constructed a quantum model for the hydrogen atom by the time he was 27. Even though his model later proved to be incorrect, Bohr remained a central figure in the drive to understand the atom. He was awarded the Nobel Prize in physics in 1922.

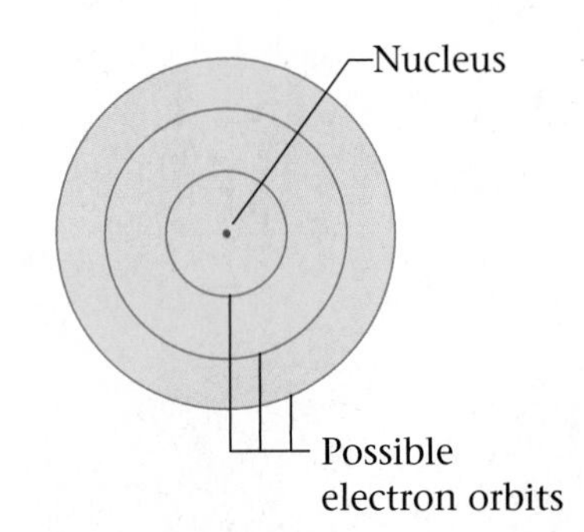

Figure 10.17
The Bohr model of the hydrogen atom represented the electron as restricted to certain circular orbits around the nucleus.

lar orbits corresponding to the various allowed energy levels. He suggested that the electron could jump to a different orbit by absorbing or emitting a photon of light with exactly the correct energy content. Thus, in the Bohr atom, the energy levels in the hydrogen atom represented certain allowed circular orbits (Figure 10.17).

At first Bohr's model appeared very promising. It fit the hydrogen atom very well. However, when this model was applied to atoms other than hydrogen, it did not work. In fact, further experiments showed that the Bohr model is fundamentally incorrect. Although the Bohr model paved the way for later theories, it is important to realize that the current theory of atomic structure is not the same as the Bohr model. Electrons do *not* move around the nucleus in circular orbits like planets orbiting the sun. Surprisingly, as we shall see later in this chapter, we do not know exactly how the electrons move in an atom.

10.6 The Wave Mechanical Model of the Atom

AIM: To understand how the electron's position is represented in the wave mechanical model.

Louis Victor de Broglie

By the mid-1920s it had become apparent that the Bohr model was incorrect. Scientists needed to pursue a totally new approach. Two young physicists, Louis Victor de Broglie from France and Erwin Schrödinger from Austria, suggested that because light seems to have both wave and particle characteristics (it behaves simultaneously as a wave and as a stream of particles), the electron might also exhibit both of these characteristics. Although everyone had assumed that the electron was a tiny particle, these scientists said it might be useful to find out whether it could be described as a wave.

When Schrödinger carried out a mathematical analysis based on this idea, he found that it led to a new model for the hydrogen atom that seemed to apply equally well to other atoms—something Bohr's model failed to do. We will now explore a general picture of this model, which is called the **wave mechanical model** of the atom.

In the Bohr model, the electron was assumed to move in circular orbits. In the wave mechanical model, on the other hand, the electron states are described by orbitals. *Orbitals are nothing like orbits.* To approximate the idea of an orbital, picture a single male firefly in a room in the center of which an open vial of female sex-attractant hormones is suspended. The room is extremely dark and there is a camera in one corner with its shutter open. Every time the firefly "flashes," the camera records a pinpoint of light and thus the firefly's position in the room at that moment. The firefly senses the sex attractant, and as you can imagine, it spends a lot of time at or close to it. However, now and then the insect flies randomly around the room.

When the film is taken out of the camera and developed, the picture will probably look like Figure 10.18. Because a picture is brightest where the

Figure 10.18
A representation of the photo of the firefly experiment. Remember that a picture is brightest where the film has been exposed to the most light. Thus the intensity of the color reflects how often the firefly visited a given point in the room. Notice that the brightest area is in the center of the room near the source of the sex attractant.

Figure 10.19
The probability map, or orbital, that describes the hydrogen electron in its lowest possible energy state. The more intense the color of a given dot, the more likely it is that the electron will be found at that point. We have no information about when the electron will be at a particular point or about how it moves. Note that the probability of the electron's presence is highest closest to the positive nucleus (located at the center of this diagram), as might be expected.

film has been exposed to the most light, the color intensity at any given point tells us how often the firefly visited a given point in the room. Notice that, as we might expect, the firefly spent the most time near the room's center.

Now suppose you are watching the firefly in the dark room. You see it flash at a given point far from the center of the room. Where do you expect to see it next? There is really no way to be sure. The firefly's flight path is not precisely predictable. However, if you had seen the time-exposure picture of the firefly's activities (Figure 10.18), you would have some idea where to look next. Your best chance would be to look more toward the center of the room. Figure 10.18 suggests there is the highest probability (the highest odds, the greatest likelihood) of finding the firefly at any particular moment near the center of the room. You *can't be sure* the firefly will fly toward the center of the room, but it *probably* will. So the time-exposure picture is a kind of "probability map" of the firefly's flight pattern.

According to the wave mechanical model, the electron in the hydrogen atom can be pictured as being something like this firefly. Schrödinger found that he could not precisely describe the electron's path. His mathematics enabled him only to predict the probabilities of finding the electron at given points in space around the nucleus. In its ground state the hydrogen electron has a probability map like that shown in Figure 10.19. The more intense the color at a particular point, the more probable that the electron will be found at that point at a given instant. The model gives *no information about when* the electron occupies a certain point in space or *how it moves.* In fact, we have good reasons to believe that we can *never know* the details of electron motion, no matter how sophisticated our models may become. But one thing we feel confident about is that the electron *does not* orbit the nucleus in circles as Bohr suggested.

10.7 The Hydrogen Orbitals

AIM: To learn about the shapes of orbitals designated by *s*, *p*, and *d*.

The probability map for the hydrogen electron shown in Figure 10.19 is called an **orbital.** Although the probability of finding the electron decreases at greater distances from the nucleus, the probability of finding it even at great distances from the nucleus never becomes exactly zero. A useful analogy might be the lack of a sharp boundary between the earth's atmosphere and "outer space." The atmosphere fades away gradually, but there are always a few molecules present. Because the edge of an orbital is "fuzzy," an orbital does not have an exactly defined size. So chemists arbitrarily define its size as the sphere that contains 90% of the total electron probability (Figure 10.20b). This means that the electron spends 90% of the time inside this surface and 10% somewhere outside this surface. (Note that we are *not* saying the electron travels only on the *surface* of the sphere.) The orbital represented in Figure 10.20 is named the **1*s* orbital,** and it describes the hydrogen electron's lowest energy state (the ground state).

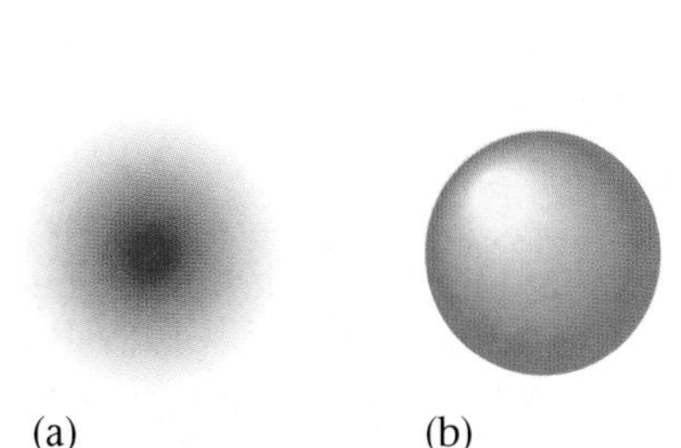

Figure 10.20
(a) The hydrogen 1*s* orbital.
(b) The size of the orbital is defined by a sphere that contains 90% of the total electron probability. That is, the electron can be found *inside* this sphere 90% of the time. The 1*s* orbital is often represented simply as a sphere. However, the most accurate picture of the orbital is the probability map represented in (a).

In Section 10.4 we saw that the hydrogen atom can absorb energy to transfer the electron to a higher energy state (an excited state). In terms of the obsolete Bohr model, this meant the electron was transferred to an orbit with a larger radius. In the wave mechanical model, these higher energy states correspond to different kinds of orbitals with different shapes.

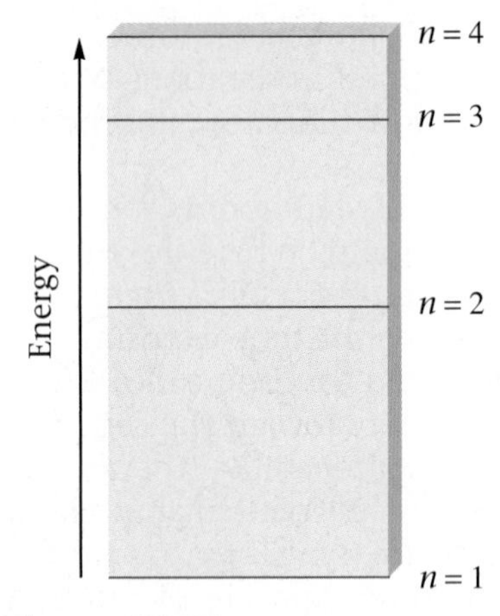

Figure 10.21
The first four principal energy levels in the hydrogen atom. Each level is assigned an integer, n.

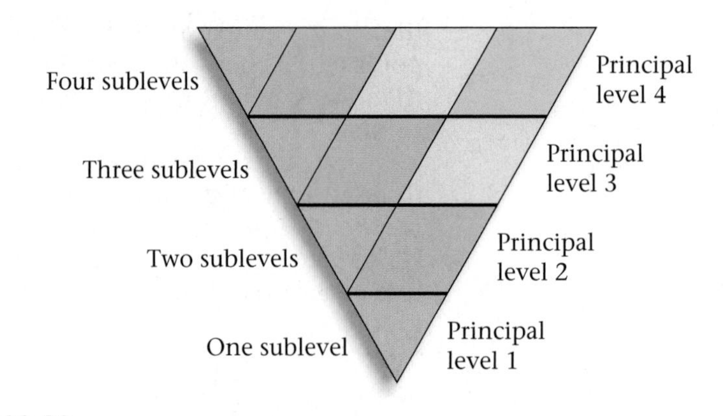

Figure 10.22
An illustration of how principal levels can be divided into sublevels.

At this point we need to stop and consider how the hydrogen atom is organized. Remember, we showed earlier that the hydrogen atom has discrete energy levels. We call these levels **principal energy levels** and label them with integers (Figure 10.21). Next we find that each of these levels is subdivided into **sublevels.** The following analogy should help you understand this. Picture an inverted triangle (Figure 10.22). We divide the principal levels into various numbers of sublevels. Principal level 1 consists of one sublevel, principal level 2 has two sublevels, principal level 3 has three sublevels, and principal level 4 has four sublevels.

Like our triangle, the principal energy levels in the hydrogen atom contain sublevels. As we will see presently, these sublevels contain spaces for the electron that we call orbitals. Principal energy level 1 consists of just one sublevel, or one type of orbital. The spherical shape of this orbital is shown in Figure 10.20. We label this orbital 1*s*. The number 1 is for the principal energy level, and *s* is a shorthand way to label a particular sublevel (type of orbital).

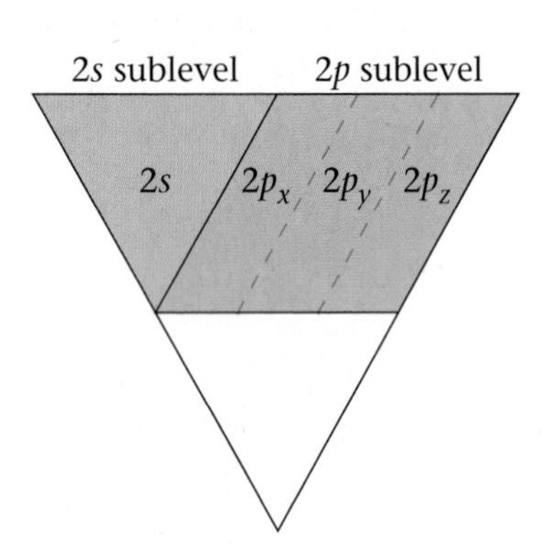

Figure 10.23
Principal level 2 shown divided into the 2*s* and 2*p* sublevels.

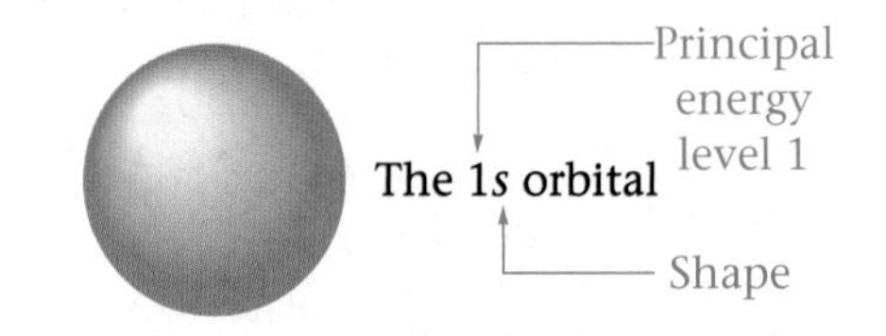

Principal energy level 2 has two sublevels. (Note the correspondence between the principal energy level number and the number of sublevels.) These sublevels are labeled 2*s* and 2*p*. The 2*s* sublevel consists of one orbital (called the 2*s*), and the 2*p* sublevel consists of three orbitals (called $2p_x$, $2p_y$, and $2p_z$). Let's return to the inverted triangle to illustrate this. Figure 10.23 shows principal level 2 divided into the sublevels 2*s* and 2*p* (which is subdivided into $2p_x$, $2p_y$, and $2p_z$). The orbitals have the shapes shown in Figures 10.24 and 10.25. The 2*s* orbital is spherical like the 1*s* orbital but larger in size (see Figure 10.24). The three 2*p* orbitals are not spherical but have two "lobes." These orbitals are shown in Figure 10.25 both as electron probability maps and as surfaces that contain 90% of the total electron probability. Notice that the label *x*, *y*, or *z* on a given 2*p* orbital tells along which axis the lobes of that orbital are directed.

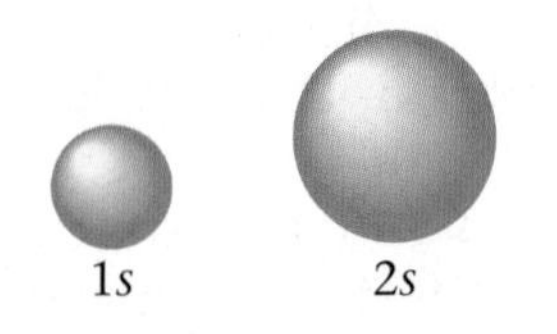

Figure 10.24
The relative sizes of the 1*s* and 2*s* orbitals of hydrogen.

What we have learned so far about the hydrogen atom is summarized in Figure 10.26. Principal energy level 1 has one sublevel, which contains the 1*s* orbital. Principal energy level 2 contains two sublevels, one of which contains the 2*s* orbital and one of which contains the 2*p* orbitals (three of them). Note that each orbital is designated by a symbol or label. We summarize the information given by this label in the following box.

Orbital Labels

1. The number tells the principal energy level.
2. The letter tells the shape. The letter *s* means a spherical orbital; the letter *p* means a two-lobed orbital. The *x*, *y*, or *z* subscript on a *p* orbital label tells along which of the coordinate axes the two lobes lie.

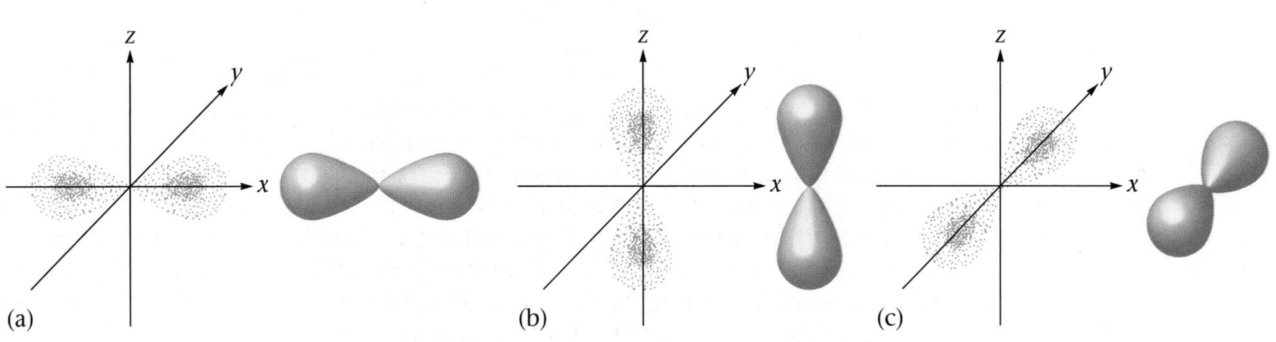

Figure 10.25
The three 2*p* orbitals: (a) $2p_x$, (b) $2p_z$, (c) $2p_y$. The *x*, *y*, or *z* label indicates along which axis the two lobes are directed. Each orbital is shown both as a probability map and as a surface that encloses 90% of the electron probability.

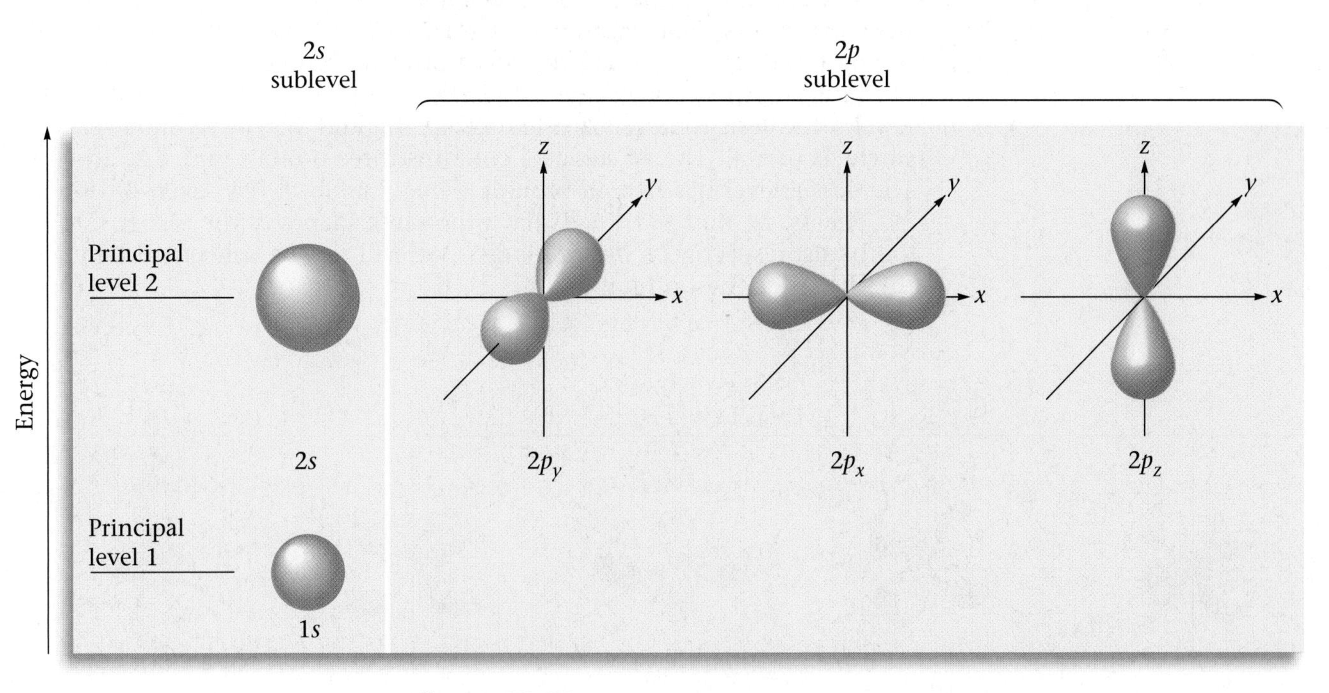

Figure 10.26
A diagram of principal energy levels 1 and 2 showing the shapes of orbitals that compose the sublevels.

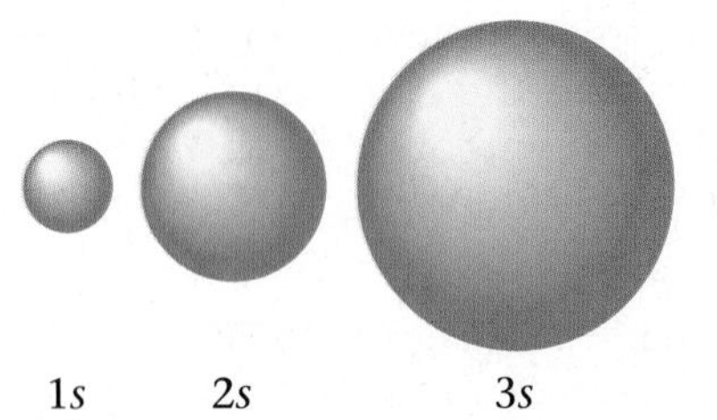

Figure 10.27
The relative sizes of the spherical 1*s*, 2*s*, and 3*s* orbitals of hydrogen.

One important characteristic of orbitals is that as the level number increases, the average distance of the electron in that orbital from the nucleus also increases. That is, when the hydrogen electron is in the 1*s* orbital (the ground state), it spends most of its time much closer to the nucleus than when it occupies the 2*s* orbital (an excited state).

You may be wondering at this point why hydrogen, which has only one electron, has more than one orbital. It is best to think of an orbital as a *potential space* for an electron. The hydrogen electron can occupy only a single orbital at a time, but the other orbitals are still available should the electron be transferred into one of them. For example, when a hydrogen atom is in its ground state (lowest possible energy state), the electron is in the 1*s* orbital. By adding the correct amount of energy (for example, a specific photon of light), we can excite the electron to the 2*s* orbital or to one of the 2*p* orbitals.

So far we have discussed only two of hydrogen's energy levels. There are many others. For example, level 3 has three sublevels (see Figure 10.22), which we label 3*s*, 3*p*, and 3*d*. The 3*s* sublevel contains a single 3*s* orbital, a spherical orbital larger than 1*s* and 2*s* (Figure 10.27). Sublevel 3*p* contains three orbitals: $3p_x$, $3p_y$, and $3p_z$, which are shaped like the 2*p* orbitals except that they are larger. The 3*d* sublevel contains five 3*d* orbitals with the shapes and labels shown in Figure 10.28. (You do not need to memorize the 3*d* orbital shapes and labels. They are shown for completeness.)

Notice as you compare levels 1, 2, and 3 that a new type of orbital (sublevel) is added in each principal energy level. (Recall that the *p* orbitals are added in level 2 and the *d* orbitals in level 3.) This makes sense because in going farther out from the nucleus, there is more space available and thus room for more orbitals.

It might help you to understand that the number of orbitals increases with the principal energy level if you think of a theater in the round. Picture a round stage with circular rows of seats surrounding it. The farther from the stage a row of seats is, the more seats it contains because the circle is larger. Orbitals divide up the space around a nucleus somewhat like the seats in this circular theater. The greater the distance from the nucleus, the more space there is and the more orbitals we find.

The pattern of increasing numbers of orbitals continues with level 4. Level 4 has four sublevels labeled 4*s*, 4*p*, 4*d*, and 4*f*. The 4*s* sublevel has a single 4*s* orbital. The 4*p* sublevel contains three orbitals ($4p_x$, $4p_y$, and $4p_z$). The 4*d* sublevel has five 4*d* orbitals. The 4*f* sublevel has seven 4*f* orbitals.

The 4*s*, 4*p*, and 4*d* orbitals have the same shapes as the earlier *s*, *p*, and *d* orbitals, respectively, but are larger. We will not be concerned here with the shapes of the *f* orbitals.

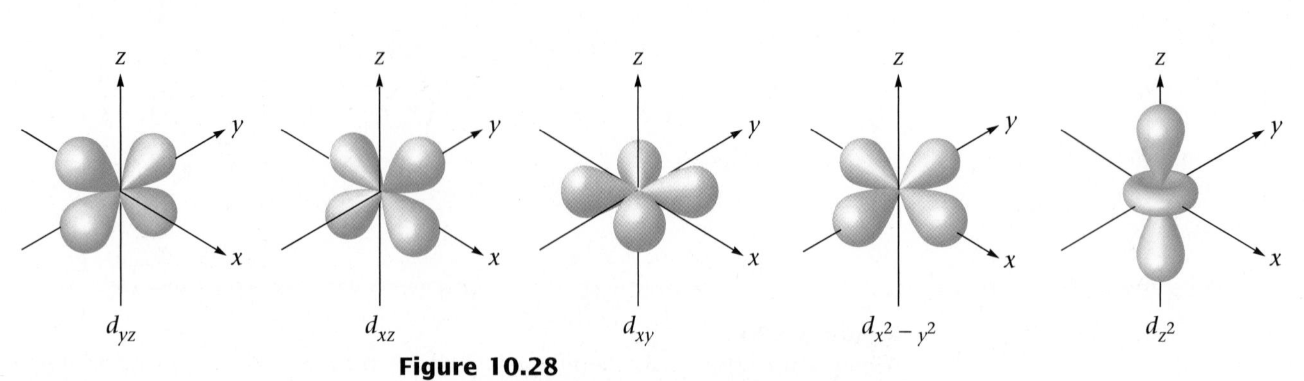

Figure 10.28
The shapes and labels of the five 3*d* orbitals.

The Wave Mechanical Model: Further Development

AIMS: To review the energy levels and orbitals of the wave mechanical model of the atom. To learn about electron spin.

A model for the atom is of little use if it does not apply to all atoms. The Bohr model was discarded because it could be applied only to hydrogen. The wave mechanical model can be applied to all atoms in basically the same form as we have just used it for hydrogen. In fact, the major triumph of this model is its ability to explain the periodic table of the elements. Recall that the elements on the periodic table are arranged in vertical groups, which contain elements that typically show similar chemical properties. The wave mechanical model of the atom allows us to explain, based on electron arrangements, why these similarities occur. We will see in due time how this is done.

Remember that an atom has as many electrons as it has protons to give it a zero overall charge. Therefore, all atoms beyond hydrogen have more than one electron. Before we can consider the atoms beyond hydrogen, we must describe one more property of electrons that determines how they can be arranged in an atom's orbitals. This property is spin. Each electron appears to be spinning as a top spins on its axis. Like the top, an electron can spin only in one of two directions. We often represent spin with an arrow: either ↑ or ↓. One arrow represents the electron spinning in the one direction, and the other represents the electron spinning in the opposite direction. For our purposes, what is most important about electron spin is that two electrons must have *opposite* spins to occupy the same orbital. That is, two electrons that have the same spin cannot occupy the same orbital. This leads to the **Pauli exclusion principle:** an atomic orbital can hold a maximum of two electrons, and those two electrons must have opposite spins.

Before we apply the wave mechanical model to atoms beyond hydrogen, we will summarize the model for convenient reference.

Principal Components of the Wave Mechanical Model of the Atom

1. Atoms have a series of energy levels called **principal energy levels,** which are designated by whole numbers symbolized by n; n can equal 1, 2, 3, 4, . . . Level 1 corresponds to $n = 1$, level 2 corresponds to $n = 2$, and so on.
2. The energy of the level increases as the value of n increases.
3. Each principal energy level contains one or more *types* of orbitals, called **sublevels.**
4. The number of sublevels present in a given principal energy level equals n. For example, level 1 contains one sublevel (1*s*); level 2 contains two sublevels (two types of orbitals), the 2*s* orbital and the three 2*p* orbitals; and so on. These are summarized in the following table. The number of each type of orbital is shown in parentheses.

(continued)

n	Sublevels (Types of Orbitals) Present
1	1*s*(1)
2	2*s*(1) 2*p*(3)
3	3*s*(1) 3*p*(3) 3*d*(5)
4	4*s*(1) 4*p*(3) 4*d*(5) 4*f*(7)

5. The *n* value is always used to label the orbitals of a given principal level and is followed by a letter that indicates the type (shape) of the orbital. For example, the designation 3*p* means an orbital in level 3 that has two lobes (a *p* orbital always has two lobes).
6. An orbital can be empty or it can contain one or two electrons, but never more than two. If two electrons occupy the same orbital, they must have opposite spins.
7. The shape of an orbital does not indicate the details of electron movement. It indicates the probability distribution for an electron residing in that orbital.

Example 10.1 Understanding the Wave Mechanical Model of the Atom

Indicate whether each of the following statements about atomic structure is true or false.

a. An *s* orbital is always spherical in shape.
b. The 2*s* orbital is the same size as the 3*s* orbital.
c. The number of lobes on a *p* orbital increases as *n* increases. That is, a 3*p* orbital has more lobes than a 2*p* orbital.
d. Level 1 has one *s* orbital, level 2 has two *s* orbitals, level 3 has three *s* orbitals, and so on.
e. The electron path is indicated by the surface of the orbital.

Solution

a. True. The size of the sphere increases as *n* increases, but the shape is always spherical.
b. False. The 3*s* orbital is larger (the electron is farther from the nucleus on average) than the 2*s* orbital.
c. False. A *p* orbital always has two lobes.
d. False. Each principal energy level has only one *s* orbital.
e. False. The electron is *somewhere inside* the orbital surface 90% of the time. The electron does not move around *on* this surface.

✔ Self-Check Exercise 10.1

Define the following terms.

a. Bohr orbits
b. orbitals
c. orbital size
d. sublevel

See Problems 10.37 through 10.44. ■

10.9 Electron Arrangements in the First Eighteen Atoms on the Periodic Table

AIMS: To understand how the principal energy levels fill with electrons in atoms beyond hydrogen. To learn about valence electrons and core electrons.

We will now describe the electron arrangements in atoms with $Z = 1$ to $Z = 18$ by placing electrons in the various orbitals in the principal energy levels, starting with $n = 1$, and then continuing with $n = 2$, $n = 3$, and so on. For the first eighteen elements, the individual sublevels fill in the following order: 1*s*, then 2*s*, then 2*p*, then 3*s*, then 3*p*.

The most attractive orbital to an electron in an atom is always the 1*s*, because in this orbital the negatively charged electron is closer to the positively charged nucleus than in any other orbital. That is, the 1*s* orbital involves the space around the nucleus that is closest to the nucleus. As *n* increases, the orbital becomes larger—the electron, on average, occupies space farther from the nucleus.

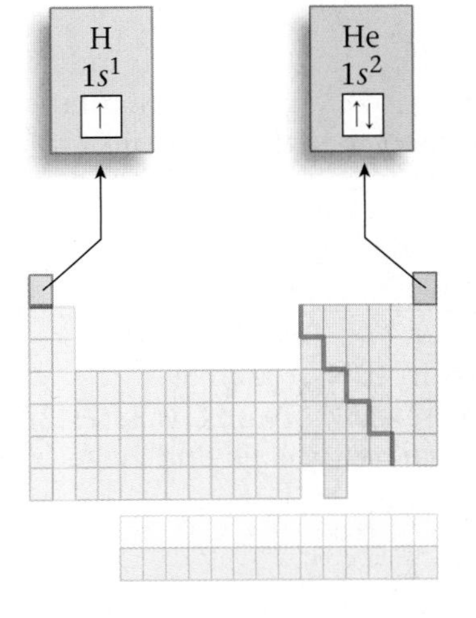

So in its ground state hydrogen has its lone electron in the 1*s* orbital. This is commonly represented in two ways. First, we say that hydrogen has the electron arrangement, or **electron configuration,** $1s^1$. This just means there is one electron in the 1*s* orbital. We can also represent this configuration by using an **orbital diagram,** also called a **box diagram,** in which orbitals are represented by boxes grouped by sublevel with small arrows indicating the electrons. For *hydrogen,* the electron configuration and box diagram are

The arrow represents an electron spinning in a particular direction. The next element is *helium,* $Z = 2$. It has two protons in its nucleus and so has two electrons. Because the 1*s* orbital is the most desirable, both electrons go there but with opposite spins. For helium, the electron configuration and box diagram are

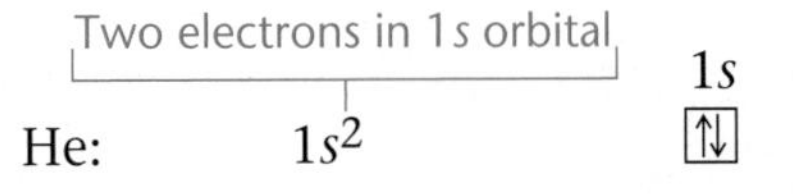

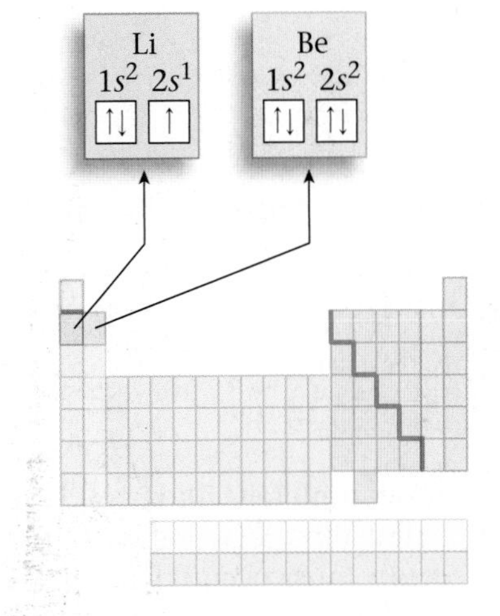

The opposite electron spins are shown by the opposing arrows in the box.

Lithium ($Z = 3$) has three electrons, two of which go into the 1*s* orbital. That is, two electrons fill that orbital. The 1*s* orbital is the only orbital for $n = 1$, so the third electron must occupy an orbital with $n = 2$—in this case the 2*s* orbital. This gives a $1s^22s^1$ configuration. The electron configuration and box diagram are

Li: $1s^22s^1$ 1s 2s

The next element, *beryllium,* has four electrons, which occupy the 1*s* and 2*s* orbitals with opposite spins.

		1*s*	2*s*
Be:	$1s^22s^2$	↑↓	↑↓

Boron has five electrons, four of which occupy the 1*s* and 2*s* orbitals. The fifth electron goes into the second type of orbital with $n = 2$, one of the 2*p* orbitals.

		1*s*	2*s*	2*p*
B:	$1s^22s^22p^1$	↑↓	↑↓	↑ | |

Because all the 2*p* orbitals have the same energy, it does not matter which 2*p* orbital the electron occupies.

Carbon, the next element, has six electrons: two electrons occupy the 1*s* orbital, two occupy the 2*s* orbital, and two occupy 2*p* orbitals. There are three 2*p* orbitals, so each of the mutually repulsive electrons occupies a different 2*p* orbital. For reasons we will not consider, in the separate 2*p* orbitals the electrons have the same spin.

B Group 3
C Group 4
N Group 5

The configuration for carbon could be written $1s^22s^22p^12p^1$ to indicate that the electrons occupy separate 2*p* orbitals. However, the configuration is usually given as $1s^22s^22p^2$, and it is understood that the electrons are in different 2*p* orbitals.

		1*s*	2*s*	2*p*
C:	$1s^22s^22p^2$	↑↓	↑↓	↑ | ↑ |

Note the like spins for the unpaired electrons in the 2*p* orbitals.

The configuration for *nitrogen,* which has seven electrons, is $1s^22s^22p^3$. The three electrons in 2*p* orbitals occupy separate orbitals and have like spins.

		1*s*	2*s*	2*p*
N:	$1s^22s^22p^3$	↑↓	↑↓	↑ | ↑ | ↑

The configuration for *oxygen,* which has eight electrons, is $1s^22s^22p^4$. One of the 2*p* orbitals is now occupied by a pair of electrons with opposite spins, as required by the Pauli exclusion principle.

		1*s*	2*s*	2*p*
O:	$1s^22s^22p^4$	↑↓	↑↓	↑↓ | ↑ | ↑

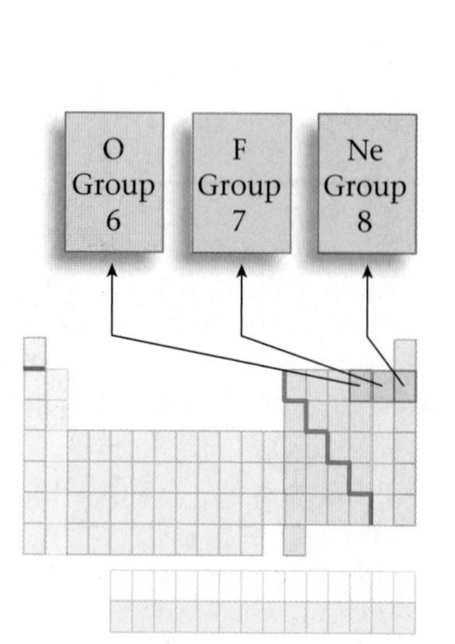

The electron configurations and orbital diagrams for *fluorine* (nine electrons) and *neon* (ten electrons) are

		1*s*	2*s*	2*p*
F:	$1s^22s^22p^5$	↑↓	↑↓	↑↓ | ↑↓ | ↑
Ne:	$1s^22s^22p^6$	↑↓	↑↓	↑↓ | ↑↓ | ↑↓

With neon, the orbitals with $n = 1$ and $n = 2$ are completely filled.

For *sodium,* which has eleven electrons, the first ten electrons occupy the 1*s*, 2*s*, and 2*p* orbitals, and the eleventh electron must occupy the first orbital with $n = 3$, the 3*s* orbital. The electron configuration for sodium is $1s^22s^22p^63s^1$. To avoid writing the inner-level electrons, we often abbreviate the configuration $1s^22s^22p^63s^1$ as $[Ne]3s^1$, where [Ne] represents the electron configuration of neon, $1s^22s^22p^6$.

H $1s^1$							He $1s^2$
Li $2s^1$	Be $2s^2$	B $2p^1$	C $2p^2$	N $2p^3$	O $2p^4$	F $2p^5$	Ne $2p^6$
Na $3s^1$	Mg $3s^2$	Al $3p^1$	Si $3p^2$	P $3p^3$	S $3p^4$	Cl $3p^5$	Ar $3p^6$

Figure 10.29
The electron configurations in the sublevel last occupied for the first eighteen elements.

The orbital diagram for sodium is

$$\begin{array}{cccc} 1s & 2s & 2p & 3s \\ \boxed{\uparrow\downarrow} & \boxed{\uparrow\downarrow} & \boxed{\uparrow\downarrow}\boxed{\uparrow\downarrow}\boxed{\uparrow\downarrow} & \boxed{\uparrow} \end{array}$$

The next element, *magnesium*, $Z = 12$, has the electron configuration $1s^22s^22p^63s^2$, or $[Ne]3s^2$.

The next six elements, *aluminum* through *argon*, have electron configurations obtained by filling the $3p$ orbitals one electron at a time. Figure 10.29 summarizes the electron configurations of the first eighteen elements by giving the number of electrons in the type of orbital (sublevel) occupied last.

Example 10.2 Writing Orbital Diagrams

Write the orbital diagram for magnesium.

Solution

Magnesium ($Z = 12$) has twelve electrons that are placed successively in the $1s$, $2s$, $2p$, and $3s$ orbitals to give the electron configuration $1s^22s^22p^63s^2$. The orbital diagram is

$$\begin{array}{cccc} 1s & 2s & 2p & 3s \\ \boxed{\uparrow\downarrow} & \boxed{\uparrow\downarrow} & \boxed{\uparrow\downarrow}\boxed{\uparrow\downarrow}\boxed{\uparrow\downarrow} & \boxed{\uparrow\downarrow} \end{array}$$

Only occupied orbitals are shown here.

✔ Self-Check Exercise 10.2

Write the complete electron configuration and the orbital diagram for each of the elements aluminum through argon.

See Problems 10.49 through 10.54. ■

At this point it is useful to introduce the concept of **valence electrons**—that is, *the electrons in the outermost (highest) principal energy level of an atom.* For example, nitrogen, which has the electron configuration $1s^22s^22p^3$, has electrons in principal levels 1 and 2. Therefore, level 2 (which has $2s$ and $2p$ sublevels) is the valence level of nitrogen, and the $2s$ and $2p$ electrons are the valence electrons. For the sodium atom (electron configuration $1s^22s^22p^63s^1$, or $[Ne]3s^1$) the valence electron is the electron in the

CHEMISTRY IN FOCUS

A Magnetic Moment

An anesthetized frog lies in the hollow core of an electromagnet. As the current in the coils of the magnet is increased, the frog magically rises and floats in midair (see photo). How can this happen? Is the electromagnet an antigravity machine? In fact, there is no magic going on here. This phenomenon demonstrates the magnetic properties of all matter. We know that iron magnets attract and repel each other depending on their relative orientations. Is a frog magnetic like a piece of iron? If a frog lands on a steel manhole cover, will it be trapped there by magnetic attractions? Of course not. The magnetism of the frog, as with most objects, shows up only in the presence of a strong inducing magnetic field. In other words, the powerful electromagnet surrounding the frog in the experiment described above *induces* a magnetic field in the frog that opposes the inducing field. The opposing magnetic field in the frog repels the inducing field, and the frog lifts up until the magnetic force is balanced by the gravitational pull on its body. The frog then "floats" in air.

A live frog levitated in a magnetic field.

How can a frog be magnetic if it is not made of iron? It's the electrons. Frogs are composed of cells containing many kinds of molecules. Of course, these molecules are made of atoms—carbon atoms, nitrogen atoms, oxygen atoms, and other types. Each of these atoms contains electrons that are moving around the atomic nuclei. When these electrons sense a strong magnetic field, they respond by moving in a fashion that produces magnetic fields aligned to oppose the inducing field. This phenomenon is called *diamagnetism*.

All substances, animate and inanimate, because they are made of atoms, exhibit diamagnetism. Andre Geim and his colleagues at the University of Nijmegan, the Netherlands, have levitated frogs, grasshoppers, plants, and water droplets, among other objects. Geim says that, given a large enough electromagnet, even humans can be levitated. He notes, however, that constructing a magnet strong enough to float a human would be very expensive, and he sees no point in it. Geim does point out that inducing weightlessness with magnetic fields may be a good way to pretest experiments on weightlessness intended as research for future space flights—to see if the ideas fly as well as the objects.

3*s* orbital, because in this case principal energy level 3 is the outermost level that contains an electron. The valence electrons are the most important electrons to chemists because, being the outermost electrons, they are the ones involved when atoms attach to each other (form bonds), as we will see in the next chapter. The inner electrons, which are known as **core electrons,** are not involved in bonding atoms to each other.

Note in Figure 10.29 that a very important pattern is developing: except for helium, *the atoms of elements in the same group (vertical column of the periodic table) have the same number of electrons in a given type of orbital* (sublevel), except that the orbitals are in different principal energy levels. Remember that the elements were originally organized into groups on the periodic table on the basis of similarities in chemical properties. Now we understand the reason behind these groupings. Elements with the same valence electron arrangement show very similar chemical behavior.

Electron Configurations and the Periodic Table

AIM: To learn about the electron configurations of atoms with *Z* greater than 18.

In the previous section we saw that we can describe the atoms beyond hydrogen by simply filling the atomic orbitals starting with level $n = 1$ and working outward in order. This works fine until we reach the element *potassium* ($Z = 19$), which is the next element after argon. Because the $3p$ orbitals are fully occupied in argon, we might expect the next electron to go into a $3d$ orbital (recall that for $n = 3$ the sublevels are $3s$, $3p$, and $3d$). However, experiments show that the chemical properties of potassium are very similar to those of lithium and sodium. Because we have learned to associate similar chemical properties with similar valence-electron arrangements, we predict that the valence-electron configuration for potassium is $4s^1$, resembling sodium ($3s^1$) and lithium ($2s^1$). That is, we expect the last electron in potassium to occupy the $4s$ orbital instead of one of the $3d$ orbitals. This means that principal energy level 4 begins to fill before level 3 has been completed. This conclusion is confirmed by many types of experiments. So the electron configuration of potassium is

$$\text{K: } 1s^2 2s^2 2p^6 3s^2 3p^6 4s^1, \text{ or } [\text{Ar}]4s^1$$

The next element is *calcium,* with an additional electron that also occupies the $4s$ orbital.

$$\text{Ca: } 1s^2 2s^2 2p^6 3s^2 3p^6 4s^2, \text{ or } [\text{Ar}]4s^2$$

The $4s$ orbital is now full.

After calcium the next electrons go into the $3d$ orbitals to complete principal energy level 3. The elements that correspond to filling the $3d$ orbitals are called transition metals. Then the $4p$ orbitals fill. Figure 10.30 gives partial electron configurations for the elements potassium through krypton.

Note from Figure 10.30 that all of the transition metals have the general configuration $[\text{Ar}]4s^2 3d^n$ except chromium ($4s^1 3d^5$) and copper ($4s^1 3d^{10}$). The reasons for these exceptions are complex and will not be discussed here.

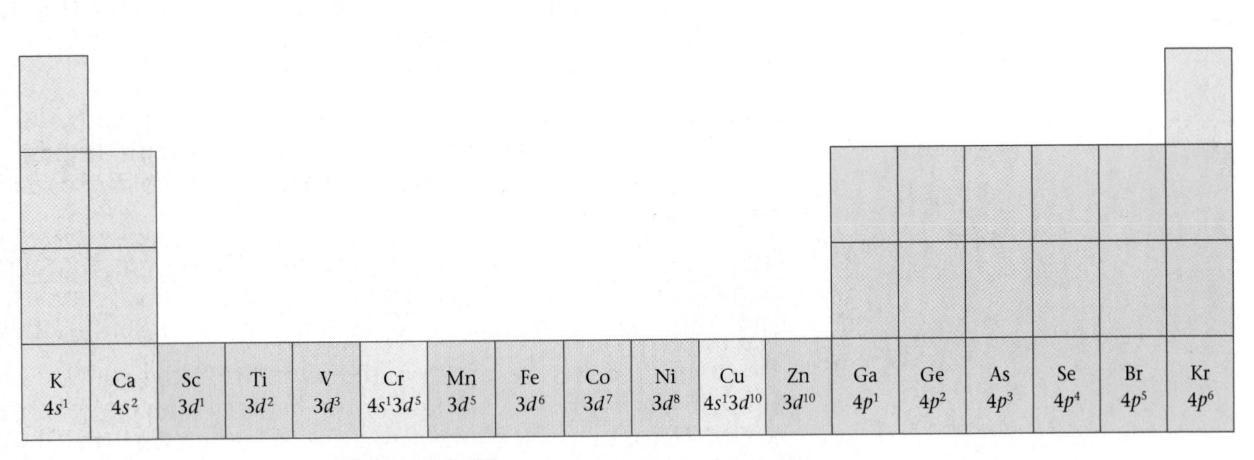

Figure 10.30
Partial electron configurations for the elements potassium through krypton. The transition metals shown in green (scandium through zinc) have the general configuration $[\text{Ar}]4s^2 3d^n$, except for chromium and copper.

CHEMISTRY IN FOCUS

The Chemistry of Bohrium

One of the best uses of the periodic table is to predict the properties of newly discovered elements. For example, the artificially synthesized element bohrium ($Z = 107$) is found in the same family as manganese, technecium, and rhenium and is expected to show chemistry similar to these elements. The problem, of course, is that only a few atoms of bohrium can be made at a time and the atoms exist for only a very short time (about 17 seconds). It's a real challenge to study the chemistry of an element under these conditions. However, a team of nuclear chemists led by Heinz W. Gaggeler of the University of Bern in Switzerland isolated six atoms of ^{267}Bh and prepared the compound BhO_3Cl. Analysis of the decay products of this compound helped define the thermochemical properties of BhO_3Cl and showed that bohrium seems to behave as might be predicted from its position in the periodic table.

Instead of continuing to consider the elements individually, we will now look at the overall relationship between the periodic table and orbital filling. Figure 10.31 shows which type of orbital is filling in each area of the periodic table. Note the points in the box below.

To help you further understand the connection between orbital filling and the periodic table, Figure 10.32 shows the orbitals in the order in which they fill.

A periodic table is almost always available to you. If you understand the relationship between the electron configuration of an element and its position on the periodic table, you can figure out the expected electron configuration of any atom.

Orbital Filling

1. In a principal energy level that has *d* orbitals, the *s* orbital from the *next* level fills before the *d* orbitals in the current level. That is, the $(n + 1)s$ orbitals always fill before the *nd* orbitals. For example, the 5*s* orbitals fill for rubidium and strontium before the 4*d* orbitals fill for the second row of transition metals (yttrium through cadmium).
2. After lanthanum, which has the electron configuration $[Xe]6s^2 5d^1$, a group of fourteen elements called the **lanthanide series,** or the lanthanides, occurs. This series of elements corresponds to the filling of the seven 4*f* orbitals.
3. After actinium, which has the configuration $[Rn]7s^2 6d^1$, a group of fourteen elements called the **actinide series,** or the actinides, occurs. This series corresponds to the filling of the seven 5*f* orbitals.
4. Except for helium, the group numbers indicate the sum of electrons in the *ns* and *np* orbitals in the highest principal energy level that contains electrons (where *n* is the number that indicates a particular principal energy level). These electrons are the valence electrons, the electrons in the outermost principal energy level of a given atom.

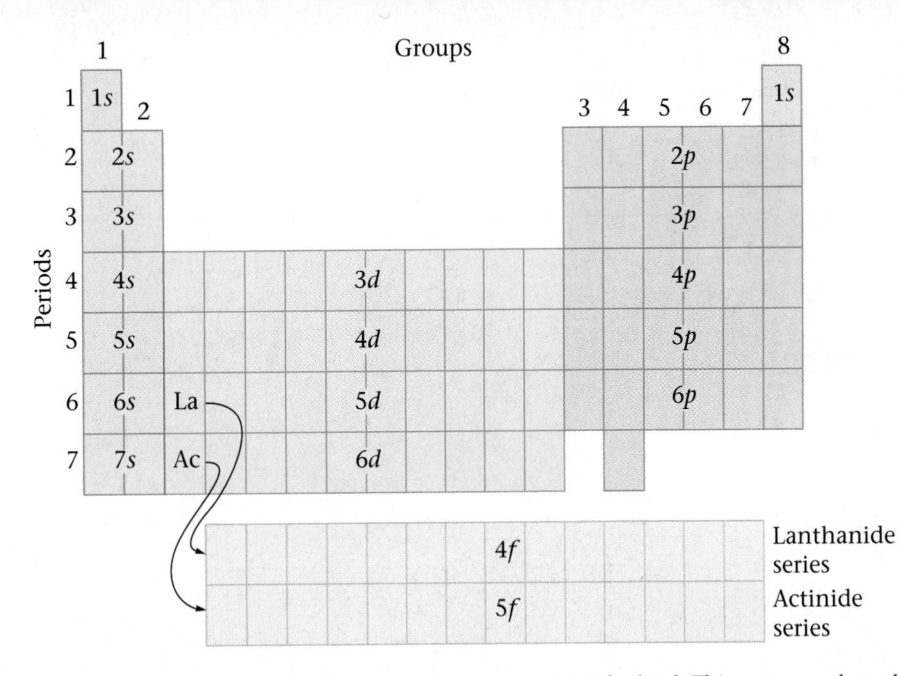

Figure 10.31
The orbitals being filled for elements in various parts of the periodic table. Note that in going along a horizontal row (a period), the $(n + 1)s$ orbital fills before the nd orbital. The group label indicates the number of valence electrons (the number of s plus the number of p electrons in the highest occupied principal energy level) for the elements in each group.

* After the 6s orbital is full, one electron goes into a 5d orbital. This corresponds to the element lanthanum ($[Xe]6s^2 5d^1$). After lanthanum, the 4f orbitals fill with electrons.

** After the 7s orbital is full, one electron goes into 6d. This is actinium ($[Rn]7s^2 6d^1$). The 5f orbitals then fill.

Order of filling of orbitals: 6d, 5f**, 7s, 6p, 5d, 4f*, 6s, 5p, 4d, 5s, 4p, 3d, 4s, 3p, 3s, 2p, 2s, 1s

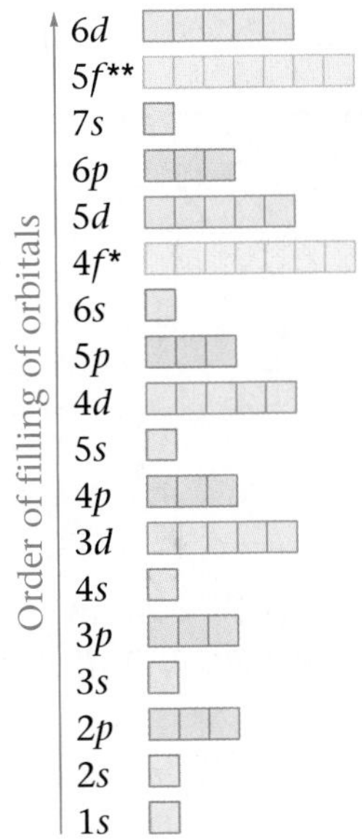

Figure 10.32
A box diagram showing the order in which orbitals fill to produce the atoms in the periodic table. Each box can hold two electrons.

Example 10.3 Determining Electron Configurations

Using the periodic table inside the front cover of the text, give the electron configurations for sulfur (S), gallium (Ga), hafnium (Hf), and radium (Ra).

Solution

Sulfur is element 16 and resides in Period 3, where the 3*p* orbitals are being filled (see Figure 10.33). Because sulfur is the fourth among the "3*p* elements," it must have four 3*p* electrons. Sulfur's electron configuration is

$$\text{S: } 1s^2 2s^2 2p^6 3s^2 3p^4, \text{ or } [\text{Ne}]3s^2 3p^4$$

Gallium is element 31 in Period 4 just after the transition metals (see Figure 10.33). It is the first element in the "4*p* series" and has a $4p^1$ arrangement. Gallium's electron configuration is

$$\text{Ga: } 1s^2 2s^2 2p^6 3s^2 3p^6 4s^2 3d^{10} 4p^1, \text{ or } [\text{Ar}]4s^2 3d^{10} 4p^1$$

Hafnium is element 72 and is found in Period 6, as shown in Figure 10.33. Note that it occurs just after the lanthanide series (see Figure 10.31). Thus the 4*f* orbitals are already filled. Hafnium is the second member of the 5*d* transition series and has two 5*d* electrons. Its electron configuration is

$$\text{Hf: } 1s^2 2s^2 2p^6 3s^2 3p^6 4s^2 3d^{10} 4p^6 5s^2 4d^{10} 5p^6 6s^2 4f^{14} 5d^2, \text{ or } [\text{Xe}]6s^2 4f^{14} 5d^2$$

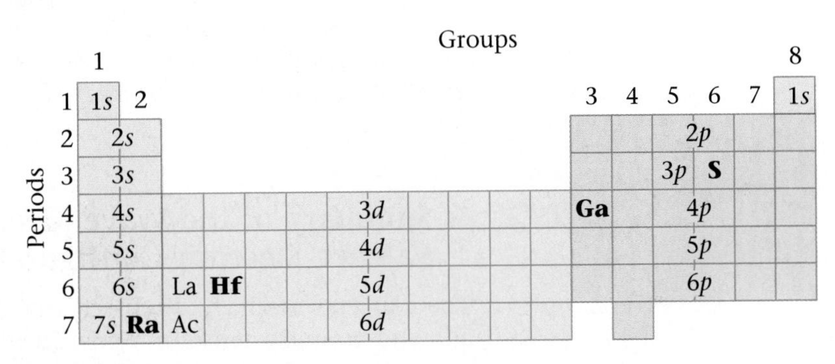

Figure 10.33
The positions of the elements considered in Example 10.3.

Representative Elements | *d*-Transition Elements | Representative Elements | Noble Gases

Group numbers

Period number, highest occupied electron level

Period	1A ns^1	2A ns^2											3A ns^2np^1	4A ns^2np^2	5A ns^2np^3	6A ns^2np^4	7A ns^2np^5	8A ns^2np^6
1	1 H $1s^1$																	2 He $1s^2$
2	3 Li $2s^1$	4 Be $2s^2$											5 B $2s^22p^1$	6 C $2s^22p^2$	7 N $2s^22p^3$	8 O $2s^22p^4$	9 F $2s^22p^5$	10 Ne $2s^22p^6$
3	11 Na $3s^1$	12 Mg $3s^2$											13 Al $3s^23p^1$	14 Si $3s^23p^2$	15 P $3s^23p^3$	16 S $3s^23p^4$	17 Cl $3s^23p^5$	18 Ar $3s^23p^6$
4	19 K $4s^1$	20 Ca $4s^2$	21 Sc $4s^23d^1$	22 Ti $4s^23d^2$	23 V $4s^23d^3$	24 Cr $4s^13d^5$	25 Mn $4s^23d^5$	26 Fe $4s^23d^6$	27 Co $4s^23d^7$	28 Ni $4s^23d^8$	29 Cu $4s^13d^{10}$	30 Zn $4s^23d^{10}$	31 Ga $4s^24p^1$	32 Ge $4s^24p^2$	33 As $4s^24p^3$	34 Se $4s^24p^4$	35 Br $4s^24p^5$	36 Kr $4s^24p^6$
5	37 Rb $5s^1$	38 Sr $5s^2$	39 Y $5s^24d^1$	40 Zr $5s^24d^2$	41 Nb $5s^14d^4$	42 Mo $5s^14d^5$	43 Tc $5s^14d^6$	44 Ru $5s^14d^7$	45 Rh $5s^14d^8$	46 Pd $4d^{10}$	47 Ag $5s^14d^{10}$	48 Cd $5s^24d^{10}$	49 In $5s^25p^1$	50 Sn $5s^25p^2$	51 Sb $5s^25p^3$	52 Te $5s^25p^4$	53 I $5s^25p^5$	54 Xe $5s^25p^6$
6	55 Cs $6s^1$	56 Ba $6s^2$	57 La* $6s^25d^1$	72 Hf $4f^{14}6s^25d^2$	73 Ta $6s^25d^3$	74 W $6s^25d^4$	75 Re $6s^25d^5$	76 Os $6s^25d^6$	77 Ir $6s^25d^7$	78 Pt $6s^15d^9$	79 Au $6s^15d^{10}$	80 Hg $6s^25d^{10}$	81 Tl $6s^26p^1$	82 Pb $6s^26p^2$	83 Bi $6s^26p^3$	84 Po $6s^26p^4$	85 At $6s^26p^5$	86 Rn $6s^26p^6$
7	87 Fr $7s^1$	88 Ra $7s^2$	89 Ac** $7s^26d^1$	104 Rf $7s^26d^2$	105 Db $7s^26d^3$	106 Sg $7s^26d^4$	107 Bh $7s^26d^5$	108 Hs $7s^26d^6$	109 Mt $7s^26d^7$	110 Uun $7s^26d^8$	111 Uuu $7s^16d^{10}$	112 Uub $7s^26d^{10}$		114 Uuq $7s^27p^2$				

f-Transition Elements

*Lanthanides	58 Ce $6s^24f^15d^1$	59 Pr $6s^24f^35d^0$	60 Nd $6s^24f^45d^0$	61 Pm $6s^24f^55d^0$	62 Sm $6s^24f^65d^0$	63 Eu $6s^24f^75d^0$	64 Gd $6s^24f^75d^1$	65 Tb $6s^24f^95d^0$	66 Dy $6s^24f^{10}5d^0$	67 Ho $6s^24f^{11}5d^0$	68 Er $6s^24f^{12}5d^0$	69 Tm $6s^24f^{13}5d^0$	70 Yb $6s^24f^{14}5d^0$	71 Lu $6s^24f^{14}5d^1$
**Actinides	90 Th $7s^25f^06d^2$	91 Pa $7s^25f^26d^1$	92 U $7s^25f^36d^1$	93 Np $7s^25f^46d^1$	94 Pu $7s^25f^66d^0$	95 Am $7s^25f^76d^0$	96 Cm $7s^25f^76d^1$	97 Bk $7s^25f^96d^0$	98 Cf $7s^25f^{10}6d^0$	99 Es $7s^25f^{11}6d^0$	100 Fm $7s^25f^{12}6d^0$	101 Md $7s^25f^{13}6d^0$	102 No $7s^25f^{14}6d^0$	103 Lr $7s^25f^{14}6d^1$

Figure 10.34
The periodic table with atomic symbols, atomic numbers, and partial electron configurations. (The configurations for elements 108, 110, 111, and 112 are not yet known.)

Radium is element 88 and is in Period 7 (and Group 2), as shown in Figure 10.33. Thus radium has two electrons in the 7*s* orbital, and its electron configuration is

$$\text{Ra: } 1s^22s^22p^63s^23p^64s^23d^{10}4p^65s^24d^{10}5p^66s^24f^{14}5d^{10}6p^67s^2, \text{ or } [\text{Rn}]7s^2$$

✔ Self-Check Exercise 10.3

Using the periodic table inside the front cover of the text, predict the electron configurations for fluorine, silicon, cesium, lead, and iodine. If you have trouble, use Figure 10.31.

See Problems 10.59 through 10.68. ■

Summary of the Wave Mechanical Model and Valence-Electron Configurations

The concepts we have discussed in this chapter are very important. They allow us to make sense of a good deal of chemistry. When it was first observed that elements with similar properties occur periodically as the atomic number increases, chemists wondered why. Now we have an explanation. The wave mechanical model pictures the electrons in an atom as arranged in orbitals, with each orbital capable of holding two electrons. As we build up the atoms, the same types of orbitals recur in going from one principal energy level to another. This means that particular valence-electron configurations recur periodically. For reasons we will explore in the next chapter, elements with a particular type of valence configuration all show very similar chemical behavior. Thus groups of elements, such as the alkali metals, show similar chemistry because all the elements in that group have the same type of valence-electron arrangement. This concept, which explains so much chemistry, is the greatest contribution of the wave mechanical model to modern chemistry.

For reference, the valence-electron configurations for all the elements are shown on the periodic table in Figure 10.34. Note the following points:

1. The group labels for Groups 1, 2, 3, 4, 5, 6, 7, and 8 indicate the *total number* of valence electrons for the atoms in these groups. For example, all the elements in Group 5 have the configuration ns^2np^3. (Any d electrons present are always in the next lower principal energy level than the valence electrons and so are not counted as valence electrons.)

2. The elements in Groups 1, 2, 3, 4, 5, 6, 7, and 8 are often called the **main-group elements,** or **representative elements.** Remember that every member of a given group (except for helium) has the same valence-electron configuration, except that the electrons are in different principal energy levels.

3. We will not be concerned in this text with the configurations for the f transition elements (lanthanides and actinides), although they are included in Figure 10.34.

Atomic Properties and the Periodic Table

AIM: To understand the general trends in atomic properties in the periodic table.

With all of this talk about electron probability and orbitals, we must not lose sight of the fact that chemistry is still fundamentally a science based on the observed properties of substances. We know that wood burns, steel rusts, plants grow, sugar tastes sweet, and so on because we *observe* these phenomena. The atomic theory is an attempt to help us understand why these things occur. If we understand why, we can hope to better control the chemical events that are so crucial in our daily lives.

In the next chapter we will see how our ideas about atomic structure help us understand how and why atoms combine to form compounds. As we explore this, and as we use theories to explain other types of chemical behavior later in the text, it is important that we distinguish the observation (steel rusts) from the attempts to explain why the observed event occurs (theories). The observations remain the same over the decades, but the theories (our explanations) change as we gain a clearer understanding of how nature operates. A good example of this is the replacement of the Bohr model for atoms by the wave mechanical model.

Because the observed behavior of matter lies at the heart of chemistry, you need to understand thoroughly the characteristic properties of the various elements and the trends (systematic variations) that occur in those properties. To that end, we will now consider some especially important properties of atoms and see how they vary, horizontally and vertically, on the periodic table.

Metals and Nonmetals

The most fundamental classification of the chemical elements is into metals and nonmetals. **Metals** typically have the following physical properties: a lustrous appearance, the ability to change shape without breaking (they can be pulled into a wire or pounded into a thin sheet), and excellent conductivity of heat and electricity. **Nonmetals** typically do not have these physical properties, although there are some exceptions. (For example, solid iodine is lustrous; the graphite form of carbon is an excellent conductor of electricity; and the diamond form of carbon is an excellent conductor of heat.) However, it is the *chemical* differences between metals and nonmetals that interest us the most: *metals tend to lose electrons to form positive ions, and nonmetals tend to gain electrons to form negative ions.* When a metal and a nonmetal react, a transfer of one or more electrons from the metal to the nonmetal often occurs.

Most of the elements are classified as metals, as is shown in Figure 10.35. Note that the metals are found on the left side and at the center of the periodic table. The relatively few nonmetals are in the upper-right corner of the table. A few elements exhibit both metallic and nonmetallic behavior; they are classified as **metalloids** or semimetals.

It is important to understand that simply being classified as a metal does not mean that an element behaves exactly like all other metals. For example, some metals can lose one or more electrons much more easily than others. In particular, cesium can give up its outermost electron (a 6*s* electron) more easily than can lithium

Gold leaf being applied to the dome of the courthouse in Huntington, West Virginia.

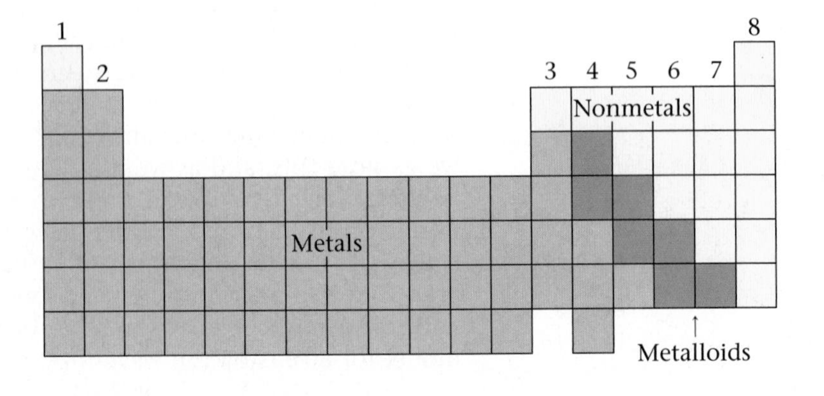

Figure 10.35
The classification of elements as metals, nonmetals, and metalloids.

CHEMISTRY IN FOCUS

Fireworks

The art of using mixtures of chemicals to produce explosives is an ancient one. Black powder—a mixture of potassium nitrate, charcoal, and sulfur—was being used in China well before A.D. 1000, and it has been used through the centuries in military explosives, in construction blasting, and for fireworks.

Before the nineteenth century, fireworks were confined mainly to rockets and loud bangs. Orange and yellow colors came from the presence of charcoal and iron filings. However, with the great advances in chemistry in the nineteenth century, new compounds found their way into fireworks. Salts of copper, strontium, and barium added brilliant colors. Magnesium and aluminum metals gave a dazzling white light.

These brightly colored fireworks are the result of complex mixtures of chemicals.

How do fireworks produce their brilliant colors and loud bangs? Actually, only a handful of different chemicals are responsible for most of the spectacular effects. To produce the noise and flashes, an oxidizer (something with a strong affinity for electrons) is reacted with a metal such as magnesium or aluminum mixed with sulfur. The resulting reaction produces a brilliant flash, which is due to the aluminum or magnesium burning, and a loud report is produced by the rapidly expanding gases. For a color effect, an element with a colored flame is included.

Yellow colors in fireworks are due to sodium. Strontium salts give the red color familiar from highway safety flares. Barium salts give a green color.

Although you might think that the chemistry of fireworks is simple, achieving the vivid white flashes and the brilliant colors requires complex combinations of chemicals. For example, because the white flashes produce high flame temperatures, the colors tend to wash out. Another problem arises from the use of sodium salts. Because sodium produces an extremely bright yellow color, sodium salts cannot be used when other colors are desired. In short, the manufacture of fireworks that produce the desired effects and are also safe to handle requires very careful selection of chemicals.*

*The chemical mixtures in fireworks are very dangerous. *Do not* experiment with chemicals on your own.

(a 2*s* electron). In fact, for the alkali metals (Group 1) the ease of giving up an electron varies as follows:

$$\text{Cs} > \text{Rb} > \text{K} > \text{Na} > \text{Li}$$

Loses an electron most easily

Note that as we go down the group, the metals become more likely to lose an electron. This makes sense because as we go down the group, the electron being removed resides, on average, farther and farther from the nucleus. That is, the 6*s* electron lost from Cs is much farther from the attractive positive nucleus—and so is easier to remove—than the 2*s* electron that must be removed from a lithium atom.

The same trend is also seen in the Group 2 metals (alkaline earth metals): the farther down in the group the metal resides, the more likely it is to lose an electron.

Just as metals vary somewhat in their properties, so do nonmetals. In general, the elements that can most effectively pull electrons from metals occur in the upper-right corner of the periodic table.

As a general rule, we can say that the most chemically active metals appear in the lower-left region of the periodic table, whereas the most chemically active nonmetals appear in the upper-right region. The properties of the semimetals, or metalloids, lie between the metals and the nonmetals, as might be expected.

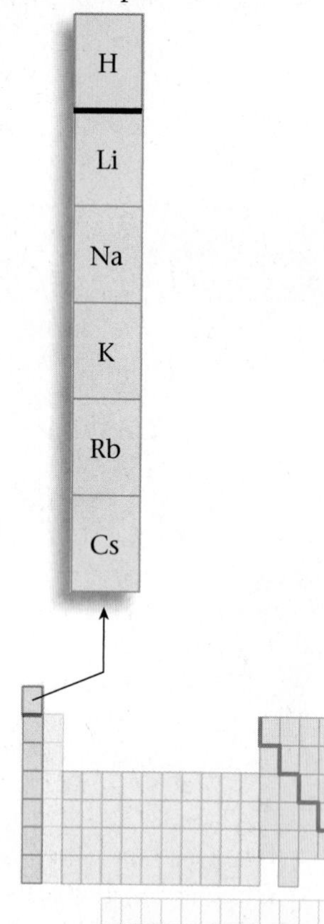

Ionization Energies

The **ionization energy** of an atom is the energy required to remove an electron from an individual atom in the gas phase:

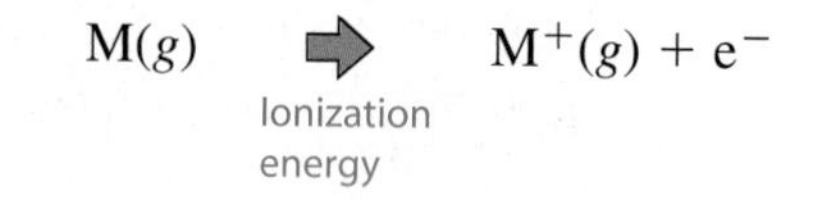

As we have noted, the most characteristic chemical property of a metal atom is losing electrons to nonmetals. Another way of saying this is to say that *metals have relatively low ionization energies*—a relatively small amount of energy is needed to remove an electron from a typical metal.

Recall that metals at the bottom of a group lose electrons more easily than those at the top. In other words, ionization energies tend to decrease in going from the top to the bottom of a group.

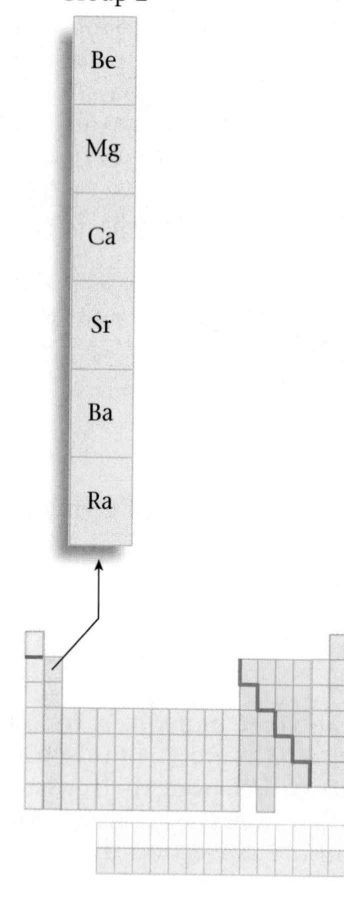

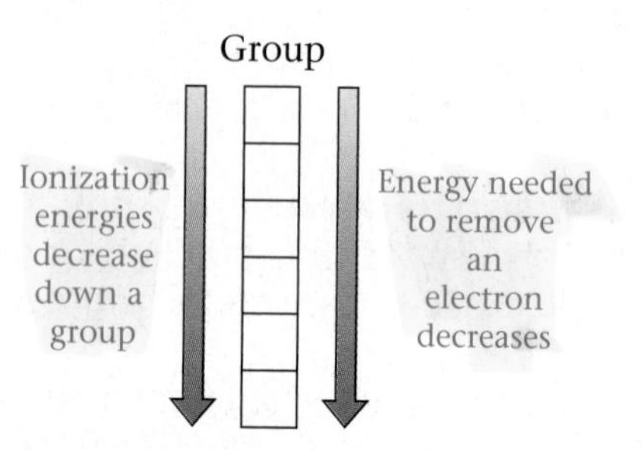

In contrast to metals, nonmetals have relatively large ionization energies. Nonmetals tend to gain, not lose, electrons. Recall that metals appear on the left side of the periodic table and nonmetals appear on the right. Thus it is not surprising that ionization energies tend to increase from left to right across a given period on the periodic table.

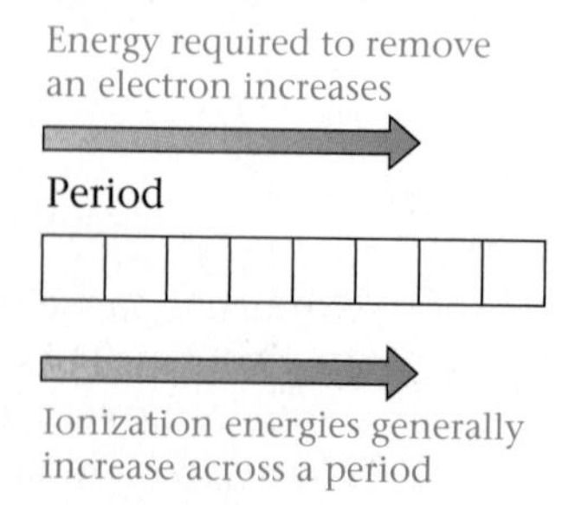

In general, the elements that appear in the lower-left region of the periodic table have the lowest ionization energies (and are therefore the most chemically active metals). On the other hand, the elements with the highest ionization energies (the most chemically active nonmetals) occur in the upper-right region of the periodic table.

Atomic Size

The sizes of atoms vary as shown in Figure 10.36. Notice that atoms get larger as we go down a group on the periodic table and that they get smaller as we go from left to right across a period.

We can understand the increase in size that we observe as we go down a group by remembering that as the principal energy level increases, the average distance of the electrons from the nucleus also increases. So atoms get bigger as electrons are added to larger principal energy levels.

Explaining the decrease in **atomic size** across a period requires a little thought about the atoms in a given row (period) of the periodic table. Recall that the atoms in a particular period all have their outermost electrons in a given principal energy level. That is, the atoms in Period 1 have their outer electrons in the 1*s* orbital (principal energy level 1), the atoms in Period 2 have their outermost electrons in principal energy level 2 (2*s* and 2*p* orbitals), and so on (see Figure 10.31). Because all the orbitals in a given principal energy level are expected to be the same size, we might expect the atoms in a given period to be the same size. However, remember that the number of protons in the nucleus increases as we move from atom to atom in the period. The resulting increase in positive charge on the nucleus tends to pull the electrons closer to the nucleus. So instead of remaining the same size across a period as electrons are added in a given principal energy level, the atoms get smaller as the electron "cloud" is drawn in by the increasing nuclear charge.

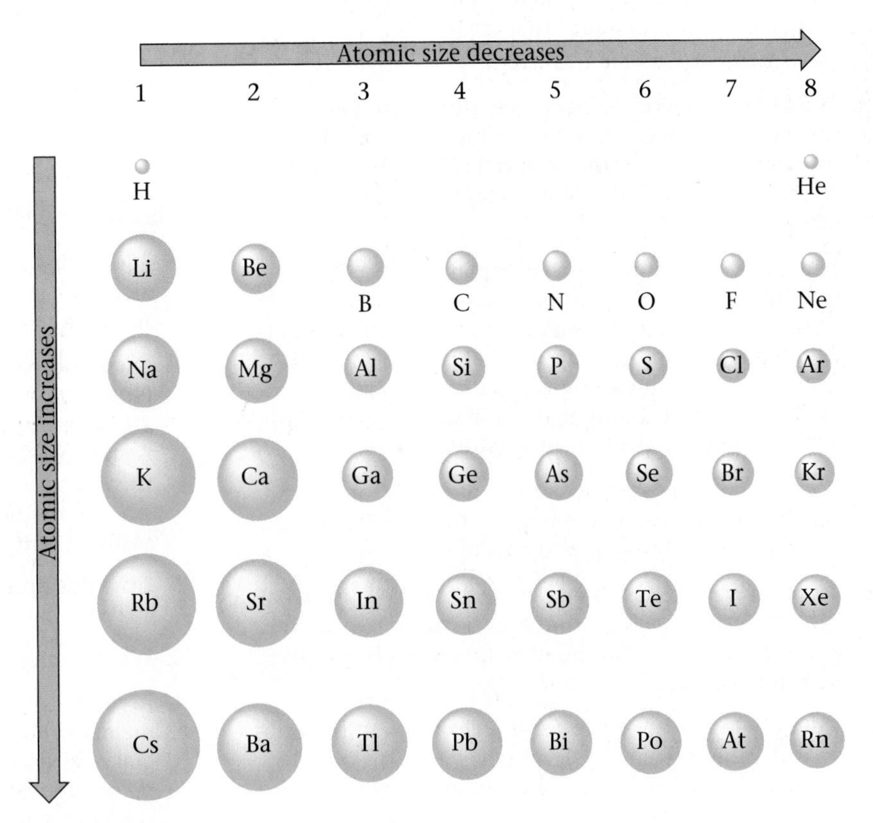

Figure 10.36
Relative atomic sizes for selected atoms. Note that atomic size increases down a group and decreases across a period.

CHAPTER 10 REVIEW

KEY TERMS

electromagnetic radiation (10.2)
wavelength (10.2)
frequency (10.2)
photon (10.2)
quantized energy levels (10.4)
wave mechanical model (10.6)
orbital (10.7)
principal energy levels (10.7)
sublevels (10.7)
Pauli exclusion principle (10.8)
electron configuration (10.9)
orbital (box) diagram (10.9)
valence electrons (10.9)
core electrons (10.9)
lanthanide series (10.10)
actinide series (10.10)
main-group (representative) elements (10.10)
metals (10.11)
nonmetals (10.11)
metalloids (10.11)
ionization energy (10.11)
atomic size (10.11)

SUMMARY

1. Energy travels through space by electromagnetic radiation ("light"), which can be characterized by the wavelength and frequency of the waves. Light can also be thought of as packets of energy called photons. Atoms can gain energy by absorbing a photon and can lose energy by emitting a photon.
2. The emissions of energy from hydrogen atoms produce only certain energies as hydrogen changes from a higher to a lower energy. This shows that the energy levels of hydrogen are quantized.
3. The Bohr model of the hydrogen atom postulated that the electron moved in circular orbits corresponding to the various allowed energy levels. Though it worked well for hydrogen, the Bohr model did not work for other atoms.
4. The wave mechanical model explains atoms by postulating that the electron has both wave and particle characteristics. Electron states are described by orbitals, which are probability maps indicating how likely it is to find the electron at a given point in space. The orbital size can be thought of as a surface containing 90% of the total electron probability.
5. According to the Pauli exclusion principle, an atomic orbital can hold a maximum of two electrons, and those electrons must have opposite spins.
6. Atoms have a series of energy levels, called principal energy levels (n), which contain one or more sublevels (types of orbitals). The number of sublevels increases with increasing n.
7. Valence electrons are the s and p electrons in the outermost principal energy level of an atom. Core electrons are the inner electrons of an atom.
8. Metals are found at the left and center of the periodic table. The most chemically active metals are found in the lower-left corner of the periodic table. The most chemically active nonmetals are located in the upper-right corner.
9. Ionization energy, the energy required to remove an electron from a gaseous atom, decreases going down a group and increases going from left to right across a period.
10. For the representative elements, atomic size increases going down a group but decreases going from left to right across a period.

IN-CLASS DISCUSSION QUESTIONS

These questions are designed to be considered by groups of students in class. Often these questions work well for introducing a particular topic in class.

1. On graph paper draw three waves—one with a wavelength of 40 small boxes, one with a wavelength of 20 boxes, and one with a wavelength of 10 boxes. Assuming the time it takes each of these waves to travel across the sheet of paper is one second, calculate the frequency for each wavelength. How do they compare? How do their energies compare? Why?
2. How does probability fit into the description of the atom?
3. What is meant by an *orbital?*
4. Account for the fact that the line that separates the metals from the nonmetals on the periodic table is diagonal downward to the right instead of horizontal or vertical.

5. Consider the following statements: "The ionization energy for the potassium atom is negative because when K loses an electron to become K^+, it achieves a noble gas electron configuration." Indicate everything that is correct in this statement. Indicate everything that is incorrect. Correct the mistaken information and explain the error.

6. In going across a row of the periodic table, protons and electrons are added and ionization energy generally increases. In going down a column of the periodic table, protons and electrons are also being added but ionization energy generally decreases. Explain.

7. Which is larger, the H 1*s* orbital or the Li 1*s* orbital? Why? Which has the larger radius, the H atom or the Li atom? Why?

8. True or false? The hydrogen atom has a 3*s* orbital. Explain.

9. Differentiate among the terms *energy level, sublevel,* and *orbital.*

10. Make sense of the fact that metals tend to lose electrons and nonmetals tend to gain electrons. Use the periodic table to support your answer.

11. Show how using the periodic table helps you find the expected electron configuration of any element.

For Questions 12–14, you will need to consider ionizations beyond the first ionization energy. For example, the second ionization energy is the energy to remove a second electron from an element.

12. Compare the first ionization energy of helium to its second ionization energy, remembering that both electrons come from the 1*s* orbital.

13. Which would you expect to have a larger second ionization energy, lithium or beryllium? Why?

14. The first four ionization energies for elements X and Y are shown below. The units are not kJ/mol.

	X	Y
first	170	200
second	350	400
third	1800	3500
fourth	2500	5000

Identify the elements X and Y. There may be more than one answer, so explain completely.

QUESTIONS AND PROBLEMS

All even-numbered exercises have answers in the back of this book and solutions in the *Solutions Guide.*

10.1 Rutherford's Atom

QUESTIONS

1. What is meant by the expression "nuclear atom"?

2. Although Rutherford's experiments confirmed the existence of the positively charged nucleus, they left many questions about the structure of the atom unanswered. Discuss.

10.2 Electromagnetic Radiation

QUESTIONS

3. What is *electromagnetic radiation?* At what speed does electromagnetic radiation travel?

4. How are the different types of electromagnetic radiation similar? How do they differ?

5. The ________ of electromagnetic radiation represents the distance between consecutive peaks in the wave, whereas the ________ of electromagnetic radiation measures the number of peaks passing a given point in space per time unit.

6. What do we mean by the *speed* of electromagnetic radiation? How do the *frequency* and the *speed* of electromagnetic radiation differ?

7. What is a "packet" of electromagnetic energy called?

8. What term is used to describe the fact that electromagnetic radiation can be thought of as either a continuous wave or as a stream of individual packets of energy?

10.3 Emission of Energy by Atoms

QUESTIONS

9. What is observed when salts containing metal ions such as Na^+, Cu^{2+}, or Li^+ are heated in a flame?

10. How does the energy of a photon of visible light emitted by an atom compare to the energy change within the atom itself?

10.4 The Energy Levels of Hydrogen

QUESTIONS

11. An atom in its lowest possible energy state is said to be in the ________ state.

12. When an atom in an excited state returns to its ground state, what happens to the excess energy of the atom?

13. How is the energy carried per photon of light related to the wavelength of the light? Does short-wavelength light carry more energy or less energy than long-wavelength light?

14. When an atom absorbs energy, it goes into a(n) ________ state.

15. Describe briefly why the study of electromagnetic radiation has been important to our understanding of the arrangement of electrons in atoms.

16. What does it mean to say that the hydrogen atom has *discrete energy levels?* How is this fact reflected in the radiation that excited hydrogen atoms emit?

17. Because a given element's atoms emit only certain photons of light, only certain ________ are occurring in those particular atoms.

18. How does the energy possessed by an emitted photon compare to the difference in energy levels that gave rise to the emission of the photon?

19. What experimental evidence do scientists have that the energy levels of hydrogen are *quantized?*

20. What is meant by the *ground state* of an atom?

10.5 The Bohr Model of the Atom

QUESTIONS

21. What are the essential points of Bohr's theory of the structure of the hydrogen atom?

22. According to Bohr's theory, what happens to the electron in an atom when the atom absorbs radiation or when it emits radiation?

23. How does the Bohr theory account for the observed phenomenon of the emission of discrete wavelengths of light by excited atoms?

24. Why was Bohr's theory for the hydrogen atom initially accepted, and why was it ultimately discarded?

10.6 The Wave Mechanical Model of the Atom

QUESTIONS

25. What major assumption (that was analogous to what had already been demonstrated for electromagnetic radiation) did de Broglie and Schrödinger make about the motion of tiny particles?

26. Discuss briefly the difference between an orbit (as described by Bohr for hydrogen) and an orbital (as described by the more modern, wave mechanical picture of the atom).

27. Why was Schrödinger not able to describe exactly the pathway an electron takes as it moves through the space of an atom?

28. Explain why we cannot *exactly* specify the location of an electron in an atom but can only discuss where an electron is *most likely* to be at any given time.

10.7 The Hydrogen Orbitals

QUESTIONS

29. Why are the orbitals of the hydrogen atom described as "probability maps"? Why are the edges of the hydrogen orbitals sometimes drawn to appear "fuzzy"?

30. When students first see a drawing of the p orbitals, they often question how the electron is able to jump through the nucleus to get from one lobe of the p orbital to the other. How would you explain this?

31. What are the differences between the $2s$ orbital and the $1s$ orbital of hydrogen? How are they similar?

32. What overall shape do the $2p$ and $3p$ orbitals have? How do the $2p$ orbitals differ from the $3p$ orbitals? How are they similar?

33. The higher the principal energy level, n, the (closer to/farther from) the nucleus is the electron.

34. When the electron in hydrogen is in the n = ________ principal energy level, the atom is in its ground state.

35. Although a hydrogen atom has only one electron, the hydrogen atom possesses a complete set of available orbitals. What purpose do these additional orbitals serve?

36. What designations are given to the sublevels in the fourth principal energy level? In the fifth principal energy level?

10.8 The Wave Mechanical Model: Further Development

QUESTIONS

37. When describing the electrons in an orbital, we use arrows pointing upward and downward (↑ and ↓) to indicate what property?

38. What is the *Pauli exclusion principle?* How many electrons can occupy an orbital, according to this principle? Why?

39. How does the *energy* of a principal energy level depend on the value of n? Does a higher value of n mean a higher or lower energy?

40. The number of sublevels in a principal energy level (increases/decreases) as n increases.

41. According to the Pauli exclusion principle, a given orbital can contain only ________ electrons.

42. According to the Pauli exclusion principle, the electrons within a given orbital must have ________ spins.

43. Which of the following orbital designations is (are) possible?
a. $1p$ c. $4d$
b. $2p$ d. $3d$

44. Which of the following orbital designations is (are) not correct?
a. $1d$ c. $4d$
b. $2d$ d. $5f$

10.9 Electron Arrangements in the First Eighteen Atoms on the Periodic Table

QUESTIONS

45. Which orbital is the *first* to be filled in any atom? Why?

46. When a hydrogen atom is in its ground state, in which orbital is its electron found? Why?

47. Where are the *valence electrons* found in an atom, and why are these particular electrons most important to the chemical properties of the atom?

48. How are the electron arrangements in a given group (vertical column) of the periodic table related? How is this relationship manifested in the properties of the elements in the given group?

PROBLEMS

49. Write the full electron configuration ($1s^22s^2$, etc.) for each of the following elements.
a. helium, $Z = 2$
b. neon, $Z = 10$
c. argon, $Z = 18$
d. krypton, $Z = 36$

50. Write the full electron configuration ($1s^22s^2$, etc.) for each of the following elements.
a. strontium, $Z = 38$
b. zinc, $Z = 30$
c. helium, $Z = 2$
d. bromine, $Z = 35$

51. Write the full electron configuration ($1s^22s^2$, etc.) for each of the following elements.
a. sodium, $Z = 11$
b. cesium, $Z = 55$
c. nitrogen, $Z = 7$
d. beryllium, $Z = 4$

52. Write the full electron configuration ($1s^22s^2$, etc.) for each of the following elements.
a. potassium, $Z = 19$
b. chlorine, $Z = 17$
c. magnesium, $Z = 12$
d. carbon, $Z = 6$

53. Write the complete orbital diagram for each of the following elements, using boxes to represent orbitals and arrows to represent electrons.
a. helium, $Z = 2$
b. neon, $Z = 10$
c. krypton, $Z = 36$
d. xenon, $Z = 54$

54. Write the complete orbital diagram for each of the following elements, using boxes to represent orbitals and arrows to represent electrons.
a. aluminum, $Z = 13$
b. phosphorus, $Z = 15$
c. bromine, $Z = 35$
d. argon, $Z = 18$

55. How many valence electrons does each of the following atoms possess?
a. lithium, $Z = 3$ c. argon, $Z = 18$
b. aluminum, $Z = 13$ d. phosphorus, $Z = 15$

56. How many valence electrons does each of the following atoms possess?
a. cesium, $Z = 55$ c. krypton, $Z = 36$
b. boron, $Z = 5$ d. magnesium, $Z = 12$

10.10 Electron Configurations and the Periodic Table

QUESTIONS

57. Why do we believe that the valence electrons of calcium and potassium reside in the $4s$ orbital rather than in the $3d$ orbital?

58. Would you expect the valence electrons of rubidium and strontium to reside in the $5s$ or the $4d$ orbitals? Why?

PROBLEMS

59. Using the symbol of the previous noble gas to indicate the core electrons, write the electron configuration for each of the following elements.
   a. zirconium, $Z = 40$
   b. vanadium, $Z = 23$
   c. bromine, $Z = 35$
   d. silicon, $Z = 14$

60. Using the symbol of the previous noble gas to indicate the core electrons, write the valence shell electron configuration for each of the following elements.
   a. calcium, $Z = 20$
   b. francium, $Z = 87$
   c. yttrium, $Z = 39$
   d. cerium, $Z = 58$

61. Using the symbol of the previous noble gas to indicate the core electrons, write the electron configuration for each of the following elements.
   a. scandium, $Z = 21$
   b. yttrium, $Z = 39$
   c. lanthanum, $Z = 57$
   d. actinium, $Z = 89$

62. Using the symbol of the previous noble gas to indicate the core electrons, write the valence shell electron configuration for each of the following elements.
   a. phosphorus, $Z = 15$
   b. chlorine, $Z = 17$
   c. magnesium, $Z = 12$
   d. zinc, $Z = 30$

63. How many 3*d* electrons are found in each of the following elements?
   a. scandium, $Z = 21$
   b. chromium, $Z = 24$
   c. zinc, $Z = 30$
   d. titanium, $Z = 22$

64. How many 4*d* electrons are found in each of the following elements?
   a. yttrium, $Z = 39$
   b. zirconium, $Z = 40$
   c. strontium, $Z = 38$
   d. cadmium, $Z = 48$

65. For each of the following elements, indicate which set of orbitals is filled last.
   a. uranium, $Z = 92$
   b. polonium, $Z = 84$
   c. silver, $Z = 47$
   d. zirconium, $Z = 40$

66. For each of the following elements, indicate which set of orbitals is being filled last.
   a. plutonium, $Z = 94$
   b. nobelium, $Z = 102$
   c. praeseodymium, $Z = 59$
   d. radon, $Z = 86$

67. Write the valence shell electron configuration of each of the following elements, basing your answer on the element's location on the periodic table.
   a. hafnium, $Z = 72$
   b. radium, $Z = 88$
   c. antimony, $Z = 51$
   d. lead, $Z = 82$

68. Write the shorthand electron configuration for each of the following elements, basing your answer on the location of the element in the periodic table.
   a. palladium, $Z = 46$
   b. neptunium, $Z = 93$
   c. ruthenium, $Z = 44$
   d. gold, $Z = 79$

10.11 Atomic Properties and the Periodic Table

QUESTIONS

69. What are some of the physical properties that distinguish the metallic elements from the nonmetals? Are these properties absolute, or do some nonmetallic elements exhibit some metallic properties (and vice versa)?

70. What types of ions do the metals and the nonmetallic elements form? Do the metals lose or gain electrons in doing this? Do the nonmetallic elements gain or lose electrons in doing this?

71. Give some similarities that exist among the elements of Group 1.

72. Give some similarities that exist among the elements of Group 7.

73. Which element in Group 1 most easily loses electrons? Why?

74. Which elements in a given period (horizontal row) of the periodic table lose electrons most easily? Why?

75. Where are the most nonmetallic elements located on the periodic table? Why do these elements pull electrons from metallic elements so effectively during a reaction?

76. Why do the metallic elements of a given period (horizontal row) typically have much lower ionization energies than do the nonmetallic elements of the same period?

77. Which element in Group 2 has the largest-sized atoms? Why?

78. Though all the elements in a given period (horizontal row) of the periodic table have their valence electrons in the same types of orbitals, the sizes of the atoms decrease from left to right within a period. Explain why.

PROBLEMS

79. In each of the following groups, which element is least reactive?
   a. Group 1
   b. Group 7
   c. Group 2
   d. Group 6

80. In each of the following sets of elements, which element would be expected to have the highest ionization energy?
   a. Cs, K, Li
   b. Ba, Sr, Ca
   c. I, Br, Cl
   d. Mg, Si, S

81. Arrange the following sets of elements in order of increasing atomic size.
   a. Sn, Xe, Rb, Sr
   b. Rn, He, Xe, Kr
   c. Pb, Ba, Cs, At

82. In each of the following sets of elements, indicate which element has the smallest atomic size.
   a. Na, K, Rb
   b. Na, Si, S
   c. N, P, As
   d. N, O, F

ADDITIONAL PROBLEMS

83. The distance in meters between two consecutive peaks (or troughs) in a wave is called the ________.

84. The speed at which electromagnetic radiation moves through a vacuum is called the ________.

85. The portion of the electromagnetic spectrum between wavelengths of approximately 400 and 700 nanometers is called the ________ region.

86. A beam of light can be thought of as consisting of a stream of light particles called ________.

87. The lowest possible energy state of an atom is called the ________ state.

88. The energy levels of hydrogen (and other atoms) are ________, which means that only certain values of energy are allowed.

89. According to Bohr, the electron in the hydrogen atom moved around the nucleus in circular paths called ________.

90. In the modern theory of the atom, a(n) ________ represents a region of space in which there is a high probability of finding an electron.

91. Electrons found in the outermost principal energy level of an atom are referred to as ________ electrons.

92. An element with partially filled *d* orbitals is called a(n) ________.

93. The ________ of electromagnetic radiation represents the number of waves passing a given point in space each second.

94. Only two electrons can occupy a given orbital in an atom, and to be in the same orbital, they must have opposite ________.

95. One bit of evidence that the present theory of atomic structure is "correct" lies in the magnetic properties of matter. Atoms with *unpaired* electrons are attracted by magnetic fields and thus are said to exhibit *paramagnetism*. The degree to which this effect is observed is directly related to the *number* of unpaired electrons present in the atom. On the basis of the electron orbital diagrams for the following elements, indicate which atoms would be expected to be paramagnetic, and tell how many unpaired electrons each atom contains.
   a. phosphorus, $Z = 15$
   b. iodine, $Z = 53$
   c. germanium, $Z = 32$

96. Without referring to your textbook or a periodic table, write the full electron configuration, the orbital box diagram, and the noble gas shorthand configuration for the elements with the following atomic numbers.
   a. $Z = 19$
   b. $Z = 22$
   c. $Z = 14$
   d. $Z = 26$
   e. $Z = 30$

97. Without referring to your textbook or a periodic table, write the full electron configuration, the orbital box diagram, and the noble gas shorthand configuration for the elements with the following atomic numbers.
   a. $Z = 21$
   b. $Z = 15$
   c. $Z = 36$
   d. $Z = 38$
   e. $Z = 30$

98. Write the general valence configuration (for example, ns^1 for Group 1) for the group in which each of the following elements is found.
a. barium, $Z = 56$
b. bromine, $Z = 35$
c. tellurium, $Z = 52$
d. potassium, $Z = 19$
e. sulfur, $Z = 16$

99. How many valence electrons does each of the following atoms have?
a. titanium, $Z = 22$
b. iodine, $Z = 53$
c. radium, $Z = 88$
d. manganese, $Z = 25$

100. In the text (Section 10.6) it was mentioned that current theories of atomic structure suggest that all matter and all energy demonstrate both particle-like and wave-like properties under the appropriate conditions, although the wave-like nature of matter becomes apparent only in very small and very fast-moving particles. The relationship between wavelength (λ) observed for a particle and the mass and velocity of that particle is called the de Broglie relationship. It is

$$\lambda = h/mv$$

in which h is Planck's constant (6.63×10^{-34} J · s),* m represents the mass of the particle in kilograms, and v represents the velocity of the particle in meters per second. Calculate the "de Broglie wavelength" for each of the following, and use your numerical answers to explain why macroscopic (large) objects are not ordinarily discussed in terms of their "wave-like" properties.
a. an electron moving at 0.90 times the speed of light
b. a 150-g ball moving at a speed of 10. m/s
c. a 75-kg person walking at a speed of 2 km/h

101. Light waves move through space at a speed of ________ meters per second.

* Note that s is the abbreviation for "seconds."

102. How do we know that the energy levels of the hydrogen atom are not *continuous*, as physicists originally assumed?

103. How does the attractive force that the nucleus exerts on an electron change with the principal energy level of the electron?

104. Into how many sublevels is the third principal energy level of hydrogen divided? What are the names of the orbitals that constitute these sublevels? What are the general shapes of these orbitals?

105. A student writes the electron configuration of carbon ($Z = 6$) as $1s^3 2s^3$. Explain to him what is *wrong* with this configuration.

106. Which of the following orbital designations is (are) *not* correct?
a. $1p$
b. $3d$
c. $3f$
d. $2p$
e. $5f$
f. $6s$

107. Why do the two electrons in the $2p$ sublevel of carbon occupy *different* $2p$ orbitals?

108. Write the full electron configuration ($1s^2 2s^2$, etc.) for each of the following elements.
a. bromine, $Z = 35$
b. xenon, $Z = 54$
c. barium, $Z = 56$
d. selenium, $Z = 34$

109. Write the complete orbital diagram for each of the following elements, using boxes to represent orbitals and arrows to represent electrons.
a. scandium, $Z = 21$
b. sulfur, $Z = 16$
c. potassium, $Z = 19$
d. nitrogen, $Z = 7$

110. How many valence electrons does each of the following atoms have?
a. nitrogen, $Z = 7$
b. chlorine, $Z = 17$
c. sodium, $Z = 11$
d. aluminum, $Z = 13$

111. What name is given to the series of ten elements in which the electrons are filling the $3d$ sublevel?

112. Using the symbol of the previous noble gas to indicate the core electrons, write the valence shell electron configuration for each of the following elements.
a. zirconium, $Z = 40$
b. iodine, $Z = 53$
c. germanium, $Z = 32$
d. cesium, $Z = 55$

113. Using the symbol of the previous noble gas to indicate core electrons, write the valence shell electron configuration for each of the following elements.
a. titanium, $Z = 22$
b. selenium, $Z = 34$
c. antimony, $Z = 51$
d. strontium, $Z = 38$

114. For each of the following elements, indicate which set of orbitals is filled last.
a. chromium, $Z = 24$
b. silver, $Z = 47$
c. uranium, $Z = 92$
d. germanium, $Z = 32$

115. Write the shorthand valence shell electron configuration of each of the following elements, basing your answer on the element's location on the periodic table.
a. nickel, $Z = 28$
b. niobium, $Z = 41$
c. hafnium, $Z = 72$
d. astatine, $Z = 85$

116. Metals have relatively (low/high) ionization energies, whereas nonmetals have relatively (high/low) ionization energies.

117. In each of the following sets of elements, indicate which element shows the most active chemical behavior.
a. B, Al, In
b. Na, Al, S
c. B, C, F

118. In each of the following sets of elements, indicate which element has the smallest atomic size.
a. Ba, Ca, Ra
b. P, Si, Al
c. Rb, Cs, K

11

Chemical Bonding

Natural rock formations in Bryce Canyon, Utah.

The world around us is composed almost entirely of compounds and mixtures of compounds. Rocks, coal, soil, petroleum, trees, and human beings are all complex mixtures of chemical compounds in which different kinds of atoms are bound together. Most of the pure elements found in the earth's crust also contain many atoms bound together. In a gold nugget each gold atom is bound to many other gold atoms, and in a diamond many carbon atoms are bonded very strongly to each other. Substances composed of unbound atoms do exist in nature, but they are very rare. (Examples include the argon atoms in the atmosphere and the helium atoms found in natural gas reserves.)

The manner in which atoms are bound together has a profound effect on the chemical and physical properties of substances. For example, both graphite and diamond are composed solely of carbon atoms. However, graphite is a soft, slippery material used as a lubricant in locks, and diamond is one of the hardest materials known, valuable both as a gemstone and in industrial cutting tools. Why do these materials, both composed solely of carbon atoms, have such different properties? The answer lies in the different ways in which the carbon atoms are bound to each other in these substances.

Molecular bonding and structure play the central role in determining the course of chemical reactions, many of which are vital to our survival. Most reactions in biological systems are very sensitive to the structures of the participating molecules; in fact, very subtle differences in shape sometimes serve to channel the chemical reaction one way rather than another. Molecules that act as drugs must have exactly the right structure to perform their functions correctly. Structure also plays a central role in our senses of smell and taste. Substances have a particular odor because they fit into the specially shaped receptors in our nasal passages. Taste is also dependent on molecular shape, as we discuss in the "Chemistry in Focus" on page 340.

Diamond, composed of carbon atoms bonded together to produce one of the hardest materials known, makes a beautiful gemstone.

To understand the behavior of natural materials, we must understand the nature of chemical bonding and the factors that control the structures of compounds. In this chapter, we will present various classes of compounds that illustrate the different types of bonds. We will then develop models to describe the structure and bonding that characterize the materials found in nature.

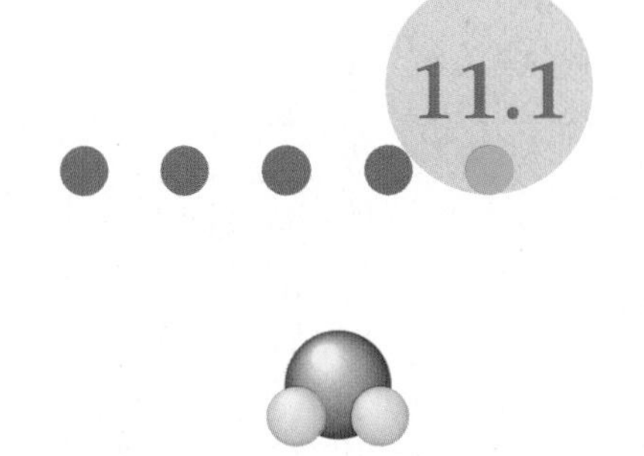

11.1 Types of Chemical Bonds

AIMS: To learn about ionic and covalent bonds and explain how they are formed. To learn about the polar covalent bond.

A water molecule.

What is a chemical bond? Although there are several possible ways to answer this question, we will define a **bond** as a force that holds groups of two or more atoms together and makes them function as a unit. For example, in water the fundamental unit is the H—O—H molecule, which we describe as being held together by the two O—H bonds. We can obtain information about the strength of a bond by measuring the energy required to break the bond, the **bond energy.**

Atoms can interact with one another in several ways to form aggregates. We will consider specific examples to illustrate the various types of chemical bonds.

In Chapter 7 we saw that when solid sodium chloride is dissolved in water, the resulting solution conducts electricity, a fact that convinces chemists that sodium chloride is composed of Na^+ and Cl^- ions. Thus, when sodium and chlorine react to form sodium chloride, electrons are transferred from the sodium atoms to the chlorine atoms to form Na^+ and Cl^- ions, which then aggregate to form solid sodium chloride. The resulting solid sodium chloride is a very sturdy material; it has a melting point of approximately 800 °C. The strong bonding forces present in sodium chloride result from the attractions among the closely packed, oppositely charged ions. This is an example of **ionic bonding.** Ionic substances are formed when an atom that loses electrons relatively easily reacts with an atom that has a high affinity for electrons. In other words, an **ionic compound** results when a metal reacts with a nonmetal.

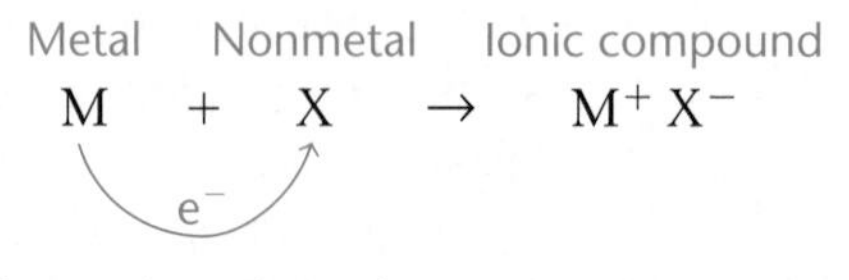

We have seen that a bonding force develops when two very different types of atoms react to form oppositely charged ions. But how does a bonding force develop between two identical atoms? Let's explore this situation by considering what happens when two hydrogen atoms are brought close together, as shown in Figure 11.1. When hydrogen atoms are close together, the two electrons are simultaneously attracted to both nuclei. Note in Figure 11.1b how the electron probability increases between the two nuclei indicating that the electrons are shared by the two nuclei.

The type of bonding we encounter in the hydrogen molecule and in many other molecules where *electrons are shared by nuclei* is called **covalent bonding.** Note that in the H_2 molecule the electrons reside primarily in the space between the two nuclei, where they are attracted simultaneously by both protons. Although we will not go into detail about it here, the increased attractive forces in this area lead to the formation of the H_2 molecule from the two separated hydrogen atoms. When we say that a bond is formed between the hydrogen atoms, we mean that the H_2 molecule is more stable than two separated hydrogen atoms by a certain quantity of energy (the bond energy).

Ionic and covalent bonds are the extreme bond types.

So far we have considered two extreme types of bonding. In ionic bonding, the participating atoms are so different that one or more electrons are

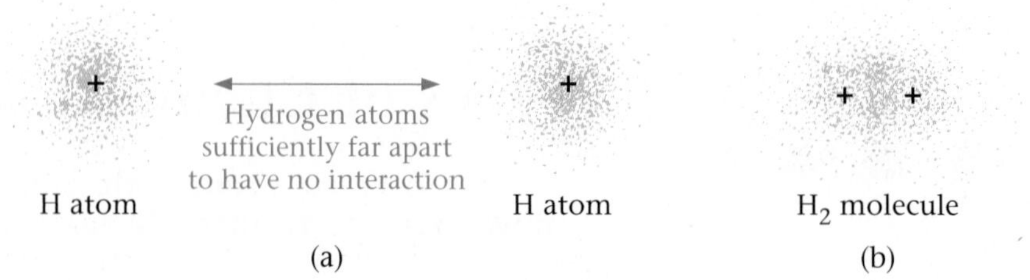

Figure 11.1
The formation of a bond between two hydrogen atoms. (a) Two separate hydrogen atoms. (b) When two hydrogen atoms come close together, the two electrons are attracted simultaneously by both nuclei. This produces the bond. Note the relatively large electron probability between the nuclei indicating sharing of the electrons.

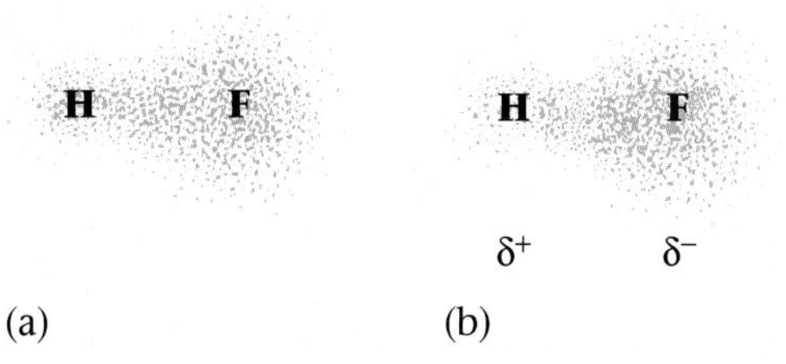

Figure 11.2
Probability representations of the electron sharing in HF. (a) What the probability map would look like if the two electrons in the H—F bond were shared equally. (b) The actual situation, where the shared pair spends more time close to the fluorine atom than to the hydrogen atom. This gives fluorine a slight excess of negative charge and the hydrogen a slight deficit of negative charge (a slight positive charge).

transferred to form oppositely charged ions. The bonding results from the attractions among these ions. In covalent bonding, two identical atoms share electrons equally. The bonding results from the mutual attraction of the two nuclei for the shared electrons. Between these extremes are intermediate cases in which the atoms are not so different that electrons are completely transferred but are different enough so that unequal sharing of electrons results, forming what is called a **polar covalent bond.** The hydrogen fluoride (HF) molecule contains this type of bond, which produces the following charge distribution,

$$\underset{\delta^+}{\text{H}}\text{—}\underset{\delta^-}{\text{F}}$$

where δ (delta) is used to indicate a partial or fractional charge.

The most logical explanation for the development of *bond polarity* (the partial positive and negative charges on the atoms in such molecules as HF) is that the electrons in the bonds are not shared equally. For example, we can account for the polarity of the HF molecule by assuming that the fluorine atom has a stronger attraction than the hydrogen atom for the shared electrons (Figure 11.2). Because bond polarity has important chemical implications, we find it useful to assign a number that indicates an atom's ability to attract shared electrons. In the next section we show how this is done.

11.2 Electronegativity

AIM: To understand the nature of bonds and their relationship to electronegativity.

We saw in the previous section that when a metal and a nonmetal react, one or more electrons are transferred from the metal to the nonmetal to give ionic bonding. On the other hand, two identical atoms react to form a covalent bond in which electrons are shared equally. When *different* nonmetals react, a bond forms in which electrons are shared *unequally,* giving a polar covalent bond. The unequal sharing of electrons between two atoms is described by a property called **electronegativity:** *the relative ability of an atom in a molecule to attract shared electrons to itself.*

Increasing electronegativity

Decreasing electronegativity

				H 2.1												
Li 1.0	Be 1.5											B 2.0	C 2.5	N 3.0	O 3.5	F 4.0
Na 0.9	Mg 1.2											Al 1.5	Si 1.8	P 2.1	S 2.5	Cl 3.0
K 0.8	Ca 1.0	Sc 1.3	Ti 1.5	V 1.6	Cr 1.6	Mn 1.5	Fe 1.8	Co 1.9	Ni 1.9	Cu 1.9	Zn 1.6	Ga 1.6	Ge 1.8	As 2.0	Se 2.4	Br 2.8
Rb 0.8	Sr 1.0	Y 1.2	Zr 1.4	Nb 1.6	Mo 1.8	Tc 1.9	Ru 2.2	Rh 2.2	Pd 2.2	Ag 1.9	Cd 1.7	In 1.7	Sn 1.8	Sb 1.9	Te 2.1	I 2.5
Cs 0.7	Ba 0.9	La-Lu 1.0-1.2	Hf 1.3	Ta 1.5	W 1.7	Re 1.9	Os 2.2	Ir 2.2	Pt 2.2	Au 2.4	Hg 1.9	Tl 1.8	Pb 1.9	Bi 1.9	Po 2.0	At 2.2
Fr 0.7	Ra 0.9	Ac 1.1	Th 1.3	Pa 1.4	U 1.4	Np-No 1.4-1.3										

Key

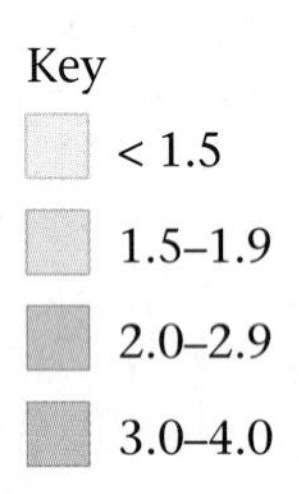

Figure 11.3
Electronegativity values for selected elements. Note that electronegativity generally increases across a period and decreases down a group. Note also that metals have relatively low electronegativity values and that nonmetals have relatively high values.

Chemists determine electronegativity values for the elements (Figure 11.3) by measuring the polarities of the bonds between various atoms. Note that electronegativity generally increases going from left to right across a period and decreases going down a group for the representative elements. The range of electronegativity values is from 4.0 for fluorine to 0.7 for cesium and francium. Remember, the higher the atom's electronegativity value, the closer the shared electrons tend to be to that atom when it forms a bond.

The polarity of a bond depends on the *difference* between the electronegativity values of the atoms forming the bond. If the atoms have very similar electronegativities, the electrons are shared almost equally and the bond shows little polarity. If the atoms have very different electronegativity values, a very polar bond is formed. In extreme cases one or more electrons are actually transferred, forming ions and an ionic bond. For example, when an element from Group 1 (electronegativity values of about 0.8) reacts with an element from Group 7 (electronegativity values of about 3), ions are formed and an ionic substance results.

The relationship between electronegativity and bond type is shown in Table 11.1. The various types of bonds are summarized in Figure 11.4.

Table 11.1 The Relationship Between Electronegativity and Bond Type

Electronegativity Difference Between the Bonding Atoms	Bond Type	Covalent Character	Ionic Character
Zero	Covalent	Decreases ↓	Increases ↓
↓	↓		
Intermediate	Polar covalent		
↓	↓		
Large	Ionic		

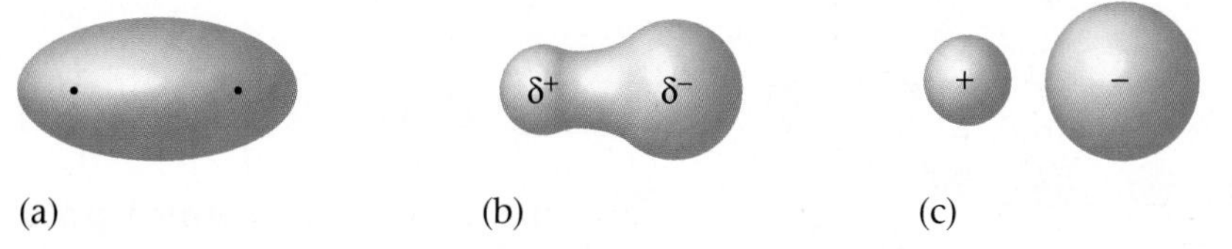

Figure 11.4
The three possible types of bonds: (a) a covalent bond formed between identical atoms; (b) a polar covalent bond, with both ionic and covalent components; and (c) an ionic bond, with no electron sharing.

Example 11.1 Using Electronegativity to Determine Bond Polarity

Using the electronegativity values given in Figure 11.3, arrange the following bonds in order of increasing polarity: H—H, O—H, Cl—H, S—H, and F—H.

Solution

The polarity of the bond increases as the difference in electronegativity increases. From the electronegativity values in Figure 11.3, the following variation in bond polarity is expected (the electronegativity value appears in parentheses below each element).

Bond	Electronegativity Values	Difference in Electronegativity Values	Bond Type	Polarity
H—H	(2.1)(2.1)	2.1 − 2.1 = 0	Covalent	Increasing ↓
S—H	(2.5)(2.1)	2.5 − 2.1 = 0.4	Polar covalent	
Cl—H	(3.0)(2.1)	3.0 − 2.1 = 0.9	Polar covalent	
O—H	(3.5)(2.1)	3.5 − 2.1 = 1.4	Polar covalent	
F—H	(4.0)(2.1)	4.0 − 2.1 = 1.9	Polar covalent	

Therefore, in order of increasing polarity, we have

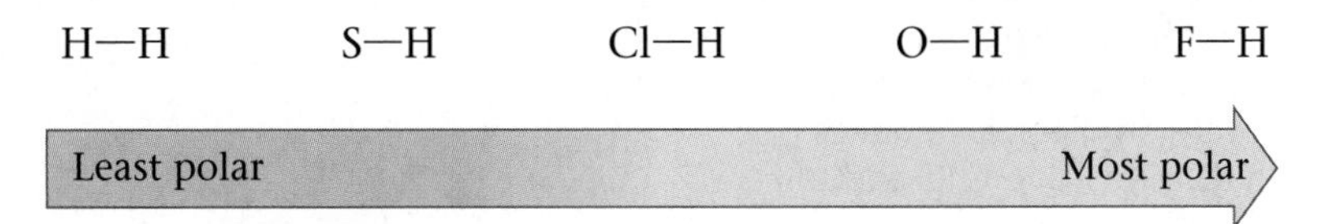

✔ Self-Check Exercise 11.1

For each of the following pairs of bonds, choose the bond that will be more polar.

a. H—P, H—C
b. O—F, O—I
c. N—O, S—O
d. N—H, Si—H

See Problems 11.17 through 11.20. ■

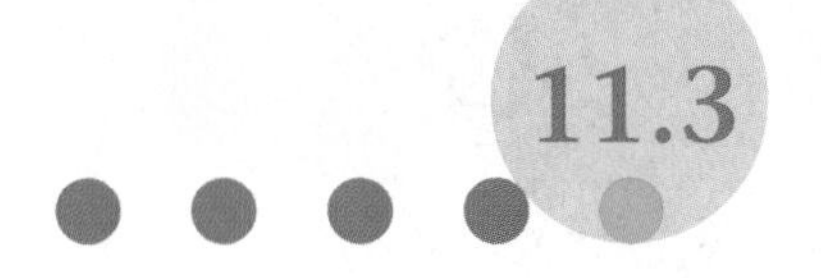

Bond Polarity and Dipole Moments

AIM: To understand bond polarity and how it is related to molecular polarity.

We saw in Section 11.1 that hydrogen fluoride has a positive end and a negative end. A molecule such as HF that has a center of positive charge and a center of negative charge is said to have a **dipole moment.** The dipolar character of a molecule is often represented by an arrow. This arrow points toward the negative charge center, and its tail indicates the positive center of charge:

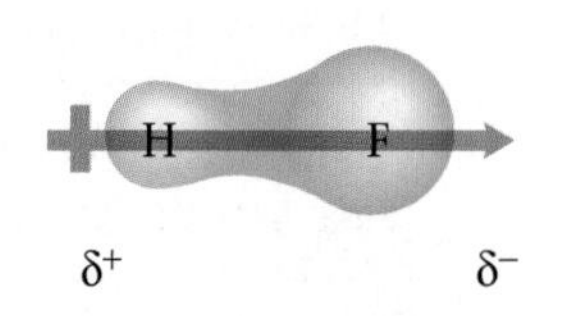

Any diatomic (two-atom) molecule that has a polar bond has a dipole moment. Some polyatomic (more than two atoms) molecules also have dipole moments. For example, because the oxygen atom in the water molecule has a greater electronegativity than the hydrogen atoms, the electrons are not shared equally. This results in a charge distribution (Figure 11.5) that causes the molecule to behave as though it had two centers of charge—one positive and one negative. So the water molecule has a dipole moment.

The fact that the water molecule is polar (has a dipole moment) has a profound impact on its properties. In fact, it is not overly dramatic to state that the polarity of the water molecule is crucial to life as we know it on earth. Because water molecules are polar, they can surround and attract both positive and negative ions (Figure 11.6). These attractions allow ionic materials to dissolve in water. Also, the polarity of water molecules causes them to attract each other strongly (Figure 11.7). This means that much energy is required to change water from a liquid to a gas (the molecules must be separated from each other to undergo this change of state). Therefore, it is the polarity of the water molecule that causes water to remain a liquid at

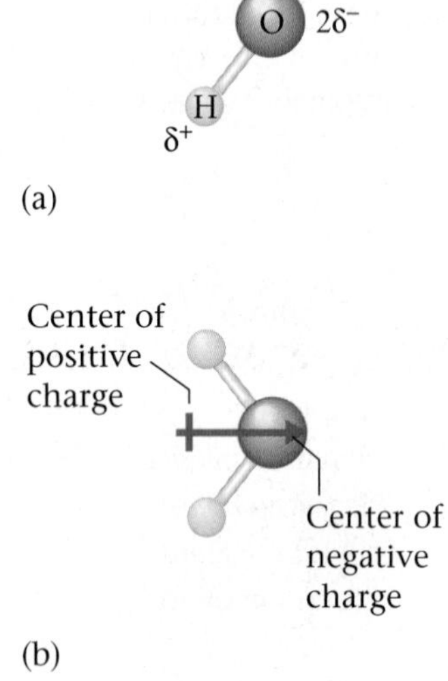

Figure 11.5
(a) The charge distribution in the water molecule. The oxygen has a charge of $2\delta^-$ because it pulls δ^- of charge from each hydrogen atom ($\delta^- + \delta^- = 2\delta^-$). (b) The water molecule behaves as if it had a positive end and a negative end, as indicated by the arrow.

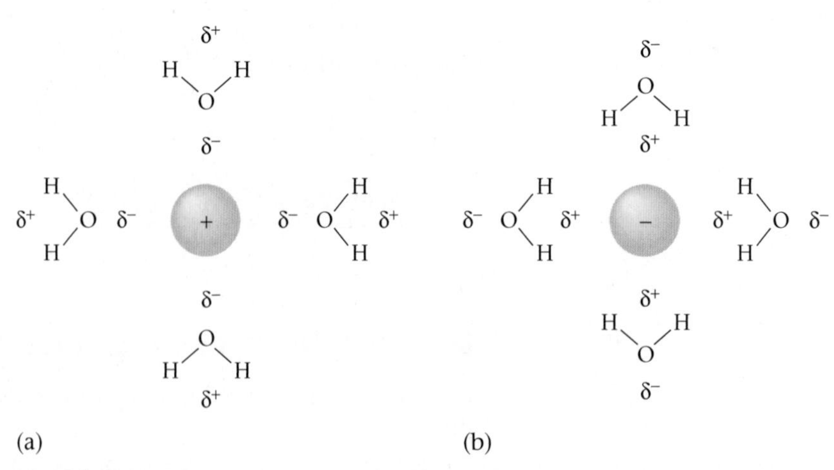

Figure 11.6
(a) Polar water molecules are strongly attracted to positive ions by their negative ends. (b) They are also strongly attracted to negative ions by their positive ends.

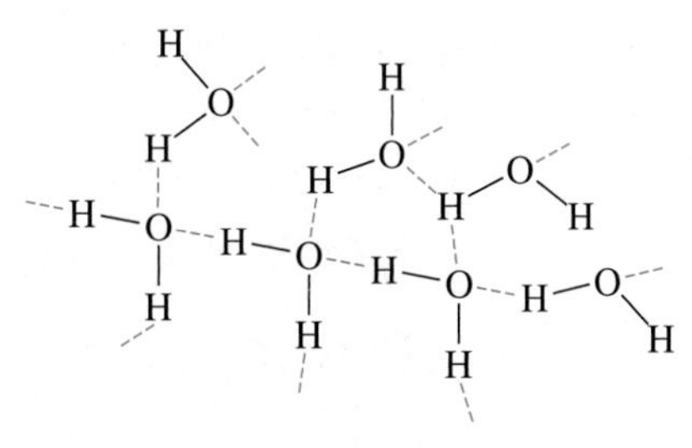

Figure 11.7
Polar water molecules are strongly attracted to each other.

the temperatures on the earth's surface. If it were nonpolar, water would be a gas and the oceans would be empty.

Stable Electron Configurations and Charges on Ions

AIMS: To learn about stable electron configurations. To learn to predict the formulas of ionic compounds.

We have seen many times that when a metal and a nonmetal react to form an ionic compound, the metal atom loses one or more electrons to the nonmetal. In Chapter 5, where binary ionic compounds were introduced, we saw that in these reactions, Group 1 metals always form 1+ cations, Group 2 metals always form 2+ cations, and aluminum in Group 3 always forms a 3+ cation. For the nonmetals, the Group 7 elements always form 1− anions, and the Group 6 elements always form 2− anions. This is further illustrated in Table 11.2.

Notice something very interesting about the ions in Table 11.2: they all have the electron configuration of neon, a noble gas. That is, sodium loses its one valence electron (the 3*s*) to form Na^+, which has an [Ne] electron configuration. Likewise, Mg loses its two valence electrons to form Mg^{2+}, which also has an [Ne] electron configuration. On the other hand, the nonmetal atoms gain just the number of electrons needed for them to achieve the noble gas electron configuration. The O atom gains two electrons and the F atom gains one electron to give O^{2-} and F^-, respectively, both of which

Table 11.2 The Formation of Ions by Metals and Nonmetals

Group	Ion Formation	Electron Configuration: Atom		Electron Configuration: Ion
1	$Na \rightarrow Na^+ + e^-$	$[Ne]3s^1$	$\xrightarrow{e^- \text{ lost}}$	[Ne]
2	$Mg \rightarrow Mg^{2+} + 2e^-$	$[Ne]3s^2$	$\xrightarrow{2e^- \text{ lost}}$	[Ne]
3	$Al \rightarrow Al^{3+} + 3e^-$	$[Ne]3s^23p^1$	$\xrightarrow{3e^- \text{ lost}}$	[Ne]
6	$O + 2e^- \rightarrow O^{2-}$	$[He]2s^22p^4 + 2e^-$	$\rightarrow [He]2s^22p^6 =$	[Ne]
7	$F + e^- \rightarrow F^-$	$[He]2s^22p^5 + e^-$	$\rightarrow [He]2s^22p^6 =$	[Ne]

have the [Ne] electron configuration. We can summarize these observations as follows:

Electron Configurations of Ions

1. Representative (main-group) metals form ions by losing enough electrons to achieve the configuration of the previous noble gas (that is, the noble gas that occurs before the metal in question on the periodic table). For example, note from the periodic table inside the front cover of the text that neon is the noble gas previous to sodium and magnesium. Similarly, helium is the noble gas previous to lithium and beryllium.
2. Nonmetals form ions by gaining enough electrons to achieve the configuration of the next noble gas (that is, the noble gas that follows the element in question on the periodic table). For example, note that neon is the noble gas that follows oxygen and fluorine, and argon is the noble gas that follows sulfur and chlorine.

Atoms in stable compounds almost always have a noble gas electron configuration.

This brings us to an important general principle. In observing millions of stable compounds, chemists have learned that **in almost all stable chemical compounds of the representative elements, all of the atoms have achieved a noble gas electron configuration.** The importance of this observation cannot be overstated. It forms the basis for all of our fundamental ideas about why and how atoms bond to each other.

We have already seen this principle operating in the formation of ions (see Table 11.2). We can summarize this behavior as follows: when representative metals and nonmetals react, they transfer electrons in such a way that both the cation and the anion have noble gas electron configurations.

On the other hand, when nonmetals react with each other, they share electrons in ways that lead to a noble gas electron configuration for each atom in the resulting molecule. For example, oxygen ($[He]2s^22p^4$), which needs two more electrons to achieve an [Ne] configuration, can get these electrons by combining with two H atoms (each of which has one electron),

		2s	2p
O:	[He]	↑↓	↑↓ ↑↓ ↑↓
			H H

to form water, H_2O. This fills the valence orbitals of oxygen.

In addition, each H shares two electrons with the oxygen atom,

$$\mathrm{H} \overset{\uparrow\downarrow}{\ } \mathrm{O} \overset{\uparrow\downarrow}{\ } \mathrm{H}$$

which fills the H 1*s* orbital, giving it a $1s^2$ or [He] electron configuration. We will have much more to say about covalent bonding in Section 11.6.

At this point let's summarize the ideas we have introduced so far.

Electron Configurations and Bonding

1. When a *nonmetal and a Group 1, 2, or 3 metal* react to form a binary ionic compound, the ions form in such a way that the valence-electron configuration of the *nonmetal* is *completed* to

achieve the configuration of the *next* noble gas, and the valence orbitals of the *metal* are *emptied* to achieve the configuration of the *previous* noble gas. In this way both ions achieve noble gas electron configurations.

2. When *two nonmetals* react to form a covalent bond, they share electrons in a way that completes the valence-electron configurations of both atoms. That is, both nonmetals attain noble gas electron configurations by sharing electrons.

Predicting Formulas of Ionic Compounds

To show how to predict what ions form when a metal reacts with a nonmetal, we will consider the formation of an ionic compound from calcium and oxygen. We can predict what compound will form by considering the valence electron configurations of the following two atoms:

$$\text{Ca: } [\text{Ar}]4s^2$$
$$\text{O: } [\text{He}]2s^22p^4$$

Now that we know something about the electron configurations of atoms, we can explain *why* these various ions are formed.

From Figure 11.3 we see that the electronegativity of oxygen (3.5) is much greater than that of calcium (1.0), giving a difference of 2.5. Because of this large difference, electrons are transferred from calcium to oxygen to form an oxygen anion and a calcium cation. How many electrons are transferred? We can base our prediction on the observation that noble gas configurations are the most stable. Note that oxygen needs two electrons to fill its valence orbitals (2s and 2p) and achieve the configuration of neon ($1s^22s^22p^6$), which is the next noble gas.

$$\text{O} + 2\text{e}^- \rightarrow \text{O}^{2-}$$
$$[\text{He}]2s^22p^4 + 2\text{e}^- \rightarrow [\text{He}]2s^22p^6, \text{ or } [\text{Ne}]$$

And by losing two electrons, calcium can achieve the configuration of argon (the previous noble gas).

$$\text{Ca} \rightarrow \text{Ca}^{2+} + 2\text{e}^-$$
$$[\text{Ar}]4s^2 \rightarrow [\text{Ar}] + 2\text{e}^-$$

Two electrons are therefore transferred as follows:

$$\underbrace{\text{Ca} + \text{O}}_{2\text{e}^-} \rightarrow \text{Ca}^{2+} + \text{O}^{2-}$$

To predict the formula of the ionic compound, we use the fact that chemical compounds are always electrically neutral—they have the same total quantities of positive and negative charges. In this case we must have equal numbers of Ca^{2+} and O^{2-} ions, and the empirical formula of the compound is CaO.

The same principles can be applied to many other cases. For example, consider the compound formed from aluminum and oxygen. Aluminum has the electron configuration $[\text{Ne}]3s^23p^1$. To achieve the neon configuration, aluminum must lose three electrons, forming the Al^{3+} ion.

$$\text{Al} \rightarrow \text{Al}^{3+} + 3\text{e}^-$$
$$[\text{Ne}]3s^23p^1 \rightarrow [\text{Ne}] + 3\text{e}^-$$

$3 \times (2-)$ balances $2 \times (3+)$.

Therefore, the ions will be Al^{3+} and O^{2-}. Because the compound must be electrically neutral, there will be three O^{2-} ions for every two Al^{3+} ions, and the compound has the empirical formula Al_2O_3.

CHEMISTRY IN FOCUS

Composite Cars

In designing fuel-efficient vehicles, weight is the enemy. The more mass a vehicle contains, the more energy will be required to move it. The problem is that saving weight almost always means higher cost. When the latest Corvette (called the C5 by autophiles) was being developed, chief engineer Dave Hill used a $10 per kilogram rule: spending an extra $10 for a part was acceptable if it saved a kilogram of mass.

A new material that is likely to be a boon to auto designers is aluminum metal foam. Metal foams are a new class of material, consisting of a sandwich of porous foamed metal between metal skins. They are 50% lighter and ten times stiffer than the same part made from steel. They are also fireproof, good thermal insulators, and excellent energy absorbers, crushing progressively on impact.

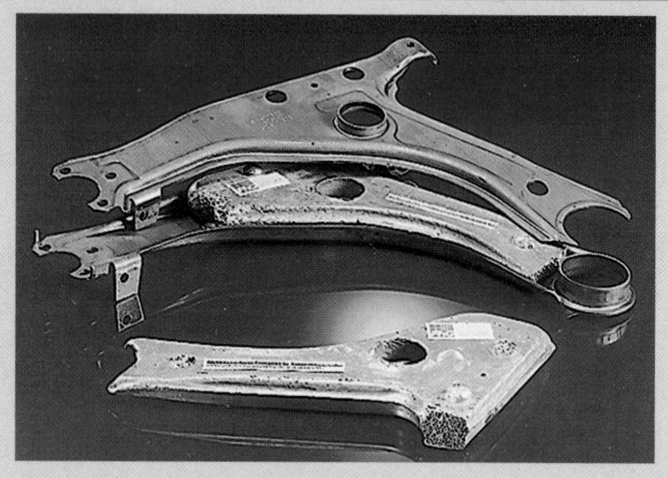

An aluminum foam part with its mold.

Aluminum metal foam was developed by the German automotive supplier Wilhelm Karmann—perhaps most associated in the United States with the Volkswagen Karmann Ghia of the 1960s. The material starts as two aluminum sheets sandwiching an aluminum powder containing a titanium hydride propellant. This assembly is crushed at high pressures into a single flat sheet that, like regular sheet metal, can be formed into a variety of three-dimensional shapes. After shaping, the part is placed in an 1150 °F oven for two minutes, where the aluminum powder melts, releasing hydrogen gas from the titanium hydride. The foaming caused by the $H_2(g)$ increases the material's thickness by a factor of 6, producing an aluminum foam between the aluminum skins. The resulting material has such a low density that it floats on water, but it is ten times stiffer than steel. The material is ideal for automotive floorpans, firewalls, roof panels, and luggage compartment walls. It is projected that as much as 20% of a typical auto could be constructed from the new metal foam. Besides being lightweight and stiff, the new foam also increases the crash-worthiness of a car due to its energy-absorbing abilities. Aluminum foam sounds like a miracle.

Table 11.3 shows common elements that form ions with noble gas electron configurations in ionic compounds.

Notice that our discussion in this section refers to metals in Groups 1, 2, and 3 (the representative metals). The transition metals exhibit more complicated behavior (they form a variety of ions), which we will not be concerned with in this text.

Table 11.3 Common Ions with Noble Gas Configurations in Ionic Compounds

Group 1	Group 2	Group 3	Group 6	Group 7	Electron Configuration
Li^+	Be^{2+}				[He]
Na^+	Mg^{2+}	Al^{3+}	O^{2-}	F^-	[Ne]
K^+	Ca^{2+}		S^{2-}	Cl^-	[Ar]
Rb^+	Sr^{2+}		Se^{2-}	Br^-	[Kr]
Cs^+	Ba^{2+}		Te^{2-}	I^-	[Xe]

11.5 Ionic Bonding and Structures of Ionic Compounds

AIMS: To learn about ionic structures. To understand factors governing ionic size.

When metals and nonmetals react, the resulting ionic compounds are very stable; large amounts of energy are required to "take them apart." For example, the melting point of sodium chloride is approximately 800 °C. The strong bonding in these ionic compounds results from the attractions among the oppositely charged cations and anions.

We write the formula of an ionic compound such as lithium fluoride simply as LiF, but this is really the empirical, or simplest, formula. The actual solid contains huge and equal numbers of Li^+ and F^- ions packed together in a way that maximizes the attractions of the oppositely charged ions. A representative part of the lithium fluoride structure is shown in Figure 11.8a. In this structure the larger F^- ions are packed together like hard spheres, and the much smaller Li^+ ions are interspersed regularly among the F^- ions. The structure shown in Figure 11.8b represents only a tiny part of the actual structure, which continues in all three dimensions with the same pattern as that shown.

When spheres are packed together, they do not fill up all of the space. The spaces (holes) that are left can be occupied by smaller spheres.

The structures of virtually all binary ionic compounds can be explained by a model that involves packing the ions as though they were hard spheres. The larger spheres (usually the anions) are packed together, and the small ions occupy the interstices (spaces or holes) among them.

To understand the packing of ions it helps to realize that *a cation is always smaller than the parent atom, and an anion is always larger than the parent atom.* This makes sense because when a metal loses all of its valence electrons to form a cation, it gets much smaller. On the other hand, in forming an anion, a nonmetal gains enough electrons to achieve the next noble gas electron configuration and so becomes much larger. The relative sizes of the Group 1 and Group 7 atoms and their ions are shown in Figure 11.9.

Ionic Compounds Containing Polyatomic Ions

So far in this chapter we have discussed only binary ionic compounds, which contain ions derived from single atoms. However, many compounds contain polyatomic ions: charged species composed of several atoms. For example, ammonium nitrate contains the NH_4^+ and NO_3^- ions. These ions with their opposite charges attract each other in the same way as do the simple ions in binary ionic compounds. However, the *individual* polyatomic ions are held together by covalent bonds, with all of the atoms behaving as a unit. For example, in the ammonium ion, NH_4^+, there are four N—H covalent

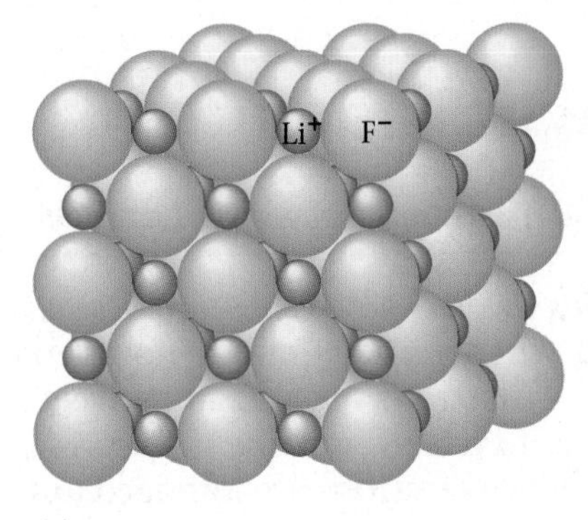

(a)

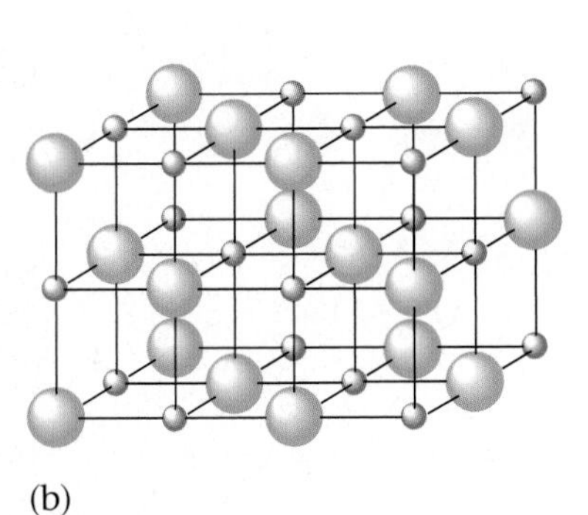
(b)

Figure 11.8
The structure of lithium fluoride. (a) This structure represents the ions as packed spheres. (b) This structure shows the positions (centers) of the ions. The spherical ions are packed in the way that maximizes the ionic attractions.

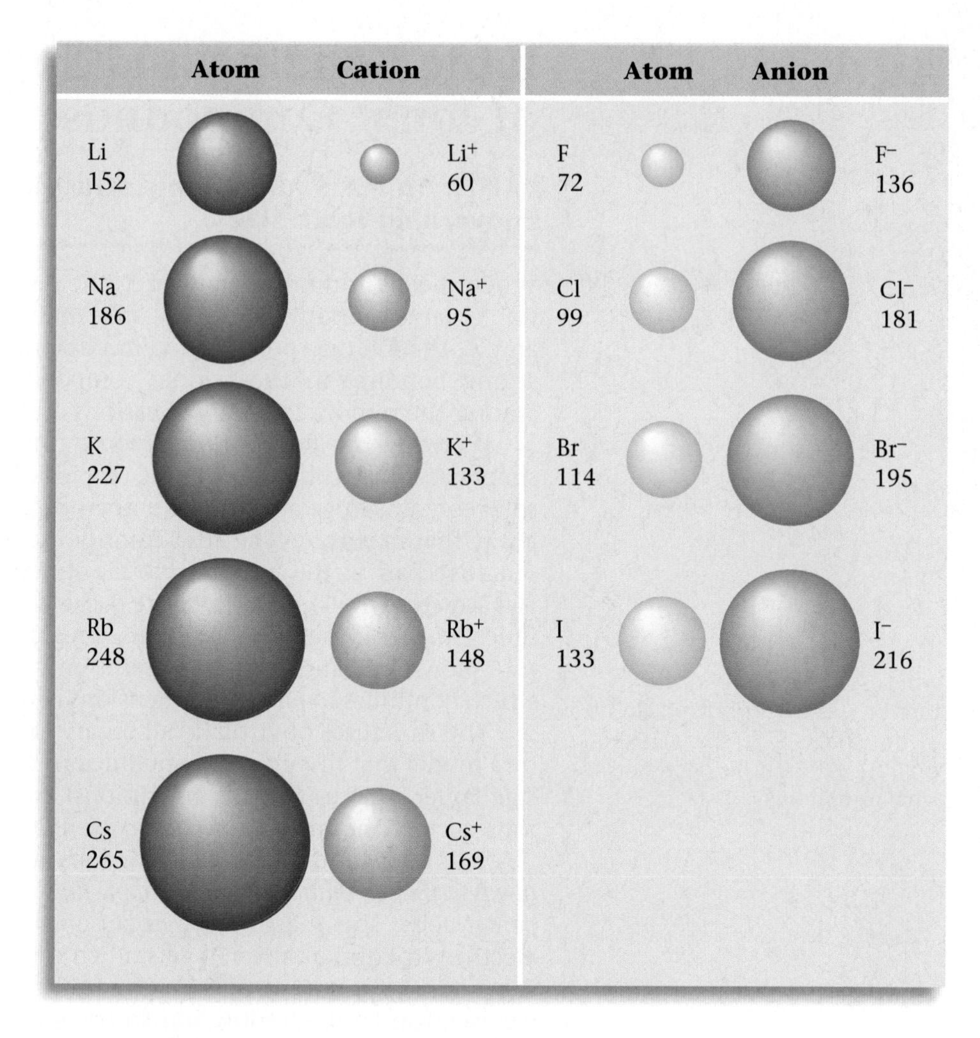

Figure 11.9
Relative sizes of some ions and their parent atoms. Note that cations are smaller and anions are larger than their parent atoms. The sizes (radii) are given in units of picometers (1 pm = 10^{-12} m).

bonds. Likewise the nitrate ion, NO_3^-, contains three covalent N—O bonds. Thus, although ammonium nitrate is an ionic compound because it contains the NH_4^+ and NO_3^- ions, it also contains covalent bonds in the individual polyatomic ions. When ammonium nitrate is dissolved in water, it behaves as a strong electrolyte like the binary ionic compounds sodium chloride and potassium bromide. As we saw in Chapter 7, this occurs because when an ionic solid dissolves, the ions are freed to move independently and can conduct an electric current.

The common polyatomic ions, which are listed in Table 5.4, are all held together by covalent bonds.

Lewis Structures

AIM: To learn to write Lewis structures.

Bonding involves just the valence electrons of atoms. Valence electrons are transferred when a metal and a nonmetal react to form an ionic compound. Valence electrons are shared between nonmetals in covalent bonds.

Remember that the electrons in the highest principal energy level of an atom are called the valence electrons.

The **Lewis structure** is a representation of a molecule that shows how the valence electrons are arranged among the atoms in the molecule. These representations are named after G. N. Lewis, who conceived the idea while lecturing to a class of general chemistry students in 1902. The rules for writ-

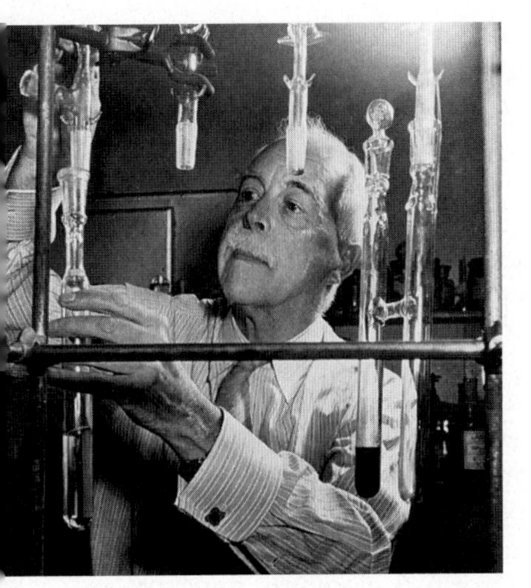

G. N. Lewis in his lab.

ing Lewis structures are based on observations of many molecules from which chemists have learned that the *most important requirement for the formation of a stable compound is that the atoms achieve noble gas electron configurations.*

We have already seen this rule operate in the reaction of metals and nonmetals to form binary ionic compounds. An example is the formation of KBr, where the K^+ ion has the [Ar] electron configuration and the Br^- ion has the [Kr] electron configuration. In writing Lewis structures, *we include only the valence electrons.* Using dots to represent valence electrons, we write the Lewis structure for KBr as follows:

K^+	$[:\ddot{\underset{..}{Br}}:]^-$
Noble gas configuration [Ar]	Noble gas configuration [Kr]

No dots are shown on the K^+ ion because it has lost its only valence electron (the 4*s* electron). The Br^- ion is shown with eight electrons because it has a filled valence shell.

Next we will consider Lewis structures for molecules with covalent bonds, involving nonmetals in the first and second periods. The principle of achieving a noble gas electron configuration applies to these elements as follows:

1. Hydrogen forms stable molecules where it shares two electrons. That is, it follows a **duet rule.** For example, when two hydrogen atoms, each with one electron, combine to form the H_2 molecule, we have

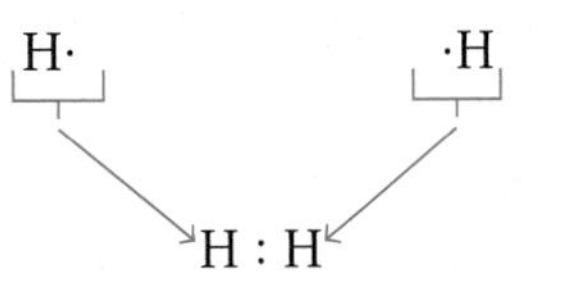

By sharing electrons, each hydrogen in H_2 has, in effect, two electrons; that is, each hydrogen has a filled valence shell.

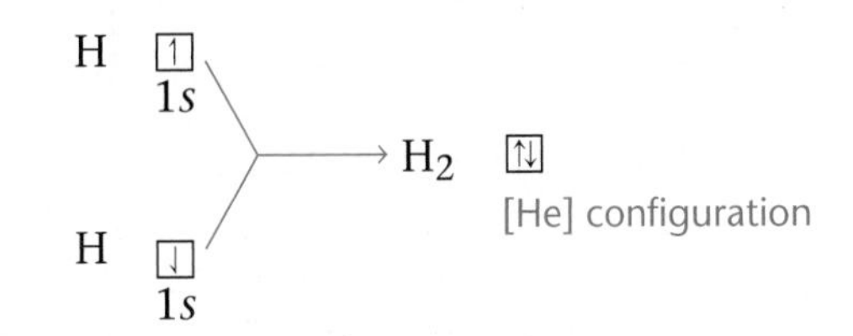

2. Helium does not form bonds because its valence orbital is already filled; it is a noble gas. Helium has the electron configuration $1s^2$ and can be represented by the Lewis structure

He:
[He] configuration

3. The second-row nonmetals carbon through fluorine form stable molecules when they are surrounded by enough electrons to fill the valence orbitals—that is, the one 2*s* and the three 2*p* orbitals. Eight electrons are required to fill these orbitals, so these elements typically obey the **octet rule;** they are surrounded by eight electrons. An example is the F_2 molecule, which has the following Lewis structure:

Carbon, nitrogen, oxygen, and fluorine almost always obey the octet rule in stable molecules.

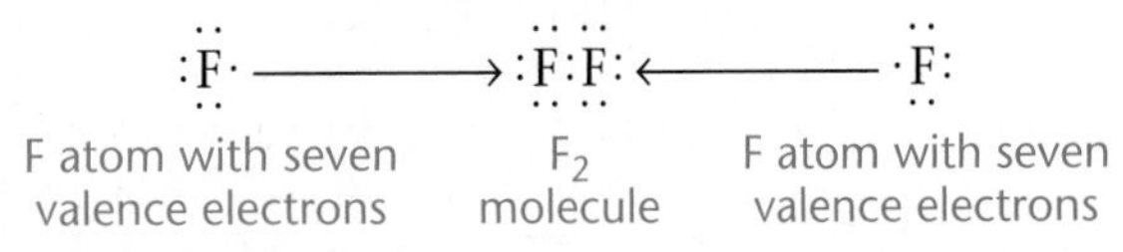

Note that each fluorine atom in F_2 is, in effect, surrounded by eight valence electrons, two of which are shared with the other atom. This is a **bonding pair** of electrons, as we discussed earlier. Each fluorine atom also has three pairs of electrons that are not involved in bonding. These are called **lone pairs** or **unshared pairs.**

4. Neon does not form bonds because it already has an octet of valence electrons (it is a noble gas). The Lewis structure is

$$:\ddot{\underset{..}{Ne}}:$$

Note that only the valence electrons ($2s^22p^6$) of the neon atom are represented by the Lewis structure. The $1s^2$ electrons are core electrons and are not shown.

Lewis structures show only valence electrons.

Next we want to develop some general procedures for writing Lewis structures for molecules. Remember that Lewis structures involve only the valence electrons on atoms, so before we proceed, we will review the relationship of an element's position on the periodic table to the number of valence electrons it has. Recall that the group number gives the total number of valence electrons. For example, all Group 6 elements have six valence electrons (valence configuration ns^2np^4).

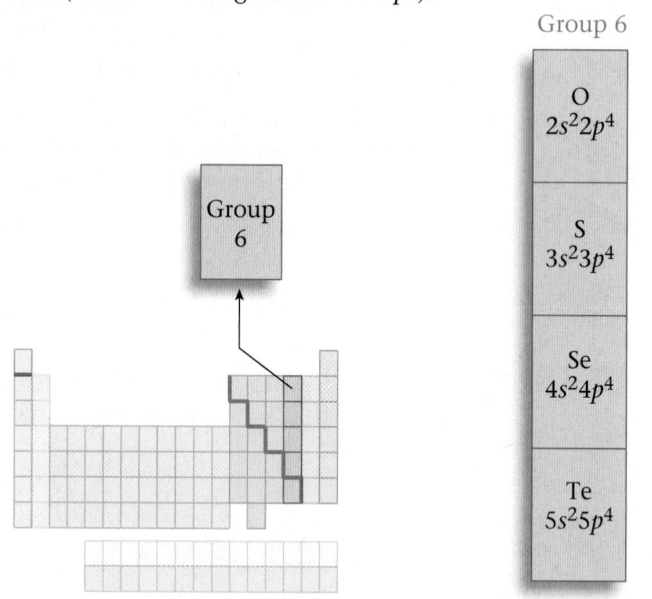

Similarly, all Group 7 elements have seven valence electrons (valence configuration ns^2np^5).

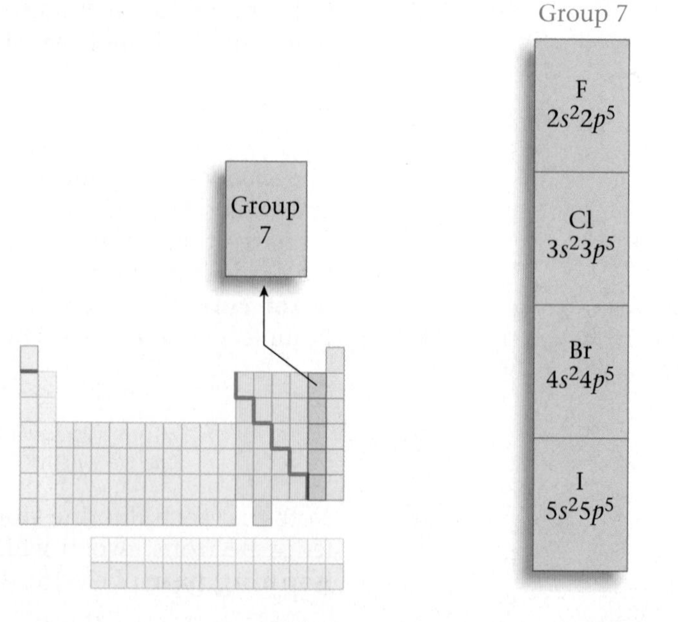

In writing the Lewis structure for a molecule, we need to keep the following things in mind:

1. We must include all the valence electrons from all atoms. The total number of electrons available is the sum of all the valence electrons from all the atoms in the molecule.
2. Atoms that are bonded to each other share one or more pairs of electrons.
3. The electrons are arranged so that each atom is surrounded by enough electrons to fill the valence orbitals of that atom. This means two electrons for hydrogen and eight electrons for second-row nonmetals.

The best way to make sure we arrive at the correct Lewis structure for a molecule is to use a systematic approach. We will use the approach summarized by the following rules.

Steps for Writing Lewis Structures

STEP 1 Obtain the sum of the valence electrons from all of the atoms. Do not worry about keeping track of which electrons come from which atoms. It is the *total* number of valence electrons that is important.

STEP 2 Use one pair of electrons to form a bond between each pair of bound atoms. For convenience, a line (instead of a pair of dots) is often used to indicate each pair of bonding electrons.

STEP 3 Arrange the remaining electrons to satisfy the duet rule for hydrogen and the octet rule for each second-row element.

To see how these rules are applied, we will write the Lewis structures of several molecules.

Example 11.2 Writing Lewis Structures: Simple Molecules

Write the Lewis structure of the water molecule.

Solution

We will follow the steps listed above.

STEP 1 Find the sum of the *valence* electrons for H_2O.

$$\underset{\substack{\text{H}\\ \text{(Group 1)}}}{\overset{1}{\uparrow}} + \underset{\substack{\text{H}\\ \text{(Group 1)}}}{\overset{1}{\uparrow}} + \underset{\substack{\text{O}\\ \text{(Group 6)}}}{\overset{6}{\uparrow}} = 8 \text{ valence electrons}$$

STEP 2 Using a pair of electrons per bond, we draw in the two O—H bonds, using a line to indicate each pair of bonding electrons.

$$\text{H—O—H}$$

Note that

$$\text{H—O—H} \quad \text{represents} \quad \text{H:O:H}$$

STEP 3 We arrange the remaining electrons around the atoms to achieve a noble gas electron configuration for each atom. Four electrons have been used in forming the two bonds, so four electrons (8 − 4) remain to be distributed. Each hydrogen is satisfied with two electrons (duet rule), but oxygen needs eight electrons to have a noble gas electron configuration. So the remaining four electrons are added to oxygen as two lone pairs. Dots are used to represent the lone pairs.

H—Ö—H (with two lone pairs on O) might also be drawn as H:Ö:H

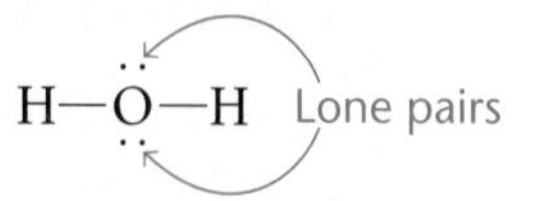

This is the correct Lewis structure for the water molecule. Each hydrogen shares two electrons, and the oxygen has four electrons and shares four to give a total of eight.

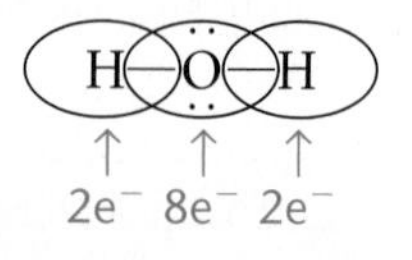

Note that a line is used to represent a shared pair of electrons (bonding electrons) and dots are used to represent unshared pairs.

Self-Check Exercise 11.2

Write the Lewis structure for HCl.

See Problems 11.59 through 11.62. ■

Lewis Structures of Molecules with Multiple Bonds

AIM: To learn how to write Lewis structures for molecules with multiple bonds.

Now let's write the Lewis structure for carbon dioxide.

STEP 1 Summing the valence electrons gives

4	+	6	+	6	= 16
↑		↑		↑	
C		O		O	
(Group 4)		(Group 6)		(Group 6)	

STEP 2 Form a bond between the carbon and each oxygen:

O—C—O

O—C—O represents O:C:O

STEP 3 Next, distribute the remaining electrons to achieve noble gas electron configurations on each atom. In this case twelve electrons (16 − 4) remain after the bonds are drawn. The distribution of these electrons is determined by a trial-and-error process. We have six pairs of electrons to distribute. Suppose we try three pairs on each oxygen to give

:Ö—C—Ö:

:Ö—C—Ö: represents :Ö:C:Ö:

Is this correct? To answer this question we need to check two things:

1. The total number of electrons. There are sixteen valence electrons in this structure, which is the correct number.
2. The octet rule for each atom. Each oxygen has eight electrons around it, but the carbon has only four. This cannot be the correct Lewis structure.

How can we arrange the sixteen available electrons to achieve an octet for each atom? Suppose we place two shared pairs between the carbon and each oxygen:

O=C=O represents O::C::O

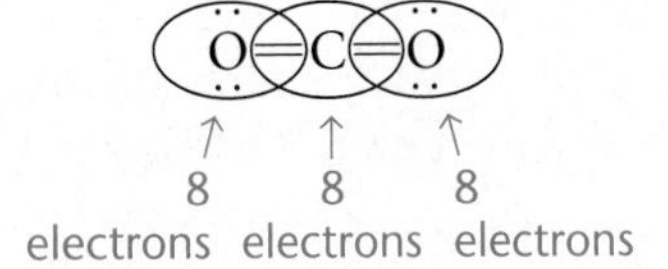

Now each atom is surrounded by eight electrons, and the total number of electrons is sixteen, as required. This is the correct Lewis structure for carbon dioxide, which has two *double* bonds. A **single bond** involves two atoms sharing one electron pair. A **double bond** involves two atoms sharing two pairs of electrons.

:O≡C—O: represents :O:::C:O:

In considering the Lewis structure for CO_2, you may have come up with

:O≡C—O: or :O—C≡O:

Note that both of these structures have the required sixteen electrons and that both have octets of electrons around each atom (verify this for yourself). Both of these structures have a **triple bond** in which three electron pairs are shared. Are these valid Lewis structures for CO_2? Yes. So there really are three Lewis structures for CO_2:

:O—C≡O: O=C=O :O≡C—O:

This brings us to a new term, **resonance.** A molecule shows resonance when *more than one Lewis structure can be drawn for the molecule.* In such a case we call the various Lewis structures **resonance structures.**

Of the three resonance structures for CO_2 shown above, the one in the center with two double bonds most closely fits our experimental information about the CO_2 molecule. In this text we will not be concerned about how to choose which resonance structure for a molecule gives the "best" description of that molecule's properties.

Next let's consider the Lewis structure of the CN^- (cyanide) ion.

STEP 1 Summing the valence electrons, we have

$$CN^-$$
$$4 + 5 + 1 = 10$$

Note that the negative charge means an extra electron must be added.

STEP 2 Draw a single bond (C—N).

STEP 3 Next, we distribute the remaining electrons to achieve a noble gas configuration for each atom. Eight electrons remain to be distributed. We can try various possibilities, such as

C—N or :C—N: or :C—N:

These structures are incorrect. To show why none is a valid Lewis structure, count the electrons around the C and N atoms. In the left structure, neither

CHEMISTRY IN FOCUS

Broccoli—Miracle Food?

Eating the right foods is critical to our health. In particular, certain vegetables, although they do not enjoy a very jazzy image, seem especially important. A case in point is broccoli, a vegetable with a humble reputation that packs a powerful chemistry wallop.

Broccoli contains a chemical called sulforaphane, which has the following Lewis structure:

$$CH_3{-}\underset{\underset{:\ddot{O}:}{\|}}{S}{-}(CH_2)_4{-}\ddot{N}{=}C{=}\ddot{S}:$$

Experiments indicate that sulforaphane furnishes protection against certain cancers and bacteria. For example, among the most common harmful bacteria in humans is *Helicobacter pylori (H. pylori)*, which has been implicated in the development of several diseases of the stomach, including inflammation, cancer, and ulcers. Antibiotics are clearly the best treatment for *H. pylori* infections. However, especially in developing countries, where *H. pylori* is rampant, antibiotics are often too expensive to be available to the general population. In addition, the bacteria sometimes evade antibiotics by "hiding" in cells on the stomach walls and then reemerging after treatment ends.

Studies at Johns Hopkins in Baltimore and Vandoeuvre-les Nancy in France have shown that sulforaphane kills *H. pylori* (even when it has taken refuge in stomach-wall cells) at concentrations that are achievable by eating broccoli. The scientists at Johns Hopkins also found that sulforaphane seems to inhibit stomach cancer in mice. Although there are no guarantees that broccoli will keep you healthy, it might not hurt to add it to your diet.

atom satisfies the octet rule. In the center structure, C has eight electrons but N has only four. In the right structure, the opposite is true. Remember that both atoms must simultaneously satisfy the octet rule. Therefore, the correct arrangement is

$:C{\equiv}N:$

represents

$:C{:::}N:$

$$:C{\equiv}N:$$

(Satisfy yourself that both carbon and nitrogen have eight electrons.) In this case we have a triple bond between C and N, in which three electron pairs are shared. Because this is an anion, we indicate the charge outside of square brackets around the Lewis structure.

$$[:C{\equiv}N:]^-$$

In summary, sometimes we need double or triple bonds to satisfy the octet rule. Writing Lewis structures is a trial-and-error process. Start with single bonds between the bonded atoms and add multiple bonds as needed.

We will write the Lewis structure for NO_2^- in Example 11.3 to make sure the procedures for writing Lewis structures are clear.

Example 11.3 Writing Lewis Structures: Resonance Structures

Write the Lewis structure for the NO_2^- anion.

Solution

STEP 1 Sum the valence electrons for NO_2^-.

$$\text{Valence electrons: } \underset{O}{6} + \underset{N}{5} + \underset{O}{6} + \underset{\substack{-1\\ \text{charge}}}{1} = 18 \text{ electrons}$$

STEP 2 Put in single bonds.

$$O—N—O$$

STEP 3 Satisfy the octet rule. In placing the electrons, we find there are two Lewis structures that satisfy the octet rule:

$$[\ddot{O}{=}\ddot{N}{-}\ddot{\underset{..}{O}}:]^- \quad \text{and} \quad [:\ddot{\underset{..}{O}}{-}\ddot{N}{=}\ddot{O}]^-$$

Verify that each atom in these structures is surrounded by an octet of electrons. Try some other arrangements to see whether other structures exist in which the eighteen electrons can be used to satisfy the octet rule. It turns out that these are the only two that work. Note that this is another case where resonance occurs; there are two valid Lewis structures.

✔ Self-Check Exercise 11.3

Ozone is a very important constituent of the atmosphere. At upper levels it protects us by absorbing high-energy radiation from the sun. Near the earth's surface it produces harmful air pollution. Write the Lewis structure for ozone, O_3.

See Problems 11.63 through 11.68. ■

Now let's consider a few more cases in Example 11.4.

Example 11.4 Writing Lewis Structures: Summary

Give the Lewis structure for each of the following:

a. HF
b. N_2
c. NH_3
d. CH_4
e. CF_4
f. NO^+
g. NO_3^-

Solution

In each case we apply the three steps for writing Lewis structures. Recall that lines are used to indicate shared electron pairs and that dots are used to indicate nonbonding pairs (lone pairs). The table on page 336 summarizes our results.

✔ Self-Check Exercise 11.4

Write the Lewis structures for the following molecules:

a. NF_3
b. O_2
c. CO
d. PH_3
e. H_2S
f. SO_4^{2-}
g. NH_4^+
h. ClO_3^-
i. SO_2

See Problems 11.55 through 11.68. ■

You may wonder how to decide which atom is the central atom in molecules of binary compounds. In cases where there is one atom of a given element and several atoms of a second element, the single atom is almost always the central atom of the molecule.

Molecule or Ion	Total Valence Electrons	Draw Single Bonds	Calculate Number of Electrons Remaining	Use Remaining Electrons to Achieve Noble Gas Configurations	Check	
					Atom	Electrons
a. HF	1 + 7 = 8	H—F	8 − 2 = 6	H—F⁝ (with three lone pairs on F)	H F	2 8
b. N_2	5 + 5 = 10	N—N	10 − 2 = 8	:N≡N:	N	8
c. NH_3	5 + 3(1) = 8	H—N—H, with H below N	8 − 6 = 2	H—N̈—H, with H below N	H N	2 8
d. CH_4	4 + 4(1) = 8	H—C—H, with H above and below C	8 − 8 = 0	H—C—H, with H above and below C	H C	2 8
e. CF_4	4 + 4(7) = 32	F—C—F, with F above and below C	32 − 8 = 24	:F̤̈—C—F̤̈:, with :F̤̈: above and below C	F C	8 8
f. NO^+	5 + 6−1 = 10	N—O	10 − 2 = 8	[:N≡O:]$^+$	N O	8 8
g. NO_3^-	5 + 3(6)+1 = 24	[O above N, bonded to two O below]	24 − 6 = 18	[:Ö: single-bonded to N; N double-bonded to one lower O and single-bonded to the other]$^-$	N O	8 8
				[:Ö: single-bonded to N; N single-bonded to one lower O and double-bonded to the other]$^-$	N O	8 8
				[:O: double-bonded to N; N single-bonded to both lower O]$^-$	N O	8 8

NO_3^- shows resonance

Remember, when writing Lewis structures, you don't have to worry about which electrons come from which atoms in a molecule. It is best to think of a molecule as a new entity that uses all the available valence electrons from the various atoms to achieve the strongest possible bonds. Think of the valence electrons as belonging to the molecule, rather than to the individual atoms. Simply distribute all the valence electrons so that noble gas electron configurations are obtained for each atom, without regard to the origin of each particular electron.

Some Exceptions to the Octet Rule

The idea that covalent bonding can be predicted by achieving noble gas electron configurations for all atoms is a simple and very successful idea. The rules we have used for Lewis structures describe correctly the bonding in most molecules. However, with such a simple model, some exceptions are inevitable. Boron, for example, tends to form compounds in which the boron atom has fewer than eight electrons around it—that is, it does not have a complete octet. Boron trifluoride, BF_3, a gas at normal temperatures and pressures, reacts very energetically with molecules such as water and ammonia that have unshared electron pairs (lone pairs).

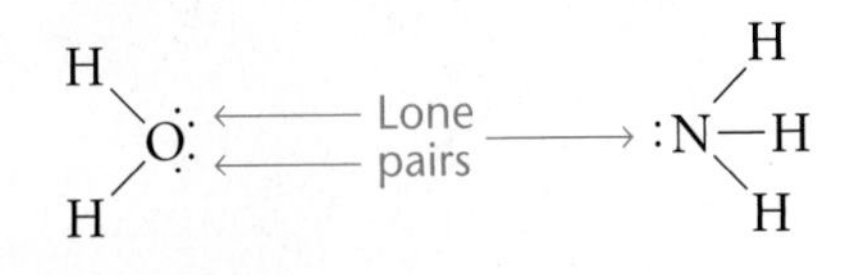

The violent reactivity of BF_3 with electron-rich molecules arises because the boron atom is electron-deficient. The Lewis structure that seems most consistent with the properties of BF_3 (twenty-four valence electrons) is

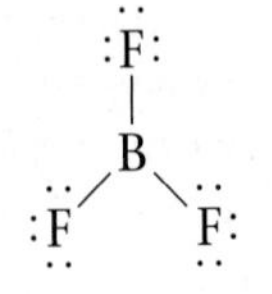

Note that in this structure the boron atom has only six electrons around it. The octet rule for boron could be satisfied by drawing a structure with a double bond between the boron and one of the fluorines. However, experiments indicate that each B—F bond is a single bond in accordance with the above Lewis structure. This structure is also consistent with the reactivity of BF_3 with electron-rich molecules. For example, BF_3 reacts vigorously with NH_3 to form H_3NBF_3.

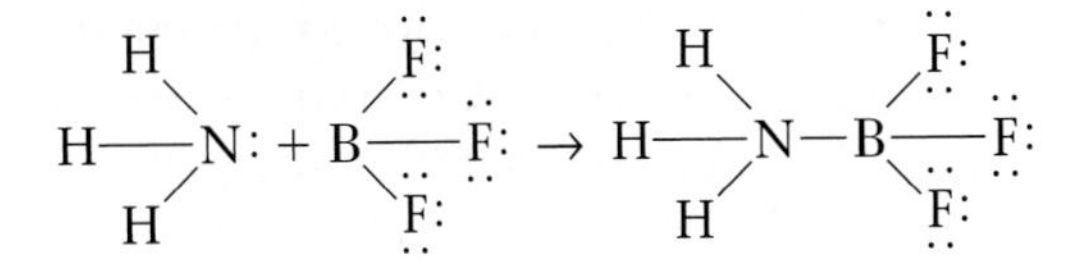

Note that in the product H_3NBF_3, which is very stable, boron has an octet of electrons.

It is also characteristic of beryllium to form molecules where the beryllium atom is electron-deficient.

The compounds containing the elements carbon, nitrogen, oxygen, and fluorine are accurately described by Lewis structures in the vast majority of cases. However, there are a few exceptions. One important example is the oxygen molecule, O_2. The following Lewis structure that satisfies the octet rule can be drawn for O_2 (see Self-Check Exercise 11.4).

$$\ddot{O}=\ddot{O}$$

Paramagnetic substances have unpaired electrons and are drawn toward the space between a magnet's poles.

However, this structure does not agree with the *observed behavior* of oxygen. For example, the photo in Figure 11.10 shows that when liquid oxygen is poured between the poles of a strong magnet, it "sticks" there until it boils away. This provides clear evidence that oxygen is paramagnetic—that is, it contains unpaired electrons. However, the above Lewis structure shows only pairs of electrons. That is, no unpaired electrons are shown. There is no simple Lewis structure that satisfactorily explains the paramagnetism of the O_2 molecule.

Any molecule that contains an odd number of electrons does not conform to our rules for Lewis structures. For example, NO and NO_2 have eleven and seventeen valence electrons, respectively, and conventional Lewis structures cannot be drawn for these cases.

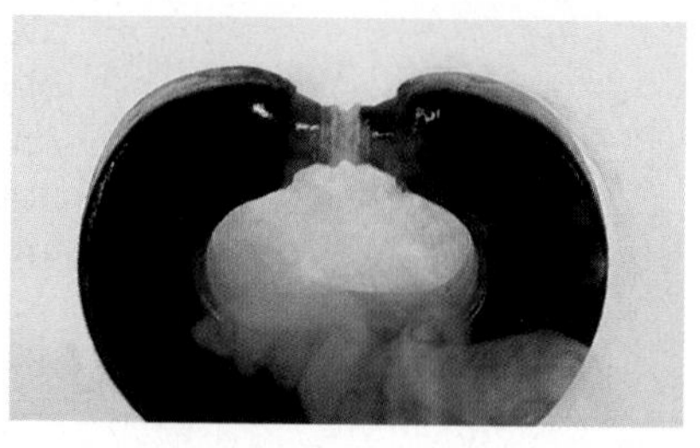

Figure 11.10
When liquid oxygen is poured between the poles of a magnet, it "sticks" until it boils away. This shows that the O_2 molecule has unpaired electrons (is paramagnetic).

Even though there are exceptions, most molecules can be described by Lewis structures in which all the atoms have noble gas electron configurations, and this is a very useful model for chemists.

Molecular Structure

AIM: To understand molecular structure and bond angles.

So far in this chapter we have considered the Lewis structures of molecules. These structures represent the arrangement of the *valence electrons* in a molecule. We use the word *structure* in another way when we talk about the **molecular structure** or **geometric structure** of a molecule. These terms refer to the three-dimensional arrangement of the *atoms* in a molecule. For example, the water molecule is known to have the molecular structure

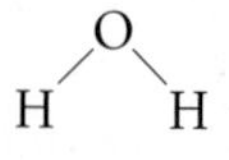

which is often called "bent" or "V-shaped." To describe the structure more precisely, we often specify the **bond angle.** For the H_2O molecule the bond angle is about 105°.

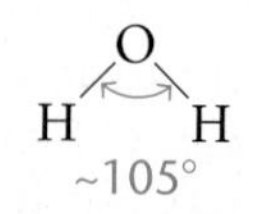

(a) (b) (c)

Computer graphics of (a) a linear molecule containing three atoms, (b) a trigonal planar molecule, and (c) a tetrahedral molecule.

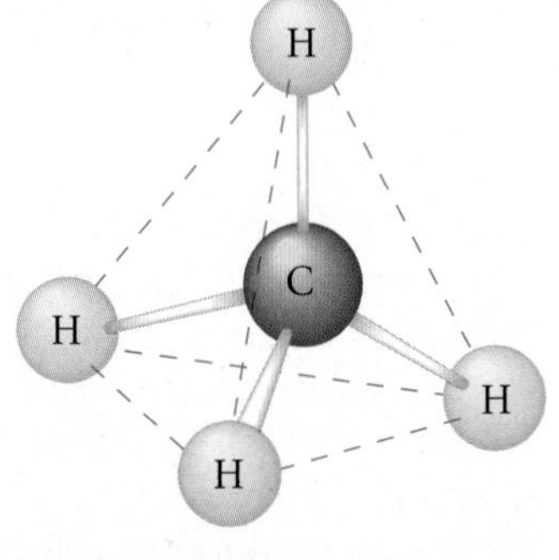

Figure 11.11
The tetrahedral molecular structure of methane. This representation is called a ball-and-stick model; the atoms are represented by balls and the bonds by sticks. The dashed lines show the outline of the tetrahedron.

On the other hand, some molecules exhibit a **linear structure** (all atoms in a line). An example is the CO_2 molecule.

O—C—O
180°

Note that a linear molecule has a 180° bond angle.

A third type of molecular structure is illustrated by BF_3, which is planar or flat (all four atoms in the same plane) with 120° bond angles.

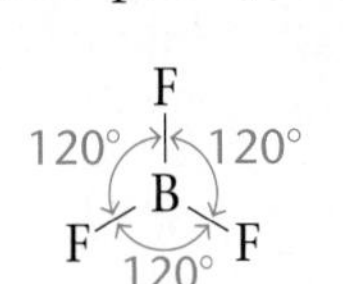

The name usually given to this structure is **trigonal planar structure,** although triangular might seem to make more sense.

Another type of molecular structure is illustrated by methane, CH_4. This molecule has the molecular structure shown in Figure 11.11, which is called a **tetrahedral structure** or a **tetrahedron.** The dashed lines shown connecting the H atoms define the four identical triangular faces of the tetrahedron.

In the next section we will discuss these various molecular structures in more detail. In that section we will learn how to predict the molecular structure of a molecule by looking at the molecule's Lewis structure.

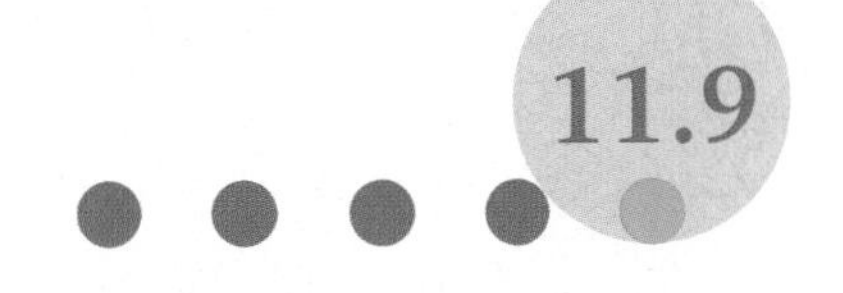

Molecular Structure: The VSEPR Model

AIM: To learn to predict molecular geometry from the number of electron pairs.

The structures of molecules play a very important role in determining their properties. For example, as we will see in the "Chemistry in Focus" on page 340, taste is directly related to molecular structure. Structure is particularly important for biological molecules; a slight change in the structure of a large biomolecule can completely destroy its usefulness to a cell and may even change the cell from a normal one to a cancerous one.

Many experimental methods now exist for determining the molecular structure of a molecule—that is, the three-dimensional arrangement of the atoms. These methods must be used when accurate information about the structure is required. However, it is often useful to be able to predict the *approximate* molecular structure of a molecule. In this section we consider a simple model that allows us to do this. This model, called the **valence shell electron pair repulsion (VSEPR) model,** is useful for predicting the molecular structures of molecules formed from nonmetals. The main idea of this model is that *the structure around a given atom is determined by minimizing repulsions between electron pairs.* This means that the bonding and nonbonding electron pairs (lone pairs) around a given atom are positioned *as far apart as possible.* To see how this model works, we will first consider the molecule $BeCl_2$, which has the following Lewis structure (it is an exception to the octet rule):

$$:\ddot{\underset{..}{Cl}}—Be—\ddot{\underset{..}{Cl}}:$$

CHEMISTRY IN FOCUS

Taste—It's the Structure That Counts

Why do certain substances taste sweet, sour, bitter, or salty? Of course, it has to do with the taste buds on our tongues. But how do these taste buds work? For example, why does sugar taste sweet to us? The answer to this question remains elusive, but it does seem clear that sweet taste depends on how certain molecules fit the "sweet receptors" in our taste buds.

One of the mysteries of the sweet taste sensation is the wide variety of molecules that taste sweet. For example, the many types of sugars include glucose and sucrose (table sugar). The first artificial sweetener was probably the Romans' sapa (see "Chemistry in Focus: Sugar of Lead" in Chapter 5), made by boiling wine in lead vessels to produce a syrup that contained lead acetate, $Pb(C_2H_3O_2)_2$, called sugar of lead because of its sweet taste. Other widely used modern artificial sweeteners include saccharin, sodium cyclamate, and aspartame, whose structures are shown in the accompanying figure. Note the great disparity of structures for these sweet-tasting molecules. It's certainly not obvious which structural features trigger a sweet sensation when these molecules interact with the taste buds.

The pioneers in relating structure to sweet taste were two chemists, Robert S. Shallenberger and Terry E. Acree of Cornell University, who almost thirty years ago suggested that all sweet-tasting substances must contain a common feature they called a glycophore. They postulated that a glycophore always contains an atom or group of atoms that have available electrons located near a hydrogen atom attached to a relatively electronegative atom. Murray Goodman, a chemist at the University of California at San Diego, expanded the definition of a glycophore to include a hydrophobic ("water-hating") region. Goodman finds that a "sweet molecule" tends to be L-shaped with positively and negatively charged regions on the upright of the L and a hydrophobic region on the base of the L. To be sweet the L must be planar. If it is twisted in one direction, it gives a bitter taste. Twisting it in the other direction makes it tasteless. Goodman reports that by using his model he can design sweeteners, but these molecules remain too expensive for commercial use.

So the search goes on for a better artificial sweetener. One thing for sure, it all has to do with molecular structure.

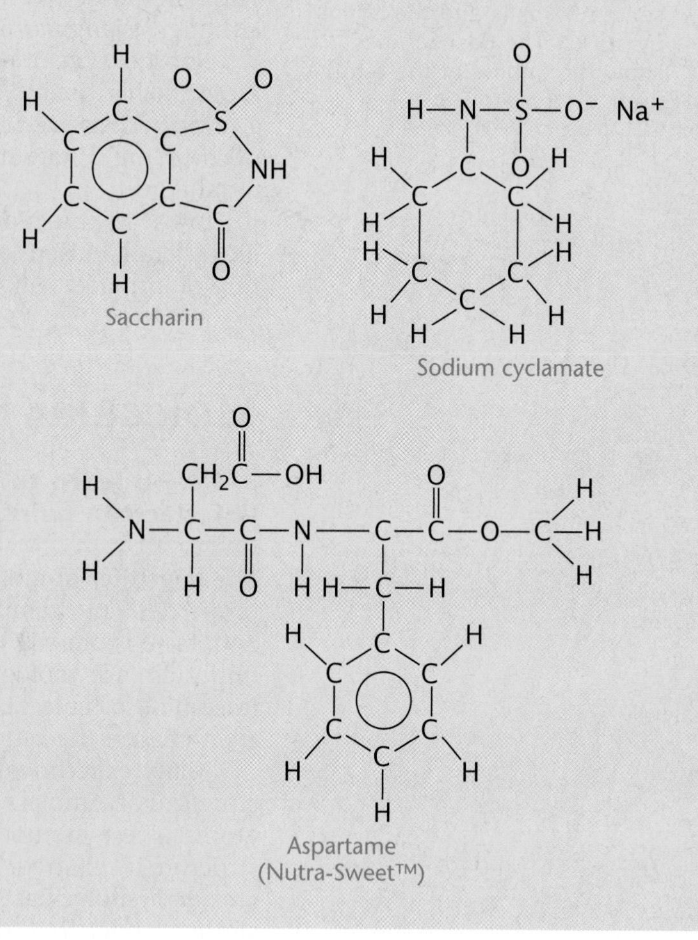

Saccharin

Sodium cyclamate

Aspartame (Nutra-Sweet™)

Note that there are two pairs of electrons around the beryllium atom. What arrangement of these electron pairs allows them to be as far apart as possible to minimize the repulsions? The best arrangement places the pairs on opposite sides of the beryllium atom at 180° from each other.

—Be—
180°

This is the maximum possible separation for two electron pairs. Now that we have determined the optimal arrangement of the electron pairs around the central atom, we can specify the molecular structure of $BeCl_2$—that is, the positions of the atoms. Because each electron pair on beryllium is shared with a chlorine atom, the molecule has a **linear structure** with a 180° bond angle.

:C̤̈l—Be—C̤̈l:
180°

Whenever two pairs of electrons are present around an atom, they should always be placed at an angle of 180° to each other to give a linear arrangement.

Next let's consider BF_3, which has the following Lewis structure (it is another exception to the octet rule):

:F̤̈:
|
:F̤̈—B—F̤̈:

Here the boron atom is surrounded by three pairs of electrons. What arrangement minimizes the repulsions among three pairs of electrons? Here the greatest distance between electron pairs is achieved by angles of 120°.

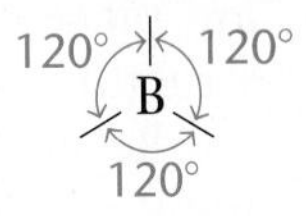

Because each of the electron pairs is shared with a fluorine atom, the molecular structure is

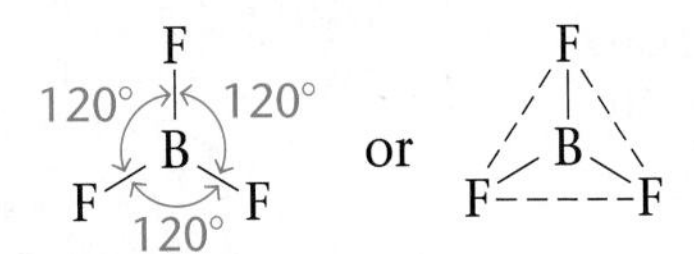

This is a planar (flat) molecule with a triangular arrangement of F atoms, commonly described as a trigonal planar structure. *Whenever three pairs of electrons are present around an atom, they should always be placed at the corners of a triangle (in a plane at angles of 120° to each other).*

Next let's consider the methane molecule, which has the Lewis structure

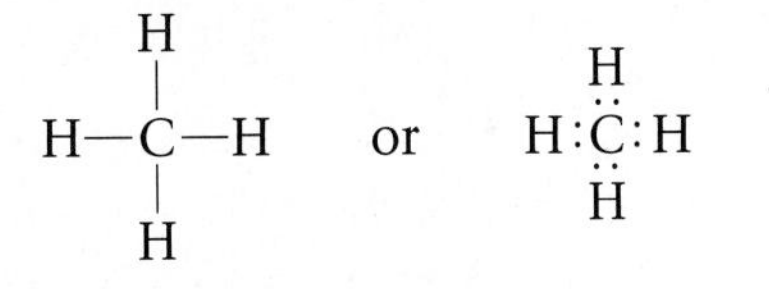

There are four pairs of electrons around the central carbon atom. What arrangement of these electron pairs best minimizes the repulsions? First we try a square planar arrangement:

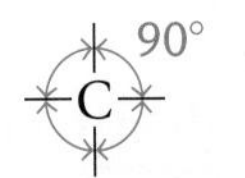

The carbon atom and the electron pairs are all in a plane represented by the surface of the paper, and the angles between the pairs are all 90°.

Is there another arrangement with angles greater than 90° that would put the electron pairs even farther away from each other? The answer is yes. We can get larger angles than 90° by using the following three-dimensional structure, which has angles of approximately 109.5°.

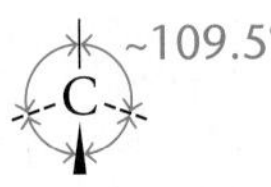

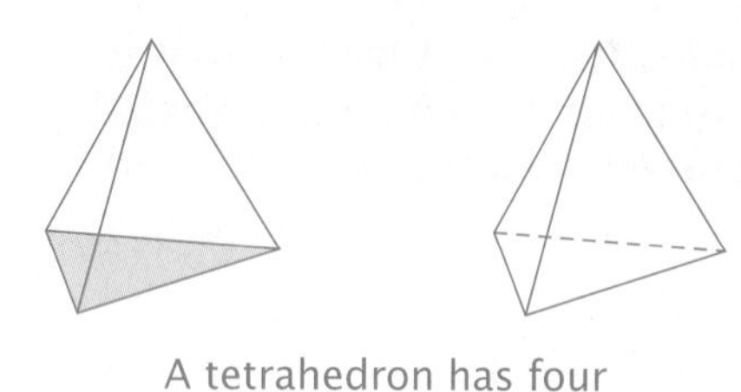

A tetrahedron has four equal triangular faces.

In this drawing the wedge indicates a position above the surface of the paper and the dashed lines indicate positions behind that surface. The solid line indicates a position on the surface of the page. The figure formed by connecting the lines is called a tetrahedron, so we call this arrangement of electron pairs the **tetrahedral arrangement.**

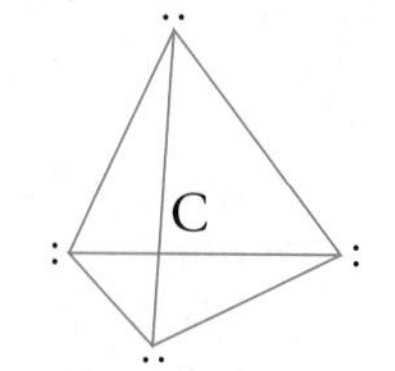

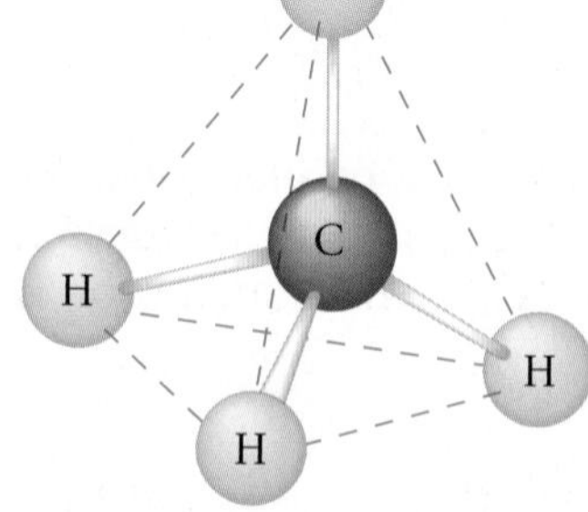

Figure 11.12
The molecular structure of methane. The tetrahedral arrangement of electron pairs produces a tetrahedral arrangement of hydrogen atoms.

This is the maximum possible separation of four pairs around a given atom. *Whenever four pairs of electrons are present around an atom, they should always be placed at the corners of a tetrahedron (the tetrahedral arrangement).*

Now that we have the arrangement of electron pairs that gives the least repulsion, we can determine the positions of the atoms and thus the molecular structure of CH_4. In methane each of the four electron pairs is shared between the carbon atom and a hydrogen atom. Thus the hydrogen atoms are placed as shown in Figure 11.12, and the molecule has a tetrahedral structure with the carbon atom at the center.

Recall that the main idea of the VSEPR model is to find the arrangement of electron pairs around the central atom that minimizes the repulsions. Then we can determine the *molecular structure* by knowing how the electron pairs are shared with the peripheral atoms. A systematic procedure for using the VSEPR model to predict the structure of a molecule is outlined below.

Steps for Predicting Molecular Structure Using the VSEPR Model

STEP 1 Draw the Lewis structure for the molecule.

STEP 2 Count the electron pairs and arrange them in the way that minimizes repulsion (that is, put the pairs as far apart as possible).

STEP 3 Determine the positions of the atoms from the way the electron pairs are shared.

STEP 4 Determine the name of the molecular structure from the positions of the *atoms.*

Example 11.5 Predicting Molecular Structure Using the VSEPR Model, I

Ammonia, NH_3, is used as a fertilizer (injected into the soil) and as a household cleaner (in aqueous solution). Predict the structure of ammonia using the VSEPR model.

Solution

STEP 1 Draw the Lewis structure.

$$\begin{array}{c} \mathrm{H{-}\ddot{N}{-}H} \\ | \\ \mathrm{H} \end{array}$$

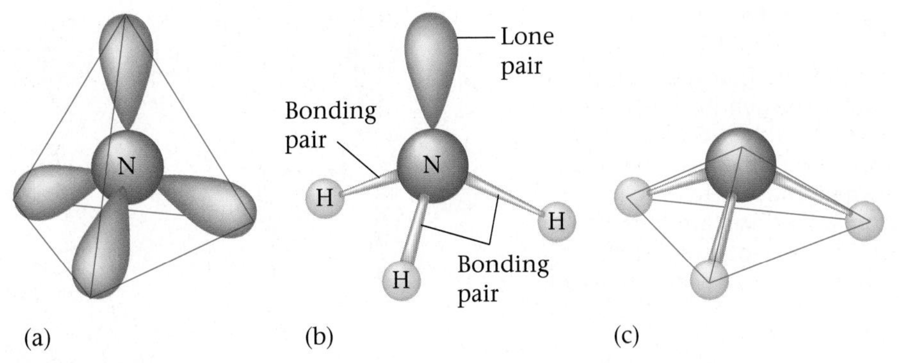

Figure 11.13
(a) The tetrahedral arrangement of electron pairs around the nitrogen atom in the ammonia molecule. (b) Three of the electron pairs around nitrogen are shared with hydrogen atoms as shown, and one is a lone pair. Although the arrangement of *electron pairs* is tetrahedral, as in the methane molecule, the hydrogen atoms in the ammonia molecule occupy only three corners of the tetrahedron. A lone pair occupies the fourth corner. (c) The NH_3 molecule has the trigonal pyramid structure (a pyramid with a triangle as a base).

STEP 2 Count the pairs of electrons and arrange them to minimize repulsions. The NH_3 molecule has four pairs of electrons around the N atom: three bonding pairs and one nonbonding pair. From the discussion of the methane molecule, we know that the best arrangement of four electron pairs is the tetrahedral structure shown in Figure 11.13a.

STEP 3 Determine the positions of the atoms. The three H atoms share electron pairs as shown in Figure 11.13b.

STEP 4 Name the molecular structure. It is very important to recognize that the name of the molecular structure is always based on the *positions of the atoms. The placement of the electron pairs determines the structure, but the name is based on the positions of the atoms.* Thus it is incorrect to say that the NH_3 molecule is tetrahedral. It has a tetrahedral arrangement of electron pairs but *not* a tetrahedral arrangement of atoms. The molecular structure of ammonia is a **trigonal pyramid** (one side is different from the other three) rather than a tetrahedron. ■

Example 11.6 Predicting Molecular Structure Using the VSEPR Model, II

Describe the molecular structure of the water molecule.

Solution

STEP 1 The Lewis structure for water is

$$\mathrm{H{-}\ddot{\underset{\cdot\cdot}{O}}{-}H}$$

STEP 2 There are four pairs of electrons: two bonding pairs and two nonbonding pairs. To minimize repulsions, these are best arranged in a tetrahedral structure as shown in Figure 11.14a.

STEP 3 Although H_2O has a tetrahedral arrangement of *electron pairs,* it is *not a tetrahedral molecule.* The *atoms* in the H_2O molecule form a V shape, as shown in Figure 11.14b and c.

STEP 4 The molecular structure is called V-shaped or bent.

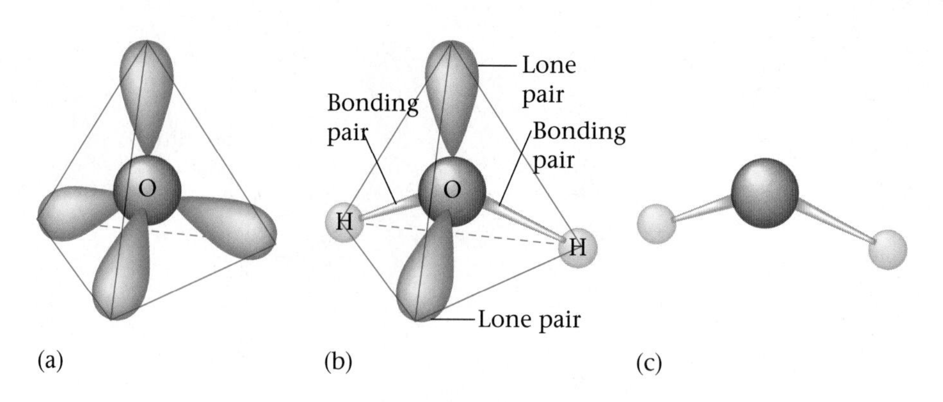

Figure 11.14
(a) The tetrahedral arrangement of the four electron pairs around oxygen in the water molecule. (b) Two of the electron pairs are shared between oxygen and the hydrogen atoms, and two are lone pairs. (c) The V-shaped molecular structure of the water molecule.

Self-Check Exercise 11.5

Predict the arrangement of electron pairs around the central atom. Then sketch and name the molecular structure for each of the following molecules or ions.

a. NH_4^+
b. SO_4^{2-}
c. NF_3
d. H_2S
e. ClO_3^-
f. BeF_2

See Problems 11.81 through 11.84. ■

The various cases we have considered are summarized in Table 11.4 on the following page. Note the following general rules.

Rules for Predicting Molecular Structure Using the VSEPR Model

1. Two pairs of electrons on a central atom in a molecule are always placed 180° apart. This is a linear arrangement of pairs.
2. Three pairs of electrons on a central atom in a molecule are always placed 120° apart in the same plane as the central atom. This is a trigonal planar (triangular) arrangement of pairs.
3. Four pairs of electrons on a central atom in a molecule are always placed 109.5° apart. This is a tetrahedral arrangement of electron pairs.
4. When *every pair* of electrons on the central atom is *shared* with another atom, the molecular structure has the same name as the arrangement of electron pairs.

Number of Pairs	Name of Arrangement
2	linear
3	trigonal planar
4	tetrahedral

5. When one or more of the electron pairs around a central atom are unshared (lone pairs), the name for the molecular structure is *different* from that for the arrangement of electron pairs (see cases 4 and 5 in Table 11.4).

Table 11.4 Arrangements of Electron Pairs and the Resulting Molecular Structures for Two, Three, and Four Electron Pairs

Case	Number of Electron Pairs	Bonds	Electron Pair Arrangement	Ball-and-Stick Model	Angle Between Pairs	Molecular Structure	Partial Lewis Structure	Ball-and-Stick Model	Example
1	2	2	Linear		180°	Linear	A—B—A		BeF_2
2	3	3	Trigonal planar (triangular)		120°	Trigonal planar (triangular)	BA_3 (trigonal)		
3	4	4	Tetrahedral		109.5°	Tetrahedral	BA_4 (A—B—A, with A above and below B)		CH_4
4	4	3	Tetrahedral		109.5°	Trigonal pyramid	A—$\ddot{B}$—A, with A below B		NH_3
5	4	2	Tetrahedral		109.5°	Bent or V-shaped	A—$\underset{..}{\ddot{B}}$—A		H_2O

11.10 Molecular Structure: Molecules with Double Bonds

AIM: To learn to apply the VSEPR model to molecules with double bonds.

Up to this point we have applied the VSEPR model only to molecules (and ions) that contain single bonds. In this section we will show that this model applies equally well to species with one or more double bonds. We will develop the procedures for dealing with molecules with double bonds by considering examples whose structures are known.

First we will examine the structure of carbon dioxide, a substance that may be contributing to the warming of the earth. The carbon dioxide molecule has the Lewis structure

$$\underset{..}{\ddot{O}}=C=\underset{..}{\ddot{O}}$$

CHEMISTRY IN FOCUS

Minimotor Molecule

Our modern society is characterized by a continual quest for miniaturization. Our computers, cell phones, portable music players, calculators, and many other devices have been greatly downsized over the last several years. The ultimate in miniaturization—machines made of single molecules. Although this idea sounds like an impossible dream, recent advances place us on the doorstep of such devices. For example, Hermann E. Gaub and his coworkers at the Center for Nanoscience at Ludwig-Maximilians University in Munich have just reported a single molecule that can do simple work.

Gaub and his associates constructed a polymer about 75 nanometers long by hooking together many light-sensitive molecules called azobenzenes:

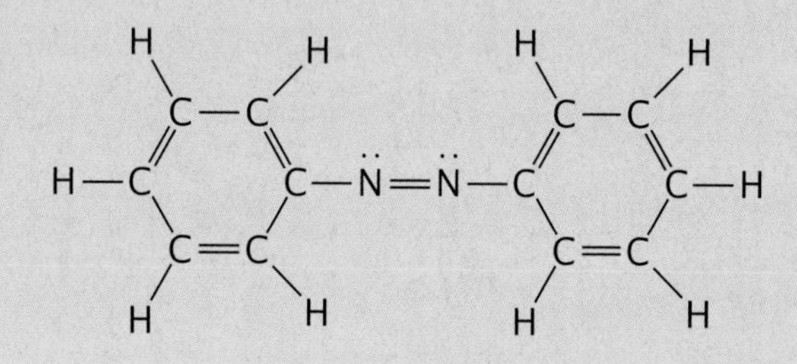

Azobenzene is ideal for this application because its bonds are sensitive to specific wavelengths of light. When azobenzene absorbs light of 420 nm, it becomes extended; light at 365 nm causes the molecule to contract.

To make their tiny machine, the German scientists attached one end of the azobenzene polymer to a tiny, bendable lever similar to the tip of an atomic-force microscope. The other end of the polymer was attached to a glass surface. Flashes of 365-nm light caused the molecule to contract, bending the lever down and storing mechanical energy. Pulses of 420-nm radiation then extended the molecule, causing the lever to rise and releasing the stored energy. Eventually, one can imagine having the lever operate some part of a nanoscale machine. It seems we are getting close to the ultimate in miniature machines.

as discussed in Section 11.7. Carbon dioxide is known by experiment to be a linear molecule. That is, it has a 180° bond angle.

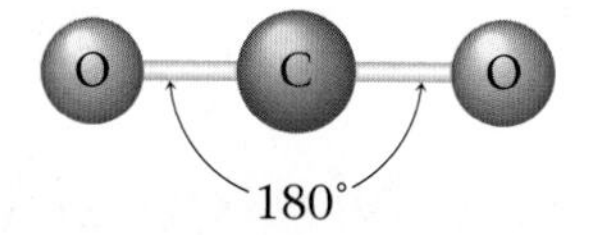

Recall from Section 11.9 that two electron pairs around a central atom can minimize their mutual repulsions by taking positions on opposite sides of the atom (at 180° from each other). This causes a molecule like $BeCl_2$, which has the Lewis structure

$$:\ddot{\underset{\cdot\cdot}{Cl}}-Be-\ddot{\underset{\cdot\cdot}{Cl}}:$$

to have a linear structure. Now recall that CO_2 has two double bonds and is known to be linear, so the double bonds must be at 180° from each other. Therefore, we conclude that each double bond in this molecule acts *effectively* as one repulsive unit. This conclusion makes sense if we think of a bond in terms of an electron density "cloud" between two atoms. For example, we can picture the single bonds in $BeCl_2$ as follows:

The minimum repulsion between these two electron density clouds occurs when they are on opposite sides of the Be atom (180° angle between them).

Each double bond in CO_2 involves the sharing of four electrons between the carbon atom and an oxygen atom. Thus we might expect the bonding cloud to be "fatter" than for a single bond:

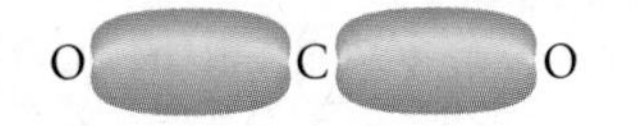

However, the repulsive effects of these two clouds produce the same result as for single bonds; the bonding clouds have minimum repulsions when they are positioned on opposite sides of the carbon. The bond angle is 180°, and so the molecule is linear:

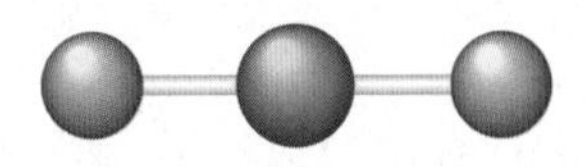

In summary, examination of CO_2 leads us to the conclusion that in using the VSEPR model for molecules with double bonds, each double bond should be treated the same as a single bond. In other words, although a double bond involves four electrons, these electrons are restricted to the space between a given pair of atoms. Therefore, these four electrons do not function as two independent pairs but are "tied together" to form one effective repulsive unit.

We reach this same conclusion by considering the known structures of other molecules that contain double bonds. For example, consider the ozone molecule, which has eighteen valence electrons and exhibits two resonance structures:

$$:\ddot{\underset{\cdot\cdot}{O}}-\ddot{O}=\ddot{O}: \longleftrightarrow \ddot{O}=\ddot{O}-\ddot{\underset{\cdot\cdot}{O}}:$$

The ozone molecule is known to have a bond angle close to 120°. Recall that 120° angles represent the minimum repulsion for three pairs of electrons.

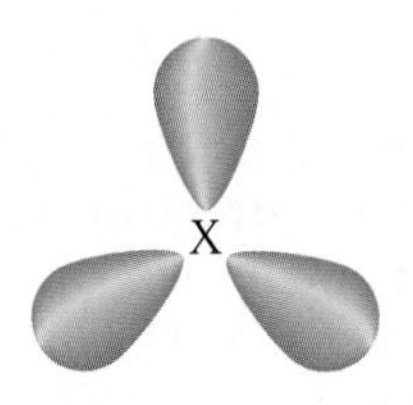

This indicates that the double bond in the ozone molecule is behaving as one effective repulsive unit:

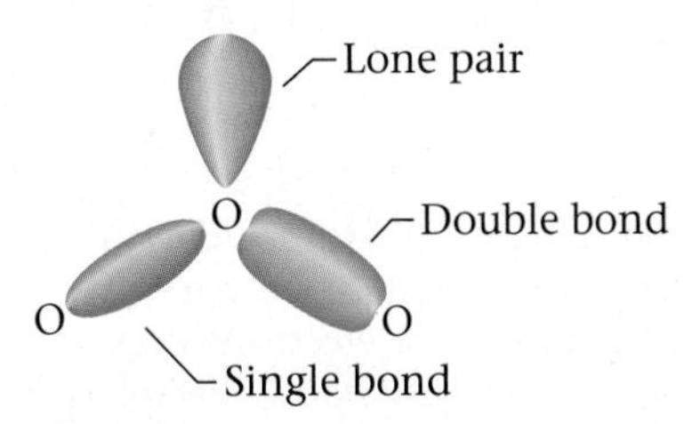

These and other examples lead us to the following rule: *When using the VSEPR model to predict the molecular geometry of a molecule, a double bond is counted the same as a single electron pair.*

Thus CO_2 has two "effective pairs" that lead to its linear structure, whereas O_3 has three "effective pairs" that lead to its bent structure with a 120° bond angle. Therefore, to use the VSEPR model for molecules (or ions) that have double bonds, we use the same steps as those given in Section 11.9, but we count any double bond the same as a single electron pair. Although we have not shown it here, triple bonds also count as one repulsive unit in applying the VSEPR model.

Example 11.7 Predicting Molecular Structure Using the VSEPR Model, III

Predict the structure of the nitrate ion.

Solution

STEP 1 The Lewis structures for NO_3^- are

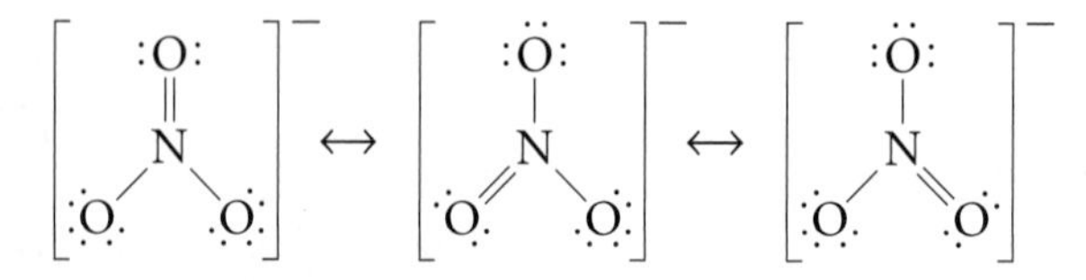

STEP 2 In each resonance structure there are effectively three pairs of electrons: the two single bonds and the double bond (which counts as one pair). These three "effective pairs" will require a trigonal planar arrangement (120° angles).

STEP 3 The atoms are all in a plane, with the nitrogen at the center and the three oxygens at the corners of a triangle (trigonal planar arrangement).

STEP 4 The NO_3^- ion has a trigonal planar structure. ■

CHAPTER 11 REVIEW

KEY TERMS

bond (11.1)
bond energy (11.1)
ionic bonding (11.1)
ionic compound (11.1)
covalent bonding (11.1)
polar covalent bond (11.1)
electronegativity (11.2)
dipole moment (11.3)
Lewis structure (11.6)
duet rule (11.6)
octet rule (11.6)
bonding pair (11.6)
lone pair (11.6)
single bond (11.7)
double bond (11.7)
triple bond (11.7)
resonance (11.7)
resonance structure (11.7)
molecular (geometric) structure (11.8)
bond angle (11.8)
linear structure (11.8)
trigonal planar structure (11.8)
tetrahedral structure (11.8)
valence shell electron pair repulsion (VSEPR) model (11.9)
trigonal pyramid (11.9)

SUMMARY

1. Chemical bonds hold groups of atoms together. They can be classified into several types. An ionic bond is formed when a transfer of electrons occurs to form ions; in a purely covalent bond, electrons are shared equally between identical atoms. Between these extremes lies the polar covalent bond, in which electrons are shared unequally between atoms with different electronegativities.

2. Electronegativity is defined as the relative ability of an atom in a molecule to attract the electrons shared in a bond. The difference in electronegativity values between the atoms involved in a bond determines the polarity of that bond.

3. In stable chemical compounds, the atoms tend to achieve a noble gas electron configuration. In the formation of a binary ionic compound involving representative elements, the valence-electron configuration of the nonmetal is completed: it achieves the configuration of the next noble gas. The valence orbitals of the metal are emptied to give the electron configuration of the previous noble gas. Two nonmetals share the valence electrons so that both atoms have completed valence-electron configurations (noble gas configurations).

4. Lewis structures are drawn to represent the arrangement of the valence electrons in a molecule. The rules for drawing Lewis structures are based on the observation that nonmetal atoms tend to achieve noble gas electron configurations by sharing electrons. This leads to a duet rule for hydrogen and to an octet rule for many other atoms.
5. Some molecules have more than one valid Lewis structure, a property called resonance. Although Lewis structures in which the atoms have noble gas electron configurations correctly describe most molecules, there are some notable exceptions, including O_2, NO, NO_2, and the molecules that contain Be and B.
6. The molecular structure of a molecule describes how the atoms are arranged in space.
7. The molecular structure of a molecule can be predicted by using the valence shell electron pair repulsion (VSEPR) model. This model bases its prediction on minimizing the repulsions among the electron pairs around an atom, which means arranging the electron pairs as far apart as possible.

In-Class Discussion Questions

These questions are designed to be considered by groups of students in class. Often these questions work well for introducing a particular topic in class.

1. Using only the periodic table, predict the most stable ion for Na, Mg, Al, S, Cl, K, Ca, and Ga. Arrange these elements from largest to smallest radius and explain why the radius varies as it does.
2. Write the proper charges so that an alkali metal, a noble gas, and a halogen have the same electron configurations. What is the number of protons in each? The number of electrons in each? Arrange them from smallest to largest radii and explain your ordering rationale.
3. What is meant by a *chemical bond?*
4. Why do atoms form bonds with one another? What can make a molecule favored compared with the lone atoms?
5. How does a bond between Na and Cl differ from a bond between C and O? What about a bond between N and N?
6. In your own words, what is meant by the term *electronegativity?* What are the trends across and down the periodic table for electronegativity? Explain them, and describe how they are consistent with trends of ionization energy and atomic radii.
7. Why are some bonds ionic and some covalent?
8. True or false? In general, a larger atom has a smaller electronegativity. Explain.
9. Why is there an octet rule (and what does *octet* mean) in writing Lewis structures?
10. Does a Lewis structure tell which electrons came from which atoms? Explain.

Questions and Problems

All even-numbered exercises have answers in the back of this book and solutions in the *Solutions Guide.*

11.1 Types of Chemical Bonds

QUESTIONS

1. Define a chemical bond.
2. What do we mean by the *bond energy* of a chemical bond? How is the *strength* of a chemical bond related to the bond energy?
3. What types of elements react to form *ionic* compounds? Give an example of the formation of an ionic compound from its elements.
4. What is *covalent* bonding? Give an example of a molecule whose atoms are held together by covalent bonding. What is *polar covalent* bonding? Give an example of a molecule whose atoms are held together by polar covalent bonding.
5. Describe the type of bonding that exists in the $Cl_2(g)$ molecule. How does this type of bonding differ from that found in the $HCl(g)$ molecule? How is it similar?
6. Compare and contrast the bonding found in the $H_2(g)$ and $HF(g)$ molecules with that found in $NaF(s)$.

11.2 Electronegativity

QUESTIONS

7. What do chemists mean by the term *electronegativity?* What does its electronegativity tell us about the atom?
8. What does it mean to say that a bond is *polar?* Give two examples of molecules with *polar* bonds. Indicate in your examples the direction of the polarity.
9. A molecule is said to possess a(n) ________ moment if the centers of positive and negative charge in the molecule do not coincide.
10. What factor determines the relative level of polarity of a polar covalent bond?

PROBLEMS

11. For each of the following sets of elements, arrange the elements in order of increasing electronegativity.
    a. Li, F, C
    b. I, Cl, F
    c. Li, Rb, Cs
12. For each of the following sets of elements, arrange the elements in order of increasing electronegativity.

a. K, Cr, Br
b. Mg, Ra, Ca
c. Cl, Na, Al

13. On the basis of the electronegativity values given in Figure 11.3, indicate whether each of the following bonds would be expected to be ionic, covalent, or polar covalent.
a. K—N c. C—N
b. Cs—O d. O—F

14. On the basis of the electronegativity values given in Figure 11.3, indicate whether each of the following bonds would be expected to be ionic, covalent, or polar covalent.
a. Na—Cl c. Cl—O
b. Cl—F d. Ca—Cl

15. Which of the following molecules contain polar covalent bonds?
a. phosphorus, P_4 c. ozone, O_3
b. oxygen, O_2 d. hydrogen fluoride, HF

16. Which of the following molecules contain polar covalent bonds?
a. nitrogen, N_2
b. astatine, At_2
c. carbon monoxide, CO
d. hydrogen fluoride, HF

17. On the basis of the electronegativity values given in Figure 11.3, indicate which is the more polar bond in each of the following pairs.
a. H—F or H—Cl c. H—Br or H—Cl
b. H—Cl or H—I d. H—I or H—Br

18. On the basis of the electronegativity values given in Figure 11.3, indicate which is the more polar bond in each of the following pairs.
a. H—O or H—N c. H—O or H—F
b. H—N or H—F d. H—O or H—Cl

19. On the basis of the electronegativity values given in Figure 11.3, indicate which is the more polar bond in each of the following pairs.
a. H—S or H—F c. N—S or N—Cl
b. O—S or O—F d. C—S or C—Cl

20. On the basis of the electronegativity values given in Figure 11.3, indicate which bond of the following pairs has more ionic character.
a. Mg—Cl or Ca—Cl c. Cu—F or Cu—I
b. Ca—F or K—F d. Na—Br or S—Br

11.3 Bond Polarity and Dipole Moments

QUESTIONS

21. What is a *dipole moment?* Give four examples of molecules that possess dipole moments, and draw the direction of the dipole as shown in this section.

22. Why is the presence of a dipole moment in the water molecule so important? What are some properties of water that are determined by its polarity?

PROBLEMS

23. In each of the following diatomic molecules, which end of the molecule is negative relative to the other end?
a. hydrogen chloride, HCl
b. carbon monoxide, CO
c. bromine monofluoride, BrF

24. In each of the following diatomic molecules, which end of the molecule is positive relative to the other end?
a. hydrogen fluoride, HF
b. chlorine monofluoride, ClF
c. iodine monochloride, ICl

25. For each of the following bonds, draw a figure indicating the direction of the bond dipole, including which end of the bond is positive and which is negative.
a. C—F c. C—O
b. Si—C d. B—C

26. For each of the following bonds, draw a figure indicating the direction of the bond dipole, including which end of the bond is positive and which is negative.
a. P—F c. P—C
b. P—O d. P—H

27. For each of the following bonds, draw a figure indicating the direction of the bond dipole, including which end of the bond is positive and which is negative.
a. Si—H c. S—H
b. P—H d. Cl—H

28. For each of the following bonds, draw a figure indicating the direction of the bond dipole, including which end of the bond is positive and which is negative.
a. I—I c. I—F
b. I—Br d. At—I

11.4 Stable Electron Configurations and Charges on Ions

QUESTIONS

29. In virtually every stable compound, each of the atoms has achieved an electron configuration analogous to that of the ______ elements.

30. The ______ elements achieve an electron configuration analogous to the previous noble gas by losing electrons from their valence shells.

31. Nonmetals form negative ions by (losing/gaining) enough electrons to achieve the electron configuration of the next noble gas.

32. Explain how the atoms in *covalent* molecules achieve configurations similar to those of the noble gases. How does this differ from the situation in ionic compounds?

PROBLEMS

33. Write the electron configuration for each of the following atoms and for the simple ion that the element most commonly forms. In each case, indicate which noble gas has the same electron configuration as the ion.
    a. sodium, $Z = 11$
    b. iodine, $Z = 53$
    c. calcium, $Z = 20$
    d. nitrogen, $Z = 7$
    e. fluorine, $Z = 9$

34. Write the electron configuration for the simple ion that each of the following elements most commonly forms, and indicate which noble gas has the same electron configuration as the ion.
    a. chlorine, $Z = 17$
    b. oxygen, $Z = 8$
    c. aluminum, $Z = 13$
    d. barium, $Z = 56$
    e. selenium, $Z = 34$

35. What simple ion does each of the following elements most commonly form?
    a. magnesium, $Z = 12$
    b. aluminum, $Z = 13$
    c. iodine, $Z = 53$
    d. calcium, $Z = 20$

36. What simple ion does each of the following elements most commonly form?
    a. strontium, $Z = 38$
    b. fluorine, $Z = 9$
    c. lithium, $Z = 3$
    d. oxygen, $Z = 8$

37. On the basis of their electron configurations, predict the formula of the simple binary ionic compounds likely to form when the following pairs of elements react with each other.
    a. aluminum, Al, and sulfur, S
    b. radium, Ra, and oxygen, O
    c. calcium, Ca, and fluorine, F
    d. cesium, Cs, and nitrogen, N
    e. rubidium, Rb, and phosphorus, P

38. On the basis of their electron configurations, predict the formula of the simple binary ionic compound likely to form when the following pairs of elements react with each other.
    a. sodium, Na, and sulfur, S
    b. barium, Ba, and selenium, Se
    c. magnesium, Mg, and bromine, Br
    d. lithium, Li, and nitrogen, N
    e. potassium, K, and hydrogen, H

39. Name the noble gas atom that has the same electron configuration as each of the ions in the following compounds.
    a. barium sulfide, BaS
    b. strontium fluoride, SrF_2
    c. magnesium oxide, MgO
    d. aluminum sulfide, Al_2S_3

40. Name the noble gas atom that has the same electron configuration as each of the ions in the following compounds.
    a. calcium nitride, Ca_3N_2
    b. magnesium sulfide, MgS
    c. aluminum fluoride, AlF_3
    d. radium bromide, $RaBr_2$
    e. cesium phosphide, Cs_3P

11.5 Ionic Bonding and Structures of Ionic Compounds

QUESTIONS

41. Is the formula we write for an ionic compound the *molecular* formula or the *empirical* formula? Why?

42. Describe in general terms the structure of ionic solids such as NaCl. How are the ions packed in the crystal?

43. Why are cations always smaller than the atoms from which they are formed?

44. Why are anions always larger than the atoms from which they are formed?

PROBLEMS

45. For each of the following pairs, indicate which species is smaller. Explain your reasoning in terms of the electron structure of each species.
    a. Na or Na^+
    b. F or F^-
    c. Mg or Mg^{2+}
    d. S or S^{2-}

46. For each of the following pairs, indicate which species is larger. Explain your reasoning in terms of the electron structure of each species.
    a. Li^+ or F^-
    b. Na^+ or Cl^-
    c. Ca^{2+} or Ca
    d. Cs^+ or I^-

47. For each of the following pairs, indicate which is smaller.
    a. Fe or Fe^{3+}
    b. Cl or Cl^-
    c. Al^{3+} or Na^+

48. For each of the following pairs, indicate which is larger.
    a. Cl^- or I^-
    b. Cl^- or Na^+
    c. Cl^- or Cl
    d. Cl^- or S^{2-}

11.6 and 11.7 Lewis Structures

QUESTIONS

49. Why are the *valence* electrons of an atom the only electrons likely to be involved in bonding to other atoms?

50. Explain what the "duet" and "octet" rules are and how they are used to describe the arrangement of electrons in a molecule.

51. What type of structure must each atom in a compound usually exhibit for the compound to be stable?

52. When elements in the second and third periods occur in compounds, what number of electrons in the valence shell represents the most stable electron arrangement? Why?

PROBLEMS

53. How many electrons are involved when two atoms in a molecule are connected by a "double bond"? Write the Lewis structure of a molecule containing a double bond.

54. What does it mean when two atoms in a molecule are connected by a "triple bond"? Write the Lewis structure of a molecule containing a triple bond.

55. Write the Lewis structure for each of the following atoms.
a. Na ($Z = 11$)
b. S ($Z = 16$)
c. B ($Z = 5$)
d. Be ($Z = 4$)
e. Br ($Z = 35$)
f. Ba ($Z = 56$)

56. Write the Lewis structure for each of the following atoms.
a. Rb ($Z = 37$)
b. Cl ($Z = 17$)
c. Kr ($Z = 36$)
d. Ba ($Z = 56$)
e. P ($Z = 15$)
f. At ($Z = 85$)

57. Give the *total* number of valence electrons in each of the following molecules.
a. N_2O
b. B_2H_6
c. C_3H_8
d. NCl_3

58. Give the *total* number of *valence* electrons in each of the following molecules.
a. CBr_4
b. NO_2
c. C_6H_6
d. H_2O_2

59. Write a Lewis structure for each of the following simple molecules. Show all bonding valence electron pairs as lines and all nonbonding valence electron pairs as dots.
a. PH_3
b. SF_2
c. HBr
d. CCl_4

60. Write a Lewis structure for each of the following simple molecules. Show all bonding valence electron pairs as lines and all nonbonding valence electron pairs as dots.
a. H_2
b. HCl
c. CF_4
d. C_2F_6

61. Write a Lewis structure for each of the following simple molecules. Show all bonding valence electron pairs as lines and all nonbonding valence electron pairs as dots.
a. C_2H_6
b. NF_3
c. C_4H_{10}
d. $SiCl_4$

62. Write a Lewis structure for each of the following simple molecules. Show all bonding valence electron pairs as lines and all nonbonding valence electron pairs as dots.
a. H_2S
b. SiF_4
c. C_2H_4
d. C_3H_8

63. Write a Lewis structure for each of the following simple molecules. Show all bonding valence electron pairs as lines and all nonbonding valence electron pairs as dots. For those molecules that exhibit resonance, draw the various possible resonance forms.
a. Cl_2O
b. CO_2
c. SO_3

64. Write a Lewis structure for each of the following simple molecules. Show all bonding valence electron pairs as lines and all nonbonding valence electron pairs as dots. For those molecules that exhibit resonance, draw the various possible resonance forms.
a. NO_2
b. H_2SO_4
c. N_2O_4

65. Write a Lewis structure for each of the following polyatomic ions. Show all bonding valence electron pairs as lines and all nonbonding valence electron pairs as dots. For those ions that exhibit resonance, draw the various possible resonance forms.
a. sulfate ion, SO_4^{2-}
b. phosphate ion, PO_4^{3-}
c. sulfite ion, SO_3^{2-}

66. Write a Lewis structure for each of the following polyatomic ions. Show all bonding valence electron pairs as lines and all nonbonding valence electron pairs as dots. For those ions that exhibit resonance, draw the various possible resonance forms.
a. chlorate ion, ClO_3^-
b. peroxide ion, O_2^{2-}
c. acetate ion, $C_2H_3O_2^-$

67. Write a Lewis structure for each of the following polyatomic ions. Show all bonding valence electron pairs as lines and all nonbonding valence electron pairs as dots. For those ions that exhibit resonance, draw the various possible resonance forms.
a. nitrite ion
b. hydrogen carbonate ion
c. hydroxide ion

68. Write a Lewis structure for each of the following polyatomic ions. Show all bonding valence electron pairs as lines and all nonbonding valence electron pairs as dots. For those ions that exhibit resonance, draw the various possible resonance forms.
a. hydrogen phosphate ion, HPO_4^{2-}
b. dihydrogen phosphate ion, $H_2PO_4^-$
c. phosphate ion, PO_4^{3-}

11.8 Molecular Structure

QUESTIONS

69. What is the geometric structure of the water molecule? How many pairs of valence electrons are there on the oxygen atom in the water molecule? What is the approximate H—O—H bond angle in water?

70. What is the geometric structure of the ammonia molecule? How many pairs of electrons surround the nitrogen atom in NH_3? What is the approximate H—N—H bond angle in ammonia?

71. What is the geometric structure of the boron trifluoride molecule, BF_3? How many pairs of valence elec-

trons are present on the boron atom in BF_3? What are the approximate F—B—F bond angles in BF_3?

72. What is the geometric structure of the SiF_4 molecule? How many pairs of valence electrons are present on the silicon atom of SiF_4? What are the approximate F—Si—F bond angles in SiF_4?

11.9 Molecular Structure: The VSEPR Model

QUESTIONS

73. Why is the geometric structure of a molecule important, especially for biological molecules?

74. What general principles determine the molecular structure (shape) of a molecule?

75. How is the structure around a given atom related to repulsion between valence electron pairs on the atom?

76. Why are all diatomic molecules *linear,* regardless of the number of valence electron pairs on the atoms involved?

77. Although the valence electron pairs in ammonia have a tetrahedral arrangement, the overall geometric structure of the ammonia molecule is *not* described as being tetrahedral. Explain.

78. Although both the BF_3 and NF_3 molecules contain the same number of atoms, the BF_3 molecule is flat, whereas the NF_3 molecule is trigonal pyramidal. Explain.

PROBLEMS

79. For the indicated atom in each of the following molecules or ions, give the number and arrangement of the electron pairs around that atom.
a. P in PO_4^{3-}
b. S in SO_4^{2-}
c. S in H_2S

80. For the indicated atom in each of the following molecules or ions, give the number and arrangement of the electron pairs around that atom.
a. P in PH_3
b. Cl in ClO_4^-
c. O in H_2O

81. Using the VSEPR theory, predict the molecular structure of each of the following molecules.
a. NCl_3
b. H_2Se
c. $SiCl_4$

82. Using the VSEPR theory, predict the molecular structure of each of the following molecules.
a. CCl_4
b. H_2S
c. GeI_4

83. Using the VSEPR theory, predict the molecular structure of each of the following polyatomic ions.
a. sulfate ion, SO_4^{2-}
b. phosphate ion, PO_4^{3-}
c. ammonium ion, NH_4^+

84. Using the VSEPR theory, predict the molecular structure of each of the following polyatomic ions.
a. dihydrogen phosphate ion, $H_2PO_4^-$
b. perchlorate ion, ClO_4^-
c. sulfite ion, SO_3^{2-}

85. For each of the following molecules or ions, indicate the bond angle expected between the central atom and any two adjacent hydrogen atoms.
a. H_2O c. NH_4^+
b. NH_3 d. CH_4

86. For each of the following molecules or ions, indicate the bond angle expected between the central atom and any two adjacent chlorine atoms.
a. Cl_2O c. CCl_4
b. NCl_3 d. C_2Cl_4

87. Predict the geometric structure of the carbonate ion, CO_3^{2-}. What are the bond angles in this molecule?

88. Predict the geometric structure of the acetylene molecule, C_2H_2. What are the bond angles in this molecule?

ADDITIONAL PROBLEMS

89. What is *resonance?* Give three examples of molecules or ions that exhibit resonance, and draw Lewis structures for each of the possible resonance forms.

90. When two atoms share two pairs of electrons, a(n) ________ bond is said to exist between them.

91. The geometric arrangement of electron pairs around a given atom is determined principally by the tendency to minimize ________ between the electron pairs.

92. In each case, which of the following pairs of bonded elements forms the more polar bond?
a. S—F or S—Cl
b. N—O or P—O
c. C—H or Si—H

93. For each case, which of the following pairs of bonded elements forms the more polar bond?
a. Br—Cl or Br—F
b. As—S or As—O
c. Pb—C or Pb—Si

94. What do we mean by the *bond energy* of a chemical bond?

95. A(n) ________ chemical bond represents the equal sharing of a pair of electrons between two nuclei.

96. For each of the following pairs of elements, identify which element would be expected to be more

electronegative. It should not be necessary to look at a table of actual electronegativity values.
a. Be or Ba
b. N or P
c. F or Cl

97. On the basis of the electronegativity values given in Figure 11.3, indicate whether each of the following bonds would be expected to be ionic, covalent, or polar covalent.
a. H—O
b. O—O
c. H—H
d. H—Cl

98. Which of the following molecules contain polar covalent bonds?
a. carbon monoxide, CO
b. chlorine, Cl_2
c. iodine monochloride, ICl
d. phosphorus, P_4

99. On the basis of the electronegativity values given in Figure 11.3, indicate which is the more polar bond in each of the following pairs.
a. N—P or N—O
b. N—C or N—O
c. N—S or N—C
d. N—F or N—S

100. In each of the following molecules, which end of the molecule is negative relative to the other end?
a. carbon monoxide, CO
b. iodine monobromide, IBr
c. hydrogen iodide, HI

101. For each of the following bonds, draw a figure indicating the direction of the bond dipole, including which end of the bond is positive and which is negative.
a. N—Cl
b. N—P
c. N—S
d. N—C

102. Write the electron configuration for each of the following atoms and for the simple ion that the element most commonly forms. In each case, indicate which noble gas has the same electron configuration as the ion.
a. aluminum, $Z = 13$
b. bromine, $Z = 35$
c. calcium, $Z = 20$
d. lithium, $Z = 3$
e. fluorine, $Z = 9$

103. What simple ion does each of the following elements most commonly form?
a. sodium
b. iodine
c. potassium
d. calcium
e. sulfur
f. magnesium
g. aluminum
h. nitrogen

104. On the basis of their electron configurations, predict the formula of the simple binary ionic compound likely to form when the following pairs of elements react with each other.
a. sodium, Na, and selenium, Se
b. rubidium, Rb, and fluorine, F
c. potassium, K, and tellurium, Te
d. barium, Ba, and selenium, Se
e. potassium, K, and astatine, At
f. francium, Fr, and chlorine, Cl

105. Which noble gas has the same electron configuration as each of the ions in the following compounds?
a. calcium bromide, $CaBr_2$
b. aluminum selenide, Al_2Se_3
c. strontium oxide, SrO
d. potassium sulfide, K_2S

106. For each of the following pairs, indicate which is smaller.
a. Rb^+ or Na^+
b. Mg^{2+} or Al^{3+}
c. F^- or I^-
d. Na^+ or K^+

107. Write the Lewis structure for each of the following atoms.
a. He ($Z = 2$)
b. Br ($Z = 35$)
c. Sr ($Z = 38$)
d. Ne ($Z = 10$)
e. I ($Z = 53$)
f. Ra ($Z = 88$)

108. What is the *total* number of *valence* electrons in each of the following molecules?
a. HNO_3
b. H_2SO_4
c. H_3PO_4
d. $HClO_4$

109. Write a Lewis structure for each of the following simple molecules. Show all bonding valence electron pairs as lines and all nonbonding valence electron pairs as dots.
a. GeH_4
b. ICl
c. NI_3
d. PF_3

110. Write a Lewis structure for each of the following simple molecules. Show all bonding valence electron pairs as lines and all nonbonding valence electron pairs as dots.
a. N_2H_4
b. C_2H_6
c. NCl_3
d. $SiCl_4$

111. Write a Lewis structure for each of the following simple molecules. Show all bonding valence electron pairs as lines and all nonbonding valence electron pairs as dots. For those molecules that exhibit resonance, draw the various possible resonance forms.
a. SO_2
b. N_2O (N in center)
c. O_3

112. Write a Lewis structure for each of the following polyatomic ions. Show all bonding valence electron pairs as lines and all nonbonding valence electron pairs as dots. For those ions that exhibit resonance, draw the various possible resonance forms.
a. nitrate ion
b. carbonate ion
c. ammonium ion

113. Why is the molecular structure of H_2O nonlinear, whereas that of BeF_2 is linear, even though both molecules consist of three atoms?

114. For the indicated atom in each of the following molecules, give the number and the arrangement of the electron pairs around that atom.
a. C in CCl_4
b. Ge in GeH_4
c. B in BF_3

115. Using the VSEPR theory, predict the molecular structure of each of the following molecules.
a. Cl_2O
b. OF_2
c. $SiCl_4$

116. Using the VSEPR theory, predict the molecular structure of each of the following polyatomic ions.
a. chlorate ion
b. chlorite ion
c. perchlorate ion

117. For each of the following molecules, indicate the bond angle expected between the central atom and any two adjacent chlorine atoms.
a. Cl_2O c. $BeCl_2$
b. CCl_4 d. BCl_3

118. Using the VSEPR theory, predict the molecular structure of each of the following molecules or ions containing multiple bonds.
a. SO_2
b. SO_3
c. HCO_3^- (hydrogen is bonded to oxygen)
d. HCN

119. Using the VSEPR theory, predict the molecular structure of each of the following molecules or ions containing multiple bonds.
a. CO_3^{2-}
b. HNO_3 (hydrogen is bonded to oxygen)
c. NO_2^-
d. C_2H_2

120. Explain briefly how substances with ionic bonding differ in properties from substances with covalent bonding.

121. Explain the difference between a covalent bond formed between two atoms of the same element and a covalent bond formed between atoms of two different elements.

Cumulative Review for Chapters 10–11

QUESTIONS

1. What is *electromagnetic radiation?* Give some examples of such radiation. Explain what the *wavelength* (λ) and *frequency* (ν) of electromagnetic radiation represent. Sketch a representation of a wave and indicate on your drawing one wavelength of the wave. At what speed does electromagnetic radiation move through space? How is this speed related to λ and ν?

2. Explain what it means for an atom to be in an *excited state* and what it means for an atom to be in its *ground state.* How does an excited atom *return* to its ground state? What is a *photon?* How is the wavelength (color) of light related to the energy of the photons being emitted by an atom? How is the energy of the photons being *emitted* by an atom related to the energy changes taking place *within* the atom?

3. Do atoms in excited states emit radiation randomly, at any wavelength? Why? What does it mean to say that the hydrogen atom has only certain *discrete energy levels* available? How do we know this? Why was the quantization of energy levels surprising to scientists when it was first discovered?

4. Describe Bohr's model of the hydrogen atom. How did Bohr envision the relationship between the electron and the nucleus of the hydrogen atom? How did Bohr's model explain the emission of only discrete wavelengths of light by excited hydrogen atoms? Why did Bohr's model not stand up as more experiments were performed using elements other than hydrogen?

5. Schrödinger and de Broglie suggested a "wave–particle duality" for small particles—that is, if electromagnetic radiation showed some particle-like properties, then perhaps small particles might exhibit some wave-like properties. Explain. How does the wave mechanical picture of the atom fundamentally differ from the Bohr model? How do wave mechanical *orbitals* differ from Bohr's *orbits?* What does it mean to say that an orbital represents a probability map for an electron?

6. Describe the general characteristics of the first (lowest-energy) hydrogen atomic orbital. How is this orbital designated symbolically? Does this orbital have a sharp "edge"? Does the orbital represent a surface upon which the electron travels at all times?

7. Use the wave mechanical picture of the hydrogen atom to describe what happens when the atom absorbs energy and moves to an "excited" state. What do the *principal energy levels* and their sublevels represent for a hydrogen atom? How do we designate specific principal energy levels and sublevels in hydrogen?

8. Describe the sublevels and orbitals that constitute the third and fourth principal energy levels of hydrogen. How is each of the orbitals designated and what are the general shapes of their probability maps?

9. Describe *electron spin.* How does electron spin affect the total number of electrons that can be accommodated in a given orbital? What does the *Pauli exclusion principle* tell us about electrons and their spins?

10. Summarize the postulates of the wave mechanical model of the atom.

11. List the *order* in which the orbitals are filled as the atoms beyond hydrogen are built up. How many electrons overall can be accommodated in the first and second principal energy levels? How many electrons can be placed in a given *s* subshell? In a given *p* subshell? In a specific *p* orbital? Why do we assign unpaired electrons in the 2*p* orbitals of carbon, nitrogen, and oxygen?

12. Define the *valence electrons* in an atom. Define the *core electrons* in an atom. Why are the valence electrons more important to the atom's chemical properties than the core electrons? How is the number of valence electrons in an atom related to the atom's position on the periodic table?

13. Sketch the overall shape of the periodic table and indicate the *general regions* of the table that represent the various *s, p, d,* and *f* orbitals being filled.

14. Using the general periodic table you developed in Question 13, show how the valence-electron configuration of most of the elements can be written just by knowing the relative *location* of the element on the table. Give specific examples.

15. What are the *representative elements?* In what region(s) of the periodic table are these elements found? In what general area of the periodic table are the *metallic* elements found? In what general area of the table are the *nonmetals* found? Where in the table are the *metalloids* located?

16. You have learned how the properties of the elements vary *systematically,* corresponding to the electron structures of the elements being considered. Discuss how the *ionization energies* and *atomic sizes* of elements vary, both within a vertical group (family) of the periodic table and within a horizontal row (period).

17. In general, what do we mean by a *chemical bond?* What does the *bond energy* tell us about the strength of a chemical bond? Name the principal types of chemical bonds.

18. What do we mean by *ionic* bonding? Give an example of a substance whose particles are held together by ionic bonding. What experimental evidence do we have for the existence of ionic bonding? In general,

what types of substances react to produce compounds having ionic bonding?

19. What do we mean by *covalent* bonding and *polar covalent* bonding? How are these two bonding types similar and how do they differ? What circumstance must exist for a bond to be purely covalent? How does a polar covalent bond differ from an ionic bond?

20. Define *electronegativity.* How does the polarity of a bond depend on the *difference* in electronegativity of the two atoms participating in the bond? If two atoms have exactly the *same* electronegativity, what type of bond will exist between the atoms? If two atoms have vastly different electronegativities, what type of bond will exist between them?

21. What does it mean to say that a molecule has a dipole moment? What is the *difference* between a polar bond and a polar molecule (one that has a dipole moment)? Give an example of a molecule that has polar bonds and that has a dipole moment. Give an example of a molecule that has polar bonds, but that does *not* have a dipole moment. What are some implications of the fact that water has a dipole moment?

22. How is the attainment of a noble gas electron configuration important to our ideas of how atoms bond to each other? When atoms of a metal react with atoms of a nonmetal, what type of electron configurations do the resulting ions attain? Explain how the atoms in a covalently bonded compound can attain noble gas electron configurations.

23. Give evidence that ionic bonds are very strong. Does an ionic substance contain discrete molecules? With what general type of structure do ionic compounds occur? Sketch a representation of a general structure for an ionic compound. Why is a cation always smaller and an anion always larger than the respective parent atom? Describe the bonding in an ionic compound containing polyatomic ions.

24. Why does a Lewis structure for a molecule show only the valence electrons? What is the most important factor for the formation of a stable compound? How do we use this requirement when writing Lewis structures?

25. In writing Lewis structures for molecules, what is meant by the *duet rule?* To which element does the duet rule apply? What do we mean by the *octet rule?* Why is attaining an octet of electrons important for an atom when it forms bonds to other atoms? What is a bonding *pair* of electrons? What is a nonbonding (or *lone*) pair of electrons?

26. For three simple molecules of your own choice, *apply* the rules for writing Lewis structures. Write your discussion as if you are explaining the method to someone who is *not* familiar with Lewis structures.

27. What does a *double* bond between two atoms represent in terms of the number of electrons shared? What does a *triple* bond represent? When writing a Lewis structure, explain how we recognize when a molecule must contain double or triple bonds. What are *resonance structures?*

28. Although many simple molecules fulfill the octet rule, some common molecules are exceptions to this rule. Give three examples of molecules whose Lewis structures are exceptions to the octet rule.

29. What do we mean by the *geometric structure* of a molecule? Draw the geometric structures of at least four simple molecules of your choosing and indicate the bond angles in the structures. Explain the main ideas of the *valence shell electron pair repulsion (VSEPR) theory.* Using several examples, explain how you would *apply* the VSEPR theory to predict their geometric structures.

30. What bond angle results when there are only two valence electron pairs around an atom? What bond angle results when there are three valence pairs? What bond angle results when there are four pairs of valence electrons around the central atom in a molecule? Give examples of molecules containing these bond angles.

31. How do we predict the geometric structure of a molecule whose Lewis structure indicates that the molecule contains a double or triple bond? Give an example of such a molecule, write its Lewis structure, and show how the geometric shape is derived.

32. Write the electron configuration for each of the following atoms or ions. For the *ions* in the list, tell with which noble gas the ion is isoelectronic. For the *atoms* in the list, tell which simple ion they would be most likely to form.

a. Mg	e. Ba	h. O^{2-}
b. Na^+	f. I^-	i. Al^{3+}
c. Cl	g. Zn	j. P^{3-}
d. F^-		

33. For the following molecules and ions, draw the Lewis structure, tell how many valence electron pairs are found around the element given in boldface, and predict the geometric structure of the molecule or ion.

a. $\mathbf{N}F_3$	c. $\mathbf{B}F_3$	e. $\mathbf{P}O_4^{3-}$
b. $\mathbf{C}F_4$	d. $\mathbf{S}O_4^{2-}$	f. $H_2\mathbf{O}$

12 Gases

A skydiver leaps into the earth's atmosphere, which is composed of a mixture of gases.

We live immersed in a gaseous solution. The earth's atmosphere is a mixture of gases that consists mainly of elemental nitrogen, N_2, and oxygen, O_2. The atmosphere both supports life and acts as a waste receptacle for the exhaust gases that accompany many industrial processes. The chemical reactions of these waste gases in the atmosphere lead to various types of pollution, including smog and acid rain. The two main sources of pollution are transportation and the production of electricity. The combustion of fuel in vehicles produces CO, CO_2, NO, and NO_2, along with unburned fragments of the petroleum used as fuel. The combustion of coal and petroleum in power plants produces NO_2 and SO_2 in the exhaust gases. These mixtures of chemicals can be activated by absorbing light to produce the photochemical smog that afflicts most large cities. The SO_2 in the air reacts with oxygen to produce SO_3 gas, which combines with water in the air to produce droplets of sulfuric acid (H_2SO_4), a major component of acid rain.

The gases in the atmosphere also shield us from harmful radiation from the sun and keep the earth warm by reflecting heat radiation back toward the earth. In fact, there is now great concern that an increase in atmospheric carbon dioxide, a product of the combustion of fossil fuels, is causing a dangerous warming of the earth. (See "Chemistry in Focus: Atmospheric Effects," in Chapter 10.)

In this chapter we will look carefully at the properties of gases. First we will see how measurements of gas properties lead to various types of laws—statements that show how the properties are related to each other. Then we will construct a model to explain why gases behave as they do. This model will show how the behavior of the individual particles of a gas leads to the observed properties of the gas itself (a collection of many, many particles).

Steve Fossett flies his balloon *Solo Spirit*, over the east coast of Australia during his attempt to make the first solo balloon flight around the world.

The study of gases provides an excellent example of the scientific method in action. It illustrates how observations lead to natural laws, which in turn can be accounted for by models.

12.1 Pressure

AIMS: To learn about atmospheric pressure and how barometers work. To learn the various units of pressure.

A gas uniformly fills any container, is easily compressed, and mixes completely with any other gas (see Section 3.1). One of the most obvious properties of a gas is that it exerts pressure on its surroundings. For example, when you blow up a balloon, the air inside pushes against the elastic sides of the balloon and keeps it firm.

Dry air (air from which the water vapor has been removed) is 78.1% N_2 molecules, 20.9% O_2 molecules, 0.9% Ar atoms, and 0.03% CO_2 molecules, along with smaller amounts of Ne, He, CH_4, Kr, and other trace components.

As a gas, water occupies 1200 times as much space as it does as a liquid at 25 °C and atmospheric pressure.

Soon after Torricelli died, a German physicist named Otto von Guericke invented an air pump. In a famous demonstration for the King of Prussia in 1683, Guericke placed two hemispheres together, pumped the air out of the resulting sphere through a valve, and showed that teams of horses could not pull the hemispheres apart. Then, after secretly opening the air valve, Guericke easily separated the hemispheres by hand. The King of Prussia was so impressed that he awarded Guericke a lifetime pension!

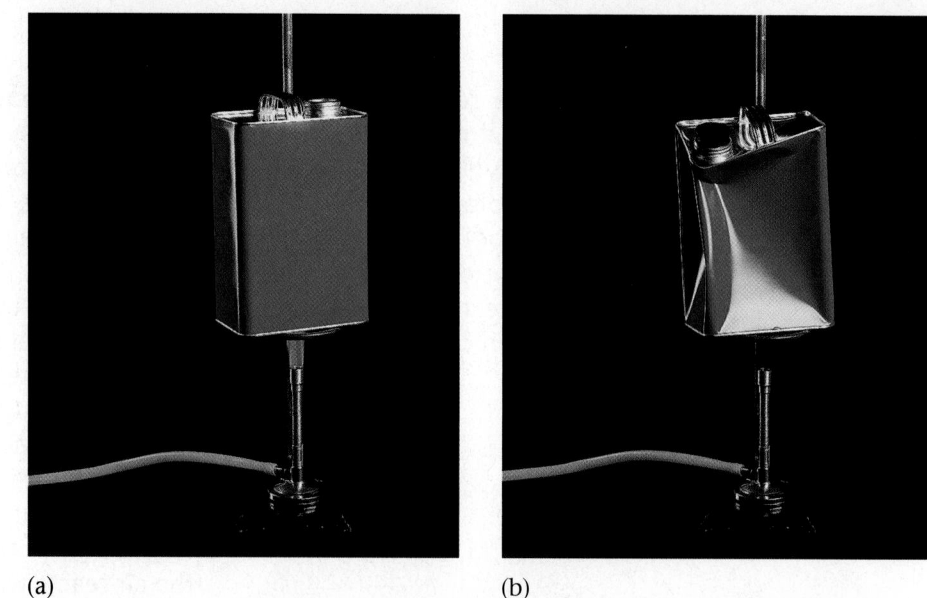

Figure 12.1
The pressure exerted by the gases in the atmosphere can be demonstrated by boiling water in a can (a), and then turning off the heat and sealing the can. As the can cools, the water vapor condenses, lowering the gas pressure inside the can. This causes the can to crumple (b).

The gases most familiar to us form the earth's atmosphere. The pressure exerted by this gaseous mixture that we call air can be dramatically demonstrated by the experiment shown in Figure 12.1. A small volume of water is placed in a metal can and the water is boiled, which fills the can with steam. The can is then sealed and allowed to cool. Why does the can collapse as it cools? It is the atmospheric pressure that crumples the can. When the can is cooled after being sealed so that no air can flow in, the water vapor (steam) inside the can condenses to a very small volume of liquid water. As a gas, the water vapor filled the can, but when it is condensed to a liquid, the liquid does not come close to filling the can. The H_2O molecules formerly present as a gas are now collected in a much smaller volume of liquid, and there are very few molecules of gas left to exert pressure outward and counteract the air pressure. As a result, the pressure exerted by the gas molecules in the atmosphere smashes the can.

A device that measures atmospheric pressure, the **barometer,** was invented in 1643 by an Italian scientist named Evangelista Torricelli (1608–1647), who had been a student of the famous astronomer Galileo. Torricelli's barometer is constructed by filling a glass tube with liquid mercury and inverting it in a dish of mercury, as shown in Figure 12.2. Notice that a large quantity of mercury stays in the tube. In fact, at sea level the height of this column of mercury averages 760 mm. Why does this mercury stay in the tube, seemingly in defiance of gravity? Figure 12.2 illustrates how the pressure exerted by the atmospheric gases on the surface of mercury in the dish keeps the mercury in the tube.

Atmospheric pressure results from the mass of the air being pulled toward the center of the earth by gravity—in other words, it results from the weight of the air. Changing weather conditions cause the atmospheric pressure to vary, so the height of the column of Hg supported by the atmosphere at sea level varies; it is not always 760 mm. The meteorologist who says a "low" is approaching means that the atmospheric pressure is going to decrease. This condition often occurs in conjunction with a storm.

Atmospheric pressure also varies with altitude. For example, when Torricelli's experiment is done in Breckenridge, Colorado (elevation 9600 feet),

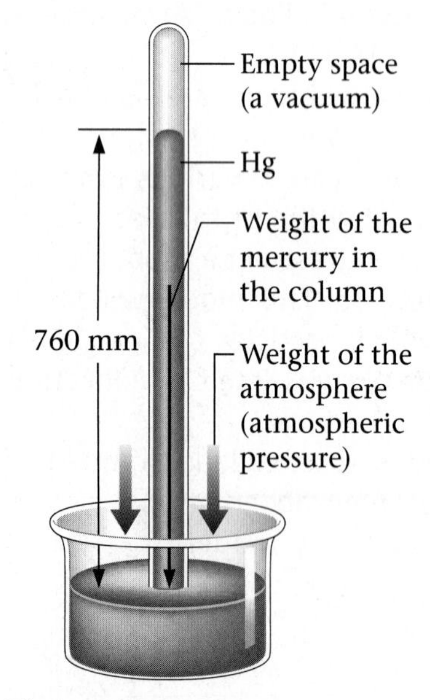

Figure 12.2
When a glass tube is filled with mercury and inverted in a dish of mercury at sea level, the mercury flows out of the tube until a column approximately 760 mm high remains (the height varies with atmospheric conditions). Note that the pressure of the atmosphere balances the weight of the column of mercury in the tube.

the atmosphere supports a column of mercury only about 520 mm high because the air is "thinner." That is, there is less air pushing down on the earth's surface at Breckenridge than at sea level.

Units of Pressure

Mercury is used to measure pressure because of its high density. By way of comparison, the column of water required to measure a given pressure would be 13.6 times as high as a mercury column used for the same purpose.

Because instruments used for measuring pressure (see Figure 12.3) often contain mercury, the most commonly used units for pressure are based on the height of the mercury column (in millimeters) that the gas pressure can support. The unit **mm Hg** (millimeters of mercury) is often called the **torr** in honor of Torricelli. The terms *torr* and *mm Hg* are used interchangeably by chemists. A related unit for pressure is the **standard atmosphere** (abbreviated atm).

$$1 \text{ standard atmosphere} = 1.000 \text{ atm} = 760.0 \text{ mm Hg} = 760.0 \text{ torr}$$

The SI unit for pressure is the **pascal** (abbreviated Pa).

$$1 \text{ standard atmosphere} = 101{,}325 \text{ Pa}$$

Thus 1 atmosphere is about 100,000 or 10^5 pascals. Because the pascal is so small we will use it sparingly in this book. A unit of pressure that is employed in the engineering sciences and that we use for measuring tire pressure is pounds per square inch, abbreviated psi.

$$1.000 \text{ atm} = 14.69 \text{ psi}$$

1.000 atm
760.0 mm Hg
760.0 torr
14.69 psi
101,325 Pa

Sometimes we need to convert from one unit of pressure to another. We do this by using conversion factors. The process is illustrated in Example 12.1.

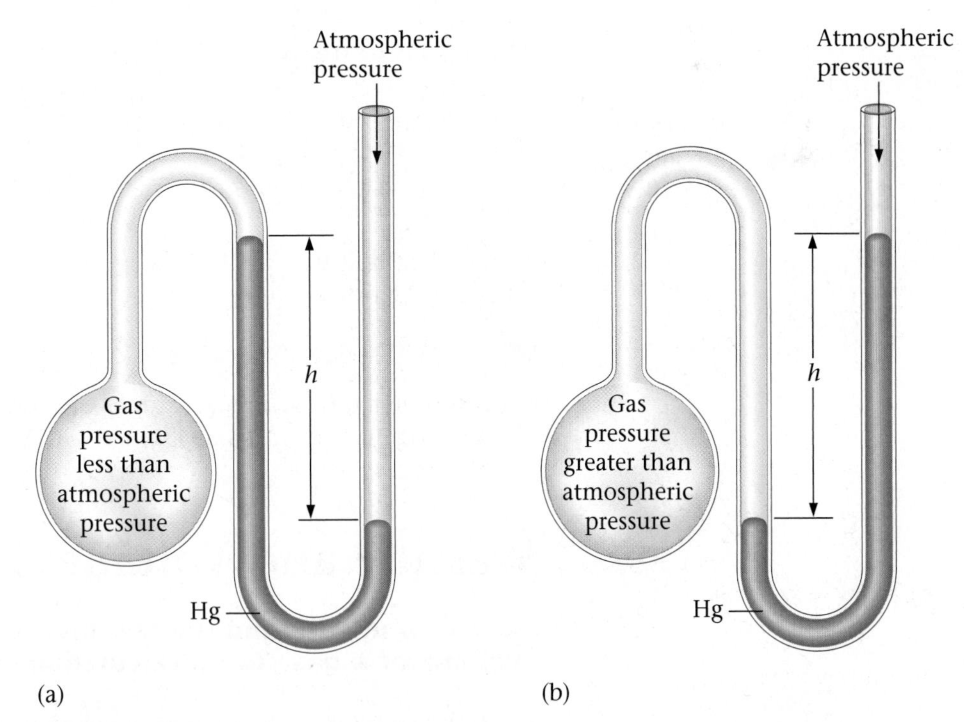

Figure 12.3
A device (called a manometer) for measuring the pressure of a gas in a container. The pressure of the gas is equal to h (the difference in mercury levels) in units of torr (equivalent to mm Hg). (a) Gas pressure = atmospheric pressure − h. (b) Gas pressure = atmospheric pressure + h.

Example 12.1 Pressure Unit Conversions

The pressure of the air in a tire is measured to be 28 psi. Represent this pressure in atmospheres, torr, and pascals.

Checking the air pressure in a tire.

Solution

To convert from pounds per square inch to atmospheres, we need the equivalence statement

$$1.000 \text{ atm} = 14.69 \text{ psi}$$

which leads to the conversion factor

$$\frac{1.000 \text{ atm}}{14.69 \text{ psi}}$$

$$28 \cancel{\text{psi}} \times \frac{1.000 \text{ atm}}{14.69 \cancel{\text{psi}}} = 1.9 \text{ atm}$$

To convert from atmospheres to torr, we use the equivalence statement

$$1.000 \text{ atm} = 760.0 \text{ torr}$$

which leads to the conversion factor

$$\frac{760.0 \text{ torr}}{1.000 \text{ atm}}$$

$$1.9 \cancel{\text{atm}} \times \frac{760.0 \text{ torr}}{1.000 \cancel{\text{atm}}} = 1.4 \times 10^3 \text{ torr}$$

$1.9 \times 760.0 = 1444$
1444 ➡ $1400 = 1.4 \times 10^3$
Round off

To change from torr to pascals, we need the equivalence statement

$$1.000 \text{ atm} = 101{,}325 \text{ Pa}$$

which leads to the conversion factor

$$\frac{101{,}325 \text{ Pa}}{1.000 \text{ atm}}$$

$$1.9 \cancel{\text{atm}} \times \frac{101{,}325 \text{ Pa}}{1.000 \cancel{\text{atm}}} = 1.9 \times 10^5 \text{ Pa}$$

$1.9 \times 101{,}325 = 192{,}517.5$
192,517.5 ➡ $190{,}000 = 1.9 \times 10^5$
Round off

NOTE: The best way to check a problem like this is to make sure the final units are the ones required.

✔ Self-Check Exercise 12.1

On a summer day in Breckenridge, Colorado, the atmospheric pressure is 525 mm Hg. What is this air pressure in atmospheres?

See Problems 12.7 through 12.12. ■

12.2 Pressure and Volume: Boyle's Law

AIMS: To understand the law that relates the pressure and volume of a gas. To do calculations involving this law.

The first careful experiments on gases were performed by the Irish scientist Robert Boyle (1627–1691). Using a J-shaped tube closed at one end (Figure 12.4), which he reportedly set up in the multi-story entryway of his house, Boyle studied the relationship between the pressure of the trapped gas and its volume. Representative values from Boyle's experiments are given in

For Boyle's law to hold, the amount of gas (moles) must not be changed. The temperature must also be constant.

The fact that the constant is sometimes 1.40×10^3 instead of 1.41×10^3 is due to experimental error (uncertainties in measuring the values of P and V).

Table 12.1 A Sample of Boyle's Observations (moles of gas and temperature both constant)

Experiment	Pressure (in Hg)	Volume (in.^{3})	Pressure × Volume (in Hg) × (in.^{3}) Actual	Rounded*
1	29.1	48.0	1396.8	1.40×10^3
2	35.3	40.0	1412.0	1.41×10^3
3	44.2	32.0	1414.4	1.41×10^3
4	58.2	24.0	1396.8	1.40×10^3
5	70.7	20.0	1414.0	1.41×10^3
6	87.2	16.0	1395.2	1.40×10^3
7	117.5	12.0	1410.0	1.41×10^3

*Three significant figures are allowed in the product because both of the numbers that are multiplied together have three significant figures.

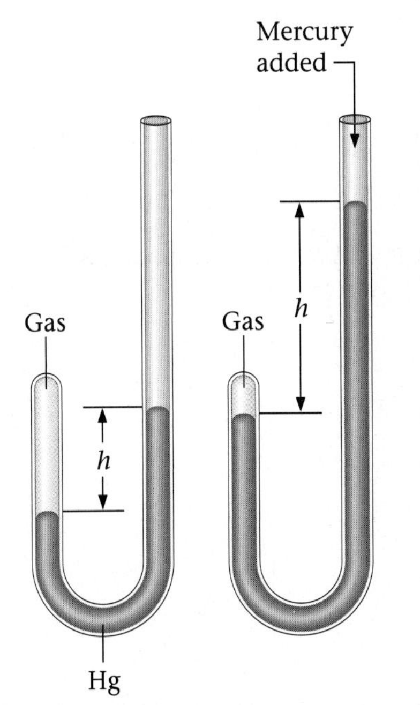

Figure 12.4
A J-tube similar to the one used by Boyle. The pressure on the trapped gas can be changed by adding or withdrawing mercury.

Table 12.1. The units given for the volume (cubic inches) and pressure (inches of mercury) are the ones Boyle used. Keep in mind that the metric system was not in use at this time.

First let's examine Boyle's observations (Table 12.1) for general trends. Note that as the pressure increases, the volume of the trapped gas decreases. In fact, if you compare the data from experiments 1 and 4, you can see that as the pressure is doubled (from 29.1 to 58.2), the volume of the gas is halved (from 48.0 to 24.0). The same relationship can be seen in experiments 2 and 5 and in experiments 3 and 6 (approximately).

We can see the relationship between the volume of a gas and its pressure more clearly by looking at the product of the values of these two properties ($P \times V$) using Boyle's observations. This product is shown in the last column of Table 12.1. Note that for all the experiments,

$$P \times V = 1.4 \times 10^3 \text{ (in Hg)} \times \text{in.}^3$$

with only a slight variation due to experimental error. Other similar measurements on gases show the same behavior. This means that the relationship of the pressure and volume of a gas can be expressed in words as

pressure times volume equals a constant

or in terms of an equation as

$$PV = k$$

which is called **Boyle's law,** where k is a constant at a specific temperature for a given amount of gas. For the data we used from Boyle's experiment, $k = 1.41 \times 10^3$ (in Hg) $\times$ in.^{3}

It is often easier to visualize the relationships between two properties if we make a graph. Figure 12.5 uses the data given in Table 12.1 to show how pressure is related to volume. This relationship, called a plot or a graph, shows that V decreases as P increases. When this type of relationship exists, we say that volume and pressure are inversely related or *inversely proportional;* when one increases, the other decreases. Boyle's law is illustrated by the gas samples in Figure 12.6.

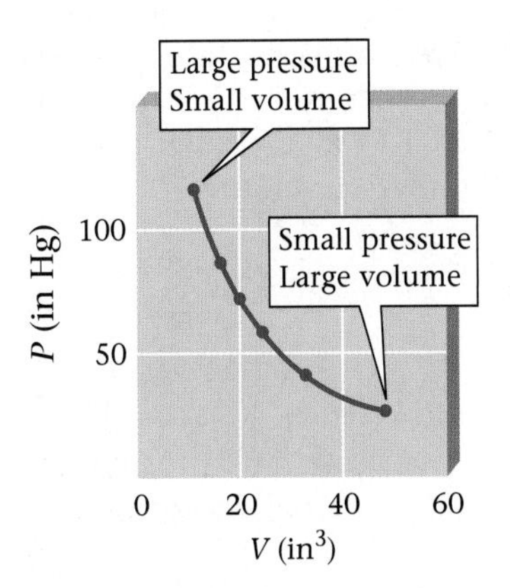

Figure 12.5
A plot of P versus V from Boyle's data in Table 12.1.

Boyle's law means that if we know the volume of a gas at a given pressure, we can predict the new volume if the pressure is changed, *provided that neither the temperature nor the amount of gas is changed.* For example, if we represent the original pressure and volumes as P_1 and V_1 and the final values as P_2 and V_2, using Boyle's law we can write

P = 1 atm

V = 1 L
T = 298 K

P = 2 atm

V = 0.50 L
T = 298 K

Figure 12.6
Illustration of Boyle's law. These three containers contain the same number of molecules. At 298 K, $P \times V = 1$ L atm in all three containers.

$$P_1V_1 = k$$

and

$$P_2V_2 = k$$

We can also say

$$P_1V_1 = k = P_2V_2$$

or simply

$$P_1V_1 = P_2V_2$$

This is really another way to write Boyle's law. We can solve for the final volume (V_2) by dividing both sides of the equation by P_2.

$$\frac{P_1V_1}{P_2} = \frac{P_2V_2}{P_2}$$

Canceling the P_2 terms on the right gives

$$\frac{P_1}{P_2} \times V_1 = V_2$$

or

$$V_2 = V_1 \times \frac{P_1}{P_2}$$

This equation tells us that we can calculate the new gas volume (V_2) by multiplying the original volume (V_1) by the ratio of the original pressure to the final pressure (P_1/P_2), as illustrated in Example 12.2.

Example 12.2 Calculating Volume Using Boyle's Law

Freon-12 (the common name for the compound CCl_2F_2) was widely used in refrigeration systems, but has now been replaced by other compounds that do not lead to the breakdown of the protective ozone in the upper atmosphere. Consider a 1.5-L sample of gaseous CCl_2F_2 at a pressure of 56 torr. If pressure is changed to 150 torr at a constant temperature,

a. Will the volume of the gas increase or decrease?

b. What will be the new volume of the gas?

Solution

a. As the first step in a gas law problem, always write down the information given, in the form of a table showing the initial and final conditions.

Initial Conditions	Final Conditions
$P_1 = 56$ torr	$P_2 = 150$ torr
$V_1 = 1.5$ L	$V_2 = ?$

Drawing a picture also is often helpful. Notice that the pressure is increased from 56 torr to 150 torr, so the volume must decrease:

$P_1V_1 \Rightarrow P_2V_2$

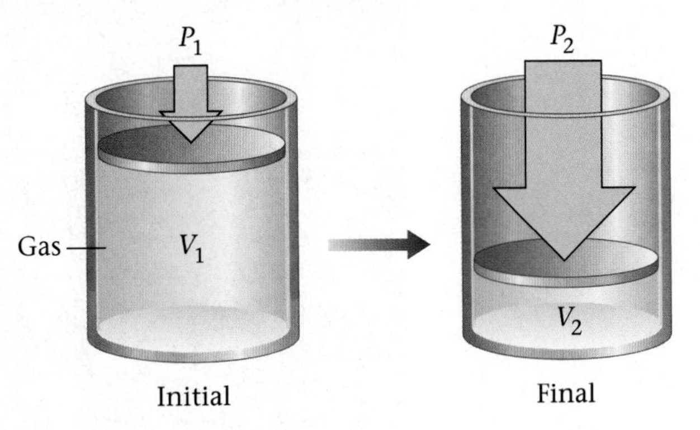

We can verify this by using Boyle's law in the form

$$V_2 = V_1 \times \frac{P_1}{P_2}$$

Note that V_2 is obtained by "correcting" V_1 using the ratio P_1/P_2. Because P_1 is less than P_2, the ratio P_1/P_2 is a fraction that is less than 1. Thus V_2 must be a fraction of (smaller than) V_1; the volume decreases.

b. We calculate V_2 as follows:

$$V_2 = V_1 \times \frac{P_1}{P_2} = \underset{\uparrow V_1}{1.5\ \text{L}} \times \frac{\overset{P_1 \downarrow}{56\ \cancel{\text{torr}}}}{\underset{\uparrow P_2}{150\ \cancel{\text{torr}}}} = 0.56\ \text{L}$$

The fact that the volume decreases in Example 12.2 makes sense because the pressure was increased. *To help catch errors, make it a habit to check whether an answer to a problem makes physical sense.*

The volume of the gas decreases from 1.5 to 0.56 L. This change is in the expected direction.

Self-Check Exercise 12.2

A sample of neon to be used in a neon sign has a volume of 1.51 L at a pressure of 635 torr. Calculate the volume of the gas after it is pumped into the glass tubes of the sign, where it shows a pressure of 785 torr.

See Problems 12.21 and 12.22. ■

Example 12.3 Calculating Pressure Using Boyle's Law

In an automobile engine the gaseous fuel–air mixture enters the cylinder and is compressed by a moving piston before it is ignited. In a certain engine the initial cylinder volume is 0.725 L. After the piston moves up, the volume is 0.075 L. The fuel–air mixture initially has a pressure of 1.00 atm. Calculate the pressure of the compressed fuel–air mixture, assuming that both the temperature and the amount of gas remain constant.

$$P_1V_1 = P_2V_2$$
$$\frac{P_1V_1}{V_2} = \frac{P_2\cancel{V_2}}{\cancel{V_2}}$$
$$P_1 \times \frac{V_1}{V_2} = P_2$$
$$\frac{0.725}{0.075} = 9.666\ldots$$

$9.666 \Rightarrow 9.7$
Round off

Solution

We summarize the given information in the following table:

Initial Conditions	Final Conditions
$P_1 = 1.00$ atm	$P_2 = ?$
$V_1 = 0.725$ L	$V_2 = 0.075$ L

Neon signs in Hong Kong.

Then we solve Boyle's law in the form $P_1V_1 = P_2V_2$ for P_2 by dividing both sides by V_2 to give the equation

$$P_2 = P_1 \times \frac{V_1}{V_2} = 1.00 \text{ atm} \times \frac{0.725 \cancel{\text{L}}}{0.075 \cancel{\text{L}}} = 9.7 \text{ atm}$$

Note that the pressure must increase because the volume gets smaller. Pressure and volume are inversely related. ■

12.3 Volume and Temperature: Charles's Law

AIMS: To learn about absolute zero. To learn about the law relating the volume and temperature of a sample of gas at constant moles and pressure, and to do calculations involving that law.

In the century following Boyle's findings, scientists continued to study the properties of gases. The French physicist Jacques Charles (1746–1823), who was the first person to fill a balloon with hydrogen gas and who made the first solo balloon flight, showed that the volume of a given amount of gas (at constant pressure) increases with the temperature of the gas. That is, the volume increases when the temperature increases. A plot of the volume of a given sample of gas (at constant pressure) versus its temperature (in Celsius degrees) gives a straight line. This type of relationship is called *linear,* and this behavior is shown for several gases in Figure 12.7.

The solid lines in Figure 12.7 are based on actual measurements of temperature and volume for the gases listed. As we cool the gases they eventually liquefy, so we cannot determine any experimental points below this temperature. However, when we extend each straight line (which is called *extrapolation* and is shown here by a dashed line), something very interesting

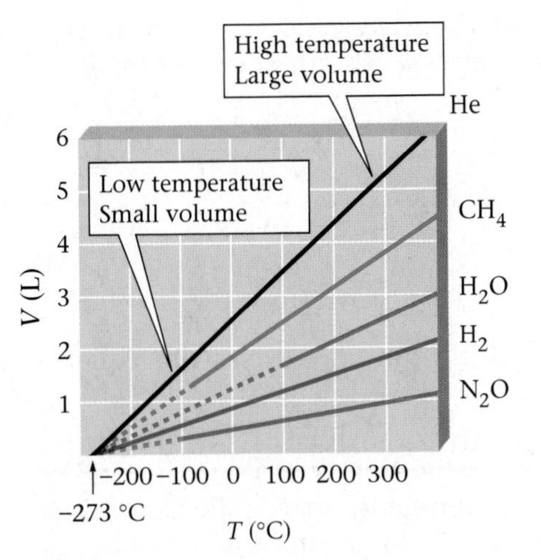

Figure 12.7
Plots of V (L) versus T (°C) for several gases. Note that each sample of gas contains a different number of moles to spread out the plots.

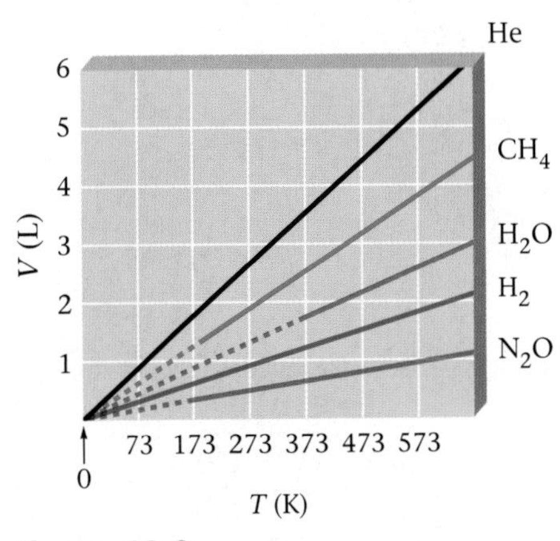

Figure 12.8
Plots of V versus T as in Figure 12.7, except that here the Kelvin scale is used for temperature.

Temperatures such as 0.00000002 K have been obtained in the laboratory, but 0 K has never been reached.

The air in a balloon expands when it is heated. This means that some of the air escapes from the balloon, lowering the air density inside and thus making the balloon buoyant.

From Figure 12.8 for Helium

V (L)	T (K)	b
0.7	73	0.01
1.7	173	0.01
2.7	273	0.01
3.7	373	0.01
5.7	573	0.01

happens. *All* of the lines extrapolate to zero volume at the same temperature: −273 °C. This suggests that −273 °C is the lowest possible temperature, because a negative volume is physically impossible. In fact, experiments have shown that matter cannot be cooled to temperatures lower than −273 °C. Therefore, this temperature is defined as **absolute zero** on the Kelvin scale.

When the volumes of the gases shown in Figure 12.7 are plotted against temperature on the Kelvin scale rather than the Celsius scale, the plots shown in Figure 12.8 result. These plots show that the volume of each gas is *directly proportional to the temperature* (in kelvins) and extrapolates to zero when the temperature is 0 K. Let's illustrate this statement with an example. Suppose we have 1 L of gas at 300 K. When we double the temperature of this gas to 600 K (without changing its pressure), the volume also doubles, to 2 L. Verify this type of behavior by looking carefully at the lines for various gases shown in Figure 12.8.

The direct proportionality between volume and temperature (in kelvins) is represented by the equation known as **Charles's law:**

$$V = bT$$

where T is in kelvins and b is the proportionality constant. Charles's law holds for a given sample of gas at constant pressure. It tells us that (for a given amount of gas at a given pressure) the volume of the gas is directly proportional to the temperature on the Kelvin scale:

$$V = bT \quad \text{or} \quad \frac{V}{T} = b = \text{constant}$$

Notice that in the second form, this equation states that the *ratio* of V to T (in kelvins) must be constant. (This is shown for helium in the margin.) Thus, when we triple the temperature (in kelvins) of a sample of gas, the volume of the gas triples also.

$$\frac{V}{T} = \frac{3 \times V}{3 \times T} = b = \text{constant}$$

We can also write Charles's law in terms of V_1 and T_1 (the initial conditions) and V_2 and T_2 (the final conditions).

$$\frac{V_1}{T_1} = b \quad \text{and} \quad \frac{V_2}{T_2} = b$$

Charles's law in the form $V_1/T_1 = V_2/T_2$ applies only when both the amount of gas (moles) and the pressure are constant.

Thus

$$\frac{V_1}{T_1} = \frac{V_2}{T_2}$$

We will illustrate the use of this equation in Examples 12.4 and 12.5.

Example 12.4 Calculating Volume Using Charles's Law, I

A 2.0-L sample of air is collected at 298 K and then cooled to 278 K. The pressure is held constant at 1.0 atm.

a. Does the volume increase or decrease?

b. Calculate the volume of the air at 278 K.

Solution

a. Because the gas is cooled, the volume of the gas must decrease:

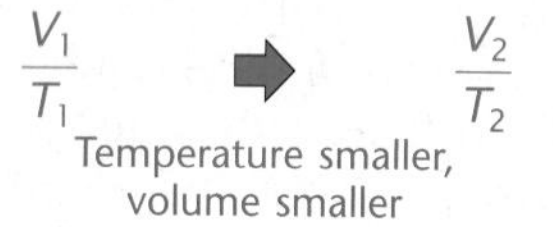

Temperature smaller, volume smaller

$$\frac{V}{T} = \text{constant}$$

T is decreased, so V must decrease to maintain a constant ratio.

b. To calculate the new volume, V_2, we will use Charles's law in the form

$$\frac{V_1}{T_1} = \frac{V_2}{T_2}$$

We are given the following information:

Initial Conditions	**Final Conditions**
$T_1 = 298$ K	$T_2 = 278$ K
$V_1 = 2.0$ L	$V_2 = ?$

We want to solve the equation

$$\frac{V_1}{T_1} = \frac{V_2}{T_2}$$

for V_2. We can do this by multiplying both sides by T_2 and canceling.

$$T_2 \times \frac{V_1}{T_1} = \frac{V_2}{\cancel{T_2}} \times \cancel{T_2} = V_2$$

Thus

$$V_2 = T_2 \times \frac{V_1}{T_1} = 278\ \cancel{\text{K}} \times \frac{2.0\ \text{L}}{298\ \cancel{\text{K}}} = 1.9\ \text{L}$$

Note that the volume gets smaller when the temperature decreases, just as we predicted. ■

Example 12.5 Calculating Volume Using Charles's Law, II

A sample of gas at 15 °C (at 1 atm) has a volume of 2.58 L. The temperature is then raised to 38 °C (at 1 atm).

a. Does the volume of the gas increase or decrease?

b. Calculate the new volume.

Solution

a. In this case we have a given sample (constant amount) of gas that is heated from 15 °C to 38 °C *while the pressure is held constant.* We know from Charles's law that the volume of a given sample of gas is directly proportional to the temperature (at constant pressure). So the increase in temperature will *increase* the volume; the new volume will be greater than 2.58 L.

b. To calculate the new volume, we use Charles's law in the form

$$\frac{V_1}{T_1} = \frac{V_2}{T_2}$$

We are given the following information:

Initial Conditions	**Final Conditions**
$T_1 = 15\ °C$	$T_2 = 38\ °C$
$V_1 = 2.58\ L$	$V_2 = ?$

As is often the case, the temperatures are given in Celsius degrees. However, for us to use Charles's law, the temperature *must be in kelvins.* Thus we must convert by adding 273 to each temperature.

Initial Conditions	**Final Conditions**
$T_1 = 15\ °C = 15 + 273 = 288\ K$	$T_2 = 38\ °C = 38 + 273 = 311\ K$
$V_1 = 2.58\ L$	$V_2 = ?$

Solving for V_2 gives

$$V_2 = V_1 \times \frac{T_2}{T_1} = 2.58\ L\left(\frac{311\ \cancel{K}}{288\ \cancel{K}}\right) = 2.79\ L$$

The new volume (2.79 L) is greater than the initial volume (2.58 L), as we expected.

Researchers take samples from a steaming volcanic vent at Mount Baker in Washington.

✔ Self-Check Exercise 12.3

A child blows a bubble that contains air at 28 °C and has a volume of 23 cm^3 at 1 atm. As the bubble rises, it encounters a pocket of cold air (temperature 18 °C). If there is no change in pressure, will the bubble get larger or smaller as the air inside cools to 18 °C? Calculate the new volume of the bubble.

See Problems 12.29 and 12.30. ■

Notice from Example 12.5 that we adjust the volume of a gas for a temperature change by multiplying the original volume by the ratio of the Kelvin temperatures—final (T_2) over initial (T_1). Remember to check whether your answer makes sense. When the temperature increases (at constant pressure), the volume must increase, and vice versa.

Example 12.6 Calculating Temperature Using Charles's Law

In former times, gas volume was used as a way to measure temperature using devices called gas thermometers. Consider a gas that has a volume of 0.675 L at 35 °C and 1 atm pressure. What is the temperature (in units of °C) of a room where this gas has a volume of 0.535 L at 1 atm pressure?

Solution

The information given in the problem is

Initial Conditions	Final Conditions
$T_1 = 35\ °C = 35 + 273 = 308\ K$	$T_2 = ?$
$V_1 = 0.0675\ L$	$V_2 = 0.535\ L$
$P_1 = 1\ atm$	$P_2 = 1\ atm$

The pressure remains constant, so we can use Charles's law in the form

$$\frac{V_1}{T_1} = \frac{V_2}{T_2}$$

and solve for T_2. First we multiply both sides by T_2.

$$T_2 \times \frac{V_1}{T_1} = \frac{V_2}{\cancel{T_2}} \times \cancel{T_2} = V_2$$

Next we multiply both sides by T_1.

$$\cancel{T_1} \times T_2 \times \frac{V_1}{\cancel{T_1}} = T_1 \times V_2$$

This gives

$$T_2 \times V_1 = T_1 \times V_2$$

Now we divide both sides by V_1 (multiply by $1/V_1$),

$$\frac{1}{\cancel{V_1}} \times T_2 \times \cancel{V_1} = \frac{1}{V_1} \times T_1 \times V_2$$

and obtain

$$T_2 = T_1 \times \frac{V_2}{V_1}$$

We have now isolated T_2 on one side of the equation, and we can do the calculation.

$$T_2 = T_1 \times \frac{V_2}{V_1} = (308 \text{ K}) \times \frac{0.535 \not{L}}{0.675 \not{L}} = 244 \text{ K}$$

To convert from units of K to units of °C, we subtract 273 from the Kelvin temperature.

$$T_{°C} = T_K - 273 = 244 - 273 = -29 \text{ °C}$$

The room is very cold; the new temperature is −29 °C. ■

12.4 Volume and Moles: Avogadro's Law

AIM: To understand the law relating the volume and the number of moles of a sample of gas at constant temperature and pressure, and to do calculations involving this law.

What is the relationship between the volume of a gas and the number of molecules present in the gas sample? Experiments show that when the number of moles of gas is doubled (at constant temperature and pressure), the volume doubles. In other words, the volume of a gas is directly proportional to the number of moles if temperature and pressure remain constant. Figure 12.9 illustrates this relationship, which can also be represented by the equation

$$V = an \quad \text{or} \quad \frac{V}{n} = a$$

where V is the volume of the gas, n is the number of moles, and a is the proportionality constant. Note that this equation means that the ratio of V to n is constant as long as the temperature and pressure remain constant. Thus, when the number of moles of gas is increased by a factor of 5, the volume also increases by a factor of 5,

$$\frac{V}{n} = \frac{\not{5} \times V}{\not{5} \times n} = a = \text{constant}$$

and so on. In words, this equation means that *for a gas at constant temperature and pressure, the volume is directly proportional to the number of moles of gas.* This relationship is called **Avogadro's law** after the Italian scientist Amadeo Avogadro, who first postulated it in 1811.

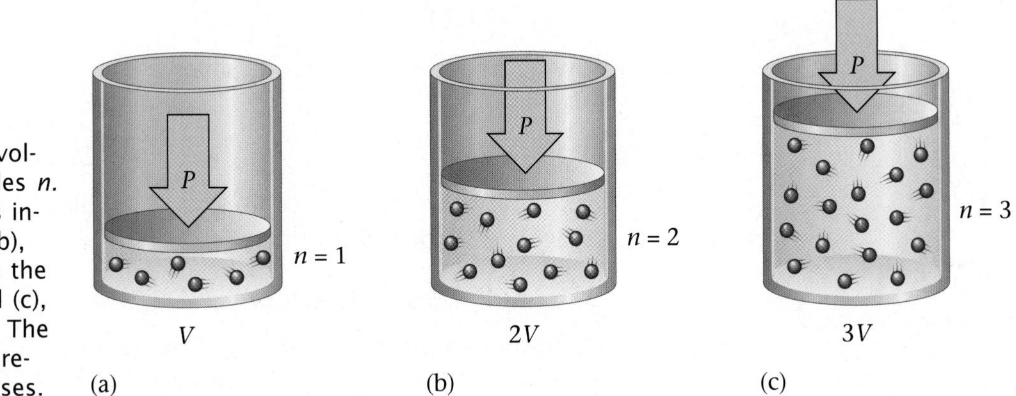

Figure 12.9 The relationship between volume V and number of moles n. As the number of moles is increased from 1 to 2 (a to b), the volume doubles. When the number of moles is tripled (c), the volume is also tripled. The temperature and pressure remain the same in these cases.

For cases where the number of moles of gas is changed from an initial amount to another amount (at constant T and P), we can represent Avogadro's law as

$$\underset{\text{Initial amount}}{\frac{V_1}{n_1}} = a = \underset{\text{Final amount}}{\frac{V_2}{n_2}}$$

or

$$\frac{V_1}{n_1} = \frac{V_2}{n_2}$$

We will illustrate the use of this equation in Example 12.7.

Example 12.7 Using Avogadro's Law in Calculations

Suppose we have a 12.2-L sample containing 0.50 mol of oxygen gas, O_2, at a pressure of 1 atm and a temperature of 25 °C. If all of this O_2 is converted to ozone, O_3, at the same temperature and pressure, what will be the volume of the ozone formed?

Solution

To do this problem we need to compare the moles of gas originally present to the moles of gas present after the reaction. We know that 0.50 mol of O_2 is present initially. To find out how many moles of O_3 will be present after the reaction, we need to use the balanced equation for the reaction.

$$3O_2(g) \rightarrow 2O_3(g)$$

We calculate the moles of O_3 produced by using the appropriate mole ratio from the balanced equation.

$$0.50 \text{ mol } O_2 \times \frac{2 \text{ mol } O_3}{3 \text{ mol } O_2} = 0.33 \text{ mol } O_3$$

Avogadro's law states that

$$\frac{V_1}{n_1} = \frac{V_2}{n_2}$$

where V_1 is the volume of n_1 moles of O_2 gas and V_2 is the volume of n_2 moles of O_3 gas. In this case we have

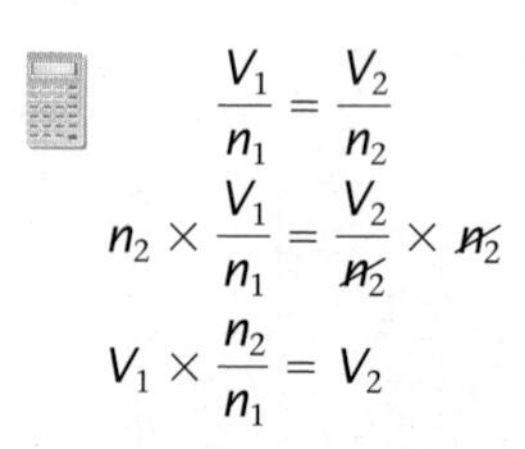

Initial Conditions	**Final Conditions**
$n_1 = 0.50$ mol	$n_2 = 0.33$ mol
$V_1 = 12.2$ L	$V_2 = ?$

Solving Avogadro's law for V_2 gives

$$V_2 = V_1 \times \frac{n_2}{n_1} = 12.2 \text{ L}\left(\frac{0.33 \text{ mol}}{0.50 \text{ mol}}\right) = 8.1 \text{ L}$$

Note that the volume decreases, as it should, because fewer molecules are present in the gas after O_2 is converted to O_3.

Self-Check Exercise 12.4

Consider two samples of nitrogen gas (composed of N_2 molecules). Sample 1 contains 1.5 mol of N_2 and has a volume of 36.7 L at 25 °C and 1 atm.

Sample 2 has a volume of 16.5 L at 25 °C and 1 atm. Calculate the number of moles of N_2 in Sample 2.

See Problems 12.41 through 12.44. ■

12.5 The Ideal Gas Law

AIM: To understand the ideal gas law and use it in calculations.

We have considered three laws that describe the behavior of gases as it is revealed by experimental observations.

Constant n means a constant number of moles of gas.

Boyle's law: $PV = k$ or $V = \frac{k}{P}$ (at constant T and n)

Charles's law: $V = bT$ (at constant P and n)

Avogadro's law: $V = an$ (at constant T and P)

These relationships, which show how the volume of a gas depends on pressure, temperature, and number of moles of gas present, can be combined as follows:

$$V = R\left(\frac{Tn}{P}\right)$$

$R = 0.08206 \frac{\text{L atm}}{\text{K mol}}$

where R is the combined proportionality constant and is called the **universal gas constant.** When the pressure is expressed in atmospheres and the volume is in liters, R always has the value 0.08206 L atm/K mol. We can rearrange the above equation by multiplying both sides by P,

$$P \times V = P \times R\left(\frac{Tn}{P}\right)$$

to obtain the **ideal gas law** written in its usual form,

$$PV = nRT$$

The ideal gas law involves all the important characteristics of a gas: its pressure (P), volume (V), number of moles (n), and temperature (T). Knowledge of any three of these properties is enough to define completely the condition of the gas, because the fourth property can be determined from the ideal gas law.

It is important to recognize that the ideal gas law is based on experimental measurements of the properties of gases. A gas that obeys this equation is said to behave *ideally.* That is, this equation defines the behavior of an **ideal gas.** Most gases obey this equation closely at pressures of approximately 1 atm or lower, when the temperature is approximately 0 °C or higher. You should assume ideal gas behavior when working problems involving gases in this text.

The ideal gas law can be used to solve a variety of problems. Example 12.8 demonstrates one type, where you are asked to find one property characterizing the condition of a gas given the other three properties.

Example 12.8 Using the Ideal Gas Law in Calculations

A sample of hydrogen gas, H_2, has a volume of 8.56 L at a temperature of 0 °C and a pressure of 1.5 atm. Calculate the number of moles of H_2 present in this gas sample. (Assume that the gas behaves ideally.)

Solution

In this problem we are given the pressure, volume, and temperature of the gas: $P = 1.5$ atm, $V = 8.56$ L, and $T = 0$ °C. Remember that the temperature must be changed to the Kelvin scale.

$$T = 0\ °\text{C} = 0 + 273 = 273\ \text{K}$$

We can calculate the number of moles of gas present by using the ideal gas law, $PV = nRT$. We solve for n by dividing both sides by RT:

$$\frac{PV}{RT} = n\frac{\cancel{RT}}{\cancel{RT}}$$

to give

$$\frac{PV}{RT} = n$$

Thus

$$n = \frac{PV}{RT} = \frac{(1.5\ \cancel{\text{atm}})(8.56\ \cancel{\text{L}})}{\left(0.08206\ \dfrac{\cancel{\text{L}}\ \cancel{\text{atm}}}{\cancel{\text{K}}\ \text{mol}}\right)(273\ \cancel{\text{K}})} = 0.57\ \text{mol}$$

✔ Self-Check Exercise 12.5

A weather balloon contains 1.10×10^5 mol of helium and has a volume of 2.70×10^6 L at 1.00 atm pressure. Calculate the temperature of the helium in the balloon in kelvins and in Celsius degrees.

See Problems 12.53 through 12.60. ■

Example 12.9 Ideal Gas Law Calculations Involving Conversion of Units

What volume is occupied by 0.250 mol of carbon dioxide gas at 25 °C and 371 torr?

Solution

We can use the ideal gas law to calculate the volume, but we must first convert pressure to atmospheres and temperature to the Kelvin scale.

$$P = 371\ \text{torr} = 371\ \cancel{\text{torr}} \times \frac{1.000\ \text{atm}}{760.0\ \cancel{\text{torr}}} = 0.488\ \text{atm}$$

$$T = 25\ °\text{C} = 25 + 273 = 298\ \text{K}$$

We solve for V by dividing both sides of the ideal gas law ($PV = nRT$) by P.

$$PV = nRT$$
$$\frac{PV}{P} = \frac{nRT}{P}$$
$$V = \frac{nRT}{P}$$

$$V = \frac{nRT}{P} = \frac{(0.250\ \cancel{\text{mol}})\left(0.08206\ \dfrac{\text{L}\ \cancel{\text{atm}}}{\cancel{\text{K}}\ \cancel{\text{mol}}}\right)(298\ \cancel{\text{K}})}{0.488\ \cancel{\text{atm}}} = 12.5\ \text{L}$$

The volume of the sample of CO_2 is 12.5 L.

✔ Self-Check Exercise 12.6

Radon, a radioactive gas formed naturally in the soil, can cause lung cancer. It can pose a hazard to humans by seeping into houses, and there is concern about this problem in many areas. A 1.5-mol sample of radon gas has a volume of 21.0 L at 33 °C. What is the pressure of the gas?

See Problems 12.53 through 12.60. ■

Note that R has units of L atm/K mol. Accordingly, whenever we use the ideal gas law, we must express the volume in units of liters, the temperature in kelvins, and the pressure in atmospheres. When we are given data in other units, we must first convert to the appropriate units.

The ideal gas law can also be used to calculate the changes that will occur when the conditions of the gas are changed as illustrated in Example 12.10.

Example 12.10 Using the Ideal Gas Law Under Changing Conditions

Suppose we have a 0.240-mol sample of ammonia gas at 25 °C with a volume of 3.5 L at a pressure of 1.68 atm. The gas is compressed to a volume of 1.35 L at 25 °C. Use the ideal gas law to calculate the final pressure.

Solution

In this case we have a sample of ammonia gas in which the conditions are changed. We are given the following information:

Initial Conditions	**Final Conditions**
$V_1 = 3.5$ L	$V_2 = 1.35$ L
$P_1 = 1.68$ atm	$P_2 = ?$
$T_1 = 25$ °C $= 25 + 273 = 298$ K	$T_2 = 25$ °C $= 25 + 273 = 298$ K
$n_1 = 0.240$ mol	$n_2 = 0.240$ mol

Note that both n and T remain constant—only P and V change. Thus we could simply use Boyle's law ($P_1V_1 = P_2V_2$) to solve for P_2. However, we will use the ideal gas law to solve this problem in order to introduce the idea that one equation—the ideal gas equation—can be used to do almost any gas problem. The key idea here is that in using the ideal gas law to describe a change in conditions for a gas, we always *solve the ideal gas equation in such a way that the variables that change are on one side of the equals sign and the constant terms are on the other side.* That is, we start with the ideal gas equation in the conventional form ($PV = nRT$) and rearrange it so that all the terms that change are moved to one side and all the terms that do not change are moved to the other side. In this case the pressure and volume change, and the temperature and number of moles remain constant (as does R, by definition). So we write the ideal gas law as

$$\underset{\text{Change}}{PV} = \underset{\text{Remain constant}}{nRT}$$

Because n, R, and T remain the same in this case, we can write $P_1V_1 = nRT$ and $P_2V_2 = nRT$. Combining these gives

$$P_1V_1 = nRT = P_2V_2 \quad \text{or} \quad P_1V_1 = P_2V_2$$

and

$$P_2 = P_1 \times \frac{V_1}{V_2} = (1.68 \text{ atm})\left(\frac{3.5 \not{\text{L}}}{1.35 \not{\text{L}}}\right) = 4.4 \text{ atm}$$

CHECK: Does this answer make sense? The volume was decreased (at constant temperature and constant number of moles), which means that the pressure should increase, as the calculation indicates.

Self-Check Exercise 12.7

A sample of methane gas that has a volume of 3.8 L at 5 °C is heated to 86 °C at constant pressure. Calculate its new volume.

See Problems 12.61 and 12.62. ■

Note that in solving Example 12.10, we actually obtained Boyle's law ($P_1V_1 = P_2V_2$) from the ideal gas equation. You might well ask, "Why go to all this trouble?" The idea is to learn to use the ideal gas equation to solve all types of gas law problems. This way you will never have to ask yourself, "Is this a Boyle's law problem or a Charles's law problem?"

We continue to practice using the ideal gas law in Example 12.11. Remember, the key idea is to rearrange the equation so that the quantities that change are moved to one side of the equation and those that remain constant are moved to the other.

Example 12.11 Calculating Volume Changes Using the Ideal Gas Law

A sample of diborane gas, B_2H_6, a substance that bursts into flames when exposed to air, has a pressure of 0.454 atm at a temperature of −15 °C and a volume of 3.48 L. If conditions are changed so that the temperature is 36 °C and the pressure is 0.616 atm, what will be the new volume of the sample?

Solution

We are given the following information:

Initial Conditions	Final Conditions
$P_1 = 0.454$ atm	$P_2 = 0.616$ atm
$V_1 = 3.48$ L	$V_2 = ?$
$T_1 = -15\ °C = 273 - 15 = 258$ K	$T_2 = 36\ °C = 273 + 36 = 309$ K

$$PV = nRT$$
$$\frac{PV}{T} = \frac{nR\cancel{T}}{\cancel{T}}$$
$$\frac{PV}{T} = nR$$

Note that the value of n is not given. However, we know that n is constant (that is, $n_1 = n_2$) because no diborane gas is added or taken away. Thus, in this experiment, n is constant and P, V, and T change. Therefore, we rearrange the ideal gas equation ($PV = nRT$) by dividing both sides by T,

$$\underset{\text{Change}}{\frac{PV}{T}} = \underset{\text{Constant}}{nR}$$

which leads to the equation

$$\frac{P_1V_1}{T_1} = nR = \frac{P_2V_2}{T_2}$$

or

$$\frac{P_1V_1}{T_1} = \frac{P_2V_2}{T_2}$$

We can now solve for V_2 by dividing both sides by P_2 and multiplying both sides by T_2.

$$\frac{1}{P_2} \times \frac{P_1V_1}{T_1} = \frac{\cancel{P_2}V_2}{T_2} \times \frac{1}{\cancel{P_2}} = \frac{V_2}{T_2}$$

$$T_2 \times \frac{P_1V_1}{P_2T_1} = \frac{V_2}{\cancel{T_2}} \times \cancel{T_2} = V_2$$

CHEMISTRY IN FOCUS

Snacks Need Chemistry, Too!

Have you ever wondered what makes popcorn pop? The popping is linked with the properties of gases. What happens when a gas is heated? Charles's law tells us that if the pressure is held constant, the volume of the gas must increase as the temperature is increased. But what happens if the gas being heated is trapped at a constant volume? We can see what happens by rearranging the ideal gas law ($PV = nRT$) as follows:

$$P = \left(\frac{nR}{V}\right)T$$

When n, R, and V are held constant, the pressure of a gas is directly proportional to the temperature. Thus, as the temperature of the trapped gas increases, its pressure also increases. This is exactly what happens inside a kernel of popcorn as it is heated. The moisture inside the kernel vaporized by the heat produces increasing pressure. The pressure finally becomes so great that the kernel breaks open, allowing the starch inside to expand to about 40 times its original size.

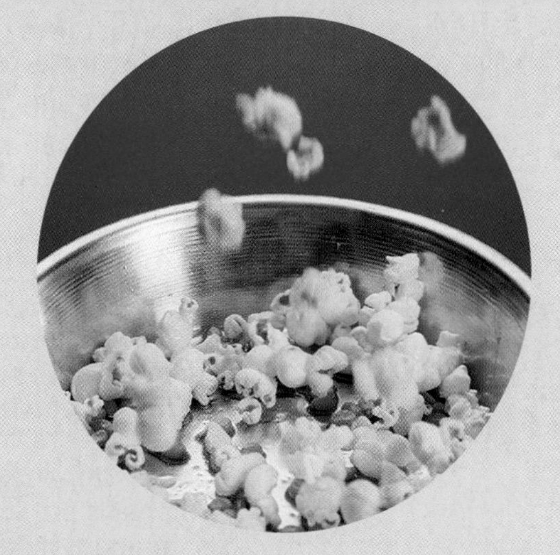

Popcorn popping.

What's special about popcorn? Why does it pop while "regular" corn doesn't? William da Silva, a biologist at the University of Campinas in Brazil, has traced the "popability" of popcorn to its outer casing, called the pericarp. The molecules in the pericarp of popcorn, which are packed in a much more orderly way than in regular corn, transfer heat unusually quickly, producing a very fast pressure jump that pops the kernel. In addition, because the pericarp of popcorn is much thicker and stronger than that of regular corn, it can withstand more pressure, leading to a more explosive pop when the moment finally comes.

That is,

$$\frac{T_2P_1V_1}{P_2T_1} = V_2$$

It is sometimes convenient to think in terms of the ratios of the initial temperature and pressure and the final temperature and pressure. That is,

Always convert the temperature to the Kelvin scale and the pressure to atmospheres when applying the ideal gas law.

$$V_2 = \frac{T_2P_1V_1}{T_1P_2} = V_1 \times \frac{T_2}{T_1} \times \frac{P_1}{P_2}$$

Substituting the information given yields

$$V_2 = \frac{309 \text{ K}}{258 \text{ K}} \times \frac{0.454 \text{ atm}}{0.616 \text{ atm}} \times 3.48 \text{ L} = 3.07 \text{ L}$$

✔ Self-Check Exercise 12.8

A sample of argon gas with a volume of 11.0 L at a temperature of 13 °C and a pressure of 0.747 atm is heated to 56 °C and a pressure of 1.18 atm. Calculate the final volume.

See Problems 12.61 and 12.62. ■

The equation obtained in Example 12.11,

$$\frac{P_1V_1}{T_1} = \frac{P_2V_2}{T_2}$$

is often called the **combined gas law** equation. It holds when the amount of gas (moles) is held constant. While it may be convenient to remember this equation, it is not necessary because you can always use the ideal gas equation.

12.6 Dalton's Law of Partial Pressures

AIM: To understand the relationship between the partial and total pressures of a gas mixture, and to use this relationship in calculations.

Many important gases contain a mixture of components. One notable example is air. Scuba divers who are going deeper than 150 feet use another important mixture, helium and oxygen. Normal air is not used because the nitrogen present dissolves in the blood in large quantities as a result of the high pressures experienced by the diver under several hundred feet of water. When the diver returns too quickly to the surface, the nitrogen bubbles out of the blood just as soda fizzes when it's opened, and the diver gets "the bends"—a very painful and potentially fatal condition. Because helium gas is only sparingly soluble in blood, it does not cause this problem.

Studies of gaseous mixtures show that each component behaves independently of the others. In other words, a given amount of oxygen exerts the same pressure in a 1.0-L vessel whether it is alone or in the presence of nitrogen (as in the air) or helium.

Among the first scientists to study mixtures of gases was John Dalton. In 1803 Dalton summarized his observations in this statement: *For a mixture of gases in a container, the total pressure exerted is the sum of the partial pressures of the gases present. The* **partial pressure** *of a gas is the pressure that the gas would exert if it were alone in the container.* This statement, known as **Dalton's law of partial pressures,** can be expressed as follows for a mixture containing three gases:

$$P_{total} = P_1 + P_2 + P_3$$

where the subscripts refer to the individual gases (gas 1, gas 2, and gas 3). The pressures P_1, P_2, and P_3 are the partial pressures; that is, each gas is responsible for only part of the total pressure (Figure 12.10).

Figure 12.10
When two gases are present, the total pressure is the sum of the partial pressures of the gases.

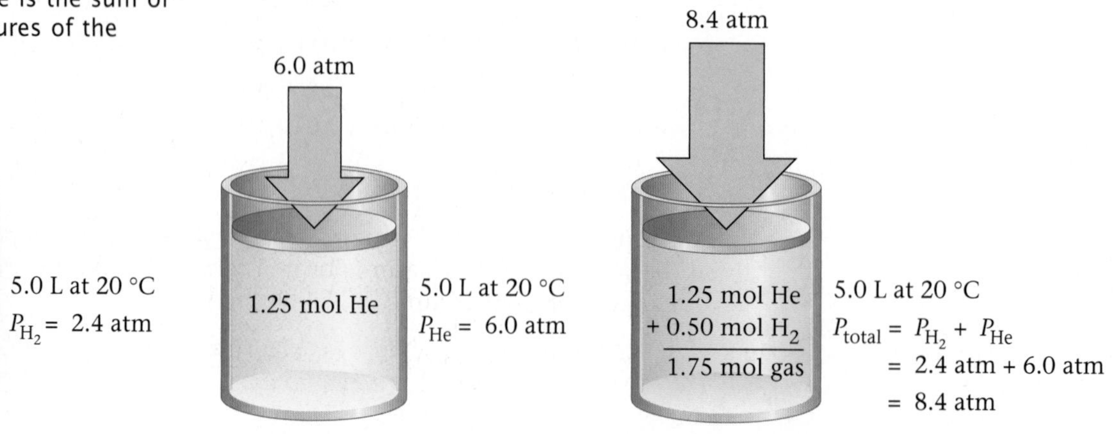

$$PV = nRT$$
$$\frac{PV\!\!\!/}{V\!\!\!/} = \frac{nRT}{V}$$
$$P = \frac{nRT}{V}$$

Assuming that each gas behaves ideally, we can calculate the partial pressure of each gas from the ideal gas law:

$$P_1 = \frac{n_1RT}{V},\ P_2 = \frac{n_2RT}{V},\ P_3 = \frac{n_3RT}{V}$$

The total pressure of the mixture, P_{total}, can be represented as

$$\begin{aligned} P_{total} = P_1 + P_2 + P_3 &= \frac{n_1RT}{V} + \frac{n_2RT}{V} + \frac{n_3RT}{V} \\ &= n_1\left(\frac{RT}{V}\right) + n_2\left(\frac{RT}{V}\right) + n_3\left(\frac{RT}{V}\right) \\ &= (n_1 + n_2 + n_3)\left(\frac{RT}{V}\right) \\ &= n_{total}\left(\frac{RT}{V}\right) \end{aligned}$$

where n_{total} is the sum of the numbers of moles of the gases in the mixture. Thus, for a mixture of ideal gases, it is the *total number of moles of particles* that is important, not the *identity* of the individual gas particles. This idea is illustrated in Figure 12.11.

The fact that the pressure exerted by an ideal gas is affected by the *number* of gas particles and is independent of the *nature* of the gas particles tells us two important things about ideal gases:

1. The volume of the individual gas particle (atom or molecule) must not be very important.
2. The forces among the particles must not be very important.

If these factors were important, the pressure of the gas would depend on the nature of the individual particles. For example, an argon atom is much larger than a helium atom. Yet 1.75 mol of argon gas in a 5.0-L container at 20 °C exerts the same pressure as 1.75 mol of helium gas in a 5.0-L container at 20 °C.

The same idea applies to the forces among the particles. Although the forces among gas particles depend on the nature of the particles, this seems to have little influence on the behavior of an ideal gas. We will see that these observations strongly influence the model that we will construct to explain ideal gas behavior.

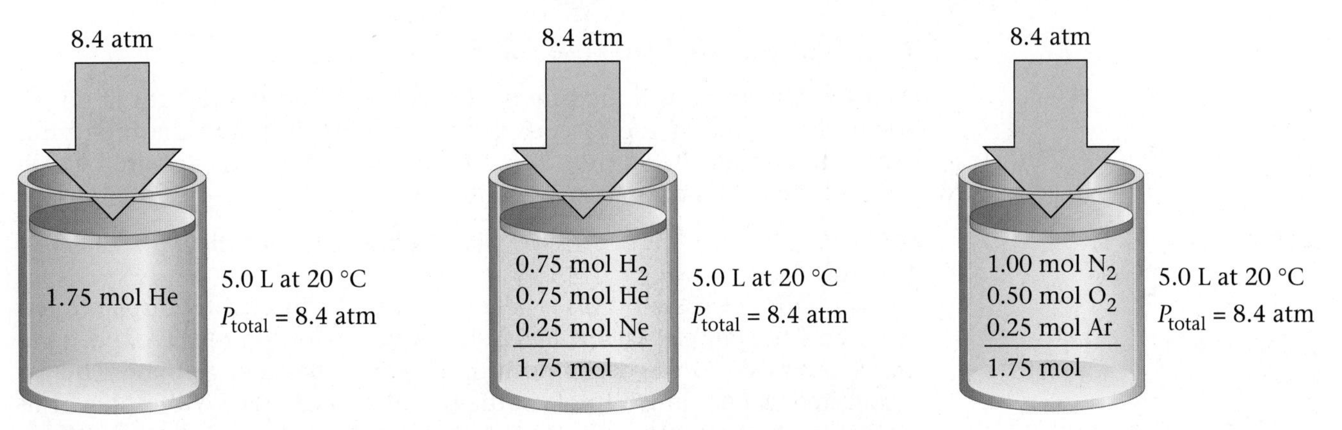

Figure 12.11
The total pressure of a mixture of gases depends on the number of moles of gas particles (atoms or molecules) present, not on the identities of the particles. Note that these three samples show the same total pressure because each contains 1.75 mol of gas. The detailed nature of the mixture is unimportant.

Example 12.12 Using Dalton's Law of Partial Pressures, I

Mixtures of helium and oxygen are used in the "air" tanks of underwater divers for deep dives. For a particular dive, 12 L of O_2 at 25 °C and 1.0 atm and 46 L of He at 25 °C and 1.0 atm were both pumped into a 5.0-L tank. Calculate the partial pressure of each gas and the total pressure in the tank at 25 °C.

Solution

$$PV = nRT$$
$$\frac{PV}{RT} = \frac{n\cancel{RT}}{\cancel{RT}}$$
$$\frac{PV}{RT} = n$$

Because the partial pressure of each gas depends on the moles of that gas present, we must first calculate the number of moles of each gas by using the ideal gas law in the form

$$n = \frac{PV}{RT}$$

From the above description we know that $P = 1.0$ atm, $V = 12$ L for O_2 and 46 L for He, and $T = 25 + 273 = 298$ K. Also, $R = 0.08206$ L atm/K mol (as always).

$$\text{Moles of } O_2 = n_{O_2} = \frac{(1.0\ \cancel{\text{atm}})(12\ \cancel{\text{L}})}{(0.08206\ \cancel{\text{L}}\ \cancel{\text{atm}}/\cancel{\text{K}}\ \text{mol})(298\ \cancel{\text{K}})} = 0.49\ \text{mol}$$

$$\text{Moles of He} = n_{He} = \frac{(1.0\ \cancel{\text{atm}})(46\ \cancel{\text{L}})}{(0.08206\ \cancel{\text{L}}\ \cancel{\text{atm}}/\cancel{\text{K}}\ \text{mol})(298\ \cancel{\text{K}})} = 1.9\ \text{mol}$$

Divers use a mixture of oxygen and helium in their breathing tanks when diving to depths greater than 150 feet.

The tank containing the mixture has a volume of 5.0 L, and the temperature is 25 °C (298 K). We can use these data and the ideal gas law to calculate the partial pressure of each gas.

$$P = \frac{nRT}{V}$$

$$P_{O_2} = \frac{(0.49\ \cancel{\text{mol}})(0.08206\ \cancel{\text{L}}\ \text{atm}/\cancel{\text{K}}\ \cancel{\text{mol}})(298\ \cancel{\text{K}})}{5.0\ \cancel{\text{L}}} = 2.4\ \text{atm}$$

$$P_{He} = \frac{(1.9\ \cancel{\text{mol}})(0.08206\ \cancel{\text{L}}\ \text{atm}/\cancel{\text{K}}\ \cancel{\text{mol}})(298\ \cancel{\text{K}})}{5.0\ \cancel{\text{L}}} = 9.3\ \text{atm}$$

The total pressure is the sum of the partial pressures.

$$P_{total} = P_{O_2} + P_{He} = 2.4\ \text{atm} + 9.3\ \text{atm} = 11.7\ \text{atm}$$

✔ Self-Check Exercise 12.9

A 2.0-L flask contains a mixture of nitrogen gas and oxygen gas at 25 °C. The total pressure of the gaseous mixture is 0.91 atm, and the mixture is known to contain 0.050 mol of N_2. Calculate the partial pressure of oxygen and the moles of oxygen present.

See Problems 12.67 through 12.70. ■

A mixture of gases occurs whenever a gas is collected by displacement of water. For example, Figure 12.12 shows the collection of the oxygen gas that is produced by the decomposition of solid potassium chlorate. The gas is collected by bubbling it into a bottle that is initially filled with water. Thus the gas in the bottle is really a mixture of water vapor and oxygen. (Water vapor is present because molecules of water escape from the surface of the liquid and collect as a gas in the space above the liquid.) Therefore, the total pressure exerted by this mixture is the sum of the partial pressure of the gas being collected and the partial pressure of the water vapor. The partial pressure of the water vapor is called the vapor pressure of water. Because wa-

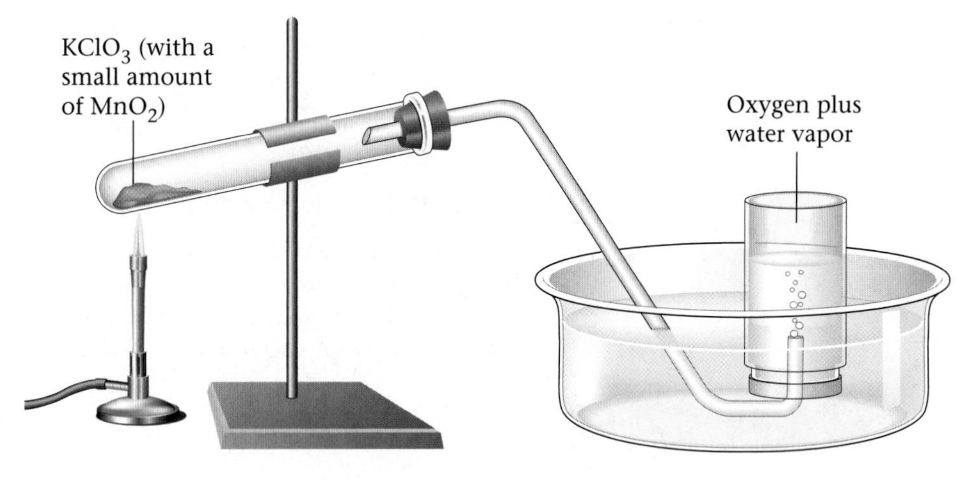

Figure 12.12
The production of oxygen by thermal decomposition of $KClO_3$. MnO_2 is mixed with the $KClO_3$ to make the reaction proceed at a more rapid rate.

ter molecules are more likely to escape from hot water than from cold water, the *vapor pressure* of water increases with temperature. This is shown by the values of vapor pressure at various temperatures shown in Table 12.2.

Example 12.13 Using Dalton's Law of Partial Pressures, II

A sample of solid potassium chlorate, $KClO_3$, was heated in a test tube (see Figure 12.12) and decomposed according to the reaction

$$2KClO_3(s) \rightarrow 2KCl(s) + 3O_2(g)$$

The oxygen produced was collected by displacement of water at 22 °C. The resulting mixture of O_2 and H_2O vapor had a total pressure of 754 torr and a volume of 0.650 L. Calculate the partial pressure of O_2 in the gas collected and the number of moles of O_2 present. The vapor pressure of water at 22 °C is 21 torr.

Table 12.2 The Vapor Pressure of Water as a Function of Temperature

T (°C)	*P* (torr)
0.0	4.579
10.0	9.209
20.0	17.535
25.0	23.756
30.0	31.824
40.0	55.324
60.0	149.4
70.0	233.7
90.0	525.8

Solution

We know the total pressure (754 torr) and the partial pressure of water (vapor pressure = 21 torr). We can find the partial pressure of O_2 from Dalton's law of partial pressures:

$$P_{\text{total}} = P_{O_2} + P_{H_2O} = P_{O_2} + 21 \text{ torr} = 754 \text{ torr}$$

or

$$P_{O_2} + 21 \text{ torr} = 754 \text{ torr}$$

We can solve for P_{O_2} by subtracting 21 torr from both sides of the equation.

$$P_{O_2} = 754 \text{ torr} - 21 \text{ torr} = 733 \text{ torr}$$

Next we solve the ideal gas law for the number of moles of O_2.

$$n_{O_2} = \frac{P_{O_2}V}{RT}$$

$$PV = nRT$$
$$\frac{PV}{RT} = \frac{n\cancel{RT}}{\cancel{RT}}$$
$$\frac{PV}{RT} = n$$

In this case, $P_{O_2} = 733$ torr. We change the pressure to atmospheres as follows:

$$\frac{733 \cancel{\text{torr}}}{760 \cancel{\text{torr}}/\text{atm}} = 0.964 \text{ atm}$$

Then,

$$V = 0.650 \text{ L}$$
$$T = 22 \text{ °C} = 22 + 273 = 295 \text{ K}$$
$$R = 0.08206 \text{ L atm/K mol}$$

so

$$n_{O_2} = \frac{(0.964 \cancel{\text{atm}})(0.650 \cancel{\text{L}})}{(0.08206 \cancel{\text{L}}\ \cancel{\text{atm}}/\cancel{\text{K}}\ \text{mol})(295 \cancel{\text{K}})} = 2.59 \times 10^{-2}\ \text{mol}$$

Self-Check Exercise 12.10

Consider a sample of hydrogen gas collected over water at 25 °C where the vapor pressure of water is 24 torr. The volume occupied by the gaseous mixture is 0.500 L, and the total pressure is 0.950 atm. Calculate the partial pressure of H_2 and the number of moles of H_2 present.

See Problems 12.71 through 12.74. ■

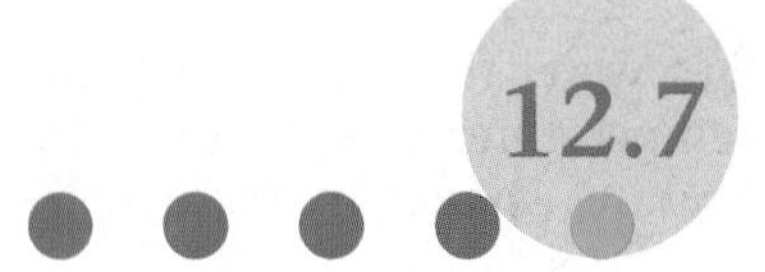

Laws and Models: A Review

AIM: To understand the relationship between laws and models (theories).

In this chapter we have considered several properties of gases and have seen how the relationships among these properties can be expressed by various laws written in the form of mathematical equations. The most useful of these is the ideal gas equation, which relates all the important gas properties. However, under certain conditions gases do not obey the ideal gas equation. For example, at high pressures and/or low temperatures, the properties of gases deviate significantly from the predictions of the ideal gas equation. On the other hand, as the pressure is lowered and/or the temperature is increased, almost all gases show close agreement with the ideal gas equation. This means that an ideal gas is really a hypothetical substance. At low pressures and/or high temperatures, real gases *approach* the behavior expected for an ideal gas.

At this point we want to build a model (a theory) to explain *why* a gas behaves as it does. We want to answer the question, *What are the characteristics of the individual gas particles that cause a gas to behave as it does?* However, before we do this let's briefly review the scientific method. Recall that a law is a generalization about behavior that has been observed in many experiments. Laws are very useful; they allow us to predict the behavior of similar systems. For example, a chemist who prepares a new gaseous compound can assume that that substance will obey the ideal gas equation (at least at low P and/or high T).

However, laws do not tell us *why* nature behaves the way it does. Scientists try to answer this question by constructing theories (building models). The models in chemistry are speculations about how individual atoms or molecules (microscopic particles) cause the behavior of macroscopic systems (collections of atoms and molecules in large enough numbers so that we can observe them).

A model is considered successful if it explains known behavior and predicts correctly the results of future experiments. But a model can never be proved absolutely true. In fact, by its very nature *any model is an approximation* and is destined to be modified, at least in part. Models range from the simple (to predict approximate behavior) to the extraordinarily complex (to account precisely for observed behavior). In this text, we use relatively simple models that fit most experimental results.

The Kinetic Molecular Theory of Gases

AIM: To understand the basic postulates of the kinetic molecular theory.

A relatively simple model that attempts to explain the behavior of an ideal gas is the **kinetic molecular theory.** This model is based on speculations about the behavior of the individual particles (atoms or molecules) in a gas. The assumptions (postulates) of the kinetic molecular theory can be stated as follows:

Postulates of the Kinetic Molecular Theory of Gases

1. Gases consist of tiny particles (atoms or molecules).
2. These particles are so small, compared with the distances between them, that the volume (size) of the individual particles can be assumed to be negligible (zero).
3. The particles are in constant random motion, colliding with the walls of the container. These collisions with the walls cause the pressure exerted by the gas.
4. The particles are assumed not to attract or to repel each other.
5. The average kinetic energy of the gas particles is directly proportional to the Kelvin temperature of the gas.

The kinetic energy referred to in postulate 5 is the energy associated with the motion of a particle. Kinetic energy (KE) is given by the equation $KE = \frac{1}{2}mv^2$, where m is the mass of the particle and v is the velocity (speed) of the particle. The greater the mass or velocity of a particle, the greater its kinetic energy. Postulate 5 means that if a gas is heated to higher temperatures, the average speed of the particles increases; therefore, their kinetic energy increases.

Although real gases do not conform exactly to the five assumptions listed here, we will see in the next section that these postulates do indeed explain *ideal* gas behavior—behavior shown by real gases at high temperatures and/or low pressures.

The Implications of the Kinetic Molecular Theory

AIMS: To understand the term temperature. To learn how the kinetic molecular theory explains the gas laws.

In this section we will discuss the *qualitative* relationships between the kinetic molecular (KM) theory and the properties of gases. That is, without going into the mathematical details, we will show how the kinetic molecular theory explains some of the observed properties of gases.

The Meaning of Temperature

In Chapter 2 we introduced temperature very practically as something we measure with a thermometer. We know that as the temperature of an object increases, the object feels "hotter" to the touch. But what does temperature really mean? How does matter change when it gets "hotter"? In Chapter 3 we introduced the idea that temperature is an index of molecular motion. The kinetic molecular theory allows us to further develop this concept. As postulate 5 of the KM theory states, the temperature of a gas reflects how rapidly, on average, its individual gas particles are moving. At high temperatures the particles move very fast and hit the walls of the container frequently, whereas at low temperatures the particles' motions are more sluggish and they collide with the walls of the container much less often. Therefore, temperature really is a measure of the motions of the gas particles. In fact, the Kelvin temperature of a gas is directly proportional to the average kinetic energy of the gas particles.

The Relationship Between Pressure and Temperature

To see how the meaning of temperature given above helps to explain gas behavior, picture a gas in a rigid container. As the gas is heated to a higher temperature, the particles move faster, hitting the walls more often. And, of course, the impacts become more forceful as the particles move faster. If the pressure is due to collisions with the walls, the gas pressure should increase as temperature is increased.

Is this what we observe when we measure the pressure of a gas as it is heated? Yes. A given sample of gas in a rigid container (if the volume is not changed) shows an increase in pressure as its temperature is increased.

The Relationship Between Volume and Temperature

Now picture the gas in a container with a movable piston. As shown in Figure 12.13a, the gas pressure P_{gas} is just balanced by an external pressure P_{ext}. What happens when we heat the gas to a higher temperature? As the temperature increases, the particles move faster, causing the gas pressure to increase. As soon as the gas pressure P_{gas} becomes greater than P_{ext} (the pressure holding the piston), the piston moves up until $P_{gas} = P_{ext}$. Therefore, the KM model predicts that the volume of the gas will increase as we raise its temperature at a constant pressure. This agrees with experimental observations (as summarized by Charles's law).

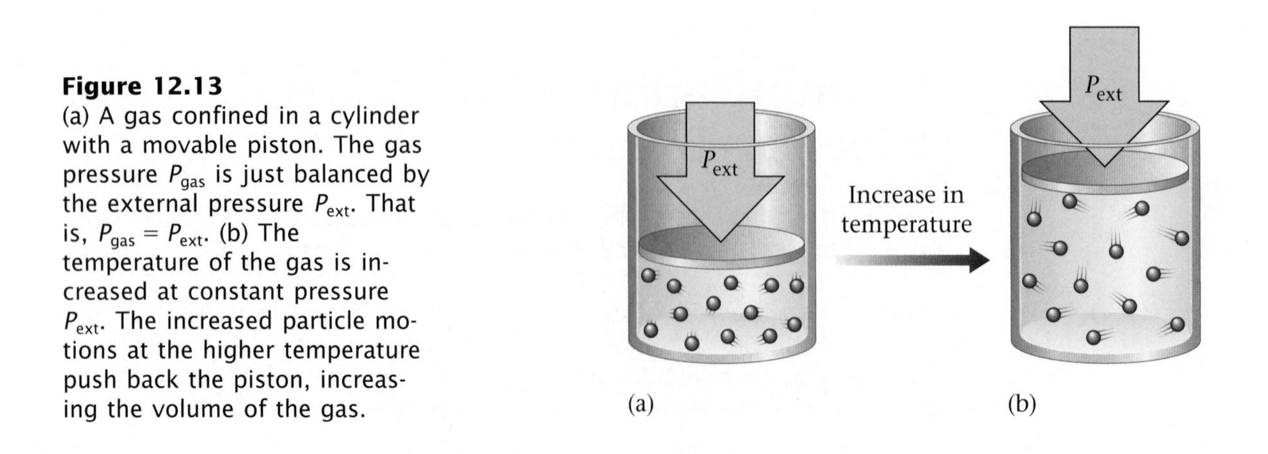

Figure 12.13
(a) A gas confined in a cylinder with a movable piston. The gas pressure P_{gas} is just balanced by the external pressure P_{ext}. That is, $P_{gas} = P_{ext}$. (b) The temperature of the gas is increased at constant pressure P_{ext}. The increased particle motions at the higher temperature push back the piston, increasing the volume of the gas.

Example 12.14 Using the Kinetic Molecular Theory to Explain Gas Law Observations

Use the KM theory to predict what will happen to the pressure of a gas when its volume is decreased (n and T constant). Does this prediction agree with the experimental observations?

Solution

When we decrease the gas's volume (make the container smaller), the particles hit the walls more often because they do not have to travel so far between the walls. This would suggest an increase in pressure. This prediction on the basis of the model is in agreement with experimental observations of gas behavior (as summarized by Boyle's law). ■

In this section we have seen that the predictions of the kinetic molecular theory generally fit the behavior observed for gases. This makes it a useful and successful model.

12.10 Gas Stoichiometry

AIMS: To understand the molar volume of an ideal gas. To learn the definition of STP. To use these concepts and the ideal gas equation.

We have seen repeatedly in this chapter just how useful the ideal gas equation is. For example, if we know the pressure, volume, and temperature for a given sample of gas, we can calculate the number of moles present: $n = PV/RT$. This fact makes it possible to do stoichiometric calculations for reactions involving gases. We will illustrate this process in Example 12.15.

Example 12.15 Gas Stoichiometry: Calculating Volume

Calculate the volume of oxygen gas produced at 1.00 atm and 25 °C by the complete decomposition of 10.5 g of potassium chlorate. The balanced equation for the reaction is

$$2KClO_3(s) \rightarrow 2KCl(s) + 3O_2(g)$$

Solution

This is a stoichiometry problem very much like the type we considered in Chapter 9. The only difference is that in this case, we want to calculate the volume of a gaseous product rather than the number of grams. To do so, we can use the relationship between moles and volume given by the ideal gas law.

We'll summarize the steps required to do this problem in the following schematic:

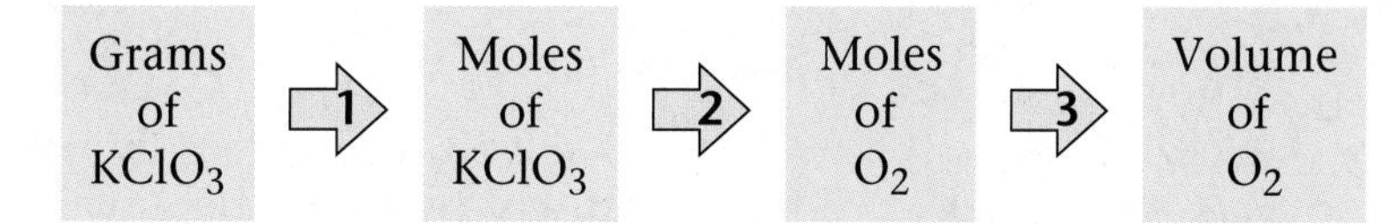

$\frac{10.5}{122.6} = 0.085644$

0.085644 ➡ 0.0856
Round off

$0.0856 = 8.56 \times 10^{-2}$

STEP 1 To find the moles of $KClO_3$ in 10.5 g, we use the molar mass of $KClO_3$ (122.6 g).

$$10.5 \text{ g } KClO_3 \times \frac{1 \text{ mol } KClO_3}{122.6 \text{ g } KClO_3} = 8.56 \times 10^{-2} \text{ mol } KClO_3$$

STEP 2 To find the moles of O_2 produced, we use the mole ratio of O_2 to $KClO_3$ derived from the balanced equation.

$$8.56 \times 10^{-2} \text{ mol } KClO_3 \times \frac{3 \text{ mol } O_2}{2 \text{ mol } KClO_3} = 1.28 \times 10^{-1} \text{ mol } O_2$$

STEP 3 To find the volume of oxygen produced, we use the ideal gas law $PV = nRT$, where

$P = 1.00$ atm
$V = ?$
$n = 1.28 \times 10^{-1}$ mol, the moles of O_2 we calculated
$R = 0.08206$ L atm/K mol
$T = 25\ °C = 25 + 273 = 298$ K

Solving the ideal gas law for V gives

$$V = \frac{nRT}{P} = \frac{(1.28 \times 10^{-1} \text{ mol})\left(0.08206 \frac{\text{L atm}}{\text{K mol}}\right)(298 \text{ K})}{1.00 \text{ atm}} = 3.13 \text{ L}$$

Thus 3.13 L of O_2 will be produced.

✔ Self-Check Exercise 12.11

Calculate the volume of hydrogen produced at 1.50 atm and 19 °C by the reaction of 26.5 g of zinc with excess hydrochloric acid according to the balanced equation

$$Zn(s) + 2HCl(aq) \rightarrow ZnCl_2(aq) + H_2(g)$$

See Problems 12.85 through 12.92. ■

In dealing with the stoichiometry of reactions involving gases, it is useful to define the volume occupied by 1 mol of a gas under certain specified conditions. For 1 mol of an ideal gas at 0 °C (273 K) and 1 atm, the volume of the gas given by the ideal gas law is

$$V = \frac{nRT}{P} = \frac{(1.00 \text{ mol})(0.08206 \text{ L atm/K mol})(273 \text{ K})}{1.00 \text{ atm}} = 22.4 \text{ L}$$

This volume of 22.4 L is called the **molar volume** of an ideal gas.

STP: 0 °C and 1 atm

The conditions 0 °C and 1 atm are called **standard temperature and pressure** (abbreviated **STP**). Properties of gases are often given under these conditions. Remember, the molar volume of an ideal gas is 22.4 L *at STP*. That is, 22.4 L contains 1 mol of an ideal gas at STP.

Example 12.16 Gas Stoichiometry: Calculations Involving Gases at STP

A sample of nitrogen gas has a volume of 1.75 L at STP. How many moles of N_2 are present?

Solution

We could solve this problem by using the ideal gas equation, but we can take a shortcut by using the molar volume of an ideal gas at STP. Because

1 mol of an ideal gas at STP has a volume of 22.4 L, a 1.75-L sample of N_2 at STP contains considerably less than 1 mol. We can find how many moles by using the equivalence statement

$$1.000 \text{ mol} = 22.4 \text{ L (STP)}$$

which leads to the conversion factor we need:

$$1.75 \cancel{\text{L N}_2} \times \frac{1.000 \text{ mol N}_2}{22.4 \cancel{\text{L N}_2}} = 7.81 \times 10^{-2} \text{ mol N}_2$$

✔ Self-Check Exercise 12.12

Ammonia is commonly used as a fertilizer to provide a source of nitrogen for plants. A sample of $NH_3(g)$ occupies a volume of 5.00 L at 25 °C and 15.0 atm. What volume will this sample occupy at STP?

See Problems 12.95 through 12.98. ■

Standard conditions (STP) and molar volume are also useful in carrying out stoichiometric calculations on reactions involving gases, as shown in Example 12.17.

Example 12.17 Gas Stoichiometry: Reactions Involving Gases at STP

Quicklime, CaO, is produced by heating calcium carbonate, $CaCO_3$. Calculate the volume of CO_2 produced at STP from the decomposition of 152 g of $CaCO_3$ according to the reaction

$$CaCO_3(s) \rightarrow CaO(s) + CO_2(g)$$

Solution

The strategy for solving this problem is summarized by the following schematic:

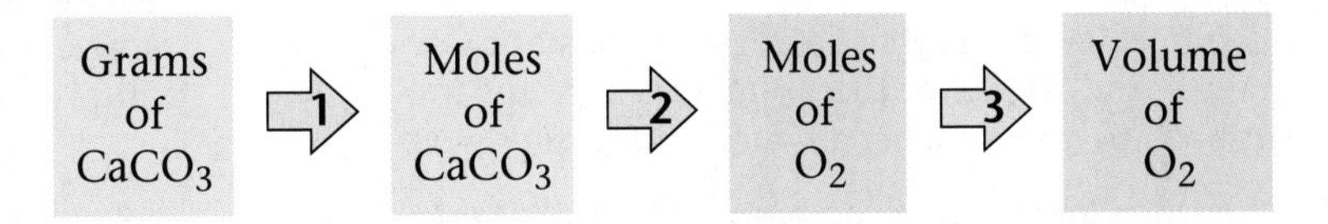

Step 1 Using the molar mass of $CaCO_3$ (100.1 g), we calculate the number of moles of $CaCO_3$.

$$152 \cancel{\text{g CaCO}_3} \times \frac{1 \text{ mol CaCO}_3}{100.0 \cancel{\text{g CaCO}_3}} = 1.52 \text{ mol CaCO}_3$$

Step 2 Each mole of $CaCO_3$ produces 1 mol of CO_2, so 1.52 mol of CO_2 will be formed.

Step 3 We can convert the moles of CO_2 to volume by using the molar volume of an ideal gas, because the conditions are STP.

$$1.52 \cancel{\text{mol CO}_2} \times \frac{22.4 \text{ L CO}_2}{1 \cancel{\text{mol CO}_2}} = 34.1 \text{ L CO}_2$$

Thus the decomposition of 152 g of $CaCO_3$ produces 34.1 L of CO_2 at STP. ■

Remember that the molar volume of an ideal gas is 22.4 L at STP.

Note that the final step in Example 12.17 involves calculating the volume of gas from the number of moles. Because the conditions were specified as STP, we were able to use the molar volume of a gas at STP. If the conditions of a problem are different from STP, we must use the ideal gas law to compute the volume, as we did in Section 12.5.

CHAPTER  12 REVIEW

KEY TERMS

barometer (12.1)
mm Hg (12.1)
torr (12.1)
standard atmosphere (12.1)
pascal (12.1)
Boyle's law (12.2)
absolute zero (12.3)
Charles's law (12.3)
Avogadro's law (12.4)
universal gas constant (12.5)
ideal gas law (12.5)
ideal gas (12.5)
combined gas law (12.5)
partial pressure (12.6)
Dalton's law of partial pressures (12.6)
kinetic molecular theory (12.8)
molar volume (12.10)
standard temperature and pressure (STP) (12.10)

SUMMARY

1. Atmospheric pressure is measured with a barometer. The most commonly used units of pressure are mm Hg (torr), atmospheres, and pascals (the SI unit).
2. Boyle's law states that the volume of a given amount of gas is inversely proportional to its pressure (at constant temperature): $PV = k$ or $P = k/V$. That is, as pressure increases, volume decreases.
3. Charles's law states that, for a given amount of gas at constant pressure, the volume is directly proportional to the temperature (in kelvins): $V = bT$. At −273 °C (0 K), the volume of a gas extrapolates to zero, and this temperature is called absolute zero.
4. Avogadro's law states that for a gas at constant temperature and pressure, the volume is directly proportional to the number of moles of gas: $V = an$.
5. These three laws can be combined into the ideal gas law, $PV = nRT$, where R is called the universal gas constant. This equation makes it possible to calculate any one of the properties—volume, pressure, temperature, or moles of gas present—given the other three. A gas that obeys this equation is said to behave ideally.
6. From the ideal gas equation we can derive the combined gas law,

$$\frac{P_1V_1}{T_1} = \frac{P_2V_2}{T_2}$$

which holds when the amount of gas (moles) remains constant.

7. The pressure of a gas mixture is described by Dalton's law of partial pressures, which states that the total pressure of the mixture of gases in a container is the sum of the partial pressures of the gases that make up the mixture.
8. The kinetic molecular theory of gases is a model that accounts for ideal gas behavior. This model assumes that a gas consists of tiny particles with negligible volumes, that there are no interactions among particles, and that the particles are in constant motion, colliding with the container walls to produce pressure.

IN-CLASS DISCUSSION QUESTIONS

These questions are designed to be considered by groups of students in class. Often these questions work well for introducing a particular topic in class.

1. As you increase the temperature of a gas in a sealed, rigid container, what happens to the density of the gas? Would the results be the same if you did the same experiment in a container with a movable piston at a constant external pressure? Explain.
2. A diagram in a chemistry book shows a magnified view of a flask of air.

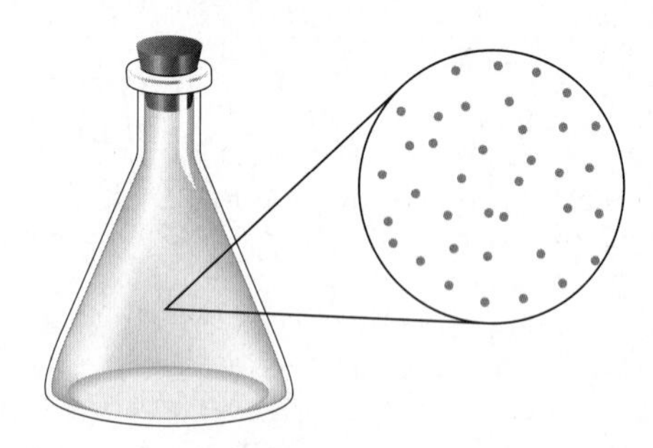

What do you suppose is between the dots (which represent air molecules)?
a. air
b. dust
c. pollutants
d. oxygen
e. nothing

3. If you put a drinking straw in water, place your finger over the opening, and lift the straw out of the water, some water stays in the straw. Explain.

4. A chemistry student relates the following story: I noticed my tires were a bit low and went to the gas station. As I was filling the tires I thought about the kinetic molecular (KM) theory. I noticed the tires because the volume was low, and I realized that I was increasing both the pressure and volume of the tires. "Hmmm," I thought, "that goes against what I learned in chemistry, where I was told pressure and volume are inversely proportional." What is the fault of the logic of the chemistry student in this situation? Explain under what conditions pressure and volume are inversely related (draw pictures and use the KM theory).

5. Chemicals X and Y (both gases) react to form the gas XY, but it takes some time for the reaction to occur. Both X and Y are placed in a container with a piston (free to move), and you note the volume. As the reaction occurs, what happens to the volume of the container? Explain your answer.

6. Which statement best explains why a hot-air balloon rises when the air in the balloon is heated?
   a. According to Charles's law, the temperature of a gas is directly related to its volume. Thus the volume of the balloon increases, decreasing the density.
   b. Hot air rises inside the balloon, which lifts the balloon.
   c. The temperature of a gas is directly related to its pressure. The pressure therefore increases, which lifts the balloon.
   d. Some of the gas escapes from the bottom of the balloon, thus decreasing the mass of gas in the balloon. This decreases the density of the gas in the balloon, which lifts the balloon.
   e. Temperature is related to the velocity of the gas molecules. Thus the molecules are moving faster, hitting the balloon more, and lifting the balloon.

   For choices you did not pick, explain what you feel is wrong with them, and justify the choice you did pick.

7. Draw a highly magnified view of a sealed, rigid container filled with a gas. Then draw what it would look like if you cooled the gas significantly, but kept the temperature above the boiling point of the substance in the container. Also draw what the container would look like if you heated the gas significantly. Finally, sketch each situation if you evacuated enough of the gas to decrease the pressure by a factor of 2.

8. If you release a helium balloon, it soars upward and eventually pops. Explain this behavior.

9. If you have any two gases in different containers that are the same size at the same pressure and same temperature, what is true about the moles of each gas? Why is this true?

10. Using postulates of the kinetic molecular theory, give a molecular interpretation of Boyle's law, Charles's law, and Dalton's law of partial pressures.

11. Rationalize the following observations.
    a. Aerosol cans will explode if heated.
    b. You can drink through a soda straw.
    c. A thin-walled can will collapse when the air inside is removed by a vacuum pump.
    d. Manufacturers produce different types of tennis balls for high and low altitudes.

12. Show how Boyle's law and Charles's law are special cases of the ideal gas law.

13. You have a balloon covering the mouth of a flask filled with air at 1 atm. You apply heat to the bottom of the flask until the volume of the balloon equals that of the flask.
    a. Which has more air in it—the balloon or the flask? Or do both have the same amount? Explain.
    b. In which is the pressure greater—the balloon or the flask? Or is the pressure the same? Explain.

14. Look at the demonstration discussed in Figure 12.1. How would this demonstration change if water was not added to the can? Explain.

15. How does Dalton's law of partial pressures help us with our model of ideal gases? That is, which postulates of the kinetic molecular theory does it support?

16. Draw molecular-level views that show the differences among solids, liquids, and gases.

17. Explain how increasing the number of moles of gas affects the pressure (assuming constant volume and temperature).

18. Explain how increasing the number of moles of gas affects the volume (assuming constant pressure and temperature).

Questions and Problems

All even-numbered exercises have answers in the back of this book and solutions in the *Solutions Guide*.

12.1 Pressure

QUESTIONS

1. The introduction to this chapter says that "we live immersed in a gaseous solution." What does that mean?

2. How are the three states of matter similar, and how do they differ?

3. What is meant by "the pressure of the atmosphere"? What causes this pressure?

4. Describe a simple mercury barometer. How is such a barometer used to measure the pressure of the atmosphere?

5. If two gases that do not react with each other are placed in the same container, they will _______ completely with each other.

6. One *standard atmosphere* of pressure is equivalent to _______ mm Hg.

PROBLEMS

7. Convert the following pressures into *atmospheres*.
   a. 109.2 kPa c. 781 mm Hg
   b. 781 torr d. 15.2 psi

8. Convert the following pressures into *atmospheres*.
   a. 105.2 kPa c. 752 mm Hg
   b. 75.2 cm Hg d. 767 torr

9. Convert the following pressures into units of *mm Hg*.
   a. 822 torr c. 1.14 atm
   b. 121.4 kPa d. 9.75 psi

10. Convert the following pressures into units of *mm Hg*.
    a. 0.9975 atm c. 99.7 kPa
    b. 225,400 Pa d. 1.078 atm

11. Convert the following pressures into pascals.
    a. 774 torr c. 112.5 kPa
    b. 0.965 atm d. 801 mm Hg

12. Convert the following pressures into units of kilopascals.
    a. 2.07×10^6 Pa c. 10.9 atm
    b. 795 mm Hg d. 659 torr

12.2 Pressure and Volume: Boyle's Law

QUESTIONS

13. Pretend that you're talking to a friend who has not yet taken any science courses, and describe how you would explain Boyle's law to her.

14. When the volume of a sample of gas is decreased, the pressure of the sample of gas _______.

15. The volume of a sample of ideal gas is inversely proportional to the _______ on the gas at constant temperature.

16. A mathematical expression that summarizes Boyle's law is _______.

PROBLEMS

17. For each of the following sets of pressure/volume data, calculate the missing quantity. Assume that the temperature and the amount of gas remain constants.
    a. V = 117 mL at 652 mm Hg; V = 125 mL at ? mm Hg
    b. V = 20.2 L at 1.02 atm; V = ? at 2.04 atm
    c. V = 64.2 mL at 755 torr; V = 1.00 L at ? mm Hg

18. For each of the following sets of pressure/volume data, calculate the missing quantity. Assume that the temperature and the mass of gas remain constant.
    a. V = 541 mL at 1.00 atm; V = ? at 699 torr
    b. V = 2.32 L at 110.2 kPa; V = ? at 0.995 atm
    c. V = 4.15 mL at 135 atm; V = 10.0 mL at ? mm Hg

19. For each of the following sets of pressure/volume data, calculate the missing quantity. Assume that the temperature and the amount of gas remain constant.
    a. V = 19.3 L at 102.1 kPa; V = 10.0 L at ? kPa
    b. V = 25.7 mL at 755 torr; V = ? at 761 mm Hg
    c. V = 51.2 L at 1.05 atm; V = ? at 112.2 kPa

20. For each of the following sets of pressure/volume data, calculate the missing quantity. Assume that the temperature and the amount of gas remain the same.
    a. V = 291 mL at 1.07 atm; V = ? at 2.14 atm
    b. V = 1.25 L at 755 mm Hg; V = ? at 3.51 atm
    c. V = 2.71 L at 101.4 kPa; V = 3.00 L at ? mm Hg

21. If the pressure on a 1.00-L sample of gas is increased from 760. mm Hg to 2.00 atm at constant temperature, what will the volume of the gas sample become?

22. If the pressure exerted on the gas in a weather balloon decreases from 1.01 atm to 0.562 atm as it rises, by what factor will the volume of the gas in the balloon increase as it rises?

23. A 1.04-L sample of gas at 759 mm Hg pressure is expanded until its volume is 2.24 L. What will be the pressure in the expanded gas sample (at constant temperature)?

24. What pressure would have to be applied to a 27.2-mL sample of gas at 25 °C and 1.00 atm to compress its volume to 1.00 mL without a change in temperature?

12.3 Volume and Temperature: Charles's Law

QUESTIONS

25. Pretend that you're talking to a friend who has not yet taken any science courses, and describe how you would explain the concept of absolute zero to him.

26. How can Charles's law be used to determine absolute zero?

27. The volume of a sample of ideal gas is _______ proportional to its temperature (K) at constant pressure.

28. A mathematical expression that summarizes Charles's law is _______.

PROBLEMS

29. A favorite demonstration in introductory chemistry is to illustrate how the volume of a gas is affected by temperature by blowing up a balloon at room temperature and then placing the balloon into a container of dry ice or liquid nitrogen (both of

which are *very* cold). Suppose a balloon containing 1.15 L of air at 25.2 °C is placed into a flask containing liquid nitrogen at −78.5 °C. What will the volume of the sample become (at constant pressure)?

30. If 525 mL of gas at 25 °C is heated to 50. °C, at constant pressure, calculate the new volume of the sample.

31. For each of the following sets of volume/temperature data, calculate the missing quantity. Assume that the pressure and the amount of gas remain constant.
a. $V = 1.14$ L at 21 °C; $V = ?$ at 42 °C
b. $V = 257$ mL at 45 °C; $V = 300$ mL at ? °C
c. $V = 2.78$ L at −50 °C; $V = 5.00$ L at ? °C

32. For each of the following sets of volume/temperature data, calculate the missing quantity. Assume that the pressure and the mass of gas remain constant.
a. $V = 73.5$ mL at 0 °C; $V = ?$ at 25 °C
b. $V = 15.2$ L at 298 K; $V = 10.0$ L at ? °C
c. $V = 1.75$ mL at 2.3 K; $V = ?$ at 0 °C

33. For each of the following sets of volume/temperature data, calculate the missing quantity. Assume that the pressure and the amount of gas remain constant.
a. $V = 25$ mL at 25 °C; $V = ?$ at 0 °C
b. $V = 10.2$ L at 100. °C; $V = ?$ at 100 K
c. $V = 551$ mL at 75 °C; $V = 1.00$ mL at ? °C

34. For each of the following sets of volume/temperature data, calculate the missing quantity. Assume that the pressure and the mass of gas remain constant.
a. $V = 2.01 \times 10^2$ L at 1150 °C; $V = 5.00$ L at ? °C
b. $V = 44.2$ mL at 298 K; $V = ?$ at 0 K
c. $V = 44.2$ mL at 298 K; $V = ?$ at 0 °C

35. If 5.00 L of an ideal gas is cooled from 24 °C to −272 °C, what will the volume of the gas become?

36. A sample of helium gas occupies a volume of 29.4 mL at 18 °C. Would the volume of this gas sample be doubled at 36 °C? At what temperature would the volume of the gas sample be half as big? Assume the pressure remains constant.

37. The label on an aerosol spray can contains a warning that the can should not be heated to over 130 °F because of the danger of explosion due to the pressure increase as it is heated. Calculate the potential volume of the gas contained in a 500.-mL aerosol can when it is heated from 25 °C to 54 °C (approximately 130 °F), assuming a constant pressure.

38. As we noted in Example 12.6, gas volume was formerly used as a way to measure temperature by applying Charles's law. Suppose a sample in a gas thermometer has a volume of 135 mL at 11 °C. Indicate what temperature would correspond to each of the following volumes, assuming that the pressure remains constant: 113 mL, 142 mL, 155 mL, 127 mL.

12.4 Volume and Moles: Avogadro's Law

QUESTIONS

39. At conditions of constant temperature and pressure, the volume of a sample of ideal gas is ________ proportional to the number of moles of gas present.

40. A mathematical expression that summarizes Avogadro's law is ________.

PROBLEMS

41. If 0.105 mol of helium gas occupies a volume of 2.35 L at a certain temperature and pressure, what volume would 0.337 mol of helium occupy under the same conditions?

42. If 2.01 g of helium gas occupies a volume of 12.0 L at 25 °C, what volume will 6.52 g of helium gas occupy under the same conditions?

43. If 3.25 mol of argon gas occupies a volume of 100. L at a particular temperature and pressure, what volume does 14.15 mol of argon occupy under the same conditions?

44. If 3.20 g of O_2 gas occupies a volume of 2.24 L at 0 °C and a pressure of 1.00 atm, what volume would be occupied by 32.00 g of O_2 gas under the same conditions?

12.5 The Ideal Gas Law

QUESTIONS

45. What do we mean by an *ideal* gas?

46. Under what conditions do *real* gases behave most ideally?

47. Show how Boyle's gas law can be derived from the ideal gas law.

48. Show how Charles's gas law can be derived from the ideal gas law.

PROBLEMS

49. Given the following sets of values for three of the gas variables, calculate the unknown quantity.
a. $P = 782.4$ mm Hg; $V = ?$; $n = 0.1021$ mol; $T =$ 26.2 °C
b. $P = ?$ mm Hg; $V = 27.5$ mL; $n = 0.007812$ mol; $T = 16.6$ °C
c. $P = 1.045$ atm; $V = 45.2$ mL; $n = 0.002241$ mol; $T = ?$ °C

50. Given each of the following sets of values for an ideal gas, calculate the unknown quantity.
a. $P = 782$ mm Hg; $V = ?$; $n = 0.210$ mol; $T = 27$ °C
b. $P = ?$ mm Hg; $V = 644$ mL; $n = 0.0921$ mol; $T =$ 303 K

c. $P = 745$ mm Hg; $V = 11.2$ L; $n = 0.401$ mol; $T =$? K

51. Given each of the following sets of values for three of the gas variables, calculate the unknown quantity.
a. $P = 7.74 \times 10^3$ Pa; $V = 12.2$ mL; $n =$? mol; $T = 298$ K
b. $P =$? mm Hg; $V = 43.0$ mL; $n = 0.421$ mol; $T = 223$ K
c. $P = 455$ mm Hg; $V =$? mL; $n = 4.4 \times 10^{-2}$ mol; $T = 331$ °C

52. Given each of the following sets of values for an ideal gas, calculate the unknown quantity.
a. $P = 1.01$ atm; $V =$?; $n = 0.00831$ mol; $T = 25$ °C
b. $P =$? atm; $V = 602$ mL; $n = 8.01 \times 10^{-3}$ mol; $T = 310$ K
c. $P = 0.998$ atm; $V = 629$ mL; $n =$? mol; $T = 35$ °C

53. What volume does 0.103 mol of N_2 gas occupy at a temperature of 27 °C and a pressure of 784 mm Hg?

54. What mass of helium is needed to pressurize an evacuated 25.0-L tank to a pressure of 120. atm at 25 °C?

55. What mass of helium gas is needed to pressurize a 100.0-L tank to 255 atm at 25 °C? What mass of oxygen gas would be needed to pressurize a similar tank to the same specifications?

56. Suppose two 200.0-L tanks are to be filled separately with the gases helium and hydrogen. What mass of each gas is needed to produce a pressure of 135 atm in its respective tank at 24 °C?

57. At what temperature will a 1.0-g sample of neon gas exert a pressure of 500. torr in a 5.0-L container?

58. What is the temperature if 30.1 g of N_2 in a 1.00-L container is at a pressure of 25.2 atm?

59. What is the pressure in a 25-L vessel containing 1.0 kg of oxygen gas at 300. K?

60. Determine the pressure in a 125-L tank containing 56.2 kg of oxygen gas at 21 °C.

61. Suppose a 24.3-mL sample of helium gas at 25 °C and 1.01 atm is heated to 50. °C and compressed to a volume of 15.2 mL. What will be the pressure of the sample?

62. What will be the new volume if 125 mL of He gas at 100 °C and 0.981 atm is cooled to 25 °C and the pressure is increased to 1.15 atm?

63. What will the volume of the sample become if 459 mL of an ideal gas at 27 °C and 1.05 atm is cooled to 15 °C and 0.997 atm?

64. If 2.51 g of H_2 is placed in an evacuated 255-mL container at 100. °C, what will be the pressure inside the container?

12.6 Dalton's Law of Partial Pressures

QUESTIONS

65. Explain why the measured properties of a mixture of gases depend only on the total number of moles of particles, not on the identity of the individual gas particles. How is this observation summarized as a law?

66. We often collect small samples of gases in the laboratory by bubbling the gas into a bottle or flask containing water. Explain why the gas becomes saturated with water vapor and how we must take the presence of water vapor into account when calculating the properties of the gas sample.

PROBLEMS

67. If a gaseous mixture is made of 2.41 g of He and 2.79 g of Ne in an evacuated 1.04-L container at 25 °C, what will be the partial pressure of each gas and the total pressure in the container?

68. A gaseous mixture consists of 6.91 g of N_2, 4.71 g of O_2, and 2.95 g of He. What volume does this mixture occupy at 28 °C and 1.05 atm pressure?

69. A tank contains a mixture of 3.0 mol of N_2, 2.0 mol of O_2, and 1.0 mol of CO_2 at 25 °C and a total pressure of 10.0 atm. Calculate the partial pressure (in torr) of each gas in the mixture.

70. A 50.0-L tank contains 5.21 kg of N_2 and 4.49 kg of O_2. What is the pressure in the tank at 24 °C?

71. A sample of oxygen gas is saturated with water vapor at 27 °C. The total pressure of the mixture is 772 torr, and the vapor pressure of water is 26.7 torr at 27 °C. What is the partial pressure of the oxygen gas?

72. A sample of oxygen gas is collected by displacement of water at 25 °C and 1.02 atm total pressure. If the vapor pressure of water is 23.756 mm Hg at 25 °C, what is the partial pressure of the oxygen gas in the sample?

73. A 500.-mL sample of O_2 gas at 24 °C was prepared by decomposing a 3% aqueous solution of hydrogen peroxide, H_2O_2, in the presence of a small amount of manganese catalyst by the reaction

$$2H_2O_2(aq) \rightarrow 2H_2O(g) + O_2(g)$$

The oxygen thus prepared was collected by displacement of water. The total pressure of gas collected was 755 mm Hg. What is the partial pressure of O_2 in the mixture? How many moles of O_2 are in the mixture? (The vapor pressure of water at 24 °C is 23 mm Hg.)

74. Small quantities of hydrogen gas can be prepared in the laboratory by the addition of aqueous hydrochloric acid to metallic zinc.

$$Zn(s) + 2HCl(aq) \rightarrow ZnCl_2(aq) + H_2(g)$$

Typically, the hydrogen gas is bubbled through water for collection and becomes saturated with water vapor. Suppose 240. mL of hydrogen gas is collected at 30. °C and has a total pressure of 1.032 atm by this process. What is the partial pressure of hydrogen gas in the sample? How many moles of hydrogen gas are present in the sample? How many grams of zinc must have reacted to produce this quantity of hydrogen? (The vapor pressure of water is 32 torr at 30 °C.)

12.7 Laws and Models: A Review

QUESTIONS

75. What is a scientific *law?* What is a *theory?* How do these concepts differ? Does a law explain a theory, or does a theory attempt to explain a law?

76. When is a scientific theory considered to be successful? Are all theories successful? Will a theory that has been successful in the past necessarily be successful in the future?

12.8 The Kinetic Molecular Theory of Gases

QUESTIONS

77. What do we assume about the volume of the actual molecules themselves in a sample of gas, compared to the bulk volume of the gas overall? Why?

78. How do chemists explain on a molecular basis the fact that gases in containers exert pressure on the walls of the container?

79. Temperature is a measure of the average ________ of the molecules in a sample of gas.

80. The kinetic molecular theory of gases suggests that gas particles exert ________ attractive or repulsive forces on each other.

12.9 The Implications of the Kinetic Molecular Theory

QUESTIONS

81. How is the phenomenon of temperature explained on the basis of the kinetic molecular theory? What microscopic property of gas molecules is reflected in the temperature measured?

82. Explain, in terms of the kinetic molecular theory, how an increase in the temperature of a gas confined to a rigid container causes an increase in the pressure of the gas.

12.10 Gas Stoichiometry

QUESTIONS

83. What is the *molar volume* of a gas? Do all gases that behave ideally have the same molar volume?

84. What conditions are considered "standard temperature and pressure" (STP) for gases? Suggest a reason why these particular conditions might have been chosen for STP.

PROBLEMS

85. When calcium carbonate is heated strongly, carbon dioxide gas is released

$$CaCO_3(s) \rightarrow CaO(s) + CO_2(g)$$

What volume of $CO_2(g)$, measured at STP, is produced if 15.2 g of $CaCO_3(s)$ is heated?

86. Consider the following reaction

$$2Al(s) + 3O_2(g) \rightarrow 2Al_2O_3(s)$$

What volume of oxygen gas at STP would be needed to react completely with 1.55 g of aluminum?

87. Consider the following *unbalanced* chemical equation for the combustion of propane.

$$C_3H_8(g) + O_2(g) \rightarrow CO_2(g) + H_2O(g)$$

What volume of oxygen gas at 25 °C and 1.04 atm is needed for the complete combustion of 5.53 g of propane?

88. Although we generally think of combustion reactions as involving oxygen gas, other rapid oxidation reactions are also referred to as combustions. For example, if magnesium metal is placed into chlorine gas, a rapid oxidation takes place, and magnesium chloride is produced.

$$Mg(s) + Cl_2(g) \rightarrow MgCl_2(s)$$

What volume of chlorine gas, measured at STP, is required to react completely with 1.02 g of magnesium?

89. Ammonia and gaseous hydrogen chloride combine to form ammonium chloride.

$$NH_3(g) + HCl(g) \rightarrow NH_4Cl(s)$$

If 4.21 L of $NH_3(g)$ at 27 °C and 1.02 atm is combined with 5.35 L of $HCl(g)$ at 26 °C and 0.998 atm, what mass of $NH_4Cl(s)$ will be produced? Which gas is the limiting reactant? Which gas is present in excess?

90. When calcium carbonate is heated strongly, carbon dioxide gas is evolved.

$$CaCO_3(s) \rightarrow CaO(s) + CO_2(g)$$

If 4.74 g of calcium carbonate is heated, what volume of $CO_2(g)$ would be produced when collected at 26 °C and 0.997 atm?

91. Many transition metal salts are hydrates: they contain a fixed number of water molecules bound per formula unit of the salt. For example, copper(II) sulfate most commonly exists as the pentahydrate, $CuSO_4 \cdot 5H_2O$. If 5.00 g of $CuSO_4 \cdot 5H_2O$ is heated strongly so as to drive off all of the waters of

hydration as water vapor, what volume will this water vapor occupy at 350. °C and a pressure of 1.04 atm?

92. If water is added to magnesium nitride, ammonia gas is produced when the mixture is heated.

$$Mg_3N_2(s) + 3H_2O(l) \rightarrow 3MgO(s) + 2NH_3(g)$$

If 10.3 g of magnesium nitride is treated with water, what volume of ammonia gas would be collected at 24 °C and 752 mm Hg?

93. What volume does a mixture of 14.2 g of He and 21.6 g of H_2 occupy at 28 °C and 0.985 atm?

94. What volume does a mixture of 26.2 g of O_2 and 35.1 g of N_2 occupy at 35 °C and 755 mm Hg?

95. An ideal gas has a volume of 50. mL at 100. °C and a pressure of 690 torr. Calculate the volume of this sample of gas at STP.

96. A sample of hydrogen gas has a volume of 145 mL when measured at 44 °C and 1.47 atm. What volume would the hydrogen sample occupy at STP?

97. A mixture contains 5.00 g *each* of O_2, N_2, CO_2, and Ne gas. Calculate the volume of this mixture at STP. Calculate the partial pressure of each gas in the mixture at STP.

98. A gaseous mixture contains 6.25 g of He and 4.97 g of Ne. What volume does the mixture occupy at STP? Calculate the partial pressure of each gas in the mixture at STP.

99. Given the following *unbalanced* chemical equation for the combination reaction of sodium metal and chlorine gas

$$Na(s) + Cl_2(g) \rightarrow NaCl(s)$$

what volume of chlorine gas, measured at STP, is necessary for the complete reaction of 4.81 g of sodium metal?

100. Consider the following reaction:

$$Ca(s) + Cl_2(g) \rightarrow CaCl_2(s)$$

What volume of chlorine gas at 25 °C, and 1.02 atm would be necessary to react completely with 4.15 g of calcium metal?

101. During the making of steel, iron(II) oxide is reduced to metallic iron by treatment with carbon monoxide gas.

$$FeO(s) + CO(g) \rightarrow Fe(s) + CO_2(g)$$

Suppose 1.45 kg of Fe reacts. What volume of $CO(g)$ is required, and what volume of $CO_2(g)$ is produced, each measured at STP?

102. Consider the following reaction:

$$Zn(s) + 2HCl(aq) \rightarrow ZnCl_2(aq) + H_2(g)$$

What mass of zinc metal should be taken so as to produce 125 mL of H_2 measured at STP when reacted with excess hydrochloric acid?

Additional Problems

103. When doing any calculation involving gas samples, we must express the temperature in terms of the ________ temperature scale.

104. Two moles of ideal gas occupy a volume that is ________ the volume of 1 mol of ideal gas under the same temperature and pressure conditions.

105. Summarize the postulates of the kinetic molecular theory for gases. How does the kinetic molecular theory account for the observed properties of temperature and pressure?

106. Give a formula or equation that represents each of the following gas laws.
   a. Boyle's law
   b. Charles's law
   c. Avogadro's law
   d. the ideal gas law
   e. the combined gas law

107. For a mixture of gases in the same container, the total pressure exerted by the mixture of gases is the ________ of the pressures that those gases would exert if they were alone in the container under the same conditions.

108. A helium tank contains 25.2 L of helium at 8.40 atm pressure. Determine how many 1.50-L balloons at 755 mm Hg can be inflated with the gas in the tank, assuming that the tank will also have to contain He at 755 mm Hg after the balloons are filled (that is, it is not possible to empty the tank completely). The temperature is 25 °C in all cases.

109. As weather balloons rise from the earth's surface, the pressure of the atmosphere becomes less, tending to cause the volume of the balloons to expand. However, the temperature is much lower in the upper atmosphere than at sea level. Would this temperature effect tend to make such a balloon expand or contract? Weather balloons do, in fact, expand as they rise. What does this tell you?

110. When ammonium carbonate is heated, three gases are produced by its decomposition.

$$(NH_4)_2CO_3(s) \rightarrow 2NH_3(g) + CO_2(g) + H_2O(g)$$

What total volume of gas is produced, measured at 453 °C and 1.04 atm, if 52.0 g of ammonium carbonate is heated?

111. Carbon dioxide gas, in the dry state, may be produced by heating calcium carbonate.

$$CaCO_3(s) \rightarrow CaO(s) + CO_2(g)$$

What volume of CO_2, collected dry at 55 °C and a pressure of 774 torr, is produced by complete thermal decomposition of 10.0 g of $CaCO_3$?

112. Carbon dioxide gas, saturated with water vapor, can be produced by the addition of aqueous acid to calcium carbonate.

$$CaCO_3(s) + 2H^+(aq) \rightarrow Ca^{2+}(aq) + H_2O(l) + CO_2(g)$$

How many moles of $CO_2(g)$, collected at 60. °C and 774 torr total pressure, are produced by the complete reaction of 10.0 g of $CaCO_3$ with acid? What volume does this wet CO_2 occupy? What volume would the CO_2 occupy at 774 torr if a desiccant (a chemical drying agent) were added to remove the water? (The vapor pressure of water at 60. °C is 149.4 mm Hg.)

113. Sulfur trioxide, SO_3, is produced in enormous quantities each year for use in the synthesis of sulfuric acid.

$$S(s) + O_2(g) \rightarrow SO_2(g)$$
$$2SO_2(g) + O_2(g) \rightarrow 2SO_3(g)$$

What volume of $O_2(g)$ at 350. °C and a pressure of 5.25 atm is needed to completely convert 5.00 g of sulfur to sulfur trioxide?

114. Calculate the volume of $O_2(g)$ produced at 25 °C and 630. torr when 50.0 g of $KClO_3(s)$ is heated in the presence of a small amount of MnO_2 catalyst.

115. If 10.0 g of liquid helium at 1.7 K is completely vaporized, what volume does the helium occupy at STP?

116. Perform the indicated pressure conversions.
a. 752 mm Hg into pascals
b. 458 kPa into atmospheres
c. 1.43 atm into mm Hg
d. 842 torr into mm Hg

117. Convert the following pressures into mm Hg.
a. 0.903 atm
b. 2.1240×10^6 Pa
c. 445 kPa
d. 342 torr

118. Convert the following pressures into pascals.
a. 645 mm Hg
b. 221 kPa
c. 0.876 atm
d. 32 torr

119. For each of the following sets of pressure/volume data, calculate the missing quantity. Assume that the temperature and the amount of gas remain constant.
a. $V = 123$ L at 4.56 atm; $V = ?$ at 1002 mm Hg
b. $V = 634$ mL at 25.2 mm Hg; $V = 166$ mL at ? atm
c. $V = 443$ L at 511 torr; $V = ?$ at 1.05 kPa

120. For each of the following sets of pressure/volume data, calculate the missing quantity. Assume that the temperature and the amount of gas remain constant.
a. $V = 255$ mL at 1.00 mm Hg; $V = ?$ at 2.00 torr
b. $V = 1.3$ L at 1.0 kPa; $V = ?$ at 1.0 atm
c. $V = 1.3$ L at 1.0 kPa; $V = ?$ at 1.0 mm Hg

121. A particular balloon is designed by its manufacturer to be inflated to a volume of no more than 2.5 L. If the balloon is filled with 2.0 L of helium at sea level, is released, and rises to an altitude at which the atmospheric pressure is only 500. mm Hg, will the balloon burst?

122. What pressure is needed to compress 1.52 L of air at 755 mm Hg to a volume of 450 mL (at constant temperature)?

123. An expandable vessel contains 729 mL of gas at 22 °C. What volume will the gas sample in the vessel have if it is placed in a boiling water bath (100. °C)?

124. For each of the following sets of volume/temperature data, calculate the missing quantity. Assume that the pressure and the amount of gas remain constant.
a. $V = 100.$ mL at 74 °C; $V = ?$ at −74 °C
b. $V = 500.$ mL at 100 °C; $V = 600.$ mL at ? °C
c. $V = 10{,}000$ L at 25 °C; $V = ?$ at 0 K

125. For each of the following sets of volume/temperature data, calculate the missing quantity. Assume that the pressure and the amount of gas remain constant.
a. $V = 22.4$ L at 0 °C; $V = 44.4$ L at ? K
b. $V = 1.0 \times 10^{-3}$ mL at −272 °C; $V = ?$ at 25 °C
c. $V = 32.3$ L at −40 °C; $V = 1000.$ L at ? °C

126. A 75.2-mL sample of helium at 12 °C is heated to 192 °C. What is the new volume of the helium (assuming constant pressure)?

127. If 5.12 g of oxygen gas occupies a volume of 6.21 L at a certain temperature and pressure, what volume will 25.0 g of oxygen gas occupy under the same conditions?

128. If 23.2 g of a given gas occupies a volume of 93.2 L at a particular temperature and pressure, what mass of the gas occupies a volume of 10.4 L under the same conditions?

129. Given each of the following sets of values for three of the gas variables, calculate the unknown quantity.
a. $P = 21.2$ atm; $V = 142$ mL; $n = 0.432$ mol; $T =$? K
b. $P = ?$ atm; $V = 1.23$ mL; $n = 0.000115$ mol; $T =$ 293 K
c. $P = 755$ mm Hg; $V = ?$ mL; $n = 0.473$ mol; $T =$ 131 °C

130. Given each of the following sets of values for three of the gas variables, calculate the unknown quantity.
a. $P = 1.034$ atm; $V = 21.2$ mL; $n = 0.00432$ mol; $T = ?$ K
b. $P = ?$ atm; $V = 1.73$ mL; $n = 0.000115$ mol; $T =$ 182 K
c. $P = 1.23$ mm Hg; $V = ?$ L; $n = 0.773$ mol; $T =$ 152 °C

131. What is the pressure inside a 10.0-L flask containing 14.2 g of N_2 at 26 °C?

132. Suppose three 100.-L tanks are to be filled separately with the gases CH_4, N_2, and CO_2, respectively. What

mass of each gas is needed to produce a pressure of 120. atm in its tank at 27 °C?

133. At what temperature does 4.00 g of helium gas have a pressure of 1.00 atm in a 22.4-L vessel?

134. What is the pressure in a 100.-mL flask containing 55 mg of oxygen gas at 26 °C?

135. A weather balloon is filled with 1.0 L of helium at 23 °C and 1.0 atm. What volume does the balloon have when it has risen to a point in the atmosphere where the pressure is 220 torr and the temperature is −31 °C?

136. At what temperature does 100. mL of N_2 at 300. K and 1.13 atm occupy a volume of 500. mL at a pressure of 1.89 atm?

137. If 1.0 mol of $N_2(g)$ is injected into a 5.0-L tank already containing 50. g of O_2 at 25 °C, what will be the total pressure in the tank?

138. A gaseous mixture contains 12.1 g of N_2 and 4.05 g of He. What is the volume of this mixture at STP?

139. A flask of hydrogen gas is collected at 1.023 atm and 35 °C by displacement of water from the flask. The vapor pressure of water at 35 °C is 42.2 mm Hg. What is the partial pressure of hydrogen gas in the flask?

140. Consider the following chemical equation:

$$N_2(g) + 3H_2(g) \rightarrow 2NH_3(g)$$

What volumes of nitrogen gas and hydrogen gas, each measured at 11 °C and 0.998 atm, are needed to produce 5.00 g of ammonia?

141. Consider the following *unbalanced* chemical equation:

$$C_6H_{12}O_6(s) + O_2(g) \rightarrow CO_2(g) + H_2O(g)$$

What volume of oxygen gas, measured at 28 °C and 0.976 atm, is needed to react with 5.00 g of $C_6H_{12}O_6$? What volume of each product is produced under the same conditions?

142. Consider the following *unbalanced* chemical equation:

$$Cu_2S(s) + O_2(g) \rightarrow Cu_2O(s) + SO_2(g)$$

What volume of oxygen gas, measured at 27.5 °C and 0.998 atm, is required to react with 25 g of copper(I) sulfide? What volume of sulfur dioxide gas is produced under the same conditions?

143. When sodium bicarbonate, $NaHCO_3(s)$, is heated, sodium carbonate is produced, with the evolution of water vapor and carbon dioxide gas.

$$2NaHCO_3(s) \rightarrow Na_2CO_3(s) + H_2O(g) + CO_2(g)$$

What total volume of gas, measured at 29 °C and 769 torr, is produced when 1.00 g of $NaHCO_3(s)$ is completely converted to $Na_2CO_3(s)$?

144. What volume does 35 moles of N_2 occupy at STP?

145. A sample of oxygen gas has a volume of 125 L at 25 °C and a pressure of 0.987 atm. Calculate the volume of this oxygen sample at STP.

146. A mixture contains 5.0 g of He, 1.0 g of Ar, and 3.5 g of Ne. Calculate the volume of this mixture at STP. Calculate the partial pressure of each gas in the mixture at STP.

147. What volume of CO_2 measured at STP is produced when 27.5 g of $CaCO_3$ is decomposed?

$$CaCO_3(s) \rightarrow CaO(s) + CO_2(g)$$

148. Concentrated hydrogen peroxide solutions are explosively decomposed by traces of transition metal ions (such as Mn or Fe):

$$2H_2O_2(aq) \rightarrow 2H_2O(l) + O_2(g)$$

What volume of pure $O_2(g)$, collected at 27 °C and 764 torr, would be generated by decomposition of 125 g of a 50.0% by mass hydrogen peroxide solution?

13 Liquids and Solids

Ice, the solid form of water, provides recreation for this ice climber.

You have only to think about water to appreciate how different the three states of matter are. Flying, swimming, and ice skating are all done in contact with water in its various states. We swim in liquid water and skate on water in its solid form (ice). Airplanes fly in an atmosphere containing water in the gaseous state (water vapor). To allow these various activities, the arrangements of the water molecules must be significantly different in their gas, liquid, and solid forms.

In Chapter 12 we saw that the particles of a gas are far apart, are in rapid random motion, and have little effect on each other. Solids are obviously very different from gases. Gases have low densities, have high compressibilities, and completely fill a container. Solids have much greater densities than gases, are compressible only to a very slight extent, and are rigid; a solid maintains its shape regardless of its container. These properties indicate that the components of a solid are close together and exert large attractive forces on each other.

The properties of liquids lie somewhere between those of solids and of gases—but not midway between, as can be seen from some of the properties of the three states of water. For example, it takes about seven times more energy to change liquid water to steam (a gas) at 100 °C than to melt ice to form liquid water at 0 °C.

$$H_2O(s) \rightarrow H_2O(l) \qquad \text{energy required} \cong 6\ \text{kJ/mol}$$
$$H_2O(l) \rightarrow H_2O(g) \qquad \text{energy required} \cong 41\ \text{kJ/mol}$$

These values indicate that going from the liquid to the gaseous state involves a much greater change than going from the solid to the liquid. Therefore, we can conclude that the solid and liquid states are more similar than the liquid and gaseous states. This is also demonstrated by the densities of the three states of water (Table 13.1). Note that water in its gaseous state is about 2000 times less dense than in the solid and liquid states and that the latter two states have very similar densities.

We find in general that the liquid and solid states show many similarities and are strikingly different from the gaseous state (see Figure 13.1). The best way to picture the solid state is in terms of closely packed, highly ordered particles in contrast to the widely spaced, randomly arranged particles of a gas. The liquid state lies in between, but its properties indicate that it much more closely resembles the solid than the gaseous state. It is useful to picture a liquid in terms of particles that are generally quite close together, but with a more disordered arrangement than for the solid state and with some empty spaces. For most substances, the solid state has a higher density than the liquid, as Figure 13.1 suggests. However, water is an exception to this rule. Ice has an unusual amount of empty space and so is less dense than liquid water, as indicated in Table 13.1.

Wind surfers use liquid water for recreation.

In this chapter we will explore the important properties of liquids and solids. We will illustrate

Table 13.1 Densities of the Three States of Water

State	Density (g/cm^3)
solid (0 °C, 1 atm)	0.9168
liquid (25 °C, 1 atm)	0.9971
gas (100 °C, 1 atm)	5.88×10^{-4}

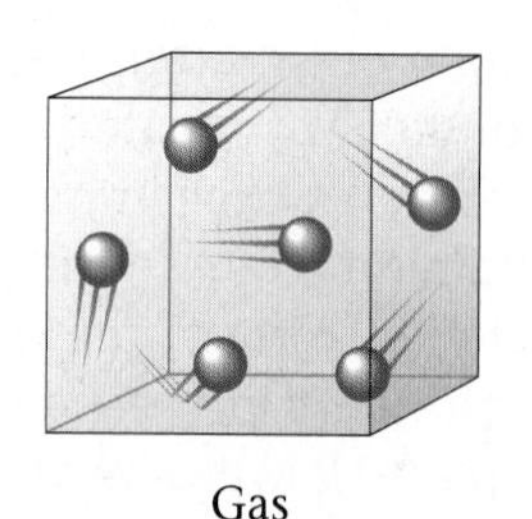
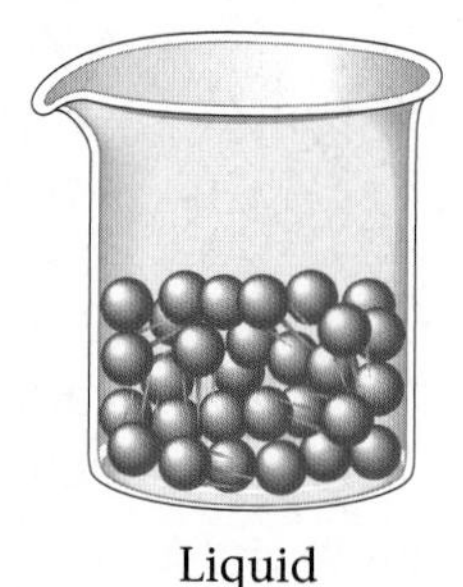
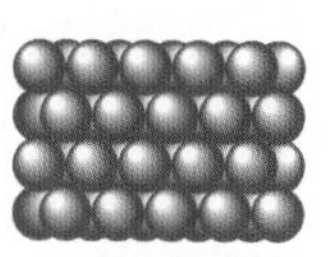

Gas Liquid Solid

Figure 13.1
Representations of the gas, liquid, and solid states.

many of these properties by considering one of the earth's most important substances: water.

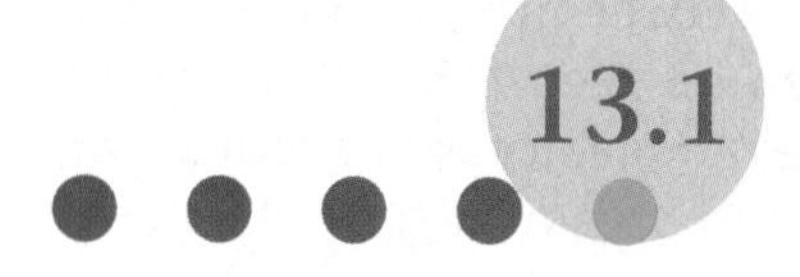

Water and Its Phase Changes

AIM: To learn some of the important features of water.

In the world around us we see many solids (soil, rocks, trees, concrete, and so on), and we are immersed in the gases of the atmosphere. But the liquid we most commonly see is water; it is virtually everywhere, covering about 70% of the earth's surface. Approximately 97% of the earth's water is found in the oceans, which are actually mixtures of water and huge quantities of dissolved salts.

Water is one of the most important substances on earth. It is crucial for sustaining the reactions within our bodies that keep us alive, but it also affects our lives in many indirect ways. The oceans help moderate the earth's temperature. Water cools automobile engines and nuclear power plants. Water provides a means of transportation on the earth's surface and acts as a medium for the growth of the myriad creatures we use as food, and much more.

The water we drink often has a taste because of the substances dissolved in it. It is not pure water.

Pure water is a colorless, tasteless substance that at 1 atm pressure freezes to form a solid at 0 °C and vaporizes completely to form a gas at 100 °C. This means that (at 1 atm pressure) the liquid range of water occurs between the temperatures 0 °C and 100 °C.

What happens when we heat liquid water? First the temperature of the water rises. Just as with gas molecules, the motions of the water molecules increase as it is heated. Eventually the temperature of the water reaches 100 °C; now bubbles develop in the interior of the liquid, float to the surface, and burst—the boiling point has been reached. An interesting thing happens at the boiling point: even though heating continues, the temperature stays at 100 °C until all the water has changed to vapor. Only when all of the water has changed to the gaseous state does the temperature begin to rise again. (We are now heating the vapor.) At 1 atm pressure, liquid water always changes to gaseous water at 100 °C, the **normal boiling point** for water.

The experiment just described is represented in Figure 13.2, which is called the **heating/cooling curve** for water. Going from left to right on this graph means energy is being added (heating). Going from right to left on the graph means that energy is being removed (cooling).

When liquid water is cooled, the temperature decreases until it reaches 0 °C, where the liquid begins to freeze (see Figure 13.2). The temperature

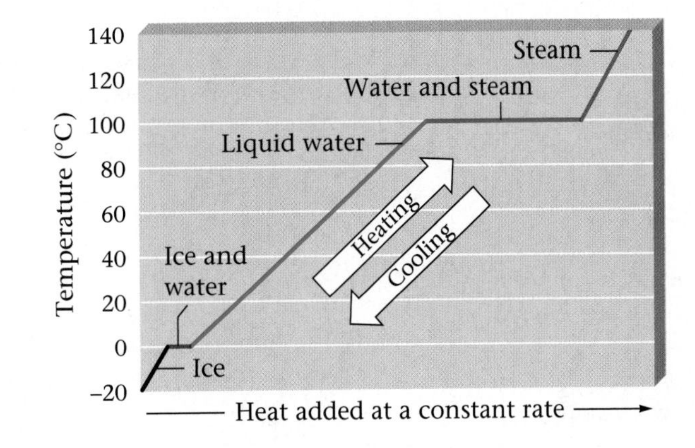

Figure 13.2
The heating/cooling curve for water heated or cooled at a constant rate. The plateau at the boiling point is longer than the plateau at the melting point, because it takes almost seven times as much energy (and thus seven times the heating time) to vaporize liquid water as to melt ice. Note that to make the diagram clear, the blue line is not drawn to scale. It actually takes more energy to melt ice and boil water than to heat water from 0 °C to 100 °C.

remains at 0 °C until all the liquid water has changed to ice and then begins to drop again as cooling continues. At 1 atm pressure, water freezes (or, in the opposite process, ice melts) at 0 °C. This is called the **normal freezing point** of water. Liquid and solid water can coexist indefinitely if the temperature is held at 0 °C. However, at temperatures below 0 °C liquid water freezes, while at temperatures above 0 °C ice melts.

Interestingly, water expands when it freezes. That is, one gram of ice at 0 °C has a greater volume than one gram of liquid water at 0 °C. This has very important practical implications. For instance, water in a confined space can break its container when it freezes and expands. This accounts for the bursting of water pipes and engine blocks that are left unprotected in freezing weather.

The expansion of water when it freezes also explains why ice cubes float. Recall that density is defined as mass/volume. When one gram of liquid water freezes, its volume becomes greater (it expands). Therefore, the *density* of one gram of ice is less than the density of one gram of water, because in the case of ice we divide by a slightly larger volume. For example, at 0 °C the density of liquid water is

$$\frac{1.00 \text{ g}}{1.00 \text{ mL}} = 1.00 \text{ g/mL}$$

and the density of ice is

$$\frac{1.00 \text{ g}}{1.09 \text{ mL}} = 0.917 \text{ g/mL}$$

The lower density of ice also means that ice floats on the surface of lakes as they freeze, providing a layer of insulation that helps to prevent lakes and rivers from freezing solid in the winter. This means that aquatic life continues to have liquid water available through the winter.

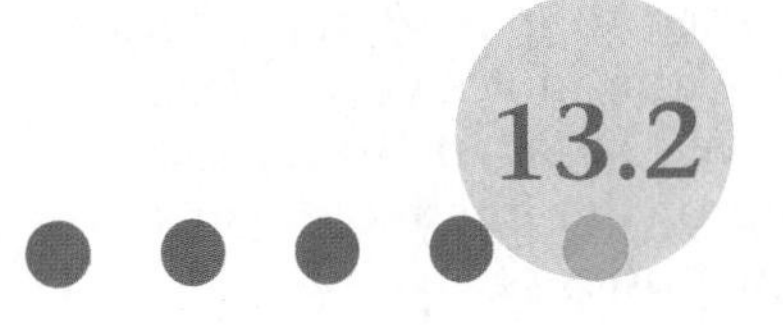

13.2 Energy Requirements for the Changes of State

AIMS: To learn about interactions among water molecules. To understand and use heat of fusion and heat of vaporization.

It is important to recognize that changes of state from solid to liquid and from liquid to gas are *physical* changes. No *chemical* bonds are broken in these processes. Ice, water, and steam all contain H_2O molecules. When

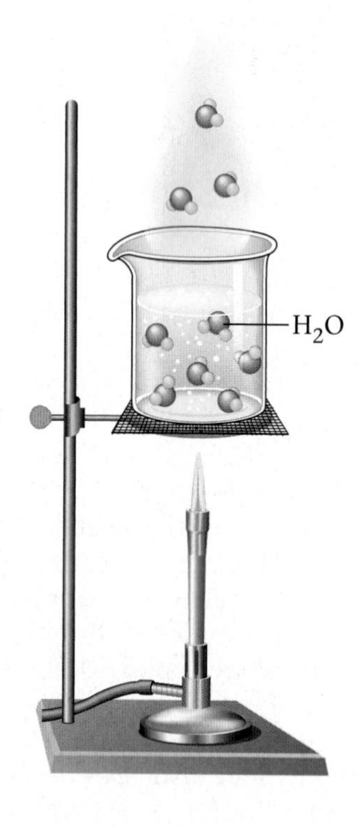

Figure 13.3
Both liquid water and gaseous water contain H_2O molecules. In liquid water the H_2O molecules are close together, whereas in the gaseous state the molecules are widely separated. The bubbles contain gaseous water.

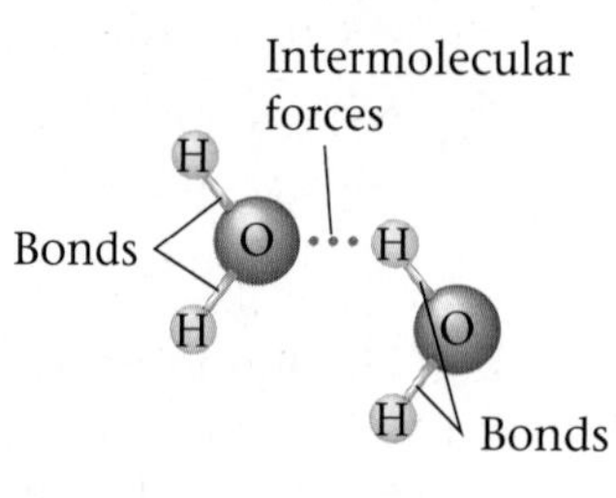

Figure 13.4
Intramolecular (bonding) forces exist between the atoms in a molecule and hold the molecule together. Intermolecular forces exist between molecules. These are the forces that cause water to condense to a liquid or form a solid at low enough temperatures. Intermolecular forces are typically much weaker than intramolecular forces.

water is boiled to form steam, water molecules are separated from each other (see Figure 13.3) but the individual molecules remain intact.

The bonding forces that hold the atoms of a molecule together are called **intramolecular** (within the molecule) **forces.** The forces that occur among molecules that cause them to aggregate to form a solid or a liquid are called **intermolecular** (between the molecules) **forces.** These two types of forces are illustrated in Figure 13.4.

Remember that temperature is a measure of the random motions (average kinetic energy) of the particles in a substance.

It takes energy to melt ice and to vaporize water, because intermolecular forces between water molecules must be overcome. In ice the molecules are virtually locked in place, although they can vibrate about their positions. When energy is added, the vibrational motions increase, and the molecules eventually achieve the greater movement and disorder characteristic of liquid water. The ice has melted. As still more energy is added, the gaseous state is eventually reached, in which the individual molecules are far apart and interact relatively little. However, the gas still consists of water molecules. It would take *much* more energy to overcome the covalent bonds and decompose the water molecules into their component atoms.

The energy required to melt 1 mol of a substance is called the **molar heat of fusion.** For ice, the molar heat of fusion is 6.02 kJ/mol. The energy required to change 1 mol of liquid to its vapor is called the **molar heat of vaporization.** For water, the molar heat of vaporization is 40.6 kJ/mol at 100 °C. Notice in Figure 13.2 that the plateau that corresponds to the vaporization of water is much longer than that for the melting of ice. This occurs because it takes much more energy (almost seven times as much) to vaporize a mole of water than to melt a mole of ice. This is consistent with our models of solids, liquids, and gases (see Figure 13.1). In liquids, the particles (molecules) are relatively close together, so most of the intermolecular forces are still present. However, when the molecules go from the liquid to the gaseous state, they must be moved far apart. To separate the molecules enough to form a gas, virtually all of the intermolecular forces must be overcome, and this requires large quantities of energy.

Example 13.1 Calculating Energy Changes: Solid to Liquid

Calculate the energy required to melt 8.5 g of ice at 0 °C. The molar heat of fusion for ice is 6.02 kJ/mol.

CHEMISTRY IN FOCUS

Whales Need Changes of State

Sperm whales are prodigious divers. They commonly dive a mile or more into the ocean, hovering at that depth in search of schools of squid or fish. To remain motionless at a given depth, the whale must have the same density as the surrounding water. Because the density of seawater increases with depth, the sperm whale has a system that automatically increases its density as it dives. This system involves the spermaceti organ found in the whale's head. Spermaceti is a waxy substance with the formula

$$CH_3\text{—}(CH_2)_{15}\text{—}O\text{—}\underset{\underset{O}{\|}}{C}\text{—}(CH_2)_{14}\text{—}CH_3$$

which is a liquid above 30 °C. At the ocean surface the spermaceti in the whale's head is a liquid, warmed by the flow of blood through the spermaceti organ. When the whale dives, this blood flow decreases and the colder water causes the spermaceti to begin freezing. Because solid spermaceti is more dense than the liquid state, the sperm whale's density increases as it dives, matching the increase in the water's density.* When the whale wants to resurface, blood flow through the spermaceti organ increases, remelting the spermaceti and making the whale more buoyant. So the sperm whale's sophisticated density-regulating mechanism is based on a simple change of state.

A sperm whale.

*For most substances, the solid state is more dense than the liquid state. Water is an important exception.

Solution

The molar heat of fusion is the energy required to melt *1 mol* of ice. In this problem we have 8.5 g of solid water. We must find out how many moles of ice this mass represents. Because the molar mass of water is 16 + 2(1) = 18, we know that 1 mol of water has a mass of 18 g, so we can convert 8.5 g of H_2O to moles of H_2O.

$$8.5\ \cancel{\text{g H}_2\text{O}} \times \frac{1\ \text{mol H}_2\text{O}}{18\ \cancel{\text{g H}_2\text{O}}} = 0.47\ \text{mol H}_2\text{O}$$

Because 6.02 kJ of energy is required to melt a mole of solid water, our sample will take about half this amount (we have approximately half a mole of ice). To calculate the exact amount of energy required, we will use the equivalence statement

$$6.02\ \text{kJ required for 1 mol of H}_2\text{O}$$

which leads to the conversion factor we need:

$$0.47\ \cancel{\text{mol H}_2\text{O}} \times \frac{6.02\ \text{kJ}}{\cancel{\text{mol H}_2\text{O}}} = 2.8\ \text{kJ}$$

This can be represented symbolically as

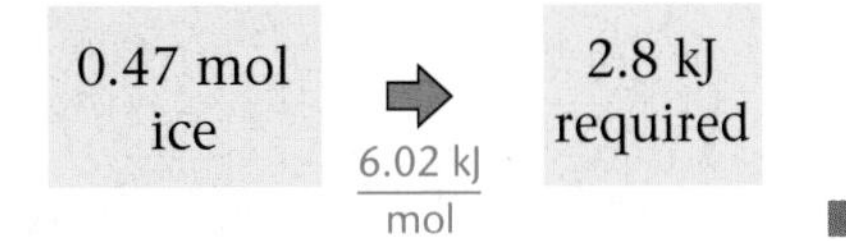

■

Example 13.2 Calculating Energy Changes: Liquid to Gas

Specific heat capacity was discussed in Section 3.7.

Calculate the energy (in kJ) required to heat 25 g of liquid water from 25 °C to 100. °C and change it to steam at 100. °C. The specific heat capacity of liquid water is 4.18 J/g °C, and the molar heat of vaporization of water is 40.6 kJ/mol.

Solution

This problem can be split into two parts: (1) heating the water to its boiling point and (2) converting the liquid water to vapor at the boiling point.

STEP 1: HEATING TO BOILING We must first supply energy to heat the liquid water from 25 °C to 100. °C. Because 4.18 J is required to heat *one* gram of water by *one* Celsius degree, we must multiply by both the mass of water (25 g) and the temperature change (100. °C − 25 °C = 75 °C),

Energy required (Q)	=	Specific heat capacity (s)	×	Mass of water (m)	×	Temperature change (ΔT)

which we can represent by the equation

$$Q = s \times m \times \Delta T$$

Thus

$$Q = 4.18 \frac{\text{J}}{\cancel{\text{g}}\ \cancel{°\text{C}}} \times 25\ \cancel{\text{g}} \times 75\ \cancel{°\text{C}} = 7.8 \times 10^3\ \text{J}$$

Energy required to heat 25 g of water from 25 °C to 100. °C; Specific heat capacity; Mass of water; Temperature change

$$= 7.8 \times 10^3\ \cancel{\text{J}} \times \frac{1\ \text{kJ}}{1000\ \cancel{\text{J}}} = 7.8\ \text{kJ}$$

STEP 2: VAPORIZATION Now we must use the molar heat of vaporization to calculate the energy required to vaporize the 25 g of water at 100. °C. The heat of vaporization is given *per mole* rather than per gram, so we must first convert the 25 g of water to moles.

$$25\ \cancel{\text{g H}_2\text{O}} \times \frac{1\ \text{mol H}_2\text{O}}{18\ \cancel{\text{g H}_2\text{O}}} = 1.4\ \text{mol H}_2\text{O}$$

We can now calculate the energy required to vaporize the water.

$$\frac{40.6\ \text{kJ}}{\cancel{\text{mol H}_2\text{O}}} \times 1.4\ \cancel{\text{mol H}_2\text{O}} = 57\ \text{kJ}$$

Molar heat of vaporization; Moles of water

The total energy is the sum of the two steps.

$$7.8\ \text{kJ} + 57\ \text{kJ} = 65\ \text{kJ}$$

Heat from 25 °C to 100. °C; Change to vapor

✔ Self-Check Exercise 13.1

Calculate the total energy required to melt 15 g of ice at 0 °C, heat the water to 100. °C, and vaporize it to steam at 100. °C.

HINT: Break the process into three steps and then take the sum.

See Problems 13.15 through 13.18. ■

13.3 Intermolecular Forces

AIMS: To learn about dipole–dipole attraction, hydrogen bonding, and London dispersion forces. To understand the effect of these forces on the properties of liquids.

We have seen that covalent bonding forces within molecules arise from the sharing of electrons, but how do intermolecular forces arise? Actually several types of intermolecular forces exist. To illustrate one type, we will consider the forces that exist among water molecules.

The polarity of a molecule was discussed in Section 11.3.

As we saw in Chapter 11, water is a polar molecule—it has a dipole moment. When molecules with dipole moments are put together, they orient themselves to take advantage of their charge distributions. Molecules with dipole moments can attract each other by lining up so that the positive and negative ends are close to each other, as shown in Figure 13.5a. This is called a **dipole–dipole attraction.** In the liquid, the dipoles find the best compromise between attraction and repulsion, as shown in Figure 13.5b.

Dipole–dipole forces are typically only about 1% as strong as covalent or ionic bonds, and they become weaker as the distance between the dipoles increases. In the gas phase, where the molecules are usually very far apart, these forces are relatively unimportant.

See Section 11.2 for a discussion of electronegativity.

Particularly strong dipole–dipole forces occur between molecules in which hydrogen is bound to a highly electronegative atom, such as nitrogen, oxygen, or fluorine. Two factors account for the strengths of these interactions: the great polarity of the bond and the close approach of the dipoles, which is made possible by the very small size of the hydrogen atom. Because dipole–dipole attractions of this type are so unusually strong, they are given a special name—**hydrogen bonding.** Figure 13.6 illustrates hydrogen bonding among water molecules.

Hydrogen bonding has a very important effect on various physical properties. For example, the boiling points for the covalent compounds of

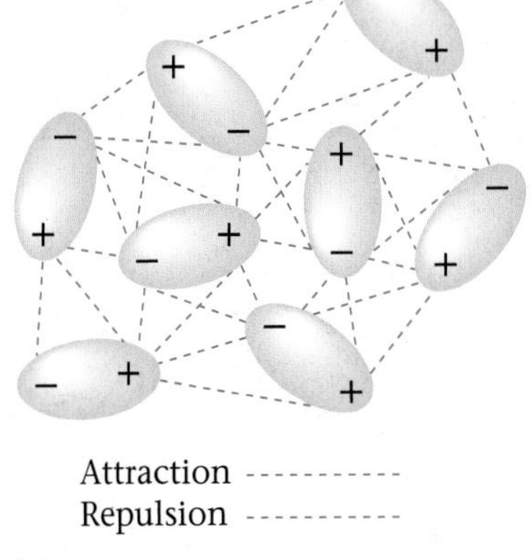

Figure 13.5
(a) The interaction of two polar molecules. (b) The interaction of many dipoles in a liquid.

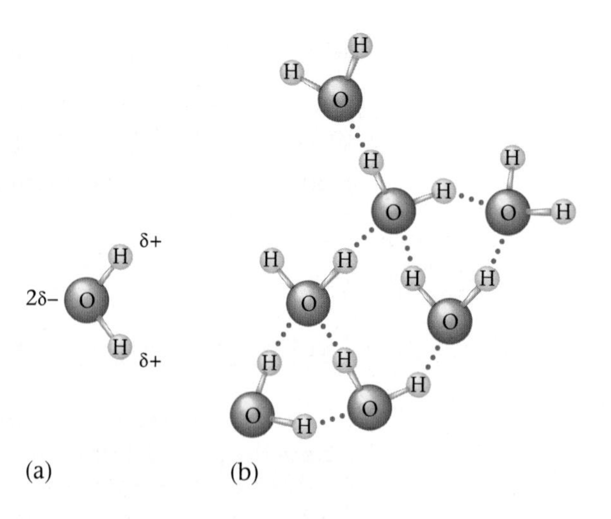

Figure 13.6
(a) The polar water molecule. (b) Hydrogen bonding among water molecules. The small size of the hydrogen atoms allows the molecules to get very close and thus to produce strong interactions.

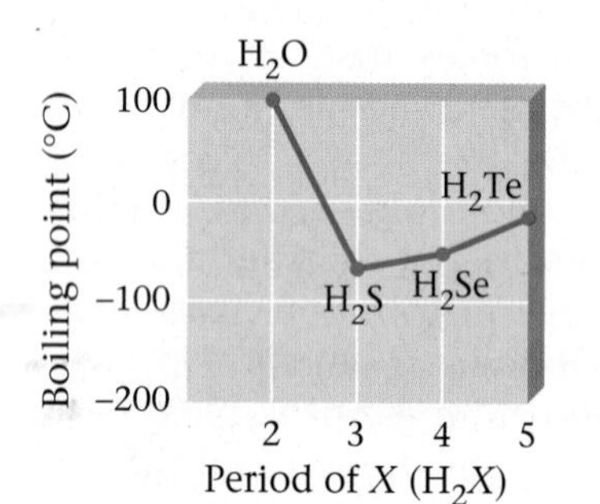

Figure 13.7
The boiling points of the covalent hydrides of elements in Group 6.

hydrogen with the elements in Group 6 are given in Figure 13.7. Note that the boiling point of water is much higher than would be expected from the trend shown by the other members of the series. Why? Because the especially large electronegativity value of the oxygen atom compared with that of the other group members causes the O—H bonds to be much more polar than the S—H, Se—H, or Te—H bonds. This leads to very strong hydrogen-bonding forces among the water molecules. An unusually large quantity of energy is required to overcome these interactions and separate the molecules to produce the gaseous state. That is, water molecules tend to remain together in the liquid state even at relatively high temperatures—hence the very high boiling point of water.

However, even molecules without dipole moments must exert forces on each other. We know this because all substances—even the noble gases—exist in the liquid and solid states at very low temperatures. There must be forces to hold the atoms or molecules as close together as they are in these condensed states. The forces that exist among noble gas atoms and nonpolar molecules are called **London dispersion forces.** To understand the origin of these forces, consider a pair of noble gas atoms. Although we usually assume that the electrons of an atom are uniformly distributed about the nucleus (see Figure 13.8a), this is apparently not true at every instant.

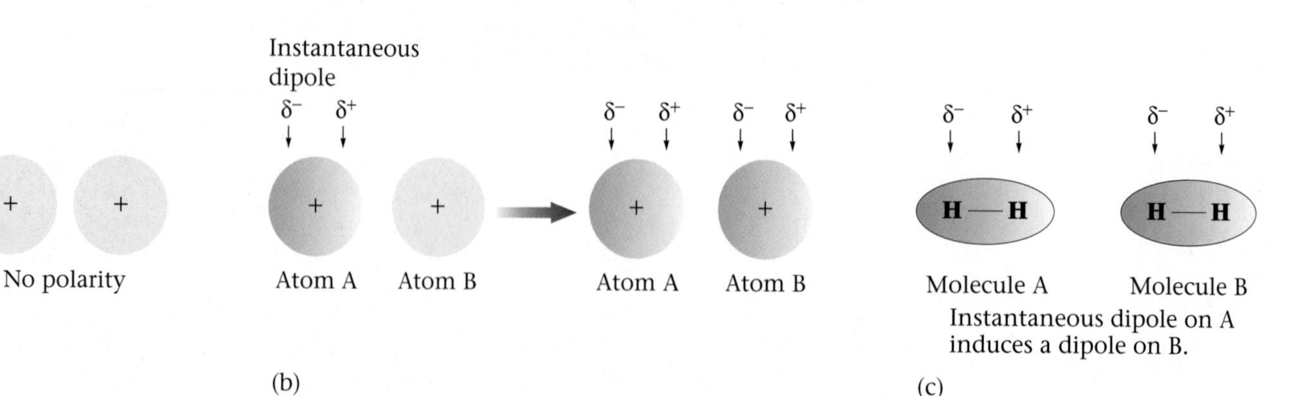

Figure 13.8
(a) Two atoms with spherical electron probability. These atoms have no polarity. (b) The atom on the left develops an instantaneous dipole when more electrons happen to congregate on the left than on the right. (c) Nonpolar molecules also interact by developing instantaneous dipoles.

Table 13.2 The Freezing Points of the Group 8 Elements

Element	Freezing Point (°C)
helium*	−272.0 (25 atm)
neon	−248.6
argon	−189.4
krypton	−157.3
xenon	−111.9

*Helium will not freeze unless the pressure is increased above 1 atm.

Atoms can develop a temporary dipolar arrangement of charge as the electrons move around the nucleus (see Figure 13.8b). This *instantaneous dipole* can then *induce* a similar dipole in a neighboring atom, as shown in Figure 13.8c. The interatomic attraction thus formed is both weak and short-lived, but it can be very significant for large atoms and large molecules, as we will see.

The motions of the atoms must be greatly slowed down before the weak London dispersion forces can lock the atoms into place to produce a solid. This explains, for instance, why the noble gas elements have such low freezing points (see Table 13.2).

Nonpolar molecules such as H_2, N_2, and I_2, none of which has a permanent dipole moment, also attract each other by London dispersion forces (see Figure 13.8d). London forces become more significant as the sizes of atoms or molecules increase. Larger size means there are more electrons available to form the dipoles.

Evaporation and Vapor Pressure

AIM: To understand the relationship among vaporization, condensation, and vapor pressure.

We all know that a liquid can evaporate from an open container. This is clear evidence that the molecules of a liquid can escape the liquid's surface and form a gas. This process, which is called **vaporization** or **evaporation,** requires energy to overcome the relatively strong intermolecular forces in the liquid.

Water is used to absorb heat from nuclear reactors. The water is then cooled in cooling towers before it is returned to the environment.

The fact that vaporization requires energy has great practical significance; in fact, one of the most important roles that water plays in our world is to act as a coolant. Because of the strong hydrogen bonding among its molecules in the liquid state, water has an unusually large heat of vaporization (41 kJ/mol). A significant portion of the sun's energy is spent evaporating water from the oceans, lakes, and rivers rather than warming the earth. The vaporization of water is also crucial to our body's temperature-control system, which relies on the evaporation of perspiration.

Vapor Pressure

Vapor, not *gas,* is the term we customarily use for the gaseous state of a substance that exists naturally as a solid or liquid at 25 °C and 1 atm.

A system at equilibrium is dynamic on the molecular level, but shows no visible changes.

When we place a given amount of liquid in a container and then close it, we observe that the amount of liquid at first decreases slightly but eventually becomes constant. The decrease occurs because there is a transfer of molecules from the liquid to the vapor phase (Figure 13.9). However, as the number of vapor molecules increases, it becomes more and more likely that some of them will return to the liquid. The process by which vapor molecules form a liquid is called **condensation.** Eventually, the same number of molecules are leaving the liquid as are returning to it: the rate of condensation equals the rate of evaporation. *At this point no further change occurs in the amounts of liquid or vapor, because the two opposite processes exactly balance each other;* the system is at *equilibrium.* Note that this system is highly *dynamic* on the molecular level—molecules are constantly escaping from and entering the liquid. However, there is no *net* change because the two opposite processes just *balance* each other. As an analogy, consider two island cities connected by a bridge. Suppose the traffic flow on the bridge is the same in both directions. There is motion—we can see the cars traveling across the bridge—but the number of cars in each city is not changing because an equal number enter and leave each one. The result is no *net* change in the number of autos in each city: an equilibrium exists.

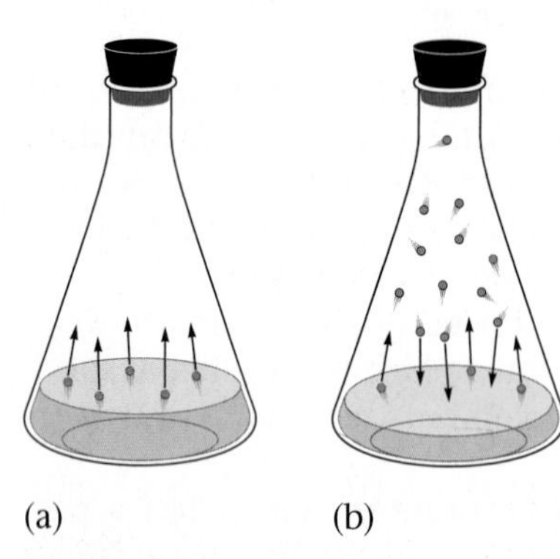

Figure 13.9
Behavior of a liquid in a closed container. (a) Net evaporation occurs at first, so the amount of liquid decreases slightly. (b) As the number of vapor molecules increases, the rate of condensation increases. Finally the rate of condensation equals the rate of evaporation. The system is at equilibrium.

The pressure of the vapor present at equilibrium with its liquid is called the *equilibrium vapor pressure* or, more commonly, the **vapor pressure** of the liquid. A simple barometer can be used to measure the vapor pressure of a liquid, as shown in Figure 13.10. Because mercury is so dense, any common liquid injected at the bottom of the column of mercury floats to the top, where it produces a vapor, and the pressure of this vapor pushes some mercury out of the tube. When the system reaches equilibrium, the vapor pressure can be determined from the change in the height of the mercury column.

In effect, we are using the space above the mercury in the tube as a closed container for each liquid. However, in this case as the liquid vaporizes, the vapor formed creates a pressure that pushes some mercury out of the tube and lowers the mercury level. The mercury level stops changing when the excess liquid floating on the mercury comes to equilibrium with the vapor. The change in the mercury level (in millimeters) from its initial

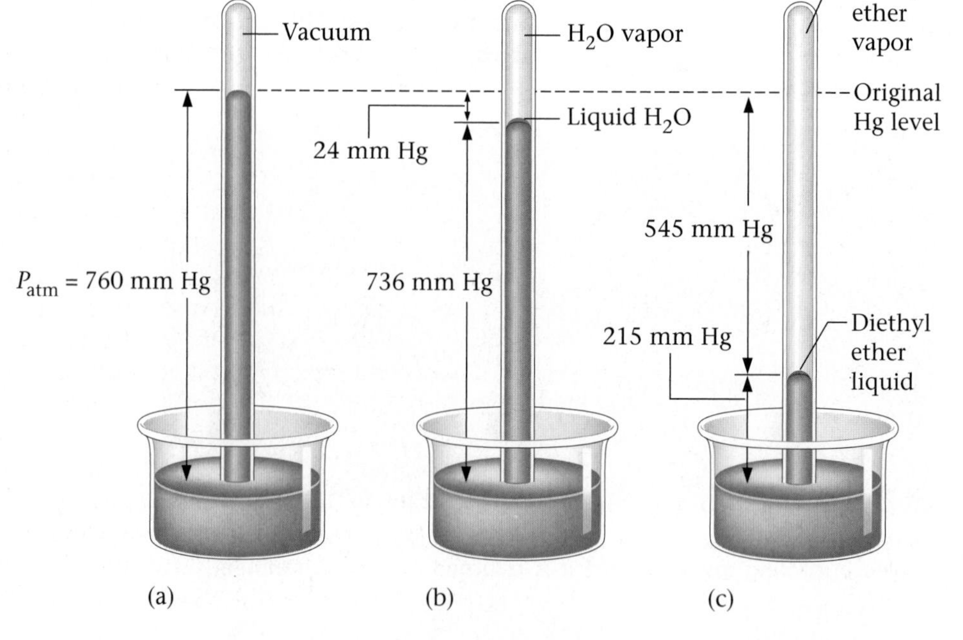

Figure 13.10
(a) It is easy to measure the vapor pressure of a liquid by using a simple barometer of the type shown here. (b) The water vapor pushed the mercury level down 24 mm (760 − 736), so the vapor pressure of water is 24 mm Hg at this temperature. (c) Diethyl ether is much more volatile than water and thus shows a higher vapor pressure. In this case, the mercury level has been pushed down 545 mm (760 − 215), so the vapor pressure of diethyl ether is 545 mm Hg at this temperature.

position (before the liquid was injected) to its final position is equal to the vapor pressure of the liquid.

The vapor pressures of liquids vary widely (see Figure 13.10). Liquids with high vapor pressures are said to be *volatile*—they evaporate rapidly.

The vapor pressure of a liquid at a given temperature is determined by the *intermolecular forces* that act among the molecules. Liquids in which the intermolecular forces are large have relatively low vapor pressures, because such molecules need high energies to escape to the vapor phase. For example, although water is a much smaller molecule than diethyl ether, $C_2H_5—O—C_2H_5$, the strong hydrogen-bonding forces in water cause its vapor pressure to be much lower than that of ether (see Figure 13.10).

Example 13.3 Using Knowledge of Intermolecular Forces to Predict Vapor Pressure

Predict which substance in each of the following pairs will show the largest vapor pressure at a given temperature.

a. $H_2O(l)$, $CH_3OH(l)$

b. $CH_3OH(l)$, $CH_3CH_2CH_2CH_2OH(l)$

Solution

a. Water contains two polar O—H bonds; methanol (CH_3OH) has only one. Therefore, the hydrogen bonding among H_2O molecules is expected to be much stronger than that among CH_3OH molecules. This gives water a lower vapor pressure than methanol.

b. Each of these molecules has one polar O—H bond. However, because $CH_3CH_2CH_2CH_2OH$ is a much larger molecule than CH_3OH, it has much greater London forces and thus is less likely to escape from its liquid. Thus $CH_3CH_2CH_2CH_2OH(l)$ has a lower vapor pressure than $CH_3OH(l)$. ■

13.5 The Solid State: Types of Solids

AIM: To learn about the various types of crystalline solids.

Solids play a very important role in our lives. The concrete we drive on, the trees that shade us, the windows we look through, the paper that holds this print, the diamond in an engagement ring, and the plastic lenses in eyeglasses are all important solids. Most solids, such as wood, paper, and glass, contain mixtures of various components. However, some natural solids, such as diamonds and table salt, are nearly pure substances.

Many substances form **crystalline solids**—those with a regular arrangement of their components. This is illustrated by the partial structure of sodium chloride shown in Figure 13.11. The highly ordered arrangement of the components in a crystalline solid produces beautiful, regularly shaped crystals such as those shown in Figure 13.12.

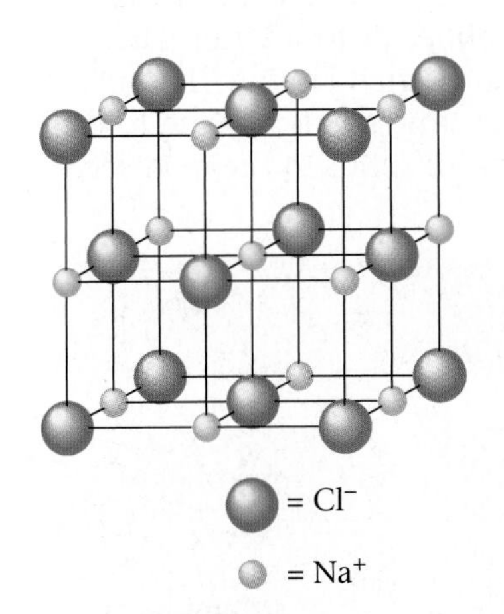

Figure 13.11
The regular arrangement of sodium and chloride ions in sodium chloride, a crystalline solid.

There are many different types of crystalline solids. For example, both sugar and salt have beautiful crystals that we can easily see. However, although both dissolve readily in water, the properties of the resulting solutions are quite different. The salt solution readily conducts an electric current; the sugar solution does not. This behavior arises from the different natures of the components in these two solids. Common salt, NaCl, is an ionic solid that contains Na^+ and Cl^- ions. When solid sodium chloride dissolves in water, sodium ions and chloride ions are distributed through-

(a) (b) (c)

Figure 13.12
Several crystalline solids: (a) quartz, SiO_2; (b) rock salt, NaCl; and (c) iron pyrite, FeS_2.

out the resulting solution. These ions are free to move through the solution to conduct an electric current. Table sugar (sucrose), on the other hand, is composed of neutral molecules that are dispersed throughout the water when the solid dissolves. No ions are present, and the resulting solution does not conduct electricity. These examples illustrate two important types of crystalline solids: **ionic solids,** represented by sodium chloride; and **molecular solids,** represented by sucrose.

A third type of crystalline solid is represented by elements such as graphite and diamond (both pure carbon), boron, silicon, and all metals. These substances, which contain atoms of only one element covalently bonded to each other, are called **atomic solids.**

We have seen that crystalline solids can be grouped conveniently into three classes as shown in Figure 13.13. Notice that the names of the three classes come from the components of the solid. An ionic solid contains ions, a molecular solid contains molecules, and an atomic solid contains atoms. Examples of the three types of solids are shown in Figure 13.14.

The internal forces in a solid determine many of the properties of the solid.

The properties of a solid are determined primarily by the nature of the forces that hold the solid together. For example, although argon, copper, and diamond are all atomic solids (their components are atoms), they have strikingly different properties. Argon has a very low melting point (−189 °C), whereas diamond and copper melt at high temperatures (about 3500 °C and 1083 °C, respectively). Copper is an excellent conductor of electricity (it is widely used for electrical wires), whereas both argon and diamond are insulators. The shape of copper can easily be changed; it is both malleable (will form thin sheets) and ductile (can be pulled into a wire). Diamond, on the other hand, is the hardest natural substance known. The marked differences in properties among these three atomic solids are due to differences in bonding. We will explore the bonding in solids in the next section.

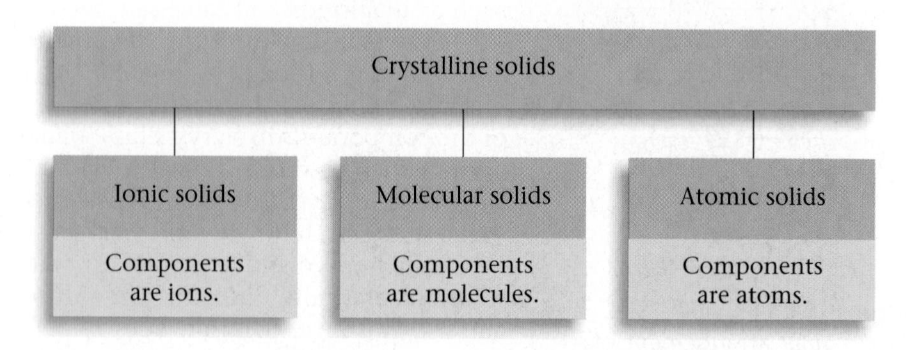

Figure 13.13
The classes of crystalline solids.

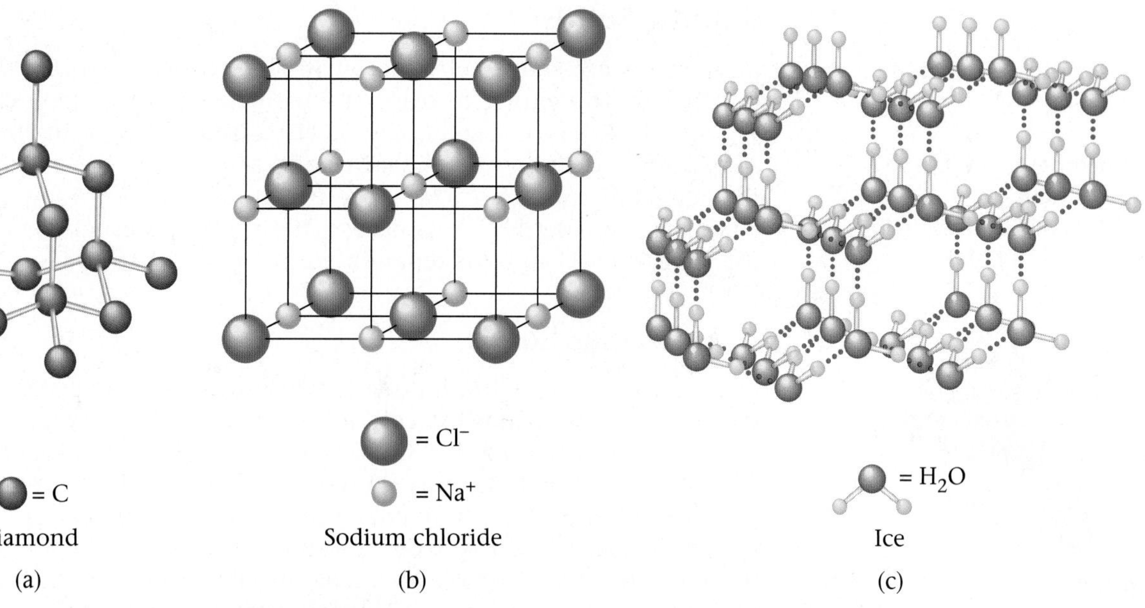

Figure 13.14
Examples of three types of crystalline solids. Only part of the structure is shown in each case. The structures continue in three dimensions with the same patterns. (a) An atomic solid. Each sphere represents a carbon atom in diamond. (b) An ionic solid. The spheres represent alternating Na^+ and Cl^- ions in solid sodium chloride. (c) A molecular solid. Each unit of three spheres represents an H_2O molecule in ice. The dashed lines show the hydrogen bonding among the polar water molecules.

Bonding in Solids

AIMS: To understand the interparticle forces in crystalline solids. To learn about how the bonding in metals determines metallic properties.

We have seen that crystalline solids can be divided into three classes, depending on the fundamental particle or unit of the solid. Ionic solids consist of oppositely charged ions packed together, molecular solids contain molecules, and atomic solids have atoms as their fundamental particles. Examples of the various types of solids are given in Table 13.3.

Table 13.3 Examples of the Various Types of Solids

Type of Solid	Examples	Fundamental Unit(s)
ionic	sodium chloride, $NaCl(s)$	Na^+, Cl^- ions
ionic	ammonium nitrate, $NH_4NO_3(s)$	NH_4^+, NO_3^- ions
molecular	dry ice, $CO_2(s)$	CO_2 molecules
molecular	ice, $H_2O(s)$	H_2O molecules
atomic	diamond, $C(s)$	C atoms
atomic	iron, $Fe(s)$	Fe atoms
atomic	argon, $Ar(s)$	Ar atoms

Ionic Solids

Ionic solids were also discussed in Section 11.5.

When spheres are packed together, there are many small empty spaces (holes) left among the spheres.

Ionic solids are stable substances with high melting points that are held together by the strong forces that exist between oppositely charged ions. The structures of ionic solids can be visualized best by thinking of the ions as spheres packed together as efficiently as possible. For example, in NaCl the larger Cl^- ions are packed together much like one would pack balls in a box. The smaller Na^+ ions occupy the small spaces ("holes") left among the spherical Cl^- ions, as represented in Figure 13.15.

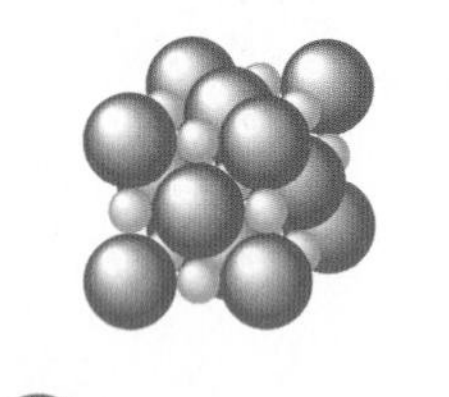

= Cl^- = Na^+

Figure 13.15
The packing of Cl^- and Na^+ ions in solid sodium chloride.

Molecular Solids

In a molecular solid the fundamental particle is a molecule. Examples of molecular solids include ice (contains H_2O molecules), dry ice (contains CO_2 molecules), sulfur (contains S_8 molecules), and white phosphorus (contains P_4 molecules). The latter two substances are shown in Figure 13.16.

Molecular solids tend to melt at relatively low temperatures because the intermolecular forces that exist among the molecules are relatively weak. If the molecule has a dipole moment, dipole–dipole forces hold the solid together. In solids with nonpolar molecules, London dispersion forces hold the solid together.

Part of the structure of solid phosphorus is represented in Figure 13.17. Note that the distances between P atoms in a given molecule are much shorter than the distances between the P_4 molecules. This is because the covalent bonds *between atoms* in the molecule are so much stronger than the London dispersion forces *between molecules*.

Atomic Solids

The properties of atomic solids vary greatly because of the different ways in which the fundamental particles, the atoms, can interact with each other. For example, the solids of the Group 8 elements have very low melting points (see Table 13.2), because these atoms, having filled valence orbitals, cannot form covalent bonds with each other. So the forces in these solids are the relatively weak London dispersion forces.

On the other hand, diamond, a form of solid carbon, is one of the hardest substances known and has an extremely high melting point (about 3500 °C). The incredible hardness of diamond arises from the very strong covalent carbon–carbon bonds in the crystal, which lead to a giant mole-

Figure 13.16
(Left) Sulfur crystals contain S_8 molecules. (Right) White phosphorus contains P_4 molecules. It is so reactive with the oxygen in air that it must be stored under water.

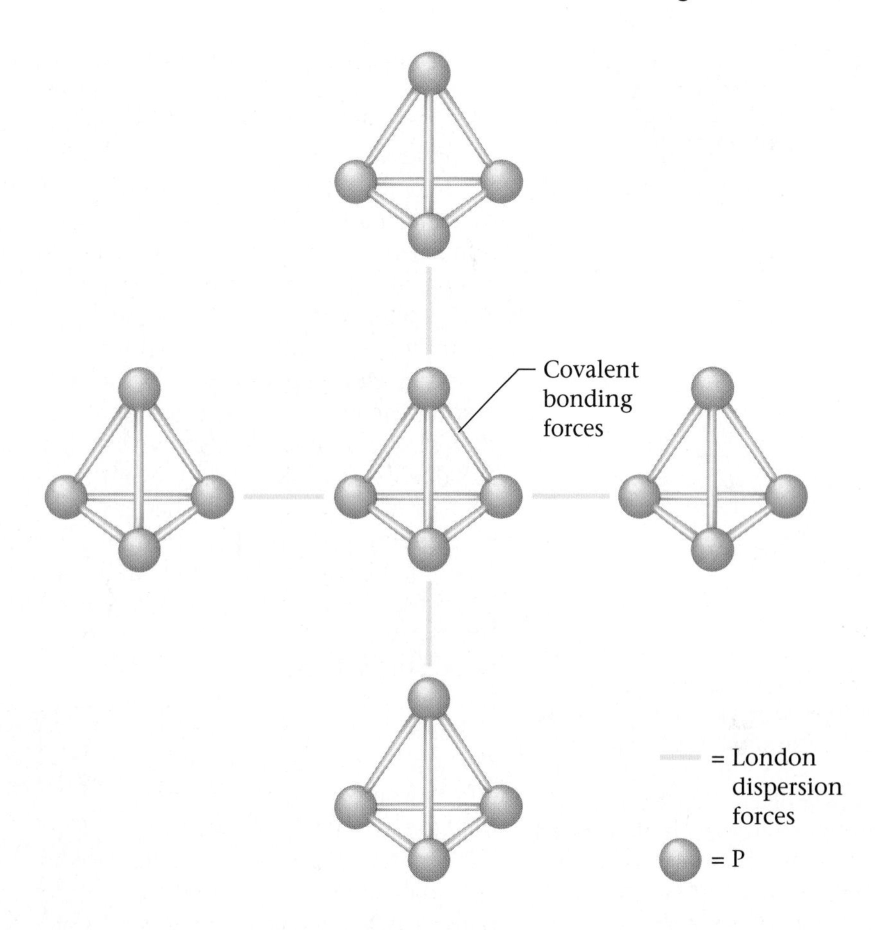

Figure 13.17
A representation of part of the structure of solid phosphorus, a molecular solid that contains P_4 molecules.

cule. In fact, the entire crystal can be viewed as one huge molecule. A small part of the diamond structure is represented in Figure 13.14. In diamond each carbon atom is bound covalently to four other carbon atoms to produce a very stable solid. Several other elements also form solids whereby the atoms join together covalently to form giant molecules. Silicon and boron are examples.

At this point you might be asking yourself, "Why aren't solids such as a crystal of diamond, which is a 'giant molecule,' classified as molecular solids?" The answer is that, by convention, a solid is classified as a molecular solid only if (like ice, dry ice, sulfur, and phosphorus) it contains small molecules. Substances like diamond that contain giant molecules are called network solids.

Bonding in Metals

Metals represent another type of atomic solid. Metals have familiar physical properties: they can be pulled into wires, can be hammered into sheets, and are efficient conductors of heat and electricity. However, although the shapes of most pure metals can be changed relatively easily, metals are also durable and have high melting points. These facts indicate that it is difficult to separate metal atoms but relatively easy to slide them past each other. In other words, the bonding in most metals is *strong* but *nondirectional.*

The simplest picture that explains these observations is the **electron sea model,** which pictures a regular array of metal atoms in a "sea" of valence electrons that are shared among the atoms in a nondirectional way and that are quite mobile in the metal crystal. The mobile electrons can conduct heat and electricity, and the atoms can be moved rather easily, as, for example, when the metal is hammered into a sheet or pulled into a wire.

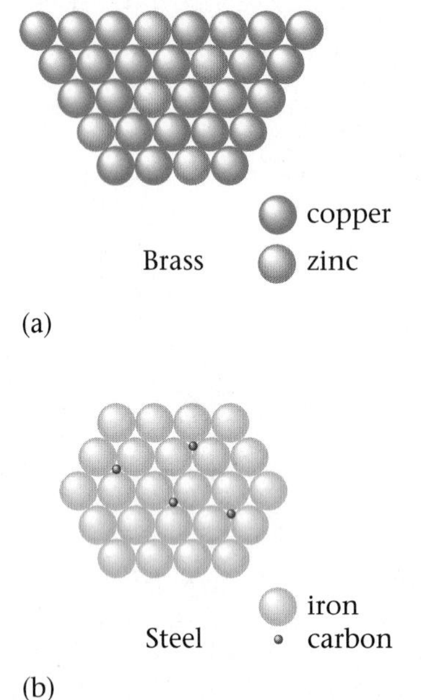

Figure 13.18
Two types of alloys. (a) Brass is a substitutional alloy in which copper atoms in the host crystal are replaced by the similarly sized zinc atoms. (b) Steel is an interstitial alloy in which carbon atoms occupy interstices (holes) among the closely packed iron atoms.

Because of the nature of the metallic crystal, other elements can be introduced relatively easily to produce substances called alloys. An **alloy** is best defined as *a substance that contains a mixture of elements and has metallic properties*. There are two common types of alloys.

In a **substitutional alloy** some of the host metal atoms are *replaced* by other metal atoms of similar sizes. For example, in brass approximately one-third of the atoms in the host copper metal have been replaced by zinc atoms, as shown in Figure 13.18a. Sterling silver (93% silver and 7% copper) and pewter (85% tin, 7% copper, 6% bismuth, and 2% antimony) are other examples of substitutional alloys.

An **interstitial alloy** is formed when some of the interstices (holes) among the closely packed metal atoms are occupied by atoms much smaller than the host atoms, as shown in Figure 13.18b. Steel, the best-known interstitial alloy, contains carbon atoms in the "holes" of an iron crystal. The presence of interstitial atoms changes the properties of the host metal. Pure iron is relatively soft, ductile, and malleable because of the absence of strong directional bonding. The spherical metal atoms can be moved rather easily with respect to each other. However, when carbon, which forms strong directional bonds, is introduced into an iron crystal, the presence of the directional carbon–iron bonds makes the resulting alloy harder, stronger, and less ductile than pure iron. The amount of carbon directly affects the properties of steel. *Mild steels* (containing less than 0.2% carbon) are still ductile and malleable and are used for nails, cables, and chains. *Medium steels* (containing 0.2–0.6% carbon) are harder than mild steels and are used in rails and structural steel beams. *High-carbon steels* (containing 0.6–1.5% carbon) are tough and hard and are used for springs, tools, and cutlery.

A steel sculpture in Chicago.

Many types of steel contain other elements in addition to iron and carbon. Such steels are often called *alloy steels* and can be viewed as being mixed interstitial (carbon) and substitutional (other metals) alloys. An example is stainless steel, which has chromium and nickel atoms substituted for some of the iron atoms. The addition of these metals greatly increases the steel's resistance to corrosion.

Example 13.4 Identifying Types of Crystalline Solids

Name the type of crystalline solid formed by each of the following substances:

a. ammonia
b. iron
c. cesium fluoride
d. argon
e. sulfur

CHEMISTRY IN FOCUS

Metal with a Memory

A distraught mother walks into the optical shop carrying her mangled pair of $400 eyeglasses. Her child had gotten into her purse, found her glasses, and twisted them into a pretzel. She hands them to the optometrist with little hope that they can be salvaged. The optometrist says not to worry and drops the glasses into a dish of warm water where the glasses magically spring back to their original shape. The optometrist hands the restored glasses to the woman and says there is no charge for repairing them.

How can the frames "remember" their original shape when placed in warm water? The answer is a nickel–titanium alloy called Nitinol that was developed in the late 1950s and early 1960s at the Naval Ordnance Laboratory in White Oak, Maryland, by William J. Buehler. (The name Nitinol comes from *Ni*ckel *Ti*tanium *N*aval *O*rdnance *L*aboratory.)

Nitinol has the amazing ability to remember a shape originally impressed in it. For example, note the accompanying photos. What causes Nitinol to behave this way? Although the details are too complicated to describe here, this phenomenon results from two different forms of solid Nitinol. When Nitinol is heated to a sufficiently high temperature, the Ni and Ti atoms arrange themselves in a way that leads to the most compact and regular pattern of the atoms—a form called austenite (A). When the alloy is cooled, its atoms rearrange slightly to a form called martensite (M). The shape desired (for example, the word *ICE*) is set into the alloy at a high temperature (A form), then the metal is cooled, causing it to assume the M form. In this process no visible change is noted. Then, if the image is deformed, it will magically return if the alloy is heated (hot water works fine) to a temperature that changes it back to the A form.

Nitinol has many medical applications, including hooks used by orthopedic surgeons to attach ligaments and tendons to bone and "baskets" to catch blood clots. In the latter case a length of Nitinol wire is shaped into a tiny basket and this shape is set at a high temperature. The wires forming the basket are then straightened so they can be inserted as a small bundle through a catheter. When the wires warm up in the blood, the basket shape springs back and acts as a filter to stop blood clots from moving to the heart.

One of the most promising consumer uses of Nitinol is for eyeglass frames. Nitinol is also now being used for braces to straighten crooked teeth.

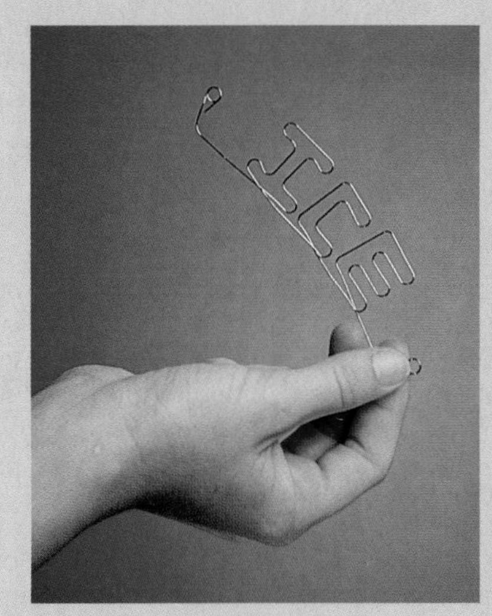

The word *ICE* is formed from Nitinol wire.

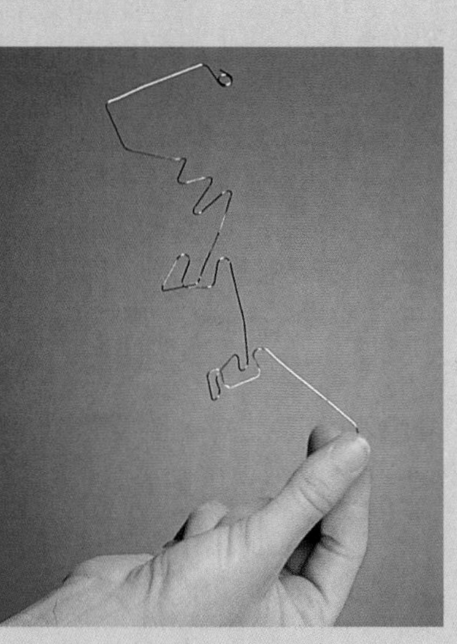

The wire is stretched to obliterate the word *ICE*.

The wire pops back to *ICE* when immersed in warm water.

Solution

a. Solid ammonia contains NH_3 molecules, so it is a molecular solid.

b. Solid iron contains iron atoms as the fundamental particles. It is an atomic solid.

c. Solid cesium fluoride contains the Cs^+ and F^- ions. It is an ionic solid.

d. Solid argon contains argon atoms, which cannot form covalent bonds to each other. It is an atomic solid.

e. Sulfur contains S_8 molecules, so it is a molecular solid.

✔ Self-Check Exercise 13.2

Name the type of crystalline solid formed by each of the following substances:

a. sulfur trioxide

b. barium oxide

c. gold

See Problems 13.41 and 13.42. ■

Chapter 13 Review

Key Terms

normal boiling point (13.1)
heating/cooling curve (13.1)
normal freezing point (13.1)
intramolecular forces (13.2)
intermolecular forces (13.2)
molar heat of fusion (13.2)
molar heat of vaporization (13.2)
dipole–dipole attraction (13.3)
hydrogen bonding (13.3)
London dispersion forces (13.3)
vaporization (evaporation) (13.4)
condensation (13.4)
vapor pressure (13.4)
crystalline solid (13.5)
ionic solid (13.5)
molecular solid (13.5)
atomic solid (13.5)
electron sea model (13.6)
alloy (13.6)
substitutional alloy (13.6)
interstitial alloy (13.6)

Summary

1. Liquids and solids exhibit some similarities and are very different from the gaseous state.

2. The temperature at which a liquid changes its state to a gas (at 1 atm pressure) is called the normal boiling point of that liquid. Similarly, the temperature at which a liquid freezes (at 1 atm pressure) is the normal freezing point. Changes of state are physical changes, not chemical changes.

3. To convert a substance from the solid to the liquid and then to the gaseous state requires the addition of energy. Forces among the molecules in a solid or a liquid must be overcome by the input of energy. The energy required to melt 1 mol of a substance is called the molar heat of fusion, and the energy required to change 1 mol of liquid to the gaseous state is called the molar heat of vaporization.

4. There are several types of intermolecular forces. Dipole–dipole interactions occur when molecules with dipole moments attract each other. A particularly strong dipole–dipole interaction called hydrogen bonding occurs in molecules that contain hydrogen bonded to a very electronegative element such as N, O, or F. London dispersion forces occur when instantaneous dipoles in atoms or nonpolar molecules lead to relatively weak attractions.

5. The change of a liquid to its vapor is called vaporization or evaporation. The process whereby vapor molecules form a liquid is called condensation. In a closed container, the pressure of the vapor over its liquid reaches a constant value called the vapor pressure of the liquid.

6. Many solids are crystalline (contain highly regular arrangements of their components). The three types of crystalline solids are ionic, molecular, and atomic solids. In ionic solids, the ions are packed together in a way that maximizes the attractions of oppositely charged ions and minimizes the repulsions among identically charged ions. Molecular solids are held

together by dipole–dipole attractions if the molecules are polar and by London dispersion forces if the molecules are nonpolar. Atomic solids are held together by covalent bonding forces or London dispersion forces, depending on the atoms present.

In-Class Discussion Questions

These questions are designed to be considered by groups of students in class. Often these questions work well for introducing a particular topic in class.

1. You seal a container half-filled with water. Which best describes what occurs in the container?
   a. Water evaporates until the air becomes saturated with water vapor; at this point, no more water evaporates.
   b. Water evaporates until the air becomes overly saturated (supersaturated) with water, and most of this water recondenses; this cycle continues until a certain amount of water vapor is present, and then the cycle ceases.
   c. The water does not evaporate because the container is sealed.
   d. Water evaporates, and then water evaporates and recondenses simultaneously and continuously.
   e. The water evaporates until it is eventually all in vapor form.

   Justify your choice and for choices you did not pick, explain what is wrong with them.
2. Explain the following: You add 100 mL of water to a 500-mL round-bottomed flask and heat the water until it is boiling. You remove the heat and stopper the flask, and the boiling stops. You then run cool water over the neck of the flask, and the boiling begins again. It seems as though you are boiling water by cooling it.
3. Is it possible for the dispersion forces in a particular substance to be stronger than hydrogen-bonding forces in another substance? Explain your answer.
4. Does the nature of intermolecular forces change when a substance goes from a solid to a liquid, or from a liquid to a gas? What causes a substance to undergo a phase change?
5. Generally, the vapor pressure of a liquid is related to (there may be more than one answer):
   a. the amount of liquid
   b. atmospheric pressure
   c. temperature
   d. intermolecular forces

   Explain your answer.
6. Why do liquids have a vapor pressure? Do all liquids have vapor pressures? Explain.
7. Do solids exhibit vapor pressure? Explain.
8. How does vapor pressure change with changing temperature? Explain.
9. What occurs when the vapor pressure of a liquid is equal to atmospheric pressure? Explain.
10. Water in an open beaker evaporates over time. As the water is evaporating, is the vapor pressure increasing, decreasing, or staying the same? Why?
11. What is the vapor pressure of water at 100 °C? How do you know?
12. How do the following physical properties depend on the strength of intermolecular forces? Explain.
    a. melting point
    b. boiling point
    c. vapor pressure
13. Look at Figure 13.2. Why doesn't temperature increase continuously over time? That is, why does the temperature stay constant for periods of time?
14. Which are stronger, intermolecular or intramolecular forces for a given molecule? What observation(s) have you made that supports this position? Explain.
15. Why does water evaporate at all?

Questions and Problems

All even-numbered exercises have answers in the back of this book and solutions in the *Solutions Guide*.

13.1 Water and Its Phase Changes

QUESTIONS

1. Summarize the *similarities* and the *differences* that exist among the three states of matter.
2. Describe some uses, both in nature and in industry, of water as a *cooling* agent.
3. Water is unusual, in that its solid form (ice) is *less* dense than its liquid form. Discuss some implications of this fact.
4. Discuss some implications of the fact that, unlike most substances, water *expands* in volume when it freezes.
5. Describe, on both a microscopic and a macroscopic basis, what happens to a sample of water as it is heated from room temperature to 50 °C above its normal boiling point.
6. Figure 13.2 presents the *cooling curve* for water. Discuss the meaning of the different portions of this curve (for example, explain what each flat section and each sloping section represents).

13.2 Energy Requirements for the Changes of State

QUESTIONS

7. Describe in detail the microscopic processes that take place when a solid melts.
8. Describe in detail the microscopic processes that take place when a liquid boils.

9. Explain the difference between *intra*molecular and *inter*molecular forces.

10. Which type of forces (*intra*molecular or *inter*molecular) must be overcome to melt a solid or vaporize a liquid?

11. Discuss the similarities and differences between the arrangements of molecules and the forces between molecules in liquid water versus steam, and in liquid water versus ice.

12. Water at 100 °C (its normal boiling point) could certainly give you a bad burn if it was spilled on the skin, but steam at 100 °C can give you a much *worse* burn. Explain.

PROBLEMS

13. The following data have been collected for substance X. Construct a heating curve for substance X. (The drawing does not need to be absolutely to scale, but it should clearly show relative differences.)

normal melting point	−15 °C
molar heat of fusion	2.5 kJ/mol
normal boiling point	134 °C
molar heat of vaporization	55.3 kJ/mol

14. Consider the data for substance X given in Problem 13. If the molar mass of substance X is 52 g/mol, what quantity of heat is required to melt 10.0 g of substance X at −15 °C? What quantity of heat is required to vaporize 25.0 g of substance X at 134 °C?

15. The molar heat of fusion of benzene is 9.92 kJ/mol. Its molar heat of vaporization is 30.7 kJ/mol. Calculate the heat required to melt 8.25 g of benzene at its normal melting point. Calculate the heat required to vaporize 8.25 g of benzene at its normal boiling point. Why is the heat of vaporization more than three times the heat of fusion?

16. The molar heats of fusion and vaporization for water are 6.02 kJ/mol and 40.6 kJ/mol, respectively, and the specific heat capacity of liquid water is 4.18 J/g °C. What quantity of heat energy is required to melt 25.0 g of ice at 0 °C? What quantity of heat is required to vaporize 37.5 g of liquid water at 100 °C? What quantity of heat is required to warm 55.2 g of liquid water from 0 °C to 100 °C?

17. Given that the specific heat capacities of ice and steam are 2.06 J/g °C and 2.03 J/g °C, respectively, and considering the information about water given in Problem 16, calculate the total quantity of heat evolved when 10.0 g of steam at 200. °C is condensed, cooled, and frozen to ice at −50. °C.

18. The heat of fusion of aluminum is 3.95 kJ/g. What is the *molar* heat of fusion of aluminum? What quantity of energy is needed to melt 10.0 g of aluminum? What quantity of energy is required to melt 10.0 mol of aluminum?

13.3 Intermolecular Forces

QUESTIONS

19. What is a *dipole–dipole* attraction? Give three examples of liquid substances in which you would expect dipole–dipole attractions to be large.

20. How is the strength of dipole–dipole interactions related to the *distance* between polar molecules? Are dipole–dipole forces short-range or long-range forces?

21. What is meant by *hydrogen bonding?* Give three examples of substances that would be expected to exhibit hydrogen bonding in the liquid state.

22. The normal boiling point of water is unusually high, compared to the boiling points of H_2S, H_2Se, and H_2Te. Explain this observation in terms of the *hydrogen bonding* that exists in water, but that does not exist in the other compounds.

23. Why are the dipole–dipole interactions between polar molecules *not* important in the vapor phase?

24. Although the noble gas elements are monatomic and could not give rise to dipole–dipole forces or hydrogen bonding, these elements still can be liquefied and solidified. Explain.

PROBLEMS

25. What type of intermolecular forces are active in the liquid state of each of the following substances?
a. HI
b. N_2
c. Ar
d. P_4

26. Discuss the types of intermolecular forces acting in the liquid state of each of the following substances.
a. Kr
b. S_8
c. NF_3
d. H_2O

27. The boiling points of the noble gas elements are listed below. Comment on the trend in the boiling points. Why do the boiling points vary in this manner?

He	−272 °C	Kr	−152.3 °C
Ne	−245.9 °C	Xe	−107.1 °C
Ar	−185.7 °C	Rn	−61.8 °C

28. The heats of fusion of three substances are listed below. Explain the trend this list reflects.

HI	2.87 kJ/mol
HBr	2.41 kJ/mol
HCl	1.99 kJ/mol

29. When dry ammonia gas (NH_3) is bubbled into a 125-mL sample of water, the volume of the sample (initially, at least) *decreases* slightly. Suggest a reason for this.

30. When 50 mL of liquid water at 25 °C is added to 50 mL of ethanol (ethyl alcohol), also at 25 °C, the combined volume of the mixture is considerably *less* than 100 mL. Give a possible explanation.

13.4 Evaporation and Vapor Pressure

QUESTIONS

31. Describe, on a microscopic basis, the processes of *evaporation* and *condensation*. Which process requires the input of energy?

32. If you've ever opened a bottle of rubbing alcohol or other solvent on a warm day, you may have heard a little "whoosh" as the vapor that had built up above the liquid escapes. Describe on a microscopic basis how a vapor pressure builds up in a closed container above a liquid. What processes in the container give rise to this phenomenon?

33. What do we mean by a *dynamic equilibrium?* Describe how the development of a vapor pressure above a liquid represents such an equilibrium.

34. Describe an experimental method that could be used to determine the vapor pressure of a volatile liquid.

PROBLEMS

35. Which substance in each pair would be expected to be less volatile? Explain your reasoning.
    a. $C_4H_{10}(l)$ or $H_2O(l)$
    b. $NH_3(l)$ or $NH_2OH(l)$
    c. $CH_3OH(l)$ or $CH_4(l)$

36. Which substance in each pair would be expected to show the largest vapor pressure at a given temperature? Explain your reasoning.
    a. $H_2O(l)$ or $HF(l)$
    b. $CH_3OCH_3(l)$ or $CH_3CH_2OH(l)$
    c. $CH_3OH(l)$ or $CH_3SH(l)$

37. Although water and ammonia differ in molar mass by only one unit, the boiling point of water is over 100 °C higher than that of ammonia. What forces in liquid water that do *not* exist in liquid ammonia could account for this observation?

38. Two molecules that contain the same number of each kind of atom but that have different molecular structures are said to be *isomers* of each other. For example, both ethyl alcohol and dimethyl ether (shown below) have the formula C_2H_6O and are isomers. Based on considerations of intermolecular forces, which substance would you expect to be more volatile? Which would you expect to have the higher boiling point? Explain.

dimethyl ether	ethyl alcohol
$CH_3—O—CH_3$	$CH_3—CH_2—OH$

13.5 The Solid State: Types of Solids

QUESTIONS

39. What are crystalline solids? What kind of microscopic structure do such solids have? How is this microscopic structure reflected in the macroscopic appearance of such solids?

40. On the basis of the smaller units that make up the crystals, cite three types of crystalline solids. For each type of crystalline solid, give an example of a substance that forms that type of solid.

13.6 Bonding in Solids

QUESTIONS

41. How do *ionic* solids differ in structure from *molecular* solids? What are the fundamental particles in each? Give two examples of each type of solid and indicate the individual particles that make up the solids in each of your examples.

42. A common prank on college campuses is to switch the salt and sugar on dining hall tables, which is usually easy because the substances look so much alike. Yet, despite the similarity in their appearance, these two substances differ greatly in their properties, since one is a molecular solid and the other is an ionic solid. How do the properties differ and why?

43. Ionic solids are generally considerably harder than most molecular solids. Explain.

44. Ionic solids typically have melting points hundreds of degrees higher than the melting points of molecular solids. Explain.

45. The forces holding together an ionic solid are much stronger than the forces between particles in a molecular solid. How are these strong forces reflected in the properties of an ionic solid?

46. Ordinary ice (solid water) melts at 0 °C, whereas dry ice (solid carbon dioxide) melts at a much lower temperature. Explain the differences in the melting points of these two substances on the basis of the intermolecular forces involved.

47. What is a *network* solid? Give an example of a network solid and describe the bonding in such a solid. How does a network solid differ from a molecular solid?

48. Ionic solids do not conduct electricity in the solid state, but are strong conductors in the liquid state and when dissolved in water. Explain.

49. What is an *alloy?* Explain the differences in structure between substitutional and interstitial alloys. Give an example of each type.

50. Although steel is made mostly from iron, its properties differ considerably from those of pure iron. For example, steel is fairly flexible and can be bent fairly easily, whereas wrought iron is quite brittle. Explain.

ADDITIONAL PROBLEMS

MATCHING

For exercises 51–60 choose one of the following terms to match the definition or description given.

a. alloy
b. specific heat
c. crystalline solid
d. dipole–dipole attraction
e. equilibrium vapor pressure
f. intermolecular
g. intramolecular
h. ionic solids
i. London dispersion forces
j. molar heat of fusion
k. molar heat of vaporization
l. molecular solids
m. normal boiling point
n. semiconductor

51. boiling point at pressure of 1 atm
52. energy required to melt 1 mol of a substance
53. forces between atoms in a molecule
54. forces between molecules in a solid
55. instantaneous dipole forces for nonpolar molecules
56. lining up of opposite charges on adjacent polar molecules
57. maximum pressure of vapor that builds up in a closed container
58. mixture of elements having metallic properties overall
59. repeating arrangement of component species in a solid
60. solids that melt at relatively low temperatures
61. Given the densities and conditions of ice, liquid water, and steam listed in Table 13.1, calculate the volume of 1.0 g of water under each of these circumstances.
62. In carbon compounds a given group of atoms can often be arranged in more than one way. This means that more than one structure may be possible for the same atoms. For example, both the molecules diethyl ether and 1-butanol have the same number of each type of atom, but they have different structures and are said to be *isomers* of one another.

diethyl ether	$CH_3—CH_2—O—CH_2—CH_3$
1-butanol	$CH_3—CH_2—CH_2—CH_2—OH$

Which substance would you expect to have the larger vapor pressure? Why?

63. Which of the substances in each of the following sets would be expected to have the highest boiling point? Explain why.
a. Ga, KBr, O_2
b. Hg, NaCl, He
c. H_2, O_2, H_2O

64. Which of the substances in each of the following sets would be expected to have the lowest melting point? Explain why.
a. H_2, N_2, O_2
b. Xe, NaCl, C (diamond)
c. Cl_2, Br_2, I_2

65. When a person has a severe fever, one therapy to reduce the fever is an "alcohol rub." Explain how the evaporation of alcohol from the person's skin removes heat energy from the body.
66. What is steel?
67. Some properties of aluminum are summarized in the following list.

normal melting point	658 °C
heat of fusion	3.95 kJ/g
normal boiling point	2467 °C
heat of vaporization	10.52 kJ/g
specific heat of the solid	0.902 J/g °C

a. Calculate the quantity of energy required to heat 1.00 mol of aluminum from 25 °C to its normal melting point.
b. Calculate the quantity of energy required to melt 1.00 mol of aluminum at 658 °C.
c. Calculate the amount of energy required to vaporize 1.00 mol of aluminum at 2467 °C.

68. What are some important uses of water, both in nature and in industry? What is the liquid range for water?
69. Describe, on both a microscopic and a macroscopic basis, what happens to a sample of water as it is cooled from room temperature to 50 °C below its normal freezing point.
70. Cake mixes and other packaged foods that require cooking often contain special directions for use at high elevations. Typically these directions indicate that the food should be cooked longer above 5000 ft. Explain why it takes longer to cook something at higher elevations.
71. Why is there no change in *intra*molecular forces when a solid is melted? Are intramolecular forces stronger or weaker than intermolecular forces?
72. What do we call the energies required, respectively, to melt and to vaporize 1 mol of a substance? Which of these energies is always larger for a given substance? Why?
73. The molar heat of vaporization of carbon disulfide, CS_2, is 28.4 kJ/mol at its normal boiling point of 46 °C. How much energy (heat) is required to vaporize 1.0 g of CS_2 at 46 °C? How much heat is evolved when 50. g of CS_2 is condensed from the vapor to the liquid form at 46 °C?
74. Which is stronger, a dipole–dipole attraction between two molecules or a covalent bond between two atoms within the same molecule? Explain.
75. In order for a liquid to boil, the intermolecular forces in the liquid must be overcome. Based on the types

of intermolecular forces present, arrange the expected boiling points of the liquid states of the following substances in order from lowest to highest: $NaCl(l)$, $He(l)$, $CO(l)$, $H_2O(l)$.

76. What are *London dispersion forces* and how do they arise in a nonpolar molecule? Are London forces typically stronger or weaker than dipole–dipole attractions between polar molecules? Are London forces stronger or weaker than covalent bonds? Explain.

77. Discuss the types of intermolecular forces acting in the liquid state of each of the following substances.
    a. N_2
    b. NH_3
    c. He
    d. CO_2 (linear, nonpolar)

78. Explain how the evaporation of water acts as a coolant for the earth.

79. What do we mean when we say a liquid is *volatile?* Do volatile liquids have large or small vapor pressures? What types of intermolecular forces occur in highly volatile liquids?

80. Although methane, CH_4, and ammonia, NH_3, differ in molar mass by only one unit, the boiling point of ammonia is over 100 °C higher than that of methane (a nonpolar molecule). Explain.

81. Which type of solid is likely to have the highest melting point—an ionic solid, a molecular solid, or an atomic solid? Explain.

82. What types of intermolecular forces exist in a crystal of ice? How do these forces differ from the types of intermolecular forces that exist in a crystal of solid oxygen?

83. Discuss the *electron sea model* for metals. How does this model account for the fact that metals are very good conductors of electricity?

14 Solutions

The colors of soap bubbles depend on the thickness of their walls. ▼

Most of the important chemistry that keeps plants, animals, and humans functioning occurs in aqueous solutions. Even the water that comes out of a tap is not pure water but a solution of various materials in water. For example, tap water may contain dissolved chlorine to disinfect it, dissolved minerals that make it "hard," and traces of many other substances that result from natural and human-initiated pollution. We encounter many other chemical solutions in our daily lives: air, shampoo, orange soda, coffee, gasoline, cough syrup, and many others.

A **solution** is a homogeneous mixture, a mixture in which the components are uniformly intermingled. This means that a sample from one part is the same as a sample from any other part. For example, the first sip of coffee is the same as the last sip.

The atmosphere that surrounds us is a gaseous solution containing $O_2(g)$, $N_2(g)$, and other gases randomly dispersed. Solutions can also be solids. For example, brass is a homogeneous mixture—a solution—of copper and zinc.

Brass, a solid solution of copper and zinc, is used to make musical instruments and many other objects.

These examples illustrate that a solution can be a gas, a liquid, or a solid (see Table 14.1). The substance present in the largest amount is called the **solvent,** and the other substance or substances are called **solutes.** For example, when we dissolve a teaspoon of sugar in a glass of water, the sugar is the solute and the water is the solvent.

Aqueous solutions are solutions with water as the solvent. Because they are so important, in this chapter we will concentrate on the properties of aqueous solutions.

14.1 Solubility

AIMS: To understand the process of dissolving. To learn why certain components dissolve in water.

What happens when you put a teaspoon of sugar in your iced tea and stir it, or when you add salt to water for cooking vegetables? Why do the sugar and salt "disappear" into the water? What does it mean when something dissolves—that is, when a solution forms?

Table 14.1 Various Types of Solutions

Example	State of Solution	Original State of Solute	State of Solvent
air, natural gas	gas	gas	gas
vodka in water, antifreeze in water	liquid	liquid	liquid
brass	solid	solid	solid
carbonated water (soda)	liquid	gas	liquid
seawater, sugar solution	liquid	solid	liquid

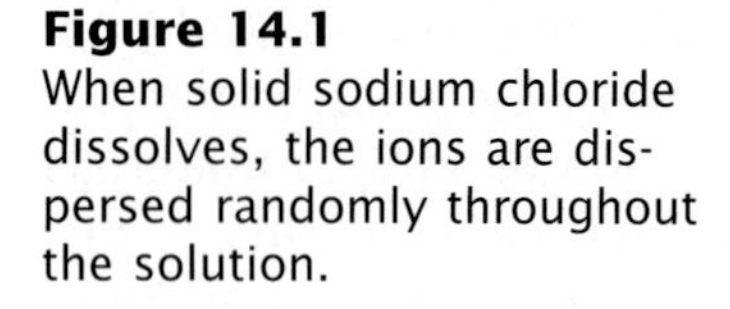

Figure 14.1
When solid sodium chloride dissolves, the ions are dispersed randomly throughout the solution.

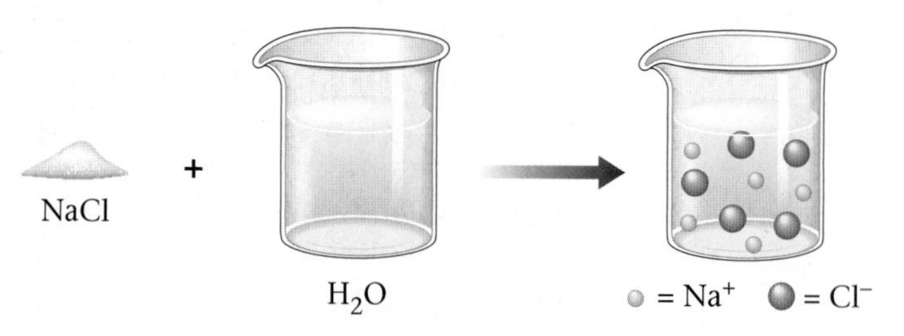

We saw in Chapter 7 that when sodium chloride dissolves in water, the resulting solution conducts an electric current. This convinces us that the solution contains *ions* that can move (this is how the electric current is conducted). The dissolving of solid sodium chloride in water is represented in Figure 14.1. Notice that in the solid state the ions are packed closely together. However, when the solid dissolves, the ions are separated and dispersed throughout the solution. The strong ionic forces that hold the sodium chloride crystal together are overcome by the strong attractions between the ions and the polar water molecules. This process is represented in Figure 14.2. Notice that each polar water molecule orients itself in a way to maximize its attraction with a Cl^- or Na^+ ion. The negative end of a water molecule is attracted to a Na^+ ion, while the positive end is attracted to a Cl^- ion. The strong forces holding the positive and negative ions in the solid are replaced by strong water–ion interactions, and the solid dissolves (the ions disperse).

Cations are positive ions. Anions are negative ions.

It is important to remember that when an ionic substance (such as a salt) dissolves in water, it breaks up into *individual* cations and anions, which are dispersed in the water. For instance, when ammonium nitrate, NH_4NO_3, dissolves in water, the resulting solution contains NH_4^+ and NO_3^- ions, which move around independently. This process can be represented as

$$NH_4NO_3(s) \xrightarrow{H_2O(l)} NH_4^+(aq) + NO_3^-(aq)$$

where (aq) indicates that the ions are surrounded by water molecules.

Figure 14.2
Polar water molecules interact with the positive and negative ions of a salt. These interactions replace the strong ionic forces holding the ions together in the undissolved solid, thus assisting in the dissolving process.

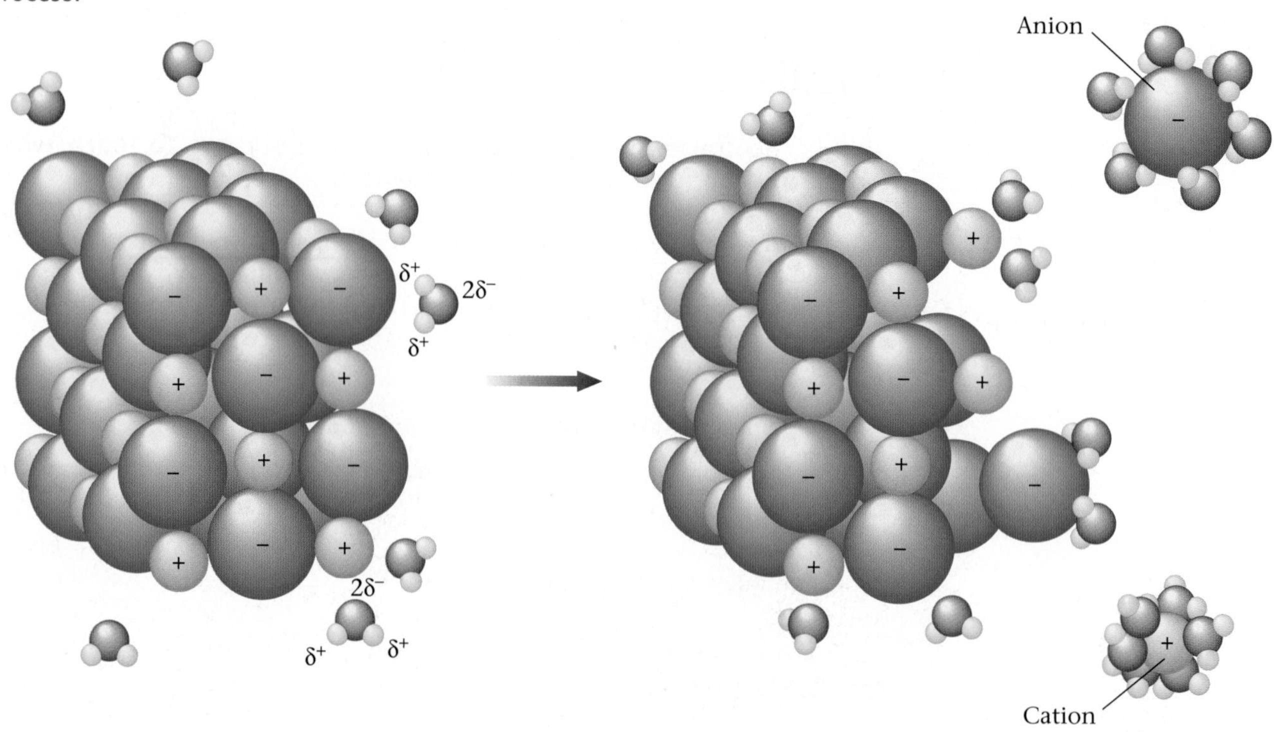

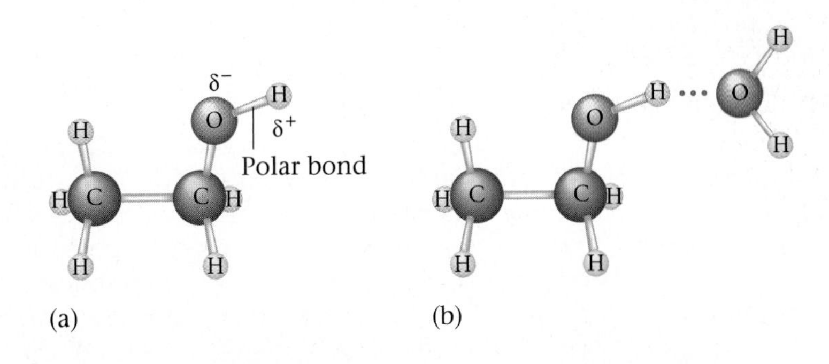

Figure 14.3
(a) The ethanol molecule contains a polar O—H bond similar to those in the water molecule. (b) The polar water molecule interacts strongly with the polar O—H bond in ethanol.

Water also dissolves many nonionic substances. Sugar is one example of a nonionic solute that is very soluble in water. Another example is ethanol, C_2H_5OH. Wine, beer, and mixed drinks are aqueous solutions of ethanol (and other substances). Why is ethanol so soluble in water? The answer lies in the structure of the ethanol molecule (Figure 14.3a). The molecule contains a polar O—H bond like those in water, which makes it very compatible with water. Just as hydrogen bonds form among water molecules in pure water (see Figure 13.6), ethanol molecules can form hydrogen bonds with water molecules in a solution of the two. This is shown in Figure 14.3b.

The sugar molecule (common table sugar has the chemical name sucrose) is shown in Figure 14.4. Notice that this molecule has many polar O—H groups, each of which can hydrogen-bond to a water molecule. Because of the attractions between sucrose and water molecules, solid sucrose is quite soluble in water.

A satellite photo of an oil spill in Tokyo Bay.

Many substances do not dissolve in water. For example, when petroleum leaks from a damaged tanker, it does not disperse uniformly in the water (does not dissolve) but rather floats on the surface because its density is less than that of water. Petroleum is a mixture of molecules like the one shown in Figure 14.5. Since carbon and hydrogen have very similar electronegativities, the bonding electrons are shared almost equally and the bonds are essentially nonpolar. The resulting molecule with its nonpolar bonds cannot form attractions to the polar water molecules and this prevents it from being soluble in water. This situation is represented in Figure 14.6.

Notice in Figure 14.6 that the water molecules in liquid water are associated with each other by hydrogen-bonding interactions. In order for a

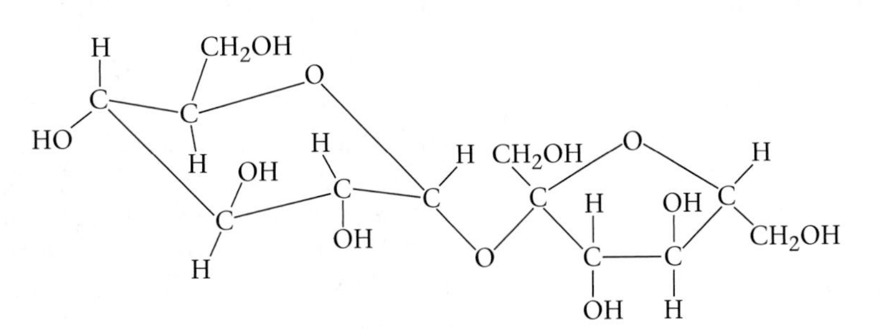

Figure 14.4
The structure of common table sugar (called sucrose). The large number of polar O—H groups in the molecule causes sucrose to be very soluble in water.

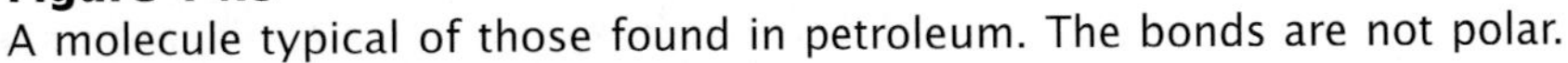

Figure 14.5
A molecule typical of those found in petroleum. The bonds are not polar.

CHEMISTRY IN FOCUS

Green Chemistry

Although some chemical industries have been culprits in the past for fouling the earth's environment, that situation is rapidly changing. In fact, a quiet revolution is sweeping through chemistry from academic labs to *Fortune* 500 companies. Chemistry is going green. *Green chemistry* means minimizing hazardous wastes, substituting water and other environmentally friendlier substances for traditional organic solvents, and manufacturing products out of recyclable materials.

A good example of green chemistry is the increasing use of carbon dioxide, one of the byproducts of the combustion of fossil fuels. For example, the Dow Chemical Company is now using CO_2 rather than chlorofluorocarbons (CFCs; substances known to catalyze the decomposition of protective stratospheric ozone) to put the "sponginess" into polystyrene egg cartons, meat trays, and burger boxes. Dow does not generate CO_2 for this process but instead uses waste gases captured from its various manufacturing processes.

Another very promising use of carbon dioxide is to replace the solvent perchloroethylene (PERC; $Cl_2C{=}CCl_2$), now used by about 80% of dry cleaners in the United States. Chronic exposure to PERC has been linked to kidney and liver damage and cancer. Although PERC is not a hazard to the general public (little PERC adheres to drycleaned garments), it represents a major concern for employees in the drycleaning industry. At high pressures CO_2 is a liquid that, when used with appropriate detergents, is a very effective solvent for the soil found on dry-clean-only fabrics. When the pressure is lowered, the CO_2 immediately changes to its gaseous form, quickly drying the clothes without the need for added heat. The gas can then be condensed and reused for the next batch of clothes.

The good news is that green chemistry makes sense economically. When all of the costs are taken into account, green chemistry is usually cheaper chemistry as well. Everybody wins.

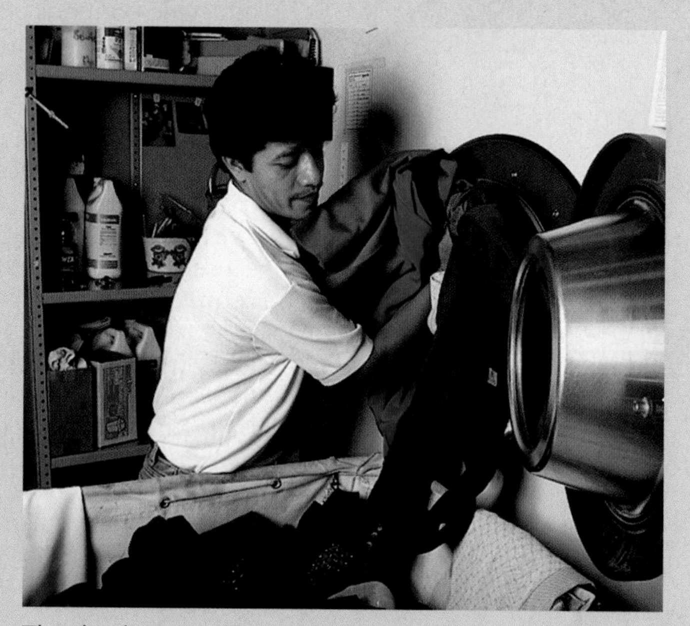

The drycleaning agent PERC is a health concern for workers in the drycleaning industry.

solute to dissolve in water, a "hole" must be made in the water structure for each solute particle. This will occur only if the lost water–water interactions are replaced by similar water–solute interactions. In the case of sodium chloride, strong interactions occur between the polar water molecules and the Na^+ and Cl^- ions. This allows the sodium chloride to dissolve. In the case of ethanol or sucrose, hydrogen-bonding interactions can occur between the O—H groups on these molecules and water molecules, making these substances soluble as well. But oil molecules are not soluble in water, because the many water–water interactions that would have to be broken to make "holes" for these large molecules are not replaced by favorable water–solute interactions.

These considerations account for the observed behavior that *"like dissolves like."* In other words, we observe that a given solvent usually dissolves solutes that have polarities similar to its own. For example, water dissolves most polar solutes, because the solute–solvent interactions formed in the

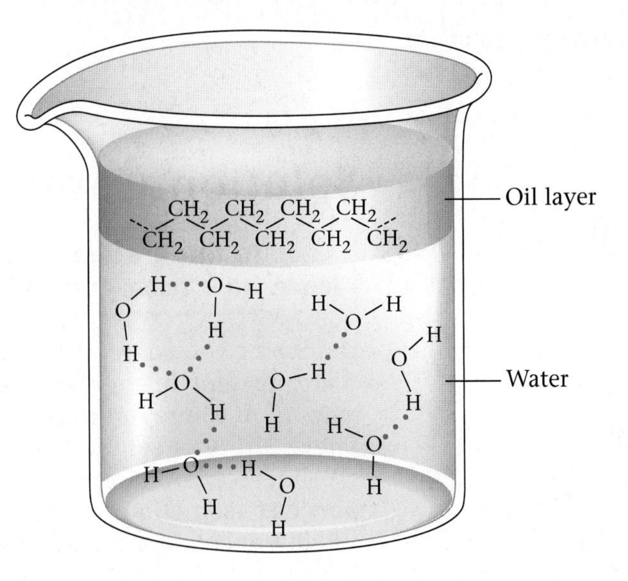

Figure 14.6
An oil layer floating on water. For a substance to dissolve, the water–water hydrogen bonds must be broken to make a "hole" for each solute particle. However, the water–water interactions will break only if they are replaced by similar strong interactions with the solute.

solution are similar to the water–water interactions present in the pure solvent. Likewise, nonpolar solvents dissolve nonpolar solutes. For example, drycleaning solvents used for removing grease stains from clothes are nonpolar liquids. "Grease" is composed of nonpolar molecules, so a nonpolar solvent is needed to remove a grease stain.

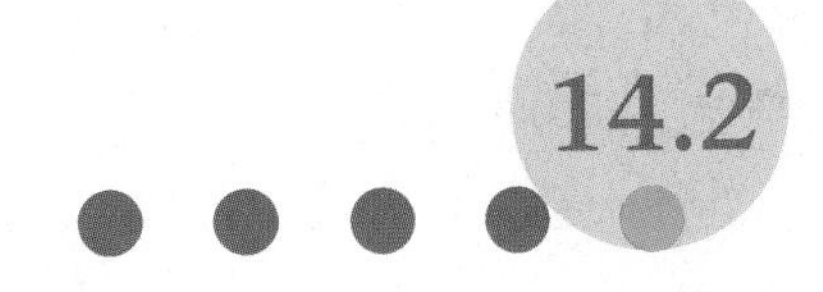

Solution Composition: An Introduction

AIM: To learn qualitative terms associated with the concentration of a solution.

Even for very soluble substances, there is a limit to how much solute can be dissolved in a given amount of solvent. For example, when you add sugar to a glass of water, the sugar rapidly disappears at first. However, as you continue to add more sugar, at some point the solid no longer dissolves but collects at the bottom of the glass. When a solution contains as much solute as will dissolve at that temperature, we say it is **saturated.** If a solid solute is added to a solution already saturated with that solute, the added solid does not dissolve. A solution that has *not* reached the limit of solute that will dissolve in it is said to be **unsaturated.** When more solute is added to an unsaturated solution, it dissolves.

Although a chemical compound always has the same composition, a solution is a mixture and the amounts of the substances present can vary in different solutions. For example, coffee can be strong or weak. Strong coffee has more coffee dissolved in a given amount of water than weak coffee. To describe a solution completely, we must specify the amounts of solvent and solute. We sometimes use the qualitative terms *concentrated* and *dilute* to describe a solution. A relatively large amount of solute is dissolved in a **concentrated** solution (strong coffee is concentrated). A relatively small amount of solute is dissolved in a **dilute** solution (weak coffee is dilute).

Although these qualitative terms serve a useful purpose, we often need to know the exact amount of solute present in a given amount of solution. In the next several sections, we will consider various ways to describe the composition of a solution.

14.3 Solution Composition: Mass Percent

AIM: To understand the concentration term *mass percent* and learn how to calculate it.

Describing the composition of a solution means specifying the amount of solute present in a given quantity of the solution. We typically give the amount of solute in terms of mass (number of grams) or in terms of moles. The quantity of solution is defined in terms of mass or volume.

One common way of describing a solution's composition is **mass percent** (sometimes called *weight percent*), which expresses the mass of solute present in a given mass of solution. The definition of mass percent is:

$$\text{Mass percent} = \frac{\text{mass of solute}}{\text{mass of solution}} \times 100\%$$
$$= \frac{\text{grams of solute}}{\text{grams of solute} + \text{grams of solvent}} \times 100\%$$

The mass of the solution is the sum of the masses of the solute and the solvent.

For example, suppose a solution is prepared by dissolving 1.0 g of sodium chloride in 48 g of water. The solution has a mass of 49 g (48 g of H_2O plus 1.0 g of NaCl), and there is 1.0 g of solute (NaCl) present. The mass percent of solute, then, is

$$\frac{1.0 \text{ g solute}}{49 \text{ g solution}} \times 100\% = 0.020 \times 100\% = 2.0\% \text{ NaCl}$$

Example 14.1 Solution Composition: Calculating Mass Percent

A solution is prepared by mixing 1.00 g of ethanol, C_2H_5OH, with 100.0 g of water. Calculate the mass percent of ethanol in this solution.

Solution

In this case we have 1.00 g of solute (ethanol) and 100.0 g of solvent (water). We now apply the definition of mass percent.

$$\text{Mass percent } C_2H_5OH = \left(\frac{\text{grams of } C_2H_5OH}{\text{grams of solution}}\right) \times 100\%$$
$$= \left(\frac{1.00 \text{ g } C_2H_5OH}{100.0 \text{ g } H_2O + 1.00 \text{ g } C_2H_5OH}\right) \times 100\%$$
$$= \frac{1.00 \text{ g}}{101.0 \text{ g}} \times 100\%$$
$$= 0.990\% \text{ } C_2H_5OH$$

✔ Self-Check Exercise 14.1

A 135-g sample of seawater is evaporated to dryness, leaving 4.73 g of solid residue (the salts formerly dissolved in the seawater). Calculate the mass percent of solute present in the original seawater.

See Problems 14.15 and 14.16. ■

Example 14.2 Solution Composition: Determining Mass of Solute

Although milk is not a true solution (it is really a suspension of tiny globules of fat, protein, and other substrates in water), it does contain a dissolved sugar called lactose. Cow's milk typically contains 4.5% by mass of lactose, $C_{12}H_{22}O_{11}$. Calculate the mass of lactose present in 175 g of milk.

Solution

We are given the following information:

$$\text{Mass of solution (milk)} = 175 \text{ g}$$
$$\text{Mass percent of solute (lactose)} = 4.5\%$$

We need to calculate the mass of solute (lactose) present in 175 g of milk. Using the definition of mass percent, we have

$$\text{Mass percent} = \frac{\text{grams of solute}}{\text{grams of solution}} \times 100\%$$

We now substitute the quantities we know:

$$\text{Mass percent} = \frac{\overbrace{\text{grams of solute}}^{\text{Mass of lactose}}}{\underbrace{175 \text{ g}}_{\text{Mass of milk}}} \times 100\% = \overbrace{4.5\%}^{\text{Mass percent}}$$

We now solve for grams of solute by multiplying both sides by 175 g,

$$\cancel{175 \text{ g}} \times \frac{\text{grams of solute}}{\cancel{175 \text{ g}}} \times 100\% = 4.5\% \times 175 \text{ g}$$

and then dividing both sides by 100%,

$$\text{Grams of solute} \times \frac{\cancel{100\%}}{\cancel{100\%}} = \frac{4.5\cancel{\%}}{100\cancel{\%}} \times 175 \text{ g}$$

to give

$$\text{Grams of solute} = 0.045 \times 175 \text{ g} = 7.9 \text{ g lactose}$$

✔ **Self-Check Exercise 14.2**

What mass of water must be added to 425 g of formaldehyde to prepare a 40.0% (by mass) solution of formaldehyde? This solution, called formalin, is used to preserve biological specimens.

HINT: Substitute the known quantities into the definition for mass percent, and then solve for the unknown quantity (mass of solvent).

See Problems 14.17 and 14.18. ■

14.4 Solution Composition: Molarity

AIMS: To understand molarity. To learn to use molarity to calculate the number of moles of solute present.

When a solution is described in terms of mass percent, the amount of solution is given in terms of its mass. However, it is often more convenient to measure the volume of a solution than to measure its mass. Because of this, chemists often describe a solution in terms of concentration. We define the *concentration* of a solution as the amount of solute in a *given volume* of solution. The most commonly used expression of concentration is **molarity (*M*).** Molarity describes the amount of solute in moles and the volume of the solution in liters. Molarity is *the number of moles of solute per volume of solution in liters.* That is

$$M = \text{molarity} = \frac{\text{moles of solute}}{\text{liters of solution}} = \frac{\text{mol}}{\text{L}}$$

A solution that is 1.0 molar (written as 1.0 M) contains 1.0 mol of solute per liter of solution.

Example 14.3 Solution Composition: Calculating Molarity, I

Calculate the molarity of a solution prepared by dissolving 11.5 g of solid NaOH in enough water to make 1.50 L of solution.

Solution

We are given the following information:

$$\text{Mass of solute} = 11.5 \text{ g NaOH}$$
$$\text{Volume of solution} = 1.50 \text{ L}$$

Because we are asked to calculate the molarity of the solution, we start by writing the definition of molarity.

$$M = \frac{\text{moles of solute}}{\text{liters of solution}}$$

We have the mass (in grams) of solute, so we need to convert the mass of solute to moles (using the molar mass of NaOH). Then we can divide the number of moles by the volume in liters.

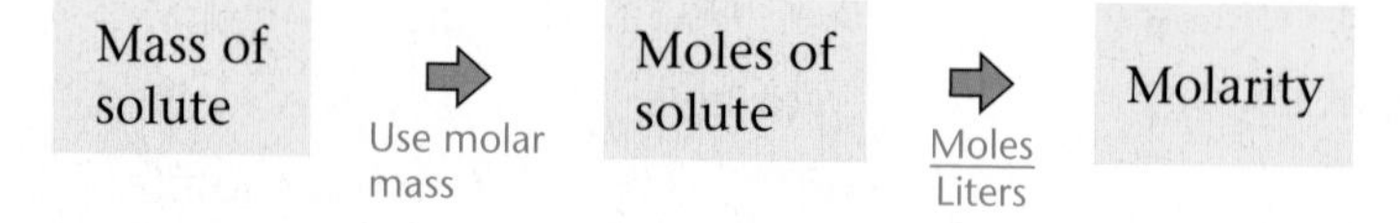

We compute the number of moles of solute, using the molar mass of NaOH (40.0 g).

$$11.5 \text{ g NaOH} \times \frac{1 \text{ mol NaOH}}{40.0 \text{ g NaOH}} = 0.288 \text{ mol NaOH}$$

Then we divide by the volume of the solution in liters.

$$\text{Molarity} = \frac{\text{moles of solute}}{\text{liters of solution}} = \frac{0.288 \text{ mol NaOH}}{1.50 \text{ L solution}} = 0.192\ M \text{ NaOH}$$

■

Example 14.4 Solution Composition: Calculating Molarity, II

Calculate the molarity of a solution prepared by dissolving 1.56 g of gaseous HCl into enough water to make 26.8 mL of solution.

Solution

We are given

$$\text{Mass of solute (HCl)} = 1.56 \text{ g}$$
$$\text{Volume of solution} = 26.8 \text{ mL}$$

Molarity is defined as

$$\frac{\text{Moles of solute}}{\text{Liters of solution}}$$

so we must change 1.56 g of HCl to moles of HCl, and then we must change 26.8 mL to liters (because molarity is defined in terms of liters). First we calculate the number of moles of HCl (molar mass = 36.5 g).

$$1.56 \text{ g HCl} \times \frac{1 \text{ mol HCl}}{36.5 \text{ g HCl}} = 0.0427 \text{ mol HCl}$$
$$= 4.27 \times 10^{-2} \text{ mol HCl}$$

Next we change the volume of the solution from milliliters to liters, using the equivalence statement 1 L = 1000 mL, which gives the appropriate conversion factor.

$$26.8 \text{ mL} \times \frac{1 \text{ L}}{1000 \text{ mL}} = 0.0268 \text{ L}$$
$$= 2.68 \times 10^{-2} \text{ L}$$

Finally, we divide the moles of solute by the liters of solution.

$$\text{Molarity} = \frac{4.27 \times 10^{-2} \text{ mol HCl}}{2.68 \times 10^{-2} \text{ L solution}} = 1.59\ M \text{ HCl}$$

✔ Self-Check Exercise 14.3

Calculate the molarity of a solution prepared by dissolving 1.00 g of ethanol, C_2H_5OH, in enough water to give a final volume of 101 mL.

See Problems 14.37 through 14.42. ■

It is important to realize that the description of a solution's composition may not accurately reflect the true chemical nature of the solute as it is present in the dissolved state. Solute concentration is always written in terms of the form of the solute *before* it dissolves. For example, describing a solution as 1.0 *M* NaCl means that the solution was prepared by dissolving 1.0 mol of solid NaCl in enough water to make 1.0 L of solution; it does not mean that the solution contains 1.0 mol of NaCl units. Actually the solution contains 1.0 mol of Na^+ ions and 1.0 mol of Cl^- ions. That is, it contains 1.0 *M* Na^+ and 1.0 *M* Cl^-.

Example 14.5 Solution Composition: Calculating Ion Concentration from Molarity

Remember, ionic compounds separate into the component ions when they dissolve in water.

$Co(NO_3)_2$

⇩

Co^{2+}

NO_3^- NO_3^-

$FeCl_3$

⇩

Fe^{3+}

Cl^- Cl^- Cl^-

Give the concentrations of all the ions in each of the following solutions:

a. 0.50 *M* $Co(NO_3)_2$

b. 1 *M* $FeCl_3$

Solution

a. When solid $Co(NO_3)_2$ dissolves, it produces ions as follows:

$$Co(NO_3)_2(s) \xrightarrow{H_2O(l)} Co^{2+}(aq) + 2NO_3^-(aq)$$

which we can represent as

$$1 \text{ mol } Co(NO_3)_2(s) \xrightarrow{H_2O(l)} 1 \text{ mol } Co^{2+}(aq) + 2 \text{ mol } NO_3^-(aq)$$

A solution of cobalt(II) nitrate.

Therefore, a solution that is 0.50 M $Co(NO_3)_2$ contains 0.50 M Co^{2+} and (2×0.50) M NO_3^-, or 1.0 M NO_3^-.

b. When solid $FeCl_3$ dissolves, it produces ions as follows:

$$FeCl_3(s) \xrightarrow{H_2O(l)} Fe^{3+}(aq) + 3Cl^-(aq)$$

or

$$1 \text{ mol } FeCl_3(s) \xrightarrow{H_2O(l)} 1 \text{ mol } Fe^{3+}(aq) + 3 \text{ mol } Cl^-(aq)$$

A solution that is 1 M $FeCl_3$ contains 1 M Fe^{3+} ions and 3 M Cl^- ions.

✔ Self-Check Exercise 14.4

Give the concentrations of the ions in each of the following solutions:

a. 0.10 M Na_2CO_3

b. 0.010 M $Al_2(SO_4)_3$

See Problems 14.49 and 14.50. ■

$$M = \frac{\text{moles of solute}}{\text{liters of solution}}$$

Liters × M ➡ Moles of solute

Often we need to determine the number of moles of solute present in a given volume of a solution of known molarity. To do this, we use the definition of molarity. When we multiply the molarity of a solution by the volume (in liters), we get the moles of solute present in that sample:

$$\text{Liters of solution} \times \text{molarity} = \cancel{\text{liters of solution}} \times \frac{\text{moles of solute}}{\cancel{\text{liters of solution}}}$$
$$= \text{moles of solute}$$

Example 14.6 Solution Composition: Calculating Number of Moles from Molarity

How many moles of Ag^+ ions are present in 25 mL of a 0.75 M $AgNO_3$ solution?

Solution

In this problem we know

Molarity of the solution = 0.75 M
Volume of the solution = 25 mL

We need to calculate the moles of Ag^+ present. To solve this problem, we must first recognize that a 0.75 M $AgNO_3$ solution contains 0.75 M Ag^+ ions and 0.75 M NO_3^- ions. Next we must express the volume in liters. That is, we must convert from mL to L.

$$25 \cancel{\text{mL}} \times \frac{1 \text{ L}}{1000 \cancel{\text{mL}}} = 0.025 \text{ L} = 2.5 \times 10^{-2} \text{ L}$$

Now we multiply the volume times the molarity.

$$2.5 \times 10^{-2} \cancel{\text{L solution}} \times \frac{0.75 \text{ mol } Ag^+}{\cancel{\text{L solution}}} = 1.9 \times 10^{-2} \text{ mol } Ag^+$$

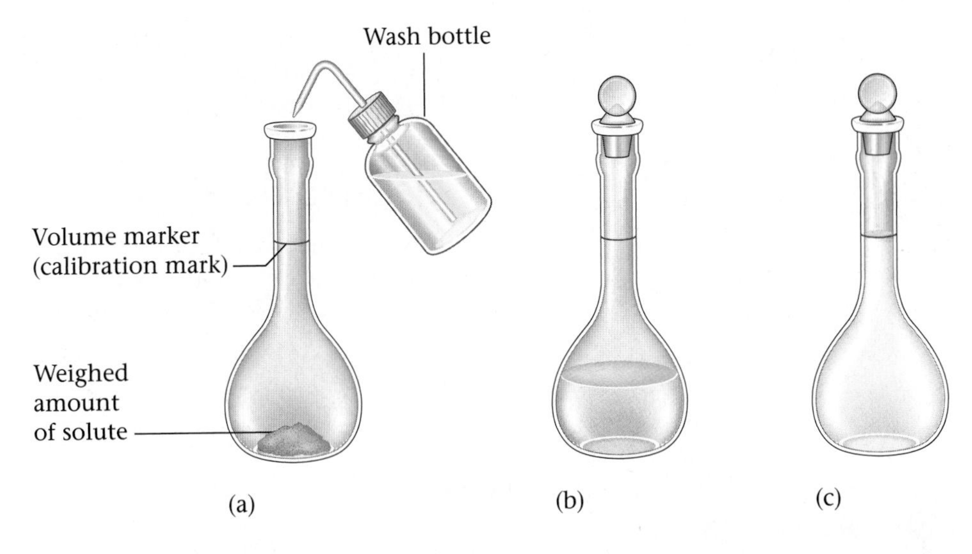

Figure 14.7
Steps involved in the preparation of a standard aqueous solution. (a) Put a weighed amount of a substance (the solute) into the volumetric flask, and add a small quantity of water. (b) Dissolve the solid in the water by gently swirling the flask (with the stopper in place). (c) Add more water (with gentle swirling) until the level of the solution just reaches the mark etched on the neck of the flask. Then mix the solution thoroughly by inverting the flask several times.

✔ Self-Check Exercise 14.5

Calculate the number of moles of Cl^- ions in 1.75 L of 1.0×10^{-3} M $AlCl_3$.

See Problems 14.49 and 14.50. ■

A **standard solution** is a solution whose *concentration is accurately known*. When the appropriate solute is available in pure form, a standard solution can be prepared by weighing out a sample of solute, transferring it completely to a *volumetric flask* (a flask of accurately known volume), and adding enough solvent to bring the volume up to the mark on the neck of the flask. This procedure is illustrated in Figure 14.7.

Example 14.7 Solution Composition: Calculating Mass from Molarity

To analyze the alcohol content of a certain wine, a chemist needs 1.00 L of an aqueous 0.200 M $K_2Cr_2O_7$ (potassium dichromate) solution. How much solid $K_2Cr_2O_7$ (molar mass = 294.2 g) must be weighed out to make this solution?

Solution

We know the following:

$$\text{Molarity of the solution} = 0.200\ M$$
$$\text{Volume of the solution} = 1.00\ \text{L}$$

We need to calculate the number of grams of solute ($K_2Cr_2O_7$) present (and thus the mass needed to make the solution). First we determine the number of moles of $K_2Cr_2O_7$ present by multiplying the volume (in liters) by the molarity.

Liters × M ⇨ Moles of solute

$$1.00\ \cancel{\text{L solution}} \times \frac{0.200\ \text{mol}\ K_2Cr_2O_7}{\cancel{\text{L solution}}} = 0.200\ \text{mol}\ K_2Cr_2O_7$$

Then we convert the moles of $K_2Cr_2O_7$ to grams, using the molar mass of $K_2Cr_2O_7$ (294.2 g).

$$0.200\ \cancel{\text{mol}\ K_2Cr_2O_7} \times \frac{294.2\ \text{g}\ K_2Cr_2O_7}{\cancel{\text{mol}\ K_2Cr_2O_7}} = 58.8\ \text{g}\ K_2Cr_2O_7$$

Therefore, to make 1.00 L of 0.200 M $K_2Cr_2O_7$, the chemist must weigh out 58.8 g of $K_2Cr_2O_7$ and dissolve it in enough water to make 1.00 L of solution. This is most easily done by using a 1.00-L volumetric flask (see Figure 14.7).

Self-Check Exercise 14.6

Formalin is an aqueous solution of formaldehyde, HCHO, used as a preservative for biological specimens. How many grams of formaldehyde must be used to prepare 2.5 L of 12.3 M formalin?

See Problems 14.51 and 14.52.

14.5 Dilution

AIM: To learn to calculate the concentration of a solution made by diluting a stock solution.

The molarities of stock solutions of the common concentrated acids are:

Sulfuric (H_2SO_4)	18 M
Nitric (HNO_3)	16 M
Hydrochloric (HCl)	12 M

To save time and space in the laboratory, solutions that are routinely used are often purchased or prepared in concentrated form (called *stock solutions*). Water (or another solvent) is then added to achieve the molarity desired for a particular solution. The process of adding more solvent to a solution is called **dilution.** For example, the common laboratory acids are purchased as concentrated solutions and diluted with water as they are needed. A typical dilution calculation involves determining how much water must be added to an amount of stock solution to achieve a solution of the desired concentration. The key to doing these calculations is to remember that *only water is added in the dilution.* The amount of solute in the final, more dilute solution is the *same* as the amount of solute in the original, concentrated stock solution. That is,

Dilution with water doesn't alter the number of moles of solute present.

Moles of solute after dilution = moles of solute before dilution

The number of moles of solute stays the same but more water is added, increasing the volume, so the molarity decreases.

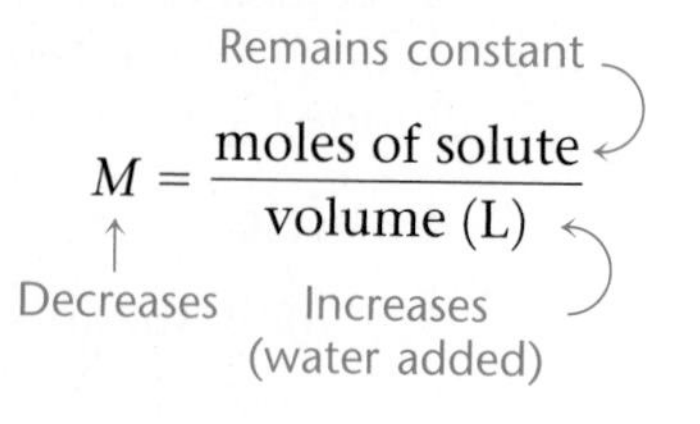

For example, suppose we want to prepare 500. mL of 1.00 M acetic acid, $HC_2H_3O_2$, from a 17.5 M stock solution of acetic acid. What volume of the stock solution is required?

The first step is to determine the number of moles of acetic acid needed in the final solution. We do this by multiplying the volume of the solution by its molarity.

$$\text{Volume of dilute solution (liters)} \times \text{molarity of dilute solution} = \text{moles of solute present}$$

The number of moles of solute present in the more dilute solution equals the number of moles of solute that must be present in the more concentrated (stock) solution, because this is the only source of acetic acid.

Because molarity is defined in terms of liters, we must first change 500. mL to liters and then multiply the volume (in liters) by the molarity.

$$\underset{\substack{V_{\text{dilute solution}} \\ \text{(in mL)}}}{500.\ \cancel{\text{mL solution}}} \times \underset{\text{Convert mL to L}}{\frac{1\ \text{L solution}}{1000\ \cancel{\text{mL solution}}}} = 0.500\ \text{L solution}$$

Liters × M ➡ Moles of solute

$$0.500\ \cancel{\text{L solution}} \times \underset{M_{\text{dilute solution}}}{\frac{1.00\ \text{mol}\ HC_2H_3O_2}{\cancel{\text{L solution}}}} = 0.500\ \text{mol}\ HC_2H_3O_2$$

Now we need to find the volume of 17.5 M acetic acid that contains 0.500 mol of $HC_2H_3O_2$. We will call this unknown volume V. Because volume × molarity = moles, we have

$$V\ \text{(in liters)} \times \frac{17.5\ \text{mol}\ HC_2H_3O_2}{\text{L solution}} = 0.500\ \text{mol}\ HC_2H_3O_2$$

Solving for $V\left(\text{by dividing both sides by } \frac{17.5\ \text{mol}}{\text{L solution}}\right)$ gives

$$V = \frac{0.500\ \cancel{\text{mol}\ HC_2H_3O_2}}{\frac{17.5\ \cancel{\text{mol}\ HC_2H_3O_2}}{\text{L solution}}} = 0.0286\ \text{L, or 28.6 mL, of solution}$$

Therefore, to make 500. mL of a 1.00 M acetic acid solution, we take 28.6 mL of 17.5 M acetic acid and dilute it to a total volume of 500. mL. This process is illustrated in Figure 14.8. Because the moles of solute remain the same before and after dilution, we can write

$$\underbrace{\underset{\substack{\text{Molarity}\\ \text{before}\\ \text{dilution}}}{M_1} \times \underset{\substack{\text{Volume}\\ \text{before}\\ \text{dilution}}}{V_1}}_{\text{Initial Conditions}} = \text{moles of solute} = \underbrace{\underset{\substack{\text{Molarity}\\ \text{after}\\ \text{dilution}}}{M_2} \times \underset{\substack{\text{Volume}\\ \text{after}\\ \text{dilution}}}{V_2}}_{\text{Final Conditions}}$$

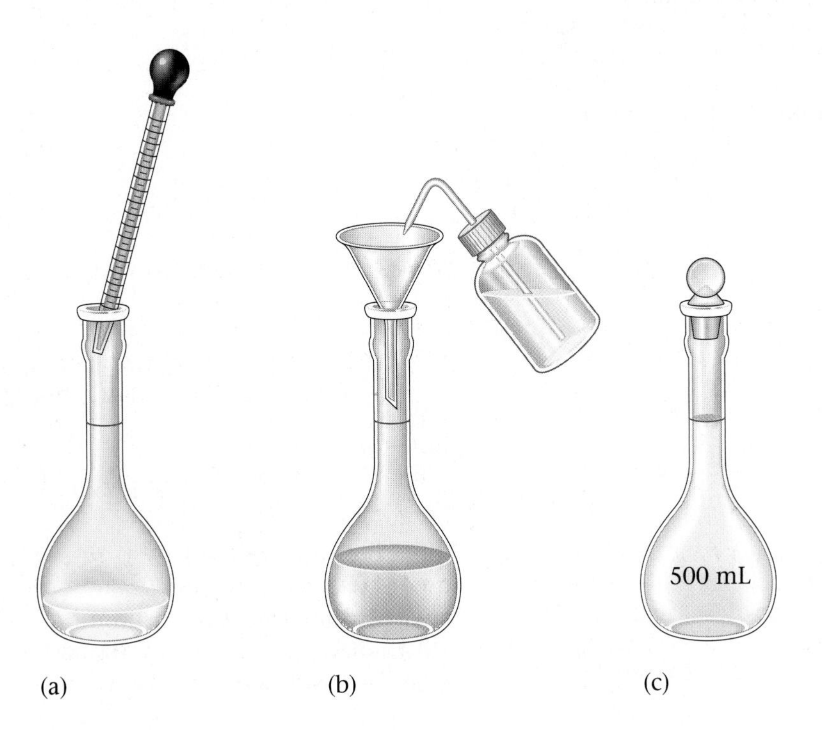

Figure 14.8
(a) 28.6 mL of 17.5 M acetic acid solution is transferred to a volumetric flask that already contains some water. (b) Water is added to the flask (with swirling) to bring the volume to the calibration mark, and the solution is mixed by inverting the flask several times. (c) The resulting solution is 1.00 M acetic acid.

We can check our calculations on acetic acid by showing that $M_1 \times V_1 = M_2 \times V_2$. In the above example, $M_1 = 17.5\ M$, $V_1 = 0.0286$ L, $V_2 = 0.500$ L, and $M_2 = 1.00\ M$, so

$$M_1 \times V_1 = 17.5\ \frac{\text{mol}}{\text{L}} \times 0.0286\ \text{L} = 0.500\ \text{mol}$$

$$M_2 \times V_2 = 1.00\ \frac{\text{mol}}{\text{L}} \times 0.500\ \text{L} = 0.500\ \text{mol}$$

and therefore

$$M_1 \times V_1 = M_2 \times V_2$$

This shows that the volume (V_2) we calculated is correct.

Example 14.8 Calculating Concentrations of Diluted Solutions

Approximate dilutions can be carried out using a calibrated beaker. Here concentrated sulfuric acid is being added to water to make a dilute solution.

What volume of 16 M sulfuric acid must be used to prepare 1.5 L of a 0.10 M H_2SO_4 solution?

Solution

We can summarize what we are given as follows:

Initial Conditions (concentrated)	**Final Conditions (dilute)**
$M_1 = 16\ \frac{\text{mol}}{\text{L}}$	$M_2 = 0.10\ \frac{\text{mol}}{\text{L}}$
$V_1 = ?$	$V_2 = 1.5$ L

We know that

$$\text{Moles of solute} = M_1 \times V_1 = M_2 \times V_2$$

and we can solve the equation

$$M_1 \times V_1 = M_2 \times V_2$$

for V_1 by dividing both sides by M_1

$$\frac{\cancel{M_1} \times V_1}{\cancel{M_1}} = \frac{M_2 \times V_2}{M_1}$$

to give

$$V_1 = \frac{M_2 \times V_2}{M_1}$$

Now we substitute the known values of M_2, V_2, and M_1.

$$V_1 = \frac{\left(0.10\ \frac{\cancel{\text{mol}}}{\cancel{\text{L}}}\right)(1.5\ \text{L})}{16\ \frac{\cancel{\text{mol}}}{\cancel{\text{L}}}} = 9.4 \times 10^{-3}\ \text{L}$$

$$9.4 \times 10^{-3}\ \cancel{\text{L}} \times \frac{1000\ \text{mL}}{1\ \cancel{\text{L}}} = 9.4\ \text{mL}$$

It is always best to add concentrated acid to water, not water to the acid. That way, if any splashing occurs accidentally, it is dilute acid that splashes.

Therefore, $V_1 = 9.4 \times 10^{-3}$ L, or 9.4 mL. To make 1.5 L of 0.10 M H_2SO_4 using 16 M H_2SO_4, we must take 9.4 mL of the concentrated acid and dilute it with water to a final volume of 1.5 L. The correct way to do this is to add the 9.4 mL of acid to about 1 L of water and then dilute to 1.5 L by adding more water.

Self-Check Exercise 14.7

What volume of 12 *M* HCl must be taken to prepare 0.75 L of 0.25 *M* HCl?

See Problems 14.57 and 14.58. ■

14.6 Stoichiometry of Solution Reactions

AIM: To understand the strategy for solving stoichiometric problems for solution reactions.

Because so many important reactions occur in solution, it is important to be able to do stoichiometric calculations for solution reactions. The principles needed to perform these calculations are very similar to those developed in Chapter 9. It is helpful to think in terms of the following steps:

Steps for Solving Stoichiometric Problems Involving Solutions

See Section 7.3 for a discussion of net ionic equations.

STEP 1 Write the balanced equation for the reaction. For reactions involving ions, it is best to write the net ionic equation.

STEP 2 Calculate the moles of reactants.

STEP 3 Determine which reactant is limiting.

STEP 4 Calculate the moles of other reactants or products, as required.

STEP 5 Convert to grams or other units, if required.

Example 14.9 Solution Stoichiometry: Calculating Mass of Reactants and Products

Calculate the mass of solid NaCl that must be added to 1.50 L of a 0.100 *M* $AgNO_3$ solution to precipitate all of the Ag^+ ions in the form of AgCl. Calculate the mass of AgCl formed.

Solution

STEP 1 *Write the balanced equation for the reaction.*
When added to the $AgNO_3$ solution (which contains Ag^+ and NO_3^- ions), the solid NaCl dissolves to yield Na^+ and Cl^- ions. Solid AgCl forms according to the following balanced net ionic reaction:

This reaction was discussed in Section 7.2.

$$Ag^+(aq) + Cl^-(aq) \rightarrow AgCl(s)$$

STEP 2 *Calculate the moles of reactants.*
In this case we must add enough Cl^- ions to just react with all the Ag^+ ions present, so we must calculate the moles of Ag^+ ions present in 1.50 L of a 0.100 *M* $AgNO_3$ solution. (Remember that a 0.100 *M* $AgNO_3$ solution contains 0.100 *M* Ag^+ ions and 0.100 *M* NO_3^- ions.)

Liters × *M* ➡ Moles of solute

$$1.50\,\cancel{L} \times \frac{0.100 \text{ mol } Ag^+}{\cancel{L}} = 0.150 \text{ mol } Ag^+$$

Moles of Ag^+ present in 1.5 L of 0.100 *M* $AgNO_3$

When aqueous sodium chloride is added to a solution of silver nitrate, a white silver chloride precipitate forms.

Step 3 *Determine which reactant is limiting.*
In this situation we want to add just enough Cl^- to react with the Ag^+ present. That is, we want to precipitate *all* the Ag^+ in the solution. Thus the Ag^+ present determines the amount of Cl^- needed.

Step 4 *Calculate the moles of Cl^- required.*
We have 0.150 mol of Ag^+ ions and, because one Ag^+ ion reacts with one Cl^- ion, we need 0.150 mol of Cl^-,

$$0.150\ \cancel{\text{mol Ag}^+} \times \frac{1\ \text{mol Cl}^-}{1\ \cancel{\text{mol Ag}^+}} = 0.150\ \text{mol Cl}^-$$

so 0.150 mol of AgCl will be formed.

$$0.150\ \text{mol Ag}^+ + 0.150\ \text{mol Cl}^- \rightarrow 0.150\ \text{mol AgCl}$$

Step 5 *Convert to grams of NaCl required.*
To produce 0.150 mol Cl^-, we need 0.150 mol NaCl. We calculate the mass of NaCl required as follows:

$$0.150\ \cancel{\text{mol NaCl}} \times \frac{58.4\ \text{g NaCl}}{\cancel{\text{mol NaCl}}} = 8.76\ \text{g NaCl}$$

Moles → Times molar mass → Mass

The mass of AgCl formed is

$$0.150\ \cancel{\text{mol AgCl}} \times \frac{143.3\ \text{g AgCl}}{\cancel{\text{mol AgCl}}} = 21.5\ \text{g AgCl}$$ ■

Example 14.10 Solution Stoichiometry: Determining Limiting Reactants and Calculating Mass of Products

See Section 7.2 for a discussion of this reaction.

When $Ba(NO_3)_2$ and K_2CrO_4 react in aqueous solution, the yellow solid $BaCrO_4$ is formed. Calculate the mass of $BaCrO_4$ that forms when 3.50×10^{-3} mol of solid $Ba(NO_3)_2$ is dissolved in 265 mL of 0.0100 *M* K_2CrO_4 solution.

Solution

Barium chromate precipitating.

Step 1 The original K_2CrO_4 solution contains the ions K^+ and CrO_4^{2-}. When the $Ba(NO_3)_2$ is dissolved in this solution, Ba^{2+} and NO_3^- ions are added. The Ba^{2+} and CrO_4^{2-} ions react to form solid $BaCrO_4$. The balanced net ionic equation is

$$Ba^{2+}(aq) + CrO_4^{2-}(aq) \rightarrow BaCrO_4(s)$$

Step 2 Next we determine the moles of reactants. We are told that 3.50×10^{-3} mol of $Ba(NO_3)_2$ is added to the K_2CrO_4 solution. Each formula unit of $Ba(NO_3)_2$ contains one Ba^{2+} ion, so 3.50×10^{-3} mol of $Ba(NO_3)_2$ gives 3.50×10^{-3} mol of Ba^{2+} ions in solution.

Because $V \times M$ = moles of solute, we can compute the moles of K_2CrO_4 in the solution from the volume and molarity of the original solution. First we must convert the volume of the solution (265 mL) to liters.

$$265\ \cancel{\text{mL}} \times \frac{1\ \text{L}}{1000\ \cancel{\text{mL}}} = 0.265\ \text{L}$$

Next we determine the number of moles of K_2CrO_4, using the molarity of the K_2CrO_4 solution (0.0100 *M*).

$$0.265\ \cancel{L} \times \frac{0.0100\ \text{mol}\ K_2CrO_4}{\cancel{L}} = 2.65 \times 10^{-3}\ \text{mol}\ K_2CrO_4$$

We know that

2.65×10^{-3} mol K_2CrO_4	➡ dissolves to give	2.65×10^{-3} mol CrO_4^{2-}

so the solution contains 2.65×10^{-3} mol of CrO_4^{2-} ions.

STEP 3 The balanced equation tells us that one Ba^{2+} ion reacts with one CrO_4^{2-} ion. Because the number of moles of CrO_4^{2-} ions (2.65×10^{-3}) is smaller than the number of moles of Ba^{2+} ions (3.50×10^{-3}), the CrO_4^{2-} will run out first.

$$Ba^{2+}(aq) \quad + \quad CrO_4^{2-}(aq) \quad \rightarrow \quad BaCrO_4(s)$$

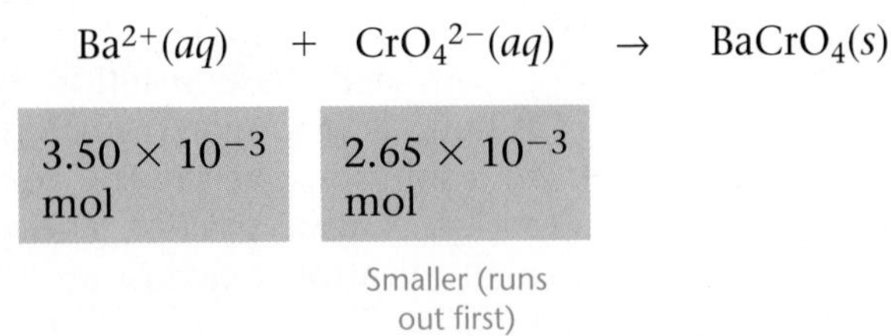

Therefore, the CrO_4^{2-} is limiting.

STEP 4 The 2.65×10^{-3} mol of CrO_4^{2-} ions will react with 2.65×10^{-3} mol of Ba^{2+} ions to form 2.65×10^{-3} mol of $BaCrO_4$.

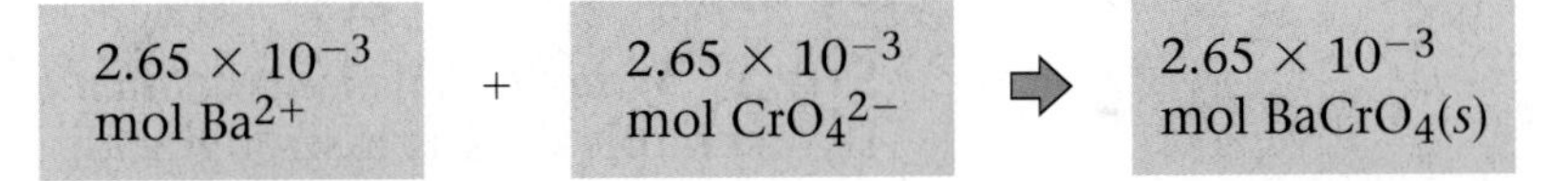

STEP 5 The mass of $BaCrO_4$ formed is obtained from its molar mass (253.3 g) as follows:

$$2.65 \times 10^{-3}\ \cancel{\text{mol}\ BaCrO_4} \times \frac{253.3\ \text{g}\ BaCrO_4}{\cancel{\text{mol}\ BaCrO_4}} = 0.671\ \text{g}\ BaCrO_4$$

✔ Self-Check Exercise 14.8

When aqueous solutions of Na_2SO_4 and $Pb(NO_3)_2$ are mixed, $PbSO_4$ precipitates. Calculate the mass of $PbSO_4$ formed when 1.25 L of 0.0500 *M* $Pb(NO_3)_2$ and 2.00 L of 0.0250 *M* Na_2SO_4 are mixed.

HINT: Calculate the moles of Pb^{2+} and SO_4^{2-} in the mixed solution, decide which ion is limiting, and calculate the moles of $PbSO_4$ formed.

See Problems 14.65 through 14.68. ■

14.7 Neutralization Reactions

AIM: To learn how to do calculations involved in acid–base reactions.

So far we have considered the stoichiometry of reactions in solution that result in the formation of a precipitate. Another common type of solution reaction occurs between an acid and a base. We introduced these reactions in Section 7.4. Recall from that discussion that an acid is a substance that furnishes H^+ ions. A strong acid, such as hydrochloric acid, HCl, dissociates (ionizes) completely in water.

$$HCl(aq) \rightarrow H^+(aq) + Cl^-(aq)$$

Strong bases are water-soluble metal hydroxides, which are completely dissociated in water. An example is NaOH, which dissolves in water to give Na^+ and OH^- ions.

$$NaOH(s) \xrightarrow{H_2O(l)} Na^+(aq) + OH^-(aq)$$

When a strong acid and strong base react, the net ionic reaction is

$$H^+(aq) + OH^-(aq) \rightarrow H_2O(l)$$

An acid–base reaction is often called a **neutralization reaction.** When just enough strong base is added to react exactly with the strong acid in a solution, we say the acid has been *neutralized.* One product of this reaction is always water. The steps in dealing with the stoichiometry of any neutralization reaction are the same as those we followed in the previous section.

Example 14.11 Solution Stoichiometry: Calculating Volume in Neutralization Reactions

What volume of a 0.100 *M* HCl solution is needed to neutralize 25.0 mL of a 0.350 *M* NaOH solution?

Solution

STEP 1 *Write the balanced equation for the reaction.*
Hydrochloric acid is a strong acid, so all the HCl molecules dissociate to produce H^+ and Cl^- ions. Also, when the strong base NaOH dissolves, the solution contains Na^+ and OH^- ions. When these two solutions are mixed, the H^+ ions from the hydrochloric acid react with the OH^- ions from the sodium hydroxide solution to form water. The balanced net ionic equation for the reaction is

$$H^+(aq) + OH^-(aq) \rightarrow H_2O(l)$$

STEP 2 *Calculate the moles of reactants.*
In this problem we are given a volume (25.0 mL) of 0.350 *M* NaOH, and we want to add just enough 0.100 *M* HCl to provide just enough H^+ ions to react with all the OH^-. Therefore, we must calculate the number of moles of OH^- ions in the 25.0-mL sample of 0.350 *M* NaOH. To do this, we first change the volume to liters and multiply by the molarity.

$$25.0\ \cancel{\text{mL NaOH}} \times \frac{1\ \cancel{\text{L}}}{1000\ \cancel{\text{mL}}} \times \frac{0.350\ \text{mol OH}^-}{\cancel{\text{L NaOH}}} = 8.75 \times 10^{-3}\ \text{mol OH}^-$$

Moles of OH^- present in 25.0 mL of 0.350 *M* NaOH

STEP 3 *Determine which reactant is limiting.*
This problem requires the addition of just enough H^+ ions to react exactly with the OH^- ions present, so the number of moles of OH^- ions present determines the number of moles of H^+ that must be added. The OH^- ions are limiting.

STEP 4 *Calculate the moles of H^+ required.*
The balanced equation tells us that the H^+ and OH^- ions react in a 1:1 ratio, so 8.75×10^{-3} mol of H^+ ions is required to neutralize (exactly react with) the 8.75×10^{-3} mol of OH^- ions present.

STEP 5 *Calculate the volume of 0.100 M HCl required.*
Next we must find the volume (V) of 0.100 M HCl required to furnish this amount of H^+ ions. Because the volume (in liters) times the molarity gives the number of moles, we have

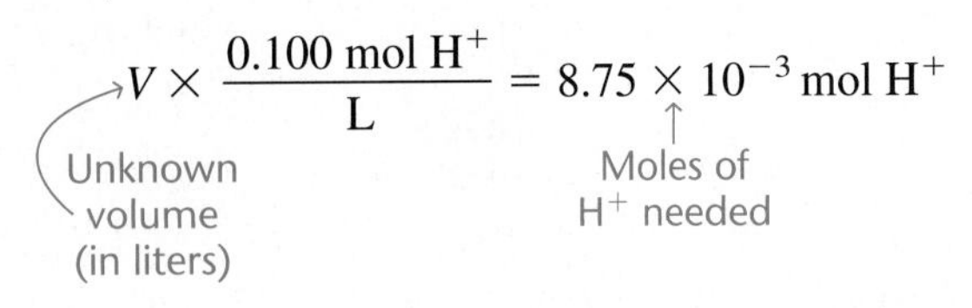

$$V \times \frac{0.100 \text{ mol H}^+}{\text{L}} = 8.75 \times 10^{-3} \text{ mol H}^+$$

Now we must solve for V by dividing both sides of the equation by 0.100.

$$V \times \frac{\cancel{0.100} \cancel{\text{mol H}^+}}{\cancel{0.100} \text{ L}} = \frac{8.75 \times 10^{-3} \cancel{\text{mol H}^+}}{0.100}$$

$$V = 8.75 \times 10^{-2} \text{ L}$$

Changing liters to milliliters, we have

$$V = 8.75 \times 10^{-2} \cancel{\text{L}} \times \frac{1000 \text{ mL}}{\cancel{\text{L}}} = 87.5 \text{ mL}$$

Therefore, 87.5 mL of 0.100 M HCl is required to neutralize 25.0 mL of 0.350 M NaOH.

Self-Check Exercise 14.9

Calculate the volume of 0.10 M HNO_3 needed to neutralize 125 mL of 0.050 M KOH.

See Problems 14.69 through 14.74. ■

14.8 Solution Composition: Normality

AIMS: To learn about normality and equivalent weight. To learn to use these concepts in stoichiometric calculations.

Normality is another unit of concentration that is sometimes used, especially when dealing with acids and bases. The use of normality focuses mainly on the H^+ and OH^- available in an acid–base reaction. Before we discuss normality, however, we need to define some terms. One **equivalent of an acid** is the *amount of that acid that can furnish 1 mol of H^+ ions.* Similarly, one **equivalent of a base** is defined as the *amount of that base that can furnish 1 mol of OH^- ions.* The **equivalent weight** of an acid or a base is the mass in grams of 1 equivalent (equiv) of that acid or base.

The common strong acids are HCl, HNO_3, and H_2SO_4. For HCl and HNO_3 each molecule of acid furnishes one H^+ ion, so 1 mol of HCl can furnish 1 mol of H^+ ions. This means that

Furnishes 1 mol of H^+

$$1 \text{ mol HCl} = 1 \text{ equiv HCl}$$

$$\text{Molar mass (HCl)} = \text{equivalent weight (HCl)}$$

Table 14.2 The Molar Masses and Equivalent Weights of the Common Strong Acids and Bases

	Molar Mass (g)	Equivalent Weight (g)
Acid		
HCl	36.5	36.5
HNO_3	63.0	63.0
H_2SO_4	98.0	$49.0 = \frac{98.0}{2}$
Base		
NaOH	40.0	40.0
KOH	56.1	56.1

Likewise, for HNO_3,

$$1 \text{ mol } HNO_3 = 1 \text{ equiv } HNO_3$$
$$\text{Molar mass } (HNO_3) = \text{equivalent weight } (HNO_3)$$

However, H_2SO_4 can furnish *two* H^+ ions per molecule, so 1 mol of H_2SO_4 can furnish *two* mol of H^+. This means that

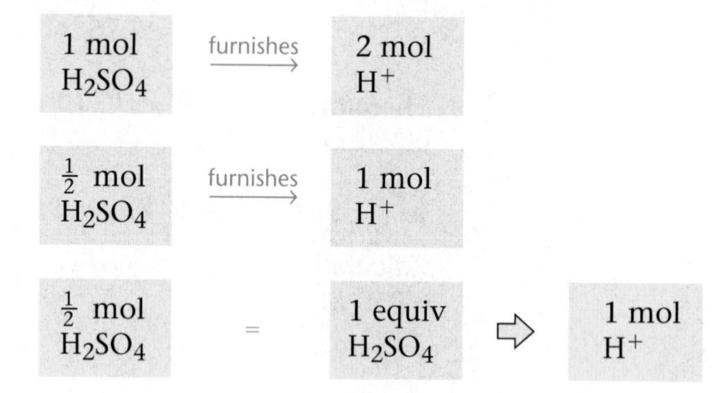

Because each mole of H_2SO_4 can furnish 2 mol of H^+, we need only to take $\frac{1}{2}$ mol of H_2SO_4 to get 1 equiv of H_2SO_4. Therefore,

$$\tfrac{1}{2} \text{ mol } H_2SO_4 = 1 \text{ equiv } H_2SO_4$$

and

$$\text{Equivalent weight } (H_2SO_4) = \tfrac{1}{2} \text{ molar mass } (H_2SO_4)$$
$$= \tfrac{1}{2}(98 \text{ g}) = 49 \text{ g}$$

The equivalent weight of H_2SO_4 is 49 g.

The common strong bases are NaOH and KOH. For NaOH and KOH, each formula unit furnishes one OH^- ion, so we can say

$$1 \text{ mol NaOH} = 1 \text{ equiv NaOH}$$
$$\text{Molar mass (NaOH)} = \text{equivalent weight (NaOH)}$$
$$1 \text{ mol KOH} = 1 \text{ equiv KOH}$$
$$\text{Molar mass (KOH)} = \text{equivalent weight (KOH)}$$

These ideas are summarized in Table 14.2.

Example 14.12 Solution Stoichiometry: Calculating Equivalent Weight

Phosphoric acid, H_3PO_4, can furnish three H^+ ions per molecule. Calculate the equivalent weight of H_3PO_4.

Solution

The key point here involves how many protons (H^+ ions) each molecule of H_3PO_4 can furnish.

H_3PO_4	⇨ furnishes	? H^+

Because each H_3PO_4 can furnish three H^+ ions, 1 mol of H_3PO_4 can furnish 3 mol of H^+ ions:

1 mol H_3PO_4	⇨ furnishes	3 mol H^+

So 1 equiv of H_3PO_4 (the amount that can furnish 1 mol of H^+) is one-third of a mole.

$\frac{1}{3}$ mol H_3PO_4	⇨ furnishes	1 mol H^+

This means the equivalent weight of H_3PO_4 is one-third its molar mass.

$$\text{Equivalent weight} = \frac{\text{Molar mass}}{3}$$

$$\text{Equivalent weight } (H_3PO_4) = \frac{\text{molar mass } (H_3PO_4)}{3}$$

$$= \frac{98.0 \text{ g}}{3} = 32.7 \text{ g}$$

■

Normality (***N***) is defined as the number of equivalents of solute per liter of solution.

$$\text{Normality} = N = \frac{\text{number of equivalents}}{\text{1 liter of solution}} = \frac{\text{equivalents}}{\text{liter}} = \frac{\text{equiv}}{\text{L}}$$

This means that a 1 N solution contains 1 equivalent of solute per liter of solution. Notice that when we multiply the volume of a solution in liters by the normality, we get the number of equivalents.

L × Normality ➡ Equiv

$$N \times V = \frac{\text{equiv}}{\cancel{\text{L}}} \times \cancel{\text{L}} = \text{equiv}$$

Example 14.13 Solution Stoichiometry: Calculating Normality

A solution of sulfuric acid contains 86 g of H_2SO_4 per liter of solution. Calculate the normality of this solution.

Whenever you need to calculate the concentration of a solution, first write the appropriate definition. Then decide how to calculate the quantities shown in the definition.

Solution

We want to calculate the normality of this solution, so we focus on the definition of normality, the number of equivalents per liter:

$$N = \frac{\text{equiv}}{\text{L}}$$

This definition leads to two questions we need to answer:

1. What is the number of equivalents?
2. What is the volume?

We know the volume; it is 1.0 L. To find the number of equivalents present, we must calculate the number of equivalents represented by 86 g of H_2SO_4. To do this calculation, we focus on the definition of the equivalent: it is the amount of acid that furnishes 1 mol of H^+. Because H_2SO_4 can furnish two H^+ ions per molecule, 1 equiv of H_2SO_4 is $\frac{1}{2}$ mol of H_2SO_4, so

$$\text{Equivalent weight } (H_2SO_4) = \frac{\text{molar mass } (H_2SO_4)}{2}$$

$$= \frac{98.0 \text{ g}}{2} = 49.0 \text{ g}$$

We have 86 g of H_2SO_4.

$$86 \cancel{\text{g } H_2SO_4} \times \frac{1 \text{ equiv } H_2SO_4}{49.0 \cancel{\text{g } H_2SO_4}} = 1.8 \text{ equiv } H_2SO_4$$

$$N = \frac{\text{equiv}}{\text{L}} = \frac{1.8 \text{ equiv } H_2SO_4}{1.0 \text{ L}} = 1.8\ N\ H_2SO_4$$

We know that 86 g is more than 1 equiv of H_2SO_4 (49 g), so this answer makes sense.

✔ Self-Check Exercise 14.10

Calculate the normality of a solution containing 23.6 g of KOH in 755 mL of solution.

See Problems 14.79 and 14.80. ■

The main advantage of using equivalents is that 1 equiv of acid contains the same number of available H^+ ions as the number of OH^- ions present in 1 equiv of base. That is,

0.75 equiv (base) will react exactly with 0.75 equiv (acid).

0.23 equiv (base) will react exactly with 0.23 equiv (acid).

And so on.

In each of these cases, the *number of* H^+ ions furnished by the sample of acid is the same as the *number of* OH^- ions furnished by the sample of base. The point is that *n equivalents of any acid will exactly neutralize n equivalents of any base.*

n equiv acid ← reacts exactly with → *n* equiv base

Because we know that equal equivalents of acid and base are required for neutralization, we can say that

$$\text{equiv (acid)} = \text{equiv (base)}$$

That is,

$$N_{acid} \times V_{acid} = \text{equiv (acid)} = \text{equiv (base)} = N_{base} \times V_{base}$$

Therefore, for any neutralization reaction, the following relationship holds:

$$N_{acid} \times V_{acid} = N_{base} \times V_{base}$$

Example 14.14 Solution Stoichiometry: Using Normality in Calculations

What volume of a 0.075 N KOH solution is required to react exactly with 0.135 L of 0.45 N H_3PO_4?

Solution

We know that for neutralization, equiv (acid) = equiv (base), or

$$N_{acid} \times V_{acid} = N_{base} \times V_{base}$$

We want to calculate the volume of base, V_{base}, so we solve for V_{base} by dividing both sides by N_{base}.

$$\frac{N_{acid} \times V_{acid}}{N_{base}} = \frac{\cancel{N_{base}} \times V_{base}}{\cancel{N_{base}}} = V_{base}$$

Now we can substitute the given values $N_{acid} = 0.45\ N$, $V_{acid} = 0.135$ L, and $N_{base} = 0.075\ N$ into the equation.

$$V_{base} = \frac{N_{acid} \times V_{acid}}{N_{base}} = \frac{\left(0.45\ \frac{\cancel{\text{equiv}}}{\cancel{\text{L}}}\right)(0.135\ \text{L})}{0.075\ \frac{\cancel{\text{equiv}}}{\cancel{\text{L}}}} = 0.81\ \text{L}$$

This gives $V_{base} = 0.81$ L, so 0.81 L of 0.075 N KOH is required to react exactly with 0.135 L of 0.45 N H_3PO_4.

✔ Self-Check Exercise 14.11

What volume of 0.50 N H_2SO_4 is required to react exactly with 0.250 L of 0.80 N KOH?

See Problems 14.85 and 14.86. ■

Chapter 14 Review

Key Terms

solution (p. 423)
solvent (p. 423)
solute (p. 423)
aqueous solution (p. 423)
saturated (14.2)
unsaturated (14.2)
concentrated (14.2)
dilute (14.2)
mass percent (14.3)
molarity (M) (14.4)
standard solution (14.4)
dilution (14.5)
neutralization reaction (14.7)
equivalent of an acid (14.8)
equivalent of a base (14.8)
equivalent weight (14.8)
normality (N) (14.8)

Summary

1. A solution is a homogeneous mixture. The solubility of a solute in a given solvent depends on the interactions between the solvent and solute particles. Water dissolves many ionic compounds and compounds with polar molecules, because strong forces occur between the solute and the polar water molecules. Nonpolar solvents tend to dissolve nonpolar solutes. "Like dissolves like."

2. Solution composition can be described in many ways. Two of the most important are in terms of mass percent of solute:

$$\text{Mass percent} = \frac{\text{mass of solute}}{\text{mass of solution}} \times 100\%$$

and molarity:

$$\text{Molarity} = \frac{\text{moles of solute}}{\text{liters of solution}}$$

3. A standard solution is one whose concentration is accurately known. Solutions are often made from a stock solution by dilution. When a solution is diluted, only solvent is added, which means that

Moles of solute after dilution = moles of solute before dilution

4. Normality is defined as the number of equivalents per liter of solution. One equivalent of acid is the amount of acid that furnishes 1 mol of H^+ ions. One equivalent of base is the amount of base that furnishes 1 mol of OH^- ions.

In-Class Discussion Questions

These questions are designed to be considered by groups of students in class. Often these questions work well for introducing a particular topic in class.

1. You have a solution of table salt in water. What happens to the salt concentration (increases, decreases, or stays the same) as the solution boils? Draw pictures to explain your answer.
2. Consider a sugar solution (solution A) with concentration x. You pour one-third of this solution into a beaker, and add an equivalent volume of water (solution B).
   a. What is the ratio of sugar in solutions A and B?
   b. Compare the volumes of solutions A and B.
   c. What is the ratio of the concentrations of sugar in solutions A and B?
3. You need to make 150.0 mL of a 0.10 *M* NaCl solution. You have solid NaCl, and your lab partner has a 2.5 *M* NaCl solution. Explain how you each independently make the solution you need.
4. You have two solutions containing solute A. To determine which solution has the highest concentration of A in molarity, which of the following must you know? (There may be more than one answer.)
   a. the mass in grams of A in each solution
   b. the molar mass of A
   c. the volume of water added to each solution
   d. the total volume of the solution
   Explain your answer.
5. Which of the following do you need to know to calculate the molarity of a salt solution? (There may be more than one answer.)
   a. the mass of salt added
   b. the molar mass of the salt
   c. the volume of water added
   d. the total volume of the solution
   Explain your answer.
6. Consider separate aqueous solutions of HCl and H_2SO_4 with the same concentrations in terms of molarity. You wish to neutralize an aqueous solution of NaOH. For which acid solution would you need to add more volume (in mL) to neutralize the base?
   a. The HCl solution.
   b. The H_2SO_4 solution.
   c. You need to know the acid concentrations to answer this question.
   d. You need to know the volume and concentration of the NaOH solution to answer this question.
   e. c and d
   Explain your answer.
7. Explain why oil and water do not mix. Also, why does oil mix so well with itself? That is, oils are quite viscous (thick) and have rather high boiling points. So why is an "oil molecule" so attracted to another "oil molecule," but not to water?
8. Draw molecular-level pictures to differentiate between concentrated and dilute solutions.
9. Can one solution have a greater concentration than another in terms of weight percent, but a lower concentration in terms of molarity? Explain.
10. How does concentration relate to density?
11. Explain why the formula $M_1V_1 = M_2V_2$ works when solving dilution problems.

Questions and Problems

All even-numbered exercises have answers in the back of this book and solutions in the *Solutions Guide*.

14.1 Solubility

QUESTIONS

1. A solution is a ______ mixture.
2. How do the properties of a *non*homogeneous (heterogeneous) mixture differ from those of a solution? Give two examples of *non*homogeneous mixtures.
3. In a solution, the substance present in the largest amount is called the ______, whereas the other substances present are called the ______.
4. A metallic alloy, such as brass, is an example of a ______ solution.
5. In Chapter 13, you learned that the bonding forces in ionic solids such as NaCl are very strong, yet many ionic solids dissolve readily in water. Explain.
6. Why are some molecular solids (such as sugar or ethyl alcohol) soluble in water, while other molecular solids (such as petroleum) are insoluble in water? What structural feature(s) of *some* molecular solids may tend to make them soluble in water?
7. A substance such as NaCl dissolves in water because the strong ionic forces that exist in solid NaCl can be overcome by, and replaced by, forces between ______ and the ions.
8. When an ionic substance such as potassium bromide, KBr, dissolves, the resulting solution contains separate, hydrated ions that behave ______ of one another.

14.2 Solution Composition: An Introduction

QUESTIONS

9. What does it mean to say that a solution is *saturated?*
10. A solution that has not reached its limit of dissolved solute is said to be ______.
11. A solution is a homogeneous mixture and, unlike a compound, has ______ composition.

12. The label "concentrated H_2SO_4" on a bottle means that there is a relatively ______ amount of H_2SO_4 present in the solution.

14.3 Solution Composition: Mass Percent

QUESTIONS

13. How do we define the *mass percent* composition of a solution? Give an example of a solution, and explain the relative amounts of solute and solvent present in the solution in terms of the mass percent composition.

14. A solution that is 9% by mass glucose contains 9 g of glucose in every ______ g of solution.

PROBLEMS

15. Calculate the mass percent of sodium acetate in each of the following solutions.
    a. 5.00 g of sodium acetate in 25.0 g of water
    b. 10.0 g of sodium acetate in 25.0 g of water
    c. 15.0 g of sodium acetate in 25.0 g of water
    d. 20.0 g of sodium acetate in 25.0 g of water

16. Calculate the mass percent of potassium nitrate, KNO_3, in each of the following solutions.
    a. 2.51 g of KNO_3 in 47.49 g of water
    b. 4.22 kg of KNO_3 in 51.43 kg of water
    c. 6.81 mg of KNO_3 in 36.39 mg of water
    d. 124 mg of KNO_3 in 38.88 g of water

17. Calculate the mass, in grams, of NaCl present in each of the following solutions.
    a. 11.5 g of 6.25% NaCl solution
    b. 6.25 g of 11.5% NaCl solution
    c. 54.3 g of 0.91% NaCl solution
    d. 452 g of 12.3% NaCl solution

18. Consider a solution of glucose in water that is 5.25% by mass glucose. Calculate the indicated quantities.
    a. the mass of glucose in 225 g of the solution
    b. the number of grams of glucose needed to prepare 225 g of the solution
    c. the amount of solution needed to obtain 35.0 g of glucose
    d. the mass of water contained in 225 g of the solution

19. A laboratory assistant prepared a potassium chloride solution for her class by dissolving 5.34 g of KCl in 152 g of water. What is the mass percent of the solution she prepared?

20. A certain alloy is made by dissolving 5.31 g of copper and 4.03 g of zinc in 145 g of iron. Calculate the percent of each component in the alloy.

21. The soda you are drinking contains 0.5% by mass sodium benzoate as a preservative. What approximate mass of sodium benzoate is contained in 1.00 L of the solution assuming that the density of the soda is 1.00 g/mL (the approximate density of water)?

22. If 67.1 g of $CaCl_2$ is added to 275 g of water, calculate the mass percent of $CaCl_2$ in the solution.

23. A solution is to be prepared that will be 4.50% by mass calcium chloride. To prepare 175 g of the solution, what mass of calcium chloride will be needed?

24. How many grams of KBr are contained in 125 g of a 6.25% (by mass) KBr solution?

25. What mass of each solute is present in 285 g of a solution that contains 5.00% by mass NaCl and 7.50% by mass Na_2CO_3?

26. How many grams of $CuCl_2$ are required to prepare 1250. g of a 1.25% (by mass) $CuCl_2$ solution?

27. Concentrated nitric acid is sold as a 70.4% by mass solution. If this solution has a density of 1.42 g/mL, what mass of HNO_3 is contained in 1.00 L of the solution?

28. A hexane solution contains as impurities 5.2% (by mass) heptane and 2.9% (by mass) pentane. Calculate the mass of each component present in 93 g of the solution.

14.4 Solution Composition: Molarity

QUESTIONS

29. A solution that is labeled "0.105 *M* NaOH" would contain ______ mol of NaOH per liter of solution.

30. A solution is labeled "0.121 *M* $AlCl_3$." How many moles of chloride ion would be contained in 1.00 L of the solution?

31. What is a *standard* solution? Describe the steps involved in preparing a standard solution.

32. If you were to prepare exactly 1.00 L of a 5 *M* NaCl solution, you would *not* need exactly 1.00 L of water. Explain.

PROBLEMS

33. For each of the following, the number of moles of solute is given, followed by the total volume of solution prepared. Calculate the molarity.
    a. 0.426 mol of NaOH; 0.500 L
    b. 0.213 mol of NaOH; 0.250 L
    c. 1.28 mol of NaOH; 1.50 L
    d. 4.26 mol of NaOH; 5.00 L

34. For each of the following solutions, the number of moles of solute is given, followed by the total volume of solution prepared. Calculate the molarity.
    a. 0.50 mol KBr; 250 mL
    b. 0.50 mol KBr; 500. mL
    c. 0.50 mol KBr; 750 mL
    d. 0.50 mol KBr; 1.0 L

35. For each of the following solutions the mass of solute is given, followed by the total volume of solution prepared. Calculate the molarity.

a. 321 g of $CaCl_2$; 1.45 L
b. 4.21 mg of NaCl; 1.65 mL
c. 6.45 g KBr; 125 mL
d. 62.5 g NH_4NO_3; 7.25 L

36. For each of the following, the mass of the solute is given, followed by the total volume of the solution prepared. Calculate the molarity.
a. 1.25 g KNO_3; 115 mL
b. 12.5 g KNO_3; 1.15 L
c. 1.25 mg KNO_3; 1.15 mL
d. 1.25 kg KNO_3; 115 L

37. If a 45.3-g sample of potassium nitrate is dissolved in enough water to make 225 mL of solution, what will be the molarity?

38. It is desired to prepare exactly 100. mL of sodium chloride solution. If 2.71 g of NaCl is weighed out, transferred to a volumetric flask, and water added to the 100-mL mark, what is the molarity of the resulting solution?

39. Standard solutions of calcium ion used to test for water hardness are prepared by dissolving pure calcium carbonate, $CaCO_3$, in dilute hydrochloric acid. A 1.745-g sample of $CaCO_3$ is placed in a 250.0-mL volumetric flask and dissolved in HCl. Then the solution is diluted to the calibration mark of the volumetric flask. Calculate the resulting molarity of calcium ion.

40. An alcoholic iodine solution ("tincture" of iodine) is prepared by dissolving 5.15 g of iodine crystals in enough alcohol to make a volume of 225 mL. Calculate the molarity of iodine in the solution.

41. Suppose 1.01 g of $FeCl_3$ is placed in a 10.0-mL volumetric flask, water is added, the mixture is shaken to dissolve the solid, and then water is added to the calibration mark of the flask. Calculate the molarity of each ion present in the solution.

42. If 495 g of NaOH is dissolved to a final total volume of 20.0 L, what is the molarity of the solution?

43. How many *moles* of the indicated solute does each of the following solutions contain?
a. 4.25 mL of 0.105 *M* $CaCl_2$ solution
b. 11.3 mL of 0.405 *M* NaOH solution
c. 1.25 L of 12.1 *M* HCl solution
d. 27.5 mL of 1.98 *M* NaCl solution

44. Calculate the number of *moles* and the number of *grams* of the indicated solutes present in each of the following solution samples.
a. 127 mL of 0.105 *M* HNO_3
b. 155 mL of 15.1 *M* NH_3
c. 2.51 L of 2.01×10^{-3} *M* KSCN
d. 12.2 mL of 2.45 *M* HCl

45. What *mass* of the indicated solute does each of the following solutions contain?
a. 2.50 L of 13.1 *M* HCl solution
b. 15.6 mL of 0.155 *M* NaOH solution
c. 135 mL of 2.01 *M* HNO_3 solution
d. 4.21 L of 0.515 *M* $CaCl_2$ solution

46. What mass in grams of the indicated solute does each of the following solution samples contain?
a. 173 mL of 1.24 *M* KBr solution
b. 2.04 L of 12.1 *M* HCl solution
c. 25 mL of 3.0 *M* NH_3 solution
d. 125 mL of 0.552 *M* $CaCl_2$ solution

47. What mass of KNO_3 is required to prepare 150. mL of 0.250 *M* KNO_3 solution?

48. What volume of 1.0 *M* NaCl solution can be prepared from 1.0 lb of NaCl?

49. Calculate the number of moles of the indicated ion present in each of the following solutions.
a. Na^+ ion in 1.00 L of 0.251 *M* Na_2SO_4 solution
b. Cl^- ion in 5.50 L of 0.10 *M* $FeCl_3$ solution
c. NO_3^- ion in 100. mL of 0.55 *M* $Ba(NO_3)_2$ solution
d. NH_4^+ ion in 250. mL of 0.350 *M* $(NH_4)_2SO_4$ solution

50. Calculate the number of moles of *each* ion present in each of the following solutions.
a. 10.2 mL of 0.451 *M* $AlCl_3$ solution
b. 5.51 L of 0.103 *M* Na_3PO_4 solution
c. 1.75 mL of 1.25 *M* $CuCl_2$ solution
d. 25.2 mL of 0.00157 *M* $Ca(OH)_2$ solution

51. Sodium hydroxide solutions of different concentrations are needed routinely in most chemistry laboratories. Calculate the masses of NaOH required to prepare 1.00 L each of 0.100, 0.500, 3.00, and 6.00 *M* solutions.

52. Strong acid solutions may have their concentration determined by reaction with measured quantities of standard sodium carbonate solution. What mass of Na_2CO_3 is needed to prepare 250. mL of 0.0500 *M* Na_2CO_3 solution?

14.5 Dilution

QUESTIONS

53. When a concentrated stock solution is diluted to prepare a less concentrated reagent, the number of ________ is the same both before and after the dilution.

54. When the volume of a given solution is doubled (by adding water), the new concentration of solute is ________ the original concentration.

PROBLEMS

55. Calculate the new molarity that results when each of the following solutions is diluted to a final total volume of 1.00 L.
a. 425 mL of 0.105 *M* HCl
b. 10.5 mL of 12.1 *M* HCl
c. 25.2 mL of 14.9 *M* HNO_3
d. 6.25 mL of 18.0 *M* H_2SO_4

56. Calculate the new molarity that results when 250. mL of water is added to each of the following solutions.
a. 125 mL of 0.251 *M* HCl
b. 445 mL of 0.499 *M* H_2SO_4
c. 5.25 L of 0.101 *M* HNO_3
d. 11.2 mL of 14.5 *M* $HC_2H_3O_2$

57. Many laboratories keep bottles of 3.0 *M* solutions of the common acids on hand. Given the following molarities of the concentrated acids, determine how many milliliters of each concentrated acid would be required to prepare 225 mL of a 3.0 *M* solution of the acid.

Acid	Molarity of Concentrated Reagent
HCl	12.1 *M*
HNO_3	15.9 *M*
H_2SO_4	18.0 *M*
$HC_2H_3O_2$	17.5 *M*
H_3PO_4	14.9 *M*

58. For convenience, one form of sodium hydroxide that is sold commercially is the saturated solution. This solution is 19.4 *M*, which is approximately 50% by mass sodium hydroxide. What volume of this solution would be needed to prepare 3.50 L of 3.00 *M* NaOH solution?

59. A chemistry student needs 125 mL of 0.150 *M* NaOH solution for her experiment, but the only solution available in the laboratory is 3.02 *M*. Describe how the student could prepare the solution she needs.

60. If 75 mL of 0.211 *M* NaOH is diluted to a final volume of 125 mL, what is the concentration of NaOH in the diluted solution?

61. How much *water* must be added to 500. mL of 0.200 *M* HCl to produce a 0.150 *M* solution? (Assume that the volumes are additive.)

62. If 25.0 mL of 0.104 *M* $BaCl_2$ solution is pipetted into a 100-mL volumetric flask, and water added to the calibration mark, what is the concentration of the resulting dilute solution?

14.6 Stoichiometry of Solution Reactions

PROBLEMS

63. One way to determine the amount of chloride ion in a water sample is to titrate the sample with standard $AgNO_3$ solution to produce solid AgCl.

$$Ag^+(aq) + Cl^-(aq) \rightarrow AgCl(s)$$

If a 25.0-mL water sample requires 27.2 mL of 0.104 *M* $AgNO_3$ in such a titration, what is the concentration of Cl^- in the sample?

64. What volume (in mL) of 0.25 *M* Na_2SO_4 solution is needed to precipitate all the barium, as $BaSO_4(s)$, from 12.5 mL of 0.15 *M* $Ba(NO_3)_2$ solution?

$$Ba(NO_3)_2(aq) + Na_2SO_4(aq) \rightarrow BaSO_4(s) + 2NaNO_3(aq)$$

65. Many metal ions are precipitated from solution by the sulfide ion. As an example, consider treating a solution of copper(II) sulfate with sodium sulfide solution:

$$CuSO_4(aq) + Na_2S(aq) \rightarrow CuS(s) + Na_2SO_4(aq)$$

What volume of 0.105 *M* Na_2S solution would be required to precipitate all of the copper(II) ion from 27.5 mL of 0.121 *M* $CuSO_4$ solution?

66. Calcium oxalate, CaC_2O_4, is very insoluble in water. What mass of sodium oxalate, $Na_2C_2O_4$, is required to precipitate the calcium ion from 37.5 mL of 0.104 *M* $CaCl_2$ solution?

67. When aqueous solutions of lead(II) ion are treated with potassium chromate solution, a bright yellow precipitate of lead(II) chromate, $PbCrO_4$, forms. How many grams of lead chromate form when a 1.00-g sample of $Pb(NO_3)_2$ is added to 25.0 mL of 1.00 *M* K_2CrO_4 solution?

68. Aluminum ion may be precipitated from aqueous solution by addition of hydroxide ion, forming $Al(OH)_3$. A large excess of hydroxide ion must not be added, however, because the precipitate of $Al(OH)_3$ will redissolve to form a soluble aluminum/hydroxide complex. How many grams of solid NaOH should be added to 10.0 mL of 0.250 *M* $AlCl_3$ to just precipitate all the aluminum?

14.7 Neutralization Reactions

PROBLEMS

69. Before strong acids can be disposed of, they are often neutralized to reduce their corrosiveness. What volume of 0.251 *M* NaOH solution would be required to neutralize 135 mL of 0.211 *M* H_2SO_4 solution?

70. What volume of 0.175 *M* HCl solution is needed to neutralize 24.9 mL of 0.451 *M* NaOH solution?

71. The concentration of a sodium hydroxide solution is to be determined. A 50.0-mL sample of 0.104 *M* HCl solution requires 48.7 mL of the sodium hydroxide solution to reach the point of neutralization. Calculate the molarity of the NaOH solution.

72. The total acidity in water samples can be determined by neutralization with standard sodium hydroxide solution. What is the total concentration of hydrogen ion, H^+, present in a water sample if 100. mL of the sample requires 7.2 mL of 2.5×10^{-3} *M* NaOH to be neutralized?

73. What volume of 1.00 *M* NaOH is required to neutralize each of the following solutions?
a. 25.0 mL of 0.154 *M* acetic acid, $HC_2H_3O_2$
b. 35.0 mL of 0.102 *M* hydrofluoric acid, HF

c. 10.0 mL of 0.143 *M* phosphoric acid, H_3PO_4
d. 35.0 mL of 0.220 *M* sulfuric acid, H_2SO_4

74. What volume of 0.101 *M* HNO_3 is required to neutralize each of the following solutions?
a. 12.7 mL of 0.501 *M* NaOH
b. 24.9 mL of 0.00491 *M* $Ba(OH)_2$
c. 49.1 mL of 0.103 *M* NH_3
d. 1.21 L of 0.102 *M* KOH

14.8 Solution Composition: Normality

QUESTIONS

75. One equivalent of an acid is the amount of the acid required to provide ________.

76. A solution that contains 1 equivalent of acid or base per liter is said to be a ________ solution.

77. Explain why the equivalent weight of H_2SO_4 is half the molar mass of this substance. How many hydrogen ions does each H_2SO_4 molecule produce when reacting with an excess of OH^- ions?

78. How many equivalents of hydroxide ion are needed to react with 1.53 equivalents of hydrogen ion? How did you know this when no balanced chemical equation was provided for the reaction?

PROBLEMS

79. For each of the following solutions, calculate the normality.
a. 25.2 mL of 0.105 *M* HCl diluted with water to a total volume of 75.3 mL
b. 0.253 *M* H_3PO_4
c. 0.00103 *M* $Ca(OH)_2$

80. For each of the following solutions, the mass of solute taken is indicated, along with the total volume of solution prepared. Calculate the normality of each solution.
a. 0.113 g NaOH; 10.2 mL
b. 12.5 mg $Ca(OH)_2$; 100. mL
c. 12.4 g H_2SO_4; 155 mL

81. Calculate the normality of each of the following solutions.
a. 0.250 *M* HCl
b. 0.105 *M* H_2SO_4
c. 5.3×10^{-2} *M* H_3PO_4

82. Calculate the normality of each of the following solutions.
a. 0.134 *M* NaOH
b. 0.00521 *M* $Ca(OH)_2$
c. 4.42 *M* H_3PO_4

83. A solution of phosphoric acid, H_3PO_4, is found to contain 35.2 g of H_3PO_4 per liter of solution. Calculate the molarity and normality of the solution.

84. A solution of the sparingly soluble base $Ca(OH)_2$ is prepared in a volumetric flask by dissolving 5.21 mg of $Ca(OH)_2$ to a total volume of 1000. mL. Calculate the molarity and normality of the solution.

85. How many milliliters of 0.50 *N* NaOH are required to neutralize exactly 15.0 mL of 0.35 *N* H_2SO_4?

86. What volume of 0.105 *N* H_3PO_4 is required to neutralize 18.7 mL of 0.204 *M* NaOH solution?

87. What volume of 0.151 *N* NaOH is required to neutralize 24.2 mL of 0.125 *N* H_2SO_4? What volume of 0.151 *N* NaOH is required to neutralize 24.1 mL of 0.125 *M* H_2SO_4?

88. Suppose that 27.34 mL of standard 0.1021 *M* NaOH is required to neutralize 25.00 mL of unknown H_2SO_4 solution. Calculate the molarity and the normality of the unknown solution.

ADDITIONAL PROBLEMS

89. A mixture is prepared by mixing 50.0 g of ethanol, 50.0 g of water, and 5.0 g of sugar. What is the mass percent of each component in the mixture? How many grams of the mixture should one take in order to have 1.5 g of sugar? How many grams of the mixture should one take in order to have 10.0 g of ethanol?

90. Explain the difference in meaning between the following two solutions: "50. g of NaCl dissolved in 1.0 L of water" and "50. g of NaCl dissolved in enough water to make 1.0 L of solution." For which solution can the molarity be calculated directly (using the molar mass of NaCl)?

91. Suppose 50.0 mL of 0.250 *M* $CoCl_2$ solution is added to 25.0 mL of 0.350 *M* $NiCl_2$ solution. Calculate the concentration, in moles per liter, of each of the ions present after mixing. Assume that the volumes are additive.

92. If 500. g of water is added to 75 g of 25% NaCl solution, what is the percent by mass of NaCl in the diluted solution?

93. Calculate the mass of AgCl formed, and the concentration of silver ion remaining in solution, when 10.0 g of solid $AgNO_3$ is added to 50. mL of 1.0×10^{-2} *M* NaCl solution. Assume there is no volume change upon addition of the solid.

94. What mass of $BaSO_4$ will be precipitated from a large container of concentrated $Ba(NO_3)_2$ solution if 37.5 mL of 0.221 *M* H_2SO_4 is added?

95. Many metal ions form insoluble sulfide compounds when a solution of the metal ion is treated with hydrogen sulfide gas. For example, nickel(II) precipitates nearly quantitatively as NiS when H_2S gas is bubbled through a nickel ion solution. How many milliliters of gaseous H_2S at STP are needed to precipitate all the nickel ion present in 10. mL of 0.050 *M* $NiCl_2$ solution?

96. Strictly speaking, the solvent is the component of a solution that is present in the largest amount on a *mole* basis. For solutions involving water, water is almost always the solvent because there tend to be many more water molecules present than molecules of any conceivable solute. To see why this is so, calculate the number of moles of water present in 1.0 L of water. Recall that the density of water is very nearly 1.0 g/mL under most conditions.

97. Aqueous ammonia is typically sold by chemical supply houses as the saturated solution, which has a concentration of 14.5 mol/L. What volume of NH_3 at STP is required to prepare 100. mL of concentrated ammonia solution?

98. What volume of hydrogen chloride gas at STP is required to prepare 500. mL of 0.100 *M* HCl solution?

99. What do we mean when we say that "like dissolves like"? Do two molecules have to be identical to be able to form a solution in one another?

100. The concentration of a solution of HCl is 33.1% by mass, and its density was measured to be 1.147 g/mL. How many milliliters of the HCl solution are required to obtain 10.0 g of HCl?

101. An experiment calls for 1.00 g of silver nitrate, but all that is available in the laboratory is a 0.50% solution of $AgNO_3$. Assuming the density of the silver nitrate solution to be very nearly that of water because it is so dilute, determine how many milliliters of the solution should be used.

102. If 14.2 g of $CaCl_2$ is added to a 50.0-mL volumetric flask, and after dissolving the salt, water is added to the calibration mark of the flask, calculate the molarity of the solution.

103. A solution is 0.1% by mass calcium chloride. Therefore, 100. g of the solution contains ________ g of calcium chloride.

104. Calculate the mass percent of KNO_3 in each of the following solutions.
a. 5.0 g of KNO_3 in 75 g of water
b. 2.5 mg of KNO_3 in 1.0 g of water
c. 11 g of KNO_3 in 89 g of water
d. 11 g of KNO_3 in 49 g of water

105. A 15.0% (by mass) NaCl solution is available. Determine what mass of the solution should be taken to obtain the following quantities of NaCl.
a. 10.0 g
b. 25.0 g
c. 100.0 g
d. 1.00 lb

106. A certain grade of steel is made by dissolving 5.0 g of carbon and 1.5 g of nickel per 100. g of molten iron. What is the mass percent of each component in the finished steel?

107. A sugar solution is prepared in such a way that it contains 10.% dextrose by mass. What quantity of this solution do we need to obtain 25 g of dextrose?

108. How many grams of Na_2CO_3 are contained in 500. g of a 5.5% by mass Na_2CO_3 solution?

109. What mass of KNO_3 is required to prepare 125 g of 1.5% KNO_3 solution?

110. A solution contains 7.5% by mass NaCl and 2.5% by mass KBr. What mass of *each* solute is contained in 125 g of the solution?

111. How many moles of each ion are present in 11.7 mL of 0.102 *M* Na_3PO_4 solution?

112. For each of the following solutions, the number of moles of solute is given, followed by the total volume of solution prepared. Calculate the molarity.
a. 0.10 mol of $CaCl_2$; 25 mL
b. 2.5 mol of KBr; 2.5 L
c. 0.55 mol of $NaNO_3$; 755 mL
d. 4.5 mol of Na_2SO_4; 1.25 L

113. For each of the following solutions, the mass of the solute is given, followed by the total volume of solution prepared. Calculate the molarity.
a. 5.0 g of $BaCl_2$; 2.5 L
b. 3.5 g of KBr; 75 mL
c. 21.5 g of Na_2CO_3; 175 mL
d. 55 g of $CaCl_2$; 1.2 L

114. If 125 g of sucrose, $C_{12}H_{22}O_{11}$, is dissolved in enough water to make 450. mL of solution, calculate the molarity.

115. Concentrated hydrochloric acid is made by pumping hydrogen chloride gas into distilled water. If concentrated HCl contains 439 g of HCl per liter, what is the molarity?

116. If 1.5 g of NaCl is dissolved in enough water to make 1.0 L of solution, what is the molarity of NaCl in the solution?

117. How many *moles* of the indicated solute does each of the following solutions contain?
a. 1.5 L of 3.0 *M* H_2SO_4 solution
b. 35 mL of 5.4 *M* NaCl solution
c. 5.2 L of 18 *M* H_2SO_4 solution
d. 0.050 L of 1.1×10^{-3} *M* NaF solution

118. How many *moles* and how many *grams* of the indicated solute does each of the following solutions contain?
a. 4.25 L of 0.105 *M* KCl solution
b. 15.1 mL of 0.225 *M* $NaNO_3$ solution
c. 25 mL of 3.0 *M* HCl
d. 100. mL of 0.505 *M* H_2SO_4

119. If 10. g of $AgNO_3$ is available, what volume of 0.25 *M* $AgNO_3$ solution can be prepared?

120. Calculate the number of moles of *each* ion present in each of the following solutions.
a. 1.25 L of 0.250 *M* Na_3PO_4 solution
b. 3.5 mL of 6.0 *M* H_2SO_4 solution
c. 25 mL of 0.15 *M* $AlCl_3$ solution
d. 1.50 L of 1.25 *M* $BaCl_2$ solution

121. Calcium carbonate, $CaCO_3$, can be obtained in a very pure state. Standard solutions of calcium ion are usually prepared by dissolving calcium carbonate in acid. What mass of $CaCO_3$ should be taken to prepare 500. mL of 0.0200 *M* calcium ion solution?

122. Calculate the new molarity when 150. mL of water is added to each of the following solutions.
   a. 125 mL of 0.200 *M* HBr
   b. 155 mL of 0.250 *M* $Ca(C_2H_3O_2)_2$
   c. 0.500 L of 0.250 *M* H_3PO_4
   d. 15 mL of 18.0 *M* H_2SO_4

123. How many milliliters of 18.0 *M* H_2SO_4 are required to prepare 35.0 mL of 0.250 *M* solution?

124. When 50. mL of 5.4 *M* NaCl is diluted to a final volume of 300. mL, what is the concentration of NaCl in the diluted solution?

125. When 10. L of water is added to 3.0 L of 6.0 *M* H_2SO_4, what is the molarity of the resulting solution? Assume the volumes are additive.

126. How many milliliters of 0.10 *M* Na_2S solution are required to precipitate all the nickel, as NiS, from 25.0 mL of 0.20 *M* $NiCl_2$ solution?

$$NiCl_2(aq) + Na_2S(aq) \rightarrow NiS(s) + 2NaCl(aq)$$

127. How many grams of $Ba(NO_3)_2$ are required to precipitate all the sulfate ion present in 15.3 mL of 0.139 *M* H_2SO_4 solution?

$$Ba(NO_3)_2(aq) + H_2SO_4(aq) \rightarrow BaSO_4(s) + 2HNO_3(aq)$$

128. What volume of 0.150 *M* HNO_3 solution is needed to exactly neutralize 35.0 mL of 0.150 *M* NaOH solution?

129. What volume of 0.250 *M* HCl is required to neutralize each of the following solutions?
   a. 25.0 mL of 0.103 *M* sodium hydroxide, NaOH
   b. 50.0 mL of 0.00501 *M* calcium hydroxide, $Ca(OH)_2$
   c. 20.0 mL of 0.226 *M* ammonia, NH_3
   d. 15.0 mL of 0.0991 *M* potassium hydroxide, KOH

130. For each of the following solutions, the mass of solute taken is indicated, as well as the total volume of solution prepared. Calculate the normality of each solution.
   a. 15.0 g of HCl; 500. mL
   b. 49.0 g of H_2SO_4; 250. mL
   c. 10.0 g of H_3PO_4; 100. mL

131. Calculate the normality of each of the following solutions.
   a. 0.50 *M* acetic acid, $HC_2H_3O_2$
   b. 0.00250 *M* sulfuric acid, H_2SO_4
   c. 0.10 *M* potassium hydroxide, KOH

132. A sodium dihydrogen phosphate solution was prepared by dissolving 5.0 g of NaH_2PO_4 in enough water to make 500. mL of solution. What are the molarity and normality of the resulting solution?

133. How many milliliters of 0.105 *M* NaOH are required to neutralize exactly 14.2 mL of 0.141 *M* H_3PO_4?

134. If 27.5 mL of 3.5×10^{-2} *N* $Ca(OH)_2$ solution is needed to neutralize 10.0 mL of nitric acid solution of unknown concentration, what is the normality of the nitric acid?

Cumulative Review for Chapters 12–14

QUESTIONS

1. What are some of the general properties of gases that distinguish them from liquids and solids?

2. One of the most obvious properties of gaseous materials is the pressure they exert on their surroundings. In particular, the pressure exerted by the atmospheric gases is important. How does the pressure of the atmosphere arise, and how is this pressure commonly measured?

3. What is the SI unit of pressure? What units of pressure are commonly used in the United States? Why are these common units more convenient to use than the SI unit? Describe a *manometer* and explain how such a device can be used to measure the pressure of gas samples.

4. Your textbook gives several definitions and formulas for Boyle's law for gases. Write, in your *own* words, what this law really tells us about gases. Now write two mathematical expressions that describe Boyle's law. Do these two expressions tell us different things, or are they different representations of the same phenomena? Sketch the general shape of a graph of pressure versus volume for an ideal gas.

5. When using Boyle's law in solving problems in the textbook, you may have noticed that questions were often qualified by stating that "the temperature and amount of gas remain the same." Why was this qualification necessary?

6. What does Charles's law tell us about how the volume of a gas sample varies as the temperature of the sample is changed? How does this volume–temperature relationship *differ* from the volume–pressure relationship of Boyle's law? Give two mathematical expressions that describe Charles's law. For Charles's law to hold true, why must the pressure and amount of gas remain the same?

7. Explain how the concept of absolute zero came about through Charles's studies of gases. Hint: What would happen to the volume of a gas sample at absolute zero (if the gas did not liquefy first)? What temperature scale is defined with its lowest point as the absolute zero of temperature? What is absolute zero in Celsius degrees?

8. What does Avogadro's law tell us about the relationship between the volume of a sample of gas and the number of molecules the gas contains? Why must the temperature and pressure be held constant for valid comparisons using Avogadro's law? Does Avogadro's law describe a direct or an inverse relationship between the volume and the number of moles of gas?

9. What do we mean specifically by an *ideal* gas? Explain why the *ideal gas law* ($PV = nRT$) is actually a combination of Boyle's, Charles's, and Avogadro's gas laws. What is the numerical value and what are the specific units of the universal gas constant, R? Why is close attention to *units* especially important when doing ideal gas law calculations?

10. Dalton's law of partial pressures concerns the properties of mixtures of gases. What is meant by the *partial pressure* of an individual gas in a mixture? How does the *total pressure* of a gaseous mixture depend on the partial pressures of the individual gases in the mixture? How does Dalton's law help us realize that in an ideal gas sample, the volume of the individual molecules is insignificant compared with the bulk volume of the sample?

11. Many common laboratory preparations of gaseous substances involve collecting the gas produced by displacement of water from a receiving container (see Figure 12.12). What happens to a gas sample when it is collected by displacement of water? How is Dalton's law of partial pressures used in determining the partial pressure of the prepared gas in such an experiment?

12. Without consulting your textbook, list and explain the main postulates of the kinetic molecular theory for gases. How do these postulates help us account for the following bulk properties of a gas: the pressure of the gas and why the pressure of the gas increases with increased temperature; the fact that a gas fills its entire container; and the fact that the volume of a given sample of gas increases as its temperature is increased.

13. What does "STP" stand for? What conditions correspond to STP? What is the volume occupied by one mole of an ideal gas at STP?

14. In general, how do we envision the structures of solids and liquids? Explain how the densities and compressibilities of solids and liquids contrast with those properties of gaseous substances. How do we know that the structures of the solid and liquid states of a substance are more comparable to each other than to the properties of the substance in the gaseous state?

15. Describe some of the physical properties of water. Why is water one of the most important substances on earth?

16. Define the *normal* boiling point of water. Why does a sample of boiling water remain at the same temperature until all the water has been boiled? Define the normal freezing point of water. Sketch a representation of a heating/cooling curve for water, marking clearly the normal freezing and boiling points.

17. Are changes in state physical or chemical changes? Explain. What type of forces must be overcome to melt or vaporize a substance (are these forces *intra*molecular or *inter*molecular)? Define the *molar heat of fusion* and

molar heat of vaporization. Why is the molar heat of vaporization of water so much larger than its molar heat of fusion? Why does the boiling point of a liquid vary with altitude?

18. What is a *dipole–dipole attraction?* How do the strengths of dipole–dipole forces compare with the strengths of typical covalent bonds? What is *hydrogen bonding?* What conditions are necessary for hydrogen bonding to exist in a substance or mixture? What experimental evidence do we have for hydrogen bonding?

19. Define *London dispersion forces.* Draw a picture showing how London forces arise. Are London forces relatively strong or relatively weak? Explain. Although London forces exist among all molecules, for what type of molecule are they the *only* major intermolecular force?

20. Why does the process of *vaporization* require an input of energy? Why is it so important that water has a large heat of vaporization? What is *condensation?* Explain how the processes of vaporization and condensation represent an *equilibrium* in a closed container. Define the *equilibrium vapor pressure* of a liquid. Describe how this pressure arises in a closed container. Describe an experiment that demonstrates vapor pressure and enables us to measure the magnitude of that pressure. How is the magnitude of a liquid's vapor pressure related to the intermolecular forces in the liquid?

21. Define a *crystalline solid.* Describe in detail some important types of crystalline solids and name a substance that is an example of each type of solid. Explain how the particles are held together in each type of solid (the interparticle forces that exist). How do the interparticle forces in a solid influence the bulk physical properties of the solid?

22. Define the bonding that exists in metals and how this model explains some of the unique physical properties of metals. What are metal *alloys?* Identify the two main types of alloys, and describe how their structures differ. Give several examples of each type of alloy.

23. Define a *solution.* Describe how an ionic solute such as NaCl dissolves in water to form a solution. How are the strong bonding forces in a crystal of ionic solute overcome? Why do the ions in a solution not attract each other so strongly as to reconstitute the ionic solute? How does a molecular solid such as sugar dissolve in water? What forces between water molecules and the molecules of a molecular solid may help the solute dissolve? Why do some substances *not* dissolve in water at all?

24. Define a *saturated* solution. Does saturated mean the same thing as saying the solution is *concentrated?* Explain. Why does a solute dissolve only to a particular extent in water? How does formation of a saturated solution represent an equilibrium?

25. The concentration of a solution may be expressed in various ways. Two means of expressing concentration were introduced in Chapter 14—mass percent and molarity. How are these two concentration expressions the same, and how do they differ? Suppose 5.0 g of NaCl is dissolved in 15.0 g of water to give a solution volume of 16.1 mL. Explain how you would calculate the mass percent of NaCl and the molarity of NaCl in this solution. Which number did you *not* use for the mass percent calculation? Which number did you *not* use for the molarity calculation? Suppose instead of being given the volume of the solution after mixing, you had been given the density of the solution. Explain how you would calculate the molarity of this solution with this information.

26. When a solution is diluted by adding additional solvent, the *concentration* of solute changes but the *amount* of solute present does not change. Explain. Suppose 250. mL of water is added to 125 mL of 0.551 *M* NaCl solution. Explain *how* you would calculate the concentration of the solution after dilution.

27. What is one *equivalent* of an acid? What does an equivalent of a base represent? How is the equivalent weight of an acid or a base related to the substance's molar mass? Give an example of an acid and a base that have equivalent weights *equal* to their molar masses. Give an example of an acid and a base that have equivalent weights that are *not equal* to their molar masses. What is a *normal* solution of an acid or a base? How is the *normality* of an acid or a base solution related to its *molarity?* Give an example of a solution whose normality is equal to its molarity, and an example of a solution whose normality is *not* the same as its molarity.

PROBLEMS

28. For each of the following gas samples, calculate the new volume or new pressure of the gas sample if the indicated changes are made at constant temperature.
    a. 275 mL of gas; the pressure is increased from 781 mm Hg to 1034 mm Hg
    b. 45.3 L of gas at 1.02 atm; the gas is transferred to a 100.-L container
    c. any sample of gas; the pressure is cut to half its original value

29. Suppose each of the following gas samples is cooled from an initial temperature of 101 °C to the indicated temperatures. Calculate the final volume for each of the gas samples.
    a. 5.23 L, 25 °C
    b. 125 mL, −25 °C
    c. 1.58 L, −201 °C

30. Calculate the indicated quantity for each gas sample.
    a. the volume at STP of a gas sample that occupies 245 mL at 29 °C and 1.51 atm
    b. the volume occupied by 21.6 g of N_2 gas at STP
    c. the partial pressure of each gas if 1.62 g of He and 2.41 g of Ne are combined in a 1.00-L container at STP
    d. the mass of helium present in a sample that has a volume of 11.2 mL at STP

31. A 1.35-g sample of impure $KClO_3$ was heated and decomposed according to the following equation:

$$2KClO_3(s) \rightarrow 2KCl(s) + 3O_2(g)$$

The oxygen gas produced was collected by displacement of water at 775 mm Hg total pressure and 24 °C. The volume of gas collected was 158 mL. Calculate the partial pressure of oxygen gas in the sample (see your textbook for the saturation vapor pressure of water at various temperatures). Calculate the number of moles of oxygen gas collected. Calculate the number of moles of $KClO_3$ that must have decomposed to produce this amount of oxygen. Calculate the mass of $KClO_3$ that must have been originally present in the sample. Calculate the percent $KClO_3$ (by mass) in the original sample.

32. Hydrogen peroxide decomposes as follows:

$$2H_2O_2(aq) \rightarrow 2H_2O(l) + O_2(g)$$

Assuming the oxygen is collected and dried (water vapor removed), calculate the volume of oxygen gas produced at STP if 100. g of 3.11% (by mass) H_2O_2 solution decomposes. What volume would this amount of dry gas have if it were collected at 24 °C and 771 mm Hg instead of STP?

33. Calculate the total energy required to melt 55.1 g of ice at 0 °C, to warm the resulting liquid water from 0 °C to 100 °C, and to boil the water at 100 °C.

34. Calculate the indicated quantity in each of the following.
    a. the percent sodium chloride by mass in a solution made by dissolving 1.01 g of NaCl, 2.11 g of KCl, and 1.55 g of $CaCl_2$ in 151 g of water
    b. the molarity of sodium ion in the solution in part (a), if the density of the resulting solution is 1.02 g/mL
    c. the molarity of chloride ion in the solution in part (a), if the density of the resulting solution is 1.02 g/mL

35. It is desired to prepare some common standard solutions for use in the lab using 500.0-mL volumetric flasks (see Figure 14.7) to contain the solutions. If the following masses of solutes are used, calculate the resulting molarity of each solution.
    a. 4.865 g NaCl
    b. 78.91 g $AgNO_3$
    c. 121.1 g Na_2CO_3

36. Suppose each of the following solutions is diluted to a final volume of 1.00 L in a volumetric flask. Calculate the new concentration of each solution.
    a. 172 mL of 0.501 *M* HCl
    b. 385 mL of 0.115 *M* KOH
    c. 0.285 L of 0.751 *M* HNO_3

37. Calculate the volume (in mL) of each of the following acid solutions that would be required to neutralize 36.2 mL of 0.259 *M* NaOH solution.
    a. 0.271 *M* HCl
    b. 0.119 *M* H_2SO_4
    c. 0.171 *M* H_3PO_4

38. What volume of 0.242 *M* H_2SO_4 produces the same number of moles of H^+ ions as each of the following?
    a. 41.5 mL of 0.118 *M* HCl
    b. 27.1 mL of 0.121 *M* H_3PO_4

15 Acids and Bases

Gargoyles on the Notre Dame cathedral in Paris in need of restoration from decades of acid rain.

Acids are very important substances. They cause lemons to be sour, digest food in the stomach (and sometimes cause heartburn), dissolve rock to make fertilizer, dissolve your tooth enamel to form cavities, and clean the deposits out of your coffee maker. Acids are essential industrial chemicals. In fact, the chemical in first place in terms of the amount manufactured in the United States is sulfuric acid, H_2SO_4. Eighty *billion* pounds of this material are used every year in the manufacture of fertilizers, detergents, plastics, pharmaceuticals, storage batteries, and metals. The acid–base properties of substances also can be used to make interesting novelties such as the foaming chewing gum described on page 459.

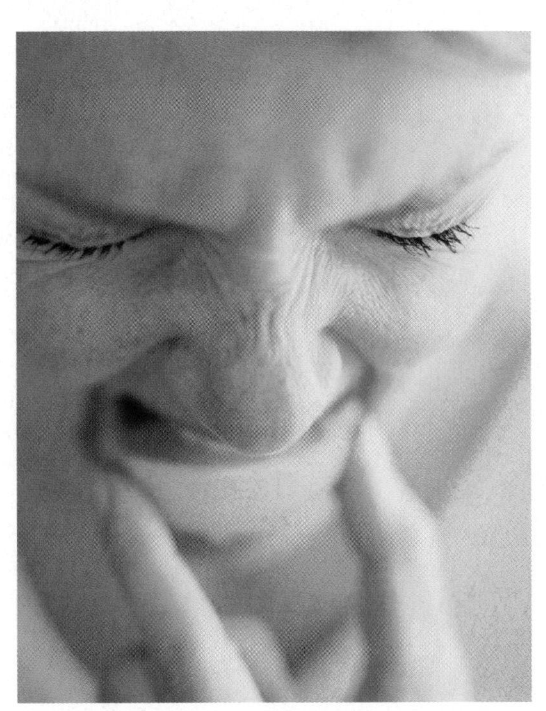
A lemon tastes sour because it contains citric acid.

In this chapter we will consider the most important properties of acids and of their opposites, the bases.

15.1 Acids and Bases

AIM: To learn about two models of acids and bases and the relationship of conjugate acid–base pairs.

Acids were first recognized as substances that taste sour. Vinegar tastes sour because it is a dilute solution of acetic acid; citric acid is responsible for the sour taste of a lemon. Bases, sometimes called *alkalis,* are characterized by their bitter taste and slippery feel. Most hand soaps and commercial preparations for unclogging drains are highly basic.

Don't taste chemical reagents!

The first person to recognize the essential nature of acids and bases was Svante Arrhenius. On the basis of his experiments with electrolytes, Arrhenius postulated that **acids** *produce hydrogen ions in aqueous solution,* whereas **bases** *produce hydroxide ions* (review Section 7.4).

For example, when hydrogen chloride gas is dissolved in water each molecule produces ions as follows:

$$HCl(g) \xrightarrow{H_2O} H^+(aq) + Cl^-(aq)$$

This solution is the strong acid known as hydrochloric acid. On the other hand, when solid sodium hydroxide is dissolved in water, its ions separate producing a solution containing Na^+ and OH^- ions.

$$NaOH(s) \xrightarrow{H_2O} Na^+(aq) + OH^-(aq)$$

This solution is called a strong base.

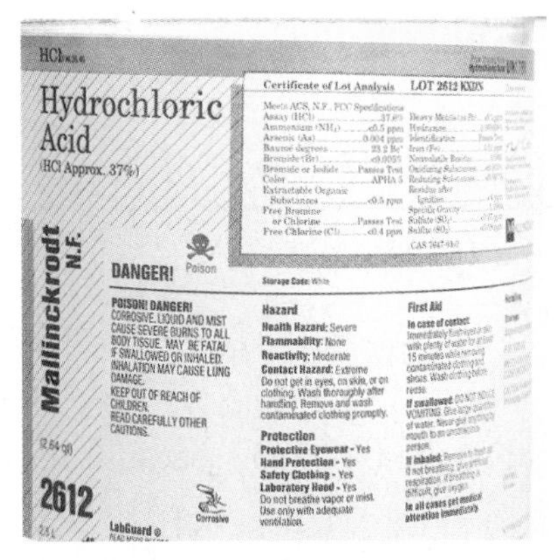

The label on a bottle of concentrated hydrochloric acid.

Although the **Arrhenius concept of acids and bases** was a major step forward in understanding acid–base chemistry, this concept is limited because it allows for only one kind of base—the hydroxide ion. A more general definition of acids and bases was suggested by the Danish chemist Johannes Brønsted and the English chemist Thomas Lowry. In the **Brønsted–Lowry model,** *an acid is a proton (H^+) donor, and a base is a proton acceptor.* According to the Brønsted–Lowry model, the general reaction that occurs when an acid is dissolved in water can best be represented as an acid (HA) donating a proton to a water molecule to form a new acid (the **conjugate acid**) and a new base (the **conjugate base**).

Recall that (*aq*) means the substance is hydrated—it has water molecules clustered around it.

$$\underset{\text{Acid}}{HA(aq)} + \underset{\text{Base}}{H_2O(l)} \rightarrow \underset{\text{Conjugate acid}}{H_3O^+(aq)} + \underset{\text{Conjugate base}}{A^-(aq)}$$

This model emphasizes the significant role of the polar water molecule in pulling the proton from the acid. Note that the conjugate base is everything that remains of the acid molecule after a proton is lost. The conjugate acid is formed when the proton is transferred to the base. A **conjugate acid–base pair** consists of two substances related to each other by the donating and accepting of a *single proton.* In the above equation there are two conjugate acid–base pairs: HA (acid) and A^- (base), and H_2O (base) and H_3O^+ (acid). For example, when hydrogen chloride is dissolved in water it behaves as an acid.

Acid–conjugate base pair

$$HCl(aq) + H_2O(l) \rightarrow H_3O^+(aq) + Cl^-(aq)$$

Base–conjugate acid pair

In this case HCl is the acid that loses an H^+ ion to form Cl^-, its conjugate base. On the other hand, H_2O (behaving as a base) gains an H^+ ion to form H_3O^+ (the conjugate acid).

How can water act as a base? Remember that the oxygen of the water molecule has two unshared electron pairs, either of which can form a covalent bond with an H^+ ion. When gaseous HCl dissolves in water, the following reaction occurs:

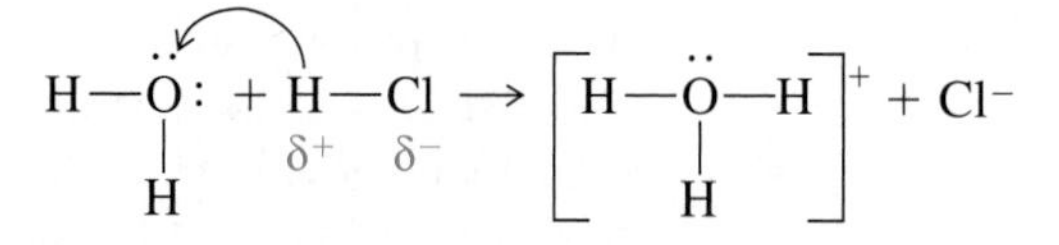

Note that an H^+ ion is transferred from the HCl molecule to the water molecule to form H_3O^+, which is called the **hydronium ion.**

Example 15.1 Identifying Conjugate Acid–Base Pairs

Which of the following represent conjugate acid–base pairs?

a. HF, F^- b. NH_4^+, NH_3 c. HCl, H_2O

Solution

a and b. HF, F^- and NH_4^+, NH_3 are conjugate acid–base pairs because the two species differ by one H^+.

$$HF \rightarrow H^+ + F^-$$

$$NH_4^+ \rightarrow H^+ + NH_3$$

CHEMISTRY IN FOCUS

Gum That Foams

Mad Dawg chewing gum is a practical joker's dream come true. It is noticeably sour when someone first starts to chew it, but the big surprise comes about ten chews later when brightly colored foam oozes from the person's mouth. Although the effect is dramatic, the cause is simple acid–base chemistry.

The foam consists of sugar and saliva churned into a bubbling mess by carbon dioxide released from the gum. The carbon dioxide is formed when sodium bicarbonate ($NaHCO_3$) present in the gum is mixed with citric acid and malic acid (also present in the gum) in the moist environment of the mouth. As $NaHCO_3$ dissolves in the water of the saliva, it separates into its ions:

$$NaHCO_3(s) \xrightarrow{H_2O} Na^+(aq) + HCO_3^-(aq)$$

The bicarbonate ion, when exposed to H^+ ions from acids, decomposes to carbon dioxide and water:*

$$H^+(aq) + HCO_3^-(aq) \rightarrow H_2O(l) + CO_2(g)$$

The acids present in the gum also cause it to be sour, stimulating extra salivation and thus extra foam.

Although the chemistry behind Mad Dawg is well understood, the development of the gum into a safe, but fun, product was not so easy. In fact, early versions of the gum exploded because the acids and the sodium bicarbonate mixed prematurely. As solids, citric and malic acids and sodium bicarbonate do not react with each other. However, the presence of water frees the ions to move and react. In the manufacture of the gum, colorings and flavorings are applied as aqueous solutions. The water caused the gum to explode in early attempts to manufacture it. The makers of Mad Dawg obviously solved the problem. Buy some Mad Dawg and cut it open to see how they did it.

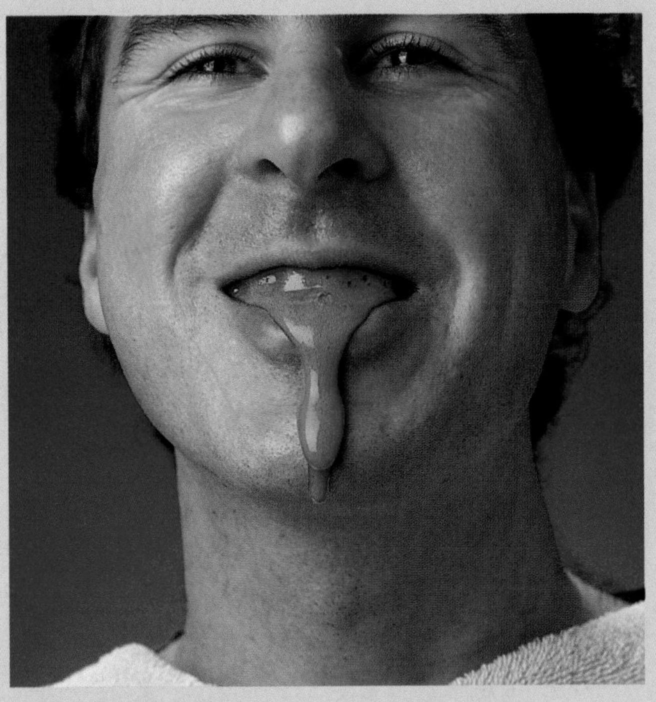

Chewing Mad Dawg gum.

*This reaction is often used to power "bottle rockets" by adding vinegar (dilute acetic acid) to baking soda (sodium bicarbonate).

c. HCl and H_2O are not a conjugate acid–base pair because they are not related by the removal or addition of one H^+. The conjugate base of HCl is Cl^-. The conjugate acid of H_2O is H_3O^+. ■

Example 15.2 Writing Conjugate Bases

Write the conjugate base for each of the following:

a. $HClO_4$ b. H_3PO_4 c. $CH_3NH_3^+$

Solution

To get the conjugate base for an acid, we must remove an H^+ ion.

a. $\underset{\text{Acid}}{HClO_4} \rightarrow H^+ + \underset{\text{Conjugate base}}{ClO_4^-}$

b. $H_3PO_4 \rightarrow H^+ + H_2PO_4^-$
Acid Conjugate base

c. $CH_3NH_3^+ \rightarrow H^+ + CH_3NH_2$
Acid Conjugate base

Self-Check Exercise 15.1

Which of the following represent conjugate acid–base pairs?

a. H_2O, H_3O^+

b. OH^-, HNO_3

c. H_2SO_4, SO_4^{2-}

d. $HC_2H_3O_2$, $C_2H_3O_2^-$

See Problems 15.7 through 15.14. ■

Acid Strength

AIMS: To understand what acid strength means. To understand the relationship between acid strength and the strength of the conjugate base.

We have seen that when an acid dissolves in water, a proton is transferred from the acid to water:

$$HA(aq) + H_2O(l) \rightarrow H_3O^+(aq) + A^-(aq)$$

In this reaction a new acid, H_3O^+ (called the conjugate acid), and a new base, A^- (the conjugate base), are formed. The conjugate acid and base can react with one another,

$$H_3O^+(aq) + A^-(aq) \rightarrow HA(aq) + H_2O(l)$$

to re-form the parent acid and a water molecule. Therefore, this reaction can occur "in both directions." The forward reaction is

$$HA(aq) + H_2O(l) \rightarrow H_3O^+(aq) + A^-(aq)$$

and the reverse reaction is

$$H_3O^+(aq) + A^-(aq) \rightarrow HA(aq) + H_2O(l)$$

Note that the products in the forward reaction are the reactants in the reverse reaction. We usually represent the situation in which the reaction can occur in both directions by double arrows:

$$HA(aq) + H_2O(l) \rightleftharpoons H_3O^+(aq) + A^-(aq)$$

This situation represents a competition for the H^+ ion between H_2O (in the forward reaction) and A^- (in the reverse reaction). If H_2O "wins" this competition—that is, if H_2O has a very high attraction for H^+ compared to A^-—then the solution will contain mostly H_3O^+ and A^-. We describe this situation by saying that the H_2O molecule is a much stronger base (more attraction for H^+) than A^-. In this case the forward reaction predominates:

$$HA(aq) + H_2O(l) \Rightarrow H_3O^+(aq) + A^-(aq)$$

We say that the acid HA is **completely ionized** or **completely dissociated.** This situation represents a **strong acid.**

The opposite situation can also occur. Sometimes A^- "wins" the competition for the H^+ ion. In this case A^- is a much stronger base than H_2O and the reverse reaction predominates:

$$HA(aq) + H_2O(l) \Leftarrow H_3O^+(aq) + A^-(aq)$$

Here A^- has a much larger attraction for H^+ than does H_2O, and most of the HA molecules remain intact. This situation represents a **weak acid.**

We can determine what is actually going on in a solution by measuring its ability to conduct an electric current. Recall from Chapter 7 that a solution can conduct a current in proportion to the number of ions that are present (see Figure 7.2). When 1 mol of solid sodium chloride is dissolved in 1 L of water, the resulting solution is an excellent conductor of an electric current because the Na^+ and Cl^- ions separate completely. We call NaCl a strong electrolyte. Similarly, when 1 mol of hydrogen chloride is dissolved in 1 L of water, the resulting solution is an excellent conductor. Therefore, hydrogen chloride is also a strong electrolyte, which means that each HCl molecule must produce H^+ and Cl^- ions. This tells us that the forward reaction predominates:

A hydrochloric acid solution readily conducts electric current, as shown by the brightness of the bulb.

$$HCl(aq) + H_2O(l) \rightleftharpoons H_3O^+(aq) + Cl^-(aq)$$

(Accordingly, the arrow pointing right is longer than the arrow pointing left.) In solution there are virtually no HCl molecules, only H^+ and Cl^- ions. This shows that Cl^- is a very poor base compared to the H_2O molecule; it has virtually no ability to attract H^+ ions in water. This aqueous solution of hydrogen chloride (called *hydrochloric acid*) is a strong acid.

In general, the strength of an acid is defined by the position of its ionization (dissociation) reaction:

$$HA(aq) + H_2O(l) \rightleftharpoons H_3O^+(aq) + A^-(aq)$$

A strong acid is one for which *the forward reaction predominates.* This means that almost all the original HA is dissociated (ionized) (see Figure 15.1a). There is an important connection between the strength of an acid and that of its conjugate base. *A strong acid contains a relatively weak conjugate base*—one that has a low attraction for protons. A strong acid can be described as an acid whose conjugate base is a much weaker base than water (Figure 15.2). In this case the water molecules win the competition for the H^+ ions.

A strong acid is completely dissociated in water. No HA molecules remain. Only H_3O^+ and A^- are present.

A strong acid has a weak conjugate base.

In contrast to hydrochloric acid, when acetic acid, $HC_2H_3O_2$, is dissolved in water, the resulting solution conducts an electric current only weakly. That is, acetic acid is a weak electrolyte, which means that only a few ions are present. In other words, for the reaction

$$HC_2H_3O_2(aq) + H_2O(l) \rightleftharpoons H_3O^+(aq) + C_2H_3O_2^-(aq)$$

the reverse reaction predominates (thus the arrow pointing left is longer). In fact, measurements show that only about one in one hundred (1%) of the $HC_2H_3O_2$ molecules is dissociated (ionized) in a 0.1 *M* solution of acetic acid. Thus acetic acid is a weak acid. When acetic acid molecules are placed in water, almost all of the molecules remain undissociated. This tells us that the acetate ion, $C_2H_3O_2^-$, is an effective base—it very successfully attracts

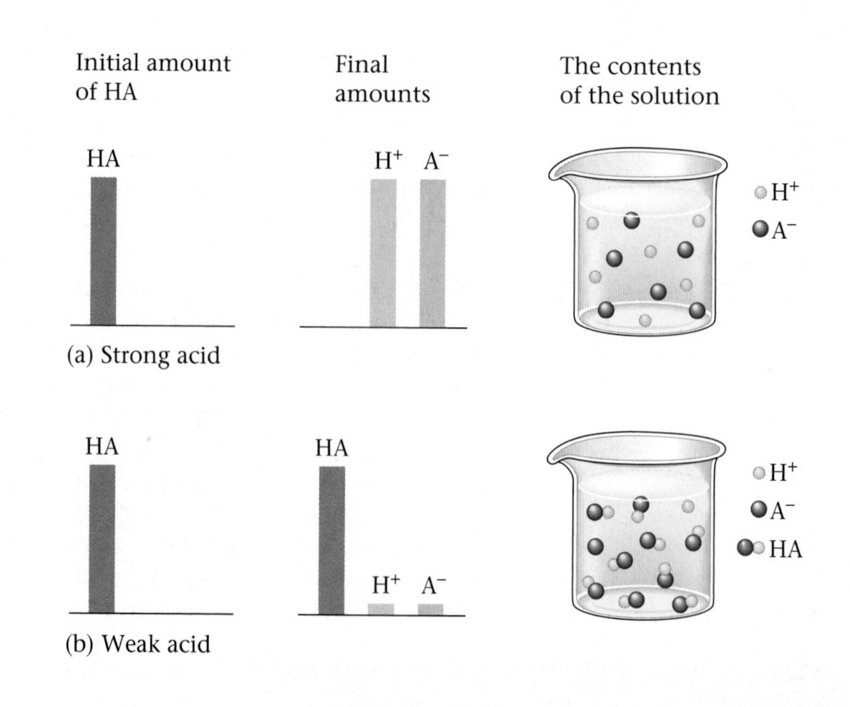

Figure 15.1
Graphical representation of the behavior of acids of different strengths in aqueous solution. (a) A strong acid is completely dissociated. (b) In contrast, only a small fraction of the molecules of a weak acid are dissociated.

A weak acid is mostly undissociated in water.

H^+ ions in water. This means that acetic acid remains largely in the form of $HC_2H_3O_2$ molecules in solution. A weak acid is one for which the *reverse reaction predominates.*

$$HA(aq) + H_2O(l) \Leftarrow H_3O^+(aq) + A^-(aq)$$

Most of the acid originally placed in the solution is still present as HA at equilibrium. That is, a weak acid dissociates (ionizes) only to a very small extent in aqueous solution (see Figure 15.1b). In contrast to a strong acid,

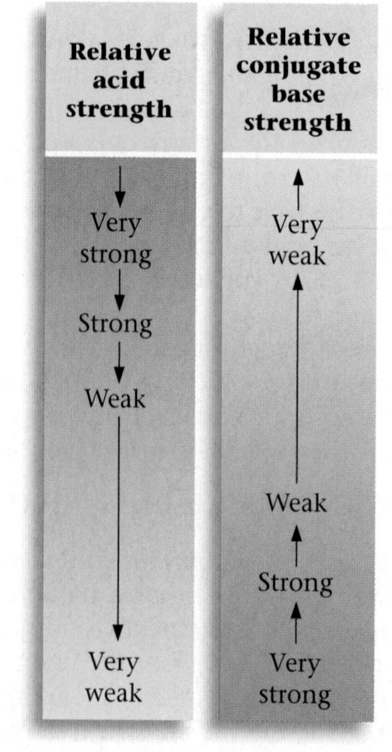

Figure 15.2
The relationship of acid strength and conjugate base strength for the dissociation reaction

$$\underset{\text{Acid}}{HA(aq)} + H_2O(l) \rightleftharpoons H_3O^+(aq) + \underset{\text{Conjugate base}}{A^-(aq)}$$

CHEMISTRY IN FOCUS

Carbonation—A Cool Trick

The sensations of taste and smell greatly affect our daily experience. For example, memories are often triggered by an odor that matches one that occurred when an event was originally stored in our memory banks. Likewise, the sense of taste has a powerful effect on our lives. For example, many people crave the intense sensation produced by the compounds found in chili peppers.

One sensation that is quite refreshing for most people is the effect of a chilled, carbonated beverage in the mouth. The sharp, tingling sensation experienced is not directly due to the bubbling of the dissolved carbon dioxide in the beverage. Rather, it arises because protons are produced as the CO_2 interacts with the water in the tissues of the mouth:

$$CO_2 + H_2O \rightleftharpoons H^+ + HCO_3^-$$

This reaction is speeded up by a biological catalyst—an enzyme—called carbonic anhydrase. The acidification of the fluids in the nerve endings in the mouth leads to the sharp sensation produced by carbonated drinks.

Carbon dioxide also stimulates nerve sites that detect "coolness" in the mouth. In fact, researchers have identified a mutual enhancement between cooling and the presence of CO_2. Studies show that at a given concentration of CO_2, a colder drink feels more "pungent" than a warmer one. When tests were conducted on drinks in which the carbon dioxide concentration was varied, the results showed that a drink felt colder as the CO_2 concentration was increased, even though the drinks were all actually at the same temperature.

Thus a beverage can seem colder if it has a higher concentration of carbon dioxide. At the same time, cooling a carbonated beverage can intensify the tingling sensation caused by the acidity induced by the CO_2. This is truly a happy synergy.

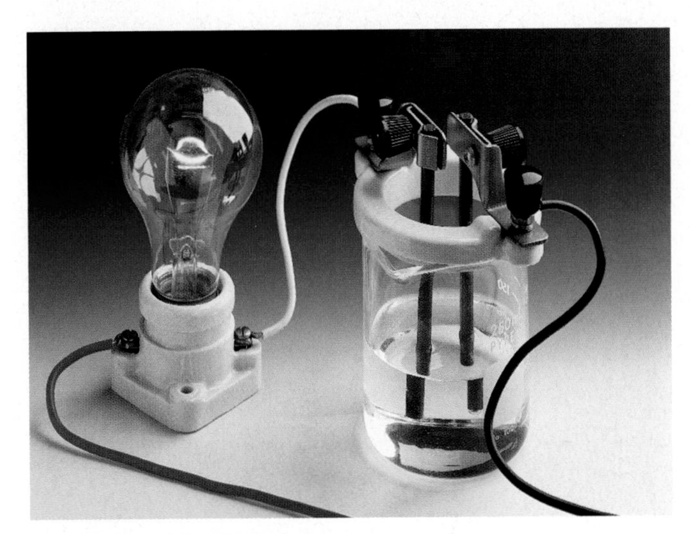

An acetic acid solution conducts only a small amount of current as shown by the dimly lit bulb.

Perchloric acid can explode when handled improperly.

a weak acid has a conjugate base that is a much stronger base than water. In this case a water molecule is not very successful in pulling an H^+ ion away from the conjugate base. *A weak acid contains a relatively strong conjugate base* (Figure 15.2).

The various ways of describing the strength of an acid are summarized in Table 15.1.

The common strong acids are sulfuric acid, $H_2SO_4(aq)$; hydrochloric acid, $HCl(aq)$; nitric acid, $HNO_3(aq)$; and perchloric acid, $HClO_4(aq)$. Sulfuric acid is actually a **diprotic acid,** an acid that can furnish two protons. The acid H_2SO_4 is a strong acid that is virtually 100% dissociated in water:

Table 15.1 Ways to Describe Acid Strength

Property	Strong Acid	Weak Acid
the acid ionization (dissociation) reaction	forward reaction predominates	reverse reaction predominates
strength of the conjugate base compared with that of water	A^- a much weaker base than H_2O	A^- a much stronger base than H_2O

CHEMISTRY IN FOCUS

Plants Fight Back

Plants sometimes do not seem to get much respect. We often think of them as rather dull life forms. We are used to animals communicating with each other, but we think of plants as mute. However, this perception is now changing. It is now becoming clear that plants communicate with other plants and also with insects. Ilya Roskin and his colleagues at Rutgers University, for example, have found that tobacco plants under attack by disease signal distress using the chemical salicylic acid, a precursor of aspirin. When a tobacco plant is infected with tobacco mosaic virus (TMV), which forms dark blisters on leaves and causes them to pucker and yellow, the sick plant produces large amounts of salicylic acid to alert its immune system to fight the virus. In addition, some of the salicylic acid is converted to methyl salicylate, a volatile compound that evaporates from the sick plant. Neighboring plants absorb this chemical and turn it back to salicylic acid, thus triggering their immune systems to protect them against the impending attack by TMV. Thus, as a tobacco plant gears up to fight an attack by TMV, it also warns its neighbors to be ready for this virus.

In another example of plant communication, a tobacco leaf under attack by a caterpillar emits a chemical signal that attracts a parasitic wasp that stings and kills the insect. Even more impressive is the ability of the plant to customize the emitted signal so that the wasp attracted will be the one that specializes in killing the particular caterpillar involved in the attack. The plant does this by changing the proportions of two chemicals emitted when a caterpillar chews on a leaf. Studies have shown that other plants, such as corn and cotton, also emit wasp-attracting chemicals when they face attack by caterpillars.

This research shows that plants can "speak up" to protect themselves. Scientists hope to learn to help them do this even more effectively.

Salicylic acid

Methyl salicylate

A wasp lays its eggs on a gypsy moth caterpillar on the leaf of a corn plant.

$$H_2SO_4(aq) \rightarrow H^+(aq) + HSO_4^-(aq)$$

The HSO_4^- ion is also an acid but it is a weak acid:

$$HSO_4^-(aq) \rightleftharpoons H^+(aq) + SO_4^{2-}(aq)$$

Most of the HSO_4^- ions remain undissociated.

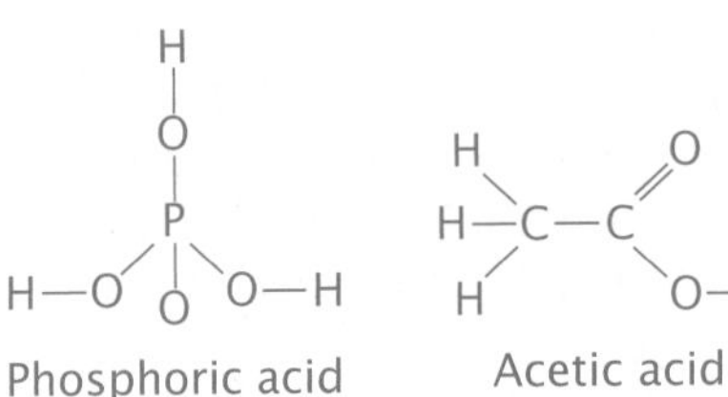
Phosphoric acid   Acetic acid

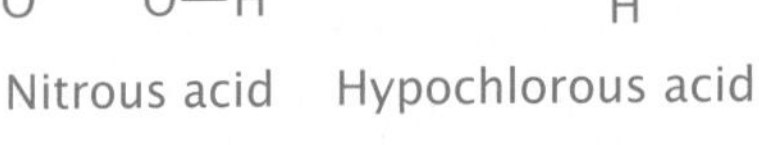
Nitrous acid   Hypochlorous acid

Most acids are **oxyacids,** in which the acidic hydrogen is attached to an oxygen atom (several oxyacids are shown in the margin). The strong acids we have mentioned, except hydrochloric acid, are typical examples. **Organic acids,** those with a carbon-atom backbone, commonly contain the **carboxyl group:**

$$-\mathrm{C}(=\mathrm{O})-\mathrm{O}-\mathrm{H}$$

Acids of this type are usually weak. An example is acetic acid, CH_3COOH, which is often written as $HC_2H_3O_2$.

There are some important acids in which the acidic proton is attached to an atom other than oxygen. The most significant of these are the hydrohalic acids HX, where X represents a halogen atom. Examples are $HCl(aq)$, a strong acid, and $HF(aq)$, a weak acid.

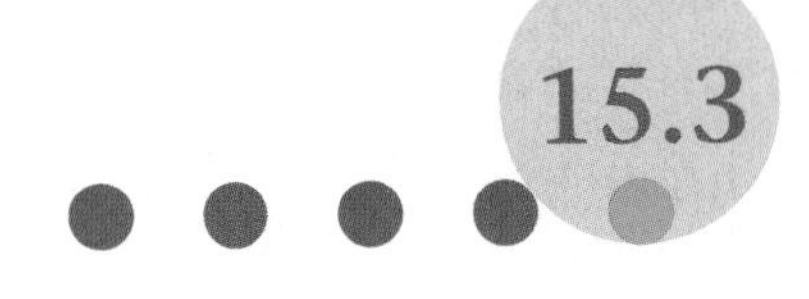

15.3 Water as an Acid and a Base

AIM: To learn about the ionization of water.

A substance is said to be *amphoteric* if it can behave either as an acid or as a base. Water is the most common **amphoteric substance.** We can see this clearly in the **ionization of water,** which involves the transfer of a proton from one water molecule to another to produce a hydroxide ion and a hydronium ion.

$$H_2O(l) + H_2O(l) \rightleftharpoons H_3O^+(aq) + OH^-(aq)$$

In this reaction one water molecule acts as an acid by furnishing a proton, and the other acts as a base by accepting the proton. The forward reaction for this process does not occur to a very great extent. That is, in pure water only a tiny amount of H_3O^+ and OH^- exist. At 25 °C the actual concentrations are

$$[H_3O^+] = [OH^-] = 1.0 \times 10^{-7}\ M$$

Notice that in pure water the concentrations of $[H_3O^+]$ and $[OH^-]$ are equal because they are produced in equal numbers in the ionization reaction.

One of the most interesting and important things about water is that the mathematical *product* of the H_3O^+ and OH^- concentrations is always constant. We can find this constant by multiplying the concentrations of H_3O^+ and OH^- at 25 °C:

$$[H_3O^+][OH^-] = (1.0 \times 10^{-7})(1.0 \times 10^{-7}) = 1.0 \times 10^{-14}$$

We call this constant K_w. Thus at 25 °C

$$[H_3O^+][OH^-] = 1.0 \times 10^{-14} = K_w$$

To simplify the notation we often write H_3O^+ as just H^+. Thus we would write the K_w expression as follows:

$$[H^+][OH^-] = 1.0 \times 10^{-14} = K_w$$

$\mathbf{K_w}$ is called the **ion-product constant** for water. The units are customarily omitted when the value of the constant is given and used.

$K_w = [H^+][OH^-]$
$= 1.0 \times 10^{-14}$

It is important to recognize the meaning of K_w. In any aqueous solution at 25 °C, *no matter what it contains,* the product of $[H^+]$ and $[OH^-]$ must always equal 1.0×10^{-14}. This means that if the $[H^+]$ goes up, the $[OH^-]$ must go down so that the product of the two is still 1.0×10^{-14}. For example, if HCl gas is dissolved in water, increasing the $[H^+]$, the $[OH^-]$ must decrease.

There are three possible situations we might encounter in an aqueous solution. If we add an acid to water (an H^+ donor), we get an *acidic solution.* In this case, because we have added a source of H^+, the $[H^+]$ will be greater than the $[OH^-]$. On the other hand, if we add a base (a source of OH^-) to water, the $[OH^-]$ will be greater than the $[H^+]$. This is a *basic solution.* Finally, we might have a situation in which $[H^+] = [OH^-]$. This is called a *neutral solution.* Pure water is automatically neutral but we can also obtain a neutral solution by adding equal amounts of H^+ and OH^-. It is very important that you understand the definitions of neutral, acidic, and basic solutions. In summary:

Remember that H^+ represents H_3O^+.

1. In a **neutral solution,** $[H^+] = [OH^-]$
2. In an **acidic solution,** $[H^+] > [OH^-]$
3. In a **basic solution,** $[OH^-] > [H^+]$

In each case, however, $K_w = [H^+][OH^-] = 1.0 \times 10^{-14}$.

Example 15.3 Calculating Ion Concentrations in Water

Calculate $[H^+]$ or $[OH^-]$ as required for each of the following solutions at 25 °C, and state whether the solution is neutral, acidic, or basic.

a. $1.0 \times 10^{-5}\ M\ OH^-$

b. $1.0 \times 10^{-7}\ M\ OH^-$

c. $10.0\ M\ H^+$

Solution

$K_w = [H^+][OH^-]$

$\frac{K_w}{[OH^-]} = [H^+]$

a. We know that $K_w = [H^+][OH^-] = 1.0 \times 10^{-14}$. We need to calculate the $[H^+]$. However, the $[OH^-]$ is given—it is $1.0 \times 10^{-5}\ M$—so we will solve for $[H^+]$ by dividing both sides by $[OH^-]$.

$$[H^+] = \frac{1.0 \times 10^{-14}}{[OH^-]} = \frac{1.0 \times 10^{-14}}{1.0 \times 10^{-5}} = 1.0 \times 10^{-9}\ M$$

Because $[OH^-] = 1.0 \times 10^{-5}\ M$ is greater than $[H^+] = 1.0 \times 10^{-9}\ M$, the solution is basic. (Remember: the more negative the exponent, the smaller the number.)

b. Again the $[OH^-]$ is given, so we solve the K_w expression for $[H^+]$.

$$[H^+] = \frac{1.0 \times 10^{-14}}{[OH^-]} = \frac{1.0 \times 10^{-14}}{1.0 \times 10^{-7}} = 1.0 \times 10^{-7}\ M$$

Here $[H^+] = [OH^-] = 1.0 \times 10^{-7}\ M$, so the solution is neutral.

$K_w = [H^+][OH^-]$

$\frac{K_w}{[H^+]} = [OH^-]$

c. In this case the $[H^+]$ is given, so we solve for $[OH^-]$.

$$[OH^-] = \frac{1.0 \times 10^{-14}}{[H^+]} = \frac{1.0 \times 10^{-14}}{10.0} = 1.0 \times 10^{-15}\ M$$

Now we compare $[H^+] = 10.0\ M$ with $[OH^-] = 1.0 \times 10^{-15}\ M$. Because $[H^+]$ is greater than $[OH^-]$, the solution is acidic.

Self-Check Exercise 15.2

Calculate $[H^+]$ in a solution in which $[OH^-] = 2.0 \times 10^{-2}$ *M*. Is this solution acidic, neutral, or basic?

See Problems 15.31 through 15.34. ■

Example 15.4 Using the Ion-Product Constant in Calculations

Is it possible for an aqueous solution at 25 °C to have $[H^+] = 0.010$ *M* and $[OH^-] = 0.010$ *M*?

Solution

The concentration 0.010 *M* can also be expressed as 1.0×10^{-2} *M*. Thus, if $[H^+] = [OH^-] = 1.0 \times 10^{-2}$ *M*, the product

$$[H^+][OH^-] = (1.0 \times 10^{-2})(1.0 \times 10^{-2}) = 1.0 \times 10^{-4}$$

This is not possible. The product of $[H^+]$ and $[OH^-]$ must always be 1.0×10^{-14} in water at 25 °C, so a solution could not have $[H^+] = [OH^-] = 0.010$ *M*. If H^+ and OH^- are added to water in these amounts, they will react with each other to form H_2O,

$$H^+ + OH^- \rightarrow H_2O$$

until the product $[H^+][OH^-] = 1.0 \times 10^{-14}$.

This is a general result. When H^+ and OH^- are added to water in amounts such that the product of their concentrations is greater than 1.0×10^{-14}, they will react to form water until enough H^+ and OH^- are consumed so that $[H^+][OH^-] = 1.0 \times 10^{-14}$. ■

The pH Scale

AIMS: To understand pH and pOH. To learn to find pOH and pH for various solutions. To learn to use a calculator in these calculations.

To express small numbers conveniently, chemists often use the "p scale," which is based on common logarithms (base 10 logs). In this system, if *N* represents some number, then

$$pN = -\log N = (-1) \times \log N$$

That is, the p means to take the log of the number that follows and multiply the result by −1. For example, to express the number 1.0×10^{-7} on the p scale, we first take the log of 1.0×10^{-7}. On most calculators this means entering the number and then pressing the log key.

1. Enter 1.0×10^{-7}.
2. Press the (log) key.

Now the calculator shows the log of the number 1.0×10^{-7}, which is

$$-7.00$$

Next we must multiply by −1. On most calculators this is done by using the (+/−) key (the (+/−) key just reverses the sign, which is what multiplying by −1 really does). In this case we get the result

$$7.00$$

Airplane Rash

Because airplanes remain in service for many years, it is important to spot corrosion that might weaken the structure at an early stage. In the past, looking for minute signs of corrosion has been very tedious and labor-intensive, especially for large planes. This situation is about to change, however, thanks to a new paint system developed by Gerald S. Frankel and Jian Zhang of Ohio State University. The paint they created turns pink in areas that are beginning to corrode, making these areas easy to spot.

The secret to the paint's magic is phenolphthalein, the common acid–base indicator that turns pink in a basic solution. The corrosion of the aluminum skin of the airplane involves a reaction that forms OH^- ions, producing a basic area at the site of the corrosion that turns the phenolphthalein pink. Because this system is highly sensitive, corrosion can be corrected before it damages the plane.

Next time you fly, if the plane has pink spots you might want to wait for a later flight!

So

$$p(1.0 \times 10^{-7}) = -\log(1.0 \times 10^{-7}) = 7.00$$

The pH scale provides a compact way to represent solution acidity.

Because the $[H^+]$ in an aqueous solution is typically quite small, using the p scale in the form of the **pH scale** provides a convenient way to represent solution acidity. The pH is defined as

$$pH = -\log[H^+]$$

To obtain the pH value of a solution, we must compute the negative log of the $[H^+]$. On a typical calculator, this involves the following steps:

Steps for Calculating pH on a Calculator

STEP 1 Enter the $[H^+]$.

STEP 2 Press the (log) key.

STEP 3 Press the (+/–) (change-of-sign) key.

In the case where $[H^+] = 1.0 \times 10^{-5}\ M$, following the above steps gives a pH value of 5.00.

To represent pH to the appropriate number of significant figures, you need to know the following rule for logarithms: *the number of decimal places for a log must be equal to the number of significant figures in the original number.* Thus

2 significant figures

$$[H^+] = 1.0 \times 10^{-5}\ M$$

and

$$pH = 5.00$$

2 decimal places

Example 15.5 Calculating pH

Calculate the pH value for each of the following solutions at 25 °C.

a. A solution in which $[H^+] = 1.0 \times 10^{-9}\ M$

b. A solution in which $[OH^-] = 1.0 \times 10^{-6}\ M$

Solution a

For this solution $[H^+] = 1.0 \times 10^{-9}$.

Step 1 Enter the number.

Step 2 Push the (log) key to give −9.00.

Step 3 Push the (+/−) key to give 9.00.

$$pH = 9.00$$

Solution b

In this case we are given the $[OH^-]$. Thus we must first calculate $[H^+]$ from the K_w expression. We solve

$$K_w = [H^+][OH^-] = 1.0 \times 10^{-14}$$

for $[H^+]$ by dividing both sides by $[OH^-]$.

$$[H^+] = \frac{1.0 \times 10^{-14}}{[OH^-]} = \frac{1.0 \times 10^{-14}}{1.0 \times 10^{-6}} = 1.0 \times 10^{-8}$$

Now that we know the $[H^+]$, we can calculate the pH by taking the three steps listed above. Doing so yields pH = 8.00.

Table 15.2 The Relationship of the H^+ Concentration of a Solution to Its pH

$[H^+]$	pH
1.0×10^{-1}	1.00
1.0×10^{-2}	2.00
1.0×10^{-3}	3.00
1.0×10^{-4}	4.00
1.0×10^{-5}	5.00
1.0×10^{-6}	6.00
1.0×10^{-7}	7.00

✔ Self-Check Exercise 15.3

Calculate the pH value for each of the following solutions at 25 °C.

a. A solution in which $[H^+] = 1.0 \times 10^{-3}\ M$

b. A solution in which $[OH^-] = 5.0 \times 10^{-5}\ M$

See Problems 15.41 through 15.44. ■

The pH decreases as $[H^+]$ increases, and vice versa.

Because the pH scale is a log scale based on 10, *the pH changes by 1 for every power-of-10 change in the $[H^+]$*. For example, a solution of pH 3 has an H^+ concentration of 10^{-3} *M*, which is 10 times that of a solution of pH 4 ($[H^+] = 10^{-4}$ *M*) and 100 times that of a solution of pH 5. This is illustrated in Table 15.2. Also note from Table 15.2 that *the pH decreases as the $[H^+]$ increases*. That is, a lower pH means a more acidic solution. The pH scale and the pH values for several common substances are shown in Figure 15.3.

We often measure the pH of a solution by using a pH meter, an electronic device with a probe that can be inserted into a solution of unknown pH. A pH meter is shown in Figure 15.4. Colored indicator paper is also commonly used to measure the pH of a solution when less accuracy is needed. A drop of the solution to be tested is placed on this special paper, which promptly turns to a color characteristic of a given pH (see Figure 15.5).

Log scales similar to the pH scale are used for representing other quantities. For example,

The symbol p means −log.

$$pOH = -\log[OH^-]$$

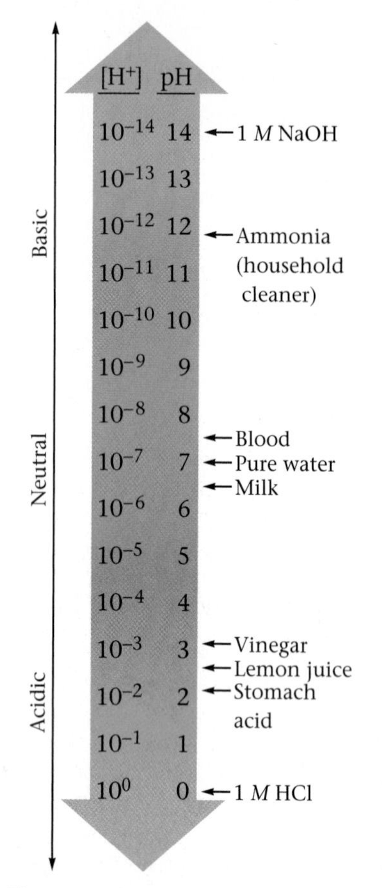

Figure 15.3
The pH scale and pH values of some common substances.

Figure 15.4
A pH meter. The electrodes on the right are placed in the solution with unknown pH. The difference between the [H$^+$] in the solution sealed into one of the electrodes and the [H$^+$] in the solution being analyzed is translated into an electrical potential and registered on the meter as a pH reading.

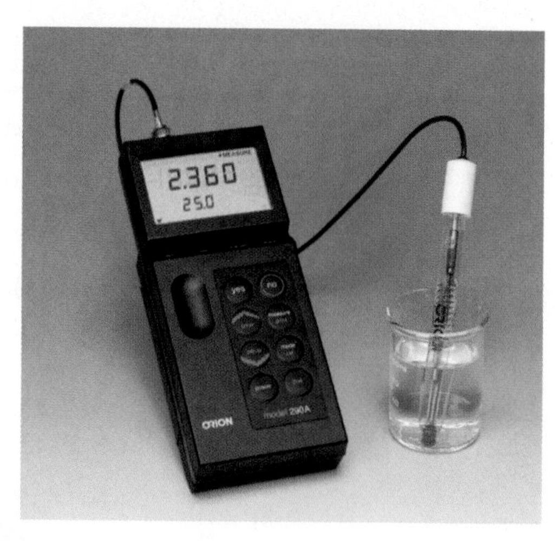

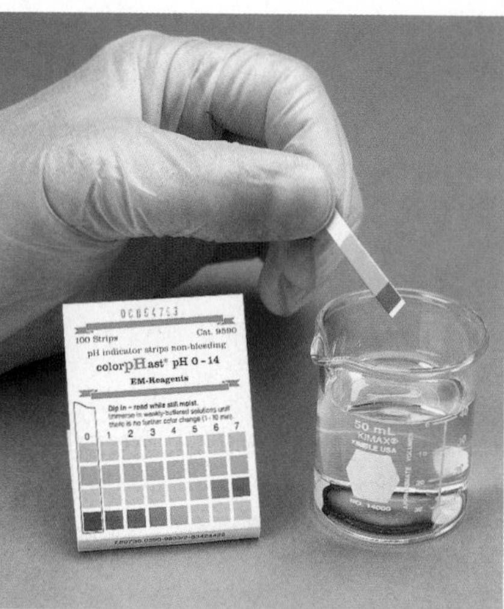

Figure 15.5
Indicator paper being used to measure the pH of a solution. The pH is determined by comparing the color that the solution turns the paper to the color chart.

Therefore, in a solution in which

$$[OH^-] = 1.0 \times 10^{-12}\,M$$

the pOH is

$$-\log[OH^-] = -\log(1.0 \times 10^{-12}) = 12.00$$

Example 15.6 Calculating pH and pOH

Calculate the pH and pOH for each of the following solutions at 25 °C.

a. $1.0 \times 10^{-3}\,M\ OH^-$

b. $1.0\,M\ H^+$

Solution

a. We are given the [OH$^-$], so we can calculate the pOH value by taking $-\log[OH^-]$.

$$pOH = -\log[OH^-] = -\log(1.0 \times 10^{-3}) = 3.00$$

To calculate the pH, we must first solve the K_w expression for [H$^+$].

$$[H^+] = \frac{K_w}{[OH^-]} = \frac{1.0 \times 10^{-14}}{1.0 \times 10^{-3}} = 1.0 \times 10^{-11}\,M$$

Now we compute the pH.

$$\text{pH} = -\log[\text{H}^+] = -\log(1.0 \times 10^{-11}) = 11.00$$

b. In this case we are given the $[H^+]$ and we can compute the pH.

$$\text{pH} = -\log[\text{H}^+] = -\log(1.0) = 0$$

We next solve the K_w expression for $[OH^-]$.

$$[\text{OH}^-] = \frac{K_w}{[\text{H}^+]} = \frac{1.0 \times 10^{-14}}{1.0} = 1.0 \times 10^{-14}\,M$$

Now we compute the pOH.

$$\text{pOH} = -\log[\text{OH}^-] = -\log(1.0 \times 10^{-14}) = 14.00 \quad ■$$

We can obtain a convenient relationship between pH and pOH by starting with the K_w expression $[H^+][OH^-] = 1.0 \times 10^{-14}$ and taking the negative log of both sides.

$$-\log([\text{H}^+][\text{OH}^-]) = -\log(1.0 \times 10^{-14})$$

Because the log of a product equals the sum of the logs of the terms—that is, $\log(A \times B) = \log A + \log B$—we have

$$\underbrace{-\log[\text{H}^+]}_{\text{pH}}\ \underbrace{-\log[\text{OH}^-]}_{\text{pOH}} = -\log(1.0 \times 10^{-14}) = 14.00$$

which gives the equation

$$\text{pH} + \text{pOH} = 14.00$$

This means that once we know either the pH or the pOH for a solution, we can calculate the other. For example, if a solution has a pH of 6.00, the pOH is calculated as follows:

$$\begin{aligned} \text{pH} + \text{pOH} &= 14.00 \\ \text{pOH} &= 14.00 - \text{pH} \\ \text{pOH} &= 14.00 - 6.00 = 8.00 \end{aligned}$$

Example 15.7 Calculating pOH from pH

The pH of blood is about 7.4. What is the pOH of blood?

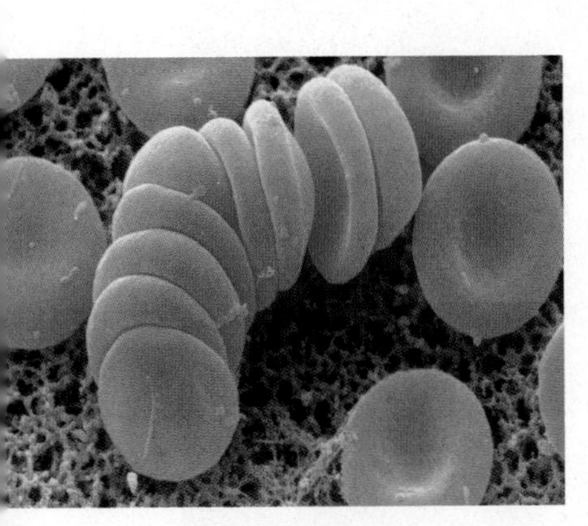

Red blood cells can exist only over a narrow range of pH.

Solution

$$\begin{aligned} \text{pH} + \text{pOH} &= 14.00 \\ \text{pOH} &= 14.00 - \text{pH} \\ &= 14.00 - 7.4 \\ &= 6.6 \end{aligned}$$

The pOH of blood is 6.6.

✔ **Self-Check Exercise 15.4**

A sample of rain in an area with severe air pollution has a pH of 3.5. What is the pOH of this rainwater?

See Problems 15.45 and 15.46. ■

It is also possible to find the $[H^+]$ or $[OH^-]$ from the pH or pOH. To find the $[H^+]$ from the pH, we must go back to the definition of pH:

$$pH = -\log[H^+]$$

or

$$-pH = \log[H^+]$$

To arrive at $[H^+]$ on the right-hand side of this equation we must "undo" the log operation. This is called taking the *antilog* or the *inverse* log.

$$\text{Inverse log }(-pH) = \text{inverse log }(\log[H^+])$$
$$\text{Inverse log }(-pH) = [H^+]$$

This operation may involve a 10^x key on some calculators.

There are different methods for carrying out the inverse log operation on various calculators. One common method is the two-key (inv) (log) sequence. (Consult the user's manual for your calculator to find out how to do the antilog or inverse log operation.) The steps in going from pH to $[H^+]$ are as follows:

Steps for Calculating $[H^+]$ from pH

STEP 1 Enter the pH.

STEP 2 Change the sign of the pH by using the (+/–) key.

STEP 3 Take the inverse log (antilog) of $-pH$ to give $[H^+]$ by using the (inv) (log) keys in that order. (Your calculator may require different keys for this operation.)

For practice, we will convert pH = 7.0 to $[H^+]$.

Step 1 pH = 7.0

Step 2 $-pH = -7.0$

Measuring the pH of the water in a river.

Step 3 (inv) (log) -7.0 gives 1×10^{-7}

$$[H^+] = 1 \times 10^{-7}\,M$$

This process is illustrated further in Example 15.8.

Example 15.8 Calculating $[H^+]$ from pH

The pH of a human blood sample was measured to be 7.41. What is the $[H^+]$ in this blood?

Solution

Step 1 $pH = 7.41$

Step 2 $-pH = -7.41$

Step 3 $[H^+]$ = inverse log of -7.41

$$\text{(inv) (log)}\ -7.41 = 3.9 \times 10^{-8}$$
$$[H^+] = 3.9 \times 10^{-8}\,M$$

Notice that because the pH has two decimal places, we need two significant figures for $[H^+]$.

✔ Self-Check Exercise 15.5

The pH of rainwater in a polluted area was found to be 3.50. What is the $[H^+]$ for this rainwater?

See Problems 15.49 and 15.50. ■

A similar procedure is used to change from pOH to $[OH^-]$, as shown in Example 15.9.

Example 15.9 Calculating $[OH^-]$ from pOH

The pOH of the water in a fish tank is found to be 6.59. What is the $[OH^-]$ for this water?

Solution

We use the same steps as for converting pH to $[H^+]$, except that we use the pOH to calculate the $[OH^-]$.

Step 1 $pOH = 6.59$

Step 2 $-pOH = -6.59$

Step 3 (inv) (log) $-6.59 = 2.6 \times 10^{-7}$

$$[OH^-] = 2.6 \times 10^{-7}\,M$$

Note that two significant figures are required.

✔ Self-Check Exercise 15.6

The pOH of a liquid drain cleaner was found to be 10.50. What is the $[OH^-]$ for this cleaner?

See Problems 15.51 and 15.52. ■

Calculating the pH of Strong Acid Solutions

AIM: To learn to calculate the pH of solutions of strong acids.

In this section we will learn to calculate the pH for a solution containing a strong acid of known concentration. For example, if we know a solution contains 1.0 *M* HCl, how can we find the pH of the solution? To answer this question we must know that when HCl dissolves in water, each molecule dissociates (ionizes) into H^+ and Cl^- ions. That is, we must know that HCl is a strong acid. Thus, although the label on the bottle says 1.0 *M* HCl, the solution contains virtually no HCl molecules. A 1.0 *M* HCl solution contains H^+ and Cl^- ions rather than HCl molecules. Typically, container labels indicate the substance(s) used to make up the solution but do not necessarily describe the solution components after dissolution. In this case,

$$1.0\,M\,\text{HCl} \rightarrow 1.0\,M\,\text{H}^+ \text{ and } 1.0\,M\,\text{Cl}^-$$

Therefore, the $[H^+]$ in the solution is 1.0 *M*. The pH is then

$$\text{pH} = -\log[\text{H}^+] = -\log(1.0) = 0$$

Example 15.10 Calculating the pH of Strong Acid Solutions

Calculate the pH of 0.10 *M* HNO_3.

Solution

HNO_3 is a strong acid, so the ions in solution are H^+ and NO_3^-. In this case,

$$0.10\,M\,\text{HNO}_3 \rightarrow 0.10\,M\,\text{H}^+ \text{ and } 0.10\,M\,\text{NO}_3^-$$

Thus

$$[\text{H}^+] = 0.10\,M \qquad \text{and} \qquad \text{pH} = -\log(0.10) = 1.00$$

✔ Self-Check Exercise 15.7

Calculate the pH of a solution of 5.0×10^{-3} *M* HCl.

See Problems 15.57 and 15.58. ■

Buffered Solutions

AIM: To understand the general characteristics of buffered solutions.

Water: pH = 7
0.01 *M* HCl: pH = 2

A **buffered solution** is one that resists a change in its pH even when a strong acid or base is added to it. For example, when 0.01 mol of HCl is added to 1 L of pure water, the pH changes from its initial value of 7 to 2, a change of 5 pH units. However, when 0.01 mol of HCl is added to a solution containing both 0.1 *M* acetic acid ($HC_2H_3O_2$) and 0.1 *M* sodium acetate ($NaC_2H_3O_2$), the pH changes from an initial value of 4.74 to 4.66, a change of only 0.08 pH units. The latter solution is buffered—it undergoes only a very slight change in pH when a strong acid or base is added to it.

Buffered solutions are vitally important to living organisms whose cells can survive only in a very narrow pH range. Many goldfish have died because their owners did not realize the importance of buffering the aquar-

For goldfish to survive, the pH of the water must be carefully controlled.

ium water at an appropriate pH. For humans to survive, the pH of the blood must be maintained between 7.35 and 7.45. This narrow range is maintained by several different buffering systems.

A solution is **buffered** by the *presence of a weak acid and its conjugate base.* An example of a buffered solution is an aqueous solution that contains acetic acid and sodium acetate. The sodium acetate is a salt that furnishes acetate ions (the conjugate base of acetic acid) when it dissolves. To see how this system acts as a buffer, we must recognize that the species present in this solution are

$$HC_2H_3O_2, \quad \underbrace{Na^+, \quad C_2H_3O_2^-}$$

When $NaC_2H_3O_2$ is dissolved, it produces the separated ions

What happens in this solution when a strong acid such as HCl is added? In pure water, the H^+ ions from the HCl would accumulate, thus lowering the pH.

$$HCl \xrightarrow{100\%} H^+ + Cl^-$$

However, this buffered solution contains $C_2H_3O_2^-$ ions, which are basic. That is, $C_2H_3O_2^-$ has a strong affinity for H^+, as evidenced by the fact that $HC_2H_3O_2$ is a weak acid. This means that the $C_2H_3O_2^-$ and H^+ ions do not exist together in large numbers. Because the $C_2H_3O_2^-$ ion has a high affinity for H^+, these two combine to form $HC_2H_3O_2$ molecules. Thus the H^+ from the added HCl does not accumulate in solution but reacts with the $C_2H_3O_2^-$ as follows:

$$H^+(aq) + C_2H_3O_2^-(aq) \rightarrow HC_2H_3O_2(aq)$$

Next consider what happens when a strong base such as sodium hydroxide is added to the buffered solution. If this base were added to pure water, the OH^- ions from the solid would accumulate and greatly change (raise) the pH.

$$NaOH \xrightarrow{100\%} Na^+ + OH^-$$

However, in the buffered solution the OH^- ion, which has a *very strong* affinity for H^+, reacts with $HC_2H_3O_2$ molecules as follows:

$$HC_2H_3O_2(aq) + OH^-(aq) \rightarrow H_2O(l) + C_2H_3O_2^-(aq)$$

This happens because, although $C_2H_3O_2^-$ has a strong affinity for H^+, OH^- has a much stronger affinity for H^+ and thus can remove H^+ ions from acetic acid molecules.

Note that the buffering materials dissolved in the solution prevent added H^+ or OH^- from building up in the solution. Any added H^+ is trapped by $C_2H_3O_2^-$ to form $HC_2H_3O_2$. Any added OH^- reacts with $HC_2H_3O_2$ to form H_2O and $C_2H_3O_2^-$.

The general properties of a buffered solution are summarized in Table 15.3.

Table 15.3 The Characteristics of a Buffer

1. The solution contains a weak acid HA and its conjugate base A^-.
2. The buffer resists changes in pH by reacting with any added H^+ or OH^- so that these ions do not accumulate.
3. Any added H^+ reacts with the base A^-.

   $H^+(aq) + A^-(aq) \rightarrow HA(aq)$

4. Any added OH^- reacts with the weak acid HA.

   $OH^-(aq) + HA(aq) \rightarrow H_2O(l) + A^-(aq)$

CHAPTER 15 REVIEW

KEY TERMS

acid (15.1)
base (15.1)
Arrhenius concept of acids and bases (15.1)
Brønsted–Lowry model (15.1)
conjugate acid (15.1)
conjugate base (15.1)
conjugate acid–base pair (15.1)
hydronium ion (15.1)
completely ionized (dissociated) (15.2)
strong acid (15.2)
weak acid (15.2)
diprotic acid (15.2)
oxyacid (15.2)
organic acid (15.2)
carboxyl group (15.2)
amphoteric substance (15.3)
ionization of water (15.3)
ion-product constant, K_w (15.3)
pH scale (15.4)
buffered solution (15.6)
buffered (15.6)

SUMMARY

1. Acids or bases in water are commonly described by two different models. Arrhenius postulated that acids produce H^+ ions in aqueous solutions and that bases produce OH^- ions. The Brønsted–Lowry model is more general: an acid is a proton donor, and a base is a proton acceptor. Water acts as a Brønsted–Lowry base when it accepts a proton from an acid to form a hydronium ion:

$$\underset{\text{Acid}}{HA(aq)} + \underset{\text{Base}}{H_2O(l)} \rightleftharpoons \underset{\text{Conjugate acid}}{H_3O^+(aq)} + \underset{\text{Conjugate base}}{A^-(aq)}$$

A conjugate base is everything that remains of the acid molecule after the proton is lost. A conjugate acid is formed when a proton is transferred to the base. Two substances related in this way are called a conjugate acid–base pair.

2. A strong acid or base is one that is completely ionized (dissociated). A weak acid is one that is ionized (dissociated) only to a slight extent. Strong acids have weak conjugate bases. Weak acids have relatively strong conjugate bases.

3. Water is an amphoteric substance—it can behave either as an acid or as a base. The ionization of water reveals this property; one water molecule transfers a proton to another water molecule to produce a hydronium ion and a hydroxide ion.

$$H_2O(l) + H_2O(l) \rightleftharpoons H_3O^+(aq) + OH^-(aq)$$

The expression

$$K_w = [H_3O^+][OH^-] = [H^+][OH^-]$$

is called the ion-product constant. It has been shown experimentally that at 25 °C,

$$[H^+] = [OH^-] = 1.0 \times 10^{-7}\ M$$

so $K_w = 1.0 \times 10^{-14}$.

4. In an acidic solution, $[H^+]$ is greater than $[OH^-]$. In a basic solution, $[OH^-]$ is greater than $[H^+]$. In a neutral solution, $[H^+] = [OH^-]$.

5. To describe $[H^+]$ in aqueous solutions, we use the pH scale.

$$pH = -\log[H^+]$$

Note that the pH decreases as $[H^+]$ (acidity) increases.

6. The pH of strong acid solutions can be calculated directly from the concentration of the acid, because 100% dissociation occurs in aqueous solution.

7. A buffered solution is one that resists a change in its pH even when a strong acid or base is added to it. A buffered solution contains a weak acid and its conjugate base.

IN-CLASS DISCUSSION QUESTIONS

These questions are designed to be considered by groups of students in class. Often these questions work well for introducing a particular topic in class.

1. You are asked for the H^+ concentration in a solution of NaOH(*aq*). Because sodium hydroxide is a strong base, can we say there is no H^+, since having H^+ would imply that the solution is acidic?
2. Explain why Cl^- does not affect the pH of an aqueous solution.
3. Write the general reaction for an acid acting in water. What is the base in this case? The conjugate acid? The conjugate base?
4. Differentiate among the terms *concentrated, dilute, weak,* and *strong* in describing acids. Use molecular-level pictures to support your answer.
5. What is meant by "pH"? True or false: A strong acid always has a lower pH than a weak acid does. Explain.
6. Consider two separate solutions: one containing a weak acid, HA, and one containing HCl. Assume that you start with 10 molecules of each.
   a. Draw a molecular-level picture of what each solution looks like.

b. Arrange the following from strongest to weakest base: Cl^-, H_2O, A^-. Explain.

7. Why is the pH of water at 25 °C equal to 7.00?

8. Can the pH of a solution be negative? Explain.

9. Stanley's grade-point average (GPA) is 3.28. What is Stanley's p(GPA)?

10. A friend asks the following: "Consider a buffered solution made up of the weak acid HA and its salt NaA. If a strong base like NaOH is added, the HA reacts with the OH^- to make A^-. Thus, the amount of acid (HA) is decreased, and the amount of base (A^-) is increased. Analogously, adding HCl to the buffered solution forms more of the acid (HA) by reacting with the base (A^-). How can we claim that a buffered solution resists changes in the pH of the solution?" How would you explain buffering to your friend?

11. Mixing together aqueous solutions of acetic acid and sodium hydroxide can make a buffered solution. Explain.

12. Could a buffered solution be made by mixing aqueous solutions of HCl and NaOH? Explain.

Questions and Problems

All even-numbered exercises have answers in the back of this book and solutions in the *Solutions Guide*.

15.1 Acids and Bases

QUESTIONS

1. What are some physical properties that historically led chemists to classify various substances as acids and bases?

2. In the Arrhenius definition, what characterizes an acid? What characterizes a base? Why are the Arrhenius definitions too restrictive?

3. What is an *acid* in the Brønsted–Lowry model? What is a *base?* What specific ion is the basis for the Brønsted–Lowry model for acid–base reactions?

4. How do the components of a conjugate acid–base pair differ from one another? Give an example of a conjugate acid–base pair to illustrate your answer.

5. Using the symbol HA to represent a general acid, write an equation showing how HA forms its conjugate Brønsted–Lowry base when dissolved in water.

6. When an acid is dissolved in water, what ion does the water form? What is the relationship of this ion to water itself?

PROBLEMS

7. Which of the following form a conjugate acid–base pair? For those pairs that are *not* conjugate acid–base pairs, write the *correct* conjugate acid–base pair for each species.
   a. H_3PO_4, PO_4^{3-} c. HCN, CN^-
   b. HBr, BrO^- d. HF, F^-

8. Which of the following represent conjugate acid–base pairs? For those pairs that are not conjugates, write the correct conjugate acid or base for each species in the pair.
   a. HSO_4^-, SO_4^{2-} c. $H_2PO_4^-$, PO_4^{3-}
   b. HBr, Br^- d. HNO_3, NO_2^-

9. In each of the following chemical equations, identify the conjugate acid–base pairs.
   a. $HF + H_2O \rightleftharpoons F^- + H_3O^+$
   b. $CN^- + H_2O \rightleftharpoons HCN + OH^-$
   c. $HCO_3^- + H_2O \rightleftharpoons H_2CO_3 + OH^-$

10. In each of the following chemical equations, identify the conjugate acid–base pairs.
   a. $NH_4^+ + H_2O \rightleftharpoons NH_3 + H_3O^+$
   b. $HC_2H_3O_2 + H_2O \rightleftharpoons C_2H_3O_2^- + H_3O^+$
   c. $CH_3NH_2 + H_2O \rightleftharpoons CH_3NH_3^+ + OH^-$

11. Write the conjugate *acid* for each of the following bases:
   a. HSO_4^- c. ClO_4^-
   b. SO_3^{2-} d. $H_2PO_4^-$

12. Write the conjugate *acid* for each of the following bases.
   a. HCO_3^- c. ClO_2^-
   b. Br^- d. CH_3NH^-

13. Write the conjugate *base* for each of the following acids:
   a. H_2S c. NH_3
   b. HS^- d. H_2SO_3

14. Write the conjugate *base* for each of the following acids.
   a. HCO_3^- c. HSO_3^-
   b. $HBrO_4$ d. $HC_2H_3O_2$

15. Write a chemical equation showing how each of the following species behaves as a *base* when dissolved in water.
   a. NH_3 c. O^{2-}
   b. NH_2^- d. F^-

16. Write a chemical equation showing how each of the following species can behave as an *acid* when dissolved in water.
   a. $HBrO_3$ c. HNO_3
   b. HI d. NH_4^+

15.2 Acid Strength

QUESTIONS

17. What does it mean to say that an acid is *strong* in aqueous solution? What does this reveal about the ability of the acid's anion to attract protons?

18. What does it mean to say that an acid is *weak* in aqueous solution? What does this reveal about the ability of the acid's anion to attract protons?

19. How is the strength of an acid related to the fact that a competition for protons exists in aqueous solution between water molecules and the anion of the acid?

20. A strong acid has a weak conjugate base, whereas a weak acid has a relatively strong conjugate base. Explain.

21. Write the formula for the *hydronium* ion. Write an equation for the formation of the hydronium ion when an acid is dissolved in water.

22. Name four strong acids. For each of these, write the equation showing the acid dissociating in water.

23. What group of atoms characterizes organic acids? Are most organic acids strong or weak?

24. Most acids are oxyacids, in which the acidic proton is bonded to an oxygen atom. Write the formulas of three acids that are oxyacids. Write the formulas of three acids that are *not* oxyacids.

25. Which of the following acids have relatively *strong* conjugate bases?
a. HCN c. $HBrO_4$
b. H_2S d. HNO_3

26. Which of the following bases have relatively *strong* conjugate acids?
a. SO_4^{2-} c. CN^-
b. Br^- d. CH_3COO^- ($C_2H_3O_2^-$)

15.3 Water as an Acid and a Base

QUESTIONS

27. What does it mean to say that a substance is *amphoteric?* Write equations illustrating how water exhibits amphoterism.

28. Anions containing hydrogen (for example, HCO_3^- and $H_2PO_4^{2-}$) show amphoteric behavior when reacting with other acids or bases. Write equations illustrating the amphoterism of these anions.

29. What is meant by the *ion-product constant* for water, K_w. What does this constant signify? Write an equation for the chemical reaction from which the constant is derived.

30. What happens to the hydroxide ion concentration in aqueous solutions when we increase the hydrogen ion concentration by adding an acid? What happens to the hydrogen ion concentration in aqueous solutions when we increase the hydroxide ion concentration by adding a base? Explain.

PROBLEMS

31. Calculate the $[H^+]$ in each of the following solutions, and indicate whether the solution is acidic or basic.
a. $[OH^-] = 5.99 \times 10^{-8}\ M$
b. $[OH^-] = 8.99 \times 10^{-6}\ M$
c. $[OH^-] = 7.00 \times 10^{-7}\ M$
d. $[OH^-] = 1.43 \times 10^{-12}\ M$

32. Calculate the $[H^+]$ in each of the following solutions, and indicate whether the solution is acidic, basic, or neutral.
a. $[OH^-] = 3.99 \times 10^{-5}\ M$
b. $[OH^-] = 2.91 \times 10^{-9}\ M$
c. $[OH^-] = 7.23 \times 10^{-2}\ M$
d. $[OH^-] = 9.11 \times 10^{-7}\ M$

33. Calculate the $[OH^-]$ in each of the following solutions, and indicate whether the solution is acidic, basic, or neutral.
a. $[H^+] = 8.89 \times 10^{-7}\ M$
b. $[H^+] = 1.19 \times 10^{-7}\ M$
c. $[H^+] = 7.00 \times 10^{-7}\ M$
d. $[H^+] = 1.00 \times 10^{-7}\ M$

34. Calculate the $[OH^-]$ in each of the following solutions, and indicate whether the solution is acidic or basic.
a. $[H^+] = 1.34 \times 10^{-2}\ M$
b. $[H^+] = 6.99 \times 10^{-7}\ M$
c. $[H^+] = 4.01 \times 10^{-9}\ M$
d. $[H^+] = 4.02 \times 10^{-13}\ M$

35. For each pair of concentrations, tell which represents the more acidic solution.
a. $[H^+] = 1.2 \times 10^{-3}\ M$ or $[H^+] = 4.5 \times 10^{-4}\ M$
b. $[H^+] = 2.6 \times 10^{-6}\ M$ or $[H^+] = 4.3 \times 10^{-8}\ M$
c. $[H^+] = 0.000010\ M$ or $[H^+] = 0.0000010\ M$

36. For each pair of concentrations, tell which represents the more basic solution.
a. $[H^+] = 1.59 \times 10^{-7}\ M$ or $[H^+] = 1.04 \times 10^{-8}\ M$
b. $[H^+] = 5.69 \times 10^{-8}\ M$ or $[OH^-] = 4.49 \times 10^{-6}\ M$
c. $[H^+] = 5.99 \times 10^{-8}\ M$ or $[OH^-] = 6.01 \times 10^{-7}\ M$

15.4 The pH Scale

QUESTIONS

37. Why do scientists tend to express the acidity of a solution in terms of its pH, rather than in terms of the molarity of hydrogen ion present? How is pH defined mathematically?

38. Using Figure 15.3, list the approximate pH of five "everyday" solutions. How do the familiar properties (such as sour taste for acids) of these solutions correspond to their pH?

39. For a hydrogen ion concentration of $2.33 \times 10^{-6}\ M$, how many *decimal places* should we give when expressing the pH of the solution?

40. As the hydrogen ion concentration of a solution *increases,* does the pH of the solution increase or decrease? Explain.

PROBLEMS

41. Calculate the pH corresponding to each of the hydrogen ion concentrations given below, and indicate whether each solution is acidic or basic.

a. $[H^+] = 4.97 \times 10^{-7}\ M$
b. $[H^+] = 1.01 \times 10^{-12}\ M$
c. $[H^+] = 2.49 \times 10^{-4}\ M$
d. $[H^+] = 1.00 \times 10^{-7}\ M$

42. Calculate the pH corresponding to each of the hydrogen ion concentrations given below. Tell whether each solution is acidic, basic, or neutral.
a. $[H^+] = 0.00100\ M$
b. $[H^+] = 2.19 \times 10^{-4}\ M$
c. $[H^+] = 9.18 \times 10^{-11}\ M$
d. $[H^+] = 4.71 \times 10^{-7}\ M$

43. Calculate the pH corresponding to each of the hydroxide ion concentrations given below. Tell whether each solution is acidic, basic, or neutral.
a. $[OH^-] = 7.42 \times 10^{-5}\ M$
b. $[OH^-] = 0.00151\ M$
c. $[OH^-] = 3.31 \times 10^{-2}\ M$
d. $[OH^-] = 9.01 \times 10^{-9}\ M$

44. Calculate the pH of each of the solutions indicated below. Tell whether each solution is acidic or basic.
a. $[H^+] = 3.99 \times 10^{-6}\ M$
b. $[OH^-] = 4.21 \times 10^{-8}\ M$
c. $[H^+] = 8.25 \times 10^{-11}\ M$
d. $[OH^-] = 9.21 \times 10^{-3}\ M$

45. Calculate the pH corresponding to each of the pOH values listed, and indicate whether each solution is acidic, basic, or neutral.
a. pOH = 4.32
b. pOH = 8.90
c. pOH = 1.81
d. pOH = 13.1

46. Calculate the pOH corresponding to each of the pH values listed, and tell whether the solutions are acidic or basic.
a. pH = 7.45
b. pH = 1.89
c. pH = 13.15
d. pH = 5.55

47. For each hydrogen ion concentration listed, calculate the pH of the solution as well as the concentration of hydroxide ion in the solution. Indicate whether the solutions are acidic or basic.
a. $[H^+] = 4.76 \times 10^{-8}\ M$
b. $[H^+] = 8.92 \times 10^{-3}\ M$
c. $[H^+] = 7.00 \times 10^{-5}\ M$
d. $[H^+] = 1.25 \times 10^{-12}\ M$

48. For each hydroxide ion concentration listed, calculate the pOH of the solution as well as the concentration of hydrogen ion in the solution. Indicate whether the solutions are acidic or basic.
a. $[OH^-] = 4.01 \times 10^{-2}\ M$
b. $[OH^-] = 9.87 \times 10^{-9}\ M$
c. $[OH^-] = 5.23 \times 10^{-5}\ M$
d. $[OH^-] = 6.29 \times 10^{-12}\ M$

49. Calculate the hydrogen ion concentration, in moles per liter, for solutions with each of the following pH values.
a. pH = 9.01
b. pH = 6.89
c. pH = 1.02
d. pH = 7.00

50. Calculate the hydrogen ion concentration, in moles per liter, for solutions with each of the following pH values.
a. pH = 1.04
b. pH = 13.1
c. pH = 5.99
d. pH = 8.62

51. Calculate the hydrogen ion concentration, in moles per liter, for solutions with each of the following pOH values.
a. pOH = 4.95
b. pOH = 7.00
c. pOH = 12.94
d. pOH = 1.02

52. Calculate the hydroxide ion concentration, in moles per liter, for solutions with each of the following pOH values.
a. pOH = 7.00
b. pOH = 12.91
c. pOH = 1.82
d. pOH = 9.41

53. Calculate the pH of each of the following solutions from the information given.
a. $[H^+] = 4.78 \times 10^{-2}\ M$
b. pOH = 4.56
c. $[OH^-] = 9.74 \times 10^{-3}\ M$
d. $[H^+] = 1.24 \times 10^{-8}\ M$

54. Calculate the pH of each of the following solutions from the information given.
a. pOH = 11.31
b. $[OH^-] = 7.22 \times 10^{-5}\ M$
c. $[H^+] = 9.93 \times 10^{-4}\ M$
d. $[OH^-] = 1.49 \times 10^{-8}\ M$

15.5 Calculating the pH of Strong Acid Solutions

QUESTIONS

55. When 1 mol of gaseous hydrogen chloride is dissolved in enough water to make 1 L of solution, approximately how many HCl molecules remain in the solution? Explain.

56. A bottle of acid solution is labeled "3 M HNO_3." What are the substances that are actually present in the solution? Are any HNO_3 molecules present? Why or why not?

PROBLEMS

57. Calculate the hydrogen ion concentration and the pH of each of the following solutions of strong acids.
a. $1.04 \times 10^{-4}\ M$ HCl
b. $0.00301\ M$ HNO_3
c. $5.41 \times 10^{-4}\ M$ $HClO_4$
d. $6.42 \times 10^{-2}\ M$ HNO_3

58. Calculate the pH of each of the following solutions of strong acids.

a. 1.21×10^{-3} *M* HNO_3
b. 0.000199 *M* $HClO_4$
c. 5.01×10^{-5} *M* HCl
d. 0.00104 *M* HBr

15.6 Buffered Solutions

QUESTIONS

59. What characteristic properties do buffered solutions possess?

60. What two components make up a buffered solution? Give an example of a combination that would serve as a buffered solution.

61. Which component of a buffered solution is capable of combining with an added strong acid? Using your example from Question 60, show how this component would react with added HCl.

62. Which component of a buffered solution consumes added strong base? Using your example from Question 60, show how this component would react with added NaOH.

PROBLEMS

63. Which of the following combinations would act as buffered solutions?
a. HCl and NaCl
b. CH_3COOH and KCH_3COO
c. H_2S and NaHS
d. H_2S and Na_2S

64. For those combinations in Question 63 that behave as buffered solutions, write equations showing how the components of the buffer consume added strong acid (HCl) or strong base (NaOH).

Additional Problems

65. The concepts of acid–base equilibria were developed in this chapter for aqueous solutions (in aqueous solutions, water is the solvent and is intimately involved in the equilibria). However, the Brønsted–Lowry acid–base theory can be extended easily to other solvents. One such solvent that has been investigated in depth is liquid ammonia, NH_3.
a. Write a chemical equation indicating how HCl behaves as an acid in liquid ammonia.
b. Write a chemical equation indicating how OH^- behaves as a base in liquid ammonia.

66. *Strong bases* are bases that completely ionize in water to produce hydroxide ion, OH^-. The strong bases include the hydroxides of the Group 1 elements. For example, if 1.0 mol of NaOH is dissolved per liter, the concentration of OH^- ion is 1.0 *M*. Calculate $[OH^-]$, pOH, and pH for each of the following strong base solutions.
a. 0.10 *M* NaOH
b. 2.0×10^{-4} *M* KOH
c. 6.2×10^{-3} *M* CsOH
d. 0.0001 *M* NaOH

67. Which of the following conditions indicate an *acidic* solution?
a. $pH < 7.0$
b. $pOH < 7.0$
c. $[H^+] > [OH^-]$
d. $[H^+] > 1.0 \times 10^{-7}$ *M*

68. Which of the following conditions indicate a *basic* solution?
a. $pOH < 7.0$
b. $pH > 7.0$
c. $[OH^-] < [H^+]$
d. $[H^+] < 1.0 \times 10^{-7}$ *M*

69. Buffered solutions are mixtures of a weak acid and its conjugate base. Explain why a mixture of a *strong* acid and its conjugate base (such as HCl and Cl^-) is not buffered.

70. Which of the following acids are classified as *strong* acids?
a. HNO_3
b. CH_3COOH ($HC_2H_3O_2$)
c. HCl
d. HF
e. $HClO_4$

71. Is it possible for a solution to have $[H^+] = 0.002$ *M* and $[OH^-] = 5.2 \times 10^{-6}$ *M* at 25 °C? Explain.

72. Despite HCl's being a strong acid, the pH of 1.00×10^{-7} *M* HCl is *not* exactly 7.00. Can you suggest a reason why?

73. According to Arrhenius, bases are species that produce ________ ion in aqueous solution.

74. According to the Brønsted–Lowry model, a base is a species that ________ protons.

75. A conjugate acid–base pair consists of two substances related by the donating and accepting of a(n) ________.

76. Acetate ion, $C_2H_3O_2^-$, has a stronger affinity for protons than does water. Therefore, when dissolved in water, acetate ion behaves as a(n) ________.

77. An acid such as HCl that strongly conducts an electric current when dissolved in water is said to be a(n) ________ acid.

78. Organic acids, which behave as typical weak acids, contain the ________ group.

79. Because of ________, even pure water contains measurable quantities of H^+ and OH^-.

80. The ion-product constant for water, K_w, has the value ________ at 25 °C.

81. The number of ________ in the logarithm of a number is equal to the number of significant figures in the number.

82. A solution with pH = 4 has a (higher/lower) hydrogen ion concentration than a solution with pOH = 4.

83. A 0.20 *M* HCl solution contains ________ *M* hydrogen ion and ________ *M* chloride ion concentrations.

84. A buffered solution is one that resists a change in ________ when either a strong acid or a strong base is added to it.

85. A solution is buffered when a weak acid and its ________ are present in comparable amounts.

86. When sodium hydroxide, NaOH, is added dropwise to a buffered solution, the ________ component of the buffer consumes the added hydroxide ion.

87. When hydrochloric acid, HCl, is added dropwise to a buffered solution, the ________ component of the buffer consumes the added hydrogen ion.

88. Which of the following represent conjugate acid–base pairs? For those pairs that are not conjugates, write the correct conjugate acid or base for each species in the pair.
a. H_2O, OH^-
b. H_2SO_4, SO_4^{2-}
c. H_3PO_4, $H_2PO_4^-$
d. $HC_2H_3O_2$, $C_2H_3O_2^-$

89. In each of the following chemical equations, identify the conjugate acid–base pairs.
a. $CH_3NH_2 + H_2O \rightleftharpoons CH_3NH_3^+ + OH^-$
b. $CH_3COOH + NH_3 \rightleftharpoons CH_3COO^- + NH_4^+$
c. $HF + NH_3 \rightleftharpoons F^- + NH_4^+$

90. Write the conjugate *acid* for each of the following:
a. NH_3
b. NH_2^-
c. H_2O
d. OH^-

91. Write the conjugate *base* for each of the following:
a. H_3PO_4
b. HCO_3^-
c. HF
d. H_2SO_4

92. Write chemical equations showing the ionization (dissociation) in water for each of the following acids.
a. CH_3CH_2COOH (Only the last H is acidic.)
b. NH_4^+
c. H_2SO_4
d. H_3PO_4

93. Which of the following bases have relatively *strong* conjugate acids?
a. F^-
b. Cl^-
c. HSO_4^-
d. NO_3^-

94. Calculate $[H^+]$ in each of the following solutions, and indicate whether the solution is acidic, basic, or neutral.
a. $[OH^-] = 4.22 \times 10^{-3}\ M$
b. $[OH^-] = 1.01 \times 10^{-13}\ M$
c. $[OH^-] = 3.05 \times 10^{-7}\ M$
d. $[OH^-] = 6.02 \times 10^{-6}\ M$

95. Calculate $[OH^-]$ in each of the following solutions, and indicate whether the solution is acidic, basic, or neutral.
a. $[H^+] = 4.21 \times 10^{-7}\ M$
b. $[H^+] = 0.00035\ M$
c. $[H^+] = 0.00000010\ M$
d. $[H^+] = 9.9 \times 10^{-6}\ M$

96. For each pair of concentrations, tell which represents the more basic solution.
a. $[H^+] = 0.000013\ M$ or $[OH^-] = 0.0000032\ M$
b. $[H^+] = 1.03 \times 10^{-6}\ M$ or $[OH^-] = 1.54 \times 10^{-8}\ M$
c. $[OH^-] = 4.02 \times 10^{-7}\ M$ or $[OH^-] = 0.0000001\ M$

97. Calculate the pH of each of the solutions indicated below. Tell whether the solution is acidic, basic, or neutral.
a. $[H^+] = 1.49 \times 10^{-3}\ M$
b. $[OH^-] = 6.54 \times 10^{-4}\ M$
c. $[H^+] = 9.81 \times 10^{-9}\ M$
d. $[OH^-] = 7.45 \times 10^{-10}\ M$

98. Calculate the pH corresponding to each of the hydroxide ion concentrations given below. Tell whether each solution is acidic, basic, or neutral.
a. $[OH^-] = 1.4 \times 10^{-6}\ M$
b. $[OH^-] = 9.35 \times 10^{-9}\ M$
c. $[OH^-] = 2.21 \times 10^{-1}\ M$
d. $[OH^-] = 7.98 \times 10^{-12}\ M$

99. Calculate the pOH corresponding to each of the pH values listed, and indicate whether each solution is acidic, basic, or neutral.
a. pH = 1.02
b. pH = 13.4
c. pH = 9.03
d. pH = 7.20

100. For each hydrogen or hydroxide ion concentration listed, calculate the concentration of the complementary ion and the pH and pOH of the solution.
a. $[H^+] = 5.72 \times 10^{-4}\ M$
b. $[OH^-] = 8.91 \times 10^{-5}\ M$
c. $[H^+] = 2.87 \times 10^{-12}\ M$
d. $[OH^-] = 7.22 \times 10^{-8}\ M$

101. Calculate the hydrogen ion concentration, in moles per liter, for solutions with each of the following pH values.
a. pH = 8.34
b. pH = 5.90
c. pH = 2.65
d. pH = 12.6

102. Calculate the hydrogen ion concentration, in moles per liter, for solutions with each of the following pH or pOH values.
a. pH = 5.41
b. pOH = 12.04
c. pH = 11.91
d. pOH = 3.89

103. Calculate the hydrogen ion concentration, in moles per liter, for solutions with each of the following pH or pOH values.
a. pOH = 0.90
b. pH = 0.90
c. pOH = 10.3
d. pH = 5.33

104. Calculate the hydrogen ion concentration and the pH of each of the following solutions of strong acids.
a. $1.4 \times 10^{-3}\ M\ HClO_4$
b. $3.0 \times 10^{-5}\ M$ HCl
c. $5.0 \times 10^{-2}\ M\ HNO_3$
d. 0.0010 M HCl

105. Write the formulas for *three* combinations of weak acid and salt that would act as buffered solutions. For each of your combinations, write chemical equations showing how the components of the buffered solution would consume added acid and base.

16 Equilibrium

Equilibrium can be analogous to traffic flowing both ways on a bridge, such as San Francisco's Golden Gate Bridge. ▼

Chemistry is mostly about reactions—processes in which groups of atoms are reorganized. So far we have learned to describe chemical reactions by using balanced equations and to calculate amounts of reactants and products. However, there are many important characteristics of reactions that we have not yet considered.

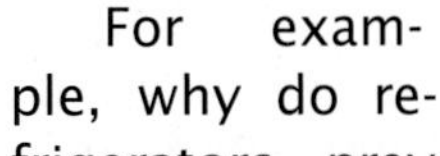

Refrigeration prevents food spoilage.

For example, why do refrigerators prevent food from spoiling? That is, why do the chemical reactions that cause food to decompose occur more slowly at lower temperatures? On the other hand, how can a chemical manufacturer speed up a chemical reaction that runs too slowly to be economical?

Another question that arises is why chemical reactions carried out in a closed vessel appear to stop at a certain point. For example, when the reaction of reddish-brown nitrogen dioxide to form colorless dinitrogen tetroxide,

$$\underset{\text{Reddish-brown}}{2NO_2(g)} \rightarrow \underset{\text{Colorless}}{N_2O_4(g)}$$

is carried out in a closed container, the reddish-brown color at first fades but stops changing after a time and then stays the same color indefinitely if left undisturbed (see Figure 16.1). We will account for all of these important observations about reactions in this chapter.

16.1 How Chemical Reactions Occur

AIM: To understand the collision model of how chemical reactions occur.

In writing the equation for a chemical reaction, we put the reactants on the left and the products on the right with an arrow between them. But how do the atoms in the reactants reorganize to form the products?

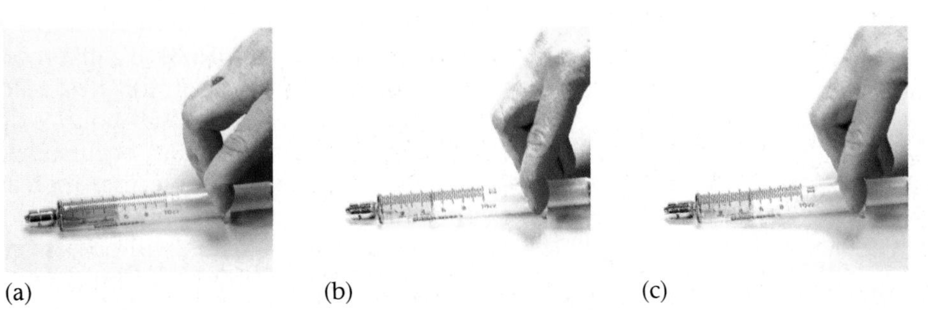

(a) (b) (c)

Figure 16.1
(a) A sample containing a large quantity of reddish-brown NO_2 gas. (b) As the reaction to form colorless N_2O_4 occurs, the sample becomes lighter brown. (c) After equilibrium is reached [$2NO_2(g) \rightleftharpoons N_2O_4(g)$], the color remains the same.

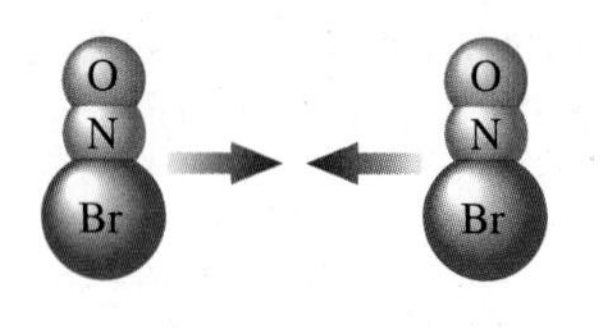

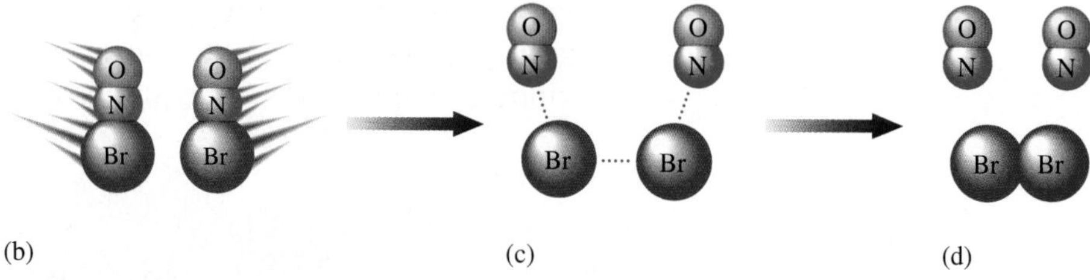

Figure 16.2
Visualizing the reaction $2BrNO(g) \rightarrow 2NO(g) + Br_2(g)$. (a) Two BrNO molecules approach each other at high speeds. (b) The collision occurs. (c) The energy of the collision causes the Br—N bonds to break and allows the Br—Br bond to form. (d) The products: one Br_2 and two NO molecules.

Chemists believe that molecules react by colliding with each other. Some collisions are violent enough to break bonds, allowing the reactants to rearrange to form the products. For example, consider the reaction

$$2BrNO(g) \rightleftharpoons 2NO(g) + Br_2(g)$$

which we picture to occur as shown in Figure 16.2. Notice that the Br—N bonds in the two BrNO molecules must be broken and a new Br—Br bond must be formed during a collision in order for the reactants to become products.

The idea that reactions occur during molecular collisions, which is called the **collision model,** explains many characteristics of chemical reactions. For example, it explains why a reaction proceeds faster if the concentrations of the reacting molecules are increased (higher concentrations lead to more collisions and therefore to more reaction events). The collision model also explains why reactions go faster at higher temperatures, as we will see in the next section.

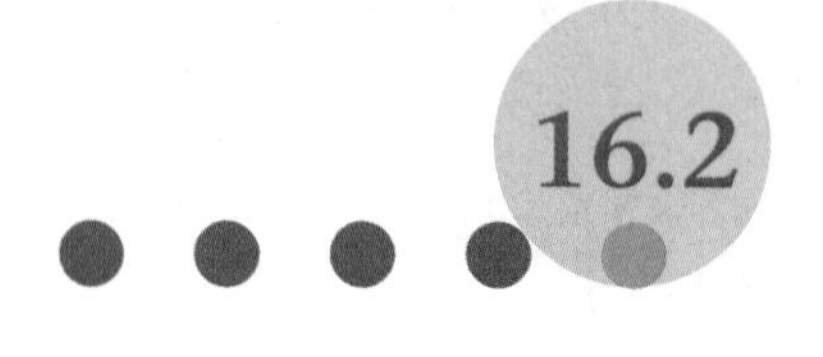

16.2 Conditions That Affect Reaction Rates

AIMS: To understand activation energy. To understand how a catalyst speeds up a reaction.

It is easy to see why reactions speed up when the *concentrations* of reacting molecules are increased: higher concentrations (more molecules per unit volume) lead to more collisions and so to more reaction events. But reactions also speed up when the *temperature* is increased. Why? The answer lies in the fact that not all collisions possess enough energy to break bonds. A minimum energy called the **activation energy (E_a)** is needed for a reaction to occur (see Figure 16.3). If a given collision possesses an energy greater than E_a, that collision can result in a reaction. If a collision has an energy less than E_a, the molecules will bounce apart unchanged.

Recall from Section 12.8 that the average kinetic energy of a collection of molecules is directly proportional to the temperature (*K*).

The reason that a reaction occurs faster as the temperature is increased is that the speeds of the molecules increase with temperature. So at higher temperatures, the average collision is more energetic. This makes it more likely that a given collision will possess enough energy to break bonds and to produce the molecular rearrangements needed for a reaction to occur.

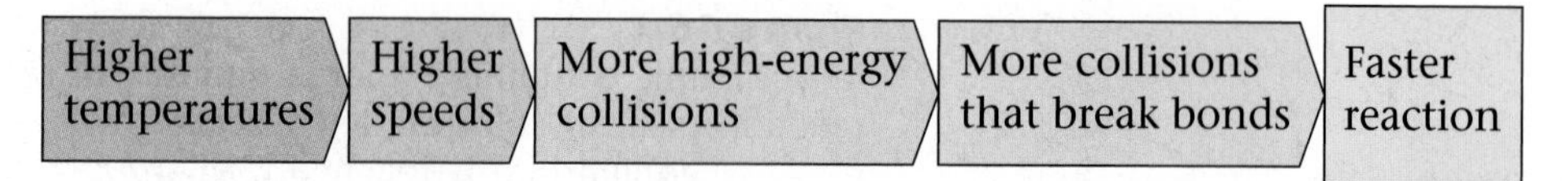

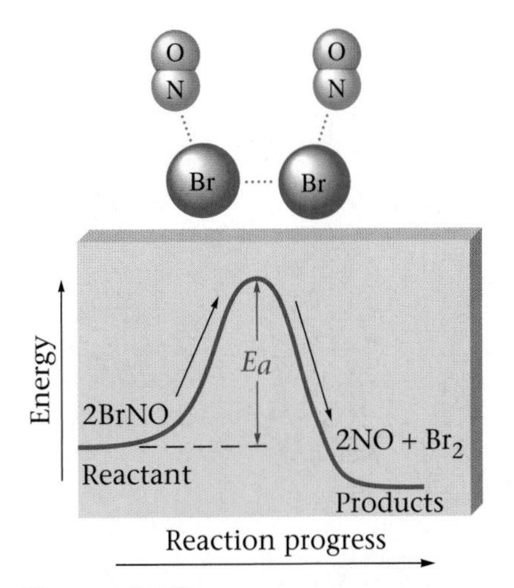

Figure 16.3
When molecules collide, a certain minimum energy called the activation energy (E_a) is needed for a reaction to occur. If the energy contained in a collision of two BrNO molecules is greater than E_a, the reaction can go "over the hump" to form products. If the collision energy is less than E_a, the colliding molecules bounce apart unchanged.

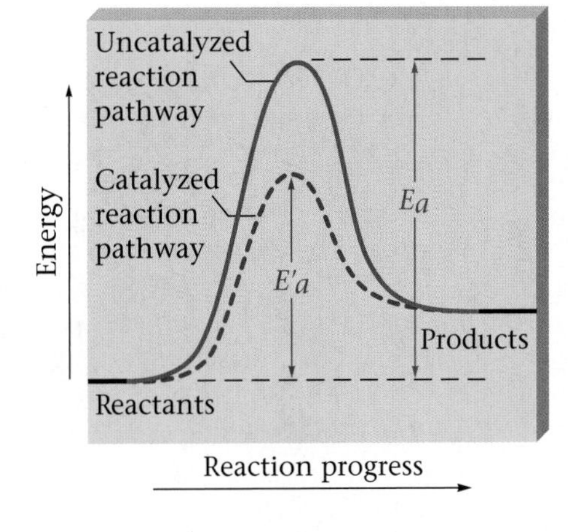

Figure 16.4
Comparison of the activation energies for an uncatalyzed reaction (E_a) and for the same reaction with a catalyst present (E'_a). Note that a catalyst works by lowering the activation energy for a reaction.

Is it possible to speed up a reaction without changing the temperature or the reactant concentrations? Yes, by using something called a **catalyst,** *a substance that speeds up a reaction without being consumed.* This may sound too good to be true, but it is a very common occurrence. In fact, you would not be alive now if your body did not contain thousands of catalysts called **enzymes.** Enzymes allow our bodies to speed up complicated reactions that would be too slow to sustain life at normal body temperatures. For example, the enzyme carbonic anhydrase speeds up the reaction between carbon dioxide and water

$$CO_2(g) + H_2O(l) \rightleftharpoons H^+(aq) + HCO_3^-(aq)$$

to help prevent an excess accumulation of carbon dioxide in our blood.

Although we cannot consider the details here, a catalyst works because it provides a new pathway for the reaction—a pathway that has a lower activation energy than the original pathway, as illustrated in Figure 16.4. Because of the lower activation energy, more collisions will have enough energy to allow a reaction. This in turn leads to a faster reaction.

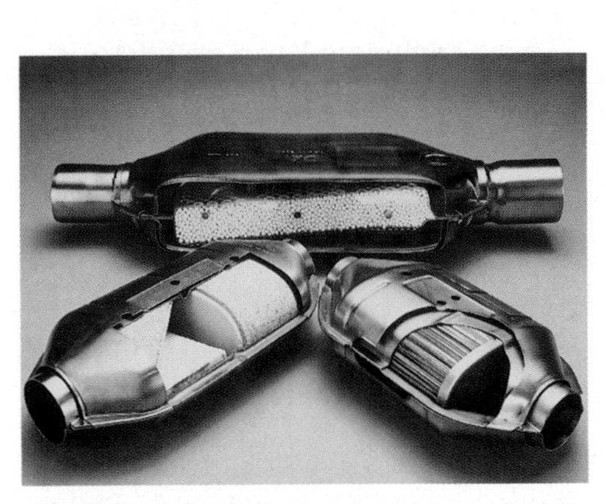

Cutaways of catalytic converters used in automobiles.

A very important example of a reaction involving a catalyst occurs in our atmosphere; it is the breakdown of ozone, O_3, catalyzed by chlorine atoms. Ozone is one constituent of the earth's upper atmosphere that is especially crucial, because it absorbs harmful high-energy radiation from the sun. There are natural processes that result in both the formation and the destruction of ozone in the upper atmosphere. The natural balance of all these opposing processes has resulted in an amount of ozone that has been relatively constant over the years. However, the ozone level now seems to be decreasing, especially over Antarctica (Figure 16.5), apparently because chlorine atoms act as catalysts for the decomposition of ozone to oxygen by the following pair of reactions:

Although O atoms are too reactive to exist near the earth's surface, they do exist in the upper atmosphere.

$$Cl + O_3 \rightarrow ClO + O_2$$
$$O + ClO \rightarrow Cl + O_2$$
$$\text{Sum: } \overline{Cl + O_3 + O + ClO \rightarrow ClO + O_2 + Cl + O_2}$$

When species that appear on both sides of the equation are canceled, the end result is the reaction

$$O + O_3 \rightarrow 2O_2$$

CHEMISTRY IN FOCUS

Protecting the Ozone

Chlorofluorocarbons (CFCs) are ideal compounds for refrigerators and air conditioners because they are nontoxic and noncorrosive. However, the chemical inertness of these substances, once thought to be their major virtue, turns out to be their fatal flaw. When these compounds leak into the atmosphere, as they inevitably do, they are so unreactive they persist there for decades. Eventually these CFCs reach altitudes where ultraviolet light causes them to decompose, producing chlorine atoms that promote the destruction of the ozone in the stratosphere (see discussion in Section 16.2). Because of this problem, the world's industrialized nations have signed an agreement (called the Montreal Protocol) that banned CFCs in 1996 (with a 10-year grace period for developing nations). So we must find substitutes for the CFCs—and fast.

In fact, the search for substitutes is now well under way. Worldwide production of CFCs has already decreased to half of the 1986 level of 1.13 million metric tons. One strategy for replacing the CFCs has been to switch to similar compounds that contain carbon and hydrogen atoms substituted for chlorine atoms. For example, the U. S. appliance industry has switched from Freon-12 (CF_2Cl_2) to the compound CH_2FCH_3 (called HFC-134a) for home refrigerators, and most of the new cars and trucks sold in the United States have air conditioners that employ HFC-134a. Converting the 140 million autos currently on the road in the United States that use CF_2Cl_2 will pose a major headache, but experience suggests that replacement of Freon-12 with HFC-134a is less expensive than was originally feared. For example, Volvo Cars of North America estimates that a Volvo can be converted from Freon-12 to HFC-134a for around $300.

A related environmental issue involves replacing the halons for firefighting applications. In particular, scientists are seeking an effective replacement for CF_3Br (halon-1301), the nontoxic "magic gas" used to flood enclosed spaces such as offices, aircraft, race cars, and military tanks in case of fire. The compound CF_3I, which appears to have a lifetime in the atmosphere of only a few days, looks like a promising candidate but much more research on the toxicology and ozone-depleting properties of CF_3I will be required before it receives government approval as a halon substitute.

The chemical industry has responded amazingly fast to the ozone depletion emergency. It is encouraging that we can act rapidly when an environmental crisis occurs. Now we need to get better at keeping the environment at a higher priority as we plan for the future.

An Amana refrigerator, one of many appliances that now use HFC-134a. This compound is replacing CFCs, which lead to the destruction of the atmospheric ozone.

Notice that a chlorine atom is used up in the first reaction but a chlorine atom is formed again by the second reaction. Therefore, the amount of chlorine does not change as the overall process occurs. This means that the chlorine atom is a true catalyst: it participates in the process but is not consumed. Estimates show that *one chlorine atom can catalyze the destruction of about one million ozone molecules per second.*

The chlorine atoms that promote this damage to the ozone layer are present because of pollution. Specifically, they come from the decomposition of compounds called Freons, such as CF_2Cl_2, which have been widely used in refrigerators and air conditioners. The Freons have leaked into the atmosphere, where they are decomposed by light to produce chlorine atoms and other substances. As a result, the manufacture of Freons was banned by

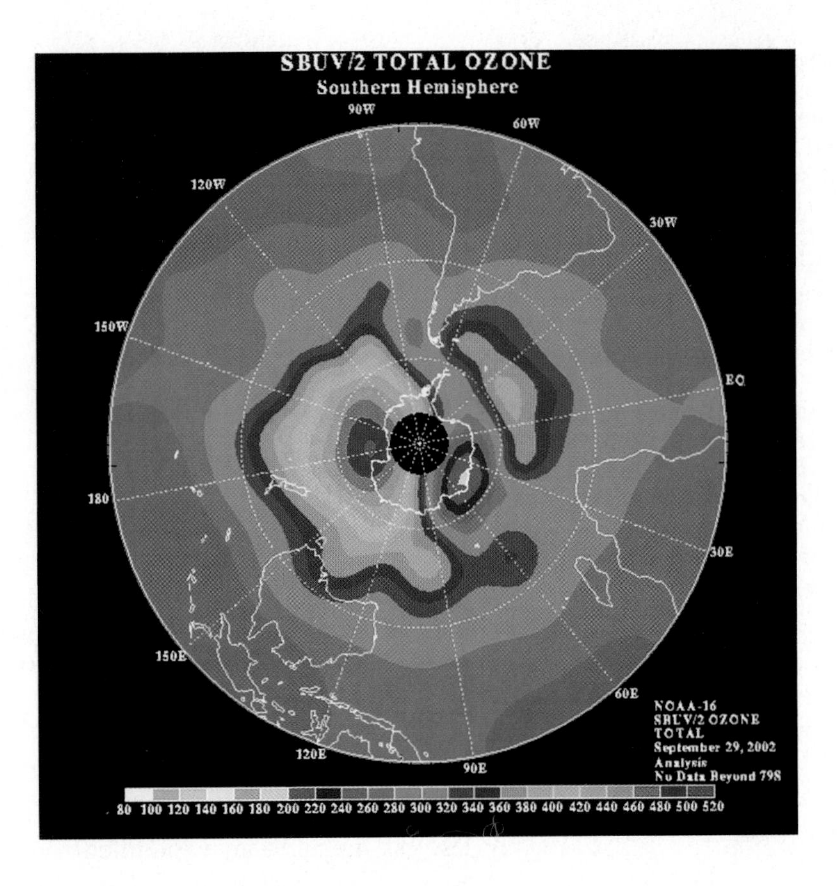

Figure 16.5
A photo showing the ozone "hole" over Antarctica.

agreement among the nations of the world as of the end of 1996. Substitute compounds are now being used in newly manufactured refrigerators and air conditioners.

The Equilibrium Condition

AIM: To learn how equilibrium is established.

Equilibrium is a word that implies balance or steadiness. When we say that someone is maintaining his or her equilibrium, we are describing a state of balance among various opposing forces. The term is used in a similar but more specific way in chemistry. Chemists define **equilibrium** as the *exact balancing of two processes, one of which is the opposite of the other.*

We first encountered the concept of equilibrium in Section 13.4, when we described the way vapor pressure develops over a liquid in a closed container (see Figure 13.9). This equilibrium process is summarized in Figure 16.6. The *equilibrium state* occurs when the rate of evaporation exactly equals the rate of condensation.

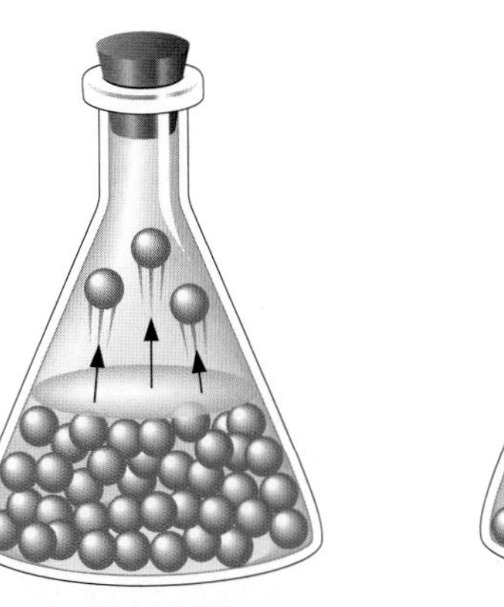
Evaporation

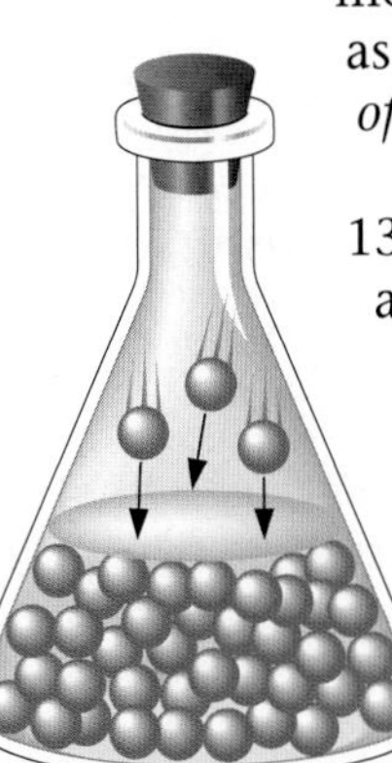
Condensation

So far in this textbook we have usually assumed that reactions proceed to completion—that is, until one of the reactants "runs out." Indeed, many reactions *do* proceed essentially to completion. For such reactions we can assume that the reactants are converted to products until the limiting reactant is completely consumed. On the

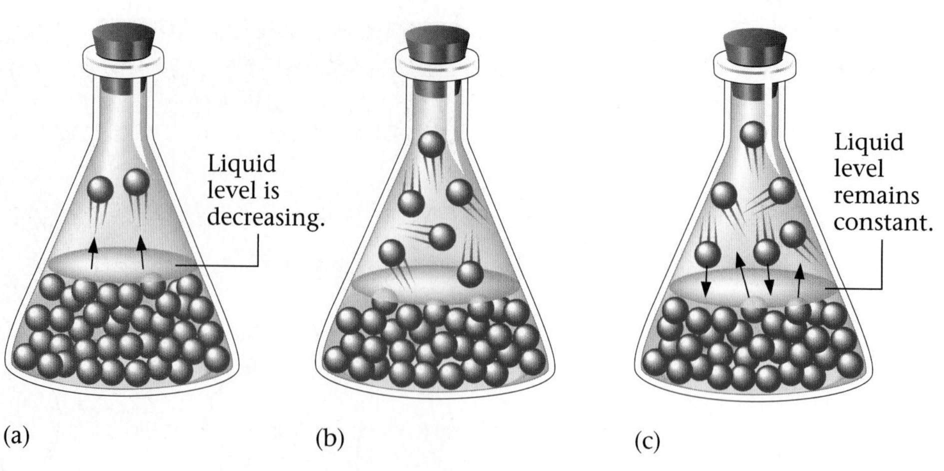

Figure 16.6
The establishment of the equilibrium vapor pressure over a liquid in a closed container. (a) At first there is a net transfer of molecules from the liquid state to the vapor state. (b) After a while, the amount of the substance in the vapor state becomes constant—both the pressure of the vapor and the level of the liquid stay the same. This is the equilibrium state. (c) The equilibrium state is very dynamic. The vapor pressure and liquid level remain constant because exactly the same number of molecules escape the liquid as return to it.

other hand, there are many chemical reactions that "stop" far short of completion when they are allowed to take place in a closed container. An example is the reaction of nitrogen dioxide to form dinitrogen tetroxide.

$$\underset{\text{Reddish-brown}}{NO_2(g) + NO_2(g)} \rightarrow \underset{\text{Colorless}}{N_2O_4(g)}$$

The reactant NO_2 is a reddish-brown gas, and the product N_2O_4 is a colorless gas. Imagine an experiment where pure NO_2 is placed in an empty, sealed glass vessel at 25 °C. The initial dark brown color will decrease in intensity as the NO_2 is converted to colorless N_2O_4 (see Figure 16.1). However, even over a long period of time, the contents of the reaction vessel do not become colorless. Instead, the intensity of the brown color eventually becomes constant, which means that the concentration of NO_2 is no longer changing. This simple observation is a clear indication that the reaction has "stopped" short of completion. In fact, the reaction has not stopped. Rather, the system has reached **chemical equilibrium,** *a dynamic state where the concentrations of all reactants and products remain constant.*

This situation is similar to the one where a liquid in a closed container develops a constant vapor pressure, except that in this case two opposite chemical reactions are involved. When pure NO_2 is first placed in the closed flask, there is no N_2O_4 present. As collisions between NO_2 molecules occur, N_2O_4 is formed and the concentration of N_2O_4 in the container increases. However, the reverse reaction can also occur. A given N_2O_4 molecule can decompose into two NO_2 molecules.

$$N_2O_4(g) \rightarrow NO_2(g) + NO_2(g)$$

That is, chemical reactions are *reversible;* they can occur in either direction. We usually indicate this fact by using double arrows.

$$2NO_2(g) \underset{\text{Reverse}}{\overset{\text{Forward}}{\rightleftharpoons}} N_2O_4(g)$$

In this case the double arrows mean that either two NO_2 molecules can combine to form an N_2O_4 molecule (the *forward* reaction) or an N_2O_4 molecule can decompose to give two NO_2 molecules (the *reverse* reaction).

Equilibrium is reached whether pure NO_2, pure N_2O_4, or a mixture of NO_2 and N_2O_4 is initially placed in a closed container. In any of these cases, conditions will eventually be reached in the container such that N_2O_4 is being formed and is decomposing at exactly the same rate. This leads to chemical equilibrium, a dynamic situation where the concentrations of reactants and products remain the same indefinitely, as long as the conditions are not changed.

Chemical Equilibrium: A Dynamic Condition

AIM: To learn about the characteristics of chemical equilibrium.

Equilibrium is a dynamic situation.

Because no changes occur in the concentrations of reactants or products in a reaction system at equilibrium, it may appear that everything has stopped. However, this is not the case. On the molecular level there is frantic activity. Equilibrium is not static but is a highly *dynamic* situation. Consider again the analogy between chemical equilibrium and two island cities connected by a single bridge. Suppose the traffic flow on the bridge is the same in both directions. It is obvious that there is motion (we can see the cars traveling across the bridge), but the number of cars in each city does not change because there is an equal flow of cars entering and leaving. The result is no *net* change in the number of cars in each of the two cities.

A double arrow ($\rightleftharpoons$) is used to show that a reaction is occurring in both directions.

To see how this concept applies to chemical reactions, let's consider the reaction between steam and carbon monoxide in a closed vessel at a high temperature where the reaction takes place rapidly.

$$H_2O(g) + CO(g) \rightleftharpoons H_2(g) + CO_2(g)$$

Assume that the same number of moles of gaseous CO and gaseous H_2O are placed in a closed vessel and allowed to react (Figure 16.7a).

When CO and H_2O, the reactants, are mixed, they immediately begin to react to form the products, H_2 and CO_2. This leads to a decrease in the concentrations of the reactants, but the concentrations of the products,

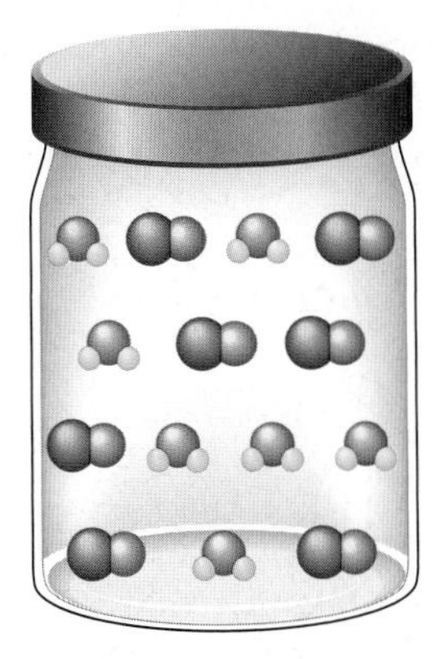

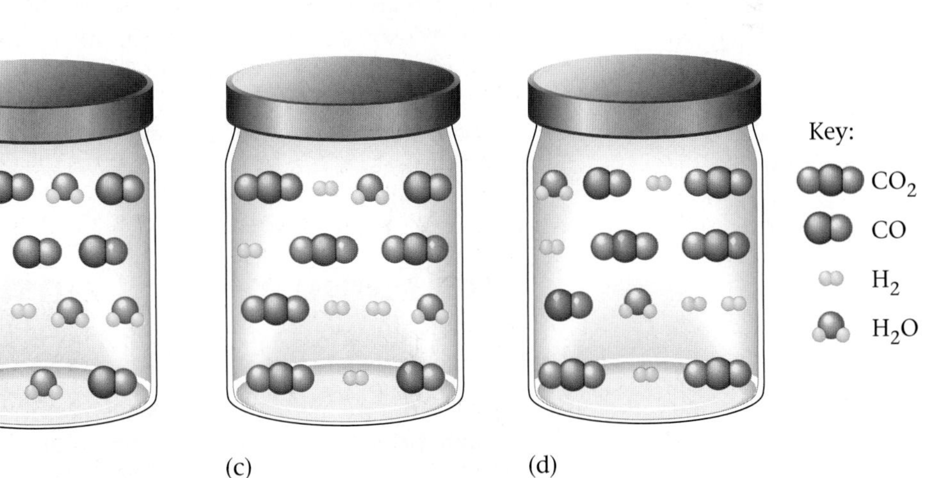

Figure 16.7
The reaction of H_2O and CO to form CO_2 and H_2 as time passes. (a) Equal numbers of moles of H_2O and CO are mixed in a closed container. (b) The reaction begins to occur, and some products (H_2 and CO_2) are formed. (c) The reaction continues as time passes and more reactants are changed to products. (d) Although time continues to pass, the numbers of reactant and product molecules are the same as in (c). No further changes are seen as time continues to pass. The system has reached equilibrium.

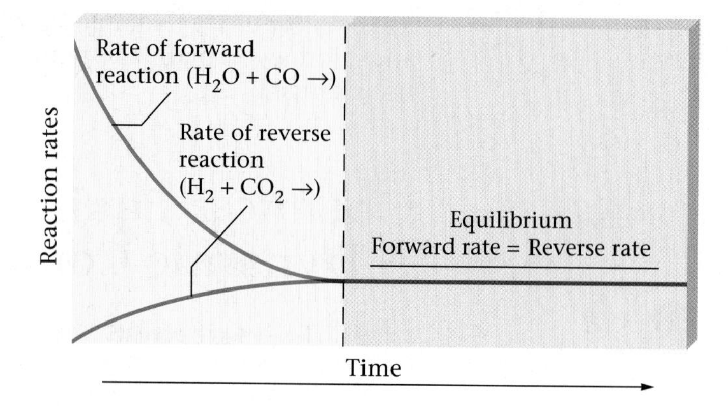

Figure 16.8
The changes with time in the rates of the forward and reverse reactions for $H_2O(g) + CO(g) \rightleftharpoons H_2(g) + CO_2(g)$ when equal numbers of moles of $H_2O(g)$ and $CO(g)$ are mixed. At first, the rate of the forward reaction decreases and the rate of the reverse reaction increases. Equilibrium is reached when the forward rate and the reverse rate become the same.

which were initially at zero, are increasing (Figure 16.7b). After a certain period of time, the concentrations of reactants and products no longer change at all—equilibrium has been reached (Figure 16.7c and d). Unless the system is somehow disturbed, no further changes in the concentrations will occur.

Why does equilibrium occur? We saw earlier in this chapter that molecules react by colliding with one another, and that the more collisions, the faster the reaction. This is why the speed of a reaction depends on concentrations. In this case the concentrations of H_2O and CO are lowered as the forward reaction occurs—that is, as products are formed.

$$H_2O + CO \rightarrow H_2 + CO_2$$

As the concentrations of the reactants decrease, the forward reaction slows down (Figure 16.8). But as in the traffic-on-the-bridge analogy, there is also movement in the reverse direction.

$$H_2 + CO_2 \rightarrow H_2O + CO$$

Initially in this experiment, no H_2 and CO_2 are present, so this reverse reaction cannot occur. However, as the forward reaction proceeds, the concentrations of H_2 and CO_2 build up and the speed (or rate) of the reverse reaction increases (Figure 16.8) as the forward reaction slows down. Eventually the concentrations reach levels at which *the rate of the forward reaction equals the rate of the reverse reaction.* The system has reached equilibrium.

The Equilibrium Constant: An Introduction

AIM: To understand the law of chemical equilibrium and to learn how to calculate values for the equilibrium constant.

Science is based on the results of experiments. The development of the equilibrium concept is typical. On the basis of their observations of many chemical reactions, two Norwegian chemists, Cato Maximilian Guldberg and Peter Waage, proposed in 1864 the **law of chemical equilibrium** (originally called the *law of mass action*) as a general description of the equilibrium condition. Guldberg and Waage postulated that for a reaction of the type

$$aA + bB \rightleftharpoons cC + dD$$

where A, B, C, and D represent chemical species and *a, b, c,* and *d* are their coefficients in the balanced equation, the law of mass action is represented by the following **equilibrium expression:**

$$K = \frac{[C]^c[D]^d}{[A]^a[B]^b}$$

Square brackets, [ ], indicate concentration units of mol/L.

The square brackets indicate the concentrations of the chemical species *at equilibrium* (in units of mol/L), and *K* is a constant called the **equilibrium constant.** Note that the equilibrium expression is a special ratio of the concentrations of the products to the concentrations of the reactants. Each concentration is raised to a power corresponding to its coefficient in the balanced equation.

The law of chemical equilibrium as proposed by Guldberg and Waage is based on experimental observations. Experiments on many reactions showed that the equilibrium condition could always be described by this special ratio, called the equilibrium expression.

To see how to construct an equilibrium expression, consider the reaction where ozone changes to oxygen:

$$\underset{\substack{\uparrow\\ \text{Reactant}}}{\overset{\substack{\text{Coefficient}\\ \downarrow}}{2O_3(g)}} \rightleftharpoons \underset{\substack{\uparrow\\ \text{Product}}}{\overset{\substack{\text{Coefficient}\\ \downarrow}}{3O_2(g)}}$$

To obtain the equilibrium expression, we place the concentration of the product in the numerator and the concentration of the reactant in the denominator.

$$\frac{[O_2]}{[O_3]} \begin{array}{l} \leftarrow \text{Product} \\ \leftarrow \text{Reactant} \end{array}$$

Then we use the coefficients as powers.

$$K = \frac{[O_2]^3}{[O_3]^2} \quad \text{Coefficients become powers}$$

Example 16.1 Writing Equilibrium Expressions

Write the equilibrium expression for the following reactions.

a. $H_2(g) + F_2(g) \rightleftharpoons 2HF(g)$

b. $N_2(g) + 3H_2(g) \rightleftharpoons 2NH_3(g)$

Solution

Applying the law of chemical equilibrium, we place products over reactants (using square brackets to denote concentrations in units of moles per liter) and raise each concentration to the power that corresponds to the coefficient in the balanced chemical equation.

a. $K = \frac{[HF]^2}{[H_2][F_2]}$ ← Product (coefficient of 2 becomes power of 2) ← Reactants (coefficients of 1 become powers of 1)

Note that when a coefficient (power) of 1 occurs, it is not written but is understood.

b. $K = \frac{[NH_3]^2}{[N_2][H_2]^3}$

✔ Self-Check Exercise 16.1

Write the equilibrium expression for the following reaction.

$$4NH_3(g) + 7O_2(g) \rightleftharpoons 4NO_2(g) + 6H_2O(g)$$

See Problems 16.15 through 16.18. ■

What does the equilibrium expression mean? It means that, for a given reaction at a given temperature, the special ratio of the concentrations of the products to reactants defined by the equilibrium expression will always be equal to the same number—namely, the equilibrium constant K. For example, consider a series of experiments on the ammonia synthesis reaction

$$N_2(g) + 3H_2(g) \rightleftharpoons 2NH_3(g)$$

carried out at 500 °C to measure the concentrations of N_2, H_2, and NH_3 present at equilibrium. The results of these experiments are shown in Table 16.1. In this table, subscript zeros next to square brackets are used to indicate *initial concentrations:* the concentrations of reactants and products originally mixed together before any reaction has occurred.

Consider the results of experiment I. One mole each of N_2 and H_2 were sealed into a 1-L vessel at 500 °C and allowed to reach chemical equilibrium. At equilibrium the concentrations in the flask were found to be $[N_2] = 0.921\ M$, $[H_2] = 0.763\ M$, and $[NH_3] = 0.157\ M$. The equilibrium expression for the reaction

$$N_2(g) + 3H_2(g) \rightleftharpoons 2NH_3(g)$$

is

$$K = \frac{[NH_3]^2}{[N_2][H_2]^3} = \frac{(0.157)^2}{(0.921)(0.763)^3}$$
$$= 0.0602 = 6.02 \times 10^{-2}$$

Similarly, as shown in Table 16.1, we can calculate for experiments II and III that K, the equilibrium constant, has the value 6.02×10^{-2}. In fact, whenever N_2, H_2, and NH_3 are mixed together at this temperature, the system *always* comes to an equilibrium position such that

$$K = 6.02 \times 10^{-2}$$

regardless of the amounts of the reactants and products that are mixed together initially.

Table 16.1 Results of Three Experiments for the Reaction $N_2(g) + 3H_2(g) \rightleftharpoons 2NH_3(g)$ at 500 °C

	Initial Concentrations			***Equilibrium Concentrations***			
Experiment	$[N_2]_0$	$[H_2]_0$	$[NH_3]_0$	$[N_2]$	$[H_2]$	$[NH_3]$	$\frac{[NH_3]^2}{[N_2][H_2]^3} = K$*
I	1.000 M	1.000 M	0	0.921 M	0.763 M	0.157 M	$\frac{(0.157)^2}{(0.921)(0.763)^3} = 0.0602$
II	0	0	1.000 M	0.399 M	1.197 M	0.203 M	$\frac{(0.203)^2}{(0.399)(1.197)^3} = 0.0602$
III	2.00 M	1.00 M	3.00 M	2.59 M	2.77 M	1.82 M	$\frac{(1.82)^2}{(2.59)(2.77)^3} = 0.0602$

*The units for K are customarily omitted.

It is important to see from Table 16.1 that the *equilibrium concentrations are not always the same.* However, even though the individual sets of equilibrium concentrations are quite different for the different situations, the *equilibrium constant, which depends on the ratio of the concentrations, remains the same.*

For a reaction at a given temperature, there are many equilibrium positions but only one value for *K*.

Each *set of equilibrium concentrations* is called an **equilibrium position.** It is essential to distinguish between the equilibrium constant and the equilibrium positions for a given reaction system. There is only *one* equilibrium constant for a particular system at a particular temperature, but there is an *infinite* number of equilibrium positions. The specific equilibrium position adopted by a system depends on the initial concentrations; the equilibrium constant does not.

Note that in the above discussion, the equilibrium constant was given without units. In certain cases the units are included when the values of equilibrium constants are given, and in other cases they are omitted. We will not discuss the reasons for this. We will omit the units in this text.

Example 16.2 Calculating Equilibrium Constants

The reaction of sulfur dioxide with oxygen in the atmosphere to form sulfur trioxide has important environmental implications because SO_3 combines with moisture to form sulfuric acid droplets, an important component of acid rain. The following results were collected for two experiments involving the reaction at 600 °C between gaseous sulfur dioxide and oxygen to form gaseous sulfur trioxide:

$$2SO_2(g) + O_2(g) \rightleftharpoons 2SO_3(g)$$

	Initial	Equilibrium
Experiment I	$[SO_2]_0 = 2.00\ M$	$[SO_2] = 1.50\ M$
	$[O_2]_0 = 1.50\ M$	$[O_2] = 1.25\ M$
	$[SO_3]_0 = 3.00\ M$	$[SO_3] = 3.50\ M$
Experiment II	$[SO_2]_0 = 0.500\ M$	$[SO_2] = 0.590\ M$
	$[O_2]_0 = 0$	$[O_2] = 0.045\ M$
	$[SO_3]_0 = 0.350\ M$	$[SO_3] = 0.260\ M$

The law of chemical equilibrium predicts that the value of *K* should be the same for both experiments. Verify this by calculating the equilibrium constant observed for each experiment.

Spruce needles turned brown by acid rain.

Solution

The balanced equation for the reaction is

$$2SO_2(g) + O_2(g) \rightleftharpoons 2SO_3(g)$$

From the law of chemical equilibrium, we can write the equilibrium expression

$$K = \frac{[SO_3]^2}{[SO_2]^2[O_2]}$$

For experiment I we calculate the value of K by substituting the observed *equilibrium* concentrations,

$$[SO_3] = 3.50\ M$$
$$[SO_2] = 1.50\ M$$
$$[O_2] = 1.25\ M$$

into the equilibrium expression:

$$K_I = \frac{(3.50)^2}{(1.50)^2(1.25)} = 4.36$$

For experiment II at equilibrium

$$[SO_3] = 0.260\ M$$
$$[SO_2] = 0.590\ M$$
$$[O_2] = 0.045\ M$$

and

$$K_{II} = \frac{(0.260)^2}{(0.590)^2(0.045)} = 4.32$$

Notice that the values calculated for K_I and K_{II} are nearly the same, as we expected. That is, the value of K is constant, within differences due to rounding off and due to experimental error. These experiments show *two different equilibrium positions* for this system, but K, the equilibrium constant, is, indeed, constant. ■

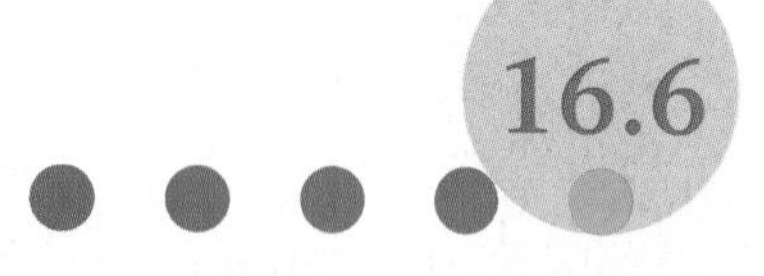

Heterogeneous Equilibria

AIM: To understand the role that liquids and solids play in constructing the equilibrium expression.

So far we have discussed equilibria only for systems in the gaseous state, where all reactants and products are gases. These are examples of **homogeneous equilibria,** in which all substances are in the same state. However, many equilibria involve more than one state and are called **heterogeneous equilibria.** For example, the thermal decomposition of calcium carbonate in the commercial preparation of lime occurs by a reaction involving solids and gases.

In terms of amount produced, lime is among the top ten chemicals manufactured in the United States.

$$CaCO_3(s) \rightleftharpoons \underset{\text{Lime}}{CaO(s)} + CO_2(g)$$

Straightforward application of the law of equilibrium leads to the equilibrium expression

$$K = \frac{[CO_2][CaO]}{[CaCO_3]}$$

However, experimental results show that the *position of a heterogeneous equilibrium does not depend on the amounts of pure solids or liquids present.* The fundamental reason for this behavior is that the concentrations of pure solids and liquids cannot change. In other words, we might say that the concentrations of pure solids and liquids are constants. Therefore, we can write the equilibrium expression for the decomposition of solid calcium carbonate as

The concentrations of pure liquids and solids are constant.

$$K' = \frac{[CO_2]C_1}{C_2}$$

where C_1 and C_2 are constants representing the concentrations of the solids CaO and $CaCO_3$, respectively. This expression can be rearranged to give

$$\frac{C_2K'}{C_1} = K = [CO_2]$$

where the constants C_2, K', and C_1 are combined into a single constant K. This leads us to the following general statement: the concentrations of pure solids or pure liquids involved in a chemical reaction *are not included in the equilibrium expression* for the reaction. This applies *only* to pure solids or liquids. It does not apply to solutions or gases, because their concentrations can vary.

For example, consider the decomposition of liquid water to gaseous hydrogen and oxygen:

$$2H_2O(l) \rightleftharpoons 2H_2(g) + O_2(g)$$

where

$$K = [H_2]^2[O_2]$$

Water is not included in the equilibrium expression because it is a pure liquid. However, when the reaction is carried out under conditions where the water is a gas rather than a liquid,

$$2H_2O(g) \rightleftharpoons 2H_2(g) + O_2(g)$$

we have

$$K = \frac{[H_2]^2[O_2]}{[H_2O]^2}$$

because the concentration of water vapor can change.

Example 16.3 Writing Equilibrium Expressions for Heterogeneous Equilibria

Write the expressions for K for the following processes.

a. Solid phosphorus pentachloride is decomposed to liquid phosphorus trichloride and chlorine gas.

b. Deep-blue solid copper(II) sulfate pentahydrate is heated to drive off water vapor and form white solid copper(II) sulfate.

Solution

a. The reaction is

$$PCl_5(s) \rightleftharpoons PCl_3(l) + Cl_2(g)$$

In this case, neither the pure solid PCl_5 nor the pure liquid PCl_3 is included in the equilibrium expression. The equilibrium expression is

$$K = [Cl_2]$$

As solid copper(II) sulfate pentahydrate, $CuSO_4 \cdot 5H_2O$, is heated, it loses H_2O, eventually forming white $CuSO_4$.

b. The reaction is

$$CuSO_4 \cdot 5H_2O(s) \rightleftharpoons CuSO_4(s) + 5H_2O(g)$$

The two solids are not included. The equilibrium expression is

$$K = [H_2O]^5$$

Self-Check Exercise 16.2

Write the equilibrium expression for each of the following reactions.

a. $2KClO_3(s) \rightleftharpoons 2KCl(s) + 3O_2(g)$

This reaction is often used to produce oxygen gas in the laboratory.

b. $NH_4NO_3(s) \rightleftharpoons N_2O(g) + 2H_2O(g)$

c. $CO_2(g) + MgO(s) \rightleftharpoons MgCO_3(s)$

d. $SO_3(g) + H_2O(l) \rightleftharpoons H_2SO_4(l)$

See Problems 16.25 through 16.28. ■

16.7 Le Châtelier's Principle

AIM: To learn to predict the changes that occur when a system at equilibrium is disturbed.

It is important to understand the factors that control the *position* of a chemical equilibrium. For example, when a chemical is manufactured, the chemists and chemical engineers in charge of production want to choose conditions that favor the desired product as much as possible. That is, they want the equilibrium to lie far to the right (toward products). When the process for the synthesis of ammonia was being developed, extensive studies were carried out to determine how the equilibrium concentration of ammonia depended on the conditions of temperature and pressure.

In this section we will explore how various changes in conditions affect the equilibrium position of a reaction system. We can predict the effects of changes in concentration, pressure, and temperature on a system at equilibrium by using **Le Châtelier's principle,** which states that when *a change is imposed on a system at equilibrium, the position of the equilibrium shifts in a direction that tends to reduce the effect of that change.*

The Effect of a Change in Concentration

Let us consider the ammonia synthesis reaction. Suppose there is an equilibrium position described by these concentrations:

$$[N_2] = 0.399\ M \qquad [H_2] = 1.197\ M \qquad [NH_3] = 0.203\ M$$

What will happen if 1.000 mol/L of N_2 is suddenly injected into the system? We can begin to answer this question by remembering that for the system at equilibrium, the rates of the forward and reverse reactions exactly balance,

$$N_2(g) + 3H_2(g) \rightleftharpoons 2NH_3(g)$$

as indicated here by arrows of the same length. When the N_2 is added, there are suddenly more collisions between N_2 and H_2 molecules. This increases the rate of the forward reaction (shown here by the greater length of the arrow pointing in that direction),

$$N_2(g) + 3H_2(g) \rightleftharpoons 2NH_3(g)$$

and the reaction produces more NH_3. As the concentration of NH_3 increases, the reverse reaction also speeds up (as more collisions between NH_3 molecules occur) and the system again comes to equilibrium. However, the new equilibrium position has more NH_3 than was present in the original position. We say that the equilibrium has shifted to the *right*—toward the products. The original and new equilibrium positions are shown below.

Equilibrium Position I		Equilibrium Position II
$[N_2] = 0.399\ M$	⇨	$[N_2] = 1.348\ M$
$[H_2] = 1.197\ M$	1.000 mol/L of N_2 added	$[H_2] = 1.044\ M$
$[NH_3] = 0.203\ M$		$[NH_3] = 0.304\ M$

Note that the equilibrium does in fact shift to the right; the concentration of H_2 decreases (from 1.197 *M* to 1.044 *M*), the concentration of NH_3 increases (from 0.203 *M* to 0.304 *M*), and, of course, because nitrogen was added, the concentration of N_2 shows an increase relative to the original amount present.

It is important to note at this point that, although the equilibrium shifted to a new position, the *value of K did not change.* We can demonstrate this by inserting the equilibrium concentrations from positions I and II into the equilibrium expression.

- Position I: $K = \dfrac{[NH_3]^2}{[N_2][H_2]^3} = \dfrac{(0.203)^2}{(0.399)(1.197)^3} = 0.0602$
- Position II: $K = \dfrac{[NH_3]^2}{[N_2][H_2]^3} = \dfrac{(0.304)^2}{(1.348)(1.044)^3} = 0.0602$

These values of *K* are the same. Therefore, although the equilibrium *position* shifted when we added more N_2, the *equilibrium constant K* remained the same.

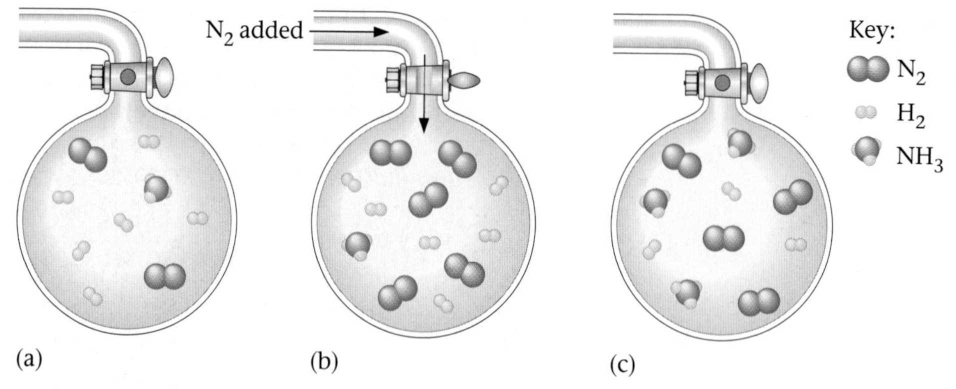

Figure 16.9
(a) The initial equilibrium mixture of N_2, H_2, and NH_3. (b) Addition of N_2. (c) The new equilibrium position for the system containing more N_2 (because of the addition of N_2), less H_2, and more NH_3 than in (a).

Could we have predicted this shift by using Le Châtelier's principle? Because the change in this case was to add nitrogen, Le Châtelier's principle predicts that the system will shift in a direction that *consumes* nitrogen. This tends to offset the original change—the addition of N_2. Therefore, Le Châtelier's principle correctly predicts that adding nitrogen will cause the equilibrium to shift to the right (Figure 16.9) as some of the added nitrogen is consumed.

A system at equilibrium shifts in the direction that compensates for any imposed change.

If ammonia had been added instead of nitrogen, the system would have shifted to the left, consuming ammonia. Another way of stating Le Châtelier's principle, then, is to say that *when a reactant or product is added to a system at equilibrium, the system shifts away from the added component.* On the other hand, if *a reactant or product is removed, the system shifts toward the removed component.* For example, if we had removed nitrogen, the system would have shifted to the left and the amount of ammonia present would have been reduced.

A real-life example that shows the importance of Le Châtelier's principle is the effect of high elevations on the oxygen supply to the body. If you have ever traveled to the mountains on vacation, you may have noticed that you felt "light-headed" and especially tired during the first few days of your visit. These feelings resulted from a decreased supply of oxygen to your body because of the lower air pressure that exists at higher elevations. For example, the oxygen supply in Leadville, Colorado (elevation ~ 10,000 ft), is only about two-thirds that found at sea level. We can understand the effects of diminished oxygen supply in terms of the following equilibrium:

$$Hb(aq) + 4O_2(g) \rightleftharpoons Hb(O_2)_4(aq)$$

where Hb represents hemoglobin, the iron-containing protein that transports O_2 from your lungs to your tissues, where it is used to support metabolism. The coefficient 4 in the equation signifies that each hemoglobin molecule picks up four O_2 molecules in the lungs. Note by Le Châtelier's principle that a lower oxygen pressure will cause this equilibrium to shift to the left, away from oxygenated hemoglobin. This leads to an inadequate oxygen supply at the tissues, which in turn results in fatigue and a "woozy" feeling.

This problem can be solved in extreme cases, such as when climbing Mt. Everest or flying in a plane at high altitudes, by supplying extra oxygen from a tank. This extra oxygen pushes the equilibrium to its normal position. However, lugging around an oxygen tank would not be very practical for people who live in the mountains. In fact, nature solves this problem in a very interesting way. The body adapts to living at high elevations by producing additional hemoglobin—the other way to shift this equilibrium to the right. Thus, people who live at high elevations have significantly higher hemoglobin levels than those living at sea level. For example, the Sherpas who live in Nepal can function in the rarefied air at the top of Mt. Everest without an auxiliary oxygen supply.

Example 16.4 Using Le Châtelier's Principle: Changes in Concentration

Arsenic, As_4, is obtained from nature by first reacting its ore with oxygen (called *roasting*) to form solid As_4O_6. (As_4O_6, a toxic compound fatal in doses of 0.1 g or more, is the "arsenic" made famous in detective stories.) The As_4O_6 is then reduced using carbon:

$$As_4O_6(s) + 6C(s) \rightleftharpoons As_4(g) + 6CO(g)$$

Predict the direction of the shift in the equilibrium position for this reaction that occurs in response to each of the following changes in conditions.

a. Addition of carbon monoxide

b. Addition or removal of $C(s)$ or $As_4O_6(s)$

c. Removal of $As_4(g)$

Solution

a. Le Châtelier's principle predicts a shift away from the substance whose concentration is increased. The equilibrium position will shift to the left when carbon monoxide is added.

b. Because the amount of a pure solid has no effect on the equilibrium position, changing the amount of carbon or tetraarsenic hexoxide will have no effect.

c. When gaseous arsenic is removed, the equilibrium position will shift to the right to form more products. In industrial processes, the desired product is often continuously removed from the reaction system to increase the yield.

When blue anhydrous $CoCl_2$ reacts with water, pink $CoCl_2 \cdot 6H_2O$ is formed.

Self-Check Exercise 16.3

Novelty devices for predicting rain contain cobalt(II) chloride and are based on the following equilibrium:

$$\underset{\text{Blue}}{CoCl_2(s)} + 6H_2O(g) \rightleftharpoons \underset{\text{Pink}}{CoCl_2 \cdot 6H_2O(s)}$$

What color will this indicator be when rain is likely due to increased water vapor in the air?

See Problems 16.33 through 16.36. ■

The Effect of a Change in Volume

When the volume of a gas is decreased (when a gas is compressed), the pressure increases. This occurs because the molecules present are now contained in a smaller space and they hit the walls of their container more often, giving a greater pressure. Therefore, when the volume of a gaseous reaction system at equilibrium is suddenly reduced, leading to a sudden increase in

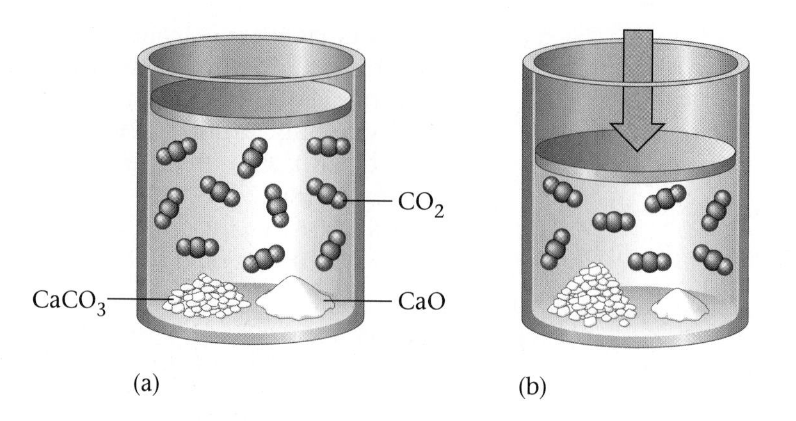

Figure 16.10
The reaction system $CaCO_3(s) \rightleftharpoons CaO(s) + CO_2(g)$. (a) The system is initially at equilibrium. (b) Then the piston is pushed in, decreasing the volume and increasing the pressure. The system shifts in the direction that consumes CO_2 molecules, thus lowering the pressure again.

pressure, by Le Châtelier's principle the system will shift in the direction that reduces the pressure.

For example, consider the reaction

$$CaCO_3(s) \rightleftharpoons CaO(s) + CO_2(g)$$

in a container with a movable piston (Figure 16.10). If the volume is suddenly decreased by pushing in the piston, the pressure of the CO_2 gas initially increases. How can the system offset this pressure increase? By shifting to the left—the direction that reduces the amount of gas present. That is, a shift to the left will use up CO_2 molecules, thus lowering the pressure. (There will then be fewer molecules present to hit the walls, because more of the CO_2 molecules have combined with CaO and thus have become part of the solid $CaCO_3$.)

Therefore, when the volume of a gaseous reaction system at equilibrium is decreased (thus increasing the pressure), *the system shifts in the direction that gives the smaller number of gas molecules.* So a decrease in the system volume leads to a shift that decreases the total number of gaseous molecules in the system.

Suppose we are running the reaction

$$N_2(g) + 3H_2(g) \rightleftharpoons 2NH_3(g)$$

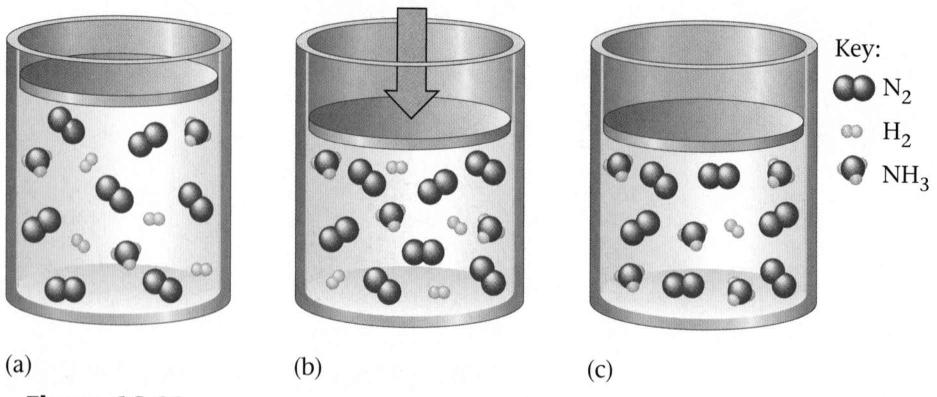

Figure 16.11
(a) A mixture of $NH_3(g)$, $N_2(g)$, and $H_2(g)$ at equilibrium. (b) The volume is suddenly decreased. (c) The new equilibrium position for the system containing more NH_3 and less N_2 and H_2. The reaction $N_2(g) + 3H_2(g) \rightleftharpoons 2NH_3(g)$ shifts to the right (toward the side with fewer molecules) when the container volume is decreased.

and we have a mixture of the gases nitrogen, hydrogen, and ammonia at equilibrium (Figure 16.11a). If we suddenly reduce the volume, what will happen to the equilibrium position? Because the decrease in volume initially increases the pressure, the system moves in the direction that lowers its pressure. The reaction system can reduce its pressure by reducing the number of gas molecules present. This means that the reaction

$$N_2(g) + 3H_2(g) \rightleftharpoons 2NH_3(g)$$

4 gaseous molecules — 2 gaseous molecules

shifts to the right, because in this direction four molecules (one of nitrogen and three of hydrogen) react to produce two molecules (of ammonia), thus *reducing the total number of gaseous molecules present.* The equilibrium position shifts to the right—toward the side of the reaction that involves the smaller number of gaseous molecules in the balanced equation.

The opposite is also true. When the container volume is increased (which lowers the pressure of the system), the system shifts so as to increase its pressure. An increase in volume in the ammonia synthesis system produces a shift to the left to increase the total number of gaseous molecules present (to increase the pressure).

Example 16.5 Using Le Châtelier's Principle: Changes in Volume

Predict the shift in equilibrium position that will occur for each of the following processes when the volume is reduced.

a. The preparation of liquid phosphorus trichloride by the reaction

$$P_4(s) + 6Cl_2(g) \rightleftharpoons 4PCl_3(l)$$

6 gaseous molecules — 0 gaseous molecules

Solution a

P_4 and PCl_3 are a pure solid and a pure liquid, respectively, so we need to consider only the effect on Cl_2. If the volume is decreased, the Cl_2 pressure will initially increase, so the position of the equilibrium will shift to the right, consuming gaseous Cl_2 and lowering the pressure (to counteract the original change).

b. The preparation of gaseous phosphorus pentachloride according to the equation

$$PCl_3(g) + Cl_2(g) \rightleftharpoons PCl_5(g)$$

2 gaseous molecules — 1 gaseous molecule

Solution b

Decreasing the volume (increasing the pressure) will shift this equilibrium to the right, because the product side contains only one gaseous molecule while the reactant side has two. That is, the system will respond to the decreased volume (increased pressure) by lowering the number of molecules present.

c. The reaction of phosphorus trichloride with ammonia:

$$PCl_3(g) + 3NH_3(g) \rightleftharpoons P(NH_2)_3(g) + 3HCl(g)$$

Solution c

Both sides of the balanced reaction equation have four gaseous molecules. A change in volume will have no effect on the equilibrium position. There is no shift in this case, because the system cannot change the number of molecules present by shifting in either direction.

Self-Check Exercise 16.4

For each of the following reactions, predict the direction the equilibrium will shift when the volume of the container is increased.

a. $H_2(g) + F_2(g) \rightleftharpoons 2HF(g)$

b. $CO(g) + 2H_2(g) \rightleftharpoons CH_3OH(g)$

c. $2SO_3(g) \rightleftharpoons 2SO_2(g) + O_2(g)$

See Problems 16.33 through 16.36. ■

The Effect of a Change in Temperature

It is important to remember that although the changes we have just discussed may alter the equilibrium *position,* they do not alter the equilibrium *constant.* For example, the addition of a reactant shifts the equilibrium position to the right but has no effect on the value of the equilibrium constant; the new equilibrium concentrations satisfy the original equilibrium constant. This was demonstrated earlier in this section for the addition of N_2 to the ammonia synthesis reaction.

The effect of temperature on equilibrium is different, however, because *the value of K changes with temperature.* We can use Le Châtelier's principle to predict the direction of the change in K.

To do this we need to classify reactions according to whether they produce heat or absorb heat. A reaction that produces heat (heat is a "product") is said to be *exothermic.* A reaction that absorbs heat is called *endothermic.* Because heat is needed for an endothermic reaction, energy (heat) can be regarded as a "reactant" in this case.

In an exothermic reaction, heat is treated as a product. For example, the synthesis of ammonia from nitrogen and hydrogen is exothermic (produces heat). We can represent this by treating energy as a product:

$$N_2(g) + 3H_2(g) \rightleftharpoons 2NH_3(g) + \underset{\text{Energy released}}{92\ \text{kJ}}$$

Le Châtelier's principle predicts that when we add energy to this system at equilibrium by heating it, the shift will be in the direction that consumes energy—that is, to the left.

On the other hand, for an endothermic reaction (one that absorbs energy), such as the decomposition of calcium carbonate,

$$CaCO_3(s) + \underset{\text{Energy needed}}{556\ \text{kJ}} \rightleftharpoons CaO(s) + CO_2(g)$$

energy is treated as a reactant. In this case an increase in temperature causes the equilibrium to shift to the right.

In summary, to use Le Châtelier's principle to describe the effect of a temperature change on a system at equilibrium, *simply treat energy as a reactant (in an endothermic process) or as a product (in an exothermic process),* and *predict the direction of the shift* in the same way you would if an actual reactant or product were being added or removed.

Example 16.6 Using Le Châtelier's Principle: Changes in Temperature

For each of the following reactions, predict how the equilibrium will shift as the temperature is increased.

a. $N_2(g) + O_2(g) \rightleftharpoons NO(g)$ (endothermic)

Solution a

This is an endothermic reaction, so energy can be viewed as a reactant.

$$N_2(g) + O_2(g) + \text{energy} \rightleftharpoons 2NO(g)$$

Thus the equilibrium will shift to the right as the temperature is increased (energy added).

b. $2SO_2(g) + O_2(g) \rightleftharpoons 2SO_3(g)$ (exothermic)

Solution b

This is an exothermic reaction, so energy can be regarded as a product.

$$2SO_2(g) + O_2(g) \rightleftharpoons 2SO_3(g) + \text{energy}$$

As the temperature is increased, the equilibrium will shift to the left.

✔ Self-Check Exercise 16.5

For the exothermic reaction

$$2SO_2(g) + O_2(g) \rightleftharpoons 2SO_3(g)$$

predict the equilibrium shift caused by each of the following changes.

a. SO_2 is added.

b. SO_3 is removed.

c. The volume is decreased.

d. The temperature is decreased.

See Problems 16.33 through 16.42. ■

We have seen how Le Châtelier's principle can be used to predict the effects of several types of changes on a system at equilibrium. To summarize these ideas, Table 16.2 shows how various changes affect the equilibrium position of the endothermic reaction $N_2O_4(g) \rightleftharpoons 2NO_2(g)$. The effect of a temperature change on this system is depicted in Figure 16.12.

Table 16.2 Shifts in the Equilibrium Position for the Reaction Energy + $N_2O_4(g) \rightleftharpoons 2NO_2(g)$

Change	Shift
addition of $N_2O_4(g)$	right
addition of $NO_2(g)$	left
removal of $N_2O_4(g)$	left
removal of $NO_2(g)$	right
decrease in container volume	left
increase in container volume	right
increase in temperature	right
decrease in temperature	left

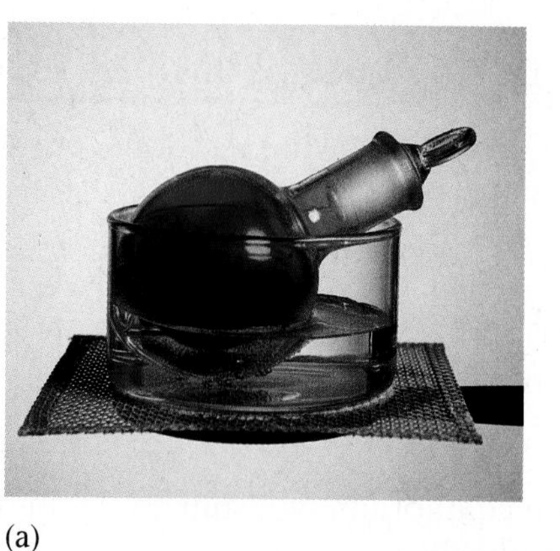

(a)

(b)

Figure 16.12
Shifting the $N_2O_4(g) \rightleftharpoons 2NO_2(g)$ equilibrium by changing the temperature. (a) At 100 °C the flask is definitely reddish-brown due to a large amount of NO_2 present. (b) At 0 °C the equilibrium is shifted toward colorless $N_2O_4(g)$.

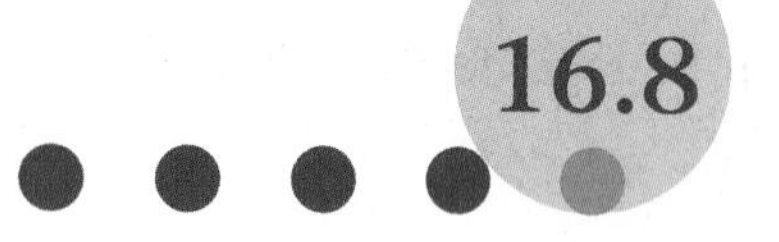

16.8 Applications Involving the Equilibrium Constant

AIM: To learn to calculate equilibrium concentrations from equilibrium constants.

Knowing the value of the equilibrium constant for a reaction allows us to do many things. For example, the size of K tells us the inherent tendency of the reaction to occur. A value of K much larger than 1 means that at equilibrium, the reaction system will consist of mostly products—the equilibrium lies to the right. For example, consider a general reaction of the type

$$A(g) \rightarrow B(g)$$

where

$$K = \frac{[B]}{[A]}$$

If K for this reaction is 10,000 (10^4), then at equilibrium,

$$\frac{[B]}{[A]} = 10{,}000 \quad \text{or} \quad \frac{[B]}{[A]} = \frac{10{,}000}{1}$$

That is, at equilibrium [B] is 10,000 times greater than [A]. This means that the reaction strongly favors the product B. Another way of saying this is that the reaction goes essentially to completion. That is, virtually all of A becomes B.

On the other hand, a small value of K means that the system at equilibrium consists largely of reactants—the equilibrium position is far to the left. The given reaction does not occur to any significant extent.

Another way we use the equilibrium constant is to calculate the equilibrium concentrations of reactants and products. For example, if we know the value of K and the concentrations of all the reactants and products except one, we can calculate the missing concentration. This is illustrated in Example 16.7.

Example 16.7 Calculating Equilibrium Concentration Using Equilibrium Expressions

Gaseous phosphorus pentachloride decomposes to chlorine gas and gaseous phosphorus trichloride. In a certain experiment, at a temperature where $K = 8.96 \times 10^{-2}$, the equilibrium concentrations of PCl_5 and PCl_3 were found to be 6.70×10^{-3} *M* and 0.300 *M*, respectively. Calculate the concentration of Cl_2 present at equilibrium.

Solution

For this reaction, the balanced equation is

$$PCl_5(g) \rightleftharpoons PCl_3(g) + Cl_2(g)$$

and the equilibrium expression is

$$K = \frac{[PCl_3][Cl_2]}{[PCl_5]} = 8.96 \times 10^{-2}$$

We know that

$$[PCl_5] = 6.70 \times 10^{-3}\, M$$
$$[PCl_3] = 0.300\, M$$

We want to calculate $[Cl_2]$. We will rearrange the equilibrium expression to solve for the concentration of Cl_2. First we divide both sides of the expression

$$K = \frac{[PCl_3][Cl_2]}{[PCl_5]}$$

by $[PCl_3]$ to give

$$\frac{K}{[PCl_3]} = \frac{\cancel{[PCl_3]}[Cl_2]}{\cancel{[PCl_3]}[PCl_5]} = \frac{[Cl_2]}{[PCl_5]}$$

Next we multiply both sides by $[PCl_5]$.

$$\frac{K[PCl_5]}{[PCl_3]} = \frac{[Cl_2]\cancel{[PCl_5]}}{\cancel{[PCl_5]}} = [Cl_2]$$

Then we can calculate $[Cl_2]$ by substituting the known information.

$$[Cl_2] = K \times \frac{[PCl_5]}{[PCl_3]} = (8.96 \times 10^{-2})\frac{(6.70 \times 10^{-3})}{(0.300)}$$
$$[Cl_2] = 2.00 \times 10^{-3}$$

The equilibrium concentration of Cl_2 is 2.00×10^{-3} *M*. ■

Solubility Equilibria

AIM: To learn to calculate the solubility product of a salt, given its solubility, and vice versa.

Solubility is a very important phenomenon. Consider the following examples.

- Because sugar and table salt dissolve readily in water, we can flavor foods easily.
- Because calcium sulfate is less soluble in hot water than in cold water, it coats tubes in boilers, reducing thermal efficiency.
- When food lodges between teeth, acids form that dissolve tooth enamel, which contains the mineral hydroxyapatite, $Ca_5(PO_4)_3OH$. Tooth decay can be reduced by adding fluoride to toothpaste. Fluoride replaces the hydroxide in hydroxyapatite to produce the corresponding fluorapatite,

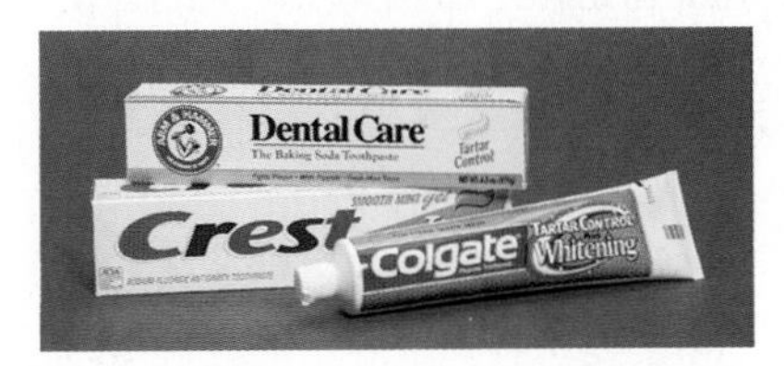

Toothpastes containing sodium fluoride, an additive that helps prevent tooth decay.

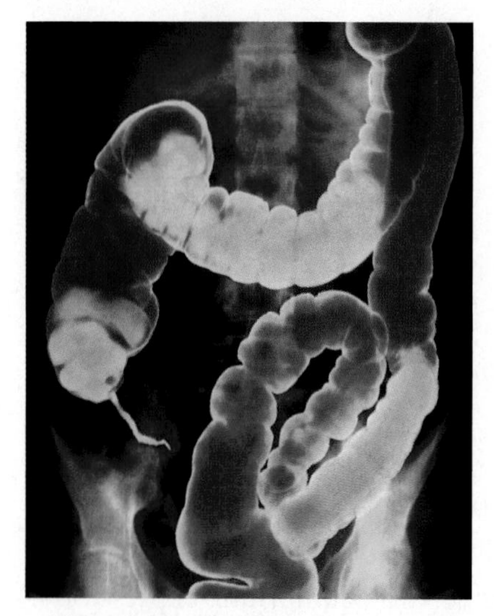

This X ray of the large intestine has been enhanced by the patient's consumption of barium sulfate.

$Ca_5(PO_4)_3F$, and calcium fluoride, CaF_2, both of which are less soluble in acids than the original enamel.

- The use of a suspension of barium sulfate improves the clarity of X rays of the digestive tract. Barium sulfate contains the toxic ion Ba^{2+}, but its very low solubility makes ingestion of solid $BaSO_4$ safe.

In this section we will consider the equilibria associated with dissolving solids in water to form aqueous solutions. When a typical ionic solid dissolves in water, it dissociates completely into separate cations and anions. For example, calcium fluoride dissolves in water as follows:

$$CaF_2(s) \xrightarrow{H_2O(l)} Ca^{2+}(aq) + 2F^-(aq)$$

When the solid salt is first added to the water, no Ca^{2+} and F^- ions are present. However, as dissolving occurs, the concentrations of Ca^{2+} and F^- increase, and it becomes more and more likely that these ions will collide and re-form the solid. Thus two opposite (competing) processes are occurring—the dissolving reaction shown above and the reverse reaction to re-form the solid:

$$Ca^{2+}(aq) + 2F^-(aq) \rightarrow CaF_2(s)$$

Ultimately, equilibrium is reached. No more solid dissolves and the solution is said to be saturated.

We can write an equilibrium expression for this process according to the law of chemical equilibrium.

$$K_{sp} = [Ca^{2+}][F^-]^2$$

where $[Ca^{2+}]$ and $[F^-]$ are expressed in mol/L. The constant K_{sp} is called the **solubility product constant,** or simply the **solubility product.**

Pure liquids and pure solids are never included in an equilibrium expression.

Because CaF_2 is a pure solid, it is not included in the equilibrium expression. It may seem strange at first that the amount of excess solid present does not affect the position of the solubility equilibrium. Surely more solid means more surface area exposed to the solvent, which would seem to result in greater solubility. This is not the case, however, because both dissolving and re-forming of the solid occur at the surface of the excess solid. When a solid dissolves, it is the ions at the surface that go into solution. And when the ions in solution re-form the solid, they do so on the surface of the solid. So doubling the surface area of the solid doubles not only the rate of dissolving but also the rate of re-formation of the solid. The amount of excess solid present therefore has no effect on the equilibrium position. Similarly, although either increasing the surface area by grinding up the solid or stirring the solution speeds up the attainment of equilibrium, neither procedure changes the *amount* of solid dissolved at equilibrium.

Example 16.8 Writing Solubility Product Expressions

Write the balanced equation describing the reaction for dissolving each of the following solids in water. Also write the K_{sp} expression for each solid.

a. $PbCl_2(s)$ b. $Ag_2CrO_4(s)$ c. $Bi_2S_3(s)$

Solution

a. $PbCl_2(s) \rightleftharpoons Pb^{2+}(aq) + 2Cl^-(aq)$; $K_{sp} = [Pb^{2+}][Cl^-]^2$

b. $Ag_2CrO_4(s) \rightleftharpoons 2Ag^+(aq) + CrO_4^{2-}(aq)$; $K_{sp} = [Ag^+]^2[CrO_4^{2-}]$

c. $Bi_2S_3(s) \rightleftharpoons 2Bi^{3+}(aq) + 3S^{2-}(aq)$; $K_{sp} = [Bi^{3+}]^2[S^{2-}]^3$

Self-Check Exercise 16.6

Write the balanced equation for the reaction describing the dissolving of each of the following solids in water. Also write the K_{sp} expression for each solid.

a. $BaSO_4(s)$ b. $Fe(OH)_3(s)$ c. $Ag_3PO_4(s)$

See Problems 16.57 and 16.58. ■

Example 16.9 Calculating Solubility Products

Copper(I) bromide, CuBr, has a measured solubility of 2.0×10^{-4} mol/L at 25 °C. That is, when excess CuBr(*s*) is placed in 1.0 L of water, we can determine that 2.0×10^{-4} mol of the solid dissolves to produce a saturated solution. Calculate the solid's K_{sp} value.

Solution

At first it is not obvious how to use the given information to solve the problem, but think about what happens when the solid dissolves. When the solid CuBr is placed in contact with water, it dissolves to form the separated Cu^+ and Br^- ions:

$$CuBr(s) \rightleftharpoons Cu^+(aq) + Br^-(aq)$$

where

$$K_{sp} = [Cu^+][Br^-]$$

Therefore, we can calculate the value of K_{sp} if we know $[Cu^+]$ and $[Br^-]$, the equilibrium concentrations of the ions. We know that the measured solubility of CuBr is 2.0×10^{-4} mol/L. This means that 2.0×10^{-4} mol of solid CuBr dissolves per 1.0 L of solution to come to equilibrium. The reaction is

$$CuBr(s) \rightarrow Cu^+(aq) + Br^-(aq)$$

so

$$2.0 \times 10^{-4}\text{ mol/L } CuBr(s) \rightarrow 2.0 \times 10^{-4}\text{ mol/L } Cu^+(aq) + 2.0 \times 10^{-4}\text{ mol/L } Br^-(aq)$$

We can now write the equilibrium concentrations

$$[Cu^+] = 2.0 \times 10^{-4}\text{ mol/L}$$

and

$$[Br^-] = 2.0 \times 10^{-4}\text{ mol/L}$$

These equilibrium concentrations allow us to calculate the value of K_{sp} for CuBr.

$$K_{sp} = [Cu^+][Br^-] = (2.0 \times 10^{-4})(2.0 \times 10^{-4}) = 4.0 \times 10^{-8}$$

The units for K_{sp} values are omitted.

Self-Check Exercise 16.7

Calculate the K_{sp} value for barium sulfate, $BaSO_4$, which has a solubility of 3.9×10^{-5} mol/L at 25 °C.

See Problems 16.59 through 16.62. ■

Solubilities must be expressed in mol/L in K_{sp} calculations.

We have seen that the known solubility of an ionic solid can be used to calculate its K_{sp} value. The reverse is also possible: the solubility of an ionic solid can be calculated if its K_{sp} value is known.

Example 16.10 Calculating Solubility from K_{sp} Values

The K_{sp} value for solid AgI(*s*) is 1.5×10^{-16} at 25 °C. Calculate the solubility of AgI(*s*) in water at 25 °C.

Solution

The solid AgI dissolves according to the equation

$$AgI(s) \rightleftharpoons Ag^+(aq) + I^-(aq)$$

and the corresponding equilibrium expression is

$$K_{sp} = 1.5 \times 10^{-16} = [Ag^+][I^-]$$

Because we do not know the solubility of this solid, we will assume that x moles per liter dissolves to reach equilibrium. Therefore,

$$x\,\frac{\text{mol}}{\text{L}}\,AgI(s) \rightarrow x\,\frac{\text{mol}}{\text{L}}\,Ag^+(aq) + x\,\frac{\text{mol}}{\text{L}}\,I^-(aq)$$

and at equilibrium,

$$[Ag^+] = x\,\frac{\text{mol}}{\text{L}}$$

$$[I^-] = x\,\frac{\text{mol}}{\text{L}}$$

Substituting these concentrations into the equilibrium expression gives

$$K_{sp} = 1.5 \times 10^{-16} = [Ag^+][I^-] = (x)(x) = x^2$$

Thus

$$x^2 = 1.5 \times 10^{-16}$$

$$x = \sqrt{1.5 \times 10^{-16}} = 1.2 \times 10^{-8}\ \text{mol/L}$$

The solubility of AgI(*s*) is 1.2×10^{-8} mol/L.

✔ Self-Check Exercise 16.8

The K_{sp} value for lead chromate, $PbCrO_4$, is 2.0×10^{-16} at 25 °C. Calculate its solubility at 25 °C.

See Problems 16.69 and 16.70. ■

Chapter 16 Review

Key Terms

collision model (16.1)
activation energy (E_a) (16.2)
catalyst (16.2)
enzyme (16.2)
equilibrium (16.3)
chemical equilibrium (16.3)
law of chemical equilibrium (16.5)
equilibrium expression (16.5)
equilibrium constant (16.5)
equilibrium position (16.5)
homogeneous equilibria (16.6)
heterogeneous equilibria (16.6)
Le Châtelier's principle (16.7)
solubility product (K_{sp}) (16.9)

Summary

1. Chemical reactions can be described by the collision model, which assumes that molecules must collide in order to react. In terms of this model, a certain threshold energy, called the activation energy (E_a), must be overcome for a collision to form products.

2. A catalyst is a substance that speeds up a reaction without being consumed. A catalyst operates by providing a lower-energy pathway for the reaction in question. Enzymes are biological catalysts.

3. When a chemical reaction is carried out in a closed vessel, the system achieves chemical equilibrium, the state where the concentrations of both reactants and products remain constant over time. Equilibrium is a highly dynamic state; reactants are converted continually into products, and vice versa, as molecules collide with each other. At equilibrium, the rates of the forward and reverse reactions are equal.

4. The law of chemical equilibrium is a general description of the equilibrium condition. It states that for a reaction of the type

$$aA + bB \rightleftharpoons cC + dD$$

the equilibrium expression is given by

$$K = \frac{[C]^c\,[D]^d}{[A]^a\,[B]^b}$$

where K is the equilibrium constant.

5. For each reaction system at a given temperature, there is only one value for the equilibrium constant, but there is an infinite number of possible equilibrium positions. An equilibrium position is defined as a particular set of equilibrium concentrations that satisfy the equilibrium expression. A specific equilibrium position depends on the initial concentrations. The amount of a pure liquid or a pure solid is never included in the equilibrium expression.

6. Le Châtelier's principle allows us to predict the effects of changes in concentration, volume, and temperature on a system at equilibrium. This principle states that when a change is imposed on a system at equilibrium, the equilibrium position will shift in a direction that tends to compensate for the imposed change.

7. The principle of equilibrium can also be applied when an excess of a solid is added to water to form a saturated solution. The solubility product (K_{sp}) is an equilibrium constant defined by the law of chemical equilibrium. Solubility is an equilibrium position, and the K_{sp} value of a solid can be determined by measuring its solubility. Conversely, the solubility of a solid can be determined if its K_{sp} value is known.

In-Class Discussion Questions

These questions are designed to be considered by groups of students in class. Often these questions work well for introducing a particular topic in class.

1. Consider an equilibrium mixture of four chemicals (A, B, C, and D, all gases) reacting in a closed flask according to the following equation:

$$A + B \rightleftharpoons C + D$$

   a. You add more A to the flask. How does the concentration of each chemical compare to its original concentration after equilibrium is reestablished? Justify your answer.
   b. You have the original set-up at equilibrium, and add more D to the flask. How does the concentration of each chemical compare to its original concentration after equilibrium is reestablished? Justify your answer.

2. The boxes shown below represent a set of initial conditions for the reaction:

$K = 25$

Draw a quantitative molecular picture that shows what this system looks like after the reactants are mixed in one of the boxes and the system reaches equilibrium. Support your answer with calculations.

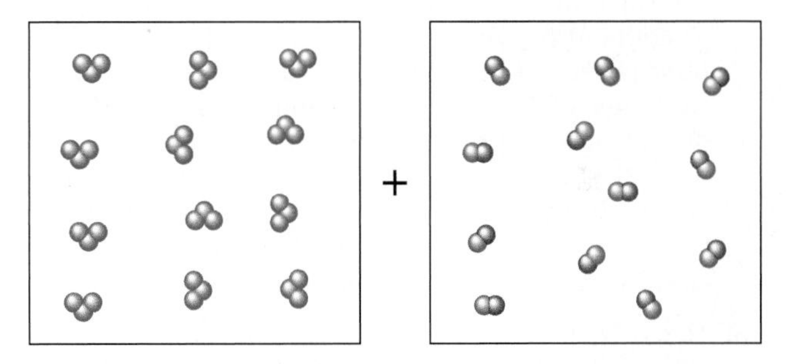

3. For the reaction $H_2 + I_2 \rightleftharpoons 2HI$, consider two possibilities: (a) you add 0.5 mol of each reactant, allow the system to come to equilibrium, and then add 1 mol of H_2, and allow the system to reach equilibrium again, or (b) you add 1.5 mol H_2 and 0.5 mol I_2 and allow the system to come to equilibrium. Will the final equilibrium mixture be different for the two procedures? Explain.

4. Given the reaction $A + B \rightleftharpoons C + D$, consider the following situations:
   a. You have 1.3 M A and 0.8 M B initially.
   b. You have 1.3 M A, 0.8 M B, and 0.2 M C initially.
   c. You have 2.0 M A and 0.8 M B initially.

   Order the preceding situations in terms of increasing equilibrium concentration of D and explain your order. Give the order in terms of increasing equilibrium concentration of B and explain.

5. Consider the reaction $A(g) + 2B(g) \rightleftharpoons C(g) + D(g)$ in a 1.0-L rigid flask. Answer the following questions for each situation (i–iv):
   i. Estimate a range (as small as possible) for the requested substance. For example, [A] could be between 95 *M* and 100 *M*.
   ii. Explain how you decided on the limits for the estimated range.
   iii. Indicate what other information would enable you to narrow your estimated range and explain.
   iv. Compare the estimated concentrations for (a) through (d), and explain any differences.
   a. If at equilibrium [A] = 1 *M*, and then 1 mol C is added, estimate the value for [A] once equilibrium is reestablished.
   b. If at equilibrium [B] = 1 *M*, and then 1 mol of C is added, estimate the value for [B] once equilibrium is reestablished.
   c. If at equilibrium [C] = 1 *M*, and then 1 mol of C is added, estimate the value for [C] once equilibrium is reestablished.
   d. If at equilibrium [D] = 1 *M*, and then 1 mol of C is added, estimate the value for [D] once equilibrium is reestablished.

6. Consider the reaction $A + B \rightleftharpoons C + D$. A friend asks the following: "I know we have been told that if a mixture of A, B, C, and D is in equilibrium and more A is added, more C and D will form. But how can more C and D form if we do not add more B?" What do you tell your friend?

7. Consider the following statements: "Consider the reaction $A(g) + B(g) \rightleftharpoons C(g)$, for which at equilibrium [A] = 2 *M*, [B] = 1 *M*, and [C] = 4 *M*. To a 1-L container of the system at equilibrium you add 3 mol of B. A possible equilibrium condition is [A] = 1 *M*, [B] = 3 *M*, and [C] = 6 *M*, because in both cases, $K = 2$." Indicate everything you think is correct in these statements, and everything that is incorrect. Correct the incorrect statements, and explain.

8. The value of the equilibrium constant, *K*, is dependent on which of the following? (There may be more than one answer.)
   a. the initial concentrations of the reactants
   b. the initial concentrations of the products
   c. the temperature of the system
   d. the nature of the reactants and products

   Explain.

9. Devise as many ways as you can to experimentally determine the K_{sp} value of a solid. Explain why each of these would work.

10. You are browsing through the *Handbook of Hypothetical Chemistry* when you come across a solid that is reported to have a K_{sp} value of zero in water at 25 °C. What does this mean?

11. A friend tells you: "The constant K_{sp} of a salt is called the solubility product constant and is calculated from the equilibrium concentrations of ions in the solution. Thus, if salt A dissolves to a greater extent than salt B, salt A must have a higher K_{sp} than salt B." Do you agree with your friend? Explain.

12. What do you suppose happens to the K_{sp} value of a solid as the temperature of the solution changes? Consider both increasing and decreasing temperatures, and explain your answer.

Questions and Problems

All even-numbered exercises have answers in the back of this book and solutions in the *Solutions Guide*.

16.1 How Chemical Reactions Occur

QUESTIONS

1. For the reaction $H_2(g) + Br_2(g) \rightarrow 2HBr(g)$, list the types of bonds that must be broken and the type of bond that must form in order for the chemical reaction to take place.

2. For the simple reaction $2CO(g) + O_2(g) \rightarrow 2CO_2(g)$, list the types of bonds that must be broken and the types of bonds that must form for the chemical reaction to take place.

16.2 Conditions That Affect Reaction Rates

QUESTIONS

3. How do chemists envision reactions taking place in terms of the *collision model* for reactions? Give an example of a simple reaction and how you might envision the reaction's taking place by means of a collision between the molecules.

4. In Figure 16.3, the height of the reaction "hill" is indicated as E_a. What does the symbol E_a stand for, and what does it represent in terms of a chemical reaction?

5. What is a *catalyst?* How does a catalyst speed up a reaction?

6. What are *enzymes* and why are they important?

16.3 The Equilibrium Condition

QUESTIONS

7. How does *equilibrium* represent the balancing of opposing processes? Give an example of an "equilibrium" encountered in everyday life, showing how the processes involved oppose each other.

8. How do chemists define a state of chemical equilibrium?

9. What does it mean to say that chemical reactions are *reversible?* Are all chemical reactions, in principle, reversible? Are some reactions more likely to occur in one direction than in the other?

10. How do chemists recognize a system that has reached a state of chemical equilibrium? When writing chemical equations, how do we indicate reactions that come to a state of chemical equilibrium?

16.4 Chemical Equilibrium: A Dynamic Condition

QUESTIONS

11. What does it mean to say that a state of chemical or physical equilibrium is *dynamic*?

12. Although a reaction that has reached a state of chemical equilibrium appears to an outside observer to have "stopped," what is still going on in the system? How do we know this?

16.5 The Equilibrium Constant: An Introduction

QUESTIONS

13. In general terms, what does the equilibrium constant for a reaction represent? What is the algebraic form of the equilibrium constant for a typical reaction? What do square brackets indicate when we write an equilibrium constant?

14. There is only one value of the equilibrium constant for a particular system at a particular temperature, but there is an infinite number of equilibrium positions. Explain.

PROBLEMS

15. Write the equilibrium expression for each of the following reactions.
    a. $4NH_3(g) + 5O_2(g) \rightleftharpoons 4NO(g) + 6H_2O(g)$
    b. $2NO(g) + O_2(g) \rightleftharpoons 2NO_2(g)$
    c. $CH_3OH(g) \rightleftharpoons CH_2O(g) + H_2(g)$

16. Write the equilibrium expression for each of the following reactions.
    a. $NO(g) + O_3(g) \rightleftharpoons NO_2(g) + O_2(g)$
    b. $SO_2(g) + NO_2(g) \rightleftharpoons SO_3(g) + NO(g)$
    c. $2Cl_2(g) + 2H_2O(g) \rightleftharpoons 4HCl(g) + O_2(g)$

17. Write the equilibrium expression for each of the following reactions.
    a. $2SO_2(g) + O_2(g) \rightleftharpoons 2SO_3(g)$
    b. $H_2(g) + F_2(g) \rightleftharpoons 2HF(g)$
    c. $2NO_2(g) + O_3(g) \rightleftharpoons N_2O_5(g) + O_2(g)$

18. Write the equilibrium expression for each of the following reactions.
    a. $CO(g) + 2H_2(g) \rightleftharpoons CH_3OH(g)$
    b. $2NO_2(g) \rightleftharpoons 2NO(g) + O_2(g)$
    c. $P_4(g) + 6Br_2(g) \rightleftharpoons 4PBr_3(g)$

19. Suppose that for the reaction

$$CH_3OH(g) \rightleftharpoons CH_2O(g) + H_2(g)$$

it is determined that, at a particular temperature, the equilibrium concentrations are $[CH_3OH(g)] = 0.00215\ M$, $[CH_2O(g)] = 0.441\ M$, and $[H_2(g)] = 0.0331\ M$. Calculate the value of K for the reaction at this temperature.

20. Suppose that for the following reaction

$$NO(g) + O_3(g) \rightleftharpoons NO_2(g) + O_2(g)$$

it is determined that, at a particular temperature, the equilibrium concentrations are as follows: $[NO(g)]$ $2.1 \times 10^{-3}\ M$, $[O_3(g)] = 4.7 \times 10^{-5}\ M$, $[NO_2(g)] = 3.5 \times 10^{-2}\ M$ and $[O_2(g)] = 7.8 \times 10^{-3}\ M$. Calculate the value of K for the reaction at this temperature.

21. At high temperatures, elemental nitrogen and oxygen react with each other to form nitrogen monoxide.

$$N_2(g) + O_2(g) \rightleftharpoons 2NO(g)$$

Suppose the system is analyzed at a particular temperature, and the equilibrium concentrations are found to be $[N_2] = 0.041\ M$, $[O_2] = 0.0078\ M$, and $[NO] = 4.7 \times 10^{-4}\ M$. Calculate the value of K for the reaction.

22. Suppose that for the following reaction

$$2NO(g) + 2H_2(g) \rightleftharpoons N_2(g) + 2H_2O(g)$$

it is determined that, at equilibrium at a particular temperature, the concentrations are as follows: $[NO(g)]$ $8.1 \times 10^{-3}\ M$, $[H_2(g)] = 4.1 \times 10^{-5}\ M$, $[N_2(g)] = 5.3 \times 10^{-2}\ M$ and $[H_2O(g)] = 2.9 \times 10^{-3}\ M$. Calculate the value of K for the reaction at this temperature.

16.6 Heterogeneous Equilibria

QUESTIONS

23. What is a *homogeneous* equilibrium system? Give an example of a homogeneous equilibrium reaction. What is a *heterogeneous* equilibrium system? Write two chemical equations that represent heterogeneous equilibria.

24. Explain why the position of a heterogeneous equilibrium does not depend on the amounts of pure solid or pure liquid reactants or products present.

PROBLEMS

25. Write the equilibrium expression for each of the following heterogeneous equilibria.
    a. $SO_3(g) + H_2O(l) \rightleftharpoons H_2SO_4(l)$
    b. $2NH_3(g) + CO_2(g) \rightleftharpoons N_2CH_4O(s) + H_2O(g)$
    c. $ZrI_4(s) \rightleftharpoons Zr(s) + 2I_2(g)$

26. Write the equilibrium expression for each of the following heterogeneous equilibria.
    a. $2H_2(g) + CO(g) \rightleftharpoons CH_3OH(l)$
    b. $2K(s) + H_2(g) \rightleftharpoons 2KH(s)$
    c. $MgO(s) + CO_2(g) \rightleftharpoons MgCO_3(s)$

27. Write the equilibrium expression for each of the following heterogeneous equilibria.

a. $2HgO(s) \rightleftharpoons 2Hg(l) + O_2(g)$
b. $2KClO_3(s) \rightleftharpoons 2KCl(s) + 3O_2(g)$
c. $C(s) + CO_2(g) \rightleftharpoons 2CO(g)$

28. Write the equilibrium expression for each of the following heterogeneous equilibria.
a. $2NBr_3(s) \rightleftharpoons N_2(g) + 3Br_2(g)$
b. $CuO(s) + H_2(g) \rightleftharpoons Cu(l) + H_2O(g)$
c. $PbCO_3(s) \rightleftharpoons PbO(s) + CO_2(g)$

16.7 Le Châtelier's Principle

QUESTIONS

29. In your own words, describe what Le Châtelier's principle tells us about how we can change the position of a reaction system at equilibrium.

30. Consider the reaction

$$2CO(g) + O_2(g) \rightleftharpoons 2CO_2(g)$$

Suppose the system is already at equilibrium, and then an additional mole of $CO(g)$ is injected into the system at constant temperature. Does the amount of $CO_2(g)$ in the system increase or decrease? Does the value of K for the reaction change?

31. For an equilibrium involving gaseous substances, what effect, in general terms, is realized when the volume of the system is decreased?

32. What is the effect on the equilibrium position if an endothermic reaction is performed at a higher temperature? Does the net amount of product increase or decrease? Does the value of the equilibrium constant change if the temperature is increased?

PROBLEMS

33. For the reaction system

$$4NH_3(g) + 5O_2(g) \rightleftharpoons 4NO(g) + 6H_2O(g)$$

which has already reached a state of equilibrium, predict the effect that each of the following changes will have on the position of the equilibrium. Tell whether the equilibrium will shift to the right, will shift to the left, or will not be affected.
a. The pressure of oxygen is increased by injecting one additional mole of oxygen into the reaction vessel.
b. A desiccant (a material that absorbs water) is added to the system.
c. The system is compressed and the ammonia liquefies.

34. Suppose the reaction system

$$2SO_2(g) + O_2(g) \rightleftharpoons 2SO_3(g)$$

has already reached equilibrium. Predict the effect of each of the following changes on the position of the equilibrium. Tell whether the equilibrium will shift to the right, will shift to the left, or will not be affected.
a. Additional $SO_2(g)$ is added to the system.
b. The $SO_3(g)$ present is liquefied and removed from the system.
c. A very efficient catalyst is used.

35. Suppose the reaction system

$$CH_4(g) + 2O_2(g) \rightleftharpoons CO_2(g) + 2H_2O(l)$$

has already reached equilibrium. Predict the effect of each of the following changes on the position of the equilibrium. Tell whether the equilibrium will shift to the right, will shift to the left, or will not be affected.
a. Any liquid water present is removed from the system.
b. CO_2 is added to the system by dropping a chunk of dry ice into the reaction vessel.
c. The reaction is performed in a metal cylinder fitted with a piston, and the piston is compressed to decrease the total volume of the system.
d. Additional $O_2(g)$ is added to the system from a cylinder of pure O_2.

36. Suppose the reaction system

$$UO_2(s) + 4HF(g) \rightleftharpoons UF_4(g) + 2H_2O(g)$$

has already reached equilibrium. Predict the effect of each of the following changes on the position of the equilibrium. Tell whether the equilibrium will shift to the right, will shift to the left, or will not be affected.
a. Additional $UO_2(s)$ is added to the system.
b. 5.0 mol of $Xe(g)$ is added to the system.
c. The reaction is performed in a glass reaction vessel; $HF(g)$ attacks and reacts with glass.
d. Water vapor is removed.
e. The size of the reaction vessel is increased.

37. Old-fashioned "smelling salts" consist of ammonium carbonate, $(NH_4)_2CO_3$. The reaction for the decomposition of ammonium carbonate

$$(NH_4)_2CO_3(s) \rightleftharpoons 2NH_3(g) + CO_2(g) + H_2O(g)$$

is endothermic. What would be the effect on the position of this equilibrium if the reaction were performed at a lower temperature?

38. The reaction $N_2(g) + 3H_2(g) \rightleftharpoons 2NH_3(g)$ is exothermic as written. For the maximum production of ammonia, should this reaction be performed at a lower or a higher temperature? Explain.

39. The reaction

$$C_2H_2(g) + 2Br_2(g) \rightleftharpoons C_2H_2Br_4(g)$$

is exothermic in the forward direction. Will an increase in temperature shift the position of the equilibrium toward reactants or products?

40. The reaction

$$4NO(g) + 6H_2O(g) \rightleftharpoons 4NH_3(g) + 5O_2(g)$$

is strongly endothermic. Will an increase in temperature shift the equilibrium position toward products or toward reactants?

41. Plants synthesize the sugar dextrose according to the following reaction by absorbing radiant energy from the sun (photosynthesis).

$$6CO_2(g) + 6H_2O(g) \rightleftharpoons C_6H_{12}O_6(s) + 6O_2(g)$$

Will an increase in temperature tend to favor or discourage the production of $C_6H_{12}O_6(s)$?

42. Chemists have expended a great deal of effort to maximize the production of ammonia from the elements

$$N_2(g) + 3H_2(g) \rightleftharpoons 2NH_3(g) + 92 \text{ kJ}$$

Suggest three changes that could be made to this equilibrium system, which would tend to *increase* the yield of ammonia.

16.8 Applications Involving the Equilibrium Constant

QUESTIONS

43. Suppose a reaction has the equilibrium constant $K = 1.3 \times 10^8$. What does the magnitude of this constant tell you about the relative concentrations of products and reactants that will be present once equilibrium is reached? Is this reaction likely to be a good source of the products?

44. Which is better for the maximum production of products, a reaction with a small equilibrium constant or a reaction with a large equilibrium constant? Explain.

PROBLEMS

45. For the reaction

$$2NBr_3(g) \rightleftharpoons N_2(g) + 3Br_2(g)$$

the system at equilibrium at a particular temperature is analyzed, and the following concentrations are found: $[NBr_3(g)] = 2.07 \times 10^{-3}$ *M*; $[N_2(g)] = 4.11 \times 10^{-2}$ *M*; $[Br_2(g)] = 1.06 \times 10^{-3}$ *M*. Calculate the value of *K* for the reaction at this temperature.

46. For the reaction

$$N_2(g) + 3Cl_2(g) \rightleftharpoons 2NCl_3(g)$$

an analysis of an equilibrium mixture is performed. It is found that $[NCl_3(g)] = 1.9 \times 10^{-1}$ *M*, $[N_2(g)] = 1.4 \times 10^{-3}$ *M*, and $[Cl_2(g)] = 4.3 \times 10^{-4}$ *M*. Calculate *K* for the reaction.

47. For the reaction

$$2CO(g) + O_2(g) \rightleftharpoons 2CO_2(g)$$

it is found at equilibrium at a certain temperature that the concentrations are $[CO(g)] = 2.7 \times 10^{-4}$ *M*, $[O_2(g)] = 1.9 \times 10^{-3}$ *M*, and $[CO_2(g)] = 1.1 \times 10^{-1}$ *M*. Calculate *K* for the reaction at this temperature.

48. For the reaction

$$CaCO_3(s) \rightleftharpoons CaO(s) + CO_2(g)$$

it is found that at equilibrium $[CO_2] = 2.1 \times 10^{-3}$ *M* at a particular temperature. Calculate *K* for the reaction at this temperature.

49. The equilibrium constant for the reaction

$$H_2(g) + F_2(g) \rightleftharpoons 2HF(g)$$

has the value 2.1×10^3 at a particular temperature. When the system is analyzed at equilibrium at this temperature, the concentrations of both $H_2(g)$ and $F_2(g)$ are found to be 0.0021 *M*. What is the concentration of $HF(g)$ in the equilibrium system under these conditions?

50. For the reaction

$$2H_2O(g) \rightleftharpoons 2H_2(g) + O_2(g)$$

$K = 2.4 \times 10^{-3}$ at a given temperature. At equilibrium it is found that $[H_2O(g)] = 1.1 \times 10^{-1}$ *M* and $[H_2(g)] = 1.9 \times 10^{-2}$ *M*. What is the concentration of $O_2(g)$ under these conditions?

51. For the reaction

$$N_2(g) + O_2(g) \rightleftharpoons 2NO(g)$$

the equilibrium constant *K* has the value 1.71×10^{-3} at a particular temperature. If the concentrations of both $N_2(g)$ and $O_2(g)$ are 0.0342 *M* in an equilibrium mixture at this temperature, what is the concentration of $NO(g)$ under these conditions?

52. For the reaction

$$N_2O_4(g) \rightleftharpoons 2NO_2(g)$$

the equilibrium constant *K* has the value 8.1×10^{-3} at a particular temperature. If the concentration of $NO_2(g)$ is found to be 0.0021 *M* in the equilibrium system, what is the concentration of $N_2O_4(g)$ under these conditions?

16.9 Solubility Equilibria

QUESTIONS

53. Explain how the dissolving of an ionic solute in water represents an equilibrium process.

54. What is the special name given to the equilibrium constant for the dissolving of an ionic solute in water?

55. Why does the amount of excess solid solute present in a solution not affect the amount of solute that ultimately dissolves in a given amount of solvent?

56. Which of the following will affect the total amount of solute that can dissolve in a given amount of solvent?
   a. The solution is stirred.
   b. The solute is ground to fine particles before dissolving.
   c. The temperature changes.

PROBLEMS

57. Write the balanced chemical equation describing the dissolving of the following solids in water. Write the expression for K_{sp} for each process.
    a. $Co_2S_3(s)$
    b. $AgI(s)$
    c. $MgCO_3(s)$
    d. $Fe(OH)_3(s)$

58. Write the balanced chemical equations describing the dissolving of the following solids in water. Write the expression for K_{sp} for each process.
    a. $PbBr_2(s)$
    b. $Ag_2S(s)$
    c. $PbCO_3(s)$
    d. $Sr_3(PO_4)_2(s)$

59. Zinc carbonate dissolves in water to the extent of 1.12×10^{-4} g/L at 25 °C. Calculate the solubility product K_{sp} for $ZnCO_3$ at 25 °C.

60. Zinc carbonate, $ZnCO_3(s)$, dissolves in water to give a solution that is 1.7×10^{-5} *M* at 22 °C. Calculate K_{sp} for $ZnCO_3(s)$ at this temperature.

61. A saturated solution of nickel(II) sulfide contains approximately 3.6×10^{-4} g of dissolved NiS per liter at 20 °C. Calculate the solubility product K_{sp} for NiS at 20 °C.

62. Copper(II) chromate, $CuCrO_4(s)$, dissolves in water to give a solution containing 1.1×10^{-5} g per liter at 23 °C. Calculate K_{sp} for $CuCrO_4(s)$.

63. The solubility product constant, K_{sp}, for calcium carbonate at room temperature is approximately 3.0×10^{-9}. Calculate the solubility of $CaCO_3$ in grams per liter under these conditions.

64. The solubility product constant, K_{sp}, for barium carbonate is 8.2×10^{-9} at a particular temperature. Calculate the solubility of $BaCO_3$ in mol/L at this temperature.

65. Approximately 1.5×10^{-3} g of iron(II) hydroxide, $Fe(OH)_2(s)$, dissolves per liter of water at 18 °C. Calculate K_{sp} for $Fe(OH)_2(s)$ at this temperature.

66. Mercuric sulfide is an extremely insoluble compound, having $K_{sp} = 3 \times 10^{-53}$ near room temperature. Calculate the solubility of HgS in mol/L at this temperature.

67. Magnesium fluoride dissolves in water to the extent of 8.0×10^{-2} g/L at 25 °C. Calculate the solubility of $MgF_2(s)$ in moles per liter, and calculate K_{sp} for MgF_2 at 25 °C.

68. Lead(II) chloride, $PbCl_2(s)$, dissolves in water to the extent of approximately 3.6×10^{-2} *M* at 20 °C. Calculate K_{sp} for $PbCl_2(s)$, and calculate its solubility in grams per liter.

69. Mercury(I) chloride, Hg_2Cl_2, was formerly administered orally as a purgative. Although we usually think of mercury compounds as highly toxic, the K_{sp} of mercury(I) chloride is small enough (1.3×10^{-18}) that the amount of mercury that dissolves and enters the bloodstream is tiny. Calculate the concentration of mercury(I) ion present in a saturated solution of Hg_2Cl_2.

70. The solubility product of iron(III) hydroxide is very small: $K_{sp} = 4 \times 10^{-38}$ at 25 °C. A classical method of analysis for unknown samples containing iron is to add NaOH or NH_3. This precipitates $Fe(OH)_3$, which can then be filtered and weighed. To demonstrate that the concentration of iron remaining in solution in such a sample is very small, calculate the solubility of $Fe(OH)_3$ in moles per liter and in grams per liter.

ADDITIONAL PROBLEMS

71. Before two molecules can react, chemists envision that the molecules must first *collide* with one another. Is collision among molecules the only consideration for the molecules to react with one another?

72. Why does an increase in temperature favor an increase in the speed of a reaction?

73. The minimum energy required for molecules to react with each other is called the ________ energy.

74. A ________ speeds up a reaction without being consumed.

75. Equilibrium may be defined as the ________ of two processes, one of which is the opposite of the other.

76. When a chemical system has reached equilibrium, the concentrations of all reactants and products remain ________ with time.

77. What does it mean to say that all chemical reactions are, to one extent or another, *reversible?*

78. What does it mean to say that chemical equilibrium is a *dynamic* process?

79. At the point of chemical equilibrium, the rate of the forward reaction ________ the rate of the reverse reaction.

80. Equilibria involving reactants or products in more than one state are said to be ________.

81. According to Le Châtelier's principle, when a large excess of a gaseous reactant is added to a reaction system at equilibrium, the amounts of products ________.

82. Addition of an inert substance (one that does not participate in the reaction) does not change the ________ of an equilibrium.

83. When the volume of a vessel containing a gaseous equilibrium system is decreased, the ________ of the gaseous substances present is initially increased.

84. Why does increasing the temperature for an exothermic process tend to favor the conversion of products back to reactants?

85. What is meant by the *solubility product* for a sparingly soluble salt? Choose a sparingly soluble salt and show how the salt ionizes when dissolved in water, and write the expression for its solubility product.

86. For a given reaction at a given temperature, the special ratio of products to reactants defined by the equilibrium constant is always equal to the same number. Explain why this is true, no matter what initial concentrations of reactants (or products) may have been taken in setting up an experiment.

87. Many sugars undergo a process called mutarotation, in which the sugar molecules interconvert between two isomeric forms, finally reaching an equilibrium between them. This is true for the simple sugar glucose, $C_6H_{12}O_6$, which exists in solution in isomeric forms called alpha-glucose and beta-glucose. If a solution of glucose at a certain temperature is analyzed, and it is found that the concentration of alpha-glucose is twice the concentration of beta-glucose, what is the value of K for the interconversion reaction?

88. Suppose $K = 4.5 \times 10^{-3}$ at a certain temperature for the reaction

$$PCl_5(g) \rightleftharpoons PCl_3(g) + Cl_2(g)$$

If it is found that the concentration of PCl_5 is twice the concentration of PCl_3, what must be the concentration of Cl_2 under these conditions?

89. For the reaction

$$CaCO_3(s) \rightleftharpoons CaO(s) + CO_2(g)$$

the equilibrium constant K has the form $K = [CO_2]$. Using a handbook to find density information about $CaCO_3(s)$ and $CaO(s)$, show that the *concentrations* of the two solids (the number of moles contained in 1 L of volume) are constant.

90. As you know from Chapter 7, most metal carbonate salts are sparingly soluble in water. Below are listed several metal carbonates along with their solubility products, K_{sp}. For each salt, write the equation showing the ionization of the salt in water, and calculate the solubility of the salt in mol/L.

Salt	K_{sp}
$BaCO_3$	5.1×10^{-9}
$CdCO_3$	5.2×10^{-12}
$CaCO_3$	2.8×10^{-9}
$CoCO_3$	1.5×10^{-13}

91. Teeth and bones are composed, to a first approximation, of calcium phosphate, $Ca_3(PO_4)_2(s)$. The K_{sp} for this salt is 1.3×10^{-32} at 25 °C. Calculate the concentration of calcium ion in a saturated solution of $Ca_3(PO_4)_2$.

92. Under what circumstances can we compare the solubilities of two salts by directly comparing the values of their solubility products?

93. How does the collision model account for the fact that a reaction proceeds faster when the concentrations of the reactants are increased?

94. How does an increase in temperature result in an increase in the number of successful collisions between reactant molecules? What does an increase in temperature mean on a molecular basis?

95. Explain why the development of a vapor pressure above a liquid in a closed container represents an equilibrium. What are the opposing processes? How do we recognize when the system has reached a state of equilibrium?

96. Write the equilibrium expression for each of the following reactions.
a. $H_2(g) + Br_2(g) \rightleftharpoons 2HBr(g)$
b. $2H_2(g) + S_2(g) \rightleftharpoons 2H_2S(g)$
c. $H_2(g) + C_2N_2(g) \rightleftharpoons 2HCN(g)$

97. Write the equilibrium expression for each of the following reactions.
a. $2O_3(g) \rightleftharpoons 3O_2(g)$
b. $CH_4(g) + 2O_2(g) \rightleftharpoons CO_2(g) + 2H_2O(g)$
c. $C_2H_4(g) + Cl_2(g) \rightleftharpoons C_2H_4Cl_2(g)$

98. At high temperatures, elemental bromine, Br_2, dissociates into individual bromine atoms.

$$Br_2(g) \rightleftharpoons 2Br(g)$$

Suppose that in an experiment at 2000 °C, it is found that $[Br_2] = 0.97\ M$ and $[Br] = 0.034\ M$ at equilibrium. Calculate the value of K.

99. Gaseous phosphorus pentachloride decomposes according to the reaction

$$PCl_5(g) \rightleftharpoons PCl_3(g) + Cl_2(g)$$

The equilibrium system was analyzed at a particular temperature, and the concentrations of the substances present were determined to be $[PCl_5] = 1.1 \times 10^{-2}\ M$, $[PCl_3] = 0.325\ M$, and $[Cl_2] = 3.9 \times 10^{-3}\ M$. Calculate the value of K for the reaction.

100. Write the equilibrium expression for each of the following heterogeneous equilibria.
a. $4Al(s) + 3O_2(g) \rightleftharpoons 2Al_2O_3(s)$
b. $NH_3(g) + HCl(g) \rightleftharpoons NH_4Cl(s)$
c. $2Mg(s) + O_2(g) \rightleftharpoons 2MgO(s)$

101. Write the equilibrium expression for each of the following heterogeneous equilibria.
a. $P_4(s) + 5O_2(g) \rightleftharpoons P_4O_{10}(s)$
b. $CO_2(g) + 2NaOH(s) \rightleftharpoons Na_2CO_3(s) + H_2O(g)$
c. $NH_4NO_3(s) \rightleftharpoons N_2O(g) + 2H_2O(g)$

102. What is the effect on the position of a reaction system at equilibrium when an exothermic reaction is performed at a higher temperature? Does the value of the equilibrium constant change in this situation?

103. Suppose the reaction system

$$2NO(g) + O_2(g) \rightleftharpoons 2NO_2(g)$$

has already reached equilibrium. Predict the effect of each of the following changes on the position of the equilibrium. Tell whether the equilibrium will shift to the right, will shift to the left, or will not be affected.
a. Additional oxygen is injected into the system.
b. NO_2 is removed from the reaction vessel.
c. 1.0 mol of helium is injected into the system.

104. The reaction

$$PCl_3(l) + Cl_2(g) \rightleftharpoons PCl_5(s)$$

liberates 124 kJ of energy per mole of PCl_3 reacted. Will an increase in temperature shift the equilibrium position toward products or toward reactants?

105. For the process

$$CO(g) + H_2O(g) \rightleftharpoons CO_2(g) + H_2(g)$$

it is found that the equilibrium concentrations at a particular temperature are $[H_2] = 1.4\ M$, $[CO_2] = 1.3\ M$, $[CO] = 0.71\ M$, and $[H_2O] = 0.66\ M$. Calculate the equilibrium constant K for the reaction under these conditions.

106. For the reaction

$$N_2(g) + 3H_2(g) \rightleftharpoons 2NH_3(g)$$

$K = 1.3 \times 10^{-2}$ at a given temperature. If the system at equilibrium is analyzed and the concentrations of both N_2 and H_2 are found to be 0.10 M, what is the concentration of NH_3 in the system?

107. The equilibrium constant for the reaction

$$2NOCl(g) \rightleftharpoons 2NO(g) + Cl_2(g)$$

has the value 9.2×10^{-6} at a particular temperature. The system is analyzed at equilibrium, and it is found that the concentrations of $NOCl(g)$ and $NO(g)$ are 0.44 M and $1.5 \times 10^{-3}\ M$, respectively. What is the concentration of $Cl_2(g)$ in the equilibrium system under these conditions?

108. As you learned in Chapter 7, most metal hydroxides are sparingly soluble in water. Write balanced chemical equations describing the dissolving of the following metal hydroxides in water. Write the expression for K_{sp} for each process.
a. $Cu(OH)_2(s)$
b. $Cr(OH)_3(s)$
c. $Ba(OH)_2(s)$
d. $Sn(OH)_2(s)$

109. The three common silver halides (AgCl, AgBr, and AgI) are all sparingly soluble salts. Given the values for K_{sp} for these salts below, calculate the concentration of silver ion, in mol/L, in a saturated solution of each salt.

Silver Halide	K_{sp}
AgCl	1.8×10^{-10}
AgBr	5.0×10^{-13}
AgI	8.3×10^{-17}

110. Approximately 9.0×10^{-4} g of silver chloride, $AgCl(s)$, dissolves per liter of water at 10 °C. Calculate K_{sp} for $AgCl(s)$ at this temperature.

111. Mercuric sulfide, HgS, is one of the least soluble salts known, with $K_{sp} = 1.6 \times 10^{-54}$ at 25 °C. Calculate the solubility of HgS in moles per liter and in grams per liter.

112. Approximately 0.14 g of nickel(II) hydroxide, $Ni(OH)_2(s)$, dissolves per liter of water at 20 °C. Calculate K_{sp} for $Ni(OH)_2(s)$ at this temperature.

113. For the reaction $N_2(g) + 3H_2(g) \rightarrow 2NH_3(g)$, list the types of bonds that must be broken and the type of bonds that must form in order for the chemical reaction to take place.

114. What does the *activation energy* for a reaction represent? How is the activation energy related to whether a collision between molecules is successful?

115. What are the catalysts in living cells called? Why are these biological catalysts necessary?

116. When a reaction system has reached chemical equilibrium, the concentrations of the reactants and products no longer changes with time. Why does the amount of product no longer increase, even though large concentrations of the reactants may still be present?

117. Ammonia, a very important industrial chemical, is produced by the direct combination of the elements under carefully controlled conditions.

$$N_2(g) + 3H_2(g) \rightleftharpoons 2NH_3(g)$$

Suppose, in an experiment, that the reaction mixture is analyzed after equilibrium is reached and it is found, at a particular temperature, that $[NH_3(g)] = 0.34\ M$, $[H_2(g)] = 2.1 \times 10^{-3}\ M$, and $[N_2(g)] = 4.9 \times 10^{-4}\ M$. Calculate the value of K at this temperature.

118. Write the equilibrium expression for each of the following heterogeneous equilibria.

a. $2LiHCO_3(s) \rightleftharpoons Li_2CO_3(s) + H_2O(g) + CO_2(g)$
b. $PbCO_3(s) \rightleftharpoons PbO(s) + CO_2(g)$
c. $4Al(s) + 3O_2(g) \rightleftharpoons 2Al_2O_3(s)$

119. Suppose a reaction has the equilibrium constant $K = 4.5 \times 10^{-6}$ at a particular temperature. If an experiment is set up with this reaction, will there be large relative concentrations of products present at equilibrium? Is this reaction useful as a means of producing the products? How might the reaction be made more useful?

Cumulative Review for Chapters 15–16

QUESTIONS

1. Compare the Arrhenius and Brønsted–Lowry definitions of an acid and a base.

2. Describe the relationship between a conjugate acid–base pair in the Brønsted–Lowry model. Write balanced chemical equations showing the following molecules/ions behaving as Brønsted–Lowry acids in water: HCl, H_2SO_4, H_3PO_4, NH_4^+. Write balanced chemical equations showing the following molecules/ions behaving as Brønsted–Lowry bases in water: NH_3, HCO_3^-, NH_2^-, $H_2PO_4^-$.

3. Acetic acid is a weak acid in water. What does this indicate about the affinity of the acetate ion for protons compared to the affinity of water molecules for protons? If a solution of sodium acetate is dissolved in water, the solution is basic. Explain. Write equilibrium reaction equations for the ionization of acetic acid in water and for the reaction of the acetate ion with water in a solution of sodium acetate.

4. How is the *strength* of an acid related to the *position* of its ionization equilibrium? Write the equations for the dissociation (ionization) of HCl, HNO_3, and $HClO_4$ in water. Since all these acids are strong acids, what does this indicate about the basicity of the Cl^-, NO_3^-, and ClO_4^- ions? Are aqueous solutions of NaCl, $NaNO_3$, or $NaClO_4$ basic?

5. Explain how water is an *amphoteric* substance. Write the chemical equation for the autoionization of water. Write the expression for the equilibrium constant, K_w, for this reaction. What values does K_w have at 25 °C? What are $[H^+]$ and $[OH^-]$ in pure water at 25 °C? How does $[H^+]$ compare to $[OH^-]$ in an acidic solution? How does $[H^+]$ compare to $[OH^-]$ in a basic solution?

6. How is the pH scale defined? What range of pH values corresponds to acidic solutions? What range corresponds to basic solutions? Why is pH = 7.00 considered *neutral?* When the pH of a solution changes by one unit, by what factor does the hydrogen ion concentration change in the solution? How is pOH defined? How are pH and pOH for a given solution related? Explain.

7. Describe a *buffered* solution. Give three examples of buffered solutions. For each of your examples, write equations and explain how the components of the buffered solution consume added strong acids or bases. Why is buffering of solutions in biological systems so important?

8. Explain the *collision model* for chemical reactions. What "collides"? Do all collisions result in the breaking of bonds and formation of products? Why? How does the collision model explain why higher concentrations and higher temperatures tend to make reactions occur faster?

9. Sketch a graph for the progress of a reaction illustrating the *activation energy* for the reaction. Define "activation energy." Explain how an increase in temperature for a reaction affects the number of collisions that possess an energy greater than E_a. Does an increase in temperature change E_a? How does a *catalyst* speed up a reaction? Does a catalyst change E_a for the reaction?

10. Explain what it means that a reaction "has reached a state of chemical equilibrium." Explain why equilibrium is a *dynamic* state: does a reaction really "stop" when the system reaches a state of equilibrium? Explain why, once a chemical system has reached equilibrium, the concentrations of all reactants remain *constant* with time. Why does this *constancy* of concentration not contradict our picture of equilibrium as being *dynamic?* What happens to the *rates* of the forward and reverse reactions as a system proceeds to equilibrium from a starting point where only reactants are present?

11. Describe how we write the equilibrium expression for a reaction. Give three examples of balanced chemical equations and the corresponding expressions for their equilibrium constants.

12. Although the equilibrium constant for a given reaction always has the same value at the same temperature, the actual *concentrations* present at equilibrium may differ from one experiment to another. Explain. What do we mean by an *equilibrium position?* Is the equilibrium position always the same for a reaction, regardless of the amounts of reactants taken?

13. Compare *homogeneous* and *heterogeneous* equilibria. Give a balanced chemical equation and write the corresponding equilibrium constant expression as an example of each of these cases. How does the fact that an equilibrium is *heterogeneous* influence the expression we write for the equilibrium constant for the reaction?

14. In your own words, paraphrase Le Châtelier's principle. Give an example (including a balanced chemical equation) of how each of the following changes can affect the position of equilibrium in favor of additional products for a system: the concentration of one of the reactants is increased; one of the products is selectively removed from the system; the reaction system is compressed to a smaller volume; the temperature is increased for an endothermic reaction; the temperature is decreased for an exothermic process.

15. Explain how dissolving a slightly soluble salt to form a saturated solution is an *equilibrium* process. Give three balanced chemical equations for solubility processes and write the expressions for K_{sp} corresponding to the reactions you have chosen. When writing expressions for K_{sp}, why is the concentration

of the sparingly soluble salt itself not included in the expression? Given the value for the solubility product for a sparingly soluble salt, explain how the molar solubility, and the solubility in g/L, may be calculated.

PROBLEMS

16. Choose 10 species that might be expected to behave as Brønsted–Lowry acids or bases in aqueous solution. For each of your choices, (a) write an equation demonstrating *how* the species behaves as an acid or base in water, and (b) write the formula of the conjugate base or acid for each of the species you have chosen.

17. Identify the Brønsted–Lowry conjugate acid–base pairs in each of the following.
    a. $HC_2H_3O_2(aq) + H_2O(l) \rightarrow C_2H_3O_2^-(aq) + H_3O^+(aq)$
    b. $SO_3^{2-}(aq) + H_2O(l) \rightarrow HSO_3^-(aq) + OH^-(aq)$
    c. $F^-(aq) + H_2O(l) \rightarrow HF(aq) + OH^-(aq)$
    d. $H_2SO_3(aq) + H_2O(l) \rightarrow HSO_3^-(aq) + H_3O^+(aq)$
    e. $S^{2-}(aq) + H_2O(l) \rightarrow HS^-(aq) + OH^-(aq)$

18. Which of the following are *not* conjugate acid–base pairs in the Brønsted–Lowry theory? Explain your reasoning.
    a. $S^{2-}(aq)$ and $H_2S(aq)$
    b. $H_3O^+(aq)$ and $OH^-(aq)$
    c. $H_2O(l)$ and $H_3O^+(aq)$
    d. $OH^-(aq)$ and $H_2O(l)$
    e. $NH_4^+(aq)$ and $NH_2^-(aq)$

19. For each of the given items, calculate the indicated quantity.
    a. $[H^+] = 4.01 \times 10^{-3}$ M, pH = ?
    b. $[OH^-] = 7.41 \times 10^{-8}$ M, pOH = ?
    c. $[H^+] = 9.61 \times 10^{-6}$ M, pOH = ?
    d. $[OH^-] = 6.62 \times 10^{-3}$ M, pH = ?
    e. pH = 6.325, $[OH^-]$ = ?
    f. pH = 9.413, $[H^+]$ = ?

20. Calculate the pH and pOH values for each of the following solutions.
    a. 0.00141 M HNO_3
    b. 2.13×10^{-3} M NaOH
    c. 0.00515 M HCl
    d. 5.65×10^{-5} M $Ca(OH)_2$

21. Write equilibrium constant expressions for each of the following reactions.
    a. $H_2(g) + Br_2(g) \rightleftharpoons 2HBr(g)$
    b. $2NO(g) + O_2(g) \rightleftharpoons 2NO_2(g)$
    c. $N_2H_4(l) + O_2(g) \rightleftharpoons N_2(g) + 2H_2O(g)$
    d. $SO_2Cl_2(g) \rightleftharpoons SO_2(g) + Cl_2(g)$
    e. $CO(g) + NO_2(g) \rightleftharpoons CO_2(g) + NO(g)$

22. For the reaction

$$2SO_2(g) + O_2(g) \rightleftharpoons 2SO_3(g)$$

at a particular temperature the equilibrium system contains $[SO_3(g)] = 0.42$ M, $[SO_2(g)] = 1.4 \times 10^{-3}$ M, and $[O_2(g)] = 4.5 \times 10^{-4}$ M. Calculate K for the process at this temperature.

23. Write expressions for K_{sp} for each of the following sparingly soluble salts.
    a. NiS
    b. $Co_2(CO_3)_3$
    c. $Cu(OH)_2$
    d. $AlPO_4$
    e. $PbCl_2$

17 Oxidation–Reduction Reactions and Electrochemistry

A teapot that was silver-plated by electrolysis.

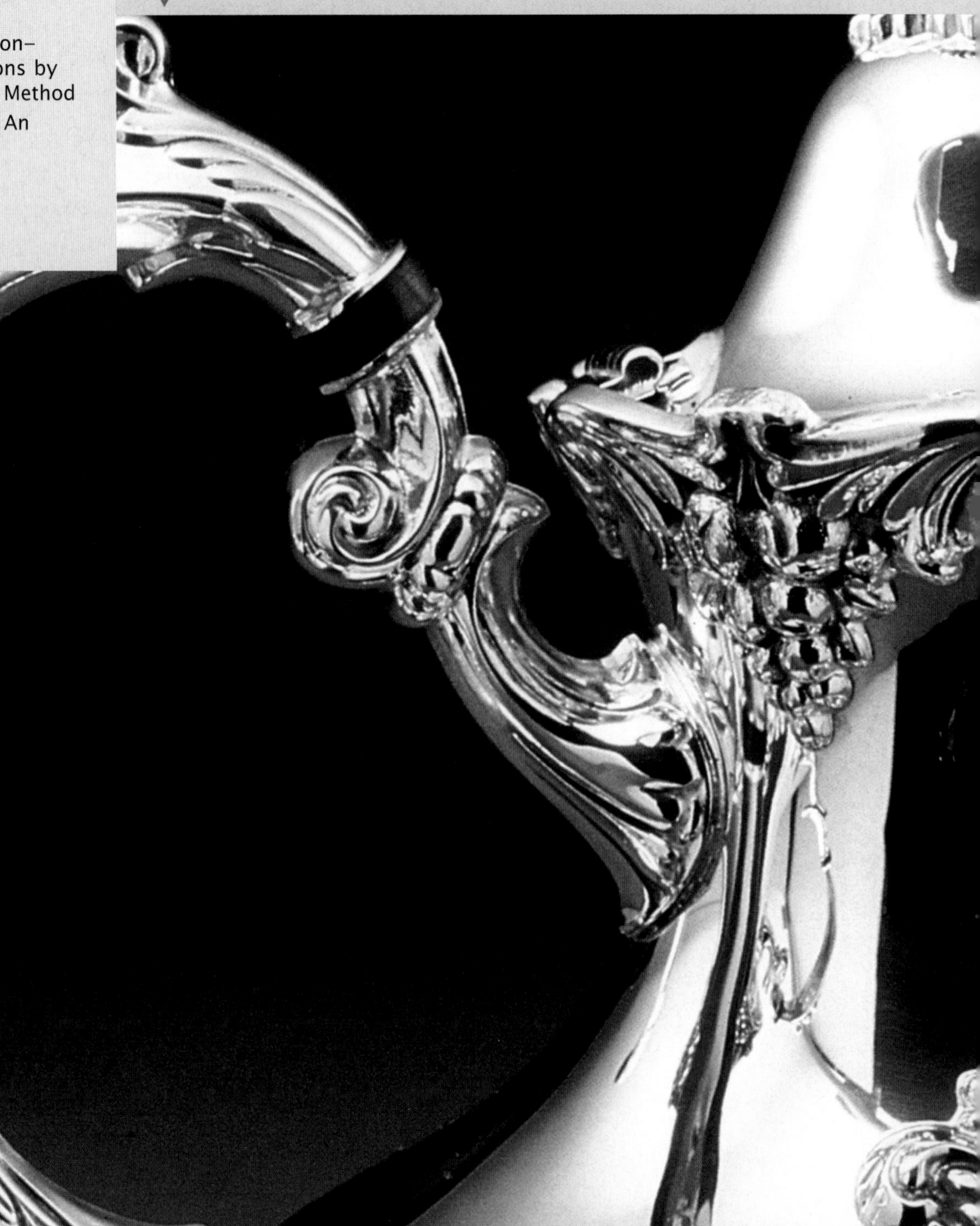

What do a forest fire, rusting steel, combustion in an automobile engine, and the metabolism of food in a human body have in common? All of these important processes involve oxidation–reduction reactions. In fact, virtually all of the processes that provide energy to heat buildings, power vehicles, and allow people to work and play depend on oxidation–reduction reactions. And every time you start your car, turn on your calculator, look at your digital watch, or listen to a radio at the beach, you are depending on an oxidation–reduction reaction to power the battery in each of these devices. In addition, because "pollution-free" vehicles have been mandated in California (and other states are soon to follow), battery-powered cars are about to become more common on U.S. roads (see "Chemistry in Focus: An Engine for the Twenty-First Century" on page 539). This will lead to increased reliance of our society on batteries and will spur the search for new, more efficient batteries.

The power generated by an alkaline AA battery and a mercury battery results from oxidation–reduction reactions.

In this chapter we will explore the properties of oxidation–reduction reactions, and we will see how these reactions are used to power batteries.

17.1 Oxidation–Reduction Reactions

AIM: To learn about metal–nonmetal oxidation–reduction reactions.

In Section 7.5 we discussed the chemical reactions between metals and nonmetals. For example, sodium chloride is formed by the reaction of elemental sodium and chlorine.

$$2Na(s) + Cl_2(g) \rightarrow 2NaCl(s)$$

Because elemental sodium and chlorine contain uncharged atoms and because sodium chloride is known to contain Na^+ and Cl^- ions, this reaction must involve a transfer of electrons from sodium atoms to chlorine atoms.

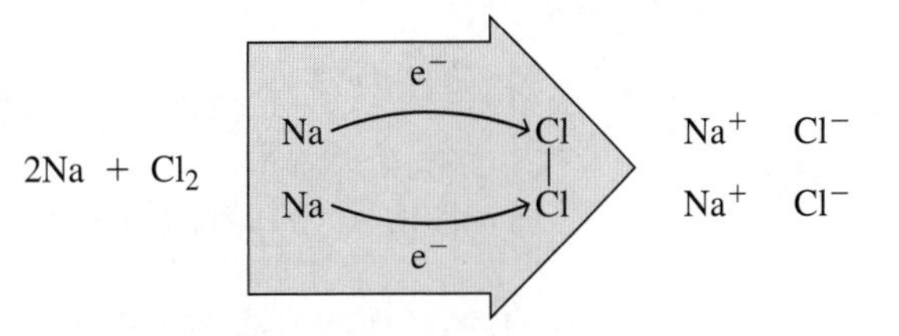

Some students use the mnemonic OIL RIG:

Oxidation Is Loss; Reduction Is Gain.

Reactions like this one, in which one or more electrons are transferred, are called **oxidation–reduction reactions,** or **redox reactions. Oxidation** is defined as a *loss of electrons.* **Reduction** is defined as a *gain of* *electrons*. In the reaction of elemental sodium and chlorine, each sodium atom loses one electron, forming a 1+ ion. Therefore, sodium is oxidized. Each chlorine atom gains one electron, forming a negative chloride ion, and is thus reduced. Whenever a metal reacts with a nonmetal to form an ionic compound, electrons are transferred from the metal to the nonmetal. So these reactions are always oxidation–reduction reactions where the metal is oxidized (loses electrons) and the nonmetal is reduced (gains electrons).

Example 17.1 Identifying Oxidation and Reduction in a Reaction

Magnesium burns in air to give a bright, white flame.

In the following reactions, identify which element is oxidized and which element is reduced.

a. $2Mg(s) + O_2(g) \rightarrow 2MgO(s)$

b. $2Al(s) + 3I_2(s) \rightarrow 2AlI_3(s)$

Solution

a. We have learned that Group 2 metals form 2+ cations and that Group 6 nonmetals form 2− anions, so we can predict that magnesium oxide contains Mg^{2+} and O^{2-} ions. This means that in the reaction given, each Mg loses two electrons to form Mg^{2+} and so is oxidized. Also each O gains two electrons to form O^{2-} and so is reduced.

b. Aluminum iodide contains the Al^{3+} and I^- ions. Thus aluminum atoms lose electrons (are oxidized). Iodine atoms gain electrons (are reduced).

✔ Self-Check Exercise 17.1

For the following reactions, identify the element oxidized and the element reduced.

a. $2Cu(s) + O_2(g) \rightarrow 2CuO(s)$

b. $2Cs(s) + F_2(g) \rightarrow 2CsF(s)$

See Problems 17.3 through 17.6. ■

Although we can identify reactions between metals and nonmetals as redox reactions, it is more difficult to decide whether a given reaction between nonmetals is a redox reaction. In fact, many of the most significant redox reactions involve only nonmetals. For example, combustion reactions such as methane burning in oxygen,

$$CH_4(g) + 2O_2(g) \rightarrow CO_2(g) + 2H_2O(g) + \text{energy}$$

are oxidation–reduction reactions. Even though none of the reactants or products in this reaction is ionic, the reaction does involve a transfer of electrons from carbon to oxygen. To explain this, we must introduce the concept of oxidation states.

17.2 Oxidation States

AIM: To learn how to assign oxidation states.

The concept of **oxidation states** (sometimes called *oxidation numbers*) lets us keep track of electrons in oxidation–reduction reactions by assigning charges to the various atoms in a compound. Sometimes these charges are quite apparent. For example, in a binary ionic compound the ions have easily identified charges: in sodium chloride, sodium is +1 and chlorine is −1;

in magnesium oxide, magnesium is +2 and oxygen is −2; and so on. In such binary ionic compounds the oxidation states are simply the charges of the ions.

Ion	Oxidation State
Na^+	+1
Cl^-	−1
Mg^{2+}	+2
O^{2-}	−2

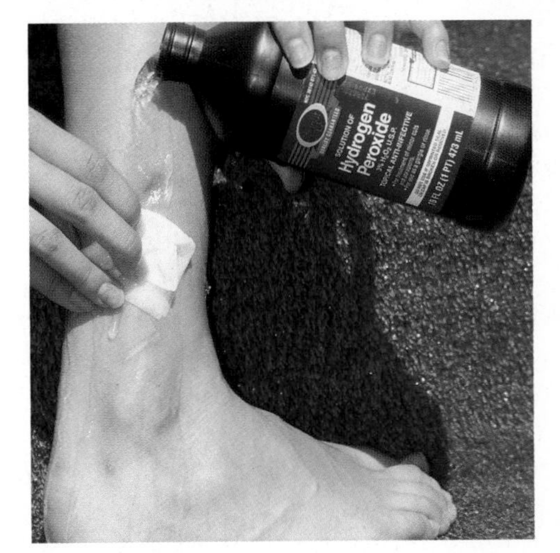

Hydrogen peroxide can be used to disinfect a wound.

In an uncombined element, all of the atoms are uncharged (neutral). For example, sodium metal contains neutral sodium atoms, and chlorine gas is made up of Cl_2 molecules, each of which contains two neutral chlorine atoms. Therefore, an atom in a pure element has no charge and is assigned an oxidation state of zero.

In a covalent compound such as water, although no ions are actually present, chemists find it useful to assign imaginary charges to the elements in the compound. The oxidation states of the elements in these compounds are equal to the imaginary charges we determine by assuming that the most electronegative atom (see Section 11.2) in a bond controls or possesses *both* of the shared electrons. For example, in the O—H bonds in water, it is assumed for purposes of assigning oxidation states that the much more electronegative oxygen atom controls both of the shared electrons in each bond. This gives the oxygen eight valence electrons.

H—Ö—H with arrows: ← $2e^-$, ← $2e^-$

In effect we say that each hydrogen has lost its single electron to the oxygen. This gives each hydrogen an oxidation state of +1 and the oxygen an oxidation state of −2 (the oxygen atom has formally gained two electrons). In virtually all covalent compounds, oxygen is assigned an oxidation state of −2 and hydrogen is assigned an oxidation state of +1.

Because fluorine is so electronegative, it is always assumed to control any shared electrons. So fluorine is always assumed to have a complete octet of electrons and is assigned an oxidation state of −1. That is, for purposes of assigning oxidation states, fluorine is always imagined to be F^- in its covalent compounds.

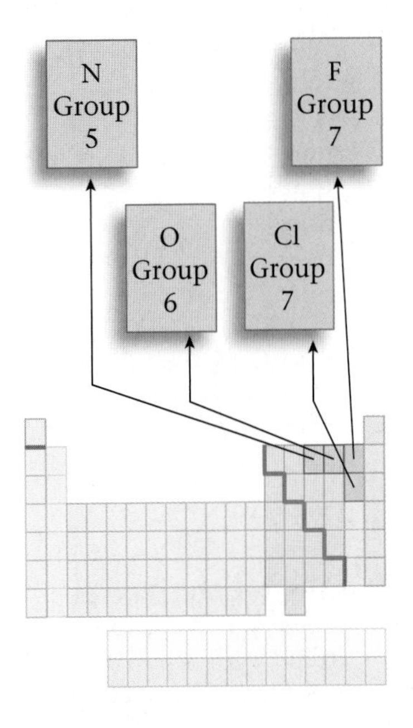

The most electronegative elements are F, O, N, and Cl. In general, we give each of these elements an oxidation state equal to its charge as an anion (fluorine is −1, chlorine is −1, oxygen is −2, and nitrogen is −3). When two of these elements are found in the same compound, we assign them in order of electronegativity, starting with the element that has the largest electronegativity.

$$F > O > N > Cl$$

Greatest electronegativity — Least electronegativity

For example, in the compound NO_2, because oxygen has a greater electronegativity than nitrogen, we assign each oxygen an oxidation state of −2. This gives a total "charge" of −4 (2 × −2) on the two oxygen atoms. Because the NO_2 molecule has zero overall charge, the N must be +4 to exactly balance the −4 on the oxygens. In NO_2, then, the oxidation state of *each* oxygen is −2 and the oxidation state of the nitrogen is +4.

The rules for assigning oxidation states are given below and are illustrated in Table 17.1. Application of these rules allows us to assign oxidation states in most compounds. The principles are illustrated by Example 17.2.

Rules for Assigning Oxidation States

1. The oxidation state of an atom in an uncombined element is 0.
2. The oxidation state of a monoatomic ion is the same as its charge.
3. Oxygen is assigned an oxidation state of −2 in most of its covalent compounds. Important exception: peroxides (compounds containing the O_2^{2-} group), in which each oxygen is assigned an oxidation state of −1.
4. In its covalent compounds with nonmetals, hydrogen is assigned an oxidation state of +1.
5. In binary compounds, the element with the greater electronegativity is assigned a negative oxidation state equal to its charge as an anion in its ionic compounds.
6. For an electrically neutral compound, the sum of the oxidation states must be zero.
7. For an ionic species, the sum of the oxidation states must equal the overall charge.

Table 17.1 Examples of Oxidation States

Substance	Oxidation States	Comments
sodium metal, Na	Na, 0	rule 1
phosphorus, P	P, 0	rule 1
sodium fluoride, NaF	Na, +1	rule 2
	F, −1	rule 2
magnesium sulfide, MgS	Mg, +2	rule 2
	S, −2	rule 2
carbon monoxide, CO	C, +2	
	O, −2	rule 3
sulfur dioxide, SO_2	S, +4	
	O, −2	rule 3
hydrogen peroxide, H_2O_2	H, +1	
	O, −1	rule 3 (exception)
ammonia, NH_3	H, +1	rule 4
	N, −3	rule 5
hydrogen sulfide, H_2S	H, +1	rule 4
	S, −2	rule 5
hydrogen iodide, HI	H, +1	rule 4
	I, −1	rule 5
sodium carbonate, Na_2CO_3	Na, +1	rule 2
	O, −2	rule 3
	C, +4	For CO_3^{2-}, the sum of the oxidation states is $+4 + 3(-2) = -2$. rule 7
ammonium chloride, NH_4Cl	N, −3	rule 5
	H, +1	rule 4 For NH_4^+, the sum of the oxidation states is $-3 + 4(+1) = +1$. rule 7
	Cl, −1	rule 2

Example 17.2 Assigning Oxidation States

Assign oxidation states to all atoms in the following molecules or ions.

a. CO_2

b. SF_6

c. NO_3^-

Solution

a. Rule 3 takes precedence here: oxygen is assigned an oxidation state of −2. We determine the oxidation state for carbon by recognizing that because CO_2 has no charge, the sum of the oxidation states for oxygen and carbon must be 0 (rule 6). Each oxygen is −2 and there are two oxygen atoms, so the carbon atom must be assigned an oxidation state of +4.

Check: $+4 + 2(-2) = 0$

b. Because fluorine has the greater electronegativity, we assign its oxidation state first. Its charge as an anion is always −1, so we assign −1 as the oxidation state of each fluorine atom (rule 5). The sulfur must then be assigned an oxidation state of +6 to balance the total of −6 from the six fluorine atoms (rule 7).

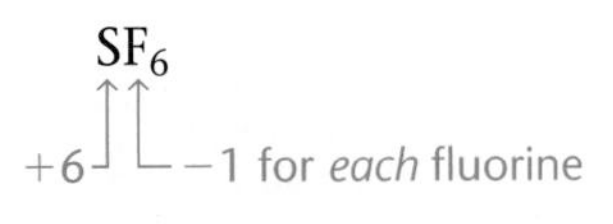

Check: $+6 + 6(-1) = 0$

c. Oxygen has a greater electronegativity than nitrogen, so we assign its oxidation state of −2 first (rule 5). Because the overall charge on NO_3^- is −1 and because the sum of the oxidation states of the three oxygens is −6, the nitrogen must have an oxidation state of +5.

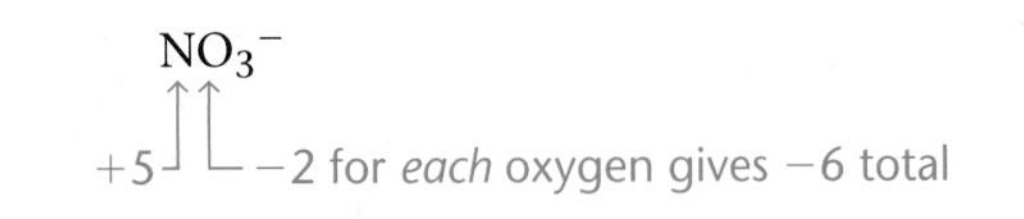

Check: $+5 + 3(-2) = -1$
This is correct; NO_3^- has a −1 charge.

✔ Self-Check Exercise 17.2

Assign oxidation states to all atoms in the following molecules or ions.

a. SO_3

b. SO_4^{2-}

c. N_2O_5

d. PF_3

e. C_2H_6

See Problems 17.13 through 17.22. ■

Oxidation–Reduction Reactions Between Nonmetals

AIMS: To understand oxidation and reduction in terms of oxidation states. To learn to identify oxidizing and reducing agents.

We have seen that oxidation–reduction reactions are characterized by a transfer of electrons. In some cases, the transfer literally occurs to form ions, such as in the reaction

$$2Na(s) + Cl_2(g) \rightarrow 2NaCl(s)$$

We can use oxidation states to verify that electron transfer has occurred.

$$2Na(s) + Cl_2(g) \rightarrow 2NaCl(s)$$

Oxidation state: 0 (element), 0 (element), +1 −1 $(Na^+)(Cl^-)$

Thus in this reaction, we represent the electron transfer as follows:

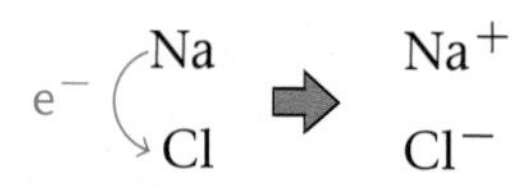

In other cases the electron transfer occurs in a different sense, such as in the combustion of methane (the oxidation state for each atom is given below each reactant and product).

$$CH_4(g) + 2O_2(g) \rightarrow CO_2(g) + 2H_2O(g)$$

Oxidation state: CH_4: −4, +1 (each H); $2O_2$: 0; CO_2: +4, −2 (each O); $2H_2O$: +1 (each H), −2

Note that the oxidation state of oxygen in O_2 is 0 because the oxygen is in elemental form. In this reaction there are no ionic compounds, but we can still describe the process in terms of the transfer of electrons. Note that carbon undergoes a change in oxidation state from −4 in CH_4 to +4 in CO_2. Such a change can be accounted for by a loss of eight electrons:

$$\underset{-4}{\text{C (in } CH_4)} \xrightarrow{\text{Loss of } 8e^-} \underset{+4}{\text{C (in } CO_2)}$$

or, in equation form,

$$\underset{-4}{CH_4} \rightarrow \underset{+4}{CO_2} + 8e^-$$

On the other hand, each oxygen changes from an oxidation state of 0 in O_2 to −2 in H_2O and CO_2, signifying a gain of two electrons per atom. Four oxygen atoms are involved, so this is a gain of eight electrons:

$$\text{4O atoms (in } 2O_2) \xrightarrow{\text{Gain of } 8e^-} 4O^{2-} \text{ (in } 2H_2O \text{ and } CO_2)$$

or, in equation form,

$$2O_2 + 8e^- \rightarrow CO_2 + 2H_2O$$

$$\underset{0}{\uparrow} \qquad 4(-2) = -8$$

Note that eight electrons are required because four oxygen atoms are going from an oxidation state of 0 to −2. So each oxygen requires two electrons. No change occurs in the oxidation state of hydrogen, and it is not involved in the electron transfer process.

Oxidation:
Loss of electrons
or
Increase in oxidation state
Reduction:
Gain of electrons
or
Decrease in oxidation state

With this background, we can now define *oxidation* and *reduction* in terms of oxidation states. **Oxidation** is an *increase* in oxidation state (a loss of electrons). **Reduction** is a *decrease* in oxidation state (a gain of electrons). Thus in the reaction

$$2Na(s) + Cl_2(g) \rightarrow 2NaCl(s)$$

Oxidizing Agent:
Accepts electrons
Contains the element reduced
Reducing Agent:
Furnishes electrons
Contains the element oxidized

sodium is oxidized and chlorine is reduced. Cl_2 is called the **oxidizing agent (electron acceptor)** and Na is called the **reducing agent (electron donor).** We can also define the *oxidizing agent* as the reactant containing the element that is reduced (gains electrons). The *reducing agent* can be defined similarly as the reactant containing the element that is oxidized (loses electrons).

Concerning the reaction

$$CH_4(g) + 2O_2(g) \rightarrow CO_2(g) + 2H_2O(g)$$

$$-4 \;\; +1 \qquad 0 \qquad +4 \;\; -2 \qquad +1 \;\; -2$$

we can say the following:

1. Carbon is oxidized because there is an increase in its oxidation state (carbon has apparently lost electrons).
2. The reactant CH_4 contains the carbon that is oxidized, so CH_4 is the reducing agent. It is the reactant that furnishes the electrons (those lost by carbon).
3. Oxygen is reduced because there has been a decrease in its oxidation state (oxygen has apparently gained electrons).
4. The reactant that contains the oxygen atoms is O_2, so O_2 is the oxidizing agent. That is, O_2 accepts the electrons.

In a redox reaction, an oxidizing agent is reduced (gains electrons) and a reducing agent is oxidized (loses electrons).

Note that when the oxidizing or reducing agent is named, the *whole compound* is specified, not just the element that undergoes the change in oxidation state.

Example 17.3 Identifying Oxidizing and Reducing Agents, I

When powdered aluminum metal is mixed with pulverized iodine crystals and a drop of water is added, the resulting reaction produces a great deal of energy. The mixture bursts into flames, and a purple smoke of I_2 vapor is produced from the excess iodine. The equation for the reaction is

$$2Al(s) + 3I_2(s) \rightarrow 2AlI_3(s)$$

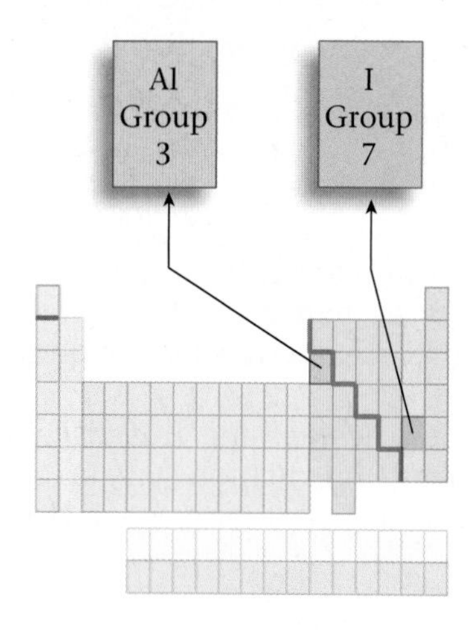

For this reaction, identify the atoms that are oxidized and those that are reduced, and specify the oxidizing and reducing agents.

Solution

The first step is to assign oxidation states.

$$\underset{\substack{\uparrow\\0\\\text{Free elements}}}{2Al(s)} + \underset{\substack{\uparrow\\0}}{3I_2(s)} \rightarrow \underset{\substack{+3 \quad -1 \text{ (each I)}\\AlI_3(s)\text{ is a salt that}\\\text{contains } Al^{3+}\text{ and } I^-\text{ ions.}}}{2AlI_3(s)}$$

Because each aluminum atom changes its oxidation state from 0 to +3 (an increase in oxidation state), aluminum is *oxidized* (loses electrons). On the other hand, the oxidation state of each iodine atom decreases from 0 to −1, and iodine is *reduced* (gains electrons). Because Al furnishes electrons for the reduction of iodine, it is the *reducing agent.* I_2 is the *oxidizing agent* (the reactant that accepts the electrons).

Example 17.4 Identifying Oxidizing and Reducing Agents, II

Metallurgy, the process of producing a metal from its ore, always involves oxidation–reduction reactions. In the metallurgy of galena (PbS), the principal lead-containing ore, the first step is the conversion of lead sulfide to its oxide (a process called *roasting*).

$$2PbS(s) + 3O_2(g) \rightarrow 2PbO(s) + 2SO_2(g)$$

The oxide is then treated with carbon monoxide to produce the free metal.

$$PbO(s) + CO(g) \rightarrow Pb(s) + CO_2(g)$$

For each reaction, identify the atoms that are oxidized and those that are reduced, and specify the oxidizing and reducing agents.

Solution

For the first reaction, we can assign the following oxidation states:

$$\underset{+2\ \ -2}{PbS(s)} + \underset{0}{3O_2(g)} \rightarrow \underset{+2\ \ -2}{2PbO(s)} + \underset{+4\ \ -2\text{ (each O)}}{2SO_2(g)}$$

The oxidation state for the sulfur atom increases from −2 to +4, so sulfur is oxidized (loses electrons). The oxidation state for each oxygen atom decreases from 0 to −2. Oxygen is reduced (gains electrons). The oxidizing agent (electron acceptor) is O_2, and the reducing agent (electron donor) is PbS.

For the second reaction, we have

$$\underset{+2\ \ -2}{PbO(s)} + \underset{+2\ \ -2}{CO(g)} \rightarrow \underset{0}{Pb(s)} + \underset{+4\ \ -2\text{ (each O)}}{CO_2(g)}$$

Lead is reduced (gains electrons; its oxidation state decreases from +2 to 0), and carbon is oxidized (loses electrons; its oxidation state increases from +2 to +4). PbO is the oxidizing agent (electron acceptor), and CO is the reducing agent (electron donor).

Self-Check Exercise 17.3

Ammonia, NH_3, which is widely used as a fertilizer, is prepared by reaction of the elements:

$$N_2(g) + 3H_2(g) \rightarrow 2NH_3(g)$$

Is this an oxidation–reduction reaction? If so, specify the oxidizing agent and the reducing agent.

See Problems 17.29 through 17.36. ■

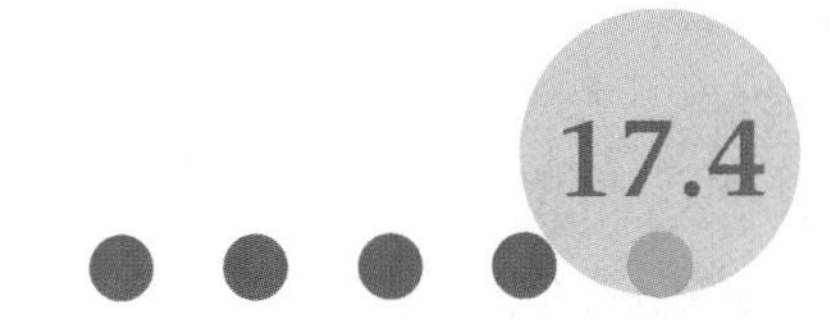

Balancing Oxidation–Reduction Reactions by the Half-Reaction Method

AIM: To learn to balance oxidation–reduction equations by using half-reactions.

Many oxidation–reduction reactions can be balanced readily by trial and error. That is, we use the procedure described in Chapter 6 to find a set of coefficients that give the same number of each type of atom on both sides of the equation.

However, the oxidation–reduction reactions that occur in aqueous solution are often so complicated that it becomes very tedious to balance them by trial and error. In this section we will develop a systematic approach for balancing the equations for these reactions.

To balance the equations for oxidation–reduction reactions that occur in aqueous solution, we separate the reaction into two half-reactions. **Half-reactions** are equations that have electrons as reactants or products. One half-reaction represents a reduction process and the other half-reaction represents an oxidation process. In a reduction half-reaction, electrons are shown on the reactant side (electrons are gained by a reactant in the equation). In an oxidation half-reaction, the electrons are shown on the product side (electrons are lost by a reactant in the equation).

For example, consider the unbalanced equation for the oxidation–reduction reaction between the cerium(IV) ion and the tin(II) ion.

$$Ce^{4+}(aq) + Sn^{2+}(aq) \rightarrow Ce^{3+}(aq) + Sn^{4+}(aq)$$

This reaction can be separated into a half-reaction involving the substance being *reduced:*

Ce^{4+} gains $1e^-$ to form Ce^{3+} and is thus reduced.

$$e^- + Ce^{4+}(aq) \rightarrow Ce^{3+}(aq) \quad \text{reduction half-reaction}$$

and a half-reaction involving the substance being *oxidized:*

Sn^{2+} loses $2e^-$ to form Sn^{4+} and is thus oxidized.

$$Sn^{2+}(aq) \rightarrow Sn^{4+}(aq) + 2e^- \quad \text{oxidation half-reaction}$$

Notice that Ce^{4+} must gain one electron to become Ce^{3+}, so one electron is shown as a reactant along with Ce^{4+} in this half-reaction. On the other hand, for Sn^{2+} to become Sn^{4+} it must lose two electrons. This means that two electrons must be shown as products in this half-reaction.

The key principle in balancing oxidation–reduction reactions is that the number of electrons lost (from the reactant that is oxidized) must equal the number of electrons gained (from the reactant that is reduced).

Number of electrons lost ⟷ (must be equal to) Number of electrons gained

In the half-reaction shown above, one electron is gained by each Ce^{4+} while two electrons are lost by each Sn^{2+}. We must equalize the number of electrons gained and lost. To do this, we first multiply the reduction half-reaction by 2.

$$2e^- + 2Ce^{4+} \rightarrow 2Ce^{3+}$$

Then we add this half-reaction to the oxidation half-reaction.

$$\begin{array}{r} 2e^- + 2Ce^{4+} \rightarrow 2Ce^{3+} \\ Sn^{2+} \rightarrow Sn^{4+} + 2e^- \\ \hline 2e^- + 2Ce^{4+} + Sn^{2+} \rightarrow 2Ce^{3+} + Sn^{4+} + 2e^- \end{array}$$

Finally, we cancel the $2e^-$ on each side to give the overall balanced equation

$$\cancel{2e^-} + 2Ce^{4+} + Sn^{2+} \rightarrow 2Ce^{3+} + Sn^{4+} + \cancel{2e^-}$$
$$2Ce^{4+} + Sn^{2+} \rightarrow 2Ce^{3+} + Sn^{4+}$$

We can now summarize what we have said about the method for balancing oxidation–reduction reactions in aqueous solution:

1. Separate the reaction into an oxidation half-reaction and a reduction half-reaction.
2. Balance the half-reactions separately.
3. Equalize the number of electrons gained and lost.
4. Add the half-reactions together and cancel electrons to give the overall balanced equation.

It turns out that most oxidation–reduction reactions occur in solutions that are distinctly basic or distinctly acidic. We will cover only the acidic case in this text, because it is the most common. The detailed procedure for balancing the equations for oxidation–reduction reactions that occur in acidic solution is given below, and Example 17.5 illustrates the use of these steps.

The Half-Reaction Method for Balancing Equations for Oxidation–Reduction Reactions Occurring in Acidic Solution

STEP 1 Identify and write the equations for the oxidation and reduction half-reactions.

STEP 2 For each half-reaction:

a. Balance all of the elements except hydrogen and oxygen.
b. Balance oxygen using H_2O.
c. Balance hydrogen using H^+.
d. Balance the charge using electrons.

STEP 3 If necessary, multiply one or both balanced half-reactions by an integer to equalize the number of electrons transferred in the two half-reactions.

STEP 4 Add the half-reactions, and cancel identical species that appear on both sides.

STEP 5 Check to be sure the elements and charges balance.

Example 17.5 Balancing Oxidation–Reduction Reactions Using the Half-Reaction Method, I

Balance the equation for the reaction between permanganate and iron(II) ions in acidic solution. The net ionic equation for this reaction is

H_2O and H^+ will be added to this equation as we balance it. We do not have to worry about them now.

$$MnO_4^-(aq) + Fe^{2+}(aq) \xrightarrow{\text{Acid}} Fe^{3+}(aq) + Mn^{2+}(aq)$$

This reaction is used to analyze iron ore for its iron content.

Solution

STEP 1 *Identify and write equations for the half-reactions.*
The oxidation states for the half-reaction involving the permanganate ion show that manganese is reduced.

Note that the left side contains oxygen but the right side does not. This will be taken care of later when we add water.

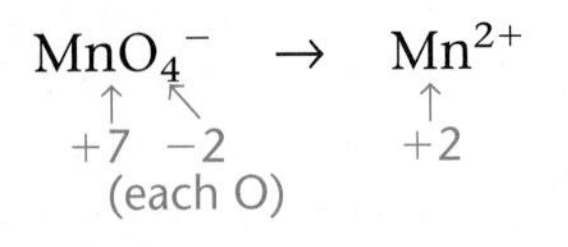

Because manganese changes from an oxidation state of +7 to +2, it is reduced. So this is the *reduction half-reaction.* It will have electrons as reactants, although we will not write them yet. The other half-reaction involves the oxidation of iron(II) to the iron(III) ion and is the *oxidation half-reaction.*

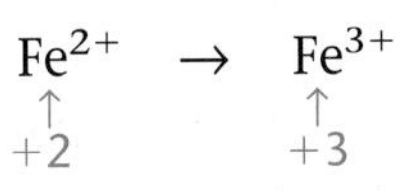

This reaction will have electrons as products, although we will not write them yet.

STEP 2 *Balance each half-reaction.*
For the reduction reaction, we have

$$MnO_4^- \rightarrow Mn^{2+}$$

a. The manganese is already balanced.

b. We balance oxygen by adding $4H_2O$ to the right side of the equation.

$$MnO_4^- \rightarrow Mn^{2+} + 4H_2O$$

The H^+ comes from the acidic solution in which the reaction is taking place.

c. Next we balance hydrogen by adding $8H^+$ to the left side.

$$8H^+ + MnO_4^- \rightarrow Mn^{2+} + 4H_2O$$

A solution containing MnO_4^- ions (*left*) and a solution containing Fe^{2+} ions (*right*).

d. All of the elements have been balanced, but we need to balance the charge using electrons. At this point we have the following charges for reactants and products in the reduction half-reaction.

$$8H^+ + MnO_4^- \rightarrow Mn^{2+} + 4H_2O$$

$$\underbrace{8+ \quad + \quad 1-}_{7+} \qquad \underbrace{2+ \quad + \quad 0}_{2+}$$

Always add electrons to the side of the half-reaction with excess positive charge.

We can equalize the charges by adding five electrons to the left side.

$$\underbrace{5e^- + 8H^+ + MnO_4^-}_{2+} \rightarrow \underbrace{Mn^{2+} + 4H_2O}_{2+}$$

Both the *elements* and the *charges* are now balanced, so this represents the balanced reduction half-reaction. The fact that five electrons appear on the reactant side of the equation makes sense, because five electrons are required to reduce MnO_4^- (in which Mn has an oxidation state of +7) to Mn^{2+} (in which Mn has an oxidation state of +2).
For the oxidation reaction,

$$Fe^{2+} \rightarrow Fe^{3+}$$

the elements are balanced, so all we have to do is balance the charge.

$$\underbrace{Fe^{2+}}_{2+} \rightarrow \underbrace{Fe^{3+}}_{3+}$$

One electron is needed on the right side to give a net 2+ charge on both sides.

$$\underbrace{Fe^{2+}}_{2+} \rightarrow \underbrace{Fe^{3+} + e^-}_{2+}$$

The number of electrons gained in the reduction half-reaction must equal the number of electrons lost in the oxidation half-reaction.

STEP 3 *Equalize the number of electrons transferred in the two half-reactions.* Because the reduction half-reaction involves a transfer of five electrons and the oxidation half-reaction involves a transfer of only one electron, the oxidation half-reaction must be multiplied by 5.

$$5Fe^{2+} \rightarrow 5Fe^{3+} + 5e^-$$

STEP 4 *Add the half-reactions and cancel identical species.*

$$5e^- + 8H^+ + MnO_4^- \rightarrow Mn^{2+} + 4H_2O$$
$$5Fe^{2+} \rightarrow 5Fe^{3+} + 5e^-$$
$$\cancel{5e^-} + 8H^+ + MnO_4^- + 5Fe^{2+} \rightarrow Mn^{2+} + 5Fe^{3+} + 4H_2O + \cancel{5e^-}$$

Note that the electrons cancel (as they must) to give the final balanced equation

$$5Fe^{2+}(aq) + MnO_4^-(aq) + 8H^+(aq) \rightarrow 5Fe^{3+}(aq) + Mn^{2+}(aq) + 4H_2O(l)$$

Note that we show the physical states of the reactants and products—(*aq*) and (*l*) in this case—only in the final balanced equation.

STEP 5 *Check to be sure that elements and charges balance.*

Elements 5Fe, 1Mn, 4O, 8H $\rightarrow$ 5Fe, 1Mn, 4O, 8H
Charges 17+ $\rightarrow$ 17+

The equation is balanced.

Example 17.6 Balancing Oxidation–Reduction Reactions Using the Half-Reaction Method, II

When an automobile engine is started, it uses the energy supplied by a lead storage battery. This battery uses an oxidation–reduction reaction between elemental lead (lead metal) and lead(IV) oxide to provide the power to start the engine. The unbalanced equation for a simplified version of the reaction is

$$Pb(s) + PbO_2(s) + H^+(aq) \rightarrow Pb^{2+}(aq) + H_2O(l)$$

Balance this equation using the half-reaction method.

Solution

STEP 1 First we identify and write the two half-reactions. One half-reaction must be

$$Pb \rightarrow Pb^{2+}$$

and the other is

$$PbO_2 \rightarrow Pb^{2+}$$

Because Pb^{2+} is the only lead-containing product, it must be the product in both half-reactions.

The first reaction involves the oxidation of Pb to Pb^{2+}. The second reaction involves the reduction of Pb^{4+} (in PbO_2) to Pb^{2+}.

STEP 2 Now we will balance each half-reaction separately.

The oxidation half-reaction

$$Pb \rightarrow Pb^{2+}$$

a–c. All the elements are balanced.

d. The charge on the left is zero and that on the right is +2, so we must add $2e^-$ to the right to give zero overall charge.

$$Pb \rightarrow Pb^{2+} + 2e^-$$

This half-reaction is balanced.

The reduction half-reaction

$$PbO_2 \rightarrow Pb^{2+}$$

a. All elements are balanced except O.

b. The left side has two oxygen atoms and the right side has none, so we add $2H_2O$ to the right side.

$$PbO_2 \rightarrow Pb^{2+} + 2H_2O$$

c. Now we balance hydrogen by adding $4H^+$ to the left.

$$4H^+ + PbO_2 \rightarrow Pb^{2+} + 2H_2O$$

d. Because the left side has a +4 overall charge and the right side has a +2 charge, we must add $2e^-$ to the left side.

$$2e^- + 4H^+ + PbO_2 \rightarrow Pb^{2+} + 2H_2O$$

The half-reaction is balanced.

Copper metal reacting with nitric acid. The solution is colored by the presence of Cu^{2+} ions. The brown gas is NO_2, which is formed when NO reacts with O_2 in the air.

STEP 3 Because each half-reaction involves $2e^-$, we can simply add the half-reactions as they are.

STEP 4

$$\begin{array}{r} Pb \rightarrow Pb^{2+} + 2e^- \\ 2e^- + 4H^+ + PbO_2 \rightarrow Pb^{2+} + 2H_2O \\ \hline 2e^- + 4H^+ + Pb + PbO_2 \rightarrow 2Pb^{2+} + 2H_2O + 2e^- \end{array}$$

Canceling electrons gives the balanced overall equation

$$Pb(s) + PbO_2(s) + 4H^+(aq) \rightarrow 2Pb^{2+}(aq) + 2H_2O(l)$$

where the appropriate states are also indicated.

STEP 5 Both the elements and the charges balance.

Elements	2Pb, 2O, 4H → 2Pb, 2O, 4H
Charges	4+ → 4+

The equation is correctly balanced.

✔ Self-Check Exercise 17.4

Copper metal reacts with nitric acid, $HNO_3(aq)$, to give aqueous copper(II) nitrate, water, and nitrogen monoxide gas as products. Write and balance the equation for this reaction.

See Problems 17.45 through 17.48. ■

17.5 Electrochemistry: An Introduction

AIMS: To understand the term electrochemistry. To learn to identify the components of an electrochemical (galvanic) cell.

Our lives would be very different without batteries. We would have to crank the engines in our cars by hand, wind our watches, and buy very long extension cords if we wanted to listen to a radio on a picnic. Indeed, our society sometimes seems to run on batteries. In this section and the next, we will find out how these important devices produce electrical energy.

A battery uses the energy from an oxidation–reduction reaction to produce an electric current. This is an important illustration of **electrochemistry,** *the study of the interchange of chemical and electrical energy.*

Electrochemistry involves two types of processes:

1. The production of an electric current from a chemical (oxidation–reduction) reaction
2. The use of an electric current to produce a chemical change

To understand how a redox reaction can be used to generate a current, let's reconsider the aqueous reaction between MnO_4^- and Fe^{2+} that we worked with in Example 17.5. We can break this redox reaction into the following half-reactions:

$$8H^+ + MnO_4^- + 5e^- \rightarrow Mn^{2+} + 4H_2O \quad \text{reduction}$$
$$Fe^{2+} \rightarrow Fe^{3+} + e^- \quad \text{oxidation}$$

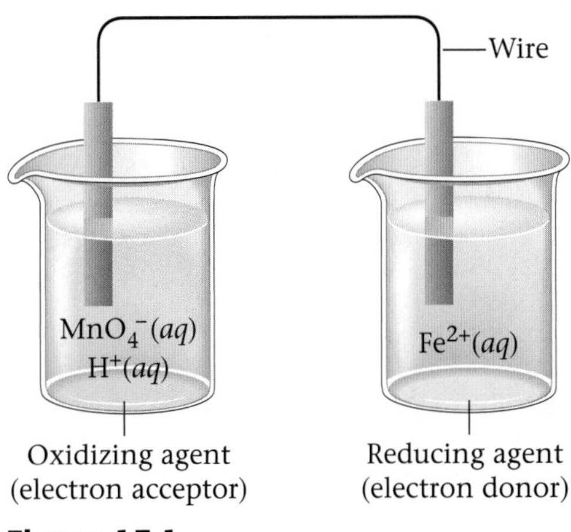

Figure 17.1
Schematic of a method for separating the oxidizing and reducing agents in a redox reaction. (The solutions also contain other ions to balance the charge.) This cell is incomplete at this point.

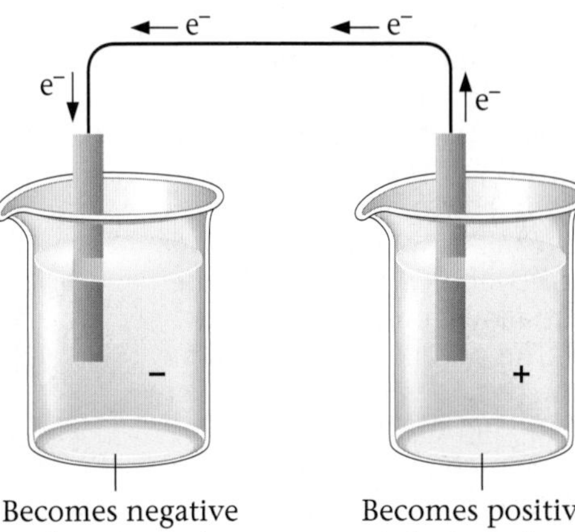

Figure 17.2
Electron flow under these conditions would lead to a build-up of negative charge on the left and positive charge on the right, which is not feasible without a huge input of energy.

The energy involved in a chemical reaction is customarily not shown in the balanced equation. In the reaction of MnO_4^- with Fe^{2+}, energy is released that can be used to do useful work.

When the reaction between MnO_4^- and Fe^{2+} occurs in solution, electrons are transferred directly as the reactants collide. No useful work is obtained from the chemical energy involved in the reaction. How can we harness this energy? The key is to *separate the oxidizing agent (electron acceptor) from the reducing agent (electron donor),* thus requiring the electron transfer to occur through a wire. That is, to get from the reducing agent to the oxidizing agent, the electrons must travel through a wire. The current produced in the wire by this electron flow can be directed through a device, such as an electric motor, to do useful work.

For example, consider the system illustrated in Figure 17.1. If our reasoning has been correct, electrons should flow through the wire from Fe^{2+} to MnO_4^-. However, when we construct the apparatus as shown, no flow of electrons occurs. Why? The problem is that if the electrons flowed from the right to the left compartment, the left compartment would become negatively charged and the right compartment would experience a build-up of positive charge (Figure 17.2). Creating a charge separation of this type would require large amounts of energy. Therefore, electron flow does not occur under these conditions.

We can, however, solve this problem very simply. The solutions must be connected (without allowing them to mix extensively) so that *ions* can also flow to keep the net charge in each compartment zero (Figure 17.3).

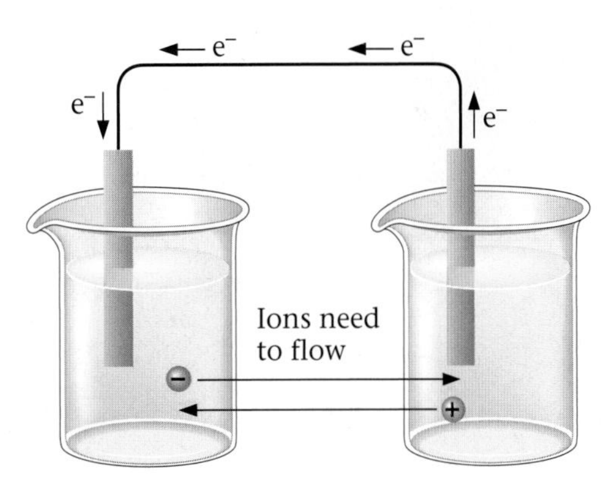

Figure 17.3
Here the ion flow between the two solutions keeps the charge neutral as electrons are transferred. This can be accomplished by having negative ions (anions) flow in the opposite direction to the electrons or by having positive ions (cations) flow in the same direction as the electrons. Both actually occur in a working battery.

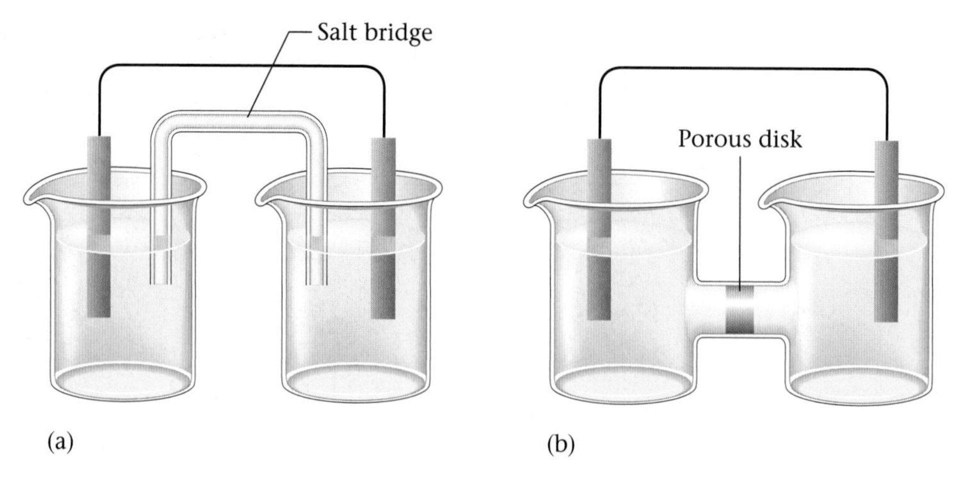

Figure 17.4
A salt bridge or a porous-disk connection allows ions to flow, completing the electric circuit. (a) The salt bridge contains a strong electrolyte either as a gel or as a solution; both ends are covered with a membrane that allows only ions to pass. (b) The porous disk allows ion flow but does not permit overall mixing of the solutions in the two compartments.

The name *galvanic cell* honors Luigi Galvani (1737–1798), an Italian scientist generally credited with the discovery of electricity. These cells are sometimes called *voltaic cells* after Alessandro Volta (1745–1827), another Italian, who first constructed cells of this type around 1800.

This can be accomplished by using a salt bridge (a U-shaped tube filled with a strong electrolyte) or a porous disk in a tube connecting the two solutions (see Figure 17.4). Either of these devices allows ion flow but prevents extensive mixing of the solutions. When we make a provision for ion flow, the circuit is complete. Electrons flow through the wire from reducing agent to oxidizing agent, and ions in the two aqueous solutions flow from one compartment to the other to keep the net charge zero.

Thus an **electrochemical battery,** also called a **galvanic cell,** is a device powered by an oxidation–reduction reaction where the oxidizing agent is separated from the reducing agent so that the electrons must travel through a wire from the reducing agent to the oxidizing agent (Figure 17.5).

Anode: The electrode where oxidation occurs. Cathode: The electrode where reduction occurs.

Notice that in a battery, the reducing agent loses electrons (which flow through the wire toward the oxidizing agent) and so is oxidized. The electrode where oxidation occurs is called the **anode.** At the other electrode, the oxidizing agent gains electrons and is thus reduced. The electrode where reduction occurs is called the **cathode.**

We have seen that an oxidation–reduction reaction can be used to generate an electric current. In fact, this type of reaction is used to produce electric currents in many space vehicles. An oxidation–reduction reaction that can be used for this purpose is hydrogen and oxygen reacting to form water.

$$2H_2(g) + O_2(g) \rightarrow 2H_2O(l)$$

	$2H_2(g)$	$O_2(g)$	$2H_2O(l)$	
Oxidation states:	0	0	+1 (each H)	−2

Alessandro Volta.

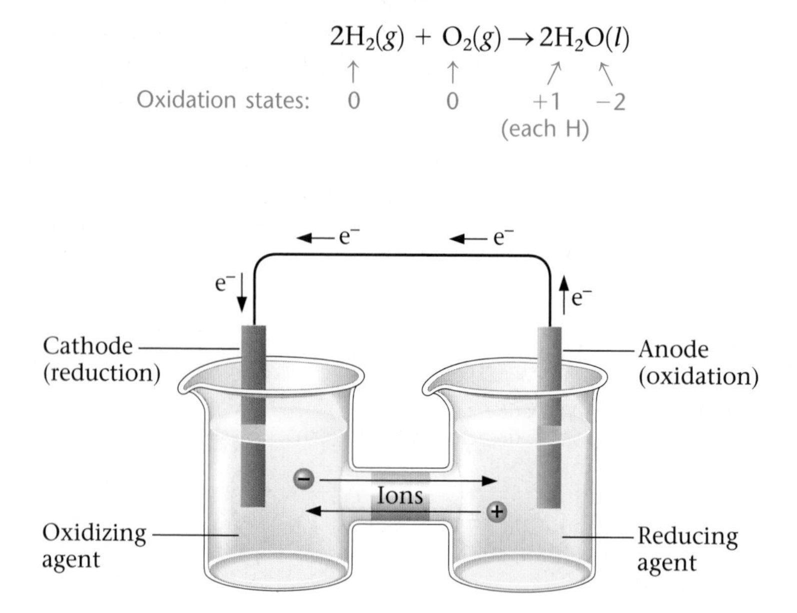

Figure 17.5
Schematic of a battery (galvanic cell).

Notice from the changes in oxidation states that in this reaction, hydrogen is oxidized and oxygen reduced. The opposite process can also occur. We can *force* a current through water to produce hydrogen and oxygen gas.

$$2H_2O(l) \xrightarrow[\text{energy}]{\text{Electrical}} 2H_2(g) + O_2(g)$$

This process, where *electrical energy is used to produce a chemical change,* is called **electrolysis.**

In the remainder of this chapter, we will discuss both types of electrochemical processes. In the next section we will concern ourselves with the practical galvanic cells we know as batteries.

17.6 Batteries

AIM: To learn about the composition and operation of commonly used batteries.

In the previous section we saw that a galvanic cell is a device that uses an oxidation–reduction reaction to generate an electric current by separating the oxidizing agent from the reducing agent. In this section we will consider several specific galvanic cells and their applications.

Lead Storage Battery

Since about 1915, when self-starters were first used in automobiles, the **lead storage battery** has been a major factor in making the automobile a practical means of transportation. This type of battery can function for several years under temperature extremes from −30 °F to 100 °F and under incessant punishment from rough roads. The fact that this same type of battery has been in use for so many years in the face of all of the changes in science and technology over that span of time attests to how well it does its job.

Remember: The oxidizing agent accepts electrons and the reducing agent furnishes electrons.

In the lead storage battery, the reducing agent is lead metal, Pb, and the oxidizing agent is lead(IV) oxide, PbO_2. We have already considered a simplified version of this reaction in Example 17.6. In an actual lead storage battery, sulfuric acid, H_2SO_4, furnishes the H^+ needed in the reaction; it also furnishes SO_4^{2-} ions that react with the Pb^{2+} ions to form solid $PbSO_4$. A schematic of one cell of the lead storage battery is shown in Figure 17.6.

In this cell the anode is constructed of lead metal, which is oxidized. In the cell reaction, lead atoms lose two electrons each to form Pb^{2+} ions, which combine with SO_4^{2-} ions present in the solution to give solid $PbSO_4$.

The cathode of this battery has lead(IV) oxide coated onto lead grids. Lead atoms in the +4 oxidation state in PbO_2 accept two electrons each (are reduced) to give Pb^{2+} ions that also form solid $PbSO_4$.

In the cell the anode and cathode are separated (so that the electrons must flow through an external wire) and bathed in sulfuric acid. The half-reactions that occur at the two electrodes and the overall cell reaction are shown below.

Anode reaction:

$$Pb + H_2SO_4 \rightarrow PbSO_4 + 2H^+ + 2e^- \quad \text{oxidation}$$

Cathode reaction:

$$PbO_2 + H_2SO_4 + 2e^- + 2H^+ \rightarrow PbSO_4 + 2H_2O \quad \text{reduction}$$

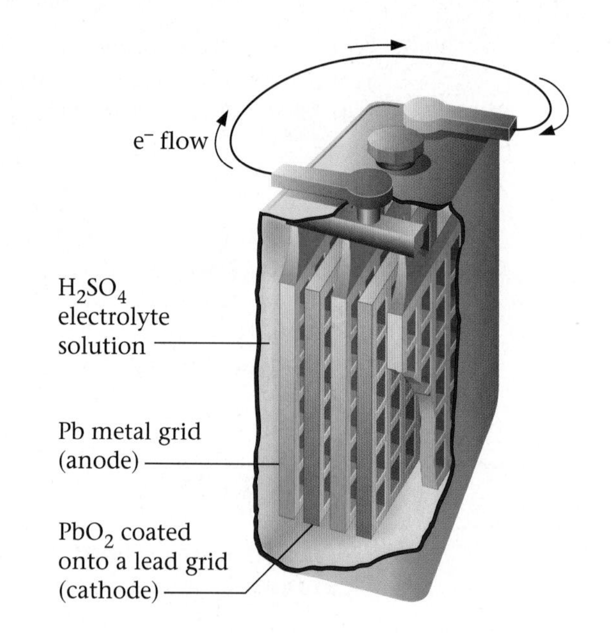

Figure 17.6
In a lead storage battery each cell consists of several lead grids that are connected by a metal bar. These lead grids furnish electrons (the lead atoms lose electrons to form Pb^{2+} ions, which combine with SO_4^{2-} ions to give solid $PbSO_4$). Because the lead is oxidized, it functions as the anode of the cell. The substance that gains electrons is PbO_2; it is coated onto lead grids, several of which are hooked together by a metal bar. The PbO_2 formally contains Pb^{4+}, which is reduced to Pb^{2+}, which in turn combines with SO_4^{2-} to form solid $PbSO_4$. The PbO_2 accepts electrons, so it functions as the cathode.

Overall reaction:

$$Pb(s) + PbO_2(s) + 2H_2SO_4(aq) \rightarrow 2PbSO_4(s) + 2H_2O(l)$$

The tendency for electrons to flow from the anode to the cathode in a battery depends on the ability of the reducing agent to release electrons and on the ability of the oxidizing agent to capture electrons. If a battery consists of a reducing agent that releases electrons readily and an oxidizing agent with a high affinity for electrons, the electrons are driven through the connecting wire with great force and can provide much electrical energy. It is useful to think of the analogy of water flowing through a pipe. The greater the pressure on the water, the more vigorously the water flows. The "pressure" on electrons to flow from one electrode to the other in a battery is called the **potential** of the battery and is measured in volts. For example, each cell in a lead storage battery produces about 2 volts of potential. In an actual automobile battery, six of these cells are connected to produce about 12 volts of potential.

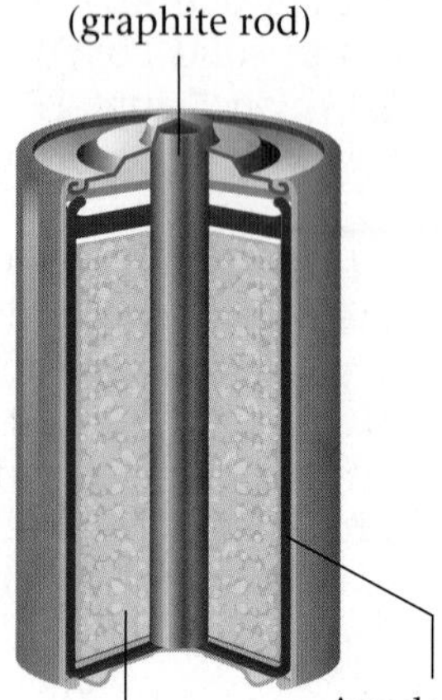

Figure 17.7
A common dry cell battery.

Dry Cell Batteries

The calculators, electronic watches, portable radios, and tape players that are so familiar to us are all powered by small, efficient **dry cell batteries.** They are called dry cells because they do not contain a liquid electrolyte. The common dry cell battery was invented more than 100 years ago by George Leclanché (1839–1882), a French chemist. In its *acid version,* the dry cell battery contains a zinc inner case that acts as the anode and a carbon (graphite) rod in contact with a moist paste of solid MnO_2, solid NH_4Cl, and carbon that acts as the cathode (Figure 17.7). The half-cell reactions are complex but can be approximated as follows:

Anode reaction: $Zn \rightarrow Zn^{2+} + 2e^-$ oxidation

Cathode reaction:

$$2NH_4^+ + 2MnO_2 + 2e^- \rightarrow Mn_2O_3 + 2NH_3 + H_2O \quad \text{reduction}$$

This cell produces a potential of about 1.5 volts.

CHEMISTRY IN FOCUS

An Engine for the Twenty-First Century

Life in the United States has the personal automobile at its very center. To preserve this way of life into the twenty-first century requires that a new power plant be developed. The current internal combustion engine makes inefficient use of energy and causes significant air pollution. One strategy under consideration involves the use of electric cars powered by storage batteries. However, the storage batteries currently available and on the near horizon have very limited capacity and require frequent charging.

The General Motors EV1, currently being tested in select cities in the United States, has a range of 80–100 miles at speeds of about 55 mph before requiring a full recharge of its lead storage batteries. Although other storage batteries are being developed, as yet none will give you anything close to the speed and range of a car powered by an internal combustion engine.

One inherent difficulty with a storage battery is that the supply of reactants is limited. When these reactants are consumed, the battery must be recharged. A different kind of battery, called a fuel cell, is one in which the reactants are continuously supplied from an external source. The best-known fuel cell is the H_2/O_2 unit used by NASA to provide electric power for the space shuttle. In this fuel cell, electric current is produced as the electrons flow from H_2 (the reducing agent) to O_2 (the oxidizing agent). This unit has never been applied to automobiles because of the difficulties in storing hydrogen.

New research, however, indicates that gasoline might serve as a good source of H_2 in an H_2/O_2 fuel cell for automobiles. Jeffrey Bentley, a mechanical engineer working for Arthur D. Little, Inc., in Cambridge, Massachusetts, found that careful heating of gasoline mixed with steam in an oxygen-poor environment results in a mixture of carbon dioxide, carbon monoxide, and hydrogen (contaminated with various sulfur compounds). Using metal catalysts developed by Nick Vanderborgh, a chemical engineer at Los Alamos National Laboratory, this mixture is then "cleaned up" to produce relatively pure H_2 that can be used in an H_2/O_2 fuel cell to produce electricity to run electric motors and power an automobile. The energy efficiency of this system is about twice that of a conventional internal combustion engine, so projected gas mileage should be about double those for current cars. In addition, the system can handle alternative fuels, such as natural gas and alcohol, without any major adjustments. As petroleum supplies dwindle, this consideration will become increasingly important. These new developments make H_2/O_2 fuel cells for automobiles much more likely in the twenty-first century.

A group of cars, powered by fuel cells outside Los Angeles Coliseum.

In the *alkaline version* of the dry cell battery, the NH_4Cl is replaced with KOH or NaOH. In this case the half-reactions can be approximated as follows:

Anode reaction: $Zn + 2OH^- \rightarrow ZnO(s) + H_2O + 2e^-$ oxidation

Cathode reaction: $2MnO_2 + H_2O + 2e^- \rightarrow Mn_2O_3 + 2OH^-$ reduction

The alkaline dry cell lasts longer, mainly because the zinc anode corrodes less rapidly under basic conditions than under acidic conditions.

Other types of dry cell batteries include the *silver cell,* which has a Zn anode and a cathode that employs Ag_2O as the oxidizing agent in a basic environment. *Mercury cells,* often used in calculators, have a Zn anode and

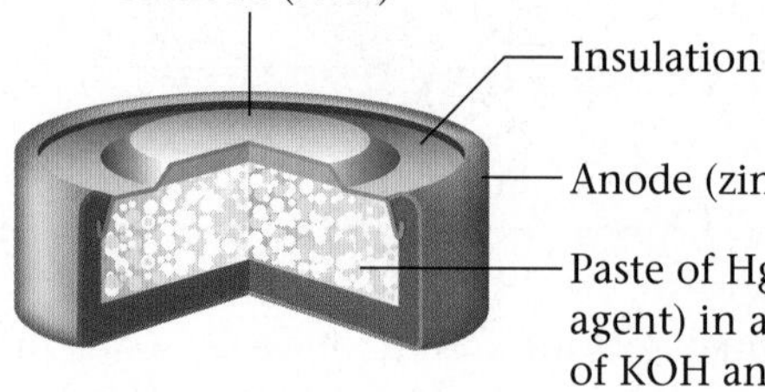

Figure 17.8
A mercury battery of the type used in small calculators.

a cathode involving HgO as the oxidizing agent in a basic medium (see Figure 17.8).

An especially important type of dry cell is the *nickel–cadmium battery,* in which the electrode reactions are

Anode reaction: $Cd + 2OH^- \rightarrow Cd(OH)_2 + 2e^-$ oxidation

Cathode reaction: $NiO_2 + 2H_2O + 2e^- \rightarrow Ni(OH)_2 + 2OH^-$ reduction

In this cell, as in the lead storage battery, the products adhere to the electrodes. Therefore, a nickel–cadmium battery can be recharged an indefinite number of times, because the products can be turned back into reactants by the use of an external source of current.

17.7 Corrosion

AIM: To understand the electrochemical nature of corrosion and to learn some ways of preventing it.

Most metals are found in nature in compounds with nonmetals such as oxygen and sulfur. For example, iron exists as iron ore (which contains Fe_2O_3 and other oxides of iron).

Corrosion can be viewed as the process of returning metals to their natural state—the ores from which they were originally obtained. Corrosion involves oxidation of the metal. Because corroded metal often loses its strength and attractiveness, this process causes great economic loss. For example, approximately one-fifth of the iron and steel produced annually is used to replace rusted metal.

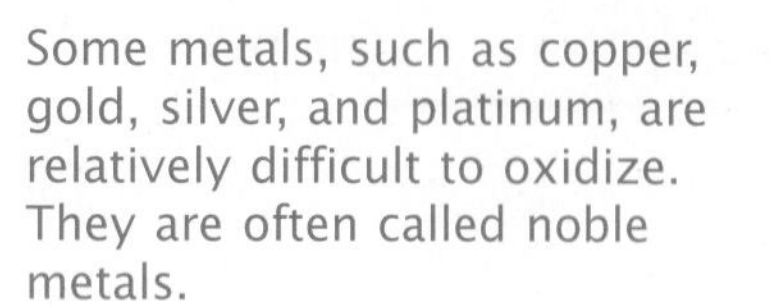
Some metals, such as copper, gold, silver, and platinum, are relatively difficult to oxidize. They are often called noble metals.

Because most metals react with O_2, we might expect them to corrode so fast in air that they wouldn't be useful. It is surprising, therefore, that the problem of corrosion does not virtually prevent the use of metals in air. Part of the explanation is that most metals develop a thin oxide coating, which tends to protect their internal atoms against further oxidation. The best example of this is aluminum. Aluminum readily loses electrons, so it should be very easily oxidized by O_2. Given this fact, why is aluminum so useful for building airplanes, bicycle frames, and so on? Aluminum is such a valuable structural material because it forms a thin adherent layer of aluminum oxide, Al_2O_3, which greatly inhibits further corrosion. Thus aluminum protects itself with this tough oxide coat. Many other metals, such as chromium, nickel, and tin, do the same thing.

Iron can also form a protective oxide coating. However, this oxide is not a very effective shield against corrosion, because it scales off easily, exposing new metal surfaces to oxidation. Under normal atmospheric conditions, copper forms an external layer of greenish copper sulfate or

CHEMISTRY IN FOCUS

Stainless Steel: It's the Pits

One of New York's giants, the Chrysler Building, boasts a much admired art-deco stainless steel pinnacle that has successfully resisted corrosion since it was built in 1929. Stainless steel is the nobility among steels. Consisting of iron, chromium (at least 13%), and nickel (with molybdenum and titanium added to more expensive types), stainless steel is highly resistant to the rusting that consumes regular steel. However, the cheaper grades of stainless steel have an Achilles heel—pit corrosion. In certain environments, pit corrosion can penetrate several millimeters in a matter of weeks.

Metallurgy, the science of producing useful metallic materials, almost always requires some kind of compromise. In the case of stainless steel, inclusions of MnS make the steel easier to machine into useful parts, but such inclusions are also the source of pit corrosion. Recently a group of British researchers analyzed stainless steel using a high-energy beam of ions that blasted atoms loose from the steel surface. Studies of the resultant atom vapor revealed the source of the problem. It turns out that when the stainless steel is cooling, the MnS inclusions "suck" chromium atoms from the surrounding area, leaving behind a chromium-deficient region. The corrosion occurs this region, as illustrated in the accompanying diagram. The essential problem is that to resist corrosion steel must contain at least 13% Cr atoms. The low-chromium region around the inclusion is not stainless steel—so it corrodes just like regular steel. This corrosion leads to a pit that causes deterioration of the steel surface.

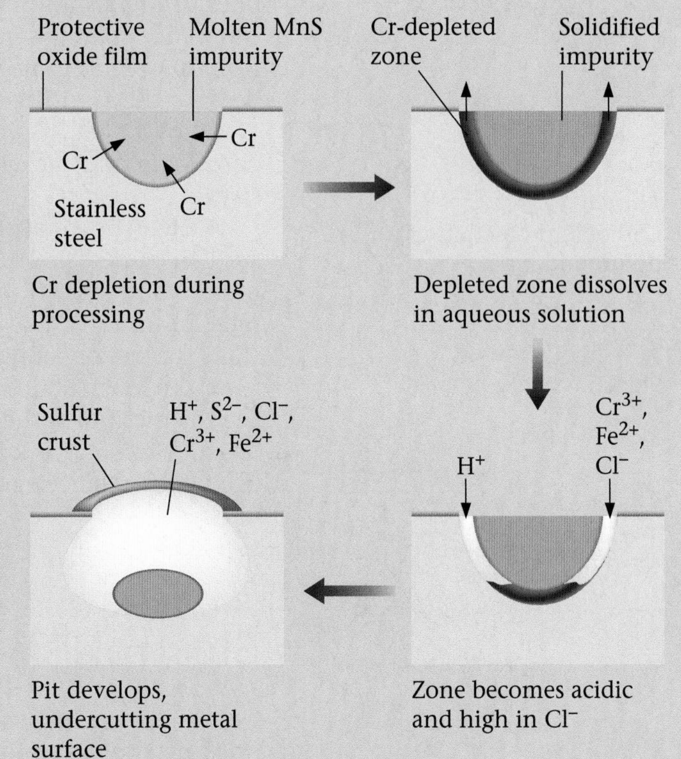

In corrosion, chromium depletion triggers pitting.

Now that the reason for the pit corrosion is understood, metallurgists should be able to develop methods of stainless steel formulation that avoid this problem. One British scientist, Mary P. Ryan, suggests that heat treatment of the stainless steel may solve the problem by allowing Cr atoms to diffuse from the inclusion back into the surrounding area. Because corrosion of regular steel is such an important issue, finding ways to make cheaper stainless steel will have a significant economic impact. We need stainless without the pits.

Built in New York in 1929, the Chrysler Building's stainless steel pinnacle has been cleaned only a few times. Despite the urban setting, the material shows few signs of corrosion.

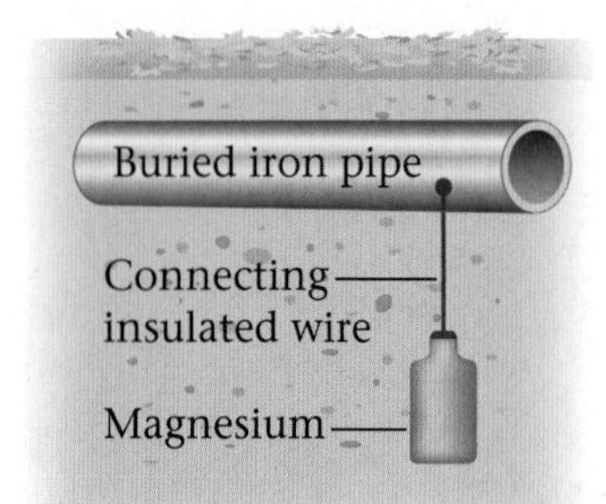

Figure 17.9
Cathodic protection of an underground pipe.

carbonate called *patina*. *Silver tarnish* is silver sulfide, Ag_2S, which in thin layers gives the silver surface a richer appearance. Gold shows no appreciable corrosion in air.

Preventing corrosion is an important way of conserving our natural supplies of metals and energy. The primary means of protection is the application of a coating—most often paint or metal plating—to protect the metal from oxygen and moisture. Chromium and tin are often used to plate steel because they oxidize to form a durable, effective oxide coating.

Alloying is also used to prevent corrosion. *Stainless steel* contains chromium and nickel, both of which form oxide coatings that protect the steel.

An alloy is a mixture of elements with metallic properties.

Cathodic protection is the method most often employed to protect steel in buried fuel tanks and pipelines. A metal that furnishes electrons more easily than iron, such as magnesium, is connected by a wire to the pipeline or tank that is to be protected (Figure 17.9). Because the magnesium is a better reducing agent than iron, electrons flow through the wire from the magnesium to the iron pipe. Thus the electrons are furnished by the magnesium rather than by the iron, keeping the iron from being oxidized. As oxidation of the magnesium occurs, the magnesium dissolves, so it must be replaced periodically.

Electrolysis

AIM: To understand the process of electrolysis and learn about the commercial preparation of aluminum.

Unless it is recharged, a battery "runs down" because the substances in it that furnish and accept electrons (to produce the electron flow) are consumed. For example, in the lead storage battery (see Section 17.6), PbO_2 and Pb are consumed to form $PbSO_4$ as the battery runs.

$$PbO_2(s) + Pb(s) + 2H_2SO_4(aq) \rightarrow 2PbSO_4(s) + 2H_2O(l)$$

However, one of the most useful characteristics of the lead storage battery is that it can be recharged. *Forcing* current through the battery in the direction opposite to the normal direction reverses the oxidation–reduction reaction. That is, $PbSO_4$ is consumed and PbO_2 and Pb are formed in the charging process. This recharging is done continuously by the automobile's alternator, which is powered by the engine.

The process of **electrolysis** involves *forcing a current through a cell to produce a chemical change that would not otherwise occur.*

One important example of this type of process is the electrolysis of water. Water is a very stable substance that can be broken down into its elements by using an electric current (Figure 17.10).

$$2H_2O(l) \xrightarrow[\text{current}]{\text{Forced electric}} 2H_2(g) + O_2(g)$$

An electrolytic cell uses electrical energy to produce a chemical change that would not otherwise occur.

The electrolysis of water to produce hydrogen and oxygen occurs whenever a current is forced through an aqueous solution. Thus, when the lead storage battery is charged, or "jumped," potentially explosive mixtures of

Figure 17.10
The electrolysis of water produces hydrogen gas at the cathode (on the left) and oxygen gas at the anode (on the right). A nonreacting strong electrolyte such as Na_2SO_4 is needed to furnish ions to allow the flow of current.

H_2 and O_2 are produced by the current flow through the solution in the battery. This is why it is very important not to produce a spark near the battery during these operations.

Another important use of electrolysis is in the production of metals from their ores. The metal produced in the greatest quantities by electrolysis is aluminum.

Aluminum is one of the most abundant elements on earth, ranking third behind oxygen and silicon. Because aluminum is a very reactive metal, it is found in nature as its oxide in an ore called *bauxite* (named after Les Baux, France, where it was discovered in 1821). Production of aluminum metal from its ore proved to be more difficult than the production of most other metals. In 1782 Lavoisier, the pioneering French chemist, recognized aluminum as a metal "whose affinity for oxygen is so strong that it cannot be overcome by any known reducing agent." As a result, pure aluminum metal remained unknown. Finally, in 1854, a process was found for producing metallic aluminum by using sodium, but aluminum remained a very expensive rarity. In fact, it is said that Napoleon III served his most honored guests with aluminum forks and spoons, while the others had to settle for gold and silver utensils!

The breakthrough came in 1886 when two men, Charles M. Hall in the United States (Figure 17.11) and Paul Heroult in France, almost simultaneously discovered a practical electrolytic process for producing aluminum, which greatly increased the availability of aluminum for many purposes. Table 17.2 shows how dramatically the price of aluminum dropped after this discovery. The effect of the electrolysis process is to reduce Al^{3+} ions to neutral Al atoms that form aluminum metal. The aluminum produced in this electrolytic process is 99.5% pure. To be useful as a structural material, aluminum is alloyed with metals such as zinc (for trailer and aircraft construction) and manganese (for cooking utensils, storage tanks, and highway signs). The production of aluminum consumes about 4.5% of all electricity used in the United States.

Table 17.2 The Price of Aluminum, 1855–1990

Date	Price of Aluminum ($/lb)*
1855	$100,000
1885	100
1890	2
1895	0.50
1970	0.30
1980	0.80
1990	0.74

*Note the precipitous drop in price after the discovery of the Hall–Heroult process in 1886.

Figure 17.11
Charles Martin Hall (1863 – 1914) was a student at Oberlin College in Ohio when he first became interested in aluminum. One of his professors commented that anyone who could find a way to manufacture aluminum cheaply would make a fortune, and Hall decided to give it a try. The 21-year-old worked in a wooden shed near his house with an iron frying pan as a container, a blacksmith's forge as a heat source, and galvanic cells constructed from fruit jars. Using these crude galvanic cells, Hall found that he could produce aluminum by passing a current through a molten mixture of Al_2O_3 and Na_3AlF_6. By a strange coincidence, Paul Heroult, a French chemist who was born and died in the same years as Hall, made the same discovery at about the same time.

Chapter 17 Review

Key Terms

oxidation–reduction (redox) reactions (17.1)
oxidation (17.1)
reduction (17.1)
oxidation states (17.2)
oxidizing agent (electron acceptor) (17.3)
reducing agent (electron donor) (17.3)
half-reactions (17.4)
electrochemistry (17.5)
electrochemical battery (galvanic cell) (17.5)
anode (17.5)
cathode (17.5)
electrolysis (17.5)
lead storage battery (17.6)
potential (17.6)
dry cell battery (17.6)
corrosion (17.7)
cathodic protection (17.7)

Summary

1. Oxidation–reduction reactions involve a transfer of electrons. Oxidation states provide a way to keep track of electrons in these reactions. A set of rules is used to assign oxidation states.

2. Oxidation is an increase in oxidation state (a loss of electrons); reduction is a decrease in oxidation state (a gain of electrons). An oxidizing agent accepts electrons, and a reducing agent donates electrons. Oxidation and reduction always occur together.

3. Oxidation–reduction equations can be balanced by inspection or by the half-reaction method. This method involves splitting a reaction into two parts (the oxidation half-reaction and the reduction half-reaction).

4. Electrochemistry is the study of the interchange of chemical and electrical energy that occurs through oxidation–reduction reactions.

5. When an oxidation–reduction reaction occurs with the reactants in the same solution, the electrons are transferred directly, and no useful work can be obtained. However, when the oxidizing agent is separated from the reducing agent, so that the electrons must flow through a wire from one to the other, chemical energy is transformed into electrical energy. The opposite process, in which electrical energy is used to produce chemical change, is called electrolysis.

6. A galvanic (electrochemical) cell is a device in which chemical energy is transformed into useful electrical energy. Oxidation occurs at the anode of a cell; reduction occurs at the cathode.

7. A battery is a galvanic cell, or group of cells, that serves as a source of electric current. The lead storage battery has a lead anode and a cathode of lead coated with PbO_2, both immersed in a solution of sulfuric acid. Dry cell batteries do not have liquid electrolytes but contain a moist paste instead.

8. Corrosion involves the oxidation of metals to form mainly oxides and sulfides. Some metals, such as aluminum, form a thin protective oxide coating that inhibits their further corrosion. Corrosion of iron can be prevented by a coating (such as paint), by alloying, and by cathodic protection.

In-Class Discussion Questions

These questions are designed to be considered by groups of students in class. Often these questions work well for introducing a particular topic in class.

1. Sketch a galvanic cell, and explain how it works. Look at Figures 17.1 and 17.5. Explain what is occurring in each container and why the cell in Figure 17.5 "works," but the one in Figure 17.1 does not.

2. Order the following molecules from lowest to highest oxidation state of the nitrogen atom: HNO_3, NH_4Cl, N_2O, NO_2, $NaNO_2$. Provide support for your answer.

3. Which of the following are oxidation–reduction reactions? Explain.
   a. $PCl_3 + Cl_2 \rightarrow PCl_5$
   b. $Cu + 2AgNO_3 \rightarrow Cu(NO_3)_2 + 2Ag$
   c. $CO_2 + 2LiOH \rightarrow Li_2CO_3 + H_2O$
   d. $FeCl_2 + 2NaOH \rightarrow Fe(OH)_2 + 2NaCl$
   e. $MnO_2 + 4HCl \rightarrow Cl_2 + 2H_2O + MnCl_2$

4. Which of the following statements is (are) true? Explain. (There may be more than one answer.)
   a. Oxidation and reduction cannot occur independently of each other.
   b. Oxidation and reduction accompany all chemical changes.
   c. Oxidation and reduction describe the loss and gain of electron(s), respectively.

5. Why do we say that when something gains electrons it is reduced? What is being reduced?

Questions and Problems

All even-numbered exercises have answers in the back of this book and solutions in the *Solutions Guide.*

17.1 Oxidation–Reduction Reactions

QUESTIONS

1. Give some examples of how we make good use of oxidation–reduction reactions in everyday life.

2. How do chemists define the processes of *oxidation* and *reduction?* Write a simple equation illustrating each of your definitions.

3. For each of the following oxidation–reduction reactions, identify which element is being oxidized and which is being reduced.
   a. $2Fe(s) + 3F_2(g) \rightarrow 2FeF_3(s)$
   b. $O_2(g) + 2Cu(s) \rightarrow 2CuO(s)$
   c. $F_2(g) + 2KI(aq) \rightarrow 2KF(aq) + I_2(s)$
   d. $2Al(s) + 3H_2(g) \rightarrow 2AlH_3(s)$

4. For each of the following oxidation–reduction reactions of metals with nonmetals, identify which element is oxidized and which is reduced.
   a. $6Na(s) + N_2(g) \rightarrow 2Na_3N(s)$
   b. $Mg(s) + Cl_2(g) \rightarrow MgCl_2(s)$
   c. $2Al(s) + 3Br_2(l) \rightarrow 2AlBr_3(s)$
   d. $4Fe(s) + 3O_2(g) \rightarrow 2Fe_2O_3(s)$

5. For each of the following oxidation–reduction reactions of metals with nonmetals, identify which element is oxidized and which is reduced.
   a. $Cl_2(g) + Cu(s) \rightarrow CuCl_2(s)$
   b. $O_2(g) + 2Ni(s) \rightarrow 2NiO(s)$
   c. $S(s) + 2Hg(l) \rightarrow Hg_2S(s)$
   d. $2K(s) + I_2(s) \rightarrow 2KI(s)$

6. For each of the following oxidation–reduction reactions, identify which element is being oxidized and which is being reduced.
   a. $2K(s) + O_2(g) \rightarrow K_2O(s)$
   b. $H_2(g) + S(s) \rightarrow H_2S(g)$
   c. $N_2(g) + 3H_2(g) \rightarrow 2NH_3(g)$
   d. $2Hg(l) + Cl_2(g) \rightarrow Hg_2Cl_2(s)$

17.2 Oxidation States

QUESTIONS

7. What is an oxidation state? Why do we define such a concept?

8. What is the oxidation state of the atoms in an uncombined element? Does it depend on whether the element occurs as a diatomic molecule (O_2, N_2) or as a larger molecule (P_4, S_8)?

9. Explain why, although it is not an ionic compound, we still assign oxygen an oxidation state of -2 in water, H_2O. Give an example of a compound in which oxygen is *not* in the -2 oxidation state.

10. Why is fluorine always assigned an oxidation state of -1? What oxidation number is *usually* assigned to the other halogen elements when they occur in compounds? In an *interhalogen* compound involving fluorine (such as ClF), which atom has a negative oxidation state?

11. Why must the sum of all the oxidation states of the atoms in a neutral molecule be zero?

12. What must be the sum of the oxidation states of all the atoms in a polyatomic ion?

PROBLEMS

13. Assign oxidation states to all of the atoms in each of the following:
   a. NF_3 c. HOCl
   b. NO_2 d. S_8

14. Assign oxidation states to all of the atoms in each of the following:
   a. P_4 c. CH_4
   b. HClO d. $H_2PO_4^-$

15. Assign oxidation states to all of the atoms in each of the following:
   a. O_2 c. $FeCl_3$
   b. O_3 d. $FeCl_2$

16. Assign oxidation states to all of the atoms in each of the following:
   a. Cl_2 c. $CoCl_3$
   b. $CoCl_2$ d. CO_2

17. Assign oxidation states to all of the atoms in each of the following:
   a. Na_3PO_4 c. Na_2HPO_4
   b. NaH_2PO_4 d. Na_3P

18. Assign oxidation states to all of the atoms in each of the following:
   a. H_2SO_4 c. NO_3^-
   b. MnO_4^- d. K_3PO_4

19. Assign oxidation states to all of the atoms in each of the following:
   a. Na_2CrO_4 c. $CrCl_3$
   b. $Na_2Cr_2O_7$ d. Cr_2O_3

20. Assign oxidation states to all of the atoms in each of the following:
   a. $CuCl_2$ c. $HCrO_4^-$
   b. $CrCl_3$ d. Cr_2O_3

21. Assign oxidation states to all of the atoms in each of the following:
   a. $NaHSO_4$ c. $KMnO_4$
   b. $CaCO_3$ d. MnO_2

22. Assign oxidation states to all of the atoms in each of the following:
   a. $MnCl_2$ c. B_2H_6
   b. $HBrO_3$ d. CF_4

17.3 Oxidation–Reduction Reactions Between Nonmetals

QUESTIONS

23. Oxidation can be defined as a loss of electrons or as an increase in oxidation state. Explain why the two

definitions mean the same thing, and give an example to support your explanation.

24. Reduction can be defined as a gain of electrons or as a decrease in oxidation state. Explain why the two definitions mean the same thing, and give an example to support your explanation.

25. What is an oxidizing agent? What is a reducing agent?

26. Give an example of a simple oxidation–reduction equation. Identify the species being oxidized and the species being reduced. Identify the oxidizing agent and the reducing agent in your example.

27. Does an oxidizing agent donate or accept electrons? Does a reducing agent donate or accept electrons?

28. Does an oxidizing agent increase or decrease its own oxidation state when it acts on another atom? Does a reducing agent increase or decrease its own oxidation state when it acts on another substance?

PROBLEMS

29. In each of the following reactions, identify which element is oxidized and which is reduced by assigning oxidation numbers.
a. $2HNO_3(aq) + 3H_2S(g) \rightarrow 2NO(g) + 4H_2O(g) + 3S(s)$
b. $2H_2O_2(aq) \rightarrow 2H_2O(l) + O_2(g)$
c. $2ZnS(s) + 3O_2(g) \rightarrow 2ZnO(s) + 2SO_2(g)$
d. $CH_4(g) + Cl_2(g) \rightarrow CH_3Cl(g) + HCl(g)$

30. In each of the following reactions, identify which element is oxidized and which is reduced by assigning oxidation numbers.
a. $Zn(s) + 2HNO_3(aq) \rightarrow Zn(NO_3)_2(aq) + H_2(g)$
b. $H_2(g) + CuSO_4(aq) \rightarrow Cu(s) + H_2SO_4(aq)$
c. $N_2(g) + 3Br_2(l) \rightarrow 2NBr_3(g)$
d. $2KBr(aq) + Cl_2(g) \rightarrow 2KCl(aq) + Br_2(l)$

31. In each of the following reactions, identify which element is oxidized and which is reduced by assigning oxidation states.
a. $2Cu(s) + S(s) \rightarrow Cu_2S(s)$
b. $2Cu_2O(s) + O_2(g) \rightarrow 4CuO(s)$
c. $4B(s) + 3O_2(g) \rightarrow 2B_2O_3(s)$
d. $6Na(s) + N_2(g) \rightarrow 2Na_3N(s)$

32. In each of the following reactions, identify which element is oxidized and which is reduced by assigning oxidation numbers.
a. $Fe_2O_3(s) + 3C(s) \rightarrow 2Fe(s) + 3CO(g)$
b. $Mg(s) + 2HBr(aq) \rightarrow MgBr_2(aq) + H_2(g)$
c. $2CoS(s) + S(s) \rightarrow Co_2S_3(s)$
d. $2Ag(s) + 2HNO_3(aq) \rightarrow 2AgNO_3(aq) + H_2(g)$

33. Pennies in the United States consist of a zinc core that is electroplated with a thin coating of copper. Zinc dissolves in hydrochloric acid, but copper does not. If a small scratch is made on the surface of a penny, it is possible to dissolve away the zinc core, leaving only the thin shell of copper. Identify which element is oxidized and which is reduced in the reaction for the dissolving of the zinc by the acid.

$$Zn(s) + 2HCl(aq) \rightarrow ZnCl_2(aq) + H_2(g)$$

34. Iron ores, usually oxides of iron, are converted to the pure metal by reaction in a blast furnace with carbon (coke). The carbon is first reacted with air to form carbon monoxide, which in turn reacts with the iron oxides as follows:

$$Fe_2O_3(s) + 3CO(g) \rightarrow 2Fe(l) + 3CO_2(g)$$

Identify the atoms that are oxidized and reduced, and specify the oxidizing and reducing agents.

35. Although magnesium metal does not react with water at room temperature, it does react vigorously with steam at higher temperatures, releasing elemental hydrogen gas from the water.

$$Mg(s) + 2H_2O(g) \rightarrow Mg(OH)_2(s) + H_2(g)$$

Identify which element is being oxidized and which is being reduced.

36. Elemental bromine can be prepared by treatment of seawater with chlorine gas.

$$Cl_2(g) + 2NaBr(aq) \rightarrow Br_2(aq) + 2NaCl(aq)$$

Identify the atoms that are oxidized and reduced, and specify the oxidizing and reducing agents.

17.4 Balancing Oxidation–Reduction Reactions by the Half-Reaction Method

QUESTIONS

37. In what *two* respects must oxidation–reduction reactions be balanced?

38. Why is a systematic method for balancing oxidation–reduction reactions necessary? Why can't these equations be balanced readily by inspection?

39. What is a half-reaction? What does each of the two half-reactions that make up an overall process represent?

40. Why must the number of electrons lost in the oxidation equal the number of electrons gained in the reduction? Is it possible to have "leftover" electrons in a reaction?

PROBLEMS

41. Balance each of the following half-reactions.
a. $Al \rightarrow Al^{3+}$
b. $I^- \rightarrow I_2$
c. $Co^{3+} \rightarrow Co^{2+}$
d. $P^{3-} \rightarrow P_4$

42. Balance each of the following half-reactions.
a. $N_2(g) \rightarrow N^{3-}(s)$
b. $O_2^{2-}(aq) \rightarrow O_2(g)$
c. $Zn(s) \rightarrow Zn^{2+}(aq)$
d. $F_2(g) \rightarrow F^-(aq)$

43. Balance each of the following half-reactions, which take place in acidic solution.
a. $NO_3^-(aq) \rightarrow NO(g)$
b. $NO_3^-(aq) \rightarrow NO_2(g)$
c. $H_2SO_4(l) \rightarrow SO_2(g)$
d. $H_2O_2(aq) \rightarrow H_2O(l)$

44. Balance each of the following half-reactions, which take place in acidic solution.
a. $HBr(aq) \rightarrow Br^-(aq) + H_2(g)$
b. $ClO_4^-(aq) \rightarrow ClO_3^-(aq)$
c. $PbSO_4(s) \rightarrow Pb(s) + HSO_4^-(aq)$
d. $HNO_2(aq) \rightarrow NO(g)$

45. Balance each of the following oxidation–reduction reactions, which take place in acidic solution, by using the "half-reaction" method.
a. $Mg(s) + Hg^{2+}(aq) \rightarrow Mg^{2+}(aq) + Hg_2^{2+}(aq)$
b. $NO_3^-(aq) + Br^-(aq) \rightarrow NO(g) + Br_2(l)$
c. $Ni(s) + NO_3^-(aq) \rightarrow Ni^{2+}(aq) + NO_2(g)$
d. $ClO_4^-(aq) + Cl^-(aq) \rightarrow ClO_3^-(aq) + Cl_2(g)$

46. Balance each of the following oxidation–reduction reactions, which take place in acidic solution, by using the "half-reaction" method.
a. $Al(s) + H^+(aq) \rightarrow Al^{3+}(aq) + H_2(g)$
b. $S^{2-}(aq) + NO_3^-(aq) \rightarrow S(s) + NO(g)$
c. $I_2(aq) + Cl_2(aq) \rightarrow IO_3^-(aq) + HCl(g)$
d. $AsO_4^-(aq) + S^{2-}(aq) \rightarrow AsO_3^-(aq) + S(s)$

47. Iodide ion, I^-, is one of the most easily oxidized species. Balance each of the following oxidation–reduction reactions, which take place in acidic solution, by using the "half-reaction" method.
a. $IO_3^-(aq) + I^-(aq) \rightarrow I_2(aq)$
b. $Cr_2O_7^{2-}(aq) + I^-(aq) \rightarrow Cr^{3+}(aq) + I_2(aq)$
c. $Cu^{2+}(aq) + I^-(aq) \rightarrow CuI(s) + I_2(aq)$

48. A solution of chlorine gas in water is used in chemical analysis to test for the presence of Br^- and I^- ions in solution. Cl_2 is able to oxidize easily these two ions to the elemental forms Br_2 and I_2, which may then be identified by their colors. Write a balanced oxidation–reduction equation for each of these processes.

17.5 Electrochemistry: An Introduction

QUESTIONS

49. How is an oxidation–reduction reaction set up as a galvanic cell (battery)? How is the transfer of electrons between reducing agent and oxidizing agent made useful?

50. What is a salt bridge? Why is a salt bridge necessary in a galvanic cell? Can some other method be used in place of the salt bridge?

51. In which direction do electrons flow in a galvanic cell, anode to cathode or vice versa?

52. Which electrode in a galvanic cell is the anode? Which is the cathode?

PROBLEMS

53. Consider the oxidation–reduction reaction

$$Al(s) + Ni^{2+}(aq) \rightarrow Al^{3+}(aq) + Ni(s)$$

Sketch a galvanic cell that makes use of this reaction. Which metal ion is reduced? Which metal is oxidized? What half-reaction takes place at the anode in the cell? What half-reaction takes place at the cathode?

54. Consider the oxidation–reduction reaction

$$Zn(s) + Pb^{2+}(aq) \rightarrow Zn^{2+}(aq) + Pb(s)$$

Sketch a galvanic cell that uses this reaction. Which metal ion is reduced? Which metal is oxidized? What half-reaction takes place at the anode in the cell? What half-reaction takes place at the cathode?

17.6 Batteries

QUESTIONS

55. Write the chemical equation for the overall cell reaction that occurs in a lead storage automobile battery. What species is oxidized in such a battery? What species is reduced? Why can such a battery be "recharged"?

56. Why does an alkaline dry cell battery typically last longer than a normal dry cell? Write the chemical equation for the overall cell reaction in an alkaline dry cell.

17.7 Corrosion

QUESTIONS

57. What process is represented by the *corrosion* of a metal? Why is corrosion undesirable?

58. Explain how some metals, notably aluminum, naturally resist complete oxidation by the atmosphere.

59. Pure iron ordinarily rusts quickly, but steel does not corrode nearly as fast. How does steel resist corrosion?

60. What is *cathodic protection,* and how is it applied to prevent oxidation of steel tanks and pipes?

17.8 Electrolysis

QUESTIONS

61. What is the difference between a *galvanic cell* and an *electrolysis cell?* Give an example of each.

62. What reactions go on during the recharging of an automobile battery?

63. Although aluminum is one of the most abundant metals on the earth, most people make an effort to recycle aluminum cans. How might this be related to the process by which aluminum metal is produced?

64. Jewelry is often manufactured by plating an expensive metal such as gold over a cheaper metal. How might such a process be set up as an electrolysis reaction?

Additional Problems

65. Reactions in which one or more ______ are transferred between species are called oxidation–reduction reactions.

66. Oxidation may be described as a(n) ______ of electrons or as an increase in ______.

67. Reduction may be described as a(n) ______ of electrons or as a decrease in ______.

68. In assigning oxidation states for a covalently bonded molecule, we assume that the more ______ element controls both electrons of the covalent bond.

69. The sum of the oxidation states of the atoms in a polyatomic ion is equal to the overall ______ of the ion.

70. What is an *oxidizing agent?* Is an oxidizing agent itself oxidized or reduced when it acts on another species?

71. An oxidizing agent causes the (oxidation/reduction) of another species, and the oxidizing agent itself is (oxidized/reduced).

72. In order to function as a good reducing agent, a species must ______ electrons easily.

73. When we balance an oxidation–reduction equation, the number of electrons lost by the reducing agent must ______ the number of electrons gained by the oxidizing agent.

74. In order to obtain useful electrical energy from an oxidation–reduction process, we must set up the reaction in such a way that the oxidation half-reaction and the reduction half-reaction are physically ______ one another.

75. An electrochemical cell that produces a current from an oxidation–reduction reaction is often called a(n) ______ cell.

76. Which process (oxidation/reduction) takes place at the anode of a galvanic cell?

77. At which electrode (anode/cathode) do species gain electrons in a galvanic cell?

78. What is an *electrolysis* reaction? Give an example of an electrolysis reaction.

79. The "pressure" on electrons to flow from one electrode to the other in a battery is called the ______ of the battery.

80. "Jump-starting" a dead automobile battery can be dangerous if precautions are not taken, because of the production of an explosive mixture of ______ and ______ gases in the battery.

81. The common acid dry cell battery typically contains an inner casing made of ______ metal, which functions as the anode.

82. Corrosion of a metal represents its ______ by species present in the atmosphere.

83. Although aluminum is a reactive metal, pure aluminum ordinarily does not corrode severely in air because a protective layer of ______ builds up on the metal's surface.

84. For each of the following unbalanced oxidation–reduction chemical equations, balance the equation by inspection, and identify which species is undergoing oxidation and which is undergoing reduction.
   a. $Fe(s) + O_2(g) \rightarrow Fe_2O_3(s)$
   b. $Al(s) + Cl_2(g) \rightarrow AlCl_3(s)$
   c. $Mg(s) + P_4(s) \rightarrow Mg_3P_2(s)$

85. In each of the following reactions, identify which element is oxidized and which is reduced.
   a. $Zn(s) + 2HCl(aq) \rightarrow ZnCl_2(aq) + H_2(g)$
   b. $2CuI(s) \rightarrow CuI_2(s) + Cu(s)$
   c. $6Fe^{2+}(aq) + Cr_2O_7^{2-}(aq) + 14H^+(aq) \rightarrow 6Fe^{3+}(aq) + 2Cr^{3+}(aq) + 7H_2O(l)$

86. In each of the following reactions, identify which element is oxidized and which is reduced.
   a. $2Al(s) + 6HCl(aq) \rightarrow 2AlCl_3(aq) + 3H_2(g)$
   b. $2HI(g) \rightarrow H_2(g) + I_2(s)$
   c. $Cu(s) + H_2SO_4(aq) \rightarrow CuSO_4(aq) + H_2(g)$

87. Carbon compounds containing double bonds (such compounds are called alkenes) react readily with many other reagents. In each of the following reactions, identify which atoms are oxidized and which are reduced, and specify the oxidizing and reducing agents.
   a. $CH_2{=}CH_2(g) + Cl_2(g) \rightarrow ClCH_2{-}CH_2Cl(l)$
   b. $CH_2{=}CH_2(g) + Br_2(g) \rightarrow BrCH_2{-}CH_2Br(l)$
   c. $CH_2{=}CH_2(g) + HBr(g) \rightarrow CH_3{-}CH_2Br(l)$
   d. $CH_2{=}CH_2(g) + H_2(g) \rightarrow CH_3{-}CH_3(g)$

88. Balance each of the following oxidation–reduction reactions by inspection.
   a. $C_3H_8(g) + O_2(g) \rightarrow CO_2(g) + H_2O(g)$
   b. $CO(g) + H_2(g) \rightarrow CH_3OH(l)$
   c. $SnO_2(s) + C(s) \rightarrow Sn(s) + CO(g)$
   d. $C_2H_5OH(l) + O_2(g) \rightarrow CO_2(g) + H_2O(g)$

89. Balance each of the following oxidation–reduction reactions, which take place in acidic solution.
   a. $MnO_4^-(aq) + H_2O_2(aq) \rightarrow Mn^{2+}(aq) + O_2(g)$
   b. $BrO_3^-(aq) + Cu^+(aq) \rightarrow Br^-(aq) + Cu^{2+}(aq)$
   c. $HNO_2(aq) + I^-(aq) \rightarrow NO(g) + I_2(aq)$

90. For each of the following oxidation–reduction reactions of metals with nonmetals, identify which element is oxidized and which is reduced.
   a. $4Na(s) + O_2(g) \rightarrow 2Na_2O(s)$
   b. $Fe(s) + H_2SO_4(aq) \rightarrow FeSO_4(aq) + H_2(g)$

c. $2Al_2O_3(s) \rightarrow 4Al(s) + 3O_2(g)$
d. $3Mg(s) + N_2(g) \rightarrow Mg_3N_2(s)$

91. For each of the following oxidation–reduction reactions of metals with nonmetals, identify which element is oxidized and which is reduced.
a. $3Zn(s) + N_2(g) \rightarrow Zn_3N_2(s)$
b. $Co(s) + S(s) \rightarrow CoS(s)$
c. $4K(s) + O_2(g) \rightarrow 2K_2O(s)$
d. $4Ag(s) + O_2(g) \rightarrow 2Ag_2O(s)$

92. Assign oxidation states to all of the atoms in each of the following:
a. NH_3 c. CO_2
b. CO d. NF_3

93. Assign oxidation states to all of the atoms in each of the following:
a. PBr_3 c. $KMnO_4$
b. C_3H_8 d. CH_3COOH

94. Assign oxidation states to all of the atoms in each of the following:
a. MnO_2 c. H_2SO_3
b. $BaCrO_4$ d. $Ca_3(PO_4)_2$

95. Assign oxidation states to all of the atoms in each of the following:
a. $CrCl_3$ c. $K_2Cr_2O_7$
b. K_2CrO_4 d. $Cr(C_2H_3O_2)_2$

96. Assign oxidation states to all of the atoms in each of the following:
a. BiO^+ c. NO_2^-
b. PO_4^{3-} d. Hg_2^{2+}

97. In each of the following reactions, identify which element is oxidized and which is reduced by assigning oxidation states.
a. $C(s) + O_2(g) \rightarrow CO_2(g)$
b. $2CO(g) + O_2(g) \rightarrow 2CO_2(g)$
c. $CH_4(g) + 2O_2(g) \rightarrow CO_2(g) + 2H_2O(g)$
d. $C_2H_2(g) + 2H_2(g) \rightarrow C_2H_6(g)$

98. In each of the following reactions, identify which element is oxidized and which is reduced by assigning oxidation states.
a. $2B_2O_3(s) + 6Cl_2(g) \rightarrow 4BCl_3(l) + 3O_2(g)$
b. $GeH_4(g) + O_2(g) \rightarrow Ge(s) + 2H_2O(g)$
c. $C_2H_4(g) + Cl_2(g) \rightarrow C_2H_4Cl_2(l)$
d. $O_2(g) + 2F_2(g) \rightarrow 2OF_2(g)$

99. Balance each of the following half-reactions.
a. $I^-(aq) \rightarrow I_2(s)$
b. $O_2(g) \rightarrow O^{2-}(s)$
c. $P_4(s) \rightarrow P^{3-}(s)$
d. $Cl_2(g) \rightarrow Cl^-(aq)$

100. Balance each of the following half-reactions, which take place in acidic solution.
a. $SiO_2(s) \rightarrow Si(s)$
b. $S(s) \rightarrow H_2S(aq)$
c. $NO_3^-(aq) \rightarrow HNO_2(aq)$
d. $NO_3^-(aq) \rightarrow NO(g)$

101. Balance each of the following oxidation–reduction reactions, which take place in acidic solution, by using the "half-reaction" method.
a. $I^-(aq) + MnO_4^-(aq) \rightarrow I_2(aq) + Mn^{2+}(aq)$
b. $S_2O_8^{2-}(aq) + Cr^{3+}(aq) \rightarrow SO_4^{2-}(aq) + Cr_2O_7^{2-}(aq)$
c. $BiO_3^-(aq) + Mn^{2+}(aq) \rightarrow Bi^{3+}(aq) + MnO_4^-(aq)$

102. Potassium permanganate, $KMnO_4$, is one of the most widely used oxidizing agents. Balance each of the following oxidation–reduction reactions of the permanganate ion in acidic solution by using the "half-reaction" method.
a. $MnO_4^-(aq) + C_2O_4^{2-}(aq) \rightarrow Mn^{2+}(aq) + CO_2(g)$
b. $MnO_4^-(aq) + Fe^{2+}(aq) \rightarrow Mn^{2+}(aq) + Fe^{3+}(aq)$
c. $MnO_4^-(aq) + Cl^-(aq) \rightarrow Mn^{2+}(aq) + Cl_2(g)$

103. Consider the oxidation–reduction reaction

$$Mg(s) + Cu^{2+}(aq) \rightarrow Mg^{2+}(aq) + Cu(s)$$

Sketch a galvanic cell that uses this reaction. Which metal ion is reduced? Which metal is oxidized? What half-reaction takes place at the anode in the cell? What half-reaction takes place at the cathode?

18 Radioactivity and Nuclear Energy

Computer simulation of a 15 solar mass star exploding as a supernova.

Because the chemistry of an atom is determined by the number and arrangement of its electrons, the properties of the nucleus do not strongly affect the chemical behavior of an atom. Therefore, you might be wondering why there is a chapter on the nucleus in a chemistry textbook. The reason for this chapter is that the nucleus is very important to all of us—a quick reading of any daily newspaper will testify to that. Nuclear processes can be used to detect explosives in airline luggage (see "Chemistry in Focus: Measurement: Past, Present, and Future," page 21), to generate electric power, and to establish the ages of very old objects such as human artifacts, rocks, and diamonds (see "Chemistry in Focus: Dating Diamonds," page 560). This chapter considers aspects of the nucleus and its properties that you should know about.

Several facts about the nucleus are immediately impressive: its very small size, its very large density, and the energy that holds it together. The radius of a typical nucleus is about 10^{-13} cm, only a hundred-thousandth the radius of a typical atom. In fact, if the nucleus of the hydrogen atom were the size of a Ping-Pong ball, the electron in the 1*s* orbital would be, on the average, 0.5 km (0.3 mile) away. The density of the nucleus is equally impressive; it is approximately 1.6×10^{14} g/cm^3. A sphere of nuclear material the size of a Ping-Pong ball would have a mass of 2.5 *billion tons!* Finally, the energies involved in nuclear processes are typically millions of times larger than those associated with normal chemical reactions, a fact that makes nuclear processes potentially attractive for generating energy.

The nucleus is believed to be made of particles called **nucleons** (**neutrons** and **protons**).

Wooden artifacts such as this dragon figurehead from a Viking Ship can be dated from their carbon-14 content.

Recall from Chapter 4 that the number of protons in a nucleus is called the **atomic number** (Z) and that the sum of the numbers of neutrons and protons is the **mass number** (A). Atoms that have identical atomic numbers but different mass numbers are called **isotopes.** The general term **nuclide** is applied to each unique atom, and we represent it as follows:

$$\begin{array}{c}\text{Mass number}\\ \downarrow \\ {}^{A}_{Z}\text{X} \leftarrow \text{Element symbol}\\ \uparrow \\ \text{Atomic number}\end{array}$$

The atomic number (Z) represents the number of protons in a nucleus; the mass number (A) represents the sum of the numbers of protons and neutrons in a nucleus.

where X represents the symbol for a particular element. For example, the following nuclides constitute the common isotopes of carbon: carbon-12, ${}^{12}_{6}\text{C}$; carbon-13, ${}^{13}_{6}\text{C}$; and carbon-14, ${}^{14}_{6}\text{C}$. Notice that all the carbon nuclides have six protons ($Z = 6$) and that they have six, seven, and eight neutrons, respectively.

Radioactive Decay

AIMS: To learn the types of radioactive decay. To learn to write nuclear equations that describe radioactive decay.

Many nuclei are **radioactive;** that is, they spontaneously decompose, forming a different nucleus and producing one or more particles. An example is carbon-14, which decays as shown in the equation

$$^{14}_{6}\text{C} \rightarrow {}^{14}_{7}\text{N} + {}^{0}_{-1}\text{e}$$

where ${}^{0}_{-1}\text{e}$ represents an electron, which in nuclear terminology is called a **beta particle,** or **β particle.** This **nuclear equation,** which is typical of those representing radioactive decay, is quite different from the chemical equations we have written before. Recall that in a balanced chemical equation the atoms must be conserved. In a nuclear equation *both the atomic number (Z) and the mass number (A) must be conserved.* That is, the sums of the Z values on both sides of the arrow must be equal, and the same restriction applies to the A values. For example, in the above equation, the sum of the Z values is 6 on both sides of the arrow (6 and 7 − 1), and the sum of the A values is 14 on both sides of the arrow (14 and 14 + 0). Notice that the mass number for the β particle is zero; the mass of the electron is so small that it can be neglected here. Of the approximately 2000 known nuclides, only 279 do not undergo radioactive decay. Tin has the largest number of nonradioactive isotopes—ten.

Over 85% of all known nuclides are radioactive.

Types of Radioactive Decay

There are several different types of radioactive decay. One frequently observed decay process involves production of an **alpha (α) particle,** which is a helium nucleus (${}^{4}_{2}\text{He}$). **Alpha-particle production** is a very common mode of decay for heavy radioactive nuclides. For example, ${}^{222}_{88}\text{Ra}$, radium-222, decays by α-particle production to give radon-218.

$$^{222}_{88}\text{Ra} \rightarrow {}^{4}_{2}\text{He} + {}^{218}_{86}\text{Rn}$$

Notice in this equation that the mass number is conserved (222 = 4 + 218) and the atomic number is conserved (88 = 2 + 86). Another α-particle producer is ${}^{230}_{90}\text{Th}$:

$$^{230}_{90}\text{Th} \rightarrow {}^{4}_{2}\text{He} + {}^{226}_{88}\text{Ra}$$

Notice that the production of an α particle results in a loss of 4 in mass number (A) and a loss of 2 in atomic number (Z).

β-particle production is another common decay process. For example, the thorium-234 nuclide produces a β particle as it changes to protactinium-234.

$$^{234}_{90}\text{Th} \rightarrow {}^{234}_{91}\text{Pa} + {}^{0}_{-1}\text{e}$$

Iodine-131 is also a β-particle producer:

$$^{131}_{53}\text{I} \rightarrow {}^{0}_{-1}\text{e} + {}^{131}_{54}\text{Xe}$$

Notice that both Z and A balance in each of these nuclear equations.

Recall that the β particle is assigned mass number 0 because its mass is tiny compared with that of a proton or neutron. The value of Z is −1 for the β particle, so the atomic number for the new nuclide is greater by 1 than the atomic number for the original nuclide. Therefore, *the net effect of β-particle production is to change a neutron to a proton.*

Table 18.1 Various Types of Radioactive Processes

Process	Example
β-particle (electron) production	${}^{227}_{89}\text{Ac} \rightarrow {}^{227}_{90}\text{Th} + {}^{0}_{-1}\text{e}$
positron production	${}^{13}_{7}\text{N} \rightarrow {}^{13}_{6}\text{C} + {}^{0}_{1}\text{e}$
electron capture	${}^{73}_{33}\text{As} + {}^{0}_{-1}\text{e} \rightarrow {}^{73}_{32}\text{Ge}$
α-particle production	${}^{210}_{84}\text{Po} \rightarrow {}^{206}_{82}\text{Pb} + {}^{4}_{2}\text{He}$
γ-ray production	excited nucleus $\rightarrow$ ground-state nucleus + ${}^{0}_{0}\gamma$ excess energy — lower energy

Production of a β particle results in no change in mass number (A) and an increase of 1 in atomic number (Z).

A gamma ray is a high-energy photon produced in connection with nuclear decay.

The ${}^{0}_{0}\gamma$ notation indicates $Z = 0$ and $A = 0$ for a γ ray. A gamma ray is often simply indicated by γ.

A **gamma ray,** or **γ ray,** is a high-energy photon of light. A nuclide in an excited nuclear energy state can release excess energy by producing a gamma ray, and γ-ray production often accompanies nuclear decays of various types. For example, in the α-particle decay of ${}^{238}_{92}\text{U}$,

$${}^{238}_{92}\text{U} \rightarrow {}^{4}_{2}\text{He} + {}^{234}_{90}\text{Th} + 2{}^{0}_{0}\gamma$$

two γ rays of different energies are produced in addition to the α particle (${}^{4}_{2}\text{He}$). Gamma rays are photons of light and so have zero charge and zero mass number.

Production of a γ ray results in no change in mass number (A) and no change in atomic number (Z).

The **positron** is a particle with the same mass as the electron but opposite charge. An example of a nuclide that decays by **positron production** is sodium-22:

$${}^{22}_{11}\text{Na} \rightarrow {}^{0}_{1}\text{e} + {}^{22}_{10}\text{Ne}$$

Note that the production of a positron appears to change a proton into a neutron.

Production of a positron results in no change in mass number (A) and a decrease of 1 in atomic number (Z).

Electron capture is a process in which one of the inner-orbital electrons is captured by the nucleus, as illustrated by the process

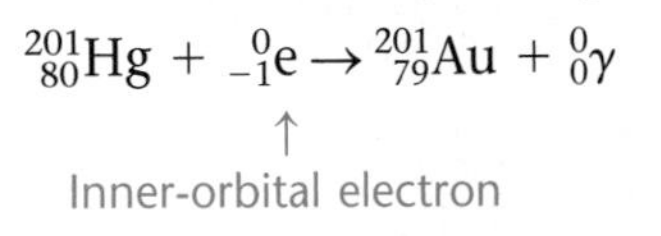

$${}^{201}_{80}\text{Hg} + {}^{0}_{-1}\text{e} \rightarrow {}^{201}_{79}\text{Au} + {}^{0}_{0}\gamma$$

This reaction would have been of great interest to the alchemists, but unfortunately, it does not occur often enough to make it a practical means of changing mercury to gold. Gamma rays are always produced along with electron capture.

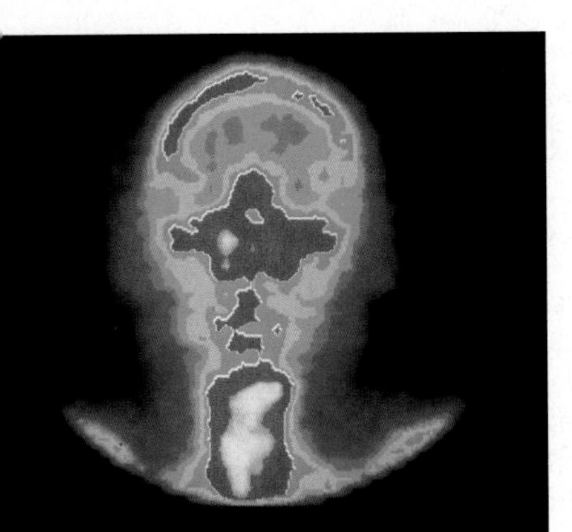

Bone scintigraph of a patient's cranium following administration of the radiopharmaceutical Technetium-99.

Table 18.1 lists the common types of radioactive decay, with examples.

Often a radioactive nucleus cannot achieve a stable (nonradioactive) state through a single decay process. In such a case, a **decay series** occurs until a stable nuclide is formed. A well-known example is the decay series that starts with ${}^{238}_{92}\text{U}$ and ends with ${}^{206}_{82}\text{Pb}$, as shown in Figure 18.1. Similar series exist for ${}^{235}_{92}\text{U}$:

$${}^{235}_{92}\text{U} \xrightarrow[\text{decays}]{\text{Series of}} {}^{207}_{82}\text{Pb}$$

and for ${}^{232}_{90}\text{Th}$:

$${}^{232}_{90}\text{Th} \xrightarrow[\text{decays}]{\text{Series of}} {}^{208}_{82}\text{Pb}$$

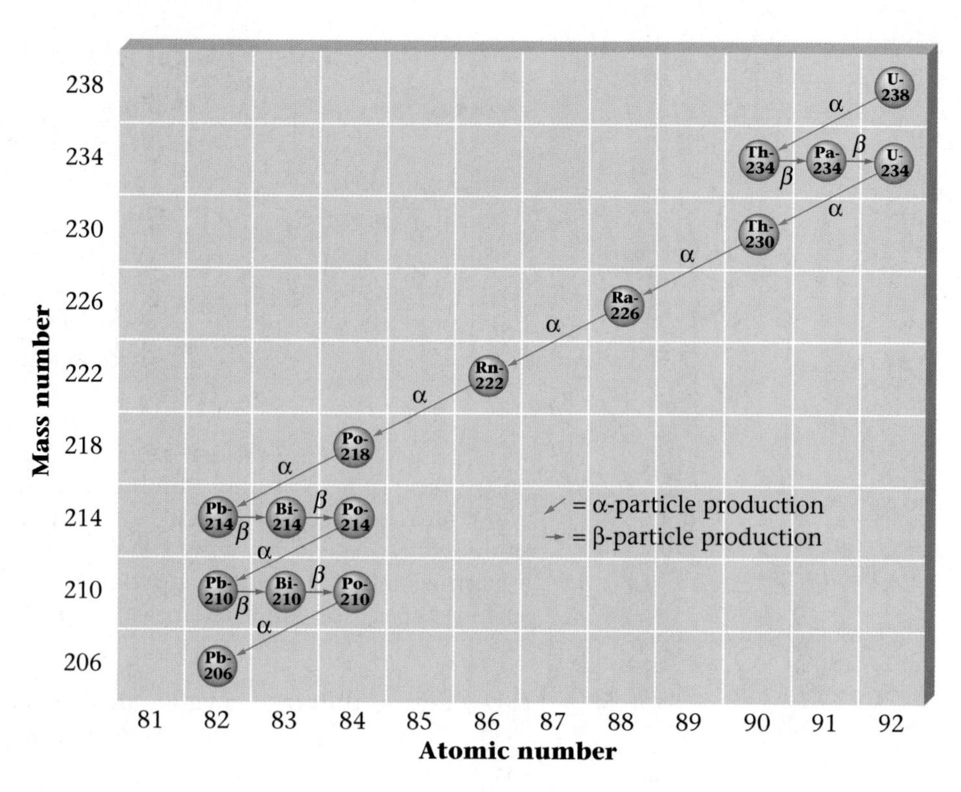

Figure 18.1
The decay series from $^{238}_{92}U$ to $^{206}_{82}Pb$. Each nuclide in the series except $^{206}_{82}Pb$ is radioactive, and the successive transformations (shown by the arrows) continue until $^{206}_{82}Pb$ is finally formed. The horizontal red arrows indicate β-particle production (Z increases by 1 and A is unchanged). The diagonal blue arrows signify α-particle production (both A and Z decrease).

Example 18.1 Writing Nuclear Equations, I

Write balanced nuclear equations for each of the following processes.

a. $^{11}_{6}C$ produces a positron.

b. $^{214}_{83}Bi$ produces a β particle.

c. $^{237}_{93}Np$ produces an α particle.

Solution

a. We must find the product nuclide represented by $^{A}_{Z}X$ in the following equation:

$$^{11}_{6}C \rightarrow {}^{0}_{1}e + {}^{A}_{Z}X$$

↑
Positron

The key to solving this problem is to recognize that both A and Z must be conserved. That is, we can find the identity of $^{A}_{Z}X$ by recognizing that the sums of the Z and A values must be the same on both sides of the equation. Thus, for X, Z must be 5 because $Z + 1 = 6$. A must be 11 because $11 + 0 = 11$. Therefore, $^{A}_{Z}X$ is $^{11}_{5}B$. (The fact that Z is 5 tells us that the nuclide is boron. See the periodic table on the inside front cover of the book.) So the balanced equation is $^{11}_{6}C \rightarrow {}^{0}_{1}e + {}^{11}_{5}B$.

CHECK:

Left Side		Right Side
$Z = 6$	$\rightarrow$	$Z = 5 + 1 = 6$
$A = 11$		$A = 11 + 0 = 11$

b. Knowing that a β particle is represented by $^{0}_{-1}e$, we can write

$$^{214}_{83}Bi \rightarrow {}^{0}_{-1}e + {}^{A}_{Z}X$$

where $Z - 1 = 83$ and $A + 0 = 214$. This means that $Z = 84$ and $A = 214$. We can now write

$$^{214}_{83}\text{Bi} \rightarrow {}^{0}_{-1}\text{e} + {}^{214}_{84}\text{X}$$

Using the periodic table, we find that $Z = 84$ for the element polonium, so $^{214}_{84}\text{X}$ must be $^{214}_{84}\text{Po}$.

CHECK:	Left Side		Right Side
	$Z = 83$	$\rightarrow$	$Z = 84 - 1 = 83$
	$A = 214$		$A = 214 + 0 = 214$

c. Because an α particle is represented by $^{4}_{2}\text{He}$, we can write

$$^{237}_{93}\text{Np} \rightarrow {}^{4}_{2}\text{He} + {}^{A}_{Z}\text{X}$$

where $A + 4 = 237$ or $A = 237 - 4 = 233$, and $Z + 2 = 93$ or $Z = 93 - 2 = 91$. Thus $A = 233$, $Z = 91$, and the balanced equation must be

$$^{237}_{93}\text{Np} \rightarrow {}^{4}_{2}\text{He} + {}^{233}_{91}\text{Pa}$$

CHECK:	Left Side		Right Side
	$Z = 93$	$\rightarrow$	$Z = 91 + 2 = 93$
	$A = 237$		$A = 233 + 4 = 237$

✔ Self-Check Exercise 18.1

The decay series for $^{238}_{92}\text{U}$ is represented in Figure 18.1. Write the balanced nuclear equation for each of the following radioactive decays.

a. Alpha-particle production by $^{226}_{88}\text{Ra}$

b. Beta-particle production by $^{214}_{82}\text{Pb}$

See Problems 18.25 through 18.28. ■

Example 18.2 Writing Nuclear Equations, II

In each of the following nuclear reactions, supply the missing particle.

a. $^{195}_{79}\text{Au} + ? \rightarrow {}^{195}_{78}\text{Pt}$

b. $^{38}_{19}\text{K} \rightarrow {}^{38}_{18}\text{Ar} + ?$

Solution

a. A does not change and Z for Pt is 1 lower than Z for Au, so the missing particle must be an electron.

$$^{195}_{79}\text{Au} + {}^{0}_{-1}\text{e} \rightarrow {}^{195}_{78}\text{Pt}$$

CHECK:	Left Side		Right Side
	$Z = 79 - 1 = 78$	$\rightarrow$	$Z = 78$
	$A = 195 + 0 = 195$		$A = 195$

This is an example of electron capture.

b. For Z and A to be conserved, the missing particle must be a positron.

$$^{38}_{19}\text{K} \rightarrow {}^{38}_{18}\text{Ar} + {}^{0}_{1}\text{e}$$

CHECK:	Left Side		Right Side
	$Z = 19$	$\rightarrow$	$Z = 18 + 1 = 19$
	$A = 38$		$A = 38 + 0 = 38$

Potassium-38 decays by positron production.

✔ Self-Check Exercise 18.2

Supply the missing species in each of the following nuclear equations.

a. ${}^{222}_{86}\text{Rn} \rightarrow {}^{218}_{84}\text{Po} + ?$

b. ${}^{15}_{8}\text{O} \rightarrow ? + {}^{0}_{1}\text{e}$

See Problems 18.21 through 18.24. ■

18.2 Nuclear Transformations

AIM: To learn how one element may be changed into another by particle bombardment.

Irene Curie and Frederick Joliot.

In 1919 Lord Rutherford observed the first **nuclear transformation,** *the change of one element into another.* He found that bombarding ${}^{14}_{7}\text{N}$ with α particles produced the nuclide ${}^{17}_{8}\text{O}$:

$${}^{14}_{7}\text{N} + {}^{4}_{2}\text{He} \rightarrow {}^{17}_{8}\text{O} + {}^{1}_{1}\text{H}$$

with a proton (${}^{1}_{1}\text{H}$) as another product. Fourteen years later, Irene Curie and her husband Frederick Joliot observed a similar transformation from aluminum to phosphorus:

$${}^{27}_{13}\text{Al} + {}^{4}_{2}\text{He} \rightarrow {}^{30}_{15}\text{P} + {}^{1}_{0}\text{n}$$

where ${}^{1}_{0}\text{n}$ represents a neutron that is produced in the process.

Notice that in both these cases the bombarding particle is a helium nucleus (an α particle). Other small nuclei, such as ${}^{12}_{6}\text{C}$ and ${}^{15}_{7}\text{N}$, also can be used to bombard heavier nuclei and cause transformations. However, because these positive bombarding ions are repelled by the positive charge of the target nucleus, the bombarding particle must be moving at a very high speed to penetrate the target. These high speeds are achieved in various types of *particle accelerators.*

Neutrons are also employed as bombarding particles to effect nuclear transformations. However, because neutrons are uncharged (and thus not repelled by a target nucleus), they are readily absorbed by many nuclei, producing new nuclides. The most common source of neutrons for this purpose is a fission reactor (see Section 18.8).

By using neutron and positive-ion bombardment, scientists have been able to extend the periodic table—that is, to produce chemical elements that are not present naturally. Prior to 1940, the heaviest known element was uranium ($Z = 92$), but in 1940, neptunium ($Z = 93$) was produced by neutron bombardment of ${}^{238}_{92}\text{U}$. The process initially gives ${}^{239}_{92}\text{U}$, which decays to ${}^{239}_{93}\text{Np}$ by β-particle production:

$${}^{238}_{92}\text{U} + {}^{1}_{0}\text{n} \rightarrow {}^{239}_{92}\text{U} \rightarrow {}^{239}_{93}\text{Np} + {}^{0}_{-1}\text{e}$$

Table 18.2 Syntheses of Some of the Transuranium Elements

Neutron Bombardment	neptunium ($Z = 93$)	${}^{238}_{92}\text{U} + {}^{1}_{0}\text{n} \rightarrow {}^{239}_{92}\text{U} \rightarrow {}^{239}_{93}\text{Np} + {}^{0}_{-1}\text{e}$
	americium ($Z = 95$)	${}^{239}_{94}\text{Pu} + 2\,{}^{1}_{0}\text{n} \rightarrow {}^{241}_{94}\text{Pu} \rightarrow {}^{241}_{95}\text{Am} + {}^{0}_{-1}\text{e}$
Positive-Ion Bombardment	curium ($Z = 96$)	${}^{239}_{94}\text{Pu} + {}^{4}_{2}\text{He} \rightarrow {}^{242}_{96}\text{Cm} + {}^{1}_{0}\text{n}$
	californium ($Z = 98$)	${}^{242}_{96}\text{Cm} + {}^{4}_{2}\text{He} \rightarrow {}^{245}_{98}\text{Cf} + {}^{1}_{0}\text{n}$ or
		${}^{238}_{92}\text{U} + {}^{12}_{6}\text{C} \rightarrow {}^{246}_{98}\text{Cf} + 4\,{}^{1}_{0}\text{n}$
	rutherfordium ($Z = 104$)	${}^{249}_{98}\text{Cf} + {}^{12}_{6}\text{C} \rightarrow {}^{257}_{104}\text{Rf} + 4\,{}^{1}_{0}\text{n}$
	dubnium ($Z = 105$)	${}^{249}_{98}\text{Cf} + {}^{15}_{7}\text{N} \rightarrow {}^{260}_{105}\text{Db} + 4\,{}^{1}_{0}\text{n}$
	seaborgium ($Z = 106$)	${}^{249}_{98}\text{Cf} + {}^{18}_{8}\text{O} \rightarrow {}^{263}_{106}\text{Sg} + 4\,{}^{1}_{0}\text{n}$

In the years since 1940, the elements with atomic numbers 93 through 112, called the **transuranium elements,** have been synthesized. In addition, preliminary reports in early 1999 indicated that element 114 also has been produced. Table 18.2 gives some examples of these processes.

Detection of Radioactivity and the Concept of Half-life

AIMS: To learn about radiation detection instruments. To understand half-life.

Geiger counters are commonly called *survey meters.*

The most familiar instrument for measuring radioactivity levels is the **Geiger–Müller counter,** or **Geiger counter** (Figure 18.2). High-energy particles from radioactive decay produce ions when they travel through matter. The probe of the Geiger counter contains argon gas. The argon atoms have no charge, but they can be ionized by a rapidly moving particle.

$$\text{Ar}(g) \xrightarrow[\text{particle}]{\text{High-energy}} \text{Ar}^{+}(g) + \text{e}^{-}$$

That is, the fast-moving particle "knocks" electrons off some of the argon atoms. Although a sample of uncharged argon atoms does not conduct a current, the ions and electrons formed by the high-energy particle allow a current to flow momentarily, so a "pulse" of current flows every time a particle enters the probe. The Geiger counter detects each pulse of current, and these events are counted.

A **scintillation counter** is another instrument often employed to detect radioactivity. This device uses a substance, such as sodium iodide, that

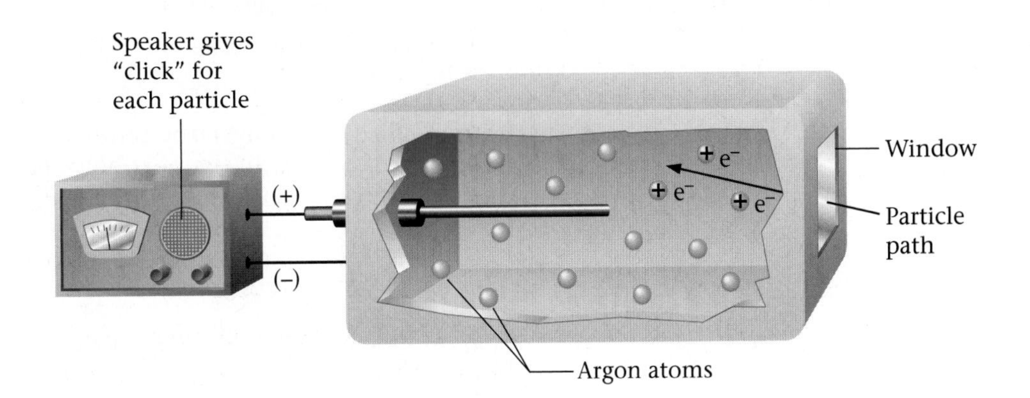

Figure 18.2
A schematic representation of a Geiger–Müller counter. The high-speed particle knocks electrons off argon atoms to form ions,

$$\text{Ar} \xrightarrow{\text{Particle}} \text{Ar}^{+} + \text{e}^{-}$$

and a pulse of current flows.

gives off light when it is struck by a high-energy particle. A detector senses the flashes of light and thus counts the decay events.

One important characteristic of a given type of radioactive nuclide is its half-life. The **half-life** is the *time required for half the original sample of nuclei to decay*. For example, if a certain radioactive sample contains 1000 nuclei at a given time and 500 nuclei (half of the original number) 7.5 days later, this radioactive nuclide has a half-life of 7.5 days.

A given type of radioactive nuclide always has the same half-life. However, the various radioactive nuclides have half-lives that cover a tremendous range. For example, $^{234}_{91}Pa$, protactinium-234, has a half-life of 1.2 minutes, and $^{238}_{92}U$, uranium-238, has a half-life of 4.5×10^9 (4.5 *billion*) years. This means that a sample containing 100 million $^{234}_{91}Pa$ nuclei will have only 50 million $^{234}_{91}Pa$ nuclei in it (half of 100 million) after 1.2 minutes have passed. In another 1.2 minutes, the number of nuclei will decrease to half of 50 million, or 25 million nuclei.

This means that a sample of $^{234}_{91}Pa$ with 100 million nuclei will show 50 million decay events (50 million $^{234}_{91}Pa$ nuclei will decay) over a time of 1.2 minutes. By contrast, a sample containing 100 million $^{238}_{92}U$ nuclei will undergo 50 million decay events over 4.5 billion years. Therefore, $^{234}_{91}Pa$ shows much greater activity than $^{238}_{92}U$. We sometimes say that $^{234}_{91}Pa$ is "hotter" than $^{238}_{92}U$.

Thus, at a given moment, a radioactive nucleus with a short half-life is much more likely to decay than one with a long half-life.

Example 18.3 Understanding Half-life

Table 18.3 The Half-lives for Some of the Radioactive Nuclides of Radium

Nuclide	Half-life
$^{223}_{88}Ra$	12 days
$^{224}_{88}Ra$	3.6 days
$^{225}_{88}Ra$	15 days
$^{226}_{88}Ra$	1600 years
$^{228}_{88}Ra$	6.7 years

Table 18.3 lists various radioactive nuclides of radium.

a. Order these nuclides in terms of activity (from most decays per day to least).

b. How long will it take for a sample containing 1.00 mol of $^{223}_{88}Ra$ to reach a point where it contains only 0.25 mol of $^{223}_{88}Ra$?

Solution

a. The shortest half-life indicates the greatest activity (the most decays over a given period of time). Therefore, the order is

Most activity (shortest half-life) — Least activity (longest half-life)

$$^{224}_{88}Ra > ^{223}_{88}Ra > ^{225}_{88}Ra > ^{228}_{88}Ra > ^{226}_{88}Ra$$

3.6 days, 12 days, 15 days, 6.7 years, 1600 years

b. In one half-life (12 days), the sample will decay from 1.00 mol of $^{223}_{88}Ra$ to 0.50 mol of $^{223}_{88}Ra$. In the next half-life (another 12 days), it will decay from 0.50 mol of $^{223}_{88}Ra$ to 0.25 mol of $^{223}_{88}Ra$.

$$1.00 \text{ mol } ^{223}_{88}Ra \xrightarrow{12 \text{ days}} 0.50 \text{ mol } ^{223}_{88}Ra \xrightarrow{12 \text{ days}} 0.25 \text{ mol } ^{223}_{88}Ra$$

Therefore, it will take 24 days (two half-lives) for the sample to change from 1.00 mol of $^{223}_{88}Ra$ to 0.25 mol of $^{223}_{88}Ra$.

A watch dial with radium paint.

✔ Self-Check Exercise 18.3

Watches with numerals that "glow in the dark" formerly were made by including radioactive radium in the paint used to letter the watch faces. Assume that to make the numeral 3 on a given watch, a sample of paint containing 8.0×10^{-7} mol of $^{228}_{88}Ra$ was used. This watch was then put in a drawer and forgotten. Many years later someone finds the watch and wishes to know when it was made. Analyzing the paint, this person finds 1.0×10^{-7} mol of $^{228}_{88}Ra$ in the numeral 3. How much time elapsed between the making of the watch and the finding of the watch?

HINT: Use the half-life of $^{228}_{88}Ra$ from Table 18.3.

See Problems 18.37 through 18.42. ■

18.4 Dating by Radioactivity

AIM: To learn how objects can be dated by radioactivity.

Archaeologists, geologists, and others involved in reconstructing the ancient history of the earth rely heavily on the half-lives of radioactive nuclei to provide accurate dates for artifacts and rocks. A method for dating ancient articles made from wood or cloth is **radiocarbon dating,** or **carbon-14 dating,** a technique originated in the 1940s by Willard Libby, an American chemist who received the Nobel Prize for his efforts.

Radiocarbon dating is based on the radioactivity of $^{14}_{6}C$, which decays by β-particle production.

$$^{14}_{6}C \rightarrow {}^{0}_{-1}e + {}^{14}_{7}N$$

Carbon-14 is continuously produced in the atmosphere when high-energy neutrons from space collide with nitrogen-14.

$$^{14}_{7}N + {}^{1}_{0}n \rightarrow {}^{14}_{6}C + {}^{1}_{1}H$$

Brigham Young researcher Scott Woodward taking a bone sample for carbon-14 dating at an archeological site in Egypt.

Just as carbon-14 is produced continuously by this process, it decomposes continuously through β-particle production. Over the years, these two opposing processes have come into balance, causing the amount of $^{14}_{6}C$ present in the atmosphere to remain approximately constant.

Carbon-14 can be used to date wood and cloth artifacts because the $^{14}_{6}C$, along with the other carbon isotopes in the atmosphere, reacts with oxygen to form carbon dioxide. A living plant consumes this carbon dioxide in the photosynthesis process and incorporates the carbon, including $^{14}_{6}C$, into its molecules. As long as the plant lives, the $^{14}_{6}C$ content in its molecules remains the same as in the atmosphere because of the plant's continuous uptake of carbon. However, as soon as a tree is cut to make a wooden bowl or a flax plant is harvested to make linen, it stops taking in carbon. There is no longer a source of $^{14}_{6}C$ to replace that lost to radioactive decay, so the material's $^{14}_{6}C$ content begins to decrease.

Because the half-life of $^{14}_{6}C$ is known to be 5730 years, a wooden bowl found in an archaeological dig that shows a $^{14}_{6}C$ content of half that found in currently living trees is approximately 5730 years old. That is, because half the $^{14}_{6}C$ present when the tree was cut has disappeared, the tree must have been cut one half-life of $^{14}_{6}C$ ago.

CHEMISTRY IN FOCUS

Dating Diamonds

While connoisseurs of gems value the clearest possible diamonds, geologists learn the most from impure diamonds. Diamonds are formed in the earth's crust at depths of about 200 kilometers, where the high pressures and temperatures favor the most dense form of carbon. As the diamond is formed, impurities are sometimes trapped, and these can be used to determine the diamond's date of "birth." One valuable dating impurity is $^{238}_{92}U$, which is radioactive and decays in a series of steps to $^{206}_{82}Pb$, which is stable (nonradioactive). Because the rate at which $^{238}_{92}U$ decays is known, determining how much $^{238}_{92}U$ has been converted to $^{206}_{82}Pb$ tells scientists the amount of time that has elapsed since the $^{238}_{92}U$ was trapped in the diamond as it was formed.

Using these dating techniques, Peter D. Kinney of Curtin University of Technology in Perth, Australia, and Henry O. A. Meyer of Purdue University in West Lafayette, Indiana, recently have identified the youngest diamond ever found. Discovered in Mbuji Mayi, Zaire, the diamond is 628 million years old, far younger than all previously dated diamonds, which range from 2.4 to 3.2 billion years old.

The great age of all previously dated diamonds had caused some geologists to speculate that all diamond formation occurred billions of years ago. However, this "youngster" suggests that diamonds have formed throughout geologic time and are probably being formed right now in the earth's crust. We won't see these diamonds for a long time, because diamonds typically remain deeply buried in the earth's crust for millions of years until they are brought to the surface by volcanic blasts called kimberlite eruptions.

It's good to know that eons from now there will be plenty of diamonds to mark the engagements of future couples.

The Hope Diamond.

18.5 Medical Applications of Radioactivity

AIM: To discuss the use of radiotracers in medicine.

Nuclides used as radiotracers have short half-lives so that they disappear rapidly from the body.

Although we owe the rapid advances of the medical sciences in recent decades to many causes, one of the most important has been the discovery and use of **radiotracers**—radioactive nuclides that can be introduced into organisms in food or drugs and subsequently *traced* by monitoring their radioactivity. For example, the incorporation of nuclides such as $^{14}_{6}C$ and $^{32}_{15}P$ into nutrients has yielded important information about how these nutrients are used to provide energy for the body.

Iodine-131 has proved very useful in the diagnosis and treatment of illnesses of the thyroid gland. Patients drink a solution containing a small amount of NaI that includes ^{131}I, and the uptake of the iodine by the thyroid gland is monitored with a scanner (Figure 18.3).

Thallium-201 can be used to assess the damage to the heart muscle in a person who has suffered a heart attack because thallium becomes concentrated in healthy muscle tissue. Technetium-99, which is also taken up by normal heart tissue, is used for damage assessment in a similar way.

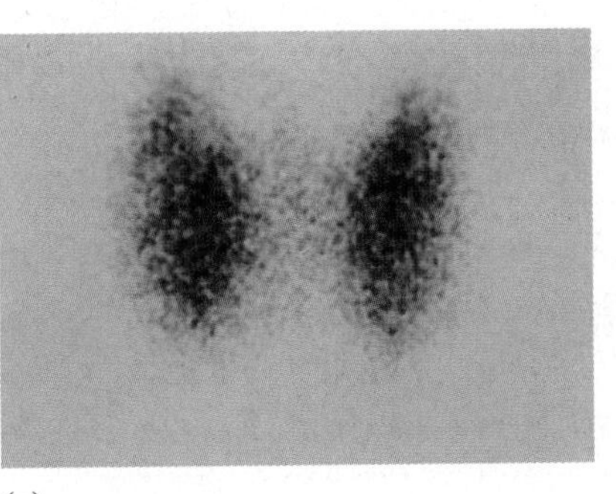

(a)

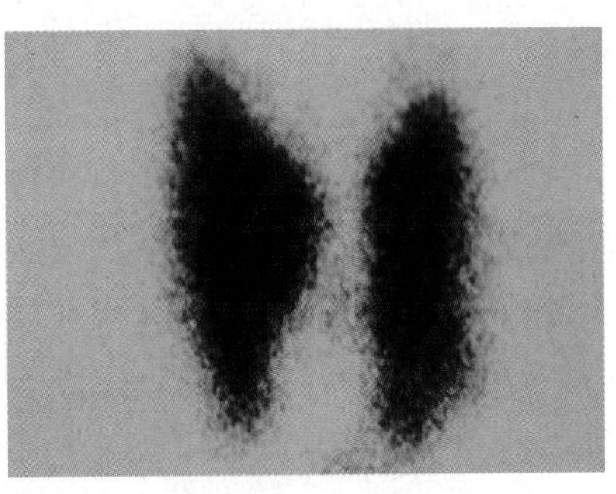

(b)

Figure 18.3
After consumption of $Na^{131}I$, the patient's thyroid is scanned for radioactivity levels to determine the efficiency of iodine absorption. (a) Scan of radioactive iodine in a normal thyroid. (b) Scan of an enlarged thyroid.

Table 18.4 Some Radioactive Nuclides, Their Half-lives, and Their Medical Applications as Radiotracers*

Nuclide	Half-life	Area of the Body Studied
^{131}I	8.1 days	thyroid
^{59}Fe	45.1 days	red blood cells
^{99}Mo	67 hours	metabolism
^{32}P	14.3 days	eyes, liver, tumors
^{51}Cr	27.8 days	red blood cells
^{87}Sr	2.8 hours	bones
^{99}Tc	6.0 hours	heart, bones, liver, lungs
^{133}Xe	5.3 days	lungs
^{24}Na	14.8 hours	circulatory system

*Z is sometimes not written when listing nuclides.

Radiotracers provide sensitive and nonsurgical methods for learning about biologic systems, for detecting disease, and for monitoring the action and effectiveness of drugs. Some useful radiotracers are listed in Table 18.4.

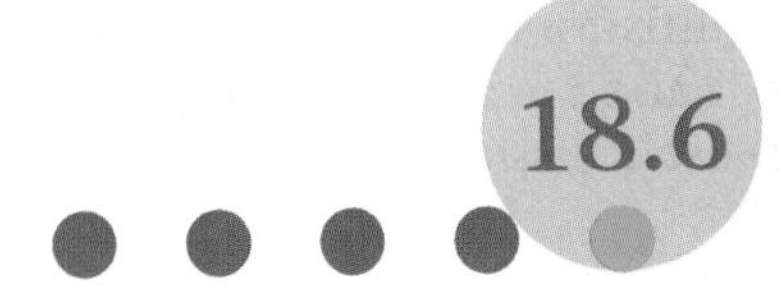

18.6 Nuclear Energy

AIM: To introduce fusion and fission as producers of nuclear energy.

The protons and the neutrons in atomic nuclei are bound together with forces that are much greater than the forces that bind atoms together to form molecules. In fact, the energies associated with nuclear processes are more than a million times those associated with chemical reactions. This potentially makes the nucleus a very attractive source of energy.

Because medium-sized nuclei contain the strongest binding forces ($^{56}_{26}Fe$ has the strongest binding forces of all), there are two types of nuclear processes that produce energy:

1. Combining two light nuclei to form a heavier nucleus. This process is called **fusion.**
2. Splitting a heavy nucleus into two nuclei with smaller mass numbers. This process is called **fission.**

As we will see in the next several sections, these two processes can supply amazing quantities of energy with relatively small masses of materials consumed.

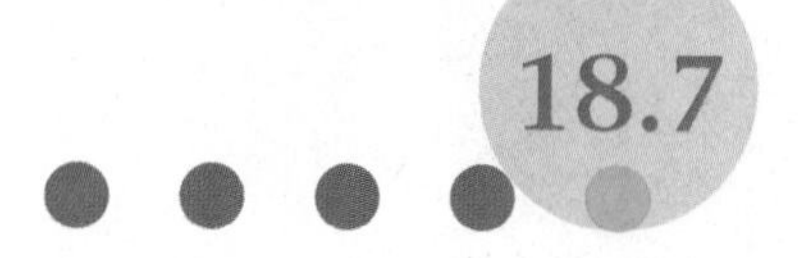

Nuclear Fission

AIM: To learn about nuclear fission.

Nuclear fission was discovered in the late 1930s when $^{235}_{92}U$ nuclides bombarded with neutrons were observed to split into two lighter elements.

$$^{1}_{0}n + ^{235}_{92}U \rightarrow ^{141}_{56}Ba + ^{92}_{36}Kr + 3\,^{1}_{0}n$$

This process, shown schematically in Figure 18.4, releases 2.1×10^{13} joules of energy per mole of $^{235}_{92}U$. Compared with what we get from typical fuels, this is a huge amount of energy. For example, the fission of 1 mol of $^{235}_{92}U$ produces about *26 million times* as much energy as the combustion of 1 mol of methane.

The process shown in Figure 18.4 is only one of the many fission reactions that $^{235}_{92}U$ can undergo. In fact, over 200 different isotopes of 35 different elements have been observed among the fission products of $^{235}_{92}U$.

In addition to the product nuclides, neutrons are produced in the fission reactions of $^{235}_{92}U$. As these neutrons fly through the solid sample of uranium, they may collide with other $^{235}_{92}U$ nuclei, producing additional fission events. Each of these fission events produces more neutrons that can, in turn, produce the fission of more $^{235}_{92}U$ nuclei. Because each fission event produces neutrons, the process can be self-sustaining. We call it a **chain reaction** (Figure 18.5). In order for the fission process to be self-sustaining, at least one neutron from each fission event must go on to split another nucleus. If, on average, *less than one* neutron causes another fission event, the process dies out. If *exactly one* neutron from each fission event causes another fission event, the process sustains itself at the same level and is said to be *critical.* If *more than one* neutron from each fission event causes another fission event, the process rapidly escalates and the heat buildup causes a violent explosion.

To achieve the critical state, a certain mass of fissionable material, called the **critical mass,** is needed. If the sample is too small, too many neutrons escape before they have a chance to cause a fission event, and the process stops.

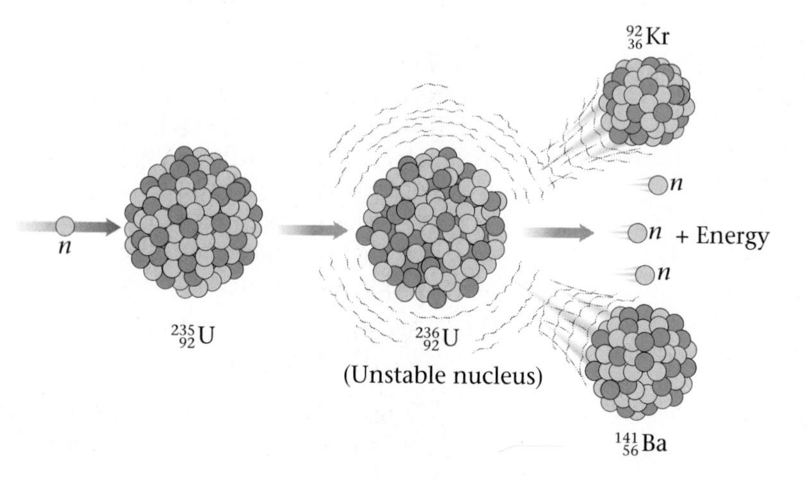

Figure 18.4
Upon capturing a neutron, the $^{235}_{92}U$ nucleus undergoes fission to produce two lighter nuclides, more neutrons (typically three), and a large amount of energy.

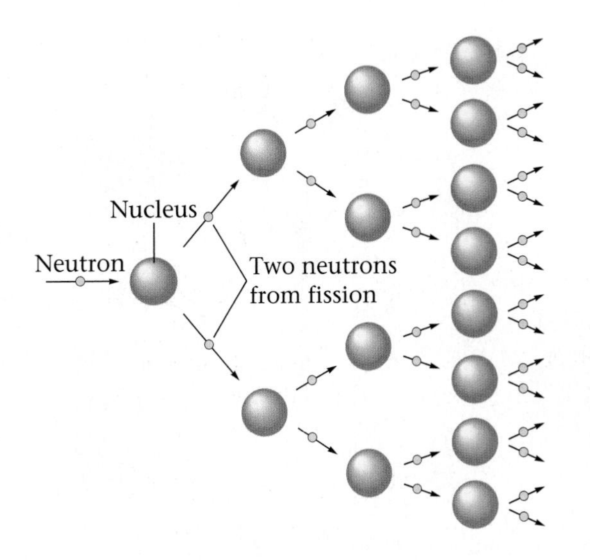

Figure 18.5
Representation of a fission process in which each event produces two neutrons that can go on to split other nuclei, leading to a self-sustaining chain reaction.

During World War II, the United States carried out an intense research effort called the Manhattan Project to build a bomb based on the principles of nuclear fission. This program produced the fission bomb, which was used with devastating effect on the cities of Hiroshima and Nagasaki in 1945. Basically, a fission bomb operates by suddenly combining two subcritical masses, which results in rapidly escalating fission events that produce an explosion of incredible intensity.

18.8 Nuclear Reactors

AIM: To understand how a nuclear reactor works.

Natural uranium consists mostly of $^{238}_{92}U$.

Because of the tremendous energies involved, fission has been developed as an energy source to produce electricity in reactors where controlled fission can occur. The resulting energy is used to heat water to produce steam that runs turbine generators, in much the same way that a coal-burning power plant generates energy by heating water to produce steam. A schematic diagram of a nuclear power plant is shown in Figure 18.6.

In the reactor core (Figure 18.7), uranium that has been enriched to approximately 3% $^{235}_{92}U$ (natural uranium contains only 0.7% $^{235}_{92}U$) is housed in metal cylinders. A *moderator* surrounding the cylinders slows the neutrons down so that the uranium fuel can capture them more efficiently. *Control rods,* composed of substances (such as cadmium) that absorb neutrons, are used to regulate the power level of the reactor. The reactor is designed so that if a malfunction occurs, the control rods are automatically inserted into the core to absorb neutrons and stop the reaction. A liquid (usually water) is circulated through the core to extract the heat generated by the energy of fission. This heat energy is then used to change water to steam, which runs turbines that in turn run electrical generators.

The core of a nuclear power plant.

Although the concentration of $^{235}_{92}U$ in the fuel elements is not great enough to allow an explosion such as that which occurs in a fission bomb, a failure of the cooling system can lead to temperatures high enough to melt the reactor core. This means that the building housing the core must be designed to contain the core even in the event of such a "meltdown." A great deal of controversy now exists about the efficiency of the safety systems in nuclear power plants. Accidents such as the one at the Three

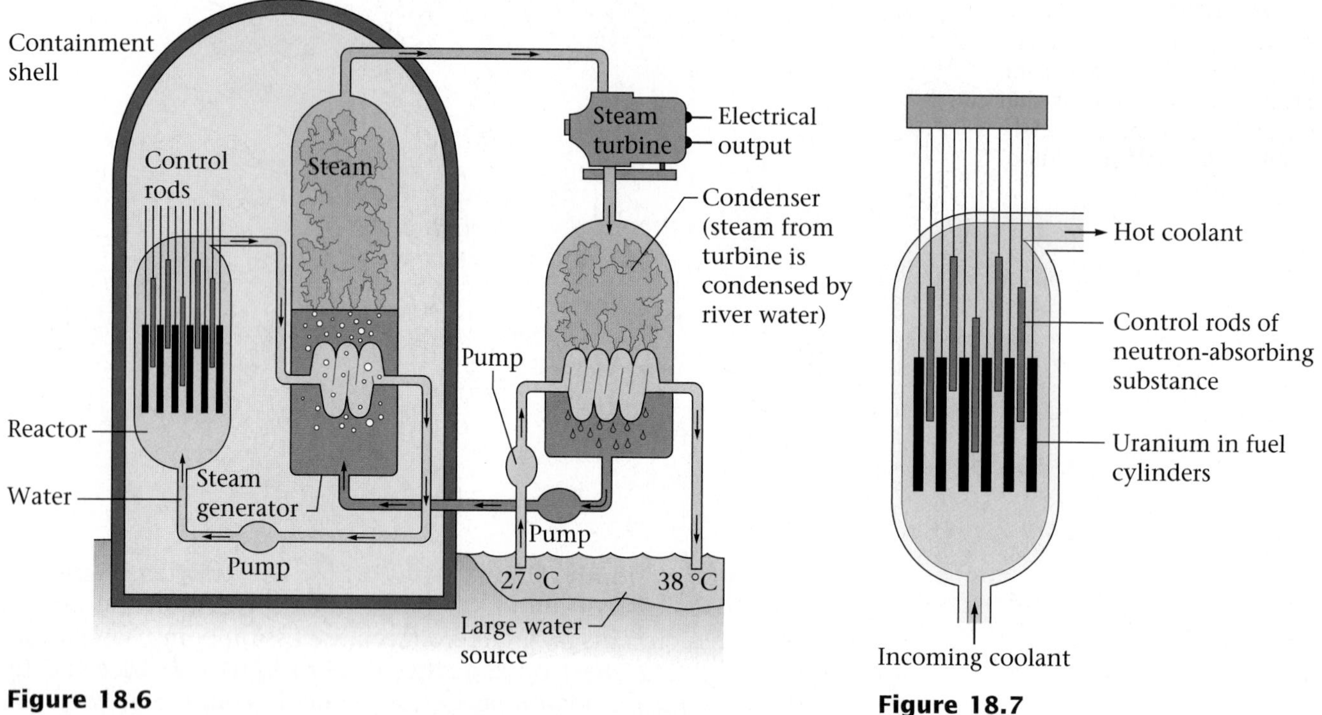

Figure 18.6
A schematic diagram of a nuclear power plant. The energy from the fission process is used to boil water, producing steam for use in a turbine-driven generator. Cooling water from a lake or river is used to condense the steam after it leaves the turbine.

Figure 18.7
A schematic of a reactor core.

Mile Island facility in Pennsylvania in 1979 and the one at Chernobyl in the Soviet Union in 1986 have led many people to question the wisdom of continuing to build fission-based power plants.

Breeder Reactors

One potential problem facing the nuclear power industry is the limited supply of $^{235}_{92}U$. Some scientists believe that we have nearly depleted the uranium deposits that are rich enough in $^{235}_{92}U$ to make the production of fissionable fuel economically feasible. Because of this possibility, reactors have been developed in which fissionable fuel is actually produced while the reactor runs. In these **breeder reactors,** the major component of natural uranium, nonfissionable $^{238}_{92}U$, is changed to fissionable $^{239}_{94}Pu$. The reaction involves absorption of a neutron, followed by production of two β particles.

$$^{1}_{0}n + ^{238}_{92}U \rightarrow ^{239}_{92}U$$
$$^{239}_{92}U \rightarrow ^{239}_{93}Np + ^{0}_{-1}e$$
$$^{239}_{93}Np \rightarrow ^{239}_{94}Pu + ^{0}_{-1}e$$

As the reactor runs and $^{235}_{92}U$ is split, some of the excess neutrons are absorbed by $^{238}_{92}U$ to produce $^{239}_{94}Pu$. The $^{239}_{94}Pu$ is then separated out and used to fuel another reactor. Such a reactor thus "breeds" nuclear fuel as it operates.

Although breeder reactors are now used in Europe, the United States is proceeding slowly with their development because much controversy surrounds their use. One problem involves the hazards that arise in handling plutonium, which is very toxic and flames on contact with air.

CHEMISTRY IN FOCUS

Stellar Nucleosynthesis

How did all the matter around us originate? One scientific answer to this question is a theory called *stellar nucleosynthesis*—the formation of nuclei in stars.

Many scientists believe that our universe originated as a cloud of neutrons that became unstable and produced an immense explosion, giving this model its name: *the big bang theory*. The model postulates that after the initial explosion, neutrons decomposed into protons and electrons,

$$ {}_{0}^{1}n \rightarrow {}_{1}^{1}H + {}_{-1}^{0}e $$

which eventually combined to form clouds of hydrogen. Over the eons, gravitational forces caused many of these hydrogen clouds to contract and heat up sufficiently to reach temperatures at which proton fusion began to occur, releasing large quantities of energy. When the tendency to expand in response to the heat from fusion and the tendency to contract in response to the forces of gravity are balanced, a stable young star such as our sun can form.

Eventually, when the star's supply of hydrogen is exhausted, the core of the star again contracts, with further heating, until it reaches temperatures at which fusion of helium nuclei can occur, leading to the formation of ${}_{6}^{12}C$ and ${}_{8}^{16}O$ nuclei. In turn, when the supply of helium nuclei runs out, further contraction and heating occur, until the fusion of heavier nuclei takes place. This process occurs repeatedly, forming heavier and heavier nuclei, until iron nuclei are formed. Because the iron nucleus has the greatest binding forces of all, iron nuclei do not release excess energy when they fuse to form heavier nuclei. This means that there is no further fusion energy to sustain the star, and it cools to a small, dense *white dwarf*.

A portion of the Cygnus Loop supernova remnant taken by the Hubble Space Telescope.

The evolution just described is characteristic of small and medium-sized stars. Much larger stars, however, become unstable at some time during their evolution and undergo a *supernova explosion*. In this explosion, nuclei with medium mass numbers are fused to produce heavy elements. Also in the explosion, light nuclei capture neutrons. These neutron-rich nuclei then produce β particles, increasing their atomic number with each β decay. This eventually leads to nuclei with large atomic numbers. In fact, almost all nuclei beyond iron are thought to originate in supernova explosions. The debris of such an explosion contains a wide variety of elements, and it is thought that this debris might eventually form a solar system such as our own.

18.9 Nuclear Fusion

AIM: To learn about nuclear fusion.

The process of combining two light nuclei—called **nuclear fusion**—produces even more energy per mole than does nuclear fission. In fact, stars produce their energy through nuclear fusion. Our sun, which presently consists of 73% hydrogen, 26% helium, and 1% other elements, gives off vast quantities of energy from the fusion of protons to form helium. One possible scheme for this process is

$$^{1}_{1}H + ^{1}_{1}H \rightarrow ^{2}_{1}H + ^{0}_{1}e + \text{energy}$$
$$^{1}_{1}H + ^{2}_{1}H \rightarrow ^{3}_{2}He + \text{energy}$$
$$^{3}_{2}He + ^{3}_{2}He \rightarrow ^{4}_{2}He + 2\,^{1}_{1}H + \text{energy}$$
$$^{3}_{2}He + ^{1}_{1}H \rightarrow ^{4}_{2}He + ^{0}_{1}e + \text{energy}$$

$^{2}_{1}H$ particles are called *deuterons*.

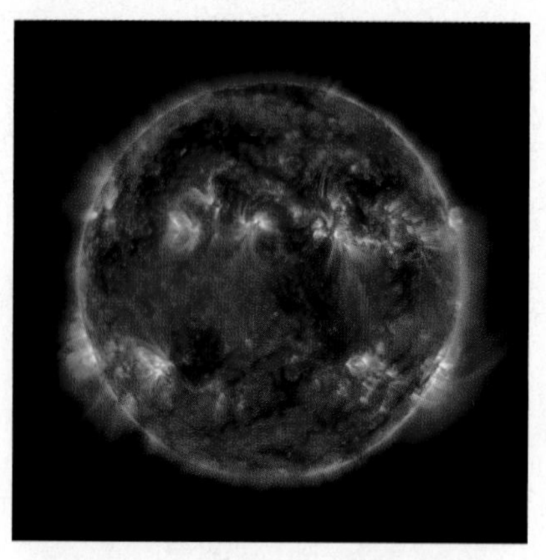

A solar flare erupts from the surface of the sun.

Intense efforts are under way to develop a feasible fusion process because of the ready availability of many light nuclides (deuterium, $^{2}_{1}H$, in seawater, for example) that can serve as fuel in fusion reactors. However, initiating the fusion process is much more difficult than initiating fission. The forces that bind nucleons together to form a nucleus become effective only at *very small* distances (approximately 10^{-13} cm), so for two protons to bind together and thereby release energy, they must get very close together. But protons, because they are identically charged, repel each other. This suggests that to get two protons (or two deuterons) close enough to bind together (the strong nuclear binding force is *not* related to charge), they must be "shot" at each other at speeds high enough to overcome their repulsion from each other. The repulsive forces between two $^{2}_{1}H$ nuclei are so great that temperatures of about 40 million K are thought to be necessary. Only at these temperatures are the nuclei moving fast enough to overcome the repulsions.

Currently, scientists are studying two types of systems to produce the extremely high temperatures required: high-powered lasers and heating by electric currents. At present, many technical problems remain to be solved, and it is not clear whether either method will prove useful.

18.10 Effects of Radiation

AIM: To see how radiation damages human tissue.

Everyone knows that being hit by a train is a catastrophic event. The energy transferred in such a collision is very large. In fact, any source of energy is potentially harmful to organisms. Energy transferred to cells can break chemical bonds and cause malfunctioning of the cell systems. This fact is behind our present concern about maintaining the ozone layer in the earth's upper atmosphere, which screens out high-energy ultraviolet radiation arriving from the sun. Radioactive elements, which are sources of high-energy particles, are also potentially hazardous. However, the effects are usually quite subtle, because even though high-energy particles are involved, the quantity of energy actually deposited in tissues *per decay event* is quite small. The resulting damage is no less real, but the effects may not be apparent for years.

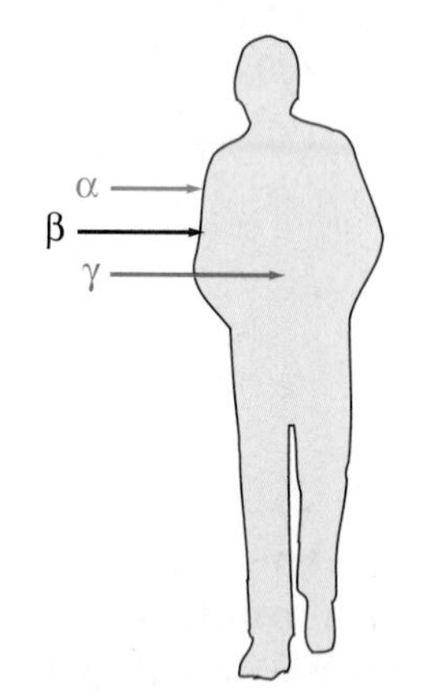

Figure 18.8
Radioactive particles and rays vary greatly in penetrating power. Gamma rays are by far the most penetrating.

Radiation damage to organisms can be classified as somatic or genetic damage. *Somatic damage* is damage to the organism itself, resulting in sickness or death. The effects may appear almost immediately if a massive dose of radiation is received; for smaller doses, damage may appear years later, usually in the form of cancer. *Genetic damage* is damage to the genetic machinery of reproductive cells, creating problems that often afflict the offspring of the organism.

The biologic effects of a particular source of radiation depend on several factors:

1. *The energy of the radiation.* The higher the energy content of the radiation, the more damage it can cause.
2. *The penetrating ability of the radiation.* The particles and rays produced in radioactive processes vary in their ability to penetrate human tissue: γ rays are highly penetrating, β particles can penetrate approximately 1 cm, and α particles are stopped by the skin (Figure 18.8).

CHEMISTRY IN FOCUS

Nuclear Waste Disposal

Our society does not have a very impressive record for safe disposal of industrial wastes. We have polluted our water and air, and some land areas have become virtually uninhabitable because of the improper burial of chemical wastes. As a result, many people are wary about the radioactive wastes from nuclear reactors. The potential threats of cancer and genetic mutations make these materials especially frightening.

Because of its controversial nature, most of the nuclear waste generated over the past 50 years has been placed in temporary storage. However, in 1982 the U.S. Congress passed the Nuclear Waste Policy Act, which established a timetable for choosing and preparing sites for the deep underground disposal of radioactive materials.

The tentative disposal plan calls for incorporation of the spent nuclear fuel into blocks of glass that will be packed in corrosion-resistant metal containers and then buried in a deep, stable rock formation indicated by the rock layers in Figure 18.9.

There are indications that this method will isolate the waste until the radioactivity decays to safe levels. Some reassuring evidence comes from a natural fission "reactor" that has been discovered at Oklo in Gabon, Africa. Spawned about 2 billion years ago when uranium in ore deposits there formed a critical mass, this "reactor" produced fission and fusion products for several thousand years. Although some of these products have migrated away from the site in the intervening 2 billion years, most have stayed in place.

Finally, more than 15 years after the Nuclear Waste Policy Act, it looks like some waste will soon be stored. In 1998 the Waste Isolation Pilot Plant (WIPP) in New Mexico was issued a license by the U.S. Environmental Protection Agency to begin receiving nuclear waste. This facility employs tunnels carved into the salt beds of an ancient ocean. Once a repository room becomes full, the salt will collapse around the waste, encapsulating it forever.

Another waste depository, under Yucca Mountain in Nevada, is being contemplated. For nearly two decades, this area has been studied to determine its suitability for storage of high-level radioactive wastes. At present, it looks to be a long time before this issue is settled.

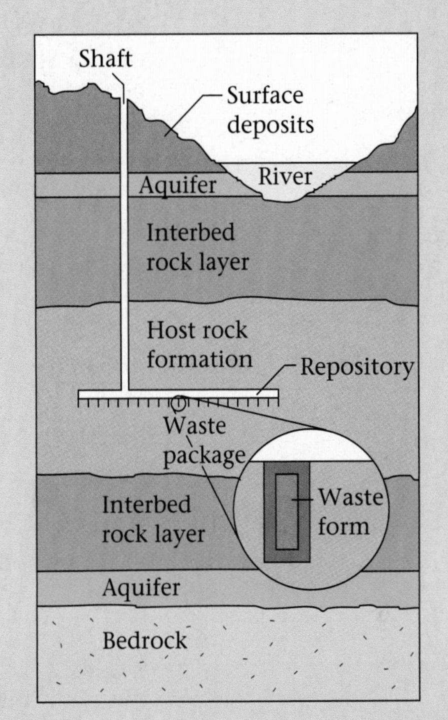

Figure 18.9
A schematic diagram for the tentative plan for deep underground isolation of nuclear waste. The disposal system would consist of a waste package buried in an underground repository. *(Reprinted with permission from* Chemical & Engineering News, *July 18, 1983. Copyright © 1983 American Chemical Society.)*

3. *The ionizing ability of the radiation.* Because ions behave quite differently from neutral molecules, radiation that removes electrons from molecules in living tissues seriously disturbs their functions. The ionizing ability of radiation varies dramatically. For example, γ rays penetrate very deeply but cause only occasional ionization. On the other hand, α particles, although they are not very penetrating, are very effective at causing ionization and produce serious damage. Therefore, the ingestion of a producer of α particles, such as plutonium, is particularly damaging.

4. *The chemical properties of the radiation source.* When a radioactive nuclide is ingested, its capacity to cause damage depends on how long it remains in the body. For example, both $^{85}_{36}Kr$ and $^{90}_{38}Sr$ are β-particle producers. Because krypton, being a noble gas, is chemically inert, it passes through the body quickly and does not have much time to do damage. Strontium, on the other hand, is chemically similar to calcium. It can collect in bones, where it may cause leukemia and bone cancer.

Because of the differences in the behavior of the particles and rays produced by radioactive decay, we have invented a unit called the **rem** that indicates the danger the radiation poses to humans.

Table 18.5 shows the physical effects of short-term exposure to various doses of radiation, and Table 18.6 gives the sources and amounts of the radiation a typical person in the United States is exposed to each year. Note that natural sources contribute about twice as much as human activities do to the total exposure. However, although the nuclear industry contributes only a small percentage of the total exposure, controversy surrounds nuclear power plants because of their *potential* for creating radiation hazards. These hazards arise mainly from two sources: accidents allowing the release of radioactive materials, and improper disposal of the radioactive products in spent fuel elements.

Table 18.5 Effects of Short-Term Exposures to Radiation

Dose (rem)	Clinical Effect
0–25	nondetectable
25–50	temporary decrease in white blood cell counts
100–200	strong decrease in white blood cell counts
500	death of half the exposed population within 30 days after exposure

Table 18.6 Typical Radiation Exposures for a Person Living in the United States (1 millirem = 10^{-3} rem)

Source	Exposure (millirems/year)
cosmic	50
from the earth	47
from building materials	3
in human tissues	21
inhalation of air	5
Total from natural sources	126
X-ray diagnosis	50
radiotherapy X rays, radioisotopes	10
internal diagnosis and therapy	1
nuclear power industry	0.2
luminous watch dials, TV tubes, industrial wastes	2
radioactive fallout	4
Total from human activities	67
Total	193 = 0.193 rems

CHAPTER 18 REVIEW

KEY TERMS

nucleons (neutrons and protons) (p. 551)
atomic number (Z) (p. 551)
mass number (A) (p. 551)
isotopes (p. 551)
nuclide (p. 551)
radioactive nuclide (18.1)
beta (β) particle (18.1)
nuclear equation (18.1)
alpha (α) particle (18.1)
alpha (α)–particle production (18.1)
beta (β)–particle production (18.1)
gamma (γ) ray (18.1)
positron (18.1)
electron capture (18.1)
decay series (18.1)
nuclear transformation (18.2)
transuranium elements (18.2)
Geiger–Müller counter (Geiger counter) (18.3)
scintillation counter (18.3)
half-life (18.3)
radiocarbon dating (carbon-14 dating) (18.4)
radiotracers (18.5)
fusion (18.6)
fission (18.6)
chain reaction (18.7)
critical mass (18.7)
breeder reactor (18.8)
rem (18.10)

SUMMARY

1. Radioactivity is the spontaneous decomposition of a nucleus to form another nucleus and produce one or more particles. We can write a nuclear equation to represent radioactive decay, in which both A (mass number) and Z (atomic number) must be conserved.

2. There are several types of radioactive decay: alpha-particle production, in which an alpha particle (helium nucleus) is produced; beta-particle (or electron) production; the production of gamma rays (high-energy photons of light); and electron capture, in which one of the inner-orbital electrons is captured by the nucleus. Often a series of decays occurs before a radioactive nucleus attains a stable state.

3. The production of new elements by nuclear transformation (the change of one element into another) is carried out by bombarding various nuclei with particles in accelerators. The transuranium elements have been synthesized in this way.

4. The half-life of a radioactive nuclide is the time required for one-half of the original sample to decay. Radiocarbon dating is based on the radioactivity of carbon-14.

5. Radiotracers—radioactive nuclides that can be introduced into organisms in food or drugs and whose pathways can be traced by monitoring their radioactivity—are used diagnostically in medicine.

6. Nuclear fusion is the process of combining two light nuclei to form a heavier, more stable nucleus. Nuclear fission involves the splitting of a heavy nucleus into two (more stable) lighter nuclei. Current nuclear reactors employ controlled fission.

7. Radiation can cause either direct damage to living tissues or damage to reproductive cells that manifests itself in the organism's offspring. The biological effects of radiation depend on the energy of the radiation, the radiation's penetrating ability and ionizing ability, and the chemical properties of the source of the radiation.

QUESTIONS AND PROBLEMS

All even-numbered exercises have answers in the back of this book and solutions in the *Solutions Guide*.

18.1 Radioactive Decay

QUESTIONS

1. What in the atom is most responsible for the atom's chemical properties?

2. How large is a typical atomic nucleus, and how does the size of the nucleus of an atom compare with the overall size of the atom?

3. Which particles in the atom are referred to as *nucleons?*

4. True or false? The *mass number* of a nucleus represents the number of neutrons present in the nucleus.

5. Give an example of what are meant by *isotopes* of an element. Give the nuclear symbols for each of the isotopes in your example. Tell how the isotopes are *similar* and how they *differ*.

6. Using Z to represent the atomic number and A to represent the mass number, give the general symbol for a nuclide of element X. Give also a specific example of the use of such symbolism.

7. Write the charge, mass number, and symbol for a *beta particle*.

8. Write the charge, mass number, and symbol for a *neutron*.

9. When an unstable nucleus produces an alpha particle, by how many units does the atomic number of the nucleus change? Does the atomic number increase or decrease?

10. When an unstable nucleus produces a beta particle, by how many units does the atomic number of the nucleus change? Does the atomic number increase or decrease?

11. What is a *decay series?*

12. What does a *gamma ray* represent? Is a gamma ray a particle? Is there a change in mass or atomic number when a nucleus produces only a gamma ray?

13. What is a *positron?* What are the mass number and charge of a positron? How do the mass number and atomic number of a nucleus change when the nucleus produces a positron?

14. What do we mean when we say a nucleus has undergone an *electron capture* process? What type of electron is captured by the nucleus in this process?

PROBLEMS

15. Neon consists primarily of two isotopes, with mass numbers 20 and 22, with a small amount of a third isotope, with mass number 21. Write the nuclear symbols for each of these isotopes. How many neutrons does each of these isotopes contain?

16. Consider your answers to Problem 15. Given the average atomic molar mass of neon (see the inside cover of this book), which of the neon isotopes predominates in nature? Explain.

17. Naturally occurring magnesium consists primarily of three isotopes, of mass numbers 24, 25, and 26. How many protons does each of these nuclides contain? How many neutrons does each of these nuclides contain? Write nuclear symbols for each of these isotopes.

18. Consider the three isotopes of magnesium discussed in Problem 17. Given that the relative natural abundances of these isotopes are 79%, 10%, and 11%, respectively, without looking at the inside cover of this book, what is the *approximate* atomic molar mass of magnesium? Explain how you made your prediction.

19. Give the nuclear symbol for each of the following.
a. a beta particle
b. an alpha particle
c. a neutron
d. a proton

20. Give the nuclear symbol for each of the following.
a. an electron
b. a positron
c. a gamma ray

21. Complete each of the following nuclear equations by supplying the missing particle.
a. ${}^{210}_{84}\text{Po} \rightarrow {}^{206}_{82}\text{Pb} + ?$
b. ${}^{140}_{57}\text{La} \rightarrow ? + 2{}^{1}_{0}\text{n}$
c. ${}^{235}_{92}\text{U} + ? \rightarrow {}^{143}_{54}\text{Xe} + {}^{90}_{38}\text{Sr} + 3{}^{1}_{0}\text{n}$

22. Complete each of the following nuclear equations by supplying the missing particle.
a. ${}^{196}_{85}\text{At} \rightarrow {}^{4}_{2}\text{He} + ?$
b. ${}^{208}_{84}\text{Po} \rightarrow {}^{4}_{2}\text{He} + ?$
c. ${}^{210}_{86}\text{Rn} \rightarrow {}^{4}_{2}\text{He} + ?$

23. Complete each of the following nuclear equations by supplying the missing particle.
a. ${}^{210}_{89}\text{Ac} \rightarrow {}^{4}_{2}\text{He} + ?$
b. ${}^{131}_{53}\text{I} \rightarrow {}^{131}_{54}\text{Xe} + ?$
c. ${}^{88}_{35}\text{Br} \rightarrow {}^{87}_{35}\text{Br} + ?$

24. Complete each of the following nuclear equations by supplying the missing particle.
a. $? \rightarrow {}^{210}_{84}\text{Po} + {}^{4}_{2}\text{He}$
b. ${}^{40}_{19}\text{K} \rightarrow {}^{40}_{18}\text{Ar} + ?$
c. ${}^{137}_{57}\text{La} + {}^{0}_{-1}\text{e} \rightarrow ?$

25. Each of the following nuclides is known to undergo radioactive decay by production of a beta particle. Write a balanced nuclear equation for each process.
a. ${}^{55}_{23}\text{V}$
b. ${}^{116}_{47}\text{Ag}$
c. ${}^{229}_{89}\text{Ac}$

26. Write a balanced nuclear equation for the decay of each of the following nuclides to produce a beta particle.
a. ${}^{136}_{53}\text{I}$
b. ${}^{133}_{51}\text{Sb}$
c. ${}^{117}_{49}\text{In}$

27. Each of the following nuclides is known to undergo radioactive decay by production of an alpha particle. Write a balanced nuclear equation for each process.
a. ${}^{227}_{90}\text{Th}$
b. ${}^{211}_{83}\text{Bi}$
c. ${}^{244}_{96}\text{Cm}$

28. The isotope ${}^{234}_{92}\text{U}$ is known to undergo a series of five successive alpha-particle emissions before ending up as the isotope ${}^{214}_{82}\text{Pb}$ (at which point the isotope undergoes a series of beta-particle emissions before finally reaching stability). Write nuclear equations for the series of alpha emissions undergone by ${}^{234}_{92}\text{U}$ as it decays to ${}^{214}_{82}\text{Pb}$.

18.2 Nuclear Transformations

QUESTIONS

29. What does a *nuclear transformation* represent? How is a nuclear transformation performed?

30. What is meant by a *nuclear bombardment* process? Give an example of such a process, and describe what the net result of the process is.

31. What are the elements with atomic numbers greater than 92 called? How have these elements been prepared?

32. Write a balanced nuclear equation for the bombardment of ${}^{27}_{13}\text{Al}$ with alpha particles to produce ${}^{30}_{15}\text{P}$ and a neutron.

18.3 Detection of Radioactivity and the Concept of Half-life

QUESTIONS

33. Describe the operation of a Geiger counter. How does a Geiger counter detect radioactive particles?

How does a scintillation counter differ from a Geiger counter?

34. What is the *half-life* of a radioactive nucleus? Does a given type of nucleus always have the same half-life? Do nuclei of different elements have the same half-life?

35. What do we mean when we say that one radioactive nucleus is "hotter" than another? Which element would have more decay events over a given period of time?

36. Consider the isotopes of radium listed in Table 18.3. Which isotope is most stable against decay? Which isotope is "hottest"?

PROBLEMS

37. The following isotopes (listed with their half-lives) have been used in the medical and biologic sciences. Arrange these isotopes in order of their relative decay activities: ^{3}H (12.2 years), ^{24}Na (15 hours), ^{131}I (8 days), ^{60}Co (5.3 years), ^{14}C (5730 years).

38. A list of several important radioactive nuclides is given in Table 18.4. Arrange these nuclides in order of their relative decay activities.

39. Silicon-31 has a half-life of approximately 2.5 hours. If we begin with a sample containing 1000 mg of Si-31, what is the approximate amount remaining after 10 hours?

40. For the first three isotopes of radium listed in Table 18.3, given a starting amount of 1000 mg, *estimate* the approximate amount of each isotope remaining after 1 month.

41. The element krypton has several radioactive isotopes. Below are listed several of these isotopes along with their half-lives. Which of the isotopes is most stable? Which of the isotopes is "hottest"? If we were to begin a half-life experiment with separate 125-μg samples of each isotope, *approximately* how much of each isotope would remain after 24 hours?

Isotope	Half-life
Kr-73	27 s
Kr-74	11.5 min
Kr-76	14.8 h
Kr-81	2.1×10^5 yr

42. Technetium-99 has been used as a radiographic agent in bone scans ($^{99}_{43}Tc$ is absorbed by bones). If $^{99}_{43}Tc$ has a half-life of 6.0 hours, what fraction of an administered dose of 100 μg of $^{99}_{43}Tc$ remains in a patient's body after 2.0 days?

18.4 Dating by Radioactivity

QUESTIONS

43. Describe in general terms how an archeological artifact is dated using carbon-14.

44. How is $^{14}_{6}C$ produced in the atmosphere? Write a balanced equation for this process.

45. Why is it assumed that the amount of $^{14}_{6}C$ in the atmosphere remains constant?

46. Why does an ancient wood or cloth artifact contain less $^{14}_{6}C$ than contemporary or more recently fabricated articles made of similar materials?

18.5 Medical Applications of Radioactivity

QUESTIONS

47. The thyroid gland is interesting in that it is practically the only place in the body where the element iodine is used. How have radiotracers been used to study and treat illnesses of the thyroid gland?

48. List four radioisotopes used in medical diagnosis or treatment, and give their nuclear symbols. Discuss why the isotopes you choose are particularly well suited to their uses.

18.6 Nuclear Energy

QUESTIONS

49. How do the forces that hold an atomic nucleus together compare in strength with the forces between atoms in a molecule?

50. Define the terms *nuclear fission* and *nuclear fusion*. Which process results in the production of a heavier nucleus? Which results in the production of smaller nuclei?

18.7 Nuclear Fission

QUESTIONS

51. How do the energies released by nuclear processes compare in magnitude with the energies of ordinary chemical processes?

52. Write an equation for the fission of $^{235}_{92}U$ by bombardment with neutrons.

53. How is it possible for the fission of $^{235}_{92}U$, once started, to lead to a chain reaction?

54. What does it mean to say that fissionable material possesses a *critical mass?* Can a chain reaction occur when a sample has less than the critical mass?

18.8 Nuclear Reactors

QUESTIONS

55. Describe the purpose of each of the major components of a nuclear reactor (moderator, control rods, containment, cooling liquid, and so on).

56. Can a nuclear explosion take place in a reactor? Is the concentration of fissionable material used in reactors large enough for this?

57. What is a *meltdown,* and how can it occur? Most nuclear reactors use water as the cooling liquid. Is there any danger of a steam explosion if the reactor core becomes overheated?

58. How does a *breeder nuclear reactor* work? Why have breeder nuclear reactors found little favor as yet in the United States?

18.9 Nuclear Fusion

QUESTIONS

59. What is the nuclear *fusion* of small nuclei? How does the energy released by fusion compare in magnitude with that released by fission?

60. What are some reasons why no practical fusion reactor has yet been developed?

61. What type of "fuel" could be used in a nuclear fusion reactor, and why is this desirable?

62. Describe some of the features of the theory of *stellar nucleosynthesis*. How does nuclear fusion play a role in this model?

18.10 Effects of Radiation

QUESTIONS

63. Although the energy transferred per event when a living creature is exposed to radiation is small, why is such exposure dangerous?

64. Explain the difference between *somatic* damage from radiation and *genetic* damage. Which type causes immediate damage to the exposed individual?

65. Describe the relative penetrating powers of alpha, beta, and gamma radiation.

66. Explain why, although gamma rays are far more penetrating than alpha particles, the latter are actually more likely to cause damage to an organism. Which radiation is more effective at causing ionization of biomolecules?

67. How do the *chemical properties* of radioactive nuclei (as opposed to the nuclear decay they undergo) influence the degree of damage they do to an organism?

68. Although nuclear processes offer the potential for an abundant source of energy, no nuclear power plants have been built in the United States for some time. In addition to the fear of a malfunction in such a plant (as happened at the Three Mile Island nuclear plant in Pennsylvania) or the threat of a terrorist attack against such a plant, there is the very practical problem of the regular disposal of the waste material from a nuclear power plant. Discuss some of the problems associated with nuclear waste and some of the proposals that have been put forth for its disposal.

Additional Problems

69. The number of protons contained in a given nucleus is called the ________ .

70. A nucleus that spontaneously decomposes is said to be ________.

71. A(n) ________, when it is produced by a nucleus at high speed, is more commonly called a beta particle.

72. In a nuclear equation, both the atomic number and the ________ number must be conserved.

73. Production of a helium nucleus from a heavy atom is referred to as ________ decay.

74. The net effect of the production of a beta particle is to convert a ________ to a ________.

75. In addition to particles, many radioactive nuclei also produce high-energy ________ rays when they decay.

76. When a nuclide decomposes through a series of steps before reaching stability, the nuclide is said to have gone through a ________ series.

77. When a nuclide produces a beta particle, the atomic number of the resulting new nuclide is one unit ________ than that of the original nuclide.

78. When a nucleus undergoes alpha decay, the ________ of the nucleus decreases by four units.

79. Machines that increase the speed of species used for nuclear bombardment processes are called ________.

80. The elements with atomic numbers of 93 or greater are referred to as the ________ elements.

81. A ________ counter contains argon gas, which is ionized by radiation, making possible the measurement of radioactive decay rates.

82. The time required for half of an original sample of a radioactive nuclide to decay is referred to as the ________ of the nuclide.

83. The radioactive nuclide that has been used in determining the age of historical wooden artifacts is ________.

84. ________ are radioactive substances that physicians introduce into the body to enable them to study the absorption and metabolism of the substance or to analyze the functioning of an organ or gland that can make use of the substance.

85. Combining two small nuclei to form a larger nucleus is referred to as the process of nuclear ________.

86. A self-sustaining nuclear process, in which the bombarding particles needed to produce the fission of further material are themselves produced as the product of the initial fission, is called a ________ reaction.

87. The most common type of nuclear reactor uses the nuclide ________ as its fissionable material.

88. A nuclear reactor that generates additional fissionable fuel (in addition to producing heat for generating electricity) is referred to as a ______ reactor.

89. The decay series from uranium-238 to lead-206 is indicated in Figure 18.1. For each *step* of the process indicated in the figure, specify what type of particle is produced by the particular nucleus involved at that point in the series.

90. The U.S. Department of Energy sells an isotope of the transuranium element californium for approximately \$10 per microgram to qualified researchers. What would a pound of the Cf nuclide cost?

91. Each of the following isotopes has been used medically for the purpose indicated. Suggest reasons why the particular element might have been chosen for this purpose.
   a. cobalt-57, for study of the body's use of vitamin B_{12}
   b. calcium-47, for study of bone metabolism
   c. iron-59, for study of red blood cell function
   d. mercury-197, for brain scans before CAT scan became available

92. The fission of $^{235}_{92}U$ releases 2.1×10^{13} joules per mole of $^{235}_{92}U$. Calculate the energy released per atom and per gram of $^{235}_{92}U$.

93. During the research that led to production of the two atomic bombs used against Japan in World War II, different mechanisms for obtaining a supercritical mass of fissionable material were investigated. In one type of bomb, what is essentially a gun was used to shoot one piece of fissionable material into a cavity containing another piece of fissionable material. In the second type of bomb, the fissionable material was surrounded with a high explosive that, when detonated, compressed the fissionable material into a smaller volume. Discuss what is meant by critical mass, and explain why the ability to achieve a critical mass is essential to sustaining a nuclear reaction.

94. Discuss some of the problems associated with the storage and disposal of the waste products of the nuclear industry.

95. The element zinc in nature consists of five isotopes with higher than 0.5% natural abundances, with mass numbers 64, 66, 67, 68, and 70. Write the nuclear symbol for each of these isotopes. How many protons does each contain? How many neutrons does each contain?

96. Aluminum exists in several isotopic forms, including $^{27}_{13}Al$, $^{28}_{13}Al$, and $^{29}_{13}Al$. Indicate the number of protons and the number of neutrons in each of these isotopes.

97. Complete each of the following nuclear equations by supplying the missing particle.
   a. $^{226}_{88}Ra \rightarrow {}^{222}_{86}Rn + ?$
   b. $^{222}_{86}Rn \rightarrow {}^{218}_{84}Po + ?$
   c. $^{2}_{1}H + {}^{3}_{1}H \rightarrow {}^{4}_{2}He + ?$

98. Complete each of the following nuclear equations by supplying the missing particle.
   a. $^{69}_{30}Zn \rightarrow {}^{69}_{31}Ga + ?$
   b. $^{74}_{35}Br \rightarrow {}^{0}_{+1}\beta + ?$
   c. $^{244}_{94}Pu \rightarrow {}^{4}_{2}He + ?$

99. Write a balanced nuclear equation for the bombardment of $^{14}_{7}N$ with alpha particles to produce $^{17}_{8}O$ and a proton.

100. The only major use of iodine in the human body is in the production of certain hormones by the thyroid gland, and iodine from the diet concentrates in this area of the body. Iodine-131 is used in the diagnosis and treatment of thyroid disease and has a half-life of 8 days. If a patient with thyroid disease consumes a sample of $Na^{131}I$ containing 10 μg of ^{131}I, how long will it take for the amount of ^{131}I to decrease to approximately 1/1000 of the original amount?

101. How have $^{131}_{53}I$ and $^{201}_{81}Tl$ been used in medical diagnosis? Why are these particular nuclides especially well suited for this purpose?

102. What is a *breeder* nuclear reactor? What difficulties with such reactors have led to their not yet being used in the United States for the generation of electricity?

Appendix

Using Your Calculator

In this section we will review how to use your calculator to perform common mathematical operations. This discussion assumes that your calculator uses the algebraic operating system, the system used by most brands.

One very important principle to keep in mind as you use your calculator is that it is not a substitute for your brain. Keep thinking as you do the calculations. Keep asking yourself, "Does the answer make sense?"

Addition, Subtraction, Multiplication, and Division

Performing these operations on a pair of numbers always involves the following steps:

1. Enter the first number, using the numbered keys and the decimal (.) key if needed.
2. Enter the operation to be performed.
3. Enter the second number.
4. Press the "equals" key to display the answer.

For example, the operation

$$15.1 + 0.32$$

is carried out as follows:

Press	***Display***
15.1	15.1
+	15.1
.32	0.32
=	15.42

The answer given by the display is 15.42. If this is the final result of a calculation, you should round it off to the correct number of significant figures (15.4), as discussed in Section 2.5. If this number is to be used in further calculations, use it exactly as it appears on the display. Round off only the final answer in the calculation.

Do the following operations for practice. The detailed procedures are given below.

a. $1.5 + 32.86$ c. 0.33×153

b. $23.5 - 0.41$ d. $\frac{9.3}{0.56}$ or $9.3 \div 0.56$

PROCEDURES

a.

Press	***Display***
1.5	1.5
+	1.5
32.86	32.86
=	34.36
Rounded:	34.4

b.

Press	***Display***
23.5	23.5
−	23.5
.41	0.41
=	23.09
Rounded:	23.1

c.

Press	***Display***
.33	0.33
×	0.33
153	153
=	50.49
Rounded:	50.

d.

Press	***Display***
9.3	9.3
÷	9.3
.56	0.56
=	16.607143
Rounded:	17

Squares, Square Roots, Reciprocals, and Logs

Now we will consider four additional operations that we often need to solve chemistry problems.

The *squaring* of a number is done with a key labeled X^2. The *square root* key is usually labeled $\sqrt{X}$. To take the *reciprocal* of a number, you need the 1/X key. The *logarithm* of a number is determined by using a key labeled log or logX.

To perform these operations, take the following steps:

1. Enter the number.
2. Press the appropriate function key.
3. The answer is displayed automatically.

For example, let's calculate the square root of 235.

Press	***Display***
235	235
$\sqrt{X}$	15.32971
Rounded:	15.3

We can obtain the log of 23 as follows:

Press	***Display***
23	23
log	1.3617278
Rounded:	1.36

Often a key on a calculator serves two functions. In this case, the first function is listed on the key and the second is shown on the calculator just above the key. For example, on some calculators the top row of keys appears as follows:

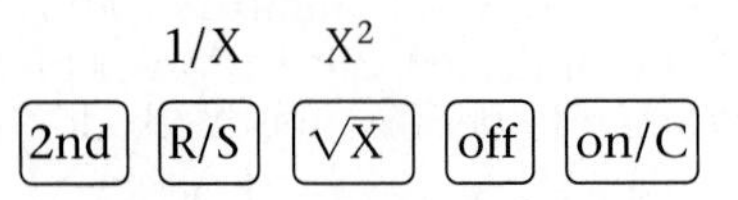

To make the calculator square a number, we must use 2nd and then $\sqrt{X}$; pressing 2nd tells the calculator we want the function that is listed *above* the key. Thus we can obtain the square of 11.56 on this calculator as follows:

Press	***Display***
11.56	11.56
2nd then $\sqrt{X}$	133.6336
Rounded:	133.6

We obtain the reciprocal of 384 (1/384) on this calculator as follows:

Press	*Display*
384	384
2nd then R/S	0.0026042
Rounded:	0.00260

Your calculator may be different. See the user's manual if you are having trouble with these operations.

Chain Calculations

In solving problems you often have to perform a series of calculations—a calculation chain. This is generally quite easy if you key in the chain as you read the numbers and operations in order. For example, to perform the calculation

$$\frac{14.68 + 1.58 - 0.87}{0.0850}$$

you should use the appropriate keys as you read it to yourself:

> 14.68 plus 1.58 equals; minus .87 equals; divided by 0.0850 equals

The details follow.

Press	*Display*
14.68	14.68
+	14.68
1.58	1.58
=	16.26
−	16.26
.87	0.87
=	15.39
÷	15.39
.0850	0.0850
=	181.05882
Rounded:	181

Note that you must press [=] after every operation to keep the calculation "up to date."

For more practice, consider the calculation

$$(0.360)(298) + \frac{(14.8)(16.0)}{1.50}$$

Here you are adding two numbers, but each must be obtained by the indicated calculations. One procedure is to calculate each number first and then add them. The first term is

$$(0.360)(298) = 107.28$$

The second term,

$$\frac{(14.8)(16.0)}{1.50}$$

can be computed easily by reading it to yourself. It "reads"

> 14.8 times 16.0 equals: divided by 1.50 equals:

and is summarized as follows:

Press	*Display*
14.8	14.8
×	14.8
16.0	16.0
=	236.8
÷	236.8
1.50	1.50
=	157.86667

Now we can keep this last number on the calculator and add it to 107.28 from the first calculation.

Press	*Display*
+	157.86667
107.28	107.28
=	265.14667
Rounded:	265

To summarize,

$$(0.360)(298) + \frac{(14.8)(16.0)}{1.50}$$

becomes

$$107.28 + 157.86667$$

and the sum is 265.14667 or, rounded to the correct number of significant figures, 265. There are other ways to do this calculation, but this is the safest way (assuming you are careful).

A common type of chain calculation involves a number of terms multiplied together in the numerator and the denominator, as in

$$\frac{(323)(.0821)(1.46)}{(4.05)(76)}$$

There are many possible sequences by which this calculation can be carried out, but the following seems the most natural.

> 323 times .0821 equals; times 1.46 equals; divided by 4.05 equals; divided by 76 equals

This sequence is summarized as follows:

Press	*Display*
323	323
×	323
.0821	0.0821
=	26.5183
×	26.5183
1.46	1.46
=	38.716718
÷	38.716718
4.05	4.05
=	9.5596835
÷	9.5596835
76	76
=	0.1257853

The answer is 0.1257853, which, when rounded to the correct number of significant figures, is 0.13. Note that when two or more numbers are multiplied in the denominator, you must divide by *each* one.

Here are some additional chain calculations (with solutions) to give you more practice.

a. $15 - (0.750)(243)$

b. $\dfrac{(13.1)(43.5)}{(1.8)(63)}$

c. $\dfrac{(85.8)(0.142)}{(16.46)(18.0)} + \dfrac{(131)(0.0156)}{10.17}$

d. $(18.1)(0.051) - \dfrac{(325)(1.87)}{(14.0)(3.81)} + \dfrac{1.56 - 0.43}{1.33}$

SOLUTIONS

a. $15 - 182 = -167$

b. 5.0

c. $0.0411 + 0.201 = 0.242$

d. $0.92 - 11.4 + 0.850 = -9.6$

In performing chain calculations, take the following steps in the order listed.

1. Perform any additions and subtractions that appear inside parentheses.
2. Complete the multiplications and divisions of individual terms.
3. Add and subtract individual terms as required.

Basic Algebra

In solving chemistry problems you will use, over and over again, relatively few mathematical procedures. In this section we review the few algebraic manipulations that you will need.

Solving an Equation

In the course of solving a chemistry problem, we often construct an algebraic equation that includes the unknown quantity (the thing we want to calculate). An example is

$$(1.5)V = (0.23)(0.08206)(298)$$

We need to "solve this equation for V." That is, we need to isolate V on one side of the equals sign with all the numbers on the other side. How can we do this? The key idea in solving an algebraic equation is that *doing the same thing on both sides of the equals sign* does not change the equality. That is, it is always "legal" to do the same thing to both sides of the equation. Here we want to solve for V, so we must get the number 1.5 on the other side of the equals sign. We can do this by dividing *both sides* by 1.5.

$$\frac{(1.5)V}{1.5} = \frac{(0.23)(0.08206)(298)}{1.5}$$

Now the 1.5 in the denominator on the left cancels the 1.5 in the numerator:

$$\frac{(\cancel{1.5})V}{\cancel{1.5}} = \frac{(0.23)(0.08206)(298)}{1.5}$$

to give

$$V = \frac{(0.23)(0.08206)(298)}{1.5}$$

Using the procedures in "Using Your Calculator" for chain calculations, we can now obtain the value for V with a calculator.

$$V = 3.7$$

Sometimes it is necessary to solve an equation that consists of symbols. For example, consider the equation

$$\frac{P_1V_1}{T_1} = \frac{P_2V_2}{T_2}$$

Let's assume we want to solve for T_2. That is, we want to isolate T_2 on one side of the equation. There are several possible ways to proceed, keeping in mind that we always do the same thing on both sides of the equals sign. First we multiply both sides by T_2.

$$T_2 \times \frac{P_1V_1}{T_1} = \frac{P_2V_2}{\cancel{T_2}} \times \cancel{T_2}$$

This cancels T_2 on the right. Next we multiply both sides by T_1.

$$T_2 \times \frac{P_1V_1}{\cancel{T_1}} \times \cancel{T_1} = P_2V_2T_1$$

This cancels T_1 on the left. Now we divide both sides by P_1V_1.

$$T_2 \times \frac{\cancel{P_1V_1}}{\cancel{P_1V_1}} = \frac{P_2V_2T_1}{P_1V_1}$$

This yields the desired equation,

$$T_2 = \frac{P_2V_2T_1}{P_1V_1}$$

For practice, solve each of the following equations for the variable indicated.

a. $PV = k$; solve for P

b. $1.5x + 6 = 3$; solve for x

c. $PV = nRT$; solve for n

d. $\dfrac{P_1V_1}{T_1} = \dfrac{P_2V_2}{T_2}$; solve for V_2

e. $\dfrac{°F - 32}{°C} = \dfrac{9}{5}$; solve for °C

f. $\dfrac{°F - 32}{°C} = \dfrac{9}{5}$; solve for °F

SOLUTIONS

a. $\dfrac{P\cancel{V}}{\cancel{V}} = \dfrac{k}{V}$

$P = \dfrac{k}{V}$

b. $1.5x + 6 - 6 = 3 - 6$

$1.5x = -3$

$\dfrac{\cancel{1.5}x}{\cancel{1.5}} = \dfrac{-3}{1.5}$

$x = -\dfrac{3}{1.5} = -2$

c. $\dfrac{PV}{RT} = \dfrac{n\cancel{RT}}{\cancel{RT}}$

$\dfrac{PV}{RT} = n$

d. $\dfrac{P_1V_1}{T_1} \times T_2 = \dfrac{P_2V_2}{\cancel{T_2}} \times \cancel{T_2}$

$\dfrac{P_1V_1T_2}{T_1P_2} = \dfrac{P_2V_2}{P_2}$

$\dfrac{P_1V_1T_2}{T_1P_2} = V_2$

e. $\dfrac{°F - 32}{\cancel{°C}} \times \cancel{°C} = \dfrac{9}{5}\,°C$

$\dfrac{5}{9}(°F - 32) = \dfrac{\cancel{5}}{\cancel{9}} \times \dfrac{\cancel{9}}{\cancel{5}}\,°C$

$\dfrac{5}{9}(°F - 32) = °C$

f. $\dfrac{°F - 32}{\cancel{°C}} \times \cancel{°C} = \dfrac{9}{5}\,°C$

$°F - \cancel{32} + \cancel{32} = \dfrac{9}{5}\,°C + 32$

$°F = \dfrac{9}{5}\,°C + 32$

Scientific (Exponential) Notation

The numbers we must work with in scientific measurements are often very large or very small; thus it is convenient to express them using powers of 10. For example, the number 1,300,000 can be expressed as 1.3×10^6, which means multiply 1.3 by 10 six times, or

$$1.3 \times 10^6 = 1.3 \times \underbrace{10 \times 10 \times 10 \times 10 \times 10 \times 10}_{10^6\ =\ 1\text{ million}}$$

A number written in scientific notation always has the form:

A number (between 1 and 10) times
the appropriate power of 10

To represent a large number such as 20,500 in scientific notation, we must move the decimal point in such a way as to achieve a number between 1 and 10 and then multiply the result by a power of 10 to compensate for moving the decimal point. In this case, we must move the decimal point four places to the left.

2 0 5 0 0

4 3 2 1

to give a number between 1 and 10:

2.05

where we retain only the significant figures (the number 20,500 has three significant figures). To compensate for moving the decimal point four places to the left, we must multiply by 10^4. Thus

$$20{,}500 = 2.05 \times 10^4$$

As another example, the number 1985 can be expressed as 1.985×10^3. To end up with the number 1.985, which is between 1 and 10, we had to move the decimal point three places to the left. To compensate for that, we must multiply by 10^3. Some other examples are given in the accompanying list.

Number	***Exponential Notation***
5.6	5.6×10^0 or 5.6×1
39	3.9×10^1
943	9.43×10^2
1126	1.126×10^3

So far, we have considered numbers greater than 1. How do we represent a number such as 0.0034 in exponential notation? First, to achieve a number between 1 and 10, we start with 0.0034 and move the decimal point three places to the right.

0 . 0 0 3 4

1 2 3

This yields 3.4. Then, to compensate for moving the decimal point to the right, we must multiply by a power of 10 with a negative exponent—in this case, 10^{-3}. Thus

$$0.0034 = 3.4 \times 10^{-3}$$

In a similar way, the number 0.00000014 can be written as 1.4×10^{-7}, because going from 0.00000014 to 1.4 requires that we move the decimal point seven places to the right.

Mathematical Operations with Exponentials

We next consider how various mathematical operations are performed using exponential numbers. First we cover the various rules for these operations; then we consider how to perform them on your calculator.

Multiplication and Division

When two numbers expressed in exponential notation are multiplied, the initial numbers are multiplied and the exponents of 10 are *added*.

$$(M \times 10^m)(N \times 10^n) = (MN) \times 10^{m+n}$$

For example (to two significant figures, as required),

$$(3.2 \times 10^4)(2.8 \times 10^3) = 9.0 \times 10^7$$

When the numbers are multiplied, if a result greater than 10 is obtained for the initial number, the decimal point is moved one place to the left and the exponent of 10 is increased by 1.

$$\begin{aligned}(5.8 \times 10^2)(4.3 \times 10^8) &= 24.9 \times 10^{10} \\ &= 2.49 \times 10^{11} \\ &= 2.5 \times 10^{11} \text{ (two significant figures)}\end{aligned}$$

Division of two numbers expressed in exponential notation involves normal division of the initial numbers and *subtraction* of the exponent of the divisor from that of the dividend. For example,

$$\frac{4.8 \times 10^8}{\underbrace{2.1 \times 10^3}_{\text{Divisor}}} = \frac{4.8}{2.1} \times 10^{(8-3)} = 2.3 \times 10^5$$

If the initial number resulting from the division is less than 1, the decimal point is moved one place to the right and the exponent of 10 is decreased by 1. For example,

$$\begin{aligned}\frac{6.4 \times 10^3}{8.3 \times 10^5} = \frac{6.4}{8.3} \times 10^{(3-5)} &= 0.77 \times 10^{-2} \\ &= 7.7 \times 10^{-3}\end{aligned}$$

Addition and Subtraction

In order for us to add or subtract numbers expressed in exponential notation, *the exponents of the numbers must be the same.* For example, to add 1.31×10^5 and 4.2×10^4, we must rewrite one number so that the exponents of both are the same. The number 1.31×10^5 can be written 13.1×10^4: decreasing the exponent by 1 compensates for moving the decimal point one place to the right. Now we can add the numbers.

$$\begin{array}{r} 13.1 \times 10^4 \\ +\ 4.2 \times 10^4 \\ \hline 17.3 \times 10^4 \end{array}$$

In correct exponential notation, the result is expressed as 1.73×10^5.

To perform addition or subtraction with numbers expressed in exponential notation, we add or subtract only the initial numbers. The exponent of the result is the same as the exponents of the numbers being added or subtracted. To subtract 1.8×10^2 from 8.99×10^3, we first convert 1.8×10^2 to 0.18×10^3 so that both numbers have the same exponent. Then we subtract.

$$\begin{array}{r} 8.99 \times 10^3 \\ -0.18 \times 10^3 \\ \hline 8.81 \times 10^3 \end{array}$$

Powers and Roots

When a number expressed in exponential notation is taken to some power, the initial number is taken to the appropriate power and the exponent of 10 is *multiplied* by that power.

$$(N \times 10^n)^m = N^m \times 10^{m \times n}$$

For example,

$$\begin{aligned}(7.5 \times 10^2)^2 &= (7.5)^2 \times 10^{2 \times 2} \\ &= 56. \times 10^4 \\ &= 5.6 \times 10^5\end{aligned}$$

When a root is taken of a number expressed in exponential notation, the root of the initial number is taken and the exponent of 10 is divided by the number representing the root. For example, we take the square root of a number as follows:

$$\sqrt{N \times 10^n} = (N \times 10^n)^{1/2} = \sqrt{N} \times 10^{n/2}$$

For example,

$$\begin{aligned}(2.9 \times 10^6)^{1/2} &= \sqrt{2.9} \times 10^{6/2} \\ &= 1.7 \times 10^3\end{aligned}$$

Using a Calculator to Perform Mathematical Operations on Exponentials

In dealing with exponential numbers, you must first learn to enter them into your calculator. First the number is keyed in and then the exponent. There is a special key that must be pressed just before the exponent is entered. This key is often labeled EE or exp. For example, the number 1.56×10^6 is entered as follows:

Press	***Display***
1.56	1.56
EE or exp	1.56 00
6	1.56 06

To enter a number with a negative exponent, use the change-of-sign key +/− after entering the exponent number. For example, the number 7.54×10^{-3} is entered as follows:

Press	***Display***
7.54	7.54
EE or exp	7.54 00
3	7.54 03
+/−	7.54 −03

Once a number with an exponent is entered into your calculator, the mathematical operations are performed exactly the same as with a "regular" number. For example, the numbers 1.0×10^3 and 1.0×10^2 are multiplied as follows:

Press	***Display***
1.0	1.0
EE or exp	1.0 00
3	1.0 03
×	1 03
1.0	1.0
EE or exp	1.0 00
2	1.0 02
=	1 05

The answer is correctly represented as 1.0×10^5.

The numbers 1.50×10^5 and 1.1×10^4 are added as follows:

Press	***Display***
1.5	1.50
EE or exp	1.50 00
5	1.50 05
+	1.5 05
1.1	1.1
EE or exp	1.1 00
4	1.1 04
=	1.61 05

The answer is correctly represented as 1.61×10^5. Note that when exponential numbers are added, the calculator automatically takes into account any difference in exponents.

To take the power, root, or reciprocal of an exponential number, enter the number first, then press the appropriate key or keys. For example, the square root of 5.6×10^3 is obtained as follows:

Press	***Display***	
5.6	5.6	
EE or exp	5.6	00
3	5.6	03
$\sqrt{X}$	7.4833148	01

The answer is correctly represented as 7.5×10^1.

Practice by performing the following operations that involve exponential numbers. The answers follow the exercises.

a. $7.9 \times 10^2 \times 4.3 \times 10^4$

b. $\dfrac{5.4 \times 10^3}{4.6 \times 10^5}$

c. $1.7 \times 10^2 + 1.63 \times 10^3$

d. $4.3 \times 10^{-3} + 1 \times 10^{-4}$

e. $(8.6 \times 10^{-6})^2$

f. $\dfrac{1}{8.3 \times 10^2}$

g. $\log(1.0 \times 10^{-7})$

h. $-\log(1.3 \times 10^{-5})$

i. $\sqrt{6.7 \times 10^9}$

SOLUTIONS

a. 3.4×10^7

b. 1.2×10^{-2}

c. 1.80×10^3

d. 4.4×10^{-3}

e. 7.4×10^{-11}

f. 1.2×10^{-3}

g. -7.00

h. 4.89

i. 8.2×10^4

Graphing Functions

In interpreting the results of a scientific experiment, it is often useful to make a graph. If possible, the function to be graphed should be in a form that gives a straight line. The equation for a straight line (a *linear equation*) can be represented in the general form

$$y = mx + b$$

where y is the *dependent variable*, x is the *independent variable*, m is the *slope*, and b is the *intercept* with the y axis.

To illustrate the characteristics of a linear equation, the function $y = 3x + 4$ is plotted in Figure A.1. For this equation $m = 3$ and $b = 4$. Note that the y intercept occurs when $x = 0$. In this case the y intercept is 4, as can be seen from the equation ($b = 4$).

The slope of a straight line is defined as the ratio of the rate of change in y to that in x:

$$m = \text{slope} = \frac{\Delta y}{\Delta x}$$

For the equation $y = 3x + 4$, y changes three times as fast as x (because x has a coefficient of 3). Thus the slope in this case is 3. This can be verified from the graph. For the triangle shown in Figure A.1,

$$\Delta y = 15 - 16 = 36 \qquad \text{and} \qquad \Delta x = 15 - 3 = 12$$

Thus

$$\text{Slope} = \frac{\Delta y}{\Delta x} = \frac{36}{12} = 3$$

This example illustrates a general method for obtaining the slope of a line from the graph of that line. Simply draw a triangle with one side parallel to the y axis and the other side parallel to the x axis, as shown in Figure A.1. Then determine the lengths of the sides to get Δy and Δx, respectively, and compute the ratio $\Delta y/\Delta x$.

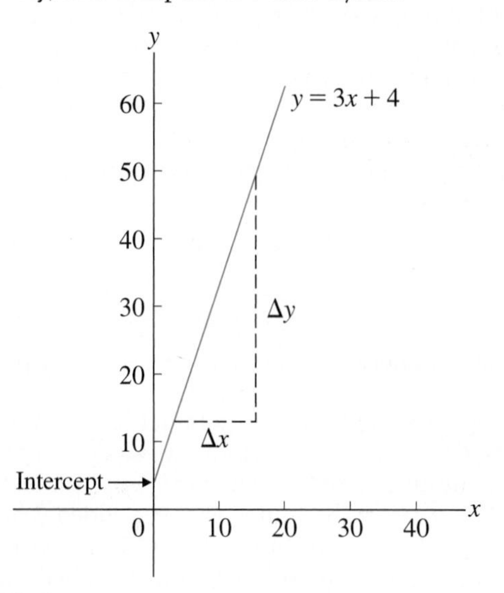

Figure A.1
Graph of the linear equation $y = 3x + 4$.

SI Units and Conversion Factors

These conversion factors are given with more significant figures than those typically used in the body of the text.

Length

SI Unit: Meter (m)

1 meter = 1.0936 yards
1 centimeter = 0.39370 inch
1 inch = 2.54 centimeters (exactly)
1 kilometer = 0.62137 mile
1 mile = 5280. feet
= 1.6093 kilometers

Volume

SI Unit: Cubic Meter (m^3)

1 liter = 10^{-3} m^3
= 1 dm^3
= 1.0567 quarts
1 gallon = 4 quarts
= 8 pints
= 3.7854 liters
1 quart = 32 fluid ounces
= 0.94635 liter

Mass

SI Unit: Kilogram (kg)

1 kilogram = 1000 grams
= 2.2046 pounds
1 pound = 453.59 grams
= 0.45359 kilogram
= 16 ounces
1 atomic mass unit = 1.66057×10^{-27} kilograms

Pressure

SI Unit: Pascal (Pa)

1 atmosphere = 101.325 kilopascals
= 760. torr (mm Hg)
= 14.70 pounds per square inch

Energy

SI Unit: Joule (J)

1 joule = 0.23901 calorie
1 calorie = 4.184 joules

Solutions
To Self-Check Exercises

Chapter 2

Self-Check Exercise 2.1

$357 = 3.57 \times 10^2$
$0.0055 = 5.5 \times 10^{-3}$

Self-Check Exercise 2.2

a. Three significant figures. The leading zeros (to the left of the 1) do not count, but the trailing zeros do.

b. Five significant figures. The one captive zero and the two trailing zeros all count.

c. This is an exact number obtained by counting the cars. It has an unlimited number of significant figures.

Self-Check Exercise 2.3

a. $12.6 \times 0.53 = 6.678 = 6.7$
   Limiting

b. $12.6 \times 0.53 = 6.7$; Limiting

$$\begin{array}{r} 6.7 \\ -4.59 \\ \hline 2.11 \end{array} = 2.1$$

(6.7: Limiting)

c. $$\begin{array}{r} 25.36 \\ -4.15 \\ \hline 21.21 \end{array} \qquad \frac{21.21}{2.317} = 9.15408 = 9.154$$

Self-Check Exercise 2.4

$$0.750\ \cancel{L} \times \frac{1.06\text{ qt}}{1\ \cancel{L}} = 0.795\text{ qt}$$

Self-Check Exercise 2.5

$$225\ \frac{\cancel{\text{mi}}}{\text{h}} \times \frac{1760\ \cancel{\text{yd}}}{1\ \cancel{\text{mi}}} \times \frac{1\ \cancel{\text{m}}}{1.094\ \cancel{\text{yd}}} \times \frac{1\text{ km}}{1000\ \cancel{\text{m}}} = 362\ \frac{\text{km}}{\text{h}}$$

Self-Check Exercise 2.6

The best way to solve this problem is to convert 172 K to Celsius degrees. To do this we will use the formula $T_{°C} = T_K - 273$.
In this case

$$T_{°C} = T_K - 273 = 172 - 273 = -101$$

So 172 K = −101 °C, which is a lower temperature than −75 °C. Thus 172 K is colder than −75 °C.

Self-Check Exercise 2.7

The problem is 41 °C = ? °F.
Using the formula

$$T_{°F} = 1.80(T_{°C}) + 32$$

we have

$$T_{°F} = ?\ °F = 1.80(41) + 32 = 74 + 32 = 106$$

That is, 41 °C = 106 °F.

Self-Check Exercise 2.8

This problem can be stated as 239 °F = ? °C.
Using the formula

$$T_{°C} = \frac{T_{°F} - 32}{1.80}$$

we have in this case

$$T_{°C} = ?\ °C = \frac{239 - 32}{1.80} = \frac{207}{1.80} = 115$$

That is, 239 °F = 115 °C.

Self-Check Exercise 2.9

We obtain the density of the cleaner by dividing its mass by its volume.

$$\text{Density} = \frac{\text{mass}}{\text{volume}} = \frac{28.1\text{ g}}{35.8\text{ mL}} = 0.785\text{ g/mL}$$

This density identifies the liquid as isopropyl alcohol.

Chapter 3

Self-Check Exercise 3.1

Items (a) and (c) are physical properties. When the solid gallium melts, it forms liquid gallium. There is no change in composition. Items (b) and (d) reflect the ability to change composition and are thus chemical properties. Statement (b) means that platinum does not react with oxygen to form some new substance. Statement (d) means that copper does react in the air to form a new substance, which is green.

Self-Check Exercise 3.2

a. Milk turns sour because new substances are formed. This is a chemical change.

b. Melting the wax is a physical change (a change of state). When the wax burns, new substances are formed. This is a chemical change.

Self-Check Exercise 3.3

a. Wine is a homogeneous mixture of alcohol and other dissolved substances dispersed uniformly in water.

b. Helium and oxygen form a homogeneous mixture.

c. Oil and vinegar salad dressing is a heterogeneous mixture. (Note the two distinct layers the next time you look at a bottle of dressing.)

d. Common salt is a pure substance (sodium chloride), so it always has the same composition. (Note that other substances such as iodine are often added to commercial preparations of table salt, which is mostly sodium chloride. Thus commercial table salt is a homogeneous mixture.)

Self-Check Exercise 3.4

The conversion factor needed is $\frac{1 \text{ cal}}{4.184 \text{ J}}$, and the conversion is

$$28.4 \text{ J} \times \frac{1 \text{ cal}}{4.184 \text{ J}} = 6.79 \text{ cal}$$

Self-Check Exercise 3.5

We know that it takes 4.184 J of energy to change the temperature of each gram of water by 1 °C, so we must multiply 4.184 by the mass of water (454 g) and the temperature change (98.6 °C − 5.4 °C = 93.2 °C).

$$4.184 \frac{\text{J}}{\text{g °C}} \times 454 \text{ g} \times 93.2 \text{ °C} = 1.77 \times 10^5 \text{ J}$$

Self-Check Exercise 3.6

From Table 3.2, the specific heat capacity for solid gold is 0.13 J/g °C. Because it takes 0.13 J to change the temperature of *one* gram of gold by *one* Celsius degree, we must multiply 0.13 by the sample size (5.63 g) and the change in temperature (32 °C − 21 °C = 11 °C).

$$0.13 \frac{\text{J}}{\text{g °C}} \times 5.63 \text{ g} \times 11 \text{ °C} = 8.1 \text{ J}$$

We can change this energy to units of calories as follows:

$$8.1 \text{ J} \times \frac{1 \text{ cal}}{4.184 \text{ J}} = 1.9 \text{ cal}$$

Self-Check Exercise 3.7

Table 3.2 lists the specific heat capacities of several metals. We want to calculate the specific heat capacity (s) for this metal and then use Table 3.2 to identify the metal. Using the equation

$$Q = s \times m \times \Delta T$$

we can solve for s by dividing both sides by m (the mass of the sample) and by ΔT:

$$\frac{Q}{m \times \Delta T} = s$$

In this case.

Q = energy (heat) required = 10.1 J
m = 2.8 g
ΔT = temperature change = 36 °C − 21 °C = 15 °C

so

$$s = \frac{Q}{m \times \Delta T} = \frac{10.1 \text{ J}}{(2.8 \text{ g})(15 \text{ °C})} = 0.24 \text{ J/g °C}$$

Table 3.2 shows that silver has a specific heat capacity of 0.24 J/g °C. The metal is silver.

Chapter 4

Self-Check Exercise 4.1

a. P_4O_{10} b. UF_6 c. $AlCl_3$

Self-Check Exercise 4.2

In the symbol $^{90}_{38}Sr$, the number 38 is the atomic number, which represents the number of protons in the nucleus of a strontium atom. Because the atom is neutral overall, it must also have 38 electrons. The number 90 (the mass number) represents the number of protons plus the number of neutrons. Thus the number of neutrons is $A - Z = 90 - 38 = 52$.

Self-Check Exercise 4.3

The atom $^{201}_{80}Hg$ has 80 protons, 80 electrons, and 201 − 80 = 121 neutrons.

Self-Check Exercise 4.4

The atomic number for phosphorus is 15 and the mass number is 15 + 17 = 32. Thus the symbol for the atom is $^{32}_{15}P$.

Self-Check Exercise 4.5

Element	Symbol	Atomic Number	Metal or Nonmetal	Family Name
a. argon	Ar	18	nonmetal	noble gas
b. chlorine	Cl	17	nonmetal	halogen
c. barium	Ba	56	metal	alkaline earth metals
d. cesium	Cs	55	metal	alkali metals

Self-Check Exercise 4.6

a. KI
$(1+) + (1-) = 0$

b. Mg_3N_2
$3(2+) + 2(3-) = (6+) + (6-) = 0$

c. Al_2O_3
$2(3+) + 3(2-) = 0$

Chapter 5

Self-Check Exercise 5.1

a. rubidium oxide
b. strontium iodide
c. potassium sulfide

Self-Check Exercise 5.2

a. The compound $PbBr_2$ must contain Pb^{2+}—named lead(II)—to balance the charges of the two Br^- ions. Thus the name is lead(II) bromide. The compound $PbBr_4$ must contain Pb^{4+}—named lead(IV)—to balance the charges of the four Br^- ions. The name is therefore lead(IV) bromide.

b. The compound FeS contains the S^{2-} ion (sulfide) and thus the iron cation present must be Fe^{2+}, iron(II). The name is iron(II) sulfide. The compound Fe_2S_3 contains three S^{2-} ions and two iron cations of unknown charge. We can determine the iron charge from the following:

$$2(?+) + 3(2-) = 0$$

↑ Iron charge ↑ S^{2-} charge

In this case, ? must represent 3 because

$$2(3+) + 3(2-) = 0$$

Thus Fe_2S_3 contains Fe^{3+} and S^{2-}, and its name is iron(III) sulfide.

c. The compound $AlBr_3$ contains Al^{3+} and Br^-. Because aluminum forms only one ion (Al^{3+}), no Roman numeral is required. The name is aluminum bromide.

d. The compound Na_2S contains Na^+ and S^{2-} ions. The name is sodium sulfide. (Because sodium forms only Na^+, no Roman numeral is needed.)

e. The compound $CoCl_3$ contains three Cl^- ions. Thus the cobalt cation must be Co^{3+}, which is named cobalt(III) because cobalt is a transition metal and can form more than one type of cation. Thus the name of $CoCl_3$ is cobalt(III) chloride.

Self-Check Exercise 5.3

Compound	*Individual Names*	*Prefixes*	*Name*
a. CCl_4	carbon chloride	none *tetra-*	carbon tetrachloride
b. NO_2	nitrogen oxide	none *di-*	nitrogen dioxide
c. IF_5	iodine fluoride	none *penta-*	iodine pentafluoride

Self-Check Exercise 5.4

a. silicon dioxide

b. dioxygen difluoride

c. xenon hexafluoride

Self-Check Exercise 5.5

a. chlorine trifluoride

b. vanadium(V) fluoride

c. copper(I) chloride

d. manganese(IV) oxide

e. magnesium oxide

f. water

Self-Check Exercise 5.6

a. calcium hydroxide

b. sodium phosphate

c. potassium permanganate

d. ammonium dichromate

e. cobalt(II) perchlorate (Perchlorate has a 1− charge, so the cation must be Co^{2+} to balance the two ClO_4^- ions.)

f. potassium chlorate

g. copper(II) nitrite (This compound contains two NO_2^- (nitrite) ions and thus must contain a Cu^{2+} cation.)

Self-Check Exercise 5.7

Compound — *Name*

a. $NaHCO_3$ — sodium hydrogen carbonate
Contains Na^+ and HCO_3^-; often called sodium bicarbonate (common name).

b. $BaSO_4$ — barium sulfate
Contains Ba^{2+} and SO_4^{2-}.

c. $CsClO_4$ — cesium perchlorate
Contains Cs^+ and ClO_4^-.

d. BrF_5 — bromine pentafluoride
Both nonmetals (Type III binary).

e. $NaBr$ — sodium bromide
Contains Na^+ and Br^- (Type I binary).

f. $KOCl$ — potassium hypochlorite
Contains K^+ and OCl^-.

g. $Zn_3(PO_4)_2$ — zinc(II) phosphate
Contains Zn^{2+} and PO_4^{3-}; Zn is a transition metal and officially requires a Roman numeral. However, because Zn forms only the Zn^{2+} cation, the II is usually left out. Thus the name of the compound is usually given as zinc phosphate.

Self-Check Exercise 5.8

Name — *Chemical Formula*

a. ammonium sulfate — $(NH_4)_2SO_4$
Two ammonium ions (NH_4^+) are required for each sulfate ion (SO_4^{2-}) to achieve charge balance.

b. vanadium(V) fluoride — VF_5
The compound contains V^{5+} ions and requires five F^- ions for charge balance.

c. disulfur dichloride — S_2Cl_2
The prefix *di-* indicates two of each atom.

d. rubidium peroxide — Rb_2O_2
Because rubidium is in Group 1, it forms only 1+ ions. Thus two Rb^+ ions are needed to balance the 2− charge on the peroxide ion (O_2^{2-}).

e. aluminum oxide — Al_2O_3
Aluminum forms only 3+ ions. Two Al^{3+} ions are required to balance the charge on three O^{2-} ions.

Chapter 6

Self-Check Exercise 6.1

a. $Mg(s) + H_2O(l) \rightarrow Mg(OH)_2(s) + H_2(g)$
Note that magnesium (which is in Group 2) always forms the Mg^{2+} cation and thus requires two OH^- anions for a zero net charge.

b. Ammonium dichromate contains the polyatomic ions NH_4^+ and $Cr_2O_7^{2-}$ (you should have these memorized). Because NH_4^+ has a 1+ charge, two NH_4^+ cations are required for each $Cr_2O_7^{2-}$, with its 2− charge, to give the formula $(NH_4)_2Cr_2O_7$. Chromium(III) oxide contains Cr^{3+} ions—signified by chromium(III)—and O^{2-} (the oxide ion). To achieve a net charge of zero, the solid must contain two Cr^{3+} ions for every three O^{2-} ions, so the formula is Cr_2O_3. Nitrogen gas contains diatomic molecules and is written $N_2(g)$, and gaseous water is written $H_2O(g)$. Thus the unbalanced equation for the decomposition of ammonium dichromate is

$$(NH_4)_2Cr_2O_7(s) \rightarrow Cr_2O_3(s) + N_2(g) + H_2O(g)$$

c. Gaseous ammonia, $NH_3(g)$, and gaseous oxygen, $O_2(g)$, react to form nitrogen monoxide gas, $NO(g)$, plus gaseous water, $H_2O(g)$. The unbalanced equation is

$$NH_3(g) + O_2(g) \rightarrow NO(g) + H_2O(g)$$

Self-Check Exercise 6.2

STEP 1 The reactants are propane, $C_3H_8(g)$, and oxygen, $O_2(g)$; the products are carbon dioxide, $CO_2(g)$, and water, $H_2O(g)$. All are in the gaseous state.

STEP 2 The unbalanced equation for the reaction is

$$C_3H_8(g) + O_2(g) \rightarrow CO_2(g) + H_2O(g)$$

STEP 3 We start with C_3H_8 because it is the most complicated molecule. C_3H_8 contains three carbon atoms per molecule, so a coefficient of 3 is needed for CO_2.

$$C_3H_8(g) + O_2(g) \rightarrow 3CO_2(g) + H_2O(g)$$

Also, each C_3H_8 molecule contains eight hydrogen atoms, so a coefficient of 4 is required for H_2O.

$$C_3H_8(g) + O_2(g) \rightarrow 3CO_2(g) + 4H_2O(g)$$

The final element to be balanced is oxygen. Note that the left side of the equation now has two oxygen atoms, and the right side has ten. We can balance the oxygen by using a coefficient of 5 for O_2.

$$C_3H_8(g) + 5O_2(g) \rightarrow 3CO_2(g) + 4H_2O(g)$$

STEP 4 ***Check:***

3 C, 8 H, 10 O → 3 C, 8 H, 10 O

Reactant atoms — Product atoms

We cannot divide all coefficients by a given integer to give smaller integer coefficients.

Self-Check Exercise 6.3

a. $NH_4NO_2(s) \rightarrow N_2(g) + H_2O(g)$ (unbalanced)
   $NH_4NO_2(s) \rightarrow N_2(g) + 2H_2O(g)$ (balanced)

b. $NO(g) \rightarrow N_2O(g) + NO_2(g)$ (unbalanced)
   $3NO(g) \rightarrow N_2O(g) + NO_2(g)$ (balanced)

c. $HNO_3(l) \rightarrow NO_2(g) + H_2O(l) + O_2(g)$ (unbalanced)
   $4HNO_3(l) \rightarrow 4NO_2(g) + 2H_2O(l) + O_2(g)$ (balanced)

Chapter 7

Self-Check Exercise 7.1

a. The ions present are

$$Ba^{2+}(aq) + 2NO_3^-(aq) + Na^+(aq) + Cl^-(aq) \rightarrow$$

Ions in $Ba(NO_3)_2(aq)$ — Ions in $NaCl(aq)$

Exchanging the anions gives the possible solid products $BaCl_2$ and $NaNO_3$. Using Table 7.1, we see that both substances are very soluble (rules 1, 2, and 3). Thus no solid forms.

b. The ions present in the mixed solution before any reaction occurs are

$$2Na^+(aq) + S^{2-}(aq) + Cu^{2+}(aq) + 2NO_3^-(aq) \rightarrow$$

Ions in $Na_2S(aq)$ — Ions in $Cu(NO_3)_2(aq)$

Exchanging the anions gives the possible solid products CuS and $NaNO_3$. According to rules 1 and 2 in Table 7.1, $NaNO_3$ is soluble, and by rule 6, CuS should be insoluble. Thus CuS will precipitate. The balanced equation is

$$Na_2S(aq) + Cu(NO_3)_2(aq) \rightarrow CuS(s) + 2NaNO_3(aq)$$

c. The ions present are

$$NH_4^+(aq) + Cl^-(aq) + Pb^{2+}(aq) + 2NO_3^-(aq) \rightarrow$$

Ions in $NH_4Cl(aq)$ — Ions in $Pb(NO_3)_2(aq)$

Exchanging the anions gives the possible solid products NH_4NO_3 and $PbCl_2$. NH_4NO_3 is soluble (rules 1 and 2) and $PbCl_2$ is insoluble (rule 3). Thus $PbCl_2$ will precipitate. The balanced equation is

$$2NH_4Cl(aq) + Pb(NO_3)_2(aq) \rightarrow PbCl_2(s) + 2NH_4NO_3(aq)$$

Self-Check Exercise 7.2

a. *Molecular equation:*

$$Na_2S(aq) + Cu(NO_3)_2(aq) \rightarrow CuS(s) + 2NaNO_3(aq)$$

Complete ionic equation:

$$2Na^+(aq) + S^{2-}(aq) + Cu^{2+}(aq) + 2NO_3^-(aq) \rightarrow CuS(s) + 2Na^+(aq) + 2NO_3^-(aq)$$

Net ionic equation:

$$S^{2-}(aq) + Cu^{2+}(aq) \rightarrow CuS(s)$$

b. *Molecular equation:*

$$2NH_4Cl(aq) + Pb(NO_3)_2(aq) \rightarrow PbCl_2(s) + 2NH_4NO_3(aq)$$

Complete ionic equation:

$$2NH_4^+(aq) + 2Cl^-(aq) + Pb^{2+}(aq) + 2NO_3^-(aq) \rightarrow PbCl_2(s) + 2NH_4^+(aq) + 2NO_3^-(aq)$$

Net ionic equation:

$$2Cl^-(aq) + Pb^{2+}(aq) \rightarrow PbCl_2(s)$$

Self-Check Exercise 7.3

a. The compound NaBr contains the ions Na^+ and Br^-. Thus each sodium atom loses one electron ($Na \rightarrow Na^+ + e^-$), and each bromine atom gains one electron ($Br + e^- \rightarrow Br^-$).

$$Na + Na + Br - Br \rightarrow (Na^+Br^-) + (Na^+Br^-)$$

(e⁻ transferred from each Na to each Br)

b. The compound CaO contains the Ca^{2+} and O^{2-} ions. Thus each calcium atom loses two electrons ($Ca \rightarrow Ca^{2+} + 2e^-$), and each oxygen atom gains two electrons ($O + 2e^- \rightarrow O^{2-}$).

$$Ca + Ca + O - O \rightarrow (Ca^{2+}O^{2-}) + (Ca^{2+}O^{2-})$$

($2e^-$ transferred from each Ca to each O)

Self-Check Exercise 7.4

a. oxidation–reduction reaction; combustion reaction
b. synthesis reaction; oxidation–reduction reaction; combustion reaction
c. synthesis reaction; oxidation–reduction reaction
d. decomposition reaction; oxidation–reduction reaction
e. precipitation reaction (and double displacement)
f. synthesis reaction; oxidation–reduction reaction
g. acid–base reaction (and double displacement)
h. combustion reaction; oxidation–reduction reaction

Chapter 8

Self-Check Exercise 8.1

The average mass of nitrogen is 14.01 amu. The appropriate equivalence statement is 1 N atom = 14.01 amu, which yields the conversion factor we need:

$$23\ \cancel{\text{N atoms}} \times \frac{14.01\ \text{amu}}{\cancel{\text{N atom}}} = 322.2\ \text{amu}$$

(exact)

Self-Check Exercise 8.2

The average mass of oxygen is 16.00 amu, which gives the equivalence statement 1 O atom = 16.00 amu. The number of oxygen atoms present is

$$288\ \cancel{\text{amu}} \times \frac{1\ \text{O atom}}{16.00\ \cancel{\text{amu}}} = 18.0\ \text{O atoms}$$

Self-Check Exercise 8.3

Note that the sample of 5.00×10^{20} atoms of chromium is less than 1 mol (6.022×10^{23} atoms) of chromium. What fraction of a mole it represents can be determined as follows:

$$5.00 \times 10^{20}\ \cancel{\text{atoms Cr}} \times \frac{1\ \text{mol Cr}}{6.022 \times 10^{23}\ \cancel{\text{atoms Cr}}} = 8.30 \times 10^{-4}\ \text{mol Cr}$$

Because the mass of 1 mol of chromium atoms is 52.00 g, the mass of 5.00×10^{20} atoms can be determined as follows:

$$8.30 \times 10^{-4}\ \cancel{\text{mol Cr}} \times \frac{52.00\ \text{g Cr}}{1\ \cancel{\text{mol Cr}}} = 4.32 \times 10^{-2}\ \text{g Cr}$$

Self-Check Exercise 8.4

Each molecule of C_2H_3Cl contains two carbon atoms, three hydrogen atoms, and one chlorine atom, so 1 mol of C_2H_3Cl molecules contains 2 mol of C atoms, 3 mol of H atoms, and 1 mol of Cl atoms.

$$\begin{aligned} &\text{Mass of 2 mol of C atoms: } 2 \times 12.01 = 24.02\ \text{g} \\ &\text{Mass of 3 mol of H atoms: } 3 \times 1.008 = 3.024\ \text{g} \\ &\text{Mass of 1 mol of Cl atoms: } 1 \times 35.45 = \underline{35.45\ \text{g}} \\ &\qquad\qquad\qquad\qquad\qquad\qquad\quad 62.494\ \text{g} \end{aligned}$$

The molar mass of C_2H_3Cl is 62.49 g (rounding to the correct number of significant figures).

Self-Check Exercise 8.5

The formula for sodium sulfate is Na_2SO_4. One mole of Na_2SO_4 contains 2 mol of sodium ions and 1 mol of sulfate ions.

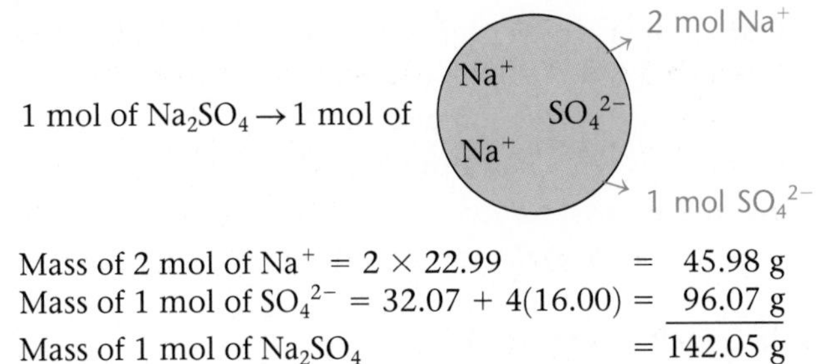

$$\begin{aligned} &\text{Mass of 2 mol of } Na^+ = 2 \times 22.99 &&= 45.98\ \text{g} \\ &\text{Mass of 1 mol of } SO_4^{2-} = 32.07 + 4(16.00) &&= \underline{96.07\ \text{g}} \\ &\text{Mass of 1 mol of } Na_2SO_4 &&= 142.05\ \text{g} \end{aligned}$$

The molar mass for sodium sulfate is 142.05 g.

A sample of sodium sulfate with a mass of 300.0 g represents more than 1 mol. (Compare 300.0 g to the molar mass of Na_2SO_4.) We calculate the number of moles of Na_2SO_4 present in 300.0 g as follows:

$$300.0\ \cancel{\text{g } Na_2SO_4} \times \frac{1\ \text{mol } Na_2SO_4}{142.05\ \cancel{\text{g } Na_2SO_4}} = 2.112\ \text{mol } Na_2SO_4$$

Self-Check Exercise 8.6

First we must compute the mass of 1 mol of C_2F_4 molecules (the molar mass). Because 1 mol of C_2F_4 contains 2 mol of C atoms and 4 mol of F atoms, we have:

$$2\ \cancel{\text{mol}}\ \text{C} \times \frac{12.01\ \text{g}}{\cancel{\text{mol}}} = 24.02\ \text{g C}$$

$$4\ \cancel{\text{mol}}\ \text{F} \times \frac{19.00\ \text{g}}{\cancel{\text{mol}}} = 76.00\ \text{g F}$$

Mass of 1 mol of C_2F_4: 100.02 g = molar mass

Using the equivalence statement 100.02 g C_2F_4 = 1 mol C_2F_4, we calculate the moles of C_2F_4 units in 135 g of Teflon.

$$135\ \cancel{\text{g } C_2F_4}\ \text{units} \times \frac{1\ \text{mol } C_2F_4}{100.02\ \cancel{\text{g } C_2F_4}} = 1.35\ \text{mol } C_2F_4\ \text{units}$$

Next, using the equivalence statement 1 mol = 6.022×10^{23} units, we calculate the number in C_2F_4 units in 135 g of Teflon.

$$135\ \cancel{\text{mol}}\ C_2F_4 \times \frac{6.022 \times 10^{23}\ \text{units}}{1\ \cancel{\text{mol}}} = 8.13 \times 10^{23}\ C_2F_4\ \text{units}$$

Self-Check Exercise 8.7

The molar mass of penicillin F is computed as follows:

$$\text{C: } 14\ \cancel{\text{mol}} \times 12.01\ \frac{\text{g}}{\cancel{\text{mol}}} = 168.1\ \text{g}$$

$$\text{H: } 20\ \cancel{\text{mol}} \times 1.008\ \frac{\text{g}}{\cancel{\text{mol}}} = 20.16\ \text{g}$$

$$\text{N: } 2\ \cancel{\text{mol}} \times 14.01\ \frac{\text{g}}{\cancel{\text{mol}}} = 28.02\ \text{g}$$

$$\text{S: } 1\ \cancel{\text{mol}} \times 32.07\ \frac{\text{g}}{\cancel{\text{mol}}} = 32.07\ \text{g}$$

$$\text{O: } 4\ \cancel{\text{mol}} \times 16.00\ \frac{\text{g}}{\cancel{\text{mol}}} = 64.00\ \text{g}$$

Mass of 1 mol of $C_{14}H_{20}N_2SO_4$ = 312.39 g = 312.4 g

$$\text{Mass percent of C} = \frac{168.1\ \text{g C}}{312.4\ \text{g } C_{14}H_{20}N_2SO_4} \times 100\% = 53.81\%$$

$$\text{Mass percent of H} = \frac{20.16\ \text{g H}}{312.4\ \text{g } C_{14}H_{20}N_2SO_4} \times 100\% = 6.453\%$$

$$\text{Mass percent of N} = \frac{28.02\ \text{g N}}{312.4\ \text{g } C_{14}H_{20}N_2SO_4} \times 100\% = 8.969\%$$

$$\text{Mass percent of S} = \frac{32.07\ \text{g S}}{312.4\ \text{g } C_{14}H_{20}N_2SO_4} \times 100\% = 10.27\%$$

$$\text{Mass percent of O} = \frac{64.00\ \text{g O}}{312.4\ \text{g } C_{14}H_{20}N_2SO_4} \times 100\% = 20.49\%$$

Check: The percentages add up to 99.99%.

Self-Check Exercise 8.8

STEP 1 0.6884 g lead and 0.2356 g chlorine

STEP 2 $0.6884 \text{ g Pb} \times \dfrac{1 \text{ mol Pb}}{207.2 \text{ g Pb}} = 0.003322 \text{ mol Pb}$

$0.2356 \text{ g Cl} \times \dfrac{1 \text{ mol Cl}}{35.45 \text{ g Cl}} = 0.006646 \text{ mol Cl}$

STEP 3 $\dfrac{0.003322 \text{ mol Pb}}{0.003322} = 1.000 \text{ mol Pb}$

$\dfrac{0.006646 \text{ mol Cl}}{0.003322} = 2.001 \text{ mol Cl}$

These numbers are very close to integers, so step 4 is unnecessary. The empirical formula is $PbCl_2$.

Self-Check Exercise 8.9

STEP 1 0.8007 g C, 0.9333 g N, 0.2016 g H, and 2.133 g O

STEP 2 $0.8007 \text{ g C} \times \dfrac{1 \text{ mol C}}{12.01 \text{ g C}} = 0.06667 \text{ mol C}$

$0.9333 \text{ g N} \times \dfrac{1 \text{ mol N}}{14.01 \text{ g N}} = 0.06662 \text{ mol N}$

$0.2016 \text{ g H} \times \dfrac{1 \text{ mol H}}{1.008 \text{ g H}} = 0.2000 \text{ mol H}$

$2.133 \text{ g O} \times \dfrac{1 \text{ mol O}}{16.00 \text{ g O}} = 0.1333 \text{ mol O}$

STEP 3 $\dfrac{0.06667 \text{ mol C}}{0.06667} = 1.001 \text{ mol C}$

$\dfrac{0.06662 \text{ mol N}}{0.06667} = 1.000 \text{ mol N}$

$\dfrac{0.2000 \text{ mol H}}{0.06662} = 3.002 \text{ mol H}$

$\dfrac{0.1333 \text{ mol O}}{0.06662} = 2.001 \text{ mol O}$

The empirical formula is CNH_3O_2.

Self-Check Exercise 8.10

STEP 1 In 100.00 g of Nylon-6 the masses of elements present are 63.68 g C, 12.38 g N, 9.80 g H, and 14.14 g O.

STEP 2 $63.68 \text{ g C} \times \dfrac{1 \text{ mol C}}{12.01 \text{ g C}} = 5.302 \text{ mol C}$

$12.38 \text{ g N} \times \dfrac{1 \text{ mol N}}{14.01 \text{ g N}} = 0.8837 \text{ mol N}$

$9.80 \text{ g H} \times \dfrac{1 \text{ mol H}}{1.008 \text{ g H}} = 9.72 \text{ mol H}$

$14.14 \text{ g O} \times \dfrac{1 \text{ mol O}}{16.00 \text{ g O}} = 0.8838 \text{ mol O}$

STEP 3 $\dfrac{5.302 \text{ mol C}}{0.8836} = 6.000 \text{ mol C}$

$\dfrac{0.8837 \text{ mol N}}{0.8837} = 1.000 \text{ mol N}$

$\dfrac{9.72 \text{ mol H}}{0.8837} = 11.0 \text{ mol H}$

$\dfrac{0.8838 \text{ mol O}}{0.8837} = 1.000 \text{ mol O}$

The empirical formula for Nylon-6 is $C_6NH_{11}O$.

Self-Check Exercise 8.11

STEP 1 First we convert the mass percents to mass in grams. In 100.0 g of the compound, there are 71.65 g of chlorine, 24.27 g of carbon, and 4.07 g of hydrogen.

STEP 2 We use these masses to compute the moles of atoms present.

$71.65 \text{ g Cl} \times \dfrac{1 \text{ mol Cl}}{35.45 \text{ g Cl}} = 2.021 \text{ mol Cl}$

$24.27 \text{ g C} \times \dfrac{1 \text{ mol C}}{12.01 \text{ g C}} = 2.021 \text{ mol C}$

$4.07 \text{ g H} \times \dfrac{1 \text{ mol H}}{1.008 \text{ g H}} = 4.04 \text{ mol H}$

STEP 3 Dividing each mole value by 2.021 (the smallest number of moles present), we obtain the empirical formula $ClCH_2$.

To determine the molecular formula, we must compare the empirical formula mass to the molar mass. The empirical formula mass is 49.48.

Cl: 35.45
C: 12.01
2 H: $2 \times (1.008)$
$ClCH_2$: 49.48 = empirical formula mass

The molar mass is known to be 98.96. We know that

Molar mass = $n \times$ (empirical formula mass)

So we can obtain the value of n as follows:

$$\frac{\text{Molar mass}}{\text{Empirical formula mass}} = \frac{98.96}{49.48} = 2$$

Molecular formula = $(ClCH_2)_2 = Cl_2C_2H_4$

This substance is composed of molecules with the formula $Cl_2C_2H_4$.

Chapter 9

Self-Check Exercise 9.1

The problem can be stated as follows:

$$4.30 \text{ mol } C_3H_8 \xrightarrow{} ? \text{ mol } CO_2$$

yields

From the balanced equation

$$C_3H_8(g) + 5O_2(g) \rightarrow 3CO_2(g) + 4H_2O(g)$$

we derive the equivalence statement

$$1 \text{ mol } C_3H_8 = 3 \text{ mol } CO_2$$

The appropriate conversion factor (moles of C_3H_8 must cancel) is $3 \text{ mol } CO_2/1 \text{ mol } C_3H_8$, and the calculation is

$$4.30 \text{ mol } C_3H_8 \times \frac{3 \text{ mol } CO_2}{1 \text{ mol } C_3H_8} = 12.9 \text{ mol } CO_2$$

Thus we can say

4.30 mol C_3H_8 yields 12.9 mol CO_2

Self-Check Exercise 9.2

The problem can be sketched as follows:

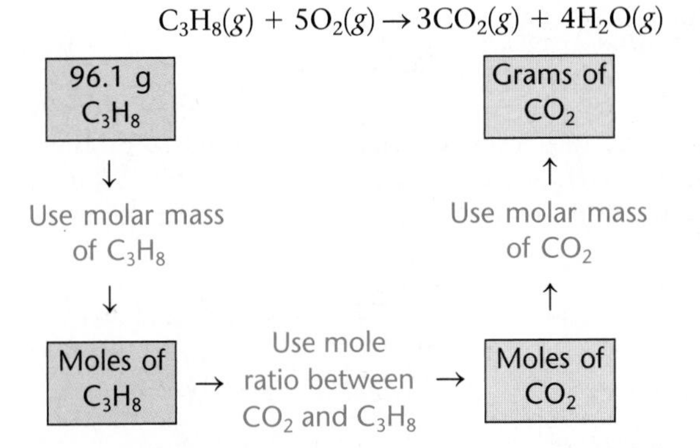

We have already done the first step in Example 9.4.

$$96.1 \text{ g } C_3H_8 \rightarrow \frac{1 \text{ mol}}{44.09 \text{ g}} \rightarrow 2.18 \text{ mol } C_3H_8$$

To find out how many moles of CO_2 can be produced from 2.18 mol of C_3H_8, we see from the balanced equation that 3 mol of CO_2 is produced for each mole of C_3H_8 reacted. The mole ratio we need is 3 mol CO_2/1 mol C_3H_8. The conversion is therefore

$$2.18 \cancel{\text{mol } C_3H_8} \times \frac{3 \text{ mol } CO_2}{1 \cancel{\text{mol } C_3H_8}} = 6.54 \text{ mol } CO_2$$

Next, using the molar mass of CO_2, which is 12.01 + 32.00 = 44.01 g, we calculate the mass of CO_2 produced.

$$6.54 \cancel{\text{mol } CO_2} \times \frac{44.01 \text{ g } CO_2}{1 \cancel{\text{mol } CO_2}} = 288 \text{ g } CO_2$$

The sequence of steps we took to find the mass of carbon dioxide produced from 96.1 g of propane is summarized in the following diagram.

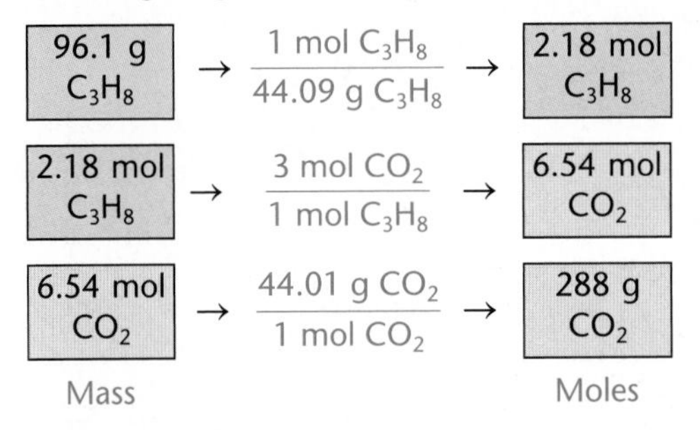

Self-Check Exercise 9.3

We sketch the problem as follows:

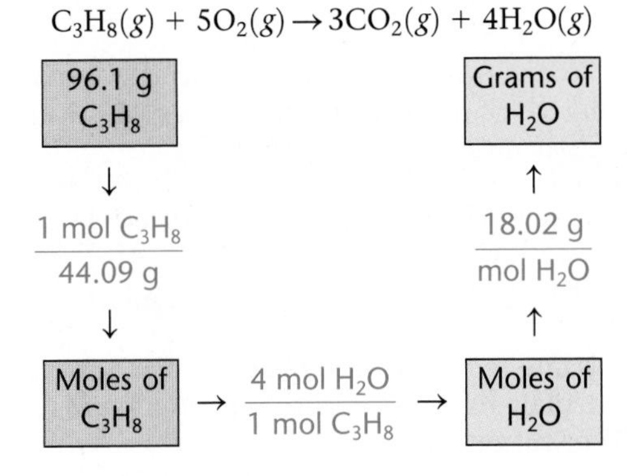

Then we do the calculations.

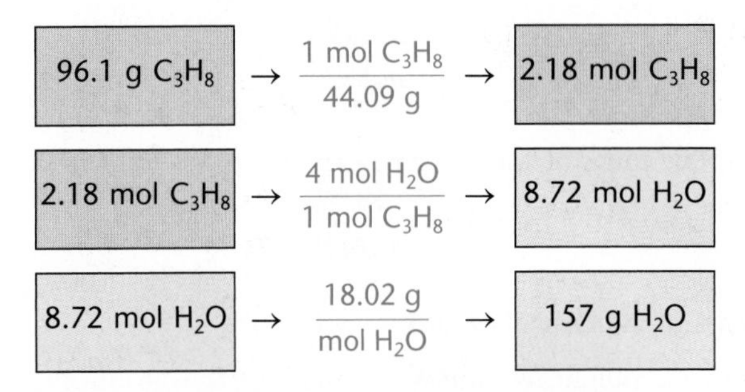

Therefore, 157 g of H_2O is produced from 96.1 g C_3H_8.

Self-Check Exercise 9.4

a. We first write the balanced equation.

$$SiO_2(s) + 4HF(aq) \rightarrow SiF_4(g) + 2H_2O(l)$$

The map of the steps required is

$$SiO_2(s) + 4HF(aq) \rightarrow SiF_4(g) + 2H_2O(l)$$

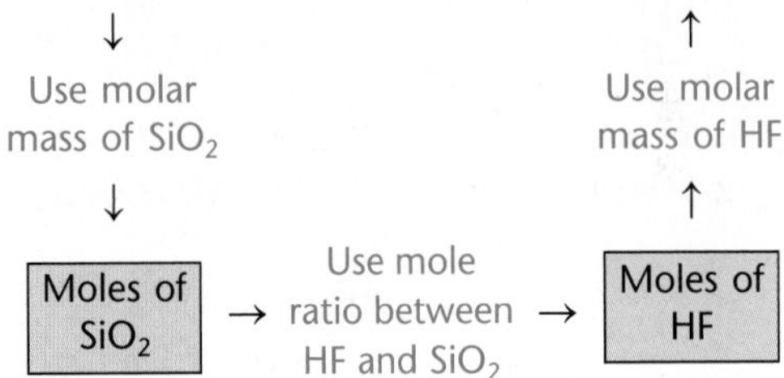

We convert 5.68 g of SiO_2 to moles as follows:

$$5.68 \cancel{\text{g } SiO_2} \times \frac{1 \text{ mol } SiO_2}{60.09 \cancel{\text{g } SiO_2}} = 9.45 \times 10^{-2} \text{ mol } SiO_2$$

Using the balanced equation, we obtain the appropriate mole ratio and convert to moles of HF.

$$9.45 \times 10^{-2} \cancel{\text{mol } SiO_2} \times \frac{4 \text{ mol HF}}{1 \cancel{\text{mol } SiO_2}} = 3.78 \times 10^{-1} \text{ mol HF}$$

Finally, we calculate the mass of HF by using its molar mass.

$$3.78 \times 10^{-1} \cancel{\text{mol HF}} \times \frac{20.01 \text{ g HF}}{\cancel{\text{mol HF}}} = 7.56 \text{ g HF}$$

b. The map for this problem is

$$SiO_2(s) + 4HF(aq) \rightarrow SiF_4(g) + 2H_2O(l)$$

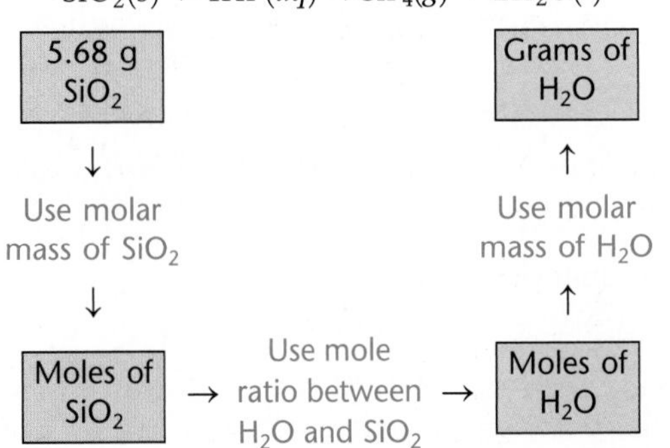

We have already accomplished the first conversion in part a. Using the balanced equation, we obtain moles of H_2O as follows:

$$9.45 \times 10^{-2}\ \cancel{\text{mol SiO}_2} \times \frac{2\text{ mol H}_2\text{O}}{1\ \cancel{\text{mol SiO}_2}} = 1.89 \times 10^{-1}\text{ mol H}_2\text{O}$$

The mass of water formed is

$$1.89 \times 10^{-1}\ \cancel{\text{mol H}_2\text{O}} \times \frac{18.02\text{ g H}_2\text{O}}{\cancel{\text{mol H}_2\text{O}}} = 3.41\text{ g H}_2\text{O}$$

Self-Check Exercise 9.5

In this problem, we know the mass of the product to be formed by the reaction

$$\text{CO}(g) + 2\text{H}_2(g) \rightarrow \text{CH}_3\text{OH}(l)$$

and we want to find the masses of reactants needed. The procedure is the same one we have been following. We must first convert the mass of CH_3OH to moles, then use the balanced equation to obtain moles of H_2 and CO needed, and then convert these moles to masses. Using the molar mass of CH_3OH (32.04 g/mol), we convert to moles of CH_3OH.

First we convert kilograms to grams.

$$6.0\ \cancel{\text{kg}}\text{ CH}_3\text{OH} \times \frac{1000\text{ g}}{\cancel{\text{kg}}} = 6.0 \times 10^3\text{ g CH}_3\text{OH}$$

Next we convert 6.0×10^3 g CH_3OH to moles of CH_3OH, using the conversion factor 1 mol CH_3OH/32.04 g CH_3OH.

$$6.0 \times 10^3\ \cancel{\text{g CH}_3\text{OH}} \times \frac{1\text{ mol CH}_3\text{OH}}{32.04\ \cancel{\text{g CH}_3\text{OH}}} = 1.9 \times 10^2\text{ mol CH}_3\text{OH}$$

Then we have two questions to answer:

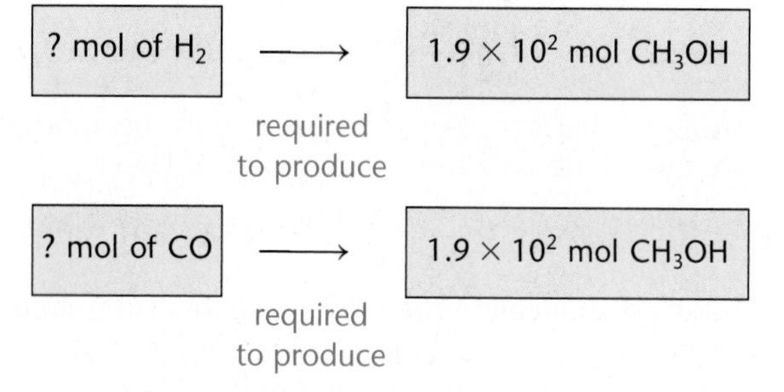

To answer these questions, we use the balanced equation

$$\text{CO}(g) + 2\text{H}_2(g) \rightarrow \text{CH}_3\text{OH}(l)$$

to obtain mole ratios between the reactants and the products. In the balanced equation the coefficients for both CO and CH_3OH are 1, so we can write the equivalence statement

$$1\text{ mol CO} = 1\text{ mol CH}_3\text{OH}$$

Using the mole ratio 1 mol CO/1 mol CH_3OH, we can now convert from moles of CH_3OH to moles of CO.

$$1.9 \times 10^2\ \cancel{\text{mol CH}_3\text{OH}} \times \frac{1\text{ mol CO}}{1\ \cancel{\text{mol CH}_3\text{OH}}} = 1.9 \times 10^2\text{ mol CO}$$

To calculate the moles of H_2 required, we construct the equivalence statement between CH_3OH and H_2, using the coefficients in the balanced equation.

$$2\text{ mol H}_2 = 1\text{ mol CH}_3\text{OH}$$

Using the mole ratio 2 mol H_2/1 mol CH_3OH, we can convert moles of CH_3OH to moles of H_2.

$$1.9 \times 10^2\ \cancel{\text{mol CH}_3\text{OH}} \times \frac{2\text{ mol H}_2}{1\ \cancel{\text{mol CH}_3\text{OH}}} = 3.8 \times 10^2\text{ mol H}_2$$

We now have the moles of reactants required to produce 6.0 kg of CH_3OH. Since we need the masses of reactants, we must use the molar masses to convert from moles to mass.

$$1.9 \times 10^2\ \cancel{\text{mol CO}} \times \frac{28.01\text{ g CO}}{1\ \cancel{\text{mol CO}}} = 5.3 \times 10^3\text{ g CO}$$

$$3.8 \times 10^2\ \cancel{\text{mol H}_2} \times \frac{2.016\text{ g H}_2}{1\ \cancel{\text{mol H}_2}} = 7.7 \times 10^2\text{ H}_2$$

Therefore, we need 5.3×10^3 g CO to react with 7.7×10^2 g H_2 to form 6.0×10^3 g (6.0 kg) of CH_3OH. This whole process is mapped in the following diagram.

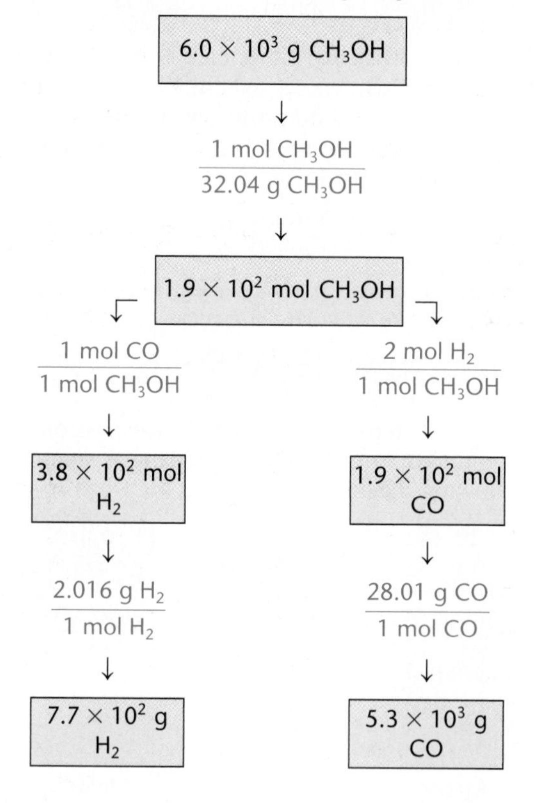

Self-Check Exercise 9.6

STEP 1 The balanced equation for the reaction is

$$6\text{Li}(s) + \text{N}_2(g) \rightarrow 2\text{Li}_3\text{N}(s)$$

STEP 2 To determine the limiting reactant, we must convert the masses of lithium (atomic mass = 6.941 g) and nitrogen (molar mass = 28.02 g) to moles.

$$56.0\ \cancel{\text{g Li}} \times \frac{1\text{ mol Li}}{6.941\ \cancel{\text{g Li}}} = 8.07\text{ mol Li}$$

$$56.0\ \cancel{\text{g N}_2} \times \frac{1\text{ mol N}_2}{28.02\ \cancel{\text{g N}_2}} = 2.00\text{ mol N}_2$$

STEP 3 Using the mole ratio from the balanced equation, we can calculate the moles of lithium required to react with 2.00 mol of nitrogen.

$$2.00\ \cancel{\text{mol N}_2} \times \frac{6\text{ mol Li}}{1\ \cancel{\text{mol N}_2}} = 12.0\text{ mol Li}$$

Therefore, 12.0 mol of Li is required to react with 2.00 mol of N_2. However, we have only 8.07 mol of Li, so lithium is limiting. It will be consumed before the nitrogen runs out.

STEP 4 Because lithium is the limiting reactant, we must use the 8.07 mol of Li to determine how many moles of Li_3N can be formed.

$$8.07\ \cancel{\text{mol Li}} \times \frac{2\ \text{mol Li}_3\text{N}}{6\ \cancel{\text{mol Li}}} = 2.69\ \text{mol Li}_3\text{N}$$

STEP 5 We can now use the molar mass of Li_3N (34.83 g) to calculate the mass of Li_3N formed.

$$2.69\ \cancel{\text{mol Li}_3\text{N}} \times \frac{34.83\ \text{g Li}_3\text{N}}{1\ \cancel{\text{mol Li}_3\text{N}}} = 93.7\ \text{g Li}_3\text{N}$$

Self-Check Exercise 9.7

a. **STEP 1** The balanced equation is

$$\text{TiCl}_4(g) + \text{O}_2(g) \rightarrow \text{TiO}_2(s) + 2\text{Cl}_2(g)$$

STEP 2 The numbers of moles of reactants are

$$6.71 \times 10^3\ \cancel{\text{g TiCl}_4} \times \frac{1\ \text{mol TiCl}_4}{189.68\ \cancel{\text{g TiCl}_4}} = 3.54 \times 10^1\ \text{mol TiCl}_4$$

$$2.45 \times 10^3\ \cancel{\text{g O}_2} \times \frac{1\ \text{mol O}_2}{32.00\ \cancel{\text{g O}_2}} = 7.66 \times 10^1\ \text{mol O}_2$$

STEP 3 In the balanced equation both $TiCl_4$ and O_2 have coefficients of 1, so

$$1\ \text{mol TiCl}_4 = 1\ \text{mol O}_2$$

and

$$3.54 \times 10^1\ \cancel{\text{mol TiCl}_4} \times \frac{1\ \text{mol O}_2}{1\ \cancel{\text{mol TiCl}_4}} = 3.54 \times 10^1\ \text{mol O}_2\ \text{required}$$

We have 7.66×10^1 mol of O_2, so the O_2 is in excess and the $TiCl_4$ is limiting. This makes sense. $TiCl_4$ and O_2 react in a 1:1 mole ratio, so the $TiCl_4$ is limiting because fewer moles of $TiCl_4$ are present than moles of O_2.

STEP 4 We will now use the moles of $TiCl_4$ (the limiting reactant) to determine the moles of TiO_2 that would form if the reaction produced 100% of the expected yield (the theoretical yield).

$$3.54 \times 10^1\ \cancel{\text{mol TiCl}_4} \times \frac{1\ \text{mol TiO}_2}{1\ \cancel{\text{mol TiCl}_4}} = 3.54 \times 10^1\ \text{mol TiO}_2$$

The mass of TiO_2 expected for 100% yield is

$$3.54 \times 10^1\ \cancel{\text{mol TiO}_2} \times \frac{79.88\ \text{g TiO}_2}{\cancel{\text{mol TiO}_2}} = 2.83 \times 10^3\ \text{g TiO}_2$$

This amount represents the theoretical yield.

b. Because the reaction is said to give only a 75.0% yield of TiO_2, we use the definition of percent yield,

$$\frac{\text{Actual yield}}{\text{Theoretical yield}} \times 100\% = \%\ \text{yield}$$

to write the equation

$$\frac{\text{Actual yield}}{2.83 \times 10^3\ \text{g TiO}_2} \times 100\% = 75.0\%\ \text{yield}$$

We now want to solve for the actual yield. First we divide both sides by 100%.

$$\frac{\text{Actual yield}}{2.83 \times 10^3\ \text{g TiO}_2} \times \frac{100\%}{100\%} = \frac{75.0}{100} = 0.750$$

Then we multiply both sides by 2.83×10^3 g TiO_2.

$$2.83 \times 10^3\ \text{g TiO}_2 \times \frac{\text{Actual yield}}{2.83 \times 10^3\ \text{g TiO}_2} = 0.750 \times 2.83 \times 10^3\ \text{g TiO}_2$$

$$\text{Actual yield} = 0.750 \times 2.83 \times 10^3\ \text{g TiO}_2 = 2.12 \times 10^3\ \text{g TiO}_2$$

Thus 2.12×10^3 g of $TiO_2(s)$ is actually obtained in this reaction.

Chapter 10

Self-Check Exercise 10.1

a. Circular pathways for electrons in the Bohr model.

b. Three-dimensional probability maps that represent the likelihood that the electron will occupy a given point in space.

c. The surface that contains 90% of the total electron probability.

d. A set of orbitals of a given type of orbital within a principal energy level. For example, there are three sublevels in principal energy level 3 (*s*, *p*, *d*).

Self-Check Exercise 10.2

Element	*Electron Configuration*	*Orbital Diagram* 1s	2s	2p	3s	3p
Al	$1s^22s^22p^63s^23p^1$ $[Ne]3s^23p^1$	↑↓	↑↓	↑↓ ↑↓ ↑↓	↑↓	↑ ☐ ☐
Si	$[Ne]3s^23p^2$	↑↓	↑↓	↑↓ ↑↓ ↑↓	↑↓	↑ ↑ ☐
P	$[Ne]3s^23p^3$	↑↓	↑↓	↑↓ ↑↓ ↑↓	↑↓	↑ ↑ ↑
S	$[Ne]3s^23p^4$	↑↓	↑↓	↑↓ ↑↓ ↑↓	↑↓	↑↓ ↑ ↑
Cl	$[Ne]3s^23p^5$	↑↓	↑↓	↑↓ ↑↓ ↑↓	↑↓	↑↓ ↑↓ ↑
Ar	$[Ne]3s^23p^6$	↑↓	↑↓	↑↓ ↑↓ ↑↓	↑↓	↑↓ ↑↓ ↑↓

Self-Check Exercise 10.3

F: $1s^22s^22p^5$ or $[\text{He}]2s^22p^5$

Si: $1s^22s^22p^63s^23p^2$ or $[\text{Ne}]3s^23p^2$

Cs: $1s^22s^22p^63s^23p^64s^23d^{10}4p^65s^24d^{10}5p^66s^1$ or $[\text{Xe}]6s^1$

Pb: $1s^22s^22p^63s^23p^64s^23d^{10}4p^65s^24d^{10}5p^66s^24f^{14}5d^{10}6p^2$ or $[\text{Xe}]6s^24f^{14}5d^{10}6p^2$

I: $1s^22s^22p^63s^23p^64s^23d^{10}4p^65s^24d^{10}5p^5$ or $[\text{Kr}]5s^24d^{10}5p^5$

Silicon (Si): In Group 4 and Period 3, it is the second of the "$3p$ elements." The configuration is $1s^22s^22p^63s^23p^2$, or $[\text{Ne}]3s^23p^2$.

Cesium (Cs): In Group 1 and Period 6, it is the first of the "$6s$ elements." The configuration is $1s^22s^22p^63s^23p^64s^23d^{10}4p^65s^24d^{10}5p^66s^1$, or $[\text{Xe}]6s^1$.

Lead (Pb): In Group 4 and Period 6, it is the second of the "$6p$ elements." The configuration is $[\text{Xe}]6s^24f^{14}5d^{10}6p^2$.

Iodine (I): In Group 7 and Period 5, it is the fifth of the "$5p$ elements." The configuration is $[\text{Kr}]5s^24d^{10}5p^5$.

Chapter 11

Self-Check Exercise 11.1

Using the electronegativity values given in Figure 11.3, we choose the bond in which the atoms exhibit the largest difference in electronegativity. (Electronegativity values are shown in parentheses.)

a. H—C (2.1)(2.5) > H—P (2.1)(2.1)

b. O—I (3.5)(2.5) > O—F (3.5)(4.0)

c. S—O (2.5)(3.5) > N—O (3.0)(3.5)

d. N—H (3.0)(2.1) > Si—H (1.8)(2.1)

Self-Check Exercise 11.2

H has one electron, and Cl has seven valence electrons. This gives a total of eight valence electrons. We first draw in the bonding pair:

H—Cl, which could be drawn as H:Cl

We have six electrons yet to place. The H already has two electrons, so we place three lone pairs around the chlorine to satisfy the octet rule.

$$\text{H—}\ddot{\underset{\cdot\cdot}{\text{Cl}}}\text{:} \quad \text{or} \quad \text{H:}\ddot{\underset{\cdot\cdot}{\text{Cl}}}\text{:}$$

Self-Check Exercise 11.3

STEP 1 O_3: 3(6) = 18 valence electrons

STEP 2 O—O—O

STEP 3 $\ddot{\text{O}}\text{=}\ddot{\text{O}}\text{—}\ddot{\underset{\cdot\cdot}{\text{O}}}\text{:}$ and $\text{:}\ddot{\underset{\cdot\cdot}{\text{O}}}\text{—}\ddot{\text{O}}\text{=}\ddot{\text{O}}$

This molecule shows resonance (it has two valid Lewis structures).

Self-Check Exercise 11.4

See table on top of page A5.

Self-Check Exercise 11.5

a. NH_4^+

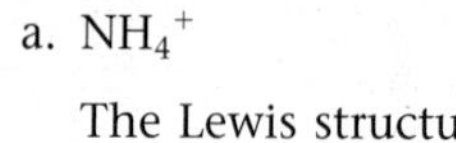

The Lewis structure is
$$\left[\begin{array}{ccc} & \text{H} & \\ & | & \\ \text{H—} & \text{N} & \text{—H} \\ & | & \\ & \text{H} & \end{array}\right]^+$$

(See Self-Check Exercise 11.4.) There are four pairs of electrons around the nitrogen. This requires a tetrahedral arrangement of electron pairs. The NH_4^+ ion has a tetrahedral molecular structure (case 3 in Table 11.4), because all electron pairs are shared.

b. SO_4^{2-}

The Lewis structure is
$$\left[\begin{array}{ccc} & \text{:}\ddot{\text{O}}\text{:} & \\ & | & \\ \text{:}\ddot{\underset{\cdot\cdot}{\text{O}}}\text{—} & \text{S} & \text{—}\ddot{\underset{\cdot\cdot}{\text{O}}}\text{:} \\ & | & \\ & \text{:}\underset{\cdot\cdot}{\text{O}}\text{:} & \end{array}\right]^{2-}$$

(See Self-Check Exercise 11.4.) The four electron pairs around the sulfur require a tetrahedral arrangement. The SO_4^{2-} has a tetrahedral molecular structure (case 3 in Table 11.4).

c. NF_3

The Lewis structure is
$$\begin{array}{ccc} & \ddot{\text{N}} & \\ \text{:}\ddot{\underset{\cdot\cdot}{\text{F}}} & | & \ddot{\underset{\cdot\cdot}{\text{F}}}\text{:} \\ & \text{:}\underset{\cdot\cdot}{\text{F}}\text{:} & \end{array}$$

(See Self-Check Exercise 11.4.) The four pairs of electrons on the nitrogen require a tetrahedral arrangement. In this case only three of the pairs are shared with the fluorine atoms, leaving one lone pair. Thus the molecular structure is a trigonal pyramid (case 4 in Table 11.4).

d. H_2S

The Lewis structure is $\text{H—}\ddot{\underset{\cdot\cdot}{\text{S}}}\text{—H}$

(See Self-Check Exercise 11.4.) The four pairs of electrons around the sulfur require a tetrahedral arrangement. In this case two pairs are shared with hydrogen atoms, leaving two lone pairs. Thus the molecular structure is bent or V-shaped (case 5 in Table 11.4).

e. ClO_3^-

The Lewis structure is
$$\left[\begin{array}{ccc} \text{:}\ddot{\underset{\cdot\cdot}{\text{O}}}\text{—} & \ddot{\text{Cl}} & \text{—}\ddot{\underset{\cdot\cdot}{\text{O}}}\text{:} \\ & | & \\ & \text{:}\underset{\cdot\cdot}{\text{O}}\text{:} & \end{array}\right]^-$$

(See Self-Check Exercise 11.4.) The four pairs of electrons require a tetrahedral arrangement. In this case, three pairs are shared with oxygen atoms, leaving one lone pair. Thus the molecular structure is a trigonal pyramid (case 4 in Table 11.4).

f. BeF_2

The Lewis structure is $\text{:}\ddot{\underset{\cdot\cdot}{\text{F}}}\text{—Be—}\ddot{\underset{\cdot\cdot}{\text{F}}}\text{:}$

The two electron pairs on beryllium require a linear arrangement. Because both pairs are shared by fluorine atoms, the molecular structure is also linear (case 1 in Table 11.4).

Chapter 12

Self-Check Exercise 12.1

We know that 1.000 atm = 760.0 mm Hg. So

$$525 \cancel{\text{mm Hg}} \times \frac{1.000 \text{ atm}}{760.0 \cancel{\text{mm Hg}}} = 0.691 \text{ atm}$$

Self-Check Exercise 12.2

Initial Conditions	*Final Conditions*
$P_1 = 635$ torr	$P_2 = 785$ torr
$V_1 = 1.51$ L	$V_2 = ?$

Molecule or Ion	Total Valence Electrons	Draw Single Bonds	Calculate Number of Electrons Remaining	Use Remaining Electrons to Achieve Noble Gas Configurations	Check	
					Atom	Electrons
a. NF_3	5 + 3(7) = 26	F—N(—F)—F	26 − 6 = 20	:F̤—N̈—F̤: with :F̤: below N	N F	8 8
b. O_2	2(6) = 12	O—O	12 − 2 = 10	:Ö=Ö:	O	8
c. CO	4 + 6 = 10	C—O	10 − 2 = 8	:C≡O:	C O	8 8
d. PH_3	5 + 3(1) = 8	H—P(—H)—H	8 − 6 = 2	H—P̈—H with H below P	P H	8 2
e. H_2S	2(1) + 6 = 8	H—S—H	8 − 4 = 4	H—S̈—H (two lone pairs on S)	S H	8 2
f. SO_4^{2-}	6+4(6) + 2 = 32	O—S(—O)(—O)—O	32 − 8 = 24	[:Ö—S(—Ö:)(—Ö:)—Ö:]$^{2-}$ (each O with three lone pairs)	S O	8 8
g. NH_4^+	5+4(1) − 1 = 8	H—N(—H)(—H)—H	8 − 8 = 0	[H—N(—H)(—H)—H]$^+$	N H	8 2
h. ClO_3^-	7 + 3(6) + 1 = 26	O—Cl(—O)—O	26 − 6 = 20	[:Ö—C̈l—Ö: with :Ö: below Cl]$^-$	Cl O	8 8
i. SO_2	6 + 2(6) = 18	O—S—O	18 − 4 = 14	Ö=S̈—Ö: and :Ö—S̈=Ö	S O	8 8

Answer to Self-Check Exercise 11.4.

Solving Boyle's law ($P_1V_1 = P_2V_2$) for V_2 gives

$$V_2 = V_1 \times \frac{P_1}{P_2}$$

$$= 1.51\ \text{L} \times \frac{635\ \cancel{\text{torr}}}{785\ \cancel{\text{torr}}} = 1.22\ \text{L}$$

Note that the volume decreased, as the increase in pressure led us to expect.

Self-Check Exercise 12.3

Because the temperature of the gas inside the bubble decreases (at constant pressure), the bubble gets smaller. The conditions are

Initial Conditions

$T_1 = 28\ °\text{C} = 28 + 273 = 301\ \text{K}$
$V_1 = 23\ \text{cm}^3$

Final Conditions

$T_2 = 18\ °\text{C} = 18 + 273 = 291\ \text{K}$
$V_2 = ?$

Solving Charles's law,

$$\frac{V_1}{T_1} = \frac{V_2}{T_2}$$

for V_2 gives

$$V_2 = V_1 \times \frac{T_2}{T_1} = 23\ \text{cm}^3 \times \frac{291\ \cancel{\text{K}}}{301\ \cancel{\text{K}}} = 22\ \text{cm}^3$$

Self-Check Exercise 12.4

Because the temperature and pressure of the two samples are the same, we can use Avogadro's law in the form

$$\frac{V_1}{n_1} = \frac{V_2}{n_2}$$

The following information is given:

Sample 1	*Sample 2*
$V_1 = 36.7$ L	$V_2 = 16.5$ L
$n_1 = 1.5$ mol	$n_2 = ?$

We can now solve Avogadro's law for the value of n_2 (the moles of N_2 in Sample 2):

$$n_2 = n_1 \times \frac{V_2}{V_1} = 1.5 \text{ mol} \times \frac{16.5 \not{L}}{36.7 \not{L}} = 0.67 \text{ mol}$$

Here n_2 is smaller than n_1, which makes sense in view of the fact that V_2 is smaller than V_1.

Note: We isolate n_2 from Avogadro's law as given above by multiplying both sides of the equation by n_2 and then by n_1/V_1,

$$\left(n_2 \times \frac{n_1}{V_1}\right)\frac{V_1}{n_1} = \left(n_2 \times \frac{n_1}{V_1}\right)\frac{V_2}{n_2}$$

to give $n_2 = n_1 \times V_2/V_1$.

Self-Check Exercise 12.5

We are given the following information:

$P = 1.00$ atm
$V = 2.70 \times 10^6$ L
$n = 1.10 \times 10^5$ mol

We solve for T by dividing both sides of the ideal gas law by nR:

$$\frac{PV}{nR} = \frac{\not{n}\not{R}T}{\not{n}\not{R}}$$

to give

$$T = \frac{PV}{nR} = \frac{(1.00 \text{ atm})(2.70 \times 10^6 \text{ L})}{(1.10 \times 10^5 \text{ mol})\left(0.08206 \frac{\text{L atm}}{\text{K mol}}\right)}$$

$$= 299 \text{ K}$$

The temperature of the helium is 299 K, or 299 − 273 = 26 °C.

Self-Check Exercise 12.6

We are given the following information about the radon sample:

$n = 1.5$ mol
$V = 21.0$ L
$T = 33\ °C = 33 + 273 = 306$ K
$P = ?$

We solve the ideal gas law ($PV = nRT$) for P by dividing both sides of the equation by V:

$$P = \frac{nRT}{V} = \frac{(1.5 \text{ mol})\left(0.08206 \frac{\text{L atm}}{\text{K mol}}\right)(306 \text{ K})}{21.0 \text{ L}}$$

$$= 1.8 \text{ atm}$$

Self-Check Exercise 12.7

To solve this problem we take the ideal gas law and separate those quantities that change from those that remain constant (on opposite sides of the equation). In this case volume and temperature change, and number of moles and pressure (and, of course, R) remain constant. So $PV = nRT$ becomes $V/T = nR/P$, which leads to

$$\frac{V_1}{T_1} = \frac{nR}{P} \quad \text{and} \quad \frac{V_2}{T_2} = \frac{nR}{P}$$

Combining these gives

$$\frac{V_1}{T_1} = \frac{nR}{P} = \frac{V_2}{T_2} \quad \text{or} \quad \frac{V_1}{T_1} = \frac{V_2}{T_2}$$

We are given

Initial Conditions

$T_1 = 5\ °C = 5 + 273 = 278$ K
$V_1 = 3.8$ L

Final Conditions

$T_2 = 86\ °C = 86 + 273 = 359$ K
$V_2 = ?$

Thus

$$V_2 = \frac{T_2V_1}{T_1} = \frac{(359 \text{ K})(3.8 \text{ L})}{278 \text{ K}} = 4.9 \text{ L}$$

Check: Is the answer sensible? In this case the temperature was increased (at constant pressure), so the volume should increase. The answer makes sense.

Note that this problem could be described as a "Charles's law problem." The real advantage of using the ideal gas law is that you need to remember only *one* equation to do virtually any problem involving gases.

Self-Check Exercise 12.8

We are given the following information:

Initial Conditions

$P_1 = 0.747$ atm
$T_1 = 13\ °C = 13 + 273 = 286$ K
$V_1 = 11.0$ L

Final Conditions

$P_2 = 1.18$ atm
$T_2 = 56\ °C = 56 + 273 = 329$ K
$V_2 = ?$

In this case the number of moles remains constant. Thus we can say

$$\frac{P_1V_1}{T_1} = nR \quad \text{and} \quad \frac{P_2V_2}{T_2} = nR$$

or

$$\frac{P_1V_1}{T_1} = \frac{P_2V_2}{T_2}$$

Solving for V_2 gives

$$V_2 = V_1 \times \frac{T_2}{T_1} \times \frac{P_1}{P_2} = (11.0 \text{ L})\left(\frac{329 \text{ K}}{286 \text{ K}}\right)\left(\frac{0.747 \text{ atm}}{1.18 \text{ atm}}\right)$$

$$= 8.01 \text{ L}$$

Self-Check Exercise 12.9

As usual when dealing with gases, we can use the ideal gas equation $PV = nRT$. First consider the information given:

$P = 0.91 \text{ atm} = P_{total}$
$V = 2.0$ L
$T = 25\ °C = 25 + 273 = 298$ K

Given this information, we can calculate the number of moles of gas in the mixture:

$n_{total} = n_{N_2} + n_{O_2}$. Solving for n in the ideal gas equation gives

$$n_{total} = \frac{P_{total}V}{RT} = \frac{(0.91\ \text{atm})(2.0\ \text{L})}{\left(0.08206\ \frac{\text{L atm}}{\text{K mol}}\right)(298\ \text{K})} = 0.074\ \text{mol}$$

We also know that 0.050 mol of N_2 is present. Because

$$n_{total} = \underset{(0.050\ \text{mol})}{\underset{\uparrow}{n_{N_2}}} + n_{O_2} = 0.074\ \text{mol}$$

we can calculate the moles of O_2 present.

$$0.050\ \text{mol} + n_{O_2} = 0.074\ \text{mol}$$
$$n_{O_2} = 0.074\ \text{mol} - 0.050\ \text{mol} = 0.024\ \text{mol}$$

Now that we know the moles of oxygen present, we can calculate the partial pressure of oxygen from the ideal gas equation.

$$P_{O_2} = \frac{n_{O_2}RT}{V} = \frac{(0.024\ \text{mol})\left(0.08206\ \frac{\text{L atm}}{\text{K mol}}\right)(298\ \text{K})}{2.0\ \text{L}}$$
$$= 0.29\ \text{atm}$$

Although it is not requested, note that the partial pressure of the N_2 must be 0.62 atm, because

$$\underbrace{0.62\ \text{atm}}_{P_{N_2}} + \underbrace{0.29\ \text{atm}}_{P_{O_2}} = \underbrace{0.91\ \text{atm}}_{P_{total}}$$

Self-Check Exercise 12.10

The volume is 0.500 L, the temperature is 25 °C (or 25 + 273 = 298 K), and the total pressure is given as 0.950 atm. Of this total pressure, 24 torr is due to the water vapor. We can calculate the partial pressure of the H_2 because we know that

$$P_{total} = P_{H_2} + \underset{24\ \text{torr}}{\underset{\uparrow}{P_{H_2O}}} = 0.950\ \text{atm}$$

Before we carry out the calculation, however, we must convert the pressures to the same units. Converting P_{H_2O} to atmospheres gives

$$24\ \text{torr} \times \frac{1.000\ \text{atm}}{760.0\ \text{torr}} = 0.032\ \text{atm}$$

Thus

$$P_{total} = P_{H_2} + P_{H_2O} = 0.950\ \text{atm} = P_{H_2} + 0.032\ \text{atm}$$

and

$$P_{H_2} = 0.950\ \text{atm} - 0.032\ \text{atm} = 0.918\ \text{atm}$$

Now that we know the partial pressure of the hydrogen gas, we can use the ideal gas equation to calculate the moles of H_2.

$$n_{H_2} = \frac{P_{H_2}V}{RT} = \frac{(0.918\ \text{atm})(0.500\ \text{L})}{\left(0.08206\ \frac{\text{L atm}}{\text{K mol}}\right)(298\ \text{K})}$$
$$= 0.0188\ \text{mol} = 1.88 \times 10^{-2}\ \text{mol}$$

The sample of gas contains 1.88×10^{-2} mol of H_2, which exerts a partial pressure of 0.918 atm.

Self-Check Exercise 12.11

We will solve this problem by taking the following steps:

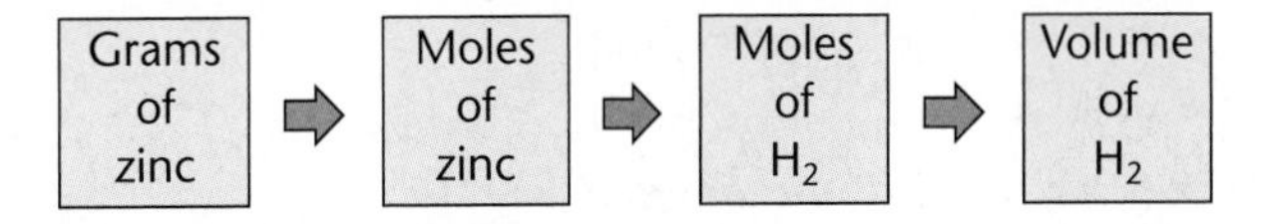

STEP 1 Using the atomic mass of zinc (65.38), we calculate the moles of zinc in 26.5 g.

$$26.5\ \text{g Zn} \times \frac{1\ \text{mol Zn}}{65.38\ \text{g Zn}} = 0.405\ \text{mol Zn}$$

STEP 2 Using the balanced equation, we next calculate the moles of H_2 produced.

$$0.405\ \text{mol Zn} \times \frac{1\ \text{mol}\ H_2}{1\ \text{mol Zn}} = 0.405\ \text{mol}\ H_2$$

STEP 3 Now that we know the moles of H_2, we can compute the volume of H_2 by using the ideal gas law, where

$P = 1.50$ atm
$V = ?$
$n = 0.405$ mol
$R = 0.08206$ L atm/K mol
$T = 19\ °C = 19 + 273 = 292$ K

$$V = \frac{nRT}{P} = \frac{(0.405\ \text{mol})\left(0.08206\ \frac{\text{L atm}}{\text{K mol}}\right)(292\ \text{K})}{1.50\ \text{atm}}$$
$$= 6.47\ \text{L of}\ H_2$$

Self-Check Exercise 12.12

Although there are several possible ways to do this problem, the most convenient method involves using the molar volume at STP. First we use the ideal gas equation to calculate the moles of NH_3 present:

$$n = \frac{PV}{RT}$$

where $P = 15.0$ atm, $V = 5.00$ L, and $T = 25\ °C + 273 = 298$ K.

$$n = \frac{(15.0\ \text{atm})(5.00\ \text{L})}{\left(0.08206\ \frac{\text{L atm}}{\text{K mol}}\right)(298\ \text{K})} = 3.07\ \text{mol}$$

We know that at STP each mole of gas occupies 22.4 L. Therefore, 3.07 mol has the volume

$$3.07\ \text{mol} \times \frac{22.4\ \text{L}}{1\ \text{mol}} = 68.8\ \text{L}$$

The volume of the ammonia at STP is 68.8 L.

Chapter 13

Self-Check Exercise 13.1

Energy to melt the ice:

$$15\ \text{g}\ H_2O \times \frac{1\ \text{mol}\ H_2O}{18\ \text{g}\ H_2O} = 0.83\ \text{mol}\ H_2O$$

$$0.83\ \text{mol}\ H_2O \times 6.02\ \frac{\text{kJ}}{\text{mol}\ H_2O} = 5.0\ \text{kJ}$$

Energy to heat the water from 0 °C to 100 °C:

$$4.18 \frac{\text{J}}{\text{g}\,^\circ\text{C}} \times 15 \text{ g} \times 100\ ^\circ\text{C} = 6300 \text{ J}$$

$$6300 \text{ J} \times \frac{1 \text{ kJ}}{1000 \text{ J}} = 6.3 \text{ kJ}$$

Energy to vaporize the water at 100 °C:

$$0.83 \text{ mol } H_2O \times 40.6 \frac{\text{kJ}}{\text{mol } H_2O} = 34 \text{ kJ}$$

Total energy required:

$$5.0 \text{ kJ} + 6.3 \text{ kJ} + 34 \text{ kJ} = 45 \text{ kJ}$$

Self-Check Exercise 13.2

a. Contains SO_3 molecules—a molecular solid.

b. Contains Ba^{2+} and O^{2-} ions—an ionic solid.

c. Contains Au atoms—an atomic solid.

Chapter 14

Self-Check Exercise 14.1

$$\text{Mass percent} = \frac{\text{mass of solute}}{\text{mass of solution}} \times 100\%$$

For this sample, the mass of solution is 135 g and the mass of the solute is 4.73 g, so

$$\text{Mass percent} = \frac{4.73 \text{ g solute}}{135 \text{ g solution}} \times 100\% = 3.50\%$$

Self-Check Exercise 14.2

Using the definition of mass percent, we have

$$\frac{\text{Mass of solute}}{\text{Mass of solution}} = \frac{\text{grams of solute}}{\text{grams of solute} + \text{grams of solvent}} \times 100\% = 40.0\%$$

There are 425 grams of solute (formaldehyde). Substituting, we have

$$\frac{425 \text{ g}}{425 \text{ g} + \text{grams of solvent}} \times 100\% = 40.0\%$$

We must now solve for grams of solvent (water). This will take some patience, but we can do it if we proceed step by step. First we divide both sides by 100%.

$$\frac{425 \text{ g}}{425 \text{ g} + \text{grams of solvent}} \times \frac{100\%}{100\%} = \frac{40.0\%}{100\%} = 0.400$$

Now we have

$$\frac{425 \text{ g}}{425 \text{ g} + \text{grams of solvent}} = 0.400$$

Next we multiply both sides by (425 g + grams of solvent).

$$(425 \text{ g} + \text{grams of solvent}) \times \frac{425 \text{ g}}{425 \text{ g} + \text{grams of solvent}} = 0.400 \times (425 \text{ g} + \text{grams of solvent})$$

This gives

$$425 \text{ g} = 0.400 \times (425 \text{ g} + \text{grams of solvent})$$

Carrying out the multiplication gives

$$425 \text{ g} = 170. \text{ g} + 0.400 \text{ (grams of solvent)}$$

Now we subtract 170. g from both sides,

$$425 \text{ g} - 170. \text{ g} = 170. \text{ g} - 170. \text{ g} + 0.400 \text{ (grams of solvent)}$$

$$255 \text{ g} = 0.400 \text{ (grams of solvent)}$$

and divide both sides by 0.400.

$$\frac{255 \text{ g}}{0.400} = \frac{0.400}{0.400} \text{ (grams of solvent)}$$

We finally have the answer:

$$\frac{255 \text{ g}}{0.400} = 638 \text{ g} = \text{grams of solvent} = \text{mass of water needed}$$

Self-Check Exercise 14.3

The moles of ethanol can be obtained from its molar mass (46.1).

$$1.00 \text{ g } C_2H_5OH \times \frac{1 \text{ mol } C_2H_5OH}{46.1 \text{ g } C_2H_5OH} = 2.17 \times 10^{-2} \text{ mol } C_2H_5OH$$

$$\text{Volume in liters} = 101 \text{ mL} \times \frac{1 \text{ L}}{1000 \text{ mL}} = 0.101 \text{ L}$$

$$\text{Molarity of } C_2H_5OH = \frac{\text{moles of } C_2H_5OH}{\text{liters of solution}} = \frac{2.17 \times 10^{-2} \text{ mol}}{0.101 \text{ L}} = 0.215\ M$$

Self-Check Exercise 14.4

When Na_2CO_3 and $Al_2(SO_4)_3$ dissolve in water, they produce ions as follows:

$$Na_2CO_3(s) \xrightarrow{H_2O(l)} 2Na^+(aq) + CO_3^{2-}(aq)$$

$$Al_2(SO_4)_3(s) \xrightarrow{H_2O(l)} 2Al^{3+}(aq) + 3SO_4^{2-}(aq)$$

Therefore, in a 0.10 M Na_2CO_3 solution, the concentration of Na^+ ions is $2 \times 0.10\ M = 0.20\ M$ and the concentration of CO_3^{2-} ions is 0.10 M. In a 0.010 M $Al_2(SO_4)_3$ solution, the concentration of Al^{3+} ions is $2 \times 0.010\ M = 0.020\ M$ and the concentration of SO_4^{2-} ions is $3 \times 0.010\ M = 0.030\ M$.

Self-Check Exercise 14.5

When solid $AlCl_3$ dissolves, it produces ions as follows:

$$AlCl_3(s) \xrightarrow{H_2O(l)} Al^{3+}(aq) + 3Cl^-(aq)$$

so a $1.0 \times 10^{-3}\ M$ $AlCl_3$ solution contains $1.0 \times 10^{-3}\ M$ Al^{3+} ions and $3.0 \times 10^{-3}\ M$ Cl^- ions.

To calculate the moles of Cl^- ions in 1.75 L of the $1.0 \times 10^{-3}\ M$ $AlCl_3$ solution, we must multiply the volume by the molarity.

$$1.75 \text{ L solution} \times 3.0 \times 10^{-3}\ M\ Cl^-$$

$$= 1.75 \text{ L solution} \times \frac{3.0 \times 10^{-3} \text{ mol } Cl^-}{\text{L solution}}$$

$$= 5.25 \times 10^{-3} \text{ mol } Cl^- = 5.3 \times 10^{-3} \text{ mol } Cl^-$$

Self-Check Exercise 14.6

We must first determine the number of moles of formaldehyde in 2.5 L of 12.3 M formalin. Remember that volume of solution (in liters) times molarity gives moles of solute.

In this case, the volume of solution is 2.5 L and the molarity is 12.3 mol of HCHO per liter of solution.

$$2.5 \text{ L solution} \times \frac{12.3 \text{ mol HCHO}}{\text{L solution}} = 31 \text{ mol HCHO}$$

Next, using the molar mass of HCHO (30.0 g), we convert 31 mol of HCHO to grams.

$$31 \text{ mol HCHO} \times \frac{30.0 \text{ g HCHO}}{1 \text{ mol HCHO}} = 9.3 \times 10^2 \text{ g HCHO}$$

Therefore, 2.5 L of 12.3 *M* formalin contains 9.3×10^2 g of formaldehyde. We must weigh out 930 g of formaldehyde and dissolve it in enough water to make 2.5 L of solution.

Self-Check Exercise 14.7

We are given the following information:

$$M_1 = 12 \frac{\text{mol}}{\text{L}} \qquad M_2 = 0.25 \frac{\text{mol}}{\text{L}}$$

$$V_1 = ? \text{ (what we need to find)} \qquad V_2 = 0.75 \text{ L}$$

Using the fact that the moles of solute do not change upon dilution, we know that

$$M_1 \times V_1 = M_2 \times V_2$$

Solving for V_1 by dividing both sides by M_1 gives

$$V_1 = \frac{M_2 \times V_2}{M_1} = \frac{0.25 \frac{\text{mol}}{\text{L}} \times 0.75 \text{ L}}{12 \frac{\text{mol}}{\text{L}}}$$

and

$$V_1 = 0.016 \text{ L} = 16 \text{ mL}$$

Self-Check Exercise 14.8

STEP 1 When the aqueous solutions of Na_2SO_4 (containing Na^+ and SO_4^{2-} ions) and $Pb(NO_3)_2$ (containing Pb^{2+} and NO_3^- ions) are mixed, solid $PbSO_4$ is formed.

$$Pb^{2+}(aq) + SO_4^{2-}(aq) \rightarrow PbSO_4(s)$$

STEP 2 We must first determine whether Pb^{2+} or SO_4^{2-} is the limiting reactant by calculating the moles of Pb^{2+} and SO_4^{2-} ions present. Because 0.0500 *M* $Pb(NO_3)_2$ contains 0.0500 *M* Pb^{2+} ions, we can calculate the moles of Pb^{2+} ions in 1.25 L of this solution as follows:

$$1.25 \text{ L} \times \frac{0.0500 \text{ mol } Pb^{2+}}{\text{L}} = 0.0625 \text{ mol } Pb^{2+}$$

The 0.0250 *M* Na_2SO_4 solution contains 0.0250 *M* SO_4^{2-} ions, and the number of moles of SO_4^{2-} ions in 2.00 L of this solution is

$$2.00 \text{ L} \times \frac{0.0250 \text{ mol } SO_4^{2-}}{\text{L}} = 0.0500 \text{ mol } SO_4^{2-}$$

STEP 3 Pb^{2+} and SO_4^{2-} react in a 1:1 ratio, so the amount of SO_4^{2-} ions is limiting because SO_4^{2-} is present in the smaller number of moles.

STEP 4 The Pb^{2+} ions are present in excess, and only 0.0500 mol of solid $PbSO_4$ will be formed.

STEP 5 We calculate the mass of $PbSO_4$ by using the molar mass of $PbSO_4$ (303.3 g).

$$0.0500 \text{ mol } PbSO_4 \times \frac{303.3 \text{ g } PbSO_4}{1 \text{ mol } PbSO_4} = 15.2 \text{ g } PbSO_4$$

Self-Check Exercise 14.9

STEP 1 Because nitric acid is a strong acid, the nitric acid solution contains H^+ and NO_3^- ions. The KOH solution contains K^+ and OH^- ions. When these solutions are mixed, the H^+ and OH^- react to form water.

$$H^+(aq) + OH^-(aq) \rightarrow H_2O(l)$$

STEP 2 The number of moles of OH^- present in 125 mL of 0.050 *M* KOH is

$$125 \text{ mL} \times \frac{1 \text{ L}}{1000 \text{ mL}} \times \frac{0.050 \text{ mol } OH^-}{\text{L}} = 6.3 \times 10^{-3} \text{ mol } OH^-$$

STEP 3 H^+ and OH^- react in a 1:1 ratio, so we need 6.3×10^{-3} mol of H^+ from the 0.100 *M* HNO_3.

STEP 4 6.3×10^{-3} mol of OH^- requires 6.3×10^{-3} mol of H^+ to form 6.3×10^{-3} mol of H_2O.
Therefore,

$$V \times \frac{0.100 \text{ mol } H^+}{\text{L}} = 6.3 \times 10^{-3} \text{ mol } H^+$$

where *V* represents the volume in liters of 0.100 *M* HNO_3 required. Solving for *V*, we have

$$V = \frac{6.3 \times 10^{-3} \text{ mol } H^+}{\frac{0.100 \text{ mol } H^+}{\text{L}}} = 6.3 \times 10^{-2} \text{ L}$$

$$= 6.3 \times 10^{-2} \text{ L} \times \frac{1000 \text{ mL}}{\text{L}} = 63 \text{ mL}$$

Self-Check Exercise 14.10

From the definition of normality, $N = \text{equiv/L}$, we need to calculate (1) the equivalents of KOH and (2) the volume of the solution in liters. To find the number of equivalents, we use the equivalent weight of KOH, which is 56.1 g (see Table 14.2).

$$23.6 \text{ g KOH} \times \frac{1 \text{ equiv KOH}}{56.1 \text{ g KOH}} = 0.421 \text{ equiv KOH}$$

Next we convert the volume to liters.

$$755 \text{ mL} \times \frac{1 \text{ L}}{1000 \text{ mL}} = 0.755 \text{ L}$$

Finally, we substitute these values into the equation that defines normality.

$$\text{Normality} = \frac{\text{equiv}}{\text{L}} = \frac{0.421 \text{ equiv}}{0.755 \text{ L}} = 0.558\, N$$

Self-Check Exercise 14.11

To solve this problem, we use the relationship

$$N_{acid} \times V_{acid} = N_{base} \times V_{base}$$

where

$$N_{acid} = 0.50 \frac{\text{equiv}}{\text{L}}$$

$$V_{acid} = ?$$

$$N_{base} = 0.80 \frac{\text{equiv}}{\text{L}}$$

$$V_{base} = 0.250 \text{ L}$$

We solve the equation

$$N_{acid} \times V_{acid} = N_{base} \times V_{base}$$

for V_{acid} by dividing both sides by N_{acid}.

$$\frac{N_{acid} \times V_{acid}}{N_{acid}} = \frac{N_{base} \times V_{base}}{N_{acid}}$$

$$V_{acid} = \frac{N_{base} \times V_{base}}{N_{acid}} = \frac{\left(0.80 \frac{\text{equiv}}{\text{L}}\right) \times (0.250\ \text{L})}{0.50 \frac{\text{equiv}}{\text{L}}}$$

$$V_{acid} = 0.40\ \text{L}$$

Therefore, 0.40 L of 0.50 N H_2SO_4 is required to neutralize 0.250 L of 0.80 N KOH.

Chapter 15

Self-Check Exercise 15.1

The conjugate acid–base pairs are

H_2O,	H_3O^+
Base	Conjugate acid

and

$HC_2H_3O_2$,	$C_2H_3O_2^-$
Acid	Conjugate base

The members of both pairs differ by one H^+.

Self-Check Exercise 15.2

Because $[H^+][OH^-] = 1.0 \times 10^{-14}$, we can solve for $[H^+]$.

$$[H^-] = \frac{1.0 \times 10^{-14}}{[OH^-]} = \frac{1.0 \times 10^{-14}}{2.0 \times 10^{-2}} = 5.0 \times 10^{-13}\ M$$

This solution is basic: $[OH^-] = 2.0 \times 10^{-2}\ M$ is greater than $[H^+] = 5.0 \times 10^{-13}\ M$.

Self-Check Exercise 15.3

a. Because $[H^+] = 1.0 \times 10^{-3}\ M$, we get pH = 3.00 by using the regular steps.

b. Because $[OH^-] = 5.0 \times 10^{-5}\ M$, we can find $[H^+]$ from the K_w expression.

$$[H^+] = \frac{K_w}{[OH^-]} = \frac{1.0 \times 10^{-14}}{5.0 \times 10^{-5}} = 2.0 \times 10^{-10}\ M$$

Then we follow the regular steps to get pH from $[H^+]$.

1. Enter 2.0×10^{-10}.
2. Push [log].
3. Push [+/−].

pH = 9.70

Self-Check Exercise 15.4

$$\text{pOH} + \text{pH} = 14.00$$
$$\text{pOH} = 14.00 - \text{pH} = 14.00 - 3.5$$
$$\text{pOH} = 10.5$$

Self-Check Exercise 15.5

STEP 1 pH = 3.50

STEP 2 −pH = −3.50

STEP 3 [inv] [log] $-3.50 = 3.2 \times 10^{-4}$

$[H^+] = 3.2 \times 10^{-4}\ M$

Self-Check Exercise 15.6

STEP 1 pOH = 10.50

STEP 2 −pOH = −10.50

STEP 3 [inv] [log] $-10.50 = 3.2 \times 10^{-11}$

$[OH^-] = 3.2 \times 10^{-11}\ M$

Self-Check Exercise 15.7

Because HCl is a strong acid, it is completely dissociated:

$$5.0 \times 10^{-3}\ M\ \text{HCl} \rightarrow 5.0 \times 10^{-3}\ M\ H^+ \text{ and } 5.0 \times 10^{-3}\ M\ Cl^-$$

so $[H^+] = 5.0 \times 10^{-3}\ M$.

$$\text{pH} = -\log(5.0 \times 10^{-3}) = 2.30$$

Chapter 16

Self-Check Exercise 16.1

Applying the law of chemical equilibrium gives

$$K = \frac{[NO_2]^4[H_2O]^6}{[NH_3]^4[O_2]^7}$$

Coefficient of NO_2 ↓; Coefficient of H_2O ↓; ← Coefficient of O_2; ↑ Coefficient of NH_3

Self-Check Exercise 16.2

a. $K = [O_2]^3$ — The solids are not included.

b. $K = [N_2O][H_2O]^2$ — The solid is not included. Water is gaseous in this reaction, so it is included.

c. $K = \frac{1}{[CO_2]}$ — The solids are not included.

d. $K = \frac{1}{[SO_3]}$ — Water and H_2SO_4 are pure liquids and so are not included.

Self-Check Exercise 16.3

When rain is imminent, the concentration of water vapor in the air increases. This shifts the equilibrium to the right, forming $CoCl_2 \cdot 6H_2O(s)$, which is pink.

Self-Check Exercise 16.4

a. No change. Both sides of the equation contain the same number of gaseous components. The system cannot change its pressure by shifting its equilibrium position.

b. Shifts to the left. The system can increase the number of gaseous components present, and so increase the pressure, by shifting to the left.

c. Shifts to the right to increase the number of gaseous components and thus its pressure.

Self-Check Exercise 16.5

a. Shifts to the right away from added SO_2.

b. Shifts to the right to replace removed SO_3.

c. Shifts to the right to decrease its pressure.

d. Shifts to the right. Energy is a product in this case, so a decrease in temperature favors the forward reaction (which produces energy).

Self-Check Exercise 16.6

a. $BaSO_4(s) \rightleftharpoons Ba^{2+}(aq) + SO_4^{2-}(aq)$; $K_{sp} = [Ba^{2+}][SO_4^{2-}]$

b. $Fe(OH)_3(s) \rightleftharpoons Fe^{3+}(aq) + 3OH^-(aq)$;
$K_{sp} = [Fe^{3+}][OH^-]^3$

c. $Ag_3PO_4(s) \rightleftharpoons 3Ag^+(aq) + PO_4^{3-}(aq)$;
$K_{sp} = [Ag^+]^3[PO_4^{3-}]$

Self-Check Exercise 16.7

$(3.9 \times 10^{-5})^2 = 1.5 \times 10^{-9} = K_{sp}$

Self-Check Exercise 16.8

$PbCrO_4(s) \rightleftharpoons Pb^{2+}(aq) + CrO_4^{2-}(aq)$
$K_{sp} = [Pb^{2+}][CrO_4^{2-}] = 2.0 \times 10^{-16}$
$[Pb^{2+}] = x$
$[CrO_4^{2-}] = x$
$K_{sp} = 2.0 \times 10^{-16} = x^2$
$x = [Pb^{2+}] = [CrO_4^{2-}] = 1.4 \times 10^{-8}$

Chapter 17

Self-Check Exercise 17.1

a. CuO contains Cu^{2+} and O^{2-} ions, so copper is oxidized ($Cu \rightarrow Cu^{2+} + 2e^-$) and oxygen is reduced ($O + 2e^- \rightarrow O^{2-}$).

b. CsF contains Cs^+ and F^- ions. Thus cesium is oxidized ($Cs \rightarrow Cs^+ + e^-$) and fluorine is reduced ($F + e^- \rightarrow F^-$).

Self-Check Exercise 17.2

a. SO_3
We assign oxygen first. Each O is assigned an oxidation state of −2, giving a total of −6 (3 × −2) for the three oxygen atoms. Because the molecule has zero charge overall, the sulfur must have an oxidation state of +6.
Check: $+6 + 3(-2) = 0$

b. SO_4^{2-}
As in part a, each oxygen is assigned an oxidation state of −2, giving a total of −8 (4 × −2) on the four oxygen atoms. The anion has a net charge of −2, so the sulfur must have an oxidation state of +6.
Check: $+6 + 4(-2) = -2$
SO_4^{2-} has a charge of −2, so this is correct.

c. N_2O_5
We assign oxygen before nitrogen because oxygen is more electronegative. Thus each O is assigned an oxidation state of −2, giving a total of −10 (5 × −2) on the five oxygen atoms. Therefore, the oxidation states of the *two* nitrogen atoms must total +10, because N_2O_5 has no overall charge. Each N is assigned an oxidation state of +5.
Check: $2(+5) + 5(-2) = 0$

d. PF_3
First we assign the fluorine an oxidation state of −1, giving a total of −3 (3 × −1) on the three fluorine atoms. Thus P must have an oxidation state of +3.
Check: $+3 + 3(-1) = 0$

e. C_2H_6
In this case it is best to recognize that hydrogen is always +1 in compounds with nonmetals. Thus each H is assigned an oxidation state of +1, which means that the six H atoms account for a total of +6 (6 × +1). Therefore, the *two* carbon atoms must account for −6, and each carbon is assigned an oxidation state of −3.
Check: $2(-3) + 6(+1) = 0$

Self-Check Exercise 17.3

We can tell whether this is an oxidation–reduction reaction by comparing the oxidation states of the elements in the reactants and products:

$$N_2 + 3H_2 \rightarrow 2NH_3$$

Oxidation states: 0 0 −3 +1 (each H)

Nitrogen goes from 0 to −3. Thus it gains three electrons and is reduced. Each hydrogen atom goes from 0 to +1 and is thus oxidized, so this is an oxidation–reduction reaction. The oxidizing agent is N_2 (it takes electrons from H_2). The reducing agent is H_2 (it gives electrons to N_2).

$$N_2 + 3H_2 \rightarrow 2NH_3$$

$6e^-$

Self-Check Exercise 17.4

The unbalanced equation for this reaction is

$$Cu(s) + HNO_3(aq) \rightarrow Cu(NO_3)_2(aq) + H_2O(l) + NO(g)$$

Copper metal | Nitric acid | Aqueous copper(II) nitrate (contains Cu^{2+}) | Water | Nitrogen monoxide

STEP 1 The oxidation half-reaction is

$$Cu + HNO_3 \rightarrow Cu(NO_3)_2$$

Oxidation states: 0 +1 +5 −2 (each O) +2 +5 −2 (each O)

The copper goes from 0 to +2 and thus is oxidized. This reduction reaction is

$$HNO_3 \rightarrow NO$$

Oxidation states: +1 +5 −2 (each O) +2 −2

In this case nitrogen goes from +5 in HNO_3 to +2 in NO and so is reduced. Notice two things about these reactions:

1. The HNO_3 must be included in the oxidation half-reaction to supply NO_3^- in the product $Cu(NO_3)_2$.
2. Although water is a product in the overall reaction, it does not need to be included in either half-reaction at the beginning. It will appear later as we balance the equation.

STEP 2 *Balance the oxidation half-reaction.*

$$Cu + HNO_3 \rightarrow Cu(NO_3)_2$$

a. Balance nitrogen first.

$$Cu + 2HNO_3 \rightarrow Cu(NO_3)_2$$

b. Balancing nitrogen also caused oxygen to balance.

c. Balance hydrogen using H^+.

$$Cu + 2HNO_3 \rightarrow Cu(NO_3)_2 + 2H^+$$

d. Balance the charge using e^-.

$$Cu + 2HNO_3 \rightarrow Cu(NO_3)_2 + 2H^+ + 2e^-$$

This is the balanced oxidation half-reaction.

Balance the reduction half-reaction.

$$HNO_3 \rightarrow NO$$

a. All elements are balanced except hydrogen and oxygen.

b. Balance oxygen using H_2O.

$$HNO_3 \rightarrow NO + 2H_2O$$

c. Balance hydrogen using H^+.

$$3H^+ + HNO_3 \rightarrow NO + 2H_2O$$

d. Balance the charge using e^-.

$$3e^- + 3H^+ + HNO_3 \rightarrow NO + 2H_2O$$

This is the balanced reduction half-reaction.

STEP 3 We equalize electrons by multiplying the oxidation half-reaction by 3:

$$3 \times (Cu + 2HNO_3 \rightarrow Cu(NO_3)_2 + 2H^+ + 2e^-)$$

gives

$$3Cu + 6HNO_3 \rightarrow 3Cu(NO_3)_2 + 6H^+ + 6e^-$$

Multiplying the reduction half-reaction by 2:

$$2 \times (3e^- + 3H^+ + HNO_3 \rightarrow NO + 2H_2O)$$

gives

$$6e^- + 6H^+ + 2HNO_3 \rightarrow 2NO + 4H_2O$$

STEP 4 We can now add the balanced half-reactions, which both involve a six-electron change.

$$3Cu + 6HNO_3 \rightarrow 3Cu(NO_3)_2 + 6H^+ + 6e^-$$
$$6e^- + 6H^+ + 2HNO_3 \rightarrow 2NO + 4H_2O$$
$$\cancel{6e^-} + \cancel{6H^+} + 3Cu + 8HNO_3 \rightarrow 3Cu(NO_3)_2 + 2NO + 4H_2O + \cancel{6H^+} + \cancel{6e^-}$$

Canceling species common to both sides gives the balanced overall equation:

$$3Cu(s) + 8HNO_3(aq) \rightarrow 3Cu(NO_3)_2(aq) + 2NO(g) + 4H_2O(l)$$

STEP 5 Check the elements and charges.

Elements	3Cu, 8H, 8N, 24O → 3Cu, 8H, 8N, 24O
Charges	0 → 0

Chapter 18

Self-Check Exercise 18.1

a. An alpha particle is a helium nucleus, ${}^{4}_{2}He$. We can initially represent the production of an α particle by ${}^{226}_{88}Ra$ as follows:

$${}^{226}_{88}Ra \rightarrow {}^{4}_{2}He + {}^{A}_{Z}X$$

Because we know that both A and Z are conserved, we can write

$$A + 4 = 226 \quad \text{and} \quad Z + 2 = 88$$

Solving for A gives 222 and for Z gives 86, so ${}^{A}_{Z}X$ is ${}^{222}_{86}X$. Because Rn has $Z = 86$, ${}^{A}_{Z}X$ is ${}^{222}_{86}Rn$. The overall balanced equation is

$${}^{226}_{88}Ra \rightarrow {}^{4}_{2}He + {}^{222}_{86}Rn$$

Check: $Z = 88 \rightarrow Z = 86 + 2 = 88$

$A = 226 \qquad A = 222 + 4 = 226$

b. Using a similar strategy, we have

$${}^{214}_{82}Pb \rightarrow {}^{0}_{-1}e + {}^{A}_{Z}X$$

Because $Z - 1 = 82$, $Z = 83$, and because $A + 0 = 214$, $A = 214$. Therefore, ${}^{A}_{Z}X = {}^{214}_{83}Bi$. The balanced equation is

$${}^{214}_{82}Pb \rightarrow {}^{0}_{-1}e + {}^{214}_{83}Bi$$

Check: $Z = 82 \rightarrow Z = 83 - 1 = 82$

$A = 214 \qquad A = 214 + 0 = 214$

Self-Check Exercise 18.2

a. The missing particle must be ${}^{4}_{2}H$ (an α particle), because

$${}^{222}_{86}Rn \rightarrow {}^{218}_{84}Po + {}^{4}_{2}He$$

is a balanced equation.

Check: $Z = 86 \rightarrow Z = 84 + 2 = 86$

$A = 222 \qquad A = 218 + 4 = 222$

b. The missing species must be ${}^{15}_{7}X$ or ${}^{15}_{7}N$, because the balanced equation is

$${}^{15}_{8}O \rightarrow {}^{15}_{7}N + {}^{0}_{1}e$$

Check: $Z = 8 \rightarrow Z = 7 + 1 = 8$

$A = 15 \qquad A = 15 + 0 = 15$

Self-Check Exercise 18.3

Let's do this problem by thinking about the number of half-lives required to go from 8.0×10^{-7} mol to 1.0×10^{-7} mol of ${}^{228}_{88}Ra$.

8.0×10^{-7} mol $\xrightarrow{\text{First half-life}}$ 4.0×10^{-7} mol $\xrightarrow{\text{Second half-life}}$ 2.0×10^{-7} mol $\xrightarrow{\text{Third half-life}}$ 1.0×10^{-7} mol

It takes three half-lives, then, for the sample to go from 8.0×10^{-7} mol of ${}^{228}_{88}Ra$ to 1.0×10^{-7} mol of ${}^{228}_{88}Ra$. From Table 18.3, we know that the half-life of ${}^{228}_{88}Ra$ is 6.7 years. Therefore, the elapsed time is 3(6.7 years) = 20.1 years, or 2.0×10^1 years when we use the correct number of significant figures.

Answers to Even-Numbered End-of-Chapter Questions and Exercises

Chapter 1

2. The answer depends on the student's experiences.
4. Unfortunately, many examples exist. Chemical and biological weapons are produced in some countries. Although the development of new plastics has been a boon in many endeavors, their use also depletes fossil fuel reserves and increases our solid waste problems. Although biotechnology has produced many exciting new drugs and treatments, the testing procedures that have been developed for determining whether a person has a genetic *likelihood* of developing a particular disease may also be used to make it impossible or difficult for that person to obtain health or life insurance.
6. Answers will depend on the student's choices.
8. Recognize the problem and state it clearly; propose possible solutions or explanations; decide which solution/explanation is best through experiments.
10. (a) quantitative: (b) qualitative: (c) qualitative: (d) quantitative: (e) qualitative: (f) quantitative
12. A natural law is a *summary of observed, measurable behavior* that occurs repeatedly and consistently. A theory is an attempt to *explain* such behavior.
14. Most applications of chemistry are oriented toward the interpretation of observations and the solving of problems. Although memorization of some facts may *aid* in these endeavors, it is the ability to combine, relate, and synthesize information that is most important in the study of chemistry.
16. In real-life situations, the problems and applications likely to be encountered are not simple textbook examples. You must be able to observe an event, hypothesize a cause, and then test this hypothesis. You must be able to carry what has been learned in class forward to new, different situations.

Chapter 2

2. Any number can be expressed as the product of a number between 1 and 10 and a power of 10 (either positive or negative). The power of 10 depends on the number of places to the left (positive) or right (negative) that the decimal point is moved.
4. positive; negative
6. (a) positive: (b) positive: (c) negative: (d) negative
8. (a) 9.367421×10^6: (b) 7.241×10^3: (c) 5.519×10^{-4}: (d) 5.408×10^0: (e) 6.24×10^2: (f) 6.319×10^1: (g) 7.215×10^{-9}: (h) 7.21×10^{-1}
10. (a) 0.0007327: (b) 151: (c) 1: (d) 0.0005399: (e) 221: (f) 0.0783: (g) 0.1218: (h) 0.0002918: (i) 7251: (j) 0.000000001911: (k) 995.1: (l) 0.09951
12. (a) 8.714×10^5: (b) 6.591×10^4: (c) 2.31×10^2: (d) 1.519×10^{-14}: (e) 1.294×10^{-2}: (f) 1.921×10^{-4}: (g) 1×10^0: (h) 7.354211×10^{14}
14. (a) 3.1×103: (b) 1×10^6: (c) 1 or 1×10^0: (d) 1.8×10^{-5}: (e) 1×10^7: (f) 1.00×10^6: (g) 1.00×10^{-7}: (h) 1×10^1
16. grams
18. (a) mega: (b) milli: (c) nano: (d) mega: (e) centi: (f) micro
20. slightly more than a pound
22. approximately 1 cup (about half a pint)
24. kilogram
26. the woman
28. (a) centimeter: (b) meter: (c) kilometer
30. d
32. 40 quarters
34. uncertainty
36. The scale of the ruler is marked to the nearest tenth of a centimeter. Writing 2.850 would imply that the scale was marked to the nearest hundredth of a centimeter (and that the zero in the thousandths place had been estimated).
38. (a) probably 2: (b) infinite (definition): (c) infinite (definition): (d) probably 1: (e) 3 (exact quantity)
40. final
42. (a) 4.23×10^{-1}: (b) 7.12×10^6: (c) 4.45×10^{-4}: (d) 2.30×10^{-4}: (e) 9.72×10^5
44. (a) 3.42×10^{-4}: (b) 1.034×10^4: (c) 1.7992×10^1: (d) 3.37×10^5
46. decimal
48. Most calculators would say 0.66666666. If the 2 and 3 were experimentally determined numbers, this quotient would imply far too many significant figures.
50. none
52. (a) 641.0: (b) 1.327: (c) 77.34: (d) 3215
54. (a) one: (b) four: (c) two: (d) three
56. (a) 2.045: (b) 3.8×10^3: (c) 5.19×10^{-5}: (d) 3.8418×10^{-7}
58. an infinite number, a definition
60. (1000 mL/1 L); (1 L/1000 mL)
62. $\frac{1 \text{ lb}}{\$0.79}$
64. (a) 2.44 yd: (b) 42.2 m: (c) 115 in.: (d) 2238 cm: (e) 648 mi: (f) 716.9 km: (g) 0.0362 km: (h) 5.01×10^4 cm
66. (a) 0.2543 kg: (b) 2.75×10^3 g: (c) 6.06 lb: (d) 97.0 oz: (e) 1.177 lb: (f) 794 g: (g) 2.5×10^2 g: (h) 1.62 oz
68. 3.1×10^2 km; 3.1×10^5 m; 1.0×10^6 ft
70. 1×10^{-8} cm; 4×10^{-9} in.; 0.1 nm
72. freezing/melting
74. 273
76. Fahrenheit (F)
78. (a) 63 K: (b) 2 °C: (b) 505 °C: (d) 1051 K
80. (a) 173 °F: (b) 104 °F: (c) −459 °F: (d) 90. °F
82. (a) 2 °C: (b) 28 °C: (c) −5.8 °F (−6 °F): (d) −40 °C (−40 is where both temperature scales have the same value)
84. g/cm^3 (g/mL)
86. 100 $in.^3$
88. Density is a characteristic property of a pure substance.
90. copper
92. (a) 22 g/cm^3: (b) 0.034 g/cm^3: (c) 0.962 g/cm^3: (d) 2.1×10^{-5} g/cm^3
94. 1.70×10^3 g
96. float
98. 11.7 mL
100. (a) 966 g: (b) 394 g: (c) 567 g: (d) 135 g
102. (a) 301,100,000,000,000,000,000,000: (b) 5,091,000,000: (c) 720: (d) 123,400: (e) 0.000432002: (f) 0.03001: (g) 0.00000029901: (h) 0.42
104. (a) cm: (b) m: (c) km: (d) cm: (e) mm
106. (a) 5.07×10^4 kryll: (b) 0.12 blim: (c) 3.70×10^{-5} $blim^2$
108. 20. in.
110. $1
112. °X = 1.26 °C + 14

114. 3.50 g/L (3.50×10^{-3} g/cm^3)
116. 959 g
118. (a) negative: (b) negative: (c) positive: (d) zero: (e) negative
120. (a) 2, positive: (b) 11, negative: (c) 3, positive: (d) 5, negative: (e) 5, positive: (f) 0, zero: (g) 1, negative: (h) 7, negative
122. (a) 1, positive: (b) 3, negative: (c) 0, zero: (d) 3, positive: (e) 9, negative
124. (a) 0.0000298: (b) 4,358,000,000: (c) 0.0000019928: (d) 602,000,000,000,000,000,000,000: (e) 0.101: (f) 0.00787: (g) 98,700,000: (h) 378.99: (i) 0.1093: (j) 2.9004: (k) 0.00039: (l) 0.00000001904
126. (a) 1×10^{-2}: (b) 1×10^2: (c) 5.5×10^{-2}: (d) 3.1×10^9: (e) 1×10^3: (f) 1×10^8: (g) 2.9×10^2: (h) 3.453×10^4
128. kelvin, K
130. centimeter
132. 0.105 m
134. 1 kg
136. 10
138. 2.8 (the hundredths place is estimated)
140. (a) 0.000426: (b) 4.02×10^{-5}: (c) 5,990,000: (d) 400.: (e) 0.00600
142. (a) 2149.6: (b) 5.37×10^3: (c) 3.83×10^{-2}: (d) -8.64×10^5
144. (a) 7.6166×10^6: (b) 7.24×10^3: (c) 1.92×10^{-5}: (d) 2.4482×10^{-3}
146. 1 yr/12 mo; 12 mo/1 yr
148. (a) 25.7 kg: (b) 3.38 gal: (c) 0.132 qt: (d) 1.09×10^4 mL: (e) 2.03×10^3 g: (f) 0.58 qt
150. for exactly 6 gross, 864 pencils
152. (a) 352 K: (b) −18 °C: (c) −43 °C: (d) 257 °F
154. 78.2 g
156. 0.59 g/cm^3
158. (a) 23 °F: (b) 32 °F: (c) −321 °F: (d) −459 °F: (e) 187 °F: (f) −459 °F

Chapter 3

2. solid, liquid, gas (vapor)
4. liquids
6. gaseous
8. The stronger the interparticle forces, the more rigid the substance.
10. Because gases are mostly empty space, they can be *compressed* easily to smaller volumes. In solids and liquids, most of the sample's bulk volume is filled with the molecules, leaving little empty space.
12. chemical
14. The sulfate ion of nickel sulfate reacts with the barium ion of barium chloride to form a new substance, barium sulfate.
16. Changes in state: solid → liquid; liquid → gas; solid → gas, etc.
18. (a) chemical: (b) physical: (c) chemical: (d) chemical: (e) chemical: (f) chemical: (g) chemical: (h) chemical: (i) chemical: (j) physical: (k) chemical
20. element
22. compounds
24. In general, the properties of a compound are very different from the properties of its constituent elements. For example, the properties of water are altogether different from the properties of the elements (hydrogen gas and oxygen gas) that make it up.
26. Mixtures have variable compositions; a pure substance has a fixed composition.
28. Heterogeneous mixtures: salad dressing, jelly beans, the change in my pocket; solutions: window cleaner, shampoo, rubbing alcohol
30. (a) mixture: (b) mixture: (c) mixture: (d) pure substances (ideally)
32. (a) homogeneous: (b) heterogeneous: (c) heterogeneous: (d) homogeneous: (e) the paper itself is basically homogeneous
34. Consider a mixture of salt (sodium chloride) and sand. Salt is soluble in water; sand is not. The mixture is added to water and stirred to dissolve the salt, and is then filtered. The salt solution passes through the filter; the sand remains on the filter. The water can then be evaporated from the salt.
36. The solution is heated to vaporize (boil) the water. The water vapor is then cooled so that it condenses back to the liquid state, and the liquid is collected. After all the water is vaporized from the original sample, pure sodium chloride remains. The process consists of physical changes.
38. 1 cal = 4.184 J, so the calorie is about 4 times larger than the joule.
40. As the steam is cooled from 150 °C to 100 °C, the molecules of vapor gradually slow down as they lose kinetic energy. At 100 °C, the steam condenses into liquid water, and the temperature remains at 100 °C until all the steam has condensed. As the liquid water cools, the molecules in the liquid move more and more slowly as they lose kinetic energy. At 0 °C, the liquid water freezes.
42. temperature
44. 1700 J
46. (a) 110.5 kcal: (b) 4.369 kcal: (c) 0.2424 kcal: (d) 45.53 kcal
48. (a) 1.230×10^4 cal: (b) 2.904×10^5 cal: (c) 9.4×10^8 cal: (d) 4.201×10^6 cal
50. (a) 0.3202 kJ: (b) 1.870 kcal: (c) 117.0 cal: (d) 3.819 kcal
52. 7.11 J/g °C
54. 50 g
56. 13,580 J = 14 kJ
58. 0.031 cal/g °C
60. gold, 65 J; mercury, 70. J; carbon, 360 J
62. 1.42 J/g °C
64. compound
66. physical
68. 380,000 J (3.8×10^5 J)
70. Aluminum will lose more heat because it has the higher specific heat capacity.
72. 24.2 °C
74. 22 °C
76. far apart
78. chemical
80. chemical
82. electrolysis
84. (a) heterogeneous: (b) heterogeneous: (c) homogeneous (if no lumps!): (d) heterogeneous (although it may appear homogeneous): (e) heterogeneous
86. 9.0 J
88. (a) 185.0 J: (b) 679.5 J: (c) 1.557×10^4 J: (d) 6.117×10^5 J
90. (a) 54.42 kJ: (b) 1.301×10^4 cal: (c) 1475 kJ: (d) 4.120 kcal
92. 4.4×10^3 J
94. $H_2O(l)$, 7.03×10^3 J; $H_2O(s)$, 3.41×10^3 J; $H_2O(g)$, 3.4×10^3 J; Al, 1.5×10^3 J; Fe, 7.6×10^2 J; Hg, 2.4×10^2 J; C, 1.2×10^3 J; Ag, 4.0×10^2 J; Au, 2.2×10^2 J
96. 0.124 J/g °C

Chapter 4

2. Robert Boyle
4. 112 elements are known; 88 occur naturally; 24 are manmade. Table 4.1 lists the most common elements on the earth.
6. The four most abundant elements in living creatures are oxygen, carbon, hydrogen, and nitrogen (see Table 4.2). In the non-

living world, the most abundant elements are oxygen, silicon, aluminum, and iron (see Table 4.1).
8. Sb (antimony); Cu (copper); Au (gold); Pb (lead); Hg (mercury); K (potassium); Ag (silver); Na (sodium); Sn (tin); W (tungsten); Fe (iron)
10. (a) 4: (b) 6: (c) 9: (d) 1: (e) 5: (f) 7: (g) 12: (h) 2: (i) 14; (j) 13
12. ruthenium; Tl; nobelium; Th; Pu; osmium
14. (a) bromine: (b) boron: (c) beryllium: (d) bismuth: (e) barium
16. A given compound always contains the same relative amount of each constituent element. For example, in carbon dioxide, each CO_2 molecule always contains one carbon atom and two oxygen atoms. If a molecule does not contain twice as many oxygen atoms as carbon atoms, then it is not carbon dioxide.
18. According to Dalton, all atoms of the same element are *identical;* in particular, every atom of a given element has the same *mass* as every other atom of that element. If a given compound always contains the *same relative numbers* of atoms of each kind, and those atoms always have the same *masses,* then the compound made from those elements always contains the same relative masses of its elements.
20. (a) C_6H_6: (b) N_2O_4: (c) $CaCl_2$: (d) $FeBr_3$: (e) $NaNO_3$: (f) Ca_3N_2
22. (a) False; Rutherford's bombardment experiments with metal foil suggested that the α particles were being deflected by coming near a *dense, positively charged* atomic nucleus. (b) False; the proton and the electron have opposite charges, but the mass of the electron is *much smaller* than the mass of the proton. (c) True
24. The protons and neutrons are found in the nucleus. The protons are positively charged; the neutrons have no charge. The protons and neutrons each weigh approximately the same.
26. neutron; electron
28. The electrons; outside the nucleus
30. false
32. mass
34. Atoms of the same element (atoms with the same number of protons in the nucleus) may have different numbers of neutrons, and so will have different masses.
36. (a) 32: (b) 30: (c) 24: (d) 74: (e) 38: (f) 27: (g) 4: (h) 3
38. (a) $^{58}_{27}Co$; (b) $^{10}_{5}B$; (c) $^{23}_{12}Mg$; (d) $^{132}_{53}I$; (e) $^{19}_{9}F$; (f) $^{65}_{29}Cu$
40. (a) 92 protons, 143 neutrons, 92 electrons: (b) 6 protons, 7 neutrons, 6 electrons: (c) 26 protons, 31 neutrons, 26 electrons: (d) 82 protons, 126 neutrons, 82 electrons: (e) 37 protons, 49 neutrons, 37 electrons: (f) 20 protons, 21 neutrons, 20 electrons
42.

Name	*Neutrons*	*Atomic Number*	*Mass Number*	*Symbol*
nitrogen	6	7	13	$^{13}_{7}N$
nitrogen	7	7	14	$^{14}_{7}N$
lead	124	82	206	$^{206}_{82}Pb$
iron	31	26	57	$^{57}_{26}Fe$
krypton	48	36	84	$^{84}_{36}Kr$

44. vertical, groups
46. Metallic elements are found toward the *left* and *bottom* of the periodic table; there are far more metallic elements than nonmetals.
48. nonmetallic gaseous elements: oxygen, nitrogen, fluorine, chlorine, hydrogen, and the noble gases; no metallic gaseous elements at room conditions
50. A metalloid is an element that has some properties common to both metallic and nonmetallic elements. The metalloids are found in the "stair-step" region marked on most periodic tables.
52. (a) 1A, alkali metals: (b) 3A: (c) 8A, noble gases: (d) 4A: (e) 6A: (f) 8A, noble gases: (g) 6A
54. (a) Li, 3, Group 1A, metal: (b) As, 33, Group 5A, metalloid: (c) Rn, 86, Group 8A, nonmetal: (d) Ra, 88, Group 2A, metal: (e) Ge, 32, Group 4A, metalloid
56. Most elements are too reactive to be found in the uncombined form in nature and are found only in compounds.
58. These elements are found uncombined in nature and do not readily react with other elements. Although these elements were once thought to form no compounds, this now has been shown to be untrue.
60. diatomic gases: H_2, N_2, O_2, F_2, Cl_2; monoatomic gases: He, Ne, Kr, Xe, Rn, and Ar
62. chlorine
64. diamond
66. Positive ions are produced when an atom loses one or more electrons; negative ions are produced when an atom gains one or more electrons. No.
68. positively
70. *-ide*
72. nonmetallic
74. (a) Fe_2S_3: (b) FeS: (c) Al_2O_3: (d) AlI_3: (e) Rb_2S: (f) AlP: (g) MnS: (h) SnS_2
76. (a) 2: (b) 3: (c) 2: (d) 4: (e) 1: (f) 2
78. (a) 3+: (b) 2−: (c) 2+: (d) 3−: (e) 1+: (f) 1+
80. Sodium chloride is an *ionic* compound, consisting of Na^+ and Cl^- *ions.* When NaCl is dissolved in water, these ions are *set free* and can move independently to conduct the electric current. Sugar crystals, although they may *appear* similar visually, contain *no* ions. When sugar is dissolved in water, it dissolves as uncharged *molecules.* No electrically charged species are present in a sugar solution to carry the electric current.
82. The total number of positive charges must equal the total number of negative charges so that the crystals of an ionic compound have *no net charge.* A macroscopic sample of compound ordinarily has no net charge.
84. (a) CoO: (b) Co_2S_3: (c) $AlCl_3$: (d) Ba_3N_2: (e) Ca_2C: (f) K_3N
86. (a) 7, halogens: (b) 8, noble gases: (c) 2, alkaline earth elements: (d) 2, alkaline earth elements: (e) 4: (f) 6: (g) 8, noble gases: (h) 1, alkali metals
88. Group 3: boron, B, 5; aluminum, Al, 13; gallium, Ga, 31; indium, In, 49; Group 5: nitrogen, N, 7; phosphorus, P, 15; arsenic, As, 33; antimony, Sb, 51; Group 8: helium, He, 2; neon, Ne, 10; argon, Ar, 18; krypton, Kr, 36
90. Most of an atom's mass is concentrated in the nucleus: the *protons* and *neutrons* that constitute the nucleus have similar masses and are each nearly 2000 times more massive than electrons. The chemical properties of an atom depend on the number and location of the *electrons* it possesses. Electrons are found in the outer regions of the atom and are involved in interactions between atoms.
92. $C_6H_{12}O_6$
94. (a) 29 electrons, 34 neutrons, 29 electrons: (b) 35 protons, 45 neutrons, 35 electrons: (c) 12 protons, 12 neutrons, 12 electrons
96. The chief use of gold in ancient times was as *ornamentation,* whether in statuary or in jewelry. Gold possesses an especially beautiful luster; since it is relatively soft and malleable, it can be worked finely by artisans. Among the metals, gold is inert to attack by most substances in the environment.
98. (a) I: (b) Si: (c) W: (d) Fe: (e) Cu: (f) Co
100. (a) Br: (b) Bi: (c) Hg: (d) V: (e) F: (f) Ca
102. (a) osmium: (b) zirconium: (c) rubidium: (d) radon: (e) uranium: (f) manganese: (g) nickel: (h) bromine
104. (a) CO_2: (b) $AlCl_3$: (c) $HClO_4$: (d) SCl_6
106. (a) $^{13}_{6}C$: (b) $^{13}_{6}C$: (c) $^{13}_{6}C$: (d) $^{44}_{19}K$: (e) $^{41}_{20}Ca$: (f) $^{35}_{19}K$
108. (a) $^{41}_{20}Ca$, 20, 21, 41; $^{55}_{25}Mn$, 25, 30, 55; $^{109}_{47}Ag$, 47, 62, 109; $^{45}_{21}Sc$, 21, 24, 45

Chapter 5

2. A binary chemical compound contains only two elements; the major types are ionic (compounds of a metal and a nonmetal) and nonionic or molecular (compounds between two nonmetals). Answers depend on student responses.
4. cation (positive ion)
6. Some substances do not contain molecules; the formula we write reflects only the relative number of each type of atom present.
8. Roman numeral
10. (a) sodium bromide: (b) calcium sulfide: (c) aluminum iodide: (d) cesium oxide: (e) magnesium fluoride: (f) strontium chloride: (g) lithium oxide: (h) barium iodide
12. (a) incorrect, Ag_2O: (b) correct: (c) incorrect, BaO: (d) correct: (e) incorrect, LiBr
14. (a) tin(II) oxide: (b) tin(IV) oxide: (c) copper(II) sulfide: (d) copper(I) sulfide: (e) iron(III) chloride: (f) iron(II) iodide
16. (a) stannous sulfide: (b) stannic sulfide: (c) cobaltic chloride: (d) cobaltous iodide: (e) mercuric bromide: (f) mercurous bromide
18. (a) xenon hexafluoride: (b) silicon tetrachloride: (c) diphosphorus pent(a)oxide: (d) water: (e) arsenic tribromide: (f) carbon tetraiodide
20. (a) diboron hexahydride, nonionic: (b) calcium nitride, ionic: (c) carbon tetrabromide, nonionic: (d) silver sulfide [silver(I) sulfide], ionic: (e) copper(II) chloride (cupric chloride), ionic: (f) chlorine monofluoride, nonionic
22. (a) barium fluoride: (b) radium oxide: (c) dinitrogen oxide: (d) rubidium oxide: (e) diarsenic pent(a)oxide: (f) calcium nitride
24. An oxyanion is an anion containing a particular element plus oxygen. If there are two oxyanions of an element, the higher oxidation state is given the *-ate* ending and the lower oxidation state is given the *-ite* ending. If there are four oxyanions of an element, the highest oxidation state is indicated with the prefix *per-* and the lowest oxidation state with the prefix *hypo-*. For example: ClO^- (hypochlorite), ClO_2^- (chlorite), ClO_3^- (chlorate), ClO_4^- (perchlorate).
26. *hypo-* (fewest); *per-* (most)
28. BrO^- (hypobromite); BrO_2^- (bromite); BrO_3^- (bromate); BrO_4^- (perbromate)
30. (a) NO_3^-: (b) NO_2^-: (c) NH_4^+: (d) CN^-
32. (a) CO_3^{2-}: (b) HCO_3^-: (c) $C_2H_3O_2^-$: (d) CN^-
34. (a) ammonium ion: (b) dihydrogen phosphate ion: (c) sulfate ion: (d) hydrogen sulfite ion (bisulfite ion): (e) perchlorate ion: (f) iodate ion
36. (a) ammonium sulfate: (b) potassium perchlorate: (c) iron(III) sulfate: (d) calcium phosphate: (e) calcium hydroxide: (f) potassium carbonate
38. oxygen
40. (a) hypochlorous acid: (b) sulfurous acid: (c) bromic acid: (d) hypoiodous acid: (e) perbromic acid: (f) hydrosulfuric acid: (g) hydroselenic acid: (h) phosphorous acid
42. (a) PbO_2: (b) $SnBr_2$: (c) CuS: (d) CuI: (e) Hg_2Cl_2: (f) CrF_3
44. (a) N_2O: (b) NO_2: (c) N_2O_4: (d) SF_6: (e) PBr_3: (f) CI_4: (g) OCl_2
46. (a) $BaSO_3$: (b) $Ca(H_2PO_4)_2$: (c) NH_4ClO_4: (d) $NaMnO_4$: (e) $Fe_2(SO_4)_3$: (f) $CoCO_3$: (g) $Ni(OH)_2$: (h) $ZnCrO_4$
48. (a) HCN: (b) HNO_3: (c) H_2SO_4: (d) H_3PO_4: (e) HClO or HOCl: (f) HBr: (g) $HBrO_2$: (h) HF
50. (a) $Mg(HSO_4)_2$: (b) $CsClO_4$: (c) FeO: (d) $H_2Te(aq)$: (e) $Sr(NO_3)_2$: (f) $Sn(C_2H_3O_2)_4$: (g) $MnSO_4$: (h) N_2O_4: (i) Na_2HPO_4: (j) Li_2O_2: (k) HNO_2: (l) $Co(NO_3)_3$
52. A moist paste of NaCl would contain Na^+ and Cl^- ions in solution and would serve as a *conductor* of electrical impulses.
54. $H \rightarrow H^+$ (hydrogen ion) $+ e^-$; $H + e^- \rightarrow H^-$ (hydride ion)
56. ClO_4^-, $HClO_4$; IO_3^-, HIO_3; ClO^-, $HClO$; BrO_2^-, $HBrO_2$; ClO_2^-, $HClO_2$
58. (a) gold(III) bromide (auric bromide): (b) cobalt(III) cyanide (cobaltic cyanide): (c) magnesium hydrogen phosphate: (d) diboron hexahydride (common name diborane): (e) ammonia: (f) silver(I) sulfate (usually called silver sulfate): (g) beryllium hydroxide
60. (a) ammonium carbonate: (b) ammonium hydrogen carbonate, ammonium bicarbonate: (c) calcium phosphate: (d) sulfurous acid: (e) manganese(IV) oxide: (f) iodic acid: (g) potassium hydride
62. (a) $M(C_2H_3O_2)_4$: (b) $M(MnO_4)_4$: (c) MO_2: (d) $M(HPO_4)_2$: (e) $M(OH)_4$: (f) $M(NO_2)_4$
64. M^+ compounds: MD, M_2E, M_3F; M^{2+} compounds: MD_2, ME, M_3F_2; M^{3+} compounds: MD_3, M_2E_3, MF
66. $Ca(NO_3)_2$, $CaSO_4$, $Ca(HSO_4)_2$, $Ca(H_2PO_4)_2$, CaO, $CaCl_2$
$Sr(NO_3)_2$, $SrSO_4$, $Sr(HSO_4)_2$, $Sr(H_2PO_4)_2$, SrO, $SrCl_2$
NH_4NO_3, $(NH_4)_2SO_4$, NH_4HSO_4, $NH_4H_2PO_4$, $(NH_4)_2O$, NH_4Cl
$Al(NO_3)_3$, $Al_2(SO_4)_3$, $Al(HSO_4)_3$, $Al(H_2PO_4)_3$, Al_2O_3, $AlCl_3$
$Fe(NO_3)_3$, $Fe_2(SO_4)_3$, $Fe(HSO_4)_3$, $Fe(H_2PO_4)_3$, Fe_2O_3, $FeCl_3$
$Ni(NO_3)_2$, $NiSO_4$, $Ni(HSO_4)_2$, $Ni(H_2PO_4)_2$, NiO, $NiCl_2$
$AgNO_3$, Ag_2SO_4, $AgHSO_4$, AgH_2PO_4, Ag_2O, AgCl $Au(NO_3)_3$, $Au_2(SO_4)_3$, $Au(HSO_4)_3$, $Au(H_2PO_4)_3$, Au_2O_3, $AuCl_3$
KNO_3, K_2SO_4, $KHSO_4$, KH_2PO_4, K_2O, KCl
$Hg(NO_3)_2$, $HgSO_4$, $Hg(HSO_4)_2$, $Hg(H_2PO_4)_2$, HgO, $HgCl_2$
$Ba(NO_3)_2$, $BaSO_4$, $Ba(HSO_4)_2$, $Ba(H_2PO_4)_2$, BaO, $BaCl_2$
68. helium
70. F_2, Cl_2 (gas); Br_2 (liquid); I_2, At_2 (solid)
72. 1−
74. 1−
76. (a) $Al(13e) \rightarrow Al^{3+}(10e) + 3e^-$: (b) $S(16e) + 2e^- \rightarrow S^{2-}(18e)$: (c) $Cu(29e) \rightarrow Cu^+(28e) + e^-$: (d) $F(9e) + e^- \rightarrow F^-(10e)$: (e) $Zn(30e) \rightarrow Zn^{2+}(28e) + 2e^-$: (f) $P(15e) + 3e^- \rightarrow P^{3-}(18e)$
78. (a) Na_2S: (b) KCl: (c) BaO: (d) MgSe: (e) $CuBr_2$: (f) AlI_3: (g) Al_2O_3: (h) Ca_3N_2
80. (a) incorrect: (b) incorrect: (c) incorrect: (d) correct: (e) incorrect
82. (a) cobaltic bromide: (b) plumbic iodide: (c) ferric oxide: (d) ferrous sulfide: (e) stannic chloride: (f) stannous oxide
84. (a) iron(III) acetate: (b) bromine monofluoride: (c) potassium peroxide: (d) silicon tetrabromide: (e) copper(II) permanganate: (f) calcium chromate
86. (a) CO_3^{2-}: (b) HCO_3^-: (c) $C_2H_3O_2^-$: (d) CN^-
88. (a) carbonate: (b) chlorate: (c) sulfate: (d) phosphate: (e) perchlorate: (f) permanganate
90. (a) $CaCl_2$: (b) Ag_2O: (c) Al_2S_3: (d) $BeBr_2$: (e) H_2S: (f) KH: (g) MgI_2: (h) CsF
92. (a) NaH_2PO_4: (b) $LiClO_4$; (c) $Cu(HCO_3)_2$: (d) $KC_2H_3O_2$: (e) BaO_2: (f) Cs_2SO_3

Chapter 6

2. The air pockets inside the biscuit are a result of the evolution of CO_2 by the baking powder.
4. Bubbling takes place as the hydrogen peroxide chemically decomposes into water and oxygen gas.
6. change in odor as acetic acid is produced
8. atoms
10. the same
12. water
14. $C_3H_8(g) + O_2(g) \rightarrow CO_2(g) + H_2O(g)$
16. $BaCl_2(aq) + Na_2SO_4(aq) \rightarrow BaSO_4(s) + NaCl(aq)$
18. $C_3H_8(g) + O_2(g) \rightarrow CO_2(g) + H_2O(g)$; $C_3H_8(g) + O_2(g) \rightarrow CO(g) + H_2O(g)$
20. $CaCO_3(s) + HCl(aq) \rightarrow CaCl_2(aq) + H_2O(l) + CO_2(g)$
22. $SiO_2(s) + C(s) \rightarrow Si(s) + CO(g)$
24. $H_2S(g) + O_2(g) \rightarrow SO_2(g) + H_2O(g)$
26. $Fe_2O_3(s) + CO(g) \rightarrow Fe(l) + CO_2(g)$

28. $O_2(g) \rightarrow O_3(g)$
30. $NH_3(g) + HNO_3(aq) \rightarrow NH_4NO_3(aq)$
32. $Xe(g) + F_2(g) \rightarrow XeF_4(s)$
34. $BaO_2(s) + H_2SO_4(aq) \rightarrow BaSO_4(s) + H_2O_2(aq)$
36. whole numbers
38. (a) $2H_2O_2(aq) \rightarrow 2H_2O(l) + O_2(g)$;
(b) $2Ag(s) + H_2S(g) \rightarrow Ag_2S(s) + H_2(g)$;
(c) $2FeO(s) + C(s) \rightarrow 2Fe(l) + CO_2(g)$;
(d) $Cl_2(g) + 2KI(aq) \rightarrow 2KCl(aq) + I_2(s)$;
(e) $Na_2B_4O_7(s) + H_2SO_4(aq) + 5H_2O(l) \rightarrow 4H_3BO_3(s) + Na_2SO_4(aq)$;
(f) $CaC_2(s) + 2H_2O(l) \rightarrow Ca(OH)_2(s) + C_2H_2(g)$;
(g) $2NaCl(s) + H_2SO_4(l) \rightarrow 2HCl(g) + Na_2SO_4(s)$;
(h) $SiO_2(s) + 2C(s) \rightarrow S(l) + 2CO(g)$
40. (a) $CaF_2(s) + H_2SO_4(l) \rightarrow CaSO_4(s) + 2HF(g)$;
(b) $3KBr(s) + H_3PO_4(aq) \rightarrow K_3PO_4(aq) + 3HBr(g)$;
(c) $TiCl_4(l) + 4Na(s) \rightarrow 4NaCl(s) + Ti(s)$;
(d) $K_2CO_3(s) \rightarrow K_2O(s) + CO_2(g)$;
(e) $4KO_2(s) + 2H_2O(l) \rightarrow 4KOH(aq) + 3O_2(g)$;
(f) $2Na_2O_2(s) + 2H_2O(g) + 4CO_2(g) \rightarrow 4NaHCO_3(s) + O_2(g)$;
(g) $2KNO_2(s) + 2C(s) \rightarrow K_2CO_3(s) + CO(g) + N_2(g)$;
(h) $3BaO(s) + 2Al(s) \rightarrow Al_2O_3(s) + 3Ba(s)$
42. (a) $2Ag_2S(s) + 2H_2O(l) \rightarrow 4Ag(s) + 2H_2S(g) + O_2(g)$;
(b) $CaO(s) + SO_3(g) \rightarrow CaSO_4(s)$;
(c) $CS_2(g) + 3Cl_2(g) \rightarrow CCl_4(l) + S_2Cl_2(g)$;
(d) $2Al(s) + 3F_2(g) \rightarrow 2AlF_3(s)$;
(e) $4NH_3(g) + 5O_2(g) \rightarrow 4NO(g) + 6H_2O(l)$;
(f) $PI_3(s) + 3H_2O(l) \rightarrow H_3PO_3(aq) + 3HI(g)$;
(g) $4C(s) + S_8(s) \rightarrow 4CS_2(l)$;
(h) $CaH_2(s) + 2H_2O(l) \rightarrow Ca(OH)_2(aq) + 2H_2(g)$
44. (a) $Ba(NO_3)_2(aq) + Na_2CrO_4(aq) \rightarrow BaCrO_4(s) + 2NaNO_3(aq)$;
(b) $PbCl_2(aq) + K_2SO_4(aq) \rightarrow PbSO_4(s) + 2KCl(aq)$;
(c) $C_2H_5OH(l) + 3O_2(g) \rightarrow 2CO_2(g) + 3H_2O(l)$;
(d) $CaC_2(s) + 2H_2O(l) \rightarrow Ca(OH)_2(s) + C_2H_2(g)$;
(e) $Sr(s) + 2HNO_3(aq) \rightarrow Sr(NO_3)_2(aq) + H_2(g)$;
(f) $BaO_2(s) + H_2SO_4(aq) \rightarrow BaSO_4(s) + H_2O_2(aq)$;
(g) $2AsI_3(s) \rightarrow 2As(s) + 3I_2(s)$;
(h) $2CuSO_4(aq) + 4KI(s) \rightarrow 2CuI(s) + I_2(s) + 2K_2SO_4(aq)$
46. $Al(s) + O_2(g) \rightarrow Al_2O_3(s)$
48. $C_{12}H_{22}O_{11}(aq) + H_2O(l) \rightarrow 4C_2H_5OH(aq) + 4CO_2(g)$
50. $2Al_2O_3(s) + 3C(s) \rightarrow 4Al(s) + 3CO_2(g)$
52. $2Li(s) + S(s) \rightarrow Li_2S(s)$; $2Na(s) + S(s) \rightarrow Na_2S(s)$; $2K(s) + S(s) \rightarrow K_2S(s)$; $2Rb(s) + S(s) \rightarrow Rb_2S(s)$; $2Cs(s) + S(s) \rightarrow Cs_2S(s)$; $2Fr(s) + S(s) \rightarrow Fr_2S(s)$
54. $BaO_2(s) + H_2O(l) \rightarrow BaO(s) + H_2O_2(aq)$
56. $2KClO_3(s) \rightarrow 2KCl(s) + 3O_2(g)$
58. $NH_3(g) + HCl(g) \rightarrow NH_4Cl(s)$
60. Charring represents the formation of elemental carbon from the sugars/starches in the muffin.
62. $Fe(s) + S(s) \rightarrow FeS(s)$
64. $K_2CrO_4(aq) + BaCl_2(aq) \rightarrow BaCrO_4(s) + 2KCl(aq)$
66. $2NaCl(aq) + 2H_2O(l) \rightarrow Cl_2(g) + H_2(g) + 2NaOH(aq, s)$
$2NaBr(aq) + 2H_2O(l) \rightarrow Br_2(l) + H_2(g) + 2NaOH(aq, s)$
$2NaI(aq) + 2H_2O(l) \rightarrow I_2(s) + H_2(g) + 2NaOH(aq, s)$
68. $CaC_2(s) + 2H_2O(l) \rightarrow Ca(OH)_2(s) + C_2H_2(g)$
70. $CuO(s) + H_2SO_4(aq) \rightarrow CuSO_4(aq) + H_2O(l)$
72. $Na_2SO_3(aq) + S(s) \rightarrow Na_2S_2O_3(aq)$
74. (a) $ZnCl_2(aq) + Na_2CO_3(aq) \rightarrow ZnCO_3(s) + 2NaCl(aq)$;
(b) $2Al(s) + 3H_2SO_4(aq) \rightarrow Al_2(SO_4)_3(aq) + 3H_2(g)$;
(c) $Mn(s) + 2S(s) \rightarrow MnS_2(s)$;
(d) $C_5H_{12}(l) + 8O_2(g) \rightarrow 5CO_2(g) + 6H_2O(g)$;
(e) $H_2O(l) + Br_2(l) \rightarrow HBr(aq) + HOBr(aq)$;
(f) $MnS_2(s) + 3O_2(g) \rightarrow MnO_2(s) + 2SO_2(g)$;
(g) $PbCl_2(aq) + K_2CrO_4(aq) \rightarrow PbCrO_4(s) + 2KCl(aq)$;
(h) $2AgNO_3(aq) + H_2SO_4(aq) \rightarrow Ag_2SO_4(s) + 2HNO_3(aq)$
76. (a) $Pb(NO_3)_2(aq) + K_2CrO_4(aq) \rightarrow PbCrO_4(s) + 2KNO_3(aq)$;
(b) $BaCl_2(aq) + Na_2SO_4(aq) \rightarrow BaSO_4(s) + 2NaCl(aq)$;
(c) $2CH_3OH(l) + 3O_2(g) \rightarrow 2CO_2(g) + 4H_2O(g)$;
(d) $Na_2CO_3(aq) + S(s) + SO_2(g) \rightarrow CO_2(g) + Na_2S_2O_3(aq)$;
(e) $Cu(s) + 2H_2SO_4(aq) \rightarrow CuSO_4(aq) + SO_2(g) + 2H_2O(l)$;
(f) $MnO_2(s) + 4HCl(aq) \rightarrow MnCl_2(aq) + Cl_2(g) + 2H_2O(l)$;
(g) $As_2O_3(s) + 6KI(aq) + 6HCl(aq) \rightarrow 2AsI_3(s) + 6KCl(aq) + 3H_2O(l)$;
(h) $2Na_2S_2O_3(aq) + I_2(aq) \rightarrow Na_2S_4O_6(aq) + 2NaI(aq)$

Chapter 7

2. Driving forces are types of *changes* in a system that pull a reaction in the *direction of product formation;* driving forces include formation of a *solid,* formation of *water,* formation of a *gas,* and transfer of electrons.
4. The net charge of a precipitate must be *zero.* The total number of positive charges equals the total number of negative charges.
6. ions
8. The simplest evidence is that solutions of ionic substances conduct electricity.
10. "Insoluble" and "slightly soluble" have roughly the same meanings. However, if a substance is highly toxic and is found in a water supply, for example, the difference between "insoluble" and "slightly soluble" could be crucial.
12. (a) soluble: (b) soluble: (c) insoluble: (d) soluble: (e) insoluble: (f) insoluble: (g) insoluble: (h) soluble
14. (a) Rule 4: (b) Rule 6: (c) Rule 5: (d) Rule 6
16. (a) calcium sulfate: (b) silver iodide: (c) lead(II) phosphate: (d) iron(III) hydroxide: (e) no precipitate: all potassium and sodium salts are soluble: (f) barium carbonate
18. (a) $Na_2S(aq) + CuCl_2(aq) \rightarrow \underline{CuS(s)} + 2NaCl(aq)$: (b) $K_3PO_4(aq) + AlCl_3(aq) \rightarrow \underline{AlPO_4(s)} + 3KCl(aq)$: (c) $H_2SO_4(aq) + BaCl_2(aq) \rightarrow \underline{BaSO_4(s)} + 2HCl(aq)$: (d) $3NaOH(aq) + FeCl_3(aq) \rightarrow \underline{Fe(OH)_3(s)} + 3NaCl(aq)$: (e) $2NaCl(aq) + Hg_2(NO_3)_2(aq) \rightarrow \underline{Hg_2Cl_2(s)} + 2NaNO_3(aq)$: (f) $3K_2CO_3(aq) + 2Cr(C_2H_3O_2)_3(aq) \rightarrow \underline{Cr_2(CO_3)_3(s)} + 6KC_2H_3O_2(aq)$
20. (a) $2AgNO_3(aq) + H_2SO_4(aq) \rightarrow Ag_2SO_4(s) + 2HNO_3(aq)$;
(b) $Ca(NO_3)_2(aq) + H_2SO_4(aq) \rightarrow CaSO_4(s) + 2HNO_3(aq)$;
(c) $Pb(NO_3)_2(aq) + H_2SO_4(aq) \rightarrow PbSO_4(s) + 2HNO_3(aq)$
22. (a) $(NH_4)_2S(aq) + CoCl_2(aq) \rightarrow CoS(s) + 2NH_4Cl(aq)$;
(b) $FeCl_3(aq) + 3NaOH(aq) \rightarrow Fe(OH)_3(s) + 3NaCl(aq)$;
(c) $CuSO_4(aq) + Na_2CO_3(aq) \rightarrow CuCO_3(s) + Na_2SO_4(aq)$
24. spectator
26. (a) no reaction: (b) $K_2S(aq) + Ca(NO_3)_2(aq) \rightarrow CaS(s) + 2KNO_3(aq)$: (c) $NaOH(aq) + AgNO_3(aq) \rightarrow AgOH(s) + NaNO_3(aq)$: (d) $3Na_2CO_3(aq) + 2FeCl_3(aq) \rightarrow Fe_2(CO_3)_3(s) + 6NaCl(aq)$: (e) $(NH_4)_3PO_4(aq) + AlCl_3(aq) \rightarrow AlPO_4(s) + 3NH_4Cl(aq)$: (f) no reaction
28. $Ag^+(aq) + Cl^-(aq) \rightarrow AgCl(s)$; $Pb^{2+}(aq) + 2Cl^-(aq) \rightarrow PbCl_2(s)$; $Hg_2{}^{2+}(aq) + 2Cl^-(aq) \rightarrow Hg_2Cl_2(s)$
30. $Co^{2+}(aq) + S^{2-}(aq) \rightarrow CoS(s)$; $2Co^{3+}(aq) + 3S^{2-}(aq) \rightarrow Co_2S_3(s)$; $Fe^{2+}(aq) + S^{2-}(aq) \rightarrow FeS(s)$; $2Fe^{3+}(aq) + 3S^{2-}(aq) \rightarrow Fe_2S_3(s)$
32. The strong bases are those hydroxide compounds that dissociate fully when dissolved in water. The strong bases that are highly soluble in water (NaOH, KOH) are also strong electrolytes.
34. acids: HCl (hydrochloric), HNO_3 (nitric), H_2SO_4 (sulfuric); bases: hydroxides of Group 1A elements: NaOH, KOH, RbOH, CsOH
36. salt
38. $RbOH(s) \rightarrow Rb^+(aq) + OH^-(aq)$; $CsOH(s) \rightarrow Cs^+(aq) + OH^-(aq)$
40. (a) $HCl(aq) + KOH(aq) \rightarrow H_2O(l) + KCl(aq)$: (b) $HClO_4(aq) + NaOH(aq) \rightarrow NaClO_4(aq) + H_2O(l)$: (c) $CsOH(aq) + HNO_3(aq) \rightarrow CsNO_3(aq) + H_2O(l)$: (d) $2KOH(aq) + H_2SO_4(aq) \rightarrow 2H_2O(l) + K_2SO_4(aq)$
42. transfer
44. The metal loses electrons, the nonmetal gains electrons.
46. 3−; three; six
48. $AlBr_3$ is made up of Al^{3+} ions and Br^- ions. Aluminum atoms each lose 3 electrons to become Al^{3+} ions. Bromine atoms each gain 1 electron to become Br^- ions (so each Br_2 molecule gains 2 electrons to become 2 Br^- ions).

50. (a) $2Fe(s) + 3S(s) \rightarrow Fe_2S_3(s)$: (b) $Zn(s) + 2HNO_3(aq) \rightarrow Zn(NO_3)_2(aq) + H_2(g)$: (c) $2Sn(s) + O_2(g) \rightarrow 2SnO(s)$: (d) $2K(s) + H_2(g) \rightarrow 2KH(s)$: (e) $2Cs(s) + 2H_2O(l) \rightarrow 2CsOH(s) + H_2(g)$

52. Examples of formation of water: $HCl(aq) + NaOH(aq) \rightarrow H_2O(l) + NaCl(aq)$; $H_2SO_4(aq) + 2KOH(aq) \rightarrow 2H_2O(l) + K_2SO_4(aq)$. Examples of formation of a gaseous product: $Mg(s) + 2HCl(aq) \rightarrow MgCl_2(aq) + H_2(g)$; $2KClO_3(s) \rightarrow 2KCl(s) + 3O_2(g)$.

54. (a) oxidation–reduction: (b) oxidation–reduction: (c) acid–base: (d) acid–base, precipitation: (e) precipitation: (f) precipitation: (g) oxidation–reduction: (h) oxidation–reduction: (i) acid–base

56. oxidation–reduction

58. A decomposition reaction is one in which a given compound is broken down into simpler compounds or constituent elements. The reactions $CaCO_3(s) \rightarrow CaO(s) + CO_2(g)$ and $2HgO(s) \rightarrow 2Hg(l) + O_2(g)$ represent decomposition reactions. Such reactions often may be classified in other ways. For example, the reaction of $HgO(s)$ is also an oxidation–reduction reaction.

60. (a) $C_2H_5OH(l) + 3O_2(g) \rightarrow 2CO_2(g) + 3H_2O(g)$; (b) $2C_6H_{14}(l) + 19O_2(g) \rightarrow 12CO_2(g) + 14H_2O(g)$; (c) $C_6H_{12}(l) + 9O_2(g) \rightarrow 6CO_2(g) + 6H_2O(g)$

62. Answer depends on student selection.

64. (a) $2Co(s) + 3S(s) \rightarrow Co_2S_3(s)$: (b) $2NO(g) + O_2(g) \rightarrow 2NO_2(g)$: (c) $FeO(s) + CO_2(g) \rightarrow FeCO_3(s)$: (d) $2Al(s) + 3F_2(g) \rightarrow 2AlF_3(s)$: (e) $2NH_3(g) + H_2CO_3(aq) \rightarrow (NH_4)_2CO_3(s)$

66. (a) $2NI_3(s) \rightarrow N_2(g) + 3I_2(s)$: (b) $BaCO_3(s) \rightarrow BaO(s) + CO_2(g)$: (c) $C_6H_{12}O_6(s) \rightarrow 6C(s) + 6H_2O(g)$: (d) $Cu(NH_3)_4SO_4(s) \rightarrow CuSO_4(s) + 4NH_3(g)$: (e) $3NaN_3(s) \rightarrow Na_3N(s) + 4N_2(g)$

68. (a) silver ion: $Ag^+(aq) + Cl^-(aq) \rightarrow AgCl(s)$ lead(II) ion: $Pb^{2+}(aq) + 2Cl^-(aq) \rightarrow PbCl_2(s)$ mercury(I) ion: $Hg_2^{2+}(aq) + 2Cl^-(aq) \rightarrow Hg_2Cl_2(s)$
(b) sulfate ion: $Ca^{2+}(aq) + SO_4^{2-}(aq) \rightarrow CaSO_4(s)$ carbonate ion: $Ca^{2+}(aq) + CO_3^{2-}(aq) \rightarrow CaCO_3(s)$ phosphate ion: $3Ca^{2+}(aq) + 2PO_4^{3-}(aq) \rightarrow Ca_3(PO_4)_2(s)$
(c) hydroxide ion: $Fe^{3+}(aq) + 3OH^-(aq) \rightarrow Fe(OH)_3(s)$ sulfide ion: $2Fe^{3+}(aq) + 3S^{2-}(aq) \rightarrow Fe_2S_3(s)$ phosphate ion: $Fe^{3+}(aq) + PO_4^{3-}(aq) \rightarrow FePO_4(s)$
(d) barium ion: $Ba^{2+}(aq) + SO_4^{2-}(aq) \rightarrow BaSO_4(s)$ calcium ion: $Ca^{2+}(aq) + SO_4^{2-}(aq) \rightarrow CaSO_4(s)$ lead(II) ion: $Pb^{2+}(aq) + SO_4^{2-}(aq) \rightarrow PbSO_4(s)$
(e) chloride ion: $Hg_2^{2+}(aq) + 2Cl^-(aq) \rightarrow Hg_2Cl_2(s)$ sulfide ion: $Hg_2^{2+}(aq) + S^{2-}(aq) \rightarrow Hg_2S(s)$ carbonate: $Hg_2^{2+}(aq) + CO_3^{2-}(aq) \rightarrow Hg_2CO_3(s)$
(f) chloride ion: $Ag^+(aq) + Cl^-(aq) \rightarrow AgCl(s)$ hydroxide ion: $Ag^+(aq) + OH^-(aq) \rightarrow AgOH(s)$ carbonate ion: $2Ag^+(aq) + CO_3^{2-}(aq) \rightarrow Ag_2CO_3(s)$

70. (a) $HNO_3(aq) + KOH(aq) \rightarrow H_2O(l) + \underline{KNO_3}(aq)$;
(b) $H_2SO_4(aq) + Ba(OH)_2(aq) \rightarrow \underline{BaSO_4}(s) + 2H_2O(l)$;
(c) $HClO_4(aq) + NaOH(aq) \rightarrow H_2O(l) + \underline{NaClO_4}(aq)$;
(d) $2HCl(aq) + Ca(OH)_2(aq) \rightarrow \underline{CaCl_2}(aq) + H_2O(l)$

72. (a) soluble (Rule 2: most potassium salts are soluble): (b) soluble (Rule 2: most ammonium salts are soluble): (c) insoluble (Rule 6: most carbonate salts are only slightly soluble): (d) insoluble (Rule 6: most phosphate salts are only slightly soluble): (e) soluble (Rule 2: most sodium salts are soluble): (f) insoluble (Rule 6: most carbonate salts are only slightly soluble): (g) soluble (Rule 3: most chloride salts are soluble)

74. (a) $AgNO_3(aq) + HCl(aq) \rightarrow \underline{AgCl}(s) + HNO_3(aq)$:
(b) $CuSO_4(aq) + (NH_4)_2CO_3(aq) \rightarrow \underline{CuCo_3}(s) + (NH_4)_2SO_4(aq)$:
(c) $FeSO_4(aq) + K_2CO_3(aq) \rightarrow \underline{FeCO_3}(s) + K_2SO_4(aq)$: (d) no reaction: (e) $Pb(NO_3)_2(aq) + Li_2CO_3(aq) \rightarrow \underline{PbCO_3}(s) + 2LiNO_3(aq)$:
(f) $SnCl_4(aq) + 4NaOH(aq) \rightarrow \underline{Sn(OH)_4}(s) + 4NaCl(aq)$

76. $Fe^{2+}(aq) + S^{2-}(aq) \rightarrow FeS(s)$; $2Cr^{3+}(aq) + 3S^{2-}(aq) \rightarrow Cr_2S_3(s)$; $Ni^{2+}(aq) + S^{2-}(aq) \rightarrow NiS(s)$

78. These anions tend to form insoluble precipitates with many metal ions. The following are illustrative for cobalt(II) chloride, tin(II) chloride, and copper(II) nitrate reacting with sodium salts of the given anions.
(a) $CoCl_2(aq) + Na_2S(aq) \rightarrow CoS(s) + 2NaCl(aq)$; $SnCl_2(aq) + Na_2S(aq) \rightarrow SnS(s) + 2NaCl(aq)$; $Cu(NO_3)_2(aq) + Na_2S(aq) \rightarrow CuS(s) + 2NaNO_3(aq)$;
(b) $CoCl_2(aq) + Na_2CO_3(aq) \rightarrow CoCO_3(s) + 2NaCl(aq)$; $SnCl_2(aq) + Na_2CO_3(aq) \rightarrow SnCO_3(s) + 2NaCl(aq)$; $Cu(NO_3)_2(aq) + Na_2CO_3(aq) \rightarrow CuCO_3(s) + 2NaNO_3(aq)$;
(c) $CoCl_2(aq) + 2NaOH(aq) \rightarrow Co(OH)_2(s) + 2NaCl(aq)$; $SnCl_2(aq) + 2NaOH(aq) \rightarrow Sn(OH)_2(s) + 2NaCl(aq)$; $Cu(NO_3)_2(aq) + 2NaOH(aq) \rightarrow Cu(OH)_2(s) + 2NaNO_3(aq)$
(d) $3CoCl_2(aq) + 2Na_3PO_4(aq) \rightarrow Co_3(PO_4)_2(s) + 6NaCl(aq)$ $3SnCl_2(aq) + 2Na_3PO_4(aq) \rightarrow Sn_3(PO_4)_2(s) + 6NaCl(aq)$ $3Cu(NO_3)_2(aq) + 2Na_3PO_4(aq) \rightarrow Cu_3(PO_4)_2(s) + 6NaNO_3(aq)$

80. (a) $2Na(s) + O_2(g) \rightarrow Na_2O_2(s)$: (b) $Fe(s) + H_2SO_4(aq) \rightarrow FeSO_4(aq) + H_2(g)$: (c) $2Al_2O_3(s) \rightarrow 4Al(s) + 3O_2(g)$: (d) $2Fe(s) + 3Br_2(l) \rightarrow 2FeBr_3(s)$: (e) $Zn(s) + 2HNO_3(aq) \rightarrow Zn(NO_3)_2(aq) + H_2(g)$

82. (a) $2C_4H_{10}(l) + 13O_2(g) \rightarrow 8CO_2(g) + 10H_2O(g)$: (b) $C_4H_{10}O(l) + 6O_2(g) \rightarrow 4CO_2(g) + 5H_2O(g)$: (c) $2C_4H_{10}O_2(l) + 11O_2(g) \rightarrow 8CO_2(g) + 10H_2O(g)$

84. (a) $2NaHCO_3(s) \rightarrow Na_2CO_3(s) + H_2O(g) + CO_2(g)$: (b) $2NaClO_3(s) \rightarrow 2NaCl(s) + 3O_2(g)$: (c) $2HgO(s) \rightarrow 2Hg(l) + O_2(g)$: (d) $C_{12}H_{22}O_{11}(s) \rightarrow 12C(s) + 11H_2O(g)$: (e) $2H_2O_2(l) \rightarrow 2H_2O(l) + O_2(g)$

86. $Fe(s) + H_2SO_4(aq) \rightarrow FeSO_4(aq) + H_2(g)$; $Zn(s) + H_2SO_4(aq) \rightarrow ZnSO_4(aq) + H_2(g)$; $Mg(s) + H_2SO_4(aq) \rightarrow MgSO_4(aq) + H_2(g)$; $Co(s) + H_2SO_4(aq) \rightarrow CoSO_4(aq) + H_2(g)$; $Ni(s) + H_2SO_4(aq) \rightarrow NiSO_4(aq) + H_2(g)$

88. (a) one: (b) one: (c) two: (d) two: (e) three

90. The reaction $C(s) + O_2(g) \rightarrow CO_2(g)$ is such an example.

92. (a) $2C_3H_8O(l) + 9O_2(g) \rightarrow 6CO_2(g) + 8H_2O(g)$; oxidation–reduction, combustion: (b) $HCl(aq) + AgC_2H_3O_2(aq) \rightarrow AgCl(s) + HC_2H_3O_2(aq)$, precipitation, double-displacement: (c) $3HCl(aq) + Al(OH)_3(s) \rightarrow AlCl_3(aq) + 3H_2O(l)$, acid–base, double-displacement: (d) $2H_2O_2(aq) \rightarrow 2H_2O(l) + O_2(g)$, oxidation–reduction, decomposition: (e) $N_2H_4(l) + O_2(g) \rightarrow N_2(g) + 2H_2O(g)$, oxidation–reduction, combustion

94. $2Na(s) + Cl_2(g) \rightarrow 2NaCl(s)$; $2Al(s) + 3Cl_2(g) \rightarrow 2AlCl_3(s)$; $Zn(s) + Cl_2(g) \rightarrow ZnCl_2(s)$; $Ca(s) + Cl_2(g) \rightarrow CaCl_2(s)$; $2Fe(s) + 3Cl_2(g) \rightarrow 2FeCl_3(s)$

Chapter 8

2. 307 corks; 116 stoppers; 2640 g (2.64×10^2 g)

4. The average atomic mass is a *weighted* average including a contribution based on the mass of the individual isotopes of an element and their abundance in nature.

6. (a) 1: (b) 10: (c) 50: (d) 100: (e) 100

8. 126.9 amu; 555 atoms; 5.72×10^4 amu

10. 2.000 mol ($2 \times 6.022 \times 10^{23}$ atoms $= 1.204 \times 10^{24}$ atoms)

12. 55.85 g

14. 177 g

16. 2.657×10^{-23} g

18. 0.50 mol O

20. (a) 1.58×10^{-4} mol: (b) 1.000×10^{-2} mol: (c) 0.9705 mol: (d) 0.1000 mol: (e) 0.5403 mol: (f) 4.684×10^2 mol

22. (a) 2.34×10^{-1} g: (b) 0.252 g: (c) 1.10×10^6 g: (d) 2.80×10^{-5} g: (e) 0.413 g: (f) 105 g

24. (a) 1.05×10^{19} atoms: (b) 6.20×10^{20} atoms: (c) 0.0467 mol cobalt: (d) 0.00995 mol cobalt: (e) 249 g: (f) 2.55×10^{24} atoms: (g) 4.32×10^{22} atoms

26. adding (summing)

28. (a) nitrogen dioxide, 46.01 g: (b) dinitrogen monoxide, 44.02 g: (c) xenon tetrafluoride, 207.3 g: (d) sodium hypochlorite, 74.44 g: (e) nitric acid, 63.02 g: (f) sodium acetate, 82.03 g
30. (a) 68.15 g: (b) 162.99 g: (c) 139.3 g: (d) 136.09 g: (e) 174.18 g: (f) 212.27 g
32. (a) 1.14 mol: (b) 3.29×10^{-5} mol: (c) 22.5 mol: (d) 0.0174 mol: (e) 0.0217 mol: (f) 0.0976 mol
34. (a) 4.10×10^{-5} mol: (b) 5.26 mol: (c) 2704 mol: (d) 0.0697 mol: (e) 467 mol
36. (a) 0.0132 g: (b) 5.31×10^{-4} g: (c) 3.70×10^{4} g: (d) 0.633 g: (e) 0.115 g: (f) 112 g
38. (a) 0.0559 g CO_2: (b) 4.96×10^{5} g NCl_3: (c) 0.361 g NH_4NO_3: (d) 324 g H_2O: (e) 1.00×10^{4} g $CuSO_4$
40. (a) 3.84×10^{24} molecules: (b) 1.37×10^{23} molecules: (c) 8.76×10^{16} molecules: (d) 1.58×10^{18} molecules: (e) 4.03×10^{22} molecules
42. (a) 0.0141 mol S: (b) 0.0159 mol S: (c) 0.0258 mol S: (d) 0.0254 mol S
44. less
46. (a) 58.91% Na, 41.09% S: (b) 43.75% N, 6.295% H, 49.96% O: (c) 35.00% N, 5.037% H, 59.96% O: (d) 27.22% N, 3.916% H, 68.87% Cl: (e) 91.10% P, 8.896% H; (f) 3.688% H, 37.77% P, 58.54% O
48. (a) 28.45% Cu: (b) 44.33% Cu: (c) 44.06% Fe: (d) 34.43% Fe: (e) 18.84% Co: (f) 13.40% Co: (g) 88.12% Sn: (h) 78.77% Sn
50. (a) 34.43% Fe: (b) 29.63% O: (c) 92.25% C: (d) 11.92% N: (e) 93.10% Ag: (f) 45.39% Co: (g) 30.45% N: (h) 43.66% Mn
52. (a) 33.73% NH_4^+: (b) 39.81% Cu^{2+}: (c) 64.94% Au^{3+}: (d) 63.51% Ag^+
54. The empirical formula represents the smallest whole-number ratio of the elements present in a compound. The molecular formula indicates the actual number of atoms of each element found in a molecule of the substance.
56. a, c
58. H_3PO_4
60. $BaSO_4$
62. $N_2H_8CO_3$ [actually $(NH_4)_2CO_3$]
64. Co_2S_3
66. CuO
68. AlF_3
70. Li_3N
72. $Al_2S_3O_{12}$ [actually $Al_2(SO_4)_3$]
74. PCl_3, PCl_5
76. molar mass
78. C_6H_6
80. $C_6H_{24}O_6$
82. empirical formula = NaCN; molecular formula = NaCN
84. 5.00 g Al, 0.185 mol, 1.12×10^{23} atoms; 0.140 g Fe, 0.00250 mol, 1.51×10^{21} atoms; 2.7×10^{2} g Cu, 4.3 mol, 2.6×10^{24} atoms; 0.00250 g Mg, 1.03×10^{-4} mol, 6.19×10^{19} atoms; 0.062 g Na, 2.7×10^{-3} mol, 1.6×10^{21} atoms; 3.95×10^{-18} g U, 1.66×10^{-20} mol, 1.00×10^{4} atoms
86. 24.8% X, 17.4% Y, 57.8% Z. If the molecular formula were actually $X_4Y_2Z_6$, the percentage composition would be the same: the *relative* mass of each element present would not change. The molecular formula is always a whole-number multiple of the empirical formula.
88. Cu_2O, CuO
90. (a) 2.82×10^{23} H atoms, 1.41×10^{23} O atoms: (b) 9.32×10^{22} C atoms, 1.86×10^{23} O atoms: (c) 1.02×10^{19} C atoms and H atoms: (d) 1.63×10^{25} C atoms, 2.99×10^{25} H atoms, 1.50×10^{25} O atoms
92. (a) 4.141 g C, 52.96% C, 2.076×10^{23} atoms C: (b) 0.0305 g C, 42.88% C, 1.53×10^{21} atoms C: (c) 14.4 g C, 76.6% C, 7.23×10^{23} atoms C
94. 2.12 g Fe
96. 7.86 g Hg
98. 2.554×10^{-22} g
100. (a) 0.9331 g N: (b) 1.388 g N: (c) 0.8537 g N: (d) 1.522 g N
102. MgN_2O_6 [$Mg(NO_3)_2$]
104. The average mass takes into account not only the exact masses of the isotopes of an element, but also the relative abundance of the isotopes in nature.
106. 8.61×10^{11} sodium atoms; 6.92×10^{24} amu
108. (a) 2.0×10^{2} g K: (b) 0.0612 g Hg: (c) 1.27×10^{-3} g Mn: (d) 325 g P: (e) 2.7×10^{6} g Fe: (f) 868 g Li: (g) 0.2290 g F
110. (a) 151.9 g: (b) 454.4 g: (c) 150.7 g: (d) 129.8 g: (e) 187.6 g
112. (a) 0.311 mol: (b) 0.270 mol: (c) 0.0501 mol: (d) 2.8 mol: (e) 6.2 mol
114. (a) 4.2 g: (b) 3.05×10^{5} g: (c) 0.533 g: (d) 1.99×10^{3} g: (e) 4.18×10^{3} g
116. (a) 1.15×10^{22} molecules: (b) 2.08×10^{24} molecules: (c) 4.95×10^{22} molecules: (d) 2.18×10^{22} molecules: (e) 6.32×10^{20} formula units (substance is ionic)
118. (a) 38.76% Ca, 19.97% P, 41.27% O: (b) 53.91% Cd, 15.38% S, 30.70% O: (c) 27.93% Fe, 24.06% S, 48.01% O: (d) 43.66% Mn, 56.34% Cl: (e) 29.16% N, 8.392% H, 12.50% C, 49.95% O: (f) 27.37% Na, 1.200% H, 14.30% C, 57.14% O: (g) 27.29% C, 72.71% O: (h) 63.51% Ag, 8.246% N, 28.25% O
120. (a) 36.76% Fe: (b) 93.10% Ag: (c) 55.28% Sr: (d) 55.80% C: (e) 37.48% C: (f) 52.92% Al: (g) 36.70% K: (h) 52.45% K
122. $C_3H_7N_2O$
124. HgO
126. $BaCl_2$

Chapter 9

2. The coefficients of the balanced chemical equation indicate the *relative numbers of molecules* (or moles) of each reactant that combine, as well as the number of molecules (or moles) of each product formed.
4. Balanced chemical equations tell us in what proportions *on a mole basis* substances combine; since the molar masses of $C(s)$ and $O_2(g)$ are different, 1 g of O_2 could not represent the same number of moles as 1 g of C.
6. (a) $3MnO_2(s) + 4Al(s) \rightarrow 3Mn(s) + 2Al_2O_3(s)$. Three formula units of manganese(IV) oxide react with four aluminum atoms, producing three manganese atoms and two formula units of aluminum oxide. Three moles of solid manganese(IV) oxide react with four moles of solid aluminum, to produce three moles of solid manganese and two moles of solid aluminum oxide. (b) $B_2O_3(s) + 3CaF_2(s) \rightarrow 2BF_3(g) + 3CaO(s)$. One diboron trioxide molecule reacts with three formula units of calcium fluoride, producing two molecules of boron trifluoride and three formula units of calcium oxide. One mole of solid diboron trioxide reacts with three moles of solid calcium fluoride, to give two moles of gaseous boron trifluoride and three moles of solid calcium oxide. (c) $3NO_2(g) + H_2O(l) \rightarrow 2HNO_3(aq) + NO(g)$. Three molecules of nitrogen dioxide react with one molecule of water, to produce two molecules of nitric acid and one molecule of nitrogen monoxide. Three moles of gaseous nitrogen dioxide react with one mole of liquid water, to produce two moles of aqueous nitric acid and one mole of nitrogen monoxide gas. (d) $C_6H_6(g) + 3H_2(g) \rightarrow C_6H_{12}(g)$. One molecule of C_6H_6 (*benzene*) reacts with three molecules of hydrogen, producing just one molecule of C_6H_{12} (*cyclohexane*). One mole of gaseous benzene reacts with three moles of hydrogen gas, giving one mole of gaseous cyclohexane.

8. False; the amounts of reactants must be on a *mole* basis.
10. $CaH_2(s) + 2H_2O(l) \rightarrow Ca(OH)_2(aq) + 2H_2(g)$;for $Ca(OH)_2$, $\frac{2 \text{ mol } H_2O}{1 \text{ mol } Ca(OH)_2}$; for H_2, $\frac{2 \text{ mol } H_2O}{2 \text{ mol } H_2}$
12. (a) 3.00 mol C, 2.75 mol H_2O: (b) 0.750 mol SO_2, 0.500 mol H_2O: (c) 0.500 mol Bi, 0.375 mol O_2: (d) 0.250 mol SO_2, 0.250 mol H_2O
14. (a) 0.50 mol NH_4Cl (27 g): (b) 0.13 mol CS_2 (9.5 g); 0.25 mol H_2S (8.5 g): (c) 0.50 mol H_3PO_3 (41 g); 1.5 mol HCl (55 g): (d) 0.50 mol $NaHCO_3$ (42 g)
16. (a) 0.469 mol O_2: (b) 0.938 mol Se: (c) 0.625 mol CH_3CHO: (d) 1.25 mol Fe
18. Stoichiometry is the process of using a chemical equation to calculate the relative masses of reactants and products involved in a reaction.
20. (a) 7.65×10^{-6} mol: (b) 0.0374 mol: (c) 3.68×10^{-3} mol: (d) 0.564 mol: (e) 0.0953 mol: (f) 1.17 mol: (g) 25.6 mol
22. (a) 119 g: (b) 678 g: (c) 0.0437 g: (d) 256 g: (e) 0.0206 g; (f) 170. g: (g) 2.11×10^{-4} g
24. (a) 1.03 mol $CuCl_2$: (b) 0.0736 mol $NiCl_2$: (c) 0.240 mol NaOH: (d) 0.250 mol HCl
26. (a) 1.38 g B, 14.0 g HCl: (b) 13.5 g Cu_2O, 6.04 g SO_2: (c) 35.9 g Cu, 6.04 g SO_2: (d) 29.0 g $CaSiO_3$, 11.0 g CO_2
28. 4.11×10^4 g O_2 (90.7 lb)
30. 9.13 g I_2
32. 47.8 g CO_2
34. 4.64 g CoO
36. 0.631 g S
38. 2.75 g CO_2
40. 2.07 g MgO
42. To determine the limiting reactant, first calculate the number of moles of each reactant present. Then determine how these numbers of moles correspond to the stoichiometric ratio indicated by the balanced chemical equation for the reaction. For each reactant, use the stoichiometric ratios from the balanced chemical equation to calculate how much of the *other* reactants would be required to react completely.
44. A reactant is present in excess if there is more of that reactant present than is required to react with the limiting reactant. The limiting reactant, by definition, cannot be present in excess. No.
46. (a) H_2SO_4 is limiting, 4.90 g SO_2, 0.918 g H_2O: (b) H_2SO_4 is limiting, 6.30 g $Mn(SO_4)_2$, 0.918 g H_2O: (c) O_2 is limiting, 6.67 g SO_2, 1.88 g H_2O: (d) $AgNO_3$ is limiting, 3.18 g Ag, 2.09 g $Al(NO_3)_3$
48. (a) O_2 is limiting, 0.458 g CO_2: (b) CO_2 is limiting, 0.409 g H_2O: (c) MnO_2 is limiting, 0.207 g H_2O: (d) I_2 is limiting, 1.28 g ICl
50. (a) CO is limiting reactant; 11.4 mg CH_3OH: (b) I_2 is limiting reactant; 10.7 mg AlI_3: (c) HBr is limiting reactant; 12.4 mg $CaBr_2$; 2.23 mg H_2O: (d) H_3PO_4 is limiting reactant; 15.0 mg $CrPO_4$; 0.309 mg H_2
52. 136 g urea
54. 1.79 g Fe_2O_3
56. 0.626 g CuI; 0.690 g KI_3; 0.573 g K_2SO_4
58. 0.67 kg SiC
60. If the reaction occurs in a solvent, the product may have a substantial solubility in the solvent; the reaction may come to equilibrium before the full yield of product is achieved (see Chapter 16); loss of product may occur through operator error.
62. 94.60% yield
64. $2LiOH(s) + CO_2(g) \rightarrow Li_2CO_3(s) + H_2O(g)$. 142 g of CO_2 can be ultimately absorbed; 102 g is 71.8% of the canister's capacity.
66. theoretical, 2.72 g $BaSO_4$; percent, 74.3%
68. 28.6 g $NaHCO_3$
70. $C_6H_{12}O_6 + 6O_2 \rightarrow 6CO_2 + 6H_2O$; 1.47 g CO_2
72. at least 325 mg
74. (a) $UO_2(s) + 4HF(aq) \rightarrow UF_4(aq) + 2H_2O(l)$. One formula unit of uranium(IV) oxide combines with four molecules of hydrofluoric acid, producing one uranium(IV) fluoride molecule and two water molecules. One mole of uranium(IV) oxide combines with four moles of hydrofluoric acid to produce one mole of uranium(IV) fluoride and two moles of water. (b) $2NaC_2H_3O_2(aq) + H_2SO_4(aq) \rightarrow Na_2SO_4(aq) + 2HC_2H_3O_2(aq)$. Two molecules (formula units) of sodium acetate react exactly with one molecule of sulfuric acid, producing one molecule (formula unit) of sodium acetate and two molecules of acetic acid. Two moles of sodium acetate combine with one mole of sulfuric acid, producing one mole of sodium sulfate and two moles of acetic acid. (c) $Mg(s) + 2HCl(aq) \rightarrow MgCl_2(aq) + H_2(g)$. One magnesium atom reacts with two hydrochloric acid molecules (formula units) to produce one molecule (formula unit) of magnesium chloride and one molecule of hydrogen gas. One mole of magnesium combines with two moles of hydrochloric acid, producing one mole of magnesium chloride and one mole of gaseous hydrogen. (d) $B_2O_3(s) + 3H_2O(l) \rightarrow 2B(OH)_3(s)$. One molecule (formula unit) of diboron trioxide reacts exactly with three molecules of water, producing two molecules of boron trihydroxide (boric acid). One mole of diboron trioxide combines with three moles of water to produce two moles of boron trihydroxide (boric acid).
76. for O_2, 5 mol O_2/1 mol C_3H_8; for CO_2, 3 mol CO_2/1 mol C_3H_8; for H_2O, 4 mol H_2O/1 mol C_3H_8
78. (a) 0.0587 mol NH_4Cl; (b) 0.0178 mol $CaCO_3$: (c) 0.0217 mol Na_2O: (d) 0.0323 mol PCl_3
80. (a) 3.2×10^2 g HNO_3: (b) 0.0612 g Hg: (c) 4.49×10^{-3} g K_2CrO_4: (d) 1.40×10^3 g $AlCl_3$: (e) 7.2×10^6 g SF_6: (f) 2.13×10^3 g NH_3: (g) 0.9397 g Na_2O_2
82. 1.9×10^2 kg SO_3
84. 0.667 g O_2
86. 0.0771 g H_2
88. (a) Br_2 is limiting reactant, 6.4 g NaBr: (b) $CuSO_4$ is limiting reactant, 5.1 g $ZnSO_4$, 2.0 g Cu: (c) NH_4Cl is limiting reactant, 1.6 g NH_3, 1.7 g H_2O, 5.5 g NaCl: (d) Fe_2O_3 is limiting reactant, 3.5 g Fe, 4.1 g CO_2
90. 0.624 mol N_2, 17.5 g N_2; 1.25 mol H_2O, 22.5 g H_2O
92. 5.0 g

Chapter 10

2. Rutherford's experiments did not indicate how the electrons of an atom are arranged or how they move through space. Rutherford suggested that the electrons might revolve around the nucleus like the planets move around the sun, but he could not explain why the negative electrons were not completely attracted into the positively charged nucleus.
4. The different forms of electromagnetic radiation all exhibit the same wavelike behavior and are propagated through space at the same speed (the "speed of light"). The types of electromagnetic radiation differ in their frequency (and wavelength) and in the resulting amount of energy carried per photon.
6. The *speed* of electromagnetic radiation represents how fast energy is transferred through space, whereas the *frequency* of the radiation tells us how many waves pass a given point per time unit.
8. wave-particle nature (duality)
10. The energy of the photon is exactly the same as the energy change within the atom.
12. It is emitted as a photon.
14. excited
16. When excited hydrogen atoms emit their excess energy, the photons of radiation emitted always have exactly the same wavelength and energy. This means that the hydrogen atom possesses only certain allowed energy states, and that the photons emitted correspond to the electron changing from one of these allowed energy states to another allowed energy state. The energy of the photon emitted corresponds to the energy difference between the

allowed states. If the hydrogen atom did *not* possess discrete energy levels, then the photons emitted would have random wavelengths and energies.
18. They are identical.
20. The ground state of an atom is its lowest possible energy state.
22. It moves to an orbit farther from or closer to the nucleus, respectively.
24. Bohr's theory explained the experimentally observed line spectrum of hydrogen exactly. The theory was discarded because the calculated properties did not correspond closely to experimental measurements for atoms other than hydrogen.
26. An orbit refers to a definite, exact circular pathway around the nucleus, in which Bohr postulated that an electron would be found. An orbital represents a region of space in which there is a high probability of finding the electron.
28. Any experiment to measure the exact location of an electron (such as shooting a beam of light at it) would cause the electron to move. Any measurement would involve the application or removal of energy, which would disturb the electron.
30. The drawing is a contour map, indicating a 90% probability that the electron is within the region of space bounded by that region. The electron may be anywhere within this region.
32. two-lobed ("dumbbell"-shaped); lower in energy and closer to the nucleus; similar shape
34. 1
36. fourth: 4*s*, 4*p*, 4*d*, 4*f*, fifth: 5*s*, 5*p*, 5*d*, 5*f*, (5*g*)
38. The Pauli exclusion principle states that an orbital can hold a maximum of two electrons, and those two electrons must have opposite spin.
40. increases
42. paired (opposite spin)
44. a and b are incorrect
46. For a hydrogen atom in its ground state, the electron is in the 1*s* orbital. The 1*s* orbital has the lowest energy of all hydrogen orbitals.
48. similar type of orbitals being filled in the same way; chemical properties of the members of the group are similar
50. (a) $1s^22s^22p^63s^23p^64s^23d^{10}4p^65s^2$: (b) $1s^22s^22p^63s^23p^64s^23d^{10}$: (c) $1s^2$: (d) $1s^22s^22p^63s^23p^64s^23d^{10}4p^5$
52. (a) $1s^22s^22p^63s^23p^64s^1$: (b) $1s^22s^22p^63s^23p^5$: (c) $1s^22s^22p^63s^2$: (d) $1s^22s^22p^2$

54. (a) 1*s* [↑↓] 2*s* [↑↓] 2*p* [↑↓|↑↓|↑↓] 3*s* [↑↓] 3*p* [↑| |]
(b) 1*s* [↑↓] 2*s* [↑↓] 2*p* [↑↓|↑↓|↑↓] 3*s* [↑↓] 3*p* [↑|↑|↑]
(c) 1*s* [↑↓] 2*s* [↑↓] 2*p* [↑↓|↑↓|↑↓] 3*s* [↑↓] 3*p* [↑↓|↑↓|↑↓] 4*s* [↑↓]
3*d* [↑↓|↑↓|↑↓|↑↓|↑↓] 4*p* [↑↓|↑↓|↑]
(d) 1*s* [↑↓] 2*s* [↑↓] 2*p* [↑↓|↑↓|↑↓] 3*s* [↑↓] 3*p* [↑↓|↑↓|↑↓]

56. (a) 1: (b) 3: (c) 8: (d) 2
58. The properties of Rb and Sr suggest that they are members of Groups 1 and 2, respectively, and so must be filling the 5*s* orbital. The 5*s* orbital is lower in energy than (and fills before) the 4*d* orbitals.
60. (a) $[Ar]4s^2$: (b) $[Rn]7s^1$: (c) $[Kr]5s^24d^1$: (d) $[Xe]6s^24f^15d^1$
62. (a) $[Ne]3s^23p^3$: (b) $[Ne]3s^23p^5$: (c) $[Ne]3s^2$: (d) $[Ar]4s^23d^{10}$
64. (a) 1: (b) 2: (c) 0: (d) 10
66. (a) 5*f*: (b) 5*f*: (c) 4*f*: (d) 6*p*
68. (a) $[Kr]5s^24d^8$: (b) $[Rn]7s^25f^46d^1$: (c) $[Kr]5s^14d^7$ (d) $[Xe]6s^14f^{14}5d^{10}$
70. The metallic elements lose electrons and form positive ions (cations); the nonmetallic elements gain electrons and form negative ions (anions).
72. All exist as diatomic molecules (F_2, Cl_2, Br_2, I_2); are nonmetals; have relatively high electronegativities; and form 1− ions in reacting with metallic elements.
74. Elements at the left of a period (horizontal row) lose electrons most readily; at the left of a period (given principal energy level) the nuclear charge is the smallest and the electrons are least tightly held.
76. The elements of a given period (horizontal row) have valence electrons in the same subshells, but nuclear charge increases across a period going from left to right. Atoms at the left side have smaller nuclear charges and bind their valence electrons less tightly.
78. The nuclear charge increases from left to right within a period, pulling progressively more tightly on the valence electrons.
80. (a) Li: (b) Ca: (c) Cl: (d) S
82. (a) Na: (b) S: (c) N: (d) F
84. speed of light
86. photons
88. quantized
90. orbital
92. transition metal
94. spins
96. (a) $1s^22s^22p^63s^23p^64s^1$; $[Ar]4s^1$;
1*s* [↑↓] 2*s* [↑↓] 2*p* [↑↓|↑↓|↑↓] 3*s* [↑↓] 3*p* [↑↓|↑↓|↑↓] 4*s* [↑]
(b) $1s^22s^22p^63s^23p^64s^23d^2$; $[Ar]4s^23d^2$;
1*s* [↑↓] 2*s* [↑↓] 2*p* [↑↓|↑↓|↑↓] 3*s* [↑↓] 3*p* [↑↓|↑↓|↑↓] 4*s* [↑↓]
3*d* [↑|↑| | |]
(c) $1s^22s^22p^63s^23p^2$; $[Ne]3s^23p^2$;
1*s* [↑↓] 2*s* [↑↓] 2*p* [↑↓|↑↓|↑↓] 3*s* [↑↓] 3*p* [↑|↑|]
(d) $1s^22s^22p^63s^23p^64s^23d^6$; $[Ar]4s^23d^6$;
1*s* [↑↓] 2*s* [↑↓] 2*p* [↑↓|↑↓|↑↓] 3*s* [↑↓] 3*p* [↑↓|↑↓|↑↓] 4*s* [↑↓]
3*d* [↑↓|↑|↑|↑|↑]
(e) $1s^22s^22p^63s^23p^64s^23d^{10}$; $[Ar]4s^23d^{10}$;
1*s* [↑↓] 2*s* [↑↓] 2*p* [↑↓|↑↓|↑↓] 3*s* [↑↓] 3*p* [↑↓|↑↓|↑↓] 4*s* [↑↓]
3*d* [↑↓|↑↓|↑↓|↑↓|↑↓]
98. (a) ns^2: (b) ns^2np^5: (c) ns^2np^4: (d) ns^1: (e) ns^2np^4
100. (a) 2.7×10^{-12} m: (b) 4.4×10^{-34} m: (c) 2×10^{-35} m. The wavelengths for the ball and the person are infinitesimally small; the wavelength for the electron is nearly the same order of magnitude as the diameter of an atom.
102. Light is emitted from the hydrogen atom only at certain fixed wavelengths. If the energy levels of hydrogen were continuous, a hydrogen atom would emit energy at all possible wavelengths.
104. The third principal energy level of hydrogen is divided into three sublevels (3*s*, 3*p*, and 3*d*); there is a single 3*s* orbital, a set of three 3*p* orbitals, and a set of five 3*d* orbitals. See Figures 10.25–10.28 for the shapes of these orbitals.
106. a, c
108. (a) $1s^22s^22p^63s^23p^64s^23d^{10}4p^5$
(b) $1s^22s^22p^63s^23p^64s^23d^{10}4p^65s^24d^{10}5p^6$
(c) $1s^22s^22p^63s^23p^64s^23d^{10}4p^65s^24d^{10}5p^66s^2$
(d) $1s^22s^22p^63s^23p^64s^23d^{10}4p^4$
110. (a) five (2*s*, 2*p*): (b) seven (3*s*, 3*p*): (c) one (3*s*): (d) three (3*s*, 3*p*)
112. (a) $[Kr]5s^24d^2$: (b) $[Kr]5s^24d^{10}5p^5$: (c) $[Ar]4s^23d^{10}4p^2$: (d) $[Xe]6s^1$

114. The position of the element (in terms of both the vertical column and the horizontal row) indicates which set of orbitals is being filled last (Fig. 10.31). (a) $3d$: (b) $4d$: (c) $5f$: (d) $4p$
116. metals, low; nonmetals, high
118. (a) Ca: (b) P: (c) K

Chapter 11

2. The *bond energy* is the amount of energy required to *break* a chemical bond. The larger the bond energy, the stronger the bond.
4. Covalent bonding exists when two atoms mutually complete their valence shells by equal sharing of one or more electron pairs. H_2 is an example. Polar covalent bonding also involves the sharing of one or more electron pairs between two atoms, but the pairs are not shared equally by the two atoms; this typically happens when the two atoms are not identical and have different electronegativities. HF is an example.
6. In H_2 and HF, the bonding is covalent in nature, with an electron pair being shared between the atoms. In H_2, the two atoms are identical and so the sharing is equal; in HF, the two atoms are different and so the bonding is polar covalent. Both of these are in marked contrast to the situation in NaF: NaF is an ionic compound, and an electron is completely transferred from sodium to fluorine, thereby producing the separate ions.
8. A bond is polar if the centers of positive and negative charge do not coincide at the same point. The bond has a negative end and a positive end. Any molecule in which the atoms in the bonds are not identical will have polar bonds (although the molecule as a whole may not be polar). Two simple examples are HCl and HF.
10. the difference in electronegativity between the atoms in the bond
12. (a) K < Cr < Br: (b) Ra < Ca < Mg: (c) Na < Al < Cl
14. (a) ionic: (b) polar covalent: (c) polar covalent: (d) ionic
16. c, d
18. The difference in electronegativities is given in parentheses. (a) H—O (1.4), H—N (0.9), H—O bond is more polar; (b) H—N (0.9), H—F (1.9), H—F bond is more polar; (c) H—O (1.4), H—F (1.9), H—F bond is more polar; (d) H—O (1.4), H—Cl (0.9), H—O bond is more polar.
20. (a) Ca—Cl: (b) K—F: (c) Cu—F: (d) Na—Br
22. The presence of strong bond dipoles and a large overall dipole moment makes water a very polar substance. Properties of water that are dependent on its dipole moment involve its freezing point, melting point, vapor pressure, and its ability to dissolve many substances.
24. (a) H: (b) Cl: (c) I
26. (a) $^{\delta+}P \rightarrow F^{\delta-}$: (b) $^{\delta+}P \rightarrow O^{\delta-}$: (c) $^{\delta+}P \rightarrow C^{\delta-}$:
(d) P—H (H and P have essentially the same electronegativities)
28. (a) nonpolar: (b) $^{\delta+}I \rightarrow Br^{\delta-}$: (c) $^{\delta+}I \rightarrow F^{\delta-}$: (d) $^{\delta+}At \rightarrow I^{\delta-}$
30. metallic
32. Atoms in covalent molecules gain a configuration like that of a noble gas by sharing one or more pairs of electrons between atoms: such shared pairs of electrons "belong" to each of the bonding atoms at the same time. In ionic bonding, one atom completely donates one or more electrons to another atom, and then the resulting ions behave independently of one another (they are not "attached" to one another, although they are mutually attracted).
34. (a) $1s^22s^22p^63s^23p^6$, [Ar]: (b) $1s^22s^22p^6$, [Ne]:
(c) $1s^22s^22p^6$, [Ne]: (d) $1s^22s^22p^63s^23p^64s^23d^{10}4p^6\,5s^24d^{10}5p^6$, [Xe]: (e) $1s^22s^22p^63s^23p^64s^23d^{10}4p^6$, [Kr]
36. (a) Sr^{2+}: (b) F^-: (c) Li^+: (d) O^{2-}
38. (a) Na_2S: (b) BaSe: (c) $MgBr_2$: (d) Li_3N: (e) KH
40. (a) Ca^{2+}[Ar], N^{3-}[Ne]: (b) Mg^{2+}[Ne], S^{2-}[Ar];
(c) Al^{3+}[Ne], F^-[Ne]: (d) Ra^{2+}[Rn], Br^-[Kr];
(e) Cs^+[Xe], P^{3-}[Ar]
42. An ionic solid such as NaCl consists of an array of alternating positively and negatively charged ions: that is, each positive ion has as its nearest neighbors a group of negative ions, and each negative ion has a group of positive ions surrounding it. In most ionic solids, the ions are packed as tightly as possible.
44. In forming an anion, an atom gains additional electrons in its outermost (valence) shell. Having additional electrons in the valence shell increases the repulsive forces between electrons, and the outermost shell becomes larger to accommodate this.
46. (a) F^- is larger than Li^+. The F^- ion has a filled $n = 2$ shell. A lithium atom has *lost* the electron from its $n = 2$ shell, leaving the $n = 1$ shell as its outermost. (b) Cl^- is larger than Na^+, since its valence electrons are in the $n = 3$ shell (Na^+ has lost its $3s$ electron). (c) Ca is larger than Ca^{2+}. Positive ions are always smaller than the atoms from which they are formed. (d) I^- is larger. Both Cs^+ and I^- have the same electron configuration (isoelectronic with Xe) and have their valence electrons in the same shell. However, Cs^+ has two more positive charges in its nucleus than does I^-; this charge causes the $n = 5$ shell of Cs^+ to be smaller than that of I^- (the electrons are pulled in closer to the nucleus by the positive charge).
48. (a) I^-: (b) Cl^-: (c) Cl^-: (d) S^{2-}
50. When atoms form covalent bonds, they try to attain a valence-electron configuration similar to that of the following noble gas element. When the elements in the first few horizontal rows of the periodic table form covalent bonds, they attempt to achieve the configurations of the noble gases helium (two valence electrons, duet rule) andneon and argon (eight valence electrons, octet rule).
52. These elements attain a total of eight valence electrons, giving the valence-electron configurations of the noble gases Ne and Ar.
54. Two atoms in a molecule are connected by a triple bond if the atoms share three pairs of electrons (six electrons) to complete their outermost shells. A simple molecule containing a triple bond is acetylene, C_2H_2 (H : C : : : C : H).

56. (a) Rb· (b) :C̈l̤·

(c) :K̈r̤: (d) Ba:

(e) ·Ṗ· (f) :Ät·

58. (a) 32: (b) 17: (c) 30

60. (a) H—H (b) H—C̈l̤:

(c)
```
     ..
    :F:
     |
 ..  |  ..
:F—C—F:
 ..  |  ..
    :F:
     ..
```
(d)
```
    .. ..
   :F::F:
    |  |
 .. |  |  ..
:F—C—C—F:
 .. |  |  ..
   :F::F:
    .. ..
```

62. (a) H_2S H—S̈—H

(b) SiF_4
```
      ..
     :F:
 ..   |   ..
:F—Si—F:
 ..   |   ..
     :F:
      ..
```
(c) C_2H_4
```
H   H
|   |
C = C
|   |
H   H
```

(d) C_3H_8
```
  H H H
  | | |
H—C—C—C—H
  | | |
  H H H
```

64. (a) NO_2 Ö=N̈—Ö· ↔ ·Ö—N̈=Ö ↔

Ö=N—Ö: ↔ :Ö—N=Ö

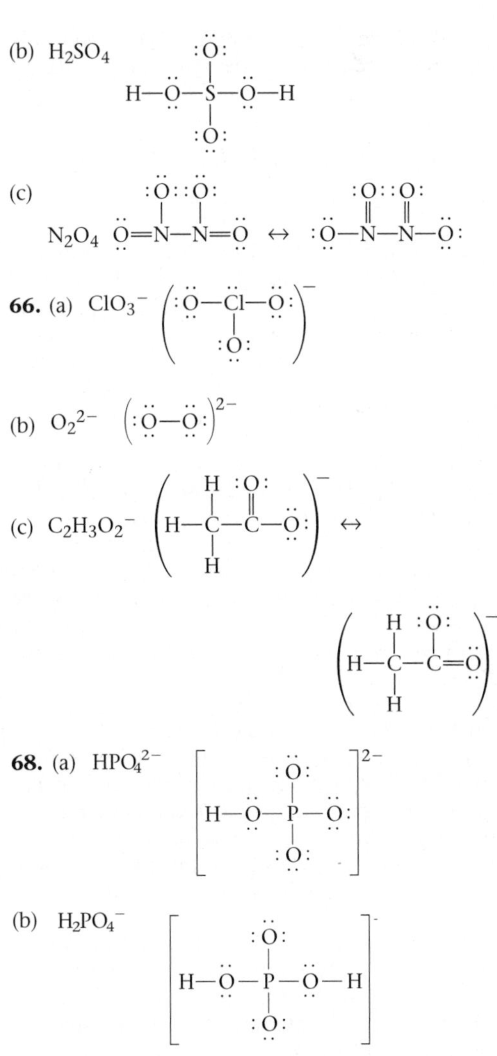

(c) PO_4^{3-} $\left[:\ddot{O}-P(:\ddot{O}:)_2-\ddot{O}:\right]^{3-}$

70. The geometric structure of NH_3 is that of a trigonal pyramid. The nitrogen atom of NH_3 is surrounded by four electron pairs (three are bonding, one is a lone pair). The H—N—H bond angle is somewhat less than 109.5° (because of the presence of the lone pair).
72. SiF_4 has a tetrahedral geometric structure; eight pairs of electrons on Si; ~109.5°
74. The general molecular structure of a molecule is determined by how many electron pairs surround the central atom in the molecule, and by which of those pairs are used for bonding to the other atoms of the molecule.
76. Geometry shows that only two points in space are needed to indicate a straight line. A diatomic molecule represents two points in space.
78. In NF_3, the nitrogen atom has *four* pairs of valence electrons; in BF_3, only *three* pairs of valence electrons surround the boron atom. The nonbonding pair on nitrogen in NF_3 pushes the three F atoms out of the plane of the N atom.
80. Each of the indicated atoms is surrounded by four pairs of electrons arranged tetrahedrally. Although the electron pairs are tetrahedral, the overall geometric shape of the molecules may *not* be tetrahedral.
82. (a) tetrahedral: (b) bent (nonlinear): (c) tetrahedral
84. (a) basically tetrahedral arrangement of the oxygens around the phosphorus: (b) tetrahedral: (c) trigonal pyramid
86. (a) approximately tetrahedral (a little less than 109.5°): (b) approximately tetrahedral (a little less than 109.5°): (c) tetrahedral (109.5°): (d) trigonal planar (120°) because of the double bond
88. The acetylene molecule would be linear, with bond angles of 180°, because of the presence of the triple bond between the carbon atoms.
90. double
92. (a) S—F: (b) P—O: (c) C—H
94. The bond energy is the energy required to break the bond.
96. (a) Be: (b) N: (c) F
98. a, c
100. (a) O: (b) Br: (c) I
102. (a) Al: $1s^22s^22p^63s^23p^1$; Al^{3+}: $1s^22s^22p^6$; Ne has the same configuration as Al^{3+}. (b) Br: $1s^22s^22p^63s^23p^64s^23d^{10}4p^5$; Br^-: $1s^22s^22p^63s^23p^64s^2 3d^{10}4p^6$; Kr has the same configuration as Br^-. (c) Ca: $1s^22s^22p^63s^23p^64s^2$; Ca^{2+}: $1s^22s^22p^63s^23p^6$; Ar has the same configuration as Ca^{2+}. (d) Li: $1s^22s^1$; Li^+: $1s^2$; He has the same configuration as Li^+. (e) F: $1s^22s^22p^5$; F^-: $1s^22s^22p^6$; Ne has the same configuration as F^-.
104. (a) Na_2Se: (b) RbF: (c) K_2Te: (d) BaSe: (e) KAt: (f) FrCl
106. (a) Na^+: (b) Al^{3+}: (c) F^-: (d) Na^+
108. (a) 24: (b) 32: (c) 32: (d) 32
110. (a) N_2H_4 H—N̈(H)—N̈(H)—H

(b) C_2H_6 H—C(H)(H)—C(H)(H)—H

(c) NCl_3 :C̈l—N̈(—C̈l:)—C̈l:

(d) $SiCl_4$:C̈l—Si(—C̈l:)(—C̈l:)—C̈l:

112. (a) NO_3^- [Ö=N(—Ö:)—Ö:]⁻ ↔ [:Ö—N(—Ö:)=Ö]⁻ ↔ [:Ö—N(=Ö)—Ö:]⁻

(b) CO_3^{2-} [Ö=C(—Ö:)—Ö:]²⁻ ↔ [:Ö—C(—Ö:)=Ö]²⁻ ↔ [:Ö—C(=Ö)—Ö:]²⁻

(c) NH_4^+ [H—N(H)(H)—H]⁺

114. (a) four pairs arranged tetrahedrally: (b) four pairs arranged tetrahedrally: (c) three pairs arranged trigonally (planar)
116. (a) trigonal pyramid: (b) nonlinear (V-shaped): (c) tetrahedral
118. (a) nonlinear (V-shaped): (b) trigonal planar: (c) basically trigonal planar around C, distorted somewhat by H: (d) linear
120. Ionic compounds tend to be hard, crystalline substances with relatively high melting and boiling points. Covalently bonded substances tend to be gases, liquids, or relatively soft solids, with much lower melting and boiling points.

Chapter 12

2. Solids are rigid and incompressible and have definite shapes and volumes. Liquids are less rigid than solids; although they have definite volumes, liquids take the shape of their containers. Gases have no fixed volume or shape; they take the volume and shape of their container, and are affected more by changes in their pressure and temperature than are solids or liquids.
4. A mercury barometer consists of a tube filled with mercury that is then inverted over a reservoir of mercury, the surface of which is open to the atmosphere. The pressure of the atmosphere is reflected in the height to which the column of mercury in the tube is supported.
6. 760 mm Hg
8. (a) 1.038 atm: (b) 0.989 atm: (c) 0.989 atm: (d) 1.01 atm
10. (a) 758.1 mm Hg: (b) 1691 mm Hg: (c) 748 mm Hg: (d) 819.3 mm Hg
12. (a) 2.07×10^3 kPa: (b) 106 kPa: (c) 1.10×10^3 kPa: (d) 87.9 kPa
14. increases
16. $PV = k$; $P_1V_1 = P_2V_2$
18. (a) 588 mL: (b) 2.54 L: (c) 4.26×10^4 mm Hg
20. (a) 146 mL: (b) 0.354 L: (c) 687 mm Hg
22. factor $(1.01/0.562) = 1.80$
24. 27.2 atm
26. Absolute zero can be determined using Charles's law, which states that the volume of an ideal gas sample is directly proportional to its temperature. If we measure the volume of a gas sample at several different temperatures and then plot these data, it is possible to extrapolate a straight line mathematically. At the point on the extrapolated line where the volume of the gas sample would be zero, the temperature is absolute zero.
28. $V = bT$; $V_1/T_1 = V_2/T_2$
30. 569 mL
32. (a) 80.2 mL: (b) −77 °C (196 K): (c) 208 mL (2.1×10^2 mL)
34. (a) 35.4 K = −238 °C: (b) 0 mL (absolute zero; a real gas would condense to a solid or liquid): (c) 40.5 mL
36. no (the temperature is doubled in Celsius); −128 °C
38. 113 mL (238 K, −35 °C); 142 mL (299 K, 26 °C); 155 mL (326 K, 53 °C); 127 mL (267 K, −6 °C)
40. $V = an$; $V_1/n_1 = V_2/n_2$
42. 39.0 L
44. 22.4 L
46. Real gases behave most ideally at relatively high temperatures and relatively low pressures. We usually assume that a real gas's behavior approaches ideal behavior if the temperature is over 0 °C (273 K) and the pressure is 1 atm or lower.
48. For an ideal gas, $PV = nRT$ is true under any conditions. Consider a particular sample of gas (n remains constant) at a particular fixed pressure (P remains constant). Suppose that at temperature T_1 the volume of the gas sample is V_1. For this set of conditions, the ideal gas equation would be given by $PV_1 = nRT_1$. If the temperature of the gas sample changes to a new temperature, T_2, then the volume of the gas sample changes to a new volume, V_2. For this new set of conditions, the ideal gas equation would be given by $PV_2 = nRT_2$. If we make a *ratio* of these two expressions for the ideal gas equation for this gas sample, and cancel out terms that are constant for this situation (P, n, and R), we get

$$\frac{PV_1}{PV_2} = \frac{nRT_1}{nRT_2}, \text{ or } \frac{V_1}{V_2} = \frac{T_1}{T_2},$$

which can be rearranged to the familiar form of Charles's law,

$$\frac{V_1}{T_1} = \frac{V_2}{T_2}.$$

50. (a) 5.02 L: (b) 3.56 atm = 2.70×10^3 mm Hg: (c) 334 K
52. (a) 0.201 L: (b) 0.338 atm: (c) 2.48×10^{-2} mol
54. 491 g He
56. 4.44×10^3 g He; 2.24×10^3 g H_2
58. 286 K (13 °C)
60. 340 atm
62. 85.2 mL
64. 150. atm
66. As a gas is bubbled through water, the bubbles of gas become saturated with water vapor, thus forming a gaseous mixture. The total pressure for a sample of gas that has been collected by bubbling through water is made up of two components: the pressure of the sample gas and the pressure of water vapor. The partial pressure of the gas equals the total pressure of the sample minus the vapor pressure of water.
68. 26.6 L
70. 159 atm
72. 751 mm Hg
74. $P_{hydrogen} = 0.990$ atm; 9.55×10^{-3} mol H_2; 0.625 g Zn
76. A theory is successful if it explains known experimental observations. Theories that have been successful in the past may not be successful in the future (for example, as technology evolves, more sophisticated experiments may be possible in the future).
78. The phenomenon of pressure arises from the molecules of gas colliding with the walls of the container.
80. no
82. If the temperature of a sample of gas is increased, the average kinetic energy of the particles of gas increases. This means that the speeds of the particles increase. If the particles have a higher speed, they hit the walls of the container more frequently and with greater force, thereby increasing the pressure.
84. STP = 0 °C, 1 atm pressure. These conditions were chosen because they are easy to attain and reproduce *experimentally*. The barometric pressure within a laboratory will usually be near 1 atm, and 0 °C can be attained with a simple ice bath.
86. 1.93 L
88. 0.940 L
90. 1.17 L
92. 5.03 L (dry volume)
94. 52.7 L
96. 184 mL
98. 40.5 L; $P_{He} = 0.864$ atm; $P_{Ne} = 0.136$ atm
100. 2.48 L
102. 0.365 g
104. twice
106. (a) $PV = k$; $P_1V_1 = P_2V_2$: (b) $V = bT$; $V_1/T_1 = V_2/T_2$: (c) $V = an$; $V_1/n_1 = V_2/n_2$: (d) $PV = nRT$: (e) $P_1V_1/T_1 = P_2V_2/T_2$
108. 125 balloons
110. 124 L
112. 0.0999 mol CO_2; 3.32 L wet; 2.68 L dry
114. 18.1 L O_2
116. (a) 1.00×10^5 Pa: (b) 4.52 atm: (c) 1087 (1.09×10^3) mm Hg: (d) 842 mm Hg
118. (a) 8.60×10^4 Pa: (b) 2.21×10^5 Pa: (c) 8.88×10^4 Pa: (d) 4.3×10^3 Pa
120. (a) 128 mL: (b) 1.3×10^{-2} L: (c) 9.8 L
122. 2.55×10^3 mm Hg
124. (a) 57.3 mL: (b) 448 K = 175 °C: (c) zero (absolute zero; a real gas would condense to a solid or liquid)

126. 123 mL
128. 2.59 g
130. (a) 61.8 K: (b) 0.993 atm: (c) 1.66×10^4 L
132. 487 mol gas needed; 7.79 kg CH_4; 13.6 kg N_2; 21.4 kg CO_2
134. 0.42 atm
136. 2.51×10^3 K
138. 32.4 L
140. 3.43 L N_2; 10.3 L H_2
142. 5.8 L O_2; 3.9 L SO_2
144. 7.8×10^2 L
146. 32 L; $P_{He} = 0.86$ atm; $P_{Ar} = 0.017$ atm; $P_{Ne} = 0.12$ atm
148. 22.4 L O_2

Chapter 13

2. Water, as perspiration, helps cool the human body; on a hot day, you might take an extra shower to cool off. Large bodies of water (e.g., oceans) have a cooling effect on nearby land masses. Water is used as a coolant in *many* commercial situations: for example, some nuclear power plants use water to cool the reactor core; many office buildings are air-conditioned by circulating cold water combined with fan systems.
4. The expansion of water when it freezes often results in broken water pipes during cold weather. This expansion also makes ice float on liquid water. Aquatic life, and probably all life as we know it, would not be possible if ice sank in water.
6. Sloped portions of a heating/cooling curve represent *changes in temperature* as heat is applied or removed; flat portions of such curves represent equilibrium transitions between states.
8. As the temperature of a liquid is increased, the molecules begin moving more rapidly and with greater energy. As the boiling point is reached, molecules possess enough kinetic energy to escape from the surface of the liquid.
10. intermolecular
12. The steam has a greater energy content because of the heat of vaporization.
14. 0.48 kJ; 27 kJ
16. 8.35 kJ; 84.4 kJ; 23.1 kJ
18. 107 kJ/mol; 39.5 kJ; 1.07×10^3 kJ
20. Dipole–dipole forces are stronger at shorter distances; they are relatively short-range forces.
22. The hydrogen bonding that can exist when H is bonded to O (or N or F) is an additional intermolecular force, which means additional energy must be added to separate the molecules during boiling.
24. The fact that such nonpolar atoms *can* be liquefied and solidified indicates that *some* kind of intermolecular forces are possible between atoms in these substances. London dispersion forces arise when a temporary (instantaneous) dipolar arrangement of charge develops as the electrons of an atom move around its nucleus. This instantaneous dipole can induce a similar dipole in a neighboring atom, leading to a momentary attraction.
26. (a) London dispersion forces: (b) London dispersion forces; (c) dipole–dipole forces and London dispersion forces: (d) dipole–dipole (hydrogen bonding) and London dispersion forces
28. An increase in the heat of fusion is observed for an increase in the size of the halogen atom (the electron cloud of a larger atom is more easily polarized by a neighboring dipole, thus giving larger London dispersion forces).
30. For a homogeneous mixture to form, the forces between molecules of the two substances being mixed must be at least *comparable in magnitude* to the intermolecular forces within each separate substance. In the case of a water–ethanol mixture, the forces that exist when water and ethanol are mixed are stronger than water–water or ethanol–ethanol forces in the separate substances. Ethanol and water molecules can approach one another more closely in the mixture than either substance's molecules could approach a like molecule in the separate substances. Strong hydrogen bonding occurs in both ethanol and water.
32. When a liquid is placed into a closed container, a dynamic equilibrium is set up, in which vaporization of the liquid and condensation of the vapor are occurring at the same rate. Once the equilibrium has been achieved, there is a net concentration of molecules in the vapor state, which gives rise to the observed vapor pressure.
34. A method is shown in Figure 13.10. The apparatus consists of a barometer into which a volatile liquid may be injected. Because mercury is more dense than other liquids, the injected volatile liquid rises to the top of the mercury column in the tube and floats on top of the mercury. The volatile liquid evaporates into the empty space above the mercury (a virtual vacuum). As the liquid is converted to the gaseous state, the level of the mercury column drops as the pressure of vapor builds up.
36. (a) HF: Although both substances are capable of hydrogen bonding, water has two O—H bonds that can be involved in hydrogen bonding versus only one F—H bond in HF. (b) CH_3OCH_3: Because no H is attached to the O atom, no hydrogen bonding can exist. Thus, the molecule should be relatively more volatile than CH_3CH_2OH even though it contains the same number of atoms of each element: (c) CH_3SH: Hydrogen bonding is not as important for a S—H bond (because S has a lower electronegativity than O). Since there is relatively little hydrogen bonding, CH_3SH is more volatile than CH_3OH.
38. Both substances have the same molar mass. Ethyl alcohol contains a hydrogen atom directly bonded to an oxygen atom, however. Therefore, hydrogen bonding can exist in ethyl alcohol, whereas only weak dipole–dipole forces exist in dimethyl ether. Dimethyl ether is more volatile; ethyl alcohol has a higher boiling point.
40. *Ionic* solids have positive and negative ions as their fundamental particles; a simple example is sodium chloride, in which Na^+ and Cl^- ions are held together by strong electrostatic forces. *Molecular* solids have molecules as their fundamental particles, with the molecules being held together in the crystal by dipole–dipole forces, hydrogen-bonding forces, or London dispersion forces (depending on the identity of the substance); simple examples of molecular solids include ice (H_2O) and ordinary table sugar (sucrose). *Atomic* solids have simple atoms as their fundamental particles, with the atoms being held together in the crystal by either covalent bonding (as in graphite or diamond) or metallic bonding (as in copper or other metals).
42. sugar: molecular solid, relatively "soft," melts at a relatively low temperature, dissolves as molecules, does not conduct electricity when dissolved or melted; salt: ionic solid, relatively "hard," melts at a high temperature, dissolves as + and − ions, conducts electricity when dissolved or melted
44. Ionic solids consist of a crystal lattice of positively and negatively charged ions. A given ion is surrounded by several ions of the opposite charge, all of which electrostatically attract it strongly. This pattern repeats itself throughout the crystal. The existence of these strong electrostatic forces throughout the crystal means a great deal of energy must be applied to overcome the forces and melt the solid.
46. Ice contains nonlinear, highly polar water molecules, with extensive, strong hydrogen bonding. Dry ice consists of linear, nonpolar molecules, and only very weak intermolecular forces are possible.
48. Although ions exist in both the solid and liquid states, in the solid state the ions are rigidly held in place in the crystal lattice and cannot move so as to conduct an electrical current.
50. Steel is an interstitial alloy of iron, in which carbon atoms occupy the interstices in the iron crystal. Adding carbon to iron to make steel introduces new properties because of the directional nature of the iron–carbon interactions.

52. j
54. f
56. d
58. a
60. l
62. Dimethyl ether has the larger vapor pressure. No hydrogen bonding is possible because the O atom does not have a hydrogen atom attached. Hydrogen bonding can occur *only* when a hydrogen atom is *directly* attached to a strongly electronegative atom (such as N, O, or F). Hydrogen bonding is possible in ethanol (ethanol contains an —OH group).
64. (a) H_2. London dispersion forces are the only intermolecular forces present in these nonpolar molecules; typically these forces become larger with increasing atomic size (as the atoms become bigger, the edge of the electron cloud lies farther from the nucleus and becomes more easily distorted). (b) Xe. Only the relatively weak London forces exist in a crystal of Xe atoms, whereas in NaCl strong ionic forces exist, and in diamond strong covalent bonding exists between carbon atoms. (c) Cl_2. Only London forces exist among such nonpolar molecules.
66. *Steel* is a general term applied to alloys consisting primarily of iron, but with small amounts of other substances added. Whereas pure iron itself is relatively soft, malleable, and ductile, steels are typically much stronger and harder and much less subject to damage.
68. Water is the solvent in which cellular processes take place in living creatures. Water in the oceans moderates the earth's temperature. Water is used in industry as a cooling agent, and it serves as a means of transportation. The liquid range is 0 °C to 100 °C at 1 atm pressure.
70. At higher altitudes, the boiling points of liquids are lower because there is a lower atmospheric pressure above the liquid. The temperature at which food cooks is determined by the temperature to which the water in the food can be heated before it escapes as steam. Thus, food cooks at a lower temperature at high elevations where the boiling point of water is lowered.
72. Heat of fusion (melt); heat of vaporization (boil). The heat of vaporization is always larger, because virtually all of the intermolecular forces must be overcome to form a gas. In a liquid, considerable intermolecular forces remain. Going from a solid to a liquid requires less energy than going from a liquid to a gas.
74. Dipole–dipole interactions are typically 1% as strong as a covalent bond. Dipole–dipole interactions represent electrostatic attractions between portions of molecules that carry only a *partial* positive or negative charge, and such forces require the molecules that are interacting to come *near* each other.
76. London dispersion forces are relatively weak forces that arise among noble gas atoms and in nonpolar molecules. London forces arise from *instantaneous dipoles* that develop when one atom (or molecule) momentarily distorts the electron cloud of another atom (or molecule). London forces are typically weaker than either permanent dipole–dipole forces or covalent bonds.
78. For each mole of liquid water that evaporates, several kilojoules of heat must be absorbed by the water from its surroundings to overcome attractive forces among the molecules.
80. In NH_3, strong hydrogen bonding can exist. Because CH_4 molecules are nonpolar, only the relatively weak London dispersion forces exist.
82. Strong *hydrogen-bonding* forces are present in an ice crystal, while only the much weaker *London forces* exist in the crystal of a nonpolar substance like oxygen.

Chapter 14

2. A *non*homogeneous mixture may differ in composition in various places in the mixture, whereas a solution (a homogeneous mixture) has the same composition throughout. Examples of nonhomogeneous mixtures include spaghetti sauce, a jar of jelly beans, and a mixture of salt and sugar.
4. solid
6. One substance will mix with and dissolve in another substance if the intermolecular forces are similar in the two substances. When the mixture forms, the forces between particles in the mixture will be similar to the forces present in the separate substances. Both sugar and ethyl alcohol molecules contain polar —OH groups, which are comparable to the polar —OH structure in water. Sugar or ethyl alcohol molecules can hydrogen-bond with water molecules and intermingle with them freely to form a solution. Substances like petroleum (whose molecules contain only carbon and hydrogen) are very nonpolar and cannot form interactions with polar water molecules.
8. independently
10. unsaturated
12. large
14. 100.
16. (a) 5.02% KNO_3: (b) 7.58% KNO_3: (c) 15.8% KNO_3: (d) 0.318% KNO_3
18. (a) 11.8 g glucose: (b) 11.8 g glucose: (c) 667 g solution: (d) 213 g water
20. 3.44% Cu; 2.61% Zn; 94.0% Fe
22. 19.6% $CaCl_2$
24. 7.81 g KBr
26. 15.6 g $CuCl_2$
28. 4.8 g heptane; 2.7 g pentane; 86 g hexane
30. 0.363 mol Cl^-
32. If a solution has a concentration of 5 *M*, then 1 L of solution (*not* solvent) contains 5 mol of solute. To prepare such a solution, place 5 mol of NaCl in a 1-L flask, and then add the amount of water that is necessary to yield a *total* volume of 1 L after mixing. The NaCl will occupy some space, so the amount of water added will be less than 1.00 L.
34. (a) 2.0 *M*: (b) 1.0 *M*: (c) 0.67 *M*: (d) 0.50 *M*
36. (a) 0.108 *M*: (b) 0.108 *M*: (c) 0.0108 *M*: (d) 0.108 *M*
38. 0.464 *M*
40. 0.0902 *M*
42. 0.619 *M*
44. (a) 0.0133 mol; 0.838 g: (b) 2.34 mol; 39.9 g: (c) 0.00505 mol; 0.490 g: (d) 0.0299 mol; 1.09 g
46. (a) 25.6 g: (b) 901 g: (c) 1.3 g: (d) 7.66 g
48. 7.8 L
50. (a) 4.60×10^{-3} mol Al^{3+}; 1.38×10^{-2} mol Cl^-: (b) 1.70 mol Na^+; 0.568 mol PO_4^{3-}: (c) 2.19×10^{-3} mol Cu^{2+}; 4.38×10^{-3} mol Cl^-: (d) 3.96×10^{-5} mol Ca^{2+}; 7.91×10^{-5} mol OH^-
52. 1.33 g
54. half
56. (a) 0.0837 *M*: (b) 0.320 *M*: (c) 0.0964 *M*: (d) 0.622 *M*
58. 0.541 L (541 mL)
60. 0.13 *M*
62. 0.0260 *M*
64. 7.5 mL
66. 0.523 g
68. 0.300 g
70. 64.2 mL
72. 1.8×10^{-4} *M*
74. (a) 63.0 mL: (b) 2.42 mL: (c) 50.1 mL: (d) 1.22 L
76. 1 normal
78. 1.53 equivalents OH^- ion. By definition, one equivalent of OH^- ion exactly neutralizes one equivalent of H^+ ion.
80. (a) 0.277 *N*: (b) 3.37×10^{-3} *N*: (c) 1.63 *N*
82. (a) 0.134 *N*: (b) 0.0104 *N*: (c) 13.3 *N*
84. 7.03×10^{-5} *M*; 1.41×10^{-4} *N*
86. 36.3 mL
88. 0.05583 *M*; 0.1117 *N*
90. Molarity is defined as the number of moles of solute contained in 1 liter of *total* solution volume (solute plus solvent after mixing). In the first example, the total volume after mixing is *not* known and the molarity cannot be calculated. In the second ex-

ample, the final volume after mixing is known and the molarity can be calculated simply.
92. 3.3%
94. 1.93 g
96. 56 mol
98. 1.12 L HCl at STP
100. 26.3 mL
102. 2.56 *M*
104. (a) 6.3% KNO_3: (b) 0.25% KNO_3: (c) 11% KNO_3: (d) 18% KNO_3
106. 4.7% C; 1.4% Ni; 93.9% Fe
108. 28 g Na_2CO_3
110. 9.4 g NaCl; 3.1 g KBr
112. (a) 4.0 *M*: (b) 1.0 *M*: (c) 0.73 *M*: (d) 3.6 *M*
114. 0.812 *M*
116. 0.026 *M*
118. (a) 0.446 mol; 33.3 g: (b) 0.00340 mol; 0.289 g; (c) 0.075 mol; 2.7 g: (d) 0.0505 mol; 4.95 g
120. (a) 0.938 mol Na^+; 0.313 mol PO_4^{3-}: (b) 0.042 mol H^+; 0.021 mol SO_4^{2-}: (c) 0.0038 mol Al^{3+}; 0.011 mol Cl^-: (d) 1.88 mol Ba^{2+}; 3.75 mol Cl^-
122. (a) 0.0909 *M*: (b) 0.127 *M*: (c) 0.192 *M*: (d) 1.6 *M*
124. 0.90 *M*
126. 50. mL
128. 35.0 mL
130. (a) 0.822 *N* HCl: (b) 4.00 *N* H_2SO_4: (c) 3.06 *N* H_3PO_4
132. 0.083 *M* NaH_2PO_4; 0.17 *N* NaH_2PO_4
134. 9.6×10^{-2} *N* HNO_3

Chapter 15

2. In the Arrhenius definition, an acid is a substance that produces hydrogen ions (H^+) when dissolved in water, whereas a base is a substance that produces hydroxide ions (OH^-) in aqueous solution. These definitions proved to be too restrictive because the only bases permitted were compounds of hydroxide ion, and the only solvent permitted was water.
4. A conjugate acid–base pair differs by one hydrogen ion, H^+. For example, $HC_2H_3O_2$ (acetic acid) differs from its conjugate base, $C_2H_3O_2^-$ (acetate ion), by a single H^+ ion.

$$HC_2H_3O_2(aq) \rightleftharpoons C_2H_3O_2^-(aq) + H^+(aq)$$

6. When an acid is dissolved in water, the hydronium ion (H_3O^+) is formed. The hydronium ion is the conjugate acid of water (H_2O).
8. (a) a conjugate pair (HSO_4^- is the acid, SO_4^{2-} is the base): (b) not a conjugate pair (Br^- is the conjugate base of HBr; BrO^- is the conjugate base of HBrO): (c) not a conjugate pair ($H_2PO_4^-$ is the conjugate acid of HPO_4^- and also the conjugate base of H_3PO_4; HPO_4^- is the conjugate acid of PO_4^{3-}): (d) not a conjugate pair (NO_3^- is the conjugate base of HNO_3; NO_2^- is the conjugate base of HNO_2)
10. (a) NH_4^+ (acid), H_2O (base); NH_3 (base), H_3O^+ (acid); (b) $HC_2H_3O_2$ (acid), H_2O (base); $C_2H_3O_2^-$ (base), H_3O^+ (acid); (c) CH_3NH_2 (base), H_2O (acid); $CH_3NH_3^+$ (acid), OH^- (base)
12. (a) H_2CO_3: (b) HBr: (c) $HClO_2$: (d) CH_3NH_2
14. (a) CO_3^{2-}: (b) BrO_4^-: (c) SO_3^{2-}: (d) $C_2H_3O_2^-$
16. (a) $HBrO_3 + H_2O \rightarrow BrO_3^- + H_3O^+$: (b) $HI + H_2O \rightarrow I^- + H_3O^+$: (c) $HNO_3 + H_2O \rightarrow NO_3^- + H_3O^+$: (d) $NH_4^+ + H_2O \rightleftharpoons NH_3 + H_3O^+$
18. If an acid is weak in aqueous solution, it does not easily transfer protons to water (and does not fully ionize). If an acid does not lose protons easily, then the acid's anion must strongly attract protons.
20. A strong acid loses its protons easily and fully ionizes in water; the acid's conjugate base is poor at attracting and holding protons and is a relatively weak base. A weak acid resists loss of its protons and does not ionize to a great extent in water; the acid's conjugate base attracts and holds protons tightly and is a relatively strong base.
22. H_2SO_4 (sulfuric): $H_2SO_4 + H_2O \rightarrow HSO_4^- + H_3O^+$; HCl (hydrochloric): $HCl + H_2O \rightarrow Cl^- + H_3O^+$; HNO_3 (nitric): $HNO_3 + H_2O \rightarrow NO_3^- + H_3O^+$; $HClO_4$ (perchloric): $HClO_4 + H_2O \rightarrow ClO_4^- + H_3O^+$
24. oxyacids: $HClO_4$, HNO_3, H_2SO_4, $HC_2H_3O_2$; non-oxyacids: HCl, HBr, HF, HI, HCN
26. (a) HSO_4^- is a *moderately* strong acid: (b) HBr is a strong acid: (c) HCN is a weak acid: (d) $HC_2H_3O_2$ is a weak acid
28. HCO_3^- can behave as an acid if it reacts with a substance that more strongly gains protons than does HCO_3^- itself. For example, HCO_3^- would behave as an acid when reacting with hydroxide ion (a much stronger base): $HCO_3^-(aq) + OH^-(aq) \rightarrow CO_3^{2-}(aq) + H_2O(l)$. On the other hand, HCO_3^- would behave as a base when reacted with a substance that more readily loses protons than does HCO_3^- itself. For example, HCO_3^- would behave as a base when reacting with hydrochloric acid (a much stronger acid): $HCO_3^-(aq) + HCl(aq) \rightarrow H_2CO_3(aq) + Cl^-(aq)$. $H_2PO_4^- + OH^- \rightarrow HPO_4^{2-} + H_2O$ and $H_2PO_4^- + H_3O^+ \rightarrow H_3PO_4 + H_2O$.
30. The concentrations of H^+ and OH^- ions in water and in dilute aqueous solutions are *not* independent of one another. Rather, they are related by the ion product equilibrium constant, K_w. $K_w = [H^+(aq)][OH^-(aq)] = 1.00 \times 10^{-14}$ at 25 °C. If the concentration of one ion is *increased* by addition of a reagent producing H^+ or OH^-, then the concentration of the complementary ion will *decrease* so that the constant's value will hold true. If an acid is added to a solution, the concentration of hydroxide ion in the solution will decrease. Similarly, if a base is added to a solution, then the concentration of hydrogen ion will decrease.
32. (a) $[H^+(aq)] = 2.5 \times 10^{-10}$ *M;* solution is basic; (b) $[H^+(aq)] = 3.4 \times 10^{-6}$ *M;* solution is acidic; (c) $[H^+(aq)] = 1.4 \times 10^{-13}$ *M;* solution is basic; (d) $[H^+(aq)] = 1.1 \times 10^{-8}$ *M;* solution is basic
34. (a) 7.5×10^{-13} *M,* acidic: (b) 1.4×10^{-8} *M,* acidic: (c) 2.5×10^{-6} *M,* basic: (d) 2.5×10^{-2} *M,* basic
36. (a) $[H^+] = 1.04 \times 10^{-8}$ *M*: (b) $[OH^-] = 4.49 \times 10^{-6}$ *M*: (c) $[OH^-] = 6.01 \times 10^{-7}$ *M*
38. household ammonia (pH 12); blood (pH 7–8); milk (pH 6–7); vinegar (pH 3); lemon juice (pH 2–3); stomach acid (pH 2)
40. The pH of a solution is defined as the *negative* of the logarithm of the hydrogen ion concentration, $pH = -\log[H^+]$. Mathematically, the *negative* sign in the definition means the pH *decreases* as the hydrogen ion concentration *increases.*
42. (a) pH = 3.000 (acidic): (b) pH = 3.660 (acidic): (c) pH = 10.037 (basic): (d) pH = 6.327 (acidic)
44. (a) pH = 5.399, acidic: (b) pH = 6.624, acidic: (c) pH = 10.084, basic: (d) pH = 11.964, basic
46. (a) pOH = 6.55, basic: (b) pOH = 12.11, acidic: (c) pOH = 0.85, basic: (d) pOH = 8.45, acidic
48. (a) pOH = 1.397, $[H^+] = 2.5 \times 10^{-13}$ *M,* basic: (b) pOH = 8.006, $[H^+] = 1.0 \times 10^{-6}$ *M,* acidic: (c) pOH = 4.281, $[H^+] = 1.9 \times 10^{-10}$ *M,* basic: (d) pOH = 11.201, $[H^+] = 1.6 \times 10^{-3}$ *M,* acidic
50. (a) 9.1×10^{-2} *M*: (b) 7.9×10^{-14} *M*: (c) 1.0×10^{-6} *M*: (d) 2.4×10^{-9} *M*
52. (a) $[OH^-] = 1.0 \times 10^{-7}$ *M*: (b) $[OH^-] = 1.2 \times 10^{-13}$ *M*: (c) $[OH^-] = 0.015$ *M*: (d) $[OH^-] = 3.9 \times 10^{-10}$ *M*
54. (a) pH = 2.69: (b) pH = 9.859: (c) pH = 3.003: (d) pH = 6.173
56. The solution contains water molecules, H_3O^+ ions (protons), and NO_3^- ions. Because HNO_3 is a strong acid that is completely ionized in water, no HNO_3 molecules are present.
58. (a) pH = 2.917: (b) pH = 3.701: (c) pH = 4.300: (d) pH = 2.983
60. A buffered solution consists of a mixture of a weak acid and its conjugate base; one example of a buffered solution is a mixture of acetic acid ($HC_2H_3O_2$) and sodium acetate (NaC_2H
62. The weak acid component of a buffered solution is
of reacting with added strong base. For example, using th

buffered solution given as an example in Question 60, acetic acid would consume added sodium hydroxide as follows: $HC_2H_3O_2(aq) + NaOH(aq) \rightarrow NaC_2H_3O_2(aq) + H_2O(l)$. Acetic acid *neutralizes* the added NaOH and prevents it from affecting the overall pH of the solution.
64. $CH_3COO^- + HCl \rightarrow CH_3COOH + Cl^-$; $CH_3COOH + NaOH \rightarrow H_2O + NaCH_3COO$; $HS^- + HCl \rightarrow H_2S + Cl^-$; $H_2S + NaOH \rightarrow H_2O + NaHS$
66. (a) $[OH^-(aq)] = 0.10\ M$, pOH = 1.00, pH = 13.00
(b) $[OH^-(aq)] = 2.0 \times 10^{-4}\ M$, pOH = 3.70, pH = 10.30
(c) $[OH^-(aq)] = 6.2 \times 10^{-3}\ M$, pOH = 2.21, pH = 11.79
(d) $[OH^-(aq)] = 0.0001\ M$, pOH = 4.0, pH = 10.0
68. a, b, d
70. a, c, e
72. Having a concentration as small as $10^{-7}\ M$ for HCl means that the contribution to the total hydrogen ion concentration from the dissociation of water must also be considered in determining the pH of the solution.
74. accepts
76. base
78. carboxyl (—COOH)
80. 1.0×10^{-14}
82. higher
84. pH
86. weak acid
88. (a) H_2O and OH^- are a conjugate acid–base pair (H_2O is the acid, having one more proton than the base, OH^-): (b) H_2SO_4 and SO_4^{2-} are *not* a conjugate acid–base pair (they differ by *two* protons). The conjugate base of H_2SO_4 is HSO_4^-; the conjugate acid of SO_4^{2-} is also HSO_4^-: (c) H_3PO_4 and $H_2PO_4^-$ are a conjugate acid–base pair (H_3PO_4 is the acid, having one more proton than the base, $H_2PO_4^-$): (d) $HC_2H_3O_2$ and $C_2H_3O_2^-$ are a conjugate acid–base pair ($HC_2H_3O_2$ is the acid, having one more proton than the base, $C_2H_3O_2^-$)
90. (a) NH_4^+: (b) NH_3: (c) H_3O^+: (d) H_2O
92. (a) $CH_3CH_2COOH + H_2O \rightleftharpoons CH_3CH_2COO^- + H_3O^+$:
(b) $NH_4^+ + H_2O \rightleftharpoons NH_3 + H_3O^+$: (c) $H_2SO_4 + H_2O \rightarrow HSO_4^- + H_3O^+$: (d) $H_3PO_4 + H_2O \rightleftharpoons H_2PO_4^- + H_3O^+$
94. (a) $[H^+(aq)] = 2.4 \times 10^{-12}\ M$, solution is basic:
(b) $[H^+(aq)] = 9.9 \times 10^{-2}\ M$, solution is acidic: (c) $[H^+(aq)] = 3.3 \times 10^{-8}\ M$, solution is basic: (d) $[H^+(aq)] = 1.7 \times 10^{-9}\ M$, solution is basic
96. (a) $[OH^-(aq)] = 0.0000032\ M$:
(b) $[OH^-(aq)] = 1.54 \times 10^{-8}\ M$: (c) $[OH^-(aq)] = 4.02 \times 10^{-7}\ M$
98. (a) pH = 8.15; solution is basic: (b) pH = 5.97; solution is acidic: (c) pH = 13.34; solution is basic: (d) pH = 2.90; solution is acidic
100. (a) $[OH^-(aq)] = 1.8 \times 10^{-11}\ M$, pH = 3.24, pOH = 10.76;
(b) $[H^+(aq)] = 1.1 \times 10^{-10}\ M$, pH = 9.95, pOH = 4.05;
(c) $[OH^-(aq)] = 3.5 \times 10^{-3}\ M$, pH = 11.54, pH = 2.46;
(d) $[H^+(aq)] = 1.4 \times 10^{-7}\ M$, pH = 6.86, pOH = 7.14
102. (a) $[H^+] = 3.9 \times 10^{-6}\ M$: (b) $[H^+] = 1.0 \times 10^{-2}\ M$:
(c) $[H^+] = 1.2 \times 10^{-12}\ M$: (d) $[H^+] = 7.8 \times 10^{-11}\ M$
104. (a) $[H^+(aq)] = 1.4 \times 10^{-3}\ M$, pH = 2.85: (b) $[H^+(aq)] = 3.0 \times 10^{-5}\ M$, pH = 4.52: (c) $[H^+(aq)] = 5.0 \times 10^{-2}\ M$, pH = 1.30: (d) $[H^+(aq)] = 0.0010\ M$, pH = 3.00

Chapter 16

2. The carbon–oxygen bonds in two carbon monoxide molecules and the oxygen–oxygen bond in an oxygen gas molecule must break, and the carbon–oxygen bonds in two carbon dioxide molecules must form.
4. E_a stands for the "activation energy" of the reaction; it represents the minimum energy the molecules must have for a reaction to occur.
6. Enzymes are biochemical catalysts that speed up the complicated reactions that would be too slow to sustain life at normal body temperatures.
8. A state of equilibrium is attained when two opposing processes are exactly balanced. The development of a vapor pressure above a liquid in a closed container is an example of a physical equilibrium. Any chemical reaction that appears to "stop" before completion is an example of a chemical equilibrium.
10. A system has reached equilibrium when no more product forms, even though significant amounts of all the needed reactants are present. This lack of further product creation indicates that the reverse process is now occurring at the same rate as the forward process—that is, every time a product molecule forms in the system, another product molecule reacts to give back the original reactants elsewhere in the system. Reactions that come to equilibrium are indicated by a double arrow.
12. Although we recognize a state of chemical equilibrium by the fact that the concentrations of reactants and products no longer change with time, the lack of change results from the fact that two opposing processes are going on at the same time (not because the reaction has "stopped"). Further reaction in the forward direction is canceled out by an equal reaction in the reverse direction. The reaction is still proceeding, but the opposite reaction is also proceeding at the same rate.
14. The equilibrium constant is a *ratio* of the concentration of products to the concentration of reactants, all at equilibrium. Depending on the amount of reactant that was originally present, different amounts of reactants and products will be present at equilibrium, but their *ratio* will always be the same for a given reaction at a given temperature. For example, the ratios 4/2 and 6/3 involve different numbers, but each of these ratios has the value 2.
16. (a) $K = \dfrac{[NO_2][O_2]}{[NO][O_3]}$: (b) $K = \dfrac{[SO_3][NO]}{[SO_2][NO_2]}$:
(c) $K = \dfrac{[HCl]^4[O_2]}{[Cl_2]^2[H_2O]}$
18. (a) $K = \dfrac{[CH_3OH(g)]}{[CO(g)][H_2(g)]^2}$: (b) $K = \dfrac{[NO(g)]^2[O_2(g)]}{[NO_2(g)]^2}$:
(c) $K = \dfrac{[PBr_3(g)]^4}{[P_4(g)][Br_2(g)]^6}$
20. $K = 2.8 \times 10^3$
22. $K = 4.0 \times 10^6$
24. Equilibrium constants represent ratios of the *concentrations* of products and reactants present at the point of equilibrium. The *concentration* of a pure solid or a pure liquid is constant and is determined by the density of the solid or liquid.
26. (a) $K = \dfrac{1}{[H_2]^2[CO]}$: (b) $K = \dfrac{1}{[H_2]}$: (c) $K = \dfrac{1}{[CO_2]}$
28. (a) $K = [N_2(g)][Br_2(g)]^3$: (b) $K = \dfrac{[H_2O(g)]}{[H_2(g)]}$: (c) $K = [CO_2(g)]$
30. $[CO_2]$ increases; K does not change
32. If heat is applied to an endothermic reaction (the temperature is raised), the equilibrium is shifted to the right. More product will be present at equilibrium than if the temperature had not been increased. The value of K increases.
34. (a) shifts right: (b) shifts right: (c) no change (catalysts do not affect the position of equilibrium)
36. (a) no effect (UO_2 is a solid): (b) no effect (Xe is not involved in the reaction): (c) shifts to left (if HF attacks glass, it is removed from the system, causing a reaction to replace the lost HF): (d) shifts to right: (e) shifts to left (4 mol gas versus 3 mol gas)
38. lower; heat is a product of the reaction, and removing heat by lowering the temperature will favor the production of ammonia.
40. For an *endo*thermic reaction, an increase in temperature will shift the position of equilibrium to the right (toward products).
42. The reaction is *exo*thermic and should be performed at as low a temperature as possible (as long as molecules still have sufficient energy to react).

44. large; K is a ratio: the larger the value, the larger the numerator (products) or the smaller the denominator (reactants)
46. $K = 3.2 \times 10^{11}$
48. $K = 2.1 \times 10^{-3}$
50. $[O_2(g)] = 8.0 \times 10^{-2}\ M$
52. $5.4 \times 10^{-4}\ M$
54. solubility product, K_{sp}
56. only the temperature
58. (a) $PbBr_2(s) \rightleftharpoons Pb^{2+}(aq) + 2Br^-(aq)$; $K_{sp} = [Pb^{2+}(aq)][Br^-(aq)]^2$: (b) $Ag_2S(s) \rightleftharpoons 2Ag^+(aq) + S^{2-}(aq)$; $K_{sp} = [Ag^+(aq)]^2[S^{2-}(aq)]$: (c) $PbCO_3(s) \rightleftharpoons Pb^{2+}(aq) + CO_3^{2-}(aq)$; $K_{sp} = [Pb^{2+}(aq)][CO_3^{2-}(aq)]$: (d) $Sr_3(PO_4)_2(s) \rightleftharpoons 3Sr^{2+}(aq) + 2PO_4^{3-}(aq)$; $K_{sp} = [Sr^{2+}(aq)]^3[PO_4^{3-}(aq)]^2$
60. $K_{sp} = 2.9 \times 10^{-10}$
62. $K_{sp} = 3.8 \times 10^{-15}$
64. $[BaCO_3] = 9.1 \times 10^{-5}\ M$ in a saturated solution.
66. $[HgS] = 5 \times 10^{-27}\ M$ in a saturated solution (very insoluble indeed)
68. $K_{sp} = 1.9 \times 10^{-4}$; 10. g/L
70. $2 \times 10^{-10}\ M$; 2×10^{-8} g/L
72. increase in temperature increases the fraction of molecules with energy $> E_a$
74. catalyst
76. constant
78. reaction is still taking place, but in opposing directions, at the same speeds
80. heterogeneous
82. position
84. In an exothermic process, heat is a product of the reaction. So adding heat (increasing the temperature) fights against the forward process.
86. An equilibrium reaction may come to many *positions* of equilibrium, but the numerical value of the equilibrium constant is fulfilled at each possible position. If different experiments vary the amounts of reactant, the *absolute amounts* of reactants and products present at the point of equilibrium will differ from one experiment to another, but the *ratio* that defines the equilibrium constant will remain the same.
88. $9.0 \times 10^{-3}\ M$
90. $BaCO_3(s) \rightleftharpoons Ba^{2+}(aq) + CO_3^{2-}(aq)$; $7.1 \times 10^{-5}\ M$
$CdCO_3(s) \rightleftharpoons Cd^{2+}(aq) + CO_3^{2-}(aq)$; $2.3 \times 10^{-6}\ M$
$CaCO_3(s) \rightleftharpoons Ca^{2+}(aq) + CO_3^{2-}(aq)$; $5.3 \times 10^{-5}\ M$
$CoCO_3(s) \rightleftharpoons Co^{2+}(aq) + CO_3^{2-}(aq)$; $3.9 \times 10^{-7}\ M$
92. Although a small solubility product generally implies a small solubility, comparisons of solubility based directly on K_{sp} values are valid only if the salts produce the same numbers of positive and negative ions per formula when they dissolve. For example, the solubilities of AgCl(s) and NiS(s) can be compared directly using K_{sp}, since each salt produces one positive and one negative ion per formula when dissolved. AgCl(s) cannot be directly compared with a salt such as $Ca_3(PO_4)_2$, however.
94. At higher temperatures, the average kinetic energy of the reactant molecules is larger, as is the probability that a collision between molecules will be energetic enough for reaction to take place. On a molecular basis, a higher temperature means a given molecule will be moving faster.

96. (a) $K = \dfrac{[HBr(g)]^2}{[H_2(g)][Br_2(g)]}$: (b) $K = \dfrac{[H_2S(g)]^2}{[H_2(g)]^2[S_2(g)]}$: (c) $K = \dfrac{[HCN(g)]^2}{[H_2(g)][C_2N_2(g)]}$
98. $K = 1.2 \times 10^{-3}$
100. (a) $K = \dfrac{1}{[O_2(g)]^3}$: (b) $K = \dfrac{1}{[NH_3(g)][HCl(g)]}$: (c) $K = \dfrac{1}{[O_2(g)]}$
102. An *exo*thermic reaction liberates energy as heat. Increasing the temperature (adding heat) for such a reaction is fighting against the reaction's own tendency to liberate heat. The net effect of raising the temperature will be a shift to the left and a decrease in the amount of product. To increase the amount of products in an exothermic reaction, heat must be *removed* from the system. Changing the temperature *does* change the numerical value of the equilibrium constant for a reaction.
104. The reaction is *exo*thermic. An increase in temperature (addition of heat) will shift the reaction to the left (toward reactants).
106. $[NH_3(g)] = 1.1 \times 10^{-3}\ M$
108. (a) $Cu(OH)_2(s) \rightleftharpoons Cu^{2+}(aq) + 2OH^-(aq)$; $K_{sp} = [Cu^{2+}(aq)][OH^-(aq)]^2$: (b) $Cr(OH)_3(s) \rightleftharpoons Cr^{3+}(aq) + 3OH^-(aq)$; $K_{sp} = [Cr^{3+}(aq)]\ [OH^-(aq)]^3$: (c) $Ba(OH)_2(s) \rightleftharpoons Ba^{2+}(aq) + 2OH^-(aq)$; $K_{sp} = [Ba^{2+}(aq)][OH^-(aq)]^2$: (d) $Sn(OH)_2(s) \rightleftharpoons Sn^{2+}(aq) + 2OH^-(aq)$; $K_{sp} = [Sn^{2+}(aq)][OH^-(aq)]^2$
110. $K_{sp} = 3.9 \times 10^{-11}$
112. $K_{sp} = 1.4 \times 10^{-8}$
114. The activation energy is the minimum energy that two colliding molecules must possess for the collision to result in a reaction.
116. Once a system has reached equilibrium, the net concentration of product no longer increases because molecules of product already present react to form the original reactants.
118. (a) $K = [H_2O][CO_2]$: (b) $K = [CO_2]$: (c) $K = \dfrac{1}{[O_2]^3}$

Chapter 17

2. Oxidation is a loss of one or more electrons by an atom or ion. Reduction is the gaining of one or more electrons by an atom or ion. Equations depend on student responses.
4. (a) sodium is oxidized, nitrogen is reduced: (b) magnesium is oxidized, chlorine is reduced: (c) aluminum is oxidized, bromine is reduced; (d) iron is oxidized, oxygen is reduced
6. (a) potassium is oxidized, oxygen is reduced: (b) hydrogen is oxidized, sulfur is reduced: (c) hydrogen is oxidized, nitrogen is reduced: (d) mercury is oxidized, chlorine is reduced
8. The oxidation state of the atoms in an uncombined element is *zero*, regardless of whether the element occurs as single atoms or diatomic molecules or a molecular substance with more than two atoms.
10. Because fluorine is the most electronegative element, its oxidation state is always negative relative to other elements; because fluorine gains only one electron to complete its outermost shell, its oxidation number in compounds is always -1. The other halogen elements are almost always more electronegative than the atoms to which they bond, and almost always have -1 oxidation numbers. However, in an interhalogen compound involving fluorine and some other halogen, since fluorine is the most electronegative element of all, the other halogens in the compound will have positive oxidation states relative to fluorine.
12. The sum of all oxidation states of the atoms in a polyatomic ion must equal the *overall charge* on the ion. For example, the hydroxide ion (OH^-) has an overall charge of -1 because hydrogen has an oxidation state of $+1$, whereas oxygen has an oxidation state of -2 in the hydroxide ion: $(-2) + (+1) = -1$.
14. (a) P, 0: (b) H, +1; Cl, +1; O, −2: (c) C, −4; H, +1: (d) H, +1; P, +5; O, −2
16. (a) Cl, 0: (b) Co, +2; Cl, −1: (c) Co, +3; Cl, −1: (d) C, +4; O, −2
18. (a) H, +1; S, +6; O, −2: (b) Mn, +7; O, −2: (c) N, +5; O, −2: (d) K, +1; P, +5; O, −2
20. (a) Cu, +2; Cl, −1: (b) Cr, +3; Cl, −1: (c) H, +1; O, −2; Cr, +6: (d) Cr, +3; O, −2
22. (a) Mn, +2; Cl, −1: (b) H, +1; Br, +5; O, −2: (c) B, +3; H, −1 (boron is less electronegative than H): (d) C, +4; F, −1
24. Electrons are negative; when an atom gains electrons, it gains one negative charge for each electron gained. For example, in the

reduction reaction $Cl + e^- \rightarrow Cl^-$, the oxidation state of chlorine decreases from 0 to −1 as the electron is gained.
26. Answer depends on student selection of example.
28. An oxidizing agent oxidizes another species by gaining the electrons lost by the other species; therefore, an oxidizing agent itself decreases in oxidation state. A reducing agent increases its oxidation state when acting on another atom or molecule.
30. (a) Zn is oxidized ($0 \rightarrow +2$); H is reduced ($+1 \rightarrow 0$): (b) H is oxidized ($0 \rightarrow +1$); Cu is reduced ($+2 \rightarrow 0$): (c) N is reduced ($0 \rightarrow -3$); Br is oxidized ($0 \rightarrow +1$); note that N is more electronegative than Br: (d) Br is oxidized ($-1 \rightarrow 0$); Cl is reduced ($0 \rightarrow -1$)
32. (a) carbon is oxidized, iron is reduced: (b) magnesium is oxidized, hydrogen is reduced: (c) cobalt is oxidized, sulfur is reduced: (d) silver is oxidized, hydrogen is reduced.
34. Iron is reduced [+3 in $Fe_2O_3(s)$, 0 in $Fe(l)$]; carbon is oxidized [+2 in $CO(g)$, +4 in $CO_2(g)$]. $Fe_2O_3(s)$ is the oxidizing agent; $CO(g)$ is the reducing agent.
36. Chlorine is reduced [0 in $Cl_2(g)$, −1 in $NaCl(s)$]; bromine is oxidized [−1 in $NaBr(aq)$, 0 in $Br_2(l)$]. $Cl_2(g)$ is the oxidizing agent; $NaBr(aq)$ is the reducing agent.
38. Oxidation–reduction reactions are often more complicated than "regular" reactions; the coefficients necessary to balance the number of electrons transferred are often large numbers.
40. Under ordinary conditions it is impossible to have "free" electrons that are not part of some atom, ion, or molecule. Thus, the total number of electrons lost by the species being oxidized must equal the total number of electrons gained by the species being reduced.
42. (a) $6e^- + N_2 \rightarrow 2N^{3-}$: (b) $O_2^{2-} \rightarrow O_2 + 2e^-$: (c) $Zn \rightarrow Zn^{2+} + 2e^-$: (d) $2e^- + F_2 \rightarrow 2F^-$
44. (a) $2e^- + 2HBr \rightarrow 2Br^- + H_2$: (b) $ClO_4^- + 2H^+ + 2e^- \rightarrow ClO_3^- + H_2O$: (c) $2e^- + H^+ + PbSO_4 \rightarrow Pb + HSO_4^-$: (d) $H^+ + e^- + HNO_2 \rightarrow NO + H_2O$
46. (a) $2Al + 6H^+ \rightarrow 2Al^{3+} + 3H_2$: (b) $8H^+ + 2NO_3^- + 3S^{2-} \rightarrow 3S + 2NO + 4H_2O$: (c) $6H_2O + I_2 + 5Cl_2 \rightarrow 2IO_3^- + 2H^+ + 10HCl$: (d) $2H^+ + AsO_4^- + S^{2-} \rightarrow S + AsO_3^- + H_2O$
48. $Cl_2(g) + 2I^-(aq) \rightarrow 2Cl^-(aq) + I_2(s)$; $Cl_2(g) + 2Br^-(aq) \rightarrow 2Cl^-(aq) + Br_2(l)$
50. A salt bridge typically consists of a U-shaped tube filled with an inert electrolyte (one involving ions that are not part of the oxidation–reduction reaction). A salt bridge completes the electrical circuit in a cell. Any method that allows transfer of charge without allowing bulk mixing of the solutions may be used (another common method is to set up one half-cell in a porous cup, which is then placed in the beaker containing the second half-cell).
52. In a galvanic cell, the anode is the electrode where oxidation occurs; the cathode is the electrode where reduction occurs.
54.

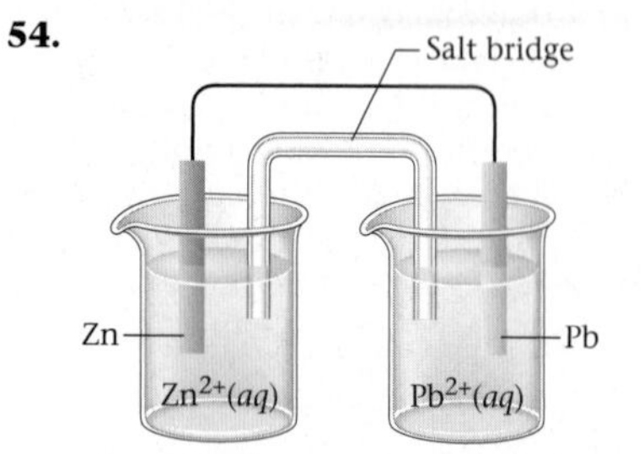

Pb^{2+} ion is reduced; Zn(s) is oxidized. The anode reaction is $Zn(s) \rightarrow Zn^{2+}(aq) + 2e^-$. The cathode reaction is $Pb^{2+}(aq) + 2e^- \rightarrow Pb(s)$.
56. Both normal and alkaline cells contain zinc as an electrode; zinc corrodes more slowly under alkaline conditions than in the highly acidic environment of a normal dry cell.
Anode: $Zn(s) + 2OH^-(aq) \rightarrow ZnO(s) + H_2O(l) + 2e^-$.
Cathode: $2MnO_2(s) + H_2O(l) + 2e^- \rightarrow Mn_2O_3(s) + 2OH^-(aq)$.
58. Aluminum is a very reactive metal when freshly isolated in the pure state. Upon standing for even a relatively short period of time, aluminum metal forms a thin coating of Al_2O_3 on its surface from reaction with atmospheric oxygen. This Al_2O_3 coating is much less reactive than the metal and protects the metal's surface from further attack.
60. In cathodic protection of steel tanks and pipes, a more reactive metal than iron is connected to the item to be protected. The active metal is then preferentially oxidized rather than the iron of the tank or pipe.
62. The main recharging reaction for the lead storage battery is $2PbSO_4(s) + 2H_2O(l) \rightarrow Pb(s) + PbO_2(s) + 2H_2SO_4(aq)$. A major side reaction is the electrolysis of water, $2H_2O(l) \rightarrow 2H_2(g) + O_2(g)$, which produces an explosive mixture of hydrogen and oxygen that accounts for many accidents during the recharging of such batteries.
64. Electrolysis is applied in electroplating by making the item to be plated the cathode in a cell containing a solution of ions of the desired plating metal.
66. loss; oxidation state
68. electronegative
70. An oxidizing agent is an atom, molecule, or ion that causes the oxidation of some other species, while itself being reduced.
72. lose
74. separate from
76. oxidation
78. In an electrolysis reaction, an ordinarily nonspontaneous reaction is forced to occur by the application of an electric current of sufficient voltage. For example, water may be electrolyzed into its elements: $2H_2O(l) \rightarrow 2H_2(g) + O_2(g)$.
80. hydrogen; oxygen
82. oxidation
84. (a) $4Fe(s) + 3O_2(g) \rightarrow 2Fe_2O_3(s)$; iron is oxidized, oxygen is reduced: (b) $2Al(s) + 3Cl_2(g) \rightarrow 2AlCl_3(s)$; aluminum is oxidized, chlorine is reduced: (c) $6Mg(s) + P_4(s) \rightarrow 2Mg_3P_2(s)$; magnesium is oxidized, phosphorus is reduced
86. (a) Al is oxidized ($0 \rightarrow +3$); H is reduced ($+1 \rightarrow 0$); (b) H is reduced ($+1 \rightarrow 0$); I is oxidized ($-1 \rightarrow 0$); (c) Cu is oxidized ($0 \rightarrow +2$); H is reduced ($+1 \rightarrow 0$)
88. (a) $C_3H_8(g) + 5O_2(g) \rightarrow 3CO_2(g) + 4H_2O(g)$: (b) $CO(g) + 2H_2(g) \rightarrow CH_3OH(l)$: (c) $SnO_2(s) + 2C(s) \rightarrow Sn(s) + 2CO(g)$: (d) $C_2H_5OH(l) + 3O_2(g) \rightarrow 2CO_2(g) + 3H_2O(g)$
90. (a) sodium is oxidized, oxygen is reduced: (b) iron is oxidized, hydrogen is reduced: (c) oxygen (O^{2-}) is oxidized, aluminum (Al^{3+}) is reduced: (d) magnesium is oxidized, nitrogen is reduced
92. (a) H, +1; N, −3: (b) C, +2; O, −2: (c) C, +4; O, −2: (d) N, +3; F, −1
94. (a) Mn, +4; O, −2: (b) Ba, +2; Cr, +6; O, −2: (c) H, +1; S, +4; O, −2: (d) Ca, +2; P, +5; O, −2
96. (a) Bi, +3; O, −2: (b) P, +5; O, −2: (c) N, +3; O, −2: (d) Hg, +1
98. (a) oxygen is oxidized, chlorine is reduced: (b) germanium is oxidized, oxygen is reduced: (c) carbon is oxidized, chlorine is reduced: (d) oxygen is oxidized, fluorine is reduced
100. (a) $SiO_2(s) + 4H^+(aq) + 4e^- \rightarrow Si(s) + 2H_2O(l)$: (b) $S(s) + 2H^+(aq) + 2e^- \rightarrow H_2S(g)$: (c) $NO_3^-(aq) + 3H^+(aq) + 2e^- \rightarrow HNO_2(aq) + H_2O(l)$: (d) $NO_3^-(aq) + 4H^+(aq) + 3e^- \rightarrow NO(g) + 2H_2O(l)$
102. (a) $16H^+(aq) + 2MnO_4^-(aq) + 5C_2O_4^{2-}(aq) \rightarrow 2Mn^{2+}(aq) + 8H_2O(l) + 10CO_2(g)$: (b) $8H^+(aq) + MnO_4^-(aq) + 5Fe^{2+}(aq) \rightarrow Mn^{2+}(aq) + 4H_2O(l) + 5Fe^{3+}(aq)$: (c) $16H^+(aq) + 2MnO_4^-(aq) + 10Cl^-(aq) \rightarrow 2Mn^{2+}(aq) + 8H_2O(l) + 5Cl_2(g)$

Chapter 18

2. The radius of a typical atomic nucleus is on the order of 10^{-13} cm, which is roughly 100,000 times smaller than the radius of an atom overall.

4. False. The mass number is the total number of protons and neutrons, not neutrons alone.
6. The atomic number (Z) is written as a left subscript, while the mass number (a) is written as a left superscript. That is, the general symbol for a nuclide is $^{A}_{Z}X$. As an example, consider the isotope of oxygen with 8 protons and 8 neutrons: its symbol would be $^{16}_{8}O$.
8. No charge, mass number = 1, $^{1}_{0}n$
10. When a nucleus produces a beta particle, the atomic number of the parent nucleus is *increased* by one unit.
12. Gamma rays are high-energy photons of electromagnetic radiation; they are not normally considered to be particles. When a nucleus produces only gamma radiation, the atomic number and mass number of the nucleus do not change.
14. Electron capture occurs when one of the inner orbital electrons is pulled into, and becomes part of, the nucleus.
16. Isotope $^{20}_{10}Ne$ predominates, which is reflected in the average atomic molar mass of 20.18 (closer to 20 than to 21 or 22).
18. Based on the predominance of Mg-24, but with significant amounts of the other isotopes, one would expect the average atomic molar mass to be slightly higher than 24 (24.31 g).
20. (a) $^{0}_{-1}e$ or $^{0}_{-1}\beta$: (b) $^{0}_{+1}e$ or $^{0}_{+1}\beta$: (c) $^{0}_{0}\gamma$
22. (a) $^{192}_{83}Bi$: (b) $^{204}_{82}Pb$: (c) $^{206}_{84}Po$
24. (a) $^{218}_{86}Rn$: (b) $^{0}_{+1}e$ (positron): (c) $^{137}_{56}Ba$
26. (a) $^{136}_{53}I \rightarrow ^{0}_{-1}e + ^{136}_{54}Xe$: (b) $^{133}_{51}Sb \rightarrow ^{0}_{-1}e + ^{133}_{52}Te$: (c) $^{117}_{49}In \rightarrow ^{0}_{-1}e + ^{117}_{50}Sn$
28. $^{234}_{92}U \rightarrow ^{230}_{90}Th \rightarrow ^{226}_{88}Ra \rightarrow ^{222}_{86}Rn \rightarrow ^{218}_{84}Po \rightarrow ^{214}_{82}Pb$
30. In a nuclear bombardment process, a target nucleus is bombarded with high-energy particles (typically subatomic particles or small atoms) from a particle accelerator. This may result in the transmutation of the target nucleus into some other element. For example, nitrogen-14 may be transmuted into oxygen-17 by bombardment with α-particles.
32. $^{27}_{13}Al + ^{4}_{2}He \rightarrow ^{30}_{15}P + ^{1}_{0}n$
34. The half-life of a nucleus is the time required for one-half of the original sample of nuclei to decay. A given isotope of an element always has the same half-life, although different isotopes of the same element may have greatly different half-lives. Nuclei of different elements have different half-lives.
36. $^{226}_{88}Ra$ is the most stable (longest half-life); $^{224}_{88}Ra$ is the "hottest" (shortest half-life).
38. highest to lowest activity

$^{87}Sr > ^{99}Tc > ^{24}Na > ^{99}Mo > ^{133}Xe > ^{131}I > ^{32}P > ^{51}Cr > ^{59}Fe$

40. For ^{223}Ra, the half-life is 12 days: After two half-lives (24 days), 250 mg remains; after three half-lives (36 days), 125 mg remains. For ^{224}Ra, the half-life is 3.6 days: One month would be approximately 8 half-life periods (29 days), and approximately 4 mg remains. For ^{225}Ra, the half-life is 15 days: One month would be two half-life periods, and 250 mg remains.
42. For an administered dose of 100 μg, 0.39 μg remains after 2 days. The fraction remaining is 0.39/100 = 0.0039; on a percentage basis, less than 0.4% of the original radioisotope remains.
44. Carbon-14 is produced in the upper atmosphere by the bombardment of nitrogen with neutrons from space:

$$^{14}_{7}N + ^{1}_{0}n \rightarrow ^{14}_{6}C + ^{1}_{1}H$$

46. We assume that the concentration of C-14 in the atmosphere is effectively constant. A living organism is constantly replenishing C-14 through the processes of either metabolism (sugars ingested in foods contain C-14) or photosynthesis (carbon dioxide contains C-14). When a plant dies, it no longer replenishes itself with C-14 from the atmosphere. As the C-14 undergoes radioactive decay, its amount decreases with time.
48. These isotopes and their uses are listed in Table 18.4.
50. Combining two light nuclei to form a heavier, more stable nucleus is called nuclear *fusion*. Splitting a heavy nucleus into nuclei with smaller mass numbers is called nuclear *fission*.
52. $^{1}_{0}n + ^{235}_{92}U \rightarrow ^{142}_{56}Ba + ^{91}_{36}Kr + 3^{1}_{0}n$
54. A critical mass of a fissionable material is the amount needed to provide a high enough internal neutron flux to sustain the chain reaction (production of enough neutrons to cause the continuous fission of further material). A sample with less than a critical mass is still radioactive, but cannot sustain a chain reaction.
56. An actual nuclear explosion, of the type produced by a nuclear weapon, cannot occur in a nuclear reactor because the concentration of the fissionable materials is not sufficient to form a supercritical mass.
58. Breeder reactors use the more common uranium-238 isotope, which is bombarded with neutrons. Through a three-step process, the uranium-238 is converted to plutonium-239, which is fissionable and can then be used to fuel another reactor. The major controversy in the United States concerns the extreme chemical toxicity of plutonium and its difficulty in handling.
60. In one type of fusion reactor, two $^{2}_{1}H$ atoms fuse to produce $^{4}_{2}He$. Because the hydrogen nuclei are positively charged, extremely high energies (temperatures of 40 million K) are needed to overcome the repulsion between the nuclei as they are shot into each other.
62. The theory of stellar nucleosynthesis says that the universe began as a cloud of neutrons that exploded (the *big bang*). After this initial explosion, neutrons supposedly decomposed into protons and electrons $^{1}_{0}n \rightarrow ^{1}_{1}H + ^{0}_{-1}e$. The products of this decomposition then combined to form large clouds of hydrogen atoms. As the hydrogen clouds became larger, gravitational forces caused these clouds to contract and heat up. Eventually the clouds of hydrogen became so dense and so hot that fusion of hydrogen nuclei into helium nuclei took place, with a great release of energy. When the tendency for the hydrogen clouds to expand from the heat of fusion was counterbalanced by the gravitational forces of the cloud, a small star formed. In addition to the fusion of hydrogen nuclei into helium, as the star's hydrogen supply was exhausted, the helium present in the star began to undergo fusion into nuclei of other elements.
64. Somatic damage is damage directly to the organism itself, causing nearly immediate sickness or death to the organism. Genetic damage is damage to the genetic machinery of the organism, which will be manifested in future generations of offspring.
66. Gamma rays penetrate long distances, but seldom cause ionization of biological molecules. Because they are much heavier, although less penetrating, alpha particles ionize biological molecules very effectively and leave a dense trail of damage in the organism. Isotopes that decay by releasing alpha particles can be ingested or breathed into the body, where the damage from the alpha particles will be more acute.
68. Most reactor waste is still in "temporary" storage. Various suggestions have been made for a more permanent solution, such as casting the spent fuel into glass bricks to contain it and then storing the bricks in corrosion-proof metal containers deep underground.
70. Radioactive
72. Mass
74. Neutron; proton
76. Radioactive decay
78. Mass number
80. Transuranium
82. Half-life
84. Radiotracers
86. Chain
88. Breeder
90. 4.5×10^{9} dollars ($4.5 billion)
92. 3.5×10^{-11} J/atom; 8.9×10^{10} J/g
94. Although nuclear waste has been generated for over 40 years, no permanent disposal plan has been implemented as yet. One proposal to dispose of such waste calls for the waste to be sealed

in blocks of glass, which would then be sealed in corrosion-proof metal drums. The drums would then be buried in deep, stable rock formations away from earthquake and other geologically active zones. In these deep storage areas, it is hoped that the waste could decay safely undisturbed until the radioactivity drops to "safe" levels.

96. $^{27}_{13}Al$: 13 protons, 14 neutrons; $^{28}_{13}Al$: 13 protons, 15 neutrons; $^{29}_{13}Al$: 13 protons, 16 neutrons

98. (a) $^{0}_{-1}e$: (b) $^{74}_{34}Se$: (c) $^{240}_{92}U$

100. For a decay of 10 μg to 1/1000 of this amount, we want to know when the remaining amount of ^{131}I is on the order of 0.01 μg.

Time (days)	*Mass (μg)*	*Time (days)*	*Mass (μg)*
0	10	48	0.156
8	5	56	0.078
16	2.5	64	0.039
24	1.25	72	0.020
32	0.625	80	0.01
40	0.313		

Approximately 80 days are required.

102. Breeder reactors convert nonfissionable ^{238}U into fissionable ^{239}Pu. The material used for fission is a combination of U-235 (which undergoes fission in a chain reaction) and the more common U-238 isotope. Excess neutrons from the U-235 fission are absorbed by the U-238, converting it to the fissionable plutonium isotope Pu-239. The chemical and physical properties of Pu-239 make it very difficult and expensive to handle and process.

Answers To Even-Numbered Cumulative Review Exercises

Chapters 1–3

2. After having covered three chapters in this book, you should have adopted an "active" approach to your study of chemistry. You can't just sit and take notes in class, or just review the solved examples in the textbook. You must learn to *interpret* problems and reduce them to simple mathematical relationships.

4. Some courses, particularly those in your major field, have obvious and immediate utility. Other courses—chemistry included—provide general *background* knowledge that will prove useful in understanding your own major, or other subjects related to your major.

6. Whenever a scientific measurement is made, we always employ the instrument or measuring device to the limits of its precision. This usually means that we *estimate* the last significant figure of the measurement. An example of the uncertainty in the last significant figure is given for measuring the length of a pin in the text in Figure 2.5. Scientists appreciate the limits of experimental techniques and instruments and always *assume* that the last digit in a number representing a measurement has been estimated. Because instruments or measuring devices always have a limit to their precision, uncertainty cannot be completely excluded from measurements.

8. Dimensional analysis is a method of problem solving that pays particular attention to the units of measurements and uses these units as if they were algebraic symbols that multiply, divide, and cancel. Consider the following example: One dozen eggs costs $1.25. Suppose we want to know how much one egg costs, and also how much three dozen eggs will cost. To solve these problems, we need two equivalence statements:

1 dozen eggs = 12 eggs

1 dozen eggs = $1.25

The calculations are:

$$\frac{\$1.25}{12 \text{ eggs}} = \$0.104 = \$0.10 \text{ as}$$

the cost of one egg and

$$\frac{\$1.25}{1 \text{ dozen}} \times 3 \text{ dozen} = \$3.75$$

as the cost of three dozen eggs. See Section 2.6 of the text for how we construct conversion factors from equivalence statements.

10. Scientists say that matter is anything that "has mass and occupies space." Matter is the "stuff" of which everything is made. It can be classified and subdivided in many ways, depending on what we are trying to demonstrate. All the types of matter we have studied are made of atoms. They differ in whether these atoms are all of one element, or are of more than one element, and also in whether these atoms are in physical mixtures or chemical combinations.

Matter can also be classified according to its physical state (solid, liquid, or gas). In addition, it can be classified as a pure substance (one type of molecule) or a mixture (more than one type of molecule).

12. An element is a fundamental substance that cannot be broken down into simpler substances by chemical methods. An element consists of atoms of only one type. Compounds, on the other hand, *can* be broken down into simpler substances. For example, both sulfur and oxygen are *elements*. When sulfur and oxygen are placed together and heated, the *compound* sulfur dioxide (SO_2) forms. Each molecule of sulfur dioxide contains one sulfur atom and two oxygen atoms. On a mass basis, SO_2 always consists of 50% each, by mass, sulfur and oxygen—that is, sulfur dioxide has a constant composition. Sulfur dioxide from any source would have the same composition (or it wouldn't be sulfur dioxide!).

14. Scientists define energy as "the capacity to do work." Energy is commonly expressed in units of *calories* or *joules* (the SI unit). One calorie is the amount of heat required to raise the temperature of one gram of water by one Celsius degree. Because 1 calorie = 4.184 joules, it takes 4.184 J to raise the temperature of one gram of water by one Celsius degree. The specific heat capacity of a particular substance is the amount of energy required to raise the temperature of one gram of that substance by one Celsius degree (e.g., the specific heat capacity of water is 1.000 cal/g °C, or 4.184 J/g °C). When energy (as heat) is added to or removed from a pure substance, the total heat flow (Q) is given by $Q = s \times m \times \Delta T$, where m is the mass of substance, s is the specific heat capacity of the substance, and ΔT is the temperature change undergone by the substance.

16. (a) 2.5 qt: (b) 1.731 lb: (c) 0.7852 kg: (d) 4.673 mi: (e) 0.252 m: (f) 0.451 L: (g) 21.2 ft: (h) 49.41 in.: (i) 0.524 ft: (j) 0.289 lb

18. (a) −40 °F: (b) 304 °C: (c) 107 °C: (d) 115 K: (e) −3.4 °C: (f) 78.6 °F

20. (a) 0.425 kJ: (b) 102 cal: (c) 0.102 kcal: (d) 3.28×10^5 J: (e) 328 kJ: (f) 7.85×10^4 cal

Chapters 4–5

2. Although you don't have to memorize all the elements, you should at least be able to give the symbol or name for the most common elements (listed in Table 4.3).

4. The main postulates of Dalton's theory are: (1) elements are made up of tiny particles called atoms; (2) all atoms of a given element are identical; (3) although all atoms of a given element are identical, these atoms are different from the atoms of all other elements; (4) atoms of one element can combine with atoms of another element to form a compound that will always have the same relative numbers and types of atoms for its composition; and (5) atoms are merely rearranged into new groupings during an ordinary chemical reaction, and no atom is ever destroyed and no new atom is ever created during such a reaction.

6. The expression "nuclear atom" indicates that the atom has a dense center of positive charge (nucleus) around which the electrons move through primarily empty space. Rutherford's experiment involved shooting a beam of α particles at a thin sheet of metal foil. According to the "plum pudding" model of the atom, these positively charged α particles should have passed through the foil. Rutherford detected that a small number of α particles bounced backward to the source of α particles or were deflected from the foil at large angles. Rutherford realized that his observations could be explained if the atoms of the metal foil had a small, dense, positively charged nucleus, with a significant amount of empty space between nuclei. The empty space between nuclei would allow most of the α particles to pass through the foil. If an α particle were to hit a nucleus head-on, it would be deflected backward. If a positively charged α particle passed *near* a positively charged nucleus, then the α particle would be deflected by the repulsive forces. Rutherford's experiment disproved the "plum pudding" model, which envisioned the atom as a uniform sphere of positive charge, with enough negatively charged electrons scattered throughout to balance out the positive charge.

8. Isotopes represent atoms of the same element that have different atomic masses. Isotopes result from the different numbers of neutrons in the nuclei of atoms of a given element. They have the same atomic number (number of protons in the nucleus) but have different mass numbers (total number of protons and neutrons in the nucleus). The different isotopes of an atom are indicated by the form $^{A}_{Z}X$, in which Z represents the atomic number and A the mass number of element X. For example, $^{13}_{6}C$ represents a nuclide of carbon with atomic number 6 (6 protons in the nucleus) and mass number 13 (6 protons plus 7 neutrons in the nucleus). The various isotopes of an element have identical *chemical* properties. The *physical* properties of the isotopes of an element may differ slightly because of the small difference in mass.

10. Most elements are too reactive to be found in nature in other than the combined form. Gold, silver, and platinum, and some of the gaseous elements (such as O_2, N_2, He, and Ar) are found in the elemental form.

12. Ionic compounds typically are hard, crystalline solids with high melting and boiling points. The ability of aqueous solutions of ionic substances to conduct electricity means that ionic substances consist of positively and negatively charged particles (ions). A sample of an ionic substance has no net electrical charge because the total number of positive charges is *balanced* by an equal number of negative charges. An ionic compound could not consist of only cations or only anions because a net charge of zero cannot be obtained when all ions have the same charge. Also, ions of like charge will repel each other.

14. When naming ionic compounds, the positive ion (cation) is named first. For simple binary Type I ionic compounds, the ending *-ide* is added to the root name of the negative ion (anion). For example, the name for K_2S would be "potassium sulf*ide*"—potassium is the cation, sulfide is the anion. Type II compounds, which involve elements that form more than one stable ion, are named by either of two systems: the "Roman numeral" system (which is preferred by most chemists) and the "*ous-ic*" system. For example, iron can react with oxygen to form either of two stable oxides, FeO or Fe_2O_3. Under the Roman numeral system, FeO would be named iron(II) oxide to show that it contains Fe^{2+} ions; Fe_2O_3 would be named iron(III) oxide to indicate that it contains Fe^{3+} ions. Under the "*ous-ic*" system, FeO is named ferr*ous* oxide and Fe_2O_3 is called ferr*ic* oxide. Type II compounds usually involve transition metals and nonmetals.

16. A polyatomic ion is an ion containing more than one atom. Some common polyatomic ions are listed in Table 5.4. Parentheses are used in writing formulas containing polyatomic ions to indicate how many polyatomic ions are present. For example, the correct formula for calcium phosphate is $Ca_3(PO_4)_2$, which indicates that three calcium ions are combined for every two phosphate ions. If we did *not* write the parentheses around the formula for the phosphate ion (that is, if we wrote Ca_3PO_{42}), people might think that 42 oxygen atoms were present!

18. Acids are substances that produce protons (H^+ ions) when dissolved in water. For acids that do *not* contain oxygen, the prefix *hydro-* and the suffix *-ic* are used with the root name of the element present in the acid (for example: HCl, *hydro*chlor*ic* acid; H_2S, *hydro*sulfur*ic* acid; HF, *hydro*fluor*ic* acid). For acids whose anions contain oxygen, a series of prefixes and suffixes is used with the name of the central atom in the anion: these prefixes and suffixes indicate the relative (not actual) number of oxygen atoms present in the anion. Most elements that form oxyanions form *two* such anions—for example, sulfur forms sulf*ite* ion (SO_3^{2-}) and sulf*ate* ion (SO_4^{2-}). For an element that forms two oxyanions, the acid containing the anions will have the ending *-ous* if the *-ite* anion is involved and the ending *-ic* if the *-ate* anion is present. For example, H_2SO_3 is sulfur*ous* acid and H_2SO_4 is sulfur*ic* acid. The Group 7 elements each form *four* oxyanions/oxyacids. The prefix *hypo-* is used for the oxyacid that contains fewer oxygen atoms than the *-ite* anion, and the prefix *per-* is used for the oxyacid that contains more oxygen atoms than the *-ate* anion. For example,

20. (a) bromine, 35, Group 7A: (b) lithium, 3, Group 1A: (c) strontium, 38, Group 2A: (d) potassium, 19, Group 1A: (e) boron, 5, Group 3A: (f) chlorine, 17, Group 7A: (g) xenon, 54, Group 8A: (h) carbon, 6, Group 4A: (i) magnesium, 12, Group 2A: (j) aluminum, 13, Group 3A

22.

	Protons	Neutrons	Electrons
(a) $^{4}_{2}He$	2	2	2
(b) $^{37}_{17}Cl$	17	20	17
(c) $^{79}_{35}Br$	35	44	35
(d) $^{41}_{20}Ca$	20	21	20
(e) $^{40}_{20}Ca$	20	20	20
(f) $^{238}_{92}U$	92	146	92
(g) $^{235}_{92}U$	92	143	92
(h) $^{1}_{1}H$	1	0	1

24. (a) 12 protons, 10 electrons: (b) 26 protons, 24 electrons: (c) 26 protons, 23 electrons: (d) 9 protons, 10 electrons: (e) 28 protons, 26 electrons: (f) 30 protons; 28 electrons: (g) 27 protons, 24 electrons: (h) 7 protons, 10 electrons: (i) 16 protons, 18 electrons: (j) 37 protons, 36 electrons: (k) 34 protons, 36 electrons: (l) 19 protons, 18 electrons

26. (a) $FeCl_3$, iron(III) chloride (ferric chloride): (b) Cu_2S, copper(I) sulfide (cuprous sulfide): (c) $CoBr_2$, cobalt(II) bromide (cobaltous bromide): (d) Fe_2O_3, iron(III) oxide (ferric oxide): (e) AuI_3, gold(III) iodide, auric iodide: (f) Cr_2S_3, chromium(III) sulfide, chromic sulfide: (g) MnO_2, manganese(IV) oxide: (h) CuO, copper(II) oxide, cupric oxide: (i) NiS, nickel(II) sulfide, nickelous sulfide

28. (a) NH_4^+, ammonium ion: (b) SO_3^{2-}, sulfite ion: (c) NO_3^-, nitrate ion: (d) SO_4^{2-}, sulfate ion: (e) NO_2^-, nitrite ion: (f) CN^-, cyanide ion: (g) OH^-, hydroxide ion: (h) ClO_4^-, perchlorate ion: (i) ClO^-, hypochlorite ion: (j) PO_4^{3-}, phosphate ion

30. (a) B_2O_3, diboron trioxide: (b) NO_2, nitrogen dioxide: (c) PCl_5, phosphorus pentachloride: (d) N_2O_4, dinitrogen tetroxide: (e) P_2O_5, diphosphorus pentoxide: (f) ICl, iodine monochloride: (g) SF_6, sulfur hexafluoride: (h) N_2O_3, dinitrogen trioxide

Chapters 6–7

2. A chemical equation indicates the substances necessary for a given chemical reaction, and the substances produced by that chemical reaction. The substances to the left of the arrow are called the *reactants;* those to the right of the arrow are called the *products*. A *balanced* equation indicates the relative numbers of molecules in the reaction.

4. Never change the *subscripts* of a *formula*: changing the subscripts changes the *identity* of a substance and makes the equation invalid. When balancing a chemical equation, we adjust only the *coefficients* in front of a formula: changing a coefficient changes the *number* of molecules being used in the reaction, *without* changing the *identity* of the substance.

6. In a precipitation reaction, a *solid* (precipitate) forms when the reactants are combined. In such a reaction, the mixture turns *cloudy* as the reactants are combined and a solid settles out on standing. For example, if you mixed solutions of barium nitrate and sodium carbonate, a precipitate of barium carbonate would form.

$$Ba(NO_3)_2(aq) + Na_2CO(aq) \rightarrow BaCO_3(s) + 2NaNO_3(aq)$$

8. Nearly all compounds containing the nitrate, sodium, potassium, and ammonium ions are soluble in water. Most salts containing the chloride and sulfate ions are soluble in water, with specific exceptions (see Table 7.1). Most compounds containing the hydroxide, sulfide, carbonate, and phosphate ions are *not* soluble in water (unless the compound also contains Na^+, K^+, or NH_4^+). For example, suppose we combine barium chloride and sulfuric acid solutions:

$BaCl_2(aq) + H_2SO_4(aq) \rightarrow BaSO_4(s) + 2HCl(aq)$
$Ba^{2+}(aq) + SO_4^{2-}(aq) \rightarrow BaSO_4(s)$ [net ionic reaction]

Because barium sulfate is not soluble in water, a precipitate of $BaSO_4(s)$ forms.

10. Acids (such as the acetic acid found in vinegar) were first noted primarily because of their sour taste, whereas bases were first characterized by their bitter taste and slippery feel on the skin. Acids and bases neutralize each other, forming water: $H^+(aq) + OH^-(aq) \rightarrow H_2O(l)$. Strong acids and bases ionize *fully* when dissolved in water, which means they are also strong electrolytes.
Strong acids: HCl, HNO_3, and H_2SO_4
Strong bases: Group 1 hydroxides (for example, NaOH and KOH)

12. Oxidation–reduction reactions are *electron-transfer* reactions. Oxidation represents a loss of electrons by an atom, molecule, or ion; reduction represents the gain of electrons. Oxidation and reduction always occur together: the electrons lost by one species must be gained by another species. For example, $Mg(s) + F_2(g) \rightarrow MgF_2(s)$ shows a simple oxidation–reduction reaction between a metal and a nonmetal. In this process, Mg atoms lose two electrons each to become Mg^{2-} ions in MgF_2: Mg is oxidized. Each atom of F_2 gains one electron to become an F ion, for a total of two electrons gained for each F_2 molecule: F_2 is reduced. $Mg \rightarrow Mg^{2+} + 2e^-$; $2(F + e^- \rightarrow F^-)$.

14. In a synthesis reaction, elements or simple compounds react to produce more complex substances. For example,

$$N_2(g) + 3H_2(g) \rightarrow 2NH_3(g)$$
$$NaOH(aq) + CO_2(g) \rightarrow NaHCO_3(s)$$

Decomposition reactions represent the breakdown of complex substances into simpler substances. For example, $2H_2O_2(aq) \rightarrow 2H_2O(l) + O_2(g)$. Synthesis and decomposition reactions are often oxidation–reduction reactions, although not always. For example, the synthesis reaction between NaOH and CO_2 does *not* represent oxidation–reduction.

16. (a) $2Na(s) + 2H_2O(l) \rightarrow 2NaOH(aq) + H_2(g)$ $2K(s) + 2H_2O(l) \rightarrow 2KOH(aq) + H_2(g)$ (b) $2Na(s) + Cl_2(g) \rightarrow 2NaCl(s)$ $2K(s) + Cl_2(g) \rightarrow 2KCl(s)$ (c) $3Na(s) + P(s) \rightarrow Na_3P(s)$ $3K(s) + P(s) \rightarrow K_3P(s)$ (d) $6Na(s) + N_2(g) \rightarrow 2Na_3N(s)$ $6K(s) + N_2(g) \rightarrow 2K_3N(s)$ (e) $2Na(s) + H_2(g) \rightarrow 2NaH(s)$ $2K(s) + H_2(g) \rightarrow 2KH(s)$

18. (a) $Ba(NO_3)_2(aq) + K_2CrO_4(aq) \rightarrow BaCrO_4(s) + 2KNO_3(aq)$ (b) $NaOH(aq) + HC_2H_3O_2(aq) \rightarrow H_2O(l) + NaC_2H_3O_2(aq)$ (then evaporate the water from the solution) (c) $AgNO_3(aq) + NaCl(aq) \rightarrow AgCl(s) + NaNO_3(aq)$ (d) $Pb(NO_3)_2(aq) + H_2SO_4(aq) \rightarrow PbSO_4(s) + 2HNO_3(aq)$ (e) $2NaOH(aq) + H_2SO_4(aq) \rightarrow Na_2SO_4(aq) + 2H_2O(l)$ (then evaporate the water from the solution) (f) $Ba(NO_3)_2(aq) + 2Na_2CO_3(aq) \rightarrow BaCO_3(s) + 2NaNO_3(aq)$

20. (a) $FeO(s) + 2HNO_3(aq) \rightarrow Fe(NO_3)_2(aq) + H_2O(l)$ acid–base; double-displacement (b) $2Mg(s) + 2CO_2(g) + O_2(g) \rightarrow 2MgCO_3(s)$ synthesis; oxidation–reduction (c) $2NaOH(s) + CuSO_4(aq) \rightarrow Cu(OH)_s(s) + Na_2SO4_4(aq)$ precipitation; double-displacement (d) $HI(aq) + KOH(aq) \rightarrow KI(aq) + H_2O(l)$ acid–base; double-displacement (e) $C_3H_8(g) + 5O_2(g) \rightarrow 3CO_2(g) + 4H_2O(g)$combustion; oxidation–reduction (f) $Co(NH_3)_6Cl_2(s) \rightarrow CoCl_2(s) + 6NH_3(g)$ decomposition (g) $2HCl(aq) + Pb(C_2H_3O_2)_2(aq) \rightarrow 2HC_2H_3O_2(aq) + PbCl_2(aq)$ precipitation; double-displacement (h) $C_{12}H_{22}O_{11}(s) \rightarrow 12C(s) + 11H_2O(g)$decomposition; oxidation–reduction (i) $2Al(s) + 6HNO_3(aq) \rightarrow 2Al(NO_3)_3(aq) + 3H_2(g)$ oxidation–reduction (j) $4B(s) + 3O_3(g) \rightarrow 2B_2O_3(s)$ synthesis; oxidation–reduction

22. Answer will depend on student examples.

Chapters 8–9

2. On a microscopic basis, one mole of a substance represents Avogadro's number (6.022×10^{23}) of individual units (atoms or molecules) of the substance. On a macroscopic basis, one mole of a substance represents the amount of substance present when the molar mass of the substance in grams is taken. Chemists have chosen these definitions so that a simple relationship will exist between measurable amounts of substances (grams) and the actual number of atoms or molecules present, and so that the number of particles present in samples of *different* substances can easily be compared.

4. The molar mass of a compound is the mass in grams of one mole of the compound and is calculated by summing the average atomic masses of all the atoms present in a molecule of the compound. For example, for H_3PO_4: molar mass $H_3PO_4 = 3(1.008 \text{ g}) + 1(30.97 \text{ g}) + 4(16.00 \text{ g}) = 97.99 \text{ g}$

6. The *empirical* formula of a compound represents the *relative* number of atoms of each type present in a molecule of the compound, whereas the *molecular* formula represents the *actual* number of atoms of each type present in a real molecule. For example, both acetylene (molecular formula C_2H_2) and benzene (molecular formula C_6H_6) have the same relative number of carbon and hydrogen atoms, and thus have the same empirical formula (CH). The molar mass of the compound must be determined before calculating the actual molecular formula. Since real molecules cannot contain fractional *parts* of atoms, the molecular formula is always a *whole-number multiple* of the empirical formula.

8. In Question 7, we chose to calculate the percentage composition of phosphoric acid, H_3PO_4: 3.085% H, 31.60% P, 65.31% O. Using these percentages, and choosing a 2.417-g sample of the compound, our new problem would be worded as follows: "A 2.417-g sample of a compound has been analyzed and was found to contain 0.07456 g H, 0.7638 g of P, and 1.579 g of O. Calculate the empirical formula of the compound."

10. For CO_2: (2 mol CO_2/1 mol C_2H_5OH). For H_2O: (3 mol H_2O/1 mol C_2H_5OH). 0.65 mol $C_2H_5OH \times$ (2 mol CO_2/1 mol C_2H_5OH) = 1.3 mol CO_2; 0.65 mol $C_2H_5OH \times$ (3 mol H_2O/1 mol C_2H_5OH) = 1.95 mol (2.0 mol) H_2O

12. When arbitrary amounts of reactants are used, one reactant will be present, stoichiometrically, in the least amount: this substance is called the *limiting reactant.* It *limits* the amount of product that can form in the experiment, because once this substance has reacted completely, the reaction must *stop.* The other reactants in the experiment are present *in excess,* which means that a portion of these reactants will be present *unchanged* after the reaction ends.

14. The *theoretical yield* for an experiment is the mass of product calculated assuming the limiting reactant for the experiment is completely consumed. The *actual yield* for an experiment is the mass of product actually collected by the experimenter. Any experiment is restricted by the skills of the experimenter and by the inherent limitations of the experimental method: for these reasons, the actual yield is often less than the theoretical yield. Although one would expect that the actual yield should never exceed the theoretical yield, in real experiments, sometimes this happens. However, an actual yield greater than a theoretical yield usually means that something is *wrong* in either the experiment (for example, impurities may be present) or the calculations.

16. (a) element; (b) 15.78% C; (c) 12.67% Al; (d) 81.71% C; (e) element; (f) 40.04% Ca; (g) 42.88% C; (h) 27.29% C; (i) 36.11% Ca (j) element

18. (a) 22.9 g Ag_2SO_4, 12.1 g $Ca(NO_3)_2$; (b) 197 g $Al(NO_3)_3$, 2.80 g H_2; (c) 41.8 g Na_3PO_4, 13.8 g H_2O; (d) 49.5 g $CaCl_2$, 8.03 g H_2O

20. 11.7 g CO; 18.3 g CO_2

Chapters 10–11

2. An atom in its *ground state* is in its lowest possible energy state. When an atom possesses more energy than in its ground state, the atom is in an *excited state.* An atom is promoted from its ground state to an excited state by absorbing energy; when the atom returns from an excited state to its ground state it emits the excess energy as electromagnetic radiation. Atoms do not gain or emit radiation randomly, but rather do so only in discrete bundles of radiation called *photons.* The photons of radiation emitted by atoms are characterized by the wavelength (color) of the radiation: longer-wavelength photons carry less energy than shorter-wavelength photons. The energy of a photon emitted by an atom corresponds *exactly* to the difference in energy between two allowed energy states in an atom.

4. Bohr pictured the electron moving in certain circular orbits around the nucleus, with each orbit being associated with a specific energy (resulting from the attraction between the nucleus and the electron and from the kinetic energy of the electron). Bohr assumed that when an atom absorbs energy, the electron moves from its ground state ($n = 1$) to an orbit farther away from the nucleus ($n = 2, 3, 4, \ldots$). Bohr postulated that when an excited atom returns to its ground state, the atom emits the excess energy as radiation. Because the Bohr orbits are located at fixed distances from the nucleus and from each other, when an electron moves from one fixed orbit to another, the energy change is of a definite amount, which corresponds to the emission of a photon with a particular characteristic wavelength and energy. When the simple Bohr model for the atom was applied to the emission spectra of other elements, however, the theory could not predict or explain the observed emission spectra of these elements.

6. The lowest-energy hydrogen atomic orbital is called the 1*s* orbital. The 1*s* orbital is spherical in shape (the electron density around the nucleus is uniform in all directions). The orbital does *not* have a sharp edge (it appears fuzzy) because the probability of finding the electron gradually decreases as distance from the nucleus increases. The orbital does *not* represent just a spherical surface on which the electron moves (this would be similar to Bohr's original theory)—instead, the 1*s* orbital represents a probability map of electron density around the nucleus for the first principal energy level.

8. The third principal energy level of hydrogen is divided into three sublevels: the 3*s*, 3*p*, and 3*d* sublevels. The 3*s* subshell consists of the single 3*s* orbital, which is spherical in shape. The 3*p* subshell consists of a set of three equal-energy 3*p* orbitals: each of these 3*p* orbitals has the same shape ("dumbbell"), but each of the 3*p* orbitals is oriented in a different direction in space. The 3*d* subshell consists of a set of five 3*d* orbitals with shapes as indicated in Figure 10.28, which are oriented in different directions around the nucleus. The fourth principal energy level of hydrogen is divided into four sublevels: the 4*s*, 4*p*, 4*d*, and 4*f* orbitals. The 4*s* subshell consists of the single 4*s* orbital. The 4*p* subshell consists of a set of three 4*p* orbitals. The 4*d* subshell consists of a set of five 4*d* orbitals. The shapes of the 4*s*, 4*p*, and 4*d* orbitals are the *same* as the shapes of the orbitals of the third principal energy level—the orbitals of the fourth principal energy level are *larger* and *farther from the nucleus* than the orbitals of the third level, however. The fourth principal energy level also contains a 4*f* subshell consisting of seven 4*f* orbitals (the shapes of the 4*f* orbitals are beyond the scope of this text).

10. Atoms have a series of *principal energy levels* indexed by the letter *n*. The $n = 1$ level is closest to the nucleus, and the energies of the levels increase as the value of *n* (and distance from the nucleus) increases. Each principal energy level is divided into *sublevels* (sets of orbitals) of different characteristic shapes designated by the letters *s, p, d,* and *f.* Each *s* subshell consists of a single *s* orbital; each *p* subshell consists of a set of three *p* orbitals; each *d* subshell consists of a set of five *d* orbitals; and so on. An orbital can be empty or it can contain one or two electrons, but never more than two electrons (if an orbital contains two electrons, then the electrons must have opposite spins). The shape of an orbital represents a probability map for finding electrons—it does not represent a trajectory or pathway for electron movements.

12. The valence electrons of an atom are found in the outermost shell of the atom that contains electrons. The core electrons are those in principal energy levels closer to the nucleus than the outermost shell. The valence electrons are the electrons gained, lost, or shared with other atoms. The arrangement of the elements on the periodic table corresponds to their valence-electron configurations (elements in the same vertical group have similar configurations). The evidence that the 4*s* subshell fills *before* the 3*d* subshell is that the elements K and Ca have properties very similar to those of the other Group 1 and Group 2 elements.

14. The general periodic table you drew for Question 13 should resemble that found in Figure 10.31. From the column and row location of an element, you should be able to determine its valence configuration. For example, the element in the third horizontal row of the second vertical column has $3s^2$ as its valence configuration. The element in the seventh vertical column of the second horizontal row has valence configuration $2s^2 2p^5$.

16. The ionization energy of an atom represents the energy required to remove an electron from the atom in the gas phase. Moving from top to bottom in a vertical group on the periodic table, the ionization energies decrease. The ionization energies increase when going from left to right within a horizontal row within the periodic table. The relative sizes of atoms also vary systematically with the location of an element on the periodic table. Within a given vertical group, the atoms become progressively larger when going from the top of the group to the bottom. Moving from left to right within a horizontal row on the periodic table, the atoms become progressively smaller.

18. To form an ionic compound, a metallic element reacts with a nonmetallic element, with the metallic element losing electrons to form a positive ion and the nonmetallic element gaining electrons to form a negative ion. The aggregate form of such a compound consists of a crystal lattice of alternating positively and negatively charged ions: a given positive ion is attracted by surrounding negatively charged ions, and a given negative ion is attracted by surrounding positively charged ions. Similar electrostatic attractions exist in three dimensions throughout the crystal of the ionic solid, leading to a very stable system (with very high melting and boiling points, for example). As evidence for the existence of ionic bonding, ionic solids do not conduct electricity (the ions are rigidly held), but melts or solutions of such substances do conduct electric current. For example, when sodium metal and chlorine gas react, a typical ionic substance (sodium chloride) results: $2Na(s) + Cl_2(g) \rightarrow 2Na^+Cl^-(s)$.

20. Electronegativity represents the relative ability of an atom in a molecule to attract shared electrons toward itself. For a bond to be polar, one of the atoms in the bond must attract the shared electron pair toward itself and away from the other atom of the bond—this happens if one atom of the bond is more electronegative than the other. If two atoms in a bond have the same electronegativity, then the two atoms pull the electron pair equally and the bond is nonpolar and covalent. If two atoms sharing a pair of electrons have *vastly* different electronegativities, the electron pair will be pulled so strongly by the more electronegative atom that a negative *ion* may be formed (as well as a positive ion for the other atom) and ionic bonding will result. If the difference in electronegativity between two atoms sharing an electron pair lies somewhere between these two extremes, then a polar covalent bond results.

22. It has been observed over many, many experiments that when an active metal like sodium or magnesium reacts with a nonmetal, the sodium atoms always form Na^+ ions and the magnesium atoms always form Mg^{2+} ions. It has also been observed that when nonmetallic elements like nitrogen, oxygen, or fluorine form simple ions, the ions are always N^{3-}, O^{2-}, and F^-, respectively. Observing that these elements always form the same ions and those ions all contain eight electrons in the outermost shell, scientists speculated that a species that has an octet of electrons (like the noble gas neon) must be very fundamentally stable. The *repeated* observation that so many elements, when reacting, tend to attain an electron configuration that is isoelectronic with a noble gas led chemists to speculate that *all* elements try to attain such a configuration for their outermost shells. Covalently and polar covalently bonded molecules also strive to attain pseudo–noble gas electron configurations. For a covalently bonded molecule like F_2, each F atom provides one electron of the pair of electrons that constitutes the covalent bond. Each F atom feels also the influence of the other F atom's electron in the shared pair, and each F atom effectively fills its outermost shell.

24. Bonding between atoms to form a molecule involves only the outermost electrons of the atoms, so only these *valence* electrons are shown in the Lewis structures of molecules. The most important requisite for the formation of a stable compound is that each atom of a molecule attain a noble gas electron configuration. In Lewis structures, arrange the bonding and nonbonding valence electrons to try to complete the octet (or duet) for as many atoms as possible.

26. You could choose practically any molecules for your discussion. Let's illustrate the method for ammonia, NH_3. First, count the total number of valence electrons available in the molecule (without regard to their source). For NH_3, since nitrogen is in Group 5, one nitrogen atom would contribute five valence electrons. Since hydrogen atoms have only one electron each, the three hydrogen atoms provide an additional three valence electrons, for a total of eight valence electrons overall. Next, write down the symbols for the atoms in the molecule, and use one pair of electrons (represented by a line) to form a bond between each pair of bound atoms.

```
H—N—H
  |
  H
```

These three bonds use six of the eight valence electrons. Because each hydrogen already has its duet and the nitrogen atom has only six electrons around it so far, the final two valence electrons must represent a lone pair on the nitrogen.

```
  ..
H—N—H
  |
  H
```

28. Boron and beryllium compounds sometimes do not fit the octet rule. For example, in BF_3, the boron atom has only six valence electrons in its outermost shell, whereas in BeF_2, the beryllium atom has only four electrons in its outermost shell. Other exceptions to the octet rule include any molecule with an odd number of valence electrons (such as NO or NO_2).

30.

Number of Valence Pairs	*Bond Angle*	*Examples*
2	180°	BeF_2, BeH_2
3	120°	BCl_3
4	109.5°	CH_4, CCl_4, GeF_4

32. (a) $1s^22s^22p^63s^2$, 2+ ion: (b) $1s^22s^22p^6$, [Ne]: (c) $1s^22s^22p^63s^23p^5$, 1− ion: (d) $1s^22s^22p^6$, [Ne]: (e) $1s^22s^22p^63s^23p^64s^23d^{10}4p^65s^24d^{10}5p^66s^2$, 2+ ion: (f) $1s^22s^22p^63s^23p^64s^23d^{10}4p^65s^24d^{10}5p^6$, [Xe]: (g) $1s^22s^22p^63s^23p^64s^23d^{10}$, 2+ ion: (h) $1s^22s^22p^6$, [Ne]: (i) $1s^22s^22p^6$, [Ne]: (j) $1s^22s^22p^63s^23p^6$, [Ar]

Chapters 12–14

2. The pressure exerted by the atmosphere results from the thick layer of gases above the surface of the earth pressing downward. Atmospheric pressure has traditionally been measured with a mercury barometer (see Figure 12.2).

4. Boyle's law says that the volume of a gas sample will decrease if you squeeze it harder (at constant temperature, for a fixed amount of gas). Two mathematical statements of Boyle's law are

$$P \times V = \text{constant}$$
$$P_1 \times V_1 = P_2 \times V_2$$

These two mathematical formulas say the same thing: if the pressure on a sample of gas is increased, the volume of the sample will decrease. A graph of Boyle's law data is given as Figure 12.5: this type of graph ($xy = k$) is known to mathematicians as a *hyperbola*.

6. Charles's law says that if you heat a sample of gas, the volume of the sample will increase (assuming the pressure and amount of gas remain the same). When the temperature is given in kelvins, Charles's law expresses a *direct* proportionality (if you *increase* T, then V *increases*), whereas Boyle's law expresses an *inverse* proportionality (if you *increase* P, then V *decreases*). Two mathematical statements of Charles's law are $V = bT$ and $(V_1/T_1) = (V_2/T_2)$. With this second formulation, we can determine volume–temperature information for a given gas sample under two sets of conditions. Charles's law holds true only if the amount of gas remains the same (the volume of a gas sample would increase if more gas were present) and also if the pressure remains the same (a change in pressure also changes the volume of a gas sample).

8. Avogadro's law says that the volume of a sample of gas is directly proportional to the number of moles (or molecules) of gas present (at constant temperature and pressure). Avogadro's law holds true only for gas samples compared under the same conditions of temperature and pressure. Avogadro's law expresses a direct proportionality: the more gas in a sample, the larger the sample's volume.

10. The "partial" pressure of an individual gas in a mixture of gases represents the pressure the gas would exert in the same container at the same temperature if it were the *only* gas present. The *total* pressure in a mixture of gases is the sum of the individual partial pressures of the gases present in the mixture. The fact that the partial pressures of the gases in a mixture are additive suggests that the total pressure in a container is a function of the *number* of molecules present, and not of the identity of the molecules or of any other property (such as the molecules' inherent atomic size).

12. The main postulates of the kinetic molecular theory for gases are: (a) gases consist of tiny particles (atoms or molecules), and the size of these particles themselves is negligible compared with the bulk volume of a gas sample: (b) the particles in a gas are in constant random motion, colliding with each other and with the walls of the container: (c) the particles in a gas sample do not assert any attractive or repulsive forces on one another; and (d) the average kinetic energy of the gas particles is directly related to the absolute temperature of the gas sample. The pressure exerted by a gas results from the molecules colliding with (and pushing on) the walls of the container; the pressure increases with temperature because, at a higher temperature, the molecules move faster and hit the walls of the container with greater force. A gas fills the volume available to it because the molecules in a gas are in constant *random* motion: the randomness of the molecules' motion means that they eventually will move out into the available volume until the distribution of molecules is uniform; at constant pressure, the volume of a gas sample increases as the temperature is increased because with each collision having greater force, the container must expand so that the molecules are farther apart if the pressure is to remain constant.

14. The molecules are much closer together in solids and liquids than in gaseous substances and interact with each other to a much

greater extent. Solids and liquids have much greater densities than do gases, and are much less compressible, because so little room exists between the molecules in the solid and liquid states (the volume of a solid or liquid is not affected very much by temperature or pressure). We know that the solid and liquid states of a substance are similar to each other in structure, since it typically takes only a few kilojoules of energy to melt 1 mol of a solid, whereas it may take 10 times more energy to convert a liquid to the vapor state.
16. The *normal* boiling point of water—that is, water's boiling point at a pressure of exactly 760 mm Hg—is 100 °C. Water remains at 100 °C while boiling, because the additional energy added to the sample is used to overcome attractive forces among the water molecules as they go from the condensed, liquid state to the gaseous state. The normal (760 mm Hg) freezing point of water is exactly 0 °C. A cooling curve for water is given in Figure 13.2.
18. Dipole–dipole forces arise when molecules with permanent dipole moments try to orient themselves so that the positive end of one polar molecule can attract the negative end of another polar molecule. Dipole–dipole forces are not nearly as strong as ionic or covalent bonding forces (only about 1% as strong as covalent bonding forces) since electrostatic attraction is related to the *magnitude* of the charges of the attracting species and drops off rapidly with distance. Hydrogen bonding is an especially strong dipole–dipole attractive force that can exist when hydrogen atoms are directly bonded to the most electronegative atoms (N, O, and F). Because the hydrogen atom is so small, dipoles involving N—H, O—H, and F—H bonds can approach each other much more closely than can other dipoles; because the magnitude of dipole–dipole forces is related to distance, unusually strong attractive forces can exist. The much higher boiling point of water than that of the other covalent hydrogen compounds of the Group 6 elements is evidence for the special strength of hydrogen bonding.
20. The vaporization of a liquid requires an input of energy to overcome the intermolecular forces that exist between the molecules in the liquid state. The large heat of vaporization of water is essential to life since much of the excess energy striking the earth from the sun is dissipated in vaporizing water. Condensation refers to the process by which molecules in the vapor state form a liquid. In a closed container containing a liquid with some empty space above the liquid, an equilibrium occurs between vaporization and condensation. When the liquid is first placed in the container, the liquid phase begins to evaporate into the empty space. As the number of molecules in the vapor phase increases, however, some of these molecules begin to reenter the liquid phase. Eventually, each time a molecule of liquid somewhere in the container enters the vapor phase, another molecule of vapor reenters the liquid phase. No further net change occurs in the amount of liquid phase. The pressure of the vapor in such an equilibrium situation is characteristic for the liquid at each temperature. A simple experiment to determine the vapor pressure of a liquid is shown in Figure 13.10. Typically, liquids with strong intermolecular forces have smaller vapor pressures (they have more difficulty in evaporating) than do liquids with very weak intermolecular forces.
22. The *electron sea model* explains many properties of metallic elements. This model pictures a regular array of metal atoms set in a "sea" of mobile valence electrons. The electrons can move easily throughout the metal to conduct heat or electricity, and the lattice of atoms and cations can be deformed with little effort, allowing the metal to be hammered into a sheet or stretched into wire. An alloy is a material that contains a mixture of elements that overall has metallic properties. *Substitutional* alloys consist of a host metal in which some of the atoms in the metal's crystalline structure are replaced by atoms of other metallic elements. For example, sterling silver is an alloy in which some silver atoms have been replaced by copper atoms. An *interstitial alloy* is formed when other, smaller atoms enter the interstices (holes) between atoms in the host metal's crystal structure. Steel is an interstitial alloy in which carbon atoms enter the interstices of a crystal of iron atoms.
24. A saturated solution contains as much solute as can dissolve at a particular temperature. Saying that a solution is *saturated* does not *necessarily* mean that the solute is present at a high concentration—for example, magnesium hydroxide dissolves only to a very small extent before the solution is saturated. A saturated solution is in equilibrium with undissolved solute: as molecules of solute dissolve from the solid in one place in the solution, dissolved molecules rejoin the solid phase in another part of the solution. Once the rates of dissolving and solid formation become equal, no further net change occurs in the concentration of the solution and the solution is saturated.
26. Adding more solvent to a solution to dilute the solution does *not* change the number of moles of solute present, but changes only the *volume* in which the solute is dispersed. If molarity is used to describe the solution's concentration, then the number of *liters* is changed when solvent is added and the number of *moles per liter* (the molarity) changes, but the actual number of *moles* of solute does *not* change. For example, 125 mL of 0.551 *M* NaCl contains 0.0689 mol of NaCl. The solution will *still* contain 0.0689 mol of NaCl after 250 mL of water is added to it. The volume and the concentration will change, but the number of moles of solute in the solution will *not* change. The 0.0689 mol of NaCl, divided by the total volume of the diluted solution in liters, gives the new molarity (0.184 *M*).
28. (a) 208 mL: (b) 0.462 atm: (c) the volume doubles
30. (a) 334 mL: (b) 17.3 L: (c) He (0.772 atm), Ne (0.228 atm): (d) 0.00200 g
32. 1.02 L O_2 at STP; 1.10 L O_2 at 24 °C, 771 mm Hg
34. (a) 0.649% NaCl: (b) 0.113 *M* Na^+: (c) 0.482 *M* Cl^-
36. (a) 0.0862 *M*: (b) 0.0443 *M*: (c) 0.214 *M*
38. (a) 10.1 mL H_2SO_4: (b) 20.3 mL H_2SO_4

Chapters 15–16

2. A conjugate acid–base pair consists of two species related to one another by the donating or accepting of a single proton, H^+. An acid has one more H^+ than its conjugate base; a base has one less H^+ than its conjugate acid.

Brønsted–Lowry acids:
$HCl(aq) + H_2O(l) \rightarrow Cl^-(aq) + H_3O^+(aq)$
$H_2SO_4(aq) + H_2O(l) \rightarrow HSO_4^-(aq) + H_3O^+(aq)$
$H_3PO_4(aq) + H_2O(l) \rightarrow H_2PO_4^-(aq) + H_3O^+(aq)$
$NH_4^+(aq) + H_2O(l) \rightarrow NH_3(aq) + H_3O^+(aq)$

Brønsted–Lowry bases:
$NH_3(aq) + H_2O(l) \rightarrow NH_4^+(aq) + OH^-(aq)$
$HCO_3^-(aq) + H_2O(l) \rightarrow H_2CO_3(aq) + OH^-(aq)$
$NH_2^-(aq) + H_2O(l) \rightarrow NH_3(aq) + OH^-(aq)$
$H_2PO_4^-(aq) + H_2O(l) \rightarrow H_3PO_4(aq) + OH^-(aq)$

4. The strength of an acid is a direct result of the position of the acid's dissociation (ionization) equilibrium. Acids whose dissociation equilibrium positions lie far to the right are called *strong* acids. Acids whose equilibrium positions lie only slightly to the right are called *weak* acids. For example, HCl, HNO_3, and $HClO_4$ are strong acids, which means that they are completely dissociated in aqueous solution (the position of equilibrium is very far to the right):

$$HCl(aq) + H_2O(l) \rightarrow Cl^-(aq) + H_3O^+(aq)$$
$$HNO_3(aq) + H_2O(l) \rightarrow NO_3^-(aq) + H_3O^+(aq)$$
$$HClO_4(aq) + H_2O(l) \rightarrow ClO_4^-(aq) + H_3O^+(aq)$$

Since these are very strong acids, their anions (Cl^-, NO_3^-, ClO_4^-) must be very *weak* bases, and solutions of their sodium salts will *not* be basic.

6. The pH of a solution is defined as $pH = -\log[H^+(aq)]$ for a solution. In pure water, the amount of $H^+(aq)$ ion present is *equal* to the amount of $OH^-(aq)$ ion—that is, pure water is *neutral*. Since $[H^+] = 1.0 \times 10^{-7}\ M$ in pure water, the pH of pure water is $-\log[1.0 \times 10^{-7}\ M] = 7.00$. Solutions in which $[H^+] > 1.0 \times 10^{-7}\ M$ ($pH < 7.00$) are acidic; solutions in which $[H^+] < 1.0 \times 10^{-7}\ M$ ($pH > 7.00$) are basic. The pH scale is logarithmic: a pH change of one unit corresponds to a change in the hydrogen ion concentration by a factor of *ten*. An analogous logarithmic expression is defined for the hydroxide ion concentration in a solution: $pOH = -\log[OH^-(aq)]$. The concentrations of hydrogen ion and hydroxide ion in water (and in aqueous solutions) are *not* independent of one another, but rather are related by the dissociation equilibrium constant for water, $K_w = [H^+][OH^-] = 1.0 \times 10^{-14}$ at 25 °C. From this expression it follows that $pH + pOH = 14.00$ for water (or an aqueous solution) at 25 °C.

8. Chemists envision that a reaction can only take place between molecules if the molecules physically *collide* with each other. Furthermore, when molecules collide, the molecules must collide with enough force for the reaction to be successful (there must be enough energy to break bonds in the reactants), and the colliding molecules must be positioned with the correct relative orientation for the products (or intermediates) to form. Reactions tend to be faster if higher concentrations are used for the reaction because if there are more molecules present per unit volume there will be more collisions between molecules in a given time period. Reactions are faster at higher temperatures because at higher temperatures the reactant molecules have a higher average kinetic energy, and the number of molecules that will collide with sufficient force to break bonds increases.

10. Chemists define equilibrium as the exact balancing of two exactly opposing processes. When a chemical reaction is begun by combining pure reactants, the only process possible initially is

$$\text{reactants} \rightarrow \text{products}$$

However, for many reactions, as the concentration of product molecules increases, it becomes more and more likely that product molecules will collide and react with each other,

$$\text{products} \rightarrow \text{reactants}$$

giving back molecules of the original reactants. At some point in the process the rates of the forward and reverse reactions become equal, and the system attains chemical equilibrium. To an outside observer the system appears to have stopped reacting. On a microscopic basis, though, both the forward and reverse processes are still going on. Every time additional molecules of the product form, however, somewhere else in the system molecules of product react to give back molecules of reactant.

Once the point is reached that product molecules are reacting at the same speed at which they are forming, there is no further net change in concentration. At the start of the reaction, the rate of the forward reaction is at its maximum, while the rate of the reverse reaction is zero. As the reaction proceeds, the rate of the forward reaction gradually decreases as the concentration of reactants decreases, whereas the rate of the reverse reaction increases as the concentration of products increases. Once the two rates have become equal, the reaction has reached a state of equilibrium.

12. The equilibrium constant for a reaction is a ratio of the concentration of products present at the point of equilibrium to the concentration of reactants still present. A *ratio* means that we have one number divided by another number (for example, the density of a substance is the ratio of a substance's mass to its volume). Since the equilibrium constant is a ratio, there are an infinite number of sets of data that can give the same ratio: for example, the ratios 8/4, 6/3, 100/50 all have the same value, 2. The actual concentrations of products and reactants will differ from one experiment to another involving a particular chemical reaction, but the ratio of the amount of product to reactant at equilibrium should be the same for each experiment.

14. Your paraphrase of Le Châtelier's principle should go something like this: "When you make any change to a system in equilibrium, this throws the system temporarily out of equilibrium, and the system responds by reacting in whatever direction it will be able to reach a new position of equilibrium." There are various changes that can be made to a system in equilibrium. Here are examples of some of them.

a. The concentration of one of the reactants is increased.
$2SO_2(g) + O_2(g) \rightleftharpoons 2SO_3(g)$
If additional SO_2 or O_2 is added to the system at equilibrium, then more SO_3 will result than if no change was made.

b. The concentration of one of the products is decreased by selectively removing it from the system.
$CH_3COOH + CH_3OH \rightleftharpoons H_2O + CH_3COOCH_3$
If H_2O were to be removed from the system by, for example, use of a drying agent, then more CH_3COOCH_3 would result than if no change was made.

c. The reaction system is compressed to a smaller volume.
$3H_2(g) + N_2(g) \rightleftharpoons 2NH_3(g)$
If this system is compressed to smaller volume, then more NH_3 would be produced than if no change was made.

d. The temperature is increased for an endothermic reaction.
$2NaHCO_3 + \text{heat} \rightleftharpoons Na_2CO_3 + H_2O + CO_2$
If heat is added to this system, then more product would be produced than if no change was made.

e. The temperature is decreased for an exothermic process.
$PCl_3 + Cl_2 \rightleftharpoons PCl_5 + \text{heat}$
If heat is removed from this system (by cooling), then more PCl_5 would be produced than if no change was made.

16. Specific answer depends on student choices. In general, for a weak acid HA and a weak base B
$HA + H_2O \rightleftharpoons A^- + H_3O^+ B + H_2O \rightleftharpoons HB^+ + OH^-$

18. a, b, and e are not conjugate acid–base pairs. They differ by more than a single H^+.

20. (a) pH = 2.851; pOH = 11.149: (b) pOH = 2.672; pH = 11.328: (c) pH = 2.288; pOH = 11.712: (d) pOH = 3.947; pH = 10.053

22. 2.0×10^8

Index/Glossary

Photo Credits

Chapter 1
xxx, Terry Donnelly/Stone/Getty Images; p. 1, PhotoDisc/Getty Images; p. 2, Courtesy, Bart Eklund; p. 4, AP Photo/Ric Risberg; p. 5, NASA; p. 10, Claude Charlier/Photo Researchers, Inc.

Chapter 2
p. 14, Richard Megna/Fundamental Photographs; p. 16, Ray Simons/Photo Researchers, Inc.; p. 19, NASA; p. 21, Ben Osborne/Stone/Getty Images; p. 22 (top) Courtesy, Mettler-Toledo; p. 22 (bottom), Andrew Lambert/Leslie Garland Picture Library/Alamy Images; p. 36, Dr. Yoshio Bando/National Institute for Materials Sciences; p. 43, Dan McCoy/Rainbow; p. 44, Thomas Pantages.

Chapter 3
p. 54, David Maisel/Stone/Getty Images; p. 56, Brian Parker/Tom Stack & Associates; p. 57 (bottom), Chip Clark; p. 59, Jim Pickerell/Stone/Getty Images; p. 62, Glenn Izett/U.S. Geological Survey; p. 65, Jim Richardson/West Light/Corbis; p. 67, ElektraVision/PictureQuest; p. 69, Neil Lucus/Nature Picture Library; p. 71, Jack Fields/Photo Researchers, Inc.

Chapter 4
p. 82, Peter Menzel/Stock Boston; p. 83 (bottom), The Granger Collection, New York; p. 85, Jeremy Woodhouse/PhotoDisc/Getty Images; p. 87, Walter Urie/West Light/Corbis; p. 88, Reproduced by permission, Manchester Literary and Philosophical Society; p. 92, Corbis-Bettmann; p. 93, E.R. Degginger; p. 97, Paul Chesley/National Geographic/Getty Images; p. 101, API/Explorer/Photo Researchers, Inc.; p. 102, Tara Piasio/IFAS/University of Florida; p. 104 (left), E.R. Degginger; p. 105 (center), Paul Silverman/Fudamental Photographs; p. 105 (bottom, right), Frank Cox; p. 110, E.R. Degginger.

Chapter 5
p. 122, John Gerlach/Tom Stack & Associates; p. 123, Bob Daemmrich/The Image Works; p. 124, Art Resource.

Chapter 6
p. 152, Gordon Garradd/SPL/Photo Researchers, Inc.; p. 153 (top), C.J. Allen/Stock Boston; p. 153 (bottom left), Stephen Derr/The Image Bank/Getty Images; p. 154 (center), Spencer Grant/PhotoEdit; p. 161, Thomas Eisner.

Chapter 7
p. 173, Dan McCoy/Rainbow; p. 186, Stephen P. Parker/Photo Researchers, Inc.; p. 193, PhotoDisc/Getty Images; p. 195, Courtesy, Morton Thiokol; p. 197, Michael Newman/PhotoEdit.

Chapter 8
p. 210, Elaine Rebman/Photo Researchers, Inc.; p. 211, AP Photo; p. 220, G.K. & Vikki Hart/The Image Bank/Getty Images; p. 225, Lyntha Scott Eiler/Library of Congress.

Chapter 9
p. 246, John Zoiner; p. 247, Corbis Images/PictureQuest; p. 249, Bill Backmann/PhotoEdit; p. 256, NASA; p. 257, AP Photo/Marc Matheny; p. 259, AP Photo/Kevin Rivoli.

Chapter 10
p. 279, Owaki-Kulla/Corbis; p. 281 (both), Kathryn E. Arnold and N.J. Marshall/University of Scotland; p. 283, Goddard Space Flight Center/NASA; p. 287, Emilio Segre Visual Archives; p. 288, The Granger Collection, New York; p. 298, Andrey K. Geim/High Field Magnet Laboratory/University of Nijmegen; p. 304, AP Photo/The Charleston Daily Mail/Chip Ellis; p. 305, Eyewire/Alamy Images.

Chapter 11
p. 316, M.L. Sinibaldi/The Stock Market/Corbis; p. 317, Tino Hammid; p. 326, Courtesy, Fraunhofer Institute for Applied Materials Research; p. 329, The Bancroft Library; p. 334, C Squared Studios/PhotoDisc/Getty Images; p. 338 (top), Donald Clegg; p. 338 (bottom), Frank Cox.

Chapter 12
p. 358, Brian Erler/Taxi/Getty Images; p. 359, AP Photo/Steve Holland; p. 362, Ken O'Donoghue; p. 366, Dave Jacobs/Stone/Getty Images; p. 367, John A. Rizzo/PhotoDisc/Getty Images; p. 369, USGA Photo by T. Casadevall; p. 380, Kurt Amsler/Vandystadt/Allsport/Getty Images.

Chapter 13
p. 398, Vandystadt/Allsport Concepts/Getty Images; p. 399, Robert Y. Ono/Corbis; p. 403, Flip Nicklin/Minden Pictures; p. 407, Photolink/PhotoDisc/Getty Images; p. 414, T.J. FlorianRainbow.

Chapter 14
p. 442, Graeme Richard White; p. 423, Lawrence Migdale/Stock Boston; p. 425, AP Photo/Science & Technology Ministry; p. 426, Michael Newman/PhotoEdit; p. 432, Tom Pantages; p. 436, Tom Pantages.

Chapter 15
p. 456, Witold Skrypczak/SuperStock; p. 457 (top), Corbis; p. 464, Agricultural Research Service/USDA; p. 471, Andrew Syred/Science Photo Library/Photo Researchers, Inc.; p. 472, David Woodfall/Stone/Getty Images; p. 475, Hans Reinhard/Bruce Coleman, Inc.

Chapter 16
p. 482, James Martin/Stone/Getty Images; p. 483 (bottom), Ken O'Donoghue; p. 485, Delphi Automotive Services; p. 486, Corutesy, Amana; p. 487, AP Photo/NOAA; p. 493, Jenny Hager/The Image Works; p. 506, Science Photo Library/Photo Researchers, Inc.

Chapter 17
p. 520, Bill Gallery/Stock Boston; p. 531 (both), Ken O'Donoghue; p. 536, Corbis-Bettmann; p. 539, *AutoWeek*/Crain Communications, Inc.; p. 541, Digital Vision/Getty Images; p. 543 (top), Runk/Schoenberger/Grant Heilman; p. 543 (bottom), The Granger Collection.

Chapter 18
p. 550, Konstantinos Kifonidis/SPL/Photo Researchers, Inc.; p. 551, Leslie Garland/Leslie Garland Picture Library/Alamy Images; p. 553, Kopal/Mediamed Publiphoto/Photo Researchers, Inc.; p. 556, Culver Pictures; p. 559 (top), Ken O'Donoghue; p. 559 (bottom), Mark W. Philbrick/BYU; p. 560, Smithsonian Institution, Natural History Museum, Department of Mineral Sciences; p. 561 (both), SIU/Visuals Unlimited; p. 563, Reinhard Janke/Peter Arnold; p. 565, NASA, p. 566, NASA.

Table 5.1 Common Simple Cations and Anions

Cation	Name	Anion	Name*
H^+	hydrogen	H^-	hydride
Li^+	lithium	F^-	fluoride
Na^+	sodium	Cl^-	chloride
K^+	potassium	Br^-	bromide
Cs^+	cesium	I^-	iodide
Be^{2+}	beryllium	O^{2-}	oxide
Mg^{2+}	magnesium	S^{2-}	sulfide
Ca^{2+}	calcium		
Ba^{2+}	barium		
Al^{3+}	aluminum		
Ag^+	silver		
Zn^{2+}	zinc		

*The root is given in color.

Table 5.2 Common Type II Cations

Ion	Systematic Name	Older Name
Fe^{3+}	iron(III)	ferric
Fe^{2+}	iron(II)	ferrous
Cu^{2+}	copper(II)	cupric
Cu^+	copper(I)	cuprous
Co^{3+}	cobalt(III)	cobaltic
Co^{2+}	cobalt(II)	cobaltous
Sn^{4+}	tin(IV)	stannic
Sn^{2+}	tin(II)	stannous
Pb^{4+}	lead(IV)	plumbic
Pb^{2+}	lead(II)	plumbous
Hg^{2+}	mercury(II)	mercuric
Hg_2^{2+}*	mercury(I)	mercurous

*Mercury(I) ions always occur bound together in pairs to form Hg_2^{2+}.

Table 5.4 Names of Common Polyatomic Ions

Ion	Name	Ion	Name
NH_4^+	ammonium	CO_3^{2-}	carbonate
NO_2^-	nitrite	HCO_3^-	hydrogen carbonate (bicarbonate is a widely used common name)
NO^{3-}	nitrate		
SO_3^{2-}	sulfite		
SO_4^{2-}	sulfate	ClO^-	hypochlorite
HSO_4^-	hydrogen sulfate (bisulfate is a widely used common name)	ClO_2^-	chlorite
		ClO_3^-	chlorate
		ClO_4^-	perchlorate
OH^-	hydroxide	$C2H_3O_2^-$	acetate
CN^-	cyanide	MnO_4^-	permanganate
PO_4^{3-}	phosphate	$Cr_2O_7^{2-}$	dichromate
HPO_4^{2-}	hydrogen phosphate	CrO_4^{2-}	chromate
$H_2PO_4^-$	dihydrogen phosphate	O_2^{2-}	peroxide